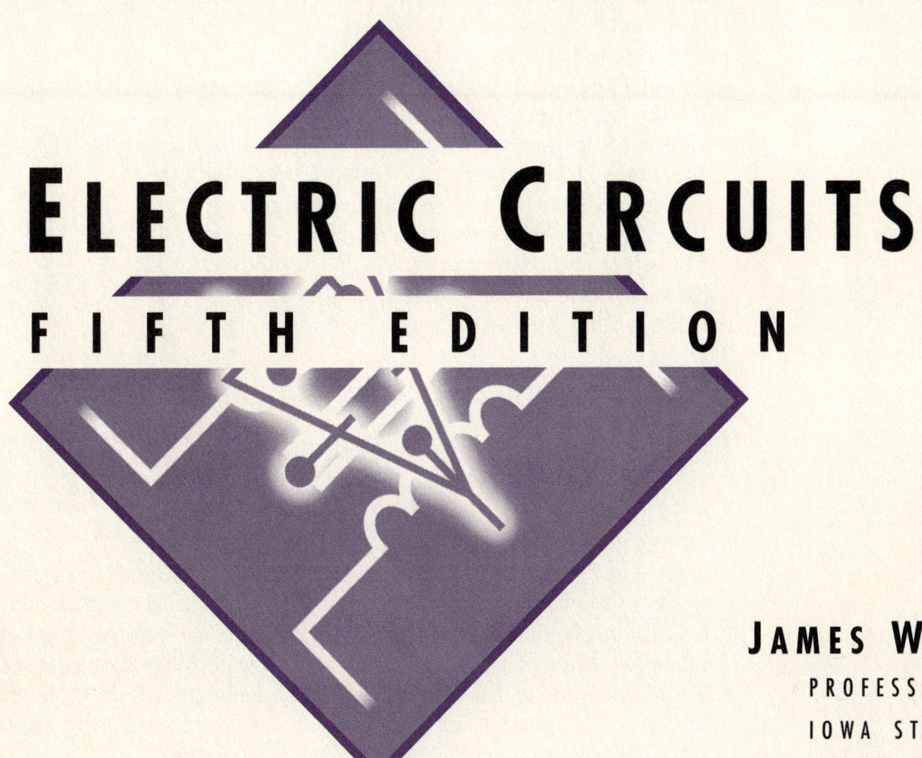

ELECTRIC CIRCUITS

FIFTH EDITION

JAMES W. NILSSON
PROFESSOR EMERITUS
IOWA STATE UNIVERSITY

SUSAN A. RIEDEL
MARQUETTE UNIVERSITY

ADDISON-WESLEY PUBLISHING COMPANY
Reading, Massachusetts • Menlo Park, California
New York • Don Mills, Ontario • Wokingham, England
Amsterdam • Bonn • Sydney • Singapore
Tokyo • Madrid • San Juan • Milan • Paris

WORLD STUDENT SERIES

Senior Sponsoring Editor: Lynne Doran Cote
Assistant Editor: Anita M. Devine
Development Editor: Laurie McGuire
Senior Production Coordinator: Kathleen A. Manley
Production Supervisor: Nancy H. Fenton
Editorial/Production Services: Lachina Publishing Services
Marketing Managers: Mary Tudor, Tom Ziolkowski
Cover Designer: Trudy Gershonov Design
Cover Design Supervisor: Eileen Hoff
Manufacturing Supervisor: Hugh Crawford
Senior Manufacturing Manager: Roy E. Logan
Compositor: Meridian Creative Group

More information on *Electric Circuits* and its accompanying supplements can be found on our World Wide Web site at http://www.aw.com/cseng/authors/nilsson/nilsson.html or via anonymous ftp at ftp://ftp.aw.com/cseng/authors/nilsson.

ISBN 0-201-40100-2
3 4 5 6 7 8 9 10—DOC—99989796

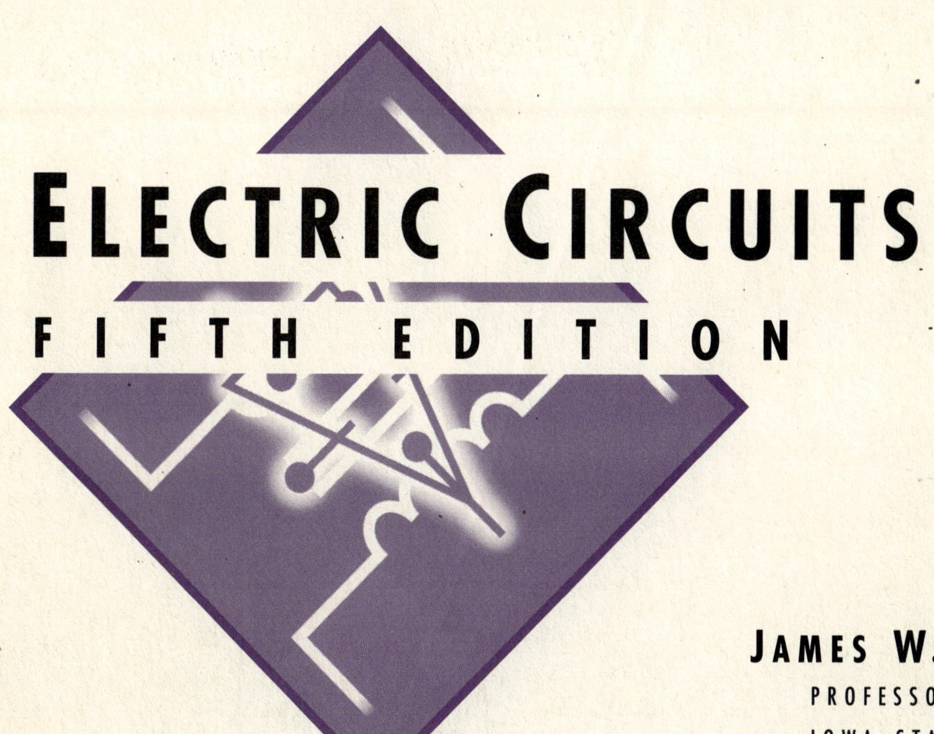

Electric Circuits

Fifth Edition

James W. Nilsson
Professor Emeritus
Iowa State University

Susan A. Riedel
Marquette University

To Anna

PREFACE

Across the engineering curriculum, factors such as course compression, computer integration, and student preparedness are affecting the way individual courses and entire disciplines are organized and taught. Even the introductory circuits course, which traditionally has been somewhat immune to change, has begun responding to such pressures, though definitive changes on a national level have yet to emerge.

Therefore, this is an exciting time to be revising the most widely used introductory circuits text of the past decade. Improvements in content, organization, and pedagogical features have been made with a mind toward reflecting course trends while preserving the time-tested strengths of the text. In addition, we are pleased and excited that Professor James Nilsson, after more than ten years of hard work in developing, writing, and revising *Electric Circuits,* is joined in the fifth edition by Professor Susan Riedel, who made valuable contributions to the third and fourth editions as well. Professor Riedel was involved in every step of the revision process and focused especially on developing and writing two of the most significant changes to this edition: the two new chapters on frequency-selective circuits and the new supplement, *Using Computer Tools for Electric Circuits.*

THE AUTHORS' AIMS

While the fifth edition of *Electric Circuits* represents a major revision, the underlying teaching approaches and philosophies, shared by both of us, remain unchanged. There are some specific aims that we wish to emphasize:

- *To build an understanding of concepts and ideas explicitly in terms of previous learning.* The learning challenges faced by students of engineering circuit analysis are prodigious; each new concept is built on a foundation of many other concepts. In *Electric Circuits,* much attention is paid to helping students recognize how new concepts and ideas fit on top of those previously learned.
- *To emphasize the relationship between conceptual understanding and problem-solving approaches.* Developing students' problem-solving skills continues to be the central challenge in this course. To address this challenge, *Electric Circuits* uses examples and simple drill exercises to demonstrate problem-solving approaches and to offer students practice opportunities. We do so not with the primary aim of giving students procedural models for solving problems; rather, we emphasize problem solving as a thought process in which one applies conceptual understanding to the solution of a practical problem. As such, in both the textual development and in the worked-out examples, we place great emphasis on a problem-solving process based on concepts rather than the use of rote procedures. Students are encouraged to think through problems before attacking them, and we often pause to consider the broader implications of a specific problem-solving situation.
- *To provide students with a strong foundation of engineering practices.* There are limited opportunities in a sophomore-year circuit analysis course to introduce students to real-world engineering experiences. We continue to emphasize the opportunities that do exist by making a strong effort to develop problems and exercises that use realistic values and represent realizable physical situations. We have included many application-type problems and exercises to help stimulate students' interest in engineering. Many of these problems require the kind of insight an engineer is expected to display when solving problems.

New to the Fifth Edition

We have come to regard each revision of *Electric Circuits* as a tremendous opportunity to make significant improvements in the book. The fifth edition has been the most extensive revision to date and has involved close collaboration between the coauthors and with reviewers, contributors, and the editorial, development, and production staff at Addison-Wesley. The goal of the fifth edition has been to better support the major teaching challenges that engineering students present. The major areas of change are as follows.

Practical Perspectives

To provide continuing motivation for students, we have introduced Practical Perspectives in Chapters 2, 3, 6, 10, 12, and 15. These sections offer examples of real-world circuits taken from real-world devices, such as telephones, hair dryers, and automobiles. Each Practical Perspective begins in the chapter opener with a brief description of a practical application of the material to follow. Once the chapter material has been presented, the chapter concludes with a quantitative analysis of the application. Several problems pertaining to the Practical Perspective are included in the chapter problems and are identified with the icon ◆. These Practical Perspectives should stimulate students' interest in applying circuit analysis to the design of useful circuits and devices. The following chart shows where to find the opening description, quantitative analysis, and related homework problems for each Practical Perspective:

DESCRIPTION	QUANTITATIVE ANALYSIS	RELATED HOMEWORK PROBLEMS
"Electric Safety" Chapter 2 Page 26	Pages 50–51	Pages 60–61
"A Rear Window Defroster" Chapter 3 Page 62	Pages 81–85	Page 103
"Proximity Switches" Chapter 6 Page 212	Pages 229–230	Pages 239–240
"Heating Appliances" Chapter 10 Page 412	Pages 441–443	Page 455
"An Ignition Circuit" Chapter 12 Page 494	Pages 530–533	Page 546
"Pushbutton Telephone Circuits" Chapter 15 Page 654	Pages 711–713	Page 722

Design Emphasis

We have in several ways increased the emphasis on the design of circuits. First, two new chapters on frequency-selective (filter) circuits replace the chapter on series and parallel resonance. Chapter 15 introduces the basic vocabulary of filter circuits through the study of the frequency response characteristics of passive first- and second-order circuits. Chapter 16 builds on this foundation and introduces many active filter circuits, all of which are based on an ideal operational amplifier. Both chapters are oriented toward the design of filter circuits.

Second, several Practical Perspective sections focus on the design aspects of the discussed circuits. The accompanying chapter problems continue these discussions of design issues.

Third, many chapter problems are explicitly labeled with the ❖ icon, enabling students and instructors to identify those problems with a design focus. Finally, the new supplement, *Using Computer Tools for Electric Circuits*, contains many examples that promote the exploration of circuit design issues using the computational support provided by computers.

Integration of Computer Tools

For electric circuits courses which use computer tools, the tool of choice for the past decade or so has been a circuit simulator, most often Spice or one of its descendants. Recently, other software tools have begun to make their way into the circuits classroom. In the fifth edition, we break new ground with *Using Computer Tools for Electric Circuits.* This supplement goes beyond circuit simulation to demonstrate the application of a wide variety of tools to the study of circuits. These tools include schematic capture and analysis tools, a matrix equation solver, a spreadsheet package, and a symbolic equation solver. The supplement is not intended to provide instruction in a particular software package. Rather, it encourages students to independently explore software tools they are familiar with and then apply these tools to support their study of circuits.

Computer tools cannot replace the traditional methods for mastering the study of electric circuits. They can, however, assist students in the learning process by providing a visual representation of a circuit's behavior, validating a calculated solution, reducing the computational burden of more complex circuits, and iterating toward a desired solution using parameter variation. This computational support is often invaluable in the design process. The supplement thus encourages students to search for new ways to apply their existing computer skills to circuit analysis.

Content and Organizational Changes

The most significant change to the fifth edition is the addition of the two new chapters on filter circuits: Chapter 15, "Introduction to Frequency-Selective Circuits," and Chapter 16, "Active Filter Circuits." We have rewritten and reorganized other topics to set the stage for these new chapters, and in response to reviewer suggestions, users of the fourth edition, and our editors, we have made these additional changes:

- Chapter 1 was rewritten with an eye toward an even greater emphasis on problem solving and placing the study of electric circuits in the context of the whole of electrical engineering.

- In Chapter 3, we have condensed the discussion of the d'Arsonval movement and the three types of meters into Section 3.5, "Measuring Voltage and Current."
- The discussion of the topological foundations for circuit analysis, which was Chapter 5 in the fourth edition, has been condensed and moved to Appendix C.
- The presentation of the differential mode of an operational amplifier and the common mode rejection ratio has been condensed and moved to Section 5.6 on the difference amplifier. The material on Bartlett's bisection theorem has been eliminated.
- The material on delta-connected three-phase loads has been eliminated, although this configuration is still explored in the problems. We have added (from the third edition) a discussion of wattmeters and the two-wattmeter method of measuring power in three-phase circuits.
- The discussion of the Laplace transform in circuit analysis has been combined with the representation of the transfer function in Chapter 14. This chapter also discusses the convolution integral and the use of the impulse function in circuit analysis.
- The material describing the construction of Bode plots is found at the end of Chapter 15, following the presentation of passive filter circuits.

Text Design and Pedagogical Features

The following features are new to the fifth edition:

- *Practical Perspective introductions* are located opposite select chapter opening pages and are highlighted with a second-color background.
- *Practical Perspective examples* at the end of the applicable chapters are set apart in an easy-to-identify separate section.
- *Practical Perspective problems* in the end-of-chapter problem sets are indicated with a ◆ icon for easy reference.
- *Key terms* are set in boldface when they are first defined. They also appear in boldface in the chapter summaries. This makes it easier for students to find the definitions of important terms.
- *Margin notes* refer students to the supplement *Using Computer Tools for Electric Circuits.* The margin notes are accompanied by a ■ icon that identifies some of the concepts in the text that could be supported with a computer tool.

- *Design problems* in the end-of-chapter problem sets are indicated with a ❖ icon for easy reference.

Examples, Drill Exercises, and Homework Problems

Solved Numerical Examples

Solved numerical examples are used extensively throughout the text to help students understand how theory is applied to circuit analysis. Because many students value such examples more than any other aspect of the text, these examples represent an important opportunity to influence the development of students' problem-solving skills. The nature and format of the examples in *Electric Circuits* are a reflection of the overall teaching approach of the text. When presenting a solution, we place great emphasis on the importance of problem solving as a thought process that applies underlying concepts, as we discussed earlier. By emphasizing this idea—even in the solution of simple problems—we hope to communicate that this approach to problem solving can help students handle the more complex problems they will encounter later on. Some characteristics of the examples include

- encouraging students to study the problem or the circuit and make initial observations before diving into a solution pathway (page 253);
- emphasizing the individual stages of the solution as part of solving the problem systematically, without suggesting that there are rote procedures for problem solving (page 48);
- exploring decision making, that is, the idea that we are often faced with choosing among many different solution approaches (page 129); and
- suggesting that students challenge their results, by emphasizing the importance of checking and testing answers based on their knowledge of circuit theory and the real world (page 43).

Drill Exercises

Drill exercises are included in the text to give students an opportunity to test their understanding of the material they have just read. The drill exercises are presented in a double-column format as a way of signaling to students that they should stop and solve the exercises before proceeding to the next section.

Homework Problems

Users of *Electric Circuits* have consistently rated the homework problems as one of the book's most attractive features. In the fifth edition, there are over 900 problems, of which 70% are either new or revised. To identify the unknowns in a problem, the circuit elements that must be calculated are in color. The problems are designed around the following objectives (in parentheses are the corresponding problem categories found in the *Instructor's Manual*):

- To give students practice in using the analytical techniques developed in the text (Practice; see Problem 4.3)
- To show students that analytical techniques are tools, not objectives (Analytical Tool; see Problem 4.4)
- To give students practice in choosing the analytical method to be used in obtaining a solution (Open Method; see Problems 4.43 and 4.44)
- To show students how the results from one solution can be used to find other information about a circuit's operation (Additional Information; see Problems 4.73 and 4.74)
- To encourage students to challenge the solution either by using an alternative method or by testing the solution to see if it makes sense in terms of known circuit behavior (Solution Check; see Problem 4.51)
- To introduce students to design-oriented problems (Design; see Problem 10.41)
- To give students practice in deriving and manipulating equations where quantities of interest are expressed as functions of circuit variables, such as R, L, C, ω, and so forth; this type of problem also supports the design process (Derivation; see Problems 9.28 and 9.30)
- To challenge students with problems that will stimulate their interest in both electrical and computer engineering (Practical; see Problem 3.65)

Every problem in every chapter has been worked out by a team of independent accuracy checkers. Special thanks to the following professors who undertook this immense task: Capt. J. D. McArthur, Capt. Stephen O'Connor, Lt. Col. Charles Smith, and Capt. Robert A. Yahn Jr., all of the United States Air Force Academy, and Curran Swift of Iowa State University.

Prerequisites

In writing the first thirteen chapters of the text, we have assumed that the reader has taken a course in elementary differential and integral calculus. We have also assumed that the reader has had an introductory physics course, at either the high school or university level, which introduced the concepts of energy, power, electric charge, electric current, electric potential, and electromagnetic fields. In writing the final six chapters, we have assumed the student has had, or is enrolled in, an introductory course in differential equations.

Course Options

The text has been designed for use in a one-semester, two-semester, or three-quarter sequence.

- *Single-semester course:* After covering Chapters 1–4 and Chapters 6–10, omitting sections 7.7 and 8.5, the instructor can choose from Chapter 5 (operational amplifiers), Chapter 11 (three-phase circuits), Chapter 12 (mutual inductance), and Chapter 17 (Fourier series) to develop the desired emphasis.
- *Two-semester sequence:* Assuming three lectures per week, the first ten chapters can be covered during the first semester, leaving Chapters 11–19 for the second semester.
- *Academic-quarter schedule:* The book can be subdivided into three parts: Chapters 1–7, Chapters 8–13, and Chapters 14–19.

The introduction of operational amplifier circuits can be omitted without interfering with the reading of the subsequent chapters. For example, if Chapter 5 is omitted, the instructor can simply skip Section 7.7, Section 8.5, Chapter 16, and those problems and drill exercises in the chapters following Chapter 5 that pertain to operational amplifiers.

There are several appendixes to help readers make effective use of their mathematical background. Appendix A reviews Cramer's method for solving simultaneous linear equations and simple matrix algebra; complex numbers are reviewed in Appendix B; Appendix C contains the topological foundations for circuit analysis; Appendix D offers a brief discussion of the decibel; Appendix E presents an abbreviated table of trigonometric identities useful in circuit analysis; and an abbreviated table of useful integrals is given in Appendix F. Appendix G provides a comprehensive list of the examples, with titles and corresponding page numbers.

SUPPLEMENTS

We have put care into the development of supplements that capitalize on the many strengths of the fifth edition. Students and professors are constantly challenged, in terms of time and energy, by the confines of the classroom and the importance of integrating new information and technologies into an electric circuits course. Through the following supplements, we believe we have succeeded in making some of these challenges more manageable.

USING COMPUTER TOOLS FOR ELECTRIC CIRCUITS

Using Computer Tools for Electric Circuits incorporates a brief introduction to MicroSim™ PSpice®, replacing the PSpice supplement from the fourth edition. It goes beyond MicroSim PSpice to provide examples of the application of a wide variety of tools to the study of circuits. These tools include MicroSim™ Schematics, a schematic capture and analysis tool (formerly part of Design Center 3); MATLAB®, a matrix equation solver; Quattro® Pro, a spreadsheet package; and Maple® V, a symbolic equation solver. Each chapter of the supplement provides five or six worked examples taken directly from the text. The disk that accompanies the supplement contains files for each example and solutions to some of the nearly 150 problems at the end of the supplement. *Using Computer Tools for Electric Circuits* is published as a separate booklet to facilitate its use at a computer, but it is bundled free with each copy of the text.

INSTRUCTOR'S MANUAL

The *Instructor's Manual* enables professors to orient themselves quickly to this text and the supplement package. For easy reference, the following information is organized for each chapter:

- a chapter overview
- problem categorizations
- problem references by chapter section
- a list of examples
- cross-references to problems in *Using Computer Tools for Electric Circuits*
- transparency master figures
- cross-references to modules in *CircuitTutor*

Solutions Manual with Transparency Masters

The *Solutions Manual* contains complete worked solutions with supporting figures to all of the end-of-chapter problems in the fifth edition. Volume I covers Chapters 1–9, and Volume II covers Chapters 10–19. These supplements, available free to all adopting faculty, were checked for accuracy by several instructors. The manuals are not available for sale to students. Transparency masters covering the chapters in each volume are located at the back of the appropriate manual and may be photocopied for students.

CircuitTutor by Tutorware, Inc., Developed by Burks Oakley II for Macintosh and Windows

CircuitTutor is an interactive software tutorial designed to help students develop the analytical skills required to master circuit problems. When working through problems, students get immediate feedback and reinforcement as well as step-by-step guidance upon request.

CircuitTutor supplements course lectures and textual material. It is available for both Macintosh and IBM-PC compatible computers. A site license is free to adopters of *Electric Circuits*. Please contact your Addison-Wesley sales representative for demonstration disks and site license information.

Acknowledgments

We would like to express our appreciation for the contributions of Norman Wittels of Worcester Polytechnic Institute. His contributions to the Practical Perspectives have greatly enhanced our original idea for this revision.

There were many people behind the scenes at Addison-Wesley Publishing Company and Lachina Publishing Services who have worked hard to produce the fifth edition. Special thanks to Lynne Doran Cote, the editor of the fifth edition, for her support in this project, and to the other members of the "Nilsson team": Anita Devine, Kathy Manley, and Laurie McGuire. They have done a superb job.

The many revisions of the text were guided by careful and thorough reviews from professors. Our heartfelt thanks to Albert L. Batten, Colorado Technical College; James G. Gottling, Ohio State University; Curran Swift, Iowa State University; Norman Wittels, Worcester Polytechnic Institute; William J. Eccles, Rose-Hulman Institute of Technology; Clifford Pollock, Cornell University;

Kwang Y. Lee, Pennsylvania State University; Dana H. Brooks, Northeastern University; Aziz S. Inan, University of Portland; A. A. (Louis) Beex, Virginia Technical Institute; Henri Merkelo, University of Illinois at Urbana; Kenneth C. Mylrea, University of Arizona; Earl Swartzlander, University of Texas at Austin; Jerry L. Prince, Johns Hopkins University; John A. Wheeldon, University of North Dakota; Thomas R. Connell, University of Akron; Larry C. Schooley, University of Arizona; Frank L. Merat, Case Western Reserve University; John D. Enderle, North Dakota State University; E. Randolph Collins Jr., Clemson University; F. Humano, California State University, Long Beach; B. A. Shenoi, Wright State University; Capt. Robert A. Yahn Jr., United States Air Force Academy; and Tim Jordanides, California State University, Long Beach.

A debt of gratitude is owed to those professors who reviewed the supplement *Using Computer Tools for Electric Circuits* at several stages of development: Mahmoud A. Abdallah, Central State University; Michael J. Batchelder, South Dakota School of Mines; Victor Demjanenko, State University of New York at Buffalo; Carl H. Durney, University of Utah; William J. Eccles, Rose-Hulman Institute of Technology; Gayle F. Miner, Brigham Young University; and Clifford Pollock, Cornell University.

First I would like to thank Professor Susan Riedel for accepting the challenge of becoming a coauthor of *Electric Circuits*. Her willingness to suggest both pedagogical and content changes and at the same time graciously accept constructive criticism when offered has made the transition to the fifth edition possible. She brings to the text an expertise in computer use and a genuine interest in and enthusiasm for teaching.

I would also like to thank Col. Alan Klayton and Lt. Col. Al Batten (Ret.) at the United States Air Force Academy for giving me the opportunity to get back in the classroom during the 1992–1994 academic years. Without this classroom experience, I would not have been inspired to develop new and modify old problems. My thanks to Tom Scott, Curran Swift, and Jim Triska at Iowa State University for their continued interest in the book.

JWN

I am extremely grateful to Professor Nilsson for having given me the opportunity to become associated with this superb book. It has been a tremendous challenge to live up to the high standards he has set. He is a wonderful teacher. Many other people have inspired my work as an engineer and an educator—you know who you are, and I thank you. I am especially grateful for the continued support and encouragement of Eileen Bernadette Moran, the editor of the fourth edition, who brought me into this project. To my children, David and Jason—you have shown patience beyond your years with my work on a book that doesn't have any really good pictures, and for that and much more, I love you!

SAR

Brief Contents

CONTENTS

CHAPTER 1 CIRCUIT VARIABLES 1

CHAPTER 2 CIRCUIT ELEMENTS 26

Chapter 3 Simple Resistive Circuits 62

Chapter 4 Techniques of Circuit Analysis 105

CHAPTER 5 THE OPERATIONAL AMPLIFIER 177

CHAPTER 6 INDUCTORS AND CAPACITORS 212

CHAPTER 7 RESPONSE OF FIRST-ORDER *RL* AND *RC* CIRCUITS 241

CHAPTER 8 NATURAL AND STEP RESPONSES OF *RLC* CIRCUITS 305

CHAPTER 9 SINUSOIDAL STEADY-STATE ANALYSIS 355

CHAPTER 10 SINUSOIDAL STEADY-STATE POWER CALCULATIONS 412

CHAPTER 11 BALANCED THREE-PHASE CIRCUITS 457

CHAPTER 12 MUTUAL INDUCTANCE 494

CHAPTER 13 INTRODUCTION TO THE LAPLACE TRANSFORM 547

CHAPTER 14 THE LAPLACE TRANSFORM IN CIRCUIT ANALYSIS 589

Chapter 15 Introduction to Frequency-Selective Circuits 654

Chapter 16 Active Filter Circuits 723

CHAPTER 17 FOURIER SERIES 779

CHAPTER 18 THE FOURIER TRANSFORM 831

CHAPTER 19 TWO-PORT CIRCUITS 869

APPENDIX A THE SOLUTION OF LINEAR SIMULTANEOUS EQUATIONS 903

APPENDIX B COMPLEX NUMBERS 929

APPENDIX C TOPOLOGY IN CIRCUIT ANALYSIS 937

APPENDIX D THE DECIBEL 951

1 CIRCUIT VARIABLES

CHAPTER CONTENTS

Electrical engineering is an exciting and challenging profession for anyone who has a genuine interest in, and aptitude for, applied science and mathematics. Over the past century and a half, electrical engineers have played a dominant role in the development of systems that have changed the way people live and work. Satellite communication links, telephones, digital computers, televisions, diagnostic and surgical medical equipment, assembly-line robots, and electrical power tools are representative components of systems that define a modern technological society. As an electrical engineer, you can participate in this ongoing technological revolution by improving and refining these existing systems and by discovering and developing new systems to meet the needs of our ever-changing society. You can thus have an impact on the evolution of our society because of the complex and intricate connections between society and electrical engineering technology.

As you embark on the study of circuit analysis, you need to gain a feel for where this study fits into the hierarchy of topics that comprise an introduction to electrical engineering. Hence we begin by presenting an overview of electrical engineering, some ideas about an engineering point of view as it relates to circuit analysis, and a review of the international system of units.

We then describe generally what circuit analysis entails. Next, we introduce the concepts of voltage and current. We follow those concepts with discussion of an ideal basic element and the need for a polarity reference system. We conclude the chapter by describing how current and voltage relate to power and energy.

1.1 Electrical Engineering: An Overview

Electrical engineering is the profession concerned with systems that produce, transmit, and measure electric signals. Electrical engineering combines the physicist's models of natural phenomena with the mathematician's tools for manipulating those models to produce systems that meet practical needs. Electrical systems pervade our lives; they are found in homes, schools, workplaces, and transportation vehicles—everywhere. We begin by presenting a few examples from each of the five major classifications of electrical systems: (1) communication systems, (2) computer systems, (3) control systems, (4) power systems, and (5) signal-processing systems. Then we describe how electrical engineers analyze and design such systems.

Communication systems are electrical systems that generate, transmit, and distribute information. Well-known examples include television equipment, such as cameras, transmitters, receivers, and VCRs; radio telescopes, which are used to explore the universe; satellite systems, which return images of other planets and our own; radar systems used to coordinate plane flights; and telephone systems.

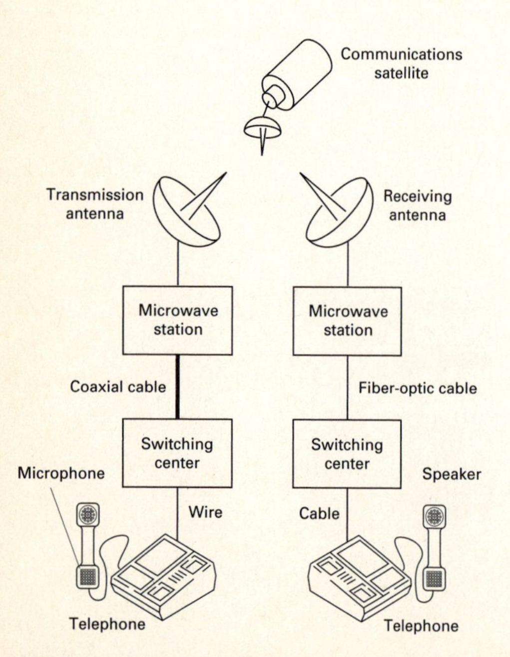

FIGURE 1.1 A telephone system.

Figure 1.1 depicts the major components of a modern telephone system. Starting at the left of the figure, inside a telephone, a microphone turns sound waves into electric signals. These signals are carried to a switching center where they are combined with the signals from tens, hundreds, or thousands of other telephones. The combined signals leave the switching center; their form depends on the distance they must travel. In our example, they are sent through wires in underground coaxial cables to a microwave transmission station. Here, the signals are transformed into microwave frequencies and broadcast from a transmission antenna through air and space, via a communications satellite, to a receiving antenna. The microwave receiving station translates the microwave signals into a form suitable for further transmission, perhaps as pulses of light to be sent through fiber-optic cable. Upon arrival at the second switching center, the combined signals are separated, and each is routed to the appropriate telephone, where an earphone acts as a speaker to convert the received electric signals back into sound waves. At each stage of the process, electric circuits operate on the signals. Imagine the challenge involved in designing, building, and operating each circuit in a way that guarantees that all of the hundreds of thousands of simultaneous calls have high-quality connections.

Computer systems use electric signals to process information ranging from word processing to mathematical computations. Systems range in size and power from pocket calculators to personal computers to supercomputers that perform such complex tasks as processing weather data and modeling chemical interactions of complex organic molecules. These systems include networks of microcircuits, or integrated circuits—postage-stamp-sized assemblies of hundreds, thousands, or millions of electrical components which often operate at speeds and power levels close to fundamental physical limits, including the speed of light and the thermodynamic laws.

Control systems use electric signals to regulate processes. Examples include the control of temperatures, pressures, and flow rates in an oil refinery; the fuel-air mixture in a fuel-injected automobile engine; mechanisms such as the motors, doors, and lights in elevators; and the locks in the Panama Canal. The autopilot and autolanding systems that help to fly and land airplanes are also familiar control systems.

Power systems generate and distribute electric power. Electric power, which is the foundation of our technology-based society, usually is generated in large quantities by nuclear, hydroelectric, and thermal (coal-, oil-, or gas-fired) generators. Power is distributed by a grid of conductors that crisscross the country. A major challenge in designing and operating such a system is to provide sufficient redundancy and control so that failure of any piece of equipment does not leave a city, state, or region completely without power.

Signal-processing systems act on electric signals which represent information. They transform the signals, and the information contained in them, to a more suitable form. There are many different ways to process the signals and their information. For example, image-processing systems gather massive quantities of data from orbiting weather satellites, reduce the amount of data to a managable level, and transform the remaining data into a video image for the evening news broadcast. A computerized tomography (CT) scan is another example of an image-processing system. It takes signals generated by a special X-ray machine and transforms them into an image such as the one in Fig. 1.2. While the original X-ray signals are of little use to a physician, once they are processed into a recognizable image the information they contain can be used in the diagnosis of disease and injury.

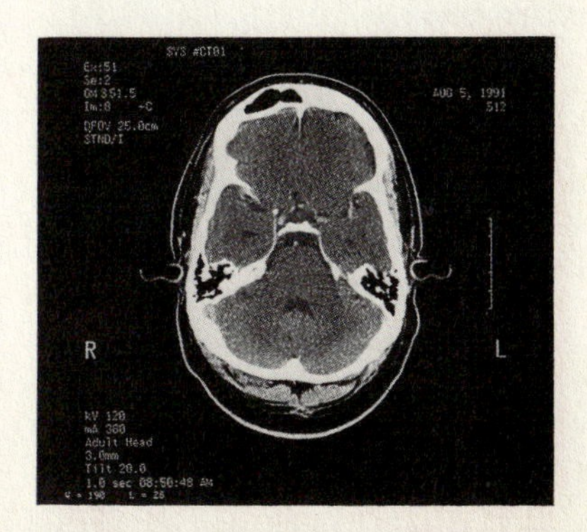

FIGURE 1.2 A CT scan of an adult head.

Considerable interaction takes place among the engineering disciplines involved in designing and operating these five classes of systems. Thus communication engineers use digital computers to control the flow of information. Computers contain control systems, and control systems contain computers. Power systems require extensive communication systems to coordinate safely and reliably the operation of components, which may be spread across a continent. A signal-processing system may involve a communication link, a computer, and a control system.

A good example of the interaction among systems is a commercial airplane, such as the one shown in Fig. 1.3. A sophisticated communication system enables the pilot and the air traffic controller to monitor the plane's location, permitting the air traffic controller to design a safe flight path for all of the nearby aircraft and enabling the pilot to keep the plane on its designated path. On the newest commercial airplanes, an onboard computer system is used for managing engine functions, implementing the navigation and flight control systems, and generating video information screens in the cockpit. A complex control system uses cockpit commands to adjust the position and speed of the airplane, producing the appropriate signals to the engines and the control surfaces (such as the wing flaps, ailerons, and rudder) to ensure the plane remains safely airborne and on the desired flight path. The plane must have its own power system to stay aloft and to provide and distribute the electric power needed to keep the cabin lights on, make the coffee, and show the movie. Signal-processing systems reduce the noise in air traffic communications and transform information about the plane's location into the more meaningful form of a video display in the cockpit. Engineering challenges abound in the design of each of these systems and their integration into a coherent whole. For example, these systems must operate in widely varying and unpredictable environmental conditions. Perhaps the most important engineering challenge is to guarantee that sufficient redundancy is incorporated in the designs to ensure that passengers arrive safely and on time at their desired destinations.

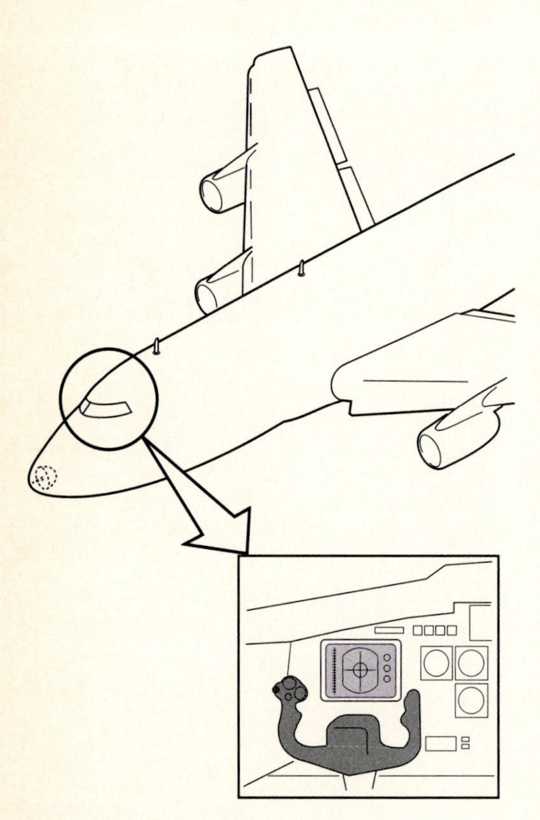

FIGURE 1.3 An airplane.

Although an electrical engineer may be interested primarily in one area, he or she must also be knowledgeable in other areas that interact with this area of interest. This interaction is part of what makes electrical engineering a challenging and exciting profession. The emphasis in engineering is on making things work, so an engineer is free to acquire and use any technique, from any field, that helps to get the job done.

CIRCUIT THEORY

In a field as diverse as electrical engineering, you might well ask whether all of its branches have anything in common. The answer is yes—electric circuits. An **electric circuit** is a mathematical model which approximates the behavior of an actual electrical system. As such, it is a foundational tool for learning—in your later courses and as a practicing engineer—the details of how to design and operate systems such as those just described. The models, the mathematical techniques, and the language of circuit theory will form the intellectual framework for your future engineering endeavors.

Note that the term *electric circuit* is commonly used to refer to an actual electrical system as well as to the model which represents it. In this text, when we talk about an electric circuit, we always mean a model, unless otherwise stated. It is the modeling aspect of circuit theory which has broad applications across engineering disciplines.

Circuit theory is a special case of electromagnetic field theory—the study of static and moving electric charges. Although generalized field theory might seem to be an appropriate starting point for investigating electric signals, its application is not only cumbersome but also requires the use of advanced mathematics. Consequently, a course in electromagnetic field theory is not a prerequisite to understanding the material in this book. We do, however, assume that you have had an introductory physics course in which electrical and magnetic phenomena were discussed.

Three basic assumptions permit us to use circuit theory, rather than electromagnetic field theory, to study a physical system represented by an electric circuit. These assumptions are as follows:

1. *Electrical effects happen instantaneously throughout a system.* We can make this assumption because we know that electric signals travel at or near the speed of light. Thus, if the system is physically small, electric signals move through it so quickly that we can consider them to affect every point in the system simultaneously. A system which is small enough so that we can make this assumption is called a **lumped-parameter system**.

2. *The net charge on every component in the system is always zero.* Thus no component can collect a net excess of charge, although some components, as you will learn later, can hold equal but opposite separated charges.

3. *There is no magnetic coupling between the components in a system.* As we demonstrate later, magnetic coupling can occur *within* a component.

That's it; there are no other assumptions. Using circuit theory provides simple solutions (of sufficient accuracy) to problems that would otherwise become hopelessly complicated if we were to use electromagnetic field theory. These benefits are so great that engineers sometimes specifically design electrical systems to ensure that these assumptions are met. The importance of assumptions 2 and 3 becomes apparent after we introduce the basic circuit elements and the rules for analyzing interconnected elements.

However, we need to take a closer look at assumption 1. The question is: "How small does a physical system have to be to qualify as a lumped-parameter system?" We can get a quantitative handle on the question by noting that electric signals propagate by wave phenomena. If the wavelength of the signal is large compared to the physical dimensions of the system, we have a lumped-parameter system. The wavelength λ is the velocity divided by the repetition rate, or **frequency**, of the signal; that is, $\lambda = c/f$. The frequency f is measured in hertz (Hz). For example, power systems in the United States operate at 60 Hz. If we use the speed of light ($c = 3 \times 10^8$ m/s) as the velocity of propagation, the wavelength is 5×10^6 m. If the power system of interest is physically smaller than this wavelength, we can represent it as a lumped-parameter system and use circuit theory to analyze its behavior. How do we define *smaller*? A good rule is the *rule of 1/10th*—if the dimension of the system is 1/10th (or smaller) of the dimension of the wavelength, you have a lumped-parameter system. Thus, so long as the physical dimension of the power system is less than 5×10^5 m, we can treat it as a lumped-parameter system.

On the other hand, the propagation frequency of radio signals is on the order of 10^9 Hz. Thus the wavelength is 0.3 m. Using the rule of 1/10th, the relevant dimensions of a communication system which sends or receives radio signals must be less than 3 cm to qualify as a lumped-parameter system. Whenever any of the pertinent physical dimensions of a system under study approach the wavelength of its signals, we must use electromagnetic field theory to analyze that system. Throughout this book we study circuits derived from lumped-parameter systems.

PROBLEM SOLVING

As a practicing engineer, you will not be asked to solve problems that have already been solved. Whether you are trying to improve the performance of an existing system or creating a new system, you will be working on unsolved problems. As a student, however, you will devote much of your attention to the discussion of problems already solved. By reading about and discussing how these problems were solved in the past, and by solving related homework and exam problems on your own, you will begin to develop the skills to successfully attack the unsolved problems you'll face as a practicing engineer.

Some general problem-solving procedures are presented here. Many of them pertain to thinking about and organizing your solution strategy *before* proceeding with calculations.

1. *Identify what's given and what's to be found.* In problem solving, you need to know your destination before you can select a route for getting there. What is the problem asking you to solve or find? Sometimes the goal of the problem is obvious; other times

you may need to paraphrase or make lists or tables of known and unknown information to see your objective.

The problem statement may contain extraneous information that you need to weed out before proceeding. On the other hand, it may offer incomplete information or more complexities than can be handled given the solution methods at your disposal. In that case, you'll need to make assumptions to fill in the missing information or simplify the problem context. Be prepared to circle back and reconsider supposedly extraneous information and/or your assumptions if your calculations get bogged down or produce an answer that doesn't seem to make sense.

2. *Sketch a circuit diagram or other visual model.* Translating a verbal problem description into a visual model is often a useful step in the solution process. If a circuit diagram is already provided, you may need to add information to it, such as labels, values, or reference directions. You may also want to redraw the circuit in a simpler, but equivalent, form. Later in this text you will learn the methods for developing such simplified equivalent circuits.

3. *Think of several solution methods and a way of choosing among them.* This course will help you build a collection of analytical tools, several of which may work on a given problem. But one method may produce fewer equations to be solved than another, or it may require only algebra instead of calculus to reach a solution. Such efficiencies, if you can anticipate them, can streamline your calculations considerably. Having an alternative method in mind also gives you a path to pursue if your first solution attempt bogs down.

4. *Calculate a solution.* Your planning up to this point should have helped you identify a good analytical method and the correct equations for the problem. Now comes the solution of those equations. Paper-and-pencil, calculator, and computer methods are all available for performing the actual calculations of circuit analysis. Efficiency and your instructor's preferences will dictate which tools you should use.

5. *Use your creativity.* If you suspect that your answer is off base or if the calculations seem to go on and on without moving you toward a solution, you should pause and consider alternatives. You may need to revisit your assumptions or select a different solution method. Or, you may need to take a less conventional problem-solving approach, such as working backward from a solution. In this text, all of the drill exercises and many of the homework problems have answers provided so that you may work backward when you get stuck. In the real world, you

won't be given answers in advance, but you may have a desired problem outcome in mind from which you can work backward. Other creative approaches include allowing yourself to see parallels with other types of problems you've successfully solved, following your intuition or hunches about how to proceed, and simply setting the problem aside temporarily and coming back to it later.

6. *Test your solution.* Ask yourself whether the solution you've obtained makes sense. Does the magnitude of the answer seem reasonable? Is the solution physically realizable? You may want to go further and rework the problem via an alternative method. Doing so will not only test the validity of your original answer, but will also help you develop your intuition about the most efficient solution methods for various kinds of problems. In the real world, safety-critical designs are always checked by several independent means. Getting into the habit of checking your answers will benefit you as a student and as a practicing engineer.

These problem-solving steps cannot be used as a recipe to solve every problem in this or any other course. You may need to skip, change the order of, or elaborate upon certain steps in order to solve a particular problem. Use these steps as a guideline to develop a problem-solving style that works for you.

1.2 International System of Units

Engineers compare theory to experiment and compare competing engineering designs using quantitative measures. Modern engineering is a multidisciplinary profession in which teams of engineers work together on projects, and they can communicate their results in a meaningful way only if they all use the same units. The International System of Units (abbreviated SI) is used by all the major engineering societies and most engineers throughout the world; hence we use it in this book.

The SI units are based on six *defined* quantities: (1) length, (2) mass, (3) time, (4) electric current, (5) thermodynamic temperature, and (6) luminous intensity. These quantities, along with the basic unit and symbol for each, are listed in Table 1.1. Although not strictly SI units, the familiar time units of minute (60 s), hour (3600 s), and so on are often used in engineering calculations. In addition, defined quantities are combined to form **derived** units. Some, such as force, energy, power, and electric charge, you already know through previous physics courses. Table 1.2 lists the derived units used in this book.

TABLE 1.1

The International System of Units (SI)

QUANTITY	BASIC UNIT	SYMBOL
Length	meter	m
Mass	kilogram	kg
Time	second	s
Electric current	ampere	A
Thermodynamic temperature	degree kelvin	K
Luminous intensity	candela	cd

In many cases the SI unit is either too small or too large to use conveniently. Standard prefixes corresponding to powers of 10, as listed in Table 1.3, are then applied to the basic unit. All of these prefixes are correct, but engineers often use only the ones for powers divisible by 3; thus centi, deci, deka, and hecto are used rarely. Also, engineers often select the prefix that places the base number in the range between 1 and 1000. Suppose that a time calculation yields a result of 10^{-5} s, that is, 0.00001 s. Most engineers would describe this quantity as 10 μs, that is, $10^{-5} = 10 \times 10^{-6}$ s, rather than as 0.01 ms or 10,000,000 ps.

TABLE 1.2

DERIVED UNITS IN SI

QUANTITY	UNIT NAME (SYMBOL)	FORMULA
Frequency	hertz (Hz)	s^{-1}
Force	newton (N)	$kg \cdot m/s^2$
Energy or work	joule (J)	$N \cdot m$
Power	watt (W)	J/s
Electric charge	coulomb (C)	$A \cdot s$
Electric potential	volt (V)	W/A
Electric resistance	ohm (Ω)	V/A
Electric conductance	siemens (S)	A/V
Electric capacitance	farad (F)	C/V
Magnetic flux	weber (Wb)	$V \cdot s$
Inductance	henry (H)	Wb/A

TABLE 1.3

STANDARDIZED PREFIXES TO SIGNIFY POWERS OF 10

PREFIX	SYMBOL	POWER	PREFIX	SYMBOL	POWER
atto	a	10^{-18}	deci	d	10^{-1}
femto	f	10^{-15}	deka	da	10
pico	p	10^{-12}	hecto	h	10^2
nano	n	10^{-9}	kilo	k	10^3
micro	μ	10^{-6}	mega	M	10^6
milli	m	10^{-3}	giga	G	10^9
centi	c	10^{-2}	tera	T	10^{12}

DRILL EXERCISES

1.1 How many dollars per millisecond would the federal government have to collect to retire a deficit of $300 billion in one year?

ANSWER: $9.51/ms.

1.2 There are approximately 142 million passenger cars registered in the United States. Assume that the average passenger-car battery stores 440 watt-hours (Wh) of energy. Estimate (in gigawatt-hours) the total energy stored in U.S. passenger cars.

ANSWER: 62.48 GWh.

1.3 A high-resolution computer display monitor has 1280 × 1024 picture elements, or pixels. Each picture element contains 24 bits of information. If a byte is defined as 8 bits, how many megabytes (MB) are required per display?

ANSWER: 3.93 MB.

1.4 Some species of bamboo can grow 250 mm/day. Assuming the individual cells in the plant are 10 μm long, how long, on average, does it take a bamboo stalk to grow a 1-cell length?

ANSWER: 3.5 s.

1.5 One liter (L) of paint covers approximately 10 m^2 of wall. How thick is the layer before it dries? (*Hint:* $1\ L = 1 \times 10^6\ mm^3$.)

ANSWER: 0.1 mm.

1.6 How long does it take for light to travel across a room which is $19'8\frac{1}{4}''$ wide?

ANSWER: 20 ns.

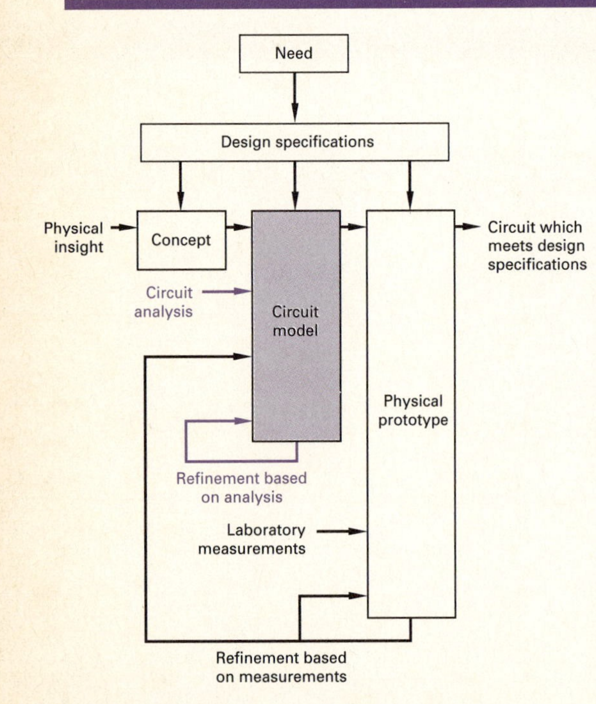

FIGURE 1.4 A conceptual model for electrical engineering design.

1.3 CIRCUIT ANALYSIS: AN OVERVIEW

Before becoming involved in the details of circuit analysis, we need to take a broad look at engineering design, specifically the design of electric circuits. The purpose of this overview is to provide you with a perspective on where circuit analysis fits within the whole of circuit design. While this book focuses on circuit analysis, we try to provide opportunities for circuit design where appropriate.

All engineering designs begin with a need, as shown in Fig. 1.4. This need may come from the desire to improve on an existing design, or it may be something brand new. A careful assessment of the need results in design specifications, which are measurable characteristics of a proposed design. Once a design is proposed, the design specifications allow us to assess whether or not the design actually meets the need.

A concept for the design comes next. The concept derives from a complete understanding of the design specifications coupled with an insight into the need, which comes from education and experience. The concept may be realized as a sketch, as a written description, or in some other form. The next step is often to translate the concept into a mathematical model. A commonly used mathematical model for electrical systems is a **circuit model**.

The elements that comprise the circuit model are called **ideal circuit components**. An ideal circuit component is a mathematical model of an actual electrical component, like a battery or a light bulb. It is important for the ideal circuit component used in a circuit model to represent the behavior of the actual electrical component to an acceptable degree of accuracy. The tools of **circuit analysis**, the focus of this book, are then applied to the circuit. Circuit analysis is based on mathematical techniques and is used to predict the behavior of the circuit model and its ideal circuit components. A comparison between the desired behavior, from the design specifications, and the predicted behavior, from circuit analysis, may lead to refinements in the circuit model and its ideal circuit elements. Once the desired and predicted behavior are in agreement, a physical prototype can be constructed.

The **physical prototype** is an actual electrical system, constructed from actual electrical components. Measurement techniques are used to determine the actual, quantitative behavior of the physical system. This actual behavior is compared with the desired behavior from the design specifications and the predicted behavior from circuit analysis. The comparisons may result in refinements to the physical prototype, the circuit model, or both. Eventually, this iterative process in which models, components, and systems are continually refined may produce a design which accurately matches the design specifications and thus meets the need.

From this description, it is clear that circuit analysis plays a very important role in the design process. Since circuit analysis is applied to circuit models, practicing engineers try to use mature circuit models so that the resulting designs will meet the design specifications in the first iteration. In this book, we use models that have been tested for between 20 and 100 years; you can assume that they are mature. The ability to model actual electrical systems with ideal circuit elements makes circuit theory extremely useful to engineers.

Saying that the interconnection of ideal circuit elements can be used to predict quantitatively the behavior of a system implies that we can describe the interconnection with mathematical equations. For the mathematical equations to be useful, we must write them in terms of measurable quantities. In the case of circuits, these quantities are voltage and current, which we discuss in Section 1.4. The study of circuit analysis involves (1) understanding the behavior of each ideal circuit element in terms of its voltage and current and (2) understanding the constraints imposed on the voltage and current as a result of interconnecting the ideal elements.

1.4 Voltage and Current

The concept of electric charge is the basis for describing all electrical phenomena. Let's review some important characteristics of electric charge. First, the charge is bipolar; that is, electrical effects are described in terms of positive and negative charges. Second, the electric charge exists in discrete quantities. Specifically, all quantities of charge are integral multiples of the electronic charge, 1.6022×10^{-19} C. Third, electrical effects are attributed to both the separation of charge and charges in motion. In circuit theory, the separation of charge creates an electric force (voltage), and the motion of charge creates an electric fluid (current).

The concepts of voltage and current are useful from an engineering point of view because they can be expressed quantitatively. Whenever positive and negative charges are separated, energy is expended. **Voltage** is the energy per unit charge that is created by the separation. We express this ratio in differential form as

$$v = \frac{dw}{dq}, \tag{1.1}$$

where

v = the voltage in volts,
w = the energy in joules, and
q = the charge in coulombs.

The electrical effects caused by charges in motion depend on the rate of charge flow. The rate of charge flow is known as the **electric current**, which is expressed as

$$i = \frac{dq}{dt}, \tag{1.2}$$

where

i = the current in amperes,
q = the charge in coulombs, and
t = the time in seconds.

Equations (1.1) and (1.2) are definitions for the magnitude of voltage and current, respectively. The bipolar nature of electric charge requires that we assign polarity references to these variables. We will do so in Section 1.5.

Although current is made up of discrete, moving electrons, we do not need to consider them individually because of the enormous number of them. Rather, we can think of electrons and their corresponding charge as one smoothly flowing entity. Thus, i is treated as a continuous variable.

One advantage of using circuit models is that we can model a component strictly in terms of the voltage and current at its terminals. Thus two physically different components could have the same relationship between the terminal voltage and terminal current. If they do, for purposes of circuit analysis, they are identical. Once we know how a component behaves at its terminals, we can analyze its behavior in a circuit. However, when developing circuit models, we are interested in a component's internal behavior. We might want to know, for example, whether charge conduction is taking place because of free electrons moving through the crystal lattice structure of a metal or whether it is because of electrons moving within the covalent bonds of a semiconductor material. However, these concerns are beyond the realm of circuit theory. In this book we use circuit models that have already been developed; we do not discuss how component models are developed.

1.5 The Ideal Basic Circuit Element

An **ideal basic circuit element** has three attributes: (1) it has only two terminals, which are points of connection to other circuit components; (2) it is described mathematically in terms of current and/or voltage; and (3) it cannot be subdivided into other elements. We use the word *ideal* to imply that a basic circuit element does not exist as a realizable physical component. However, as we discussed in Section 1.3, ideal elements can be connected in order to model actual devices and systems. We use the word *basic* to imply that the circuit element cannot be further reduced or subdivided into other elements. Thus the basic circuit elements form the building blocks for constructing circuit models, but they themselves cannot be modeled with any other type of element.

Figure 1.5 is a representation of an ideal basic circuit element. The box is blank because we are making no commitment at this time as to the type of circuit element it is. In Fig. 1.5, the voltage across the terminals of the box is denoted by v, and the current in the circuit element is denoted by i. The polarity reference for the voltage is indicated by the plus and minus signs, and the reference direction for the current is shown by the arrow placed alongside the current. The interpretation of these references given positive or negative numerical values of v and i is summarized in Table 1.4. Note that algebraically the notion of positive charge flowing in one direction is equivalent to the notion of negative charge flowing in the opposite direction.

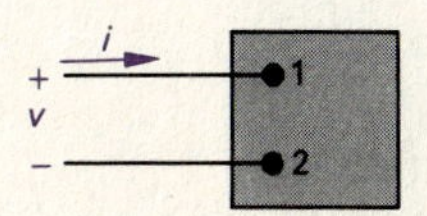

FIGURE 1.5 An ideal basic circuit element.

TABLE 1.4

INTERPRETATION OF REFERENCE DIRECTIONS IN FIG. 1.5

	POSITIVE VALUE	NEGATIVE VALUE
v	voltage drop from terminal 1 to terminal 2	voltage rise from terminal 1 to terminal 2
	or	*or*
	voltage rise from terminal 2 to terminal 1	voltage drop from terminal 2 to terminal 1
i	positive charge flowing from terminal 1 to terminal 2	positive charge flowing from terminal 2 to terminal 1
	or	*or*
	negative charge flowing from terminal 2 to terminal 1	negative charge flowing from terminal 1 to terminal 2

The assignments of the reference polarity for voltage and the reference direction for current are entirely arbitrary. However, once you have assigned the references, you must write all subsequent equations to agree with the chosen references. The most widely used sign convention applied to these references is called the **passive sign convention**, which we use throughout this book. The passive sign convention can be stated as follows:

> Whenever the reference direction for the current in an element is in the direction of the reference voltage drop across the element (as in Fig. 1.5), use a positive sign in any expression that relates the voltage to the current. Otherwise, use a negative sign.

We apply this sign convention in all the analyses that follow. Our purpose for introducing it even before we have introduced the different types of basic circuit elements is to impress on you the fact that the selection of polarity references along with the adoption of the passive sign convention is *not* a function of the basic elements nor the type of interconnections made with the basic elements. We present the application and interpretation of the passive sign convention in power calculations in Section 1.6.

DRILL EXERCISES

1.7 The current at the terminals of the element in Fig. 1.5 is

$$i = 0, \qquad t < 0;$$
$$i = 10e^{-2000t}\,\text{A}, \qquad t \geq 0.$$

Calculate the total charge (in microcoulombs) entering the element at its upper terminal.

ANSWER: 5000 μC.

1.8 The expression for the charge entering the upper terminal of Fig. 1.5 is

$$q = \frac{1}{\alpha^2} - \left(\frac{t}{\alpha} + \frac{1}{\alpha^2}\right)e^{-\alpha t} \text{ C}.$$

Find the maximum value of the current entering the terminal if $\alpha = 0.03679 \text{ s}^{-1}$.

ANSWER: 10 A.

1.6 POWER AND ENERGY

Power and energy calculations also are important in circuit analysis. One reason is that although voltage and current are useful variables in the analysis and design of electrically based systems, the useful output of the system often is nonelectrical, and this output is conveniently expressed in terms of power or energy. Another reason is that all practical devices have limitations on the amount of power that they can handle. In the design process, therefore, voltage and current calculations by themselves are not sufficient.

We now relate power and energy to voltage and current and at the same time use the power calculation to illustrate the passive sign convention. Recall from basic physics that power is the time rate of expending or absorbing energy. (A water pump rated 75 kW can deliver more liters per second than one rated 7.5 kW.) Mathematically, energy per unit time is expressed in the form of a derivative, or

$$p = \frac{dw}{dt}, \tag{1.3}$$

where

p = the power in watts,
w = the energy in joules, and
t = the time in seconds.

Thus 1 W is equivalent to 1 J/s.

The power associated with the flow of charge follows directly from the definition of voltage and current in Eqs. (1.1) and (1.2), or

$$p = \frac{dw}{dt} = \left(\frac{dw}{dq}\right)\left(\frac{dq}{dt}\right) = vi, \tag{1.4}$$

where

p = the power in watts,
v = the voltage in volts, and
i = the current in amperes.

Equation (1.4) shows that the **power** associated with a basic circuit element is simply the product of the current in the element and the voltage across the element. Therefore power is a quantity associated with a pair of terminals, and we have to be able to tell from our calculation whether power is being delivered to the pair of terminals or extracted from them. This information comes from the correct application and interpretation of the passive sign convention.

If we use the passive sign convention, Eq. (1.4) is correct if the reference direction for the current is in the direction of the reference voltage drop across the terminals. Otherwise, Eq. (1.4) must be written with a minus sign. In other words, if the current reference is in the direction of a reference voltage rise across the terminals, the expression for the power is

$$p = -vi. \tag{1.5}$$

The algebraic sign of power is based on charge movement through voltage drops and rises. As positive charges move through a drop in voltage, they lose energy, and as they move through a rise in voltage, they gain energy. Figure 1.6 summarizes the relationship between the polarity references for voltage and current and the expression for power.

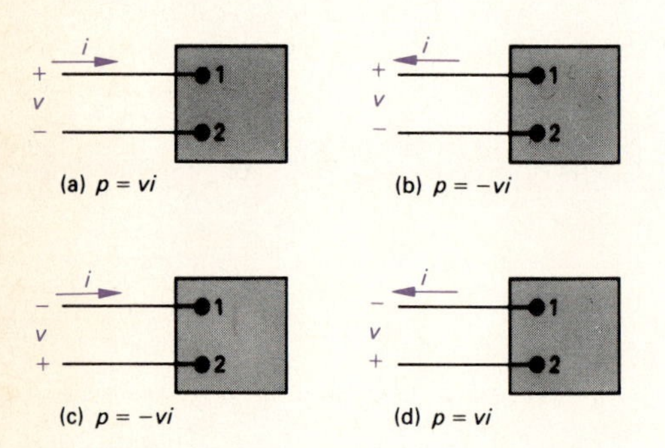

FIGURE 1.6 Polarity references and the expression for power.

We can now state the rule for interpreting the algebraic sign of power:

> If the power is positive (that is, if $p > 0$), power is being delivered to the circuit inside the box. If the power is negative (that is, if $p < 0$), power is being extracted from the circuit inside the box.

For example, suppose that we have selected the polarity references shown in Fig. 1.6(b). Assume further that our calculations for the current and voltage yield the following numerical results:

$$i = 4 \text{ A} \quad \text{and} \quad v = -10 \text{ V}.$$

Then the power associated with the terminal pair 1,2 is

$$p = -(-10)(4) = 40 \text{ W}.$$

Thus the circuit inside the box is absorbing 40 W.

To take this analysis one step further, assume that a colleague is solving the same problem but that she has chosen the reference polarities shown in Fig. 1.6(c). Her numerical values are

$$i = -4 \text{ A}, \quad v = 10 \text{ V}, \quad \text{and} \quad p = 40 \text{ W}.$$

Note that interpreting these results in terms of this reference system gives the same conclusions that we previously obtained—namely, that the circuit inside the box is absorbing 40 W. In fact, any of the reference systems in Fig. 1.6 yields this same result.

DRILL EXERCISES

1.9 Assume that a 15 V voltage drop occurs across an element from terminal 1 to terminal 2 and that a current of 5 A enters terminal 2.

a) Specify the values of v and i for the polarity references shown in Fig. 1.6(a) through (d).

b) State whether the circuit inside the box is absorbing or delivering power.

c) How much power is the circuit absorbing?

ANSWER: a) Circuit 1.6(a): $v = 15$ V, $i = -5$ A;
circuit 1.6(b): $v = 15$ V, $i = 5$ A;
circuit 1.6(c): $v = -15$ V, $i = -5$ A;
circuit 1.6(d): $v = -15$ V, $i = 5$ A;
b) delivering; c) -75 W.

1.10 Assume that the voltage at the terminals of the element in Fig. 1.5 corresponding to the current in Drill Exercise 1.7 is

$$v = 0, \quad t < 0;$$
$$v = 50e^{-2000t} \text{ V}, \quad t \geq 0.$$

Calculate the total energy (in millijoules) delivered to the circuit element.

ANSWER: 125 mJ.

1.11 A high-voltage direct-current transmission line between Celilo, Oregon, and Sylmar, California, is operating at 800 kV and carrying 1800 A, as shown. Calculate the power (in megawatts) at the Oregon end of the line and state the direction of power flow.

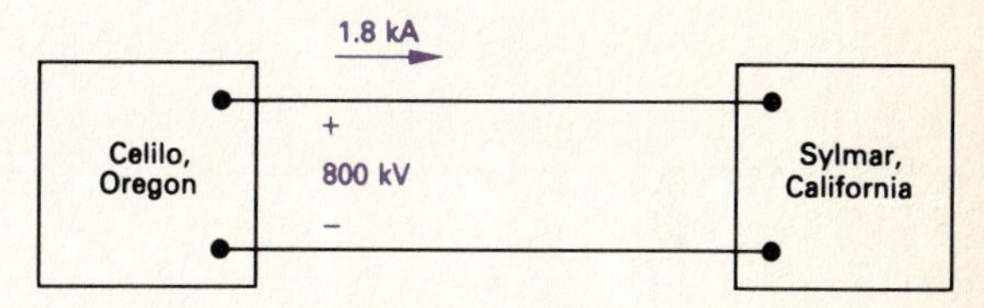

ANSWER: 1440 MW, Celilo to Sylmar.

SUMMARY

- The International System of Units (SI) enables engineers to communicate in a meaningful way about quantitative results. Table 1.1 summarizes the base SI units; Table 1.2 presents some useful derived SI units.
- Circuit analysis is based on the variables of voltage and current.
- **Voltage** is the energy per unit charge created by charge separation and has the SI unit of volt ($v = dw/dq$).
- **Current** is the rate of charge flow and has the SI unit of ampere ($i = dq/dt$).
- The **ideal basic circuit element** is a two-terminal component that cannot be subdivided; it can be described mathematically in terms of its terminal voltage and current.

- The **passive sign convention** uses a positive sign in the expression that relates the voltage and current at the terminals of an element when the reference direction for the current through the element is in the direction of the reference voltage drop across the element.
- **Power** is energy per unit of time and is equal to the product of the terminal voltage and current; it has the SI unit of watt ($p = dw/dt = vi$).
- The algebraic sign of power is interpreted as follows:
  - if $p > 0$, power is being delivered to the circuit or circuit component;
  - if $p < 0$, power is being extracted from the circuit or circuit component.

PROBLEMS

1.1 A penny is approximately 1.5 mm thick. At what average velocity does a stack of pennies have to grow in order to accumulate 421 billion dollars in one year?

1.2 Assume a telephone signal travels through a cable at one half the speed of light. If it is approximately 5 Mm across the United States, how long does it take the signal to cross the country?

1.3 Estimate the time it takes to generate 1 s of a film that uses computer-generated graphics if

a) high-resolution film recorders have a resolution of 1200×1600 picture elements (pixels) per frame;

b) each pixel requires 10 bits of data for each of the three primary colors—red, green, and blue;

c) it takes 10 floating-point calculations to determine the value of each color per pixel;

d) a motion picture runs 24 frames per second;

e) a supercomputer can perform 225 million floating-point calculations per second.

1.4 In electronic circuits it is not unusual to encounter currents in the microampere range. Assume a 5 μA current, due to the flow of electrons.

a) What is the average number of electrons per second that flow past a fixed reference cross section that is perpendicular to the direction of flow?

b) Compare the size of this number to the number of micrometers between Sydney, Australia, and San Francisco. You may assume the distance between Sydney and San Francisco is 7500 mi.

1.5 The current entering the upper terminal of Fig. 1.5 is $24 \cos 4000t$ A. Assume the charge at the upper terminal is zero at the instant the current is passing through its maximum value. Find the expression for $q(t)$.

1.6 How much energy is extracted from an electron as it flows through a 12 V battery from the negative to the positive terminal? Express your answer in attojoules.

1.7 A current of 1600 A exists in a rectangular (0.4-by-16 cm) copper bus bar. The current is due to free electrons moving through the bus bar at an average velocity of v meters/second. If the concentration of free electrons is 10^{29} electrons per cubic meter and if they are uniformly dispersed throughout the bus bar, then what is the average velocity of an electron?

1.8 The line described in Drill Exercise 1.11 is 845 mi in length. The line contains four conductors. Each conductor weighs 2526 lb per 1000 ft. How may megatons of conductor are in the line?

1.9 The references for the voltage and current at the terminal of a circuit element are as shown in Fig. 1.6(d). The numerical values for v and i are 40 V and -10 A.

a) Calculate the power at the terminals and state whether the power is being absorbed or delivered by the element in the box.

b) Given that the current is due to electron flow, state whether the electrons are entering or leaving terminal 2.

c) Do the electrons gain or lose energy as they pass through the element in the box?

1.10 Repeat Problem 1.9 with the current being 5 A.

1.11 Two electric circuits, represented by boxes A and B, are connected as shown in Fig. P1.11. The reference direction for the current i in the interconnection and the reference polarity for the voltage v across the interconnection are as shown in the figure. For each of the following sets of numerical values, calculate the power in the interconnection and state whether the power is flowing from A to B or vice versa.

a) $i = 25$ A, $v = 120$ V

b) $i = -5$ A, $v = 240$ V

c) $i = 1.5$ A, $v = -720$ V

d) $i = -20$ A, $v = -480$ V

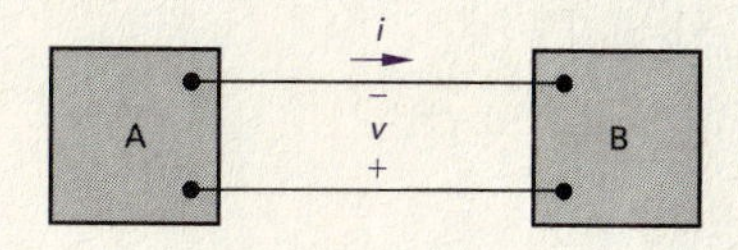

FIGURE P1.11

1.12 A 12 V battery supplies 100 mA to a radio. How much energy does the battery supply in 4 hours?

1.13 The voltage and current at the terminals of an automobile battery during a charge cycle are shown in Fig. P1.13.

a) Calculate the total charge transferred to the battery.

b) Calculate the total energy transferred to the battery.

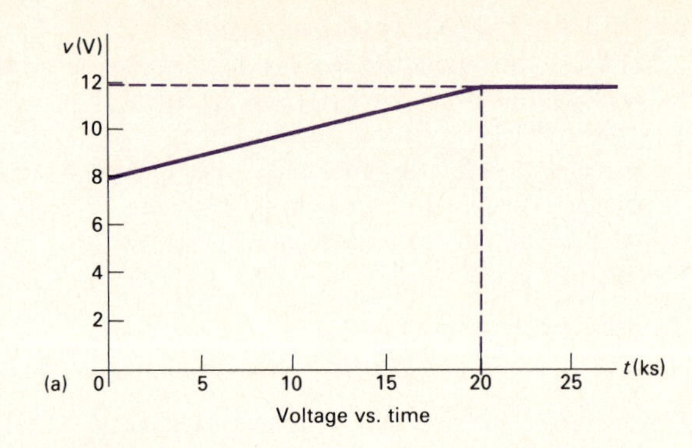

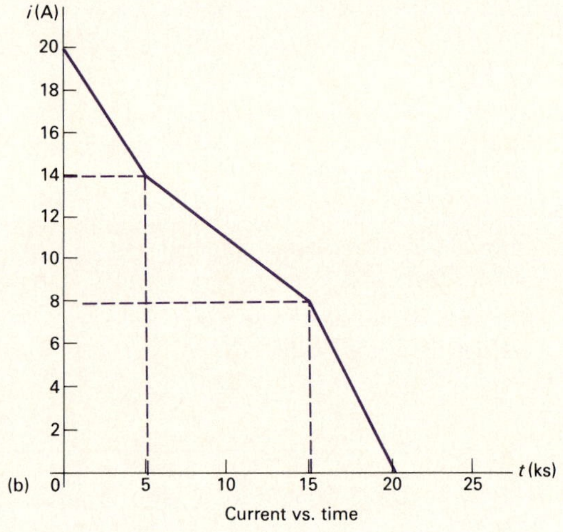

FIGURE P1.13

1.14 The voltage and current at the terminals of the circuit element in Fig. 1.5 are zero for $t < 0$. For $t \geq 0$ they are

$$v = 100e^{-500t} \text{ V}$$

and

$$i = 50 - 50e^{-500t} \text{ mA}.$$

Find the total energy delivered to the element.

1.15 The voltage and current at the terminals of the circuit element in Fig. 1.5 are zero for $t < 0$. For $t \geq 0$ they are

$$v = 80{,}000te^{-500t} \text{ V}, \quad t \geq 0;$$
$$i = 15te^{-500t} \text{ A}, \quad t \geq 0.$$

a) Find the time (in milliseconds) when the power delivered to the circuit element is maximum.

b) Find the maximum value of p in milliwatts.

c) Find the total energy delivered to the circuit element in microjoules.

1.16 The voltage and current at the terminals of the circuit element in Fig. 1.5 are zero for $t < 0$. For $t \geq 0$ they are

$$v = 20e^{-400t} - 20e^{-1600t} \text{ V}$$

and

$$i = 30 - 40e^{-400t} + 10e^{-1600t} \text{ mA.}$$

a) Find the power at $t = 625\ \mu$s.

b) How much energy is delivered to the circuit element between 0 and 625 μs?

c) Find the total energy delivered to the element.

1.17 The voltage and current at the terminals of the circuit element in Fig. 1.5 are zero for $t < 0$. For $t \geq 0$ they are

$$v = (16{,}000t + 20)e^{-800t} \text{ V}$$

and

$$i = (128t + 0.16)e^{-800t} \text{ A.}$$

a) At what instant of time is maximum power delivered to the element?

b) Find the maximum power in watts.

c) Find the total energy delivered to the element in millijoules.

1.18 The voltage and current at the terminals of the circuit element in Fig. 1.5 are shown in Fig. P1.18(a) and (b), respectively.

a) Sketch the power versus t plot for $0 \leq t \leq 50$ s.

b) Calculate the energy delivered to the circuit element at $t = 4$, 12, 36, and 50 s.

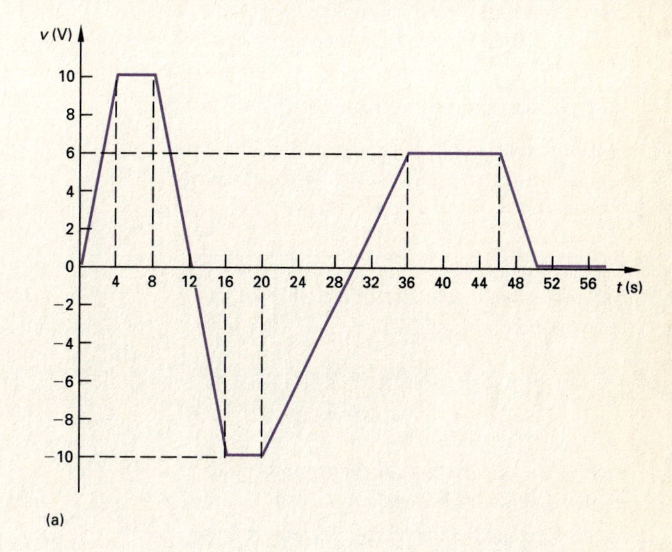

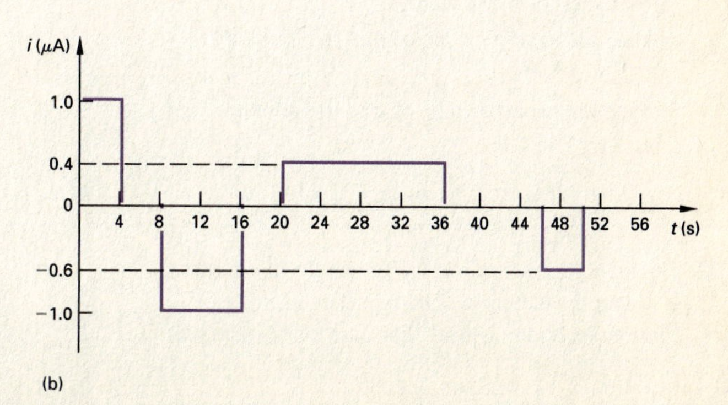

FIGURE P1.18

1.19 The voltage and current at the terminals of the circuit element in Fig. 1.5 are zero for $t < 0$ and $t > 40$ s. In the interval between 0 and 40 s the expressions are

$$v = t(1 - 0.025t) \text{ V}, \qquad 0 < t < 40 \text{ s};$$

$$i = 4 - 0.2t \text{ A}, \qquad 0 < t < 40 \text{ s}.$$

a) At what instant of time is the power being delivered to the circuit element maximum?

b) What is the power at the time found in part (a)?

c) At what instant of time is the power being extracted from the circuit element maximum?

d) What is the power at the time found in part (c)?

e) Calculate the net energy delivered to the circuit at 0, 10, 20, 30, and 40 s.

1.20 When a car has a dead battery it can often be started by connecting the battery from another car across its terminals. The positive terminals are connected together, as are the negative terminals. The connection is illustrated in Fig. P1.20. Assume the current i in Fig. P1.20 is measured and found to be -50 A.

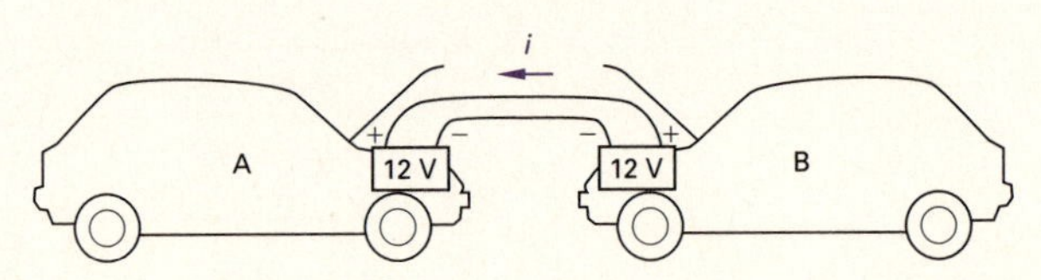

FIGURE P1.20

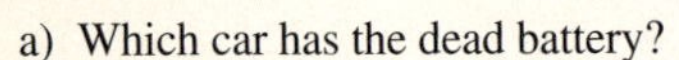

a) Which car has the dead battery?

b) If this connection is maintained for 2 min, how much energy is transferred to the dead battery?

1.21 The voltage and current at the terminals of the element in Fig. 1.5 are

$$v = 125 \cos 400\,\pi t \text{ V},$$
$$i = 32 \sin 400\,\pi t \text{ A}.$$

a) Find the maximum value of the power being delivered to the element.

b) Find the maximum value of the power being extracted from the element.

c) Find the average value of p in the interval $0 \le t \le 5$ ms.

d) Find the average value of p in the interval $0 \le t \le 6.25$ ms.

1.22 The voltage and current at the terminals of the circuit element in Fig. 1.5 are zero for $t < 0$. For $t \ge 0$ they are

$$v = 100e^{-50t} \sin 150t \text{ V}$$

and

$$i = 20e^{-50t} \sin 150t \text{ A}.$$

a) Find the power absorbed by the element at $t = 20$ ms.

b) Find the total energy (in millijoules) absorbed by the element.

1.23 The manufacturer of a 1.5 V D-cell flashlight battery says that the battery will deliver 9 mA for 40 continuous hours. During that time the voltage will drop from 1.5 V to 1.0 V. Assume the drop in voltage is linear with time. How much energy does the battery deliver in this 40 h interval?

1.24 a) In the circuit shown in Fig. P1.24, identify which elements have the voltage and current reference polarities defined using the passive sign convention.

b) The numerical values of the currents and voltages for each element are given in Table P1.24.

How much total power is absorbed and how much is delivered in this circuit?

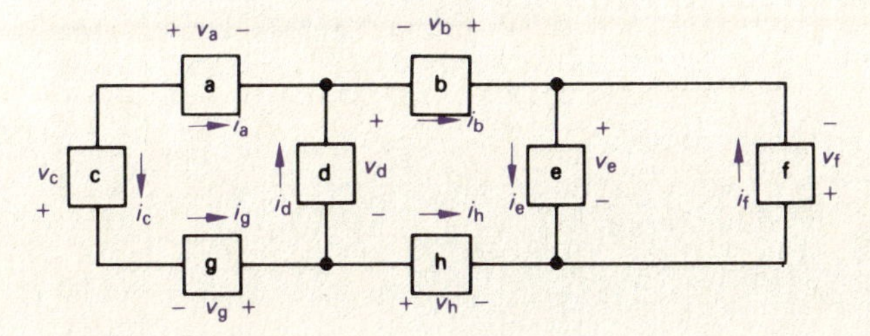

FIGURE P1.24

TABLE P1.24

ELEMENT	VOLTAGE (V)	CURRENT (A)
a	5	2
b	1	3
c	7	−2
d	−9	1
e	−20	5
f	20	2
g	−3	−2
h	−12	−3

1.25 One method of checking calculations involving interconnected circuit elements is to see that the total power delivered equals the total power absorbed (conservation-of-energy principle). With this thought in mind, check the interconnection in Fig. P1.25 and state whether it satisfies this power check. The current and voltage values for each element are given in Table P1.25.

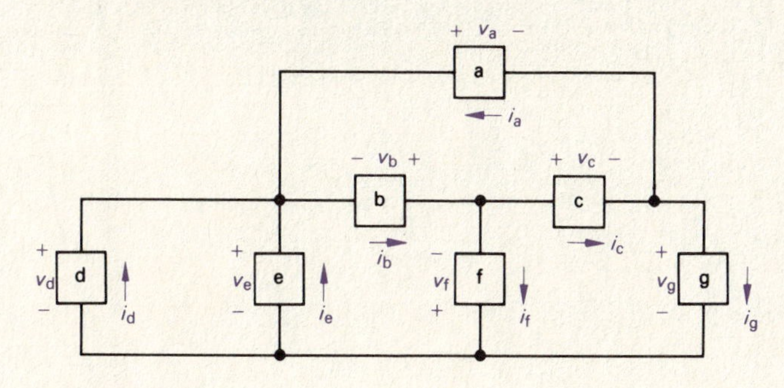

FIGURE P1.25

TABLE P1.25

ELEMENT	VOLTAGE (V)	CURRENT (A)
a	160	−10
b	−100	20
c	60	6
d	800	50
e	800	−20
f	−700	14
g	640	16

1.26 The numerical values of the voltages and currents in the interconnection seen in Fig. P1.26 are given in Table P1.26. Does the interconnection satisfy the power check?

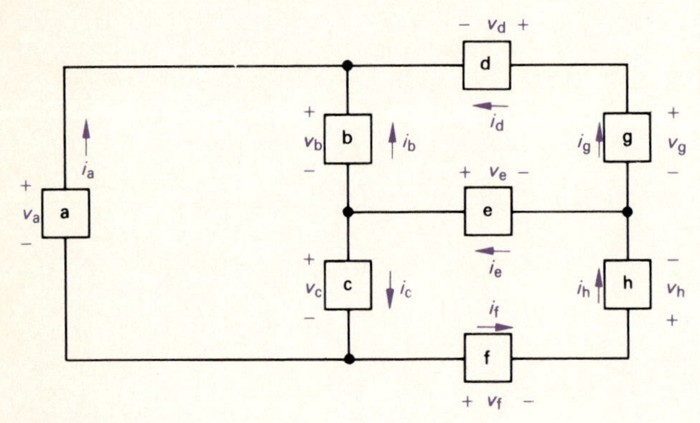

FIGURE P1.26

TABLE P1.26

ELEMENT	VOLTAGE (V)	CURRENT (A)
a	180	−45
b	120	−60
c	60	120
d	21	105
e	−24	60
f	33	165
g	117	105
h	−117	165

1.27 Assume you are an engineer in charge of a project and one of your subordinate engineers reports that the interconnection in Fig. P1.27 does not pass the power check. The data for the interconnection are given in Table P1.27.

a) Is the subordinate correct? Explain your answer.

b) If the subordinate is correct, can you find the error in the data?

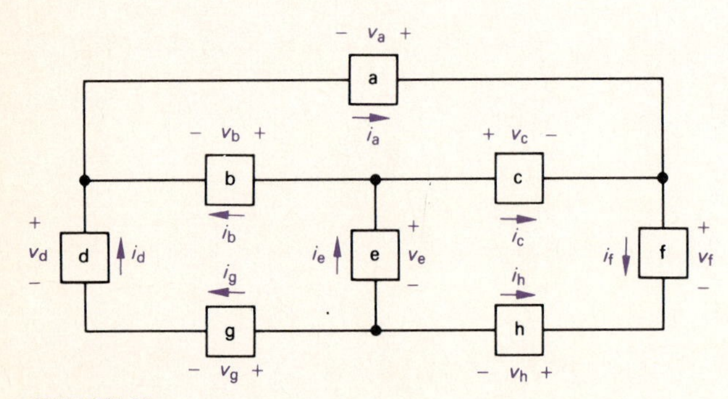

FIGURE P1.27

TABLE P1.27

ELEMENT	VOLTAGE (V)	CURRENT (A)
a	46.16	6.00
b	14.16	4.72
c	−32.0	−6.40
d	22.0	1.28
e	33.60	1.68
f	66.0	−0.40
g	2.56	1.28
h	−0.40	0.40

1.28 The numerical values for the currents and voltages in the circuit in Fig. P1.28 are given in Table P1.28. Find the total power developed in the circuit.

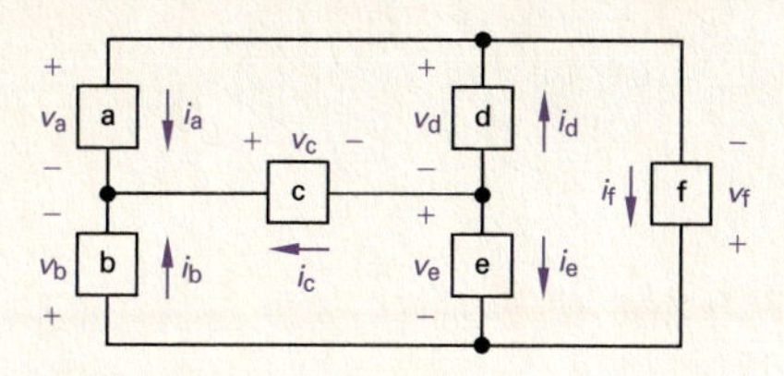

FIGURE P1.28

TABLE P1.28

ELEMENT	VOLTAGE (V)	CURRENT (A)
a	138	−51
b	−150	45
c	−18	6
d	120	−20
e	168	14
f	−288	31

PRACTICAL PERSPECTIVE

Electrical Safety

"Danger—High Voltage." This commonly seen warning is misleading. All forms of energy, including electrical energy, can be hazardous. But it's not only the voltage that harms. The static electricity shock you receive when you walk across a carpet and touch a doorknob is annoying but does not injure. Yet that spark is caused by a voltage hundreds or thousands of times larger than the voltages which can cause harm.

The thing that can actually cause injury is electrical current and how it flows through the body. Why, then, does the sign warn of high voltage? Because of the way electrical power is produced and distributed, it is easier to determine voltages than currents. Also, most electrical sources produce constant, specified voltages. So the signs warn about what is easy to measure. Determining whether and under what conditions a source can supply potentially dangerous currents, however, is more difficult, as this requires an understanding of electrical engineering.

Before we can examine this aspect of electrical safety, we have to learn how voltages and currents are produced and the relationship between them. The electrical behavior of objects, such as the human body, is quite complex and often beyond complete comprehension. To allow us to predict and control electrical phenomena, we use simplifying models in which simple mathematical relationships between voltage and current are used to approximate the actual relationships in real objects. Such models and analytical methods form the core of the electrical engineering techniques that will allow us to understand all electrical phenomena, including those relating to electrical safety.

At the end of this chapter we will use a simple electric circuit model to describe how and why people are injured by electric currents. While we may never develop a complete and accurate explanation of the electrical behavior of the human body, we can obtain a close approximation using simple circuit models to assess and improve the safety of electrical systems and devices. Developing models which provide an understanding that is imperfect but adequate for solving practical problems lies at the heart of engineering. Much of the art of electrical engineering, which you will learn with experience, is in knowing when and how to solve difficult problems by using simplifying models.

2 CIRCUIT ELEMENTS

CHAPTER CONTENTS

There are five ideal basic circuit elements:

1. voltage sources,
2. current sources,
3. resistors,
4. inductors, and
5. capacitors.

In this chapter we discuss the characteristics of voltage sources, current sources, and resistors. Although this may seem like a small number of elements with which to begin analyzing circuits, many practical systems can be modeled with just sources and resistors. They are also a useful starting point because of their relative simplicity; the mathematical relationships between voltage and current in sources and resistors are algebraic. Thus you will be able to begin learning the basic techniques of circuit analysis with only algebraic manipulations.

We will postpone introducing inductors and capacitors until Chapter 6, because their use requires that you solve integral and differential equations. However, the basic analytical techniques for solving circuits with inductors and capacitors are the same as those introduced in this chapter. So, by the time you need to begin manipulating more difficult equations, you should be very familiar with the methods of writing them.

2.1 Voltage and Current Sources

Before discussing ideal voltage and current sources, we need to consider the general nature of electrical sources. An **electrical source** is a device that is capable of converting nonelectric energy to electric energy and vice versa. A discharging battery converts chemical energy to electric energy, whereas a battery being charged converts electric energy to chemical energy. A dynamo is a machine that converts mechanical energy to electric energy and vice versa. If operating in the mechanical-to-electric mode, it is called a generator. If transforming from electric to mechanical energy, it is referred to as a motor. The important thing to remember about these sources is that they can either deliver or absorb electric power, generally maintaining either voltage or current. This behavior is of particular interest for circuit analysis and led to the creation of the ideal voltage source and the ideal current source as basic circuit elements. The challenge is to model practical sources in terms of the ideal basic circuit elements.

An **ideal voltage source** is a circuit element which maintains a prescribed voltage across its terminals regardless of the current flowing in those terminals. Similarly, an **ideal current source** is a circuit element which maintains a prescribed current through its terminals regardless of the voltage across those terminals. These circuit elements do not exist as practical devices—they are idealized models of actual voltage and current sources.

Using an ideal model for current and voltage sources places an important restriction on how we may describe them mathematically. Since an ideal voltage source provides the same voltage even if the current in the element changes, it is impossible to specify the current in an ideal voltage source as a function of its voltage. Likewise, if the only information you have about an ideal current source is the value of current supplied, it is impossible to determine the voltage across that current source. We have sacrificed our ability to relate voltage and current in a practical source for the simplicity of using ideal sources in circuit analysis.

Ideal voltage and current sources can be further described as either independent sources or dependent sources. An **independent source** establishes a voltage or current in a circuit without relying on voltages or currents elsewhere in the circuit. The value of the voltage or current supplied is specified by the value of the independent source alone. In contrast, a **dependent source** establishes a voltage or current whose value depends on the value of a voltage or current elsewhere in the circuit. You cannot specify the value of a dependent source unless you know the value of the voltage or current on which it depends.

The circuit symbols for the ideal independent sources are shown in Fig. 2.1. Note that a circle is used to represent an independent source. To completely specify an ideal independent voltage source in a circuit, you must include the value of the supplied voltage and the reference polarity, as shown in Fig. 2.1(a). Similarly, to completely specify an ideal independent current source, you must include the value of the supplied current and its reference direction, as shown in Fig. 2.1(b).

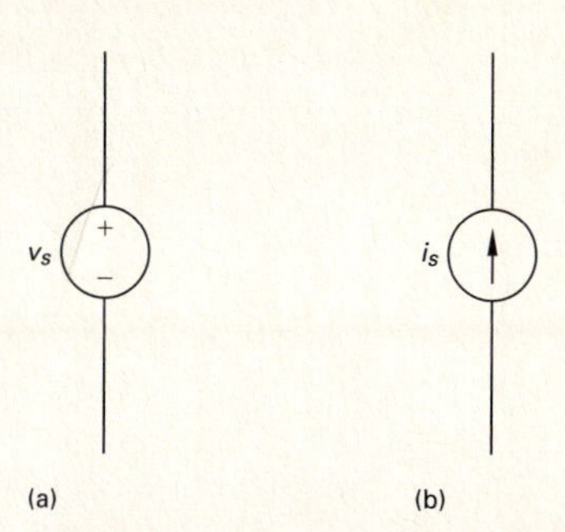

FIGURE 2.1 The circuit symbols for (a) an ideal independent voltage source and (b) an ideal independent current source.

The circuit symbols for the ideal dependent sources are shown in Fig. 2.2. A diamond is used to represent a dependent source. Both the dependent current source and the dependent voltage source may be controlled by either a voltage or a current elsewhere in the circuit, so there are a total of four variations, as indicated by the symbols in Fig. 2.2. Dependent sources are sometimes called **controlled sources**.

To completely specify an ideal dependent voltage-controlled voltage source, you must identify the controlling voltage, the equation which permits you to compute the supplied voltage from the controlling voltage, and the reference polarity for the supplied voltage. In Fig. 2.2(a), the controlling voltage is named v_x, the equation which determines the supplied voltage v_s is

$$v_s = \mu v_x,$$

and the reference polarity for v_s is as indicated. Note that μ is a multiplying constant which is dimensionless.

Similar requirements exist for completely specifying the other ideal dependent sources. In Fig. 2.2(b), the controlling current is i_x, the equation for the supplied voltage v_s is

$$v_s = \rho i_x,$$

the reference polarity is as shown, and the multiplying constant ρ has the dimension volts per ampere. In Fig. 2.2(c), the controlling voltage is v_x, the equation for the supplied current i_s is

$$i_s = \alpha v_x,$$

the reference direction is as shown, and the multiplying constant α has the dimension amperes per volt. In Fig. 2.2(d), the

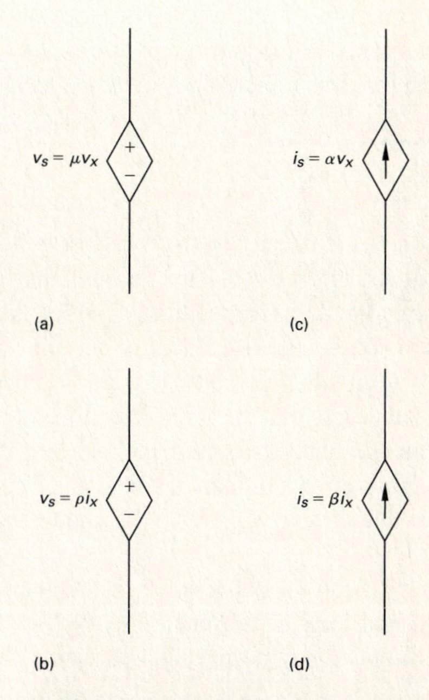

FIGURE 2.2 The circuit symbols for (a) an ideal dependent voltage-controlled voltage source, (b) an ideal dependent current-controlled voltage source, (c) an ideal dependent voltage-controlled current source, and (d) an ideal dependent current-controlled current source.

controlling current is i_x, the equation for the supplied current i_s is

$$i_s = \beta i_x,$$

the reference direction is as shown, and the multiplying constant β is dimensionless.

Finally, in our discussion of ideal sources, we note that they are examples of active circuit elements. An **active element** is one which models a device capable of generating electric energy. **Passive elements** model physical devices which cannot generate electric energy. Resistors, inductors, and capacitors are examples of passive circuit elements. Examples 2.1 and 2.2 illustrate how the characteristics of ideal independent and dependent sources limit the types of permissible interconnections of the sources.

EXAMPLE 2.1

Using the definitions of the ideal independent voltage and current sources, state which interconnections in Fig. 2.3 are permissible and which violate the constraints imposed by the ideal sources.

SOLUTION

Connection (a) is valid. Each source supplies voltage across the same pair of terminals, marked a,b. This requires that each source supply the same voltage with the same polarity, which they do.

Connection (b) is valid. Each source supplies current through the same pair of terminals, marked a,b. This requires that each source supply the same current in the same direction, which they do.

Connection (c) is not permissible. Each source supplies voltage across the same pair of terminals, marked a,b. This requires that each source supply the same voltage with the same polarity, which they do not.

Connection (d) is not permissible. Each source supplies current through the same pair of terminals, marked a,b. This requires that each source supply the same current in the same direction, which they do not.

Connection (e) is valid. The voltage source supplies voltage across the pair of terminals marked a,b. The current source supplies current through the same pair of terminals. Since an ideal voltage source supplies the same voltage regardless of the current, and an ideal current source supplies the same current regardless of the voltage, this is a permissible connection.

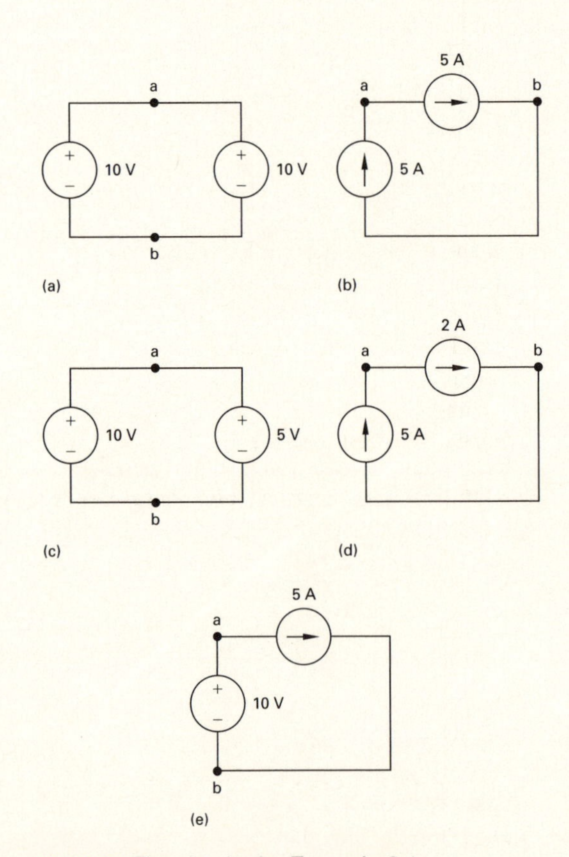

FIGURE 2.3 The circuits for Example 2.1.

EXAMPLE 2.2

Using the definitions of the ideal independent and dependent sources, state which interconnections in Fig. 2.4 are valid and which violate the constraints imposed by the ideal sources.

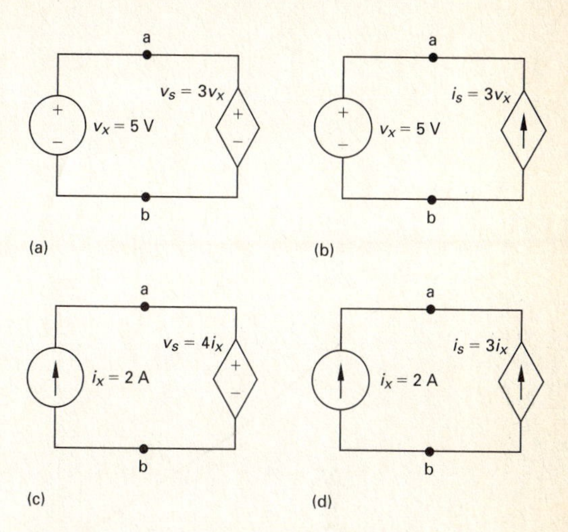

FIGURE 2.4 The circuits for Example 2.2.

SOLUTION

Connection (a) is invalid. Both the independent source and the dependent source supply voltage across the same pair of terminals, labeled a,b. This requires that each source supply the same voltage with the same polarity. The independent source supplies 5 V, but the dependent source supplies 15 V.

Connection (b) is valid. The independent voltage source supplies voltage across the pair of terminals marked a,b. The dependent current source supplies current through the same pair of terminals. Since an ideal voltage source supplies the same voltage regardless of current, and an ideal current source supplies the same current regardless of voltage, this is an allowable connection.

Connection (c) is valid. The independent current source supplies current through the pair of terminals marked a,b. The dependent voltage source supplies voltage across the same pair of terminals. Since an ideal current source supplies the same current regardless of voltage, and an ideal voltage source supplies the same voltage regardless of current, this is an allowable connection.

Connection (d) is invalid. Both the independent source and the dependent source supply current through the same pair of terminals, labeled a,b. This requires that each source supply the same current in the same reference direction. The independent source supplies 2 A, but the dependent source supplies −6 A.

2.2 ELECTRICAL RESISTANCE (OHM'S LAW)

Resistance is the capacity of materials to impede the flow of current or, more specifically, the flow of electric charge. The circuit element used to model this behavior is the **resistor**. Figure 2.5 shows the circuit symbol for the resistor, with R denoting the resistance value of the resistor.

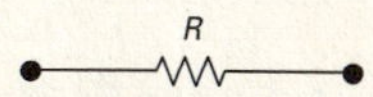

FIGURE 2.5 The circuit symbol for a resistor having a resistance R.

Conceptually, we can understand resistance if we think about the moving electrons which make up electric current interacting with and being resisted by the atomic structure of the material through which they are moving. In the course of these interactions, some amount of electric energy is converted to thermal energy and dissipated in the form of heat. This effect may be undesirable. However, many useful electrical devices take advantage of resistance heating, including stoves, toasters, irons, and space heaters.

Most materials exhibit measurable resistance to current. The amount of resistance depends upon the material. Metals such as copper and aluminum have small values of resistance, making them good choices for wiring used to conduct electric current. In fact, when represented in a circuit diagram, copper or aluminum wiring isn't usually modeled as a resistor; the resistance of the wire is so small compared to the resistance of other elements in the circuit that we can neglect the wiring resistance to simplify the diagram.

For purposes of circuit analysis, we must reference the current in the resistor to the terminal voltage. We can do so in two ways: either in the direction of the voltage drop across the resistor, or in the direction of the voltage rise across the resistor, as shown in Fig. 2.6. If we choose the former, the relationship between the voltage and current is

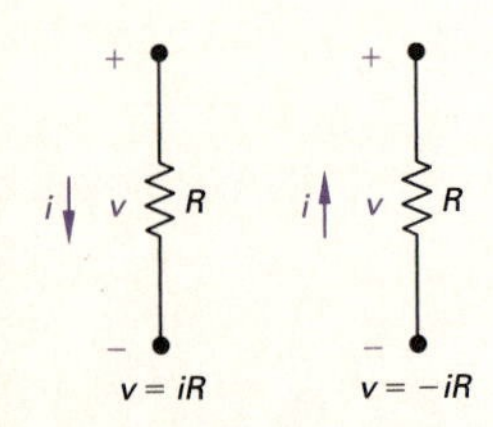

FIGURE 2.6 Two possible reference choices for the current and voltage at the terminals of a resistor, and the resulting equations.

$$v = iR, \tag{2.1}$$

where

$v =$ the voltage in volts,

$i =$ the current in amperes, and

$R =$ the resistance in ohms.

If we choose the second method, we must write

$$v = -iR, \tag{2.2}$$

where v, i, and R are, as before, measured in volts, amperes, and ohms, respectively. The algebraic signs used in Eqs. (2.1) and (2.2) are a direct consequence of the passive sign convention, which we introduced in Chapter 1. In both Eqs. (2.1) and (2.2), we assume that the resistance is positive.

Equations (2.1) and (2.2) are known as **Ohm's law** after Georg Simon Ohm, a German physicist who established its validity early in the nineteenth century. Ohm's law is the algebraic relationship between voltage and current for a resistor to which we referred in the introduction to this chapter. In SI units, resistance is measured in ohms. The Greek letter omega (Ω) is the standard symbol for an ohm. The circuit diagram symbol for an 8 Ω resistor is shown in Fig. 2.7.

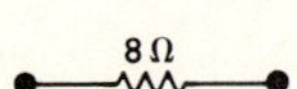

FIGURE 2.7 The circuit symbol for an 8 Ω resistor.

Ohm's law expresses the voltage as a function of the current. However, expressing the current as a function of the voltage also is convenient. Thus, from Eq. (2.1),

$$i = \frac{v}{R}, \tag{2.3}$$

or, from Eq. (2.2),

$$i = -\frac{v}{R}. \tag{2.4}$$

The reciprocal of the resistance is referred to as **conductance** is symbolized by the letter G, and is measured in siemens (S). Thus

$$G = \frac{1}{R}\ \text{S}. \tag{2.5}$$

An 8 Ω resistor has a conductance value of 0.125 S. In much of the professional literature, the unit used for conductance is the mho (ohm spelled backward), which is symbolized by an inverted omega (℧). Therefore we may also describe an 8 Ω resistor as having a conductance of 0.125 mho, or 0.125 ℧.

We use ideal resistors in circuit analysis to model the behavior of physical devices. Using the qualifier *ideal* reminds us that the resistor model makes several simplifying assumptions about the behavior of actual resistive devices. The most important of these simplifying assumptions is that the resistance of the ideal resistor is constant and its value does not vary over time. Most actual resistive devices do not have constant resistance, and their resistance does vary over time. The ideal resistor model can be used to represent a physical device whose resistance doesn't vary much from some constant value over the time period of interest in the circuit analysis. In this book we assume that the simplifying assumptions about resistance devices are valid, and we thus use ideal resistors in circuit analysis.

We may calculate the power at the terminals of a resistor in several ways. The first approach is to use the defining equation and simply calculate the product of the terminal voltage and current. For the reference systems shown in Fig. 2.6 we write

$$p = vi \tag{2.6}$$

when $v = iR$ and

$$p = -vi \tag{2.7}$$

when $v = -iR$.

A second method of expressing the power at the terminals of a resistor expresses power in terms of the current and the resistance. Substituting Eq. (2.1) into Eq. (2.6), we obtain

$$p = vi = (iR)i = i^2R. \tag{2.8}$$

Likewise, substituting Eq. (2.2) into Eq. (2.7), we have

$$p = -vi = -(-iR)i = i^2R. \tag{2.9}$$

Equations (2.8) and (2.9) are identical and demonstrate clearly that, regardless of voltage polarity and current direction, the power at the terminals of a resistor is positive. Therefore a resistor absorbs power from the circuit.

A third method of expressing the power at the terminals of a resistor is in terms of the voltage and resistance. The expression is independent of the polarity references, so

$$p = \frac{v^2}{R}. \tag{2.10}$$

Sometimes a resistor's value will be expressed as a conductance rather than as a resistance. Using the relationship between resistance and conductance given in Eq. (2.5), we may also write Eqs. (2.9) and (2.10) in terms of the conductance, or

$$p = \frac{i^2}{G} \tag{2.11}$$

and

$$p = v^2 G. \tag{2.12}$$

Equations (2.6)–(2.12) provide a variety of different methods for calculating the power absorbed by a resistor. Each yields the same answer. In analyzing a circuit, look at the information provided and choose the power equation which uses that information directly.

Example 2.3 illustrates the application of Ohm's law in conjunction with an ideal source and a resistor. Power calculations at the terminals of a resistor also are illustrated.

EXAMPLE 2.3

In each circuit in Fig. 2.8, either the value of v or i is not known.

a) Calculate the values of v and i.

b) Determine the power dissipated in each resistor.

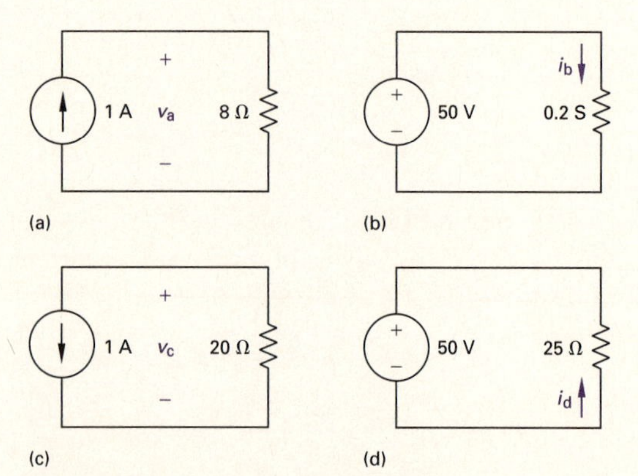

FIGURE 2.8 The circuits for Example 2.3.

SOLUTION

a) The voltage v_a in Fig. 2.8(a) is a drop in the direction of the current in the resistor. Therefore,

$$v_a = (1)(8) = 8 \text{ V}.$$

The current i_b in the resistor with a conductance of 0.2 S in Fig. 2.8(b) is in the direction of the voltage drop across the resistor. Thus

$$i_b = (50)(0.2) = 10 \text{ A}.$$

The voltage v_c in Fig. 2.8(c) is a rise in the direction of the current in the resistor. Hence

$$v_c = -(1)(20) = -20 \text{ V}.$$

The current i_d in the 25 Ω resistor in Fig. 2.8(d) is in the direction of the voltage rise across the resistor. Therefore

$$i_d = \frac{-50}{25} = -2 \text{ A}.$$

b) The power dissipated in each of the four resistors is

$$p_{8\Omega} = \frac{(8)^2}{8} = (1)^2(8) = 8 \text{ W};$$
$$p_{0.2S} = (50)^2(0.2) = 500 \text{ W};$$
$$p_{20\Omega} = \frac{(-20)^2}{20} = (1)^2(20) = 20 \text{ W};$$
$$p_{25\Omega} = \frac{(50)^2}{25} = (-2)^2(25) = 100 \text{ W}.$$

Having introduced the general characteristics of ideal sources and resistors, we next show how to use these elements to build the circuit model of a practical system.

2.3 Construction of a Circuit Model

We have already stated that one reason for an interest in the basic circuit elements is that they can be used to construct circuit models of practical systems. The skill required to develop a circuit model of a device or system is as complex as the skill required to solve the derived circuit. Although this text emphasizes the skills required to solve circuits, you also will need other skills in the practice of electrical engineering, and one of the most important is modeling.

We develop circuit models in the next two examples. In Example 2.4 we construct a circuit model based on a knowledge of the behavior of the system's components and how the components are interconnected. In Example 2.5 we create a circuit model by measuring the terminal behavior of a device.

EXAMPLE 2.4

Construct a circuit model of a flashlight.

SOLUTION

We chose the flashlight to illustrate a practical system because its components are so familiar. Figure 2.9(a) shows a photograph of a widely available flashlight. Figure 2.9(b) shows the disassembled flashlight's components.

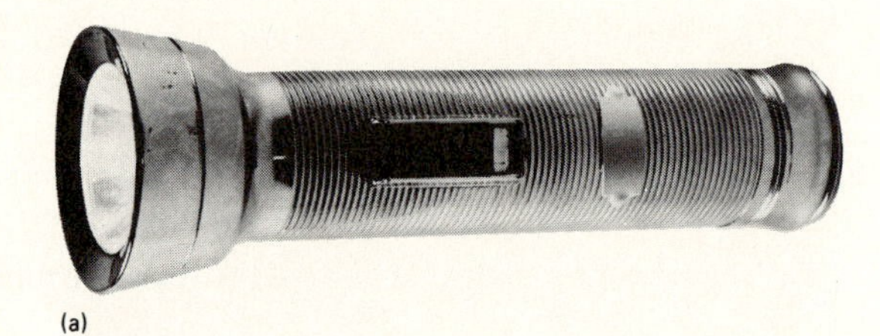

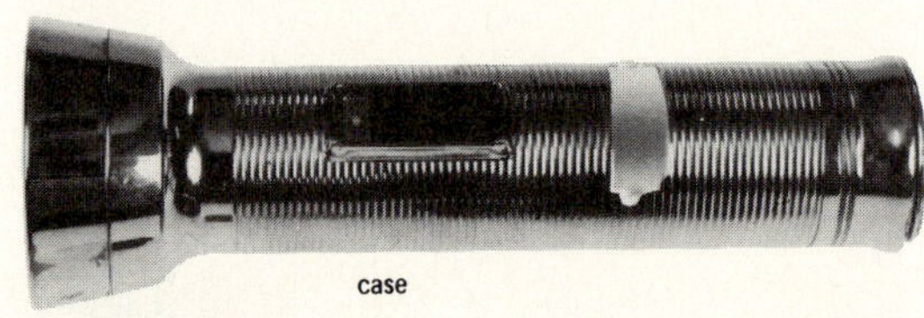

FIGURE 2.9 The flashlight viewed as an electrical system: (a) the flashlight; (b) the disassembled flashlight.

When a flashlight is regarded as an electrical system, the components of primary interest are (1) the batteries, (2) the lamp, (3) the connector, (4) the case, and (5) the switch. We now consider the circuit model for each component.

A dry-cell battery maintains a reasonably constant terminal voltage if the current demand is not excessive. Thus if the dry-cell battery is operating within its intended limits, we can model it with an ideal voltage source. The prescribed voltage then is constant and equal to the sum of two dry-cell values.

The ultimate output of the lamp is light energy, which is achieved by heating the filament in the lamp to a temperature high enough to cause radiation in the visible range. We can model the lamp with an ideal resistor. Note in this case that although the resistor accounts for the amount of electric energy converted to thermal energy, it does not predict how much of the thermal energy is converted to light energy. The resistor used to represent the lamp does predict the steady current drain on the batteries, a characteristic of the system that also is of interest. In this model, R_l symbolizes the lamp resistance.

The connector used in the flashlight serves a dual role. First, it provides an electrical conductive path between the dry cells and the case. Second, it is formed into a springy coil so that it also can apply mechanical pressure to the contact between the batteries and the lamp. The purpose of this mechanical pressure is to maintain contact between the two dry cells and between the dry cells and the lamp. Hence, in choosing the wire for the connector, we may find that its mechanical properties are more important than its electrical properties for the flashlight design. Electrically, we can model the connector with an ideal resistor, labeled R_1.

The case also serves both a mechanical and an electrical purpose. Mechanically, it contains all the other components and provides a grip for the person using it. Electrically, it provides a connection between other elements in the flashlight. If the case is metal, it conducts current between the batteries and the lamp. If it is plastic, a metal strip inside the case connects the coiled connector to the switch. Either way, an ideal resistor, which we denote R_c, models the electrical connection provided by the case.

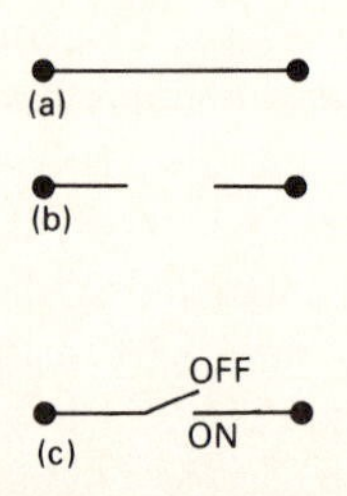

FIGURE 2.10 The circuit symbol for (a) a short circuit, (b) an open circuit, and (c) a switch.

The final component is the switch. Electrically, the switch is a two-state device. It is either ON or OFF. An ideal switch offers no resistance to the current when it is in the ON state, but it offers infinite resistance to current when it is in the OFF state. These two states represent the limiting values of a resistor—that is, the ON state corresponds to a resistor with a numerical value of zero, and the OFF state corresponds to a resistor with a numerical value of infinity. The two extreme values have the descriptive names **short circuit** ($R = 0$) and **open circuit** ($R = \infty$). Figures 2.10(a) and (b) show the graphical representation of a short circuit and an open circuit, respectively. The symbol shown in Fig. 2.10(c) represents the fact that a switch can be either a short circuit or an open circuit, depending on the position of its contacts.

We now construct the circuit model of the flashlight. Starting with the dry-cell batteries, the positive terminal of the first cell is connected to the negative terminal of the second cell, as shown in Fig. 2.11. The positive terminal of the second cell is connected to one terminal of the lamp. The other terminal of the lamp makes contact with one side of the switch, and the other side of the switch is connected to the metal case. The metal case is then connected to the negative terminal of the first dry cell by means of the metal spring. Note that the elements form a closed path or circuit. In Fig. 2.11 the dashed line depicts this closed path. Figure 2.12 shows a circuit model for the flashlight.

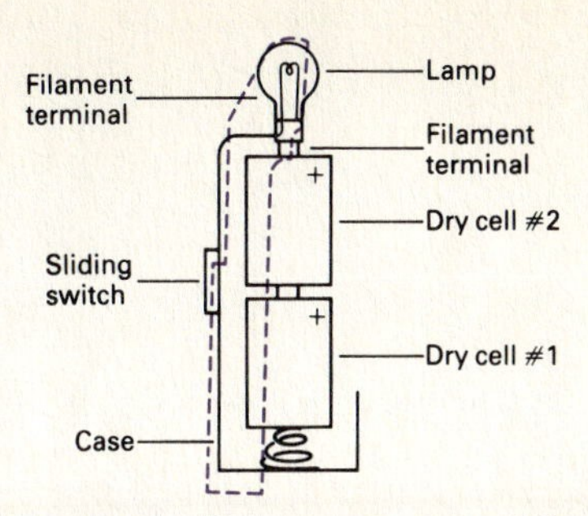

FIGURE 2.11 The arrangement of flashlight components.

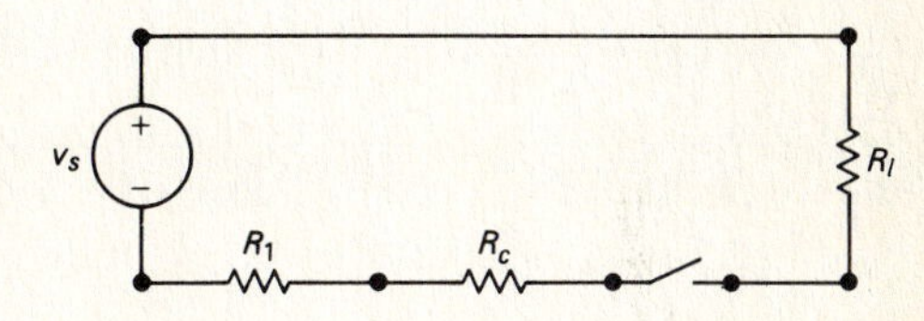

FIGURE 2.12 A circuit model for a flashlight.

We can make some general observations about modeling from our flashlight example: First, in developing a circuit model, the *electrical* behavior of each physical component is of primary interest. In the flashlight model, three very different physical components—a lamp, a coiled wire, and a metal case—are all represented by the same circuit element (a resistor) because the electrical phenomenon taking place in each is the same. Each is presenting resistance to the current flowing through the circuit.

Second, circuit models may need to account for undesired as well as desired electrical effects. For example, the heat resulting from the resistance in the lamp produces the light, a desired effect. However, the heat resulting from the resistance in the case and coil represents an unwanted or parasitic effect. It drains the dry cells and produces no useful output. Such parasitic effects must be considered or the resulting model may not adequately represent the system.

And finally, modeling requires approximation. Even for the basic system represented by the flashlight, we made simplifying assumptions in developing the circuit model. For example, we assumed an ideal switch, but in practical switches, contact resistance may be high enough to interfere with proper operation of the system. Our model does not predict this behavior. We also assumed that the coiled connector exerts enough pressure to eliminate any contact resistance between the dry cells. Our model does not predict the effect of inadequate pressure. Our use of an ideal voltage source ignores any internal dissipation of energy in the dry cells, which might be due to the parasitic heating just mentioned. We could account for this by adding an ideal resistor between the source and the lamp resistor. Our model assumes the internal loss to be negligible.

In modeling the flashlight as a circuit, we had a basic understanding of and access to the internal components of the system. However, sometimes we know only the terminal behavior of a device and must use this information in constructing the model. Example 2.5 explores such a modeling problem.

EXAMPLE 2.5

The voltage and current are measured at the terminals of the device illustrated in Fig. 2.13(a), and the values of v_t and i_t are tabulated in Fig. 2.13(b). Construct a circuit model of the device inside the box.

SOLUTION

Plotting the voltage as a function of the current yields the graph shown in Fig. 2.14. The equation of the line in this figure illustrates that the terminal voltage is directly proportional to the terminal current, $v_t = 4i_t$. In terms of Ohm's law, the device inside the box behaves like a 4 Ω resistor. Therefore, the circuit model for the device inside the box is a 4 Ω resistor.

We come back to this technique of using terminal characteristics to construct a circuit model after introducing Kirchhoff's laws and circuit analysis. (See Example 2.9, Drill Exercises 2.4–2.5, and Problems 2.16–2.19).

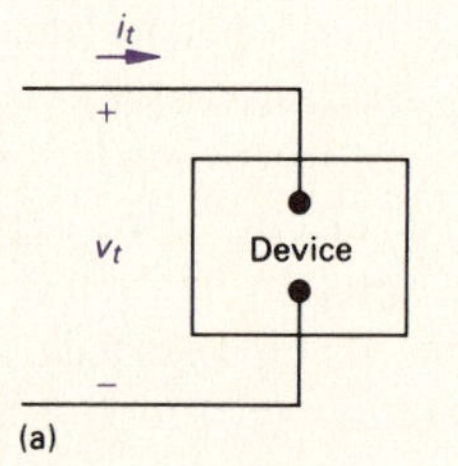

v_t(V)	i_t(A)
−40	−10
−20	−5
0	0
20	5
40	10

FIGURE 2.13 The device (a) and data (b) for Example 2.5.

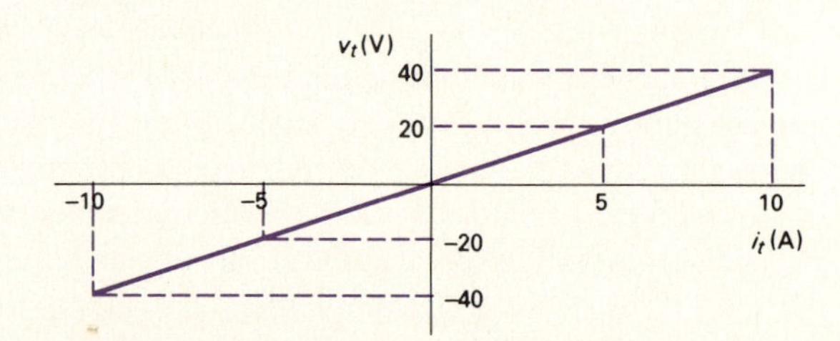

FIGURE 2.14 The values of v_t versus i_t for the device in Fig. 2.13.

2.4 KIRCHHOFF'S LAWS

A circuit is said to be solved when the voltage across and the current in every element have been determined. Ohm's law is an important equation for deriving such solutions. However, Ohm's law may not be enough to provide a complete solution. As we shall see in trying to solve the flashlight circuit from Example 2.4, we need to use two more important algebraic relationships, known as Kirchhoff's laws, to solve most circuits.

We begin by redrawing the circuit as shown in Fig. 2.15, with the switch in the ON state. Note that we have also labeled the current and voltage variables associated with each resistor, and the current associated with the voltage source. Labeling includes reference polarities, as always. For convenience, we attach the same subscript to the voltage and current labels as we do to the resistor labels. In Fig. 2.15, we have also removed some of the terminal dots of Fig. 2.12 and inserted nodes. Terminal dots are the start and end points of an individual circuit element. A **node** is a point where two or more circuit elements meet. It is necessary to identify nodes in order to use Kirchhoff's current law, as we will see in a moment. In Fig. 2.15, the nodes are labeled a, b, c, and d. Node d connects the battery and the lamp and in essence stretches all the way across the top of the diagram, though we label a single point for convenience. The dots on either side of the switch indicate its terminals, but only one is needed to represent a node, so only one is labeled c.

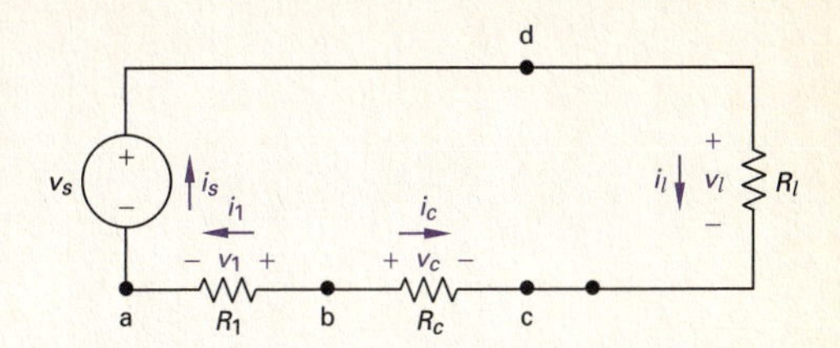

FIGURE 2.15 Circuit model of the flashlight with assigned voltage and current variables.

For the circuit shown in Fig. 2.15, we can identify seven unknowns: $i_s, i_1, i_c, i_l, v_1, v_c$, and v_l. Recall that v_s is a known voltage, as it represents the sum of the terminal voltages of the two dry cells—a constant voltage of 3 V. The problem is to find the seven unknown variables. From algebra, you know that to find n unknown quantities you must solve n simultaneous independent equations. From our discussion of Ohm's law in Section 2.2, you know that three of the necessary equations are

$$v_1 = i_1 R_1; \quad (2.13)$$

$$v_c = i_c R_c; \quad (2.14)$$

$$v_l = i_l R_l. \quad (2.15)$$

What about the other four equations?

The interconnection of circuit elements imposes constraints on the relationship between the terminal voltages and currents. These constraints are referred to as Kirchhoff's laws, after Gustav Kirchhoff, who first stated them in a paper published in 1848. The two laws that state the constraints in mathematical form are known as Kirchhoff's current law and Kirchhoff's voltage law.

We can now state **Kirchhoff's current law**:

> The algebraic sum of all the currents at any node in a circuit equals zero.

To use Kirchhoff's current law, an algebraic sign corresponding to a reference direction must be assigned to every current at the node. Assigning a positive sign to a current leaving a node requires assigning a negative sign to a current entering a node. Conversely, giving a negative sign to a current leaving a node requires giving a positive sign to a current entering a node.

Applying Kirchhoff's current law to the four nodes in the circuit shown in Fig. 2.15—using the convention that currents leaving a node are considered positive—yields four equations:

$$\text{Node a} \qquad i_s - i_1 = 0; \tag{2.16}$$

$$\text{Node b} \qquad i_1 + i_c = 0; \tag{2.17}$$

$$\text{Node c} \qquad -i_c - i_l = 0; \tag{2.18}$$

$$\text{Node d} \qquad i_l - i_s = 0. \tag{2.19}$$

Note that Eqs. (2.16)–(2.19) are not an independent set because any one of the four can be derived from the other three. In any circuit with n nodes, $n - 1$ independent current equations can be derived from Kirchhoff's current law.[1] Let's disregard Eq. (2.19) so that we have six independent equations, namely, Eqs. (2.13)–(2.18). We need one more, which we can derive from Kirchhoff's voltage law.

Before we can state Kirchhoff's voltage law, we must define a **closed path**, or **loop**. Starting at an arbitrarily selected node, we trace a closed path in a circuit through selected basic circuit elements and return to the original node without passing through any intermediate node more than once. The circuit shown in Fig. 2.15 has only one closed path or loop. For example, choosing node a as the starting point and tracing the circuit clockwise, we form the closed path by moving through nodes d, c, b, and back to a. We can now state **Kirchhoff's voltage law**:

> The algebraic sum of all the voltages around any closed path in a circuit equals zero.

To use Kirchhoff's voltage law, we must assign an algebraic sign (reference direction) to each voltage in the loop. As we trace a closed path, a voltage will appear either as a rise or a drop in the tracing direction. Assigning a positive sign to a voltage rise requires assigning a negative sign to a voltage drop. Conversely, giving a negative sign to a voltage rise requires giving a positive sign to a voltage drop.

We now apply Kirchhoff's voltage law to the circuit shown in Fig. 2.15. We elect to trace the closed path clockwise, assigning a positive algebraic sign to voltage drops. Starting at node d leads to the expression

$$v_l - v_c + v_1 - v_s = 0, \tag{2.20}$$

Matrix equation solvers can easily handle seven simultaneous equations.

which represents the seventh independent equation needed to find the seven unknown circuit variables mentioned earlier.

The thought of having to solve seven simultaneous equations to find the current delivered by a pair of dry cells to a flashlight lamp

[1] We say more about this observation in Chapter 4.

is not very appealing. Thus in the coming chapters we introduce you to analytical techniques that will enable you to solve a simple one-loop circuit by writing a single equation. However, before moving on to a discussion of these circuit techniques, we need to make several observations about the detailed analysis of the flashlight circuit. In general these observations are true and therefore are important to the discussions in subsequent chapters. They also support the contention that the flashlight circuit can be solved by defining a single unknown.

First, note that if you know the current in a resistor, you also know the voltage across the resistor, because current and voltage are directly related through Ohm's law. Thus you can associate one unknown variable with each resistor, either the current or the voltage. Choose, say, the current as the unknown variable. Then, once you solve for the unknown current in the resistor, you can find the voltage across the resistor. In general, if you know the current in a passive element, you can find the voltage across it, greatly reducing the number of simultaneous equations to be solved. For example, in the flashlight circuit, we eliminate the voltages v_c, v_l, and v_1 as unknowns. Thus at the outset we reduce the analytical task to solving four simultaneous equations rather than seven.

The second general observation relates to the consequences of connecting only two elements to form a node. According to Kirchhoff's current law, when only two elements connect to a node, if you know the current in one of the elements, you also know it in the second element. In other words, you need define only one unknown current for the two elements. When just two elements connect at a single node, the elements are said to be **in series**. The importance of this second observation is obvious when you note that each node in the circuit shown in Fig. 2.15 involves only two elements. Thus you need to define only one unknown current. The reason is that Eqs. (2.16)–(2.18) lead directly to

$$i_s = i_1 = -i_c = i_l, \tag{2.21}$$

which states that if you know any one of the element currents, you know them all. For example, choosing to use i_s as the unknown eliminates i_1, i_c, and i_l. The problem is reduced to determining one unknown, namely, i_s.

Examples 2.6 and 2.7 illustrate how to write circuit equations based on Kirchhoff's laws. Example 2.8 illustrates how to use Kirchhoff's laws and Ohm's law to find an unknown current.

EXAMPLE 2.6

Sum the currents at each node in the circuit shown in Fig. 2.16. Note that there is no connection dot (•) in the center of the diagram, where the 4 Ω branch crosses the branch containing the ideal current source i_a.

SOLUTION

In writing the equations, we use a positive sign for a current leaving a node. The four equations are

$$\begin{aligned}
&\text{node a} & i_1 + i_4 - i_2 - i_5 &= 0;\\
&\text{node b} & i_2 + i_3 - i_1 - i_b - i_a &= 0;\\
&\text{node c} & i_b - i_3 - i_4 - i_c &= 0;\\
&\text{node d} & i_5 + i_a + i_c &= 0.
\end{aligned}$$

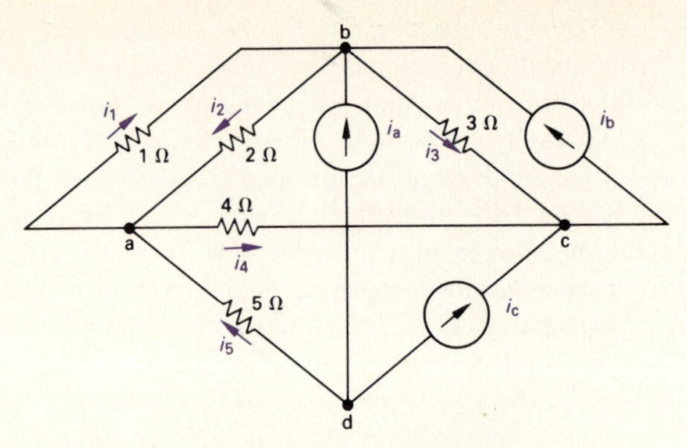

FIGURE 2.16 The circuit for Example 2.6.

EXAMPLE 2.7

Sum the voltages around each designated path in the circuit shown in Fig. 2.17.

SOLUTION

In writing the equations, we use a positive sign for a voltage drop. The four equations are

$$\begin{aligned}
&\text{path a} & -v_1 + v_2 + v_4 - v_b - v_3 &= 0;\\
&\text{path b} & -v_a + v_3 + v_5 &= 0;\\
&\text{path c} & v_b - v_4 - v_c - v_6 - v_5 &= 0;\\
&\text{path d} & -v_a - v_1 + v_2 - v_c + v_7 - v_d &= 0.
\end{aligned}$$

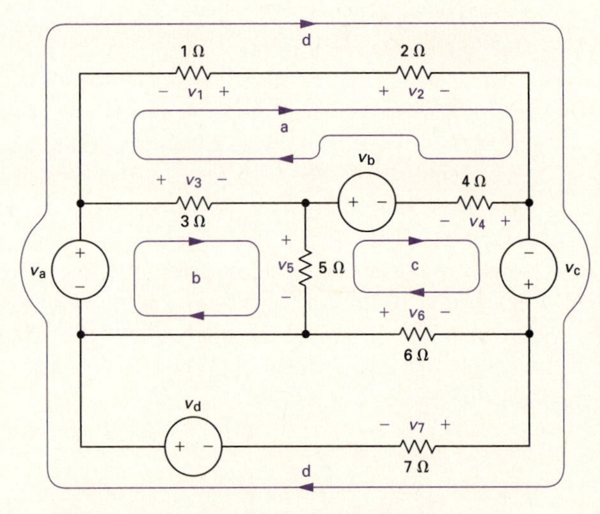

FIGURE 2.17 The circuit for Example 2.7.

EXAMPLE 2.8

a) Use Kirchhoff's laws and Ohm's law to find i_o in the circuit shown in Fig. 2.18.

b) Test the solution for i_o by verifying that the total power generated equals the total power dissipated.

SOLUTION

a) We begin by redrawing the circuit and assigning an unknown current to the 50 Ω resistor and unknown voltages across the 10 Ω and 50 Ω resistors. Figure 2.19 shows the circuit. The nodes are labeled a, b, and c to aid the discussion.

Because i_o also is the current in the 120 V source, we have two unknown currents and therefore must derive two simultaneous equations involving i_o and i_1. We obtain one of the equations by applying Kirchhoff's current law to either node b or c. Summing the currents at node b and assigning a positive sign to the currents leaving the node gives

$$i_1 - i_o - 6 = 0.$$

We obtain the second equation from Kirchhoff's voltage law in combination with Ohm's law. Noting from Ohm's law that v_o is $10i_o$ and v_1 is $50i_1$, we sum the voltages around the closed path cabc to obtain

$$-120 + 10i_o + 50i_1 = 0.$$

In writing this equation, we assigned a positive sign to voltage drops in the clockwise direction. Solving these two equations for i_o and i_1 yields

$$i_o = -3 \text{ A} \qquad \text{and} \qquad i_1 = 3 \text{ A}.$$

b) The power dissipated in the 50 Ω resistor is

$$p_{50\Omega} = (3)^2 50 = 450 \text{ W}.$$

The power dissipated in the 10 Ω resistor is

$$p_{10\Omega} = (-3)^2(10) = 90 \text{ W}.$$

The power delivered to the 120 V source is

$$p_{120V} = -120i_o = -120(-3) = 360 \text{ W}.$$

The power delivered to the 6 A source is

$$p_{6A} = -v_1(6), \qquad \text{but} \qquad v_1 = 50i_1 = 150 \text{ V}.$$

Therefore

$$p_{6A} = -150(6) = -900 \text{ W}.$$

The 6 A source is delivering 900 W, and the 120 V source is absorbing 360 W. The total power absorbed is 360 + 450 + 90 = 900 W. Therefore the solution verifies that the power delivered equals the power absorbed.

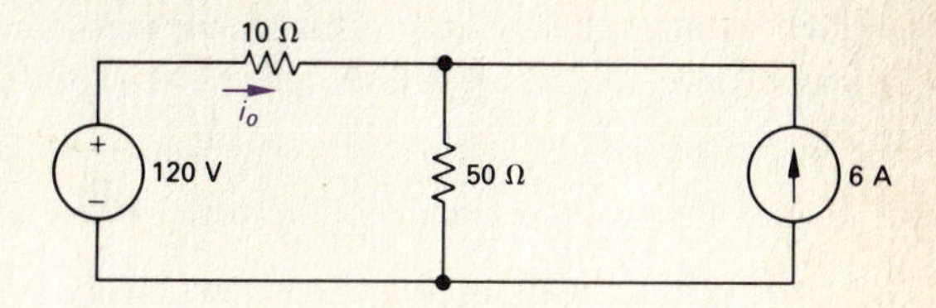

FIGURE 2.18 The circuit for Example 2.8.

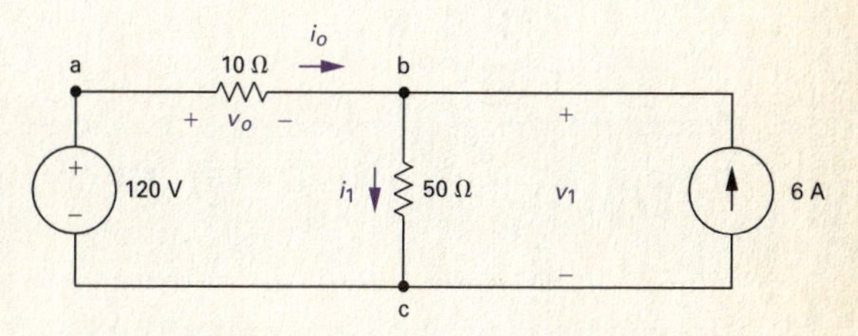

FIGURE 2.19 The circuit shown in Fig. 2.18, with the unknowns i_1, v_o, and v_1 defined.

EXAMPLE 2.9

The terminal voltage and terminal current were measured on the device shown in Fig. 2.20(a), and the values of v_t and i_t are tabulated in Fig. 2.20(b).

a) Construct a circuit model of the device inside the box.

b) Using this circuit model, predict the power this device will deliver to a 10 Ω resistor.

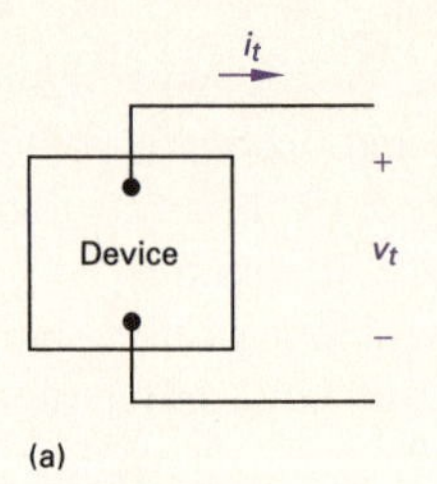

(a)

v_t (V)	i_t (A)
30	0
15	3
0	6

(b)

FIGURE 2.20 The device (a) and data (b) for Example 2.9.

SOLUTION

a) Plotting the voltage as a function of the current yields the graph shown in Fig. 2.21(a). The equation of the line plotted is

$$v_t = 30 - 5i_t.$$

Now we need to identify the components of a circuit model which will produce the same relationship between voltage and current. Kirchhoff's voltage law tells us that the voltage drops across two components in series add. From the equation, one of those components produces a 30 V drop regardless of the current. This component can be modeled as an ideal independent voltage source.

The other component produces a positive voltage drop in the direction of the current i_t. Since the voltage drop is proportional to the current, Ohm's law tells us that this component can be modeled as an ideal resistor with a value of 5 Ω. The resulting circuit model is depicted in the dashed box in Fig. 2.21(b).

Use a spreadsheet to plot the values.

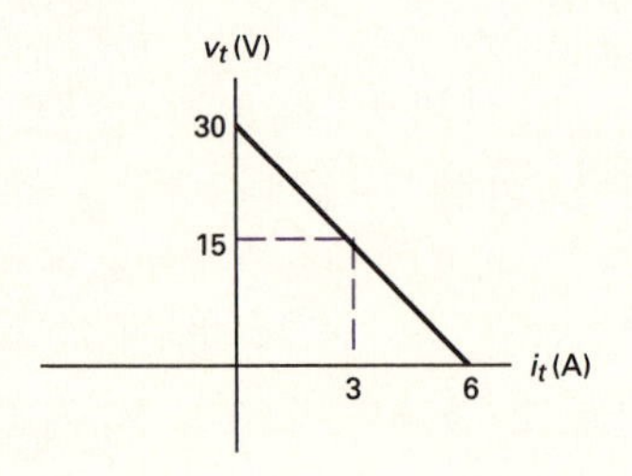

(a)

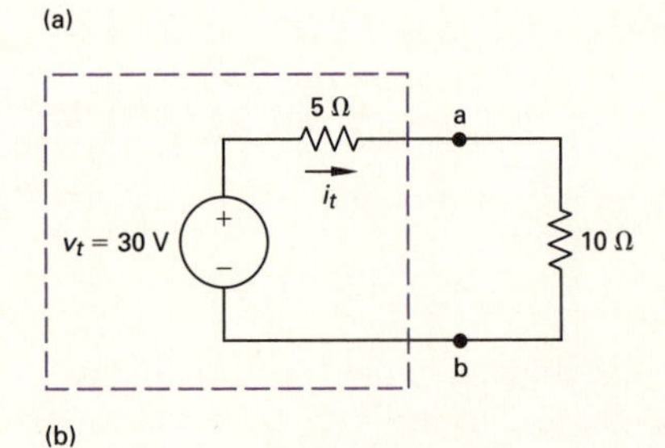

(b)

FIGURE 2.21 (a) The graph of v_t versus i_t for the device in Fig. 2.20(a); (b) the resulting circuit model for the device in Fig. 2.20(a), connected to a 10 Ω resistor.

b) Now we attach a 10 Ω resistor to the device in Fig. 2.21(b) to complete the circuit. Kirchhoff's current law tells us that the current in the 10 Ω resistor is the same as the current in the 5 Ω resistor. Using Kirchhoff's voltage law and Ohm's law, we can write the equation for the voltage drops around the circuit, starting at the voltage source and proceeding clockwise:

$$30 = 5i_t + 10i_t.$$

Solving for i_t, we get

$$i_t = 2 \text{ A}.$$

Since this is the value of current flowing in the 10 Ω resistor, we can use the power equation $p = i^2R$ to compute the power delivered to this resistor:

$$p_{10\Omega} = (2)^2(10) = 40 \text{ W}.$$

DRILL EXERCISES

2.1 a) Show that Eq. (2.20) can be written as

$$i_s R_l + i_s R_c + i_s R_1 - v_s = 0.$$

b) Write the explicit expression for i_s in terms of v_s, R_1, R_c, and R_l.

ANSWER: (a) derivation; (b) $i_s = v_s/(R_l + R_c + R_1)$.

2.2 For the circuit shown, calculate (a) i_5; (b) v_1; (c) v_2; (d) v_5; and (e) the power delivered by the 24 V source.

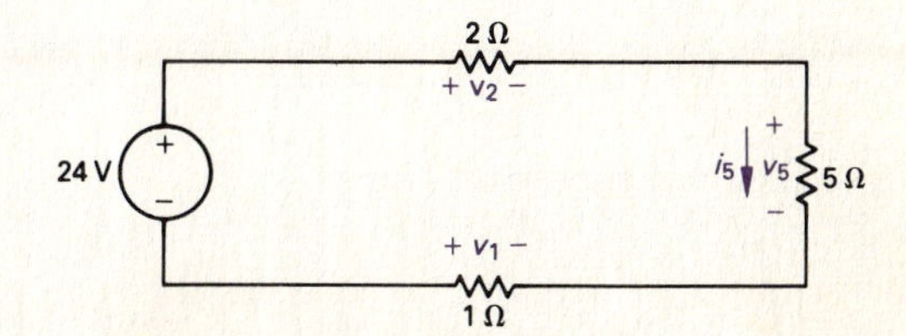

ANSWER: (a) $i_5 = 3$ A; (b) $v_1 = -3$ V; (c) $v_2 = 6$ V; (d) $v_5 = 15$ V; (e) 72 W.

2.3 Use Ohm's law and Kirchhoff's law to find the value of R in the circuit shown.

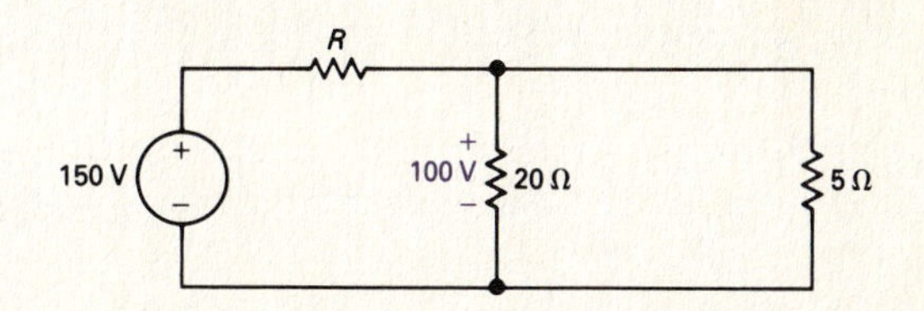

ANSWER: $R = 2\ \Omega$.

2.4 a) The terminal voltage and terminal current were measured on the device shown. The values of v_t and i_t are provided in the table. Using these values, create the straight line plot of v_t versus i_t. Compute the equation of the line and use the equation to construct a circuit model for the device using an ideal voltage source and a resistor.

b) Use the model constructed in (a) to predict the power that the device will deliver to a 25 Ω resistor.

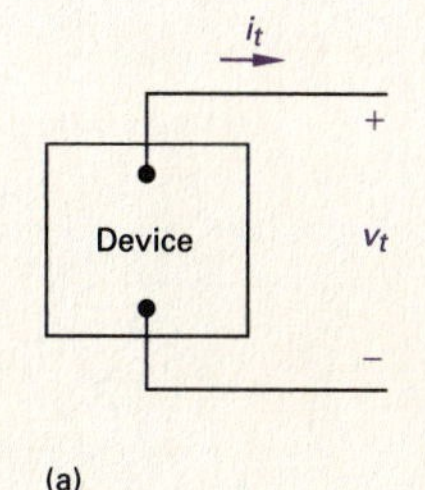

(a)

v_t (V)	i_t (A)
25	0
15	0.1
5	0.2
0	0.25

(b)

ANSWER: (a) A 25 V source in series with a 100 Ω resistor; (b) 1 W.

2.5 Repeat Drill Exercise 2.4 but use the equation of the graphed line to construct a circuit model containing an ideal current source and a resistor.

ANSWER: (a) A 1/4 A current source connected between the terminals of a 100 Ω resistor; (b) 1 W.

2.5 Analysis of a Circuit Containing a Dependent Source

We conclude this introduction to elementary circuit analysis with a discussion of a circuit that contains a dependent source, as depicted in Fig. 2.22.

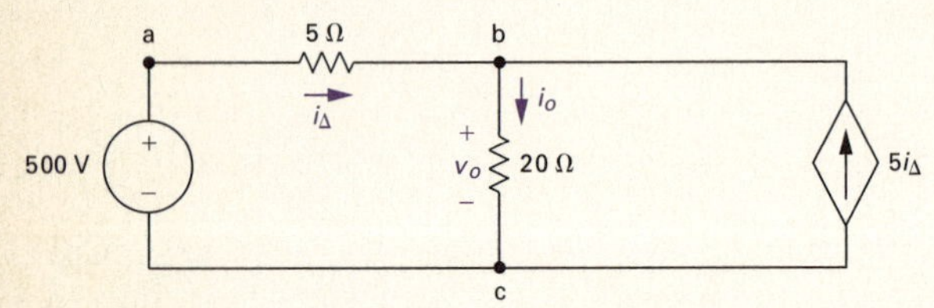

FIGURE 2.22 A circuit with a dependent source.

We want to use Kirchhoff's laws and Ohm's law to find v_o in this circuit. Before writing equations, it is good practice to examine the circuit diagram closely. This will help us identify the information that is known and the information we must calculate. It may also help us devise a strategy for solving the circuit which results in only a few calculations.

A look at the circuit in Fig. 2.22 reveals that

- once we know i_o, we can calculate v_o using Ohm's law;
- once we know i_Δ, we also know the current supplied by the dependent source $5i_\Delta$;
- the current in the 500 V source is i_Δ.

There are thus two unknown currents, i_Δ and i_o. We need to construct and solve two independent equations involving these two currents to produce a value for v_o.

From the circuit, notice the closed path containing the voltage source, the 5 Ω resistor, and the 20 Ω resistor. We can apply Kirchhoff's voltage law around this closed path. The resulting equation contains the two unknown currents:

$$500 = 5i_\Delta + 20i_o. \tag{2.22}$$

Now we need to generate a second equation containing these two currents. Consider the closed path formed by the 20 Ω resistor and the dependent current source. If we attempt to apply Kirchhoff's voltage law to this loop, we fail to develop a useful equation, because we don't know the value of the voltage across the dependent current source. In fact, the voltage across the dependent source is v_o, which is the voltage we are trying to compute. Writing an equation for this loop does not advance us toward a solution. For this same reason, we do not use the closed path containing the voltage source, the 5 Ω resistor, and the dependent source.

There are three nodes in the circuit, so we turn to Kirchhoff's current law to generate the second equation. Node a connects the voltage source and the 5 Ω resistor; as we have already observed, the current in these two elements is the same. Either node b or node c can be used to construct the second equation from Kirchhoff's current law. We select node b and produce the following equation:

$$i_o = i_\Delta + 5i_\Delta = 6i_\Delta. \tag{2.23}$$

Solving Eqs. (2.22) and (2.23) for the currents, we get

$$i_\Delta = 4 \text{ A}$$

and

$$i_o = 24 \text{ A}. \tag{2.24}$$

Using Eq. (2.24) and Ohm's law for the 20 Ω resistor, we can solve for the voltage v_o:

$$v_o = 20i_o = 480 \text{ V}.$$

Think about a circuit analysis strategy before beginning to write equations. As we have demonstrated, not every closed path provides an opportunity to write a useful equation based on Kirchhoff's voltage law. Not every node provides for a useful application of Kirchhoff's current law. Some preliminary thinking about the problem can help in selecting the most fruitful approach and the most useful analysis tools for a particular problem. Choosing a good approach and the appropriate tools will usually reduce the number and complexity of equations to be solved.

Example 2.10 illustrates another application of Ohm's law and Kirchhoff's laws to a circuit with a dependent source. Example 2.11 involves a much more complicated circuit, but with a careful choice of analysis tools, the analysis is relatively uncomplicated.

EXAMPLE 2.10

a) Use Kirchhoff's laws and Ohm's law to find the voltage v_o as shown in Fig. 2.23.

b) Show that your solution is consistent with the constraint that the total power developed in the circuit equals the total power dissipated.

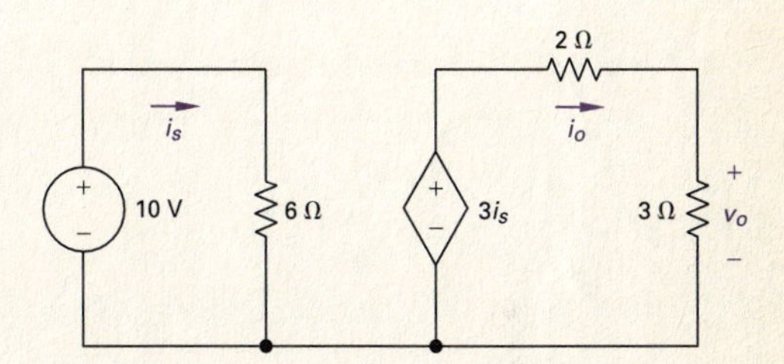

FIGURE 2.23 The circuit for Example 2.10.

SOLUTION

a) A close look at the circuit in Fig. 2.23 reveals that

- there are two closed paths, the one on the left with the current i_s, and the one on the right with the current i_o;

- once i_o is known, we can compute v_o.

We need two equations for the two currents. Since there are two closed paths and both have voltage sources, we can apply Kirchhoff's voltage law to each to give the following equations:

$$10 = 6i_s$$

and

$$3i_s = 2i_o + 3i_o.$$

Solving for the currents yields

$$i_s = 1.67 \text{ A}$$

and

$$i_o = 1 \text{ A}.$$

Applying Ohm's law to the 3 Ω resistor gives the desired voltage:

$$v_o = 3i_o = 3 \text{ V}.$$

b) To compute the power delivered to the voltage sources, we use the power equation in the form $p = vi$. The power delivered to the independent voltage source is

$$p = (10)(-1.67) = -16.7 \text{ W}.$$

The power delivered to the dependent voltage source is

$$p = (3i_s)(-i_o) = (5)(-1) = -5 \text{ W}.$$

Both sources are developing power, and the total developed power is 21.7 W.

To compute the power delivered to the resistors, we use the power equation in the form $p = i^2R$. The power delivered to the 6 Ω resistor is

$$p = (1.67)^2(6) = 16.7 \text{ W}.$$

The power delivered to the 2 Ω resistor is

$$p = (1)^2(2) = 2 \text{ W}.$$

The power delivered to the 3 Ω resistor is

$$p = (1)^2(3) = 3 \text{ W}.$$

The resistors all dissipate power, and the total power dissipated is 21.7 W, equal to the total power developed in the sources.

EXAMPLE 2.11

The circuit in Fig. 2.24 represents a common configuration encountered in the analysis and design of transistor amplifiers. Assume that the values of all the circuit elements—R_1, R_2, R_C, R_E, V_{CC}, and V_0,—are known.

a) Develop the equations needed to determine the current in each element of this circuit.

b) From these equations, devise a formula for computing i_B in terms of the circuit element values.

A symbolic equation solver can generate such a formula.

SOLUTION

A careful examination of the circuit reveals a total of six unknown currents, designated i_1, i_2, i_B, i_C, i_E, and i_{CC}. In defining these six unknown currents, we used the observation that the resistor R_C is in series with the dependent current source βi_B. We now must derive six independent equations involving these six unknowns.

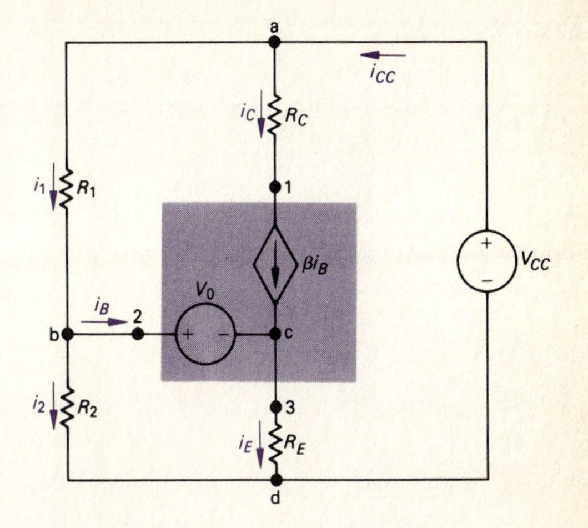

FIGURE 2.24 The circuit for Example 2.11.

a) We can derive three equations by applying Kirchhoff's current law to any three of the nodes a, b, c, and d. Let's use a, b, and c and label the currents away from the nodes as positive:

$$(1) \quad i_1 + i_C - i_{CC} = 0;$$
$$(2) \quad i_B + i_2 - i_1 = 0;$$
$$(3) \quad i_E - i_B - i_C = 0.$$

A fourth equation results from imposing the constraint presented by the series connection of R_C and the dependent source:

$$(4) \quad i_C = \beta i_B.$$

We turn to Kirchhoff's voltage law in deriving the remaining two equations. We need to select two closed paths in order to use Kirchhoff's voltage law. Note that the voltage across the dependent current source is unknown and that it cannot be determined from the source current βi_B. Therefore, we must select two closed paths which do not contain this dependent current source.

We choose the paths bcdb and badb and specify voltage drops as positive to yield

$$(5) \quad V_0 + i_E R_E - i_2 R_2 = 0;$$
$$(6) \quad -i_1 R_1 + V_{CC} - i_2 R_2 = 0.$$

b) To get a single equation for i_B in terms of the known circuit variables, you can follow these steps:

- Solve Eq. (6) for i_1, and substitute this solution for i_1 into Eq. (2).
- Solve the transformed Eq. (2) for i_2, and substitute this solution for i_2 into Eq. (5).
- Solve the transformed Eq. (5) for i_E, and substitute this solution for i_E into Eq. (3). Use Eq. (4) to eliminate i_C in Eq. (3).
- Solve the transformed Eq. (3) for i_B, and rearrange the terms to yield

$$i_B = \frac{(V_{CC}R_2)/(R_1 + R_2) - V_0}{(R_1R_2)/(R_1 + R_2) + (1 + \beta)R_E} \qquad \textbf{(2.25)}$$

Problem 2.23 asks you to verify these steps. Note that once we know i_B, we can easily obtain the remaining currents.

DRILL EXERCISES

2.6 For the circuit shown find (a) the current i_1 in microamperes and (b) the voltage v in volts.

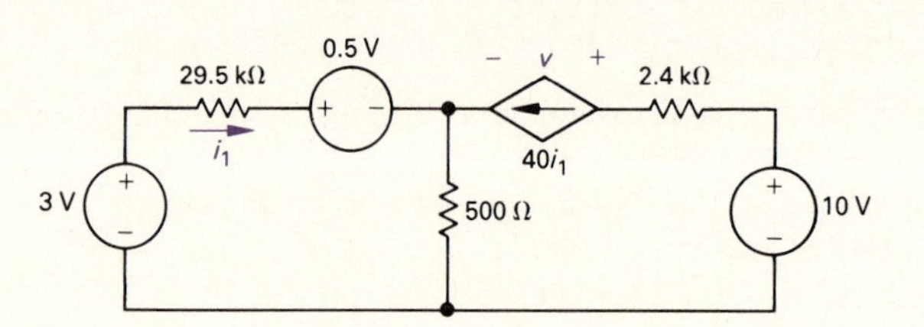

ANSWER: (a) $i_1 = 50\ \mu\text{A}$; (b) $v = 4.175$ V.

2.7 The current i_ϕ in the circuit shown is 5 A. Calculate

a) v_s;

b) the power absorbed by the independent voltage source;

c) the power delivered by the independent current source;

d) the power delivered by the controlled current source;

d) the total power dissipated in the two resistors.

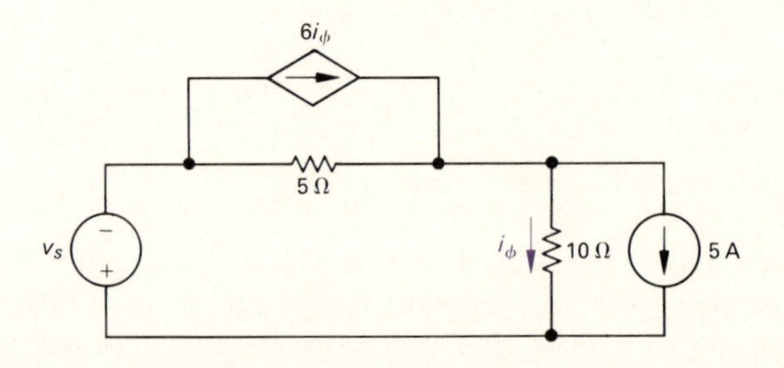

ANSWER: (a) 50 V; (b) 500 W; (c) −250 W; (d) 3000 W; (e) 2250 W.

PRACTICAL PERSPECTIVE

Electrical Safety

EXERCISE

At the beginning of this chapter, we said that current through the body can cause injury. Let's examine this aspect of electrical safety.

You might think that electrical injury is due to burns. However, that is not the case. The most common electrical injury is to the nervous system. Nerves use electrochemical signals, and electric currents can disrupt those signals. When the current path includes only skeletal muscles, the effects can include temporary paralysis (cessation of nervous signals) or involuntary muscle contractions, which are generally not life threatening. However, when the current path includes nerves and muscles which control the supply of oxygen to the brain, the problem is much more serious. Temporary paralysis of these muscles can stop a person from breathing, and a sudden muscle contraction can disrupt the signals that regulate heartbeat. The result is a halt in the flow of oxygenated blood to the brain, causing death in a few minutes unless emergency aid is given immediately. Table 2.1 shows a range of physiological reactions to various current levels. The numbers in

TABLE 2.1

PHYSIOLOGICAL REACTIONS TO CURRENT LEVELS IN HUMANS

PHYSIOLOGICAL REACTION	CURRENT
Barely perceptible	3–5 mA
Extreme pain	35–50 mA
Muscle paralysis	50–70 mA
Heart stoppage	500 mA

Note: Data taken from W. F. Cooper, *Electrical Safety Engineering*, 2d ed. (London: Butterworth, 1986); and C. D. Winburn, *Practical Electrical Safety* (Monticello, N.Y.: Marcel Dekker, 1988).

this table are approximate; they are obtained from an analysis of accidents because, obviously, it is not ethical to perform electrical experiments on people. Good electrical design will limit current to a few milliamperes or less under all possible conditions.

Now we develop a simplified electrical model of the human body. The body acts as a conductor of current, so a reasonable starting point is to model the body using resistors. Figure 2.25 shows a potentially dangerous situation. A voltage difference exists between one arm and one leg of a human being. Figure 2.25(b) shows an electrical model of the human body in Fig. 2.25(a). The arms, legs, neck, and trunk (chest and abdomen) each have a characteristic resistance. Note that the path of the current is through the trunk, which contains the heart, a potentially deadly arrangement.

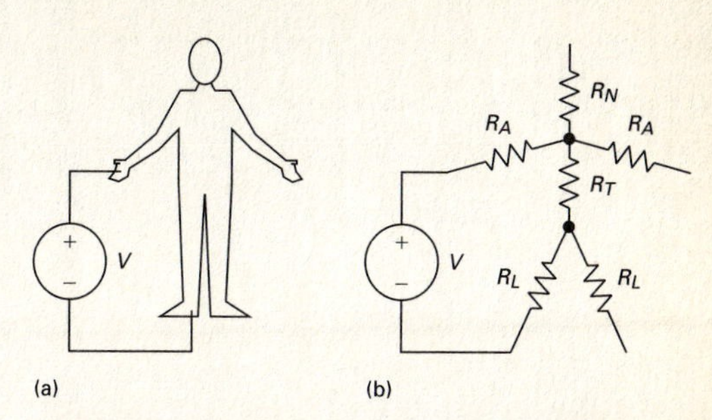

FIGURE 2.25 (a) A human body with a voltage difference between one arm and one leg; (b) a simplified model of the human body with a voltage difference between one arm and one leg.

Suppose the power company installs some equipment that could provide a 250 V shock to a human being. Is the current that results dangerous enough to warrant posting a warning sign and taking other precautions to prevent such a shock? Assume that if the source is 250 V, the resistance of the arm is 400 Ω, the resistance of the trunk is 50 Ω, and the resistance of the leg is 200 Ω.

SOLUTION

We can simplify our electrical model of the human body by noting that current will not flow through the neck or the other arm and leg. This model is shown in Fig. 2.26, with the given values of resistance included. Using Ohm's law and Kirchhoff's voltage law,

$$400i + 50i + 200i - 250 = 0.$$

Solving for the current,

$$i = \frac{250}{650} = 385 \text{ mA}.$$

Thus, the current through the region of the heart is 385 mA, nearly enough to stop the heart, according to Table 2.1. The power company must post a warning sign and take other measures to prevent someone from encountering the 250 V source.

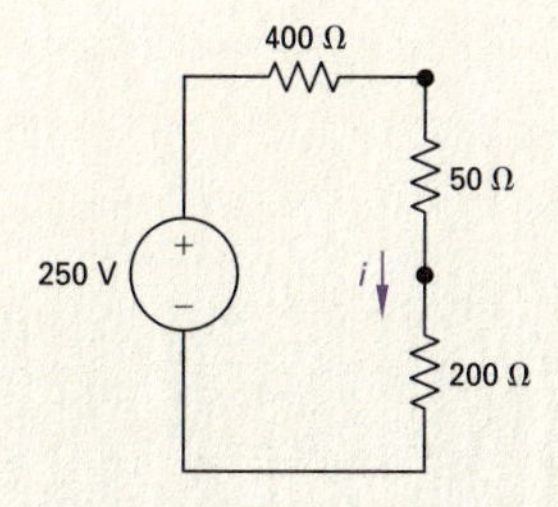

FIGURE 2.26 A model of the path of current through the human body for a 250 V shock between one arm and leg.

SUMMARY

- The three circuit elements introduced in this chapter are voltage sources, current sources, and resistors:

— An **ideal voltage source** maintains a prescribed voltage regardless of the current in the device. An **ideal current source** maintains a prescribed current regardless of the voltage across the device. Voltage and current sources are either **independent**, that is, not influenced by any other current or voltage in the circuit; or **dependent**, that is, determined by some other current or voltage in the circuit.

— A **resistor** constrains its voltage and current to be proportional to each other. The value of the proportional constant relating voltage and current in a resistor is called its **resistance** and is measured in ohms.

- **Ohm's law** establishes the proportionality of voltage and current in a resistor. Specifically,

$$v = iR \quad \textbf{(2.1)}$$

if the current flow in the resistor is in the direction of the voltage drop across it; or

$$v = -iR \quad \textbf{(2.2)}$$

if the current flow in the resistor is in the direction of the voltage rise across it.

- By combining the equation for power, $p = vi$, with Ohm's law, we can determine the power absorbed by a resistor:

$$p = i^2R. \quad \textbf{(2.8)}$$

- Circuits are described by nodes and closed paths. A **node** is a point where two or more circuit elements join. When just two elements connect to form a node, they are said to be **in series**. A **closed path** is a loop traced through connecting elements, starting and ending at the same node and encountering intermediate nodes only once each.

- The voltages and currents of interconnected circuit elements obey Kirchhoff's laws:

— **Kirchhoff's current law** states that the algebraic sum of all the currents at any node in a circuit equals zero.

— **Kirchhoff's voltage law** states that the algebraic sum of all the voltages around any closed path in a circuit equals zero.

- A circuit is solved when the voltage across and the current in every element have been determined. By combining an understanding of independent and dependent sources, Ohm's law, and Kirchhoff's laws, we can solve many simple circuits.

PROBLEMS

2.1 A pair of automotive headlamps are connected to a 12 V battery via the arrangements shown in Fig. P2.1. In the figure, the triangular symbol ▼ is used to indicate that the terminal is connected directly to the metal frame of the car.

a) Construct a circuit model using resistors and an independent voltage source.

b) Identify the correspondence between the ideal circuit element and the symbol component that it represents.

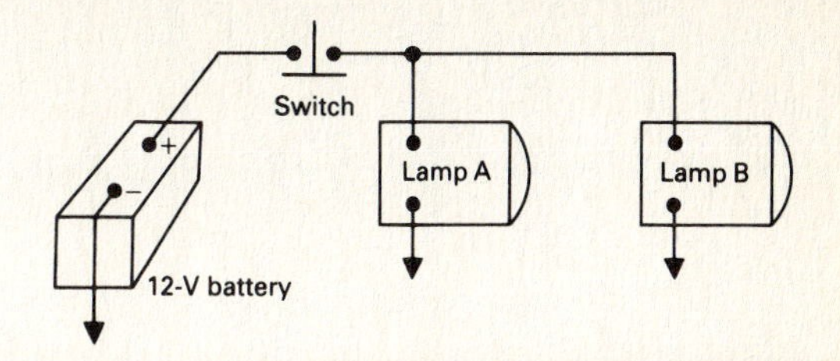

FIGURE P2.1

2.2 A simplified circuit model for a residential wiring system is shown in Fig. P2.2.

a) How many basic circuit elements are there in this model?

b) How many nodes are there in the circuit?

c) How many of the nodes connect three or more basic elements?

d) Identify the circuit elements that form a series pair.

e) What is the minimum number of unknown currents?

f) Describe seven closed paths in the circuit.

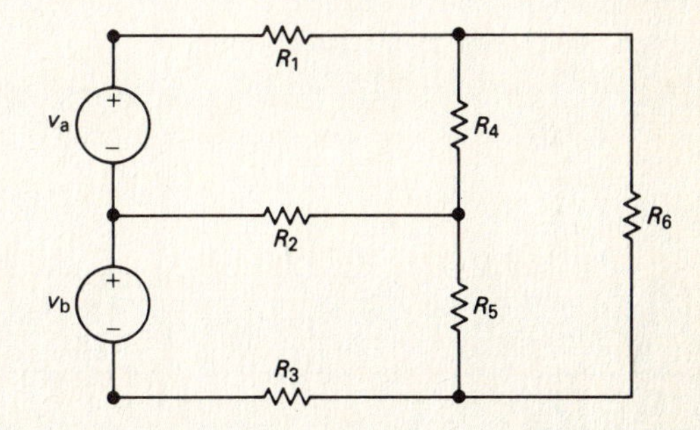

FIGURE P2.2

2.3 The current i_o in the circuit shown in Fig. P2.3 is 14 A. Find (a) i_a; (b) i_g; and (c) the power delivered by the independent current source.

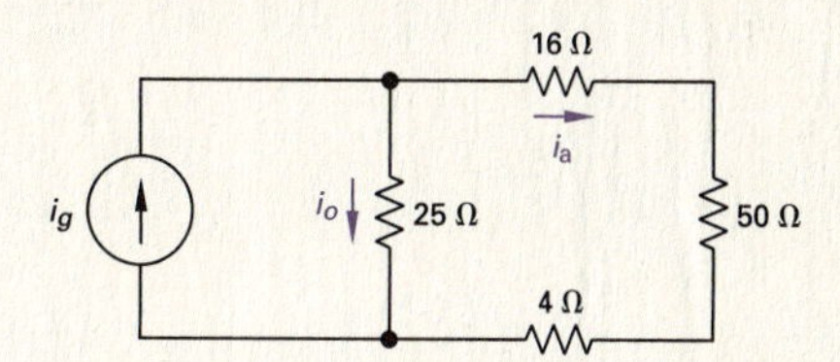

FIGURE P2.3

2.4 a) Is the interconnection of ideal sources in the circuit in Fig. P2.4 valid? Explain.

b) Identify which sources are developing power and which sources are absorbing power.

c) Verify that the total power developed in the circuit equals the total power absorbed.

d) Repeat parts (a) through (c), reversing the polarity of the 30 V source.

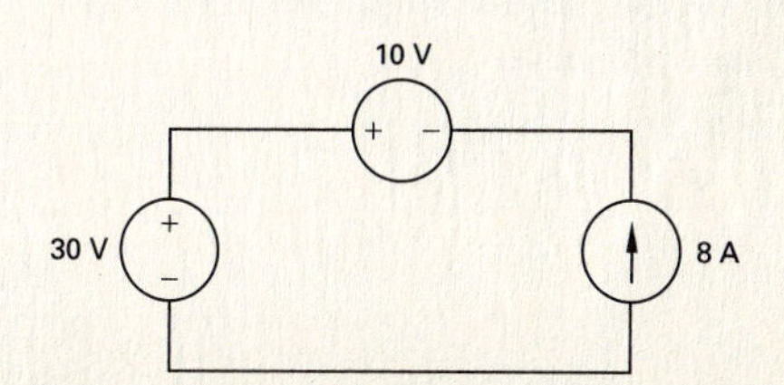

FIGURE P2.4

2.5 Is the interconnection in Fig. P2.5 valid? Explain.

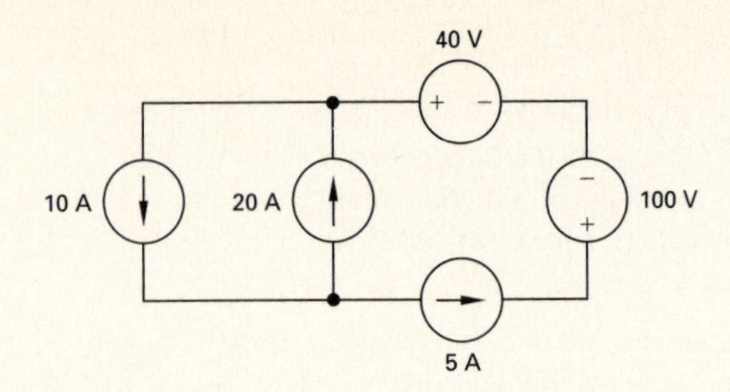

FIGURE P2.5

2.6 If the interconnection in Fig. P2.6 is valid, find the total power developed in the circuit. If the interconnection is not valid, explain why.

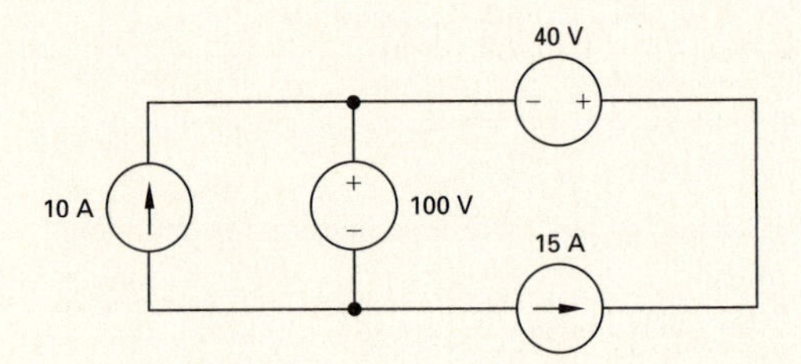

FIGURE P2.6

2.7 If the interconnection in Fig. P2.7 is valid, find the total power developed in the circuit. If the interconnection is not valid, explain why.

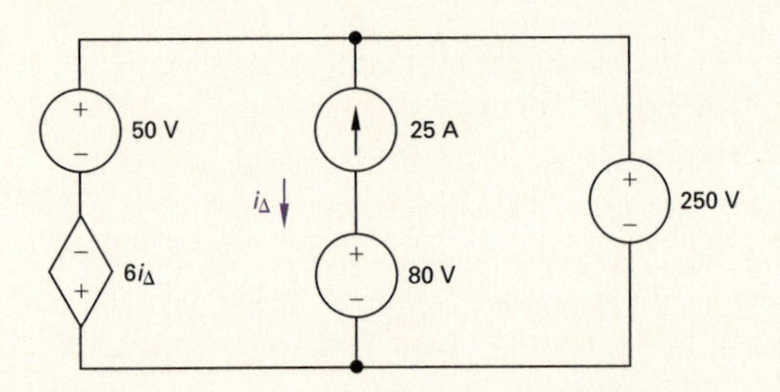

FIGURE P2.7

2.8 The interconnection of ideal sources can lead to an indeterminate solution. With this thought in mind, explain why the solutions for v_1 and v_2 in the circuit in Fig. P2.8 are not unique.

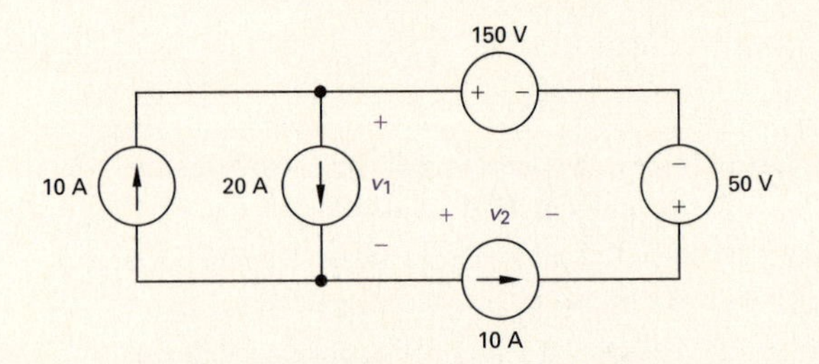

FIGURE P2.8

2.9 a) Is the interconnection in Fig. P2.9 valid? Explain.

b) Can you find the total energy developed in the circuit? Explain.

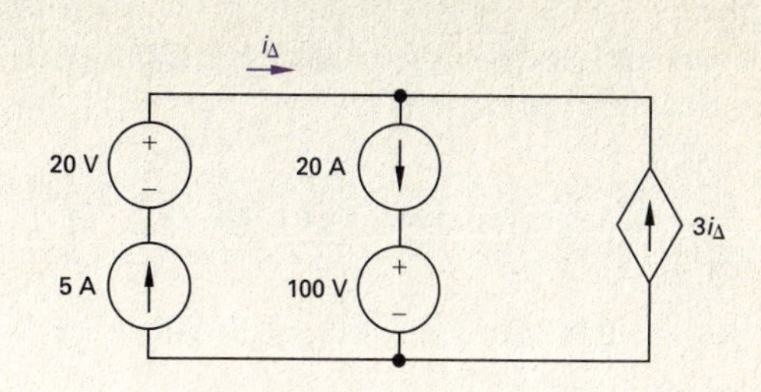

FIGURE P2.9

2.10 Find the total power developed in the circuit in Fig. P2.10 if $v_o = 125$ V.

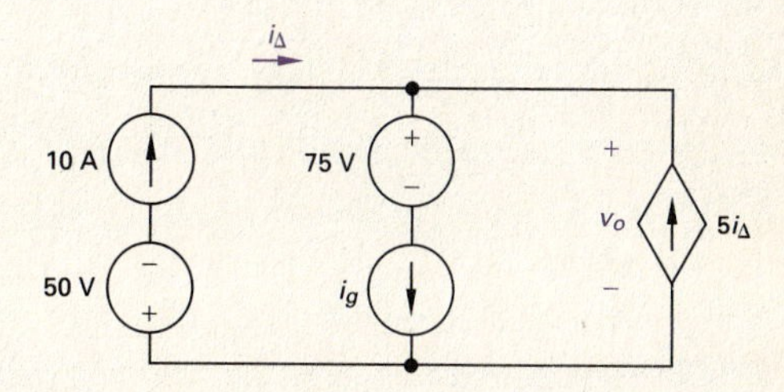

FIGURE P2.10

2.11 Given the circuit shown in Fig. P2.11, find

a) the value of i_a;

b) the value of i_b;

c) the value of v_o;

d) the power dissipated in each resistor;

e) the power delivered by the 300 V source.

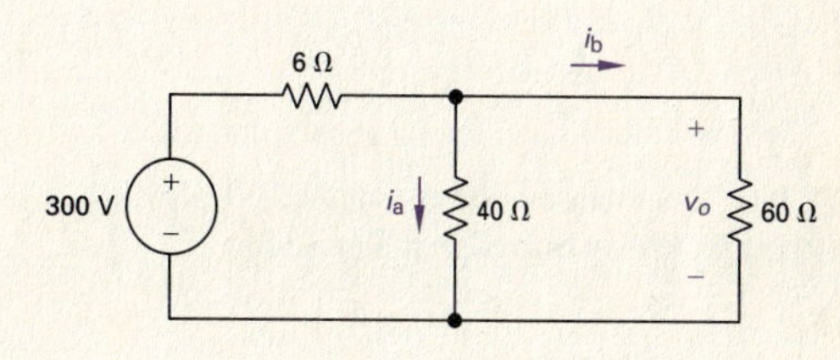

FIGURE P2.11

2.12 The current i_o in the circuit in Fig. P2.12 is 4 A.

a) Find i_1.

b) Find the power dissipated in each resistor.

c) Verify that the total power dissipated in the circuit equals the power developed by the 180 V source.

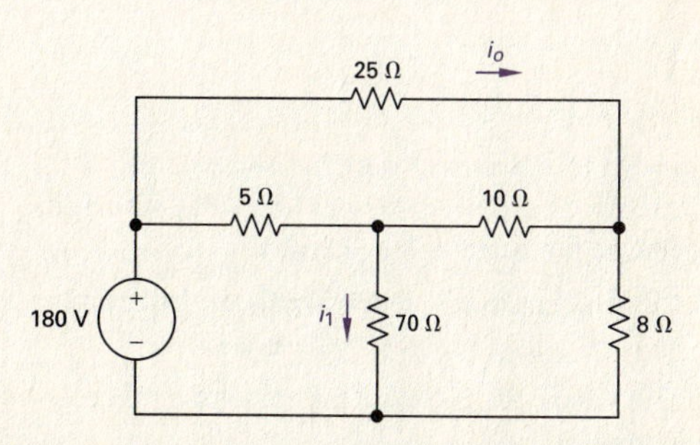

FIGURE P2.12

2.13 The currents i_a and i_b in the circuit in Fig. P2.13 are 4 A and -2 A, respectively.

a) Find i_g.

b) Find the power dissipated in each resistor.

c) Find v_g.

d) Show that the power delivered by the current source is equal to the power absorbed by all the other elements.

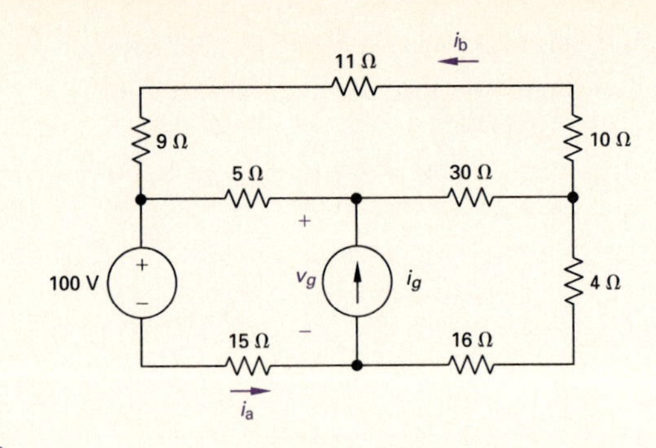

FIGURE P2.13

2.14 a) Find the currents i_a and i_b in the circuit in Fig. P2.14.

b) Find the voltage v_g.

c) Verify that the total power developed equals the total power dissipated.

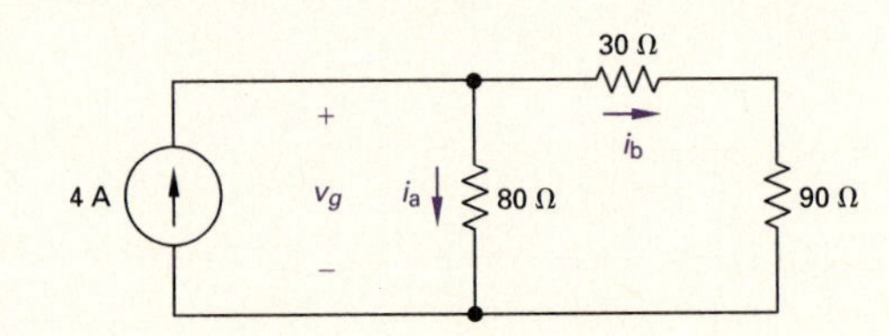

FIGURE P2.14

2.15 The currents i_1 and i_2 in the circuit in Fig. P2.15 are 20 A and 15 A, respectively.

a) Find the power supplied by each voltage source.

b) Show that the total power supplied equals the total power dissipated in the resistors.

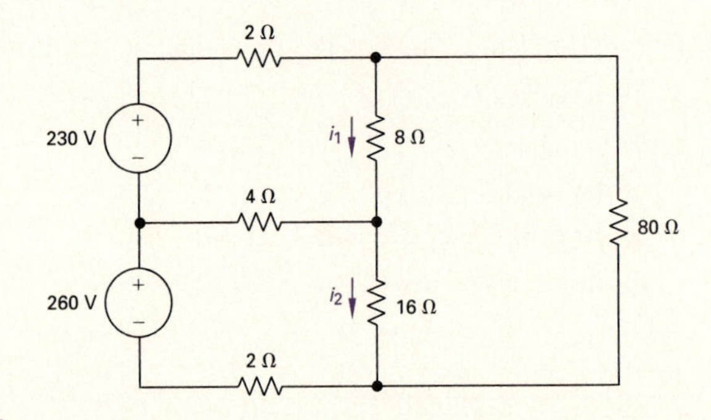

FIGURE P2.15

2.16 The voltage and current were measured at the terminals of the device shown in Fig. P2.16(a). The results are tabulated in Fig. 2.16(b).

a) Construct a circuit model for this device, using an ideal voltage source and a resistor.

b) Use the model to predict the value of i_t when v_t is zero.

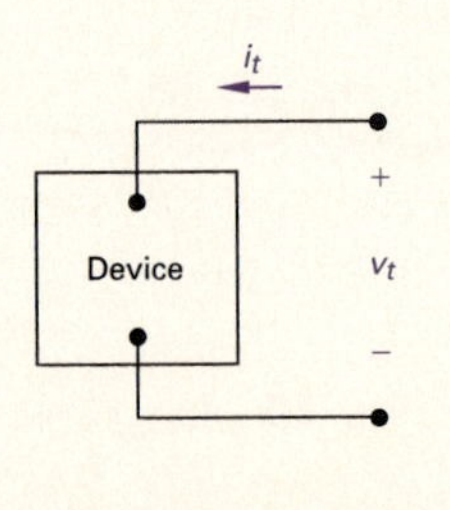

(a)

v_t (V)	i_t (A)
50	0
58	2
66	4
74	6
82	8
90	10

(b)

FIGURE P2.16

2.17 The voltage and current were measured at the terminals of the device shown in Fig. P2.17(a). The results are tabulated in Fig. P2.17(b).

a) Construct a circuit model for this device, using an ideal current source and a resistor.

b) Use the model to predict the amount of power the device will deliver to a 5 Ω resistor.

(a)

v_t (V)	i_t (A)
100	0
180	4
260	8
340	12
420	16

(b)

FIGURE P2.17

2.18 The table in Fig. P2.18(a) gives the relationship between the terminal voltage and current of the practical constant voltage source shown in Fig. P2.18(b).

a) Plot v_s versus i_s.

b) Construct a circuit model of the practical source that is valid for $0 \leq i_s \leq 24$ A, based on the equation of the line plotted in (a). (Use an ideal voltage source in series with an ideal resistor.)

c) Use your circuit model to predict the current delivered to a 1 Ω resistor connected to the terminals of the practical source.

d) Use your circuit model to predict the current delivered to a short circuit connected to the terminals of the practical source.

e) What is the actual short-circuit current?

f) Explain why the answers to parts (d) and (e) are not the same.

v_s (V)	i_s (A)
24	0
22	8
20	16
18	24
15	32
10	40
0	48

(a)

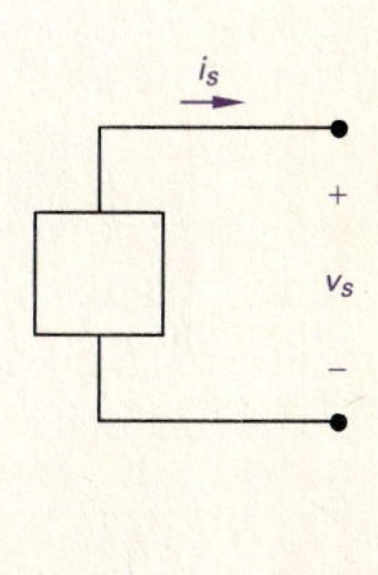

(b)

FIGURE P2.18

2.19 The table in Fig. P2.19(a) gives the relationship between the terminal current and voltage of the practical constant current source shown in Fig. P2.19(b).

a) Plot i_s versus v_s.

b) Construct a circuit model of this current source that is valid for $0 \leq v_s \leq 75$ V, based on the equation of the line plotted in (a).

c) Use your circuit model to predict the current delivered to a 2.5 kΩ resistor.

d) Use your circuit model to predict the open-circuit voltage of the current source.

e) What is the actual open-circuit voltage?

f) Explain why the answers to parts (d) and (e) are not the same.

i_s (mA)	v_s (V)
20	0
17.5	25
15	50
12.5	75
9	100
4	125
0	140

(a)

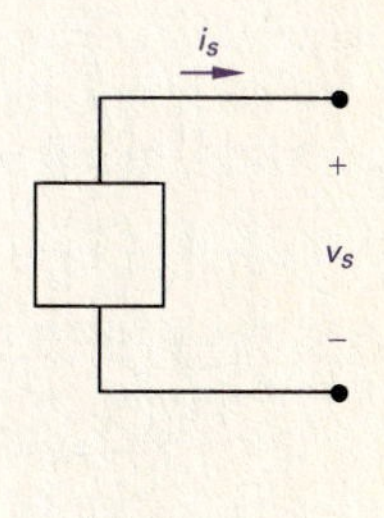

(b)

FIGURE P2.19

2.20 The variable resistor R in the circuit in Fig. P2.20 is adjusted until v_a equals 60 V. Find the value of R.

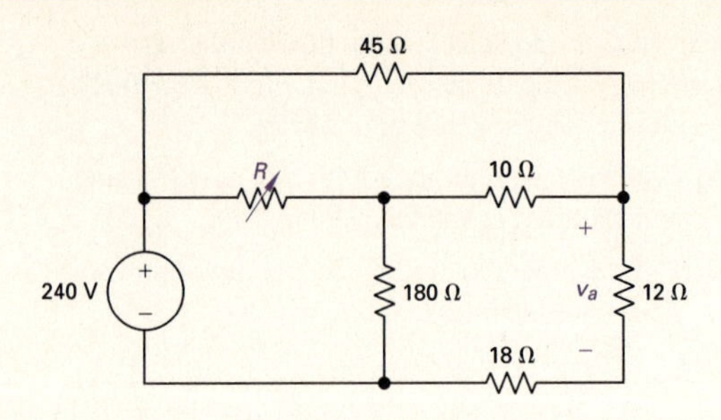

FIGURE P2.20

2.21 The voltage across the 22.5 Ω resistor in the circuit in Fig. P2.21 is 90 V, positive at the upper terminal.

a) Find the power dissipated in each resistor.

b) Find the power supplied by the 240 V ideal voltage source.

c) Verify that the power supplied equals the total power dissipated.

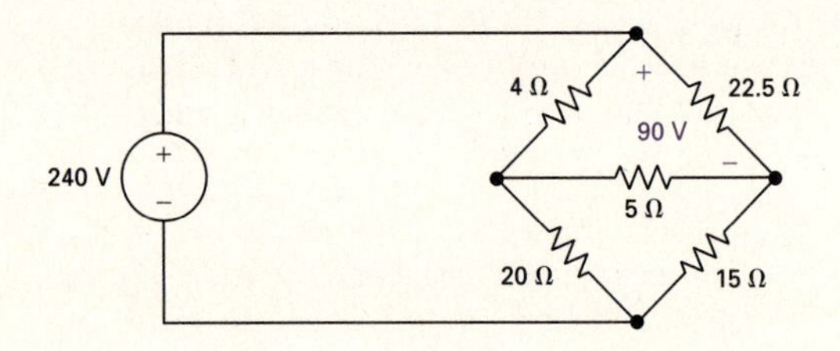

FIGURE P2.21

2.22 For the circuit shown in Fig. P2.22, find (a) R and (b) the power supplied by the 125 V source.

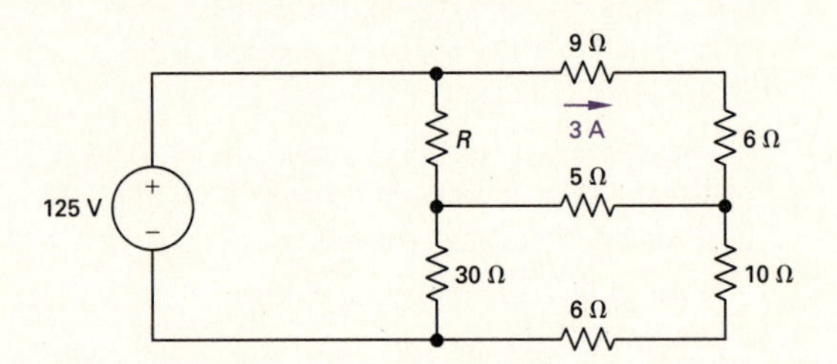

FIGURE P2.22

2.23 Derive Eq. (2.25). *Hint:* Use Eqs. (3) and (4) from Example 2.11 to express i_E as a function of i_B. Solve Eq. (2) for i_2 and substitute the result into both Eqs. (5) and (6). Solve the "new" Eq. (6) for i_1 and substitute this result into the "new" Eq. (5). Replace i_E in the "new" Eq. (5) and solve for i_B. Note that since i_{CC} appears only in Eq. (1), the solution for i_B involves the manipulation of only five equations.

2.24 For the circuit shown in Fig. 2.24, $R_1 = 40\ \text{k}\Omega$, $R_2 = 60\ \text{k}\Omega$, $R_C = 750\ \Omega$, $R_E = 120\ \Omega$, $V_{CC} = 10$ V, $V_0 = 600$ mV, and $\beta = 49$. Calculate i_B, i_C, i_E, v_{3d}, v_{bd}, i_2, i_1, v_{ab}, i_{CC}, and v_{13}. (*Note:* In the double subscript notation on voltage variables, the first subscript is positive with respect to the second subscript. See Fig. P2.24.)

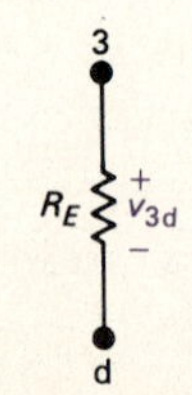

FIGURE P2.24

2.25 Find (a) i_2, (b) i_1, and (c) i_o in the circuit in Fig. P2.25.

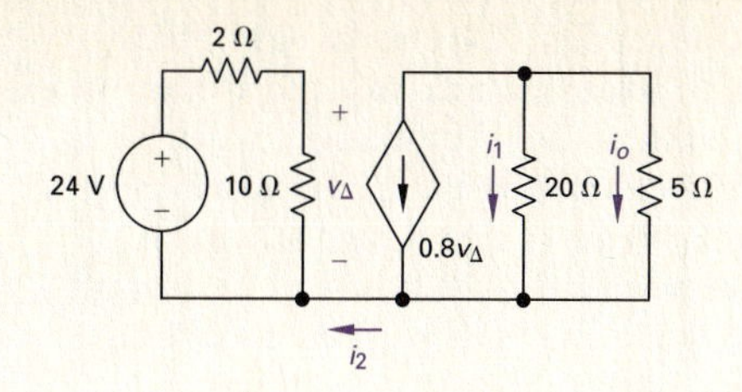

FIGURE P2.25

2.26 Find v_1 and v_g in the circuit shown in Fig. P2.26 when v_0 equals 250 mV. (*Hint:* Start at the right end of the circuit and work back toward v_g.)

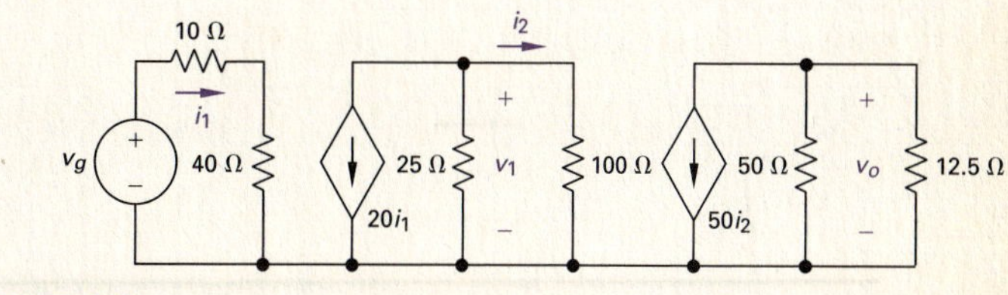

FIGURE P2.26

2.27 a) Find the voltage v_y in the circuit in Fig. P2.27.

b) Show that the total power generated in the circuit equals the total power absorbed.

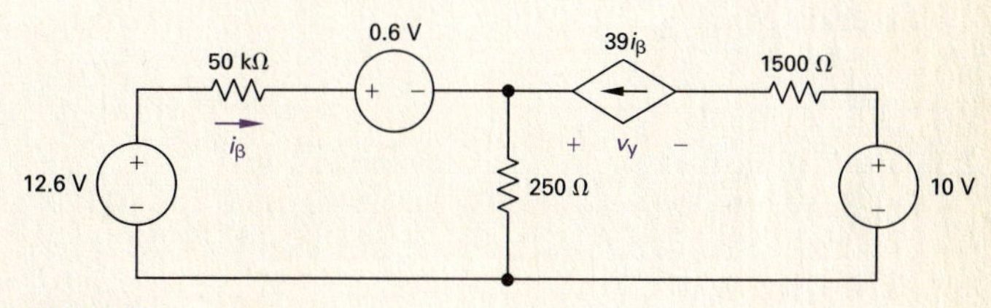

FIGURE P2.27

2.28 For the circuit shown in Fig. P2.28, calculate (a) i_Δ and v_o and (b) show that the power developed equals the power absorbed.

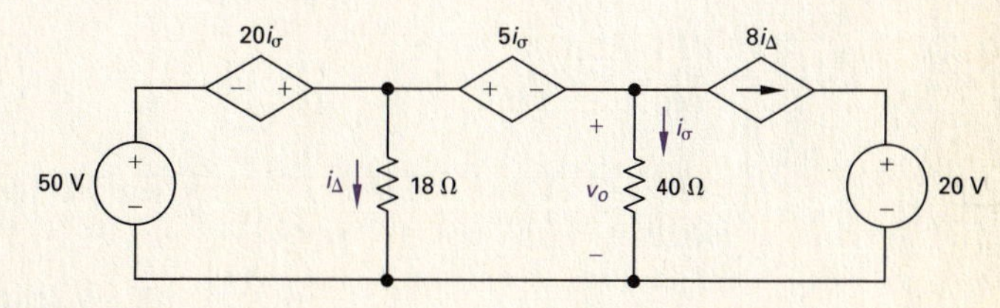

FIGURE P2.28

❖ **2.29** It is often desirable in designing an electric wiring system to be able to control a single appliance from two or more locations—for example, to control a lighting fixture from both the top and bottom of a stairwell. In home wiring systems, this type of control is implemented with three-way and four-way switches. A three-way switch is a three-terminal, two-position switch, and a four-way switch is a four-terminal, two-position switch. The switches are shown schematically in Fig. P2.29(a), which illustrates a three-way switch, and P2.29(b), which illustrates a four-way switch.

a) Show how two three-way switches can be connected between a and b in the circuit in Fig. P2.29(c) so that the lamp l can be turned ON or OFF from two locations.

b) If the lamp (appliance) is to be controlled from more than two locations, four-way switches are used in conjunction with two three-way switches. One four-way switch is required for each location in excess of two. Show how one four-way switch plus two three-way switches can be connected between a and b in Fig. P2.29(c) to control the lamp from three locations. (*Hint:* The four-way switch is placed between the three-way switches.)

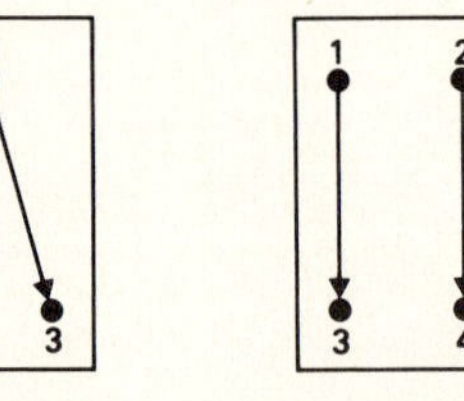

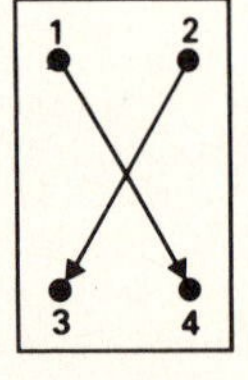

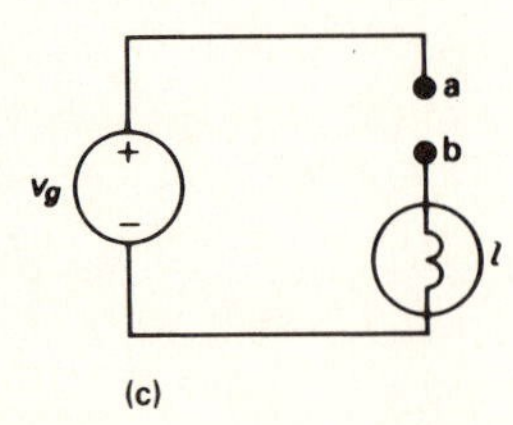

FIGURE P2.29

◆ **2.30** a) Using the values of resistance for arm, leg, and trunk provided in the Practical Perspective (Fig. 2.26), calculate the power dissipated in the arm, leg, and trunk.

b) The specific heat of water is 4.18×10^3 J/kg°C, so a mass of water M (in kilograms) heated by a power P (in watts) undergoes a rise in temperature at a rate given by

$$\frac{dT}{dt} = \frac{2.39 \times 10^{-4} P}{M} \text{°C/s}.$$

Assuming that the mass of an arm is 4 kg, the mass of a leg is 10 kg, and the mass of a trunk is 25 kg, and that the human body is mostly water, how many seconds does it take the arm, leg, and trunk to rise the 5°C that endangers living tissue?

c) How do the values you computed in part (b) compare to the few minutes it takes for oxygen starvation to injure the brain?

◆ **2.31** A person accidently grabs conductors connected to each end of a DC voltage source, one in each hand.

a) Using the resistance values for the human body provided in the Practical Perspective (Fig. 2.26), what is the minimum source voltage that can produce electrical shock sufficient to cause paralysis, preventing the person from letting go of the conductors?

b) Is there a significant risk of this type of accident occurring while servicing a personal computer, which typically has 5 V and 12 V sources?

◆ **2.32** To understand why the voltage level is not the sole determinant of potential injury due to electrical shock, consider the case of a static electricity shock mentioned in the Practical Perspective at the start of this chapter. When you shuffle your feet across a carpet, your body becomes charged. The effect of this charge is that your entire body represents a voltage potential. When you touch a metal doorknob, a voltage difference is created between you and the doorknob, and current flows—but the conduction material is air, not your body!

Suppose the model of the space between your hand and the doorknob is a 1 MΩ resistance. What voltage potential exists between your hand and the doorknob if the current causing the mild shock is 3 mA?

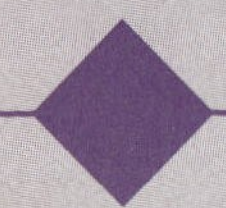

PRACTICAL PERSPECTIVE

A Rear Window Defroster

The rear window defroster grid on an automobile is an example of a resistive circuit that performs a useful function. One such grid structure is shown in part (a) of the figure here. The grid conductors can be modeled with resistors, as shown in part (b) of the figure. The number of horizontal conductors varies with the make and model of the car but typically ranges from 9 to 16.

How does this grid work to defrost the rear window? How are the properties of the grid determined? We will answer these questions in the Practical Perspective at the end of this chapter. The circuit analysis required to answer these questions arises from the goal of having uniform defrosting in both the horizontal and vertical directions.

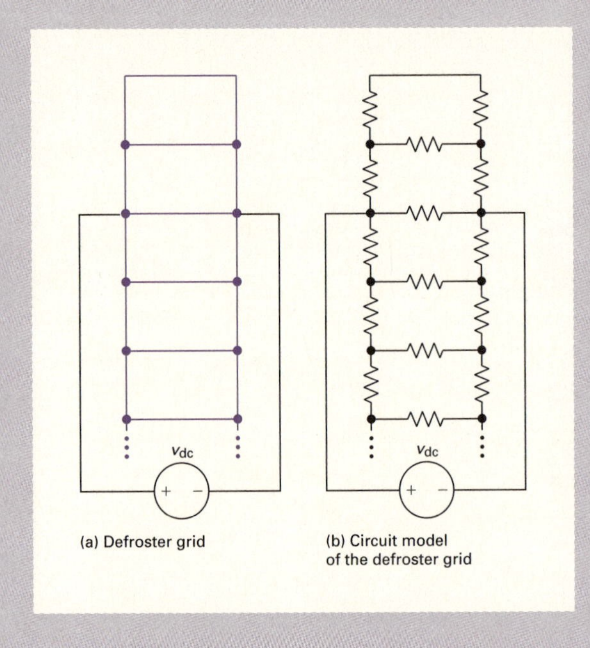

(a) Defroster grid

(b) Circuit model of the defroster grid

3 SIMPLE RESISTIVE CIRCUITS

CHAPTER CONTENTS

Our analytical toolbox now contains Ohm's law and Kirchhoff's laws. In Chapter 2 we used these tools in solving simple circuits. In this chapter we continue applying these tools, but on more complex circuits. The greater complexity lies in a greater number of elements with more complicated interconnections. This chapter focuses on reducing such circuits into simpler, equivalent circuits. We continue to focus on relatively simple circuits for two reasons: (1) it gives us a chance to acquaint ourselves thoroughly with the laws underlying more sophisticated methods; and (2) it allows us to be introduced to some circuits that have important engineering applications.

The sources in the circuits discussed in this chapter are limited to voltage and current sources that generate either constant voltages or currents; that is, voltages and currents that are invariant with time. Constant sources are often called **dc sources**. The *dc* stands for *direct current*, a description which has a historical basis but can seem misleading now. Historically, a direct current was defined as a current produced by a constant voltage. Therefore, a constant voltage became known as a direct current, or a dc voltage. The use of *dc* for *constant* stuck, and the terms *dc current* and *dc voltage* are now universally accepted in science and engineering to mean constant current and constant voltage.

3.1 Resistors in Series

In Chapter 2 we said that when just two elements connect at a single node, they are said to be in series. **Series-connected circuit elements** carry the same current. The resistors in the circuit shown in Fig. 3.1 are connected in series. We can show that these resistors carry the same current by applying Kirchhoff's current law to each node in the circuit. The series interconnection in Fig. 3.1 requires that

$$i_s = i_1 = -i_2 = i_3 = i_4 = -i_5 = -i_6 = i_7, \tag{3.1}$$

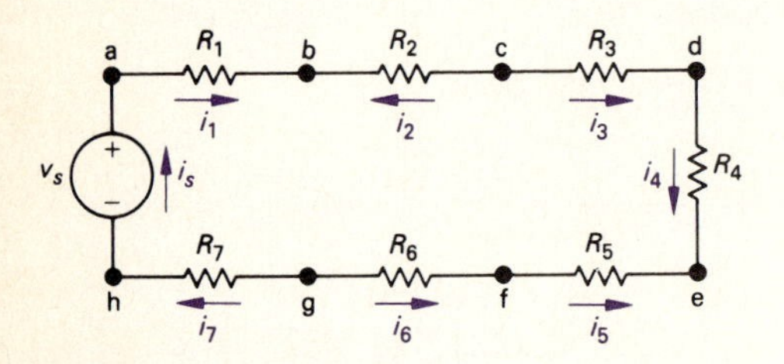

FIGURE 3.1 Resistors connected in series.

which states that if we know any one of the seven currents, we know them all. Thus we can redraw Fig. 3.1 as shown in Fig. 3.2, retaining the identity of the single current i_s.

To find i_s, we apply Kirchhoff's voltage law around the single closed loop. Defining the voltage across each resistor as a drop in the direction of i_s gives

$$-v_s + i_sR_1 + i_sR_2 + i_sR_3 + i_sR_4 + i_sR_5 + i_sR_6 + i_sR_7 = 0, \tag{3.2}$$

or

$$v_s = i_s(R_1 + R_2 + R_3 + R_4 + R_5 + R_6 + R_7). \tag{3.3}$$

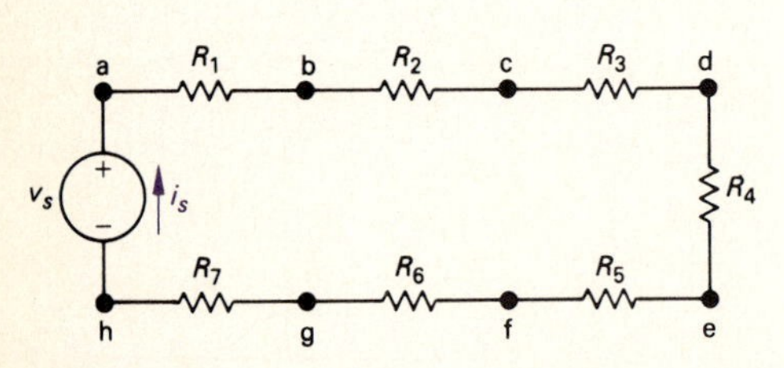

FIGURE 3.2 Series resistors with a single unknown current i_s.

The significance of Eq. (3.3) for calculating i_s is that the seven resistors can be replaced by a single resistor whose numerical value is the sum of the individual resistors—that is,

$$R_{eq} = R_1 + R_2 + R_3 + R_4 + R_5 + R_6 + R_7 \tag{3.4}$$

and

$$v_s = i_s R_{\text{eq}}. \tag{3.5}$$

Thus we can redraw Fig. 3.2 as shown in Fig. 3.3.

In general, if k resistors are connected in series, the equivalent single resistor has a resistance equal to the sum of the k resistances, or

$$R_{\text{eq}} = \sum_{i=1}^{k} R_i = R_1 + R_2 + \cdots + R_k. \tag{3.6}$$

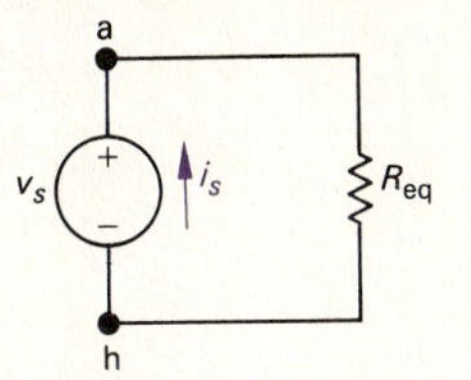

FIGURE 3.3 A simplified version of the circuit shown in Fig. 3.2.

Note that the resistance of the equivalent resistor always is larger than that of the largest resistor in the series connection.

Another way to think about this concept of an equivalent resistance is to visualize the string of resistors as being inside a black box. (An electrical engineer uses the term **black box** to imply an opaque container; that is, the contents are hidden from view. The engineer is then challenged to model the contents of the box by studying the relationship between the voltage and current at its terminals.) Determining whether the box contains k resistors or a single equivalent resistor is impossible. Figure 3.4 illustrates this method of studying the circuit shown in Fig. 3.2.

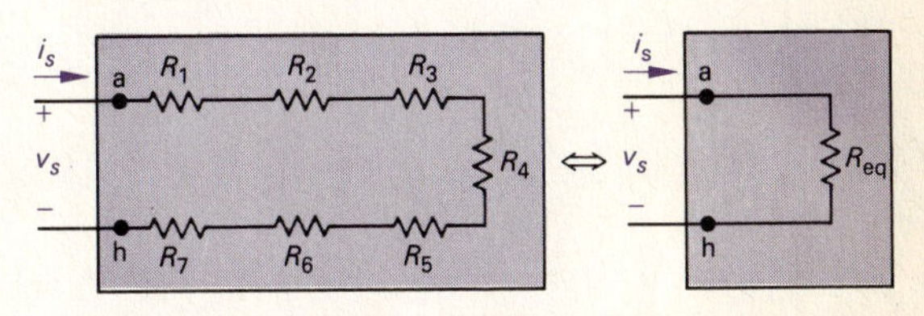

FIGURE 3.4 The black box equivalent of the circuit shown in Fig. 3.2.

3.2 Resistors in Parallel

When just two elements connect at a single node pair, they are said to be in parallel. **Parallel-connected circuit elements** have the same voltage across their terminals. The circuit shown in Fig. 3.5 illustrates resistors connected in parallel. Don't make the mistake of assuming that two elements are parallel connected merely because they are lined up in parallel in a circuit diagram. The defining characteristic of parallel-connected elements is that they have the same voltage across their terminals. In Fig. 3.6, you can see that R_1 and R_3 are not parallel connected because, between their respective terminals, another resistor dissipates some of the voltage.

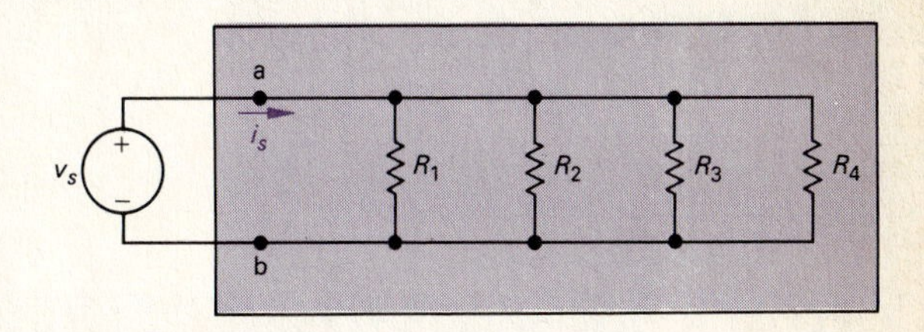

FIGURE 3.5 Resistors in parallel.

Resistors in parallel can be reduced to a single equivalent resistor using Kirchhoff's current law and Ohm's law, as we now demonstrate. In the circuit shown in Fig. 3.5, we let the currents i_1, i_2, i_3, and i_4 be the currents in the resistors R_1 through R_4, respectively. We also let the positive reference direction for each resistor current be down through the resistor, that is, from node a to node b. From Kirchhoff's current law,

$$i_s = i_1 + i_2 + i_3 + i_4. \tag{3.7}$$

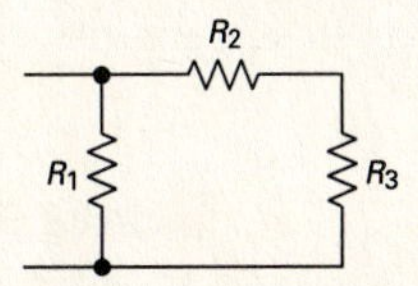

FIGURE 3.6 Nonparallel resistors.

The parallel connection of the resistors means that the voltage across each resistor must be the same. Hence, from Ohm's law,

$$i_1 R_1 = i_2 R_2 = i_3 R_3 = i_4 R_4 = v_s. \tag{3.8}$$

Therefore

$$i_1 = \frac{v_s}{R_1}, \quad i_2 = \frac{v_s}{R_2}, \quad i_3 = \frac{v_s}{R_3}, \quad \text{and} \quad i_4 = \frac{v_s}{R_4}. \tag{3.9}$$

Substituting Eq. (3.9) into Eq. (3.7) yields

$$i_s = v_s\left(\frac{1}{R_1} + \frac{1}{R_2} + \frac{1}{R_3} + \frac{1}{R_4}\right), \tag{3.10}$$

from which

$$\frac{i_s}{v_s} = \frac{1}{R_{\text{eq}}} = \frac{1}{R_1} + \frac{1}{R_2} + \frac{1}{R_3} + \frac{1}{R_4}. \tag{3.11}$$

Equation (3.11) is what we set out to show: that the four resistors in the circuit shown in Fig. 3.5 can be replaced by a single equivalent resistor. The circuit shown in Fig. 3.7 illustrates the substitution. For k resistors connected in parallel, Eq. (3.11) becomes

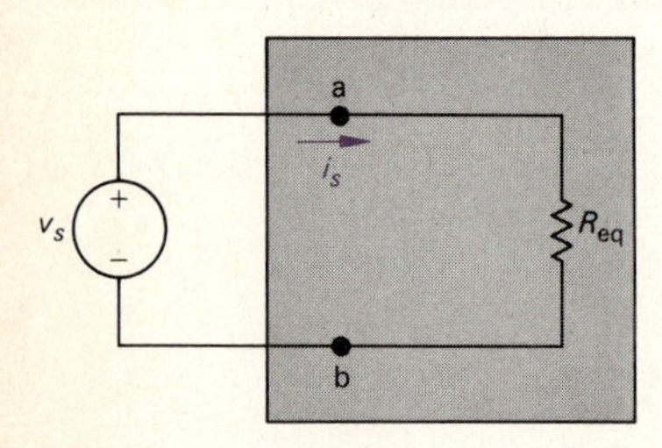

FIGURE 3.7 Replacing the four parallel resistors shown in Fig. 3.5 with a single equivalent resistor.

$$\frac{1}{R_{\text{eq}}} = \sum_{i=1}^{k} \frac{1}{R_i} = \frac{1}{R_1} + \frac{1}{R_2} + \cdots + \frac{1}{R_k}. \tag{3.12}$$

Note that the resistance of the equivalent resistor always is smaller than the resistance of the smallest resistor in the parallel connection. Sometimes, using conductance when dealing with resistors connected in parallel is more convenient. In that case, Eq. (3.12) becomes

$$G_{\text{eq}} = \sum_{i=1}^{k} G_i = G_1 + G_2 + \cdots + G_k. \tag{3.13}$$

Many times only two resistors are connected in parallel. Figure 3.8 illustrates this special case. We calculate the equivalent resistance from Eq. (3.12):

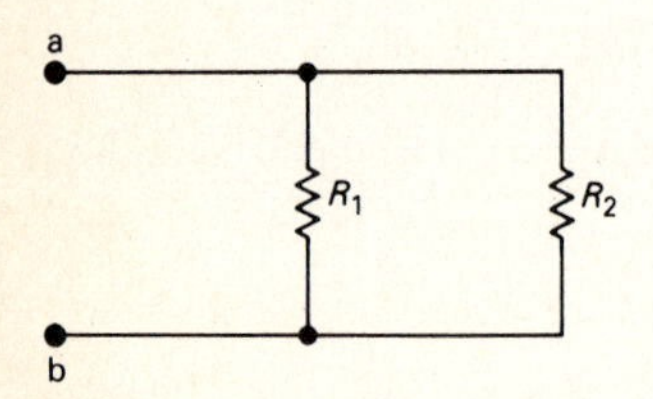

FIGURE 3.8 Two resistors connected in parallel.

$$\frac{1}{R_{\text{eq}}} = \frac{1}{R_1} + \frac{1}{R_2} = \frac{R_2 + R_1}{R_1 R_2}, \tag{3.14}$$

or

$$R_{\text{eq}} = \frac{R_1 R_2}{R_1 + R_2}. \tag{3.15}$$

Thus for just two resistors in parallel the equivalent resistance equals the product of the resistances divided by the sum of the resistances. Remember that you can only use this result in the special case of just two resistors in parallel. Example 3.1 illustrates the usefulness of these results.

EXAMPLE 3.1

Find i_s, i_1, and i_2 in the circuit shown in Fig. 3.9.

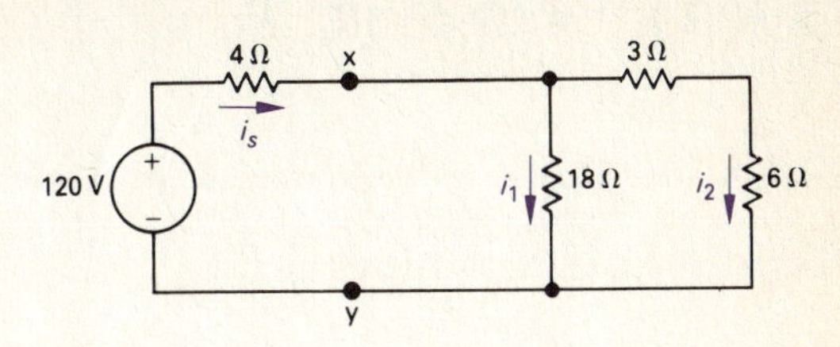

FIGURE 3.9 The circuit for Example 3.1.

SOLUTION

We begin by noting that the 3 Ω resistor is in series with the 6 Ω resistor. We therefore replace this series combination with a 9 Ω resistor, reducing the circuit to the one shown in Fig. 3.10(a). We now can replace the parallel combination of the 9 Ω and 18 Ω resistors with a single resistance of $(18 \times 9)/(18 + 9)$, or 6 Ω. Figure 3.10(b) shows this further reduction of the circuit. The nodes x and y marked on all diagrams facilitate tracing through the reduction of the circuit.

From Fig. 3.10(b) you can verify that i_s equals 120/10, or 12 A. Figure 3.11 shows the result at this point in the analysis. We added the voltage v_1 to help clarify the subsequent discussion. Using Ohm's law we compute the value of v_1:

$$v_1 = (12)(6) = 72 \text{ V}. \tag{3.16}$$

But v_1 is the voltage drop from node x to node y, so we can return to the circuit shown in Fig. 3.10(a) and again use Ohm's law to calculate i_1 and i_2. Thus

$$i_1 = \frac{v_1}{18} = \frac{72}{18} = 4 \text{ A} \tag{3.17}$$

and

$$i_2 = \frac{v_1}{9} = \frac{72}{9} = 8 \text{ A}. \tag{3.18}$$

We have found the three specified currents by using series-parallel reductions in combination with Ohm's law.

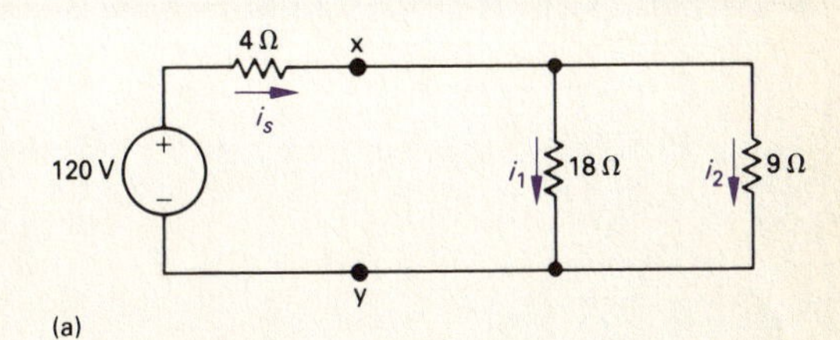

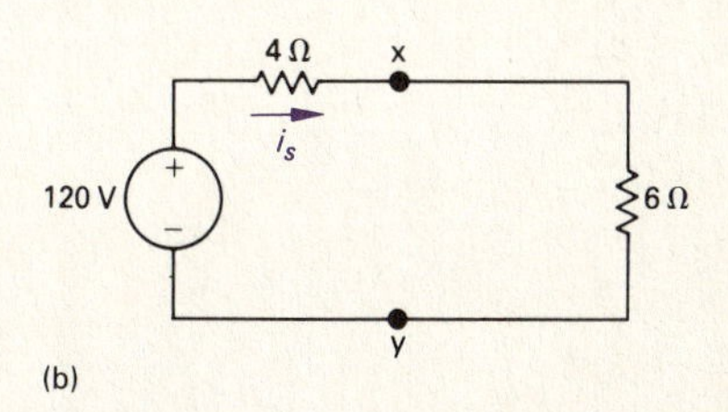

FIGURE 3.10 A simplification of the circuit shown in Fig. 3.9.

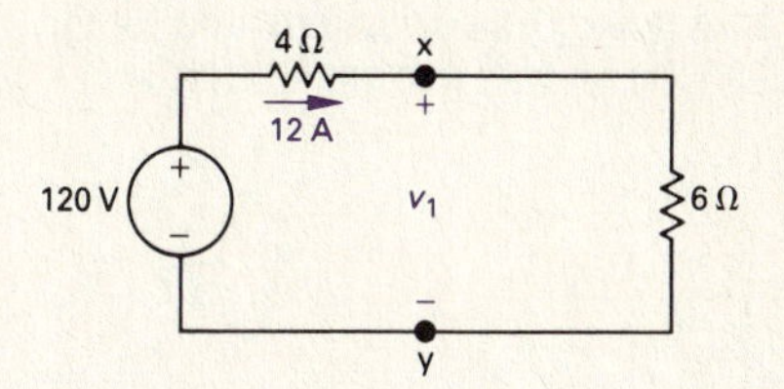

FIGURE 3.11 The circuit of Fig. 3.10(b) showing the numerical value of i_s.

Before leaving Example 3.1 completely, we suggest that you take the time to show that the solution satisfies Kirchhoff's current law at every node and Kirchhoff's voltage law around every closed path. (Note that there are three closed paths that can be tested.) Showing that the power delivered by the voltage source equals the total power dissipated in the resistors also is informative. (See Problems 3.1 and 3.2.)

DRILL EXERCISE

3.1 For the circuit shown, find (a) the voltage v; (b) the power delivered to the circuit by the current source; and (c) the power dissipated in the 10 Ω resistor.

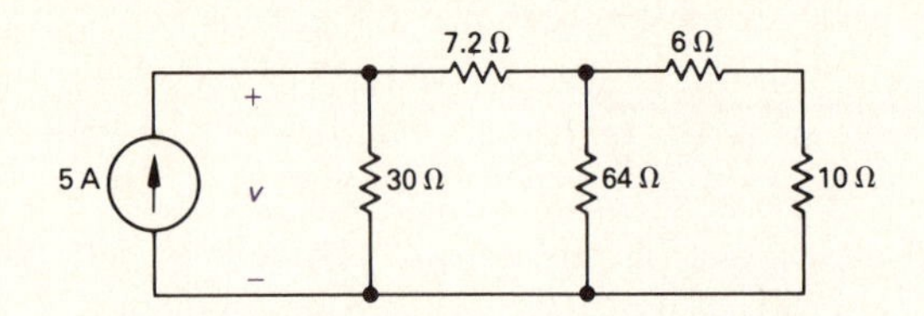

ANSWER: (a) 60 V; (b) 300 W; (c) 57.6 W.

3.3 THE VOLTAGE-DIVIDER CIRCUIT

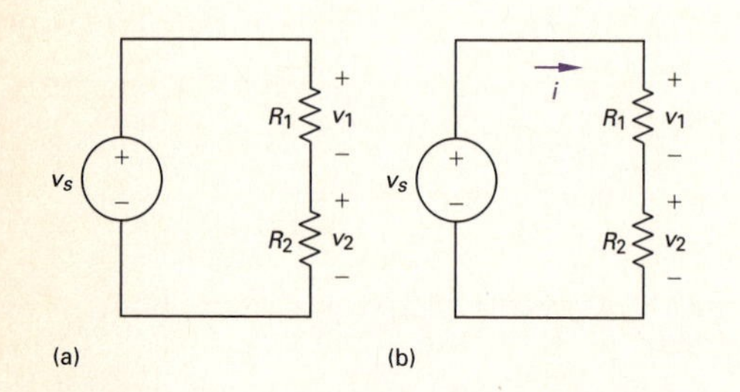

FIGURE 3.12 (a) A voltage-divider circuit and (b) the voltage-divider circuit with current i indicated.

At times—especially in electronic circuits—developing more than one voltage level from a single voltage supply is necessary. One way of doing this is by using a **voltage-divider circuit**, such as the one in Fig. 3.12.

We analyze this circuit by directly applying Ohm's law and Kirchhoff's laws. To aid the analysis, we introduce the current i as shown in Fig. 3.12(b). From Kirchhoff's current law, R_1 and R_2 carry the same current. Applying Kirchhoff's voltage law around the closed loop yields

$$v_s = iR_1 + iR_2 \tag{3.19}$$

or

$$i = \frac{v_s}{R_1 + R_2}. \tag{3.20}$$

Now we can use Ohm's law to calculate v_1 and v_2:

$$v_1 = iR_1 = v_s \frac{R_1}{R_1 + R_2}; \tag{3.21}$$

$$v_2 = iR_2 = v_s \frac{R_2}{R_1 + R_2}. \tag{3.22}$$

Equations (3.21) and (3.22) show that v_1 and v_2 are fractions of v_s. Each fraction is the ratio of the resistance across which the divided voltage is defined to the sum of the two resistances. Since this ratio is always less than 1.0, the divided voltages v_1 and v_2 are always less than the source voltage v_s.

If you desire a particular value of v_2, and v_s is specified, an infinite number of combinations of R_1 and R_2 yield the proper ratio. For example, suppose that v_s equals 15 V and v_2 is to be 5 V. Then $v_2/v_s = 1/3$ and, from Eq. (3.22), we find that this ratio is satisfied whenever $R_2 = {}^1/_2 R_1$. Other factors that may enter into the selection of R_1, and hence R_2, include the power losses that occur in dividing the source voltage and the effects of connecting the voltage-divider circuit to other circuit components.

Consider connecting a resistor R_L in parallel with R_2, as shown in Fig. 3.13. The resistor R_L acts as a load on the voltage-divider circuit. A **load** on any circuit consists of one or more circuit elements which draw power from the circuit. With the load R_L connected, the expression for the output voltage becomes

$$v_o = \frac{R_{\text{eq}}}{R_1 + R_{\text{eq}}} v_s, \tag{3.23}$$

where

$$R_{\text{eq}} = \frac{R_2 R_L}{R_2 + R_L}. \tag{3.24}$$

Substituting Eq. (3.24) into Eq. (3.23) yields

$$v_o = \frac{R_2}{R_1[1 + (R_2/R_L)] + R_2} v_s. \tag{3.25}$$

Note that Eq. (3.25) reduces to Eq. (3.22) as $R_L \to \infty$, as it should. Equation (3.25) shows that, so long as $R_L \gg R_2$, the voltage ratio v_o/v_s essentially is undisturbed by the addition of the load on the divider.

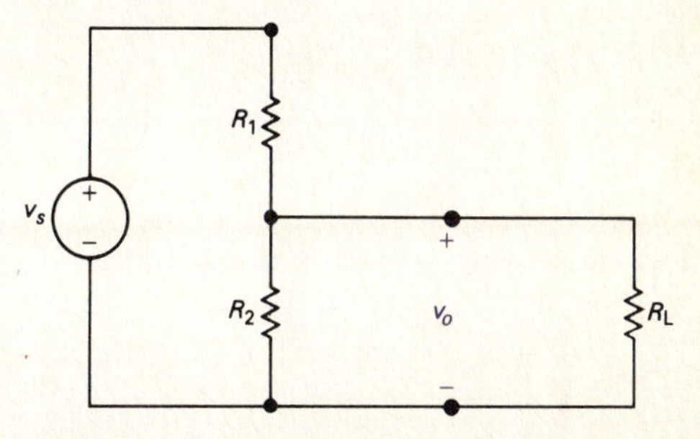

FIGURE 3.13 A voltage divider connected to a load R_L.

Another characteristic of the voltage-divider circuit of interest is the sensitivity of the divider to the tolerances of the resistors. By *tolerance* we mean a range of possible values. The resistances of commercially available resistors always vary within some percentage of their stated value. Example 3.2 illustrates the effect of resistor tolerances in a voltage-divider circuit.

Use a computer tool to explore the effect of component tolerances on circuit behavior.

EXAMPLE 3.2

The resistors used in the voltage-divider circuit shown in Fig. 3.14 have a tolerance of ±10%. Find the maximum and minimum value of v_o.

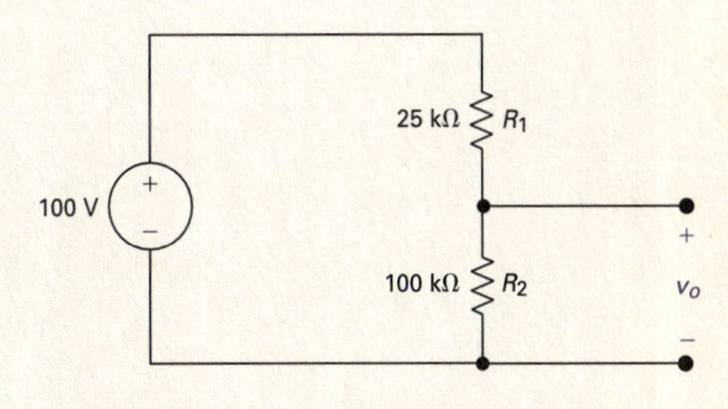

FIGURE 3.14 The circuit for Example 3.2.

SOLUTION

From Eq. (3.22), the maximum value of v_o occurs when R_2 is 10% high and R_1 is 10% low, and the minimum value of v_o occurs when R_2 is 10% low and R_1 is 10% high. Therefore

$$v_o(\text{max}) = \frac{(100)(110)}{110 + 22.5} = 83.02 \text{ V}$$

and

$$v_o(\text{min}) = \frac{(100)(90)}{90 + 27.5} = 76.60 \text{ V}.$$

Thus in making the decision to use 10% resistors in this voltage divider, we recognize that the no-load output voltage will lie between 76.60 V and 83.02 V.

DRILL EXERCISE

3.2 a) Find the no-load value of v_o in the circuit shown.

b) Find v_o when R_L is 450 kΩ.

c) How much power is dissipated in the 30 kΩ resistor if the load terminals are accidentally short-circuited?

d) What is the maximum power dissipated in the 50 kΩ resistor?

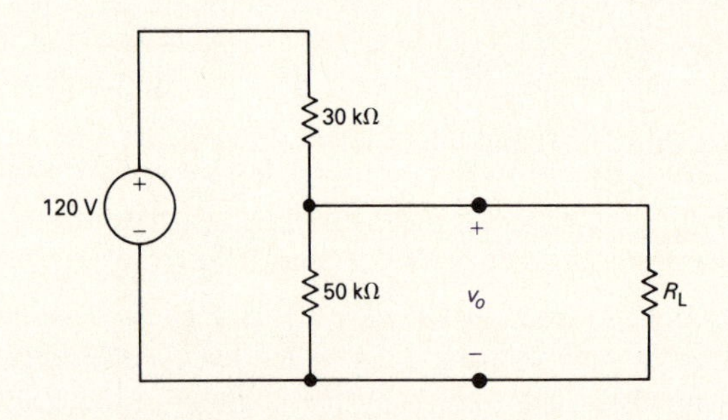

ANSWER: (a) 75 V; (b) 72 V; (c) 0.48 W; (d) 0.1125 W.

3.4 The Current-Divider Circuit

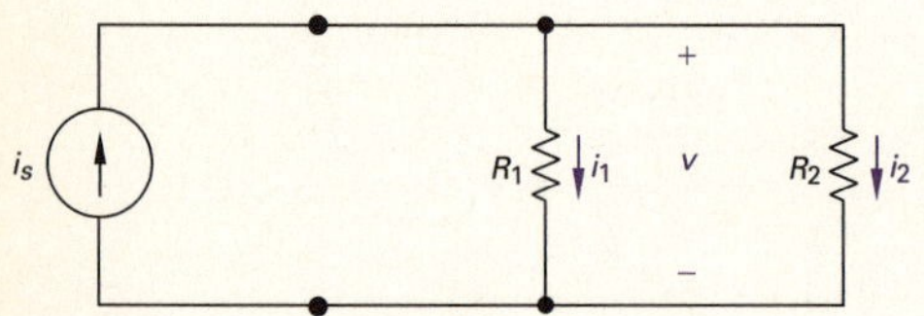

FIGURE 3.15 The current-divider circuit.

The **current-divider circuit** shown in Fig. 3.15 consists of two resistors connected in parallel across a current source. The current divider is designed to divide the current i_s between R_1 and R_2. We find the relationship between the current i_s and the current in each resistor (that is, i_1 and i_2) by directly applying Ohm's law and Kirchhoff's current law. The voltage across the parallel resistors is

$$v = i_1 R_1 = i_2 R_2 = \frac{R_1 R_2}{R_1 + R_2} i_s. \tag{3.26}$$

From Eq. (3.26),

$$i_1 = \frac{R_2}{R_1 + R_2} i_s; \tag{3.27}$$

$$i_2 = \frac{R_1}{R_1 + R_2} i_s. \tag{3.28}$$

Equations (3.27) and (3.28) show that the current divides between two resistors in parallel such that the current in one resistor equals the current entering the parallel pair multiplied by the other resistance and divided by the sum of the resistors. Example 3.3 illustrates the use of the current-divider equation.

EXAMPLE 3.3

Find the power dissipated in the 6 Ω resistor shown in Fig. 3.16.

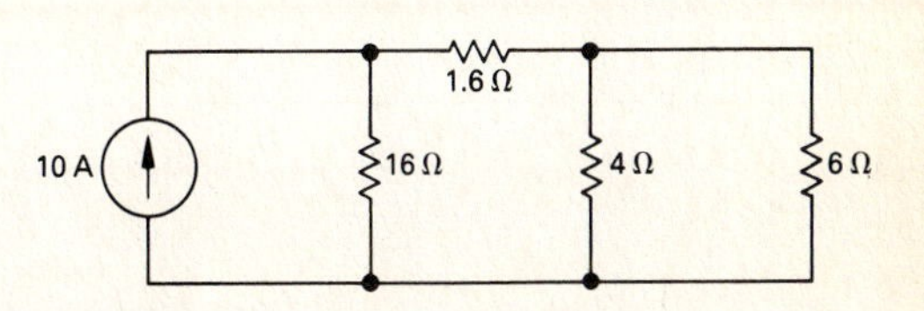

FIGURE 3.16 The circuit for Example 3.3.

SOLUTION

First, we must find the current in the resistor by simplifying the circuit with series-parallel reductions. Thus the circuit shown in Fig. 3.16 reduces to the one shown in Fig. 3.17. We find the current i_o by using the formula for current division:

$$i_o = \frac{(10)(16)}{16+4} = 8 \text{ A}.$$

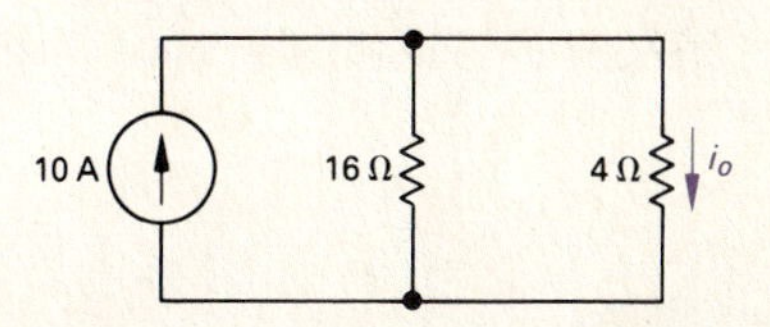

FIGURE 3.17 A simplification of the circuit shown in Fig. 3.16.

Note that i_o is the current in the 1.6 Ω resistor in Fig. 3.16. We now can further divide i_o between the 6 Ω and 4 Ω resistors. The current in the 6 Ω resistor is

$$i_6 = \frac{(8)(4)}{10} = 3.2 \text{ A},$$

and the power dissipated in the 6 Ω resistor is

$$p = (3.2)^2(6) = 61.44 \text{ W}.$$

3.5 MEASURING VOLTAGE AND CURRENT

When working with actual circuits you will often need to measure voltages and currents. We will spend some time discussing several measuring devices here and in the next section, because they are relatively simple to analyze and offer practical examples of the current- and voltage-divider configurations we have just studied.

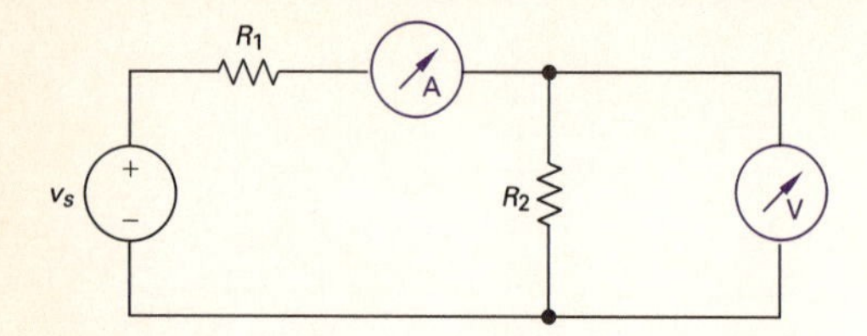

FIGURE 3.18 An ammeter connected to measure the current in R_1, and a voltmeter connected to measure the voltage across R_2.

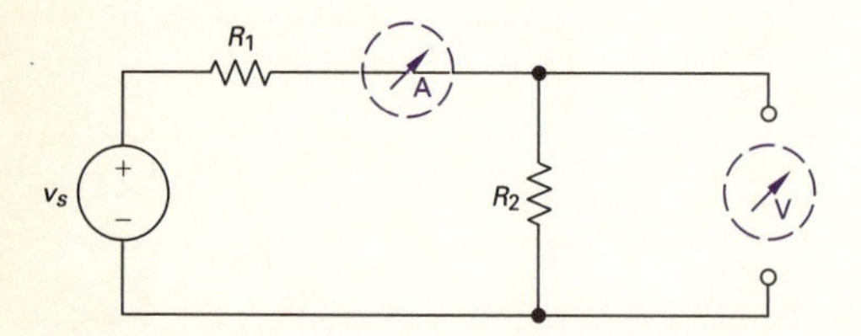

FIGURE 3.19 A short-circuit model for the ideal ammeter, and an open-circuit model for the ideal voltmeter.

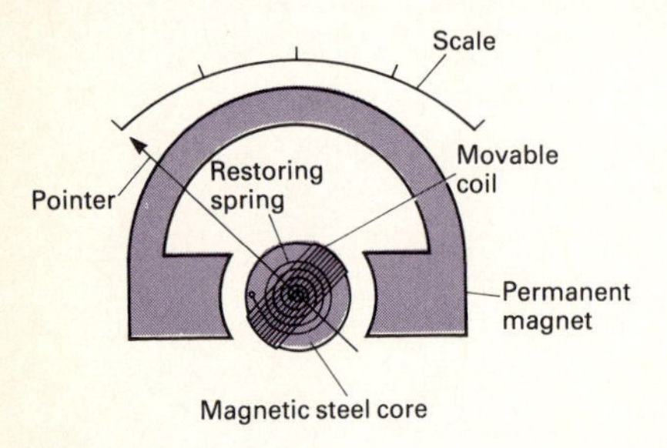

FIGURE 3.20 A schematic diagram of a d'Arsonval meter movement.

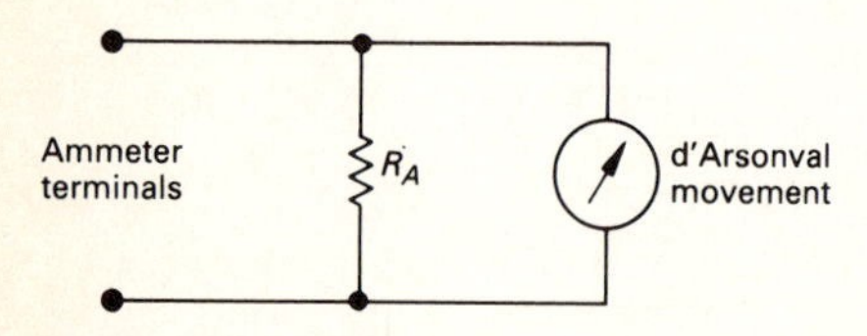

FIGURE 3.21 A direct-current ammeter circuit.

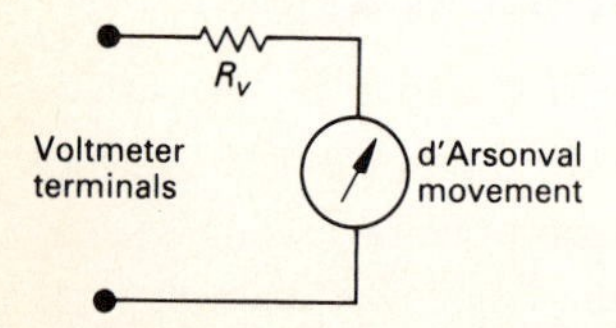

FIGURE 3.22 A direct-current voltmeter circuit.

An **ammeter** is an instrument designed to measure current; it is placed in series with the circuit element whose current is being measured. A **voltmeter** is an instrument designed to measure voltage; it is placed in parallel with the element whose voltage is being measured. An ideal ammeter or voltmeter has no effect on the circuit variable it is designed to measure. That is, an ideal ammeter has an equivalent resistance of 0 Ω and functions as a short circuit in series with the element whose current is being measured. An ideal voltmeter has an infinite equivalent resistance and thus functions as an open circuit in parallel with the element whose voltage is being measured. The configurations for an ammeter used to measure the current in R_1 and for a voltmeter used to measure the voltage in R_2 are depicted in Fig. 3.18. The ideal models for these meters in the same circuit are shown in Fig. 3.19.

There are two broad categories of meters used to measure continuous voltages and currents—digital meters and analog meters. **Digital meters** measure the continuous voltage or current signal at discrete points in time, called the sampling times. The signal is thus converted from an analog signal, which is continuous in time, to a digital signal, which exists only at discrete instants in time. A more detailed explanation of the workings of digital meters is beyond the scope of this text and course. However, you are likely to see and use digital meters in lab settings because they offer several advantages over analog meters. They introduce less resistance into the circuit to which they are connected, they are easier to connect, and the precision of the measurement is greater due to the nature of the readout mechanism.

Analog meters are based on the d'Arsonval meter movement which implements the readout mechanism. A d'Arsonval meter movement consists of a moveable coil placed in the field of a permanent magnet. When current flows in the coil, it creates a torque on the coil, causing it to rotate and move a pointer across a calibrated scale. By design, the deflection of the pointer is directly proportional to the current in the moveable coil. The coil is characterized by both a voltage rating and a current rating. For example, one commercially available meter movement is rated at 50 mV and 1 mA. This means that when the coil is carrying 1 mA, the voltage drop across the coil is 50 mV and the pointer is deflected to its full-scale position. A schematic illustration of a d'Arsonval meter movement is shown in Fig. 3.20.

An analog ammeter consists of a d'Arsonval movement in parallel with a resistor, as shown in Fig. 3.21. The purpose of the parallel resistor is to limit the amount of current in the movement's coil by shunting some of it through R_A. An analog voltmeter consists of a d'Arsonval movement in series with a resistor, as shown in Fig. 3.22. Here, the resistor is used to limit the voltage drop across the meter's coil. In both meters, the added resistor determines the full-scale reading of the meter movement.

From these descriptions we see that an actual meter is nonideal; both the added resistor and the meter movement introduce resistance in the circuit to which the meter is attached. In fact, any instrument used to make physical measurements extracts energy from the system while making measurements. The more energy extracted by the instruments, the more severely the measurement is disturbed. A real ammeter has an equivalent resistance which is not zero, and it thus effectively adds resistance to the circuit in series with the element whose current the ammeter is reading. A real voltmeter has an equivalent resistance which is not infinite, so it effectively adds resistance to the circuit in parallel with the element whose voltage is being read.

Use a circuit simulator to build meters and analyze their effect on circuits when measuring current and voltage.

How much these meters disturb the circuit being measured depends on the effective resistance of the meters compared to the resistance in the circuit. For example, using the rule of 1/10th, the effective resistance of an ammeter should be no more than 1/10th of the value of the smallest resistance in the circuit to be sure that the current being measured is nearly the same with or without the ammeter. But in an analog meter, the value of resistance is determined by the desired full-scale reading we wish to make, and it cannot be arbitrarily selected. The following examples illustrate the calculations involved in determining the resistance needed in an analog ammeter or voltmeter. The examples also consider the resulting effective resistance of the meter when it is inserted in a circuit.

EXAMPLE 3.4

a) A 50 mV, 1 mA d'Arsonval movement is to be used in an ammeter with a full-scale reading of 10 mA. Determine R_A.

b) Repeat (a) for a full-scale reading of 1 A.

c) How much resistance is added to the circuit when the 10 mA ammeter is inserted to measure current?

d) Repeat (c) for the 1 A ammeter.

SOLUTION

a) From the statement of the problem, we know that when the current at the terminals of the ammeter is 10 mA, then 1 mA is flowing through the meter coil, which means that 9 mA must be diverted through R_A. We also know that when the movement carries 1 mA, the drop across its terminals is 50 mV. Ohm's law requires that

$$9 \times 10^{-3} R_A = 50 \times 10^{-3},$$

or

$$R_A = 50/9 = 5.555\ \Omega.$$

b) When the full-scale deflection of the ammeter is 1 A, R_A must carry 999 mA when the movement carries 1 mA. In this case, then,

$$999 \times 10^{-3} R_A = 50 \times 10^{-3},$$

or

$$R_A = 50/999 \approx 50.05\ \text{m}\Omega.$$

c) Let R_m represent the equivalent resistance of the ammeter. For the 10 mA ammeter

$$R_m = \frac{50\ \text{mV}}{10\ \text{mA}} = 5\ \Omega,$$

or, alternatively,

$$R_m = \frac{(50)(50/9)}{50 + (50/9)} = 5\ \Omega.$$

d) For the 1 A ammeter

$$R_m = \frac{50 \times 10^{-3}}{1} = 0.050\ \Omega,$$

or, alternatively,

$$R_m = \frac{(50)(50/999)}{50 + (50/999)} = 0.050\ \Omega.$$

EXAMPLE 3.5

a) A 50 mV, 1 mA d'Arsonval movement is to be used in a voltmeter in which the full-scale reading is 150 V. Determine R_v.

b) Repeat (a) for a full-scale reading of 5 V.

c) How much resistance does the 150 V meter insert into the circuit?

d) Repeat (c) for the 5 V meter.

SOLUTION

a) Full-scale deflection requires 50 mV across the meter movement, and the movement has a resistance of 50 Ω. Therefore we apply

Eq. (3.22) with $R_1 = R_v$, $R_2 = 50$, $v_s = 150$, and $v_o = 50$ mV:

$$50 \times 10^{-3} = \frac{150(50)}{R_v + 50}.$$

Solving for R_v gives

$$R_v = 149{,}950\ \Omega.$$

b) For a full-scale reading of 5 V,

$$50 \times 10^{-3} = \frac{5(50)}{R_v + 50}, \qquad \text{or} \qquad R_v = 4950\ \Omega.$$

c) If we let R_m represent the equivalent resistance of the meter,

$$R_m = \frac{150}{10^{-3}} = 150{,}000\ \Omega,$$

or, alternatively,

$$R_m = 149{,}950 + 50 = 150{,}000\ \Omega.$$

d) Then,

$$R_m = \frac{5}{10^{-3}} = 5000\ \Omega,$$

or, alternatively,

$$R_m = 4950 + 50 = 5000\ \Omega.$$

DRILL EXERCISES

3.3 a) Find the current in the circuit shown.

b) If the ammeter in Example 3.4(a) is used to measure the current, what will it read?

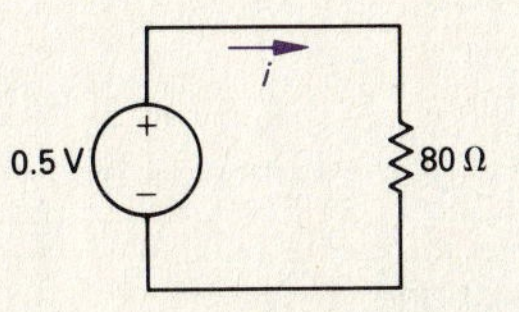

ANSWER: (a) 6.25 mA; (b) 5.88 mA.

3.4 a) Find the voltage v across the 75 kΩ resistor in the circuit shown.

b) If the 150 V voltmeter of Example 3.5(a) is used to measure the voltage, what will be the reading?

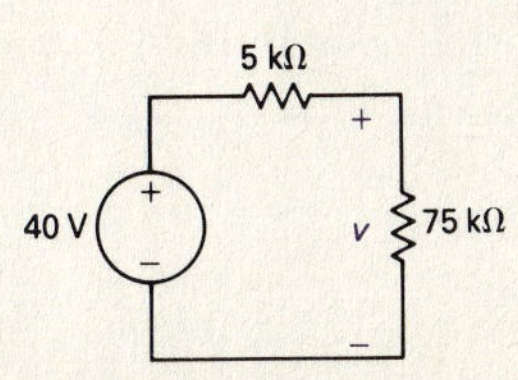

ANSWER: (a) 37.5 V; (b) 36.36 V.

3.6 The Wheatstone Bridge

There are a variety of circuit configurations used to measure resistance. Here we will focus on just one, the Wheatstone bridge. The Wheatstone bridge circuit is used to measure precisely resistances of medium values, that is, in the range of 1 Ω to 1 MΩ. In commercial models of the Wheatstone bridge, accuracies on the order of $\pm 0.1\%$ are possible. The bridge circuit consists of four resistors, a dc voltage source, and a detector. The resistance of one of the four resistors can be varied, which is indicated in Fig. 3.23 by the arrow through R_3. The dc voltage source is usually a battery, which is indicated by the battery symbol for the voltage source v in Fig. 3.23. The detector is generally a d'Arsonval movement in the microamp range and is called a galvanometer. Figure 3.23 shows the circuit arrangement of the resistances, battery, and detector where R_1, R_2, and R_3 are known resistors and R_x is the unknown resistor.

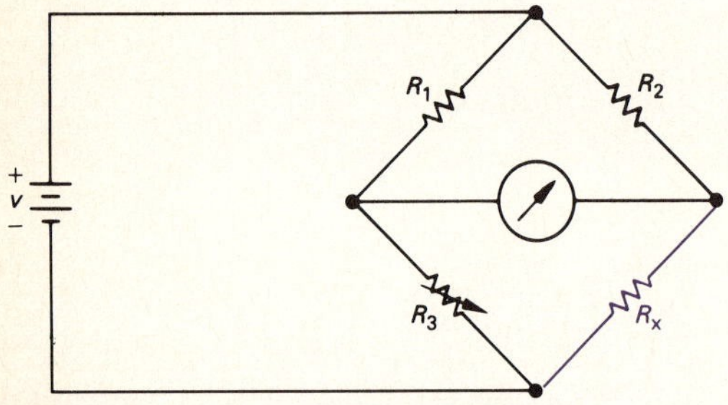

FIGURE 3.23 The Wheatstone bridge circuit.

Use a computer tool to analyze the Wheatstone bridge and monitor current as the variable resistor is adjusted.

To find the value of R_x, we adjust the variable resistor R_3 until there is no current in the galvanometer. We then calculate the unknown resistor from the simple expression

$$R_x = \frac{R_2}{R_1} R_3. \qquad \textbf{(3.29)}$$

The derivation of Eq. (3.29) follows directly from the application of Kirchhoff's laws to the bridge circuit. We redraw the bridge circuit as Fig. 3.24 to show the currents appropriate to the derivation of Eq. (3.29). When i_g is zero, that is, when the bridge is balanced, Kirchhoff's current law requires that

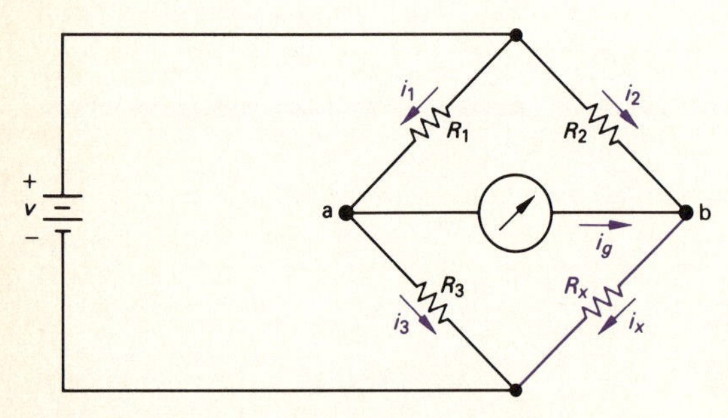

FIGURE 3.24 A balanced Wheatstone bridge ($i_g = 0$).

$$i_1 = i_3 \qquad \textbf{(3.30)}$$

and

$$i_2 = i_x. \qquad \textbf{(3.31)}$$

Now, because i_g is zero, there is no voltage drop across the detector, and therefore points a and b are at the same potential. Thus when the bridge is balanced, Kirchhoff's voltage law requires that

$$i_3 R_3 = i_x R_x \qquad \textbf{(3.32)}$$

and

$$i_1 R_1 = i_2 R_2. \qquad \textbf{(3.33)}$$

Combining Eqs. (3.30) and (3.31) with Eq. (3.32) gives

$$i_1 R_3 = i_2 R_x. \qquad \textbf{(3.34)}$$

We obtain Eq. (3.29) by first dividing Eq. (3.34) by Eq. (3.33) and then solving the resulting expression for R_x:

$$\frac{R_3}{R_1} = \frac{R_x}{R_2}, \tag{3.35}$$

from which

$$R_x = \frac{R_2}{R_1} R_3. \tag{3.36}$$

Now that we have verified the validity of Eq. (3.29), several comments about the result are in order. First, note that if the ratio R_2/R_1 is unity, the unknown resistor R_x equals R_3. In this case, the bridge resistor R_3 must vary over a range that includes the value R_x. For example, if the unknown resistance were 1000 Ω and R_3 could be varied from 0 to 100 Ω, the bridge could never be balanced. Thus to cover a wide range of unknown resistors, we must be able to vary the ratio R_2/R_1. In a commercial Wheatstone bridge, R_1 and R_2 consist of decimal values of resistances that can be switched into the bridge circuit. Normally, the decimal values are 1, 10, 100, and 1000 Ω so that the ratio R_2/R_1 can be varied from 0.001 to 1000 in decimal steps. The variable resistor R_3 is usually adjustable in integral values of resistance from 1 to 11,000 Ω.

Although Eq. (3.29) implies that R_x can vary from zero to infinity, the practical range of R_x is approximately 1 Ω to 1 MΩ. Lower resistances are difficult to measure on a standard Wheatstone bridge because of thermoelectric voltages generated at the junctions of dissimilar metals and because of thermal heating effects—that is, i^2R effects. Higher resistances are difficult to measure accurately because of leakage currents. In other words, if R_x is large, the current leakage in the electrical insulation may be comparable to the current in the branches of the bridge circuit.

DRILL EXERCISES

3.5 The bridge circuit shown is balanced when $R_1 = 100\ \Omega$, $R_2 = 1000\ \Omega$, and $R_3 = 150\ \Omega$. The bridge is energized from a 5 V dc source.

a) What is the value of R_x?

b) Suppose each bridge resistor is capable of dissipating 250 mW. Can the bridge be left in the balanced state without exceeding the power-dissipating capacity of the resistors, thereby damaging the bridge?

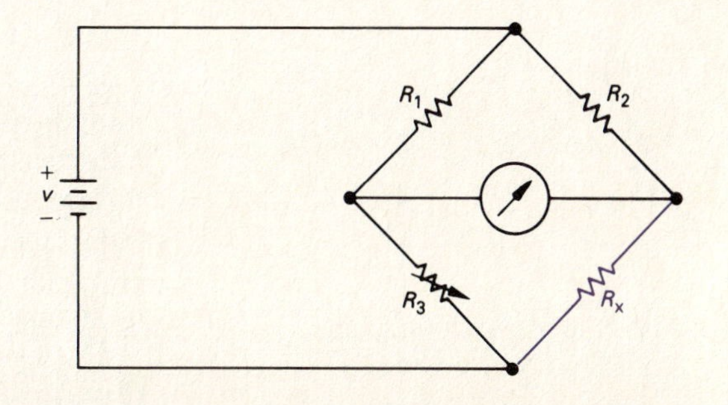

ANSWER: (a) 1500 Ω; (b) yes

3.7 DELTA-TO-WYE (OR PI-TO-TEE) EQUIVALENT CIRCUITS

The bridge configuration in Fig. 3.23 introduces an interconnection of resistances that warrants further discussion. If we replace the galvanometer with its equivalent resistance R_m, we can draw the circuit shown in Fig. 3.25. We cannot reduce the interconnected resistors of this circuit to a single equivalent resistance across the terminals of the battery if restricted to the simple series or parallel equivalent circuits introduced earlier in this chapter. The interconnected resistors can be reduced to a single equivalent resistor by means of a delta-to-wye (Δ-to-Y) or pi-to-tee (π-to-T) equivalent circuit.[1]

FIGURE 3.25 A resistive network generated by a Wheatstone bridge circuit.

The resistors R_1, R_2, and R_m (or R_3, R_m, and R_x) in the circuit shown in Fig. 3.25 are referred to as a **delta (Δ) interconnection** because the interconnection looks like the Greek letter Δ. It also is referred to as a **pi (π) interconnection** because the Δ can be shaped into a π without disturbing the electrical equivalence of the two configurations. The electrical equivalence between the Δ and π interconnections is apparent in Fig. 3.26.

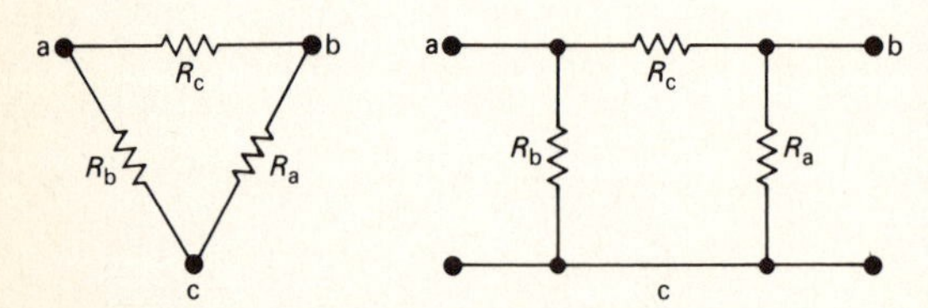

FIGURE 3.26 A Δ configuration viewed as a π configuration.

The resistors R_1, R_m, and R_3 (or R_2, R_m, and R_x) in the circuit shown in Fig. 3.25 are referred to as a **wye (Y) interconnection** because the interconnection can be shaped to look like the letter Y. It is easier to see the Y shape when the interconnection is drawn as in Fig. 3.27. The Y configuration also is referred to as a **tee (T) interconnection** because the Y structure can be shaped into a T structure without disturbing the electrical equivalence of the two structures. The electrical equivalence of the Y and the T configurations is apparent from Fig. 3.27.

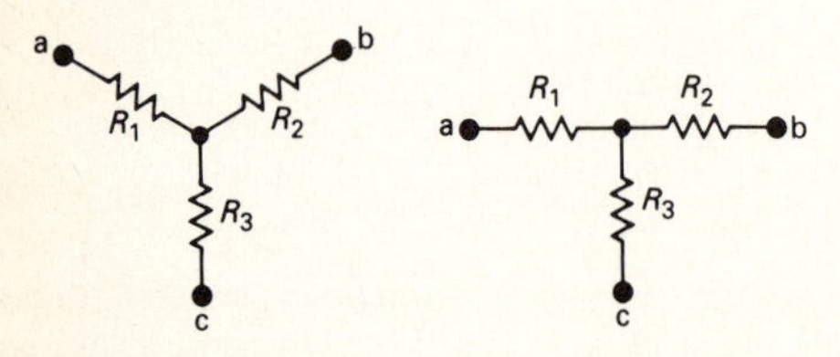

FIGURE 3.27 A Y structure viewed as a T structure.

Figure 3.28 illustrates the Δ-to-Y (or π-to-T) equivalent circuit transformation. Note that we cannot transform the Δ interconnection into the Y interconnection simply by changing the shape of the interconnections. Saying the Δ-connected circuit is equivalent to the Y-connected circuit means that the Δ configuration can be replaced with a Y configuration to make the terminal behavior of the two configurations identical. Thus if each circuit is placed in a black box, we can't tell by external measurements whether the box contains a set of Δ-connected resistors or a set of Y-connected resistors. This condition is true only if the resistance between corresponding terminal pairs is the same for each box. For example, the resistance between terminals a and b must be the same whether we use the Δ-connected set or the Y-connected set. For each pair of terminals in the Δ-connected circuit,

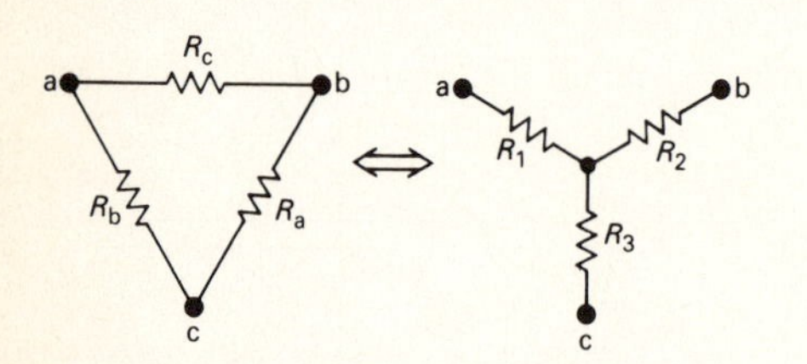

FIGURE 3.28 The Δ-to-Y transformation.

[1] Δ and Y structures are present in a variety of useful circuits, not just resistive networks. Hence the Δ-to-Y transformation is a helpful tool in circuit analysis.

the equivalent resistance can be computed using series and parallel simplifications to yield

$$R_{ab} = \frac{R_c(R_a + R_b)}{R_a + R_b + R_c} = R_1 + R_2; \quad \textbf{(3.37)}$$

$$R_{bc} = \frac{R_a(R_b + R_c)}{R_a + R_b + R_c} = R_2 + R_3; \quad \textbf{(3.38)}$$

$$R_{ca} = \frac{R_b(R_c + R_a)}{R_a + R_b + R_c} = R_1 + R_3. \quad \textbf{(3.39)}$$

Straightforward algebraic manipulation of Eqs. (3.37), (3.38), and (3.39) gives values for the Y-connected resistors in terms of the Δ-connected resistors required for the Δ-to-Y equivalent circuit:

$$R_1 = \frac{R_b R_c}{R_a + R_b + R_c}; \quad \textbf{(3.40)}$$

$$R_2 = \frac{R_c R_a}{R_a + R_b + R_c}; \quad \textbf{(3.41)}$$

$$R_3 = \frac{R_a R_b}{R_a + R_b + R_c}. \quad \textbf{(3.42)}$$

Reversing the Δ-to-Y transformation also is possible. That is, we can start with the Y structure and replace it with an equivalent Δ structure. The expressions for the three Δ-connected resistors as functions of the three Y-connected resistors are

$$R_a = \frac{R_1 R_2 + R_2 R_3 + R_3 R_1}{R_1}; \quad \textbf{(3.43)}$$

$$R_b = \frac{R_1 R_2 + R_2 R_3 + R_3 R_1}{R_2}; \quad \textbf{(3.44)}$$

$$R_c = \frac{R_1 R_2 + R_2 R_3 + R_3 R_1}{R_3}. \quad \textbf{(3.45)}$$

Example 3.6 illustrates the use of a Δ-to-Y transformation to simplify the analysis of a circuit.

EXAMPLE 3.6

Find the current and power supplied by the 40 V source in the circuit shown in Fig. 3.29.

SOLUTION

We are interested only in the current and power drain on the 40 V source, so the problem has been solved once we obtain the equivalent resistance across the terminals of the source. We can find this equivalent resistance easily after replacing either the upper Δ (100, 125, 25 Ω) or the lower Δ (40, 25, 37.5 Ω) with its equivalent Y. We choose to replace the upper Δ. We then compute the three Y resistances, defined in Fig. 3.30, from Eqs. (3.40) to (3.42). Thus

$$R_1 = \frac{100 \times 125}{250} = 50\ \Omega;$$

$$R_2 = \frac{125 \times 25}{250} = 12.5\ \Omega;$$

$$R_3 = \frac{100 \times 25}{250} = 10\ \Omega.$$

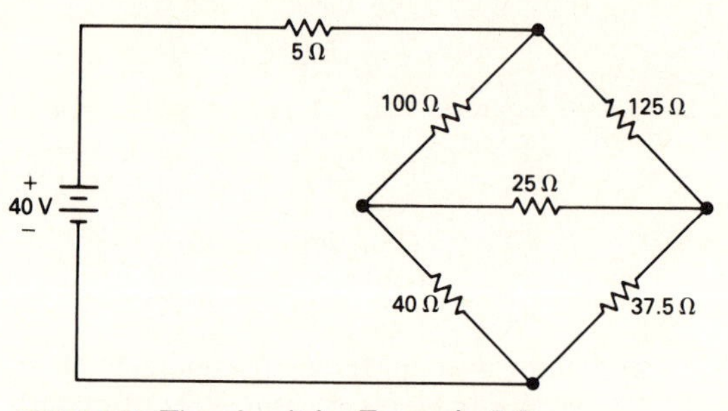

FIGURE 3.29 The circuit for Example 3.6.

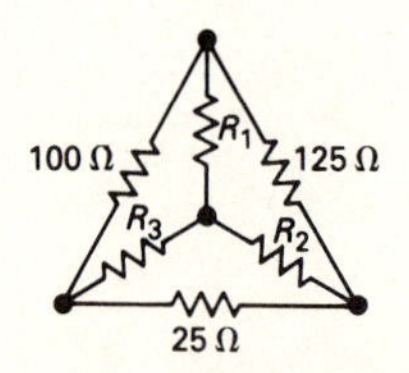

FIGURE 3.30 The equivalent Y resistors.

Substituting the Y-resistors into the circuit shown in Fig. 3.29 produces the circuit shown in Fig. 3.31. From Fig. 3.31, we can easily calculate the resistance across the terminals of the 40 V source by series-parallel simplifications:

$$R_{\text{eq}} = 55 + \frac{(50)(50)}{100} = 80\ \Omega.$$

The final step is to note that the circuit reduces to an 80 Ω resistor across a 40 V source, as shown in Fig. 3.32, from which it is apparent that the 40 V source delivers 0.5 A and 20 W to the circuit.

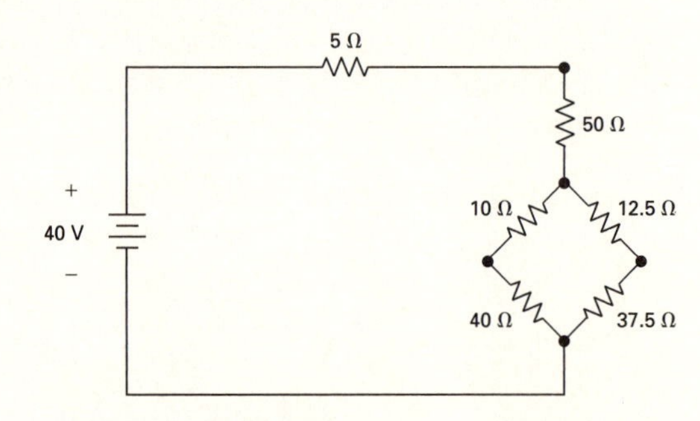

FIGURE 3.31 A simplified version of the circuit shown in Fig. 3.29.

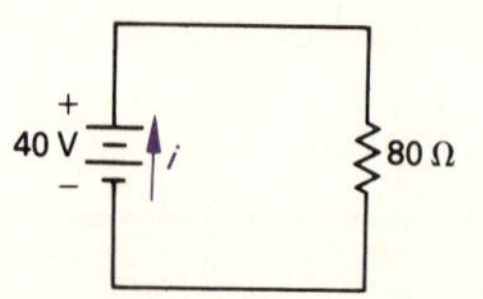

FIGURE 3.32 The final step in the simplification of the circuit shown in Fig. 3.29.

DRILL EXERCISE

3.6 a) Use a Δ-to-Y transformation to find the current i in the circuit shown.

b) Find v_1 and v_2. (*Hint:* Use the circuit that exists after the Δ-to-Y transformation.)

ANSWER: (a) 1 A; (b) $v_1 = 23.2$ V, $v_2 = 21$ V.

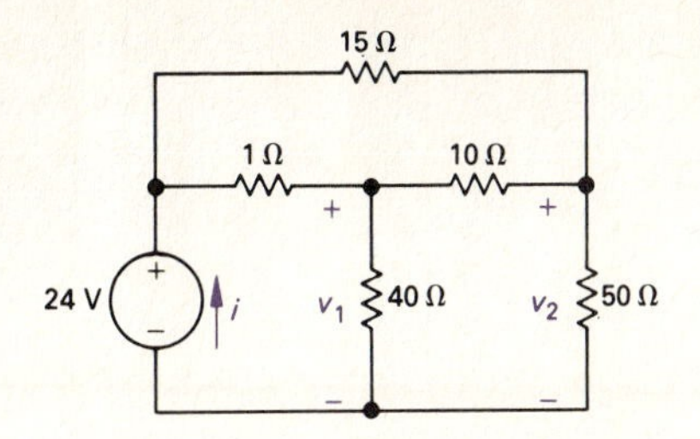

PRACTICAL PERSPECTIVE

A Rear Window Defroster

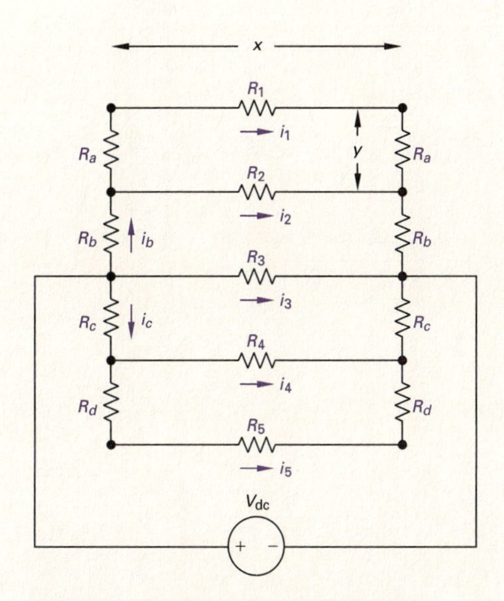

FIGURE 3.33 Model of a defroster grid.

EXAMPLE

A model of a defroster grid is shown in Fig. 3.33, where x and y denote the horizontal and vertical spacing of the grid elements. Suppose the grid structure is 1 m wide and the vertical displacement of the five horizontal grid lines is 0.025 m. Specify the numerical values of R_1 through R_5 and R_a through R_d to achieve a uniform power dissipation of 120 W/m, using a 12 V power supply.

SOLUTION

We solve the problem in general first and then compute the resistor values which satisfy the problem statement. We need to find expressions for each resistor in the grid such that the power dissipated per unit length is the same in each conductor. This will ensure uniform heating of the rear window in both the x and y directions. Thus we need to find values for the grid resistors that satisfy the following relationships:

$$i_1^2\left(\frac{R_1}{x}\right) = i_2^2\left(\frac{R_2}{x}\right) = i_3^2\left(\frac{R_3}{x}\right) = i_4^2\left(\frac{R_4}{x}\right) = i_5^2\left(\frac{R_5}{x}\right); \quad (3.46)$$

$$i_1^2\left(\frac{R_a}{y}\right) = i_1^2\left(\frac{R_1}{x}\right); \quad (3.47)$$

$$i_1^2\left(\frac{R_a}{y}\right) = i_b^2\left(\frac{R_b}{y}\right) = i_c^2\left(\frac{R_c}{y}\right) = i_5^2\left(\frac{R_d}{y}\right); \quad (3.48)$$

$$i_5^2\left(\frac{R_d}{y}\right) = i_5^2\left(\frac{R_5}{x}\right). \quad (3.49)$$

We begin the analysis of the grid by taking advantage of its symmetry. Note that if we disconnect the lower portion of the circuit (i.e., the resistors R_c, R_d, R_4, and R_5), the currents i_1, i_2, i_3, and i_b are unaffected. Thus, instead of analyzing the circuit in Fig. 3.33, we can analyze the simpler circuit in Fig. 3.34. Note further that after finding R_1, R_2, R_3, R_a, and R_b in the circuit in Fig. 3.34, we have also found the values for the remaining resistors, since

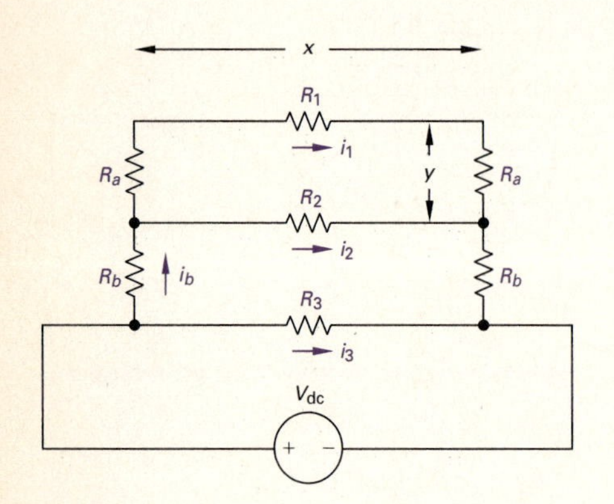

FIGURE 3.34 A simplified model of the defroster grid.

$$\begin{aligned} R_4 &= R_2; \\ R_5 &= R_1; \\ R_c &= R_b; \\ R_d &= R_a. \end{aligned} \tag{3.50}$$

Begin analysis of the simplified grid circuit in Fig. 3.34 by writing expressions for the currents i_1, i_2, i_3, and i_b. To find i_b, describe the equivalent resistance in parallel with R_3:

$$\begin{aligned} R_e &= 2R_b + \frac{R_2(R_1 + 2R_a)}{R_1 + R_2 + 2R_a} \\ &= \frac{(R_1 + 2R_a)(R_2 + 2R_b) + 2R_2R_b}{(R_1 + R_2 + 2R_a)}. \end{aligned} \tag{3.51}$$

For convenience, define the numerator of Eq. (3.51) as

$$D = (R_1 + 2R_a)(R_2 + 2R_b) + 2R_2R_b, \tag{3.52}$$

and therefore

$$R_e = \frac{D}{(R_1 + R_2 + 2R_a)}. \tag{3.53}$$

It follows directly that

$$\begin{aligned} i_b &= \frac{V_{\text{dc}}}{R_e} \\ &= \frac{V_{\text{dc}}(R_1 + R_2 + 2R_a)}{D}. \end{aligned} \tag{3.54}$$

Expressions for i_1 and i_2 can be found directly from i_b using current division. Hence

$$i_1 = \frac{i_bR_2}{R_1 + R_2 + 2R_a} = \frac{V_{\text{dc}}R_2}{D} \tag{3.55}$$

and

$$i_2 = \frac{i_b(R_1 + 2R_a)}{(R_1 + R_2 + 2R_a)} = \frac{V_{\text{dc}}(R_1 + 2R_a)}{D}. \tag{3.56}$$

The expression for i_3 is simply

$$i_3 = \frac{V_{\text{dc}}}{R_3}. \tag{3.57}$$

Now we use the constraints in Eqs. (3.46) to (3.48) to derive expressions for R_a, R_b, R_2, and R_3 as functions of R_1. From Eq. (3.47),

$$\frac{R_a}{y} = \frac{R_1}{x}$$

or

$$R_a = \frac{y}{x} R_1 = \sigma R_1, \tag{3.58}$$

where

$$\sigma = y/x.$$

Then from Eq. (3.46) we have

$$R_2 = \left(\frac{i_1}{i_2}\right)^2 R_1. \tag{3.59}$$

The ratio (i_1/i_2) is obtained directly from Eqs. (3.55) and (3.56):

$$\frac{i_1}{i_2} = \frac{R_2}{R_1 + 2R_a} = \frac{R_2}{R_1 + 2\sigma R_1}. \tag{3.60}$$

When Eq. (3.60) is substituted into Eq. (3.59), we obtain, after some algebraic manipulation (see Problem 3.62),

$$R_2 = (1 + 2\sigma)^2 R_1. \tag{3.61}$$

The expression for R_b as a function of R_1 is derived from the constraint imposed by Eq. (3.48), namely that

$$R_b = \left(\frac{i_1}{i_b}\right)^2 R_a. \tag{3.62}$$

The ratio (i_1/i_b) is derived from Eqs. (3.54) and (3.55). Thus,

$$\frac{i_1}{i_b} = \frac{R_2}{(R_1 + R_2 + 2R_a)}. \tag{3.63}$$

When Eq. (3.63) is substituted into Eq. (3.62), we obtain, after some algebraic manipulation (see Problem 3.62),

$$R_b = \frac{(1 + 2\sigma)^2 \sigma R_1}{4(1 + \sigma)^2}. \tag{3.64}$$

Finally, the expression for R_3 can be obtained from the constraint given in Eq. (3.46), or

$$R_3 = \left(\frac{i_1}{i_3}\right)^2 R_1, \tag{3.65}$$

where

$$\frac{i_1}{i_3} = \frac{R_2 R_3}{D}.$$

TABLE 3.1

SUMMARY OF RESISTANCE EQUATIONS FOR THE DEFROSTER GRID

RESISTANCE	EXPRESSION
R_a	σR_1
R_b	$\frac{(1+2\sigma)^2\sigma}{4(1+\sigma)^2} R_1$
R_2	$(1+2\sigma)^2 R_1$
R_3	$\frac{(1+2\sigma)^4}{(1+\sigma)^2} R_1$

where $\sigma = \frac{y}{x}$

Once again, after some algebraic manipulation (see Problem 3.63), the expression for R_3 can be reduced to

$$R_3 = \frac{(1+2\sigma)^4}{(1+\sigma)^2} R_1. \tag{3.66}$$

The results of our analysis are summarized in Table 3.1.

We now turn to the specific numerical example posed at the start of this Practical Perspective. From the specifications, we calculate σ and R_3. Thus,

$$\sigma = \frac{y}{x} = \frac{0.025}{1} = 0.025.$$

Since the grid is 1 m wide, the power dissipation in R_3 must be 120 W. Therefore,

$$R_3 = \frac{12^2}{120} = 1.2\ \Omega.$$

Now use the last entry in Table 3.1 to calculate R_1, since R_3 and σ are known. The result is

$$R_1 = 1.0372\ \Omega.$$

The first three entries in Table 3.1 are used to calculate R_a, R_b, and R_2. The values are

$$R_a = 0.0259\ \Omega;$$
$$R_b = 0.0068\ \Omega;$$
$$R_2 = 1.1435\ \Omega.$$

It follows from symmetry that

$$R_4 = R_2 = 1.1435\ \Omega;$$
$$R_5 = R_1 = 1.0372\ \Omega;$$
$$R_c = R_b = 0.0068\ \Omega;$$
$$R_d = R_a = 0.0259\ \Omega.$$

Now test these calculations by checking the power dissipated—it should satisfy the specification of 120 W/m. First, calculate the numerical value of D. We get

$$D = 1.2758.$$

Now use Eqs. (3.54)–(3.56) to obtain i_b, i_1, and i_2. The results are

$$i_b = 21\ \text{A};$$
$$i_1 = 10.7561\ \text{A};$$
$$i_2 = 10.2439\ \text{A}.$$

It follows that $i_b^2 R_b$ equals 3 W, and the power dissipation per meter is $3/0.025 = 120$ W/m. The value of $i_1^2 R_1$ is 120 W, or 120 W/m. The value of $i_2^2 R_2$ is 120 W, or 120 W/m. Finally, we check $i_1^2 R_a$, which also gives 3 W, or 120 W/m.

Problems 3.62–3.65 are based on this defroster grid circuit.

SUMMARY

- **Series resistors**—Resistors in series can be combined to obtain a single equivalent resistance according to the equation

$$R_{\text{eq}} = \sum_{i=1}^{k} R_i = R_1 + R_2 + \cdots + R_k.$$

- **Parallel resistors**—Resistors in parallel can be combined to obtain a single equivalent resistance according to the equation

$$\frac{1}{R_{\text{eq}}} = \sum_{i=1}^{k} \frac{1}{R_i} = \frac{1}{R_1} + \frac{1}{R_2} + \cdots + \frac{1}{R_k}.$$

  When just two resistors are in parallel, the equation for equivalent resistance can be simplified to give

$$R_{\text{eq}} = \frac{R_1 R_2}{R_1 + R_2}.$$

- **Voltage division**—When voltage is divided across series resistors, as shown in the figure, the voltage across each resistor can be found according to the equations

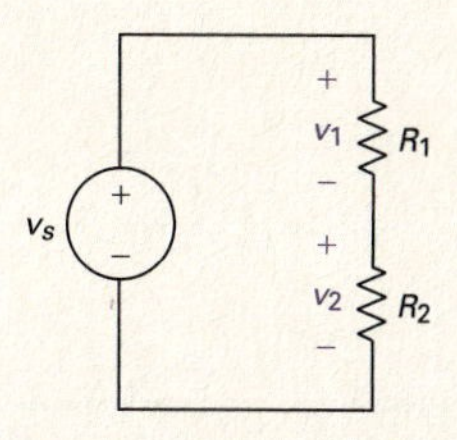

$$v_1 = \frac{R_1}{R_1 + R_2} v_s$$

  and

$$v_2 = \frac{R_2}{R_1 + R_2} v_s.$$

- **Current division**—When current is divided across parallel resistors, as shown in the figure, the current through each resistor can be found according to the equations

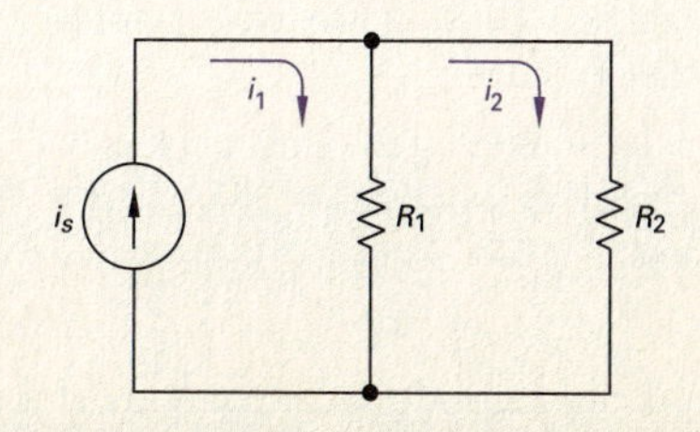

$$i_1 = i_s \frac{R_2}{R_1 + R_2}$$

and

$$i_2 = i_s \frac{R_1}{R_1 + R_2}.$$

- **Voltmeters**—A voltmeter measures voltage and must be placed in parallel with the voltage being measured. An ideal voltmeter has infinite internal resistance and thus does not alter the voltage being measured.

- **Ammeters**—An ammeter measures current and must be placed in series with the current being measured. An ideal ammeter has zero internal resistance and thus does not alter the current being measured.

- **Digital meters** and **analog meters**—These realistic meters have internal resistance, which influences the value of the circuit variable being measured. Meters based on the d'Arsonval meter movement deliberately include internal resistance as a way to limit the current in the movement's coil.

- **Wheatstone bridge**—This circuit is used to make precise measurements of a resistor's value using four resistors, a dc voltage source, and a galvanometer. A Wheatstone bridge is balanced when the resistors obey Eq. (3.29), resulting in a galvanometer reading of 0 A.

- **Delta-to-wye (pi-to-tee) transformation**—A circuit with three resistors connected in a Δ configuration (or a π-configuration) can be transformed into an equivalent circuit in which the three resistors are Y connected (or T connected). The Δ-to-Y transformation is given by Eqs. (3.40)–(3.42); the Y-to-Δ transformation is given by Eqs. (3.43)–(3.45).

PROBLEMS

3.1 a) Show that the solution of the circuit in Fig. 3.9 (see Example 3.1) satisfies Kirchhoff's current law at junctions x and y.

b) Show that the solution of the circuit in Fig. 3.9 satisfies Kirchhoff's voltage law around every closed loop.

3.2 a) Find the power dissipated in each resistor in the circuit shown in Fig. 3.9.

b) Find the power delivered by the 120 V source.

c) Show that the power delivered equals the power dissipated.

3.3 Find the equivalent resistance R_{ab} for each of the circuits in Fig. P3.3.

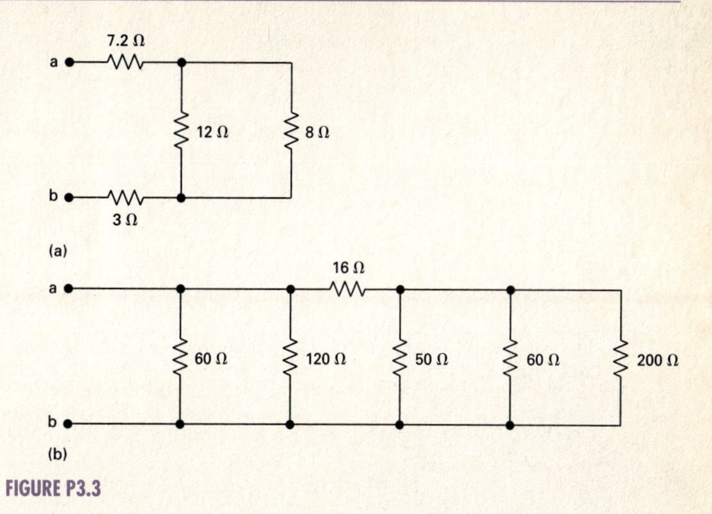

FIGURE P3.3

3.4 a) In the circuits in Fig. P3.4(a), (b), and (c), find the equivalent resistance R_{ab}.

b) For each circuit find the power delivered by the source.

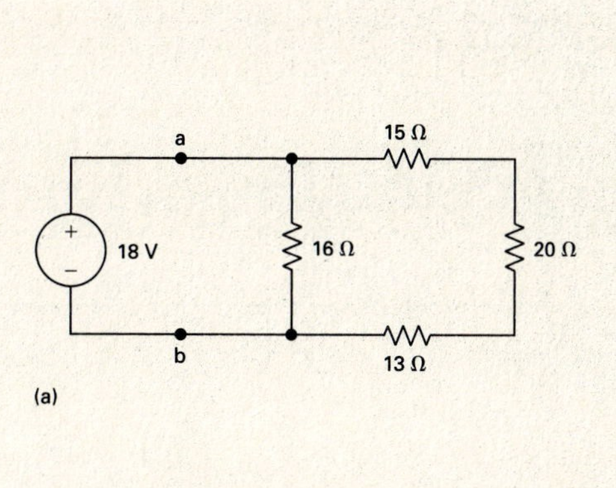

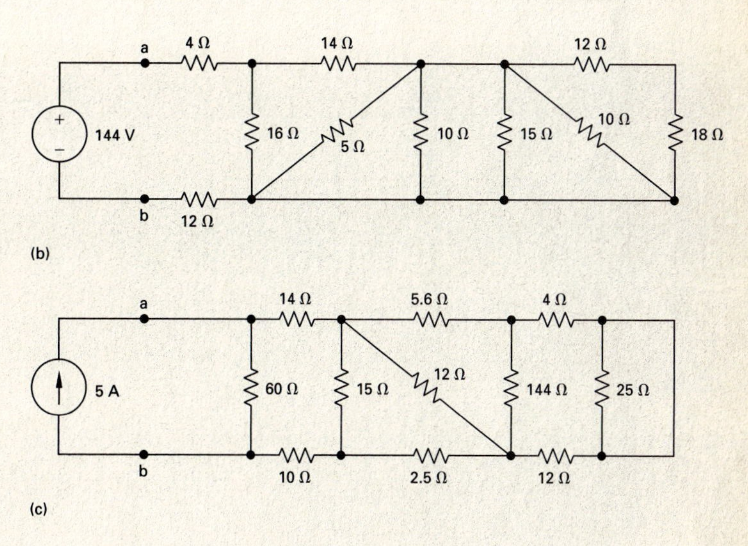

FIGURE P3.4

3.5 Find the power dissipated in the 30 Ω resistor in the circuit in Fig. P3.5.

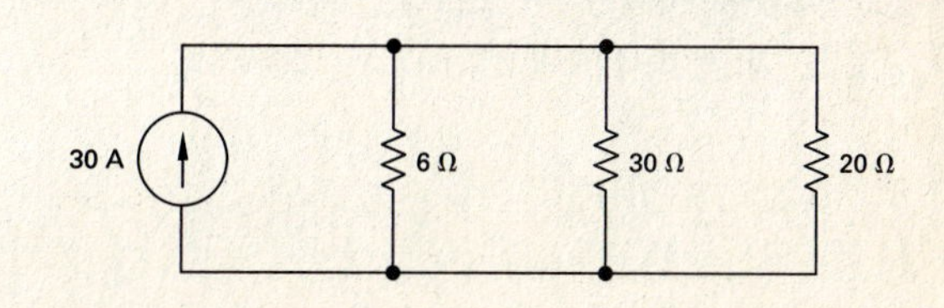

FIGURE P3.5

3.6 Find the value of v_g in the circuit in Fig. P3.6.

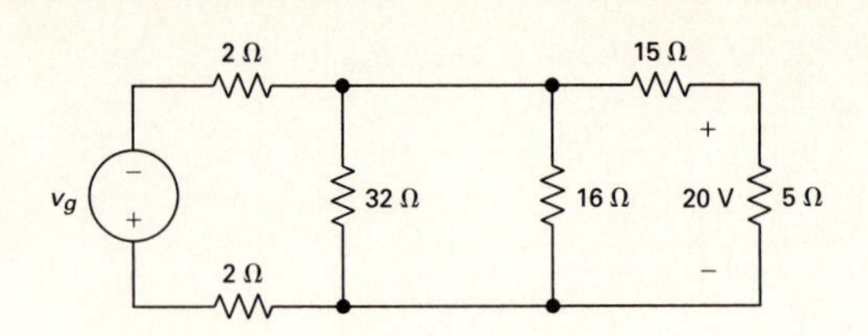

FIGURE P3.6

3.7 Find the equivalent resistance R_{ab} for each of the circuits in Fig. P3.7.

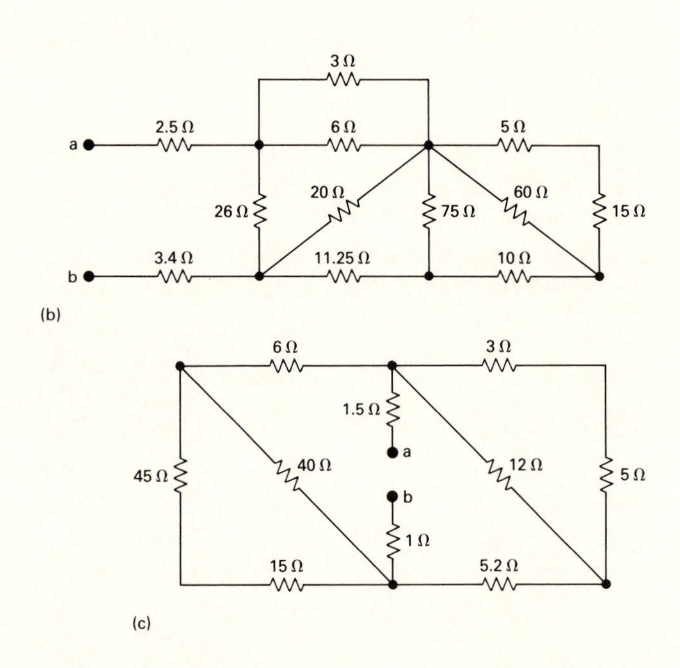

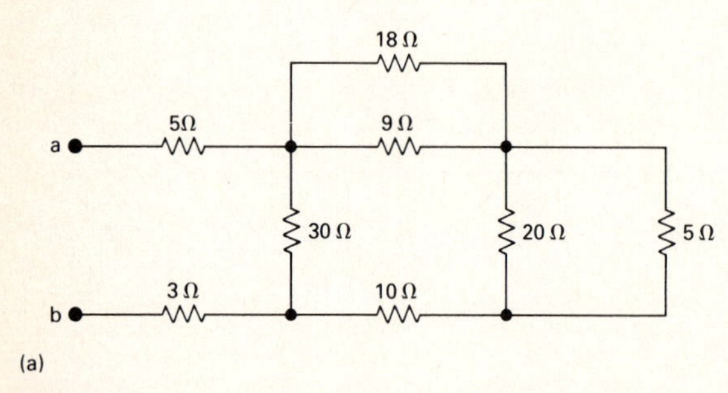

FIGURE P3.7

3.8 For the circuit in Fig. P3.8 calculate

a) v_o and i_o;

b) the power dissipated in the 12 Ω resistor;

c) the power developed by the current source.

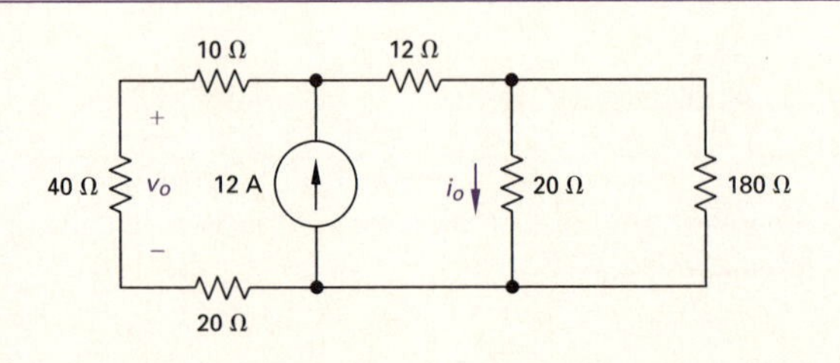

FIGURE P3.8

3.9 Find i_o and i_g in the circuit in Fig. P3.9.

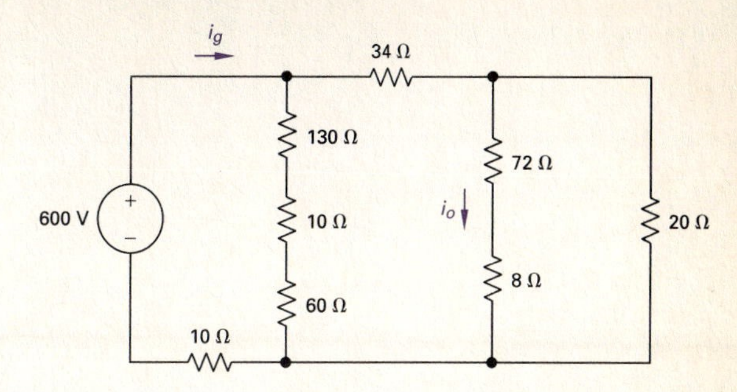

FIGURE P3.9

3.10 Find v_o and v_g in the circuit in Fig. P3.10.

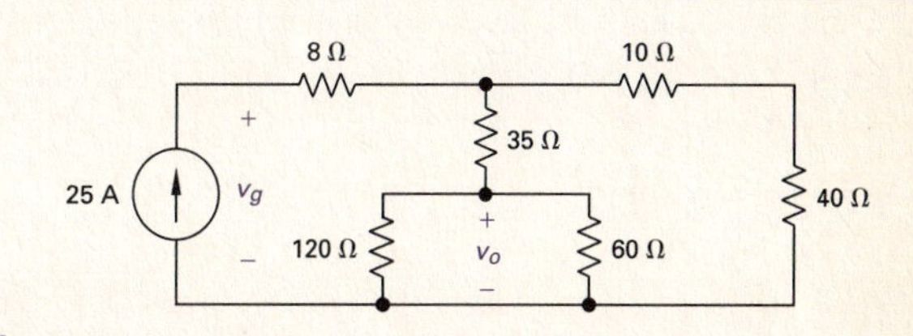

FIGURE P3.10

3.11 For the circuit in Fig. P3.11 calculate (a) i_o and (b) the power dissipated in the 15 Ω resistor.

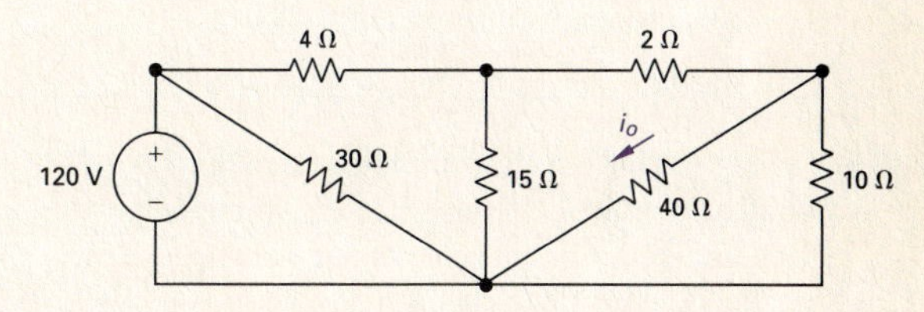

FIGURE P3.11

3.12 a) Find the voltage v_x in the circuit in Fig. P3.12.

b) Replace the 45 V source with a general voltage source equal to V_s. Assume V_s is positive at the upper terminal. Find v_x as a function of V_s.

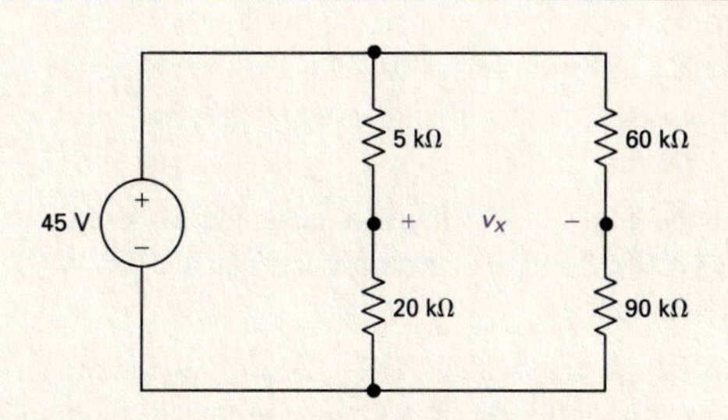

FIGURE P3.12

3.13 Find v_o in the circuit in Fig. P3.13.

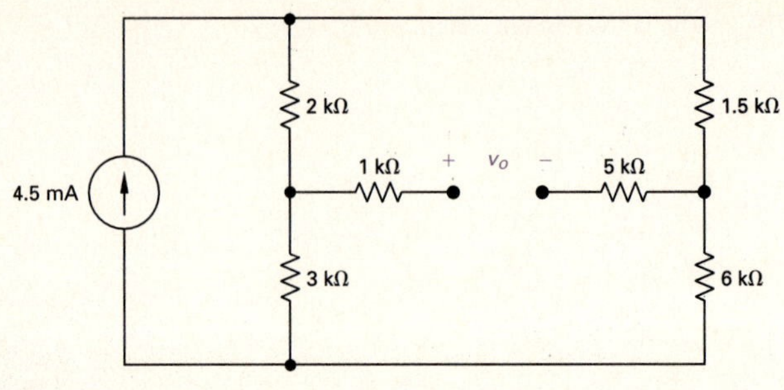

FIGURE P3.13

3.14 The current in the 12 Ω resistor in the circuit in Fig. P3.14 is 1 A, as shown.

a) Find v_g

b) Find the power dissipated in the 20 Ω resistor.

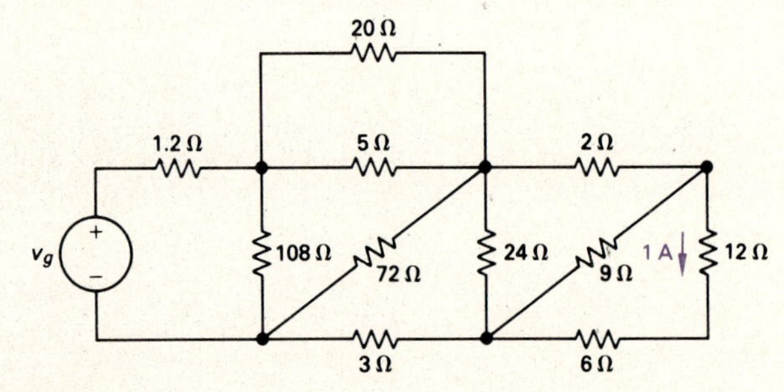

FIGURE P3.14

3.15 The high-voltage direct-current transmission line introduced in Drill Exercise 1.11 is 845 mi long. Each side of the circuit consists of two conductors in parallel. The resistance of each conductor is 0.0397 Ω/mi. The arrangement is illustrated in Fig. P3.15.

a) The voltage at the Oregon terminal of the line is 800 kV. Each conductor is carrying 1000 A, as shown in the figure. Calculate the power received at the California end of the line and the efficiency of the power transmission from Oregon to California.

b) Repeat part (a) with the voltage at Celilo raised to 1000 kV and the current remaining at 1000 A/conductor.

c) Repeat part (b) with a third conductor added to each side of the circuit and the current remaining at 1000 A/conductor.

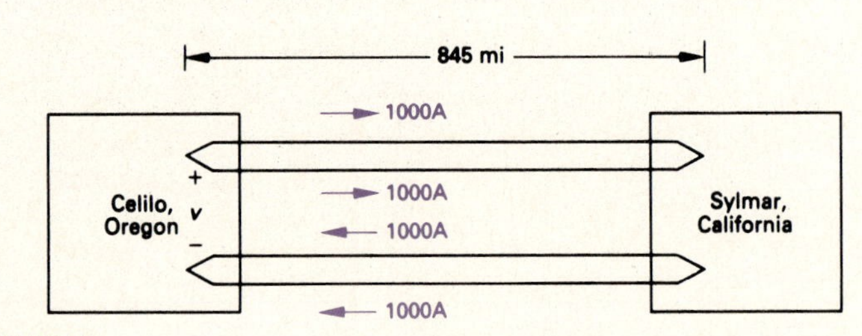

FIGURE P3.15

❖ **3.16** a) Calculate the no-load voltage v_o for the voltage-divider circuit shown in Fig. P3.16.

b) Calculate the power dissipated in R_1 and R_2.

c) Assume that only 0.25 W resistors are available. The no-load voltage is to be the same as in part (a). Specify the ohmic values of R_1 and R_2.

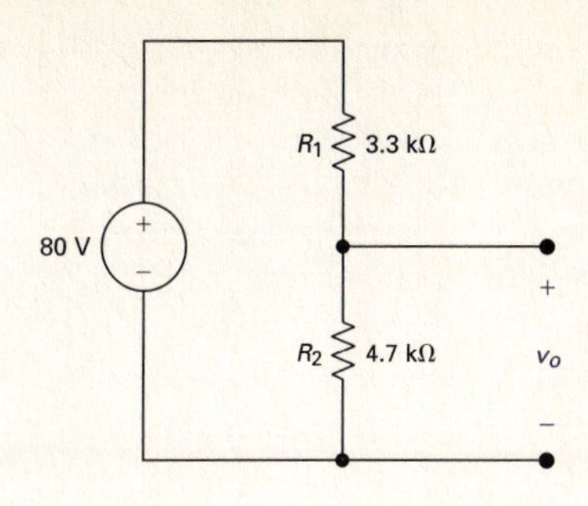

FIGURE P3.16

❖ **3.17** The no-load voltage in the voltage-divider circuit shown in Fig. P3.17 is 20 V. The smallest load resistor that is ever connected to the divider is 48 kΩ. When the divider is loaded, v_o is not to drop below 16 V.

a) Design the divider circuit to meet the specifications just mentioned. Specify the numerical value of R_1 and R_2.

b) Assume the power ratings of commercially available resistors are 1/16, 1/8, 1/4, 1, and 2 W. What power rating would you specify?

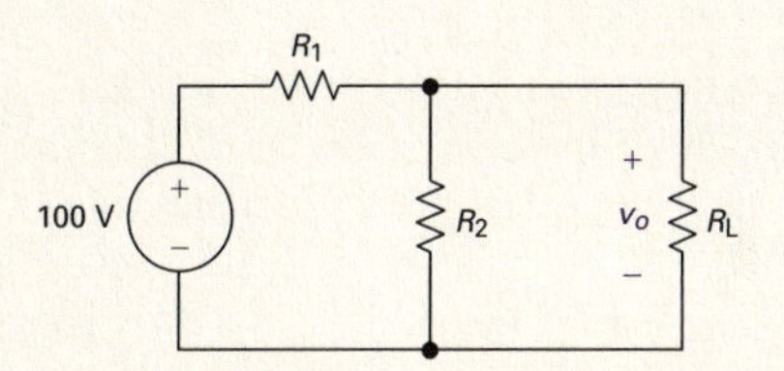

FIGURE P3.17

3.18 Assume the voltage divider in Fig. P3.17 has been constructed from 0.15 W resistors. How small can R_L be before one of the resistors in the divider is operating at its dissipation limit?

3.19 In the voltage-divider circuit shown in Fig. P3.19, the no-load value of v_o is 4 V. When the load resistance R_L is attached across the terminals a,b, v_o drops to 3 V. Find R_L.

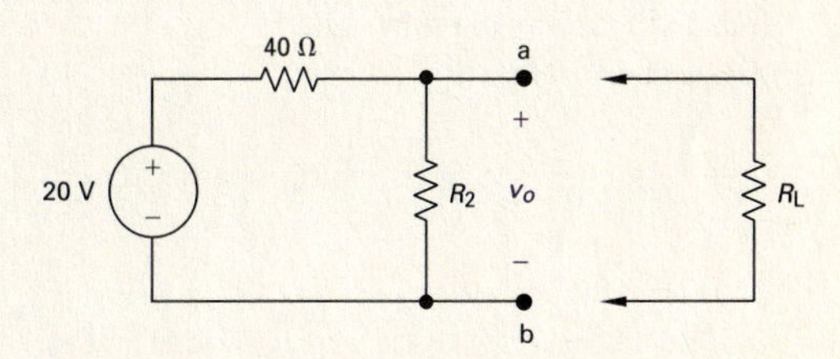

FIGURE P3.19

❖ **3.20** There is often a need to produce more than one voltage using a voltage divider. For example, the memory components of many personal computers require voltages of −12 V, 5 V, and +12 V, all with respect to a common reference terminal. Select the values of R_1, R_2, and R_3 in the circuit in Fig. P3.20 to meet the following design requirements:

1) The total power supplied to the divider circuit by the 24 V source is 80 W when the divider is unloaded.

2) The three voltages, all measured with respect to the common reference terminal, are $v_1 = 12$ V, $v_2 = 5$ V, and $v_3 = -12$ v.

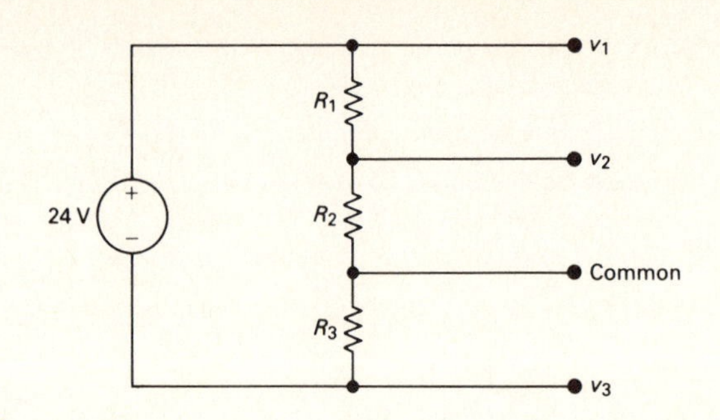

FIGURE P3.20

3.21 a) The voltage divider in Fig. P3.21(a) is loaded with the voltage divider shown in Fig. P3.21(b); that is, a is connected to a′, and b is connected to b′. Find v_o.

b) Now assume the voltage divider in Fig. P3.21(b) is connected to the voltage divider in Fig. P3.21(a) by means of a current-controlled voltage source as shown in Fig. P3.21(c). Find v_o.

c) What effect does adding the dependent-voltage source have on the operation of the voltage divider that is connected to the 480 V source?

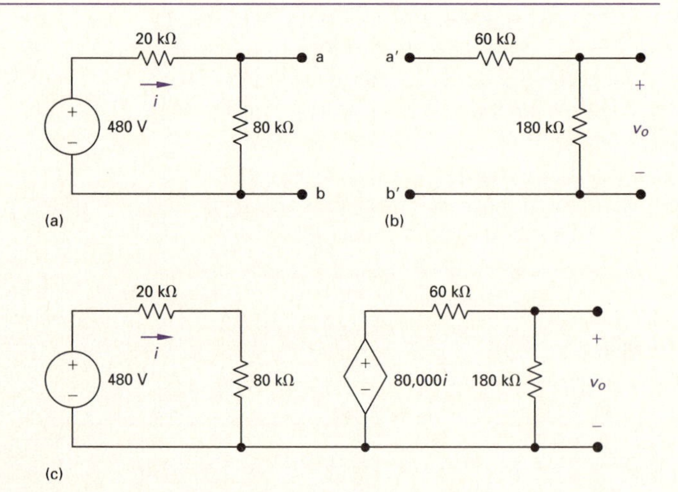

FIGURE P3.21

3.22 a) Show that the current in the kth branch of the circuit in Fig. P3.22(a) is equal to the source current i_g times the conductance of the kth branch divided by the sum of the conductances—that is,

$$i_k = \frac{i_g G_k}{[G_1 + G_2 + G_3 + \cdots + G_k + \cdots + G_N]}.$$

b) Use the result derived in part (a) to calculate the current in the 6.25 Ω resistor in the circuit in Fig. P3.22(b).

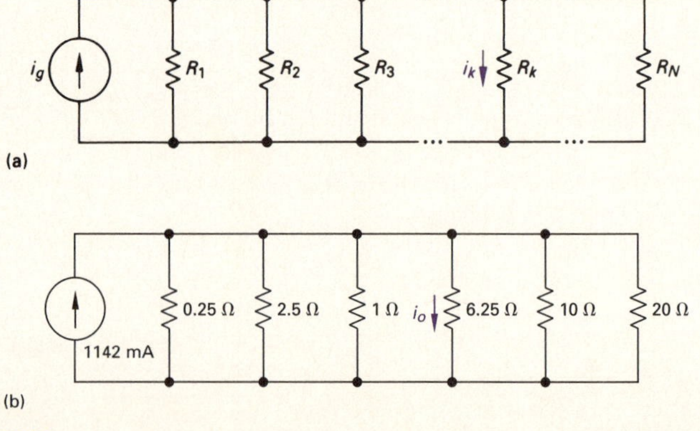

FIGURE P3.22

❖ **3.23** Specify the resistors in the circuit in Fig. P3.23 to meet the following design criteria:

$$i_g = 5 \text{ mA};\ v_g = 1 \text{ V};\ i_1 = 4i_2;\ i_2 = 8i_3;\ \text{and}\ i_3 = 5i_4.$$

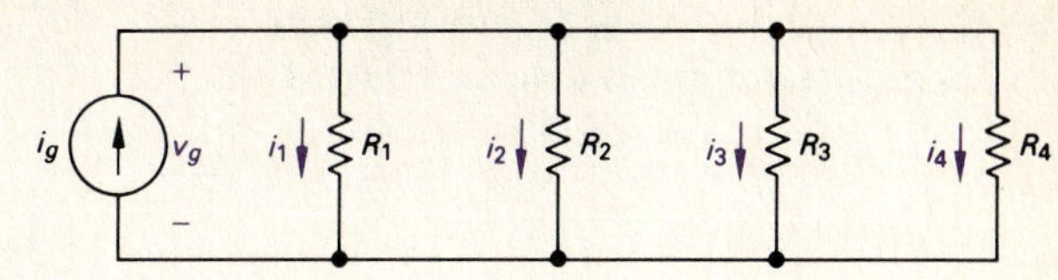

FIGURE P3.23

3.24 In the circuit in Fig. P3.24(a) the device labeled D represents a component that has the equivalent circuit shown in Fig. P3.24(b). The labels on the terminals of D show how the device is connected to the circuit. Find v_x and the power absorbed by the device.

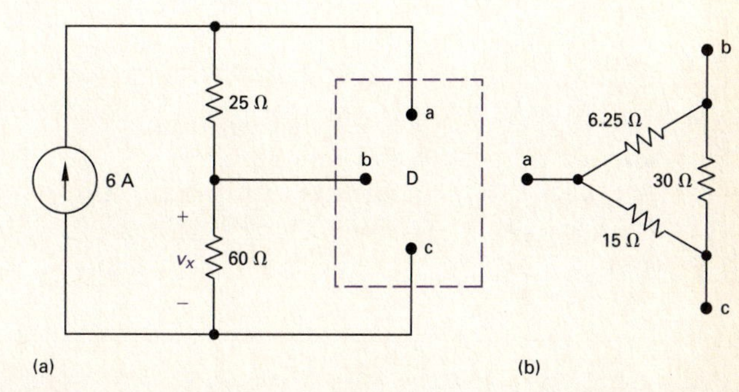

FIGURE P3.24

3.25 In Fig. P3.25(a) the box represents a field-effect transistor known as a FET. A simplified circuit model for the FET is shown in Fig. P3.25(b). Find v_o and v_{DS}.

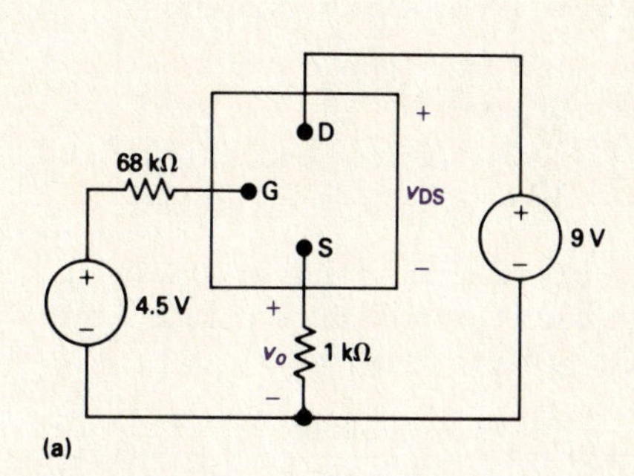

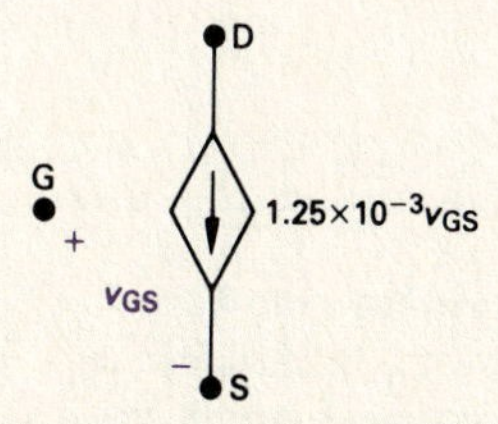

FIGURE P3.25

❖ **3.26** A d'Arsonval ammeter is shown in Fig. P3.26. Design a set of d'Arsonval ammeters to read the following full-scale current readings: (a) 5 A, (b) 2 A, (c) 1 A, and (d) 50 mA. Specify the shunt resistor R_A for each ammeter.

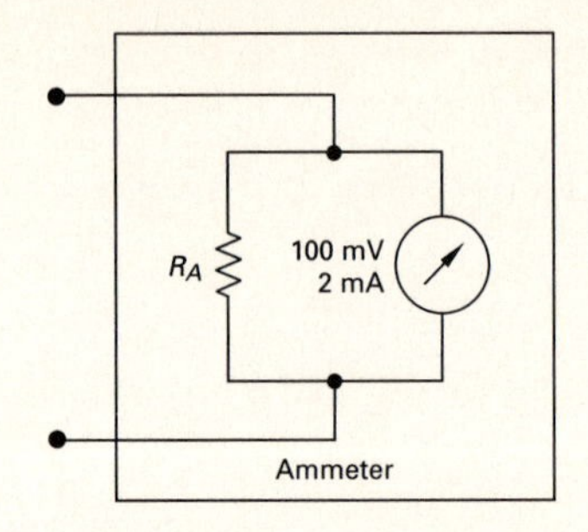

FIGURE P3.26

3.27 A shunt resistor and a 50 mV, 1 mA d'Arsonval movement are used to build a 5 A ammeter. A resistance of 0.04 Ω is placed across the terminals of the ammeter. What is the new full-scale range of the ammeter?

3.28 Two d'Arsonval ammeters are connected in parallel. Ammeter 1 uses a 2 mA, 100 mV movement and has a full-scale reading of 25 mA. Ammeter 2 uses a 1 mA, 25 mV movement and has a full-scale reading of 10 mA. What is the largest current these parallel-connected ammeters can read?

3.29 a) Show for the ammeter circuit in Fig. P3.29 that the current in the d'Arsonval movement is always 1/25th of the current being measured.

b) What would the fraction be if the 100 mV, 2 mA movement were used in a 5 A ammeter?

c) Would you expect a uniform scale on a dc d'Arsonval ammeter?

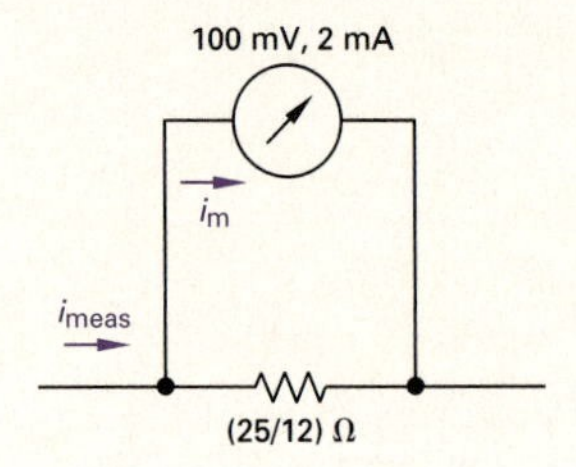

FIGURE P3.29

3.30 The ammeter in the circuit in Fig. P3.30 has a resistance of 0.5 Ω. What is the percentage of error in the reading of this ammeter if

$$\% \text{ error} = \left(\frac{\text{measured value}}{\text{true value}} - 1\right) \times 100?$$

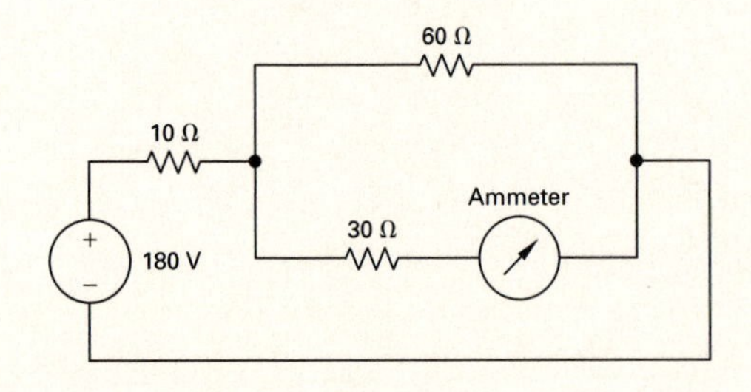

FIGURE P3.30

3.31 The paralleled ammeters described in Problem 3.28 are used to measure the current i_o in the circuit in Fig. P3.31. What is the percentage of error in the measured value?

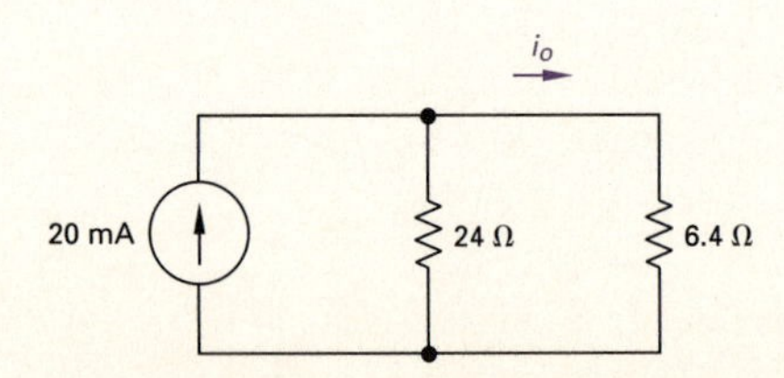

FIGURE P3.31

❖ **3.32** A d'Arsonval movement is rated at 1 mA and 50 mV. Assume 0.50 W precision resistors are available to use as shunts. What is the largest full-scale-reading ammeter that can be designed? Explain.

3.33 A d'Arsonval voltmeter is shown in Fig. P3.33. Find the value of R_V for each of the following full-scale readings: (a) 50 V, (b) 5 V, (c) 250 mV, and (d) 25 mV.

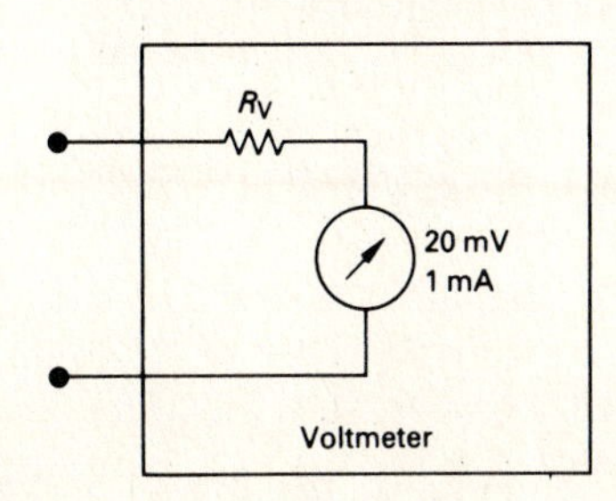

FIGURE P3.33

3.34 The voltage-divider circuit shown in Fig. P3.34 is designed so that the no-load output voltage is 8/10ths of the input voltage. A d'Arsonval voltmeter having a sensitivity of 200 Ω/V and a full-scale rating of 150 V is used to check the operation of the circuit.

a) What will the voltmeter read if it is placed across the 126 V source?

b) What will the voltmeter read if it is placed across the 60 kΩ resistor?

c) What will the voltmeter read if it is placed across the 15 kΩ resistor?

d) Will the voltmeter readings obtained in parts (b) and (c) add to the reading recorded in part (a)? Explain why or why not.

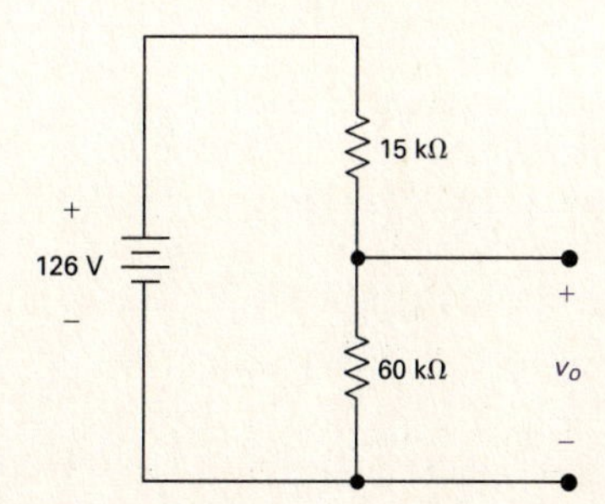

FIGURE P3.34

❖ **3.35** Design a d'Arsonval voltmeter that will have the three voltage ranges shown in Fig. P3.35.

a) Specify the values of R_1, R_2, and R_3.

b) Assume that a 750 kΩ resistor is connected between the 150 V terminal and the common terminal. The voltmeter is then connected to an unknown voltage using the common terminal and the 300 V terminal. The voltmeter reads 288 V. What is the unknown voltage?

c) What is the maximum voltage the voltmeter in part (b) can measure?

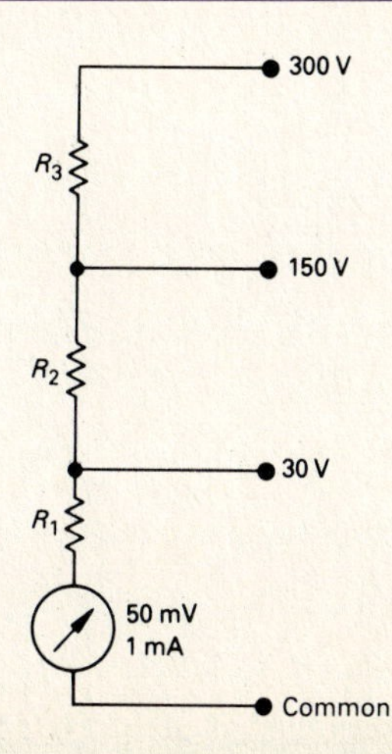

FIGURE P3.35

3.36 You have been told that the dc voltage of a power supply is about 400 V. When you go to the instrument room to get a dc voltmeter to measure the power supply voltage, you find that there are only two dc voltmeters available. One voltmeter is rated 300 V full scale and has a sensitivity of 1000 Ω/V. The second voltmeter is rated 150 V full scale and has a sensitivity of 800 Ω/V.

a) How can you use the two voltmeters to check the power supply voltage?

b) What is the maximum voltage that can be measured?

c) If the power supply voltage is 399 V, what will each voltmeter read?

3.37 Assume that in addition to the two voltmeters described in Problem 3.36, a 80 kΩ precision resistor is also available. The 80 kΩ resistor is connected in series with the series-connected voltmeters. This circuit is then connected across the terminals of the power supply. The reading on the 300 V voltmeter is 288 V, and the reading on the 150 V voltmeter is 115.2 V. What is the voltage of the power supply?

3.38 A 600 kΩ resistor is connected from the 200 V terminal to the common terminal of a dual-scale voltmeter, as shown in Fig. P3.38(a). This modified voltmeter is then used to measure the voltage across the 360 kΩ resistor in the circuit in Fig. P3.38(b).

a) What is the reading on the 500 V scale of the meter?

b) What is the percentage of error in the measured voltage?

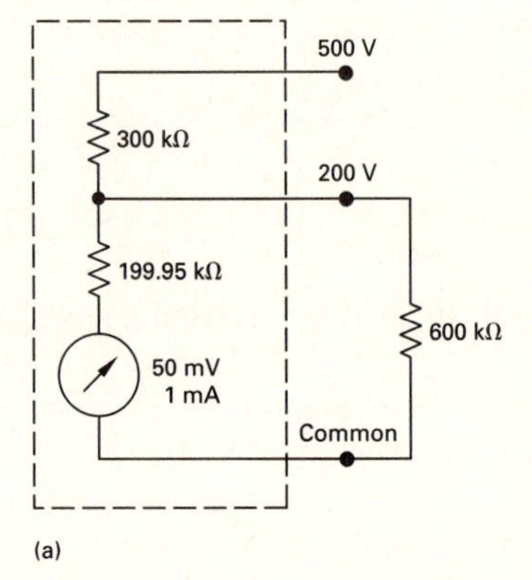

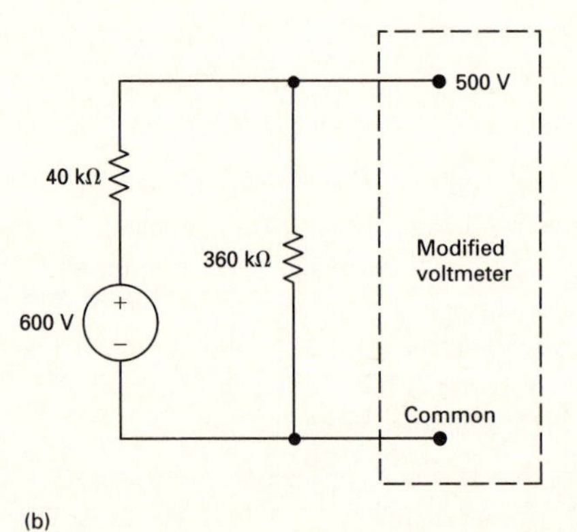

FIGURE P3.38

3.39 The voltmeter shown in Fig. P3.39(a) has a full-scale reading of 800 V. The meter movement is rated 100 mV and 1.0 mA. What is the percentage of error in the meter reading if it is used to measure the voltage v in the circuit of Fig. P3.39(b)?

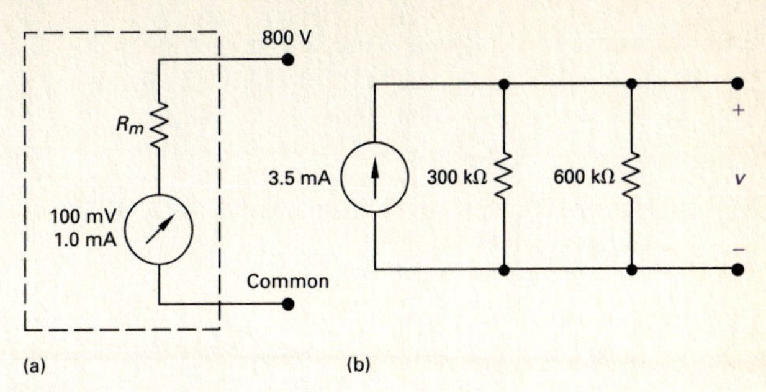

FIGURE P3.39

3.40 The elements in the circuit in Fig. 2.24 have the following values: $R_1 = 20$ kΩ, $R_2 = 80$ kΩ, $R_C = 0.82$ kΩ, $R_E = 0.20$ kΩ, $V_{CC} = 7.5$ V, $V_0 = 0.6$ V, and $\beta = 39$.

a) Calculate the value of i_B in microamperes.

b) Assume that a digital multimeter, when used as a dc ammeter, has a resistance of 1 kΩ. If the meter is inserted between terminals b and 2 to measure the current i_B, what will the meter read?

c) Using the calculated value of i_B in part (a) as the correct value, what is the percentage of error in the measurement?

3.41 The circuit model of a dc voltage source is shown in Fig. P3.41. The following voltage measurements are made at the terminals of the source: (1) With the terminals of the source open, the voltage is measured at 50 mV, and (2) with a 15 MΩ resistor connected to the terminals, the voltage is measured at 48.75 mV. All measurements are made with a digital voltmeter that has a meter resistance of 10 MΩ.

a) What is the internal voltage of the source (V_s) in millivolts?

b) What is the internal resistance of the source (R_s) in kilo-ohms?

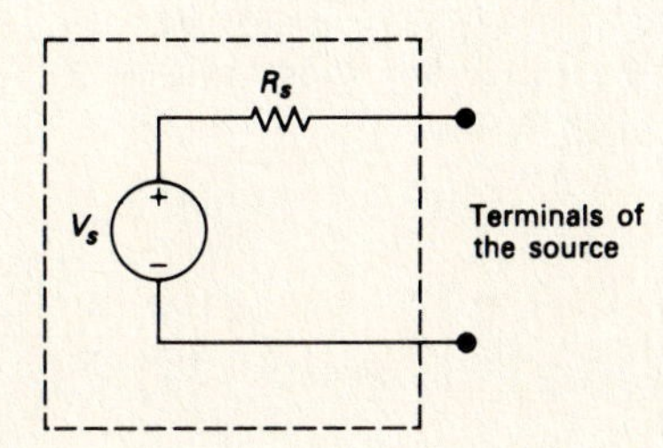

FIGURE P3.41

❖ **3.42** Assume in designing the multirange voltmeter shown in Fig. P3.42 you ignore the resistance of the meter movement.

a) Specify the values of R_1, R_2, and R_3.

b) For each of the three ranges, calculate the percentage of error that this design strategy produces.

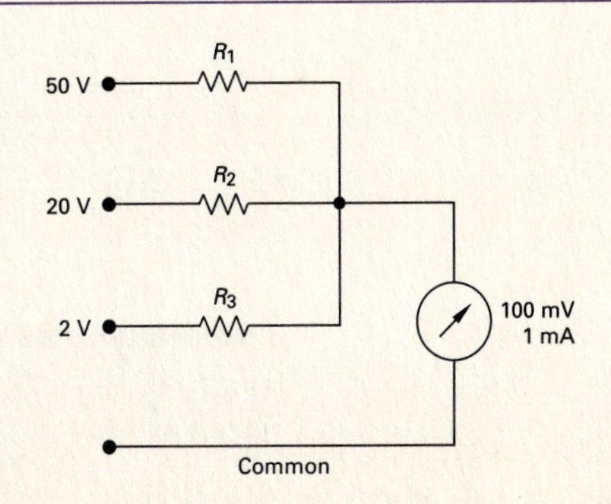

FIGURE P3.42

3.43 The bridge circuit shown in Fig. 3.23 is energized from a 9 V dc source. The bridge is balanced when $R_1 = 1000\ \Omega$, $R_2 = 4000\ \Omega$, and $R_3 = 500\ \Omega$.

a) What is the value of R_x?

b) How much current (in milliamperes) does the dc source supply?

c) Which resistor in the circuit absorbs the most power? How much power does it absorb?

d) Which resistor absorbs the least power? How much power does it absorb?

3.44 Find the power dissipated in the 9 Ω resistor in the circuit in Fig. P3.44.

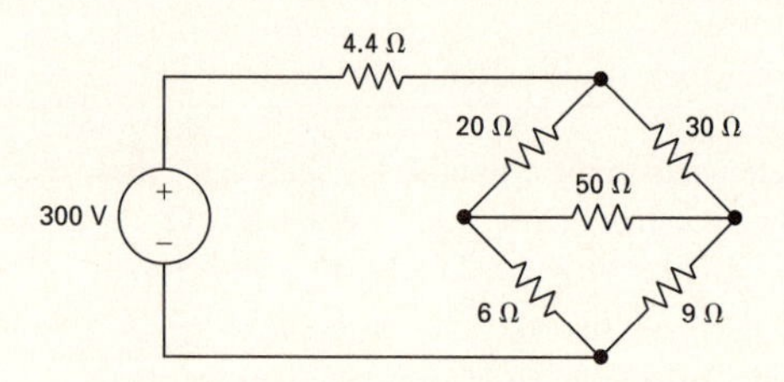

FIGURE P3.44

3.45 Find the detector current i_d in the unbalanced bridge in Fig. P3.45 if the voltage drop across the detector is negligible.

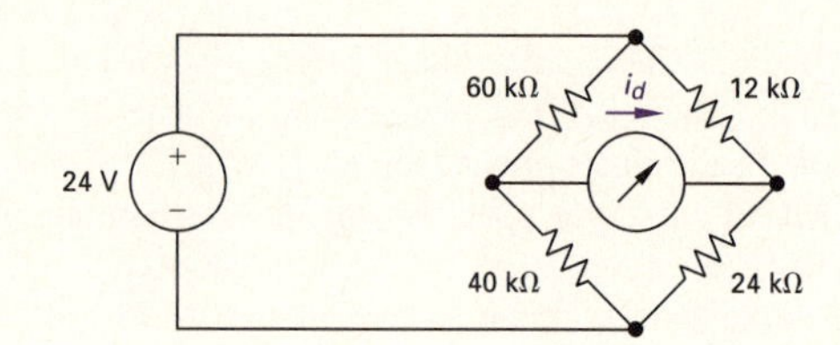

FIGURE P3.45

3.46 Assume the ideal voltage source in Fig. 3.23 is replaced by an ideal current source. Show that Eq. (3.29) is still valid.

3.47 Find R_{ab} in the circuit in Fig. P3.47.

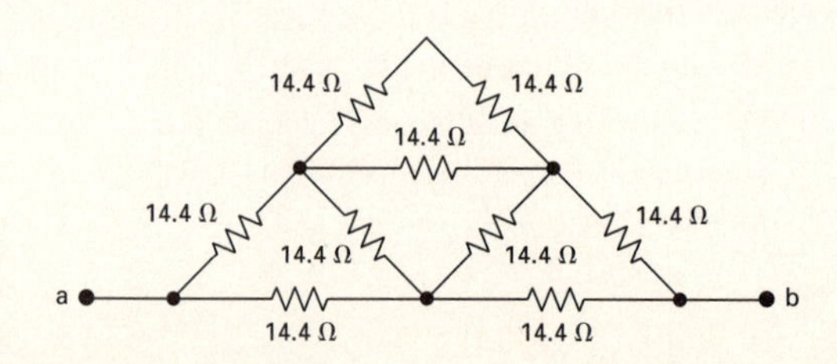

FIGURE P3.47

3.48 a) Find the equivalent resistance R_{ab} in the circuit in Fig. P3.48 by using a Δ-to-Y transformation involving the resistors R_2, R_3, and R_4.

b) Repeat part (a) using a Y-to-Δ transformation involving resistors R_2, R_4, and R_5.

c) Give two additional Δ-to-Y or Y-to-Δ transformations that could be used to find R_{ab}.

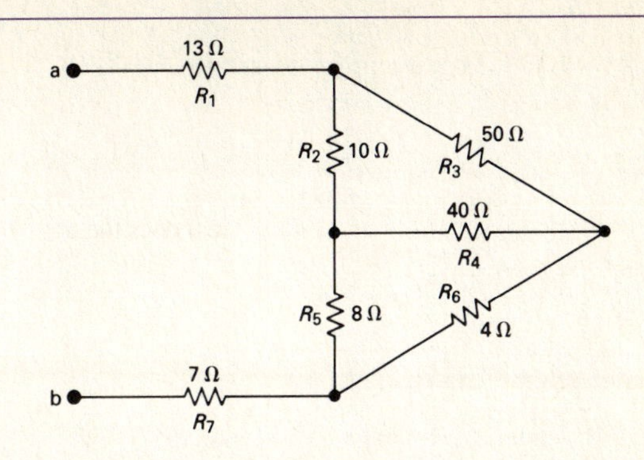

FIGURE P3.48

3.49 Use a Δ-to-Y transformation to find the current i_o in the circuit in Fig. P3.49.

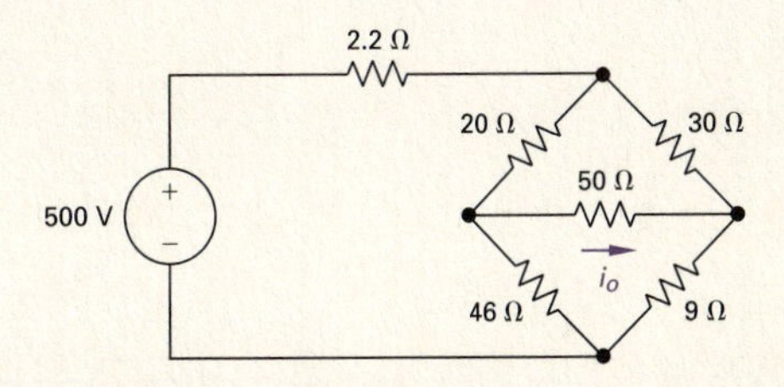

FIGURE P3.49

3.50 Find i_o and the power dissipated in the 30 Ω resistor in the circuit in Fig. P3.50.

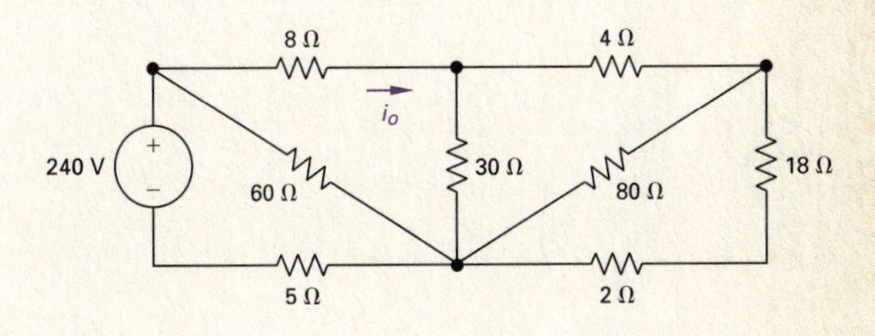

FIGURE P3.50

3.51 In the Wheatstone bridge circuit shown in Fig. 3.23, the ratio R_2/R_1 can be set to the following values: 0.001, 0.01, 0.1, 1, 10, 100, and 1000. The resistor R_3 can be varied from 1 to 11,110 Ω, in increments of 1 Ω. An unknown resistor is known to lie between 4 and 5 Ω. What should be the setting of the R_2/R_1 ratio so that the unknown resistor can be measured to four significant figures?

3.52 Derive Eqs. (3.40)–(3.45) from Eqs. (3.37)–(3.39). The following two hints should help you get started in the right direction:

1) To find R_1 as a function of R_a, R_b, and R_c, first subtract Eq. (3.38) from Eq. (3.39) and then add this result to Eq. (3.37). Use similar manipulations to find R_2 and R_3 as functions of R_a, R_b, and R_c.

2) To find R_b as a function of R_1, R_2, and R_3, take advantage of the derivations obtained by hint (1), namely, Eqs. (3.40)–(3.42). Note that these equations can be divided to obtain

$$\frac{R_2}{R_3} = \frac{R_c}{R_b}, \text{ or } R_c = \frac{R_2}{R_3}R_b,$$

and

$$\frac{R_1}{R_2} = \frac{R_b}{R_a}, \text{ or } R_a = \frac{R_2}{R_1}R_b.$$

Now use these ratios in Eq. (3.39) to eliminate R_a and R_c. Use similar manipulations to find R_a and R_c as functions of R_1, R_2, and R_3.

3.53 Show that the expressions for Δ conductances as functions of the three Y conductances are

$$G_a = \frac{G_2 G_3}{G_1 + G_2 + G_3},$$

$$G_b = \frac{G_1 G_3}{G_1 + G_2 + G_3},$$

and

$$G_c = \frac{G_1 G_2}{G_1 + G_2 + G_3},$$

where

$$G_a = \frac{1}{R_a}, \quad G_1 = \frac{1}{R_1}, \text{ etc.}$$

3.54 For the circuit shown in Fig. P3.54, find (a) i_2, (b) i_1, (c) v, and (d) the power supplied by the voltage source.

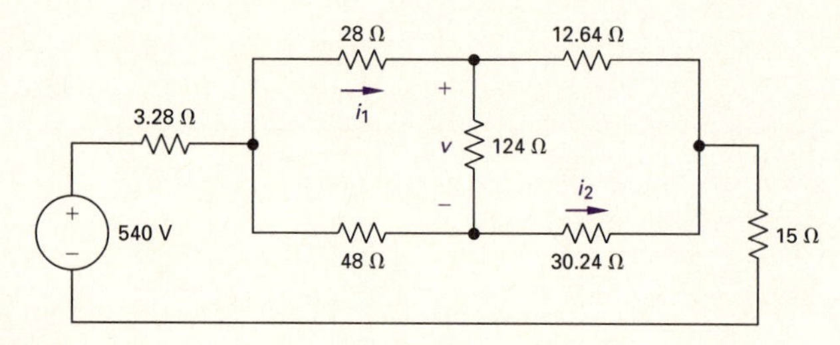

FIGURE P3.54

3.55 Find the equivalent resistance R_{ab} in the circuit in Fig. P3.55.

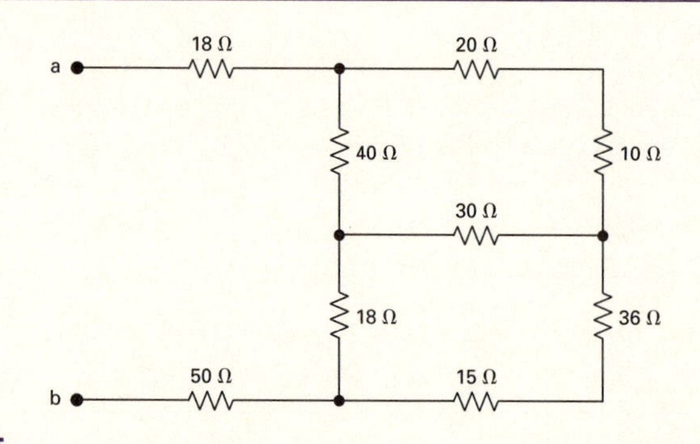

FIGURE P3.55

3.56 Use a Y-to-Δ transformation to find (a) i_o; (b) i_1; (c) i_2; and (d) the power delivered by the ideal current source in the circuit in Fig. P3.56.

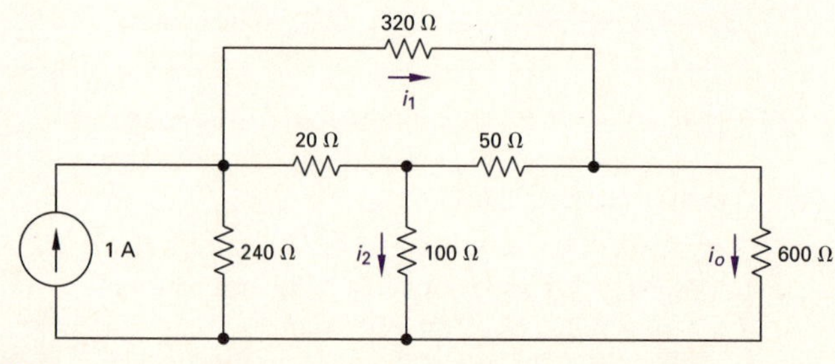

FIGURE P3.56

3.57 a) Find the resistance seen by the ideal voltage source in the circuit in Fig. P3.57.

b) If v_{ab} equals 500 V, how much power is dissipated in the 26 Ω resistor?

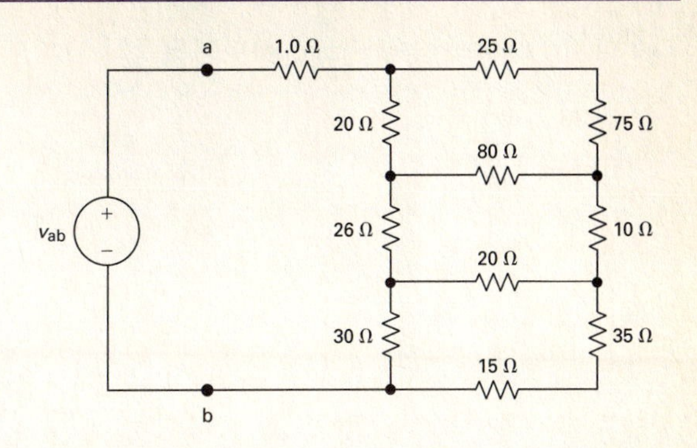

FIGURE P3.57

3.58 A voltage divider like that in Fig. 3.13 is to be designed so that $v_o = kv_s$ at no load ($R_L = \infty$) and $v_o = \alpha v_s$ at full load ($R_L = R_o$). Note that by definition $\alpha < k < 1$.

a) Show that

$$R_1 = \frac{k - \alpha}{\alpha k} R_o \quad \text{and} \quad R_2 = \frac{k - \alpha}{\alpha(1 - k)} R_o.$$

b) Specify the numerical values of R_1 and R_2 if $k = 0.90$, $\alpha = 0.75$, and $R_o = 36$ kΩ.

c) If $v_s = 120$ V, specify the maximum power that will be dissipated in R_1 and R_2.

d) Assume the load resistor is accidentally short-circuited. How much power is dissipated in R_1 and R_2?

3.59 Resistor networks are sometimes used as volume-control circuits. In this application, they are referred to as *resistance attenuators* or *pads*. A typical fixed-attenuator pad is shown in Fig. P3.59. In designing an attenuation pad, the circuit designer will select the values of R_1 and R_2 so that the ratio of v_o/v_i and the resistance seen by the input voltage source R_{ab} both have a specified value.

a) Show that if $R_{ab} = R_L$, then

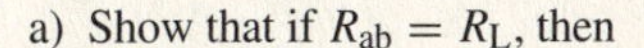

$$R_L^2 = 4R_1(R_1 + R_2)$$

and

$$\frac{v_o}{v_i} = \frac{R_2}{2R_1 + R_2 + R_L}.$$

b) Select the values of R_1 and R_2 so that $R_{ab} = R_L = 600$ Ω and $v_o/v_i = 0.6$.

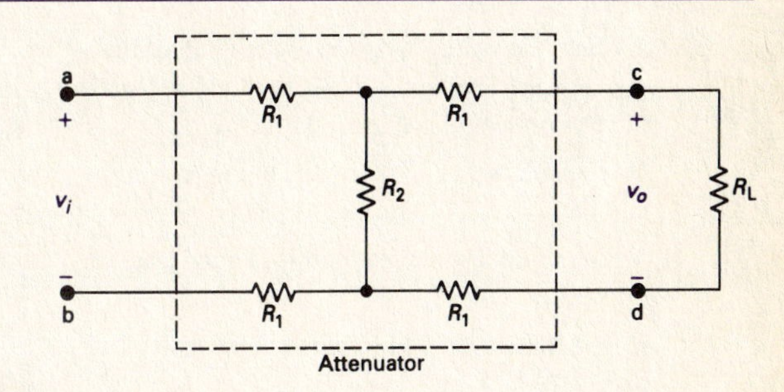

FIGURE P3.59

3.60 a) The fixed-attenuator pad shown in Fig. P3.60 is called a *bridged tee*. Use a Y-to-Δ transformation to show that $R_{ab} = R_L$ if $R = R_L$.

b) Show that when $R = R_L$, the voltage ratio v_o/v_i equals 0.50.

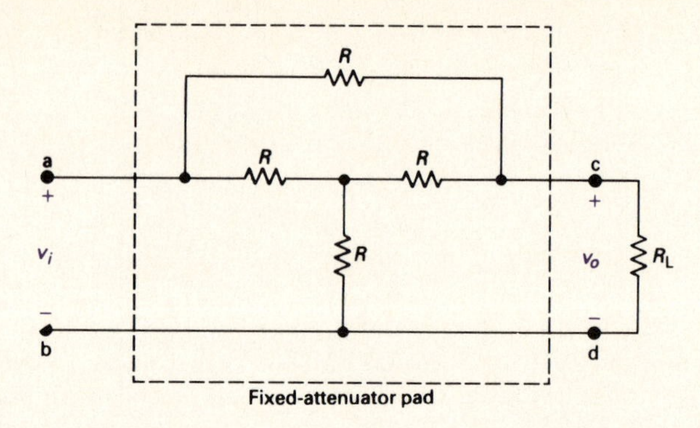

FIGURE P3.60

3.61 The design equations for the bridged-tee attenuator circuit in Fig. P3.61 are

$$R_2 = \frac{2RR_L^2}{3R^2 - R_L^2} \quad \text{and} \quad \frac{v_o}{v_i} = \frac{3R - R_L}{3R + R_L}$$

when R_2 has the value just given.

a) Design a fixed attenuator so that $v_i = 3.5v_o$ when $R_L = 300\ \Omega$.

b) Assume the voltage applied to the input of the pad designed in part (a) is 42 V. Which resistor in the pad dissipates the most power?

c) How much power is dissipated in the resistor in part (b)?

d) Which resistor in the pad dissipates the least power?

e) How much power is dissipated in the resistor in part (d)?

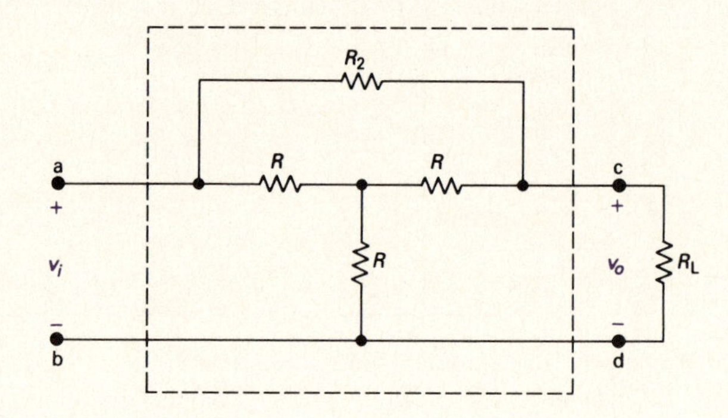

FIGURE P3.61

3.62 a) Derive Eq. (3.61).

b) Derive Eq. (3.64).

3.63 Derive Eq. (3.66).

◆ **3.64** Check the solution to the Practical Perspective example by showing that the total power dissipated equals the power developed by the 12 V source.

 **3.65** a) Design a defroster grid having five horizontal conductors to meet the following specifications: The grid is to be 1.5 m wide, the vertical separation between conductors is to be 0.03 m, and the power dissipation is to be 200 W/m when the supply voltage is 12 V.

b) Check your solution and make sure it meets the design specifications.

4 TECHNIQUES OF CIRCUIT ANALYSIS

CHAPTER CONTENTS

So far, we have analyzed relatively simple resistive circuits by applying Kirchhoff's laws in combination with Ohm's law. We can use this approach for all circuits, but as they become structurally more complicated and involve more and more elements, this direct method soon becomes cumbersome. In this chapter we introduce two powerful techniques of circuit analysis that aid in the analysis of complex circuit structures: the node-voltage method and the mesh-current method. These techniques give us two systematic methods of describing circuits with the minimum number of simultaneous equations.

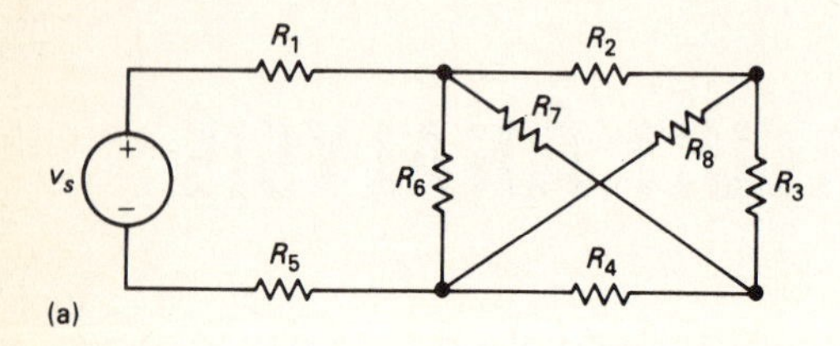

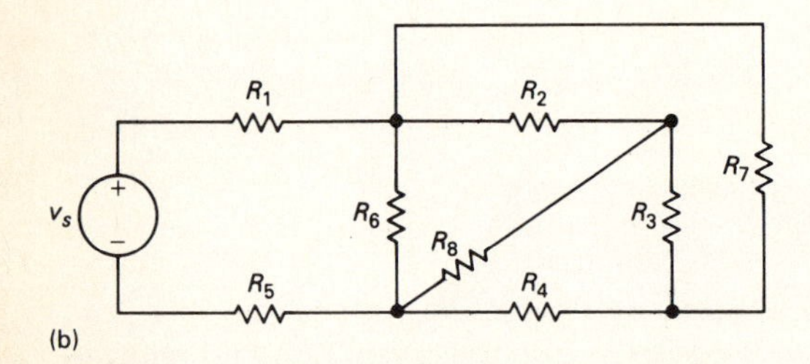

FIGURE 4.1 (a) A planar circuit; (b) the same circuit redrawn to verify that it is planar.

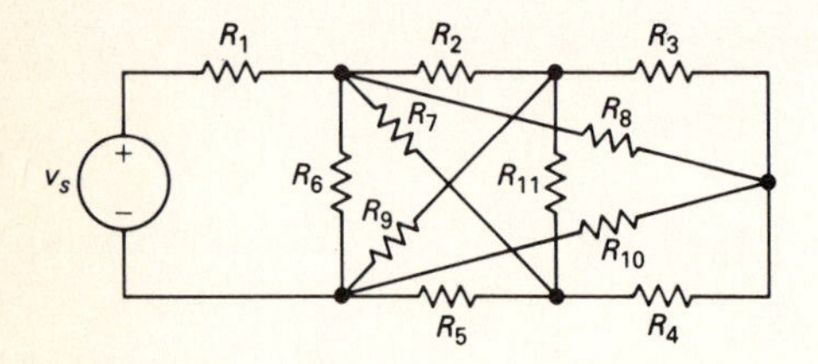

FIGURE 4.2 A nonplanar circuit.

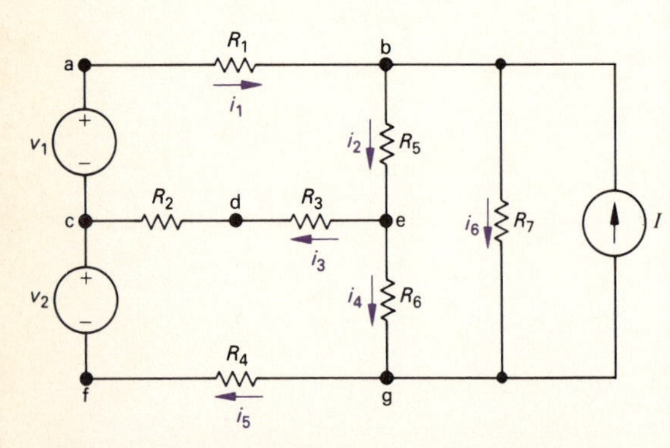

FIGURE 4.3 A circuit illustrating nodes, branches, meshes, paths, and loops.

In addition to these two general analytical methods, in this chapter we also discuss other techniques for simplifying circuits. We have already demonstrated how to use series-parallel reductions and Δ-to-Y transformations to simplify a circuit's structure. We now add source transformations and Thévenin and Norton equivalent circuits to those techniques.

We also consider two other topics which play a role in circuit analysis. One, maximum power transfer, considers the conditions necessary to ensure that the power delivered to a resistive load by a source is maximized. Thévenin equivalent circuits are used in establishing the maximum power transfer conditions. The final topic in this chapter, superposition, looks at the analysis of circuits with more than one independent source.

4.1 TERMINOLOGY

To discuss the more involved methods of circuit analysis, we must define a few basic terms necessary for a clear, concise description of important circuit features. So far, all the circuits presented have been **planar circuits**—that is, those circuits that can be drawn on a plane with no crossing branches. A circuit that is drawn with crossing branches still is considered planar if it can be redrawn with no crossover branches. For example, the circuit shown in Fig. 4.1(a) can be redrawn as Fig. 4.1(b); the circuits are equivalent because all the node connections have been maintained. Therefore, Fig. 4.1(a) is a planar circuit because it can be redrawn as one. Figure 4.2 shows a nonplanar circuit—it cannot be redrawn in such a way that all the node connections are maintained and no branches overlap. The node-voltage method is applicable to both planar and nonplanar circuits, whereas the mesh-current method is limited to planar circuits. We provide the mathematical foundation for these methods in Appendix C.

DESCRIBING A CIRCUIT—THE VOCABULARY

In Section 1.5 we defined an ideal basic circuit element. When basic circuit elements are interconnected to form a circuit, the resulting interconnection is described in terms of nodes, paths, branches, loops, and meshes. We defined both a node and a closed path, or loop, in Section 2.4. Here we restate those definitions and then define the terms *path, branch,* and *mesh.* For your convenience, all of these definitions are presented in Table 4.1, with examples of each definition taken from the circuit in Fig. 4.3.

TABLE 4.1

TERMS FOR DESCRIBING CIRCUITS

NAME	DEFINITION	EXAMPLE FROM FIG. 4.3
node	A point where two or more circuit elements join	a
essential node	A node where three or more circuit elements join	b
path	A trace of adjoining basic elements with no elements included more than once	v_1–R_1–R_5–R_6
branch	A path that connects two nodes	R_1
essential branch	A path which connects two essential nodes without passing through an essential node	v_1–R_1
loop	A path whose last node is the same as the starting node	v_1–R_1–R_5–R_6–R_4–v_2
mesh	A loop that does not enclose any other loops	v_1–R_1–R_5–R_3–R_2

EXAMPLE 4.1

For the circuit in Fig. 4.3, identify

a) all nodes;

b) all essential nodes;

c) all branches;

d) all essential branches;

e) all meshes;

f) two paths which are not loops or essential branches; and,

g) two loops which are not meshes.

SOLUTION

a) The nodes are a, b, c, d, e, f, and g.

b) The essential nodes are b, c, e, and g.

c) The branches are v_1, v_2, R_1, R_2, R_3, R_4, R_5, R_6, R_7, and I.

d) The essential branches are v_1–R_1, R_2–R_3, v_2–R_4, R_5, R_6, R_7, and I.

e) The meshes are v_1–R_1–R_5–R_3–R_2, v_2–R_2–R_3–R_6–R_4, R_5–R_7–R_6, and R_7–I.

f) R_1–R_5–R_6 is a path but it is not a loop (because it does not have the same starting and ending nodes), nor is it an essential branch (because it does not connect two essential nodes). v_2–R_2 is also a path but neither a loop nor an essential branch, for the same reasons.

g) v_1–R_1–R_5–R_6–R_4–v_2 is a loop but is not a mesh, since there are two loops within it. I–R_5–R_6 is also a loop but not a mesh.

SIMULTANEOUS EQUATIONS—HOW MANY?

The number of unknown currents in a circuit equals the number of branches, b, where the current is not known. For example, the circuit shown in Fig. 4.3 has nine branches in which the current is unknown. Recall that we must have b independent equations to solve a circuit with b unknown currents. If we let n represent the number of nodes in the circuit, we can derive $n - 1$ independent equations by applying Kirchhoff's current law to any set of $n - 1$ nodes. (Application of the current law to the nth node does not generate an independent equation, because this equation can be derived from the previous $n - 1$ equations. See Drill Exercise 4.2.) Because we need b equations to describe a given circuit and because we can obtain $n - 1$ of these equations from Kirchhoff's current law, we must apply Kirchhoff's voltage law to loops or meshes to obtain the remaining $b - (n - 1)$ equations.

Thus by counting nodes, meshes, and branches where the current is unknown, we have established a systematic method for writing the necessary number of equations to solve a circuit. Specifically, we apply Kirchhoff's current law to $n - 1$ nodes and Kirchhoff's voltage law to $b - (n - 1)$ loops (or meshes). These observations also are valid in terms of essential nodes and essential branches. Thus if we let n_e represent the number of essential nodes and b_e the number of essential branches where the current is unknown, we can apply Kirchhoff's current law at $n_e - 1$ nodes and Kirchhoff's voltage law around $b_e - (n_e - 1)$ loops or meshes. In circuits, the number of essential nodes is less than or equal to the number of nodes, and the number of essential branches is less than or equal to the number of branches. Thus it is often convenient to use essential nodes and essential branches when analyzing a circuit, since they produce fewer independent equations to solve.

A circuit may consist of disconnected parts. An example of such a circuit is examined in Drill Exercise 4.4. The statements pertaining to the number of equations that can be derived from Kirchhoff's current law, $n - 1$, and voltage law, $b - (n - 1)$, apply to connected circuits. If a circuit has n nodes and b branches and is made up of s parts, the current law can be applied $n - s$ times, and the voltage law $b - n + s$

times. Any two separate parts can be connected by a single conductor. This connection always causes two nodes to form one node. Moreover, no current exists in the single conductor, so any circuit made up of s disconnected parts can always be reduced to a connected circuit.

THE SYSTEMATIC APPROACH—AN ILLUSTRATION

We now illustrate this systematic approach by using the circuit shown in Fig. 4.4. We write the equations on the basis of essential nodes and branches. The circuit has four essential nodes and six essential branches, denoted i_1–i_6, for which the current is unknown.

We derive three of the six simultaneous equations needed by applying Kirchhoff's current law to any three of the four essential nodes. We use the nodes b, c, and e to get

$$\begin{aligned} -i_1 + i_2 + i_6 - I &= 0; \\ i_1 - i_3 - i_5 &= 0; \\ i_3 + i_4 - i_2 &= 0. \end{aligned} \tag{4.1}$$

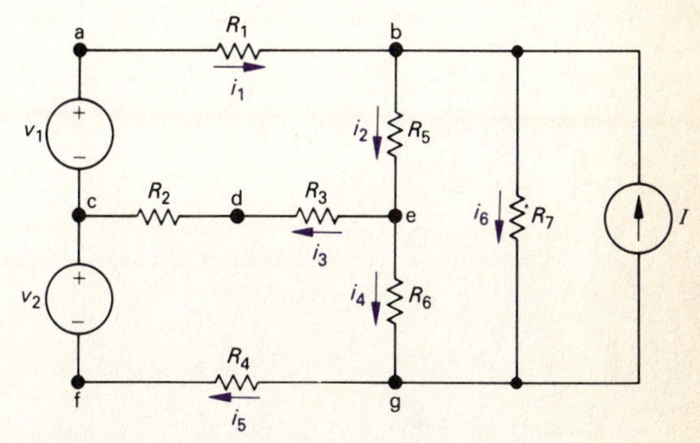

FIGURE 4.4 The circuit shown in Fig. 4.3, with six unknown branch currents defined.

We derive the remaining three equations by applying Kirchhoff's voltage law around three meshes. Because the circuit has four meshes, we need to dismiss one mesh. We choose R_7–I, because we don't know the voltage across I.[1] Using the other three meshes gives

$$\begin{aligned} R_1 i_1 + R_5 i_2 + i_3(R_2 + R_3) - v_1 &= 0; \\ -i_3(R_2 + R_3) + i_4 R_6 + i_5 R_4 - v_2 &= 0; \\ -i_2 R_5 + i_6 R_7 - i_4 R_6 &= 0. \end{aligned} \tag{4.2}$$

Rearranging Eqs. (4.1) and (4.2) to facilitate their solution yields the set

$$\begin{aligned} -i_1 + i_2 + 0i_3 + 0i_4 + 0i_5 + i_6 &= I; \\ i_1 + 0i_2 - i_3 + 0i_4 - i_5 + 0i_6 &= 0; \\ 0i_1 - i_2 + i_3 + i_4 + 0i_5 + 0i_6 &= 0; \\ R_1 i_1 + R_5 i_2 + (R_2 + R_3) i_3 + 0i_4 + 0i_5 + 0i_6 &= v_1; \\ 0i_1 + 0i_2 - (R_2 + R_3) i_3 + R_6 i_4 + R_4 i_5 + 0i_6 &= v_2; \\ 0i_1 + R_5 i_2 + 0i_3 - R_6 i_4 + 0i_5 + R_7 i_6 &= 0. \end{aligned} \tag{4.3}$$

Note that summing the current at the nth node (g in this example) gives

$$i_5 - i_4 - i_6 + I = 0. \tag{4.4}$$

Equation (4.4) is not independent, because we can derive it by summing Eqs. (4.1) and then multiplying the sum by -1. Thus Eq. (4.4) is a linear combination of Eqs. (4.1) and therefore is not independent

[1] We say more about this decision in Section 4.7.

of them. We now carry the procedure one step further. By introducing new variables, we can describe a circuit with just $n - 1$ equations or just $b - (n - 1)$ equations. Therefore these new variables allow us to obtain a solution by manipulating fewer equations, a desirable goal even if a computer is to be used to obtain a numerical solution.

Computer tools help solve circuits with multiple equations.

The new variables are known as node voltages and mesh currents. The node-voltage method enables us to describe a circuit in terms of $n_e - 1$ equations; the mesh-current method enables us to describe a circuit in terms of $b_e - (n_e - 1)$ equations. We begin in Section 4.2 with the node-voltage method.

DRILL EXERCISES

4.1 For the circuit shown, state the numerical value of the number of (a) branches, (b) branches where the current is unknown, (c) essential branches, (d) essential branches where the current is unknown, (e) nodes, (f) essential nodes, and (g) meshes.

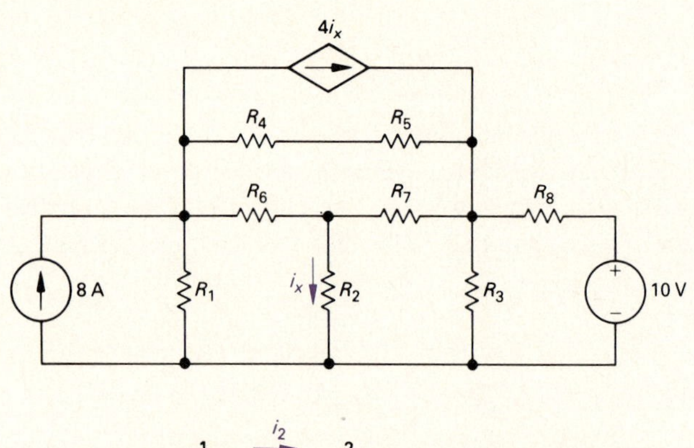

ANSWER: (a) 11; (b) 9; (c) 9; (d) 7; (e) 6; (f) 4; (g) 6.

4.2 A current leaving a node is defined as positive.

a) Sum the currents at each node in the circuit shown.

b) Show that any one of the equations in (a) can be derived from the remaining two equations.

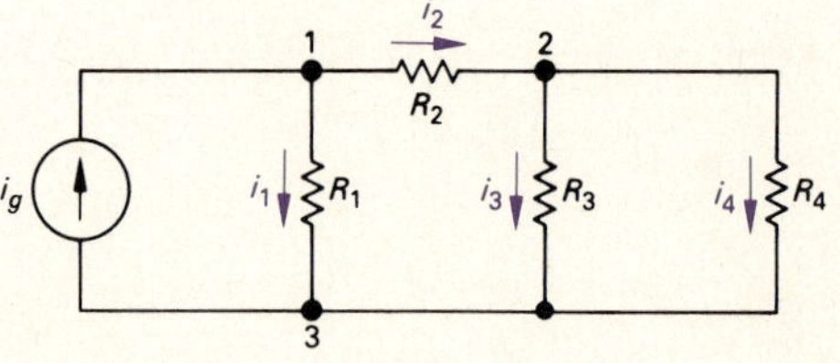

ANSWER: (a) 1: $i_1 - i_g + i_2 = 0$; 2: $i_3 + i_4 - i_2 = 0$; 3: $i_g - i_1 - i_3 - i_4 = 0$. (b) To derive any one equation from the other two equations, simply add the two equations and then multiply the resulting sum by -1.

4.3 a) If only the essential nodes and branches are identified in the circuit of Drill Exercise 4.1, how many simultaneous equations are needed to describe the circuit?

b) How many of these equations can be derived using Kirchhoff's current law?

c) How many must be derived using Kirchhoff's voltage law?

d) What two meshes should be avoided in applying the voltage law?

ANSWER: (a) 7; (b) 3; (c) 4; (d) R_4–R_5–$4i_x$ and 8 A–R_1.

4.4 a) How many separate parts does the circuit shown have?

b) How many nodes?

c) How many independent current equations can be written?

d) How many branches are there?

e) How many branches are there where the current is unknown?

f) How many equations must be written using the voltage law?

g) Assume that the lower node in each part of the circuit is joined by a single conductor. Repeat the calculations in (a) through (g).

ANSWER: (a) 2; (b) 5; (c) 3; (d) 7; (e) 6; (f) 3; (g) 1, 4, 3, 7, 6, 3.

4.2 Introduction to the Node-Voltage Method

We introduce the node-voltage method by using the essential nodes of the circuit. The first step is to make a neat layout of the circuit so that no branches cross over and to mark clearly the essential nodes on the circuit diagram, as in Fig. 4.5. This circuit has three essential nodes ($n_e = 3$); therefore we need two ($n_e - 1$) node-voltage equations to describe the circuit. The next step is to select one of the three essential nodes as a reference node. Although theoretically the choice is arbitrary, practically the choice for the reference node often is obvious. For example, the node with the most branches is usually a good choice. The optimum choice of the reference node (if one exists) will become apparent after you have gained some experience using this method. In the circuit shown in Fig. 4.5, the lower node connects the most branches, so we use it as the reference node. We flag the chosen reference node with the symbol ↓, as in Fig. 4.6.

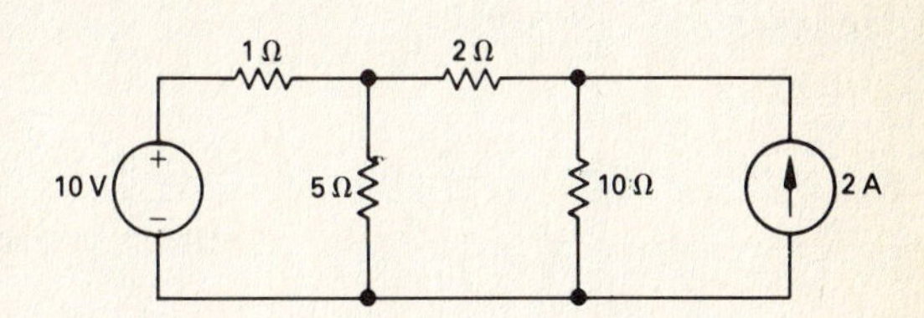

FIGURE 4.5 A circuit used to illustrate the node-voltage method of circuit analysis.

After selecting the reference node, we define the node voltages on the circuit diagram. A **node voltage** is defined as the voltage rise from the reference node to a nonreference node. For this circuit, we must define two node voltages, which are denoted v_1 and v_2 in Fig. 4.6.

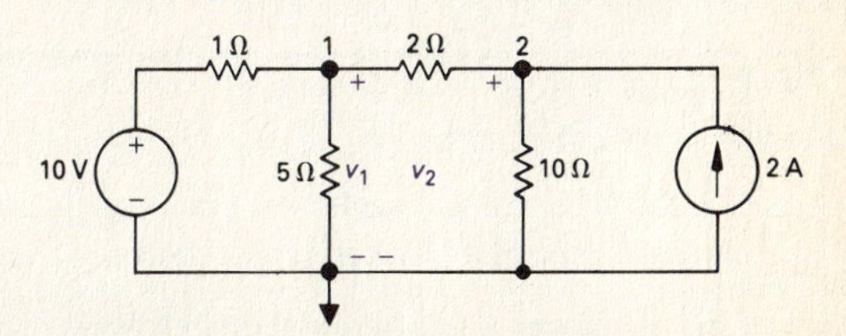

FIGURE 4.6 The circuit shown in Fig. 4.5, with a reference node and the node voltages.

We are now ready to generate the node-voltage equations. We do so by first writing the current leaving each branch connected to a nonreference node as a function of the node voltages and then summing these currents to zero in accordance with Kirchhoff's current law. For the circuit in Fig. 4.6, the current away from node 1 through the 1 Ω resistor is the voltage drop across the resistor divided by the resistance (Ohm's law). The voltage drop across the resistor, in the direction of the current away from the node, is $v_1 - 10$. Therefore the current in the 1 Ω resistor is $(v_1 - 10)/1$. Figure 4.7 depicts these observations. It shows the 10 V – 1 Ω branch, with the appropriate voltages and current.

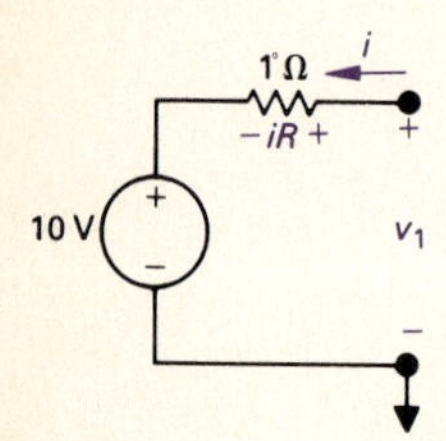

FIGURE 4.7 Computation of the branch current i.

This same reasoning yields the current in every branch where the current is unknown. Thus the current away from node 1 through the 5 Ω resistor is $v_1/5$, and the current away from node 1 through the 2 Ω resistor is $(v_1 - v_2)/2$. The sum of the three currents leaving node 1 must equal zero; therefore the node-voltage equation derived at node 1 is

$$\frac{v_1 - 10}{1} + \frac{v_1}{5} + \frac{v_1 - v_2}{2} = 0. \quad \textbf{(4.5)}$$

The node-voltage equation derived at node 2 is

$$\frac{v_2 - v_1}{2} + \frac{v_2}{10} - 2 = 0. \quad \textbf{(4.6)}$$

Note that the first term in Eq. (4.6) is the current away from node 2 through the 2 Ω resistor, the second term is the current away from node 2 through the 10 Ω resistor, and the third term is the current away from node 2 through the current source.

Use a computer tool to check these solutions.

Equations (4.5) and (4.6) are the two simultaneous equations that describe the circuit shown in Fig. 4.6 in terms of the node voltages v_1 and v_2. Solving for v_1 and v_2 yields

$$v_1 = \frac{100}{11} = 9.09 \text{ V} \quad \text{and} \quad v_2 = \frac{120}{11} = 10.91 \text{ V}.$$

Once the node voltages are known, all the branch currents can be calculated. Once these are known, the branch voltages and powers can be calculated. Example 4.2 illustrates the use of the node-voltage method.

EXAMPLE 4.2

a) Use the node-voltage method of circuit analysis to find the branch currents i_a, i_b, and i_c in the circuit shown in Fig. 4.8.

b) Find the power associated with each source, and state whether the source is delivering or absorbing power.

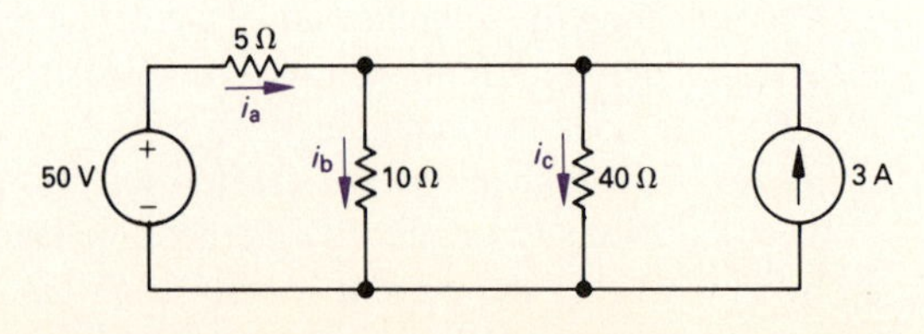

FIGURE 4.8 The circuit for Example 4.2.

SOLUTION

a) We begin by noting that the circuit has two essential nodes; thus we need to write a single node-voltage expression. We select the lower node as the reference node and define the unknown node voltage as v_1. Figure 4.9 illustrates these decisions. Summing the currents away from node 1 generates the node-voltage equation:

$$\frac{v_1 - 50}{5} + \frac{v_1}{10} + \frac{v_1}{40} - 3 = 0.$$

Solving for v_1 gives

$$v_1 = 40 \text{ V}.$$

Hence

$$i_a = \frac{50 - 40}{5} = 2 \text{ A};$$

$$i_b = \frac{40}{10} = 4 \text{ A};$$

$$i_c = \frac{40}{40} = 1 \text{ A}.$$

b) The power associated with the 50 V source is

$$p_{50\text{V}} = -50i_a = -100 \text{ W (delivering)}.$$

The power associated with the 3 A source is

$$p_{3\text{A}} = -3v_1 = -3(40) = -120 \text{ W (delivering)}.$$

We check these calculations by noting that the total delivered power is 220 W. The total power absorbed by the three resistors is $4(5) + 16(10) + 1(40)$, or 220 W, as we calculated and as it must be.

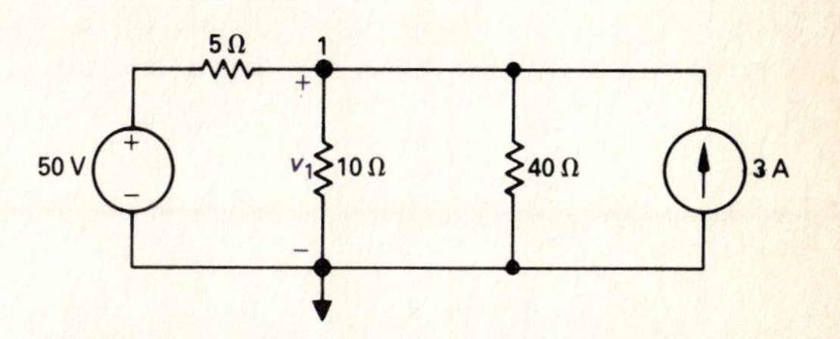

FIGURE 4.9 The circuit shown in Fig. 4.8, with a reference node and the unknown node voltage v_1.

DRILL EXERCISES

4.5 a) For the circuit shown, use the node-voltage method to find v_1, v_2, and i_1.

b) How much power is delivered to the circuit by the 12 A source?

c) Repeat (b) for the 5 A source.

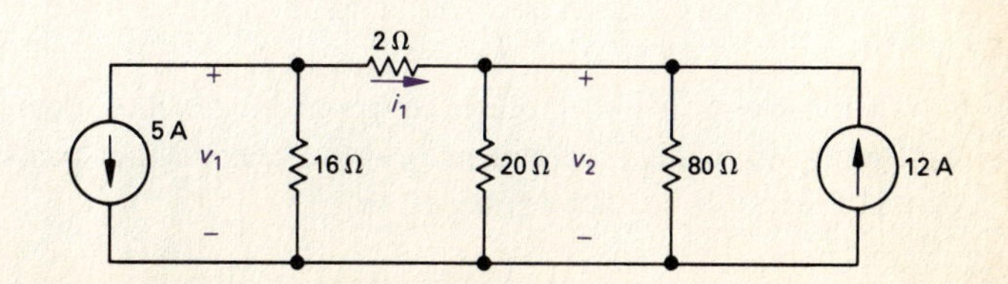

ANSWER: (a) 48 V, 64 V, −8 A; (b) 768 W; (c) −240 W.

4.6 Use the node-voltage method to find v in the circuit shown.

ANSWER: 15 V.

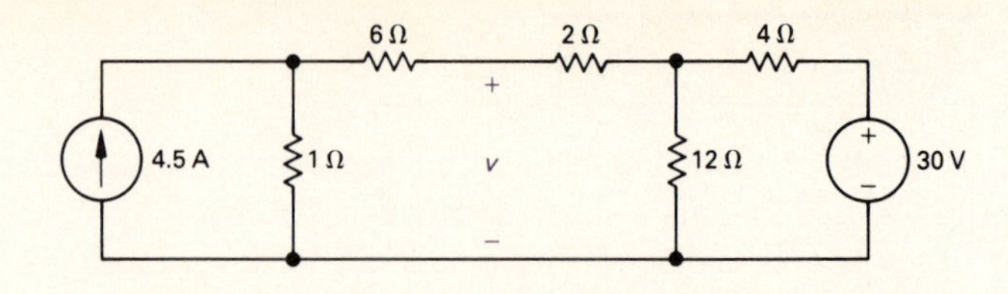

4.3 The Node-Voltage Method and Dependent Sources

If the circuit contains dependent sources, the node-voltage equations must be supplemented with the constraint equations imposed by the presence of the dependent sources. Example 4.3 illustrates the application of the node-voltage method to a circuit containing a dependent source.

EXAMPLE 4.3

Use the node-voltage method to find the power dissipated in the 5 Ω resistor in the circuit shown in Fig. 4.10.

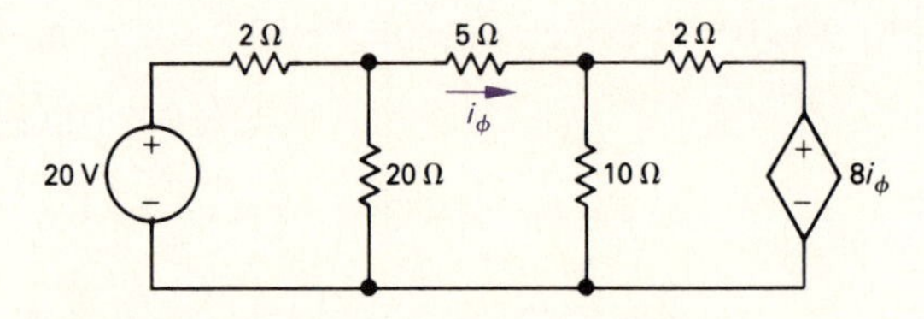

FIGURE 4.10 The circuit for Example 4.3.

SOLUTION

We begin by noting that the circuit has three essential nodes. Hence we need two node-voltage equations to describe the circuit. Four branches terminate on the lower node, so we select it as the reference node. The two unknown node voltages are defined on the circuit shown in Fig. 4.11. Summing the currents away from node 1 generates the equation

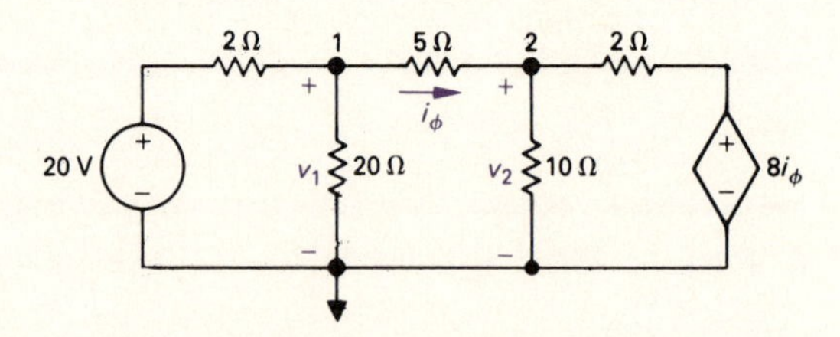

FIGURE 4.11 The circuit shown in Fig. 4.10, with a reference node and the node voltages.

$$\frac{v_1 - 20}{2} + \frac{v_1}{20} + \frac{v_1 - v_2}{5} = 0.$$

Summing the currents away from node 2 yields

$$\frac{v_2 - v_1}{5} + \frac{v_2}{10} + \frac{v_2 - 8i_\phi}{2} = 0.$$

As written, these two node-voltage equations contain three unknowns, namely, v_1, v_2, and i_ϕ. To eliminate i_ϕ we must express this controlling current in terms of the node voltages, or

$$i_\phi = \frac{v_1 - v_2}{5}.$$

Substituting this relationship into the node 2 equation simplifies the two node-voltage equations to

$$0.75v_1 - 0.2v_2 = 10$$

and

$$-v_1 + 1.6v_2 = 0.$$

Solving for v_1 and v_2 gives

$$v_1 = 16 \text{ V} \quad \text{and} \quad v_2 = 10 \text{ V}.$$

Then,

$$i_\phi = \frac{16 - 10}{5} = 1.2 \text{ A}$$

and

$$p_{5\Omega} = (1.44)(5) = 7.2 \text{ W}.$$

A good exercise to build your problem-solving intuition would be to reconsider this example, using node 2 as the reference node. Does it make the analysis easier or harder?

DRILL EXERCISE

4.7 a) Use the node-voltage method to find the power associated with each source in the circuit shown.

b) State whether the source is delivering power to the circuit or extracting power from the circuit.

ANSWER: (a) $p_{1.5\text{A}} = 15$ W, $p_{6i_2} = 1200$ W, $p_{80\text{V}} = 320$ W; (b) all sources are delivering power to the circuit.

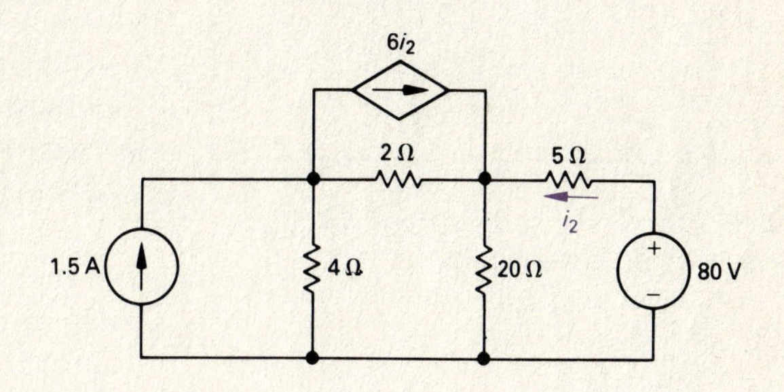

4.4 The Node-Voltage Method: Some Special Cases

When a voltage source is the only element between two essential nodes, the node-voltage method is simplified. As an example, look at the circuit in Fig. 4.12. There are three essential nodes in this circuit,

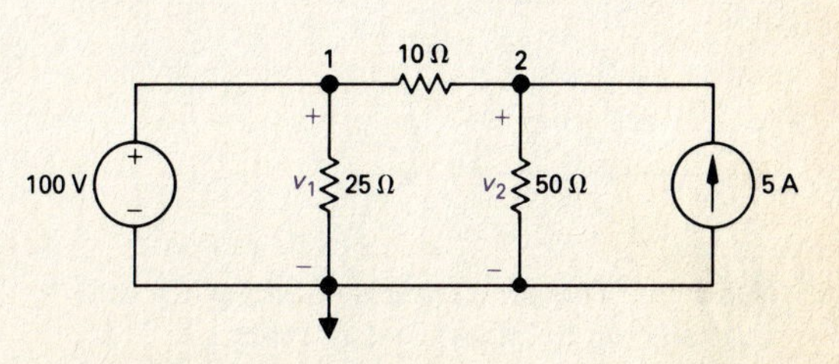

FIGURE 4.12 A circuit with a known node voltage.

which means that two simultaneous equations are needed. From these three essential nodes, a reference node has been chosen and two other nodes have been labeled. But the 100 V source constrains the voltage between node 1 and the reference node to 100 V. This means that there is only one unknown node voltage (v_2). Solution of this circuit thus involves only a single node voltage equation at node 2:

$$\frac{v_2 - v_1}{10} + \frac{v_2}{50} - 5 = 0. \tag{4.7}$$

But $v_1 = 100$ V, so Eq. (4.7) can be solved for v_2:

$$v_2 = 125 \text{ V}. \tag{4.8}$$

Knowing v_2, we can calculate the current in every branch. You should verify that the current into node 1 in the branch containing the independent voltage source is 1.5 A.

In general, when you use the node-voltage method to solve circuits which have voltage sources connected directly between essential nodes, the number of unknown node voltages is reduced. The reason is that, whenever a voltage source connects two essential nodes, it constrains the difference between the node voltages at these nodes to equal the voltage of the source. Taking the time to see if you can reduce the number of unknowns in this way will simplify circuit analysis.

Suppose that the circuit shown in Fig. 4.13 is to be analyzed using the node-voltage method. The circuit contains four essential nodes, so we anticipate writing three node-voltage equations. However, two essential nodes are connected by an independent voltage source, and two other essential nodes are connected by a current-controlled dependent voltage source. Hence, there actually is only one unknown node voltage.

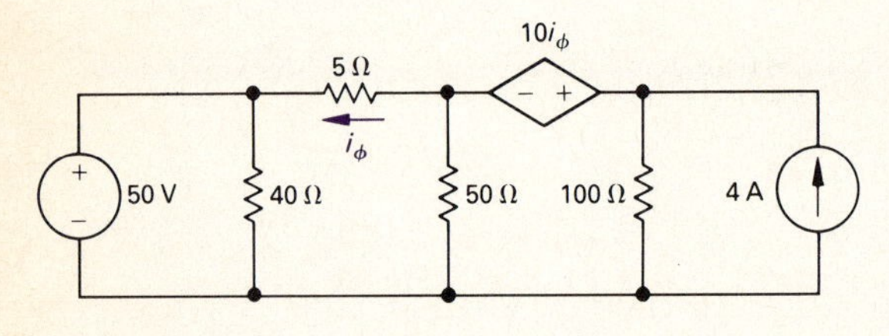

FIGURE 4.13 A circuit with a dependent voltage source connected between nodes.

Choosing which node to use as the reference node involves several possibilities. Either node on each side of the dependent voltage source looks attractive because, if chosen, one of the node voltages would be known to be either $+10i_\phi$ (left node is the reference) or $-10i_\phi$ (right node is the reference). The lower node looks even better because one node voltage is immediately known (50 V) and five branches terminate there. We therefore opt for the lower node as the reference.

Figure 4.14 shows the redrawn circuit, with the reference node flagged and the node voltages defined. Also, we introduce the current i because we cannot express the current in the dependent voltage source branch as a function of the node voltages v_2 and v_3. Thus at node 2

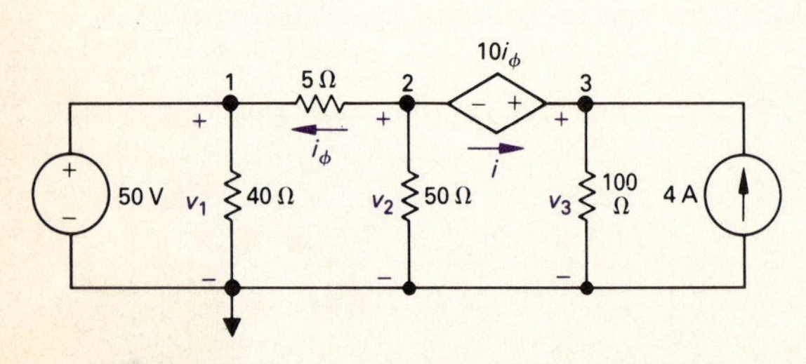

FIGURE 4.14 The circuit shown in Fig. 4.13, with the selected node voltages defined.

$$\frac{v_2 - v_1}{5} + \frac{v_2}{50} + i = 0, \tag{4.9}$$

and at node 3

$$\frac{v_3}{100} - i - 4 = 0. \tag{4.10}$$

We eliminate i simply by adding Eqs. (4.9) and (4.10) to get

$$\frac{v_2 - v_1}{5} + \frac{v_2}{50} + \frac{v_3}{100} - 4 = 0. \quad \textbf{(4.11)}$$

THE CONCEPT OF A SUPERNODE

Equation (4.11) may be written directly, without resorting to the intermediate step represented by Eqs. (4.9) and (4.10). To do so, we consider nodes 2 and 3 to be a single node and simply sum the currents away from the node in terms of the node voltages v_2 and v_3. Figure 4.15 illustrates this approach.

When a voltage source is between two essential nodes, we can combine those nodes to form a **supernode**. Obviously, Kirchhoff's current law must hold for the supernode. In Fig. 4.15, starting with the 5 Ω branch and moving counterclockwise around the supernode, we generate the equation

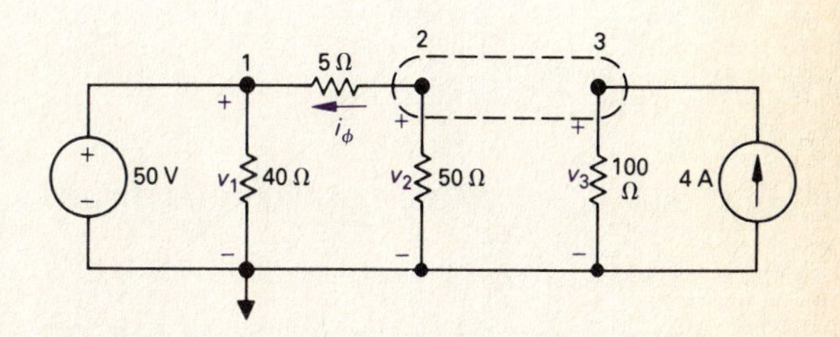

FIGURE 4.15 Considering nodes 2 and 3 to be a supernode.

$$\frac{v_2 - v_1}{5} + \frac{v_2}{50} + \frac{v_3}{100} - 4 = 0, \quad \textbf{(4.12)}$$

which is identical to Eq. (4.11). Creating a supernode at nodes 2 and 3 has made the task of analyzing this circuit easier. It is therefore always worth taking the time to look for this type of shortcut before writing any equations.

After Eq. (4.12) has been derived, the next step is to reduce the expression to a single unknown node voltage. First we eliminate v_1 from the equation because we know that $v_1 = 50$ V. Next we express v_3 as a function of v_2:

$$v_3 = v_2 + 10i_\phi. \quad \textbf{(4.13)}$$

We now express the current controlling the dependent voltage source as a function of the node voltages:

$$i_\phi = \frac{v_2 - 50}{5}. \quad \textbf{(4.14)}$$

Using Eqs. (4.13) and (4.14) and $v_1 = 50$ V reduces Eq. (4.12) to

$$v_2\left(\frac{1}{50} + \frac{1}{5} + \frac{1}{100} + \frac{10}{500}\right) = 10 + 4 + 1$$

$$v_2(0.25) = 15$$

$$v_2 = 60 \text{ V}.$$

From Eqs. (4.13) and (4.14),

$$i_\phi = \frac{60 - 50}{5} = 2 \text{ A} \quad \text{and} \quad v_3 = 60 + 20 = 80 \text{ V}.$$

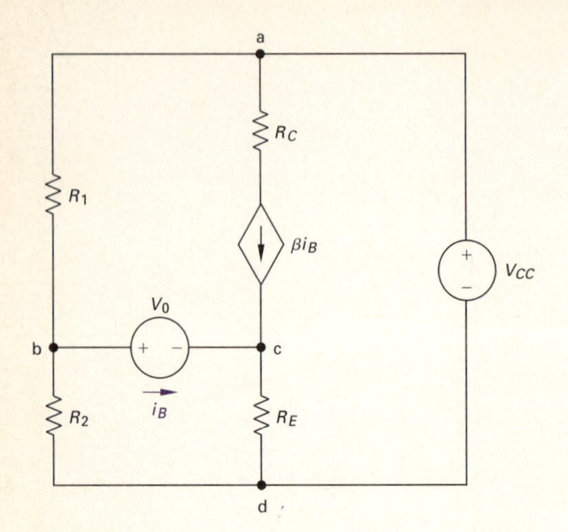

FIGURE 4.16 The transistor amplifier circuit shown in Fig. 2.24.

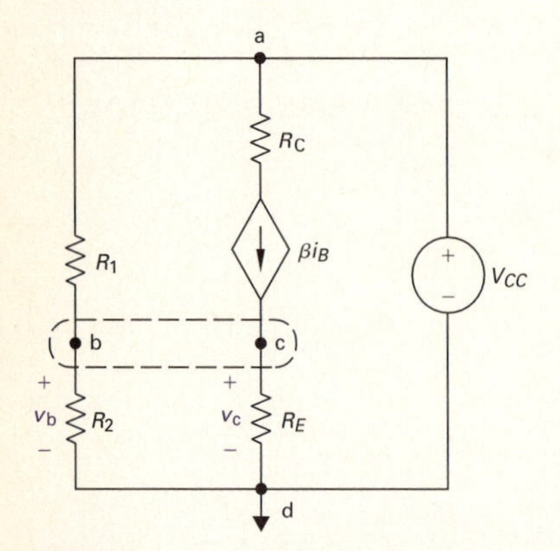

FIGURE 4.17 The circuit shown in Fig. 4.16, with voltages and the supernode identified.

Computer tools make solving large sets of simultaneous equations easier.

NODE-VOLTAGE ANALYSIS OF THE AMPLIFIER CIRCUIT

Let's use the node-voltage method to analyze the circuit first introduced in Section 2.5 and shown again in Fig. 4.16. When we used the branch-current method of analysis in Section 2.5, we faced the task of writing and solving six simultaneous equations. Here we will show how nodal analysis can simplify our task.

The circuit has four essential nodes: Nodes a and d are connected by an independent voltage source, as are nodes b and c. Therefore the problem reduces to finding a single unknown node voltage, since $(n_e - 1) - 2 = 1$. Using d as the reference node, combine nodes b and c into a supernode, label the voltage drop across R_2 as v_b, and label the voltage drop across R_E as v_c, as shown in Fig. 4.17. Then,

$$\frac{v_b}{R_2} + \frac{v_b - V_{CC}}{R_1} + \frac{v_c}{R_E} - \beta i_B = 0. \quad (4.15)$$

We now eliminate both v_c and i_B from Eq. (4.15) by noting that

$$v_c = (i_B + \beta i_B)R_E \quad (4.16)$$

and

$$v_c = v_b - V_0. \quad (4.17)$$

Substituting Eqs. (4.16) and (4.17) into Eq. (4.15) yields

$$v_b\left[\frac{1}{R_1} + \frac{1}{R_2} + \frac{1}{(1+\beta)R_E}\right] = \frac{V_{CC}}{R_1} + \frac{V_0}{(1+\beta)R_E}. \quad (4.18)$$

Solving Eq. (4.18) for v_b yields

$$v_b = \frac{V_{CC}R_2(1+\beta)R_E + V_0R_1R_2}{R_1R_2 + (1+\beta)R_E(R_1+R_2)}. \quad (4.19)$$

Using the node-voltage method to analyze this circuit reduces the problem from manipulating six simultaneous equations (see Problem 2.23) to manipulating three simultaneous equations. You should verify that, when Eq. (4.19) is combined with Eqs. (4.16) and (4.17), the solution for i_B is identical to Eq. (2.25). (See Problem 4.24.)

DRILL EXERCISES

4.8 Use the node-voltage method to find v in the circuit shown.

ANSWER: 8 V.

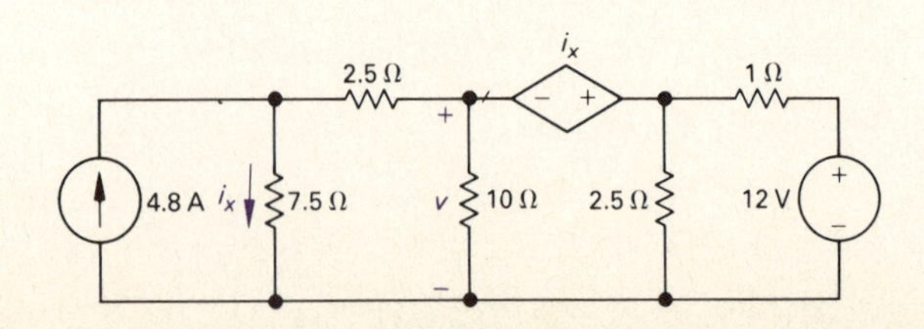

4.9 Use the node-voltage method to find v_1 in the circuit shown.

ANSWER: 120 V.

4.10 Use the node-voltage method to find v_o in the circuit shown.

ANSWER: 24 V.

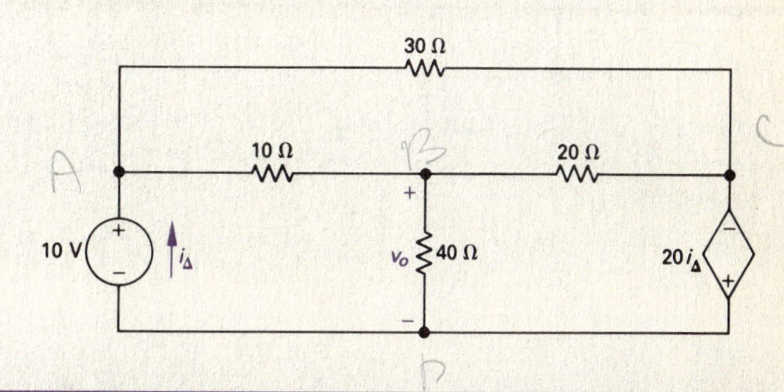

4.5 Introduction to Mesh Currents

As stated in Section 4.1, the mesh-current method of circuit analysis enables us to describe a circuit in terms of $b_e - (n_e - 1)$ equations. Recall that a mesh is a loop with no other loops inside it. The circuit in Fig. 4.1(b) is shown again in Fig. 4.18, with current arrows inside each loop to distinguish it. Recall also that the mesh-current method is applicable only to planar circuits. The circuit in Fig. 4.18 contains seven essential branches where the current is unknown, and four essential nodes. Therefore, to solve it via the mesh-current method, we must write four—7 − (4 − 1)—mesh-current equations.

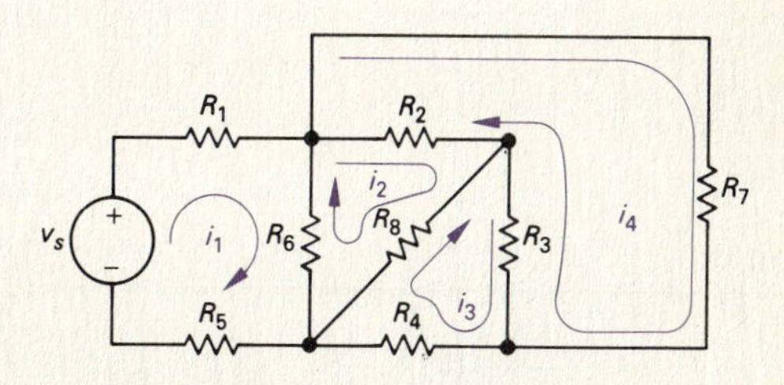

FIGURE 4.18 The circuit shown in Fig. 4.1(b), with the mesh currents defined.

A **mesh current** is the current that exists only in the perimeter of a mesh. On a circuit diagram it appears as either a closed solid line or an almost-closed solid line that follows the perimeter of the appropriate mesh. An arrowhead on the solid line indicates the reference direction for the mesh current. Figure 4.18 shows the four mesh currents that describe the circuit in Fig. 4.1(b). Note that by definition, mesh currents automatically satisfy Kirchhoff's current law. That is, at any node in the circuit, a given mesh current both enters and leaves the node.

Figure 4.18 also shows that identifying a mesh current in terms of a branch current is not always possible. For example, the mesh current i_2 is not equal to any branch current, whereas mesh currents i_1, i_3, and i_4 can be identified with branch currents. Thus measuring a mesh current is not always possible; note that there is no place where an ammeter can be inserted to measure the mesh current i_2. The fact that a mesh current can be a fictitious quantity doesn't mean that it is a useless concept. On the contrary, the mesh-current method of circuit analysis evolves quite naturally from the branch-current equations.

We can use the circuit in Fig. 4.19 to show the evolution of the mesh-current technique. We begin by using the branch currents (i_1, i_2, and i_3) to formulate the set of independent equations. For this circuit, $b_e = 3$ and $n_e = 2$. We can write only one independent current equation, so we need two independent voltage equations. Applying Kirchhoff's current law to the upper node and Kirchhoff's voltage law around the two meshes generates the following set of equations:

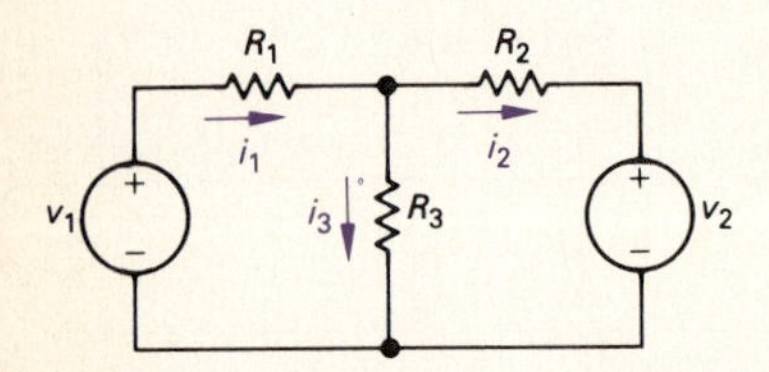

FIGURE 4.19 A circuit used to illustrate development of the mesh-current method of circuit analysis.

$$i_1 = i_2 + i_3; \quad \textbf{(4.20)}$$

$$v_1 = i_1 R_1 + i_3 R_3; \quad \textbf{(4.21)}$$

$$-v_2 = i_2 R_2 - i_3 R_3. \quad \textbf{(4.22)}$$

We reduce this set of three equations to a set of two equations by solving Eq. (4.20) for i_3 and then substituting this expression into Eqs. (4.21) and (4.22):

$$v_1 = i_1(R_1 + R_3) - i_2 R_3 \quad \textbf{(4.23)}$$

and

$$-v_2 = -i_1 R_3 + i_2(R_2 + R_3). \quad \textbf{(4.24)}$$

We can solve Eqs. (4.23) and (4.24) for i_1 and i_2 to replace the solution of three simultaneous equations with the solution of two simultaneous equations. We derived Eqs. (4.23) and (4.24) by substituting the $n_e - 1$ current equations into the $b_e - (n_e - 1)$ voltage equations. The value of the mesh-current method is that, by defining mesh currents, we automatically eliminate the $n_e - 1$ current equations. Thus the mesh-current method is equivalent to a systematic substitution of the $n_e - 1$ current equations into the $b_e - (n_e - 1)$ voltage equations. The mesh currents in Fig. 4.19 that are equivalent to eliminating the branch current i_3 from Eqs. (4.21) and (4.22) are shown in Fig. 4.20. We now apply Kirchhoff's voltage law around the two meshes, expressing all voltages across resistors in terms of the mesh currents, to get the equations

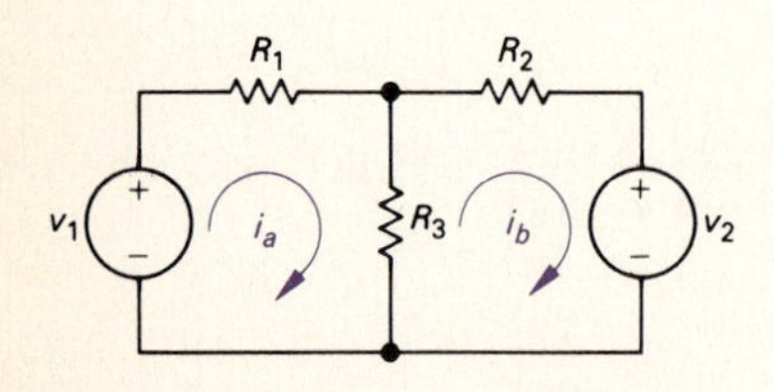

FIGURE 4.20 Mesh currents i_a and i_b.

$$v_1 = i_a R_1 + (i_a - i_b) R_3 \quad \textbf{(4.25)}$$

and

$$-v_2 = (i_b - i_a) R_3 + i_b R_2. \quad \textbf{(4.26)}$$

Collecting the coefficients of i_a and i_b in Eqs. (4.25) and (4.26) gives

$$v_1 = i_a(R_1 + R_3) - i_b R_3 \quad \textbf{(4.27)}$$

and

$$-v_2 = -i_a R_3 + i_b(R_2 + R_3). \quad \textbf{(4.28)}$$

Note that Eqs. (4.27) and (4.28) and Eqs. (4.23) and (4.24) are identical in form, with the mesh currents i_a and i_b replacing the branch currents i_1 and i_2. Note also that the branch currents shown in Fig. 4.19 can be expressed in terms of the mesh currents shown in Fig. 4.20, or

$$i_1 = i_a; \quad \textbf{(4.29)}$$

$$i_2 = i_b; \quad \textbf{(4.30)}$$

$$i_3 = i_a - i_b. \quad \textbf{(4.31)}$$

The ability to write Eqs. (4.29)–(4.31) by inspection is crucial to the mesh-current method of circuit analysis. Once you know the mesh currents, you also know the branch currents. And once you know the branch currents, you can compute any voltages or powers of interest.

Example 4.4 illustrates how the mesh-current method is used to find source powers and a branch voltage.

EXAMPLE 4.4

a) Use the mesh-current method to determine the power associated with each voltage source in the circuit shown in Fig. 4.21.

b) Calculate the voltage v_o across the 8 Ω resistor.

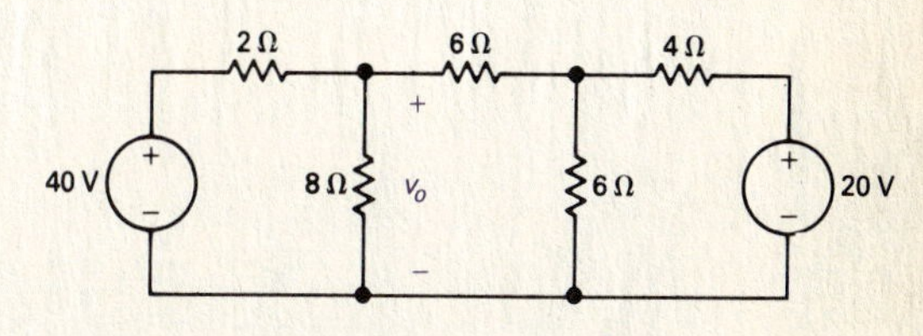

FIGURE 4.21 The circuit for Example 4.4.

SOLUTION

a) To calculate the power associated with each source, we need to know the current in each source. The circuit indicates that these source currents will be identical to mesh currents. Also, note that the circuit has seven branches where the current is unknown and five nodes. Therefore we need three—$b - (n - 1) = 7 - (5 - 1)$—mesh-current equations to describe the circuit. Figure 4.22 shows the three mesh currents used to describe the circuit in Fig. 4.21.

If we assume that the voltage drops are positive, the three mesh equations are

$$-40 + 2i_a + 8(i_a - i_b) = 0;$$

$$8(i_b - i_a) + 6i_b + 6(i_b - i_c) = 0; \quad \textbf{(4.32)}$$

$$6(i_c - i_b) + 4i_c + 20 = 0.$$

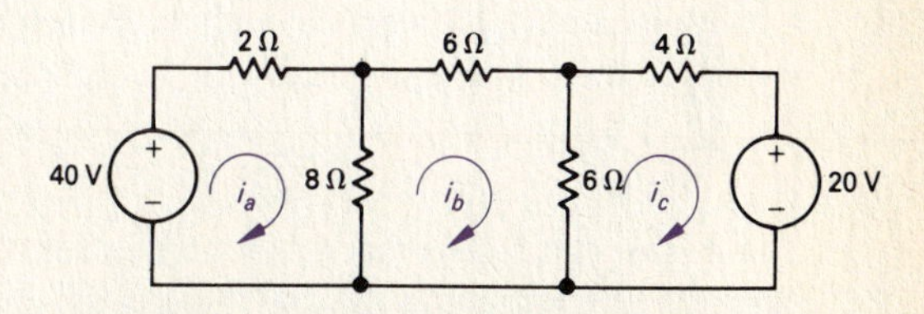

FIGURE 4.22 The three mesh currents used to analyze the circuit shown in Fig. 4.21.

Cramer's method is a useful tool when solving three or more simultaneous equations. You can review this important tool in

Computer tools using Cramer's method can help solve equations.

Appendix A. Reorganizing Eqs. (4.32) in anticipation of using Cramer's method gives

$$\begin{aligned} 10i_a - 8i_b + 0i_c &= 40; \\ -8i_a + 20i_b - 6i_c &= 0; \\ 0i_a - 6i_b + 10i_c &= -20. \end{aligned} \quad \textbf{(4.33)}$$

The characteristic determinant is

$$\Delta = \begin{vmatrix} 10 & -8 & 0 \\ -8 & 20 & -6 \\ 0 & -6 & 10 \end{vmatrix}$$

$$= 10(200 - 36) + 8(-80)$$

$$= 1640 - 640 = 1000.$$

The three mesh currents are

$$i_a = \frac{\begin{vmatrix} 40 & -8 & 0 \\ 0 & 20 & -6 \\ -20 & -6 & 10 \end{vmatrix}}{1000} = \frac{40(200 - 36) - 20(48)}{1000} = 5.6 \text{ A};$$

$$i_b = \frac{\begin{vmatrix} 10 & 40 & 0 \\ -8 & 0 & -6 \\ 0 & -20 & 10 \end{vmatrix}}{1000} = \frac{10(-120) + 8(400)}{1000} = 2.0 \text{ A};$$

$$i_c = \frac{\begin{vmatrix} 10 & -8 & 40 \\ -8 & 20 & 0 \\ 0 & -6 & -20 \end{vmatrix}}{1000} = \frac{10(-400) + 8(160 + 240)}{1000} = -0.80 \text{ A}.$$

The mesh current i_a is identical with the branch current in the 40 V source, so the power associated with this source is

$$p_{40V} = -40i_a = -224 \text{ W}.$$

The minus sign means that this source is delivering power to the network. The current in the 20 V source is identical to the mesh current i_c; therefore

$$p_{20V} = 20i_c = -16 \text{ W}.$$

The 20 V source also is delivering power to the network.

b) The branch current in the 8 Ω resistor in the direction of the voltage drop v_o is $i_a - i_b$. Therefore

$$v_o = 8(i_a - i_b) = 8(3.6) = 28.8 \text{ V}.$$

DRILL EXERCISE

4.11 Use the mesh-current method to find (a) the power delivered by the 100 V source to the circuit shown and (b) the power dissipated in the 15 Ω resistor.

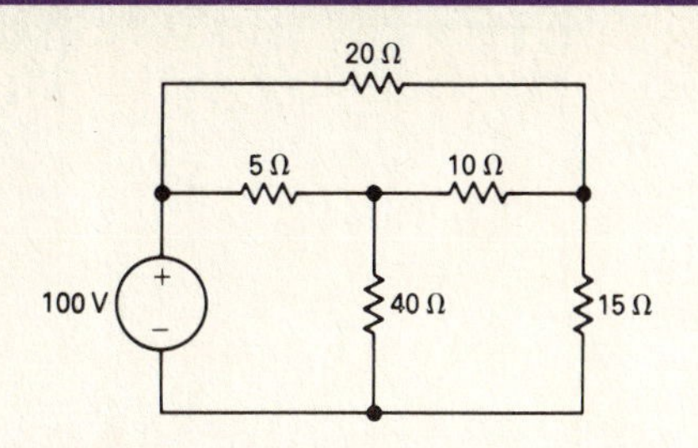

ANSWER: (a) 600 W; (b) 240 W.

4.6 THE MESH-CURRENT METHOD AND DEPENDENT SOURCES

If the circuit contains dependent sources, the mesh-current equations must be supplemented by the appropriate constraint equations. Example 4.5 illustrates the application of the mesh-current method when the circuit includes a dependent source.

EXAMPLE 4.5

Use the mesh-current method of circuit analysis to determine the power dissipated in the 4 Ω resistor in the circuit shown in Fig. 4.23.

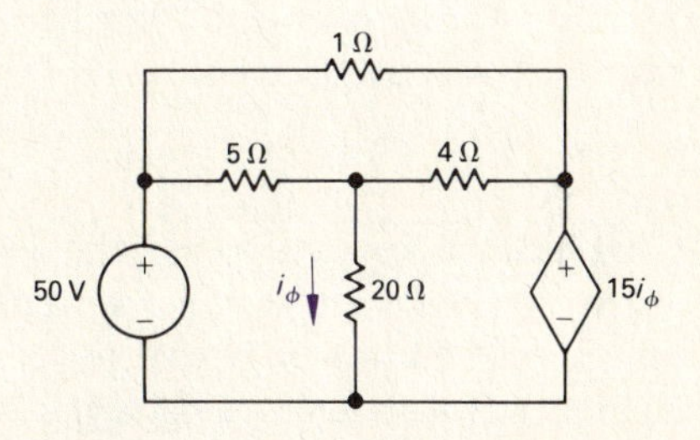

FIGURE 4.23 The circuit for Example 4.5.

SOLUTION

This circuit has six branches where the current is unknown and four nodes. Therefore we need three mesh currents to describe the circuit. They are defined on the circuit shown in Fig. 4.24. The three mesh-current equations are

Use computer tools to help solve this set of simultaneous equations.

$$50 = 5(i_1 - i_2) + 20(i_1 - i_3);$$
$$0 = 5(i_2 - i_1) + 1i_2 + 4(i_2 - i_3); \quad (4.34)$$
$$0 = 20(i_3 - i_1) + 4(i_3 - i_2) + 15i_\phi.$$

We now express the branch current controlling the dependent voltage source in terms of the mesh currents as

$$i_\phi = i_1 - i_3, \quad (4.35)$$

which is the supplemental equation imposed by the presence of the dependent source. Substituting Eq. (4.35) into Eqs. (4.34) and collecting

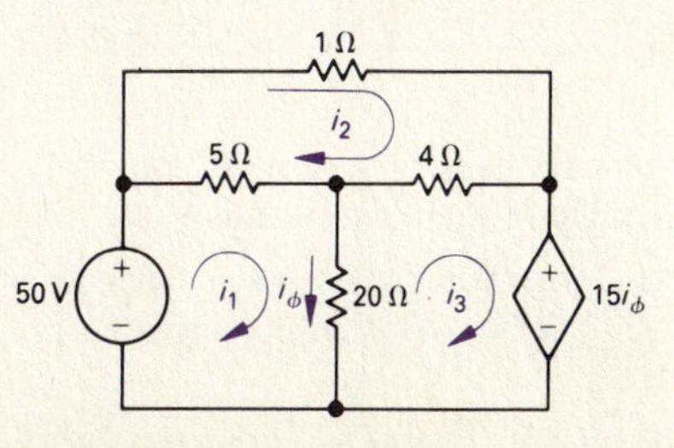

FIGURE 4.24 The circuit shown in Fig. 4.23, with the three mesh currents.

the coefficients of i_1, i_2, and i_3 in each equation generates

$$\begin{aligned} 50 &= 25i_1 - 5i_2 - 20i_3; \\ 0 &= -5i_1 + 10i_2 - 4i_3; \\ 0 &= -5i_1 - 4i_2 + 9i_3. \end{aligned}$$

The characteristic determinant is

$$\Delta = \begin{vmatrix} 25 & -5 & -20 \\ -5 & 10 & -4 \\ -5 & -4 & 9 \end{vmatrix}.$$

Expanding the characteristic determinant by the first column gives

$$\Delta = 25(90 - 16) + 5(-45 - 80) - 5(20 + 200) = 125.$$

Because we are calculating the power dissipated in the 4 Ω resistor, we compute the mesh currents i_2 and i_3:

$$i_2 = \frac{\begin{vmatrix} 25 & 50 & -20 \\ -5 & 0 & -4 \\ -5 & 0 & 9 \end{vmatrix}}{125} = \frac{-50(-45 - 20)}{125} = 26 \text{ A}; \qquad i_3 = \frac{\begin{vmatrix} 25 & -5 & 50 \\ -5 & 10 & 0 \\ -5 & -4 & 0 \end{vmatrix}}{125} = \frac{50(20 + 50)}{125} = 28 \text{ A}.$$

The current in the 4 Ω resistor oriented from left to right is $i_3 - i_2$, or 2 A. Therefore the power dissipated is

$$p_{4\Omega} = (i_3 - i_2)^2(4) = (2)^2(4) = 16 \text{ W}.$$

What if you had not been told to use the mesh-current method? Would you have chosen the node-voltage method? It reduces the problem to finding one unknown node voltage, because of the presence of two voltage sources between essential nodes. More about making such choices later.

DRILL EXERCISES

4.12 a) Determine the number of mesh-current equations needed to solve the circuit shown.

b) Use the mesh-current method to find how much power is being delivered to the dependent voltage source.

ANSWER: (a) 3; (b) −36 W.

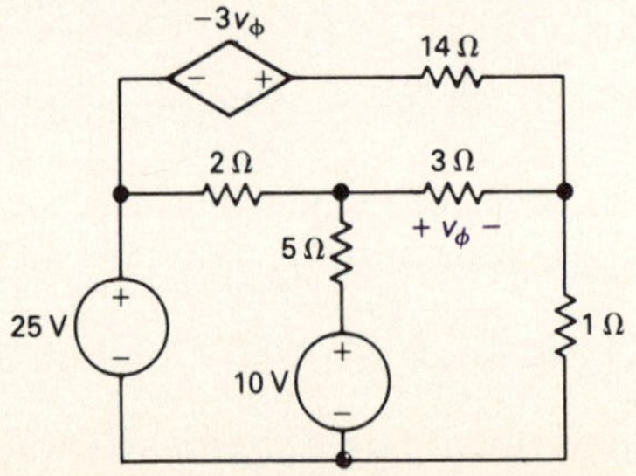

4.13 Use the mesh-current method to find v_o in the circuit shown.

ANSWER: 20 V.

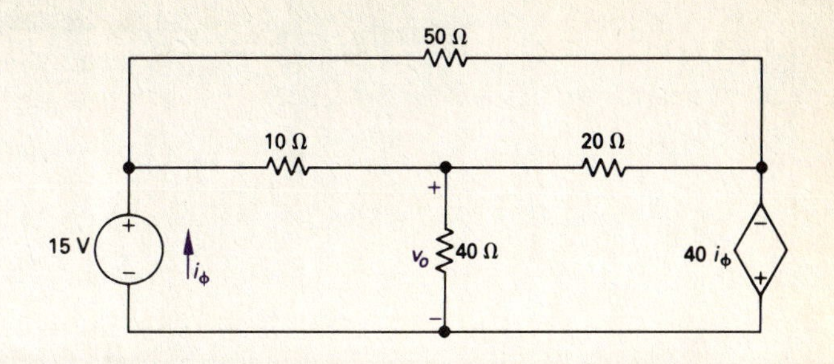

4.7 THE MESH-CURRENT METHOD: SOME SPECIAL CASES

When a branch includes a current source, the mesh-current method requires some additional manipulations. The circuit shown in Fig. 4.25 depicts the nature of the problem. We have defined the mesh currents i_a, i_b, and i_c, as well as the voltage across the 5 A current source, to aid the discussion. Note that the circuit contains five essential branches where the current is unknown and four essential nodes. Hence we need to write two—5 − (4 − 1)—mesh-current equations to solve the circuit. The presence of the current source reduces the three unknown mesh currents to two such currents because it constrains the difference between i_a and i_c to equal 5 A. Hence if we know i_a, we know i_c, and vice versa.

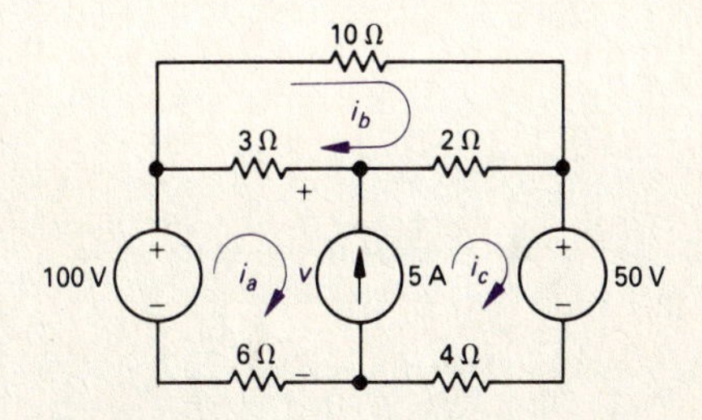

FIGURE 4.25 A circuit illustrating mesh analysis when a branch contains an independent current source.

However, when we attempt to sum the voltages around either mesh a or mesh c, we must introduce into the equations the unknown voltage across the 5 A current source. Thus for mesh a

$$100 = 3(i_a - i_b) + v + 6i_a, \quad \textbf{(4.36)}$$

and for mesh c

$$-50 = 4i_c - v + 2(i_c - i_b). \quad \textbf{(4.37)}$$

We now add Eqs. (4.36) and (4.37) to eliminate v and obtain

$$50 = 9i_a - 5i_b + 6i_c. \quad \textbf{(4.38)}$$

Summing voltages around mesh b gives

$$0 = 3(i_b - i_a) + 10i_b + 2(i_b - i_c). \quad \textbf{(4.39)}$$

We reduce Eqs. (4.38) and (4.39) to two equations and two unknowns by using the constraint that

$$i_c - i_a = 5. \quad \textbf{(4.40)}$$

Computer tools help in verifying such results.

We leave to you the verification that, when Eq. (4.40) is combined with Eqs. (4.38) and (4.39), the solutions for the three mesh currents are

$$i_a = 1.75 \text{ A}, \quad i_b = 1.25 \text{ A}, \quad \text{and} \quad i_c = 6.75 \text{ A}.$$

The Concept of a Supermesh

We can derive Eq. (4.38) without introducing the unknown voltage v, by using the concept of a supermesh. To create a supermesh, we mentally remove the current source from the circuit by simply avoiding this branch when writing the mesh-current equations. We express the voltages around the supermesh in terms of the original mesh currents. Figure 4.26 illustrates the supermesh concept. When we sum the voltages around the supermesh (denoted by the dashed line), we obtain the equation

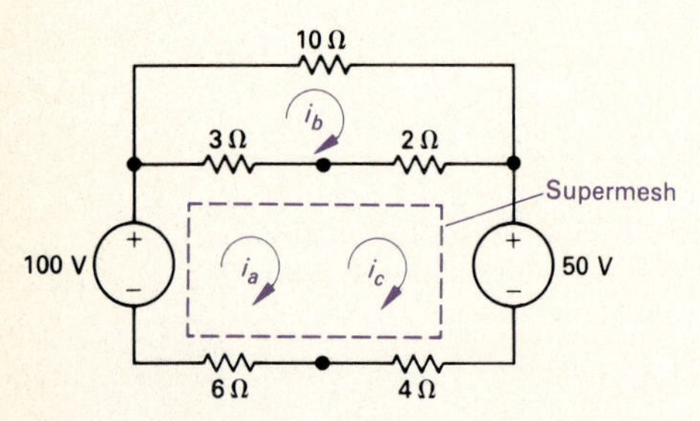

FIGURE 4.26 The circuit shown in Fig. 4.25, illustrating the concept of a supermesh.

$$-100 + 3(i_a - i_b) + 2(i_c - i_b) + 50 + 4i_c + 6i_a = 0, \tag{4.41}$$

which reduces to

$$50 = 9i_a - 5i_b + 6i_c. \tag{4.42}$$

Note that Eqs. (4.42) and (4.38) are identical. Thus the supermesh has eliminated the need for introducing the unknown voltage across the current source. Once again, taking time to look carefully at a circuit to identify a shortcut such as this provides a big payoff in simplifying the analysis.

Mesh-Current Analysis of the Amplifier Circuit

We can use the circuit first introduced in Section 2.5 (Fig. 2.24) to illustrate how the mesh-current method works when a branch contains a dependent current source. Figure 4.27 shows that circuit, with the three mesh currents denoted i_a, i_b, and i_c. This circuit has four essential nodes and five essential branches where the current is unknown. Therefore we know that the circuit can be analyzed in terms of two—$5-(4-1)$—mesh-current equations. Although we defined three mesh currents in Fig. 4.27, the dependent current source forces a constraint between mesh currents i_a and i_c, so we have only two unknown mesh currents. Using the concept of the supermesh, we redraw the circuit as shown in Fig. 4.28.

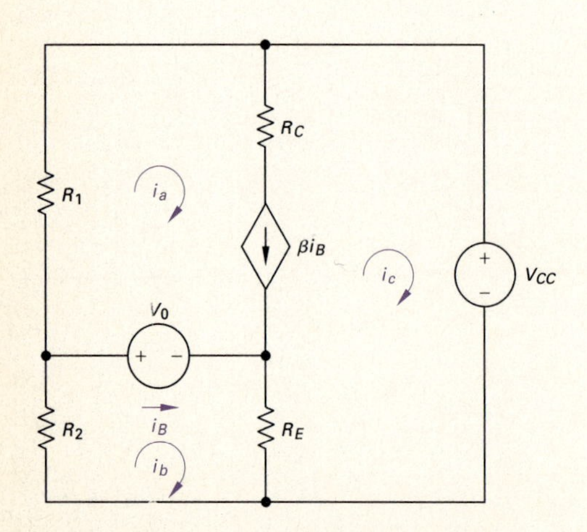

FIGURE 4.27 The circuit shown in Fig. 2.24, with the mesh currents i_a, i_b, and i_c.

We now sum the voltages around the supermesh in terms of the mesh currents i_a, i_b, and i_c to obtain

$$R_1 i_a + V_{CC} + R_E(i_c - i_b) - V_0 = 0. \tag{4.43}$$

The mesh b equation is

$$R_2 i_b + V_0 + R_E(i_b - i_c) = 0. \tag{4.44}$$

The constraint imposed by the dependent current source is

$$\beta i_b = i_a - i_c. \tag{4.45}$$

The branch current controlling the dependent current source, expressed as a function of the mesh currents, is

$$i_B = i_b - i_a. \tag{4.46}$$

From Eqs. (4.45) and (4.46),

$$i_c = (1+\beta)i_a - \beta i_b. \tag{4.47}$$

We now use Eq. (4.47) to eliminate i_c from Eqs. (4.43) and (4.44):

$$[R_1(1+\beta)R_E]i_a - (1+\beta)R_E i_b = V_0 - V_{CC}; \tag{4.48}$$

$$-(1+\beta)R_E i_a + [R_2 + (1+\beta)R_E]i_b = -V_0. \tag{4.49}$$

You should verify that the solution of Eqs. (4.48) and (4.49) for i_a and i_b gives

$$i_a = \frac{V_0 R_2 - V_{CC}R_2 - V_{CC}(1+\beta)R_E}{R_1R_2 + (1+\beta)R_E(R_1+R_2)} \tag{4.50}$$

$$i_b = \frac{-V_0 R_1 - (1+\beta)R_E V_{CC}}{R_1R_2 + (1+\beta)R_E(R_1+R_2)}. \tag{4.51}$$

We also leave you to verify that, when Eqs. (4.50) and (4.51) are used to find i_B, the result is the same as that given by Eq. (2.25).

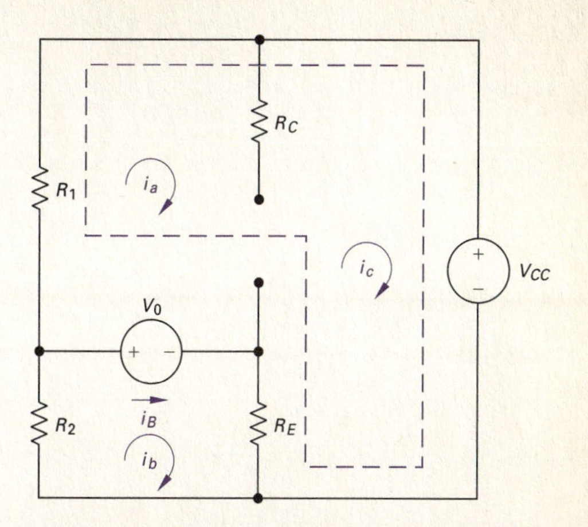

FIGURE 4.28 The circuit shown in Fig. 4.27, depicting the supermesh created by the presence of the dependent current source.

DRILL EXERCISES

4.14 Use the mesh-current method to find the power dissipated in the 2 Ω resistor in the circuit shown.

ANSWER: 72 W.

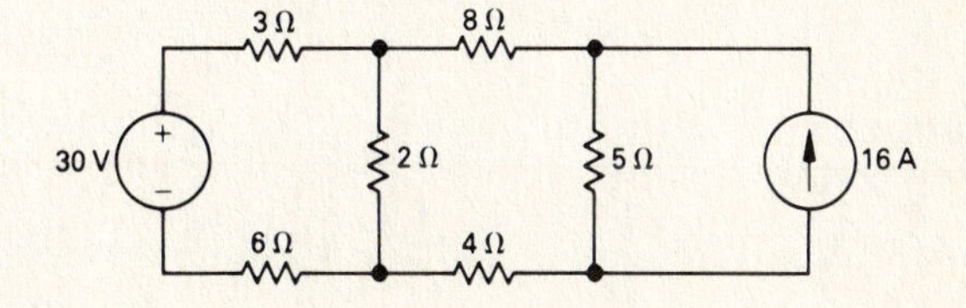

4.15 Use the mesh-current method to find the mesh current i_a in the circuit shown.

ANSWER: 10 A.

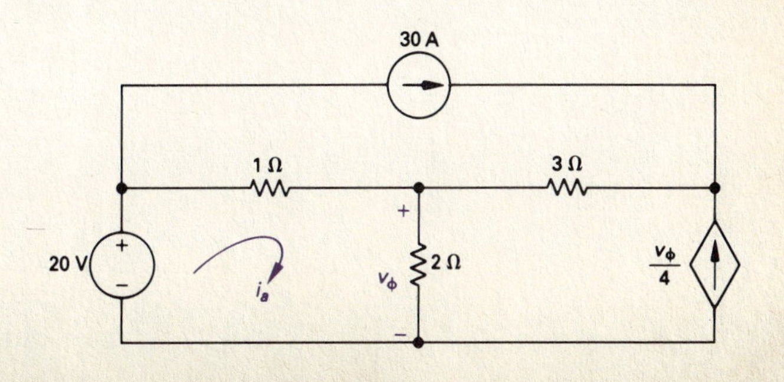

4.16 Use the mesh-current method to find the power dissipated in the 1 Ω resistor in the circuit shown.

ANSWER: 36 W.

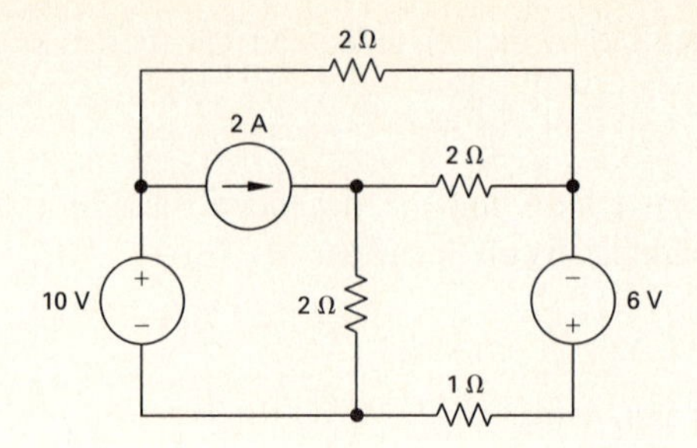

4.8 THE NODE-VOLTAGE METHOD VERSUS THE MESH-CURRENT METHOD

The greatest advantage of both the node-voltage and mesh-current methods is that they reduce the number of simultaneous equations that must be manipulated. They also require the analyst to be quite systematic in terms of organizing and writing these equations. It is natural to ask then, "When is the node-voltage method preferred to the mesh-current method, and vice versa?" As you might suspect, there is no clear-cut answer. Asking a number of questions, however, may help you identify the more efficient method before plunging into the solution process:

- Does one of the methods result in fewer simultaneous equations to solve?
- Does the circuit contain supernodes? If so, using the node-voltage method will permit you to reduce the number of equations to be solved.
- Does the circuit contain supermeshes? If so, using the mesh-current method will permit you to reduce the number of equations to be solved.
- Will solving some portion of the circuit give the requested solution? If so, which method is most efficient for solving just the pertinent portion of the circuit?

Perhaps the most important observation is that, for any situation, some time spent thinking about the problem in relation to the various analytical approaches available is time well spent. Examples 4.6 and 4.7 illustrate the process of deciding between the node-voltage and mesh-current methods.

EXAMPLE 4.6

Find the power dissipated in the 300 Ω resistor in the circuit shown in Fig. 4.29.

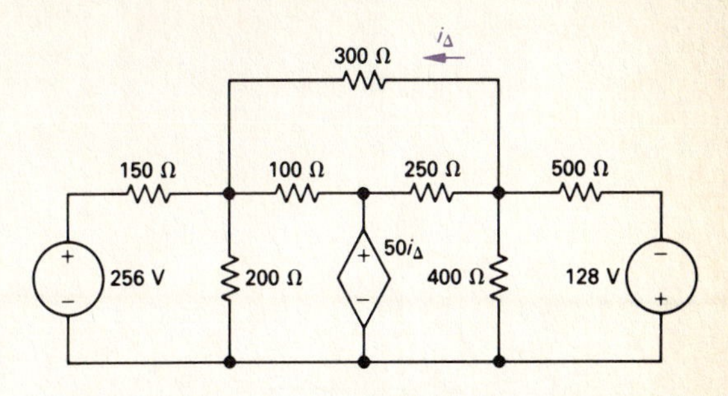

FIGURE 4.29 The circuit for Example 4.6.

SOLUTION

To find the power dissipated in the 300 Ω resistor, we need to find either the current in the resistor or the voltage across it. The mesh-current method yields the current in the resistor; this approach requires solving five simultaneous mesh equations, as depicted in Fig. 4.30. In writing the five equations, we must include the constraint $i_\Delta = -i_b$.

Before going further, let's also look at the circuit in terms of the node-voltage method. Note that, once we know the node voltages, we can calculate either the current in the 300 Ω resistor or the voltage across it. The circuit has four essential nodes, and therefore only three node-voltage equations are required to describe the circuit. Because of the dependent voltage source between two essential nodes, we have to sum the currents at only two nodes. Hence the problem is reduced to writing two node-voltage equations and a constraint equation. Because the node-voltage method requires only three simultaneous equations, it is the more attractive approach.

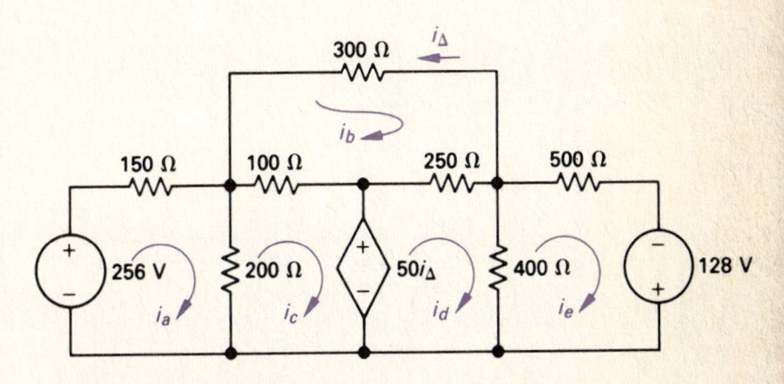

FIGURE 4.30 The circuit shown in Fig. 4.29, with the five mesh currents.

Once the decision to use the node-voltage method has been made, the next step is to select a reference node. Two essential nodes in the circuit in Fig. 4.29 merit consideration. The first is the reference node in Fig. 4.31. If this node is selected, one of the unknown node voltages is the voltage across the 300 Ω resistor, namely, v_2 in Fig. 4.31. Once we know this voltage, we calculate the power in the 300 Ω resistor by using the expression $p_{300\Omega} = v_2^2/300$. Note that, in addition to selecting the reference node, we defined the three node voltages v_1, v_2, and v_3 and indicated that nodes 1 and 3 form a supernode, because they are connected by a dependent voltage source. It is understood that a node voltage is a rise from the reference node; therefore, in Fig. 4.31, we have not placed the node voltage polarity references on the circuit diagram.

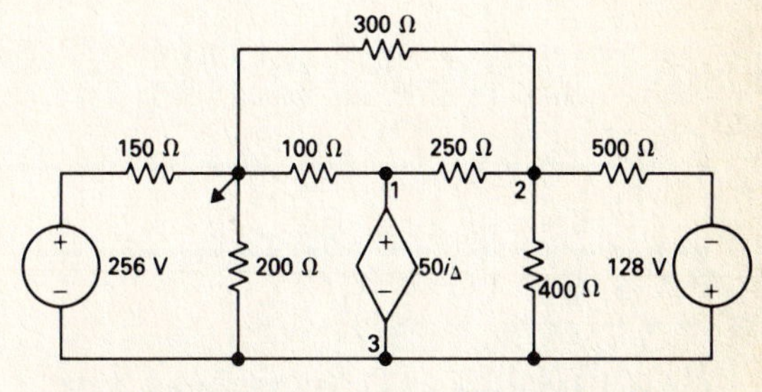

FIGURE 4.31 The circuit shown in Fig. 4.29, with a reference node.

The second node that merits consideration as the reference node is the lower node in the circuit, as shown in Fig. 4.32. It is attractive because it has the most branches connected to it and the node-voltage equations are thus easier to write. However, to find either the current in the 300 Ω resistor or the voltage across it requires an additional calculation once we know the node voltages v_a and v_c. For example, the current in the 300 Ω resistor is $(v_c - v_a)/300$, whereas the voltage across the resistor is $v_c - v_a$.

We compare these two possible reference nodes by means of the following sets of equations. The first set pertains to the circuit shown in Fig. 4.31, and the second set is based on the circuit shown in Fig. 4.32.

- Set 1 (Fig 4.31)

At the supernode,

$$\frac{v_1}{100} + \frac{v_1 - v_2}{250} + \frac{v_3}{200} + \frac{v_3 - v_2}{400} + \frac{v_3 - (v_2 + 128)}{500} + \frac{v_3 + 256}{150} = 0.$$

At v_2,

$$\frac{v_2}{300} + \frac{v_2 - v_1}{250} + \frac{v_2 - v_3}{400} + \frac{v_2 + 128 - v_3}{500} = 0.$$

From the supernode, the constraint equation is

$$v_3 = v_1 - 50i_\Delta = v_1 - \frac{v_2}{6}.$$

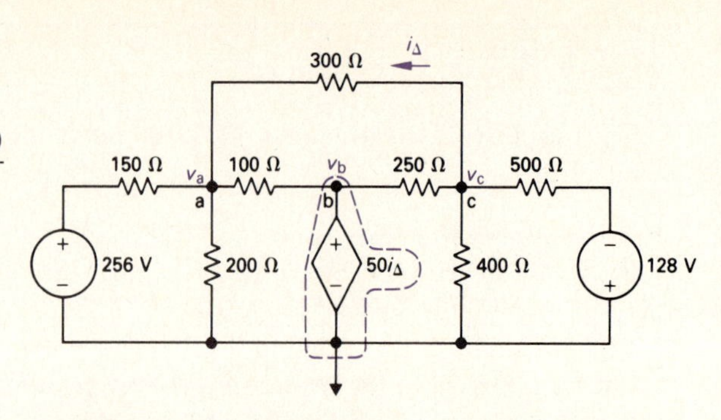

FIGURE 4.32 The circuit shown in Fig. 4.29, with an alternative reference node.

- Set 2 (Fig 4.32)

At v_a,

$$\frac{v_a}{200} + \frac{v_a - 256}{150} + \frac{v_a - v_b}{100} + \frac{v_a - v_c}{300} = 0.$$

At v_c,

$$\frac{v_c}{400} + \frac{v_c + 128}{500} + \frac{v_c - v_b}{250} + \frac{v_c - v_a}{300} = 0.$$

From the supernode, the constraint equation is

$$v_b = 50i_\Delta = \frac{50(v_c - v_a)}{300} = \frac{v_c - v_a}{6}.$$

You should verify that the solution of either set leads to a power calculation of 16.57 W dissipated in the 300 Ω resistor.

Use computer tools to assist in verifying this result.

EXAMPLE 4.7

Find the voltage v_o in the circuit shown in Fig. 4.33.

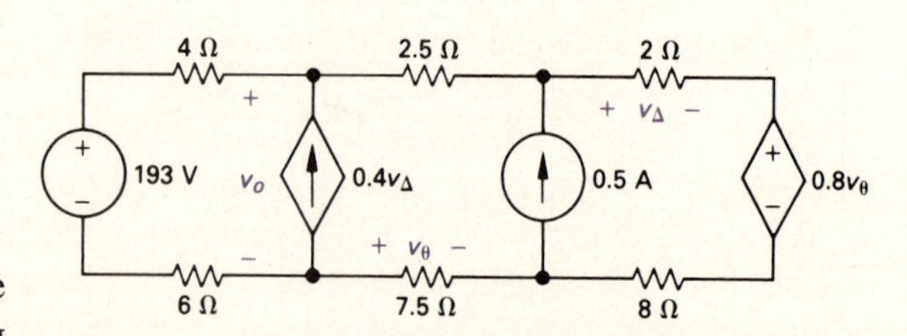

FIGURE 4.33 The circuit for Example 4.7.

SOLUTION

At first glance the node-voltage method looks appealing, because we may define the unknown voltage as a node voltage by choosing the lower terminal of the dependent current source as the reference node. The circuit has four essential nodes and two voltage-controlled dependent sources, so the node-voltage method requires manipulation of three node-voltage equations and two constraint equations.

Let's now turn to the mesh-current method for finding v_o. The circuit contains three meshes, and we can use the leftmost one to calculate v_o. If we let i_a denote the leftmost mesh current, then $v_o = 193 - 10i_a$. The presence of the two current sources reduces the problem to manipulating a single supermesh equation and two constraint equations. Hence the mesh-current method is the more attractive technique here.

To help you compare the two approaches, we summarize both methods. The mesh-current equations are based on the circuit shown in Fig. 4.34, and the node-voltage equations are based on the circuit shown in Fig. 4.35.The supermesh equation is

$$193 = 10i_a + 10i_b + 10i_c + 0.8v_\theta,$$

and the constraint equations are

$$i_b - i_a = 0.4v_\Delta = 0.8i_c; \quad v_\theta = -7.5i_b; \quad \text{and} \quad i_c - i_b = 0.5.$$

We use the constraint equations to write the supermesh equation in terms of i_a:

$$160 = 80i_a, \quad \text{or} \quad i_a = 2\text{ A};$$

$$v_o = 193 - 20 = 173\text{ V}.$$

The node-voltage equations are

$$\frac{v_o - 193}{10} - 0.4v_\Delta + \frac{v_o - v_a}{2.5} = 0;$$

$$\frac{v_a - v_o}{2.5} - 0.5 + \frac{v_a - (v_b + 0.8v_\theta)}{10} = 0;$$

$$\frac{v_b}{7.5} + 0.5 + \frac{v_b + 0.8v_\theta - v_a}{10} = 0.$$

The constraint equations are

$$v_\theta = -v_b \quad \text{and} \quad v_\Delta = \left[\frac{v_a - (v_b + 0.8v_\theta)}{10}\right]2.$$

We use the constraint equations to reduce the node-voltage equations to three simultaneous equations involving v_o, v_a, and v_b. You should verify that the node-voltage approach also gives $v_o = 173$ V.

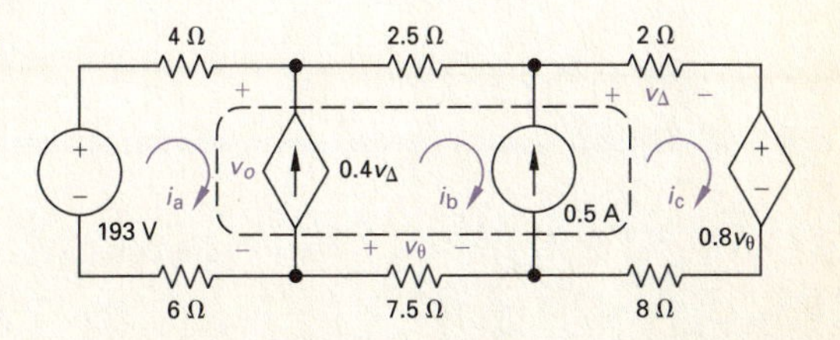

FIGURE 4.34 The circuit shown in Fig. 4.33, with the three mesh currents.

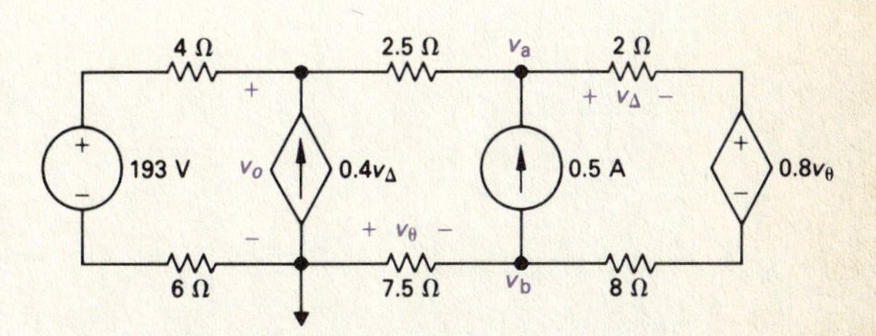

FIGURE 4.35 The circuit shown in Fig. 4.33, with node voltages.

DRILL EXERCISES

4.17 Find the power delivered by the 2 A current source in the circuit shown.

ANSWER: 70 W.

4.18 Find the power delivered by the 4 A current source in the circuit shown.

ANSWER: 40 W.

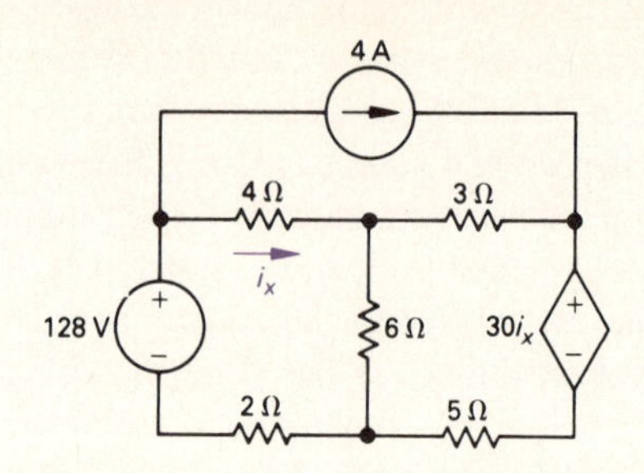

4.9 SOURCE TRANSFORMATIONS

Even though the node-voltage and mesh-current methods are powerful techniques for solving circuits, we are still interested in methods that can be used to simplify circuits. Series-parallel reductions and Δ-to-Y transformations are already on our list of simplifying techniques. We begin expanding this list with source transformations. A **source transformation**, shown in Fig. 4.36, allows a voltage source in series with a resistor to be replaced by a current source in parallel with the same resistor, or vice versa. The double-headed arrow emphasizes that a source transformation is bilateral; that is, we can start with either configuration and derive the other.

We need to find the relationship between v_s and i_s that guarantees the two configurations in Fig. 4.36 are equivalent with respect to nodes a,b. Equivalence is achieved if any resistor R_L experiences the same current flow, and thus the same voltage drop, whether connected between nodes a,b in Fig. 4.36(a) or Fig. 4.36(b).

Suppose R_L is connected between nodes a,b in Fig. 4.36(a). Using Ohm's law, the current in R_L is

$$i_L = \frac{v_s}{R + R_L}. \tag{4.52}$$

Now suppose the same resistor R_L is connected between nodes a,b in Fig. 4.36(b). Using current division, the current in R_L is

$$i_L = \frac{R}{R + R_L} i_s. \tag{4.53}$$

If the two circuits in Fig. 4.36 are equivalent, these resistor currents must be the same. Equating the right-hand sides of Eqs. (4.52) and

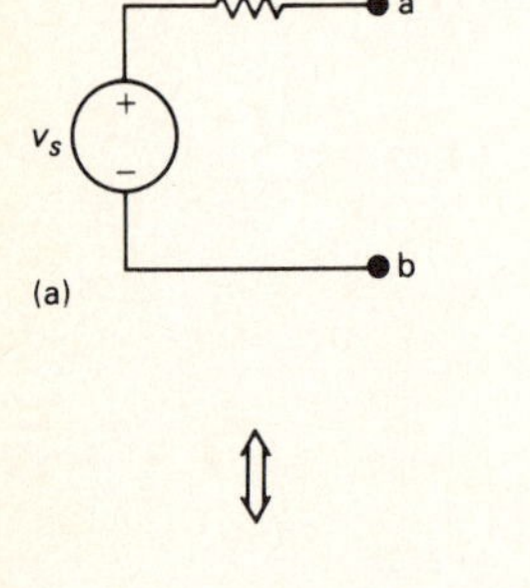

⇕

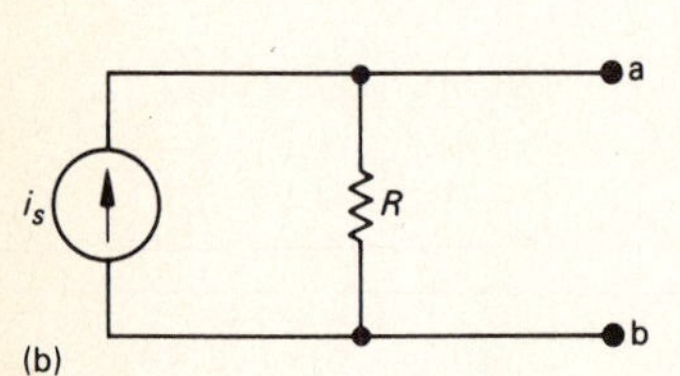

FIGURE 4.36 Source transformations.

(4.53) and simplifying,

$$i_s = \frac{v_s}{R}. \tag{4.54}$$

When Eq. (4.54) is satisfied for the circuits in Fig. 4.36, the current in R_L is the same for both circuits in the figure for all values of R_L. If the current through R_L is the same in both circuits, then the voltage drop across R_L is the same in both circuits, and the circuits are equivalent at nodes a,b.

If the polarity of v_s is reversed, the orientation of i_s must be reversed to maintain equivalence.

Example 4.8 illustrates the usefulness of making source transformations to simplify a circuit-analysis problem.

EXAMPLE 4.8

a) For the circuit shown in Fig. 4.37, find the power associated with the 6 V source.

b) State whether the 6 V source is absorbing or delivering the power calculated in (a).

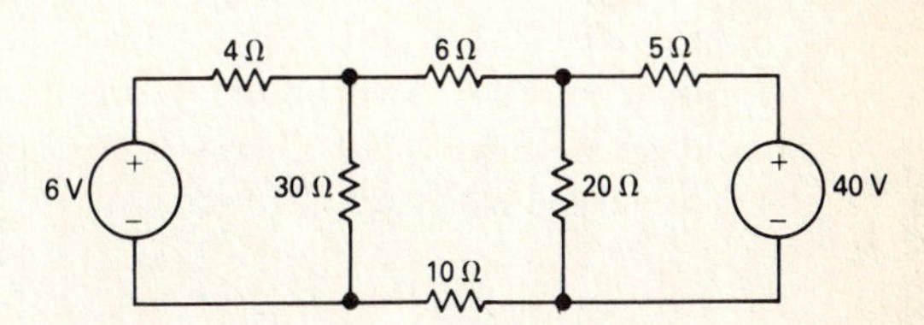

FIGURE 4.37 The circuit for Example 4.8.

SOLUTION

a) If we study the circuit shown in Fig. 4.37 knowing that the power associated with the 6 V source is of interest, several approaches come to mind. The circuit has four essential nodes and six essential branches where the current is unknown. Thus we can find the current in the branch containing the 6 V source by solving either three—6 − (4 − 1)—mesh-current equations or three—4 − 1—node-voltage equations. Choosing the mesh-current approach involves solving for the mesh current that corresponds to the branch current in the 6 V source. Choosing the node-voltage approach involves solving for the voltage across the 30 Ω resistor, from which the branch current in the 6 V source can be calculated. But by focusing on just one branch current, we can first simplify the circuit by using source transformations.

We must reduce the circuit in a way that preserves the identity of the branch containing the 6 V source. We have no reason to preserve the identity of the branch containing the 40 V source. Beginning with this branch, we can transform the 40 V source in series with the 5 Ω resistor into an 8 A current source in parallel with a 5 Ω resistor, as shown in Fig. 4.38(a). Next, we can replace

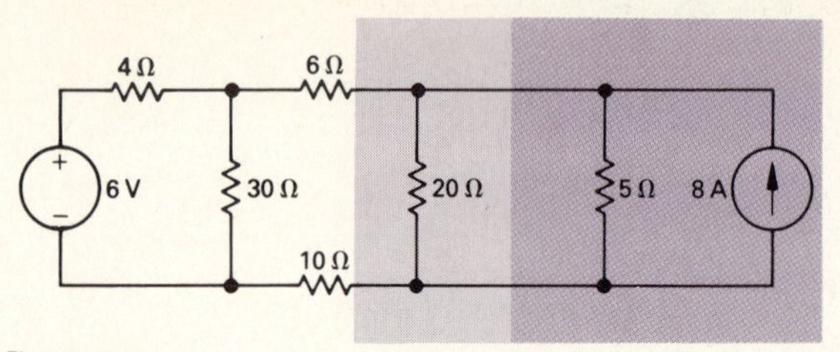

(a) First step

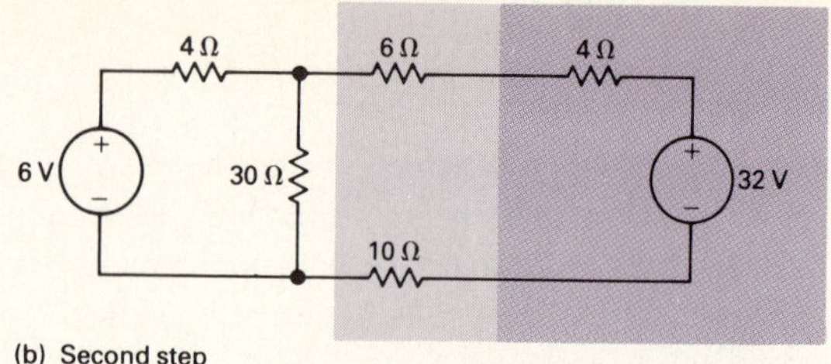

(b) Second step

the parallel combination of the 20 Ω and 5 Ω resistors with a 4 Ω resistor. This 4 Ω resistor is in parallel with the 8 A source and therefore can be replaced with a 32 V source in series with a 4 Ω resistor, as shown in Fig. 4.38(b). The 32 V source is in series with 20 Ω of resistance and, hence, can be replaced by a current source of 1.6 A in parallel with 20 Ω, as shown in Fig. 4.38(c). The 20 Ω and 30 Ω parallel resistors can be reduced to a single 12 Ω resistor. The parallel combination of the 1.6 A current source and the 12 Ω resistor transforms into a voltage source of 19.2 V in series with 12 Ω. Figure 4.38(d) shows the result of this last transformation. The current in the direction of the voltage drop across the 6 V source is (19.2 − 6)/16, or 0.825 A. Therefore the power associated with the 6 V source is

$$p_{6V} = (0.825)(6) = 4.95 \text{ W}.$$

b) The voltage source is absorbing power.

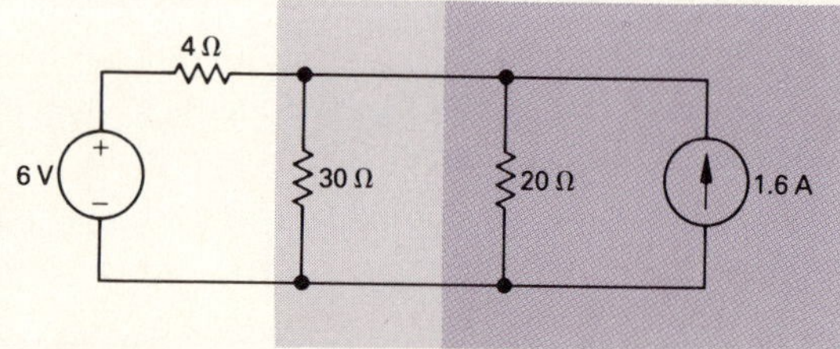

(c) Third step

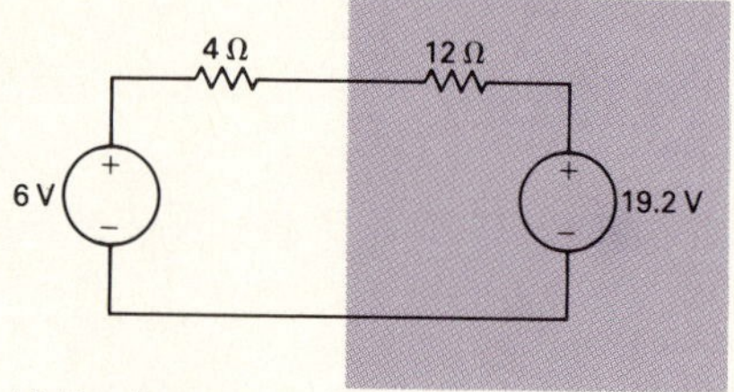

(d) Fourth step

FIGURE 4.38 Step-by-step simplification of the circuit shown in Fig. 4.37.

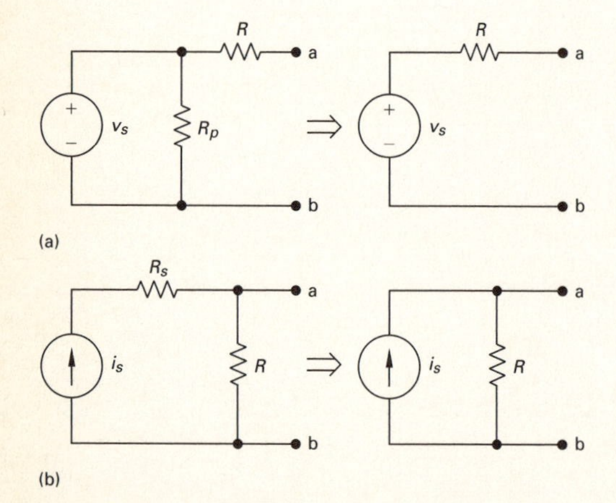

FIGURE 4.39 Equivalent circuits containing a resistance in parallel with a voltage source or in series with a current source.

A question that arises from use of the source transformation depicted in Fig. 4.38 is, "What happens if there is a resistance R_p in parallel with the voltage source, or a resistance R_s in series with the current source?" In both cases the resistance has no effect on the equivalent circuit that predicts behavior with respect to terminals a,b. Figure 4.39 summarizes this observation.

The two circuits depicted in Fig. 4.39(a) are equivalent with respect to terminals a,b because they produce the same voltage and current in any resistor R_L inserted between nodes a,b. The same can be said for the circuits in Fig. 4.39(b). Example 4.9 illustrates an application of the equivalent circuits depicted in Fig. 4.39.

EXAMPLE 4.9

a) Use source transformations to find the voltage v_o in the circuit shown in Fig. 4.40.

b) Find the power developed by the 250 V voltage source.

c) Find the power developed by the 8 A current source.

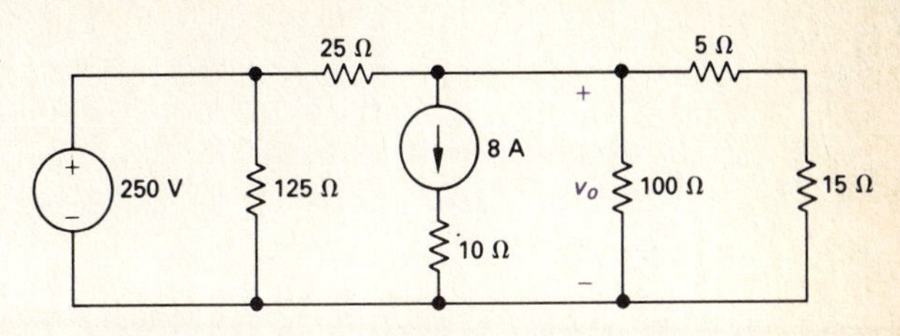

FIGURE 4.40 The circuit for Example 4.9.

SOLUTION

a) We begin by removing the 125 Ω and 10 Ω resistors, because the 125 Ω resistor is connected across the 250 V voltage source and the 10 Ω resistor is connected in series with the 8 A current source. We also combine the series-connected resistors into a single resistance of 20 Ω. Figure 4.41 shows the simplified circuit.

We now use a source transformation to replace the 250 V source and 25 Ω resistor with a 10 A source in parallel with the 25 Ω resistor, as shown in Fig. 4.42. We can now simplify the circuit shown in Fig. 4.42 by using Kirchhoff's current law to combine the parallel current sources into a single source. The parallel resistors combine into a single resistor. Figure 4.43 shows the result. Hence $v_O = 20$ V.

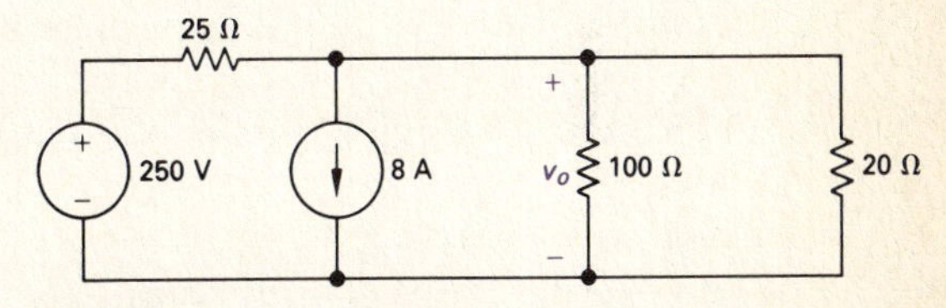

FIGURE 4.41 A simplified version of the circuit shown in Fig. 4.40.

b) The current supplied by the 250 V source equals the current in the 125 Ω resistor plus the current in the 25 Ω resistor. Thus

$$i_s = \frac{250}{125} + \frac{250 - 20}{25} = 11.2 \text{ A}.$$

Therefore the power developed by the voltage source is

$$p_{250\text{V}} \text{ (developed)} = (250)(11.2) = 2800 \text{ W}.$$

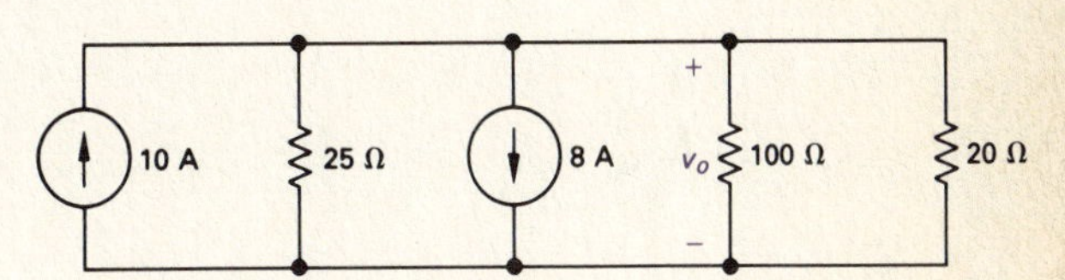

FIGURE 4.42 The circuit shown in Fig. 4.41, after a source transformation.

c) To find the power developed by the 8 A current source, we first find the voltage across the source. If we let v_s represent the voltage across the source—positive at the upper terminal of the source—we obtain

$$v_s + 8(10) = v_o = 20,$$

or

$$v_s = -60 \text{ V},$$

and the power developed by the 8 A source is 480 W. Note that the 125 Ω and 10 Ω resistors do not affect the value of v_o but do affect the power calculations.

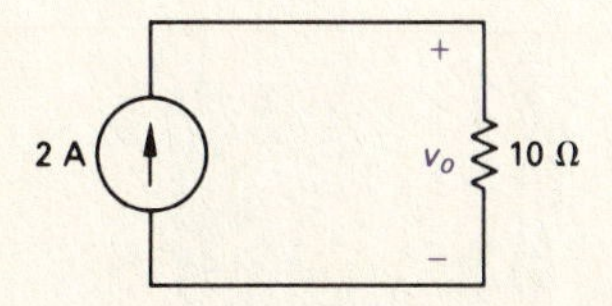

FIGURE 4.43 The circuit shown in Fig. 4.42, after combining sources and resistors.

DRILL EXERCISE

4.19 a) Use a series of source transformations to find the voltage v in the circuit shown.

b) How much power does the 120 V source deliver to the circuit?

ANSWER: (a) 48 V; (b) 374.4 W

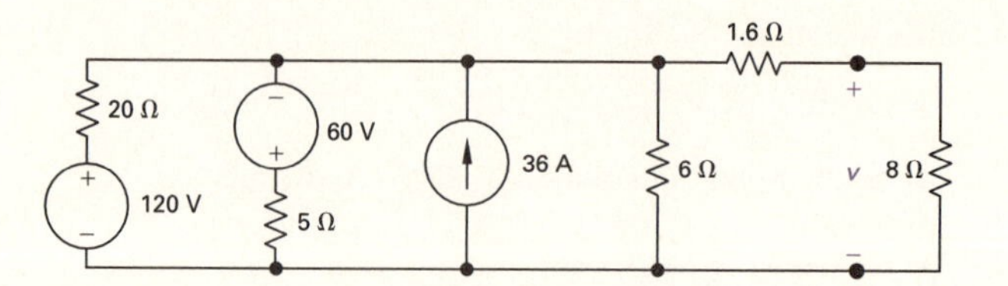

4.10 THÉVENIN AND NORTON EQUIVALENTS

At times in circuit analysis we want to concentrate on what happens at a specific pair of terminals. For example, when we plug a toaster into an outlet, we are interested primarily in the voltage and current at the terminals of the toaster. We have little or no interest in the effect that connecting the toaster has on voltages or currents elsewhere in the circuit supplying the outlet. We can expand this interest in terminal behavior to a set of appliances, each requiring a different amount of power. We then are interested in how the voltage and current delivered at the outlet change as we change appliances. In other words, we want to focus on the behavior of the circuit supplying the outlet, but only at the outlet terminals.

Thévenin and Norton equivalents are circuit simplification techniques which focus on terminal behavior and thus are extremely valuable aids in analysis. Although here we discuss them as they pertain to resistive circuits, Thévenin and Norton equivalent circuits may be used to represent any circuit made up of linear elements.

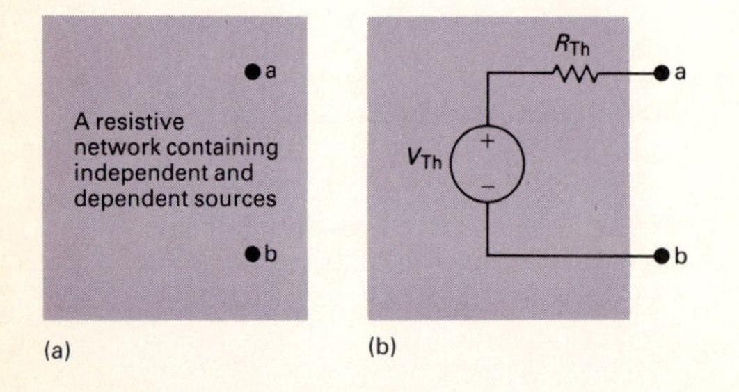

FIGURE 4.44 (a) A general circuit; (b) the Thévenin equivalent circuit.

We can best describe a Thévenin equivalent circuit by reference to Fig. 4.44, which represents any circuit made up of sources (both independent and dependent) and resistors. The letters a and b denote the pair of terminals of interest. Figure 4.44(b) shows the Thévenin equivalent. Thus, a **Thévenin equivalent circuit** is an independent voltage source V_{Th} in series with a resistor R_{Th}, which replaces an interconnection of sources and resistors. This series combination of V_{Th} and R_{Th} is equivalent to the original circuit in the sense that, if we connect the same load across the terminals a,b of each circuit, we get the same voltage and current at the terminals of the load. This equivalence holds for all possible values of load resistance.

To represent the original circuit by its Thévenin equivalent, we must be able to determine the Thévenin voltage V_{Th} and the Thévenin resistance R_{Th}. First, we note that, if the load resistance is infinitely large, we have an open-circuit condition. The open-circuit voltage at the terminals a,b in the circuit shown in Fig. 4.44(b) is V_{Th}. By hypothesis, this must be the same as the open-circuit voltage at the terminals a,b in the original circuit. Therefore to calculate the Thévenin voltage V_{Th}, we simply calculate the open-circuit voltage in the original circuit.

Reducing the load resistance to zero gives us a short-circuit condition. If we place a short circuit across the terminals a,b of the Thévenin equivalent circuit, the short-circuit current directed from a to b is

$$i_{sc} = \frac{V_{\text{Th}}}{R_{\text{Th}}}. \quad \textbf{(4.55)}$$

By hypothesis, this short-circuit current must be identical to the short-circuit current that exists in a short circuit placed across the terminals a,b of the original network. From Eq. (4.55),

$$R_{\text{Th}} = \frac{V_{\text{Th}}}{i_{sc}}. \quad \textbf{(4.56)}$$

Thus the Thévenin resistance is the ratio of the open-circuit voltage to the short-circuit current.

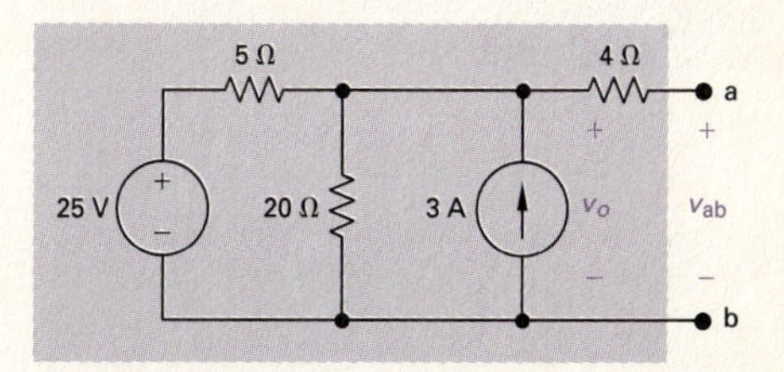

FIGURE 4.45 A circuit used to illustrate a Thévenin equivalent.

FINDING A THÉVENIN EQUIVALENT

To find the Thévenin equivalent of the circuit shown in Fig. 4.45, we first calculate the open-circuit voltage of v_{ab}. Note that when the terminals a,b are open, there is no current in the 4 Ω resistor. Therefore the open-circuit voltage v_{ab} is identical to the voltage across the 3 A current source, labeled v_o. We find the voltage v_o by solving a single node-voltage equation. Choosing the lower node as the reference node, we get

$$\frac{v_o - 25}{5} + \frac{v_o}{20} - 3 = 0. \quad \textbf{(4.57)}$$

Solving for v_o yields

$$v_o = 32 \text{ V}. \quad \textbf{(4.58)}$$

Hence the Thévenin voltage for the circuit is 32 V.

The next step is to place a short circuit across the terminals and calculate the resulting short-circuit current. Figure 4.46 shows the circuit with the short in place. Note that the short-circuit current is in the direction of the open-circuit voltage drop across the terminals a,b. If the short-circuit current is in the direction of the open-circuit voltage rise across the terminals, a minus sign must be inserted in Eq. (4.56).

The short-circuit current (i_{sc}) is easily found once v_o is known. Therefore the problem reduces to finding v_o with the short in place.

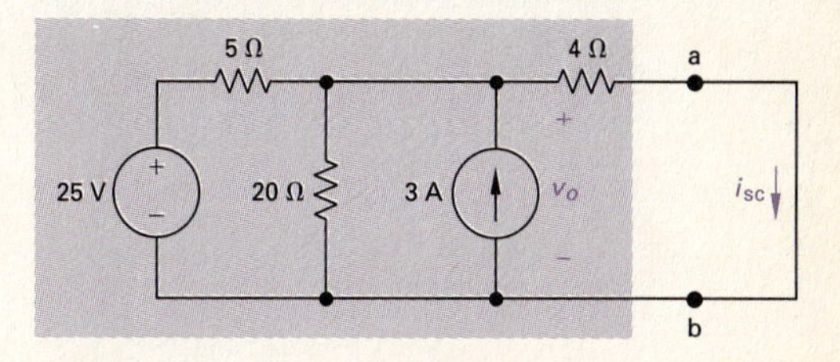

FIGURE 4.46 The circuit shown in Fig. 4.45, with terminals a and b short-circuited.

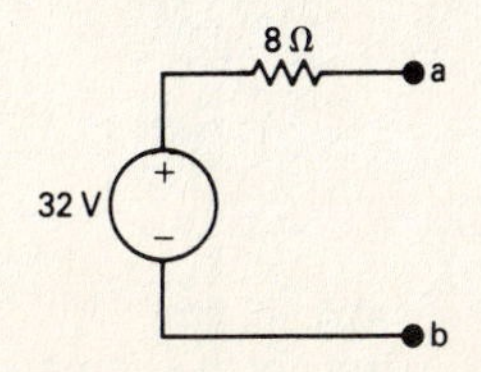

FIGURE 4.47 The Thévenin equivalent of the circuit shown in Fig. 4.45.

Again, if we use the lower node as the reference node, the equation for v_o becomes

$$\frac{v_o - 25}{5} + \frac{v_o}{20} - 3 + \frac{v_o}{4} = 0. \tag{4.59}$$

Solving Eq. (4.59) for v_o gives

$$v_o = 16 \text{ V}. \tag{4.60}$$

Hence, the short-circuit current is

$$i_{sc} = \frac{16}{4} = 4 \text{ A}. \tag{4.61}$$

We now find the Thévenin resistance by substituting the numerical results from Eqs. (4.58) and (4.61) into Eq. (4.56):

$$R_{Th} = \frac{V_{Th}}{i_{sc}} = \frac{32}{4} = 8\ \Omega. \tag{4.62}$$

Figure 4.47 shows the Thévenin equivalent for the circuit shown in Fig. 4.45.

You should verify that, if a 24 Ω resistor is connected across the terminals a,b in Fig. 4.45, the voltage across the resistor will be 24 V and the current in the resistor will be 1 A—as would be the case with the Thévenin circuit in Fig. 4.47. This same equivalence between the circuit in Figs. 4.45 and 4.47 holds for any resistor value connected between nodes a,b.

The Norton Equivalent

A **Norton equivalent circuit** consists of an independent current source in parallel with the Norton equivalent resistance. We can derive it from a Thévenin equivalent circuit simply by making a source transformation. Thus the Norton current equals the short-circuit current at the terminals of interest, and the Norton resistance is identical to the Thévenin resistance.

Using Source Transformations

Sometimes we can make effective use of source transformations to derive a Thévenin or Norton equivalent circuit. For example, we can derive the Thévenin and Norton equivalents of the circuit shown in Fig. 4.45 by making the series of source transformations shown in Fig. 4.48. This technique is most useful when the network contains only independent sources. The presence of dependent sources requires retaining the identity of the controlling voltages and/or currents, and this constraint usually prohibits continued reduction of the circuit

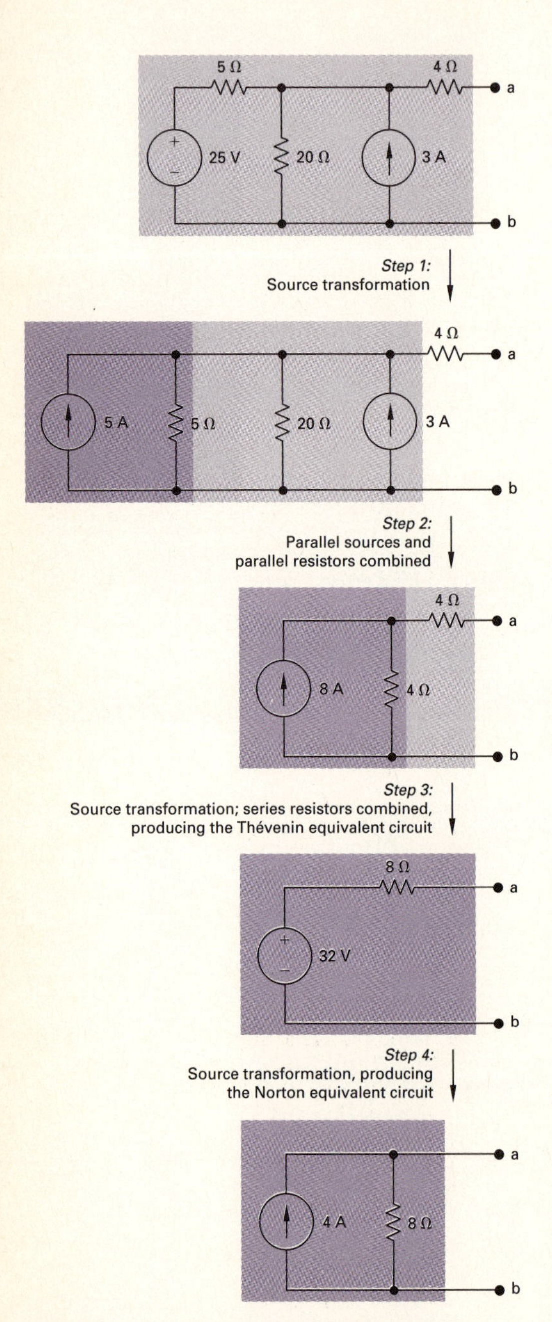

FIGURE 4.48 Step-by-step derivation of the Thévenin and Norton equivalents of the circuit shown in Fig. 4.45.

by source transformations. We discuss the problem of finding the Thévenin equivalent when a circuit contains dependent sources in Example 4.10.

EXAMPLE 4.10

Find the Thévenin equivalent for a circuit containing dependent sources, shown in Fig. 4.49.

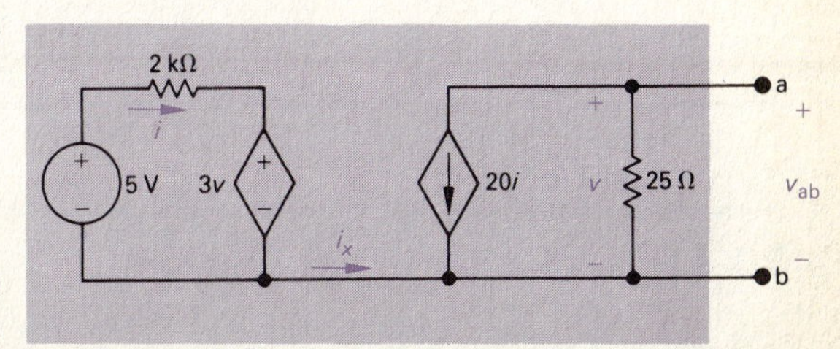

FIGURE 4.49 A circuit used to illustrate a Thévenin equivalent when the circuit contains dependent sources.

SOLUTION

The first step in analyzing the circuit in Fig. 4.49 is to recognize that the current labeled i_x must be zero. (Note the absence of a return path for i_x to enter the left-hand portion of the circuit.) The open-circuit, or Thévenin, voltage will be the voltage across the 25 Ω resistor. With $i_x = 0$,

$$V_{\mathrm{Th}} = v_{\mathrm{ab}} = (-20i)(25) = -500i.$$

The current i is

$$i = \frac{5 - 3v}{2000} = \frac{5 - 3V_{\mathrm{Th}}}{2000}.$$

In writing the equation for i, we recognize that the Thévenin voltage is identical to the control voltage. When we combine these two equations, we obtain

$$V_{\mathrm{Th}} = -5 \text{ V}.$$

To calculate the short-circuit current, we place a short circuit across a,b. When the terminals a,b are shorted together, the control voltage v is reduced to zero. Therefore, with the short in place, the circuit shown in Fig. 4.49 becomes the one shown in Fig. 4.50. With the short circuit shunting the 25 Ω resistor, all the current from the dependent current source appears in the short, so

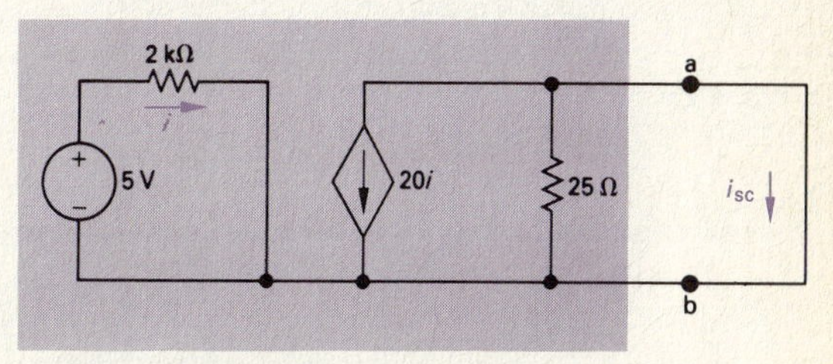

FIGURE 4.50 The circuit shown in Fig. 4.49, with terminals a and b short-circuited.

$$i_{\mathrm{sc}} = -20i.$$

As the voltage controlling the dependent voltage source has been reduced to zero, the current controlling the dependent current source is

$$i = \frac{5}{2000} = 2.5 \text{ mA}.$$

Combining these two equations yields a short-circuit current of

$$i_{\mathrm{sc}} = -20(2.5) = -50 \text{ mA}.$$

From i_{sc} and V_{Th} we get

$$R_{Th} = \frac{V_{Th}}{i_{sc}} = \frac{-5}{-50} \times 10^3 = 100\ \Omega.$$

Figure 4.51 illustrates the Thévenin equivalent for the circuit shown in Fig. 4.49. Note that the reference polarity marks on the Thévenin voltage source in Fig. 4.51 agree with the preceding equation for V_{Th}.

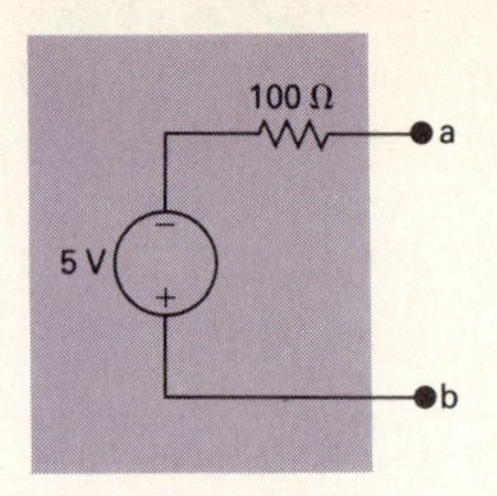

FIGURE 4.51 The Thévenin equivalent for the circuit shown in Fig. 4.49.

DRILL EXERCISES

4.20 Find the Thévenin equivalent circuit with respect to the terminals a,b for the circuit shown.

ANSWER: $V_{ab} = V_{Th} = 64.8$ V, $R_{Th} = 6\ \Omega$.

4.21 Find the Norton equivalent circuit with respect to the terminals a,b for the circuit shown.

ANSWER: $I_N = 6$ A (directed toward a), $R_N = 6\ \Omega$.

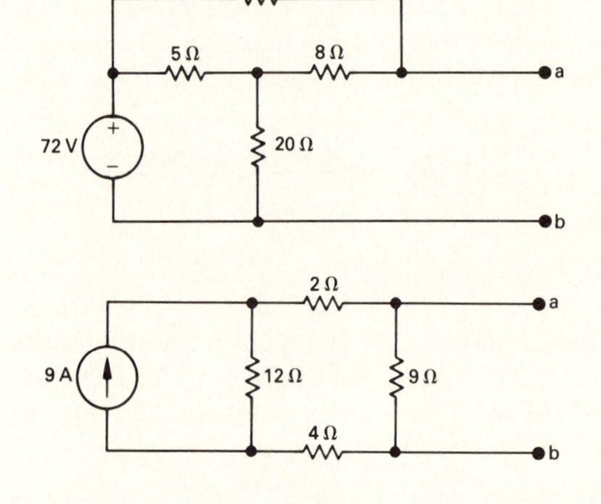

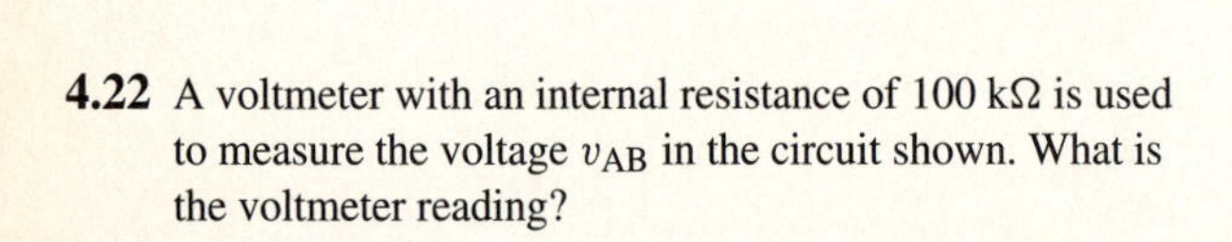

4.22 A voltmeter with an internal resistance of 100 kΩ is used to measure the voltage v_{AB} in the circuit shown. What is the voltmeter reading?

ANSWER: 120 V.

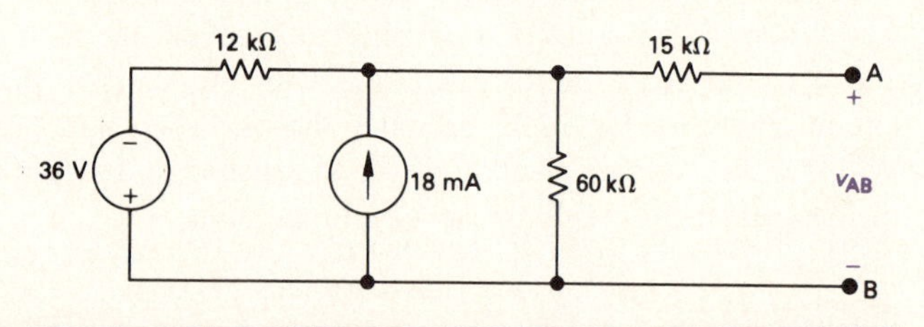

4.11 MORE ON DERIVING A THÉVENIN EQUIVALENT

The technique for determining R_{Th} that we discussed and illustrated in Section 4.10 is not always the easiest method available. Two other methods generally are simpler to use. The first is useful if the network contains only independent sources. To calculate R_{Th} for such a network, we first deactivate all independent sources and then calculate the resistance seen looking into the network at the designated terminal pair. A voltage source is deactivated by replacing it with a short circuit. A current source is deactivated by replacing it with an open circuit. For example, consider the circuit shown in Fig. 4.52. Deactivating the independent sources simplifies the circuit to the one shown in Fig. 4.53. The resistance seen looking into the terminals a,b is denoted R_{ab}, which consists of the 4 Ω resistor in series with the parallel combinations of the 5 and 20 Ω resistors. Thus

$$R_{ab} = R_{Th} = 4 + \frac{5 \times 20}{25} = 8\ \Omega. \tag{4.63}$$

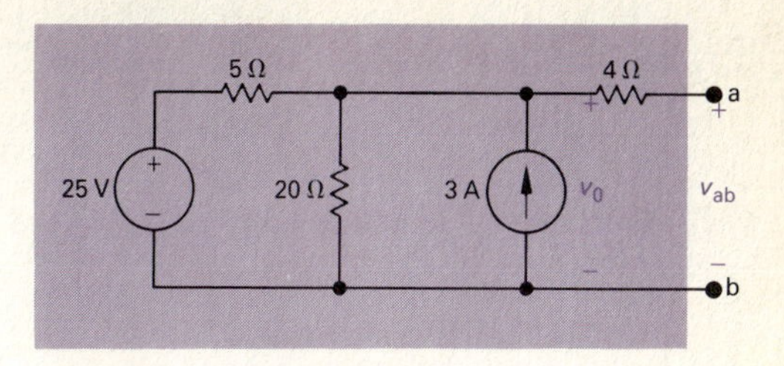

FIGURE 4.52 A circuit used to illustrate a Thévenin equivalent.

Note that the derivation of R_{Th} with Eq. (4.63) is much simpler than the same derivation with Eqs. (4.59)–(4.62).

If the circuit or network contains dependent sources, an alternative procedure for finding the Thévenin resistance R_{Th} is as follows. We first deactivate all independent sources, and we then apply either a test voltage source or a test current source to the Thévenin terminals a,b. The Thévenin resistance equals the ratio of the voltage across the test source to the current delivered by the test source. Example 4.11 illustrates this alternative procedure for finding R_{Th}, using the same circuit as Example 4.9.

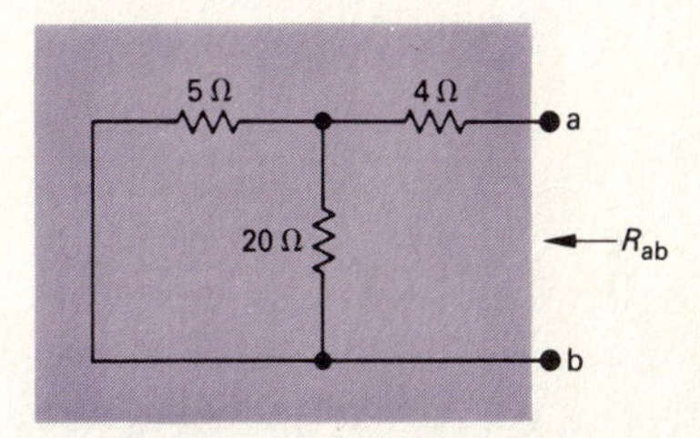

FIGURE 4.53 The circuit shown in Fig. 4.52, after deactivation of the independent sources.

EXAMPLE 4.11

Find the Thévenin resistance R_{Th} for the circuit in Fig. 4.49, using the alternative method described.

SOLUTION

We first deactivate the independent voltage source from the circuit and then excite the circuit from the terminals a,b with either a test voltage source or a test current source. If we apply a test voltage source, we will know the voltage of the dependent voltage source and hence the controlling current i. Therefore we opt for the test voltage source. Figure 4.54 shows the circuit for computing the Thévenin resistance.

The externally applied test voltage source is denoted v_T, and the current that it delivers to the circuit is labeled i_T. To find the Thévenin resistance, we simply solve the circuit shown in Fig. 4.54 for the ratio of the voltage to the current at the test source; that is, $R_{Th} = v_T/i_T$. From Fig. 4.54,

$$i_T = \frac{v_T}{25} + 20i \tag{4.64}$$

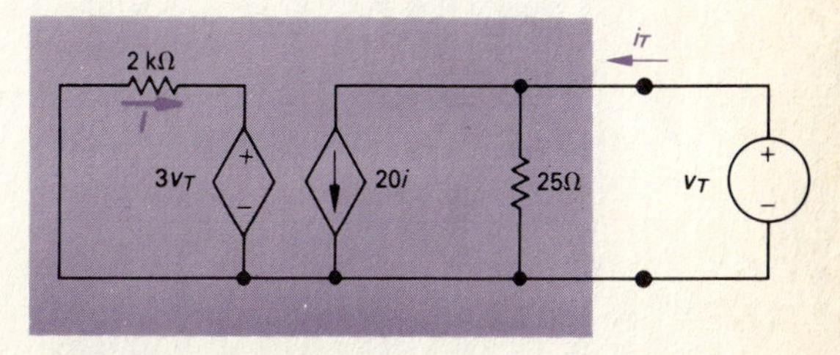

FIGURE 4.54 An alternative method for computing the Thévenin resistance.

and

$$i = \frac{-3v_T}{2} \text{ mA.} \tag{4.65}$$

We then substitute Eq. (4.65) into Eq. (4.64) and solve the resulting equation for the ratio v_T/i_T:

$$i_T = \frac{v_T}{25} - \frac{60v_T}{2000} \tag{4.66}$$

and

$$\frac{i_T}{v_T} = \frac{1}{25} - \frac{6}{200} = \frac{50}{5000} = \frac{1}{100}. \tag{4.67}$$

From Eqs. (4.66) and (4.67),

$$R_{\text{Th}} = \frac{v_T}{i_T} = 100\ \Omega. \tag{4.68}$$

In general, these computations are easier than those involved in computing the short-circuit current. Moreover, in a network containing only resistors and dependent sources, you must use the alternative method, because the ratio of the Thévenin voltage to the short-circuit current is indeterminate. That is, it is the ratio 0/0. (See Problems 4.67 and 4.68.)

DRILL EXERCISES

4.23 Find the Thévenin equivalent circuit with respect to the terminals a,b for the circuit shown.

ANSWER: $V_{\text{Th}} = v_{\text{ab}} = 20$ V, $R_{\text{Th}} = 0.625\ \Omega$.

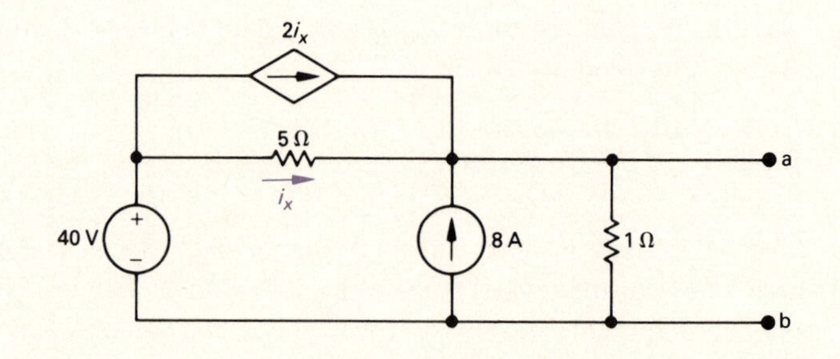

4.24 Find the Thévenin equivalent circuit with respect to the terminals a,b for the circuit shown. (Hint: define the voltage at the left-most node as v, and write two nodal equations with V_{Th} as the right node voltage.)

ANSWER: $V_{\text{Th}} = v_{\text{ab}} = 30$ V, $R_{\text{Th}} = 10\ \Omega$.

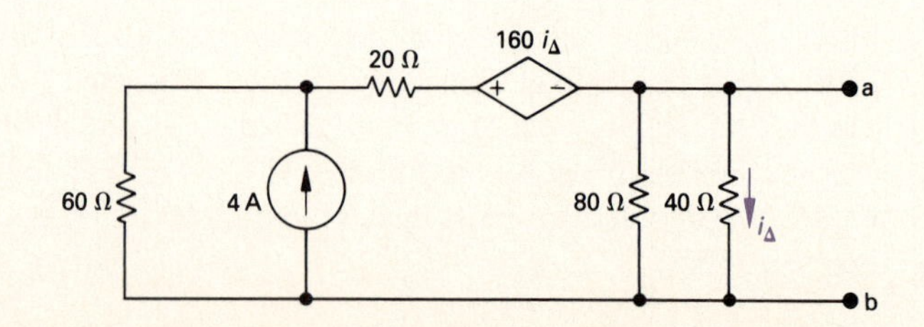

USING THE THÉVENIN EQUIVALENT IN THE AMPLIFIER CIRCUIT

At times we can use a Thévenin equivalent to reduce one portion of a circuit to greatly simplify analysis of the larger network. Let's return to the circuit first introduced in Section 2.5 and subsequently analyzed in Sections 4.4 and 4.7. To aid our discussion, we redrew the circuit and identified the branch currents of interest, as shown in Fig. 4.55.

As our previous analysis has shown, i_B is the key to finding the other branch currents. We redraw the circuit as shown in Fig. 4.56 to prepare to replace the subcircuit to the left of V_0 with its Thévenin equivalent. You should be able to determine that this modification has no effect on the branch currents i_1, i_2, i_B, and i_E.

Now we replace the circuit made up of V_{CC}, R_1, and R_2 with a Thévenin equivalent, with respect to the terminals b,d. The Thévenin voltage and resistance are

$$V_{\text{Th}} = \frac{V_{CC}R_2}{R_1 + R_2} \tag{4.69}$$

and

$$R_{\text{Th}} = \frac{R_1 R_2}{R_1 + R_2}. \tag{4.70}$$

With the Thévenin equivalent, the circuit in Fig. 4.56 becomes the one shown in Fig. 4.57.

We now derive an equation for i_B simply by summing the voltages around the left mesh. In writing this mesh equation, we recognize that $i_E = (1 + \beta)i_B$. Thus

$$V_{\text{Th}} = R_{\text{Th}}i_B + V_0 + R_E(1 + \beta)i_B, \tag{4.71}$$

from which

$$i_B = \frac{V_{\text{Th}} - V_0}{R_{\text{Th}} + (1 + \beta)R_E}. \tag{4.72}$$

When we substitute Eqs. (4.69) and (4.70) into Eq. (4.72), we get the same expression obtained in Eq. (2.25). Note that, when we have incorporated the Thévenin equivalent into the original circuit, we can obtain the solution for i_B by writing a single equation.

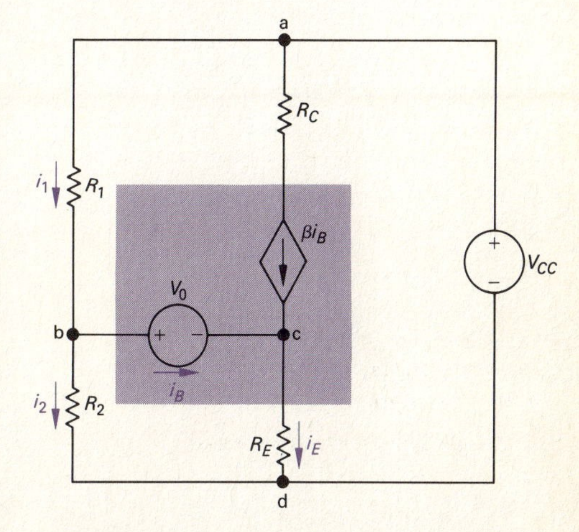

FIGURE 4.55 The application of a Thévenin equivalent in circuit analysis.

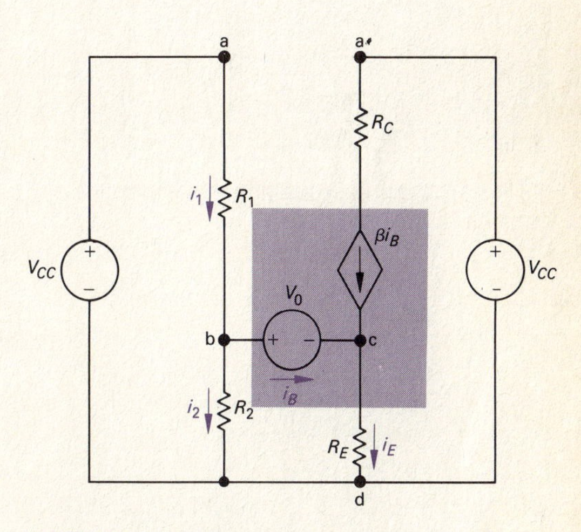

FIGURE 4.56 A modified version of the circuit shown in Fig. 4.55.

4.12 MAXIMUM POWER TRANSFER

Circuit analysis plays an important role in the analysis of systems designed to transfer power from a source to a load. We discuss power transfer in terms of two basic types of systems. The first emphasizes

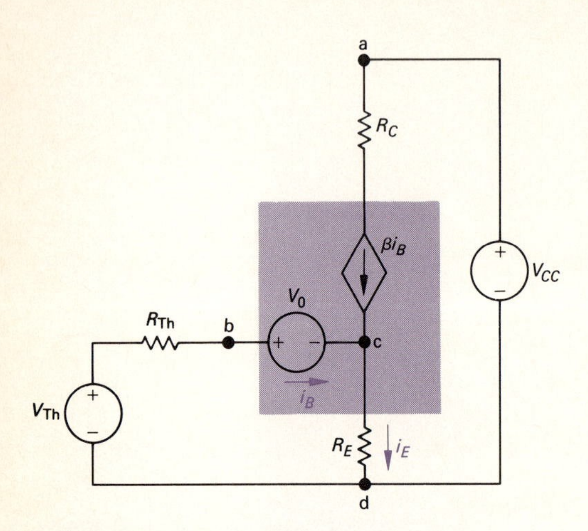

FIGURE 4.57 The circuit shown in Fig. 4.56 modified by a Thévenin equivalent.

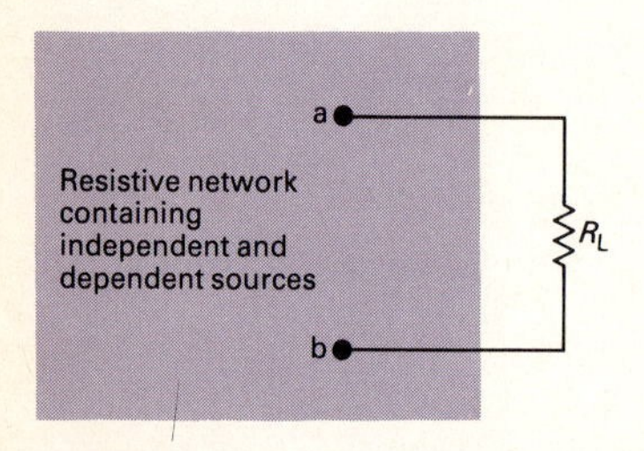

FIGURE 4.58 A circuit describing maximum power transfer.

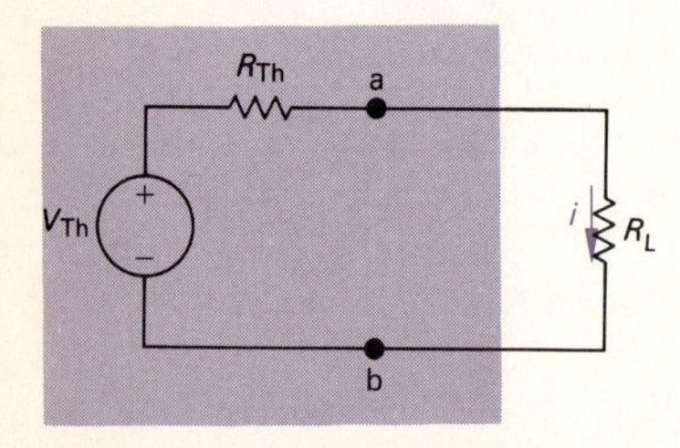

FIGURE 4.59 A circuit used to determine the value of R_L for maximum power transfer.

the efficiency of the power transfer. Power utility systems are a good example of this type because they are concerned with the generation, transmission, and distribution of large quantities of electric power. If a power utility system is inefficient, a large percentage of the power generated is lost in the transmission and distribution processes, and thus wasted.

The second basic type of system emphasizes the amount of power transferred. Communication and instrumentation systems are good examples because in the transmission of information, or data, via electric signals, the power available at the transmitter or detector is limited. Thus, transmitting as much of this power as possible to the receiver, or load, is desirable. In such applications the amount of power being transferred is small, so the efficiency of transfer is not a primary concern. We now consider maximum power transfer in systems that can be modeled by a purely resistive circuit.

Maximum power transfer can best be described with the aid of the circuit shown in Fig. 4.58. We assume a resistive network containing independent and dependent sources and a designated pair of terminals, a,b, to which a load, R_L, is to be connected. The problem is to determine the value of R_L that permits maximum power delivery to R_L. The first step in this process is to recognize that a resistive network can always be replaced by its Thévenin equivalent. Therefore we redraw the circuit shown in Fig. 4.58 as the one shown in Fig. 4.59. Replacing the original network by its Thévenin equivalent greatly simplifies the task of finding R_L. Derivation of R_L requires expressing the power dissipated in R_L as a function of the three circuit parameters V_{Th}, R_{Th}, and R_L. Thus

$$p = i^2 R_L = \left(\frac{V_{Th}}{R_{Th} + R_L} \right)^2 R_L. \qquad \textbf{(4.73)}$$

Next, we recognize that for a given circuit, V_{Th} and R_{Th} will be fixed. Therefore the power dissipated is a function of the single variable R_L. To find the value of R_L that maximizes the power, we use elementary calculus. We begin by writing an equation for the derivative of p with respect to R_L:

$$\frac{dp}{dR_L} = V_{Th}^2 \left[\frac{(R_{Th} + R_L)^2 - R_L \cdot 2(R_{Th} + R_L)}{(R_{Th} + R_L)^4} \right]. \qquad \textbf{(4.74)}$$

The derivative is zero, and p is maximized, when

$$(R_{Th} + R_L)^2 = 2R_L(R_{Th} + R_L). \qquad \textbf{(4.75)}$$

Solving Eq. (4.75) yields

$$R_L = R_{Th}. \qquad \textbf{(4.76)}$$

Thus maximum power transfer occurs when the load resistance R_L equals the Thévenin resistance R_{Th}. To find the maximum power delivered to R_L, we simply substitute Eq. (4.76) into Eq. (4.73):

$$p_{\max} = \frac{V_{\mathrm{Th}}^2 R_L}{(2R_L)^2} = \frac{V_{\mathrm{Th}}^2}{4R_L}. \tag{4.77}$$

The analysis of a circuit when the load resistor is adjusted for maximum power transfer is illustrated in Example 4.12.

EXAMPLE 4.12

a) For the circuit shown in Fig. 4.60, find the value of R_L that results in maximum power being transferred to R_L.

b) Calculate the maximum power that can be delivered to R_L.

c) When R_L is adjusted for maximum power transfer, what percentage of the power delivered by the 360 V source reaches R_L?

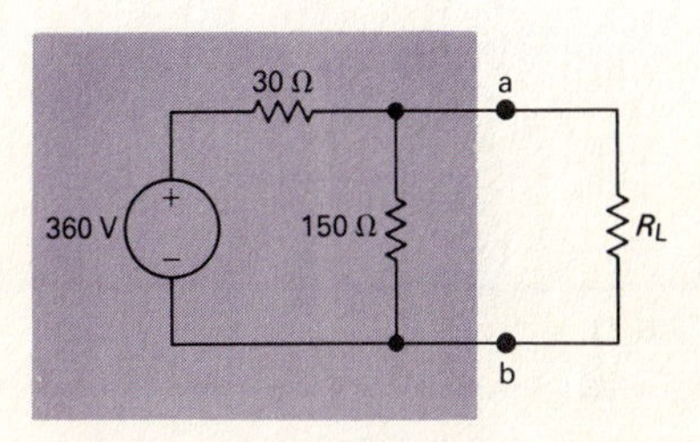

FIGURE 4.60 The circuit for Example 4.12.

SOLUTION

a) The Thévenin voltage for the circuit to the left of the terminals a,b is

$$V_{\mathrm{Th}} = \frac{360}{180} \times 150$$
$$= 300 \text{ V}.$$

The Thévenin resistance is

$$R_{\mathrm{Th}} = \frac{(150)(30)}{180}$$
$$= 25\ \Omega.$$

Replacing the circuit to the left of the terminals a,b with its Thévenin equivalent gives us the circuit shown in Fig. 4.61, which indicates that R_L must equal 25 Ω for maximum power transfer.

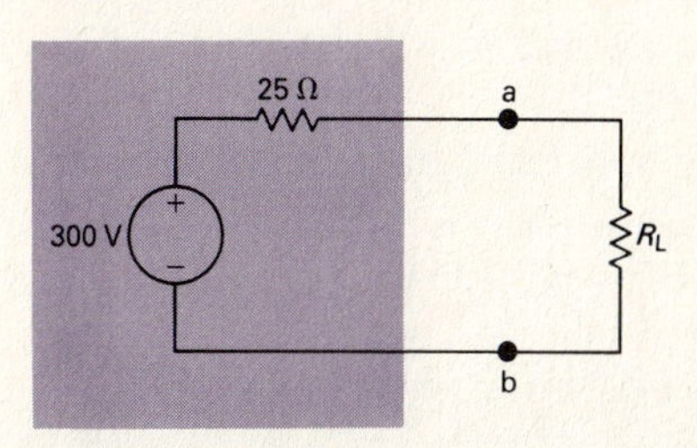

FIGURE 4.61 Reduction of the circuit shown in Fig. 4.60, by means of a Thévenin equivalent.

b) The maximum power that can be delivered to R_L is

$$p_{\max} = \left(\frac{300}{50}\right)^2 (25)$$
$$= 900 \text{ W}.$$

c) When R_L equals 25 Ω, the voltage v_{ab} is

$$v_{\mathrm{ab}} = \left(\frac{300}{50}\right)(25)$$
$$= 150 \text{ V}.$$

Use a computer tool to vary R_L and calculate the value for which maximum power is transferred.

From Fig. 4.60, when v_{ab} equals 150 V, the current in the voltage source in the direction of the voltage rise across the source is

$$i_s = \frac{360 - 150}{30} = \frac{210}{30}$$
$$= 7 \text{ A}.$$

Therefore the source is delivering 2520 W to the circuit, or

$$p_s = -i_s(360)$$
$$= -2520 \text{ W}.$$

The percentage of the source power delivered to the load is

$$\frac{900}{2520} \times 100 = 35.71\%.$$

DRILL EXERCISES

4.25 a) Find the value of R that enables the circuit shown to deliver maximum power to the terminals a,b.

b) Find the maximum power delivered to R.

ANSWER: (a) 3 Ω; (b) 1.2 kW.

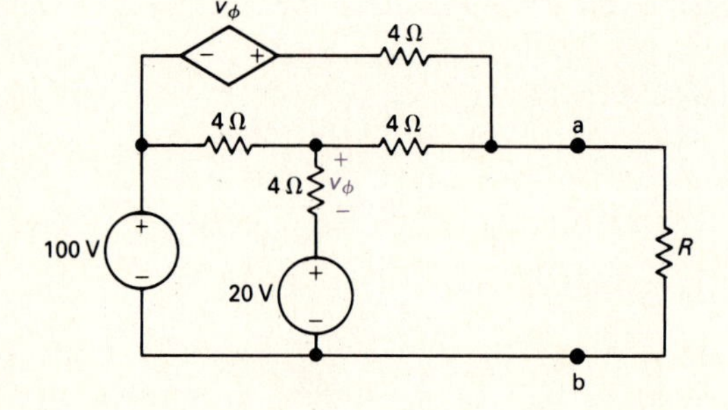

4.26 Assume that the circuit in Drill Exercise 4.25 is delivering maximum power to the load resistor R.

a) How much power is the 100 V source delivering to the network?

b) Repeat (a) for the dependent voltage source.

c) What percentage of the total power generated by these two sources is delivered to the load resistor R?

ANSWER: (a) 3000 W; (b) 800 W; (c) 31.58%.

4.13 SUPERPOSITION

A linear system obeys the principle of **superposition**, which states that whenever a linear system is excited, or driven, by more than one independent source of energy, the total response is the sum of the individual responses. An individual response is the result of an independent source acting alone. Because we are dealing with circuits made up of interconnected linear-circuit elements, we can apply the principle of superposition directly to the analysis of such circuits when they are driven by more than one independent energy source. At present, we restrict the discussion to simple resistive networks; however, the principle is applicable to any linear system.

Superposition is applied in both the analysis and the design of circuits. In analyzing a complex circuit with multiple independent voltage and current sources, there are often fewer, simpler equations to solve when the effects of the independent sources are considered one at a time. Applying superposition can thus simplify circuit analysis. Be aware, though, that sometimes applying superposition actually complicates the analysis, producing more equations to solve than with an alternative method. Superposition is required only if the independent sources in a circuit are fundamentally different. In these early chapters, all independent sources are dc sources, so superposition is not required. We introduce superposition here in anticipation of later chapters in which circuits will require superposition.

Superposition is applied in design to synthesize a desired circuit response that could not be achieved in a circuit with a single source. If the desired circuit response can be written as a sum of two or more terms, the response can be realized by including one independent source for each term of the response. This approach to the design of circuits with complex responses allows a designer to consider several simple designs instead of one complex design.

Use computer tools to manage the design of circuits with complex responses.

We demonstrate the superposition principle by using it to find the branch currents in the circuit shown in Fig. 4.62. We begin by finding the branch currents resulting from the 120 V voltage source. We denote those currents with a prime. Replacing the ideal current source with an open circuit deactivates it; Fig. 4.63 shows this. The branch currents in this circuit are the result of only the voltage source.

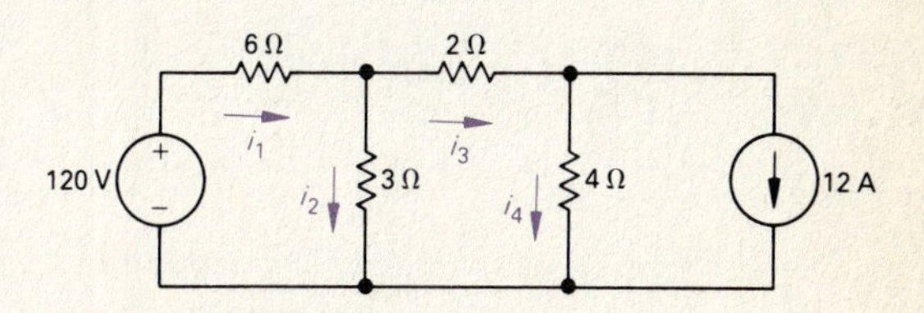

FIGURE 4.62 A circuit used to illustrate superposition.

We can easily find the branch currents in the circuit in Fig. 4.63 once we know the node voltage across the 3 Ω resistor. Denoting this voltage v_1, we write

$$\frac{v_1 - 120}{6} + \frac{v_1}{3} + \frac{v_1}{2+4} = 0, \quad \textbf{(4.78)}$$

from which

$$v_1 = 30 \text{ V}. \quad \textbf{(4.79)}$$

Now we can write the expressions for the branch currents i_1' through i_4' directly:

$$i_1' = \frac{120 - 30}{6} = 15 \text{ A}; \quad \textbf{(4.80)}$$

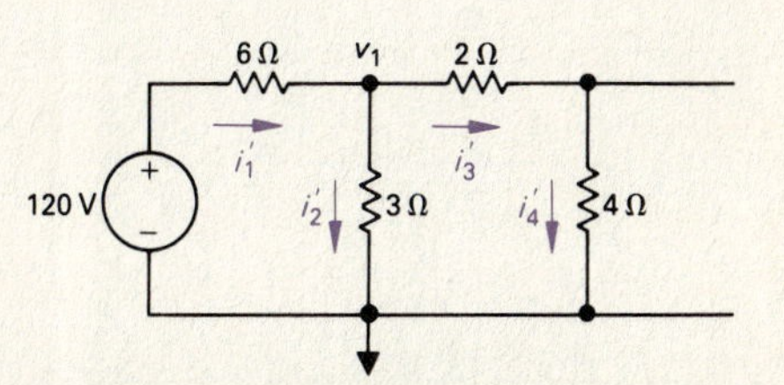

FIGURE 4.63 The circuit shown in Fig. 4.62, with the current source deactivated.

$$i'_2 = \frac{30}{3} = 10 \text{ A}; \quad \textbf{(4.81)}$$

$$i'_3 = i'_4 = \frac{30}{6} = 5 \text{ A}. \quad \textbf{(4.82)}$$

To find the component of the branch currents resulting from the current source, we deactivate the ideal voltage source and solve the circuit shown in Fig. 4.64. The double-prime notation for the currents indicates they are the components of the total current resulting from the ideal current source.

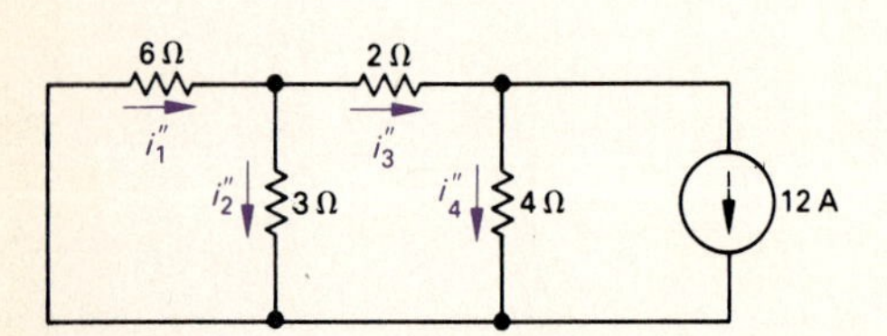

FIGURE 4.64 The circuit shown in Fig. 4.62, with the voltage source deactivated.

We determine the branch currents in the circuit shown in Fig. 4.64 by first solving for the node voltages across the 3 and 4 Ω resistors, respectively. Figure 4.65 shows the two node voltages. The two node-voltage equations that describe the circuit are

$$\frac{v_3}{3} + \frac{v_3}{6} + \frac{v_3 - v_4}{2} = 0 \quad \textbf{(4.83)}$$

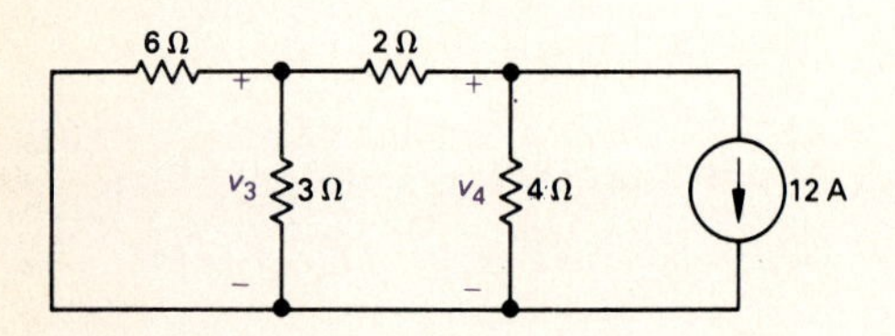

FIGURE 4.65 The circuit shown in Fig. 4.64, showing the node voltages v_3 and v_4.

and

$$\frac{v_4 - v_3}{2} + \frac{v_4}{4} + 12 = 0. \quad \textbf{(4.84)}$$

Solving Eqs. (4.83) and (4.84) for v_3 and v_4, we get

$$v_3 = -12 \text{ V} \quad \textbf{(4.85)}$$

and

$$v_4 = -24 \text{ V}. \quad \textbf{(4.86)}$$

Now we can write the branch currents i''_1 through i''_4 directly in terms of the node voltages v_3 and v_4:

$$i''_1 = \frac{-v_3}{6} = \frac{12}{6} = 2 \text{ A}; \quad \textbf{(4.87)}$$

$$i''_2 = \frac{v_3}{3} = \frac{-12}{3} = -4 \text{ A}; \quad \textbf{(4.88)}$$

$$i''_3 = \frac{v_3 - v_4}{2} = \frac{-12 + 24}{2} = 6 \text{ A}; \quad \textbf{(4.89)}$$

$$i''_4 = \frac{v_4}{4} = \frac{-24}{4} = -6 \text{ A}. \quad \textbf{(4.90)}$$

To find the branch currents in the original circuit—that is, the currents i_1, i_2, i_3, and i_4 in Fig. 4.62—we simply add the currents given by Eqs. (4.87)–(4.90) to the currents given by Eqs. (4.80)–(4.82):

$$i_1 = i'_1 + i''_1 = 15 + 2 = 17 \text{ A}; \quad \textbf{(4.91)}$$

$$i_2 = i'_2 + i''_2 = 10 - 4 = 6 \text{ A}; \quad \textbf{(4.92)}$$

$$i_3 = i'_3 + i''_3 = 5 + 6 = 11 \text{ A}; \quad \textbf{(4.93)}$$

$$i_4 = i'_4 + i''_4 = 5 - 6 = -1 \text{ A}; \quad \textbf{(4.94)}$$

You should verify that the currents given by Eqs. (4.91)–(4.94) are the correct values for the branch currents in the circuit shown in Fig. 4.62.

Use a computer tool to verify this solution.

When applying superposition to linear circuits containing both independent and dependent sources, you must recognize that the dependent sources are never deactivated. Example 4.13 illustrates the application of superposition when a circuit contains both dependent and independent sources.

EXAMPLE 4.13

Use the principle of superposition to find v_o in the circuit shown in Fig. 4.66.

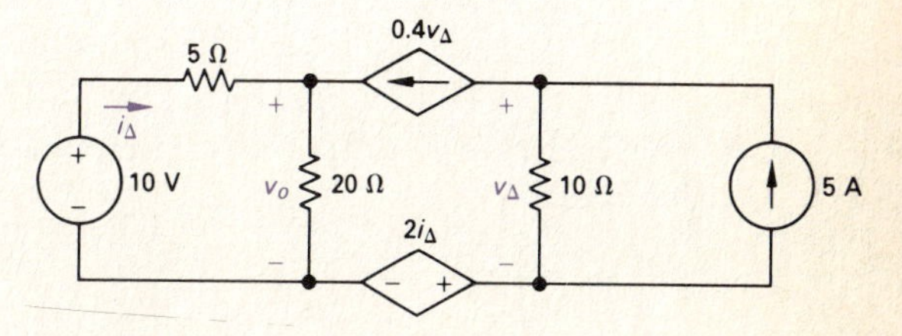

FIGURE 4.66 The circuit for Example 4.13.

SOLUTION

We begin by finding the component of v_o resulting from the 10 V source. Figure 4.67 shows the circuit. With the 5 A source deactivated, v'_Δ must equal $(-0.4v'_\Delta)(10)$. Hence v'_Δ must be zero, the branch containing the two dependent sources is open, and

$$v'_o = \frac{10}{25}(20) = 8\text{ V}.$$

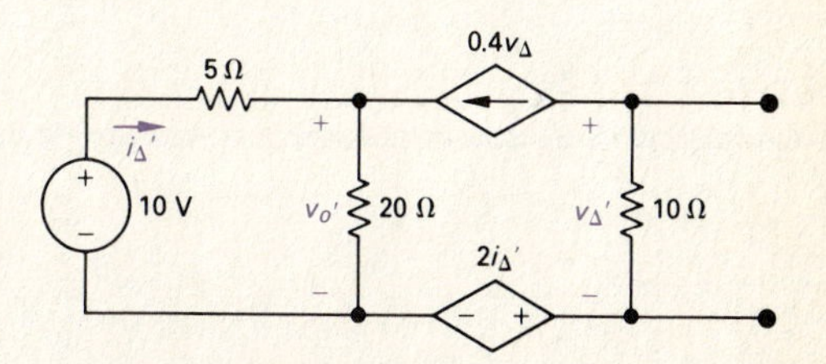

FIGURE 4.67 The circuit shown in Fig. 4.66, with the 5 A source deactivated.

When the 10 V source is deactivated, the circuit reduces to the one shown in Fig. 4.68. We have added a reference node and the node designations a, b, and c to aid the discussion. Summing the currents away from node a yields

$$\frac{v''_o}{20} + \frac{v''_o}{5} - 0.4v''_\Delta = 0, \quad \text{or} \quad 5v''_o - 8v''_\Delta = 0.$$

Summing the currents away from node b gives

$$0.4v''_\Delta + \frac{v_b - 2i''_\Delta}{10} - 5 = 0, \quad \text{or} \quad 4v''_\Delta + v_b - 2i''_\Delta = 50.$$

We now use

$$v_b = 2i''_\Delta + v''_\Delta$$

to find the value for v''_Δ. Thus

$$5v''_\Delta = 50, \quad \text{or} \quad v''_\Delta = 10\text{ V}.$$

From the node a equation,

$$5v''_o = 80, \quad \text{or} \quad v''_o = 16\text{ V}.$$

The value of v_o is the sum of v'_o and v''_o, or 24 V.

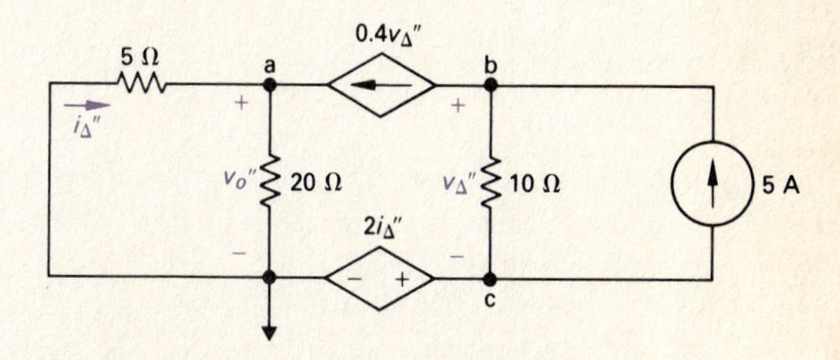

FIGURE 4.68 The circuit shown in Fig. 4.66, with the 10 V source deactivated.

DRILL EXERCISES

4.27 a) Use the principle of superposition to find the voltage v in the circuit shown.

b) Find the power dissipated in the 40 Ω resistor.

ANSWER: (a) 40 V; (b) 40 W.

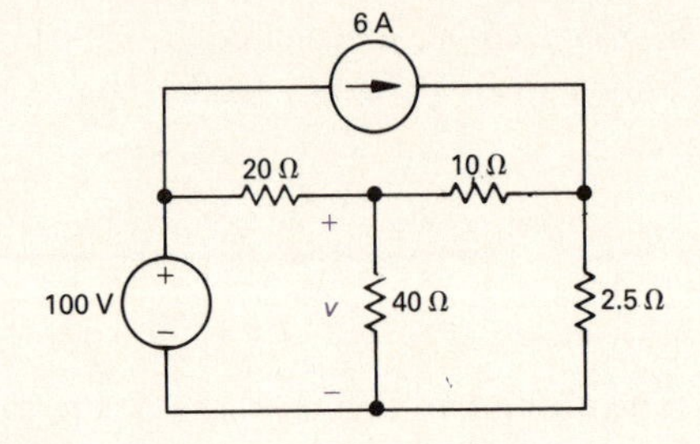

4.28 Use the principle of superposition to find the voltage v in the circuit shown.

ANSWER: 30 V.

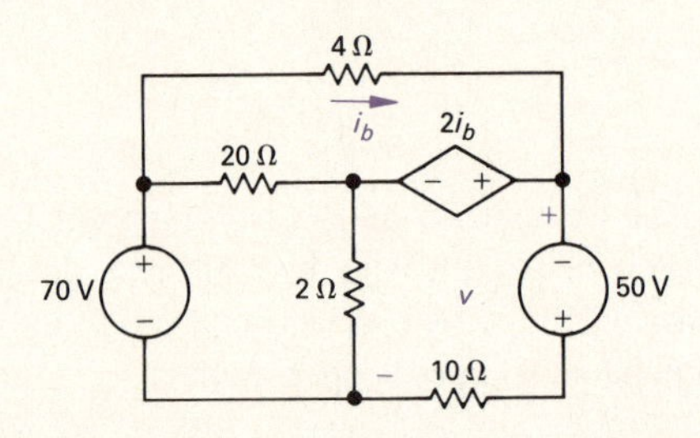

SUMMARY

- For the topics in this chapter, mastery of some basic terms, and the concepts they represent, is necessary. Those terms are **node, path, closed path, branch, mesh, essential node, essential branch,** and **planar circuit**. Table 4.1 provides definitions and examples of these terms.
- Two new circuit analysis techniques were introduced in this chapter:
  - The **node-voltage method** works with both planar and nonplanar circuits. A reference node is chosen from among the essential nodes. Voltage variables are assigned at the remaining essential nodes, and Kirchhoff's current law is used to write one equation per voltage variable. The number of equations is $n - 1$, where n is the number of essential nodes.

— The **mesh-current method** works only with planar circuits. Mesh currents are assigned to each mesh, and Kirchhoff's voltage law is used to write one equation per mesh. The number of equations is $b - (n - 1)$, where b is the number of branches in which the current is unknown, and n is the number of nodes. The mesh currents are used to find the branch currents.

- Several new circuit simplification techniques were introduced in this chapter:

  — **Source transformations** allow us to exchange a voltage source (v_s) and a series resistor (R) for a current source (i_s) and a parallel resistor (R), and vice versa. The combinations must be equivalent in terms of their terminal voltage and current. Terminal equivalence holds provided that

$$i_s = \frac{v_s}{R}.$$

  — **Thévenin equivalents** and **Norton equivalents** allow us to simplify a circuit comprised of sources and resistors into an equivalent circuit consisting of a voltage source and a series resistor (Thévenin) or a current source and a parallel resistor (Norton). The simplified circuit and the original circuit must be equivalent in terms of their terminal voltage and current. Thus keep in mind that (1) The Thévenin voltage (V_{Th}) is the open-circuit voltage across the terminals of the original circuit; (2) the Thévenin resistance (R_{Th}) is the ratio of the Thévenin voltage to the short-circuit current across the terminals of the original circuit; and (3) the Norton equivalent is obtained by performing a source transformation on a Thévenin equivalent.

- In a circuit with multiple independent sources, **superposition** allows us to activate one source at a time and sum the resulting voltages and currents to determine the voltages and currents that exist when all independent sources are active. Dependent sources are never deactivated when applying superposition.

- **Maximum power transfer** is a technique for calculating the maximum value of p which can be delivered to a load, R_L. Maximum power transfer occurs when $R_L = R_{\text{Th}}$, the Thévenin resistance as seen from the resistor R_L. The equation for the maximum power transferred is

$$p = \frac{V_{\text{Th}}^2}{4R_L}.$$

PROBLEMS

4.1 Assume the current i_g in the circuit in Fig. P4.1 is known. The resistors R_1 through R_5 are also known.

a) How many unknown currents are there?

b) How many independent equations can be written using Kirchhoff's current law (KCL)?

c) Write an independent set of KCL equations.

d) How many independent equations must be derived from Kirchhoff's voltage law (KCL)?

e) Write a set of independent KVL equations.

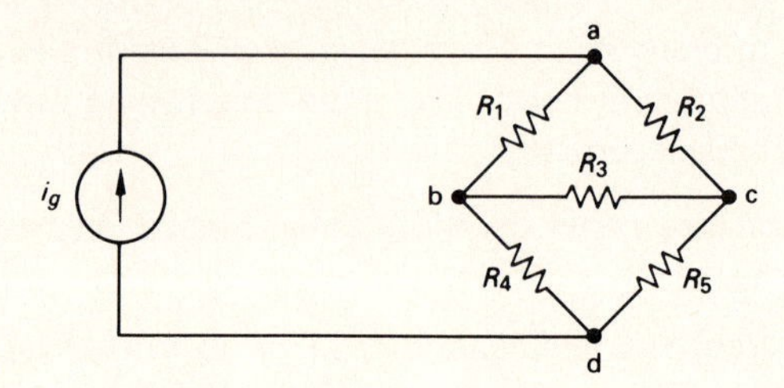

FIGURE P4.1

4.2 Use the node-voltage method to find v_o in the circuit in Fig. P4.2.

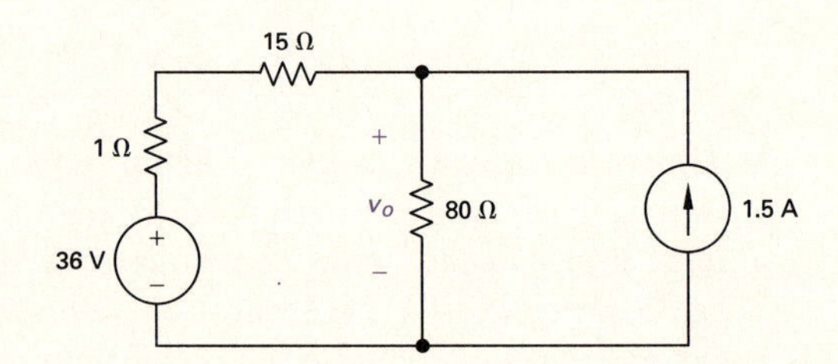

FIGURE P4.2

4.3 Use the node-voltage method to find v_1 and v_2 in the circuit shown in Fig. P4.3.

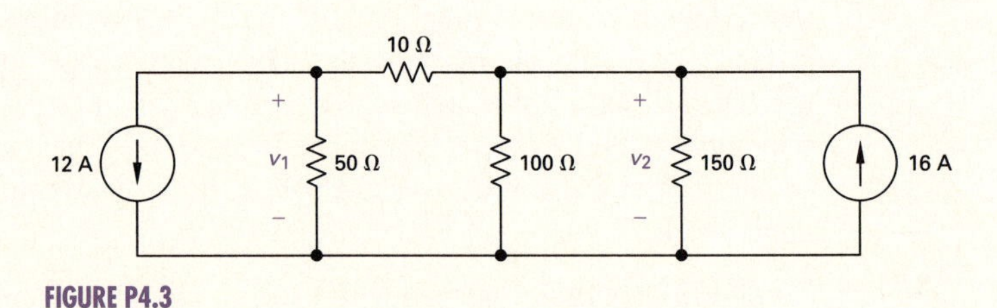

FIGURE P4.3

4.4 Use the node-voltage method to find how much power the 20 V source extracts from the circuit in Fig. P4.4.

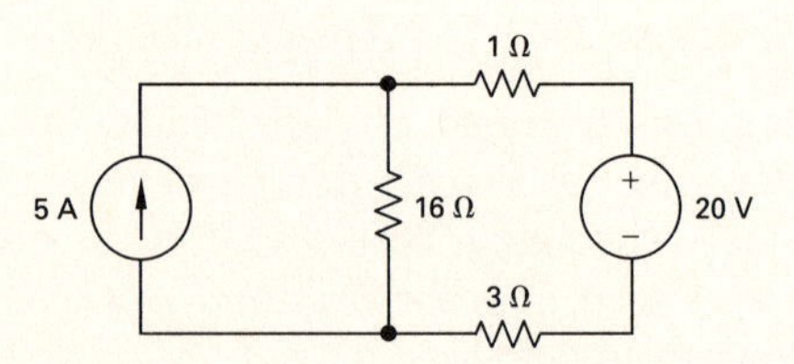

FIGURE P4.4

4.5 Use the node-voltage method to find v_1 and v_2 in the circuit in Fig. P4.5.

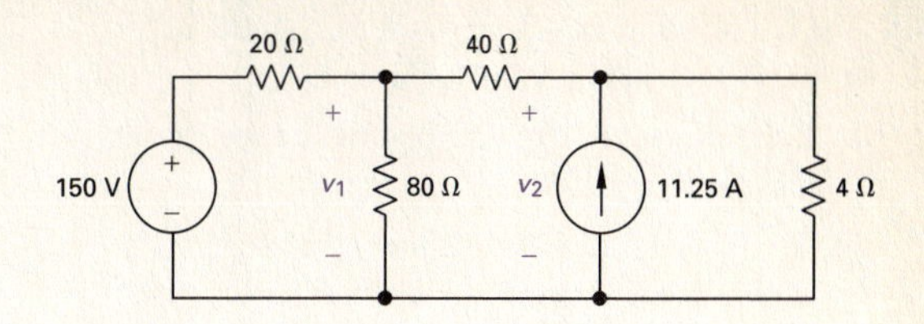

FIGURE P4.5

4.6 a) Use the node-voltage method to find v_1, v_2, and v_3 in the circuit in Fig. P4.6.

b) How much power does the 640 V voltage source deliver to the circuit?

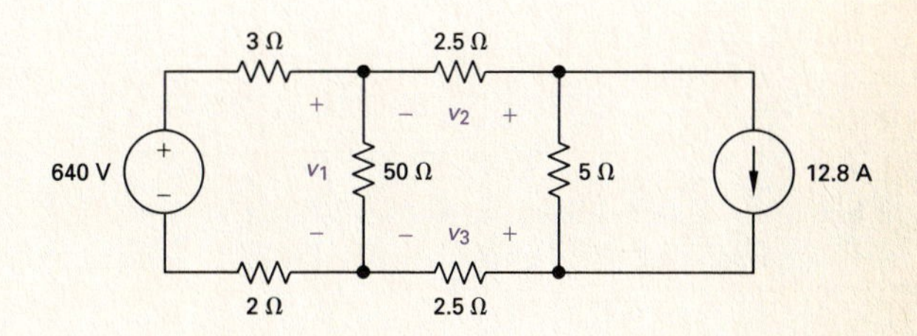

FIGURE P4.6

4.7 a) Use the node-voltage method to find the branch currents i_a through i_e in the circuit shown in Fig. P4.7.

b) Find the total power developed in the circuit.

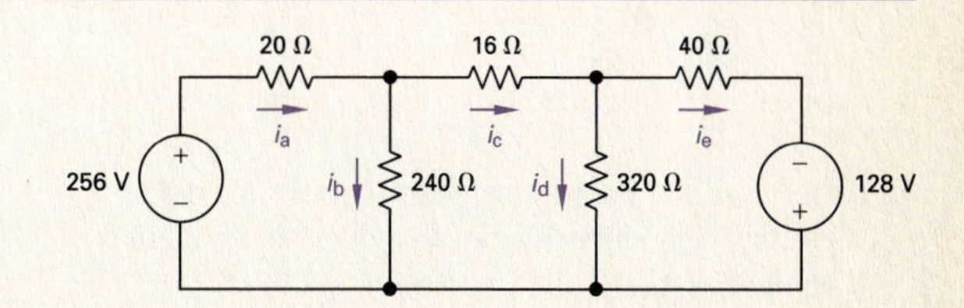

FIGURE P4.7

4.8 Use the node-voltage method to find the total power dissipated in the circuit in Fig. P4.8.

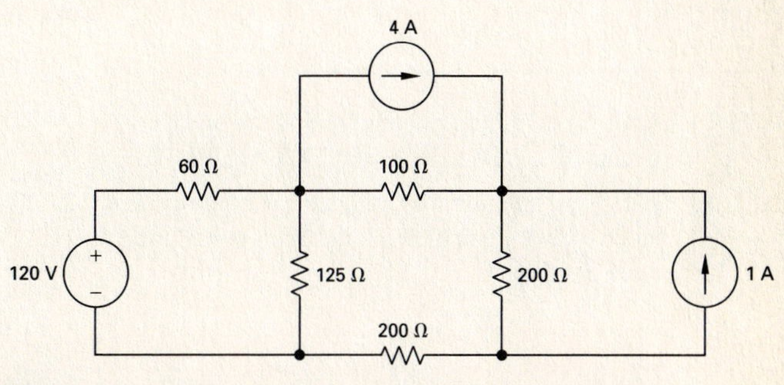

FIGURE P4.8

4.9 The circuit shown in Fig. P4.9 is a dc model of a residential power distribution circuit.

a) Use the node-voltage method to find the branch currents i_1 through i_6.

b) Test your solution for the branch currents by showing that the total power dissipated equals the total power developed.

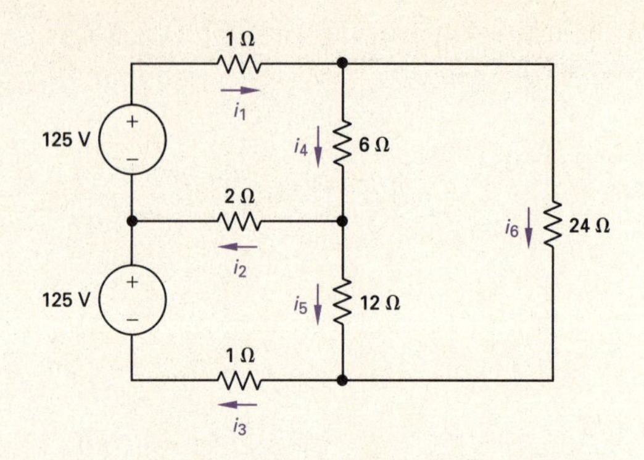

FIGURE P4.9

4.10 Use the node-voltage method to find i_o (in milliamperes) in the circuit in Fig. P4.10.

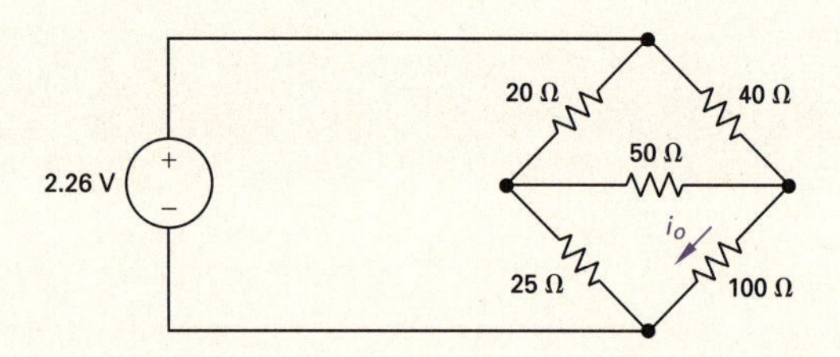

FIGURE P4.10

4.11 Use the node-voltage method to find v_1 and the power delivered by the 60 V voltage source in the circuit in Fig. P4.11.

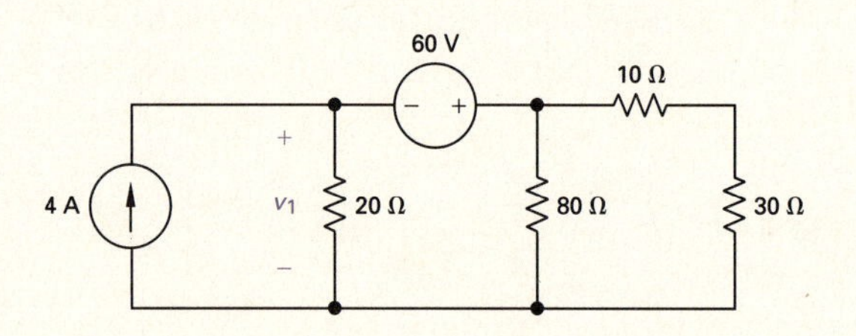

FIGURE P4.11

4.12 a) Use the node-voltage method to find the branch currents i_1, i_2, and i_3 in the circuit in Fig. P4.12.

b) Check your solution for i_1, i_2, and i_3 by showing that the power dissipated in the circuit equals the power developed.

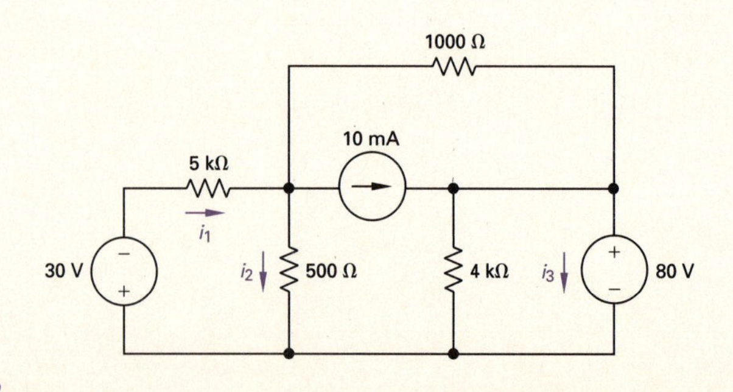

FIGURE P4.12

4.13 a) Find the power developed by the 1.5 A current source in the circuit in Fig. P4.2.

b) Find the power developed by the 36 V voltage source in the circuit in Fig. P4.2.

c) Verify that the total power developed equals the total power dissipated.

4.14 A 20 Ω resistor is connected in series with the 1.5 A current source in the circuit in Fig. P4.2.

a) Find v_o.

b) Find the power developed by the 1.5 A current source.

c) Find the power developed by the 36 V voltage source.

d) Verify that the total power developed equals the total power dissipated.

e) What effect will any finite resistance connected in series with the 1.5 A current source have on the value of v_o?

4.15 Use the node-voltage method to find the value of v_o in the circuit in Fig. P4.15.

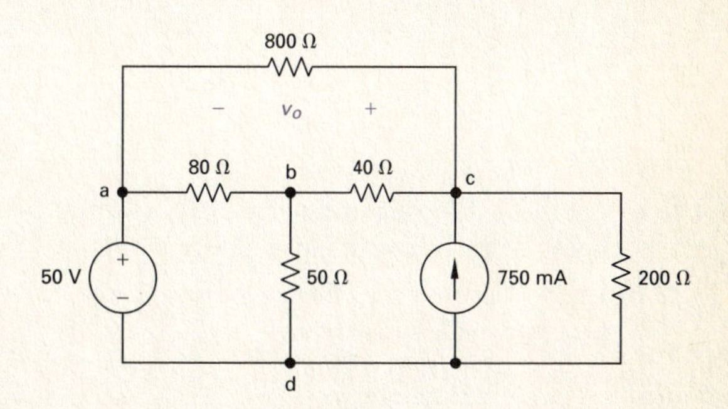

FIGURE P4.15

4.16 Check the solution for v_o in Problem 4.15 by first using a Y-to-Δ transformation to eliminate node b.

4.17 a) Use the node-voltage method to find the power dissipated in the 2 Ω resistor in the circuit in Fig. P4.17.

b) Find the power supplied by the 230 V source.

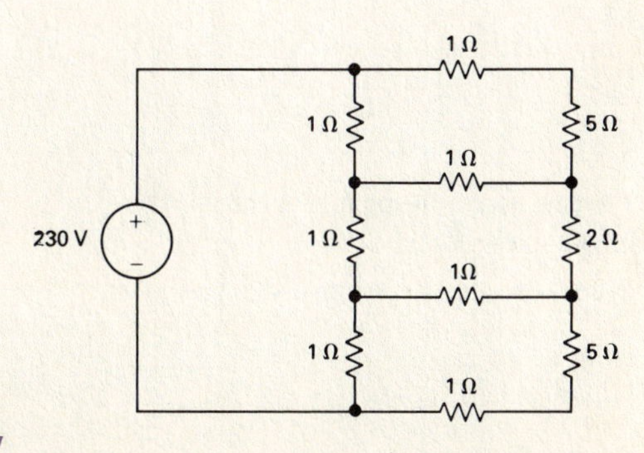

FIGURE P4.17

4.18 a) Use the node-voltage method to find v_o in the circuit in Fig. P4.18.

b) Find the power absorbed by the dependent source.

c) Find the total power developed by the independent sources.

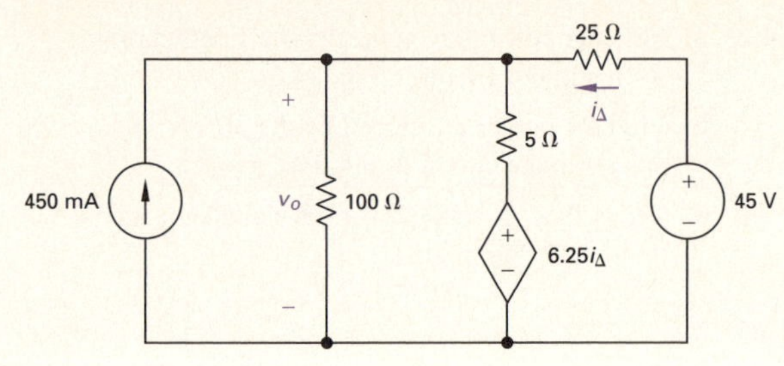

FIGURE P4.18

4.19 Use the node-voltage method to calculate the power delivered by the dependent voltage source in the circuit in Fig. P4.19.

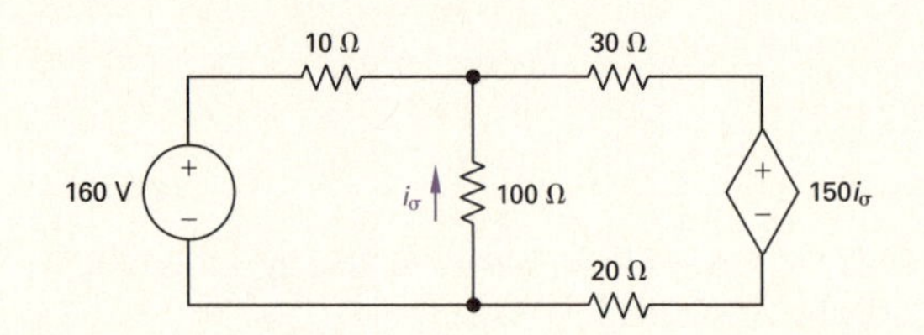

FIGURE P4.19

4.20 a) Use the node-voltage method to find the total power developed in the circuit in Fig. P4.20.

b) Check your answer by finding the total power absorbed in the circuit.

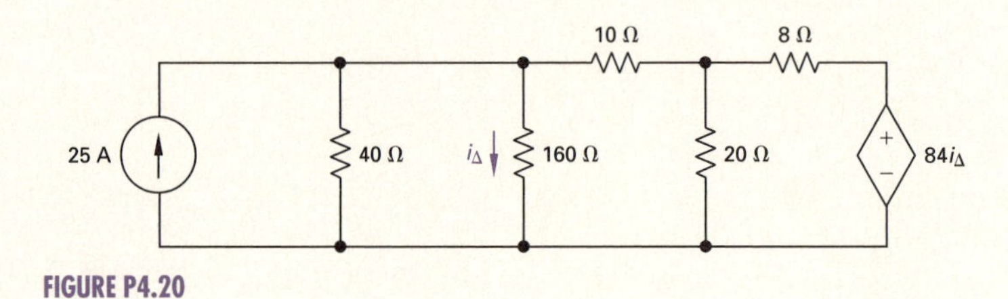

FIGURE P4.20

4.21 Use the node-voltage method to find v_Δ in the circuit in Fig. P4.21.

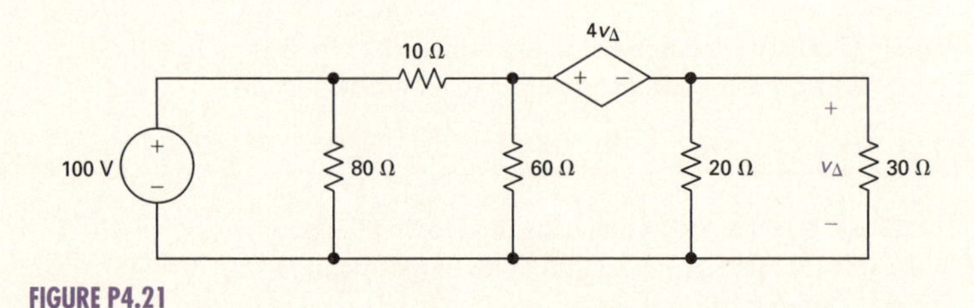

FIGURE P4.21

4.22 a) Find the node voltages v_1, v_2, and v_3 in the circuit in Fig. P4.22.

b) Find the total power dissipated in the circuit.

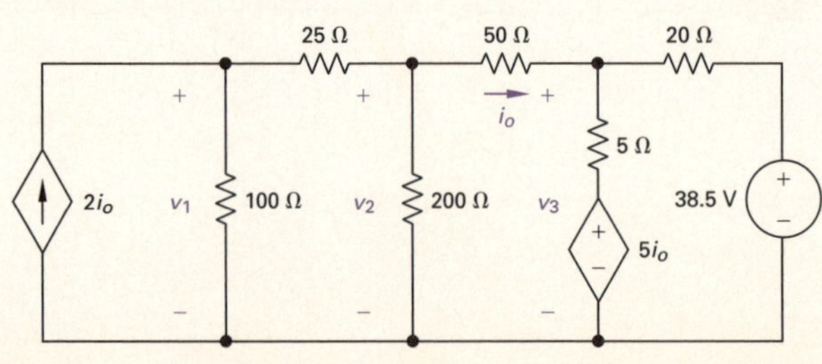

FIGURE P4.22

4.23 Assume you are a project engineer and one of your staff is assigned to analyze the circuit shown in Fig. P4.23. The reference node and node numbers given on the figure were assigned by the analyst. Her solution gives the values of v_3 and v_4 as 108 V and 81.60 V, respectively.

Test these values by checking the total power developed in the circuit against the total power dissipated. Do you agree with the solution submitted by the analyst?

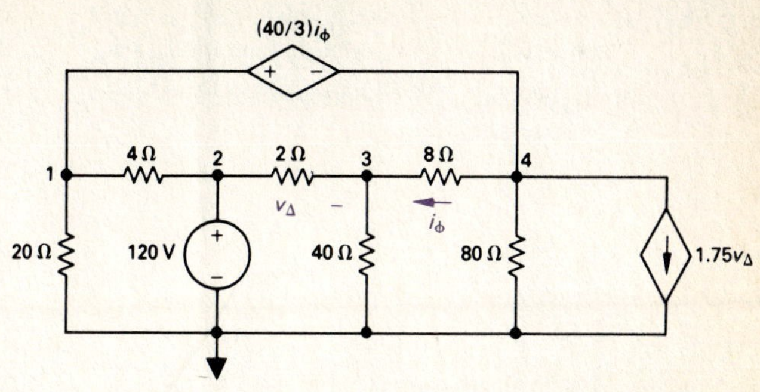

FIGURE P4.23

4.24 Show that when Eqs. (4.16), (4.17), and (4.19) are solved for i_B, the result is identical to Eq. (2.25).

4.25 Use the node-voltage method to find the power developed by the 60 V source in the circuit in Fig. P4.25.

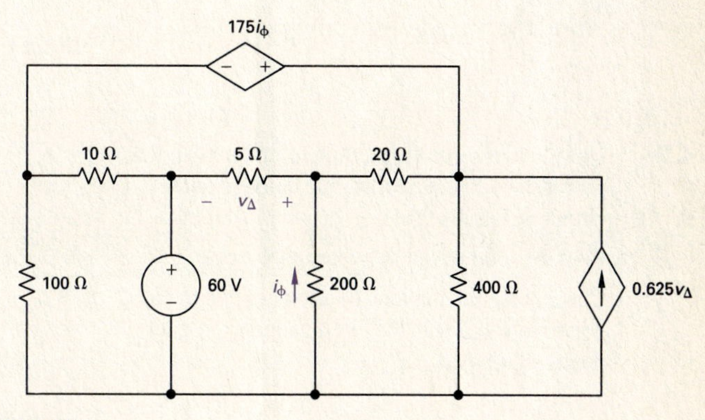

FIGURE P4.25

4.26 a) Use the node-voltage method to show that the output voltage v_o in the circuit in Fig. P4.26 is equal to the average value of the source voltages.

b) Find v_o if $v_1 = 120$ V, $v_2 = 60$ V, and $v_3 = -30$ V.

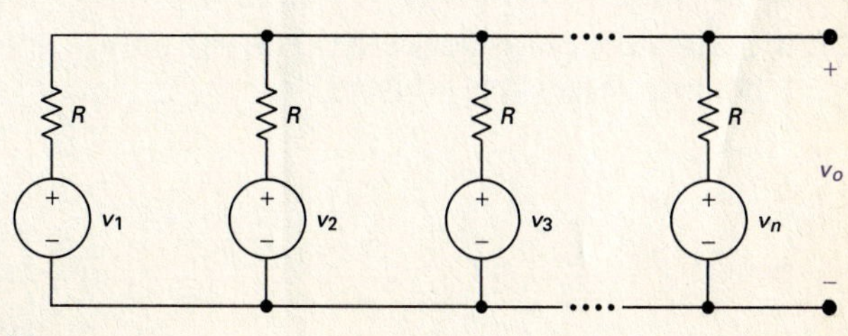

FIGURE P4.26

4.27 a) Use the mesh-current method to find the branch currents i_a, i_b, and i_c in the circuit in Fig. P4.27.

b) Repeat part (a) if the polarity of the 64 V source is reversed.

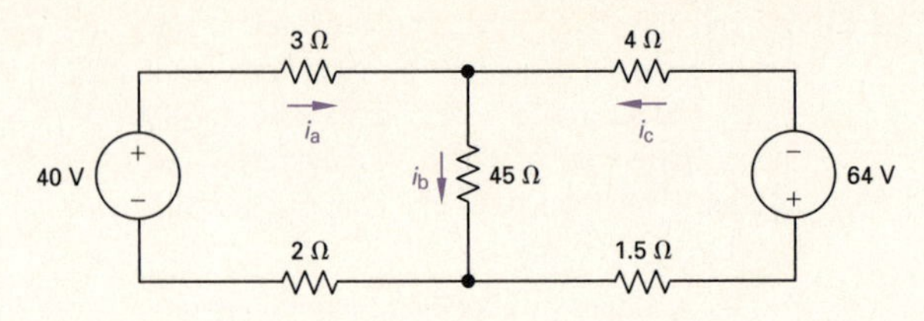

FIGURE P4.27

4.28 a) Use the mesh-current method to find the total power developed in the circuit in Fig. P4.28.

b) Check your answer by showing that the total power developed equals the total power dissipated.

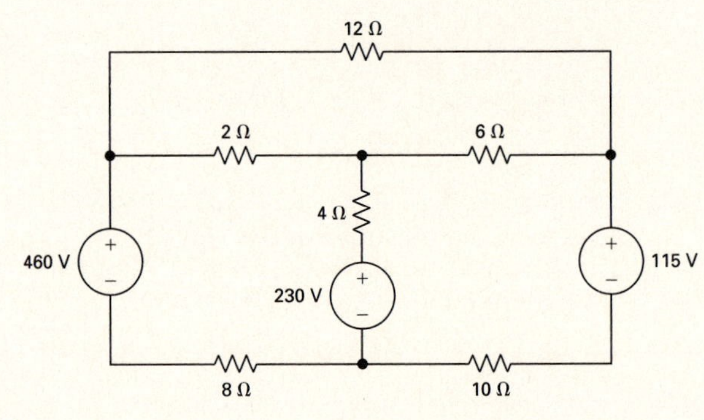

FIGURE P4.28

4.29 a) Use the mesh-current method to find how much power the 12 A current source delivers to the circuit in Fig. P4.29.

b) Find the total power delivered to the circuit.

c) Check your calculations by showing that the total power developed in the circuit equals the total power dissipated.

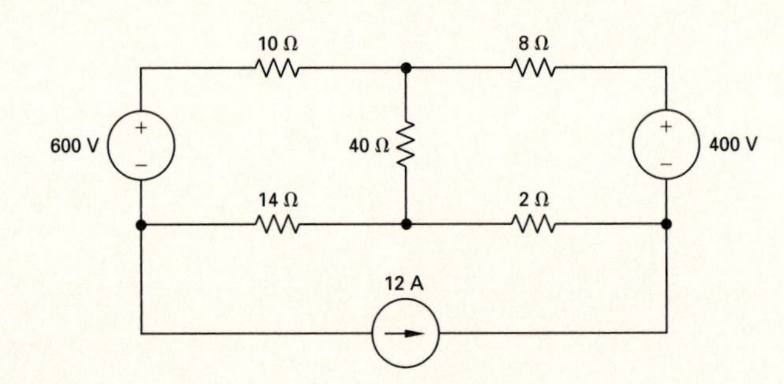

FIGURE P4.29

4.30 Use the mesh-current method to find the total power dissipated in the circuit in Fig. P4.30.

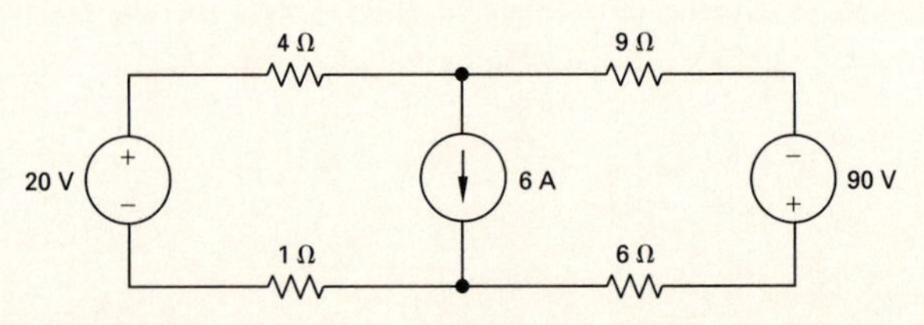

FIGURE P4.30

4.31 Assume the 20 V source in the circuit in Fig. P4.30 is increased to 120 V. Find the total power dissipated in the circuit.

4.32 a) Assume the 20 V source in the circuit in Fig. P4.30 is increased to 60 V. Find the total power dissipated in the circuit.

b) Repeat part (a) if the 6 A current source is replaced by a short circuit.

c) Explain why the answers to parts (a) and (b) are the same.

4.33 a) Use the mesh-current method to find the branch currents in i_a through i_e in the circuit in Fig. P4.33.

b) Check your solution by showing that the total power developed in the circuit equals the total power dissipated.

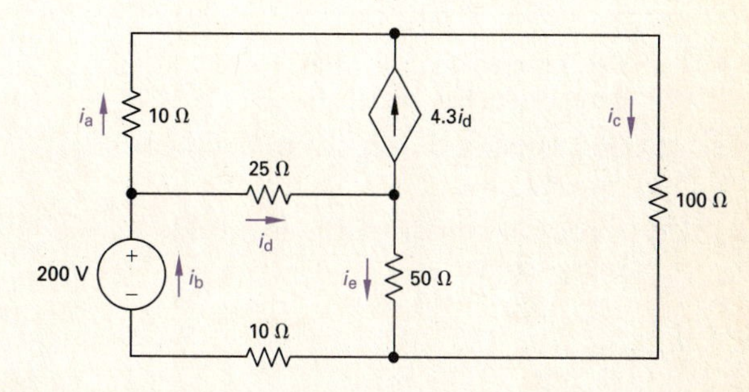

FIGURE P4.33

4.34 Use the mesh-current method to find the power developed by the 20 A source in the circuit in Fig. P4.34.

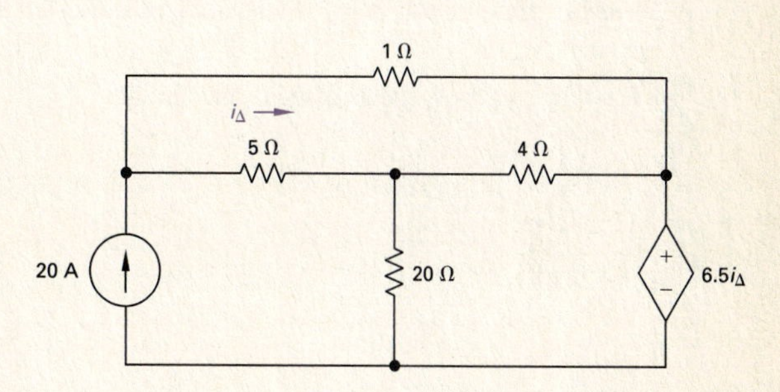

FIGURE P4.34

4.35 Use the mesh-current method to find the power delivered by the dependent voltage source in the circuit seen in Fig. P4.35.

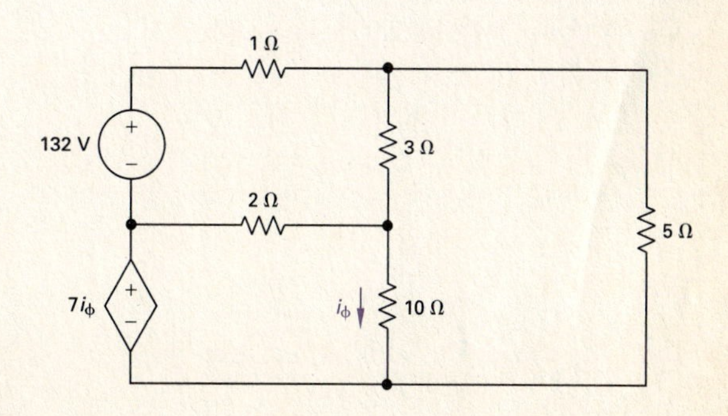

FIGURE P4.35

4.36 a) Use the mesh-current method to find v_o in the circuit in Fig. P4.36.

b) Find the power delivered by the dependent source.

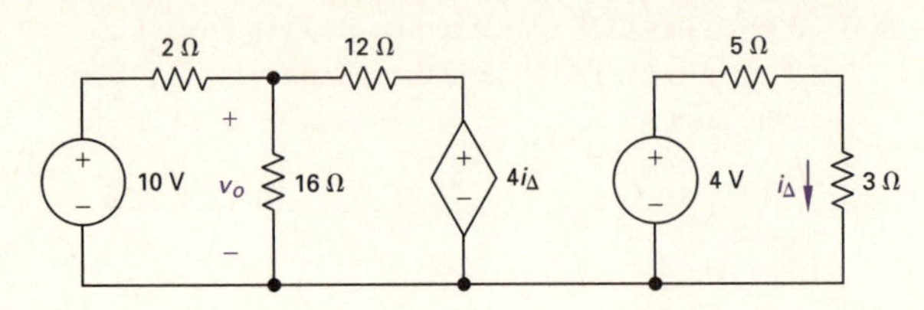

FIGURE P4.36

4.37 a) Use the mesh-current method to solve for i_Δ in the circuit in Fig. P4.37.

b) Find the power delivered by the independent current source.

c) Find the power delivered by the dependent voltage source.

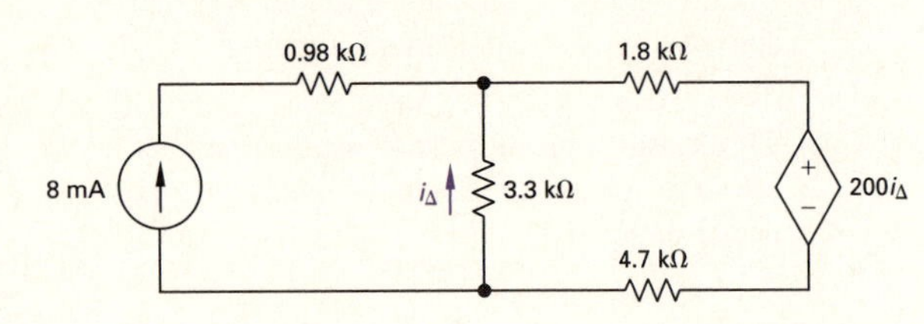

FIGURE P4.37

4.38 a) Use the mesh-current method to determine which sources in the circuit in Fig. P4.38 are generating power.

b) Find the total power dissipated in the circuit.

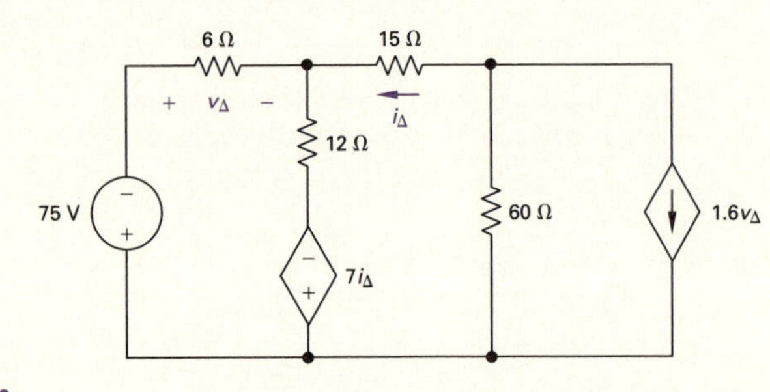

FIGURE P4.38

4.39 Use the mesh-current method to find the total power developed in the circuit in Fig. P4.39.

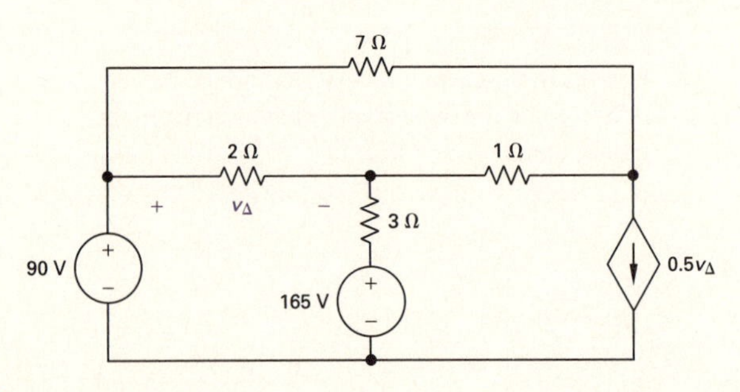

FIGURE P4.39

4.40 Use the mesh-current method to find the power dissipated in the 5 Ω resistor in the circuit in Fig. P4.40.

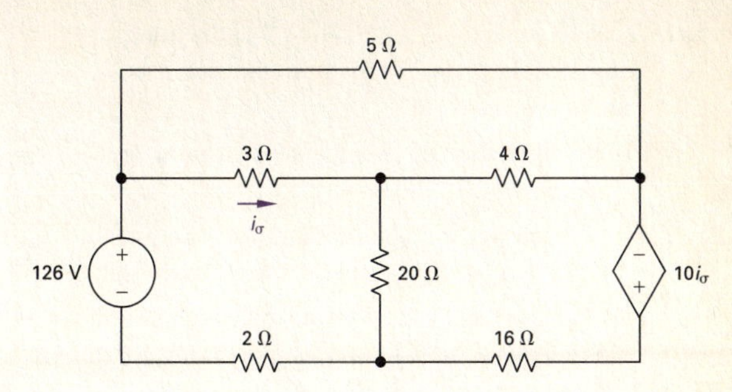

FIGURE P4.40

4.41 a) Use the mesh-current method to find the power delivered to the 2 Ω resistor in the circuit in Fig. P4.41.

b) What percentage of the total power developed in the circuit is delivered to the 2 Ω resistor?

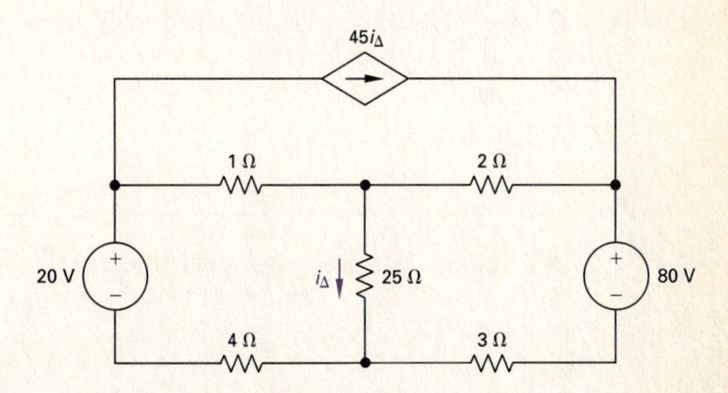

FIGURE P4.41

4.42 Use the mesh-current method to find the power developed in the dependent voltage source in the circuit in Fig. P4.42.

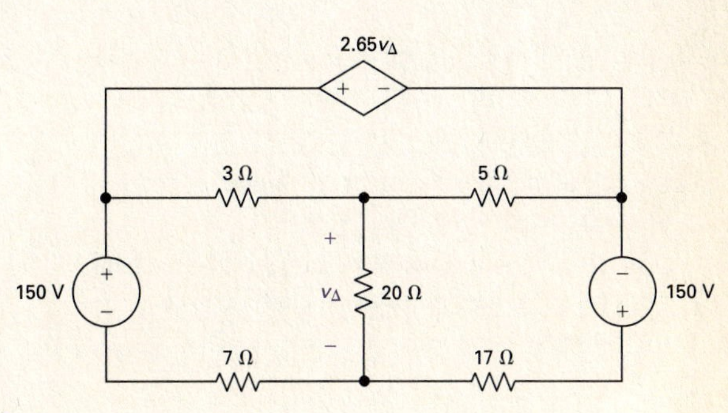

FIGURE P4.42

4.43 Assume you have been asked to find the power dissipated in the 10 Ω resistor in the circuit in Fig. P4.43.

a) Which method of circuit analysis would you recommend? Explain why.

b) Use your recommended method of analysis to find the power dissipated in the 10 Ω resistor.

c) Would you change your recommendation if the problem had been to find the power developed by the 4 A current source? Explain.

d) Find the power delivered by the 4 A current source.

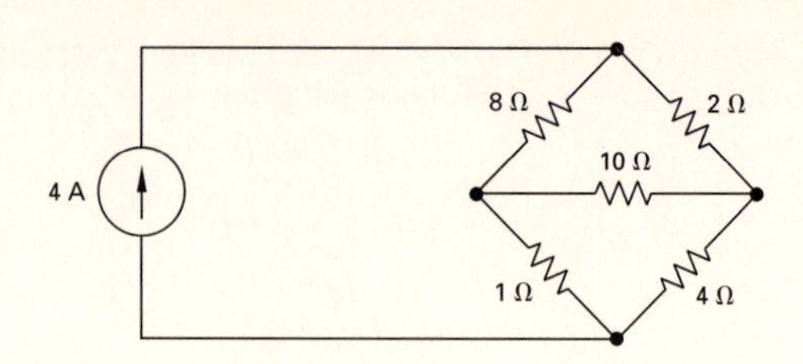

FIGURE P4.43

4.44 A 20 Ω resistor is placed in parallel with the 4 A current source in the circuit in Fig. P4.43. Assume you have been asked to calculate the power developed by the current source.

a) Which method of circuit analysis would you recommend? Explain why.

b) Find the power developed by the current source.

4.45 a) Would you use the node-voltage or mesh-current method to find the power absorbed by the 20 V source in the circuit in Fig. P4.45? Explain your choice.

b) Use the method you selected in part (a) to find the power.

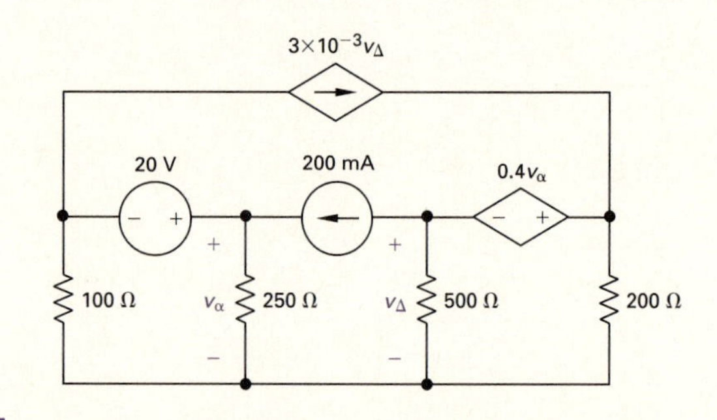

FIGURE P4.45

4.46 a) Find the branch currents i_a through i_e for the circuit shown in Fig. P4.46.

b) Check your answers by showing that the total power generated equals the total power dissipated.

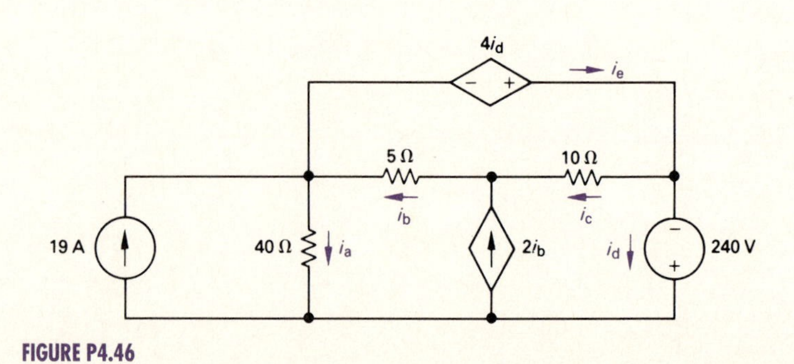

FIGURE P4.46

4.47 The circuit in Fig. P4.47 is a direct-current version of a typical three-wire distribution system. The resistors R_a, R_b, and R_c represent the resistances of the three conductors that connect the three loads R_1, R_2, and R_3 to the 125/250 V voltage supply. The resistors R_1 and R_2 represent loads connected to the 125 V circuits, and R_3 represents a load connected to the 250 V circuit.

a) Calculate v_1, v_2, and v_3.

b) Calculate the power delivered to R_1, R_2, and R_3.

c) What percentage of the total power developed by the sources is delivered to the loads?

d) The R_b branch represents the neutral conductor in the distribution circuit. What adverse effect occurs if the neutral conductor is opened? (*Hint:* Calculate v_1 and v_2 and note that appliances or loads designed for use in this circuit would have a nominal voltage rating of 125 V.)

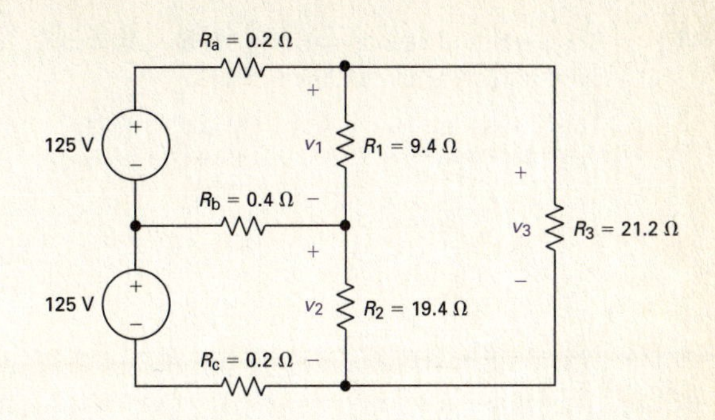

FIGURE P4.47

4.48 Show that whenever $R_1 = R_2$ in the circuit in Fig. P4.47, the current in the neutral conductor is zero. (*Hint:* Solve for the neutral conductor current as a function of R_1 and R_2).

4.49 The variable dc current source in the circuit in Fig. P4.49 is adjusted so that the power developed by the 4 A current source is zero. Find the value of the dc current.

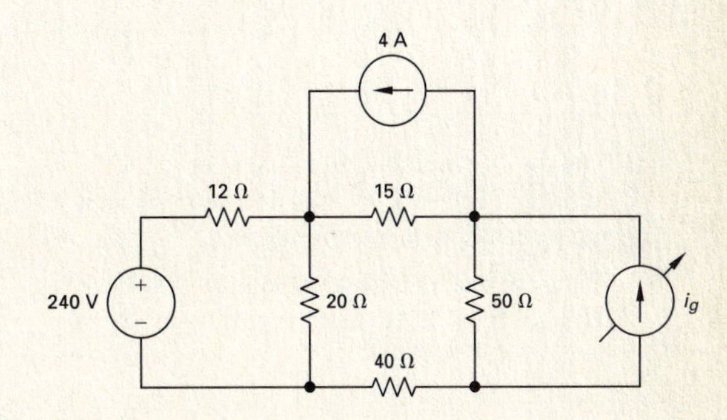

FIGURE P4.49

4.50 The variable dc voltage source in the circuit in Fig. P4.50 is adjusted so that i_o is zero. Find the value of V_{dc}.

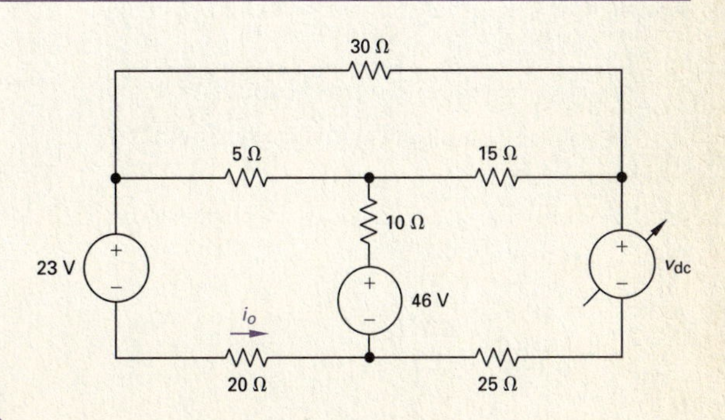

FIGURE P4.50

4.51 a) Use a series of source transformations to find the current i_o in the circuit in Fig. P4.51.

b) Verify your solution by using the node-voltage method to find i_o.

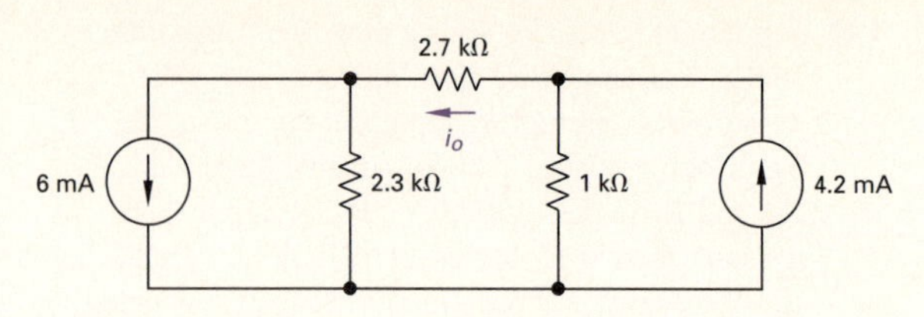

FIGURE P4.51

4.52 a) Use a series of source transformations to find i_o in the circuit in Fig. P4.52.

b) Verify your solution by using the mesh-current method to find i_o.

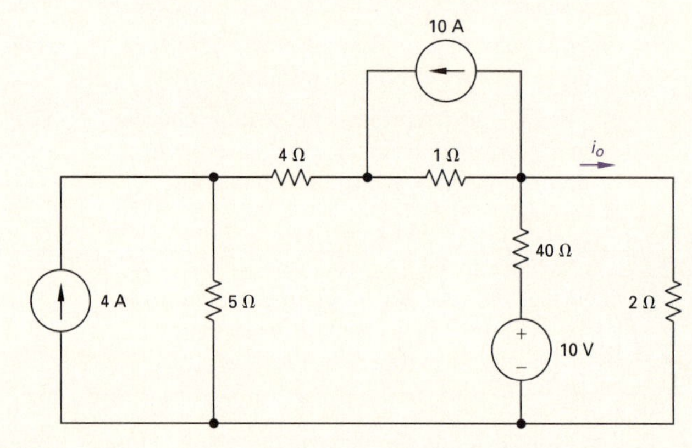

FIGURE P4.52

4.53 a) Find the current in the 10 kΩ resistor in the circuit in Fig. P4.53 by making a succession of appropriate source transformations.

b) Using the result obtained in part (a) work back through the circuit to find the power developed by the 100 V source.

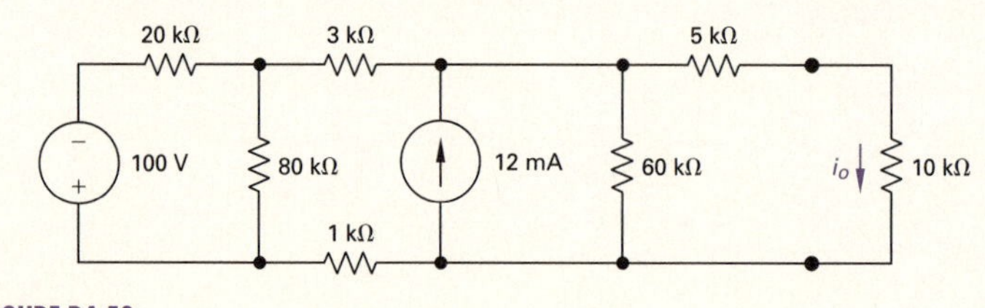

FIGURE P4.53

4.54 a) Use source transformations to find v_o in the circuit in Fig. P4.54.

b) Find the power developed by the 340 V source.

c) Find the power developed by the 5 A current source.

d) Verify that the total power developed equals the total power dissipated.

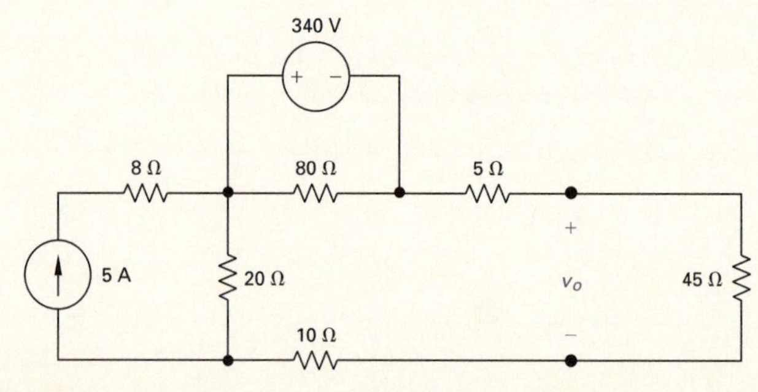

FIGURE P4.54

4.55 Find the Thévenin equivalent with respect to the terminals a,b for the circuit in Fig. P4.55.

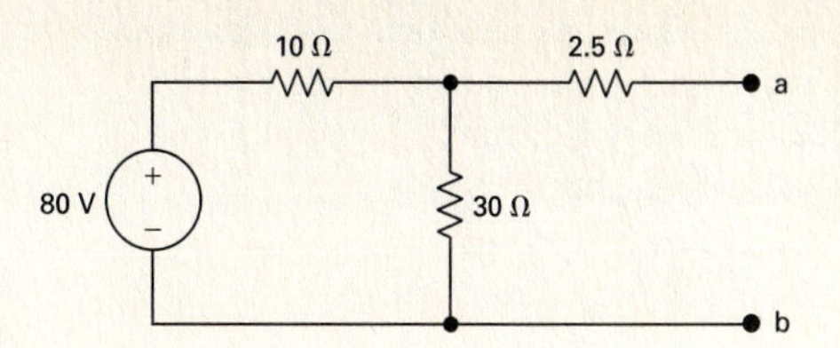

FIGURE P4.55

4.56 Find the Thévenin equivalent with respect to the terminals a,b for the circuit in Fig. P4.56.

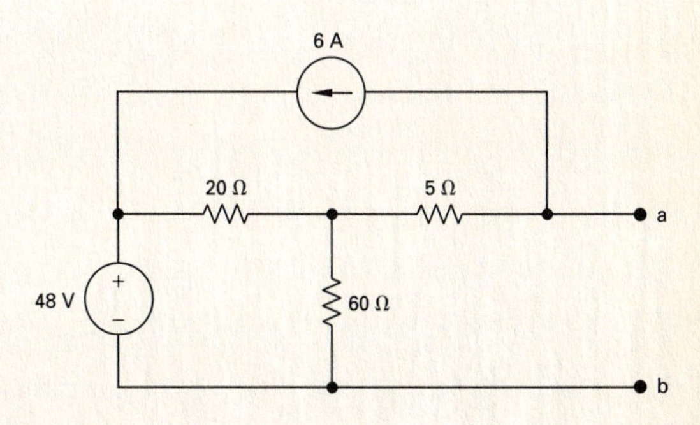

FIGURE P4.56

4.57 Find the Thévenin equivalent with respect to the terminals a,b for the circuit in Fig. P4.57.

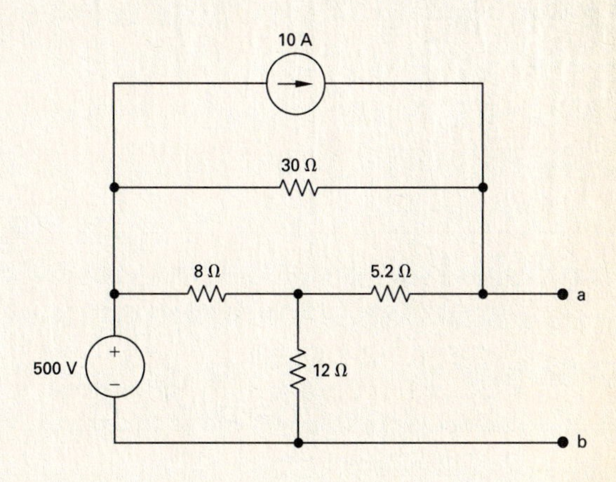

FIGURE P4.57

4.58 a) Find the Thévenin equivalent with respect to the terminals a,b for the circuit in Fig. P4.58 by finding the open-circuit voltage and the short-circuit current.

b) Solve for the Thévenin resistance by removing the independent sources. Compare your result to the Thévenin resistance found in part (a).

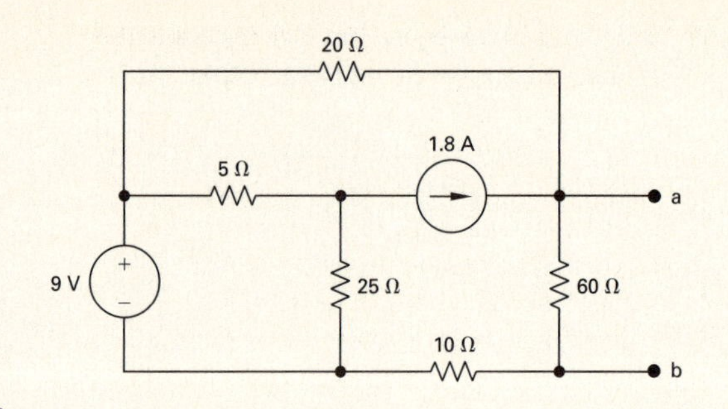

FIGURE P4.58

4.59 Find the Norton equivalent with respect to the terminals a,b in the circuit in Fig. P4.59.

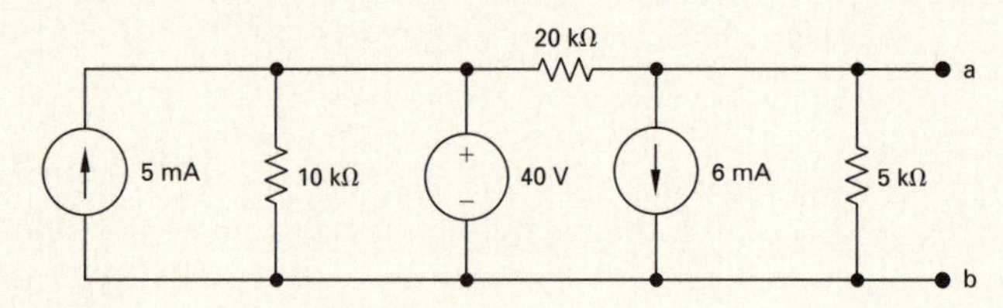

FIGURE P4.59

4.60 Determine i_o and v_o in the circuit shown in Fig. P4.60 when R_o is 0, 2, 6, 10, 15, 20, 30, 40, 50, and 70 Ω.

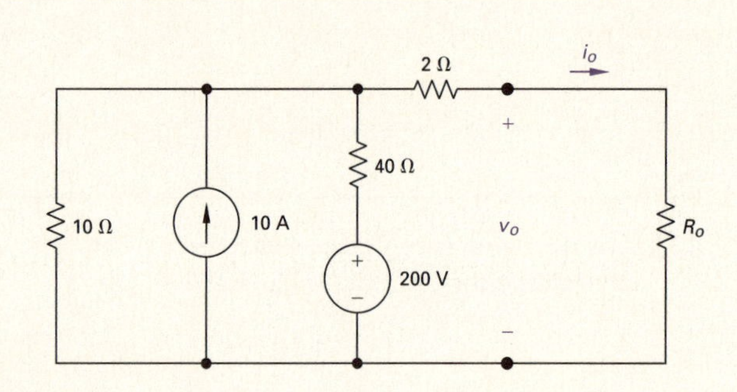

FIGURE P4.60

4.61 A voltmeter with a resistance of 60 kΩ is used to measure the voltage v_{ab} in the circuit in Fig. P4.61.

a) What is the voltmeter reading?

b) What is the percentage of error in the voltmeter reading if the percentage of error is defined as [(measured – actual)/actual] × 100?

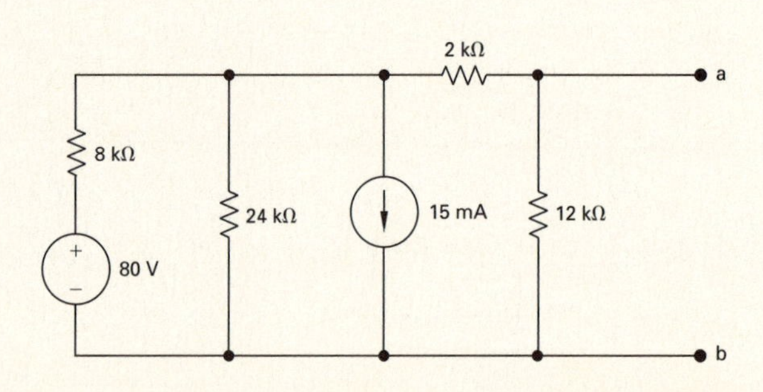

FIGURE P4.61

4.62 The Wheatstone bridge in the circuit shown in Fig. P4.62 is balanced when R_2 equals 3000 Ω. If the galvanometer has a resistance of 50 Ω, how much current will the galvanometer detect when the bridge is unbalanced by setting R_2 to 3003 Ω? (*Hint:* Find the Thévenin equivalent with respect to the galvanometer terminals when $R_2 = 3003$ Ω. Note that once we have found this Thévenin equivalent, it is easy to find the amount of unbalanced current in the galvanometer branch for different galvanometer movements.)

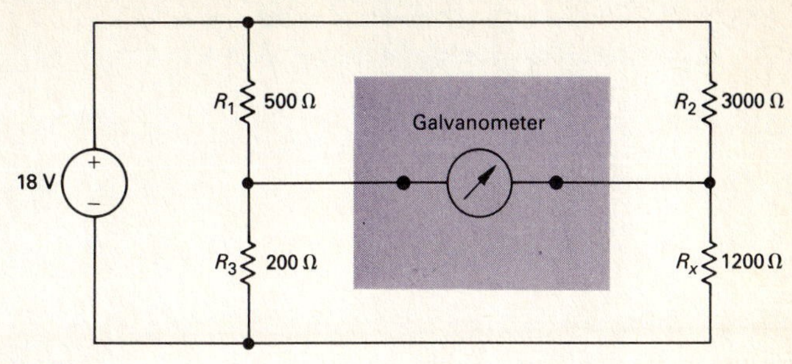

FIGURE P4.62

4.63 Determine the Thévenin equivalent with respect to the terminals a,b for the circuit shown in Fig. P4.63.

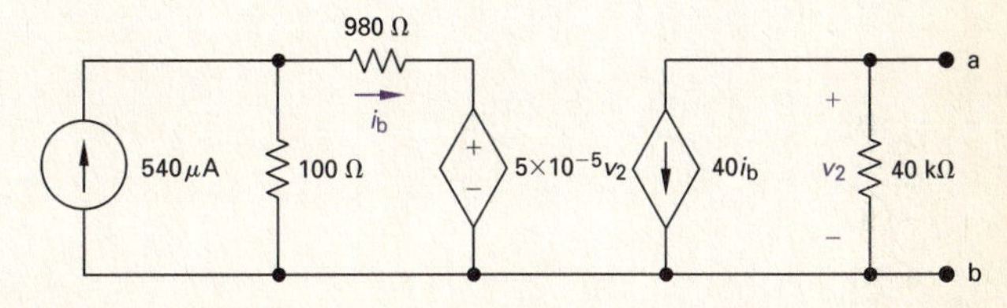

FIGURE P4.63

4.64 Find the Thévenin equivalent with respect to the terminals a,b for the circuit seen in Fig. P4.64.

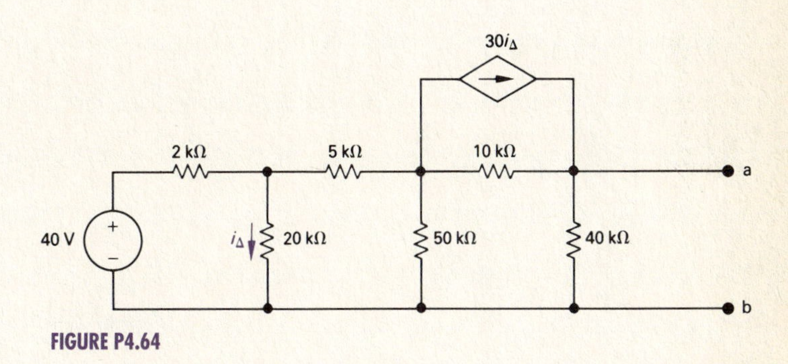

FIGURE P4.64

4.65 When a voltmeter is used to measure the voltage v_e in Fig. P4.65, it reads 5.5 V.

a) What is the resistance of the voltmeter?

b) What is the percentage of error in the voltage measurement?

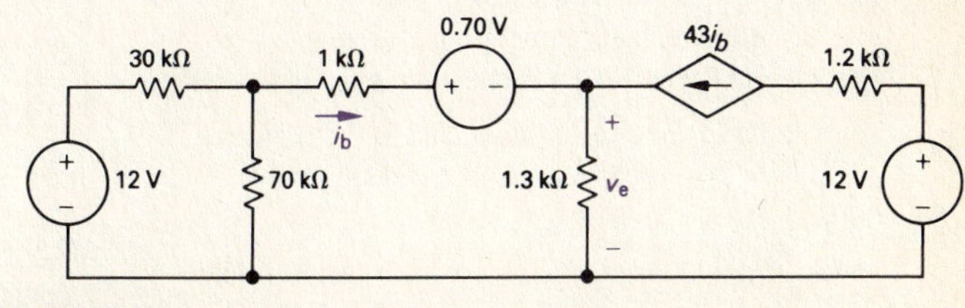

FIGURE P4.65

4.66 When an ammeter is used to measure the current i_ϕ in the circuit shown in Fig. P4.66, it reads 6 A.

a) What is the resistance of the ammeter?

b) What is the percentage of error in the current measurement?

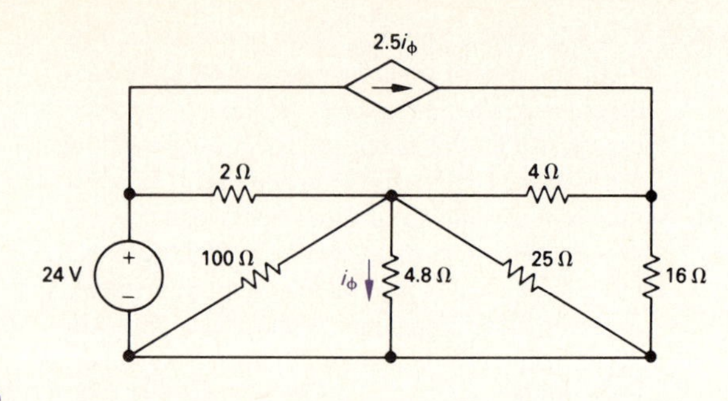

FIGURE P4.66

4.67 Find the Thévenin equivalent with respect to the terminals a,b for the circuit seen in Fig. P4.67.

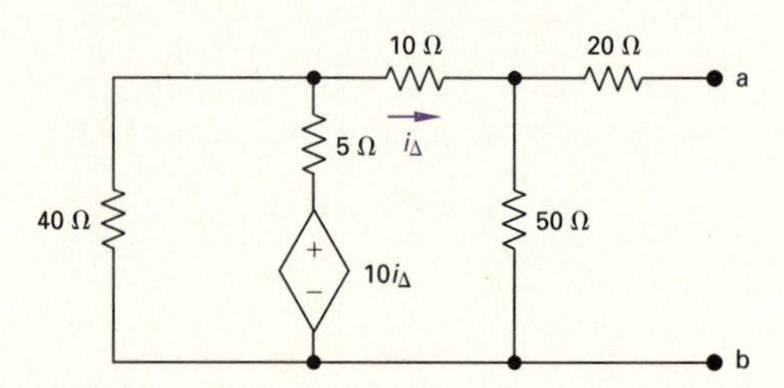

FIGURE P4.67

4.68 Find the Thévenin equivalent with respect to the terminals a,b in the circuit in Fig. P4.68.

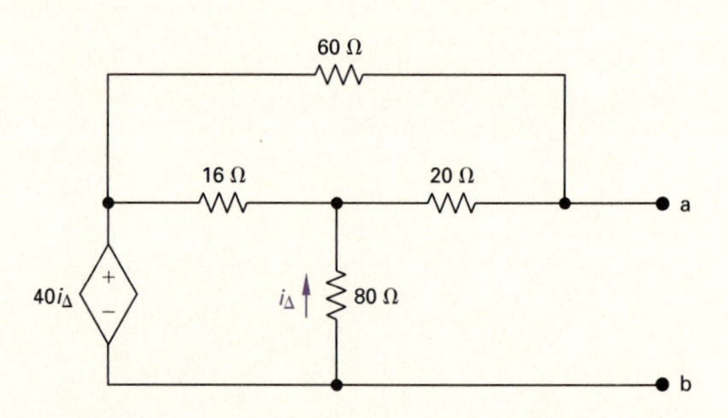

FIGURE P4.68

4.69 A Thévenin equivalent can also be determined from measurements made at the pair of terminals of interest. Assume the following measurements were made at the terminals a,b in the circuit in Fig. P4.69.

When a 20 kΩ resistor is connected to the terminals a,b, the voltage v_{ab} is measured and found to be 20 V.

When a 7.5 kΩ resistor is connected to the terminals a,b, the voltage is measured and found to be 15 V.

Find the Thévenin equivalent of the network with respect to the terminals a,b.

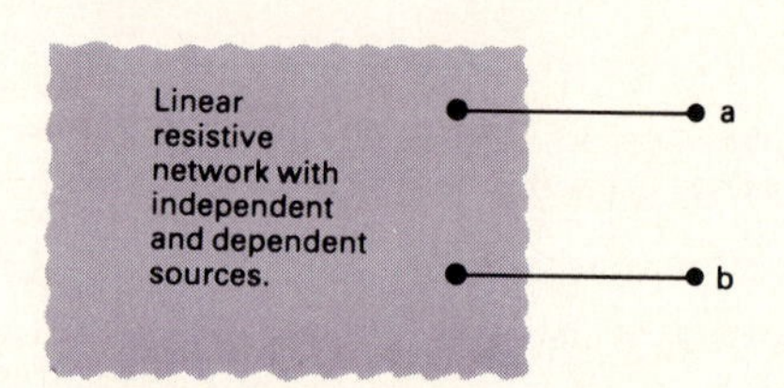

FIGURE P4.69

4.70 An automobile battery, when connected to a car radio, provides 12.72 V to the radio. When connected to a set of headlights, it provides 12 V to the headlights. Assume the radio can be modeled as a 6.36 Ω resistor and the headlights can be modeled as a 0.60 Ω resistor. What are the Thévenin and Norton equivalents for the battery?

4.71 The variable resistor (R_o) in the circuit in Fig. P4.71 is adjusted until the power dissipated in the resistor is 250 W. Find the values of R_o which satisfy this condition.

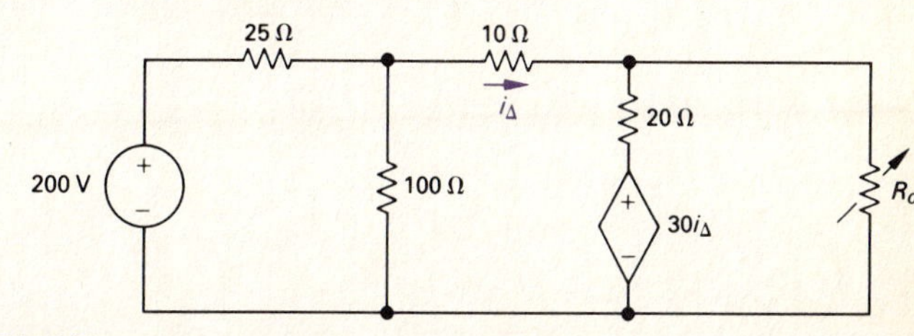

FIGURE P4.71

4.72 a) Calculate the power delivered to each resistor in Problem 4.60.

b) Plot the power delivered versus the resistance.

c) At what value of R is the power maximum?

4.73 The variable resistor in the circuit in Fig. P4.73 is adjusted for maximum power transfer to R_o.

a) Find the value of R_o.

b) Find the maximum power that can be delivered to R_o.

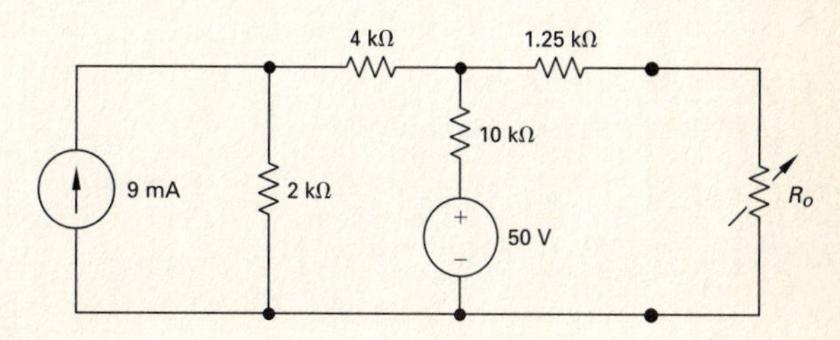

FIGURE P4.73

4.74 What percentage of the total power developed in the circuit in Fig. P4.73 is delivered to R_o when R_o is set for maximum power transfer?

4.75 The variable resistor (R_L) in the circuit in Fig. P4.75 is adjusted for maximum power transfer to R_L.

a) Find the numerical value of R_L.

b) Find the maximum power transferred to R_L.

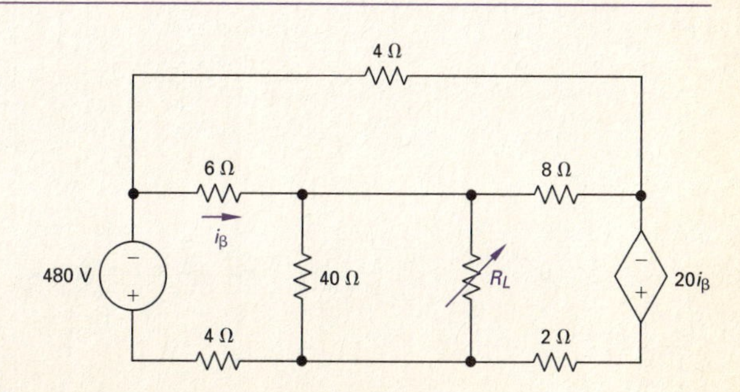

FIGURE P4.75

4.76 The variable resistor (R_0) in the circuit in Fig. P4.76 is adjusted for maximum power transfer to R_o.

a) Find the value of R_o.

b) Find the maximum power that can be delivered to R_o.

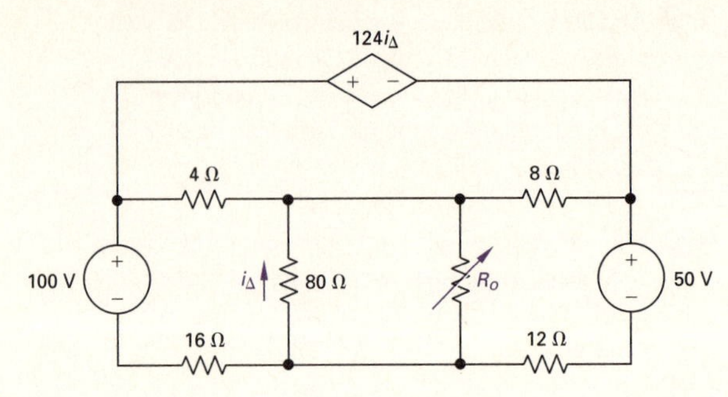

FIGURE P4.76

4.77 What percentage of the total power developed in the circuit in Fig. P4.76 is delivered to R_o?

4.78 The variable resistor (R_o) in the circuit in Fig. P4.78 is adjusted until it absorbs maximum power from the circuit.

a) Find the value of R_o.

b) Find the maximum power.

c) Find the percentage of the total power developed in the circuit that is delivered to R_o.

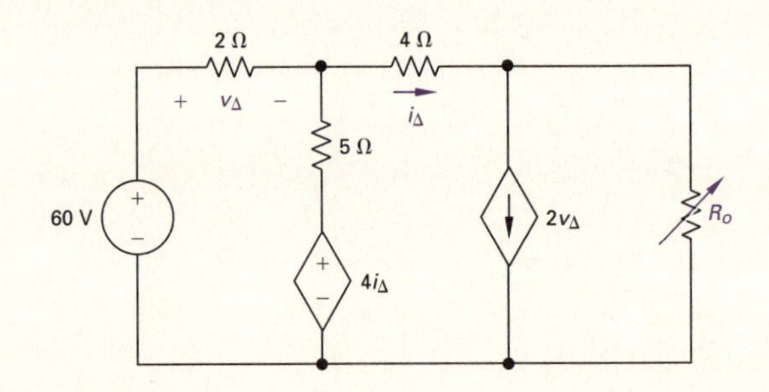

FIGURE P4.78

4.79 The variable resistor (R_o) in the circuit in Fig. P4.79 is adjusted for maximum power transfer to R_o. What percentage of the total power developed in the circuit is delivered to R_o?

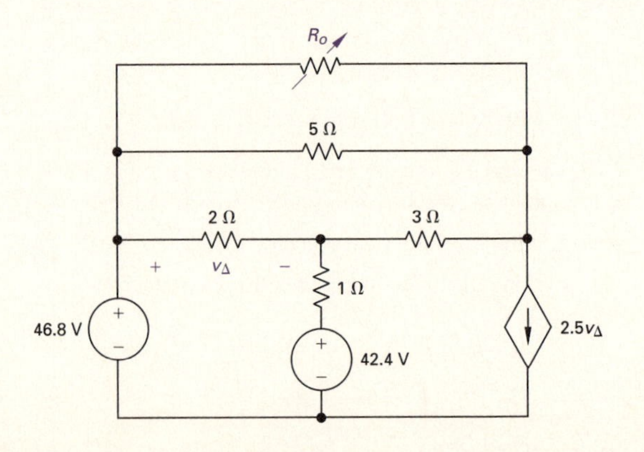

FIGURE P4.79

4.80 The variable resistor in the circuit in Fig. P4.80 is adjusted for maximum power transfer to R_o.

a) Find the numerical value of R_o.

b) Find the maximum power delivered to R_o.

c) How much power does the 280 V source deliver to the circuit when R_o is adjusted to the value found in part (a)?

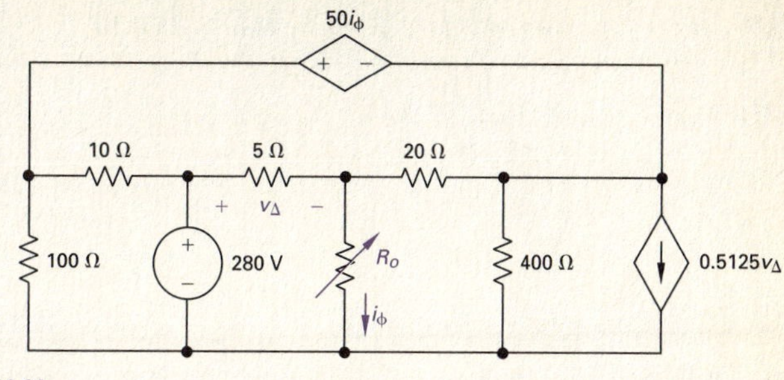

FIGURE P4.80

4.81 A variable resistor R_o is connected across the terminals a,b in the circuit in Fig. P4.64. The variable resistor is adjusted until maximum power is transferred to R_o.

a) Find the value of R_o.

b) Find the maximum power delivered to R_o.

c) Find the percentage of the total power developed in the circuit that is delivered to R_o.

4.82 a) Find the value of the variable resistor R_o in the circuit in Fig. P4.82 that will result in maximum power dissipation in the 6 Ω resistor. (*Hint:* Hasty conclusions could be hazardous to your career.)

b) What is the maximum power that can be delivered to the 6 Ω resistor?

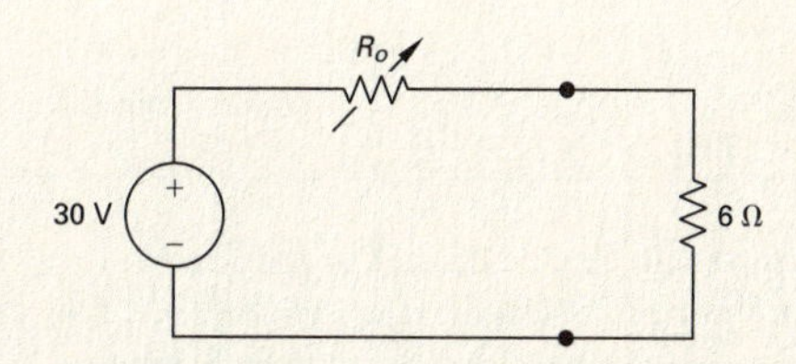

FIGURE P4.82

4.83 Use superposition to solve for i_o and v_o in the circuit in Fig. P4.83.

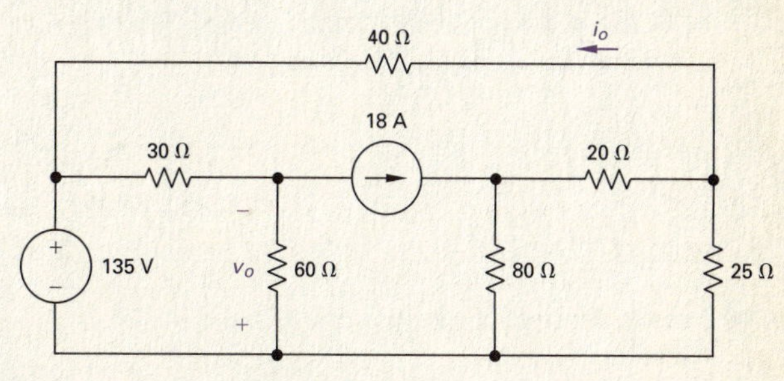

FIGURE P4.83

4.84 a) Use superposition to find the current in the 10 Ω resistor in Fig. P4.84.

b) Find the power dissipated in the 10 Ω resistor.

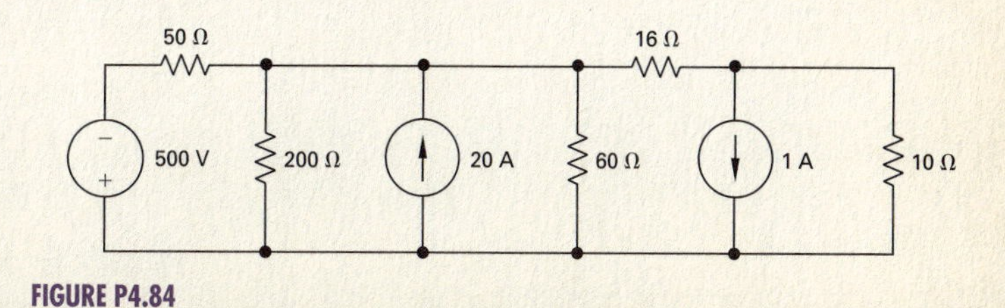

FIGURE P4.84

4.85 Use the principle of superposition to find the current i_o in the circuit shown in Fig. P4.85.

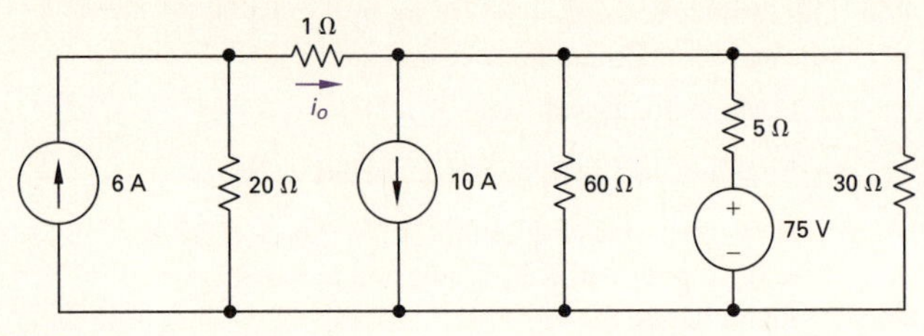

FIGURE P4.85

4.86 Use the principle of superposition to find the voltage v_o in the circuit in Fig. P4.86.

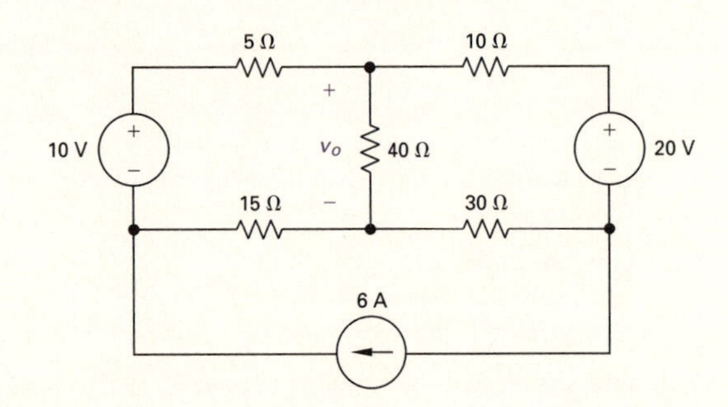

FIGURE P4.86

4.87 a) In the circuit in Fig. P4.87, before the 5 mA current source is attached to the terminals a,b, the current i_o is calculated and found to be 3.5 mA. Use superposition to find the value of i_o after the current source is attached.

b) Verify your solution by finding i_o when all three sources are acting simultaneously.

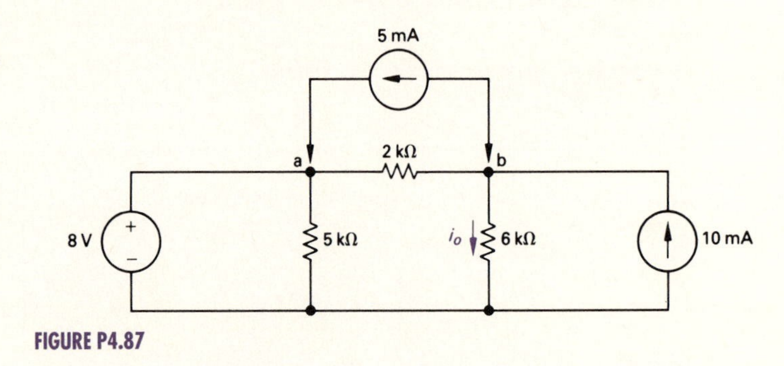

FIGURE P4.87

4.88 Use the principle of superposition to find v_o in the circuit in Fig. P4.88.

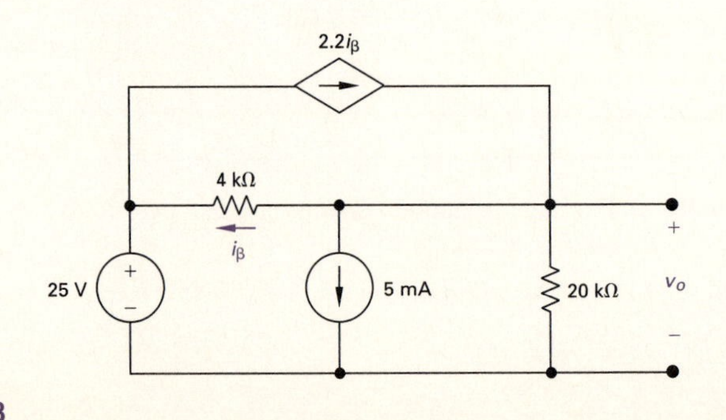

FIGURE P4.88

4.89 Use the principle of superposition to find the current leaving the positive terminal of the 165 V source in the circuit in Fig. P4.39.

4.90 Use the principle of superposition to find the voltage v_Δ in the circuit in Fig. P4.42.

4.91 Use the principle of superposition to find i_Δ in the circuit in Fig. P4.41.

4.92 Use the principle of superposition to find the voltage across the dependent current source in the circuit in Fig. P4.46. Use the upper terminal of the dependent current source as the positive reference for the voltage.

4.93 Laboratory measurements on a dc voltage source yield a terminal voltage of 75 V with no load connected to the source and 60 V when loaded with a 20 Ω resistor.

a) What is the Thévenin equivalent with respect to the terminals of the dc voltage source?

b) Show that the Thévenin resistance of the source is given by the expression

$$R_{\text{Th}} = \left(\frac{V_{\text{Th}}}{V_o} - 1\right) R_L,$$

where

$V_{\text{Th}} =$ the Thévenin voltage, and

$V_o =$ the terminal voltage corresponding to the load resistance R_L.

4.94 Two ideal dc voltage sources are connected by electrical conductors that have a resistance of r Ω/m, as shown in Fig. P4.94. A load having a resistance of R Ω moves between the two voltage sources. Let x equal the distance between the load and the source V_1, and let L equal the distance between the sources.

a) Show that

$$v = \frac{V_1 RL + R(V_2 - V_1)x}{RL + 2rLx - 2rx^2}.$$

b) Show that the voltage v will be minimum when

$$x = \frac{L}{V_2 - V_1}\left[-V_1 \pm \sqrt{V_1 V_2 - \frac{R}{2rL}(V_1 - V_2)^2}\right].$$

c) Find x when L = 16 km, $V_1 = 1000$ V, $V_2 = 1200$ V, $R = 3.9$ Ω, and $r = 5 \times 10^{-5}$ Ω/m.

d) What is the minimum value of v for the circuit of part (c)?

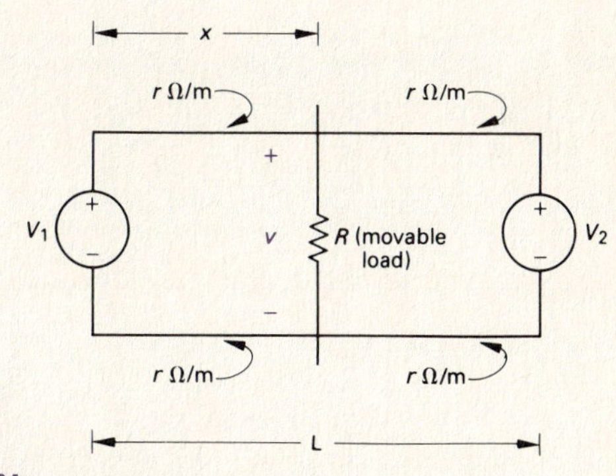

FIGURE P4.94

4.95 Assume your supervisor has asked you to determine the power developed by the 1 V source in the circuit in Fig. P4.95. Before calculating the power developed by the 1 V source, the supervisor asks you to submit a proposal describing how you plan to attack the problem. Furthermore, he asks you to explain why you have chosen your proposed method of solution.

a) Describe your plan of attack, explaining your reasoning.

b) Use the method you have outlined in (a) to find the power developed by the 1 V source.

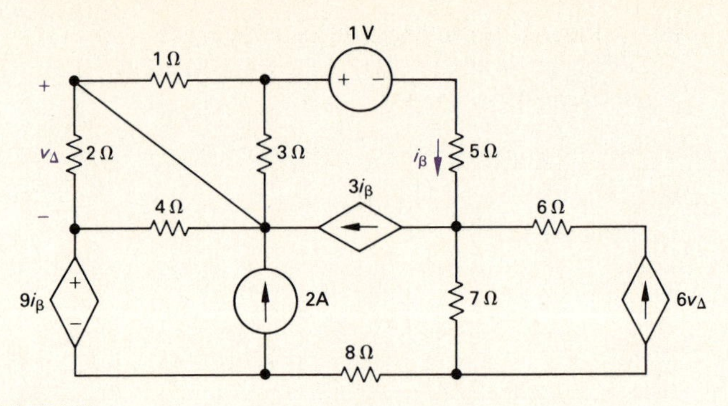

FIGURE P4.95

4.96 Find the power absorbed by the 5 A current source in the circuit in Fig. P4.96.

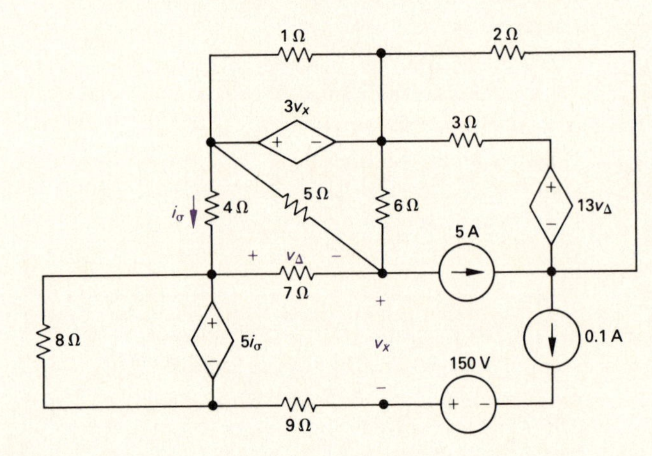

FIGURE P4.96

4.97 Find v_1, v_2, and v_3 in the circuit in Fig. P4.97.

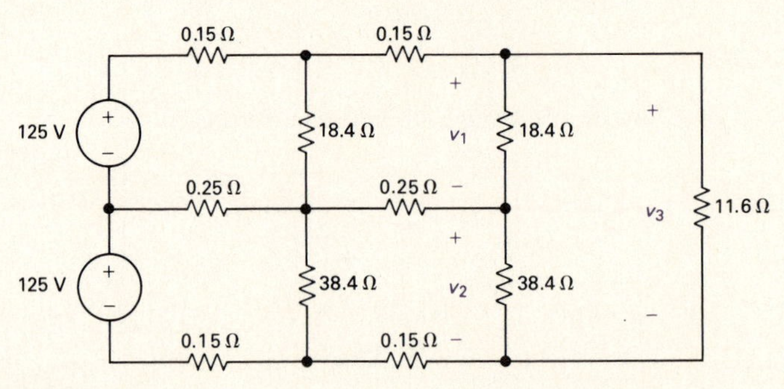

FIGURE P4.97

4.98 Find i_1 in the circuit in Fig. P4.98.

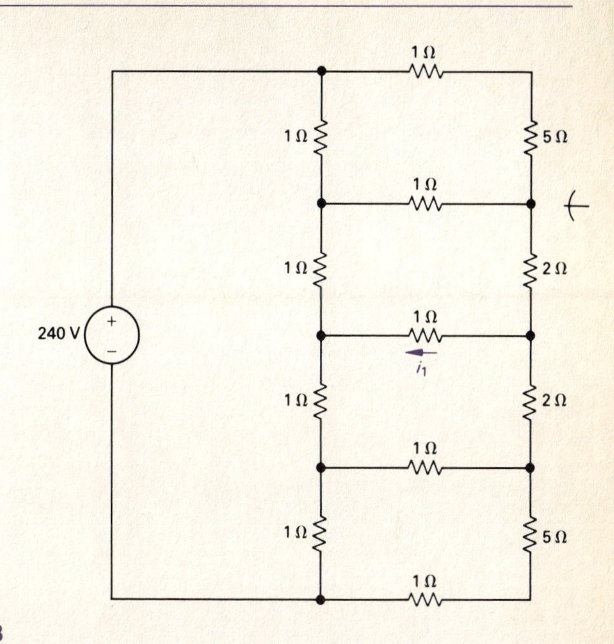

FIGURE P4.98

5 THE OPERATIONAL AMPLIFIER

CHAPTER CONTENTS

The electronic circuit known as an operational amplifier has become increasingly important. However, a detailed analysis of this circuit requires an understanding of electronic devices such as diodes and transistors. You may wonder, then, why we are introducing the circuit before discussing the circuit's electronic components. There are several reasons. First, you can develop an appreciation for how the operational amplifier can be used as a circuit building block by focusing on its terminal behavior. At an introductory level, you need not fully understand the operation of the electronic components that govern terminal behavior. Second, the circuit model of the operational amplifier requires the use of a dependent source. Thus you have a chance to use this type of source in a practical circuit rather than as an abstract circuit component. Third, you can combine the operational amplifier with resistors to perform some very useful functions, such as scaling, summing, sign changing, and subtracting. Finally, after introducing inductors and capacitors in Chapter 6, we can show you how to use the operational amplifier to design integrating and differentiating circuits.

Our focus on the terminal behavior of the operational amplifier implies taking a black box approach to its operation; that is, we are not interested in the internal structure of the amplifier, nor in the currents and voltages that exist in this structure. The important thing to remember is that the internal behavior of the amplifier accounts for the voltage and current constraints imposed at the terminals. (For now, we ask that you accept these constraints on faith.)

The operational amplifier circuit first came into existence as a basic building block in analog computers. It was referred to as *operational* because it was used to implement the mathematical operations of integration, differentiation, addition, sign changing, and scaling. In recent years, the range of application has broadened beyond implementing mathematical operations; however, the original name for the circuit persists. Engineers and technicians have a penchant for creating technical jargon; hence the operational amplifier is widely known as the op amp.

5.1 Operational Amplifier Terminals

Because we are stressing the terminal behavior of the operational amplifier, we begin by discussing the terminals on a commercially available device. In 1968, Fairchild Semiconductor introduced an op amp that has found widespread acceptance: the μA741. (The μA prefix is used by Fairchild to indicate a microcircuit fabrication of the amplifier.) This amplifier is available in several different packages. For our discussion, we assume an eight-lead DIP.[1] Figure 5.1 shows a top view of the package, with the terminal designations given alongside the terminals. The terminals of primary interest are

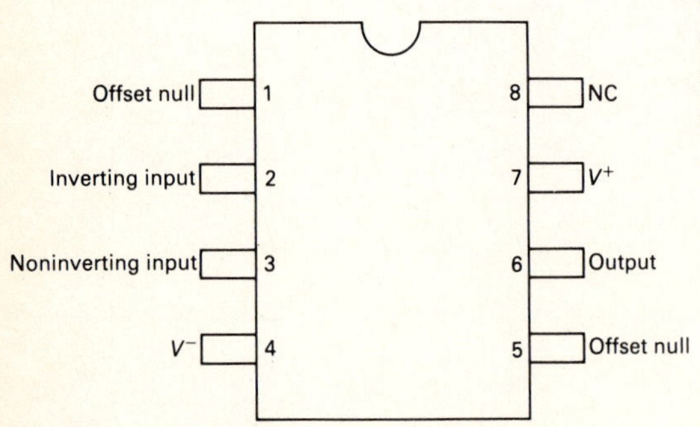

FIGURE 5.1 The eight-lead DIP package (top view).

- the inverting input,
- the noninverting input,
- the output,
- the positive power supply (V^+), and
- the negative power supply (V^-).

The remaining three terminals are of little or no concern. The offset null terminals may be used in an auxiliary circuit to compensate for

[1] DIP is an abbreviation for *dual in-line package*. This means that the terminals on each side of the package are in line, and that the terminals on opposite sides of the package also line up.

a degradation in performance because of aging and imperfections. However, the degradation in most cases is negligible, so the offset terminals often are unused and play a secondary role in circuit analysis. Terminal 8 is of no interest simply because it is an unused terminal; NC stands for no connection, which means that the terminal is not connected to the amplifier circuit.

Figure 5.2 shows a widely used circuit symbol for an op amp that contains the five terminals of primary interest.

Using word labels for the terminals is inconvenient in circuit diagrams, so we simplify the terminal designations in the following way. The noninverting input terminal is labeled plus (+), and the inverting input terminal is labeled minus (−). The power supply terminals, which are always drawn outside the triangle, are marked V^+ and V^-. The terminal at the apex of the triangular box always is understood to be the output terminal. Figure 5.3 summarizes these simplified designations.

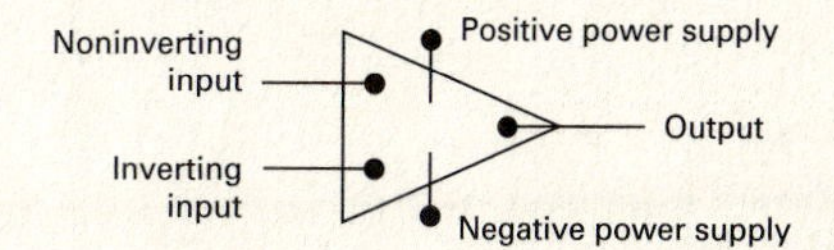

FIGURE 5.2 The circuit symbol for an operational amplifier.

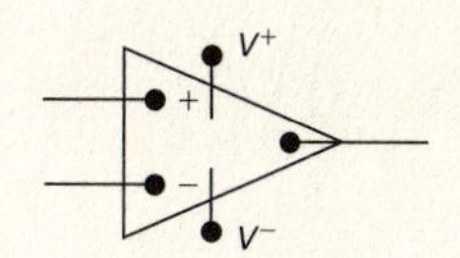

FIGURE 5.3 A simplified circuit symbol for an op amp.

5.2 TERMINAL VOLTAGES AND CURRENTS

We are now ready to introduce the terminal voltages and currents used to describe the behavior of the op amp. The voltage variables are measured from a common reference node.[2] Figure 5.4 shows the voltage variables with their reference polarities. All voltages are considered as voltage rises from the common node. This convention is the same as that used in the node-voltage method of analysis. A positive supply voltage (V_{CC}) is connected between V^+ and the common node. A negative supply voltage ($-V_{CC}$) is connected between V^- and the common node. The voltage between the inverting input terminal and the common node is denoted v_n. The voltage between the noninverting input terminal and the common node is designated as v_p. The voltage between the output terminal and the common node is denoted v_o.

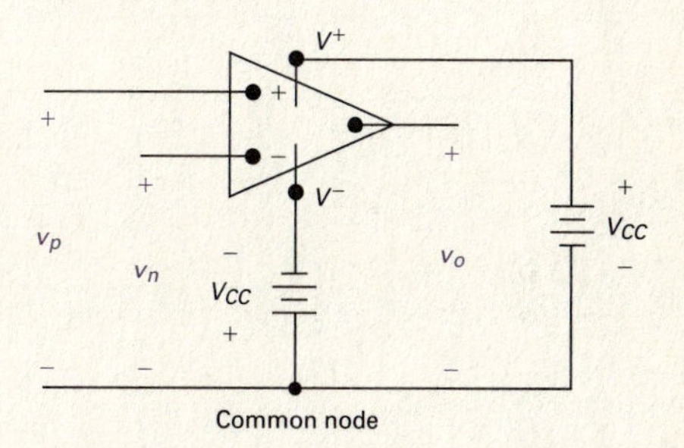

FIGURE 5.4 Terminal voltage variables.

Figure 5.5 shows the current variables with their reference directions. Note that all the current reference directions are into the terminals of the operational amplifier: i_n is the current into the inverting input terminal; i_p is the current into the noninverting input terminal; i_o is the current into the output terminal; i_{c^+} is the current into the positive power supply terminal; and i_{c^-} is the current into the negative power supply terminal.

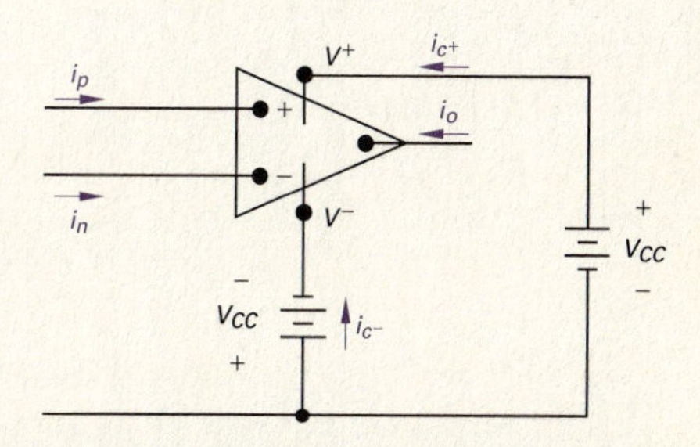

FIGURE 5.5 Terminal current variables.

The terminal behavior of the op amp as a linear circuit element is characterized by constraints on the input voltages and the input

[2] The common node is external to the op amp. It is the reference terminal of the circuit in which the op amp is embedded.

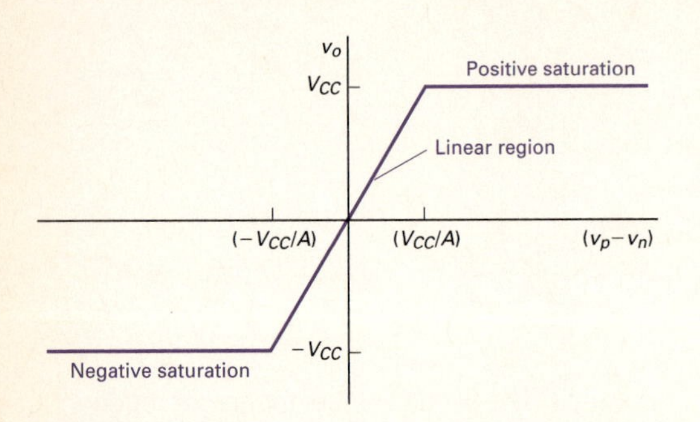

FIGURE 5.6 The voltage transfer characteristic of an op amp.

currents. The voltage constraint is derived from the voltage transfer characteristic of the op amp integrated circuit and is pictured in Fig. 5.6. The voltage transfer characteristic describes how the output voltage varies as a function of the input voltages, that is, how voltage is transferred from the input to the output. Note that for the op amp, the output voltage is a function of the difference between the input voltages, $v_p - v_n$. The equation for the voltage transfer characteristic is

$$v_o = \begin{cases} -V_{CC} & A(v_p - v_n) < -V_{CC}; \\ A(v_p - v_n) & -V_{CC} \leq A(v_p - v_n) \leq +V_{CC}; \\ +V_{CC} & A(v_p - v_n) > +V_{CC}. \end{cases} \quad \textbf{(5.1)}$$

We see from Fig. 5.6 and Eq. (5.1) that the op amp has three distinct regions of operation. When the magnitude of the input voltage difference ($|v_p - v_n|$) is small, the op amp behaves as a linear device, since the output voltage is a linear function of the input voltages. Outside this linear region, the output of the op amp saturates, and the op amp behaves as a nonlinear device, since the output voltage is no longer a linear function of the input voltages. When it is operating linearly, the op amp's output voltage is equal to the difference in its input voltages times the multiplying constant, or **gain**, A.

When we confine the op amp to its linear operating region, a constraint is imposed on the input voltages, v_p and v_n. The constraint is based on typical numerical values for V_{CC} and A in Eq. (5.1). For most op amps, the recommended dc power supply voltages seldom exceed 20 V, and the gain, A, is rarely less than 10,000, or 10^4. We see from both Fig. 5.6 and Eq. (5.1) that in the linear region, the magnitude of the input voltage difference ($|v_p - v_n|$) must be less than $20/10^4$, or 2 mV.

Typically, node voltages in the circuits we study are much larger than 2 mV, so a voltage difference of less than 2 mV means the two voltages are essentially equal. Thus, when an op amp is constrained to its linear operating region and the node voltages are much larger than 2 mV, the constraint on the input voltages of the op amp is

$$v_p = v_n. \quad \textbf{(5.2)}$$

Note that Eq. (5.2) characterizes the relationship between the input voltages for an ideal op amp, that is, an op amp whose value of A is infinite.

The input voltage constraint in Eq. (5.2) is called the *virtual short* condition at the input of the op amp. It is natural to ask how the virtual short is maintained at the input of the op amp when the op amp is embedded in a circuit, thus ensuring linear operation. The answer is that a signal is fed back from the output terminal to the inverting input terminal. This configuration is known as **negative feedback** because the signal fed back from the output subtracts from the input signal. The negative feedback causes the input voltage difference to decrease. Since the output voltage is proportional to the input voltage difference, the output voltage is also decreased, and the op amp operates in its linear region.

If a circuit containing an op amp does not provide a negative feedback path from the op amp output to the inverting input, then the op amp will normally saturate. The difference in the input signals must be extremely small to prevent saturation with no negative feedback. But even if the circuit provides a negative feedback path for the op amp, linear operation is not ensured. So how do we know whether the op amp is operating in its linear region?

The answer is, we don't! We deal with this dilemma by assuming linear operation, performing the circuit analysis, and then checking our results for contradictions. For example, suppose we assume that an op amp in a circuit is operating in its linear region, and we compute the output voltage of the op amp to be 10 V. Upon examining the circuit, we discover that V_{CC} is 6 V, resulting in a contradiction, since the op amp's output voltage can be no larger than V_{CC}. Thus our assumption of linear operation was invalid, and the op amp output must be saturated at 6 V.

We have identified a constraint on the input voltages which is based on the voltage transfer characteristic of the op amp integrated circuit, the assumption that the op amp is restricted to its linear operating region and to typical values for V_{CC} and A. Equation (5.2) represents the voltage constraint for an ideal op amp, that is, with a value of A that is infinite.

We now turn our attention to the constraint on the input currents. Analysis of the op amp integrated circuit reveals that the equivalent resistance seen by the input terminals of the op amp is very large, typically 1 MΩ or more. Ideally, the equivalent input resistance is infinite, resulting in the current constraint

$$i_p = i_n = 0. \quad \textbf{(5.3)}$$

Note that the current constraint is not based on assuming the op amp is confined to its linear operating region, as was the voltage constraint. Together, Eqs. (5.2) and (5.3) form the constraints on terminal behavior that define our ideal op amp model.

From Kirchhoff's current law we know that the sum of the currents entering the operational amplifier is zero, or

$$i_p + i_n + i_o + i_{c^+} + i_{c^-} = 0. \quad \textbf{(5.4)}$$

Substituting the constraint given by Eq. (5.3) into Eq. (5.4) gives

$$i_o = -(i_{c^+} + i_{c^-}). \quad \textbf{(5.5)}$$

The significance of Eq. (5.5) is that, even though the current at the input terminals is negligible, there may still be appreciable current at the output terminal.

Before we start analyzing circuits containing operational amplifiers, let's further simplify the circuit symbol. When we know that the amplifier is operating within its linear region, the dc voltages $\pm V_{CC}$ do not enter into the circuit equations. In this case we can remove the power supply terminals from the symbol and the dc power supplies from the circuit, as shown in Fig. 5.7. A word of caution: Because the power supply terminals have been omitted, there is a danger of inferring from the symbol that $i_p + i_n + i_o = 0$. We have already noted that such is not the case; that is, $i_p + i_n + i_o + i_{c^+} + i_{c^-} = 0$. In other words, the ideal op amp model constraint that $i_p = i_n = 0$ does not imply that $i_o = 0$.

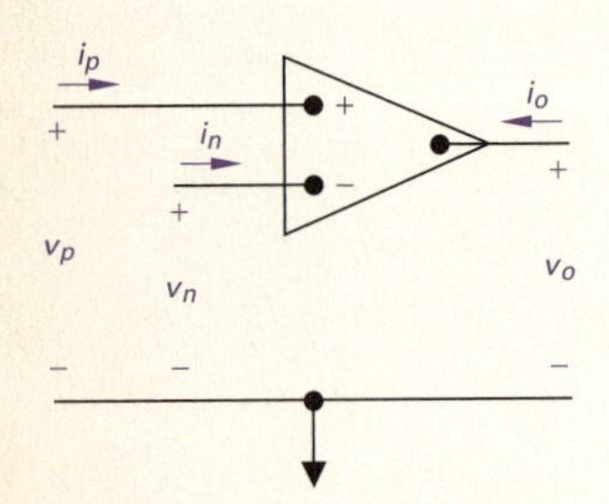

FIGURE 5.7 The op amp symbol with the power supply terminals removed.

Note that the positive and negative power supply voltages do not have to be equal in magnitude. In the linear operating region, v_o must lie between the two supply voltages. For example, if $V^+ = 15$ V and $V^- = -10$ V, then $-10 \text{ V} \leq v_o \leq 15$ V. Be aware also that the value of A is not constant under all operating conditions. For now, however, we assume that it is. A discussion of how and why the value of A can change must be delayed until after you have studied the electronic devices and components used to fabricate an amplifier.

Example 5.1 illustrates the judicious application of Eqs. (5.2) and (5.3). When we use these equations to predict the behavior of a circuit containing an op amp, in effect we are using an ideal model of the device.

EXAMPLE 5.1

The op amp in the circuit shown in Fig. 5.8 is ideal.

a) Calculate v_o if $v_a = 1$ V and $v_b = 0$ V.

b) Repeat (a) for $v_a = 1$ V and $v_b = 2$ V.

c) If $v_a = 1.5$ V, specify the range of v_b that avoids amplifier saturation.

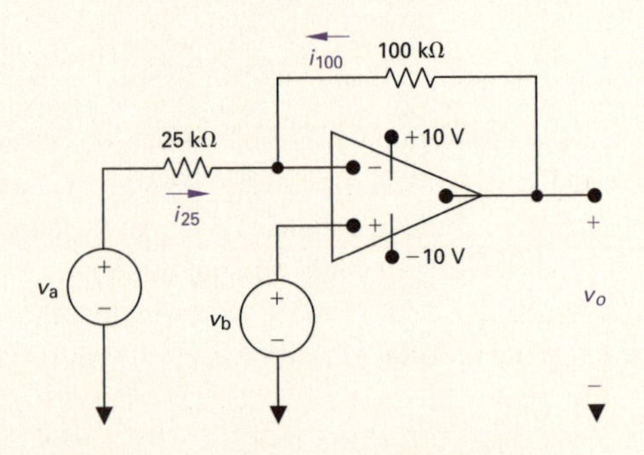

FIGURE 5.8 The circuit for Example 5.1.

SOLUTION

a) Since a negative feedback path exists from the op amp's output to its inverting input through the 100 kΩ resistor, let's assume the op amp is confined to its linear operating region. We can write a node-voltage equation at the inverting input terminal. The voltage at the inverting input terminal is 0, since $v_p = v_b = 0$ from the connected voltage source, and $v_n = v_p$ from the voltage constraint Eq. (5.2). The node-voltage equation at v_n is thus

$$i_{25} + i_{100} = i_n.$$

From Ohm's law,

$$i_{25} = (v_a - v_n)/25 = \frac{1}{25}\text{ mA}$$

and

$$i_{100} = (v_o - v_n)/100 = v_o/100 \text{ mA}.$$

The current constraint requires $i_n = 0$. Substituting the values for the three currents into the node-voltage equation, we obtain

$$\frac{1}{25} + \frac{v_o}{100} = 0.$$

Hence, v_o is -4 V. Note that because v_o lies between ± 10 V, the op amp is in its linear region of operation.

b) Using the same process as in (a), we get

$$v_p = v_b = v_n = 2 \text{ V};$$

$$i_{25} = \frac{v_a - v_n}{25} = \frac{1-2}{25} = -\frac{1}{25}\text{ mA};$$

$$i_{100} = \frac{v_o - v_n}{100} = \frac{v_o - 2}{100}\text{ mA};$$

$$i_{25} = -i_{100}.$$

Therefore $v_o = 6$ V. Again, v_o lies within ± 10 V.

c) As before, $v_n = v_p = v_b$, and $i_{25} = -i_{100}$. Because $v_a = 1.5$ V,

$$\frac{1.5 - v_b}{25} = -\frac{v_o - v_b}{100}.$$

Solving for v_b as a function of v_o gives

$$v_b = \frac{1}{5}(6 + v_o).$$

Now, if the amplifier is to be within the linear region of operation, $-10 \text{ V} \leq v_o \leq 10$ V. Substituting these limits on v_o into the expression for v_b, we see that v_b is limited to

$$-0.8 \text{ V} \leq v_b \leq 3.2 \text{ V}.$$

DRILL EXERCISE

5.1 Assume that the operational amplifier in the circuit shown is ideal.

a) Calculate v_o for the following values of v_s: 0.4, 0.72, 2.0, −0.6, −0.8, and −2.0 V.

b) Specify the range of v_s required to avoid amplifier saturation.

ANSWER: (a) −5, −9, −10, 7.5, 10, and 15 V; (b) $-1.2\text{ V} \leq v_s \leq 0.8\text{ V}$.

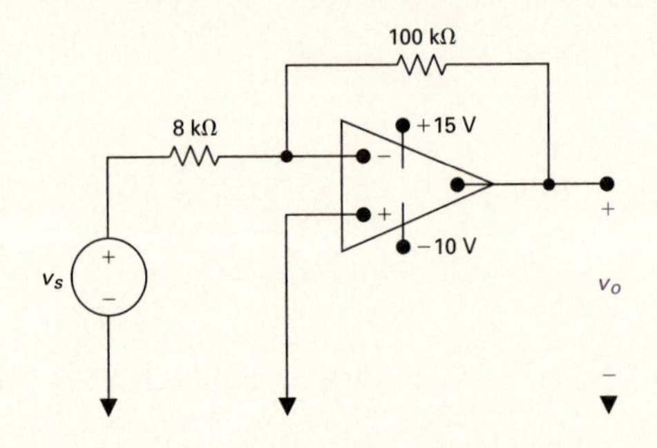

5.3 THE INVERTING-AMPLIFIER CIRCUIT

We are now ready to discuss the operation of some important op amp circuits, using Eqs. (5.2) and (5.3) to model the behavior of the device itself. Figure 5.9 shows an inverting-amplifier circuit. We assume that the op amp is operating in its linear region. Note that in addition to the op amp, the circuit consists of two resistors (R_f and R_s), a voltage signal source (v_s), and a short circuit connected between the noninverting input terminal and the common node.

We now analyze this circuit, assuming an ideal op amp. The goal is to obtain an expression for the output voltage, v_o, as a function of the source voltage, v_s. We employ a single node-voltage equation at the inverting terminal of the op amp, given as

$$i_s + i_f = i_n. \tag{5.6}$$

The voltage constraint of Eq. (5.2) sets the voltage at $v_n = 0$, since the voltage at $v_p = 0$. Therefore,

$$i_s = \frac{v_s}{R_s} \tag{5.7}$$

and

$$i_f = \frac{v_o}{R_f}. \tag{5.8}$$

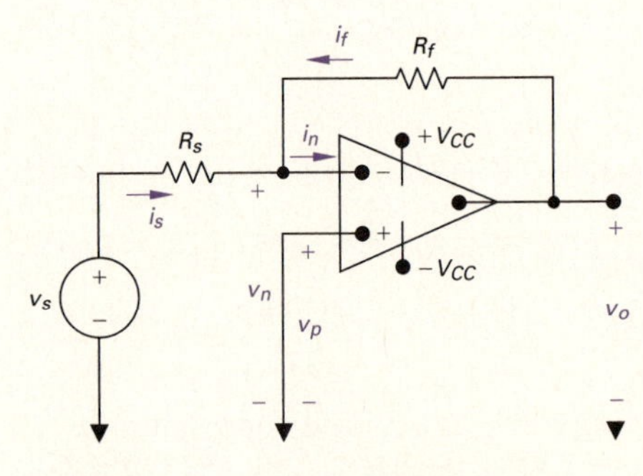

FIGURE 5.9 An inverting-amplifier circuit.

Now we invoke the constraint stated in Eq. (5.3), namely,

$$i_n = 0. \tag{5.9}$$

Substituting Eqs. (5.7), (5.8), and (5.9) into Eq. (5.6) yields the sought-after result:

$$v_o = \frac{-R_f}{R_s} v_s. \tag{5.10}$$

Note that the output voltage is an inverted, scaled replica of the input. The sign reversal from input to output is, of course, the reason for referring to the circuit as an *inverting* amplifier. The scaling factor, or gain, is the ratio R_f/R_s.

The result given by Eq. (5.10) is valid only if the op amp shown in the circuit in Fig. 5.9 is ideal, that is, if A is infinite and the input resistance is infinite. For a practical op amp, Eq. (5.10) is an approximation, usually a good one. (We say more about this later.) Equation (5.10) is important because it tells us that if the op amp gain A is large, we can specify the gain of the inverting amplifier with the external resistors R_f and R_s. The upper limit on the gain, R_f/R_s, is determined by the power supply voltages and the value of the signal voltage v_s. If we assume equal power supply voltages, that is, $V^+ = -V^- = V_{CC}$, we get

$$|v_o| < V_{CC}; \qquad \left|\frac{R_f}{R_s} v_s\right| < V_{CC}; \qquad \frac{R_f}{R_s} < \left|\frac{V_{CC}}{v_s}\right|. \tag{5.11}$$

For example, if $V_{CC} = 15$ V and $v_s = 10$ mV, the ratio R_f/R_s must be less than 1500.

In the inverting amplifier circuit shown in Fig. 5.9, the resistor R_f provides the negative feedback connection. That is, it connects the output terminal to the inverting input terminal. If R_f is removed, the feedback path is opened and the amplifier is said to be operating *open loop*. Figure 5.10 shows the open-loop operation.

Opening the feedback path drastically changes the behavior of the circuit. First, the output voltage is now

$$v_o = -Av_n, \tag{5.12}$$

assuming as before that $V^+ = -V^- = V_{CC}$; then $|v_n| < V_{CC}/A$ for linear operation. Because the inverting input current is almost zero, the voltage drop across R_s is almost zero, and the inverting input voltage nearly equals the signal voltage, v_s; that is, $v_n \approx v_s$. Hence, the op amp can operate open loop in the linear mode only if $|v_s| < V_{CC}/A$. If $|v_s| > V_{CC}/A$, the op amp simply saturates. In particular, if $v_s < -V_{CC}/A$, the op amp saturates at $+V_{CC}$, and if $v_s > V_{CC}/A$, the op amp saturates at $-V_{CC}$. Because the relationship shown in Eq. (5.12) occurs when there is no feedback path, the value of A is often called the **open-loop gain** of the op amp.

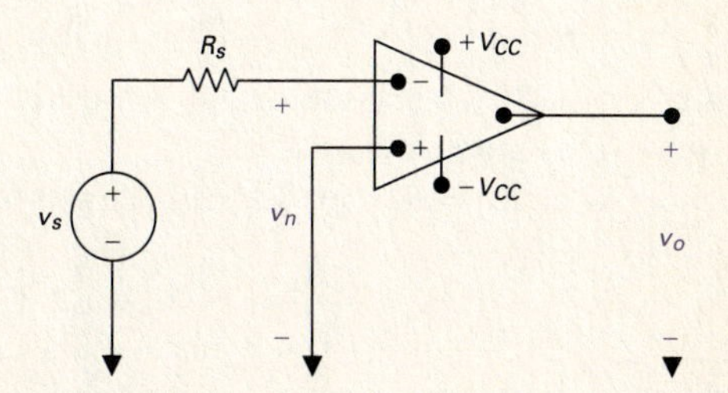

FIGURE 5.10 An inverting amplifier operating open loop.

DRILL EXERCISE

5.2 The source voltage v_s in the circuit in Drill Exercise 5.1 is -640 mV. The 100 kΩ feedback resistor is replaced by a variable resistor R_x. What range of R_x allows the inverting amplifier to operate in its linear region?

ANSWER: $0 \leq R_x \leq 187.5$ kΩ.

5.4 THE SUMMING-AMPLIFIER CIRCUIT

The output voltage of a summing amplifier is an inverted, scaled sum of the voltages applied to the input of the amplifier. Figure 5.11 shows a summing amplifier with three input voltages.

We obtain the relationship between the output voltage v_o and the three input voltages, v_a, v_b, and v_c, by summing the currents away from the inverting input terminal:

$$\frac{v_n - v_a}{R_a} + \frac{v_n - v_b}{R_b} + \frac{v_n - v_c}{R_c} + \frac{v_n - v_o}{R_f} + i_n = 0. \qquad \textbf{(5.13)}$$

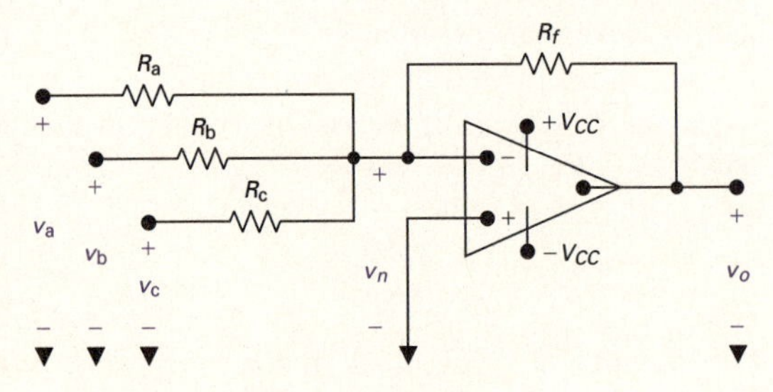

FIGURE 5.11 A summing amplifier.

Assuming an ideal operational amplifier, we can use the voltage and current constraints together with the ground imposed at v_p by the circuit to see that $v_n = v_p = 0$ and $i_n = 0$. This reduces Eq. (5.13) to

$$v_o = -\left(\frac{R_f}{R_a}v_a + \frac{R_f}{R_b}v_b + \frac{R_f}{R_c}v_c\right). \qquad \textbf{(5.14)}$$

Equation (5.14) states that the output voltage is an inverted, scaled sum of the three input voltages.

If $R_a = R_b = R_c = R_s$, then Eq. (5.14) reduces to

$$v_o = -\frac{R_f}{R_s}(v_a + v_b + v_c). \qquad \textbf{(5.15)}$$

Finally, if we make $R_f = R_s$, the output voltage is just the inverted sum of the input voltages. That is,

$$v_o = -(v_a + v_b + v_c). \qquad \textbf{(5.16)}$$

Although we illustrated the summing amplifier with just three input signals, the number of input voltages can be increased as needed. For example, you might wish to sum 16 individually recorded audio signals to form a single audio signal. The summing amplifier configuration in Fig. 5.11 could include 16 different input resistor values so that each of the input audio tracks appears in the output signal with a different amplification factor. The summing amplifier thus plays the role of an audio mixer. As with inverting-amplifier circuits, the scaling factors in summing-amplifier circuits are determined by the external resistors R_f, R_a, R_b, R_c, ..., R_n.

DRILL EXERCISE

5.3 a) Find v_o in the circuit shown if $v_a = 0.1$ V and $v_b = 0.25$ V.

b) If $v_b = 0.25$ V, how large can v_a be before the op amp saturates?

c) If $v_a = 0.10$ V, how large can v_b be before the op amp saturates?

d) Repeat (a), (b), and (c) with the polarity of v_b reversed.

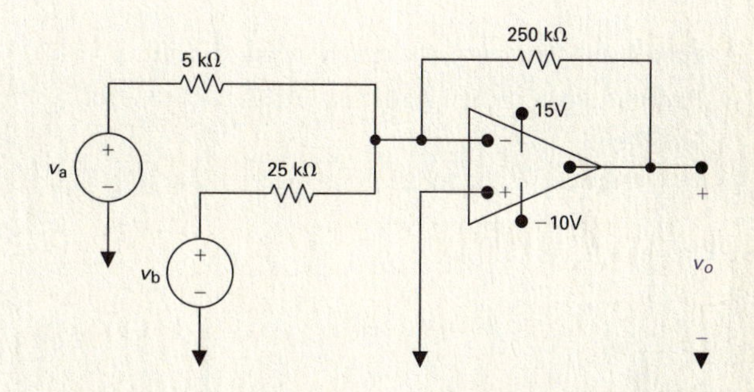

ANSWER: (a) −7.5 V; (b) 0.15 V; (c) 0.5 V; (d) −2.5, 0.25, and 2 V.

5.5 The Noninverting-Amplifier Circuit

Figure 5.12 depicts a noninverting-amplifier circuit. The signal source is represented by v_g in series with the resistor R_g. In deriving the expression for the output voltage as a function of the source voltage, we assume an ideal operational amplifier operating within its linear region. Thus, as before, we use Eqs. (5.2) and (5.3) as the basis for the derivation. Because the op amp input current is zero, we can write $v_p = v_g$ and, from Eq. (5.2), $v_n = v_g$ as well. Now, because the input current is zero ($i_n = i_p = 0$), the resistors R_f and R_s form an unloaded voltage divider across v_o. Therefore

$$v_n = v_g = \frac{v_o R_s}{R_s + R_f}. \tag{5.17}$$

Solving Eq. (5.17) for v_o gives us the sought-after expression:

$$v_o = \frac{R_s + R_f}{R_s} v_g. \tag{5.18}$$

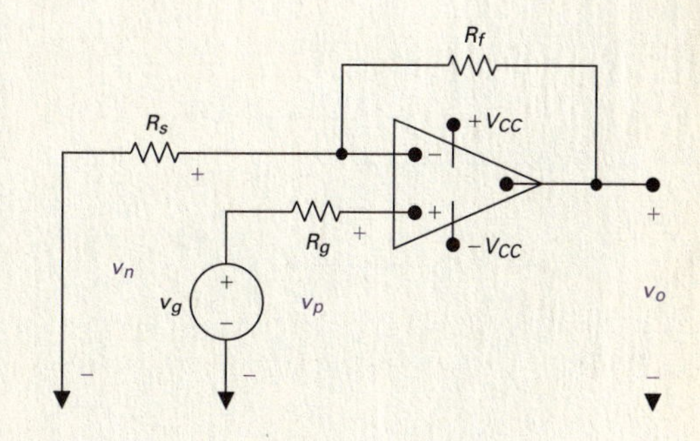

FIGURE 5.12 A noninverting amplifier.

Operation in the linear region requires that

$$\frac{R_s + R_f}{R_s} < \left|\frac{V_{CC}}{v_g}\right|.$$

Note again that, because of the ideal op amp assumption, we can express the output voltage as a function of the input voltage and the external resistors—in this case, R_s and R_f.

DRILL EXERCISE

5.4 Assume that the op amp in the circuit shown is ideal.

a) Find the output voltage when the variable resistor is set to 80 kΩ.

b) How large can R_x be before the amplifier saturates?

ANSWER: (a) 9 V; (b) 160 kΩ.

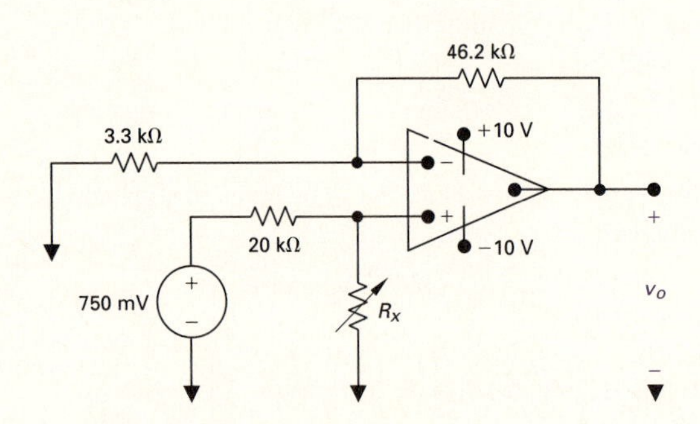

5.6 THE DIFFERENCE-AMPLIFIER CIRCUIT

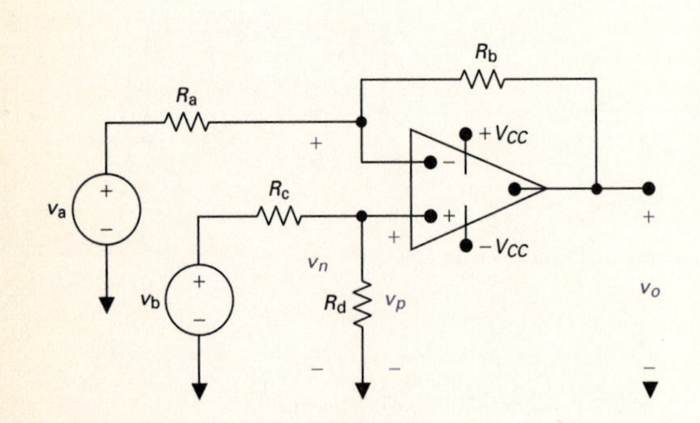

FIGURE 5.13 A difference amplifier.

The output voltage of a difference amplifier is proportional to the difference between the two input voltages. To demonstrate, we analyze the difference-amplifier circuit shown in Fig. 5.13, assuming an ideal op amp operating in its linear region. We derive the relationship between v_o and the two input voltages v_a and v_b by summing the currents away from the inverting input node:

$$\frac{v_n - v_a}{R_a} + \frac{v_n - v_o}{R_b} + i_n = 0. \quad \textbf{(5.19)}$$

Because the op amp is ideal, we use the voltage and current constraints to see that

$$i_n = i_p = 0 \quad \textbf{(5.20)}$$

and

$$v_n = v_p = \frac{R_d}{R_c + R_d} v_b. \quad \textbf{(5.21)}$$

Combining Eqs. (5.19), (5.20), and (5.21) gives the desired relationship:

$$v_o = \frac{R_d(R_a + R_b)}{R_a(R_c + R_d)} v_b - \frac{R_b}{R_a} v_a. \quad \textbf{(5.22)}$$

Equation (5.22) shows that the output voltage is proportional to the difference between a scaled replica of v_b and a scaled replica of v_a. In general the scaling factor applied to v_b is not the same as that applied to v_a. However, the scaling factor applied to each input voltage can be made equal by setting

$$\frac{R_a}{R_b} = \frac{R_c}{R_d}. \quad \textbf{(5.23)}$$

When Eq. (5.23) is satisfied, the expression for the output voltage reduces to

$$v_o = \frac{R_b}{R_a}(v_b - v_a). \quad \textbf{(5.24)}$$

Equation (5.24) indicates that the output voltage can be made a scaled replica of the difference between the input voltages v_b and v_a. As in the previous ideal amplifier circuits, the scaling is controlled by the external resistors. Furthermore, the relationship between the output voltage and the input voltages is not affected by connecting a nonzero load resistance across the output of the amplifier.

DRILL EXERCISES

5.5 a) Use the principle of superposition to derive Eq. (5.22).

b) Derive Eqs. (5.23) and (5.24).

ANSWER: (a) $v_o' = -\frac{R_b}{R_a} v_a$; $v_o'' = \frac{R_d(R_a + R_b)}{R_a(R_c + R_d)} v_b$; $v_o = v_o' + v_o''$; (b) derivation.

5.6 a) In the difference amplifier shown, $v_b = 6.0$ V. What range of values for v_a will result in linear operation?

b) Repeat (a) with the 20 kΩ resistor decreased to 5 kΩ.

ANSWER: (a) 3.5 V $\leq v_a \leq$ 8.5 V; (b) 1.25 V $\leq v_a \leq$ 6.25 V.

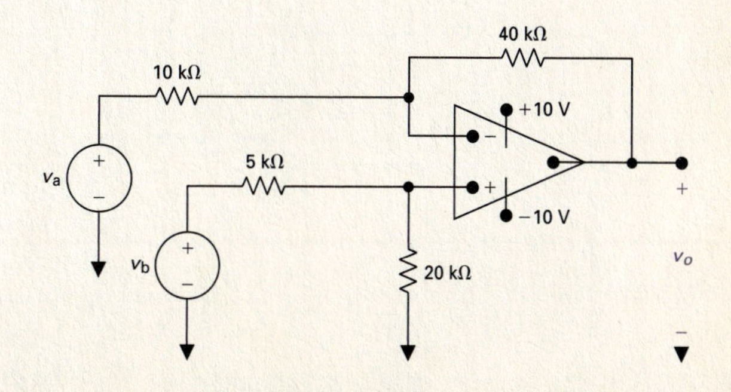

THE DIFFERENCE AMPLIFIER—ANOTHER PERSPECTIVE

We can examine the behavior of a difference amplifier more closely if we redefine its inputs in terms of two other voltages. The first is the **differential mode** input, which is the difference between the two input voltages in Fig. 5.13:

$$v_{\text{dm}} = v_{\text{b}} - v_{\text{a}}. \tag{5.25}$$

The second is the **common mode** input, which is the average of the two input voltages in Fig. 5.13:

$$v_{\text{cm}} = (v_{\text{a}} + v_{\text{b}})/2. \tag{5.26}$$

Using Eqs. (5.25) and (5.26), we can now represent the original input voltages, v_{a} and v_{b}, in terms of the differential mode and common mode voltages, v_{dm} and v_{cm}:

$$v_{\text{a}} = v_{\text{cm}} - \frac{1}{2}v_{\text{dm}} \tag{5.27}$$

and

$$v_{\text{b}} = v_{\text{cm}} + \frac{1}{2}v_{\text{dm}}. \tag{5.28}$$

Substituting Eqs. (5.27) and (5.28) into Eq. (5.22) gives the output of the difference amplifier in terms of the differential mode and common mode voltages:

$$v_o = \left[\frac{R_{\text{a}}R_{\text{d}} - R_{\text{b}}R_{\text{c}}}{R_{\text{a}}(R_{\text{c}} + R_{\text{d}})}\right]v_{\text{cm}} + \left[\frac{R_{\text{d}}(R_{\text{a}} + R_{\text{b}}) + R_{\text{b}}(R_{\text{c}} + R_{\text{d}})}{2R_{\text{a}}(R_{\text{c}} + R_{\text{d}})}\right]v_{\text{dm}} \tag{5.29}$$

$$= A_{\text{cm}}v_{\text{cm}} + A_{\text{dm}}v_{\text{dm}}, \tag{5.30}$$

where A_{cm} is the common mode gain and A_{dm} is the differential mode gain. Now, substitute $R_{\text{c}} = R_{\text{a}}$ and $R_{\text{d}} = R_{\text{b}}$, which are the values for R_{c} and R_{d} that satisfy Eq. (5.23), into Eq. (5.29):

$$v_o = (0)v_{\text{cm}} + \left(\frac{R_{\text{b}}}{R_{\text{a}}}\right)v_{\text{dm}}. \tag{5.31}$$

Thus an ideal difference amplifier has $A_{\text{cm}} = 0$, amplifies only the differential mode portion of the input voltage, and eliminates the common mode portion of the input voltage. Figure 5.14 shows a difference-amplifier circuit with differential mode and common mode input voltages in place of v_{a} and v_{b}.

Equation (5.30) provides an important perspective on the function of the difference amplifier, since in many applications it is the differential mode signal that contains the information of interest, whereas

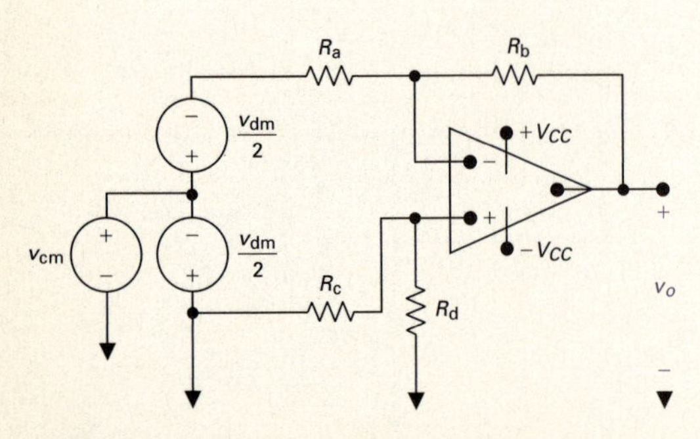

FIGURE 5.14 A difference amplifier with common mode and differential mode input voltages.

the common mode signal is the noise found in all electric signals. For example, an electrocardiograph electrode measures the voltages produced by your body to regulate your heartbeat. These voltages have very small magnitudes compared to the electrical noise which the electrode picks up from sources such as lights and electrical equipment. The noise appears as the common mode portion of the measured voltage, while the heart rate voltages comprise the differential mode portion. Thus an ideal difference amplifier would amplify only the voltage of interest and would suppress the noise.

Measuring Difference-Amplifier Performance—The Common Mode Rejection Ratio

An ideal difference amplifier has zero common mode gain and nonzero (and usually large) differential mode gain. Two factors have an influence on the ideal common mode gain—resistance mismatches (that is, Eq. [5.23] is not satisfied) or a nonideal op amp (that is, Eq. [5.20] is not satisfied). We focus here on the effect of resistance mismatches on the performance of a difference amplifier.

Suppose that resistor values are chosen which do not precisely satisfy Eq. (5.23). Instead, the relationship among the resistors R_a, R_b, R_c, and R_d is

$$\frac{R_a}{R_b} = (1-\epsilon)\frac{R_c}{R_d},$$

so

$$R_a = (1-\epsilon)R_c \quad \text{and} \quad R_b = R_d, \tag{5.32}$$

or

$$R_d = (1-\epsilon)R_b \quad \text{and} \quad R_a = R_c, \tag{5.33}$$

where ϵ is a very small number. We can see the effect of this resistance mismatch on the common mode gain of the difference amplifier by substituting Eq. (5.33) into Eq. (5.29) and simplifying the expression for A_{cm}:

$$A_{cm} = \frac{R_a(1-\epsilon)R_b - R_aR_b}{R_a[R_a + (1-\epsilon)R_b]} \tag{5.34}$$

$$= \frac{-\epsilon R_b}{R_a + (1-\epsilon)R_b} \tag{5.35}$$

$$\approx \frac{-\epsilon R_b}{R_a + R_b}. \tag{5.36}$$

We can make the approximation to give Eq. (5.36) because ϵ is very small and therefore $(1-\epsilon)$ is approximately 1 in the denominator of Eq. (5.35). Note that when the resistors in the difference amplifier satisfy Eq. (5.23), $\epsilon = 0$ and Eq. (5.36) gives $A_{cm} = 0$.

Now calculate the effect of the resistance mismatch on the differential mode gain by substituting Eq. (5.33) into Eq. (5.29) and simplifying the expression for A_{dm}:

$$A_{\text{dm}} = \frac{(1-\epsilon)R_{\text{b}}(R_{\text{a}}+R_{\text{b}}) + R_{\text{b}}[R_{\text{a}}+(1-\epsilon)R_{\text{b}}]}{2R_{\text{a}}[R_{\text{a}}+(1-\epsilon)R_{\text{b}}]} \tag{5.37}$$

$$= \frac{R_{\text{b}}}{R_{\text{a}}}\left[1 - \frac{(\epsilon/2)R_{\text{a}}}{R_{\text{a}}+(1-\epsilon)R_{\text{b}}}\right] \tag{5.38}$$

$$\approx \frac{R_{\text{b}}}{R_{\text{a}}}\left[1 - \frac{(\epsilon/2)R_{\text{a}}}{R_{\text{a}}+R_{\text{b}}}\right] \tag{5.39}$$

We use the same rationale for the approximation in Eq. (5.39) as in the computation of A_{cm}. When the resistors in the difference amplifier satisfy Eq. (5.23), $\epsilon = 0$ and Eq. (5.39) gives $A_{\text{dm}} = R_{\text{b}}/R_{\text{a}}$.

The **common mode rejection ratio (CMRR)** can be used to measure how nearly ideal a difference amplifier is. It is defined as the ratio of the differential mode gain to the common mode gain:

$$\text{CMRR} = \left|\frac{A_{\text{dm}}}{A_{\text{cm}}}\right| \tag{5.40}$$

The higher the CMRR, the more nearly ideal the difference amplifier. We can see the effect of resistance mismatch on the CMRR by substituting Eqs. (5.36) and (5.39) into Eq. (5.40):

$$\text{CMRR} = \left|\frac{\frac{R_{\text{b}}}{R_{\text{a}}}[1-(R_{\text{a}}\epsilon/2)/(R_{\text{a}}+R_{\text{b}})]}{-\epsilon R_{\text{b}}/(R_{\text{a}}+R_{\text{b}})}\right| \tag{5.41}$$

$$= \frac{R_{\text{a}}(1-\epsilon/2)+R_{\text{b}}}{\epsilon R_{\text{a}}} \tag{5.42}$$

$$\approx \frac{1+R_{\text{b}}/R_{\text{a}}}{\epsilon} \tag{5.43}$$

Use a computer tool to compute the CMRR for various resistor mismatches.

From Eq. (5.43), if the resistors in the difference amplifier are matched, $\epsilon = 0$ and $\text{CMRR} = \infty$. Even if the resistors are mismatched, we can minimize the impact of the mismatch by making the differential mode gain ($R_{\text{b}}/R_{\text{a}}$), very large, thereby making the CMRR large.

We said at the outset that another reason for nonzero common mode gain is a nonideal op amp. Note that the op amp is itself a difference amplifier, since in the linear operating region its output is proportional to the difference of its inputs; that is, $v_o = A(v_p - v_n)$. The output of a nonideal op amp is not strictly proportional to the difference between the inputs (the differential mode input) but also is comprised of a common mode signal. Internal mismatches in the components of the integrated circuit make the behavior of the op amp nonideal, in the same way that the resistor mismatches in the difference-amplifier circuit make its behavior nonideal. While a discussion of nonideal op amps is beyond the scope of this text, you may note that the CMRR is often used in assessing how nearly ideal an op amp's behavior is. In fact, it is one of the main ways of rating operational amplifiers in practice.

DRILL EXERCISES

5.7 In the difference amplifier shown, compute (a) the differential mode gain; (b) the common mode gain; and (c) the CMRR.

ANSWER: (a) 24.02; (b) –0.04; (c) 624.5.

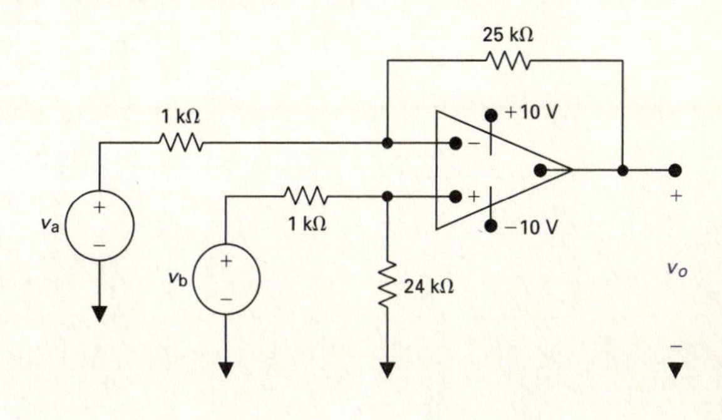

5.8 In the difference amplifier shown, what value of R_x yields a CMRR ≥ 1000?

ANSWER: 9.89 kΩ or 10.11 kΩ.

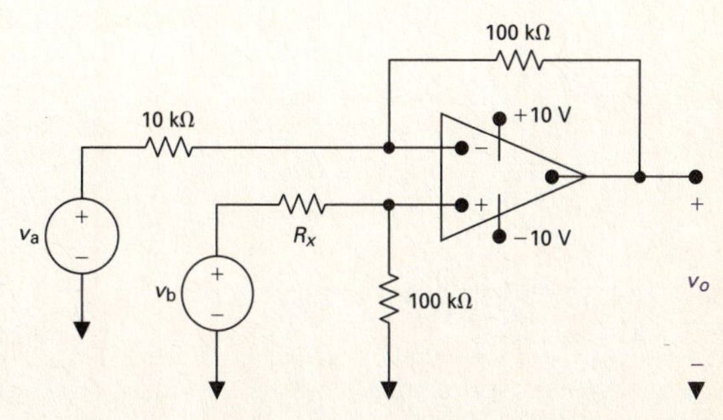

5.7 A More Realistic Model for the Operational Amplifier

We now consider a more realistic model that predicts the performance of an op amp in its linear region of operation. Such a model includes three modifications to the ideal op amp: (1) a finite input resistance, R_i; (2) a finite open-loop gain, A; and (3) a nonzero output resistance, R_o. The circuit shown in Fig. 5.15 illustrates the more realistic model.

Whenever we use the equivalent circuit shown in Fig. 5.15, we disregard the assumptions that $v_n = v_p$ (Eq. [5.2]) and $i_n = i_p = 0$ (Eq. [5.3]). Furthermore, Eq. (5.1) is no longer valid because of the presence of the nonzero output resistance, R_o. Another way to understand the circuit shown in Fig. 5.15 is to reverse our thought process. That is, we can see that the circuit reduces to the ideal model when $R_i \to \infty$, $A \to \infty$, and $R_o \to 0$. For the μA741 op amp, the typical values of R_i, A, and R_o are 2 MΩ, 10^5, and 75 Ω, respectively.

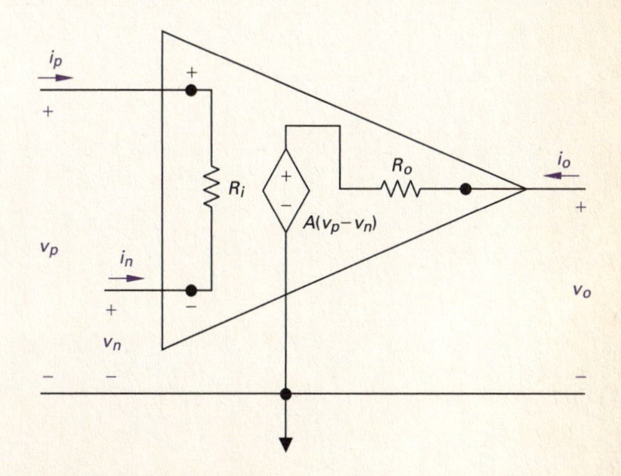

FIGURE 5.15 An equivalent circuit for an operational amplifier.

Although the presence of R_i and R_o makes the analysis of circuits containing op amps more cumbersome, such analysis remains straightforward. To illustrate, we analyze both an inverting and a noninverting amplifier, using the equivalent circuit shown in Fig. 5.15. We begin with the inverting amplifier.

Use a computer tool when analyzing nonideal op amps.

ANALYSIS OF AN INVERTING OP AMP CIRCUIT USING THE MORE REALISTIC MODEL

If we use the op amp circuit shown in Fig. 5.15, the circuit for the inverting amplifier is the one depicted in Fig. 5.16. As before, our goal is to express the output voltage, v_o, as a function of the source voltage, v_s. We obtain the desired expression by writing the two node-voltage equations that describe the circuit and then solving the resulting set of equations for v_o. In Fig. 5.16, the two nodes are labeled a and b. Also note that $v_p = 0$ by virtue of the external short-circuit connection at the noninverting input terminal. The two node-voltage equations are as follows:

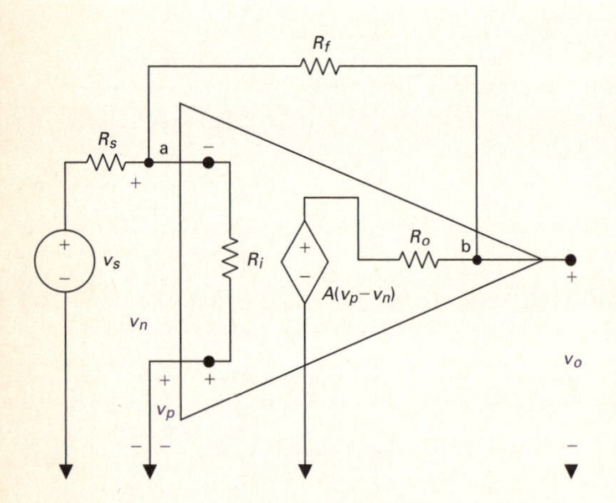

FIGURE 5.16 An inverting-amplifier circuit.

$$\text{node a:} \quad \frac{v_n - v_s}{R_s} + \frac{v_n}{R_i} + \frac{v_n - v_o}{R_f} = 0; \tag{5.44}$$

$$\text{node b:} \quad \frac{v_o - v_n}{R_f} + \frac{v_o - A(-v_n)}{R_o} = 0. \tag{5.45}$$

We rearrange Eqs. (5.44) and (5.45) so that the solution for v_o by Cramer's method becomes apparent:

$$\left(\frac{1}{R_s} + \frac{1}{R_i} + \frac{1}{R_f}\right) v_n - \frac{1}{R_f} v_o = \frac{1}{R_s} v_s; \tag{5.46}$$

$$\left(\frac{A}{R_o} - \frac{1}{R_f}\right) v_n + \left(\frac{1}{R_f} + \frac{1}{R_o}\right) v_o = 0. \tag{5.47}$$

Solving for v_o yields

$$v_o = \frac{-A + (R_o/R_f)}{\frac{R_s}{R_f}\left(1 + A + \frac{R_o}{R_i}\right) + \left(\frac{R_s}{R_i} + 1\right) + \frac{R_o}{R_f}} v_s. \tag{5.48}$$

Note that Eq. (5.48) reduces to Eq. (5.10) as $R_o \to 0$, $R_i \to \infty$, and $A \to \infty$.

If the inverting amplifier shown in Fig. 5.16 were loaded at its output terminals with a load resistance of R_L ohms, the relationship between v_o and v_s would become

$$v_o = \frac{-A + (R_o/R_f)}{\frac{R_s}{R_f}\left(1 + A + \frac{R_o}{R_i} + \frac{R_o}{R_L}\right) + \left(1 + \frac{R_o}{R_L}\right)\left(1 + \frac{R_s}{R_i}\right) + \frac{R_o}{R_f}} v_s. \tag{5.49}$$

Problems 5.32, 5.33, 5.34, and 5.36 will familiarize you with numerical calculations involving Eqs. (5.48) and (5.49).

Analysis of a Noninverting Op Amp Circuit Using the More Realistic Model

When we use the equivalent circuit shown in Fig. 5.15 to analyze a noninverting amplifier, we obtain the circuit depicted in Fig. 5.17. Here, the voltage source v_g, in series with the resistance R_g, represents the signal source. The resistor R_L denotes the load on the amplifier. Our analysis consists of deriving an expression for v_o as a function of v_g. We do so by writing the node-voltage equations at nodes a and b. At node a,

$$\frac{v_n}{R_s} + \frac{v_n - v_g}{R_g + R_i} + \frac{v_n - v_o}{R_f} = 0, \tag{5.50}$$

and at node b,

$$\frac{v_o - v_n}{R_f} + \frac{v_o}{R_L} + \frac{v_o - A(v_p - v_n)}{R_o} = 0. \tag{5.51}$$

Because the current in R_g is the same as in R_i, we have

$$\frac{v_p - v_g}{R_g} = \frac{v_n - v_g}{R_i + R_g}. \tag{5.52}$$

We use Eq. (5.52) to eliminate v_p from Eq. (5.51), giving a pair of equations involving the unknown voltages v_n and v_o. This algebraic manipulation leads to

$$v_n\left(\frac{1}{R_s} + \frac{1}{R_g + R_i} + \frac{1}{R_f}\right) - v_o\left(\frac{1}{R_f}\right) = v_g\left(\frac{1}{R_g + R_i}\right); \tag{5.53}$$

$$v_n\left[\frac{AR_i}{R_o(R_i + R_g)} - \frac{1}{R_f}\right] + v_o\left(\frac{1}{R_f} + \frac{1}{R_o} + \frac{1}{R_L}\right) = v_g\left[\frac{AR_i}{R_o(R_1 + R_g)}\right]. \tag{5.54}$$

Solving for v_o yields

$$v_o = \frac{[(R_f + R_s) + (R_s R_o / AR_i)]v_g}{R_s + \frac{R_o}{A}(1 + K_r) + \frac{R_f R_s + (R_f + R_s)(R_i + R_g)}{AR_i}}, \tag{5.55}$$

where

$$K_r = \frac{R_s + R_g}{R_i} + \frac{R_f + R_s}{R_L} + \frac{R_f R_s + R_f R_g + R_g R_s}{R_i R_L}.$$

Note that Eq. (5.55) reduces to Eq. (5.18) when $R_o \to 0$, $A \to \infty$, and $R_i \to \infty$. For the unloaded ($R_L = \infty$) noninverting amplifier,

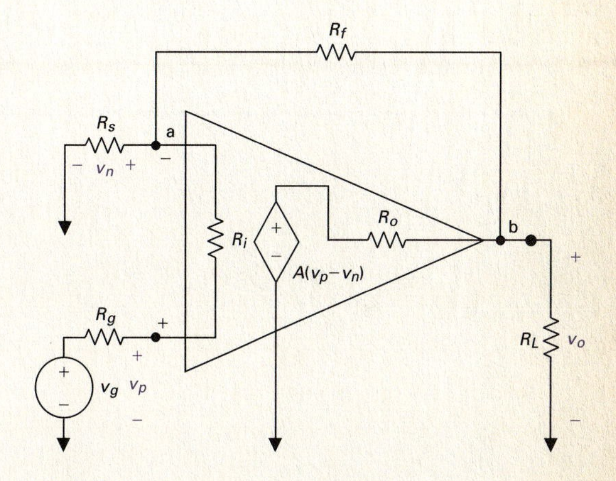

FIGURE 5.17 A noninverting-amplifier circuit.

Eq. (5.55) simplifies to

$$v_o = \frac{[(R_f + R_s) + R_s R_o / A R_i] v_g}{R_s + \frac{R_o}{A}\left(1 + \frac{R_s + R_g}{R_i}\right) + \frac{1}{AR_i}[R_f R_s + (R_f + R_s)(R_i + R_g)]}. \tag{5.56}$$

Note that, in the derivation of Eq. (5.56) from Eq. (5.55), K_r reduces to $(R_s + R_g)/R_i$. Problem 5.34 illustrates the effect of R_i, A, and R_o on the performance of a noninverting amplifier.

SUMMARY

- The equation which defines the voltage transfer characteristic of an ideal operational amplifier is

$$v_o = \begin{cases} -V_{CC} & A(v_p - v_n) < -V_{CC}; \\ A(v_p - v_n) & -V_{CC} \leq A(v_p - v_n) \leq +V_{CC}; \\ +V_{CC} & A(v_p - v_n) > +V_{CC}, \end{cases}$$

  where A is a proportionality constant known as the open-loop gain, and V_{CC} represents the power supply voltages.

- A feedback path between an op amp's output and its inverting input can constrain the op amp to its linear operating region where $v_o = A(v_p - v_n)$.

- A voltage constraint exists when the op amp is confined to its linear operating region due to typical values of V_{CC} and A. If the ideal modeling assumptions are made—meaning A is assumed to be infinite—the ideal op amp model is characterized by the voltage constraint

$$v_p = v_n.$$

- A current constraint further characterizes the ideal op amp model, because the ideal input resistance of the op amp integrated circuit is infinite. This current constraint is given by

$$i_p = i_n = 0.$$

- We considered both a simple, ideal op amp model and a more realistic model in this chapter. The differences between the two models are as follows:

Simplified Model	More Realistic Model
Infinite input resistance	Finite input resistance
Infinite open-loop gain	Finite open-loop gain
Zero output resistance	Nonzero output resistance

- An inverting amplifier is an op amp circuit producing an output voltage which is an inverted, scaled replica of the input.
- A summing amplifier is an op amp circuit producing an output voltage which is a scaled sum of the input voltages.
- A noninverting amplifier is an op amp circuit producing an output voltage which is a scaled replica of the input voltage.
- A difference amplifier is an op amp circuit producing an output voltage which is a scaled replica of the input voltage difference.
- The two voltage inputs to a difference amplifier can be used to calculate the common mode and difference mode voltage inputs, v_{cm} and v_{dm}. The output from the difference amplifier can be written in the form

$$v_o = A_{\text{cm}} v_{\text{cm}} + A_{\text{dm}} v_{\text{dm}},$$

  where A_{cm} is the common mode gain, and A_{dm} is the differential mode gain.

- In an ideal difference amplifier, $A_{\text{cm}} = 0$. To measure how nearly ideal a difference amplifier is, we use the common mode rejection ratio:

$$\text{CMRR} = \left| \frac{A_{\text{dm}}}{A_{\text{cm}}} \right|$$

  An ideal difference amplifier has an infinite CMRR.

PROBLEMS

5.1 The operational amplifier in the circuit in Fig. P5.1 is ideal.

a) Calculate v_o if $v_a = 1.0$ V and $v_b = 0$ V.

b) Calculate v_o if $v_a = 2.0$ V and $v_b = 0$ V.

c) Calculate v_o if $v_a = 2.5$ V and $v_b = 3$ V.

d) Calculate v_o if $v_a = 3.0$ V and $v_b = 2$ V.

e) Calculate v_o if $v_a = 1.5$ V and $v_b = 2.5$ V.

f) If $v_b = 4.0$ V, specify the range of v_a such that the amplifier does not saturate.

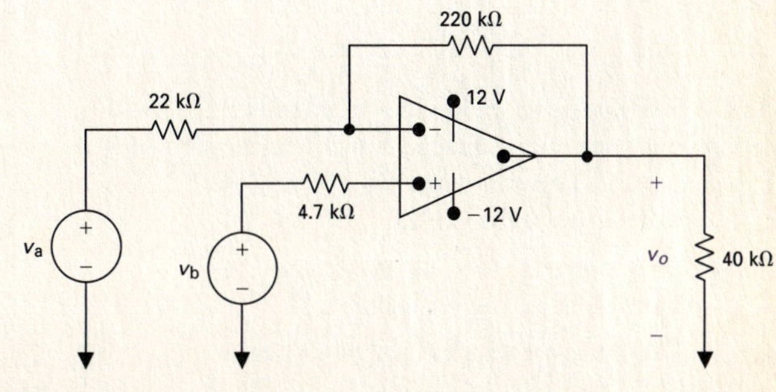

FIGURE P5.1

5.2 The operational amplifier in the circuit in Fig. P5.2 is ideal.

a) Calculate v_o.

b) Calculate i_o.

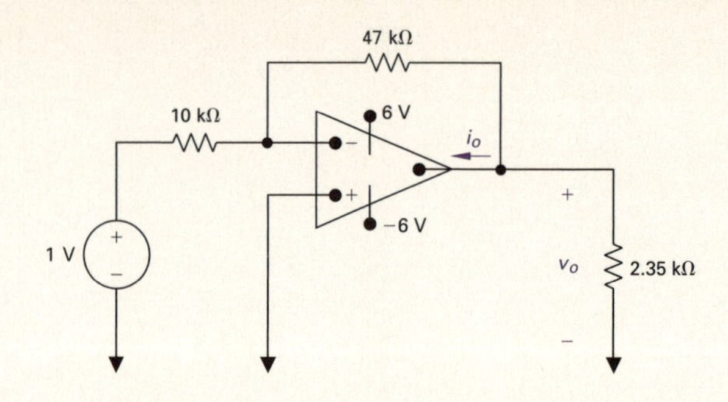

FIGURE P5.2

5.3 A voltmeter with a full-scale reading of 10 V is used to measure the output voltage in the circuit in Fig. P5.3. What is the reading of the voltmeter? Assume the operational amplifier is ideal.

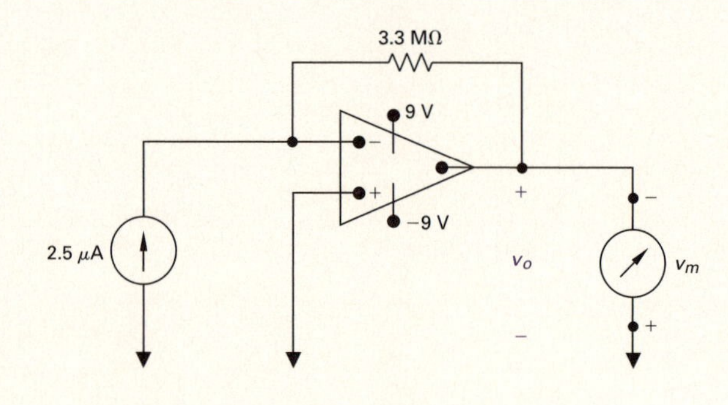

FIGURE P5.3

5.4 Find i_o in the circuit in Fig. P5.4 if the operational amplifier is ideal.

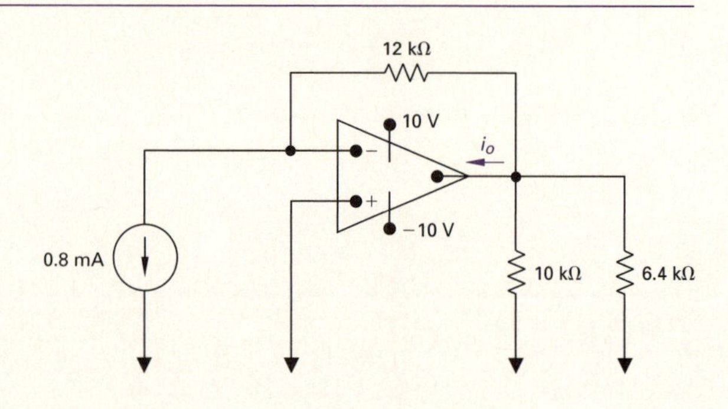

FIGURE P5.4

❖ **5.5** A circuit designer claims the circuit in Fig. P5.5 will produce an output voltage that will vary between ±9 as v_g varies between 0 and 6 V. Assume the operational amplifier is ideal.

a) Draw a graph of the output voltage v_o as a function of the input voltage v_g for $0 \leq v_g \leq 6$ V.

b) Do you agree with the designer's claim?

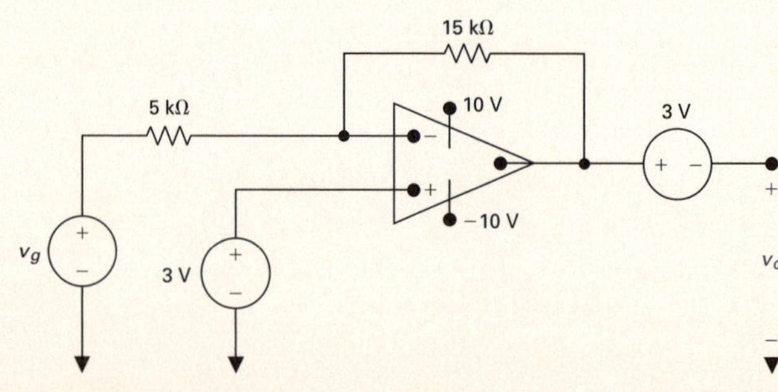

FIGURE P5.5

5.6 The operational amplifier in the circuit in Fig. P5.6 is ideal. Calculate the following:

a) v_a

b) v_o

c) i_a

d) i_o

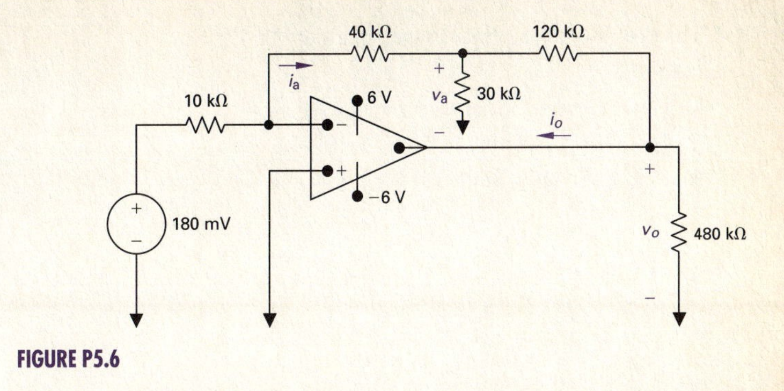

FIGURE P5.6

5.7 Find i_L (in microamperes) in the circuit in Fig. P5.7.

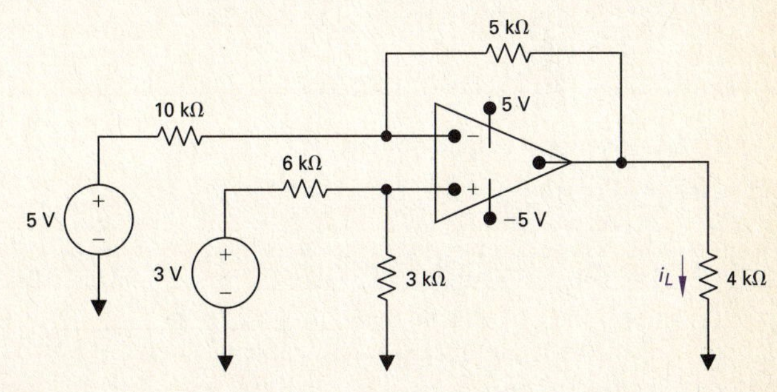

FIGURE P5.7

5.8 a) The operational amplifier in the circuit shown in Fig. P5.8 is ideal. The adjustable resistor R_Δ has a maximum value of 120 kΩ, and α is restricted to the range of $0.25 \leq \alpha \leq 0.8$. Calculate the range of v_o if $v_g = 40$ mV.

b) If α is not restricted, at what value of α will the operational amplifier saturate?

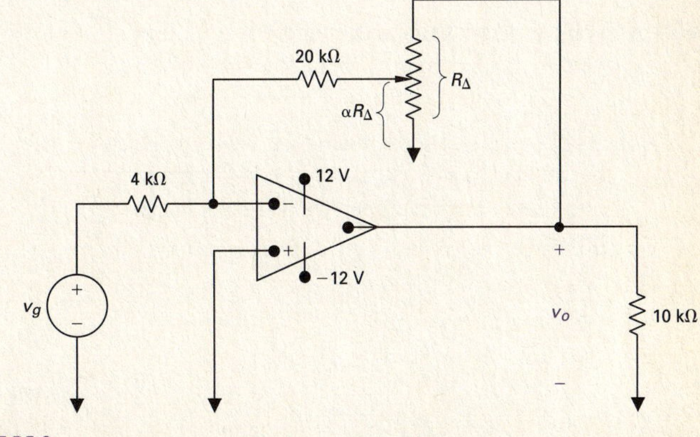

FIGURE P5.8

5.9 The operational amplifier in the circuit in Fig. P5.9 is ideal.

a) Find the range of values for σ in which the operational amplifier does not saturate.

b) Find i_o (in microamperes) when $\sigma = 0.272$.

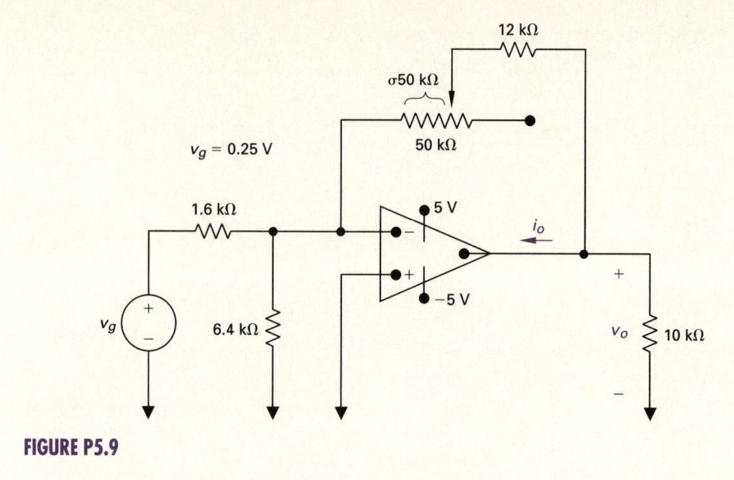

FIGURE P5.9

5.10 The operational amplifier in Fig. P5.10 is ideal.

a) Find v_o if $v_a = 1.2$ V, $v_b = -1.5$ V, and $v_c = 4$ V.

b) The voltages v_a and v_c remain at 1.2 V and 4 V, respectively. What are the limits on v_b if the operational amplifier operates within its linear region?

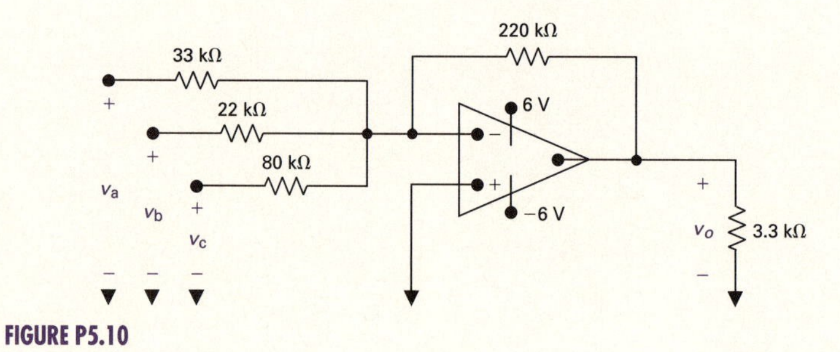

FIGURE P5.10

5.11 a) The operational amplifier in Fig. P5.11 is ideal. Find v_o if $v_a = 16$ V, $v_b = 12$ V, $v_c = -6$ V, and $v_d = 10$ V.

b) Assume v_a, v_c, and v_d retain their values as given in part (a). Specify the range of v_b such that the operational amplifier operates within its linear region.

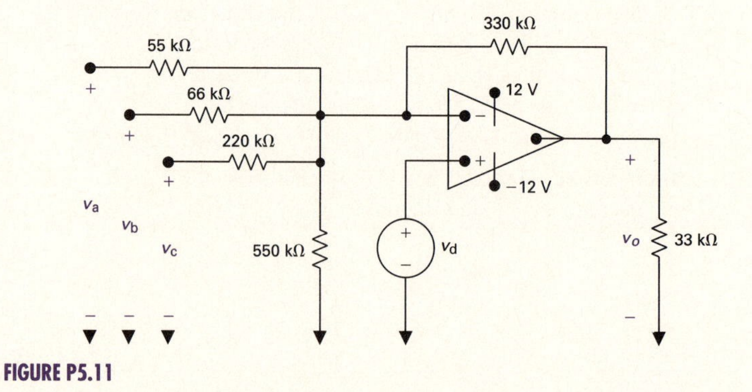

FIGURE P5.11

5.12 The 330 kΩ feedback resistor in the circuit in Fig. P5.11 is replaced by a variable resistor R_f. The voltages v_a through v_d have the same values as given in Problem 5.11(a).

a) What value of R_f will cause the operational amplifier to saturate? Note that $0 \leq R_f \leq \infty$.

b) When R_f has the value found in part (a), what is the current (in microamperes) into the output terminal of the operational amplifier?

5.13 Find i_a in the circuit in Fig. P5.13.

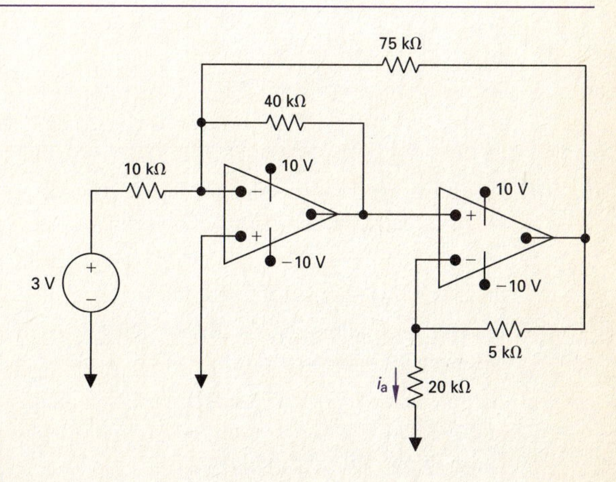

FIGURE P5.13

5.14 The operational amplifiers in the circuit in Fig. P5.14 are ideal. Find i_a.

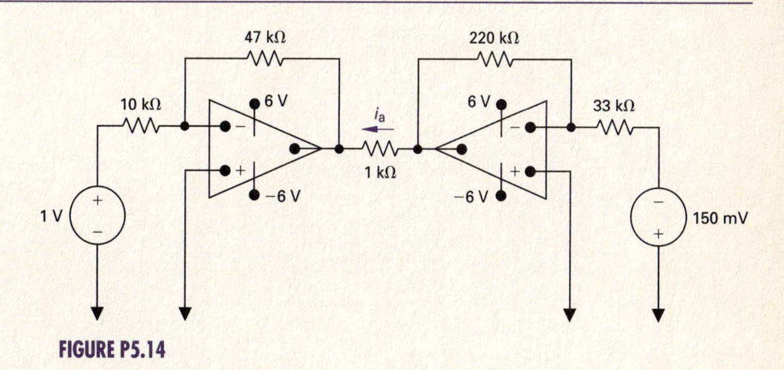

FIGURE P5.14

5.15 The variable resistor R_o in the circuit in Fig. P5.15 is adjusted until the source current i_g is zero. The operational amplifiers are ideal, and $0 \leq v_g \leq 1.2$ V.

a) What is the value of R_o?

b) If $v_g = 1.0$ V, how much power (in microwatts) is dissipated in R_o?

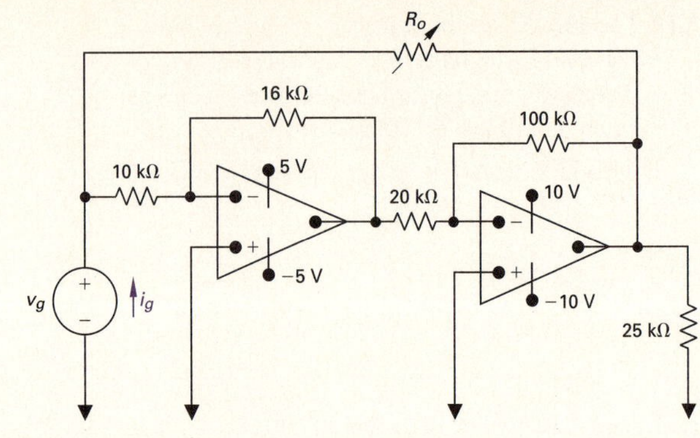

FIGURE P5.15

5.16 The circuit inside the shaded area in Fig. P5.16 is a constant current source for a limited range of values of R_L.

a) Find the value of i_L for $R_L = 1.8$ kΩ.

b) Find the maximum value for R_L for which i_L will have the value in part (a).

c) Assume that $R_L = 4.8$ kΩ. Explain the operation of the circuit. You can assume that $i_n = i_p \approx 0$ under all operating conditions.

d) Sketch i_L versus R_L for $0 \leq R_L \leq 4.8$ kΩ.

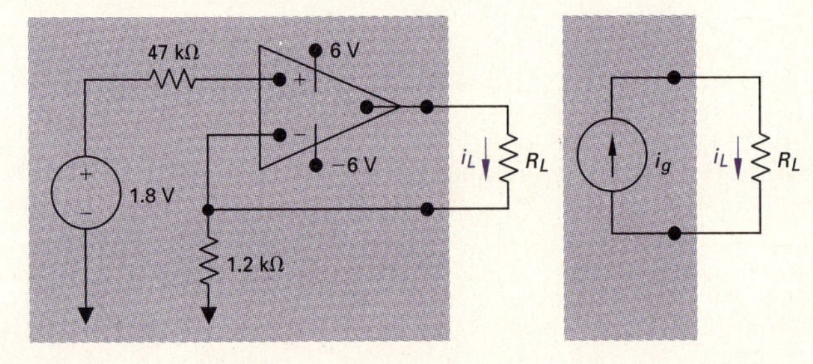

FIGURE P5.16

5.17 a) Show that when the ideal operational amplifier in Fig. P5.17 is operating in its linear region,

$$i_a = \frac{3v_g}{R}.$$

b) Show that the ideal operational amplifier will saturate when

$$R_a = \frac{R(\pm V_{CC} - 2v_g)}{3v_g}.$$

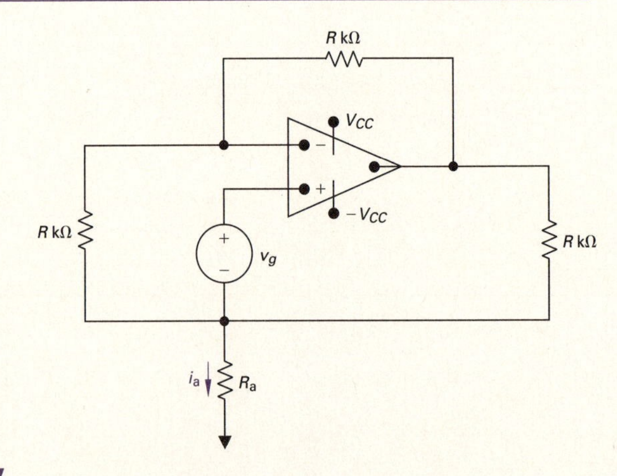

FIGURE P5.17

5.18 Refer to the circuit in Fig. 5.11, where the operational amplifier is assumed to be ideal. Given that $R_a = 4$ kΩ, $R_b = 5$ kΩ, $R_c = 20$ kΩ, $v_a = 200$ mV, $v_b = 150$ mV, $v_c = 400$ mV, and $V_{CC} = \pm 6$ V, specify the range of R_f for which the operational amplifier operates within its linear region.

❖ **5.19** Design an inverting summing amplifier so that

$$v_o = -(3v_a + 5v_b + 4v_c + 2v_d).$$

If the feedback resistor (R_f) is chosen to be 60 kΩ, draw a circuit diagram of the amplifier and specify the values of R_a, R_b, R_c, and R_d.

5.20 The operational amplifier in the circuit shown in Fig. P5.20 is ideal.

a) Calculate v_o when v_g equals 4 V.

b) Specify the range of values of v_g so that the operational amplifier operates in a linear mode.

c) Assume that v_g equals 2 V and that the 63 kΩ resistor is replaced with a variable resistor. What value of the variable resistor will cause the operational amplifier to saturate?

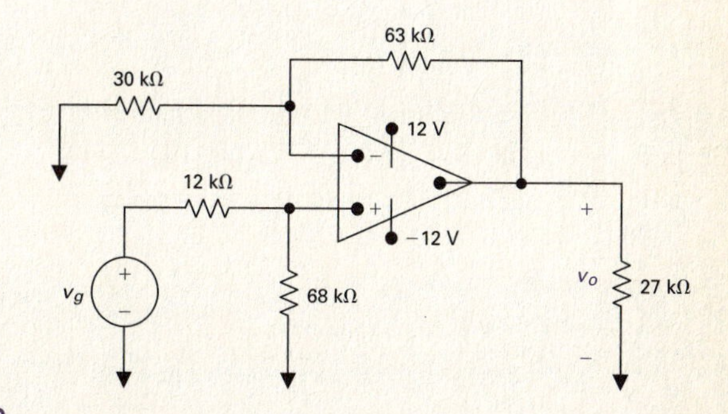

FIGURE P5.20

5.21 Assume that the ideal op amp in the circuit seen in Fig. P5.21 is operating in its linear region.

a) Show that $v_o = [(R_1 + R_2)/R_1]v_s$.

b) What happens if $R_1 \to \infty$ and $R_2 \to 0$?

c) Explain why this circuit is referred to as a voltage follower when $R_1 = \infty$ and $R_2 = 0$.

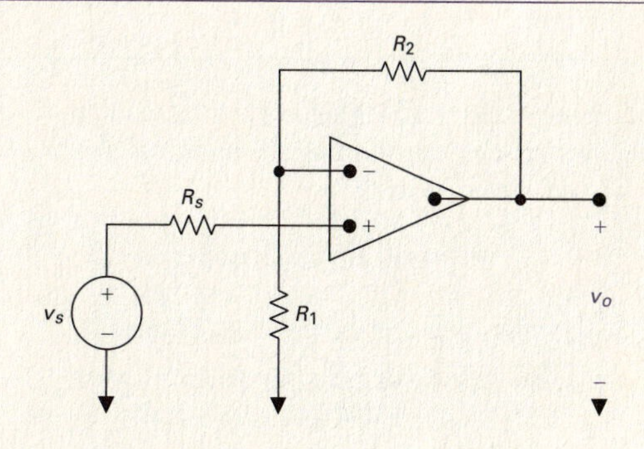

FIGURE P5.21

5.22 Assume that the ideal op amp in the circuit in Fig. P5.22 is operating in its linear region.

a) Calculate the power delivered to the 12 kΩ resistor.

b) Repeat part (a) with the operational amplifier removed from the circuit, that is, with the 12 kΩ resistor connected in the series with the voltage source and the 68 kΩ resistor.

c) Find the ratio of the power found in part (a) to that found in part (b).

d) Does the insertion of the op amp between the source and the load serve a useful purpose? Explain.

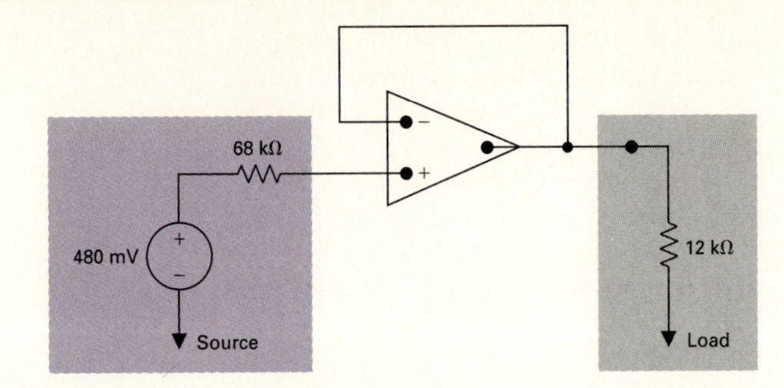

FIGURE P5.22

5.23 The op amp in the noninverting amplifier shown in Fig. P5.23 is ideal. The signal voltages v_a and v_b are 400 mV and 1200 mV, respectively.

a) Calculate v_o in volts.

b) Find i_a and i_b in microamperes.

c) What are the weighting factors associated with v_a and v_b?

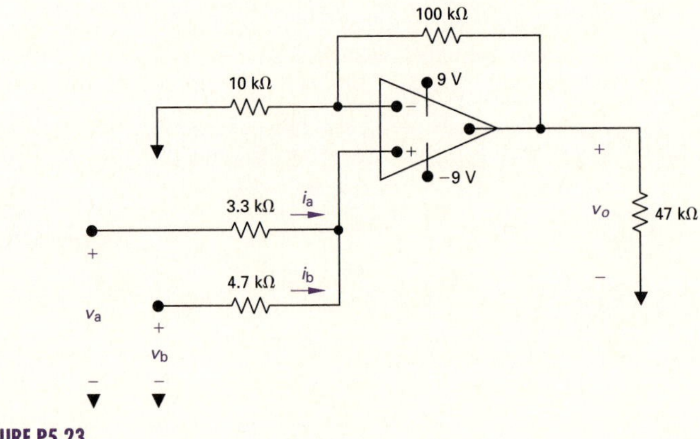

FIGURE P5.23

❖ **5.24** The circuit in Fig. P5.24 is a noninverting summing amplifier. Assume the operational amplifier is ideal. Design the circuit so that

$$v_o = 4v_a + v_b + 2v_c.$$

a) Specify the numerical values of R_b, R_c, and R_f.

b) Calculate (in microamperes) i_a, i_b, and i_c when $v_a = 0.75$ V, $v_b = 1.0$ V, and $v_c = 1.5$ V.

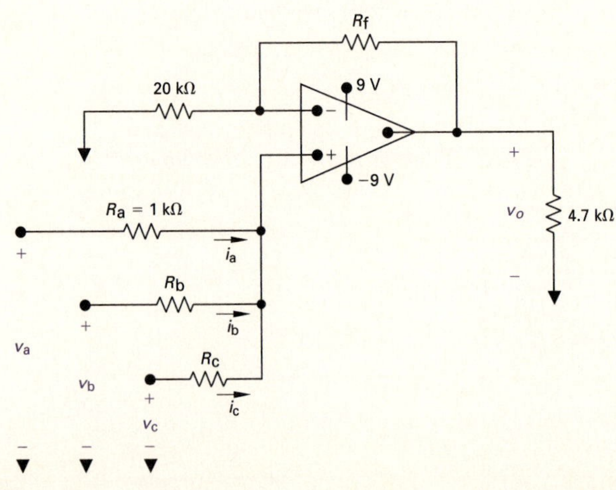

FIGURE P5.24

5.25 The operational amplifier in the noninverting summing amplifier of Fig. P5.25 is ideal.

a) Specify the values of R_f, R_b, and R_c so that

$$v_o = 3v_a + 2v_b + v_c.$$

b) Find (in microamperes) i_a, i_b, i_c, i_g, and i_s when $v_a = 0.80$ V, $v_b = 1.5$ V, and $v_c = 2.1$ V.

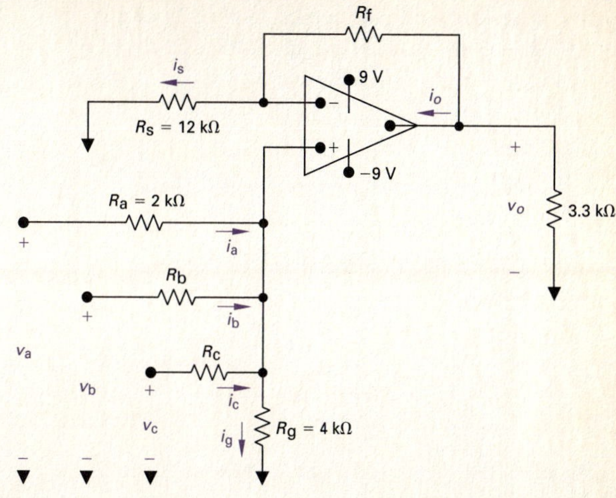

FIGURE P5.25

5.26 The operational amplifier in the circuit of Fig. P5.26 is ideal. Plot v_o versus α when $R_f = 4R_1$ and $v_g = 2$ V. Use increments of 0.1 and note by hypothesis that $0 \le \alpha \le 1.0$.

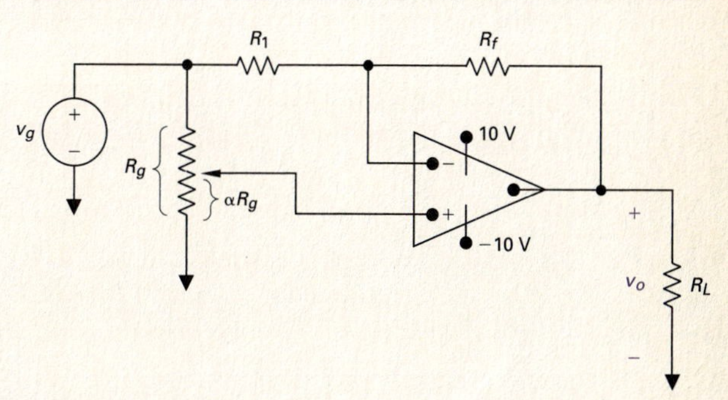

FIGURE P5.26

❖ **5.27** Design the difference-amplifier circuit in Fig. P5.27 so that $v_o = 5(v_b - v_a)$ and the voltage source v_b sees an input resistance of 1.2 MΩ. Specify the values of R_a, R_b, and R_f. Use the ideal model for the operational amplifier.

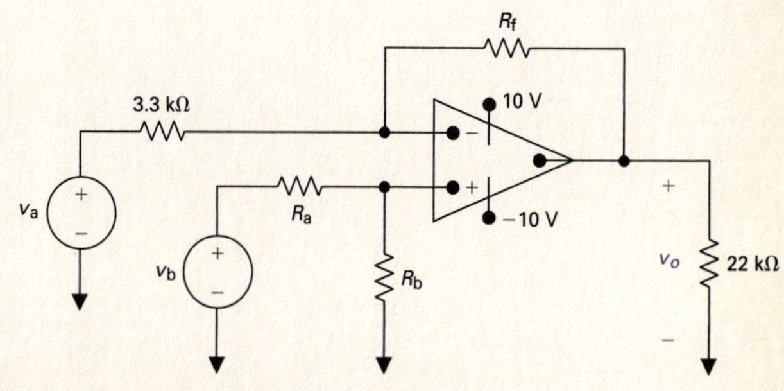

FIGURE P5.27

5.28 The resistors in the difference amplifier shown in Fig. 5.13 are $R_a = 10$ kΩ, $R_b = 100$ kΩ, $R_c = 33$ kΩ, and $R_d = 47$ kΩ. The signal voltages v_a and v_b are 0.67 and 0.8 V, respectively, and $V_{CC} = \pm 5$ V.

a) Find v_o.

b) What is the resistance seen by the signal source v_a?

c) What is the resistance seen by the signal source v_b?

❖ **5.29** Design a difference amplifier (Fig. 5.13) to meet the following criteria: $v_o = 3v_b - 4v_a$; the resistance seen by the signal source v_b is 470 kΩ; and the resistance seen by the signal source v_a is 22 kΩ when the output voltage v_o is zero. Specify the values of R_a, R_b, R_c, and R_d.

❖ **5.30** Select the values of R_b and R_f in the circuit in Fig. P5.30 so that

$$v_o = 4000(i_b - i_a).$$

The operational amplifier is ideal.

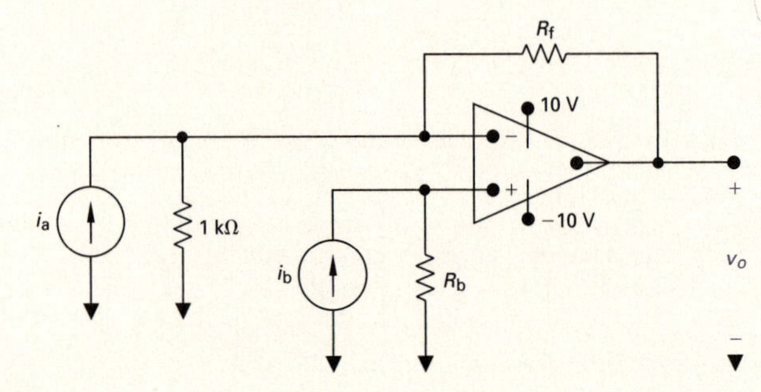

FIGURE P5.30

5.31 The operational amplifier in the adder-subtracter circuit shown in Fig. P5.31 is ideal.

a) Find v_o when $v_a = 0.4$ V, $v_b = 0.8$ V, $v_c = 0.2$ V, and $v_d = 0.6$ V.

b) If v_a, v_c, and v_d are held constant, what values of v_b will not saturate the op amp?

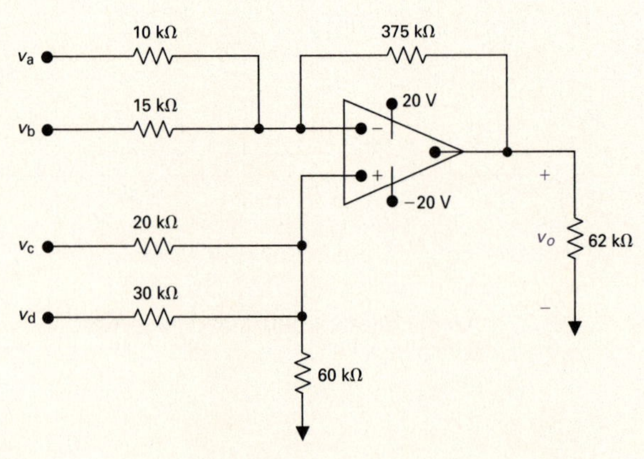

FIGURE P5.31

5.32 The inverting amplifier in the circuit in Fig. P5.32 has an input resistance of 500 kΩ, an output resistance of 5 kΩ, and an open-loop gain of 250,000. Assume that the amplifier is operating in its linear region.

a) Calculate the voltage gain (v_o/v_g) of the amplifier.

b) Calculate the value of v_n in microvolts when $v_g = 100$ mV.

c) Calculate the resistance seen by the signal source (v_g).

d) Repeat parts (a), (b), and (c) using the ideal model for the op amp.

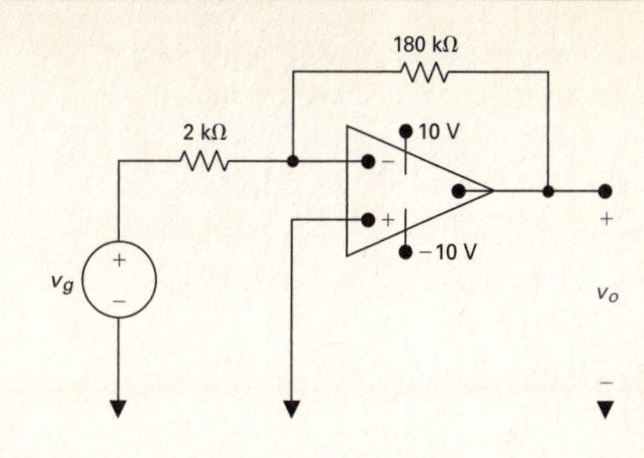

FIGURE P5.32

5.33 Repeat Problem 5.32 given that the inverting amplifier is loaded with a 1600 Ω resistor.

5.34 The operational amplifier in the noninverting amplifier circuit of Fig. P5.34 has an input resistance of 560 kΩ, an output resistance of 8 kΩ, and an open-loop gain of 50,000. Assume that the op amp is operating in its linear region.

a) Calculate the voltage gain (v_o/v_g).

b) Find the inverting and noninverting input voltages v_n and v_p (in millivolts) if $v_g = 1$ V.

c) Calculate the difference ($v_p - v_n$) in microvolts when $v_g = 1$ V.

d) Find the current drain in picoamperes on the signal source v_g when $v_g = 1$ V.

e) Repeat parts (a) through (d) assuming an ideal op amp.

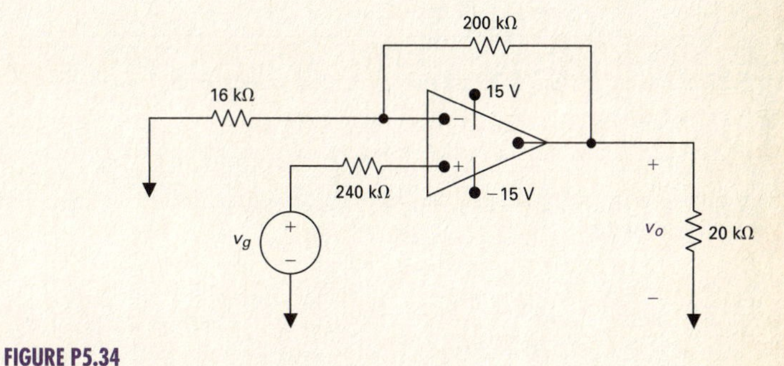

FIGURE P5.34

5.35 Assume the input resistance of the operational amplifier in Fig. P5.35 is infinite and its output resistance is zero.

a) Find v_o as a function of v_g and the open-loop gain A.

b) What is the value of v_o if $v_g = 0.4$ V and $A = 90$?

c) What is the value of v_o if $v_g = 0.4$ V and $A = \infty$?

d) How large does A have to be so that v_o is 95% of its value in part (c)?

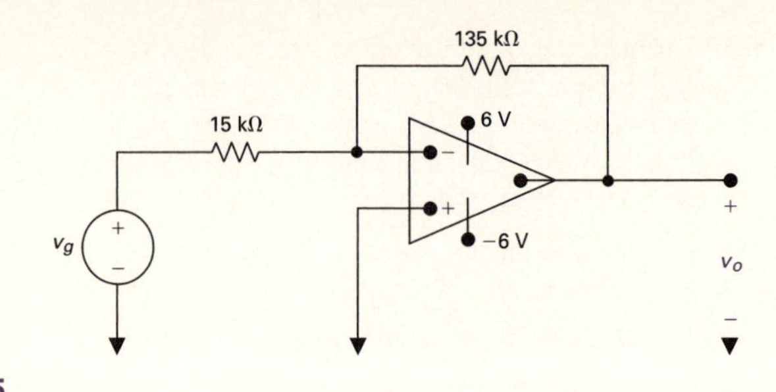

FIGURE P5.35

5.36 a) Find the Thévenin equivalent circuit with respect to the output terminals a,b for the inverting amplifier of Fig. P5.36. The dc signal source has a value of 200 mV. The operational amplifier has an input resistance of 400 kΩ, an output resistance of 800 Ω, and an open-loop gain of 10,000.

b) What is the output resistance of the inverting amplifier?

c) What is the resistance (in ohms) seen by the signal source v_s when the load at the terminals a,b is 220 Ω?

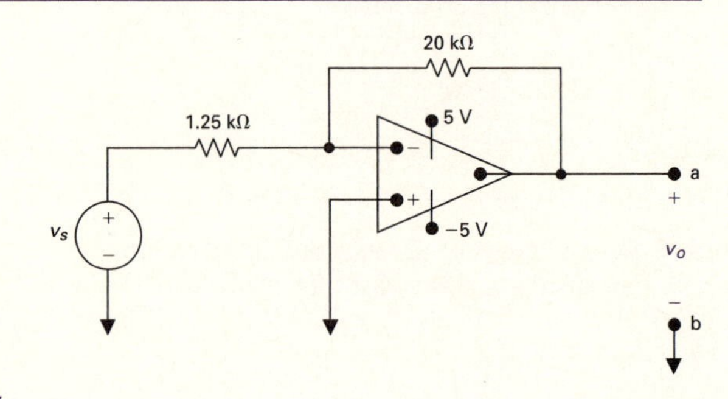

FIGURE P5.36

5.37 Repeat Problem 5.36 assuming an ideal operational amplifier.

5.38 The two operational amplifiers in the circuit in Fig. P5.38 are ideal. Calculate v_{o1} and v_{o2}.

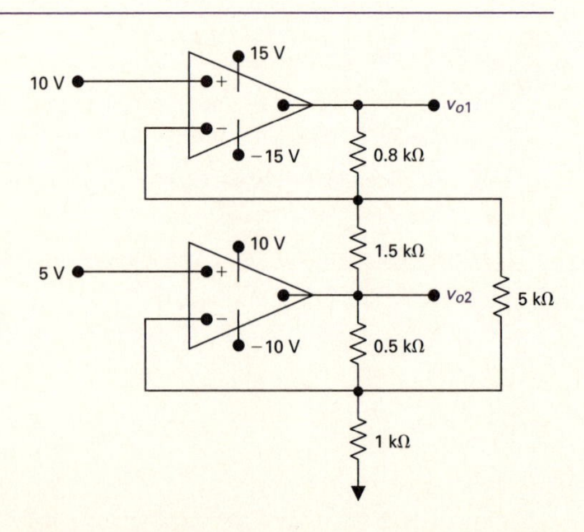

FIGURE P5.38

5.39 The resistor R_f in the circuit in Fig. P5.39 is adjusted until the ideal operational amplifier saturates. Specify R_f in kilohms.

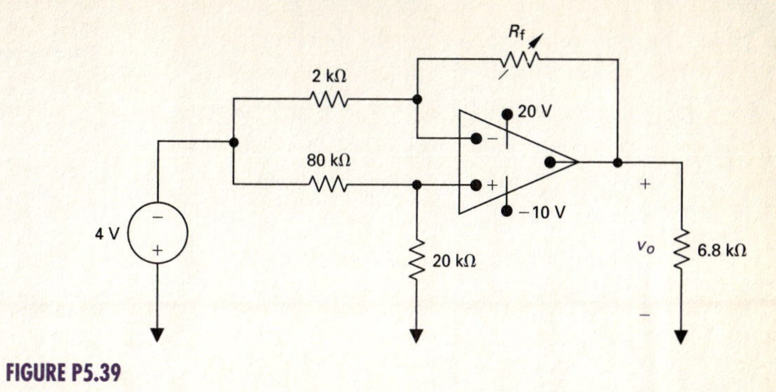

FIGURE P5.39

5.40 The operational amplifiers in the circuit in Fig. P5.40 are ideal. Find v_x, i_a, and i_o.

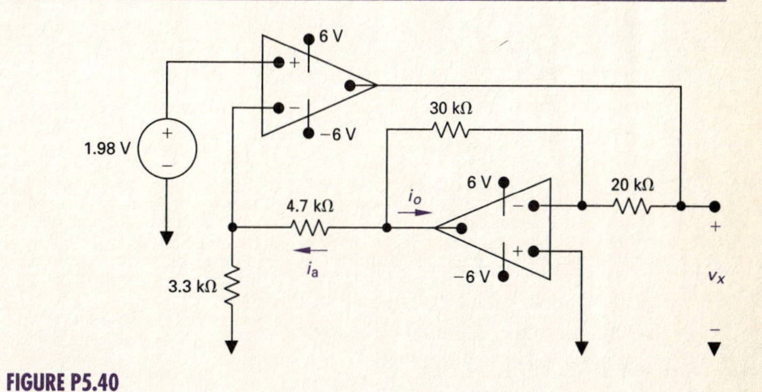

FIGURE P5.40

5.41 Find v_o amd i_o in the circuit shown in Fig. P5.41, assuming the operational amplifiers are ideal.

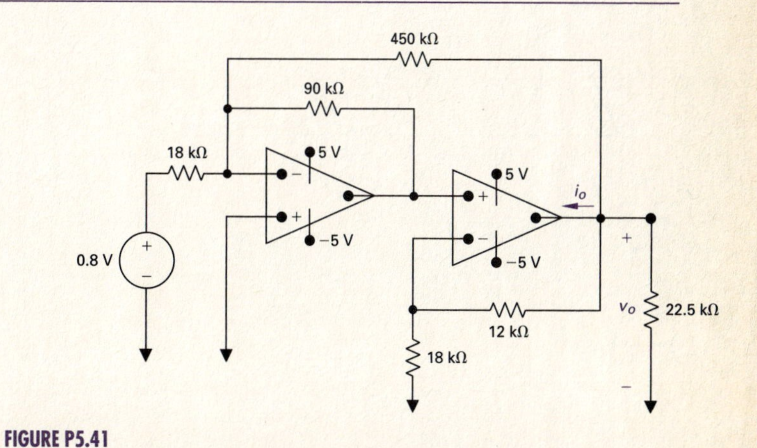

FIGURE P5.41

5.42 The operational amplifiers in the circuit shown in Fig. P5.42 are ideal.

a) Find v_o as a function of α, σ, v_{g1}, and v_{g2} when the op amps operate within their linear regions.

b) Describe the behavior of the circuit when $\alpha = \sigma = 1.0$.

c) Describe the behavior of the circuit when $\alpha = \sigma = 0$.

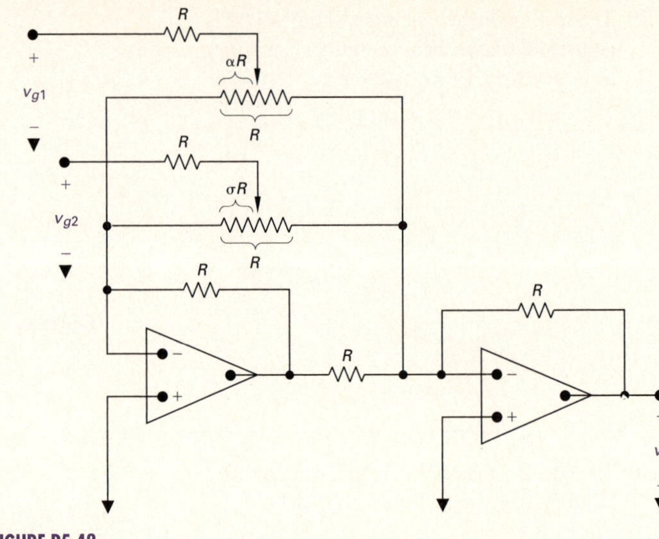

FIGURE P5.42

5.43 The voltage v_g shown in Fig. P5.43(a) is applied to the inverting amplifier shown in Fig. P5.43(b). Sketch v_o versus t, assuming the operational amplifier is ideal.

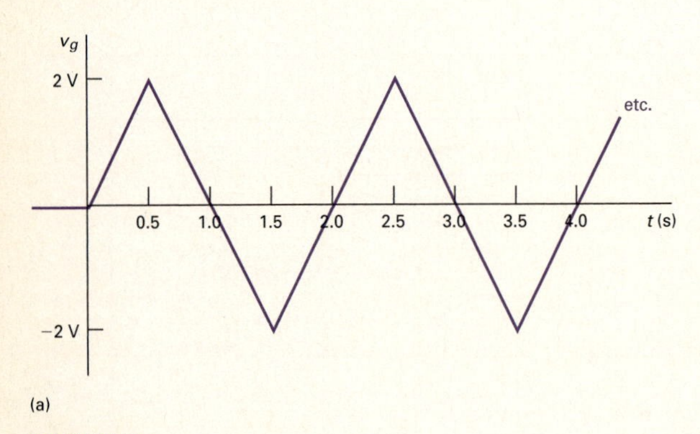

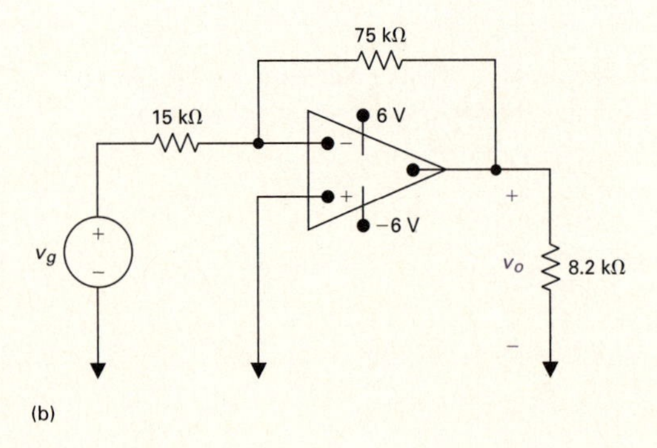

FIGURE P5.43

5.44 The signal voltage v_g in the circuit shown in Fig. P5.44 is described by the following equations:

$$v_g = 0, \quad t \leq 0,$$

$$v_g = 10\sin(\pi/3)t \text{ V}, \quad 0 \leq t \leq \infty.$$

Sketch v_o versus t, assuming the operational amplifier is ideal.

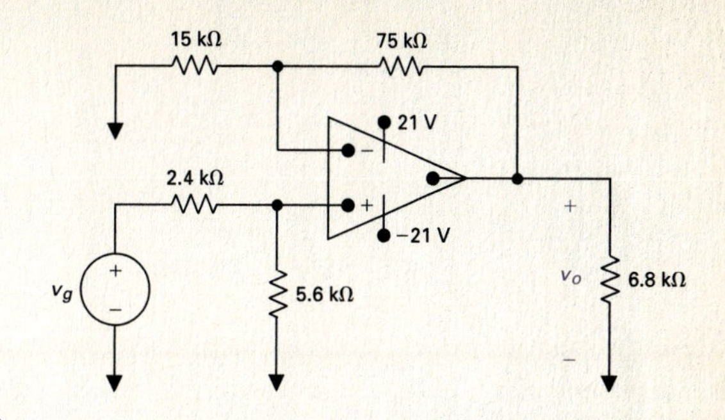

FIGURE P5.44

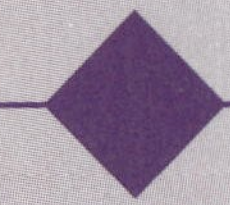

PRACTICAL PERSPECTIVE

Proximity Switches

The electrical devices we use in our daily lives contain many switches. Most switches are mechanical, such as the one used in the flashlight introduced in Chapter 2. Mechanical switches use an actuator which is pushed, pulled, slid, or rotated, causing two pieces of conducting metal to touch and create a short circuit. Sometimes designers prefer to use switches without moving parts, to increase the safety, reliability, convenience, or novelty of their products. Such switches are called proximity switches. Proximity switches can employ a variety of sensor technologies. For example, some elevator doors stay open whenever a light beam is obstructed.

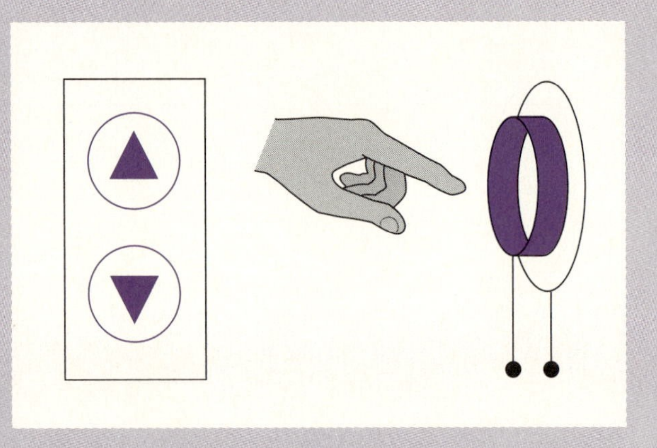

Another sensor technology used in proximity switches detects people by responding to the disruption they cause in electric fields. This type of proximity switch is used in some desk lamps which turn on and off when touched, and in elevator buttons with no moving parts (as shown in the figure). The switch is based on a capacitor. As you are about to discover in this chapter, a capacitor is a circuit element whose terminal characteristics are determined by electric fields. When you touch a capacitive proximity switch, you produce a change in the value of a capacitor, causing a voltage change which activates the switch. The design of a capacitive touch-sensitive switch is the topic of the Practical Perspective example at the end of this chapter.

6 INDUCTORS AND CAPACITORS

CHAPTER CONTENTS

We now introduce the last two ideal circuit elements mentioned in Chapter 2, namely, inductors and capacitors. Be assured that the circuit analysis techniques introduced in Chapters 3 and 4 apply to circuits containing inductors and capacitors. Therefore, once you understand the terminal behavior of these elements in terms of current and voltage, you can use Kirchhoff's laws to describe any interconnections with the other basic elements. Like other components, inductors and capacitors are easier to describe in terms of circuit variables rather than electromagnetic field variables. However, before we focus on the circuit descriptions, a brief review of the field concepts underlying these basic elements is in order.

The behavior of inductors is based on phenomena associated with magnetic fields. The source of the magnetic field is charge in motion, or current. If the current is varying with time, the magnetic field is varying with time. A time-varying magnetic field induces a voltage in any conductor linked by the field. The circuit parameter of inductance relates the induced voltage to the current. We discuss this quantitative relationship in Section 6.1.

The behavior of capacitors is based on phenomena associated with electric fields. The source of the electric field is separation of charge, or voltage. If the voltage is varying with time, the electric field is varying with time. A time-varying electric field produces a displacement current in the space occupied by the field. The circuit parameter of capacitance relates the displacement current to the voltage, where the displacement current is equal to the conduction current at the terminals of the capacitor. We discuss this quantitative relationship in Section 6.2. Section 6.3 describes techniques used to simplify circuits with series or parallel combinations of capacitors or inductors.

Energy can be stored in both magnetic and electric fields. Hence you should not be too surprised to learn that inductors and capacitors are capable of storing energy. For example, energy can be stored in an inductor and then released to fire a spark plug. Energy can be stored in a capacitor and then released to fire a flashbulb. In ideal inductors and capacitors, only as much energy can be extracted as has been stored. Because inductors and capacitors cannot generate energy, they are classified as **passive elements**.

6.1 The Inductor

Inductance is the circuit parameter used to describe an inductor. Inductance is symbolized by the letter L, is measured in henrys (H), and is represented graphically as a coiled wire—a reminder that inductance is a consequence of a conductor linking a magnetic field. Figure 6.1(a) shows an inductor. Assigning the reference direction of the current in the direction of the voltage drop across the terminals of the inductor, as shown in Fig. 6.1(b), yields

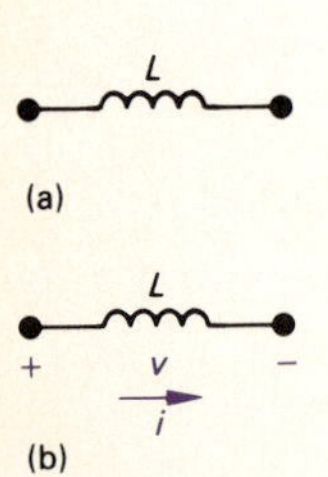

FIGURE 6.1 The graphic symbol for an inductor with an inductance of L henrys.

$$v = L\frac{di}{dt}, \tag{6.1}$$

where v is measured in volts, L in henrys, i in amperes, and t in seconds. Equation (6.1) reflects the passive sign convention shown in Fig. 6.1(b); that is, the current reference is in the direction of the voltage drop across the inductor. If the current reference is in the direction of the voltage rise, Eq. (6.1) is written with a minus sign.

Note from Eq. (6.1) that the voltage across the terminals of an inductor is proportional to the time rate of change of the current in the inductor. We can make two important observations here. First, if the current is constant, the voltage across the ideal inductor is zero. Thus the inductor behaves as a short circuit in the presence of a constant, or dc, current. Second, current cannot change instantaneously in an inductor; that is, the current cannot change by a finite amount in zero time. Equation (6.1) tells us that this change would require an infinite voltage, and infinite voltages are not possible. For example, when someone opens the switch on an inductive circuit in an actual system, the current initially continues to flow in the air across the switch, a phenomenon called *arcing*. The arc across the switch prevents the current from dropping to zero instantaneously. Switching inductive circuits is an important engineering problem because arcing and voltage surges must be controlled to prevent equipment damage. The first step to understanding the nature of this problem is to master the introductory material presented in this and the following two chapters. Example 6.1 illustrates the application of Eq. (6.1) to a simple circuit.

EXAMPLE 6.1

The independent current source in the circuit shown in Fig. 6.2 generates zero current for $t < 0$ and a pulse $10te^{-5t}$ for $t > 0$.

a) Sketch the current waveform.

b) At what instant of time is the current maximum?

c) Express the voltage across the terminals of the 100 mH inductor as a function of time.

d) Sketch the voltage waveform.

e) Is the voltage maximum when the current is maximum?

f) At what instant of time does the voltage change polarity?

g) Is there ever an instantaneous change in voltage across the inductor? If so, at what time?

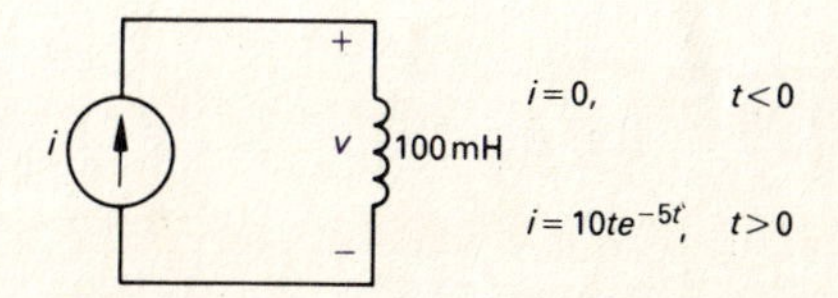

FIGURE 6.2 The circuit for Example 6.1.

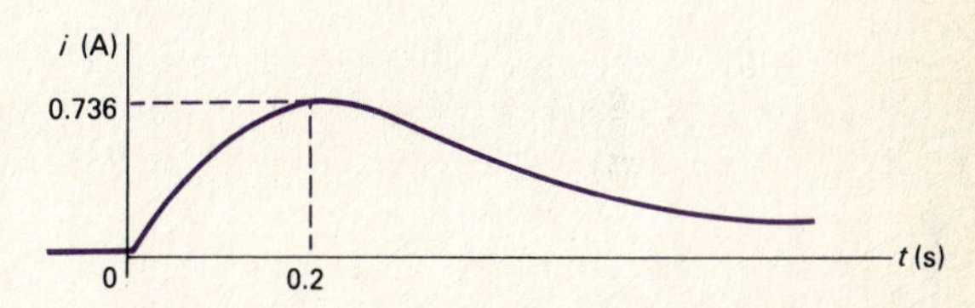

FIGURE 6.3 The current waveform for Example 6.1.

SOLUTION

a) Figure 6.3 shows the current waveform.

b) $di/dt = 10(-5te^{-5t} + e^{-5t}) = 10e^{-5t}(1 - 5t)$; $di/dt = 0$ when $t = \frac{1}{5}$ s. (See Fig. 6.3.)

c) $v = Ldi/dt = (0.1)10e^{-5t}(1 - 5t) = e^{-5t}(1 - 5t)$ V, $t > 0$; $v = 0, t < 0$.

d) Figure 6.4 shows the voltage waveform.

e) No; the voltage is proportional to di/dt, not i.

Use a computer tool to confirm your sketch.

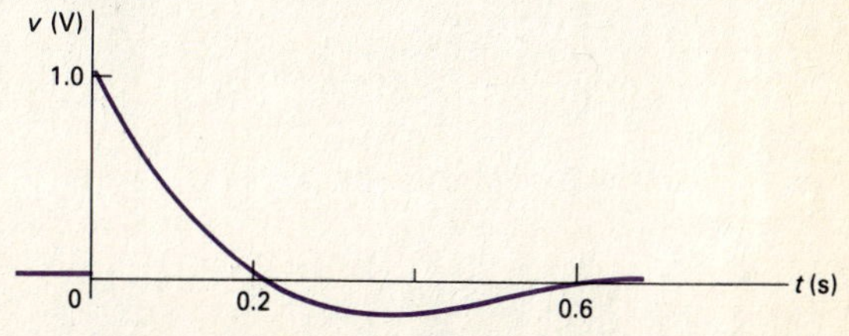

FIGURE 6.4 The voltage waveform for Example 6.1.

f) At 0.2 s, which corresponds to the moment when di/dt is passing through zero and changing sign.

g) Yes, at $t = 0$. Note that the voltage can change instantaneously across the terminals of an inductor.

Current in an Inductor in Terms of the Voltage across the Inductor

Equation (6.1) expresses the voltage across the terminals of an inductor as a function of the current in the inductor. Also desirable is the ability to express the current as a function of the voltage. To find i as a function of v, we start by multiplying both sides of Eq. (6.1) by a differential time dt:

$$v\,dt = L\left(\frac{di}{dt}\right)dt. \tag{6.2}$$

Multiplying the rate at which i varies with t by a differential change in time generates a differential change in i, so we write Eq. (6.2) as

$$v\,dt = L\,di. \tag{6.3}$$

We next integrate both sides of Eq. (6.3). For convenience, we interchange the two sides of the equation and write

$$L\int_{i(t_0)}^{i(t)} dx = \int_{t_0}^{t} v\,d\tau. \tag{6.4}$$

Note that we use x and τ as the variables of integration, while i and t become limits on the integrals. Then from Eq. (6.4),

$$i(t) = \frac{1}{L}\int_{t_0}^{t} v\,d\tau + i(t_0), \tag{6.5}$$

where $i(t)$ is the current corresponding to t, and $i(t_0)$ is the value of the inductor current when we initiate the integration, namely, t_0. In many practical applications, t_0 is zero and Eq. (6.5) becomes

$$i(t) = \frac{1}{L}\int_{0}^{t} v\,d\tau + i(0). \tag{6.6}$$

Use a symbolic equation solver to confirm this result.

Equations (6.1) and (6.5) both give the relationship between the voltage and current at the terminals of an inductor. Equation (6.1) expresses the voltage as a function of current, whereas Eq. (6.5) expresses the current as a function of voltage. In both equations the reference direction for the current is in the direction of the voltage drop across the terminals. Note that $i(t_0)$ carries its own algebraic sign. If the initial current is in the same direction as the reference direction for i, it is a positive quantity. If the initial current is in the opposite direction, it is a negative quantity. Example 6.2 illustrates the application of Eq. (6.5).

EXAMPLE 6.2

The voltage pulse applied to the 100 mH inductor shown in Fig. 6.5 is 0 for $t < 0$ and is given by the expression

$$v(t) = 20te^{-10t}$$

for $t > 0$. Also assume $i = 0$ for $t \leq 0$.

a) Sketch the voltage as a function of time.

b) Find the inductor current as a function of time.

c) Sketch the current as a function of time.

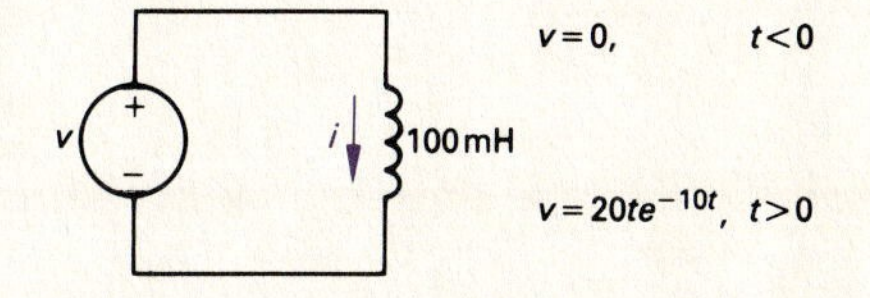

FIGURE 6.5 The circuit for Example 6.2.

SOLUTION

a) The voltage as a function of time is shown in Fig. 6.6.

b) The current in the inductor is 0 at $t = 0$. Therefore the current for $t > 0$ is

$$\begin{aligned} i &= \frac{1}{0.1}\int_0^t 20\tau e^{-10\tau} d\tau + 0 \\ &= 200\left[\frac{-e^{-10\tau}}{100}(10\tau + 1)\right]\Bigg|_0^t \\ &= 2(1 - 10te^{-10t} - e^{-10t})\text{A}, \quad t > 0. \end{aligned}$$

c) Figure 6.7 shows the current as a function of time.

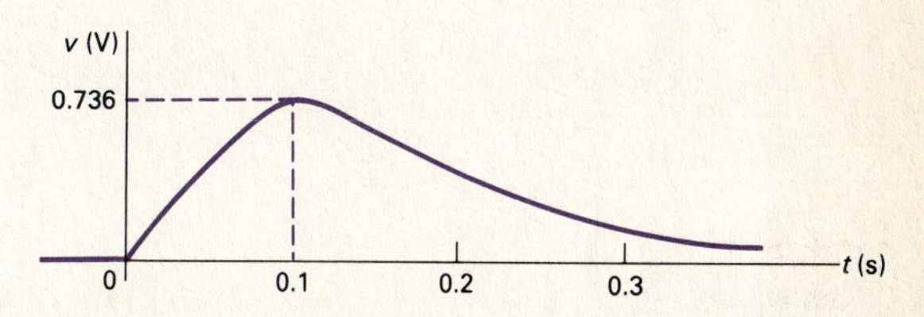

FIGURE 6.6 The voltage waveform for Example 6.2.

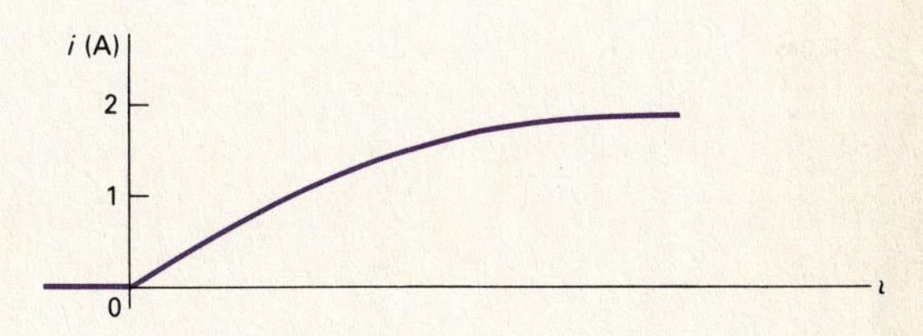

FIGURE 6.7 The current waveform for Example 6.2.

Note in Example 6.2 that i approaches a constant value of 2 A as t increases. We say more about this result after discussing the energy stored in an inductor.

POWER AND ENERGY IN THE INDUCTOR

The power and energy relationships for an inductor can be derived directly from the current and voltage relationships. If the current reference is in the direction of the voltage drop across the terminals of the inductor, the power is

$$p = vi. \tag{6.7}$$

Remember that power is in watts, voltage is in volts, and current is in amperes. If we express the inductor voltage as a function of the inductor current, Eq. (6.7) becomes

$$p = Li\frac{di}{dt}. \tag{6.8}$$

We can also express the current in terms of the voltage:

$$p = v\left[\frac{1}{L}\int_{t_0}^{t} v\,d\tau + i(t_0)\right]. \tag{6.9}$$

Equation (6.8) is most useful in expressing the energy stored in the inductor. Power is the time rate of expending energy, so

$$p = \frac{dw}{dt} = Li\frac{di}{dt}. \tag{6.10}$$

Multiplying both sides of Eq. (6.10) by a differential time gives the differential relationship

$$dw = Li\,di. \tag{6.11}$$

Both sides of Eq. (6.11) are integrated with the understanding that the reference for zero energy corresponds to zero current in the inductor. Thus

$$\int_0^w dx = L\int_0^i y\,dy;$$

$$w = \frac{1}{2}Li^2. \tag{6.12}$$

As before, we use different symbols of integration to avoid confusion with the limits placed on the integrals. In Eq. (6.12), the energy is in joules when inductance is in henrys and current is in amperes. To illustrate the application of Eqs. (6.7) and (6.12), we return to Examples 6.1 and 6.2 by means of Example 6.3.

EXAMPLE 6.3

a) For Example 6.1, plot i, v, p, and w versus time. Line up the plots vertically to allow easy assessment of each variable's behavior.

b) In what time interval is energy being stored in the inductor?

c) In what time interval is energy being extracted from the inductor?

d) What is the maximum energy stored in the inductor?

e) Evaluate the integrals

$$\int_0^{0.2} p\,dt \quad \text{and} \quad \int_{0.2}^{\infty} p\,dt$$

and comment on their significance.

f) Repeat parts (a)–(c) for Example 6.2.

g) In Example 6.2, why is there a sustained current in the inductor as the voltage approaches zero?

SOLUTION

a) The plots of i, v, p, and w follow directly from the expressions for i and v obtained in Example 6.1 and are shown in Fig. 6.8. In particular, $p = vi$ and $w = (\frac{1}{2})Li^2$.

b) An increasing energy curve indicates that energy is being stored. Thus energy is being stored in the time interval 0 to 0.2 s. Note that this corresponds to the interval when $p > 0$.

c) A decreasing energy curve indicates that energy is being extracted. Thus energy is being extracted in the time interval 0.2 s to ∞. Note that this corresponds to the interval when $p < 0$.

d) From Eq. (6.12) we see that energy is at a maximum when current is at a maximum; glancing at the graphs confirms this. From Example 6.1, maximum current = 0.736 A. Therefore, $w_{max} = 27.07$ mJ.

e) From Example 6.1,

$$i = 10te^{-5t} \text{ A} \quad \text{and} \quad v = e^{-5t}(1 - 5t) \text{ V}.$$

Therefore

$$p = vi = 10te^{-10t} - 50t^2e^{-10t} \text{ W}.$$

Thus

$$\int_0^{0.2} p\,dt = 10\left[\frac{e^{-10t}}{100}(-10t - 1)\right]_0^{0.2}$$

$$-50\left\{\frac{t^2e^{-10t}}{-10} + \frac{2}{10}\left[\frac{e^{-10t}}{100}(-10t - 1)\right]\right\}_0^{0.2}$$

$$= 0.2e^{-2} = 27.07 \text{ mJ};$$

$$\int_{0.2}^{\infty} p\,dt = 10\left[\frac{e^{-10t}}{100}(-10t - 1)\right]_{0.2}^{\infty}$$

$$-50\left\{\frac{t^2e^{-10t}}{-10} + \frac{2}{10}\left[\frac{e^{-10t}}{100}(-10t - 1)\right]\right\}_{0.2}^{\infty}$$

$$= -0.2e^{-2} = -27.07 \text{ mJ}.$$

Based on the definition of p, the area under the plot of p versus t represents the energy expended over the interval of integration. Hence the integration of the power between 0 and 0.2 s represents

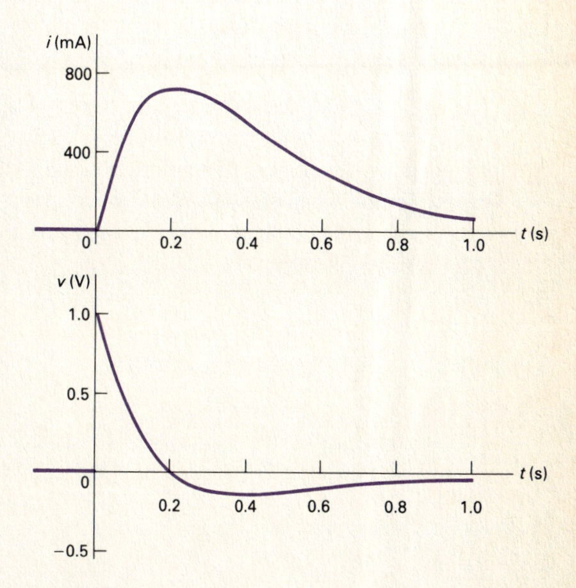

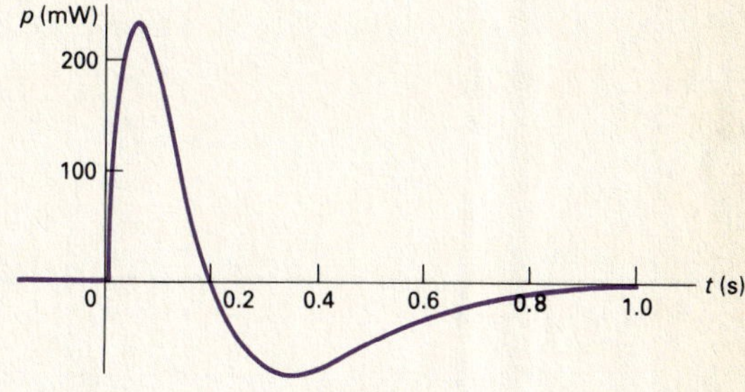

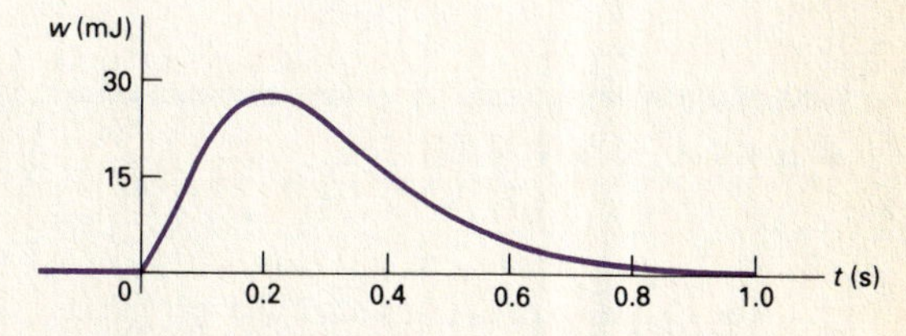

FIGURE 6.8 The variables i, v, p, and w versus t for Example 6.1.

the energy stored in the inductor during this time interval. The integral of p over the interval 0.2 s to ∞ is the energy extracted. Note that in this time interval all the energy originally stored is removed; that is, after the current peak has passed, no energy is stored in the inductor.

f) The plots of v, i, p, and w follow directly from the expressions for v and i given in Example 6.2 and are shown in Fig. 6.9. Note that in this case the power always is positive, and hence energy always is being stored during the duration of the voltage pulse.

g) The application of the voltage pulse stores energy in the inductor. Because the inductor is ideal, this energy cannot dissipate after the voltage subsides to zero. Therefore a sustained current circulates in the circuit. A lossless inductor obviously is an ideal circuit element. Practical inductors require a resistor in the circuit model. (More about this later.)

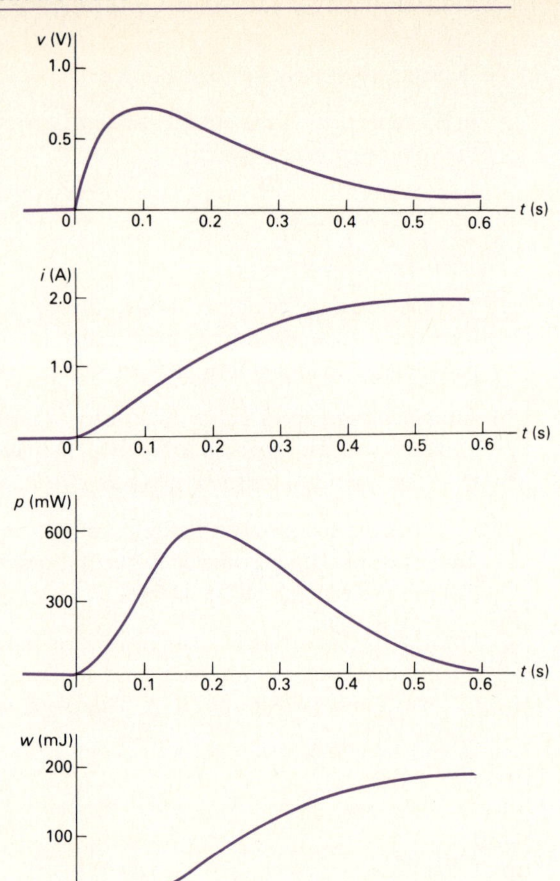

FIGURE 6.9 The variables v, i, p, and w versus t for Example 6.2.

DRILL EXERCISE

6.1 The current source in the circuit shown generates the current pulse

$$i_g(t) = 0, \quad t < 0;$$

$$i_g(t) = 5e^{-200t} - 5e^{-800t} \text{ A}, \quad t \geq 0.$$

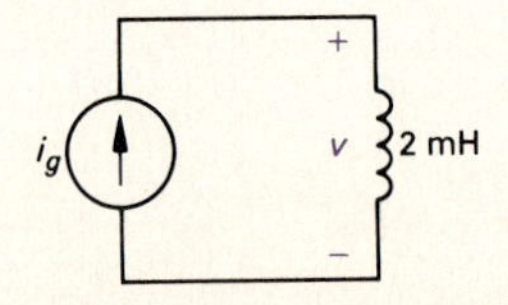

Find (a) $v(0)$; (b) the instant of time, greater than zero, when the voltage v passes through zero; (c) the expression for the power delivered to the inductor; (d) the instant when the power delivered to the inductor is maximum; (e) the maximum power; (f) the maximum energy stored in the inductor; and (g) the instant of time when the stored energy is maximum.

ANSWER: (a) 6 V; (b) 2.31 ms; (c) $50e^{-1000t} - 10e^{-400t} - 40e^{-1600t}$ W; (d) 616.58 μs; (e) 4.26 W; (f) 5.58 mJ; and (g) 2.31 ms.

6.2 The Capacitor

The circuit parameter of capacitance is represented by the letter C, is measured in farads (F), and is symbolized graphically by two short parallel conductive plates, as shown in Fig. 6.10(a). Because the farad is an extremely large quantity of capacitance, practical capacitors are based on submultiples of the farad. The most frequently encountered values lie in the picofarad (pF) to microfarad (μF) range.

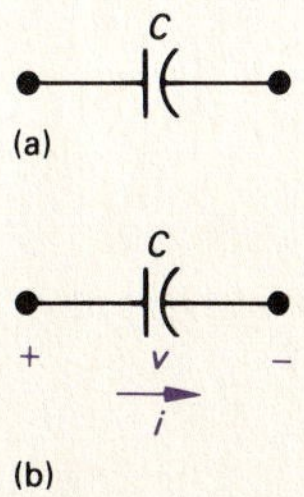

The graphic symbol for a capacitor is a reminder that capacitance occurs whenever electrical conductors are separated by a dielectric, or insulating, material. This condition implies that electric charge is not transported through the capacitor. Although applying a voltage to the terminals of the capacitor cannot move a charge through the dielectric, it can displace a charge within the dielectric. As the voltage varies with time, the displacement of charge also varies with time, causing what is known as the **displacement current**.

C

(a)

C

+ v −

i

(b)

FIGURE 6.10 The circuit symbol for a capacitor.

At the terminals, the displacement current is indistinguishable from a conduction current. The current is proportional to the rate at which the voltage across the capacitor varies with time, or, mathematically,

$$i = C\frac{dv}{dt}, \qquad \textbf{(6.13)}$$

where i is measured in amperes, C in farads, v in volts, and t in seconds.

Equation (6.13) reflects the passive sign convention shown in Fig. 6.10(b); that is, the current reference is in the direction of the voltage drop across the capacitor. If the current reference is in the direction of the voltage rise, Eq. (6.13) is written with a minus sign.

Two important observations follow from Eq. (6.13). First, voltage cannot change instantaneously across the terminals of a capacitor. Equation (6.13) indicates that such a change would produce infinite current, a physical impossibility. Second, if the voltage across the terminals is constant, the capacitor current is zero. The reason is that a conduction current cannot be established in the dielectric material of the capacitor. Only a time-varying voltage can produce a displacement current. Thus a capacitor behaves as an open circuit in the presence of a constant voltage.

Equation (6.13) gives the capacitor current as a function of the capacitor voltage. Expressing the voltage as a function of the current is also useful. To do so we multiply both sides of Eq. (6.13) by a differential time dt and then integrate the resulting differentials:

$$i\,dt = C\,dv \quad \text{or} \quad \int_{v(t_0)}^{v(t)} dx = \frac{1}{C}\int_{t_0}^{t} i\,d\tau.$$

Carrying out the integration of the left-hand side of the second equation gives

$$v(t) = \frac{1}{C}\int_{t_0}^{t} i\,d\tau + v(t_0). \tag{6.14}$$

In many practical applications of Eq. (6.14), the initial time is zero; that is, $t_0 = 0$. Thus Eq. (6.14) becomes

$$v(t) = \frac{1}{C}\int_{0}^{t} i\,d\tau + v(0). \tag{6.15}$$

We can easily derive the power and energy relationships for the capacitor. From the definition of power,

$$p = vi = Cv\frac{dv}{dt}, \tag{6.16}$$

or

$$p = i\left[\frac{1}{C}\int_{t_0}^{t} i\,d\tau + v(t_0)\right]. \tag{6.17}$$

Combining the definition of energy with Eq. (6.16) yields

$$dw = C\,v\,dv,$$

from which

$$\int_0^w dx = C\int_0^v y\,dy,$$

or

$$w = \frac{1}{2}Cv^2. \tag{6.18}$$

Use a symbolic equation solver to derive these equations.

In the derivation of Eq. (6.18), the reference for zero energy corresponds to zero voltage.

Examples 6.4 and 6.5 illustrate the application of the current, voltage, power, and energy relationships for a capacitor.

EXAMPLE 6.4

The voltage pulse described by the following equations is impressed across the terminals of a 0.5 μF capacitor:

$$\begin{aligned} v(t) &= 0, && t \leq 0; \\ v(t) &= 4t \text{ V}, && 0 \leq t \leq 1; \\ v(t) &= 4e^{-(t-1)} \text{ V}, && 1 \leq t \leq \infty. \end{aligned}$$

a) Derive the expressions for the capacitor current, power, and energy.

b) Sketch the voltage, current, power, and energy as functions of time. Line up the plots vertically.

c) Specify the interval of time when energy is being stored in the capacitor.

d) Specify the interval of time when energy is being delivered by the capacitor.

e) Evaluate the integrals

$$\int_0^1 p\,dt \quad \text{and} \quad \int_1^\infty p\,dt$$

and comment on their significance.

SOLUTION

a) From Eq. (6.13),

$$\begin{aligned} i &= (0.5 \times 10^{-6})(0) = 0, && t < 0; \\ i &= (0.5 \times 10^{-6})(4) = 2\ \mu\text{A}, && 1 < t < 1; \\ i &= (0.5 \times 10^{-6})(-4e^{-(t-1)}) = -2e^{-(t-1)}\ \mu\text{A}, && 1 < t < \infty. \end{aligned}$$

The expression for the power is derived from Eq. (6.16):

$$\begin{aligned} p &= 0, && t < 0; \\ p &= (4t)(2) = 8t\ \mu\text{W}, && 0 \leq t < 1; \\ p &= (4e^{-(t-1)})(-2e^{-(t-1)}) = -8e^{-2(t-1)}\ \mu\text{W}, && 1 < t \leq \infty. \end{aligned}$$

The energy expression follows directly from Eq. (6.18):

$$\begin{aligned} w &= 0 && t < 0; \\ w &= \tfrac{1}{2}(0.5)16t^2 = 4t^2\ \mu\text{J}, && 0 \leq t < 1, \\ w &= \tfrac{1}{2}(0.5)16e^{-2(t-1)} = 4e^{-2(t-1)}\ \mu\text{J}, && 1 \leq t \leq \infty. \end{aligned}$$

b) Figure 6.11 shows the voltage, current, power, and energy as functions of time.

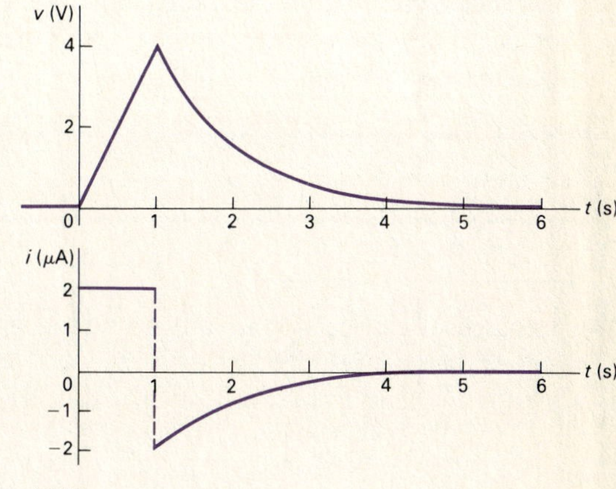

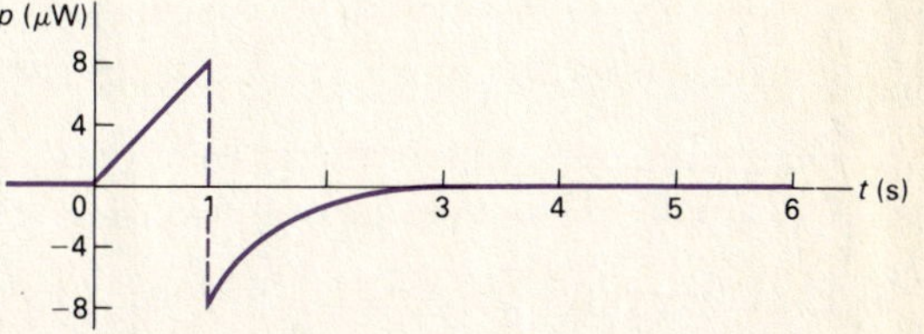

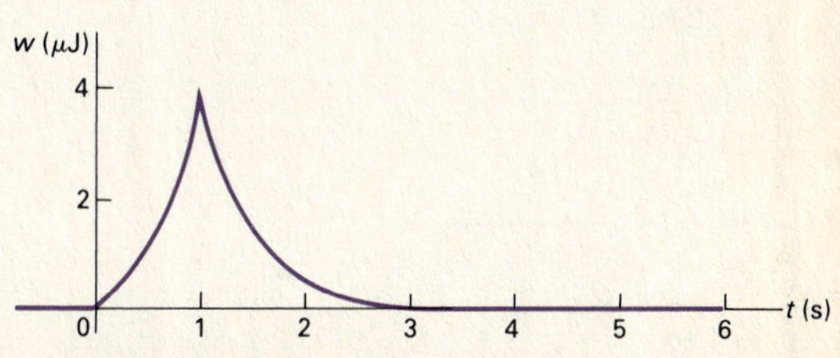

FIGURE 6.11 The variables v, i, p, and w versus t for Example 6.4.

c) Energy is being stored in the capacitor whenever the power is positive. Hence energy is being stored in the interval 0 to 1 s.

d) Energy is being delivered by the capacitor whenever the power is negative. Thus energy is being delivered for all t greater than 1 s.

e) The integral of $p\,dt$ is the energy associated with the time interval corresponding to the limits on the integral. Thus the first integral represents the energy stored in the capacitor between 0 and 1 s, whereas the second integral represents the energy returned, or delivered, by the capacitor in the interval 1 s to ∞:

$$\int_0^1 p\,dt = \int_0^1 8t\,dt = 4t^2\Big|_0^1 = 4\ \mu\text{J};$$

$$\int_1^\infty p\,dt = \int_1^\infty (-8e^{-2(t-1)})\,dt = (-8)\frac{e^{-2(t-1)}}{-2}\Big|_1^\infty = -4\ \mu\text{J}.$$

The voltage applied to the capacitor returns to zero as time increases without limit, so the energy returned by this ideal capacitor must equal the energy stored.

EXAMPLE 6.5

An uncharged 0.2 μF capacitor is driven by a triangular current pulse. The current pulse is described by

$$\begin{aligned} i(t) &= 0, && t \leq 0; \\ i(t) &= 5000t \text{ A}, && 0 \leq t \leq 20\ \mu\text{s}; \\ i(t) &= 0.2 - 5000t \text{ A}, && 20 \leq t \leq 40\ \mu\text{s}; \\ i(t) &= 0, && t \geq 40\ \mu\text{s}. \end{aligned}$$

a) Derive the expressions for the capacitor voltage, power, and energy for each of the four time intervals needed to describe the current.

b) Plot i, v, p, and w versus t. Align the plots as specified in the previous examples.

c) Why does a voltage remain on the capacitor after the current returns to zero?

SOLUTION

a) For $t \leq 0$, v, p, and w all are zero. For $0 \leq t \leq 20\ \mu$s,

$$v = 5 \times 10^6 \int_0^t 5000\tau\,d\tau + 0 = 12.5 \times 10^9 t^2 \text{ V};$$

$$p = vi = 62.5 \times 10^{12} t^3 \text{ W};$$

$$w = \frac{1}{2} C v^2 = 15.625 \times 10^{12} t^4 \text{ J}.$$

For 20 μs $\leq t \leq$ 40 μs,

$$v = 5 \times 10^6 \int_{20\mu s}^{t} (0.2 - 5000\tau)\, d\tau + 5.$$

(Note that 5 V is the voltage on the capacitor at the end of the preceding interval.) Then,

$$v = (10^6 t - 12.5 \times 10^9 t^2 - 10) \text{ V};$$

$$p = vi$$
$$= (62.5 \times 10^{12} t^3 - 7.5 \times 10^9 t^2 + 2.5 \times 10^5 t - 2) \text{ W};$$

$$w = \frac{1}{2} C v^2$$
$$= (15.625 \times 10^{12} t^4 - 2.5 \times 10^9 t^3 + 0.125 \times 10^6 t^2$$
$$-2.0 \times 10^{-6} t + 10^{-5}) \text{ J}.$$

For $t \geq 40$ μs,

$$v = 10 \text{ V};$$
$$p = vi = 0;$$
$$w = \frac{1}{2} C v^2 = 10\ \mu\text{J}.$$

b) The excitation current and the resulting voltage, power, and energy are plotted in Fig. 6.12.

c) Note that the power is always positive for the duration of the current pulse, which means that energy is continuously being stored in the capacitor. When the current returns to zero, the stored energy is trapped because the ideal capacitor offers no means for dissipating energy. Thus a voltage remains on the capacitor after i returns to zero.

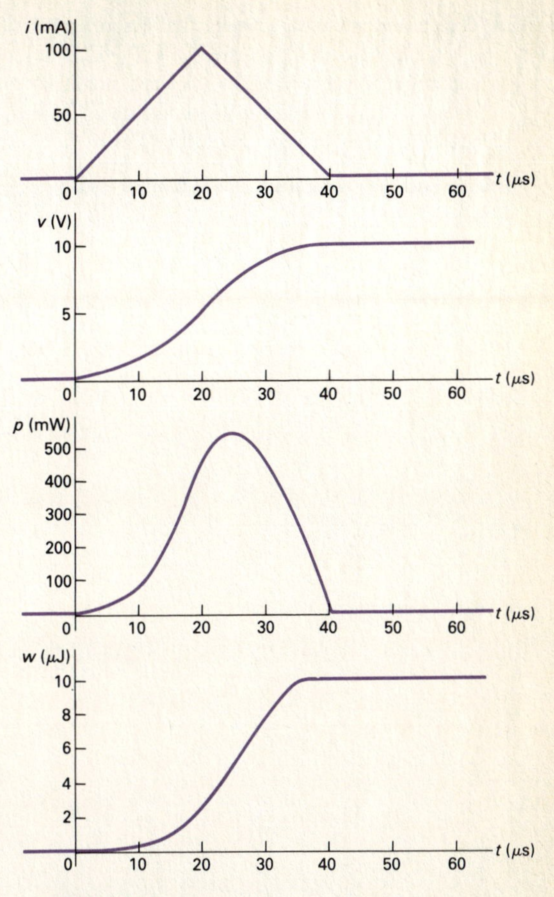

FIGURE 6.12 The variables i, v, p, and w versus t for Example 6.5.

Use a computer tool to investigate the effect of pulse shape on the current, voltage, power, and energy.

DRILL EXERCISES

6.2 The voltage at the terminals of the 0.5 μF capacitor shown in the figure is 0 for $t < 0$ and $100e^{-20,000t} \sin 40,000t$ V for $t \geq 0$. Find (a) $i(0)$; (b) the power delivered to the capacitor at $t = \pi/80$ ms; and (c) the energy stored in the capacitor at $t = \pi/80$ ms.

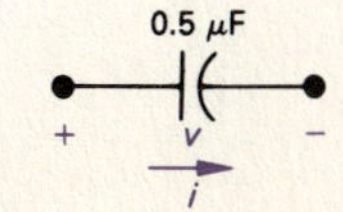

ANSWER: (a) 2 A; (b) −20.79 W; (c) 519.70 μJ.

6.3 The current in the capacitor of Drill Exercise 6.2 is 0 for $t < 0$ and $2 \cos 50{,}000t$ A for $t \geq 0$. Find (a) $v(t)$; (b) the maximum power delivered to the capacitor at any one instant of time; and (c) the maximum energy stored in the capacitor at any one instant of time.

ANSWER: (a) $80 \sin 50{,}000t$ V; (b) 80 W; (c) 1.6 mJ.

6.3 SERIES-PARALLEL COMBINATIONS OF INDUCTANCE AND CAPACITANCE

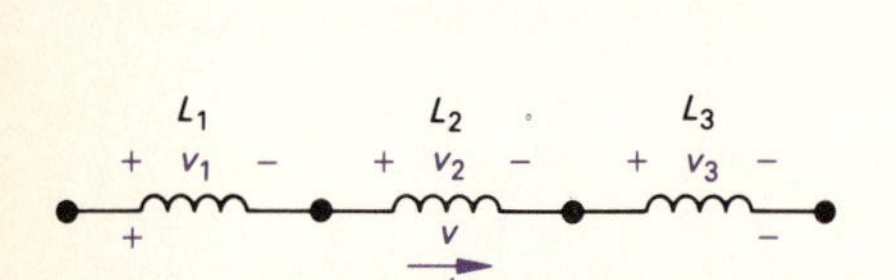

FIGURE 6.13 Inductors in series.

Just as series-parallel combinations of resistors can be reduced to a single equivalent resistor, series-parallel combinations of inductors or capacitors can be reduced to a single inductor or capacitor. Figure 6.13 shows inductors in series. Here, the inductors are forced to carry the same current; thus we define only one current for the series combination. The voltage drops across the individual inductors are

$$v_1 = L_1 \frac{di}{dt}, \quad v_2 = L_2 \frac{di}{dt}, \quad \text{and} \quad v_3 = L_3 \frac{di}{dt}.$$

The voltage across the series connection is

$$v = v_1 + v_2 + v_3 = (L_1 + L_2 + L_3)\frac{di}{dt},$$

from which it should be apparent that the equivalent inductance of series-connected inductors is the sum of the individual inductances. For n inductors in series,

$$L_{eq} = L_1 + L_2 + L_3 + \cdots + L_n. \quad \textbf{(6.19)}$$

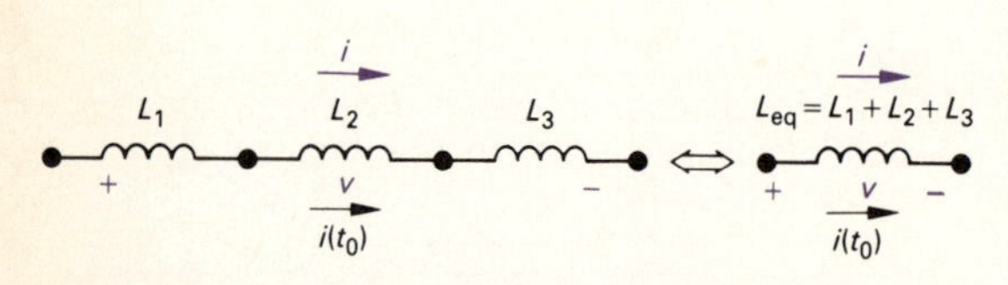

FIGURE 6.14 An equivalent circuit for inductors in series carrying an initial current $i(t_0)$.

If the original inductors carry an initial current of $i(t_0)$, the equivalent inductor carries the same initial current. Figure 6.14 shows the equivalent circuit for series inductors carrying an initial current.

Inductors in parallel have the same terminal voltage. In the equivalent circuit, the current in each inductor is a function of the terminal voltage and the initial current in the inductor. For the three inductors in parallel shown in Fig. 6.15, the currents for the individual inductors are

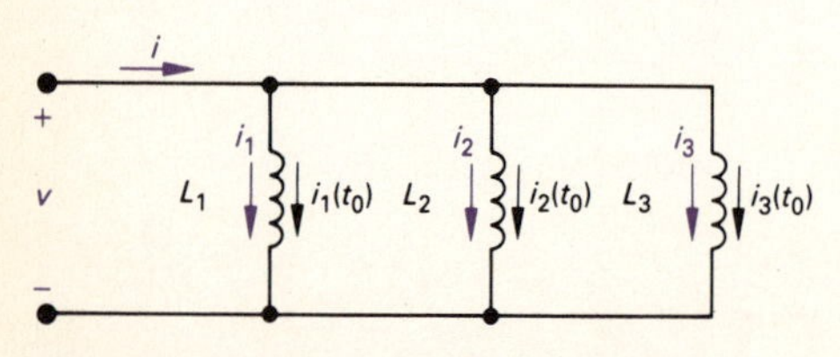

FIGURE 6.15 Three inductors in parallel.

$$i_1 = \frac{1}{L_1}\int_{t_0}^{t} v\,d\tau + i_1(t_0);$$

$$i_2 = \frac{1}{L_2}\int_{t_0}^{t} v\,d\tau + i_2(t_0); \quad \textbf{(6.20)}$$

$$i_3 = \frac{1}{L_3}\int_{t_0}^{t} v\,d\tau + i_3(t_0).$$

The current at the terminals of the three parallel inductors is the sum of the inductor currents:

$$i = i_1 + i_2 + i_3. \tag{6.21}$$

Substituting Eq. (6.20) into Eq. (6.21) yields

$$i = \left(\frac{1}{L_1} + \frac{1}{L_2} + \frac{1}{L_3}\right)\int_{t_0}^{t} v\,d\tau + i_1(t_0) + i_2(t_0) + i_3(t_0). \tag{6.22}$$

Now we can interpret Eq. (6.22) in terms of a single inductor; that is,

$$i = \frac{1}{L_{eq}}\int_{t_0}^{t} v\,d\tau + i(t_0). \tag{6.23}$$

Comparing Eq. (6.23) with (6.22) yields

$$\frac{1}{L_{eq}} = \frac{1}{L_1} + \frac{1}{L_2} + \frac{1}{L_3} \tag{6.24}$$

and

$$i(t_0) = i_1(t_0) + i_2(t_0) + i_3(t_0). \tag{6.25}$$

Figure 6.16 shows the equivalent circuit for the three parallel inductors in Fig. 6.15.

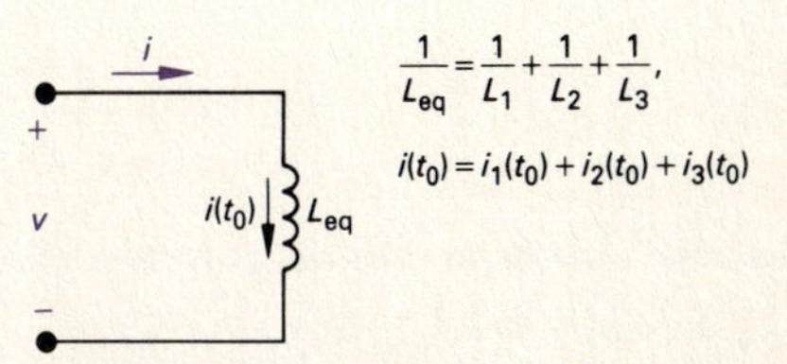

FIGURE 6.16 An equivalent circuit for three inductors in parallel.

The results expressed in Eqs. (6.24) and (6.25) can be extended to n inductors in parallel:

$$\frac{1}{L_{eq}} = \frac{1}{L_1} + \frac{1}{L_2} + \cdots + \frac{1}{L_n} \tag{6.26}$$

and

$$i(t_0) = i_1(t_0) + i_2(t_0) + \cdots + i_n(t_0). \tag{6.27}$$

Capacitors connected in series can be reduced to a single equivalent capacitor. The reciprocal of the equivalent capacitance is equal to the sum of the reciprocals of the individual capacitances. If each capacitor carries its own initial voltage, the initial voltage on the equivalent capacitor is the algebraic sum of the initial voltages on the individual capacitors. Figure 6.17 and the following equations summarize these observations:

$$\frac{1}{C_{eq}} = \frac{1}{C_1} + \frac{1}{C_2} + \cdots + \frac{1}{C_n}; \tag{6.28}$$

$$v(t_0) = v_1(t_0) + v_2(t_0) + \cdots + v_n(t_0). \tag{6.29}$$

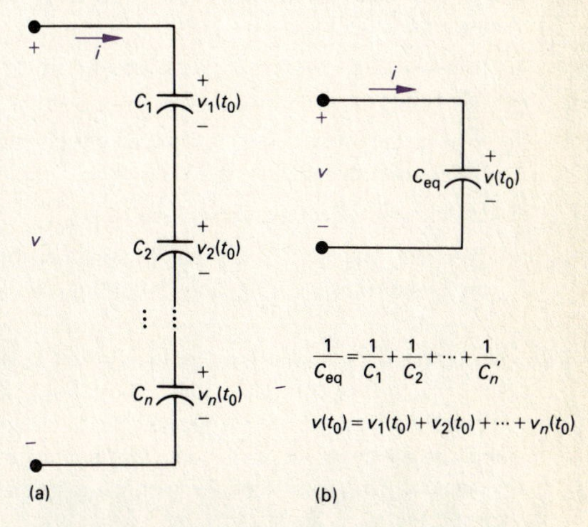

FIGURE 6.17 An equivalent circuit for capacitors connected in series: (a) the series capacitors; (b) the equivalent circuit.

We leave the derivation of the equivalent circuit for series-connected capacitors as an exercise. (See Problem 6.24.)

The equivalent capacitance of capacitors connected in parallel is simply the sum of the capacitances of the individual capacitors, as

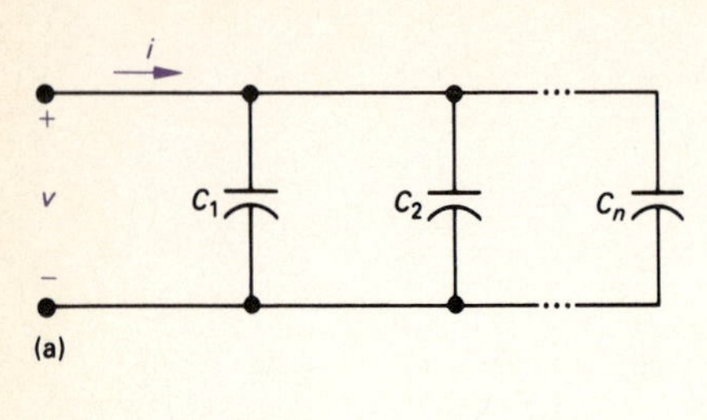

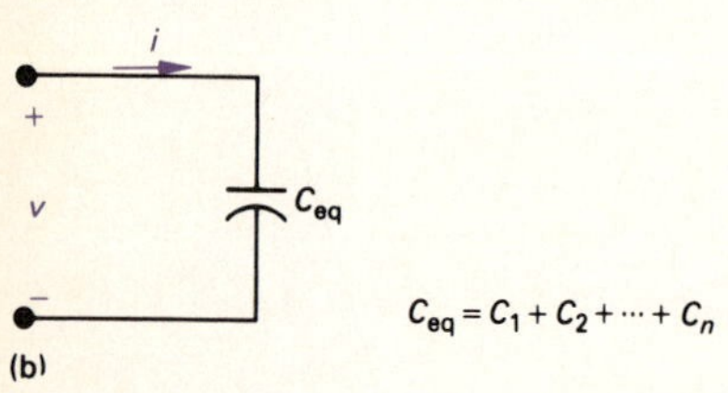

FIGURE 6.18 An equivalent circuit for capacitors connected in parallel: (a) capacitors in parallel; (b) the equivalent circuit.

Fig. 6.18 and the following equation show:

$$C_{eq} = C_1 + C_2 + \cdots + C_n. \quad \textbf{(6.30)}$$

Capacitors connected in parallel must carry the same voltage. Therefore, if there is an initial voltage across the original parallel capacitors, this same initial voltage appears across the equivalent capacitance C_{eq}. The derivation of the equivalent circuit for parallel capacitors is left as an exercise. (See Problem 6.25.)

We say more about series-parallel equivalent circuits of inductors and capacitors in Chapter 7, where we interpret results based on their use.

DRILL EXERCISES

6.4 The initial values of i_1 and i_2 in the circuit shown are -2 and $+4$ A, respectively. The voltage at the terminals of the parallel inductors for $t \geq 0$ is $-40e^{-5t}$ V.

a) If the parallel inductors are replaced by a single inductor, what is its inductance?

b) What is the initial current and its reference direction in the equivalent inductor?

c) Use the equivalent inductor to find $i(t)$.

d) Find $i_1(t)$ and $i_2(t)$. Verify that the solutions for $i_1(t)$, $i_2(t)$, and $i(t)$ satisfy Kirchhoff's current law.

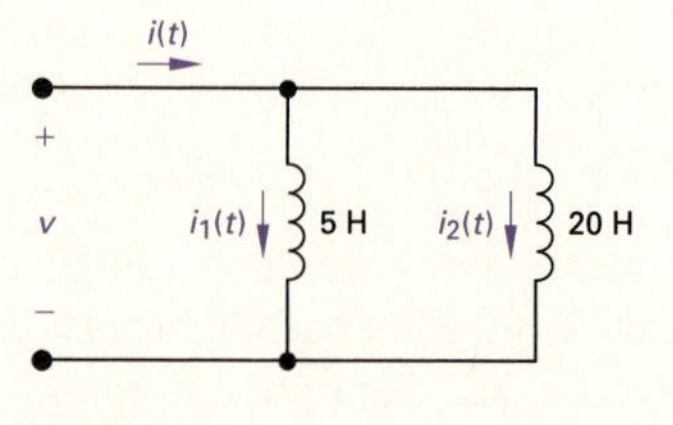

ANSWER: (a) 4 H; (b) 2 A, down; (c) $2e^{-5t}$ A; (d) $i_1(t) = 1.6e^{-5t} - 3.6$ A, $i_2(t) = 0.4e^{-5t} + 3.6$ A.

6.5 The current at the terminals of the two capacitors shown is $240e^{-10t}$ μA for $t \geq 0$. The initial values of v_1 and v_2 are -10 and -5 V, respectively. Calculate the total energy trapped in the capacitors as $t \to \infty$. (*Hint:* Don't combine the capacitors in series—find the energy trapped in each and then add.)

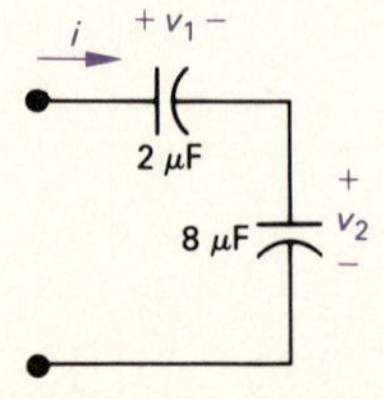

ANSWER: 20 μJ.

PRACTICAL PERSPECTIVE

Proximity Switches

At the beginning of this chapter we introduced the capacitive proximity switch. There are two forms of this switch. The one used in table lamps is based on a single-electrode switch. It is left to your investigation in Problem 6.33. In the example here, we consider the two-electrode switch used in elevator call buttons.

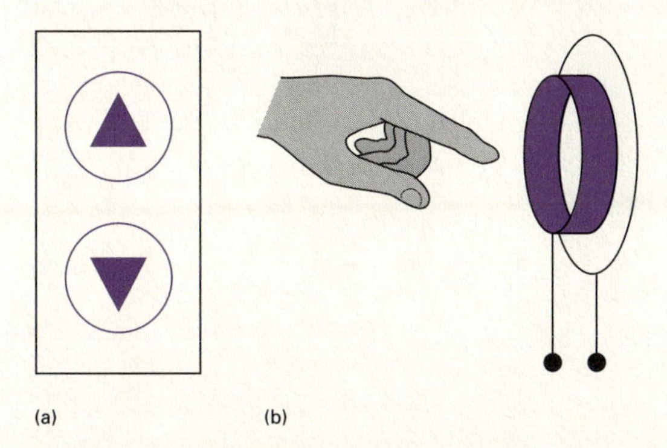

FIGURE 6.19 An elevator call button: (a) front view; (b) side view.

EXAMPLE

The elevator call button is a small cup into which the finger is inserted, as shown in Fig. 6.19. The cup is made of a metal ring electrode and a circular plate electrode which are insulated from each other. Sometimes two concentric rings embedded in insulating plastic are used instead. The electrodes are covered with an insulating layer to prevent direct contact with the metal. The resulting device can be modeled as a capacitor, as shown in Fig. 6.20.

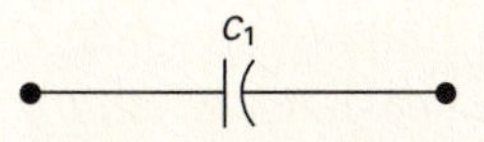

FIGURE 6.20 A capacitor model of the two-electrode proximity switch used in elevator call buttons.

Unlike most capacitors, the capacitive proximity switch permits you to insert an object, such as a finger, between the electrodes. Since your finger is much more conductive than the insulating covering surrounding the electrodes, the circuit responds as though another electrode, connected to ground, has been added. The result is a three-terminal circuit containing three capacitors, as shown in Fig. 6.21.

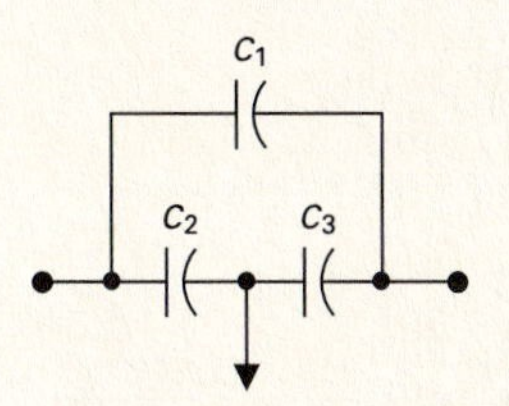

FIGURE 6.21 A circuit model of a capacitive proximity switch activated by finger touch.

The actual values of the capacitors in Figs. 6.20 and 6.21 are in the range of 10 to 50 pF, depending on the exact geometry of the switch, how the finger is inserted, whether the person is wearing gloves, and so forth. For the following problems, assume that all capacitors have the same value of 25 pF. Also assume the elevator call button is placed in the capacitive equivalent of a voltage-divider circuit, as shown in Fig. 6.22.

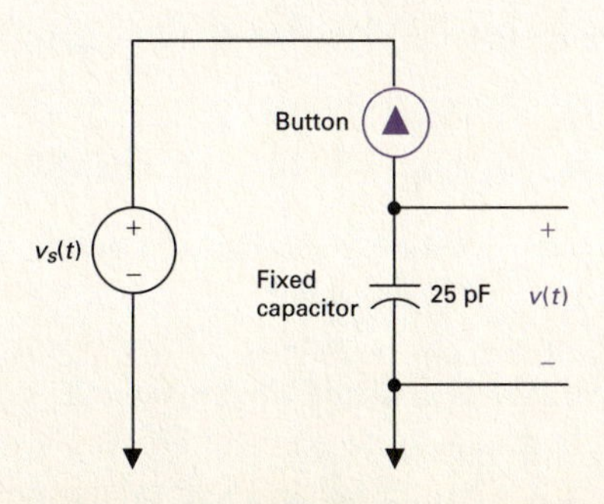

FIGURE 6.22 An elevator call button circuit.

a) Calculate the output voltage with no finger present.

b) Calculate the output voltage when a finger touches the button.

SOLUTION

a) Begin by redrawing the circuit in Fig. 6.22 with the call button replaced by its capacitive model from Fig. 6.20. The resulting circuit is shown in Fig. 6.23. Write the current equation at the single node:

$$C_1 \frac{d(v - v_s)}{dt} + C_2 \frac{dv}{dt} = 0. \quad \textbf{(6.31)}$$

Rearrange this equation to produce a differential equation for the output voltage $v(t)$:

$$\frac{dv}{dt} = \frac{C_1}{C_1 + C_2} \frac{dv_s}{dt}. \quad \textbf{(6.32)}$$

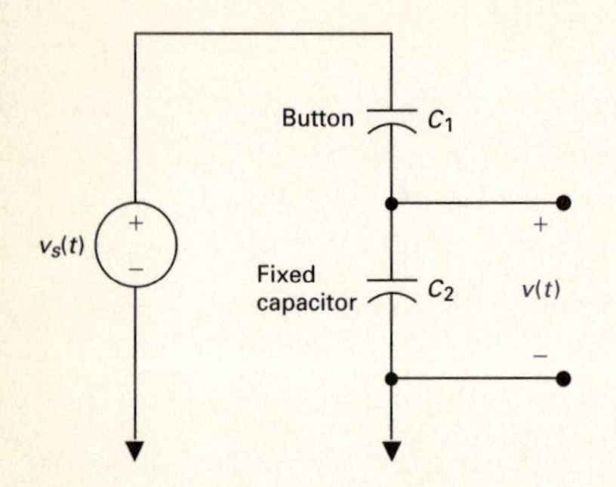

FIGURE 6.23 A model of the elevator call button circuit with no finger present.

Finally, integrate Eq. (6.32) to find the output voltage:

$$v(t) = \frac{C_1}{C_1 + C_2} v_s(t) + v(0). \tag{6.33}$$

The result in Eq. (6.33) shows that the series capacitor circuit in Fig. 6.23 forms a voltage divider just as the series resistor circuit did in Chapter 3. In both voltage-divider circuits, the output voltage does not depend on the component values, but only on their ratio. Here, $C_1 = C_2 = 25$ pF, so the capacitor ratio is $C_1/C_2 = 1$. Thus the output voltage is

$$v(t) = 0.5v_s(t) + v(0). \tag{6.34}$$

The constant term in Eq. (6.34) is due to the initial charge on the capacitor. We can assume that $v(0) = 0$ V, since the circuit which senses the output voltage eliminates the effect of the initial capacitor charge. Therefore, the sensed output voltage is

$$v(t) = 0.5v_s(t). \tag{6.35}$$

b) Now we replace the call button of Fig. 6.22 with the model of the activated switch in Fig. 6.21. The result is shown in Fig. 6.24. Again, we calculate the currents leaving the output node:

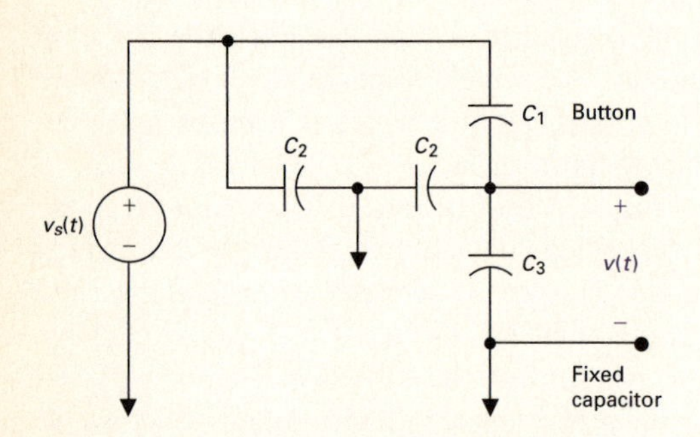

FIGURE 6.24 A model of the elevator call button circuit when activated by finger touch.

$$C_1 \frac{d(v - v_s)}{dt} + C_2 \frac{dv}{dt} + C_3 \frac{dv}{dt} = 0. \tag{6.36}$$

Rearranging to write a differential equation for $v(t)$ results in

$$\frac{dv}{dt} = \frac{C_1}{C_1 + C_2 + C_3} \frac{dv_s}{dt}. \tag{6.37}$$

Finally, solving the differential equation in Eq. (6.37), we see

$$v(t) = \frac{C_1}{C_1 + C_2 + C_3} v_s(t) + v(0). \tag{6.38}$$

If $C_1 = C_2 = C_3 = 25$ pF,

$$v(t) = 0.333v_s(t) + v(0). \tag{6.39}$$

As before, the sensing circuit eliminates $v(0)$, so the sensed output voltage is

$$v(t) = 0.333v_s(t). \tag{6.40}$$

Comparing Eqs. (6.35) and (6.40), we see that when the button is pushed, the output is one-third of the input voltage. When the button is not pushed, the output voltage is one-half of the input voltage. Any drop in output voltage is detected by the elevator's control computer and ultimately results in the elevator arriving at the appropriate floor. We leave to you the further exploration of capacitive proximity switches in Problems 6.32–6.34.

SUMMARY

- Inductance is a linear circuit parameter that relates the voltage induced by a time-varying magnetic field to the current producing the field.
- Capacitance is a linear circuit parameter that relates the current induced by a time-varying electric field to the voltage producing the field.
- Inductors and capacitors are passive elements; they can store and release energy, but they cannot generate or dissipate energy.
- The instantaneous power at the terminals of an inductor or capacitor can be positive or negative, depending on whether energy is being delivered to or extracted from the element.
- An inductor does not permit an instantaneous change in its terminal current.
- A capacitor does not permit an instantaneous change in its terminal voltage.
- An inductor does permit an instantaneous change in its terminal voltage.
- A capacitor does permit an instantaneous change in its terminal current.
- An inductor behaves as a short circuit in the presence of a constant terminal current.
- A capacitor behaves as an open circuit in the presence of a constant terminal voltage.
- Equations for voltage, current, power, and energy in ideal inductors and capacitors are given in Table 6.1.

TABLE 6.1

TERMINAL EQUATIONS FOR IDEAL INDUCTORS AND CAPACITORS*

INDUCTORS		CAPACITORS	
1. $v = L\dfrac{di}{dt}$	(V)	1. $v = \dfrac{1}{C}\int_{t_0}^{t} i\,d\tau + v(t_0)$	(V)
2. $i = \dfrac{1}{L}\int_{t_0}^{t} v\,d\tau + i(t_0)$	(A)	2. $i = C\dfrac{dv}{dt}$	(A)
3. $p = vi = Li\dfrac{di}{dt}$	(W)	3. $p = vi = Cv\dfrac{dv}{dt}$	(W)
4. $w = \frac{1}{2}Li^2$	(J)	4. $w = \frac{1}{2}Cv^2$	(J)

*The equations in this table are based on the passive sign convention.

- Inductors in series or in parallel can be replaced by an equivalent inductor. Capacitors in series or parallel can be replaced by an equivalent capacitor. The equations are summarized in Table 6.2. See Section 6.3 for a discussion on how to handle the initial conditions for series and parallel equivalent circuits involving inductors and capacitors.

TABLE 6.2

EQUATIONS FOR SERIES- AND PARALLEL-CONNECTED INDUCTORS AND CAPACITORS

	INDUCTORS	CAPACITORS
Series-connected	$L_{eq} = L_1 + L_2 + \cdots + L_n$	$\frac{1}{C_{eq}} = \frac{1}{C_1} + \frac{1}{C_2} + \cdots + \frac{1}{C_n}$
Parallel-connected	$\frac{1}{L_{eq}} = \frac{1}{L_1} + \frac{1}{L_2} + \cdots + \frac{1}{L_n}$	$C_{eq} = C_1 + C_2 + \cdots + C_n$

PROBLEMS

6.1 Evaluate the integral

$$\int_0^{\infty} p\,dt$$

for Example 6.2. Comment on the significance of the result.

6.2 The triangular current pulse shown in Fig. P6.2 is applied to a 40 mH inductor.

a) Write the expressions that describe $i(t)$ in the four intervals $t < 0$, $0 \leq t \leq 10$ ms, $10 \text{ ms} \leq t \leq 20$ ms, and $t > 20$ ms.

b) Derive the expressions for the inductor voltage, power, and energy. Use the passive sign convention.

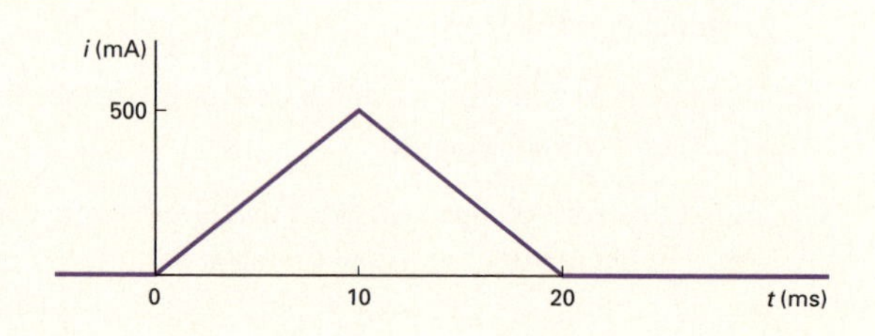

FIGURE P6.2

6.3 The voltage at the terminals of the 200 μH inductor in Fig. P6.3(a) is shown in Fig. P6.3(b). The inductor current i is known to be zero for $t \leq 0$.

a) Derive the expressions for i for $t \geq 0$.

b) Sketch i versus t for $0 \leq t \leq \infty$.

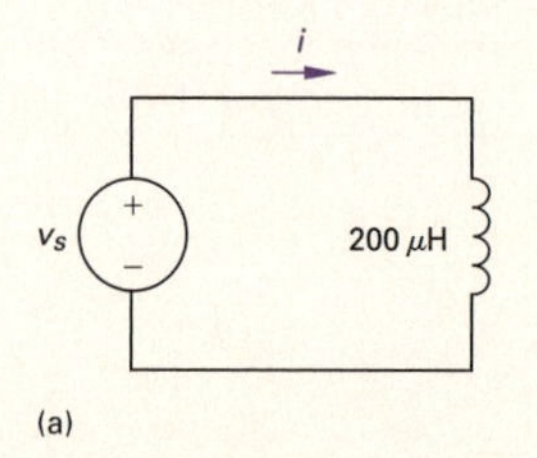

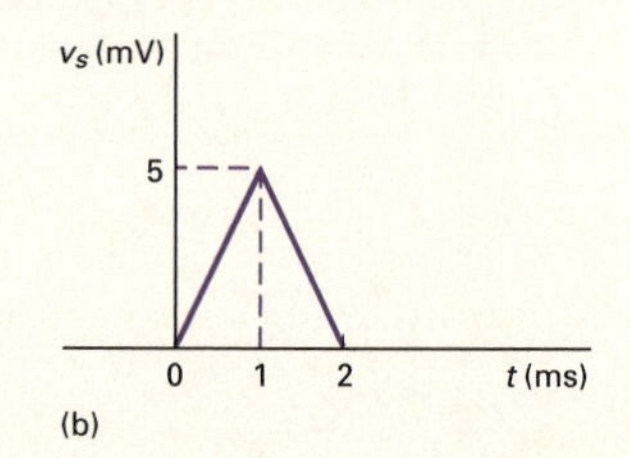

FIGURE P6.3

6.4 The current in the 2.5 mH inductor in Fig. P6.4 is known to be 6 A for $t \leq 0$. The inductor voltage for $t \geq 0^+$ is given by the expression

$$v_L(t) = 50e^{-5t} \text{ mV}, \quad 0^+ \leq t < \infty.$$

Sketch $v_L(t)$ and $i_L(t)$ for $0 \leq t \leq \infty$.

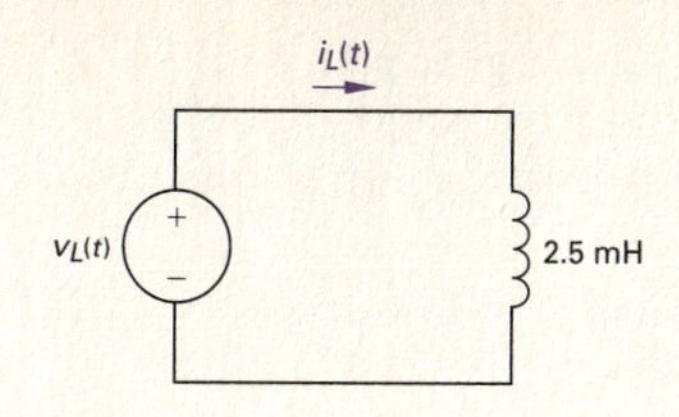

FIGURE P6.4

6.5 a) Find the inductor current in the circuit in Fig. P6.5 if $v = 250 \sin 1000t$ V, $L = 50$ mH, and $i(0) = -5$ A.

b) Sketch v, i, p, and w versus time. In making these sketches, use the format used in Fig. 6.8. Plot over one complete cycle of the voltage waveform.

c) Describe the subintervals in the time interval between 0 and 2π ms when power is being absorbed by the inductor. Repeat for the subintervals when power is being delivered by the inductor.

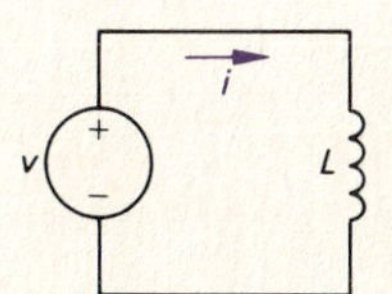

FIGURE P6.5

6.6 The current in a 20 mH inductor is known to be $7 + (15 \sin 140t - 35 \cos 140t)e^{-20t}$ mA for $t \geq 0$. Assume the passive sign convention.

a) At what instant of time is the voltage across the inductor maximum?

b) What is the maximum voltage?

6.7 The current in a 100 μH inductor is known to be

$$i_L = 20te^{-5t} \text{ A for } t \geq 0.$$

a) Find the voltage across the inductor for $t > 0$. (Assume the passive sign convention.)

b) Find the power (in microwatts) at the terminals of the inductor when $t = 100$ ms.

c) Is the inductor absorbing or delivering power at 100 ms?

d) Find the energy (in microjoules) stored in the inductor at 100 ms.

e) Find the maximum energy (in microjoules) stored in the inductor, and the time (in milliseconds) when it occurs.

6.8 The current in and the voltage across a 2.5 H inductor are known to be zero for $t \leq 0$. The voltage across the inductor is given by the graph in Fig. P6.8 for $t \geq 0$.

a) Derive the expression for the current as a function of time in the intervals $0 \leq t \leq 2$ s, $2\text{ s} \leq t \leq 6$ s, $6\text{ s} \leq t \leq 10$ s, $10\text{ s} \leq t \leq 12$ s, and $12\text{ s} \leq t \leq \infty$.

b) For $t > 0$, what is the current in the inductor when the voltage is zero?

c) Sketch i versus t for $0 \leq t \leq \infty$.

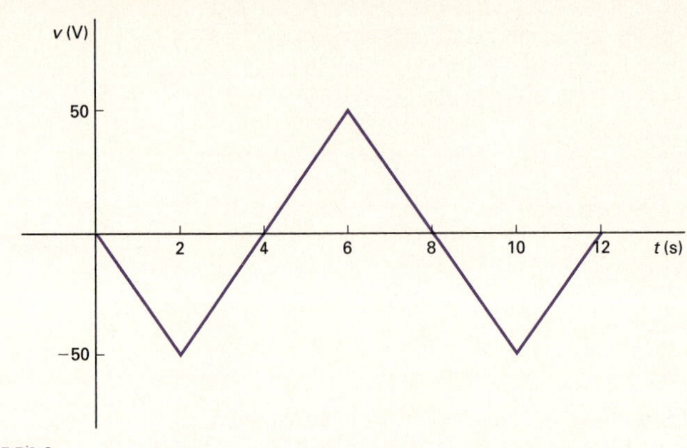

FIGURE P6.8

6.9 The current in a 2 H inductor is

$$i = 25\text{ A}, \quad t \leq 0;$$

$$i = (\mathrm{B}_1 \cos 5t + \mathrm{B}_2 \sin 5t)e^{-1t}\text{ A}, \quad t \geq 0.$$

The voltage across the inductor (passive sign convention) is 100 V at $t = 0$.
Calculate the power at the terminals of the inductor at $t = 0.5$ s. State whether the inductor is absorbing or delivering power.

6.10 The current in a 15 mH inductor is known to be

$$i = 1\text{ A}, \quad t \leq 0;$$

$$i = \mathrm{A}_1 e^{-2000t} + \mathrm{A}_2 e^{-8000t}\text{ A}, \quad t \geq 0.$$

The voltage across the inductor (passive sign convention) is 60 V at $t = 0$.

a) Find the expression for the voltage across the inductor for $t > 0$.

b) Find the time, greater than zero, when the power at the terminals of the inductor is zero.

6.11 Assume in Problem 6.10 that the value of the voltage across the inductor at $t = 0$ is -300 V instead of 60 V.

a) Find the numerical expressions for i and v for $t \geq 0$.

b) Specify the time intervals when the inductor is storing energy, and the time intervals when the inductor is delivering energy.

c) Show that the total energy extracted from the inductor is equal to the total energy stored.

6.12 Initially there was no energy stored in the 20 H inductor in the circuit in Fig. P6.12 when it was placed across the terminals of the voltmeter. At $t = 0$ the inductor was switched instantaneously to position b where it remained for 1.2 s before returning instantaneously to position a. The d'Arsonval voltmeter has a full-scale reading of 25 V and a sensitivity of 1000 Ω/V. What will the reading of the voltmeter be at the instant the switch returns to position a if the inertia of the d'Arsonval movement is negligible?

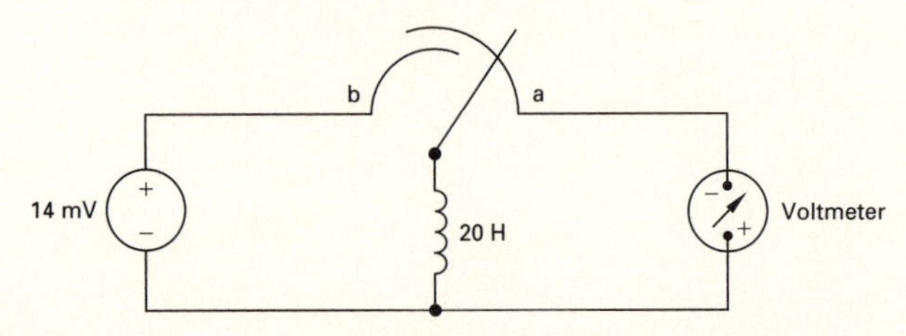

FIGURE P6.12

6.13 The voltage across the terminals of a 0.25 μF capacitor is

$$v = 40 \text{ V}, \quad t \leq 0;$$

$$v = A_1te^{-5000t} + A_2e^{-5000t} \text{ V}, \quad t \geq 0.$$

The initial current in the capacitor is 150 mA. Assume the passive sign convention.

a) What is the initial energy stored in the capacitor?

b) Evaluate the coefficients A_1 and A_2.

c) What is the expression for the capacitor current?

6.14 The voltage at the terminals of the capacitor in Fig. 6.10 is known to be

$$v = -30 \text{ V}, \quad t \leq 0;$$

$$v = 10 - 10e^{-1000t}(4 \cos 3000t + \sin 3000t) \text{ V}, \quad t \geq 0.$$

Assume $C = 0.5\ \mu$F.

a) Find the current in the capacitor for $t < 0$.

b) Find the current in the capacitor for $t > 0$.

c) Is there an instantaneous change in the voltage across the capacitor at $t = 0$?

d) Is there an instantaneous change in the current in the capacitor at $t = 0$?

e) How much energy (in microjoules) is stored in the capacitor at $t = \infty$?

6.15 The rectangular-shaped current pulse shown in Fig. P6.15 is applied to a 0.2 μF capacitor. The initial voltage on the capacitor is a 40 V drop in the reference direction of the current. Assume the passive sign convention. Derive the expression for the capacitor voltage for the time intervals in parts (a)–(c).

a) $0 \leq t \leq 100\ \mu$s

b) $100\ \mu\text{s} \leq t \leq 300\ \mu$s

c) $300\ \mu\text{s} \leq t \leq \infty$

d) Sketch $v(t)$ over the interval $-100\ \mu\text{s} \leq t \leq 500\ \mu$s.

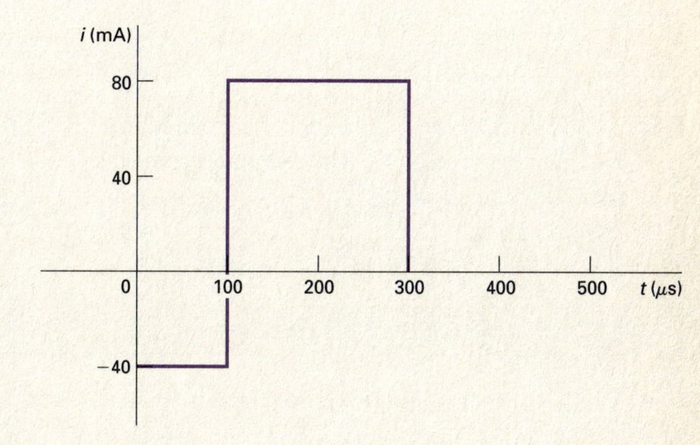

FIGURE P6.15

6.16 The current pulse shown in Fig. P6.16 is applied to a 0.25 μF capacitor. The initial voltage on the capacitor is zero.

a) Find the charge on the capacitor at $t = 15\ \mu$s.

b) Find the voltage on the capacitor at $t = 25\ \mu$s.

c) How much energy is stored in the capacitor by the current pulse?

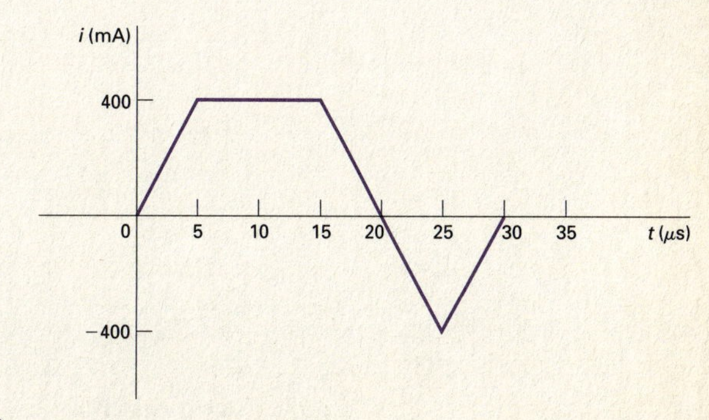

FIGURE P6.16

6.17 The initial voltage on the 0.5 μF capacitor shown in Fig. P6.17(a) is 20 V. The capacitor current has the waveform shown in Fig. P6.17(b).

a) How much energy, in microjoules, is stored in the capacitor at $t = 500\ \mu$s?

b) Repeat part (a) for $t = \infty$.

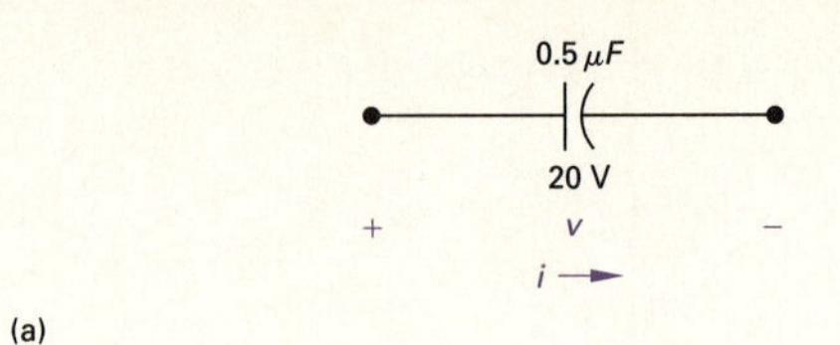

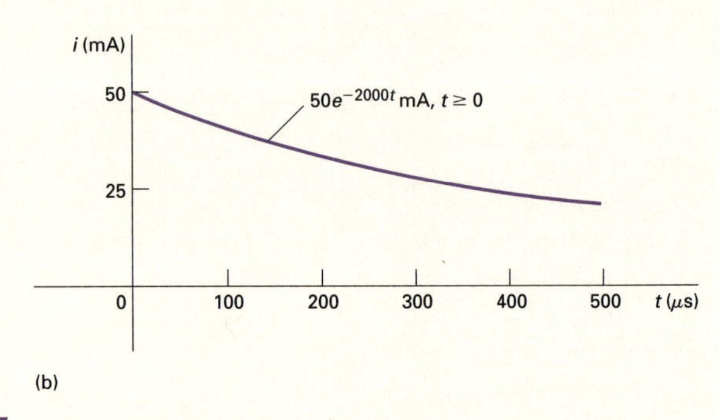

FIGURE P6.17

6.18 A 0.5 μF capacitor is subjected to a voltage pulse having a duration of 1 s. The pulse is described by the following equations:

$$v_c(t) = 40t^3 \text{ V}, \quad 0 \le t \le 1 \text{ s};$$

$$v_c(t) = 40(2-t)^3 \text{ V}, \quad 1 \text{ s} \le t \le 2 \text{ s};$$

$$v_c(t) = 0 \quad \text{elsewhere.}$$

Sketch the current pulse that exists in the capacitor during the 2 s interval.

6.19 The expressions for voltage, power, and energy derived in Example 6.5 involved both integration and manipulation of algebraic expressions. As an engineer, you cannot accept such results on faith alone. That is, you should develop the habit of asking yourself, "Do these results make sense in terms of the known behavior of the circuit they purport to describe?" With these thoughts in mind, test the expressions of Example 6.5 by performing the following checks:

a) Check the expressions to see whether the voltage is continuous in passing from one time interval to the next.

b) Check the power expression in each interval by selecting a time within the interval and seeing whether it gives the same result as the corresponding product of v and i. For example, test at 10 and 30 μs.

c) Check the energy expression within each interval by selecting a time within the interval and seeing whether the energy equation gives the same result as $\frac{1}{2}Cv^2$. Use 10 and 30 μs as test points.

6.20 Assume that the initial energy stored in the inductors of Fig. P6.20 is zero. Find the equivalent inductance with respect to the terminals a,b.

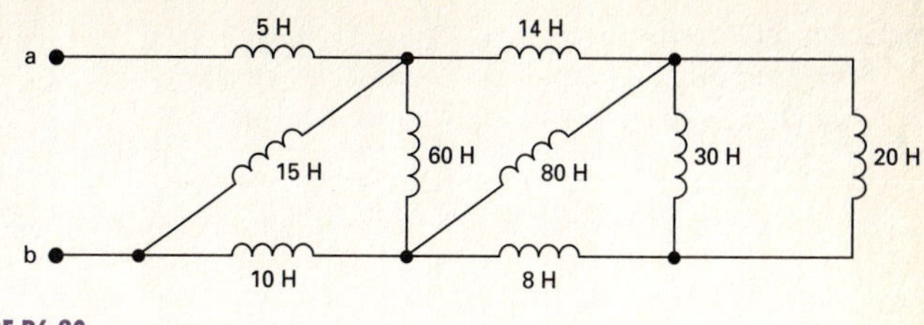

FIGURE P6.20

6.21 The two parallel inductors in Fig. P6.21 are connected across the terminals of a black box at $t = 0$. The resulting voltage v for $t \geq 0$ is known to be $800e^{-10t}$ V. It is also known that $i_1(0) = 10$A and $i_2(0) = -5$A.

a) Replace the original inductors with an equivalent inductor and find $i(t)$ for $t \geq 0$.

b) Find $i_1(t)$ for $t \geq 0$.

c) Find $i_2(t)$ for $t \geq 0$.

d) How much energy is delivered to the black box in the time interval $0 \leq t \leq \infty$?

e) How much energy was initially stored in the parallel inductors?

f) How much energy is trapped in the ideal inductors?

g) Do your solutions for i_1 and i_2 agree with the answer obtained in part (f)?

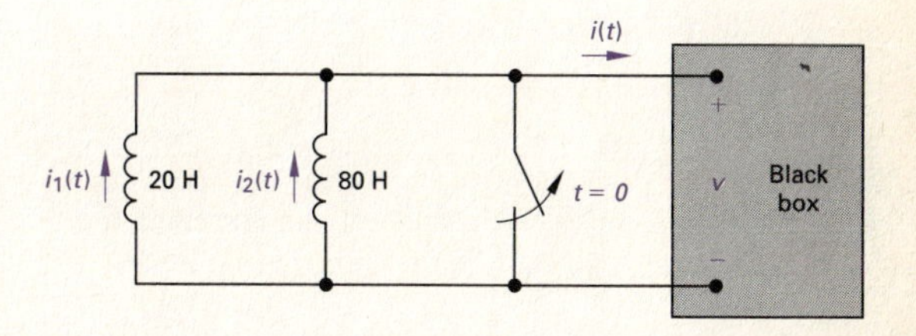

FIGURE P6.21

6.22 The three inductors in the circuit in Fig. P6.22 are connected across the terminals of a black box at $t = 0$. The resulting voltage for $t \geq 0$ is known to be

$$v_b = 100e^{-2.5t} \text{ V}.$$

If $i_1(0) = 6$ A and $i_2(0) = -4$ A, find

a) $i_o(0)$;

b) $i_o(t), t \geq 0$;

c) $i_1(t), t \geq 0$;

d) $i_2(t), t \geq 0$;

e) the initial energy stored in the three inductors;

f) the total energy delivered to the black box; and

g) the energy trapped in the ideal inductors.

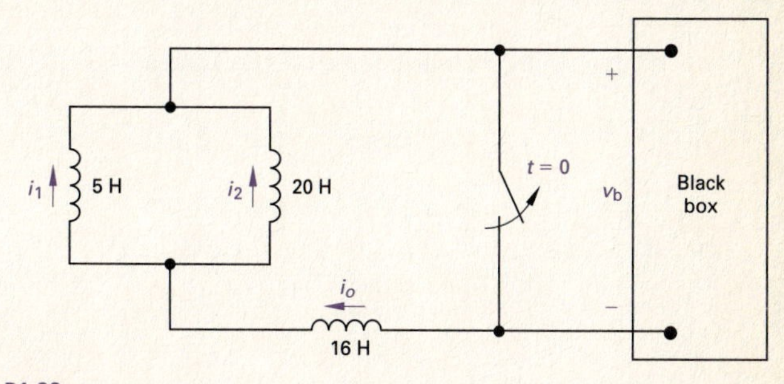

FIGURE P6.22

6.23 For the circuit shown in Fig. P6.22, how many milliseconds after the switch is opened is the energy delivered to the black box 60% of the total amount delivered?

6.24 Derive the equivalent circuit for a series connection of ideal capacitors. Assume that each capacitor has its own initial voltage. Denote these initial voltages as $v_1(t_0)$, $v_2(t_0)$, and so on. (*Hint:* Sum the voltages across the string of capacitors, recognizing that the series connection forces the current in each capacitor to be the same.)

6.25 Derive the equivalent circuit for a parallel connection of ideal capacitors. Assume that the initial voltage across the paralleled capacitors is $v(t_0)$. (*Hint:* Sum the currents into the string of capacitors, recognizing that the parallel connection forces the voltage across each capacitor to be the same.)

6.26 Find the equivalent capacitance with respect to the terminals a,b for the circuit shown in Fig. P6.26.

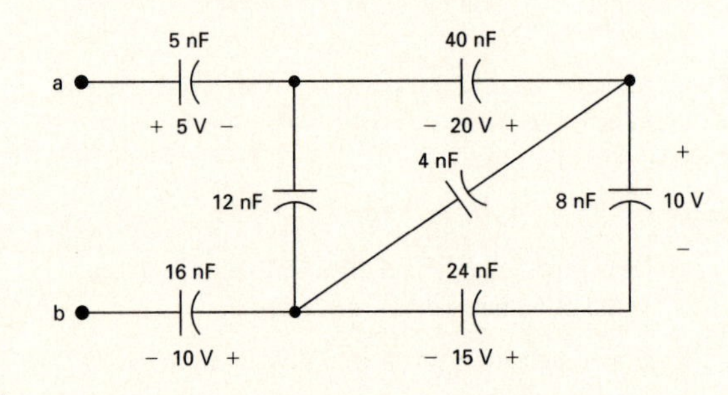

FIGURE P6.26

6.27 The four capacitors in the circuit in Fig. P6.27 are connected across the terminals of a black box at $t = 0$. The resulting current i_b for $t \geq 0$ is known to be

$$i_b = 250e^{-1000t}\ \mu\text{A}.$$

If $v_a(0) = 40$ V, $v_c(0) = 20$ V, and $v_d(0) = 80$ V, find the following for $t \geq 0$: (a) $v_b(t)$, (b) $v_a(t)$, (c) $v_c(t)$, (d) $v_d(t)$, (e) $i_1(t)$, and (f) $i_2(t)$.

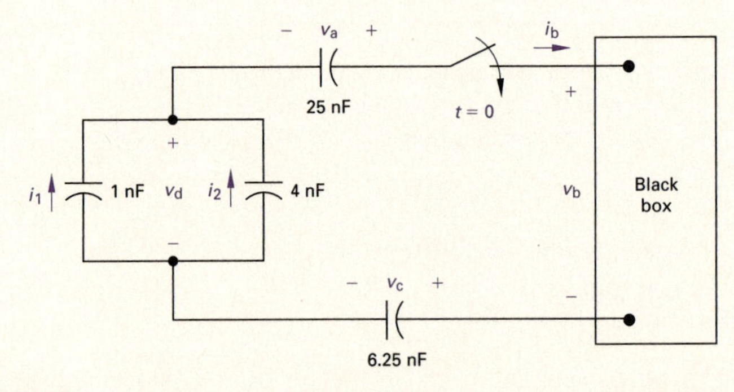

FIGURE P6.27

6.28 For the circuit in Fig. P6.27, calculate

a) the initial energy stored in the capacitors;

b) the final energy stored in the capacitors;

c) the total energy delivered to the black box;

d) the percentage of the initial energy stored that is delivered to the black box; and

e) the time, in milliseconds, it takes to deliver 10 μJ to the black box.

6.29 The two series-connected capacitors in Fig. P6.29 are connected to the terminals of a black box at $t = 0$. The resulting current $i(t)$ for $t \geq 0$ is known to be $40e^{-500t}\ \mu$A.

a) Replace the original capacitors with an equivalent capacitor and find $v_o(t)$ for $t \geq 0$.

b) Find $v_1(t)$ for $t \geq 0$.

c) Find $v_2(t)$ for $t \geq 0$.

d) How much energy is delivered to the black box in the time interval $0 \leq t \leq \infty$?

e) How much energy was initially stored in the series capacitors?

f) How much energy is trapped in the ideal capacitors?

g) Do the solutions for v_1 and v_2 agree with the answer obtained in part (f)?

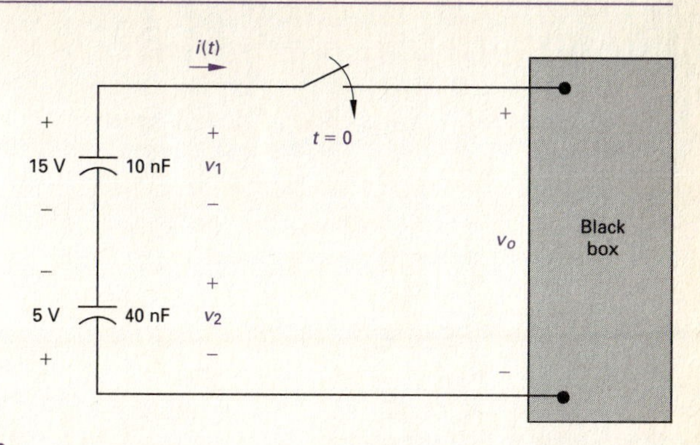

FIGURE P6.29

6.30 At $t = 0$, a series-connected capacitor and inductor are placed across the terminals of a black box, as shown in Fig. P6.30. For $t \geq 0$, it is known that

$$i_o = 1.5e^{-16{,}000t} - 0.5e^{-4000t}\ \text{A}.$$

If $v_c(0) = -50$ V find v_o for $t \geq 0$.

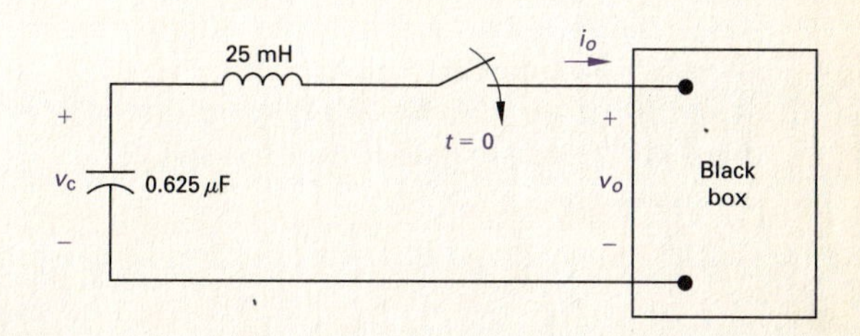

FIGURE P6.30

6.31 The current in the circuit in Fig. P6.31 is known to be

$$i_o = 50e^{-8000t}(\cos\ 6000t + 2\sin 6000t)\ \text{A}$$

for $t \geq 0^+$.

Find $v_1(0^+)$ and $v_2(0^+)$.

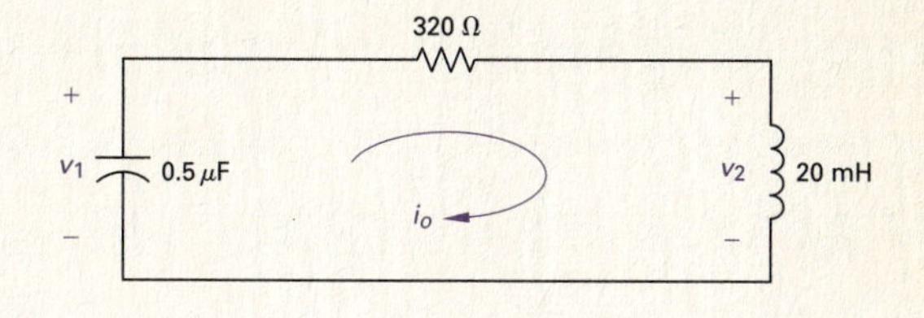

FIGURE P6.31

◆ **6.32** Rework the Practical Perspective example, except this time put the button on the bottom of the divider circuit, as shown in Fig. P6.32. Calculate the output voltage $v(t)$ when a finger is present.

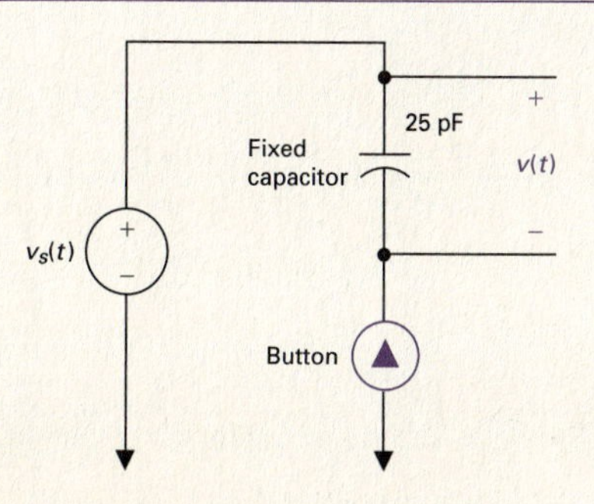

FIGURE P6.32

◆ **6.33** Some lamps are made to turn on or off when the base is touched. These use a one-terminal variation of the capacitive switch circuit discussed in the Practical Perspective. Figure P6.33 shows a circuit model of such a lamp. Calculate the change in the voltage $v(t)$ when a person touches the lamp. Assume all capacitors are initially discharged.

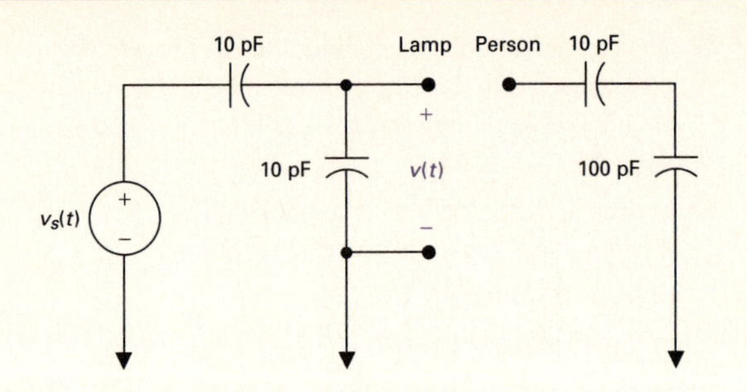

FIGURE P6.33

◆ **6.34** In the Practical Perspective example, we calculated the output voltage when the elevator button is the upper capacitor in a voltage divider. In Problem 6.33, we calculated the voltage when the button is the bottom capacitor in the divider, and we got the same result! You may wonder if this will be true for all such voltage dividers. Calculate the voltage difference (finger versus no finger) for the circuits in Figs. P6.34(a) and (b), which use two identical voltage sources.

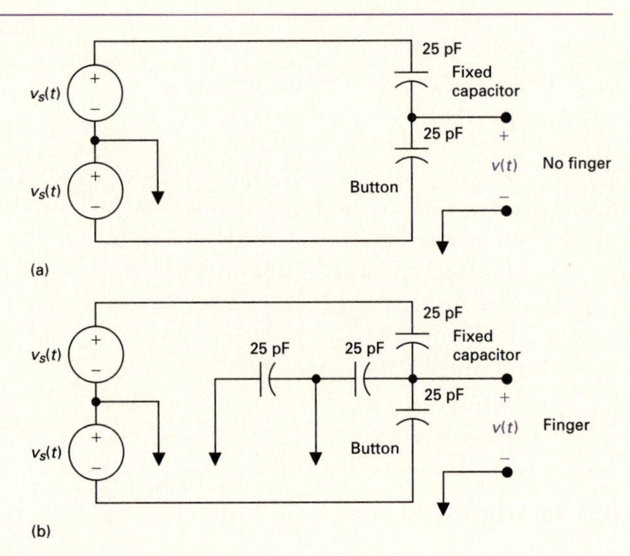

FIGURE P6.34

7 RESPONSE OF FIRST-ORDER *RL* AND *RC* CIRCUITS

CHAPTER CONTENTS

In Chapter 6 we noted that an important attribute of inductors and capacitors is their ability to store energy. We are now in a position to determine the currents and voltages that arise when energy is either released or acquired by an inductor or capacitor in response to an abrupt change in a dc voltage or current source. In this chapter, we will focus on circuits that consist only of sources, resistors, and either (but not both) inductors or capacitors. For brevity, such configurations are called ***RC*** (resistor-capacitor) and ***RL*** (resistor-inductor) **circuits**.

Our analysis of RL and RC circuits will be divided into three phases. In the first phase, we consider the currents and voltages that arise when stored energy in an inductor or capacitor is suddenly released to a resistive network. This happens when the inductor or capacitor is abruptly disconnected from its dc source. Thus we can reduce the circuit to one of the two equivalent forms shown in Fig. 7.1.

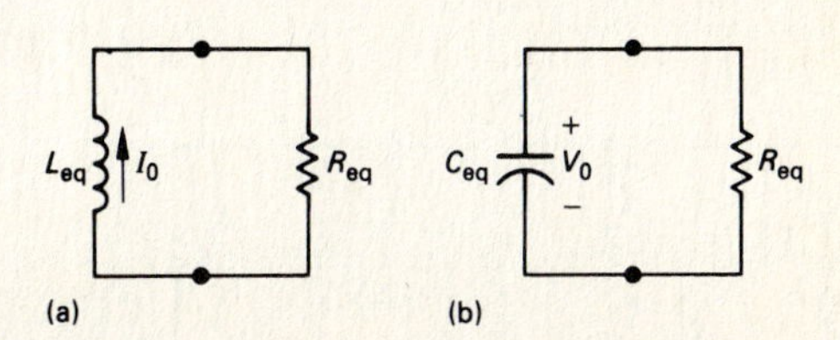

FIGURE 7.1 The two forms of the circuits for natural response: (a) an *RL* circuit; (b) an *RC* circuit.

The currents and voltages which arise in this configuration are referred to as the **natural response** of the circuit, to emphasize that the nature of the circuit itself, not external sources of excitation, determine its behavior.

In the second phase of our analysis, we consider the currents and voltages that arise when energy is being acquired by an inductor or capacitor due to the sudden application of a dc voltage or current source. This response is referred to as the **step response**. The process for finding both the natural and step responses is the same; thus, in the third phase of our analysis, we develop a general method that can be used to find the response of RL and RC circuits to any abrupt change in a dc voltage or current source.

Figure 7.2 shows the four possibilities for the general configuration of RL and RC circuits. Note that when there are no independent sources in the circuit, the Thévenin voltage or Norton current is zero, and the circuit reduces to one of those shown in Fig. 7.1; that is, we have a natural response problem.

RL and RC circuits are also known as first-order circuits, because their voltages and currents are described by first-order differential equations. No matter how complex a circuit may appear, if it can be reduced to a Thévenin or Norton equivalent connected to the terminals of an equivalent inductor or capacitor, it is a first-order circuit. (Note that if multiple inductors or capacitors exist in the original circuit, they must be interconnected so that they can be replaced by a single equivalent element.)

After introducing the techniques for analyzing the natural and step responses of first-order circuits, we discuss some special cases of interest. The first is that of sequential switching, involving circuits in which switching can take place at two or more instants in time. Next is the unbounded response. Finally, we analyze a useful circuit called the integrating amplifier.

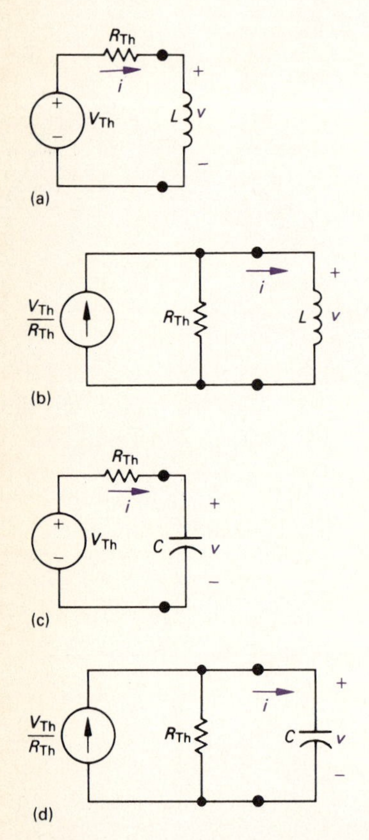

FIGURE 7.2 Four possible first-order circuits: (a) an inductor connected to a Thévenin equivalent; (b) an inductor connected to a Norton equivalent; (c) a capacitor connected to a Thévenin equivalent; (d) a capacitor connected to a Norton equivalent.

7.1 THE NATURAL RESPONSE OF AN *RL* CIRCUIT

The natural response of an RL circuit can best be described in terms of the circuit shown in Fig. 7.3. We assume that the independent current source generates a constant current of I_s amperes and that the switch has been in a closed position for a long time. We define the phrase *a long time* more accurately later in this section. For now it means that all currents and voltages have reached a constant value. Thus only constant, or dc, currents can exist in the circuit just prior to the switch's being opened, and therefore the inductor appears as a short circuit ($Ldi/dt = 0$) prior to the release of the stored energy.

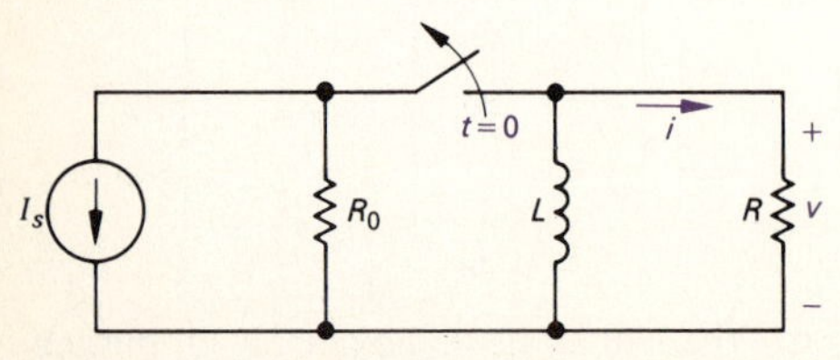

FIGURE 7.3 An RL circuit.

Since the inductor appears as a short circuit, the voltage across the inductive branch is zero, and there can be no current in either R_o or R. Therefore all the source current I_s appears in the inductive branch. Finding the natural response requires finding the voltage and current at the terminals of the resistor after the switch has been opened, that is, after the source has been disconnected and the inductor begins releasing energy. If we let $t = 0$ denote the instant when the switch is opened, the problem becomes one of finding $v(t)$ and $i(t)$ for $t \geq 0$. For $t \geq 0$, the circuit shown in Fig. 7.3 reduces to the one shown in Fig. 7.4.

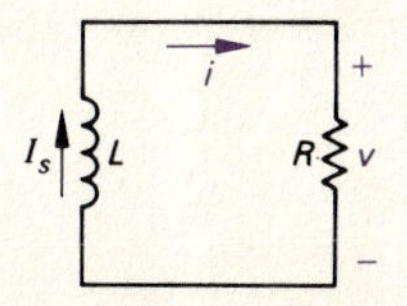

FIGURE 7.4 The circuit shown in Fig. 7.3, for $t \geq 0$.

DERIVING THE EXPRESSION FOR THE CURRENT

To find $i(t)$ we use Kirchhoff's voltage law to obtain an expression involving i, R, and L. Summing the voltages around the closed loop gives

$$L\frac{di}{dt} + Ri = 0, \quad (7.1)$$

where we use the passive sign convention. Equation (7.1) is known as a first-order ordinary differential equation, because it contains terms involving the ordinary derivative of the unknown, that is, di/dt. The highest order derivative appearing in the equation is 1; hence the term **first-order**.

You can use a symbolic equation solver to find the closed-form solution for this first-order equation.

We can go one step farther in describing this equation. The coefficients in the equation, R and L, are constants; that is, they are not functions of either the dependent variable i or the independent variable t. Thus the equation can also be described as an ordinary differential equation with constant coefficients.

To solve Eq. (7.1), we divide through by L, transpose the term involving i to the right-hand side, and then multiply both sides by a differential time dt. The result is

$$\frac{di}{dt}dt = -\frac{R}{L}i\,dt. \quad (7.2)$$

Next, we recognize the left-hand side of Eq. (7.2) as a differential change in the current i, that is, di. We now divide through by i, getting

$$\frac{di}{i} = -\frac{R}{L}dt. \quad (7.3)$$

We obtain an explicit expression for i as a function of t by integrating both sides of Eq. (7.3). Using x and y as variables of integration yields

$$\int_{i(t_0)}^{i(t)} \frac{dx}{x} = -\frac{R}{L}\int_{t_0}^{t} dy, \quad (7.4)$$

in which $i(t_0)$ is the current corresponding to time t_0, and $i(t)$ is the current corresponding to time t. Here, $t_0 = 0$. Therefore carrying out the indicated integration gives

$$\ln \frac{i(t)}{i(0)} = -\frac{R}{L}t. \quad (7.5)$$

Based on the definition of the natural logarithm,

$$i(t) = i(0)e^{-(R/L)t}. \quad (7.6)$$

Recall from Chapter 6 that an instantaneous change of current cannot occur in an inductor. Therefore, in the first instant after the switch has been opened, the current in the inductor remains unchanged. If we use 0^- to denote the time just prior to switching, and 0^+ for the time immediately following switching, then

$$i(0^-) = i(0^+) = I_0,$$

where, as in Fig. 7.1, I_0 denotes the initial current in the inductor. The initial current in the inductor is oriented in the same direction as the reference direction of i. Hence Eq. (7.6) becomes

You can use the computer to plot this equation.

$$i(t) = I_0 e^{-(R/L)t}, \quad t \geq 0, \quad (7.7)$$

which shows that the current starts from an initial value I_0 and decreases exponentially toward zero as t increases. Figure 7.5 shows this response.

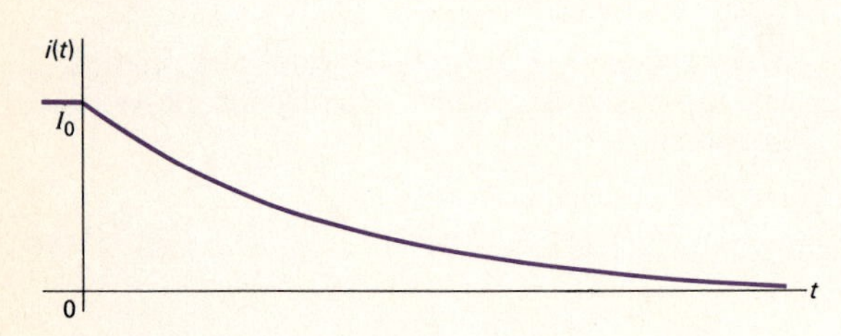

FIGURE 7.5 The current response for the circuit shown in Fig. 7.4.

We derive the voltage across the resistor in Fig. 7.4 from a direct application of Ohm's law:

$$v = iR = I_0 R e^{-(R/L)t}, \quad t \geq 0^+. \quad (7.8)$$

Note that in contrast to the expression for the current shown in Eq. (7.7), the voltage is defined only for $t > 0$, not at $t = 0$. The reason is that a step change occurs in the voltage at zero. Note that for $t < 0$, the derivative of the current is zero, so the voltage is also zero. (This result follows from $v = L\,di/dt = 0$.) Thus

$$v(0^-) = 0 \quad (7.9)$$

and

$$v(0^+) = I_0 R, \quad (7.10)$$

where $v(0^+)$ is obtained from Eq. (7.8) with $t = 0^+$.[1] With this step change at an instant in time, the value of the voltage at $t = 0$ is

[1]We can define the expressions 0^- and 0^+ more formally. The expression $x(0^-)$ refers to the limit of the variable x as $t \rightarrow 0$ from the left, or from negative time. The expression $x(0^+)$ refers to the limit of the variable x as $t \rightarrow 0$ from the right, or from positive time.

unknown. Thus we use $t \geq 0^+$ in defining the region of validity for these solutions.

We derive the power dissipated in the resistor from any of the following expressions:

$$p = vi, \quad p = i^2 R, \quad \text{or} \quad p = \frac{v^2}{R}. \tag{7.11}$$

Whichever form is used, the resulting expression can be reduced to

$$p = I_0^2 R e^{-2(R/L)t}, \quad t \geq 0^+. \tag{7.12}$$

The energy delivered to the resistor during any interval of time after the switch has been opened is

$$\begin{aligned} w &= \int_0^t p dx = \int_0^t I_0^2 R e^{-2(R/L)x} dx \\ &= \frac{1}{2(R/L)} I_0^2 R (1 - e^{-2(R/L)t}) \\ &= \frac{1}{2} L I_0^2 (1 - e^{-2(R/L)t}), \quad t \geq 0. \end{aligned} \tag{7.13}$$

Note from Eq. (7.13) that as t becomes infinite, the energy dissipated in the resistor approaches the initial energy stored in the inductor.

Use the computer to plot the voltage, current, power, and energy, and then compare.

The Significance of the Time Constant

The expressions for $i(t)$ (Eq. 7.7) and $v(t)$ (Eq. 7.8) include a term of the form $e^{-(R/L)t}$. The coefficient of t—namely, R/L—determines the rate at which the current or voltage approaches zero. The reciprocal of this ratio is the **time constant** of the circuit, denoted

$$\tau = \text{time constant} = \frac{L}{R}. \tag{7.14}$$

Using the time-constant concept, we write the expressions for current, voltage, power, and energy as

$$i(t) = I_0 e^{-t/\tau}, \quad t \geq 0; \tag{7.15}$$

$$v(t) = I_0 R e^{-t/\tau}, \quad t \geq 0^+; \tag{7.16}$$

$$p = I_0^2 R e^{-2t/\tau}, \quad t \geq 0^+; \tag{7.17}$$

$$w = \frac{1}{2} L I_0^2 (1 - e^{-2t/\tau}), \quad t \geq 0. \tag{7.18}$$

The time constant is an important parameter for first-order circuits, so mentioning several of its characteristics is worthwhile. First, it is convenient to think of the time elapsed after switching in terms of integral multiples of τ. Thus one time constant after the inductor has begun to release its stored energy to the resistor, the current has been reduced to e^{-1}, or approximately 0.37 of its initial value.

TABLE 7.1

VALUE OF $e^{-t/\tau}$ FOR t EQUAL TO INTEGRAL MULTIPLES OF τ

t	$e^{-t/\tau}$	t	$e^{-t/\tau}$
τ	3.6788×10^{-1}	6τ	2.4788×10^{-3}
2τ	1.3534×10^{-1}	7τ	9.1188×10^{-4}
3τ	4.9787×10^{-2}	8τ	3.3546×10^{-4}
4τ	1.8316×10^{-2}	9τ	1.2341×10^{-4}
5τ	6.7379×10^{-3}	10τ	4.5400×10^{-5}

Table 7.1 gives the value of $e^{-t/\tau}$ for integral multiples of τ from 1 to 10. Note that when the elapsed time exceeds five time constants, the current is less than 1% of its initial value. Thus we sometimes say that five time constants after switching has occurred, the currents and voltages have, for most practical purposes, reached their final values. For single time-constant circuits (first-order circuits) and with 1% accuracy, the phrase *a long time* implies that five or more time constants have elapsed. Thus the existence of current in the RL circuit shown in Fig. 7.1(a) is a momentary event and therefore is also referred to as the **transient response** of the circuit. The response that exists a long time after the switching has taken place is called the **steady-state response**. The phrase *a long time* then also means the time it takes the circuit to reach its steady-state value.

Any first-order circuit is characterized, in part, by the value of its time constant. If we have no method for calculating the time constant of such a circuit (perhaps because we don't know the values of its components), we can determine its value from a plot of the circuit's natural response. That's because another important characteristic of the time constant is that it gives the time required for the current to reach its final value if the current continues to change at its initial rate. To illustrate, we evaluate di/dt at 0^+ and assume that the current continues to change at this rate:

$$\frac{di}{dt}(0^+) = -\frac{R}{L}I_0 = -\frac{I_0}{\tau}. \tag{7.19}$$

Now, if i starts as I_0 and decreases at a constant rate of I_0/τ amperes per second, the expression for i becomes

$$i = I_0 - \frac{I_0}{\tau}t. \tag{7.20}$$

Equation (7.20) indicates that i would reach its final value of zero in τ seconds. Figure 7.6 shows how this graphic interpretation is useful in estimating the time constant of a circuit from a plot of its natural response. Such a plot could be generated on an oscilloscope measuring output current. Drawing the tangent to the natural response plot at $t = 0$ and reading the value at which the tangent intersects the time axis gives the value of τ.

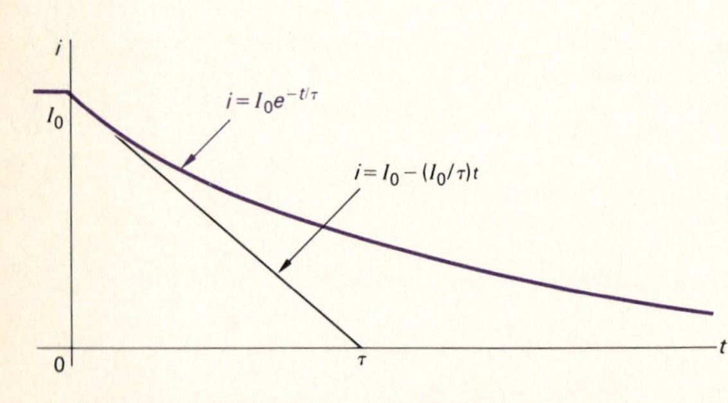

FIGURE 7.6 A graphic interpretation of the time constant of the *RL* circuit shown in Fig. 7.4.

Calculating the natural response of an RL circuit can be summarized as follows:

1. find the initial current, $i(0)$, through the inductor;
2. find the time constant of the circuit; and then
3. use Eq. (7.15) to generate $i(t)$ from $i(0)$ and τ.

All other calculations of interest follow from knowing $i(t)$. Examples 7.1 and 7.2 illustrate the numerical calculations associated with the natural response of an RL circuit.

EXAMPLE 7.1

The switch in the circuit shown in Fig. 7.7 has been closed for a long time before it is opened at $t = 0$. Find

a) $i_L(t)$ for $t \geq 0$;

b) $i_o(t)$ for $t \geq 0^+$;

c) $v_o(t)$ for $t \geq 0^+$; and

d) the percentage of the total energy stored in the 2 H inductor that is dissipated in the 10 Ω resistor.

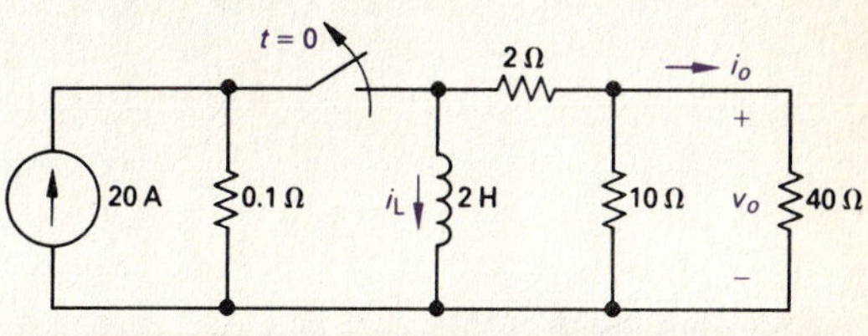

FIGURE 7.7 The circuit for Example 7.1

SOLUTION

a) The switch has been closed for a long time prior to $t = 0$, so we know the voltage across the inductor must be zero at $t = 0^-$. Therefore the initial current in the inductor is 20 A at $t = 0^-$. Hence $i_L(0^+)$ also is 20 A because an instantaneous change in the current cannot occur in an inductor. We replace the resistive circuit connected to the terminals of the inductor with a single resistor of 10 Ω:

$$R_{\text{eq}} = 2 + (40 \parallel 10) = 10\ \Omega.$$

The time constant of the circuit is L/R_{eq}, or 0.2 s, giving the expression for the inductor current as

$$i_L(t) = 20e^{-5t}\ \text{A}, \quad t \geq 0.$$

b) We find the current in the 40 Ω resistor most easily by using current division; that is,

$$i_o = -i_L \frac{10}{10 + 40}.$$

Note that this expression is valid for $t \geq 0^+$ because $i_o = 0$ at $t = 0^-$. The inductor behaves as a short circuit prior to the switch being opened, producing an instantaneous change in the current i_o. Then,

$$i_o(t) = -4e^{-5t}\ \text{A}, \quad t \geq 0^+.$$

c) We find the voltage v_o by direct application of Ohm's law:

$$v_o(t) = 40 i_o = -160e^{-5t}\ \text{V}, \quad t \geq 0^+.$$

d) The power dissipated in the 10 Ω resistor is

$$p_{10\Omega}(t) = \frac{v_o^2}{10} = 2560e^{-10t}\ \text{W}, \quad t \geq 0^+.$$

The total energy dissipated in the 10 Ω resistor is

$$w_{10\Omega}(t) = \int_0^\infty 2560e^{-10t}\,dt = 256\text{ J}.$$

The initial energy stored in the 2 H inductor is

$$w(0) = \frac{1}{2}Li^2(0) = \frac{1}{2}(2)(400) = 400\text{ J}.$$

Therefore the percentage of energy dissipated in the 10 Ω resistor is

$$\text{percentage dissipated} = \frac{256}{400}(100) = 64\%.$$

EXAMPLE 7.2

In the circuit shown in Fig. 7.8, the initial currents in inductors L_1 and L_2 have been established by sources not shown. The switch is opened at $t = 0$.

a) Find i_1, i_2, and i_3 for $t \geq 0$.

b) Calculate the initial energy stored in the parallel inductors.

c) Determine how much energy is stored in the inductors as $t \to \infty$.

d) Show that the total energy delivered to the resistive network equals the difference between the results obtained in (b) and (c).

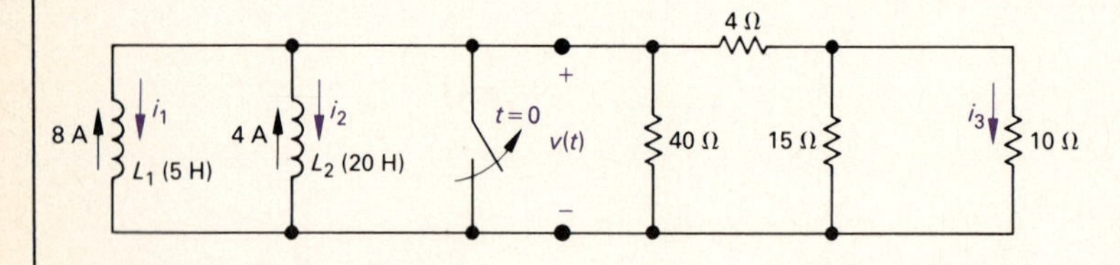

FIGURE 7.8 The circuit for Example 7.2

SOLUTION

a) The key to finding currents i_1, i_2, and i_3 lies in knowing the voltage $v(t)$. We can easily find $v(t)$ if we reduce the circuit shown in Fig. 7.8 to the equivalent form shown in Fig. 7.9. The parallel inductors simplify to an equivalent inductance of 4 H, carrying an initial current of 12 A. The resistive network reduces to a single resistance of 8 Ω. Hence the initial value of $i(t)$ is 12 A and the time constant is $\frac{4}{8}$, or 0.5 s. Therefore

$$i(t) = 12e^{-2t}\text{ A}, \qquad t \geq 0.$$

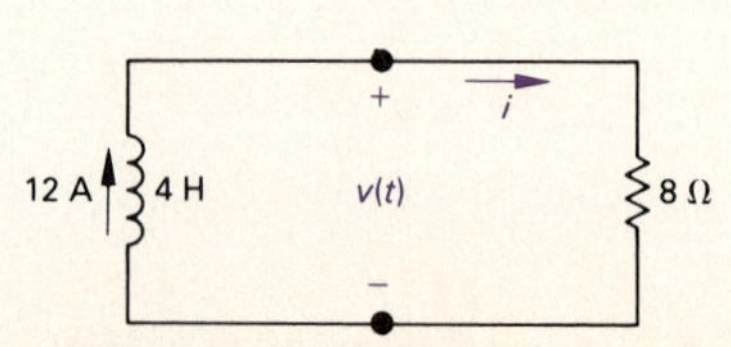

FIGURE 7.9 A simplification of the circuit shown in Fig. 7.8.

Now $v(t)$ is simply the product $8i$, so

$$v(t) = 96e^{-2t}\ \text{V}, \quad t \geq 0^+.$$

The circuit shows that $v(t) = 0$ at $t = 0^-$, so the expression for $v(t)$ is valid for $t \geq 0^+$. After obtaining $v(t)$, we can calculate i_1, i_2, and i_3:

$$i_1 = \frac{1}{5}\int_0^t 96e^{-2x}dx - 8$$

$$= 1.6 - 9.6e^{-2t}\ \text{A}, \quad t \geq 0;$$

$$i_2 = \frac{1}{20}\int_0^t 96e^{-2x}dx - 4$$

$$= -1.6 - 2.4e^{-2t}\ \text{A}, \quad t \geq 0;$$

$$i_3 = \frac{v(t)}{10}\frac{15}{25} = 5.76e^{-2t}\ \text{A}, \quad t \geq 0^+.$$

Note that the expressions for the inductor currents i_1 and i_2 are valid for $t \geq 0$, whereas the expression for the resistor current i_3 is valid for $t \geq 0^+$.

b) The initial energy stored in the inductors is

$$w = \frac{1}{2}(5)(64) + \frac{1}{2}(20)(16) = 320\ \text{J}.$$

c) As $t \to \infty$, $i \to 1.6$ A and $i_2 \to -1.6$ A. Therefore a long time after the switch has been opened, the energy stored in the two inductors is

$$w = \frac{1}{2}(5)(1.6)^2 + \frac{1}{2}(20)(-1.6)^2 = 32\ \text{J}.$$

d) We obtain the total energy delivered to the resistive network by integrating the expression for the instantaneous power from zero to infinity:

$$w = \int_0^\infty p\,dt = \int_0^\infty 1152e^{-4t}dt$$

$$= 1152\left.\frac{e^{-4t}}{-4}\right|_0^\infty = 288\ \text{J}.$$

This result is the difference between the initially stored energy (320 J) and the energy trapped in the parallel inductors (32 J). The equivalent inductor for the parallel inductors (which predicts the terminal behavior of the parallel combination) has an initial energy of 288 J; that is, the energy stored in the equivalent inductor represents the amount of energy that will be delivered to the resistive network at the terminals of the original inductors.

DRILL EXERCISES

7.1 The switch in the circuit shown has been closed for a long time and is opened at $t = 0$.

a) Calculate the initial value of i.

b) Calculate the initial energy stored in the inductor.

c) What is the time constant of the circuit for $t > 0$?

d) What is the numerical expression for $i(t)$ for $t \geq 0$?

e) What percentage of the initial energy stored has been dissipated in the 4 Ω resistor 5 ms after the switch has been opened?

ANSWER: (a) -16 A; (b) 1.28 J; (c) $t = 2.5$ ms; (d) $-16e^{-400t}$ A; (e) 98.17%.

7.2 At $t = 0$, the switch in the circuit shown moves instantaneously from position a to position b.

a) Calculate v_o for $t \geq 0^+$.

b) What percentage of the initial energy stored in the inductor is eventually dissipated in the 4 Ω resistor?

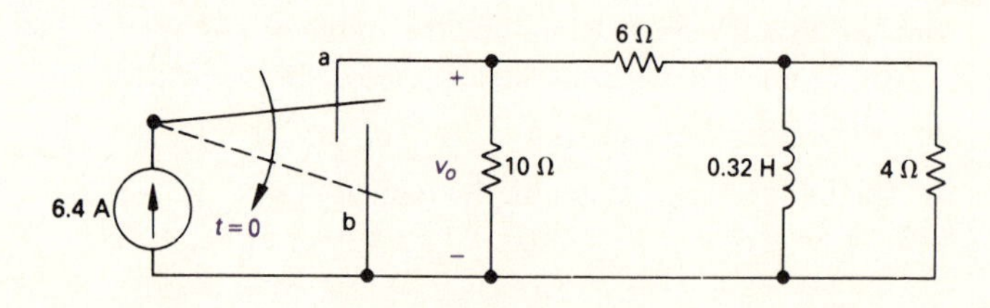

ANSWER: (a) $-8e^{-10t}$ V; (b) 80%.

7.2 THE NATURAL RESPONSE OF AN *RC* CIRCUIT

As mentioned in Section 7.1, the natural response of an RC circuit is analogous to that of an RL circuit. Consequently, we don't treat the RC circuit in the same detail as we did the RL circuit.

The natural response of an RC circuit is developed from the circuit shown in Fig. 7.10. We begin by assuming that the switch has been in position a for a long time, allowing the loop made up of the dc voltage source V_g, the resistor R_1, and the capacitor C to reach a steady-state condition. Recall from Chapter 6 that a capacitor behaves as an open circuit in the presence of a constant voltage. Thus the voltage source cannot sustain a current, and so the source voltage appears across the capacitor terminals. In Section 7.3 we will discuss how the capacitor

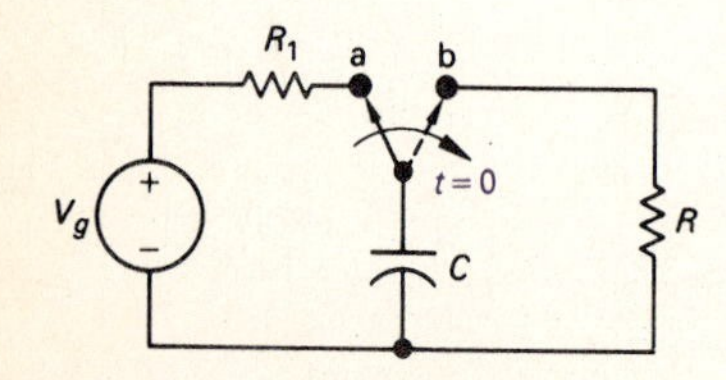

FIGURE 7.10 An *RC* circuit.

voltage actually builds to the steady-state value of the dc voltage source, but for now the important point is that when the switch is moved from position a to position b (at $t = 0$), the voltage on the capacitor is V_g. Because there can be no instantaneous change in the voltage at the terminals of a capacitor, the problem reduces to solving the circuit shown in Fig. 7.11.

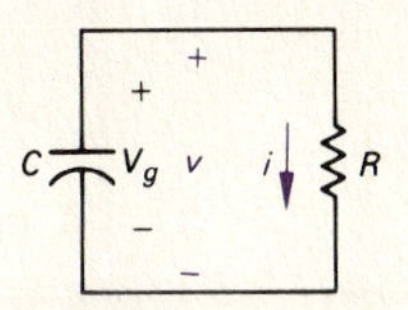

FIGURE 7.11 The circuit shown in Fig. 7.10, after switching.

DERIVING THE EXPRESSION FOR THE VOLTAGE

We can easily find the voltage $v(t)$ by thinking in terms of node voltages. Using the lower junction between R and C as the reference node and summing the currents away from the upper junction between R and C gives

$$C\frac{dv}{dt} + \frac{v}{R} = 0. \tag{7.21}$$

Comparing Eq. (7.21) with Eq. (7.1) shows that the same mathematical techniques can be used to obtain the solution for $v(t)$. We leave it to you to show that

Use a computer tool to help with this derivation.

$$v(t) = v(0)e^{-t/RC}, \quad t \geq 0. \tag{7.22}$$

As we have already noted, the initial voltage on the capacitor equals the voltage source voltage V_g, or

$$\begin{aligned} v(0^-) &= v(0) = v(0^+) \\ &= V_g = V_0, \end{aligned} \tag{7.23}$$

where V_0 denotes the initial voltage on the capacitor. The time constant for the RC circuit equals the product of the resistance and capacitance, namely,

$$\tau = RC. \tag{7.24}$$

Substituting Eqs. (7.23) and (7.24) into Eq. (7.22) yields

$$v(t) = V_0 e^{-t/\tau}, \quad t \geq 0, \tag{7.25}$$

which indicates that the natural response of an RC circuit is an exponential decay of the initial voltage. The time constant RC governs the rate of decay. Figure 7.12 shows the plot of Eq. (7.25) and the graphic interpretation of the time constant.

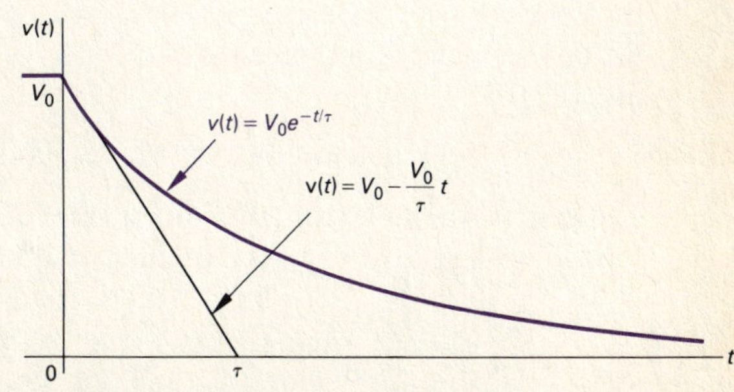

FIGURE 7.12 The natural response of an *RC* circuit.

After determining $v(t)$, we can easily derive the expressions for i, p, and w:

$$i(t) = \frac{v(t)}{R} = \frac{V_0}{R}e^{-t/\tau}, \quad t \geq 0^+; \tag{7.26}$$

$$p = vi = \frac{V_0^2}{R}e^{-2t/\tau}, \quad t \geq 0^+; \tag{7.27}$$

$$w = \int_0^t p dx = \int_0^t \frac{V_0^2}{R} e^{-2x/\tau} dx$$
$$= \frac{1}{2} C V_0^2 (1 - e^{-2t/\tau}), \quad t \geq 0. \qquad \textbf{(7.28)}$$

Calculating the natural response of an RC circuit can be summarized as follows:

1. find the initial voltage, $v(0)$, across the capacitor;
2. find the time constant of the circuit; and then
3. use Eq. (7.25) to generate $v(t)$ from $v(0)$ and τ.

All other calculations of interest follow from knowing $v(t)$. Examples 7.3 and 7.4 illustrate the numerical calculations associated with the natural response of an RC circuit.

EXAMPLE 7.3

The switch in the circuit shown in Fig. 7.13 has been in position x for a long time. At $t = 0$, the switch moves instantaneously to position y. Find

a) $v_c(t)$ for $t \geq 0$;

b) $v_o(t)$ for $t \geq 0^+$;

c) $i_o(t)$ for $t \geq 0^+$; and

d) the total energy dissipated in the 60 kΩ resistor.

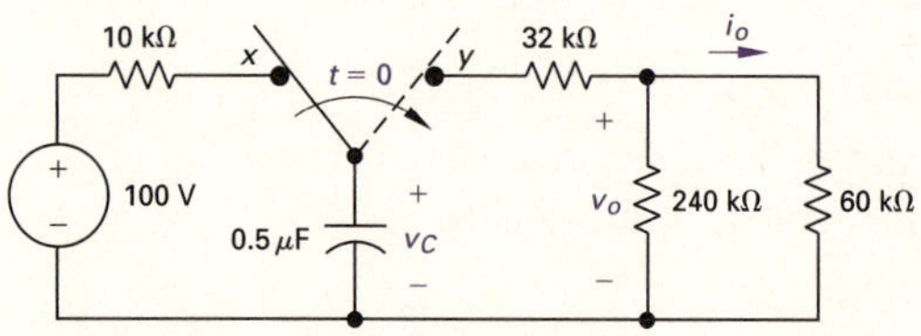

FIGURE 7.13 The circuit for Example 7.3

SOLUTION

a) Because the switch has been in position x for a long time, the 0.5 μF capacitor will charge to 100 V and be positive at the upper terminal. We can replace the resistive network connected to the capacitor at $t = 0^+$ with an equivalent resistance of 80 kΩ . Hence the time constant of the circuit is $(0.5 \times 10^{-6})(80 \times 10^3)$, or 40 ms. Then,

$$v_C(t) = 100e^{-25t} \text{ V}, \quad t \geq 0.$$

b) The easiest way to find $v_o(t)$ is to note that the resistive circuit forms a voltage divider across the terminals of the capacitor. Thus

$$v_o(t) = \frac{48}{80} v_C(t) = 60e^{-25t} \text{ V}, \quad t \geq 0^+.$$

This expression for $v_o(t)$ is valid for $t \geq 0^+$ because $v_o(0^-)$ is zero. Thus we have an instantaneous change in the voltage across the 240 kΩ resistor.

c) We find the current $i_o(t)$ from Ohm's law:

$$i_o(t) = \frac{v_o(t)}{60 \times 10^3} = e^{-25t} \text{ mA}, \quad t \geq 0^+.$$

d) The power dissipated in the 60 kΩ resistor is

$$p_{60k\Omega}(t) = i_o^2(t)(60 \times 10^3) = 60e^{-50t} \text{ mW}, \quad t \geq 0^+.$$

The total energy dissipated is

$$w_{60k\Omega} = \int_0^\infty i_o^2(t)(60 \times 10^3)dt = 1.2 \text{ mJ}.$$

EXAMPLE 7.4

The initial voltages on capacitors C_1 and C_2 in the circuit shown in Fig. 7.14 have been established by sources not shown. The switch is closed at $t = 0$.

a) Find $v_1(t)$, $v_2(t)$, and $v(t)$ for $t \geq 0$, and $i(t)$ for $t \geq 0^+$.

b) Calculate the initial energy stored in the capacitors C_1 and C_2.

c) Determine how much energy is stored in the capacitors as $t \to \infty$.

d) Show that the total energy delivered to the 250 kΩ resistor is the difference between the results obtained in (b) and (c).

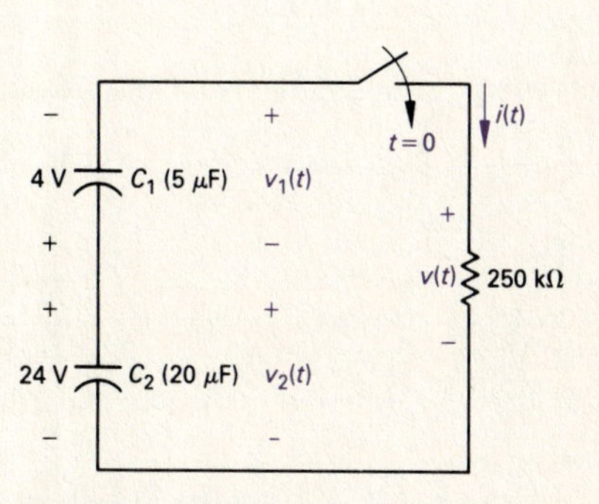

FIGURE 7.14 The circuit for Example 7.4.

SOLUTION

a) Once we know $v(t)$, we can obtain the current $i(t)$ from Ohm's law. After determining $i(t)$, we can calculate $v_1(t)$ and $v_2(t)$ because the voltage across a capacitor is a function of the capacitor current. To find $v(t)$ we replace the series-connected capacitors with an equivalent capacitor. It has a capacitance of 4 μF and is charged to a voltage of 20 V. Therefore the circuit shown in Fig. 7.14 reduces to the one shown in Fig. 7.15, which reveals that the initial value of $v(t)$ is 20 V and that the time constant of the circuit is $(4)(250) \times 10^{-3}$, or 1 s. Thus the expression for $v(t)$ is

$$v(t) = 20e^{-t} \text{ V}, \quad t \geq 0.$$

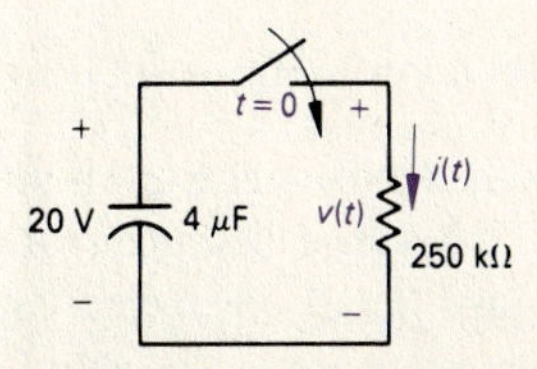

FIGURE 7.15 A simplification of the circuit shown in Fig. 7.14

The current $i(t)$ is

$$i(t) = \frac{v(t)}{250{,}000} = 80e^{-t} \ \mu\text{A}, \quad t \geq 0^+.$$

Knowing $i(t)$, we calculate the expressions for $v_1(t)$ and $v_2(t)$:

$$\begin{aligned} v_1(t) &= -\frac{10^6}{5} \int_0^t 80 \times 10^{-6} e^{-x} dx - 4 \\ &= (16e^{-t} - 20) \text{ V}, \quad t \geq 0; \end{aligned}$$

$$\begin{aligned} v_2(t) &= -\frac{10^6}{20} \int_0^t 80 \times 10^{-6} e^{-x} dx + 24 \\ &= (4e^{-t} + 20) \text{ V}, \quad t \geq 0. \end{aligned}$$

b) The initial energy stored in C_1 is

$$w_1 = \frac{1}{2}(5 \times 10^{-6})(16) = 40 \ \mu\text{J}.$$

The initial energy stored in C_2 is

$$w_2 = \frac{1}{2}(20 \times 10^{-6})(576) = 5760 \ \mu\text{J}.$$

The total energy stored in the two capacitors is

$$w_o = 40 + 5760 = 5800 \ \mu\text{J}.$$

c) As $t \to \infty$,

$$v_1 \to -20 \text{ V} \quad \text{and} \quad v_2 \to +20 \text{ V}.$$

Therefore the energy stored in the two capacitors is

$$w_\infty = \frac{1}{2}(5 + 20) \times 10^{-6}(400) = 5000 \ \mu\text{J}.$$

d) The total energy delivered to the 250 kΩ resistor is

$$w = \int_0^\infty p dt = \int_0^\infty \frac{400e^{-2t}}{250{,}000} dt = 800 \ \mu\text{J}.$$

Comparing the results obtained in (b) and (c) shows that

$$800 \ \mu\text{J} = (5800 - 5000) \ \mu\text{J}.$$

The energy stored in the equivalent capacitor in Fig. 7.15 is $\frac{1}{2}(4 \times 10^{-6})(400)$, or 800 μJ. Because this capacitor predicts the terminal behavior of the original series-connected capacitors, the energy stored in the equivalent capacitor is the energy delivered to the 250 kΩ resistor.

DRILL EXERCISES

7.3 The switch in the circuit shown has been closed for a long time and is opened at $t = 0$. Find

a) the initial value of $v(t)$;

b) the time constant for $t > 0$;

c) the numerical expression for $v(t)$ after the switch has been opened;

d) the initial energy stored in the capacitor; and

e) the length of time required to dissipate 75% of the initially stored energy.

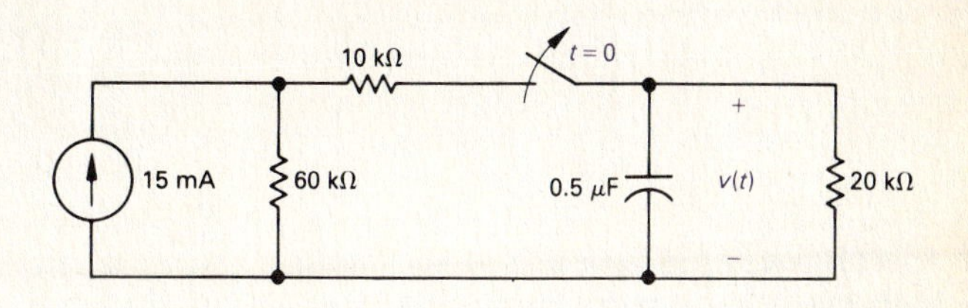

ANSWER: (a) 200 V; (b) 10 ms; (c) $200e^{-100t}$ V; (d) 10 mJ; (e) 6.93 ms.

7.4 The switch in the circuit shown has been closed for a long time before being opened at $t = 0$.

a) Find $v_o(t)$ for $t \geq 0$.

b) What percentage of the initial energy stored in the circuit has been dissipated after the switch has been open for 60 ms?

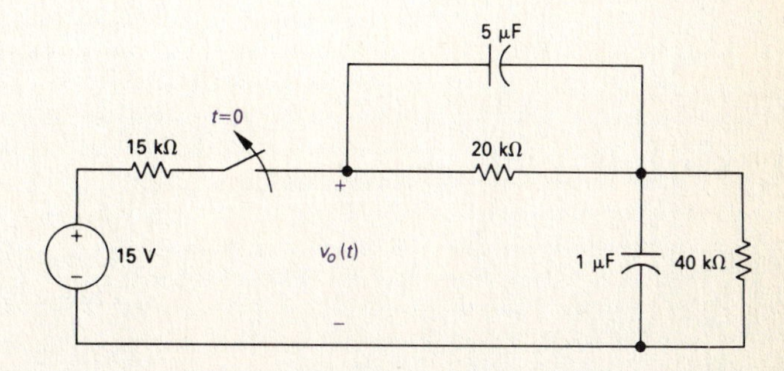

ANSWER: (a) $8e^{-25t} + 4e^{-10t}$ V; (b) 81.05%.

7.3 THE STEP RESPONSE OF *RL* AND *RC* CIRCUITS

We are now ready to discuss the problem of finding the currents and voltages generated in first-order RL or RC circuits when either dc voltage or current sources are suddenly applied. The response of a circuit to the sudden application of a constant voltage or current source is referred to as the **step response** of the circuit. In presenting the step response we show how the circuit responds when energy is being stored in the inductor or capacitor. We begin with the step response of an RL circuit.

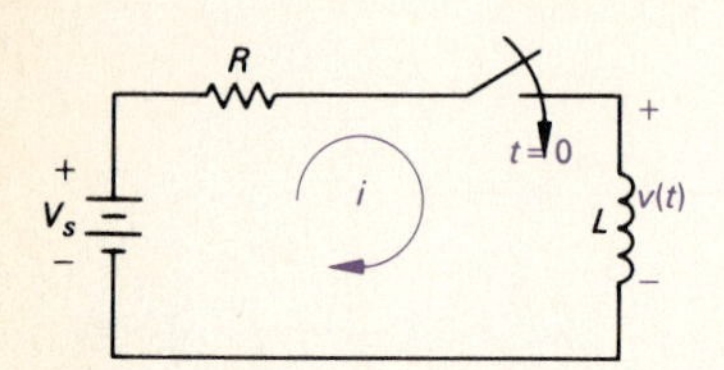

FIGURE 7.16 A circuit used to illustrate the step response of a first-order *RL* circuit.

You can use a symbolic equation solver to find the closed-form solution for this first-order equation.

THE STEP RESPONSE OF AN *RL* CIRCUIT

To begin, we modify the first-order circuit shown in Fig. 7.2(a) by adding a switch. We use the resulting circuit, shown in Fig. 7.16, in developing the step response of an RL circuit. Energy stored in the inductor at the time the switch is closed is given in terms of a nonzero initial current $i(0)$. The task is to find the expressions for the current in the circuit and for the voltage across the inductor after the switch has been closed. The procedure is the same as that used in Section 7.1; we use circuit analysis to derive the differential equation that describes the circuit in terms of the variable of interest and then use elementary calculus to solve the equation.

After the switch in Fig. 7.16 has been closed, Kirchhoff's voltage law requires that

$$V_s = Ri + L\frac{di}{dt}, \tag{7.29}$$

which can be solved for the current by separating the variables i and t and then integrating. The first step in this approach is to solve Eq. (7.29) for the derivative di/dt:

$$\frac{di}{dt} = \frac{-Ri + V_s}{L} = \frac{-R}{L}\left(i - \frac{V_s}{R}\right). \tag{7.30}$$

Next, we multiply both sides of Eq. (7.30) by a differential time dt. This step reduces the left-hand side of the equation to a differential change in the current. Thus

$$\frac{di}{dt}dt = \frac{-R}{L}\left(i - \frac{V_s}{R}\right)dt,$$

or

$$di = \frac{-R}{L}\left(i - \frac{V_s}{R}\right)dt. \tag{7.31}$$

We now separate the variables in Eq. (7.31) to get

$$\frac{di}{i - (V_s/R)} = \frac{-R}{L}dt, \tag{7.32}$$

and then integrate both sides of Eq. (7.32). Using x and y as variables for the integration, we obtain

$$\int_{I_0}^{i(t)} \frac{dx}{x - (V_s/R)} = \frac{-R}{L}\int_0^t dy, \tag{7.33}$$

where I_0 is the current at $t = 0$ and $i(t)$ is the current at any $t > 0$. Performing the integration called for in Eq. (7.33) generates the expression

$$ln\frac{i(t) - (V_s/R)}{I_0 - (V_s/R)} = \frac{-R}{L}t, \tag{7.34}$$

from which

$$\frac{i(t) - (V_s/R)}{I_0 - (V_s/R)} = e^{-(R/L)t},$$

or

$$i(t) = \frac{V_s}{R} + \left(I_0 - \frac{V_s}{R}\right)e^{-(R/L)t}. \quad \textbf{(7.35)}$$

When the initial energy in the inductor is zero, I_0 is zero. Thus Eq. (7.35) reduces to

$$i(t) = \frac{V_s}{R} - \frac{V_s}{R}e^{-(R/L)t}. \quad \textbf{(7.36)}$$

Equation (7.36) indicates that after the switch has been closed, the current increases exponentially from zero to a final value of V_s/R. The time constant of the circuit, L/R, determines the rate of increase. One time constant after the switch has been closed, the current will have reached approximately 63% of its final value, or

$$i(\tau) = \frac{V_s}{R} - \frac{V_s}{R}e^{-1} \approx 0.6321\frac{V_s}{R}. \quad \textbf{(7.37)}$$

If the current were to continue to increase at its initial rate, it would reach its final value at $t = \tau$; that is, because

$$\frac{di}{dt} = \frac{-V_s}{R}\left(\frac{-1}{\tau}\right)e^{-t/\tau} = \frac{V_s}{L}e^{-t/\tau}, \quad \textbf{(7.38)}$$

the initial rate at which $i(t)$ increases is

$$\frac{di}{dt}(0) = \frac{V_s}{L}. \quad \textbf{(7.39)}$$

If the current were to continue to increase at this rate, the expression for i would be

$$i = \frac{V_s}{L}t, \quad \textbf{(7.40)}$$

from which, at $t = \tau$,

$$i = \frac{V_s}{L}\frac{L}{R} = \frac{V_s}{R}. \quad \textbf{(7.41)}$$

Equations (7.36) and (7.40) are plotted in Fig. 7.17. The values given by Eqs. (7.37) and (7.41) are also shown in this figure.

The voltage across an inductor is Ldi/dt, so from Eq. (7.35), for $t \geq 0^+$,

$$v = L\left(\frac{-R}{L}\right)\left(I_0 - \frac{V_s}{R}\right)e^{-(R/L)t} = (V_s - I_0R)e^{-(R/L)t}. \quad \textbf{(7.42)}$$

The voltage across the inductor is zero before the switch is closed. Equation (7.42) indicates that the inductor voltage jumps to $V_s - I_0R$

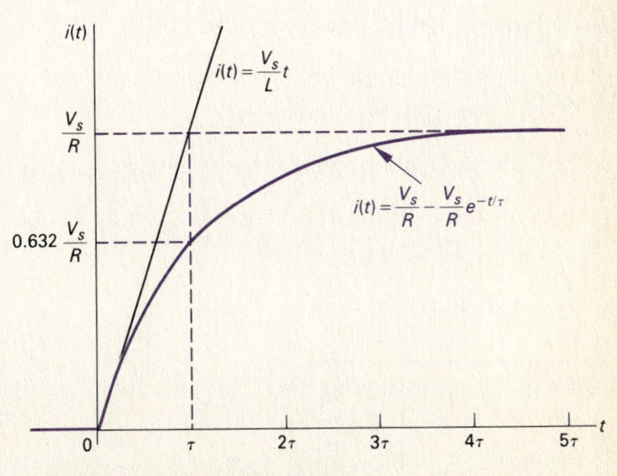

FIGURE 7.17 The step response of the *RL* circuit shown in Fig. 7.16 when $I_0 = 0$.

at the instant the switch is closed and then decays exponentially to zero.

Does the value of v at $t = 0^+$ make sense? Because the initial current is I_0 and the inductor prevents an instantaneous change in current, the current is I_0 during the instant after the switch has been closed. The voltage drop across the resistor is $I_0 R$, and the voltage impressed across the inductor is the source voltage minus this voltage, that is, $V_s - I_0 R$.

When the initial inductor current is zero, Eq. (7.42) simplifies to

$$v = V_s e^{-(R/L)t}. \tag{7.43}$$

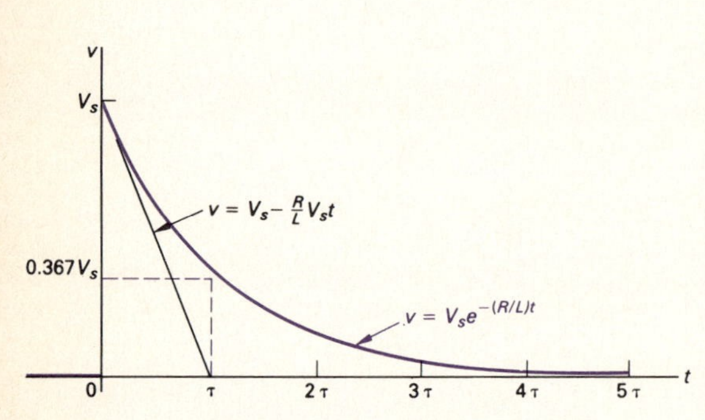

FIGURE 7.18 Inductor voltage versus time.

If the initial current is zero, the voltage across the inductor jumps to V_s. We also expect the inductor voltage to approach zero as t increases because the current in the circuit is approaching the constant value of V_s/R. Figure 7.18 shows the plot of Eq. (7.43) and the relationship between the time constant and the initial rate at which the inductor voltage is decreasing.

If there is an initial current in the inductor, Eq. (7.35) gives the solution for it. The algebraic sign of I_0 is positive if the initial current is in the same direction as i; otherwise, I_0 carries a negative sign. Example 7.5 illustrates the application of Eq. (7.35) to a specific circuit.

EXAMPLE 7.5

The switch in the circuit shown in Fig. 7.19 has been in position a for a long time. At $t = 0$, the switch moves from position a to position b. The switch is a make-before-break type; that is, the connection at position b is established before the connection at position a is broken, so there is no interruption of current through the inductor.

a) Find the expression for $i(t)$ for $t \geq 0$.

b) What is the initial voltage across the inductor just after the switch has been moved to position b?

c) Does this initial voltage make sense in terms of circuit behavior?

d) How many milliseconds after the switch has been moved does the inductor voltage equal 24 V?

e) Plot both $i(t)$ and $v(t)$ versus t.

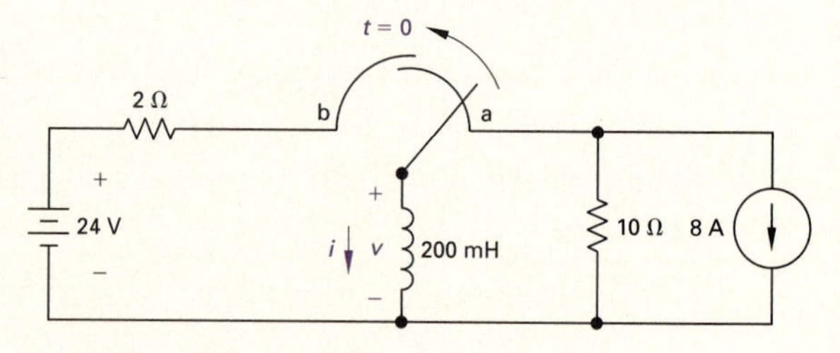

FIGURE 7.19 The circuit for Example 7.5.

SOLUTION

a) The switch has been in position a for a long time, so the 200 mH inductor is a short circuit across the 8 A current source. Therefore the inductor carries an initial current of 8 A. This current is oriented opposite to the reference direction for i; thus I_0 is -8 A. When the switch is in position b, the final value of i will be 24/2, or 12 A. The time constant of the circuit is 200/2, or 100 ms. Substituting these values into Eq. (7.35) gives

$$i = 12 + (-8 - 12)e^{-t/0.1}$$
$$= 12 - 20e^{-10t} \text{ A}, \quad t \geq 0.$$

b) The voltage across the inductor is

$$v = L\frac{di}{dt} = 0.2(200e^{-10t}) = 40e^{-10t} \text{ V}, \quad t \geq 0^+.$$

The initial inductor voltage is

$$v(0^+) = 40 \text{ V}.$$

c) Yes; in the instant after the switch has been moved to position b, the inductor sustains a current of 8 A counterclockwise around the newly formed closed path. This current causes a 16 V drop across the 2 Ω resistor. This voltage drop adds to the drop across the source, producing a 40 V drop across the inductor.

d) We find the time at which the inductor voltage equals 24 V by solving the expression

$$24 = 40e^{-10t}$$

for t:

$$t = \frac{1}{10}ln\frac{40}{24} = 51.08 \times 10^{-3} = 51.08 \text{ ms}.$$

e) Figure 7.20 shows the graphs of $i(t)$ and $v(t)$ versus t. Note that the instant of time when the current equals zero corresponds to the instant of time when the inductor voltage equals the source voltage of 24 V, as predicted by Kirchhoff's voltage law.

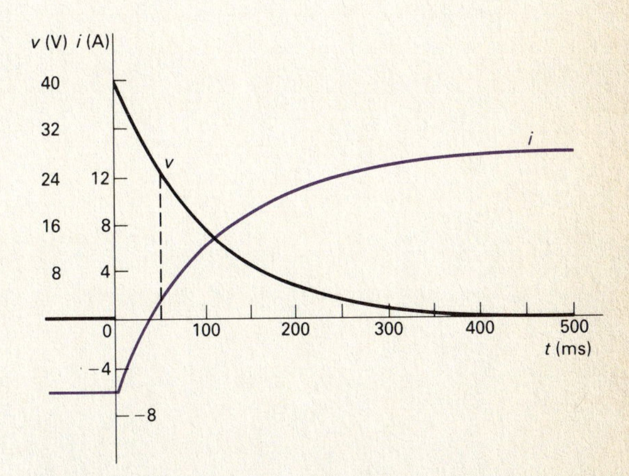

FIGURE 7.20 The current and voltage waveforms for Example 7.5.

Use a computer tool to generate these plots.

DRILL EXERCISE

7.5 Assume that the switch in the circuit shown in Fig. 7.19 has been in position b for a long time, and at $t = 0$ it moves to position a. Find (a) $i(0^+)$; (b) $v(0^+)$; (c) τ, $t > 0$; (d) $i(t)$, $t \geq 0$; and (e) $v(t)$, $t \geq 0^+$.

ANSWER: (a) 12 A; (b) −200 V; (c) 20 ms; (d) $-8 + 20e^{-50t}$ A, $t \geq 0$; (e) $-200e^{-50t}$ V, $t \geq 0^+$.

We can also describe the voltage $v(t)$ across the inductor in Fig. 7.16 directly, not just in terms of the circuit current. We begin by noting that the voltage across the resistor is the difference between the source voltage and the inductor voltage. We write

$$i(t) = \frac{V_s}{R} - \frac{v(t)}{R}, \tag{7.44}$$

where V_s is a constant. Differentiating both sides with respect to time yields

$$\frac{di}{dt} = -\frac{1}{R}\frac{dv}{dt}. \tag{7.45}$$

Then, if we multiply each side of Eq. (7.45) by the inductance L, we get an expression for the voltage across the inductor on the left-hand side, or

$$v = -\frac{L}{R}\frac{dv}{dt}. \tag{7.46}$$

Putting Eq. (7.46) into standard form yields

$$\frac{dv}{dt} + \frac{R}{L}v = 0. \tag{7.47}$$

You should verify (in Drill Exercise 7.6) that the solution to Eq. (7.47) is identical to that given in Eq. (7.42).

At this point a general observation about the step response of an RL circuit is pertinent. (This observation will prove helpful later.) When we derived the differential equation for the inductor current, we obtained Eq. (7.29). We now rewrite Eq. (7.29) as

$$\frac{di}{dt} + \frac{R}{L}i = \frac{V_s}{L}. \tag{7.48}$$

Observe that Eqs. (7.47) and (7.48) have the same form. Specifically, each equates the sum of the first derivative of the variable and a constant times the variable to a constant value. In Eq. (7.47), the constant on the right-hand side happens to be zero; hence this equation takes on the same form as the natural response equations in Section 7.1. In both Eq. (7.47) and Eq. (7.48), the constant multiplying the dependent variable is the reciprocal of the time constant, that is, $R/L = 1/\tau$. We encounter a similar situation in the derivations for the step response of an RC circuit. In Section 7.4 we will use these observations to develop a general approach to finding the natural and step responses of RL and RC circuits.

DRILL EXERCISE

7.6 a) Derive Eq. (7.47) by first converting the Thévenin equivalent in Fig. 7.16 to a Norton equivalent and then summing the currents away from the upper node, using the inductor voltage v as the variable of interest.

b) Use the separation of variables technique to find the solution to Eq. (7.47). Verify that your solution agrees with the solution given in Eq. (7.42).

ANSWER: (a) Derivation; (b) verification.

THE STEP RESPONSE OF AN *RC* CIRCUIT

We can find the step response of a first-order RC circuit by analyzing the circuit shown in Fig. 7.21. For mathematical convenience, we choose the Norton equivalent of the network connected to the equivalent capacitor. Summing the currents away from the top node in Fig. 7.21 generates the differential equation

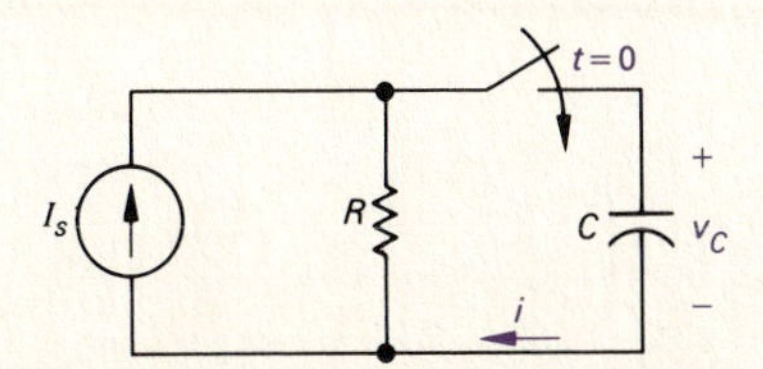

FIGURE 7.21 A circuit used to illustrate the step response of a first-order *RC* circuit.

$$C\frac{dv_C}{dt} + \frac{v_C}{R} = I_s. \tag{7.49}$$

Division of Eq. (7.49) by C gives

$$\frac{dv_C}{dt} + \frac{v_C}{RC} = \frac{I_s}{C}. \tag{7.50}$$

Comparing Eq. (7.50) with Eq. (7.48) reveals that the form of the solution for v_C is the same as that for the current in the inductive circuit, namely, Eq. (7.35). Therefore, by simply substituting the appropriate variables and coefficients, we can write the solution for v_C directly. The translation requires that I_s replace V_s, C replace L, $1/R$ replace R, and V_0 replace I_0. We get

$$v_C = I_s R + (V_0 - I_s R)e^{-t/RC}, \quad t \geq 0. \tag{7.51}$$

A similar derivation for the current in the capacitor yields the differential equation

$$\frac{di}{dt} + \frac{1}{RC}i = 0. \tag{7.52}$$

Equation (7.52) has the same form as Eq. (7.47), hence the solution for i is obtained by using the same translations used for the solution of Eq. (7.50). Thus

$$i = \left(I_s - \frac{V_0}{R}\right)e^{-t/RC}, \quad t \geq 0^+, \tag{7.53}$$

where V_0 is the initial value of v_C, the voltage across the capacitor.

We obtained Eqs. (7.51) and (7.53) by using a mathematical analogy with the solution for the step response of the inductive circuit. Let's see whether these solutions for the RC circuit make sense in terms of known circuit behavior. From Eq. (7.51), note that the initial voltage across the capacitor is V_0, the final voltage across the capacitor is $I_s R$, and the time constant of the circuit is RC. Also note that the solution for v_C is valid for $t \geq 0$. These observations are consistent with the behavior of a capacitor in parallel with a resistor when driven by a constant current source.

Equation (7.53) predicts that the current in the capacitor at $t = 0^+$ is $I_s - V_0/R$. This prediction makes sense because the capacitor voltage cannot change instantaneously and therefore the initial current in the resistor is V_0/R. The capacitor branch current changes instantaneously from zero at $t = 0^-$ to $I_s - V_0/R$ at $t = 0^+$. The capacitor current is zero at $t = \infty$. Also note that the final value of $v = I_s R$.

Example 7.6 illustrates how to use Eqs. (7.51) and (7.53) to find the step response of a first-order RC circuit.

EXAMPLE 7.6

The switch in the circuit shown in Fig. 7.22 has been in position 1 for a long time. At $t = 0$, the switch moves to position 2. Find

a) $v_o(t)$ for $t \geq 0$; and

b) $i_o(t)$ for $t \geq 0^+$.

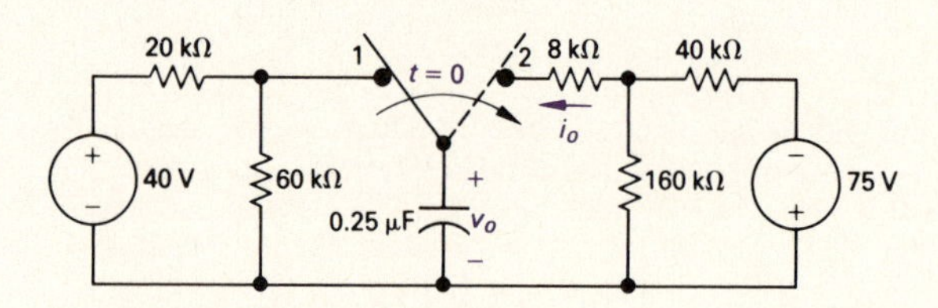

FIGURE 7.22 The circuit for Example 7.6.

SOLUTION

a) The switch has been in position 1 for a long time, so the initial value of v_o is 40(60/80), or 30 V. To take advantage of Eqs. (7.51) and (7.53), we find the Norton equivalent with respect to the terminals of the capacitor for $t \geq 0$. To do this, begin by computing the open-circuit voltage, which is given by the −75 V source divided across the 40 kΩ and 160 kΩ resistors:

$$V_{oc} = \frac{160 \times 10^3}{(40 + 160) \times 10^3}(-75) = -60 \text{ V}.$$

Next, calculate the Thévenin resistance, as seen to the right of the capacitor, by shorting the −75 V source and making series and parallel combinations of the resistors:

$$R_{\text{Th}} = 8000 + 40{,}000 \parallel 160{,}000 = 40 \text{ k}\Omega$$

The value of the Norton current source is the ratio of the open-circuit voltage to the Thévenin resistance, or $-60/(40 \times 10^3) = -1.5$ mA. The resulting Norton equivalent circuit is shown in Fig. 7.23. From Fig. 7.23, $I_s R = -60$ V and $RC = 10$ ms.

We have already noted that $v_o(0) = 30$ V, so the solution for v_o is

$$v_o = -60 + [30 - (-60)]e^{-100t}$$
$$= -60 + 90e^{-100t} \text{ V}, \quad t \geq 0.$$

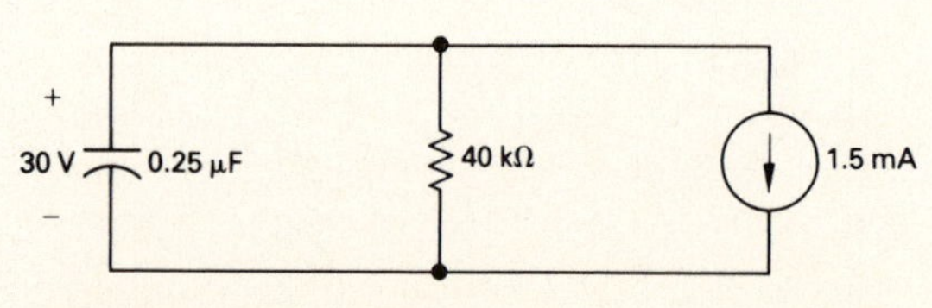

FIGURE 7.23 The equivalent circuit for $t > 0$ for the circuit shown in Fig. 7.22.

b) We write the solution for i_o directly from Eq. (7.50) by noting that $I_s = -1.5$ mA and $v_o/R = (30/40) \times 10^{-3}$, or 0.75 mA:

$$i_o = -2.25e^{-100t} \text{ mA}, \quad t \geq 0^+.$$

We check the consistency of the solutions for v_o and i_o by noting that

$$i_o = C\frac{dv_o}{dt} = (0.25 \times 10^{-6})(-9000e^{-100t}) = -2.25e^{-100t} \text{ mA}.$$

Because $dv_o(0^-)/dt = 0$, the expression for i_o clearly is valid only for $t \geq 0^+$.

DRILL EXERCISES

7.7 a) Derive Eq. (7.47) by first converting the Norton equivalent circuit shown in Fig. 7.21 to a Thévenin equivalent and then summing the voltages around the closed loop using the capacitor current i as the relevant variable.

b) Use the separation of variables technique to find the solution to Eq. (7.52). Verify that your solution agrees with that of Eq. (7.53).

ANSWER: (a) Derivation; (b) verification.

7.8 a) Find the expression for the voltage across the 160 kΩ resistor in the circuit shown in Fig. 7.22. Let this voltage be denoted v_A and assume that the reference polarity for the voltage is positive at the upper terminal of the 160 kΩ resistor.

b) Specify the interval of time for which the expression obtained in (a) is valid.

ANSWER: (a) $v_A = -60 + 72e^{-100t}$ V; (b) $t \geq 0^+$.

7.4 A GENERAL SOLUTION FOR STEP AND NATURAL RESPONSES

The general approach to finding either the natural response or the step response of the first-order RL and RC circuits shown in Fig. 7.24 is based on their differential equations being the same (compare Eqs. [7.47]–[7.52]). To generalize the solution of these four possible circuits, we let $x(t)$ represent the unknown quantity, giving $x(t)$ four possible values. It can represent the current or voltage at the terminals of an inductor, or the current or voltage at the terminals of a

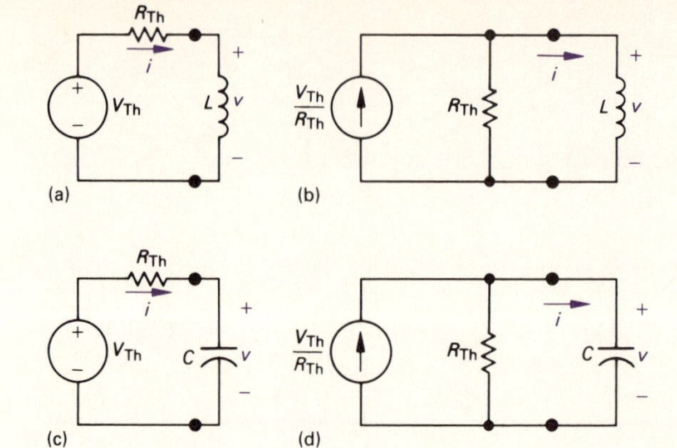

FIGURE 7.24 Four possible first-order circuits: (a) An inductor connected to a Thévenin equivalent; (b) an inductor connected to a Norton equivalent; (c) a capacitor connected to a Thévenin equivalent; (d) a capacitor connected to a Norton equivalent.

capacitor. From Eqs. (7.47), (7.48), (7.50), and (7.52), we know that the differential equation that describes any one of the four circuits in Fig. 7.24 takes the form

$$\frac{dx}{dt} + \frac{x}{\tau} = K, \tag{7.54}$$

where the value of the constant K can be zero. Because the sources in the circuit are constant voltages and/or currents, the final value of x will be constant; that is, the final value must satisfy Eq. (7.54), and, when x reaches its final value, the derivative dx/dt must be zero. Hence

$$x_f = K\tau, \tag{7.55}$$

where x_f represents the final value of the variable.

We solve Eq. (7.54) by separating the variables, beginning by solving for the first derivative:

$$\frac{dx}{dt} = \frac{-x}{\tau} + K = \frac{-(x - K\tau)}{\tau} = \frac{-(x - x_f)}{\tau}. \tag{7.56}$$

In writing Eq. (7.56), we used on Eq. (7.55) to substitute x_f for $K\tau$. We now multiply both sides of Eq. (7.56) by dt to obtain

$$\frac{dx}{x - x_f} = \frac{-1}{\tau}dt. \tag{7.57}$$

Next, we integrate Eq. (7.57). To obtain as general a solution as possible, we use time t_0 as the lower limit and t as the upper limit. Time t_0 corresponds to the time of the switching or other change. Previously we assumed that $t_0 = 0$, but this change allows the switching to take place at any time. Using u and v as symbols of integration, we get

$$\int_{x(t_0)}^{x(t)} \frac{du}{u - x_f} = -\frac{1}{\tau}\int_{t_0}^{t} dv. \tag{7.58}$$

Carrying out the integration called for in Eq. (7.58) gives

$$x(t) = x_f + [x(t_0) - x_f]e^{-(t-t_0)/\tau}. \quad \textbf{(7.59)}$$

The importance of Eq. (7.59) becomes apparent if we write it out in verbal form:

$$\begin{array}{c}\text{The unknown}\\ \text{variable as a}\\ \text{function of time}\end{array} = \begin{array}{c}\text{the final}\\ \text{value of the}\\ \text{variable}\end{array} + \left[\begin{array}{c}\text{the initial}\\ \text{value of the}\\ \text{variable}\end{array} - \begin{array}{c}\text{the final}\\ \text{value of the}\\ \text{variable}\end{array}\right] \times e^{\frac{-[t-(\text{time of switching})]}{(\text{time constant})}}. \quad \textbf{(7.60)}$$

In many cases, the time of switching—that is, t_0—is zero.

While acquiring experience with computing the step and natural responses of circuits, it may help to follow these steps:

1. Identify the variable of interest for the circuit. For RC circuits, it is most convenient to choose the capacitive voltage; for RL circuits, it is best to choose the inductive current.

2. Determine the initial value of the variable, which is its value at t_0. Note that if you choose capacitive voltage or inductive current as your variable of interest, it is not necessary to distinguish between $t = t_0^-$ and $t = t_0^+$.[2] This is because they both are continuous variables. If you make another choice of variable, you need to remember that its initial value is defined at $t = t_0^+$.

3. Calculate the final value of the variable, which is its value as $t \to \infty$.

4. Calculate the time constant for the circuit.

With these quantities, you can use Eq. (7.60) to produce the equation of the variable of interest as a function of time. You can then find equations for other circuit variables using the circuit analysis techniques introduced in Chapters 3 and 4, or by repeating the preceding steps for the other variables.

Examples 7.7, 7.8, and 7.9 illustrate how to use Eq. (7.60) to find the step response of an RC or RL circuit.

EXAMPLE 7.7

The switch in the circuit shown in Fig. 7.25 has been in position a for a long time. At $t = 0$ the switch is moved to position b.

a) What is the initial value of v_C?

b) What is the final value of v_C?

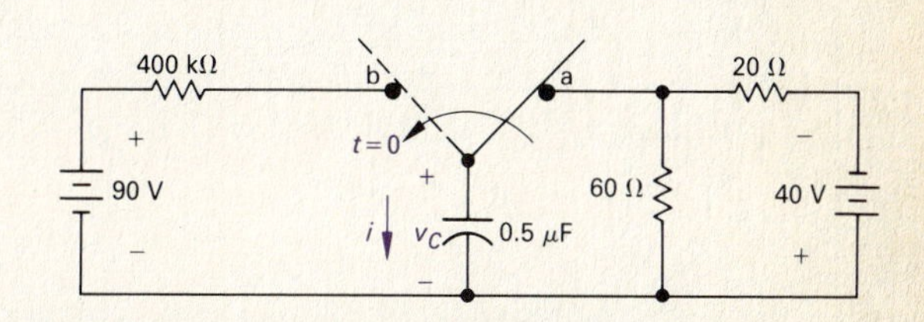

FIGURE 7.25 The circuit for Example 7.7.

[2]The expressions t_0^- and t_0^+ are analogous to 0^- and 0^+. Thus $x(t_0^-)$ is the limit of $x(t)$ as $t \to t_0$ from the left, and $x(t_0^+)$ is the limit of $x(t)$ as $t \to t_0$ from the right.

c) What is the time constant of the circuit when the switch is in position b?

d) What is the expression for $v_C(t)$ when $t \geq 0$?

e) What is the expression for $i(t)$ when $t \geq 0^+$?

f) How long after the switch is in position b does the capacitor voltage equal zero?

g) Plot $v_C(t)$ and $i(t)$ versus t.

SOLUTION

a) The switch has been in position a for a long time, so the capacitor looks like an open circuit. Therefore the voltage across the capacitor is the voltage across the 60 Ω resistor. From the voltage-divider rule, the voltage across the 60 Ω resistor is $40 \times [60/(60 + 20)]$, or 30 V. As the reference for v_C is positive at the upper terminal of the capacitor, we have $v_C(0) = -30$ V.

b) After the switch has been in position b for a long time, the capacitor will look like an open circuit in terms of the 90 V source. Thus the final value of the capacitor voltage is +90 V.

c) The time constant is

$$\tau = RC = (400 \times 10^3)(0.5 \times 10^{-6}) = 0.2 \text{ s}.$$

d) Substituting the appropriate values for v_f, $v(0)$, and t into Eq. (7.60) yields

$$v_C(t) = 90 + (-30 - 90)e^{-5t} = 90 - 120e^{-5t} \text{ V}, \quad t \geq 0.$$

e) Here the value for τ doesn't change. Thus we need to find only the initial and final values for the current in the capacitor. When obtaining the initial value, we must get the value of $i(0^+)$ because the current in the capacitor can have a step jump. This current is equal to the current in the resistor, which from Ohm's law is $[90 - (-30)]/(400 \times 10^3) = 300\ \mu$A. Note that when applying Ohm's law we recognized that the capacitor voltage cannot change instantaneously. The final value of $i(t) = 0$, so

$$i(t) = \frac{90 - (-30)}{0.4}e^{-5t} = 300e^{-5t}\ \mu\text{A}, \quad t \geq 0^+.$$

We could have obtained this solution by differentiating the solution in (d) and multiplying by the capacitance. You may want to do so for yourself. Note that this alternative approach to finding $i(t)$ also predicts the discontinuity at $t = 0$.

f) To find how long the switch must be in position b before the capacitor voltage becomes zero, we solve the equation derived in (d) for the time when $v_C(t) = 0$:

$$120e^{-5t} = 90 \quad \text{or} \quad e^{5t} = \frac{120}{90},$$

so

$$t = \frac{1}{5} ln\left(\frac{4}{3}\right) = 57.54 \text{ ms}.$$

Note that when $v_C = 0$, $i = 225$ mA and the voltage drop across the 400 kΩ resistor is 90 V.

g) Figure 7.26 shows the graphs of $v_C(t)$ and $i(t)$ versus t.

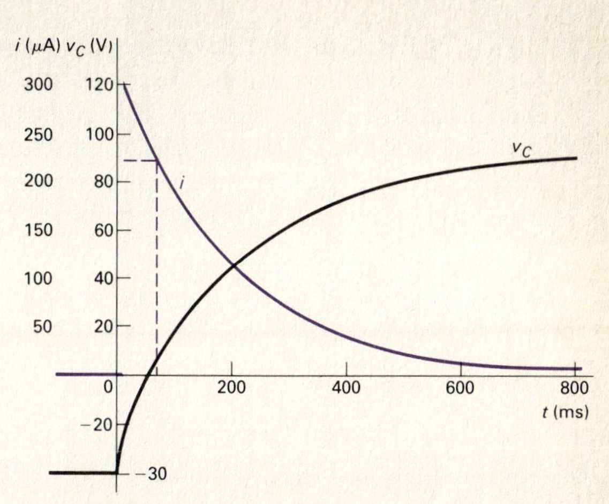

FIGURE 7.26 The current and voltage waveforms for Example 7.7.

EXAMPLE 7.8

The switch in the circuit shown in Fig. 7.27 has been open for a long time. The initial charge on the capacitor is zero. At $t = 0$, the switch is closed. Find the expression for

a) $i(t)$ for $t \geq 0^+$, and

b) $v(t)$ when $t \geq 0^+$.

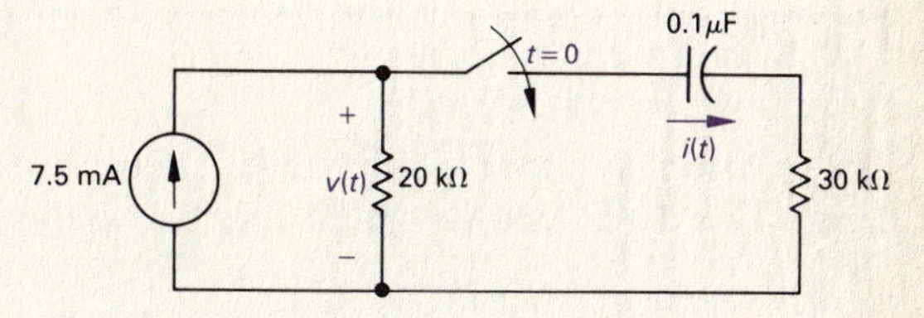

FIGURE 7.27 The circuit for Example 7.8.

SOLUTION

a) Because the initial voltage on the capacitor is zero, at the instant when the switch is closed the current in the 30 kΩ branch will be

$$i(0^+) = \frac{(7.5)(20)}{50} = 3 \text{ mA}.$$

The final value of the capacitor current will be zero because the capacitor eventually will appear as an open circuit in terms of dc current. Thus $i_f = 0$. The time constant of the circuit will equal the product of the Thévenin resistance (as seen from the capacitor) and the capacitance. Therefore $\tau = (20+30)10^3(0.1) \times 10^{-6} = 5$ ms. Substituting these values into Eq. (7.60) generates the expression

$$i(t) = 0 + (3 - 0)e^{-t/5\times10^{-3}} = 3e^{-200t} \text{ mA}, \quad t \geq 0^+.$$

b) To find $v(t)$, we note from the circuit that it equals the sum of the voltage across the capacitor and the voltage across the 30 kΩ resistor. To find the capacitor voltage (which is a drop in the direction of the current), we note that its initial value is zero and its final value is (7.5)(20), or 150 V. The time constant is the same as before, or 5 ms. Therefore we use Eq. (7.60) to write

$$\begin{aligned} v_C(t) &= 150 + (0 - 150)e^{-200t} \\ &= (150 - 150e^{-200t})\text{ V}, \quad t \geq 0. \end{aligned}$$

Hence the expression for the voltage $v(t)$ is

$$\begin{aligned} v(t) &= 150 - 150e^{-200t} + (30)(3)e^{-200t} \\ &= (150 - 60e^{-200t})\text{ V}, \quad t \geq 0^+. \end{aligned}$$

As one check on this expression, note that it predicts the initial value of the voltage across the 20 Ω resistor as 150 − 60, or 90 V. The instant the switch is closed, the current in the 20 kΩ resistor is (7.5)(30/50), or 4.5 mA. This current produces a 90 V drop across the 20 kΩ resistor, confirming the value predicted by the solution.

EXAMPLE 7.9

The switch in the circuit shown in Fig. 7.28 has been open for a long time. At $t = 0$ the switch is closed. Find the expression for

a) $v(t)$ when $t \geq 0^+$, and

b) $i(t)$ when $t \geq 0$.

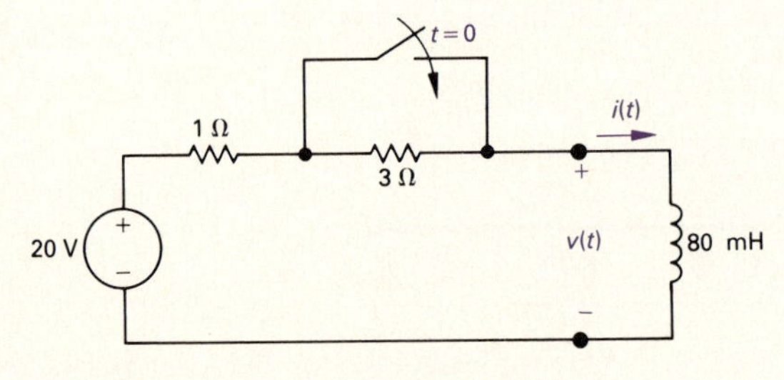

FIGURE 7.28 The circuit for Example 7.9.

SOLUTION

a) The switch has been open for a long time, so the initial current in the inductor is 5 A, oriented from top to bottom. Immediately after the switch closes, the current still is 5 A, and therefore the initial voltage across the inductor becomes 20 − 5(1), or 15 V. The final value of the inductor voltage is 0 V. With the switch closed, the time constant is 80/1, or 80 ms. We use Eq. (7.60) to write the expression for $v(t)$:

$$v(t) = 0 + (15 - 0)e^{-t/80\times10^{-3}} = 15e^{-12.5t}\text{ V}, \quad t \geq 0^+.$$

b) We have already noted that the initial value of the inductor current is 5 A. After the switch has been closed for a long time, the inductor current reaches 20/1, or 20 A. The circuit time constant is 80 ms, so the expression for $i(t)$ is

$$i(t) = 20 + (5 - 20)e^{-12.5t}$$
$$= (20 - 15e^{-12.5t}) \text{ A}, \quad t \geq 0.$$

We determine that the solutions for $v(t)$ and $i(t)$ agree by noting that

$$v(t) = L\frac{di}{dt} = 80 \times 10^{-3}[15(12.5)e^{-12.5t}]$$
$$= 15e^{-12.5t} \text{ V}, \quad t \geq 0^+.$$

DRILL EXERCISES

7.9 Assume that the switch in the circuit shown has been in position a for a long time and that at $t = 0$ it is moved to position b. Find (a) $v_C(0^+)$; (b) $v_C(\infty)$; (c) τ for $t > 0$; (d) $i(0^+)$; (e) v_C, $t \geq 0$; and (f) i, $t \geq 0^+$.

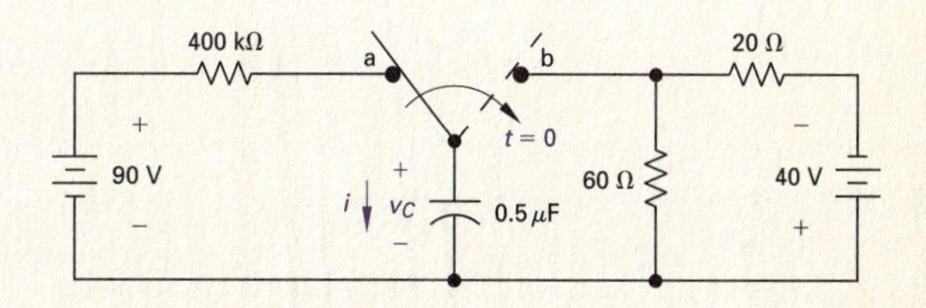

ANSWER: (a) 90 V; (b) −30 V; (c) 7.5 μs; (d) −8 A; (e) $(-30 + 120e^{-(400{,}000/3)t})$ V, $t \geq 0$; (f) $-8e^{-(400{,}000/3)t}$ A, $t \geq 0^+$.

7.10 The switch in the circuit shown has been in position a for a long time. At $t = 0$ the switch is moved to position b. Calculate (a) the initial voltage on the capacitor; (b) the final voltage on the capacitor; (c) the time constant (in microseconds) for $t > 0$; and (d) the length of time (in microseconds) required for the capacitor voltage to reach zero after the switch is moved to position b.

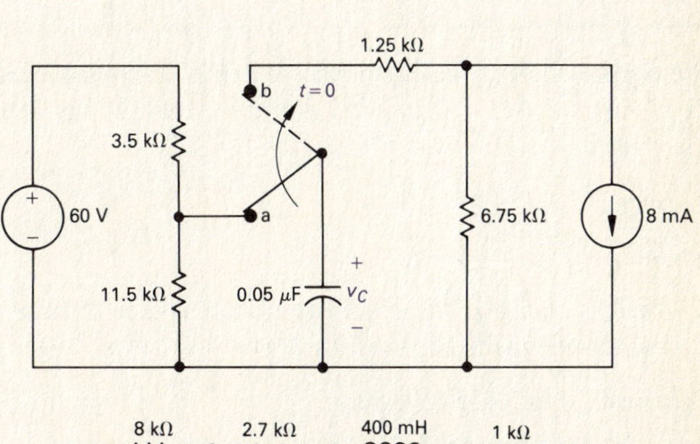

ANSWER: (a) 46 V; (b) −54 V; (c) 400 μs; (d) 246.47 μs.

7.11 After the switch in the circuit shown has been open for a long time, it is closed at $t = 0$. Calculate (a) the initial value of i; (b) the final value of i; (c) the time constant for $t \geq 0$; and (d) the numerical expression for $i(t)$ when $t \geq 0$.

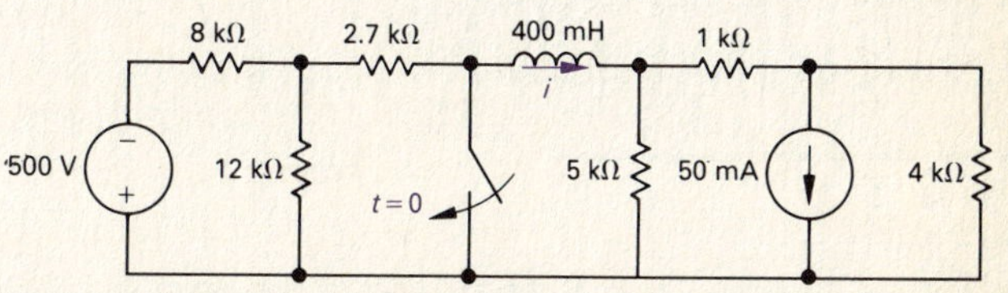

ANSWER: (a) −20 mA; (b) 40 mA; (c) 160 μs; (d) $i = (40 - 60e^{-6250t})$ mA, $t \geq 0$.

7.5 Sequential Switching

Whenever switching occurs more than once in a circuit, we have **sequential switching**. For example, a single, two-position switch may be switched back and forth, or multiple switches may be opened or closed in sequence. The time reference for all switchings cannot be $t = 0$. We determine the voltages and currents generated by a switching sequence by using the techniques described previously in this chapter. We derive the expressions for $v(t)$ and $i(t)$ for a given position of the switch or switches and then use these solutions to determine the initial conditions for the next position of the switch or switches.

With sequential switching problems, a premium is placed on obtaining the initial value $x(t_0)$. Recall that anything but inductive currents and capacitive voltages can change instantaneously at the time of switching. Thus solving first for inductive currents and capacitive voltages is even more pertinent in sequential switching problems. Drawing the circuit that pertains to each time interval in such a problem is often helpful in the solution process.

Examples 7.10 and 7.11 illustrate the analysis techniques for circuits with sequential switching. The first is a natural response problem with two switching times, and the second is a step response problem.

You can create models of switches using a circuit simulation program.

EXAMPLE 7.10

The two switches in the circuit shown in Fig. 7.29 have been closed for a long time. At $t = 0$, switch 1 is opened. Then, 35 ms later, switch 2 is opened.

a) Find $i_L(t)$ for $0 \le t \le 35$ ms.

b) Find $i_L(t)$ for $t \ge 35$ ms.

c) What percentage of the initial energy stored in the 150 mH inductor is dissipated in the 18 Ω resistor?

d) Repeat (c) for the 3 Ω resistor.

e) Repeat (c) for the 6 Ω resistor.

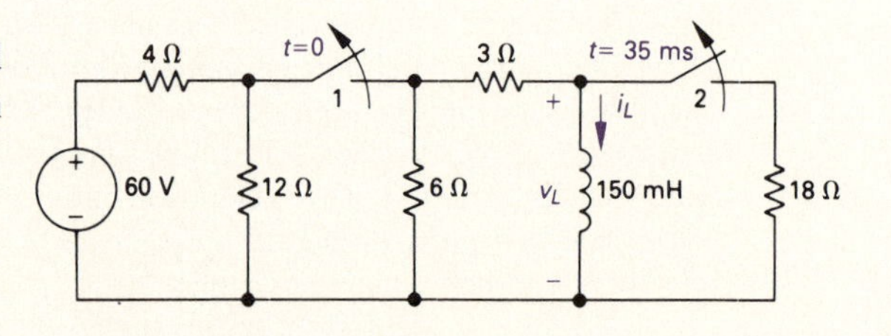

FIGURE 7.29 The circuit for Example 7.10.

SOLUTION

a) For $t < 0$ both switches are closed, causing the 150 mH inductor to short-circuit the 18 Ω resistor. The equivalent circuit is shown in Fig. 7.30. We determine the initial current in the inductor by

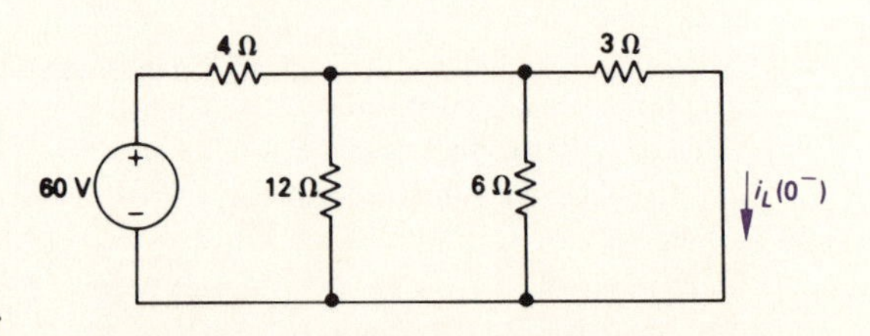

FIGURE 7.30 The circuit shown in Fig. 7.29, for $t < 0$.

solving for $i_L(0^-)$ in the circuit shown in Fig. 7.30. After making several source transformations, we find $i_L(0^-)$ to be 6 A.

For $0 \leq t \leq 35$ ms, switch 1 is open (switch 2 is closed), which disconnects the 60 V voltage source and the 4 Ω and 12 Ω resistors from the circuit. The inductor is no longer behaving as a short circuit (because the dc source is no longer in the circuit), so the 18 Ω resistor is no longer short-circuited. The equivalent circuit is shown in Fig. 7.31. Note that the equivalent resistance across the terminals of the inductor is the parallel combination of 9 Ω and 18 Ω, or 6 Ω. The time constant of the circuit is $(150/6) \times 10^{-3}$, or 25 ms. Therefore the expression for i_L is

$$i_L = 6e^{-40t} \text{ A}, \quad 0 \leq t \leq 35 \text{ ms}.$$

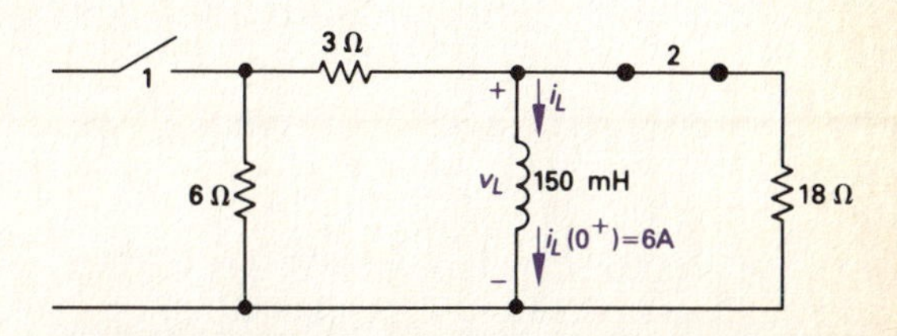

FIGURE 7.31 The circuit shown in Fig. 7.29, for $0 \leq t \leq 35$ ms.

b) When $t = 35$ ms, the value of the inductor current is

$$i_L = 6e^{-1.4} = 1.48 \text{ A}.$$

Thus when switch 2 is opened, the circuit reduces to the one shown in Fig. 7.32, and the time constant changes to $(150/9) \times 10^{-3}$, or 16.67 ms. The expression for i_L becomes

$$i_L = 1.48e^{-60(t-0.035)} \text{ A}, \quad t \geq 0.035 \text{ s}.$$

Note that the exponential function is shifted in time by 35 ms.

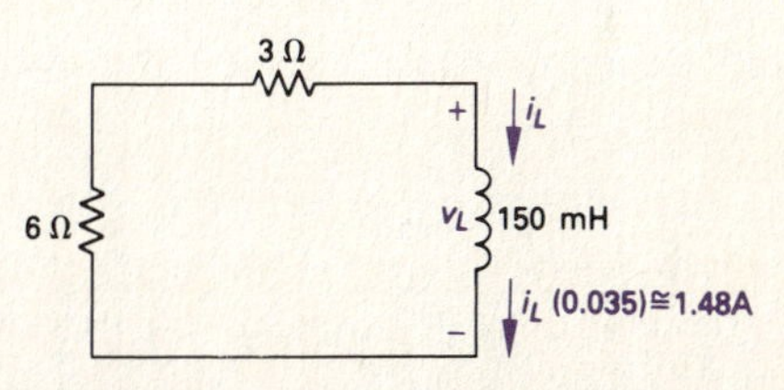

FIGURE 7.32 The circuit shown in Fig. 7.29, for $t \geq 35$ ms.

c) The 18 Ω resistor is in the circuit only during the first 35 ms of the switching sequence. During this interval, the voltage across the resistor is

$$v_L = 0.15\frac{d}{dt}(6e^{-40t})$$
$$= -36e^{-40t} \text{ V}, \quad 0 < t < 0.035 \text{ s}.$$

The power dissipated in the 18 Ω resistor is

$$p = \frac{v_L^2}{18} = 72e^{-80t} \text{ W}, \quad 0 < t < 0.035 \text{ s}.$$

Hence the energy dissipated is

$$w = \int_0^{0.035} 72e^{-80t}\,dt = \frac{72}{-80}e^{-80t}\bigg|_0^{0.035}$$
$$= 0.9(1 - e^{-2.8}) = 845.27 \text{ mJ}.$$

The initial energy stored in the 150 mH inductor is

$$w_i = \frac{1}{2}(0.15)(36) = 2.7 \text{ J} = 2700 \text{ mJ}.$$

Therefore (845.27/2700) × 100, or 31.31% of the initial energy stored in the 150 mH inductor is dissipated in the 18 Ω resistor.

d) For $0 < t < 0.035$ s, the voltage across the 3 Ω resistor is

$$v_{3\Omega} = \left(\frac{v_L}{9}\right)(3) = \frac{1}{3}v_L = -12e^{-40t} \text{ V}.$$

Therefore the energy dissipated in the 3 Ω resistor in the first 35 ms is

$$\begin{aligned} w_{3\Omega} &= \int_0^{0.035} \frac{144e^{-80t}}{3} dt \\ &= 0.6(1 - e^{-2.8}) \\ &= 563.51 \text{ mJ}. \end{aligned}$$

For $t > 0.035$ s, the current in the 3 Ω resistor is

$$i_{3\Omega} = i_L = (6e^{-1.4})e^{-60(t-0.035)} \text{ A}.$$

Hence the energy dissipated in the 3 Ω resistor for $t > 0.035$ s is

$$\begin{aligned} w_{3\Omega} &= \int_{0.035}^{\infty} i_{3\Omega}^2 \times 3dt \\ &= \int_{0.035}^{\infty} 3(36)e^{-2.8}e^{-120(t-0.035)}dt \\ &= 108e^{-2.8} \times \left.\frac{e^{-120(t-0.035)}}{-120}\right|_{0.035}^{\infty} \\ &= \frac{108}{120}e^{-2.8} = 54.73 \text{ mJ}. \end{aligned}$$

The total energy dissipated in the 3 Ω resistor is

$$\begin{aligned} w_{3\Omega}(\text{total}) &= 563.51 + 54.73 \\ &= 618.24 \text{ mJ}. \end{aligned}$$

The percentage of the initial energy stored is

$$\frac{618.24}{2700} \times 100 = 22.90\%.$$

e) Because the 6 Ω resistor is in series with the 3 Ω resistor, the energy dissipated and the percentage of the initial energy stored will be twice that of the 3 Ω resistor:

$$w_{6\Omega}(\text{total}) = 1236.48 \text{ mJ},$$

and the percentage of the initial energy stored is 45.80%. We check these calculations by observing that

$$1236.48 + 618.24 + 845.27 = 2699.99 \text{ mJ}$$

and

$$31.31 + 22.90 + 45.80 = 100.01\%.$$

The small discrepancies in the summations are the result of round-off errors.

EXAMPLE 7.11

The uncharged capacitor in the circuit shown in Fig. 7.33 is initially switched to terminal a of the three-position switch. At $t = 0$ the switch is moved to position b, where it remains for 15 ms. After the 15 ms delay, the switch is moved to position c, where it remains indefinitely.

a) Derive the numerical expression for the voltage across the capacitor.

b) Plot the capacitor voltage versus time.

c) When will the voltage on the capacitor equal 200 V?

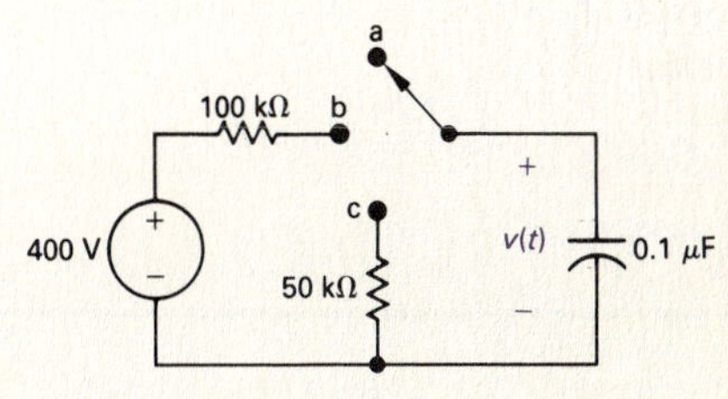

FIGURE 7.33 The circuit for Example 7.11.

SOLUTION

a) At the instant the switch is moved to position b, the initial voltage on the capacitor is zero. If the switch were to remain in position b, the capacitor would eventually charge to 400 V. The time constant of the circuit when the switch is in position b is 10 ms. Therefore we can use Eq. (7.59) with $t_0 = 0$ to write the expression for the capacitor voltage:

$$\begin{aligned} v &= 400 + (0 - 400)e^{-100t} \\ &= (400 - 400e^{-100t})\ \text{V}, \quad 0 \le t \le 15\ \text{ms}. \end{aligned}$$

Note that, because the switch remains in position b for only 15 ms, this expression is valid only for the time interval from 0 to 15 ms. After the switch has been in this position for 15 ms, the voltage on the capacitor will be

$$v(15\ \text{ms}) = 400 - 400e^{-1.5} = 310.75\ \text{V}.$$

Therefore, when the switch is moved to position c, the initial voltage on the capacitor is 310.75 V. With the switch in position c, the final value of the capacitor voltage is zero, and the time constant is 5 ms. Again, we use Eq. (7.59) to write the expression for the capacitor voltage:

$$\begin{aligned} v &= 0 + (310.75 - 0)e^{-200(t-0.015)} \\ &= 310.75e^{-200(t-0.015)}\ \text{V}, \quad 15\ \text{ms} \le t. \end{aligned}$$

In writing the expression for v, we recognized that $t_0 = 15$ ms and that this expression is valid only for $t \geq 15$ ms.

b) Figure 7.34 shows the plot of v versus t.

c) The plot in Fig. 7.34 reveals that the capacitor voltage will equal 200 V at two different times: once in the interval between 0 and 15 ms, and once after 15 ms. We find the first time by solving the expression

$$200 = 400 - 400e^{-100t_1},$$

which yields $t_1 = 6.93$ ms. We find the second time by solving the expression

$$200 = 310.75e^{-200(t_2 - 0.015)}.$$

In this case, $t_2 = 17.20$ ms.

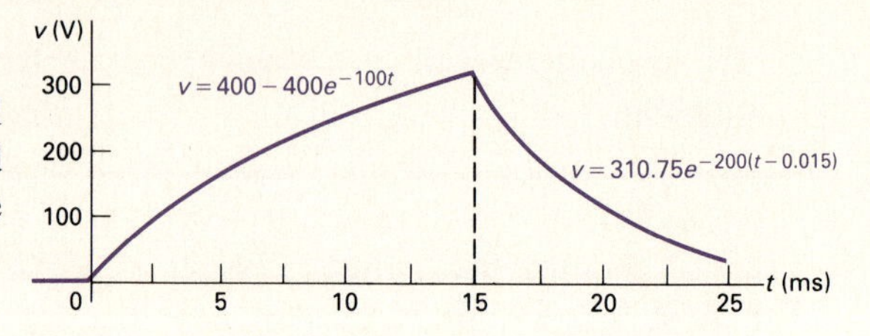

FIGURE 7.34 The capacitor voltage for Example 7.11.

DRILL EXERCISES

7.12 In the circuit shown, switch 1 has been closed and switch 2 has been open for a long time. At $t = 0$, switch 1 is opened. Then 50 ms later, switch 2 is closed. Find

a) $v_C(t)$ for $0 \leq t \leq 0.05$ s;

b) $v_C(t)$ for $t \geq 0.05$ s;

c) the total energy dissipated in the 50 kΩ resistor; and

d) the total energy dissipated in the 200 kΩ resistor.

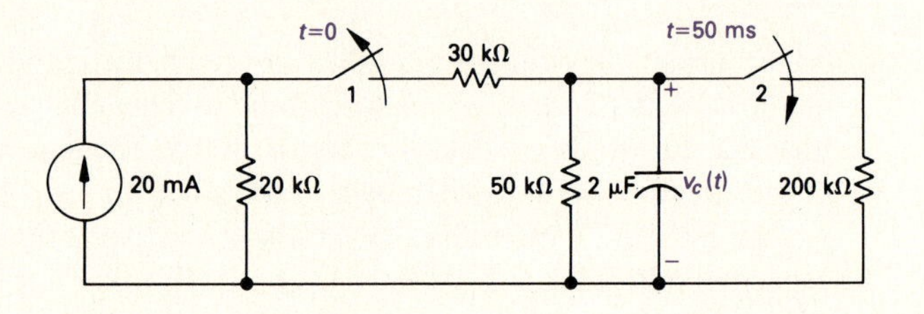

ANSWER: (a) $200e^{-10t}$ V; (b) $121.31e^{-12.5(t-0.05)}$ V; (c) 37.06 mJ; (d) 2.94 mJ.

7.13 Switch a in the circuit shown has been open for a long time, and switch b has been closed for a long time. Switch a is closed at $t = 0$ and, after remaining closed for 1 s, is opened again. Switch b is opened simultaneously, and both switches remain open indefinitely. Determine the expression for the inductor current i that is valid when (a) $0 \leq t \leq 1$ s and (b) $t \geq 1$ s.

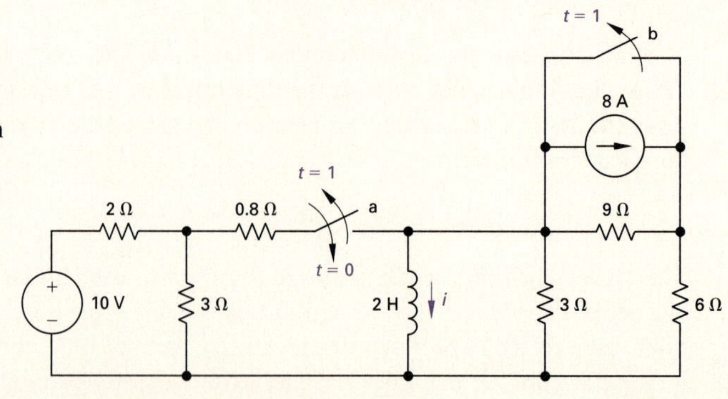

ANSWER: (a) $i(t) = (3 - 3e^{-0.5t})$ A, $0 \leq t \leq 1$ s; (b) $i(t) = (-4.8 + 5.98e^{-1.25(t-1)})$ A, $t \geq 1$ s.

7.6 Unbounded Response

A circuit response may grow, rather than decay, exponentially with time. This type of response is possible if the circuit contains dependent sources. In that case, the Thévenin equivalent resistance with respect to the terminals of either an inductor or a capacitor may be negative. This negative resistance generates a negative time constant, and the resulting currents and voltages increase without limit. In an actual circuit, the response eventually reaches a limiting value when a component breaks down or goes into a saturation state, prohibiting further increases in voltage or current.

When we consider unbounded responses, the concept of a final value is confusing. Hence, rather than using the step response solution given in Eq. (7.59), we derive the differential equation that describes the circuit containing the negative resistance and then solve it using the separation of variables technique. Example 7.12 presents an exponentially growing response in terms of the voltage across a capacitor.

EXAMPLE 7.12

a) When the switch is closed in the circuit shown in Fig. 7.35, the voltage on the capacitor is 10 V. Find the expression for v_o for $t \geq 0$.

b) Assume that the capacitor short-circuits when its terminal voltage reaches 150 V. How many milliseconds elapse before the capacitor short-circuits?

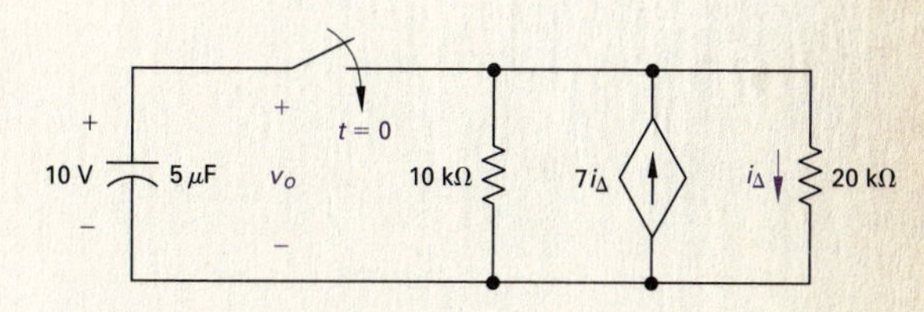

FIGURE 7.35 The circuit for Example 7.12.

SOLUTION

a) To find the Thévenin equivalent resistance with respect to the capacitor terminals, we use the test-source method described in Chapter 4. Figure 7.36 shows the resulting circuit, where v_T is the test voltage and i_T is the test current.

For v_T expressed in volts, we obtain

$$i_T = \frac{v_T}{10} - 7\left(\frac{v_T}{20}\right) + \frac{v_T}{20} \text{ mA}.$$

Solving for the ratio v_T/i_T yields the Thévenin resistance:

$$R_{\text{Th}} = \frac{v_T}{i_T} = -5 \text{ k}\Omega.$$

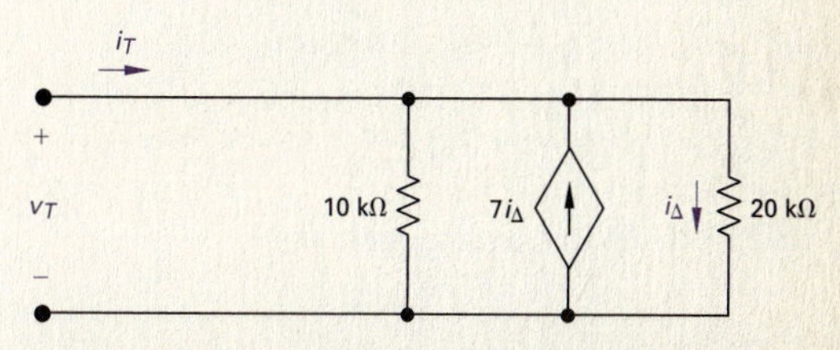

FIGURE 7.36 The test-source method used to find R_{Th}.

With this Thévenin resistance, we can simplify the circuit shown in Fig. 7.35 to the one shown in Fig. 7.37.

For $t \geq 0$, the differential equation describing the circuit shown in Fig. 7.37 is

$$(5 \times 10^{-6})\frac{dv_o}{dt} - \frac{v_o}{5} \times 10^{-3} = 0.$$

Dividing by the coefficient of the first derivative yields

$$\frac{dv_o}{dt} - 40v_o = 0.$$

We now use the separation of variables technique to find $v_o(t)$:

$$v_o(t) = 10e^{40t} \text{ V}, \quad t \geq 0.$$

b) $v_o = 150$ V when $e^{40t} = 15$. Therefore $40t = ln\, 15$, and $t = 67.70$ ms.

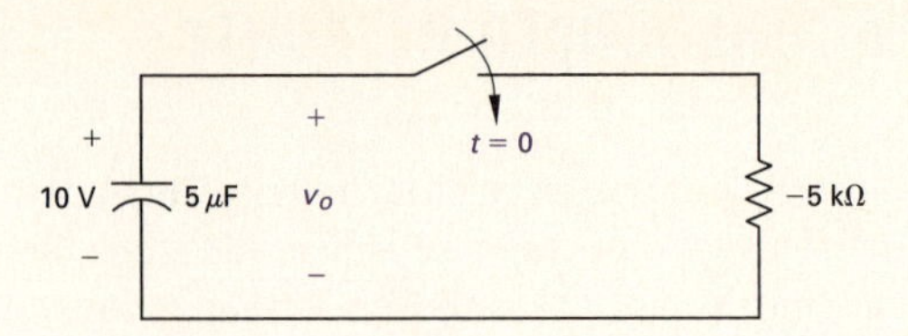

FIGURE 7.37 A simplification of the circuit shown in Fig. 7.35.

The fact that interconnected circuit elements may lead to ever-increasing currents and voltages is important to engineers. If such interconnections are unintended, the resulting circuit may experience unexpected, and potentially dangerous, component failures.

7.7 The Integrating Amplifier

Recall from the introduction to Chapter 5 that one reason for our interest in the operational amplifier is its use as an integrating amplifier. We are now ready to analyze an integrating-amplifier circuit, which is shown in Fig. 7.38. The purpose of such a circuit is to generate an output voltage proportional to the integral of the input voltage. In Fig. 7.38 we added the branch currents i_f and i_s, along with the node voltages v_n and v_p, to aid our analysis.

We assume that the operational amplifier is ideal. Thus we take advantage of the constraints

$$i_f + i_s = 0 \tag{7.61}$$

and

$$v_n = v_p. \tag{7.62}$$

Because $v_p = 0$,

$$i_s = \frac{v_s}{R_s} \tag{7.63}$$

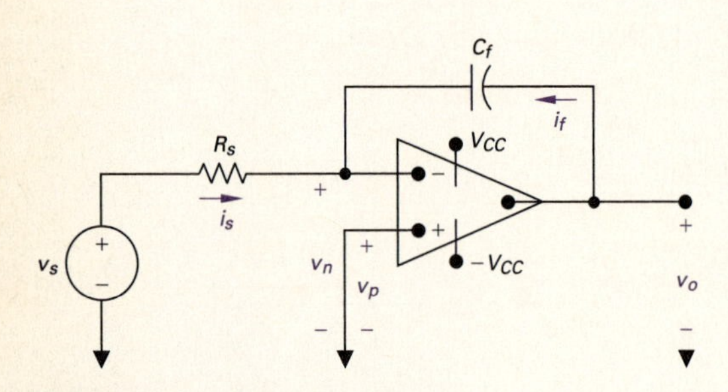

FIGURE 7.38 An integrating amplifier.

and

$$i_f = C_f \frac{dv_o}{dt}. \quad \textbf{(7.64)}$$

Hence, from Eqs. (7.61), (7.63), and (7.64),

$$\frac{dv_o}{dt} = -\frac{1}{R_s C_f} v_s. \quad \textbf{(7.65)}$$

Multiplying both sides of Eq. (7.65) by a differential time dt and then integrating from t_0 to t generates the equation

$$v_o(t) = -\frac{1}{R_s C_f} \int_{t_0}^{t} v_s dy + v_o(t_0). \quad \textbf{(7.66)}$$

In Eq. (7.66), t_0 represents the instant in time when we begin the integration. Thus $v_o(t_0)$ is the value of the output voltage at that time. Also, because $v_n = v_p = 0$, $v_o(t_0)$ is identical to the initial voltage on the feedback capacitor C_f.

Equation (7.66) states that the output voltage of an integrating amplifier equals the initial value of the voltage on the capacitor plus an inverted (minus sign), scaled ($1/R_s C_F$) replica of the integral of the input voltage. If no energy is stored in the capacitor when integration commences, Eq. (7.66) reduces to

$$v_o(t) = -\frac{1}{R_s C_f} \int_{t_0}^{t} v_s dy. \quad \textbf{(7.67)}$$

If v_s is a step change in a dc voltage level, the output voltage will vary linearly with time. For example, assume that the input voltage is the rectangular voltage pulse shown in Fig. 7.39. Assume also that the initial value of $v_o(t)$ is zero at the instant v_s steps from 0 to V_m. A direct application of Eq. (7.66) yields

$$v_o = -\frac{1}{R_s C_f} V_m t + 0, \quad 0 \le t \le t_1. \quad \textbf{(7.68)}$$

When t lies between t_1 and $2t_1$,

$$\begin{aligned} v_o &= -\frac{1}{R_s C_f} \int_{t_1}^{t} (-V_m) dy - \frac{1}{R_s C_f} V_m t_1 \\ &= \frac{V_m}{R_s C_f} t - \frac{2V_m}{R_s C_f} t_1, \quad t_1 \le t \le 2t_1. \end{aligned} \quad \textbf{(7.69)}$$

Figure 7.40 shows a sketch of $v_o(t)$ versus t. Clearly, the output voltage is an inverted, scaled replica of the integral of the input voltage.

The output voltage is proportional to the integral of the input voltage only if the operational amplifier operates within its linear range, that is, if it doesn't saturate. Examples 7.13 and 7.14 further illustrate the analysis of the integrating amplifier.

Use a symbolic equation solver to generate and plot the solution.

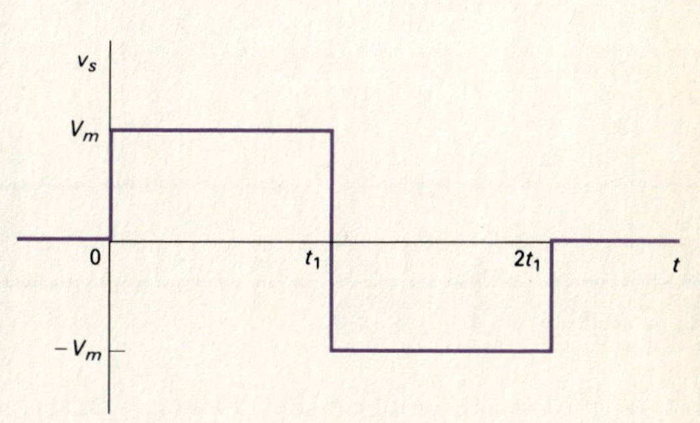

FIGURE 7.39 An input voltage signal.

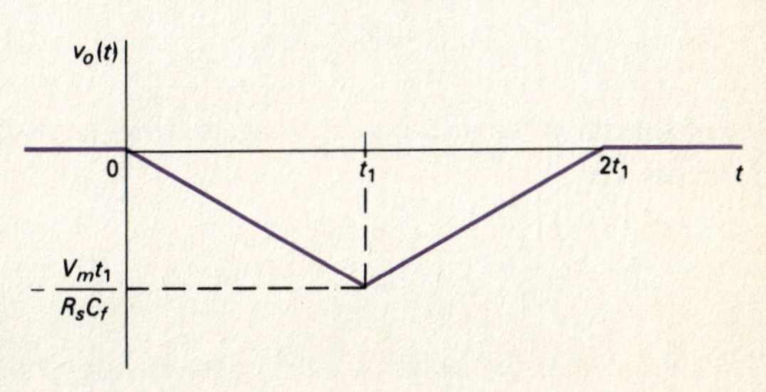

FIGURE 7.40 The output voltage of an integrating amplifier.

EXAMPLE 7.13

Assume that the numerical values for the signal voltage shown in Fig. 7.39 are $V_m = 50$ mV and $t_1 = 1$ s. This signal voltage is applied to the integrating-amplifier circuit shown in Fig. 7.38. The circuit parameters of the amplifier are $R_s = 100$ kΩ, $C_f = 0.1$ μF, and $V_{CC} = 6$ V. The initial voltage on the capacitor is zero.

a) Calculate $v_o(t)$.

b) Plot $v_o(t)$ versus t.

SOLUTION

a) For $0 \leq t \leq 1$ s,

$$v_o = \frac{-1}{(100 \times 10^3)(0.1 \times 10^{-6})} 50 \times 10^{-3} t + 0$$
$$= -5t \text{ V}, \quad 0 \leq t \leq 1 \text{ s}.$$

For $1 \leq t \leq 2$ s,

$$v_o = (5t - 10) \text{ V}.$$

b) Figure 7.41 shows a plot of $v_o(t)$ versus t.

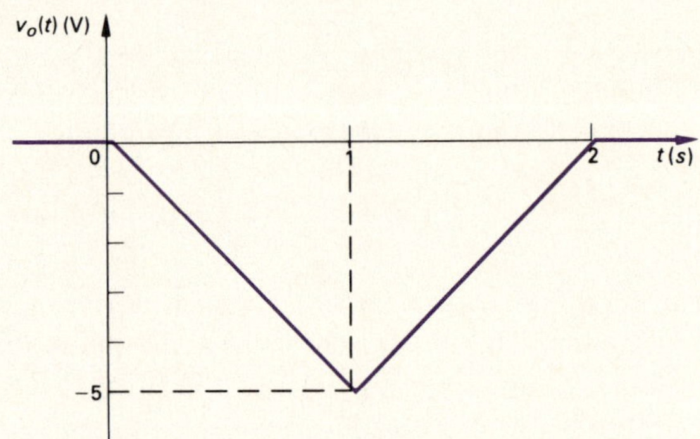

FIGURE 7.41 The output voltage for Example 7.13.

Use a computer tool to vary the value of capacitance and plot the resulting output voltage.

EXAMPLE 7.14

At the instant the switch makes contact with terminal a in the circuit shown in Fig. 7.42, the voltage on the 0.1 μF capacitor is 5 V. The switch remains at terminal a for 9 ms and then moves instantaneously to position b. How many milliseconds after making contact with terminal a does the operational amplifier saturate?

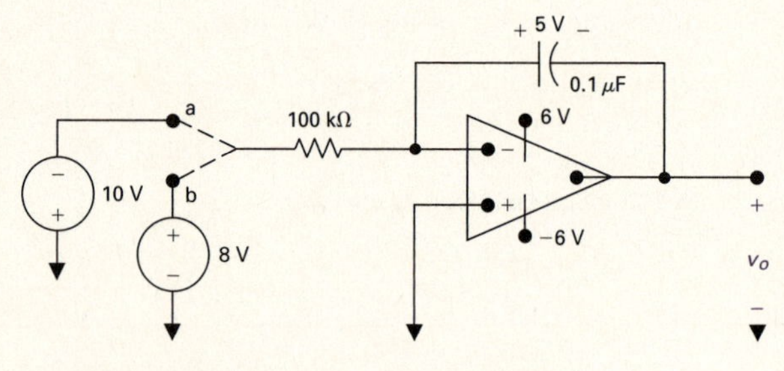

FIGURE 7.42 The circuit for Example 7.14.

SOLUTION

The expression for the output voltage during the time the switch is in position a is

$$v_o = -5 - \frac{1}{10^{-2}} \int_0^t (-10) dy = (-5 + 1000t) \text{ V}.$$

Thus 9 ms after the switch makes contact with terminal a, the output voltage is $-5 + 9$, or 4 V.

The expression for the output voltage after the switch moves to position b is

$$v_o = 4 - \frac{1}{10^{-2}} \int_{9\times 10^{-3}}^{t} 8dy$$
$$= 4 - 800(t - 9 \times 10^{-3}) = (11.2 - 800t) \text{ V}.$$

During this time interval, the voltage is decreasing, and the operational amplifier eventually saturates at −6 V. Therefore we set the expression for v_o equal to −6 V to obtain the saturation time t_s:

$$11.2 - 800t_s = -6, \quad \text{or} \quad t_s = 21.5 \text{ ms}.$$

Thus the integrating amplifier saturates 21.5 ms after making contact with terminal a.

From the examples, we see that the integrating amplifier can perform the integration function very well, but only within specified limits which avoid saturating the op amp. The op amp saturates due to the accumulation of charge on the feedback capacitor. We can prevent it from saturating by placing a resistor in parallel with the feedback capacitor. We examine such a circuit in Chapter 8.

Note that we can convert the integrating amplifier to a differentiating amplifier by interchanging the input resistance R_s and the feedback capacitor C_f. Then

$$v_o = -R_s C_f \frac{dv_s}{dt}. \quad \textbf{(7.70)}$$

We leave the derivation of Eq. (7.70) as an exercise for you. The differentiating amplifier is seldom used because in practice it is a source of unwanted or noisy signals.

Finally, we can design both integrating- and differentiating-amplifier circuits by using an inductor instead of a capacitor. However, fabricating capacitors for integrated-circuit devices is much easier, so inductors are rarely used in integrating amplifiers.

SUMMARY

- A first-order circuit may be reduced to a Thévenin (or Norton) equivalent connected to either a single equivalent inductor or capacitor.

- In the presence of a dc current, an inductor behaves like a short circuit.
- In the presence of a dc voltage, a capacitor behaves like an open circuit.
- The **time constant** of an RL circuit equals the equivalent inductance divided by the Thévenin resistance as viewed from the terminals of the equivalent inductor.
- The **time constant** of an RC circuit equals the equivalent capacitance times the Thévenin resistance as viewed from the terminals of the equivalent capacitor.
- The **natural response** is the currents and voltages that exist when stored energy is released to a circuit that contains no independent sources.
- The **step response** is the currents and voltages that result from abrupt changes in dc sources connected to a circuit. Stored energy may or may not be present at the time the abrupt changes take place.
- Capacitive voltages and inductive currents are continuous; that is, they have the same value at $t = t_0^-$ and $t = t_0^+$. Capacitive currents and inductive voltages may be discontinuous; that is, they may have different values at $t = t_0^-$ and $t = t_0^+$.
- The solution for either the natural or step response of both RL and RC circuits involves finding the initial and final value of the current or voltage of interest and the time constant of the circuit. Equations (7.59) and (7.60) summarize this approach.
- An unbounded response occurs when the Thévenin resistance is negative, which is possible when the first-order circuit contains dependent sources.
- Sequential switching in first-order circuits is analyzed by dividing the analysis into time intervals corresponding to specific switch positions. Initial values for a particular interval are determined from the solution corresponding to the immediately preceding interval.
- An integrating amplifier consists of an ideal op amp, a capacitor in the negative feedback branch, and a resistor in series with the signal source. It outputs the integral of the signal source, within specified limits that avoid saturating the op amp.

PROBLEMS

7.1 In the circuit in Fig. P7.1, the voltage and current expressions are

$$v = 200e^{-20t}\ \text{V},\quad t \geq 0^+;$$
$$i = 12.5e^{-20t}\ \text{A},\quad t \geq 0.$$

Find (a) R, (b) τ (milliseconds), (c) L, (d) the initial energy stored in the inductor, and (e) the time (in milliseconds) it takes to dissipate 40% of the initial stored energy.

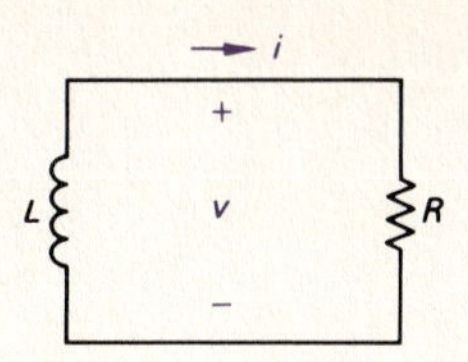

FIGURE P7.1

7.2 In the circuit shown in Fig. P7.2 the switch makes contact with position b just before breaking contact with position a. As already mentioned, this is known as a make-before-break switch and is designed so that the switch does not interrupt the current in an inductive circuit. The interval of time between "making" and "breaking" is assumed to be negligible. The switch has been in the a position for a long time. At $t = 0$ the switch is thrown from position a to position b.

a) Determine the initial current in the inductor.

b) Determine the time constant of the circuit for $t > 0$.

c) Find i, v_i, and v_2 for $t \geq 0$.

d) What percentage of the initial energy stored in the inductor is dissipated in the 40 Ω resistor 10 ms after the switch is thrown from position a to position b?

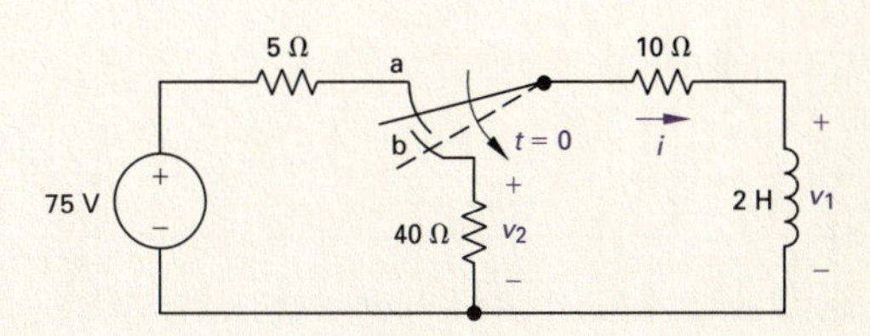

FIGURE P7.2

7.3 The switch in the circuit in Fig. P7.3 has been open for a long time. At $t = 0$ the switch is closed.

a) Determine $i_o(0^+)$ and $i_o(\infty)$.

b) Determine $i_o(t)$ for $t \geq 0^+$.

c) How many microseconds after the switch has been closed will the current in the switch equal 3.2 A?

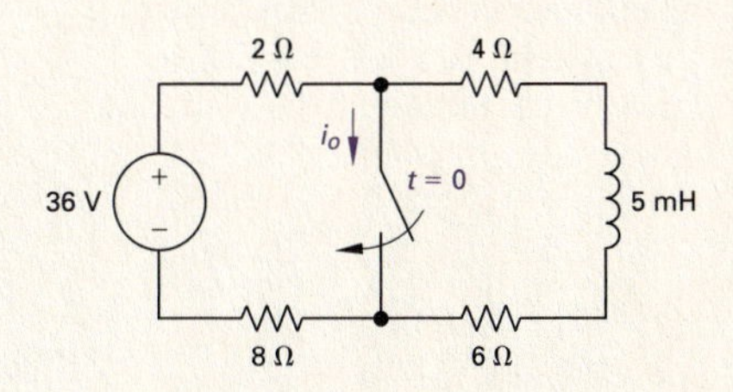

FIGURE P7.3

7.4 The switch shown in Fig. P7.4 has been open a long time before closing at $t = 0$.

a) Find $i_o(0^-)$.

b) Find $i_L(0^-)$.

c) Find $i_o(0^+)$.

d) Find $i_L(0^+)$.

e) Find $i_o(\infty)$.

f) Find $i_L(\infty)$.

g) Write the expression for $i_L(t)$ for $t \geq 0$.

h) Find $v_L(0^-)$.

i) Find $v_L(0^+)$.

j) Find $v_L(\infty)$.

k) Write the expression for $v_L(t)$ for $t \geq 0^+$.

l) Write the expression for $i_o(t)$ for $t \geq 0^+$.

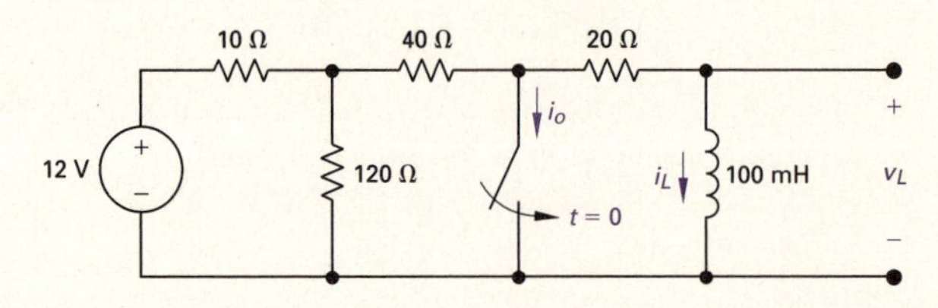

FIGURE P7.4

7.5 The switch in the circuit in Fig. P7.5 has been closed a long time. At $t = 0$ it is opened. Find $v_o(t)$ for $t \geq 0^+$.

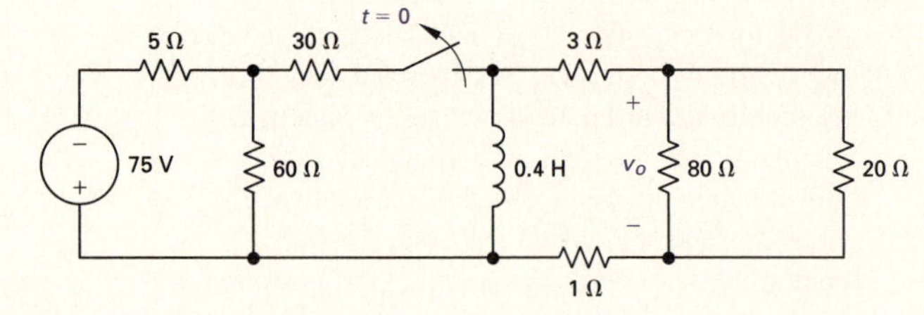

FIGURE P7.5

7.6 Assume that the switch in the circuit in Fig. P7.5 has been open for one time constant. At this instant, what percentage of the total energy stored in the 0.4 H inductor has been dissipated in the 20 Ω resistor?

7.7 The switch in the circuit in Fig. P7.7 has been in position 1 for a long time. At $t = 0$, the switch moves instantaneously to position 2. Find $v_o(t)$ for $t \geq 0^+$.

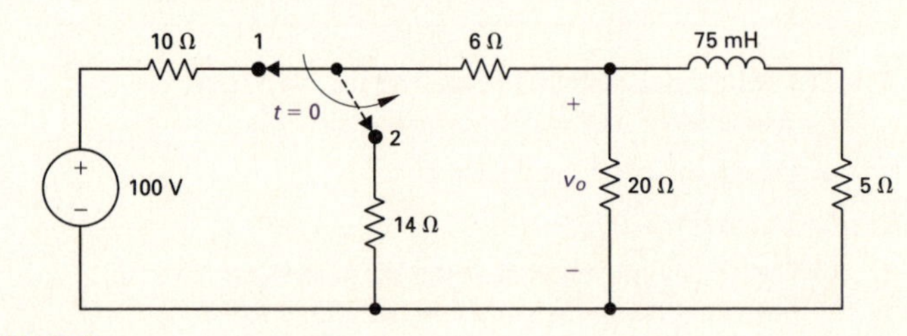

FIGURE P7.7

7.8 For the circuit of Fig. 7.7, what percentage of the initial energy stored in the inductor is eventually dissipated in the 5 Ω resistor?

7.9 In the circuit in Fig. 7.9, the switch has been closed for a long time before opening at $t = 0$.

a) Find the value of L so that $v_o(t)$ equals $0.2\, v_o(0^+)$ when $t = 2$ ms.

b) Find the percentage of the stored energy that has been dissipated in the 20 Ω resistor when $t = 2$ ms.

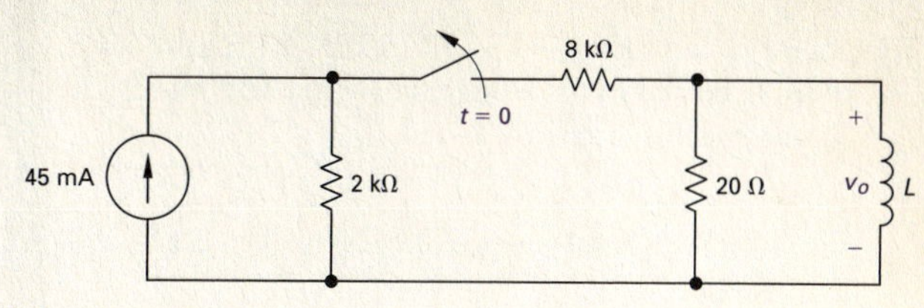

FIGURE P7.9

7.10 The switch in the circuit in Fig. P7.10 has been closed for a long time before opening at $t = 0$.

a) Find $i_1(0^-)$ and $i_2(0^-)$.

b) Find $i_1(0^+)$ and $i_2(0^+)$.

c) Find $i_1(t)$ for $t \geq 0$.

d) Find $i_2(t)$ for $t \geq 0^+$.

e) Explain why $i_2(0^-) \neq i_2(0^+)$.

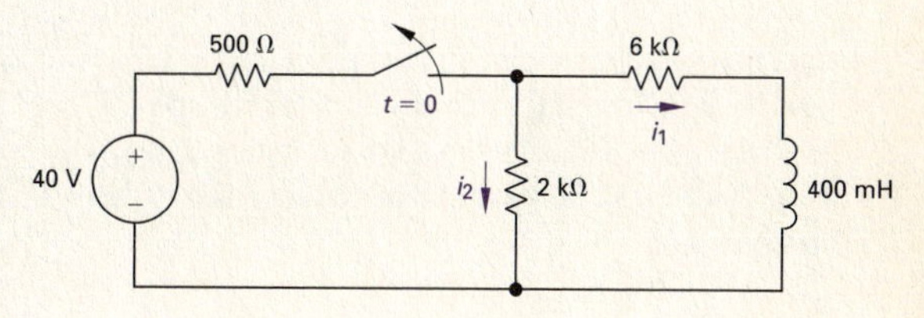

FIGURE P7.10

7.11 The switch in the circuit seen in Fig. P7.11 has been in position 1 for a long time. At $t = 0$ the switch moves instantaneously to position 2. Find the value of R so that 20% of the initial energy stored in the 15 mH inductor is dissipated in R in 12.5 μs.

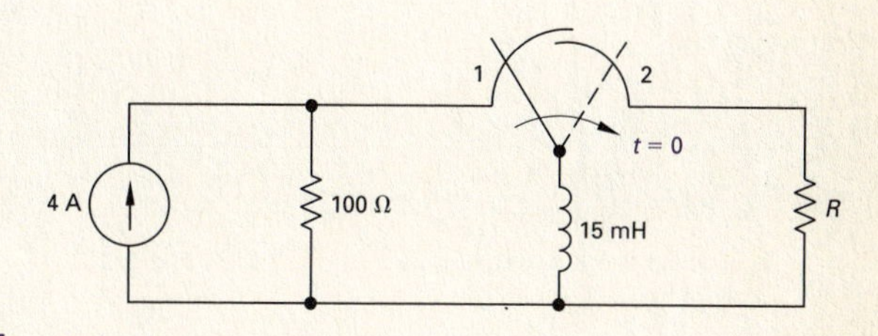

FIGURE P7.11

7.12 In the circuit in Fig. P7.11, let I_g represent the dc current source, σ represent the fraction of initial energy stored in the inductor that is dissipated in t_o seconds, and L represent the inductance.

a) Show that

$$R = \frac{L\, ln[1/(1-\sigma)]}{2t_o}$$

b) Test the expression derived in part (a) by using it to find the value of R in Problem 7.11.

7.13 In the circuit shown in Fig. P7.13, the switch has been in position a for a long time. At $t = 0$, it moves instantaneously from a to b.

a) Find $v_o(t)$ for $t \geq 0^+$.

b) What is the total energy delivered to the 5 kΩ resistor?

c) How many time constants does it take to deliver 95% of the energy found in part (b)?

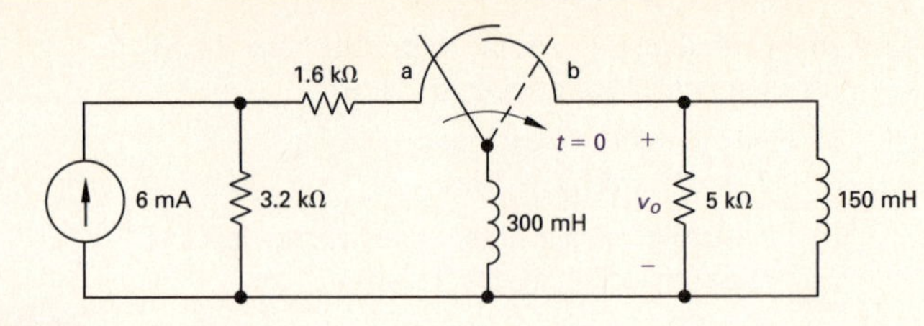

FIGURE P7.13

7.14 The two switches shown in the circuit in Fig. P7.14 operate simultaneously. Prior to $t = 0$ each switch has been in its indicated position for a long time. At $t = 0$ the two switches move instantaneously to their new positions. Find

a) $v_o(t), t \geq 0^+$; and

b) $i_o(t), t \geq 0$.

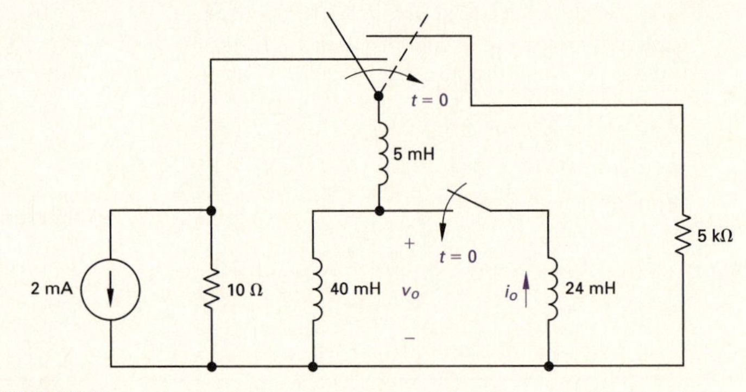

FIGURE P7.14

7.15 For the circuit seen in Fig. P7.14, find

a) the total energy dissipated in the 5 kΩ resistor; and

b) the energy trapped in the ideal inductors.

7.16 The switch in the circuit in Fig. P7.16 has been closed for a long time before opening at $t = 0$. Find $v_o(t)$ for $t \geq 0^+$.

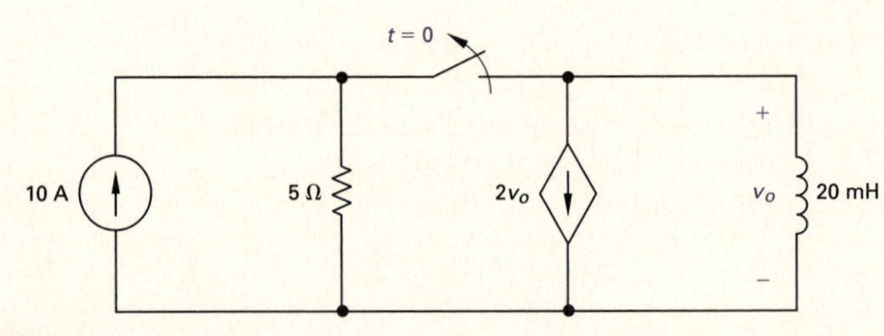

FIGURE P7.16

7.17 The switch in Fig. P7.17 has been closed for a long time before opening at $t = 0$. Find

a) $i_L(t)$, $t \geq 0$;

b) $v_L(t)$, $t \geq 0^+$; and

c) $i_\Delta(t)$, $t \geq 0^+$.

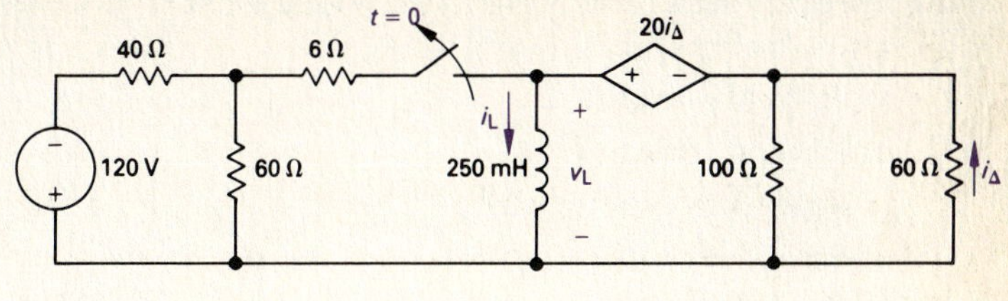

FIGURE P7.17

7.18 What percentage of the total energy dissipated in the two resistors in the circuit in Fig. P7.17 is supplied by the dependent voltage source?

7.19 The 240 V, 2 Ω source in the circuit in Fig. P7.19 is inadvertently short-circuited at its terminals a,b. At the time the fault occurs, the circuit has been in operation for a long time.

a) What is the initial value of the current i_{ab} in the short-circuit connection between terminals a,b?

b) What is the final value of the current i_{ab}?

c) How many microseconds after the short circuit has occurred is the current in the short equal to 114 A?

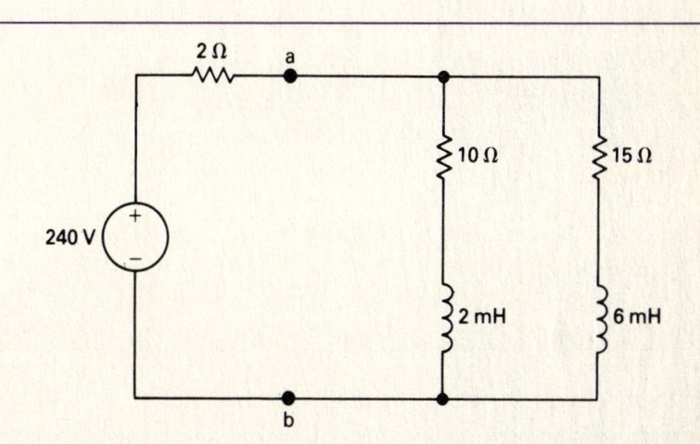

FIGURE P7.19

7.20 In the circuit in Fig. P7.20 the voltage and current expressions are

$$v = 100e^{-1000t} \text{ V}, \quad t \geq 0;$$
$$i = 5e^{-1000t} \text{ mA}, \quad t \geq 0^+.$$

Find (a) R, (b) C, (c) τ (milliseconds), (d) the initial energy stored in the capacitor, and (e) how many microseconds it takes to dissipate 80% of the initial energy stored in the capacitor.

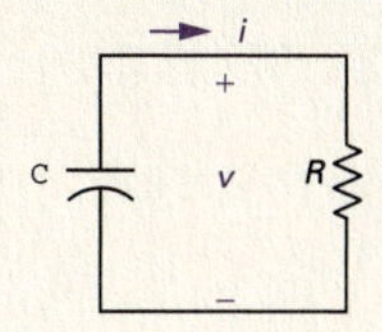

FIGURE P7.20

7.21 The switch in the circuit in Fig. P7.21 has been in position a for a long time. At $t = 0$ the switch is thrown to position b.

a) Find $i_o(t)$ for $t \geq 0^+$.

b) What percentage of the initial energy stored in the capacitor is dissipated in the 3 kΩ resistor 500 μs after the switch has been thrown?

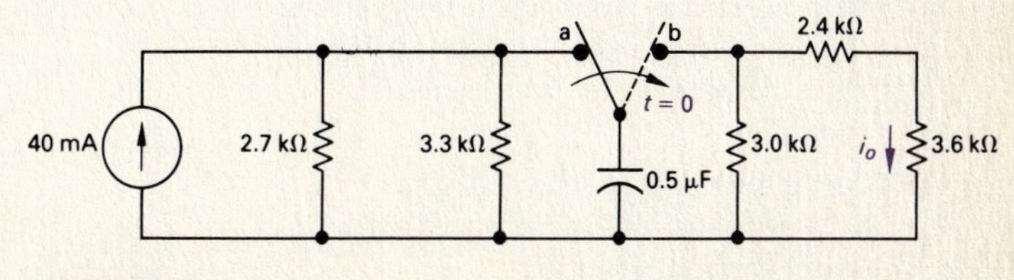

FIGURE P7.21

7.22 The switch in the circuit in Fig. P7.22 is closed at $t = 0$ after being open for a long time.

a) Find $i_1(0^-)$ and $i_2(0^-)$.

b) Find $i_1(0^+)$ and $i_2(0^+)$.

c) Explain why $i_1(0^-) = i_1(0^+)$.

d) Explain why $i_2(0^-) \neq i_2(0^+)$.

e) Find $i_1(t)$ for $t \geq 0$.

f) Find $i_2(t)$ for $t \geq 0^+$.

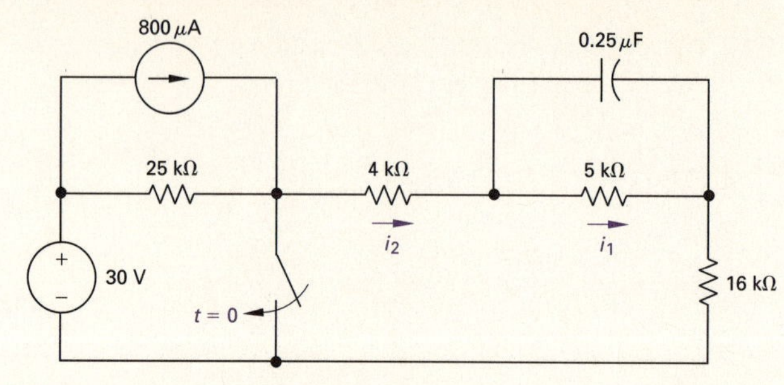

FIGURE P7.22

7.23 Both switches in the circuit in Fig. P7.23 have been closed for a long time. At $t = 0$, both switches open simultaneously.

a) Find $i_o(t)$ for $t \geq 0^+$.

b) Find $v_o(t)$ for $t \geq 0$.

c) Calculate the energy (in microjoules) trapped in the circuit.

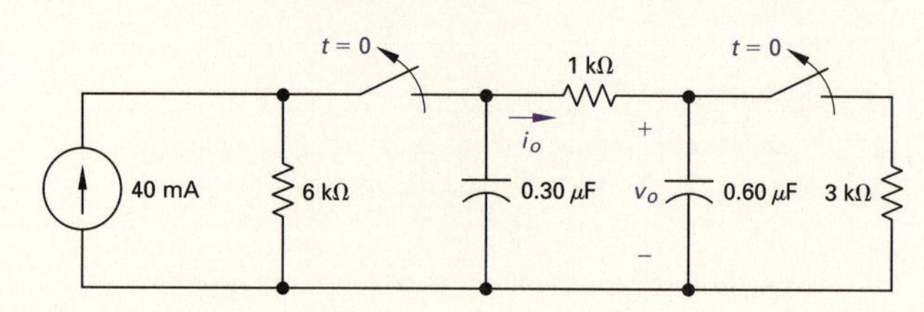

FIGURE P7.23

7.24 In the circuit shown in Fig. P7.24, both switches operate together—that is, they either open or close at the same time. The switches are closed a long time before opening at $t = 0$.

a) How many millijoules of energy have been dissipated in the 68 kΩ resistor 15 ms after the switches open?

b) How long does it take to dissipate 60% of the initially stored energy?

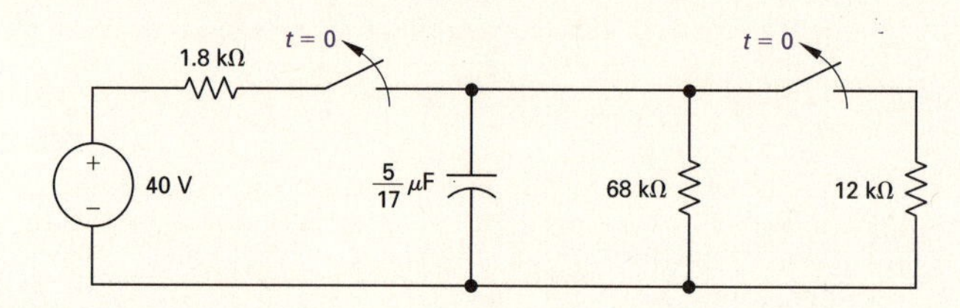

FIGURE P7.24

7.25 The two switches in the circuit seen in Fig. P7.25 are synchronized. The switches have been closed for a long time before opening at $t = 0$.

a) How many microseconds after the switches are open is the energy dissipated in the 4 kΩ resistor 10% of the initial energy stored in the 6 H inductor?

b) At the time calculated in part (a), what percentage of the total energy stored in the inductor has been dissipated?

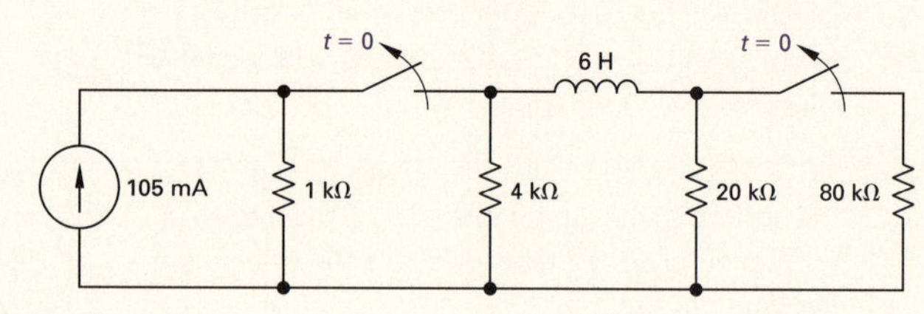

FIGURE P7.25

7.26 The switch in the circuit in Fig. P7.26 has been in position 1 for a long time before moving to position 2 at $t = 0$. Find $i_o(t)$ for $t \geq 0^+$.

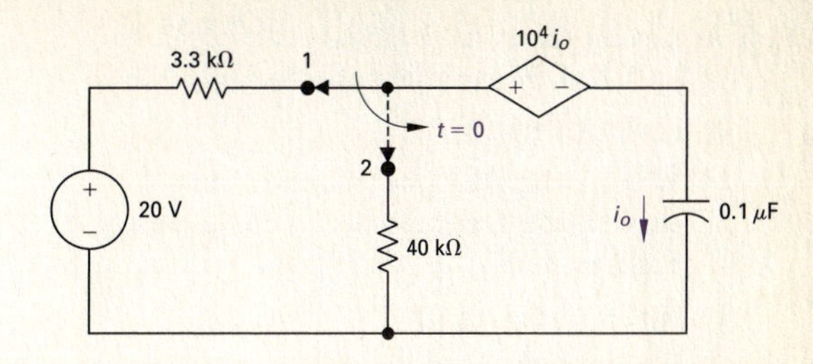

FIGURE P7.26

7.27 The switch in the circuit seen in Fig. P7.27 has been in position x for a long time. At $t = 0$ the switch moves instantaneously to position y.

a) Find α so that the time constant for $t > 0$ is 25 ms.

b) For the α found in part (a), find v_Φ.

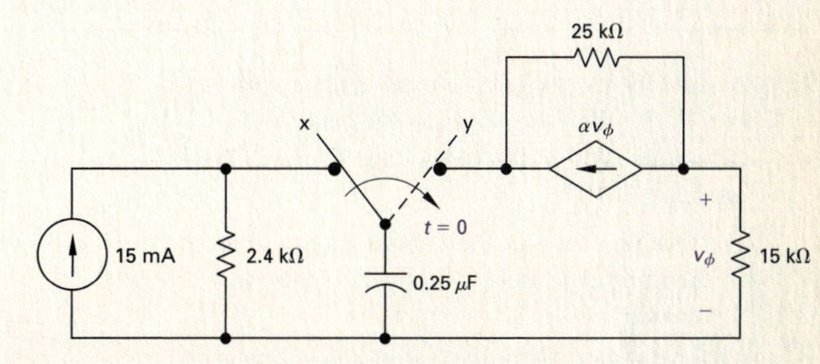

FIGURE P7.27

7.28 a) In Problem 7.27 how many microjoules of energy are generated by the dependent current source during the time the capacitor discharges to 0 V?

b) Show that for $t \geq 0$ the total energy stored and generated in the capacitive circuit equals the total energy dissipated.

7.29 The switch in the circuit in Fig. P7.29 has been in position a for a long time. At $t = 0$ the switch is thrown to position b.

a) Calculate i, v_1, and v_2 for $t \geq 0^+$.

b) Calculate the energy stored in the capacitor at $t = 0$.

c) Calculate the energy trapped in the circuit and the total energy dissipated in the 25 kΩ resistor if the switch remains in position b indefinitely.

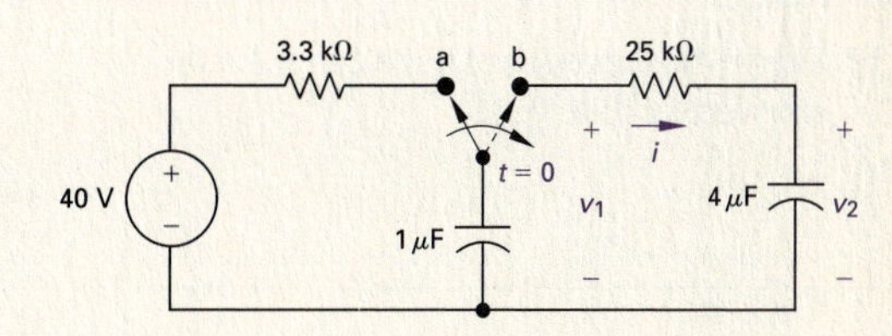

FIGURE P7.29

7.30 At the time the switch is closed in the circuit shown in Fig. P7.30, the capacitors are charged as shown.

a) Find $v_o(t)$ for $t \geq 0^+$.

b) What percentage of the total energy initially stored in the three capacitors is dissipated in the 5 kΩ resistor?

c) Find $v_1(t)$ for $t \geq 0$.

d) Find $v_2(t)$ for $t \geq 0$.

e) Find the energy (in microjoules) trapped in the ideal capacitors.

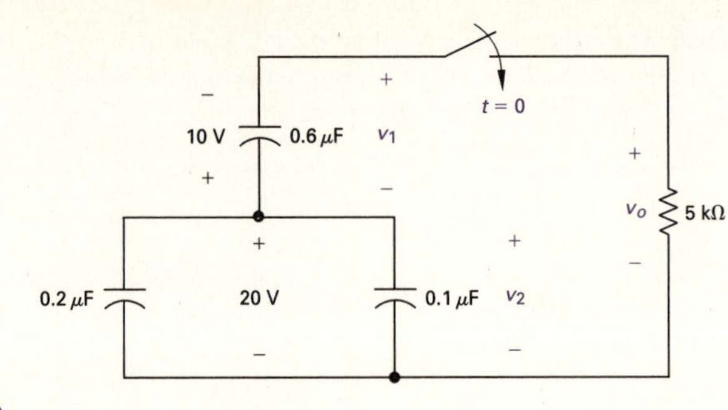

FIGURE P7.30

7.31 At the time the switch is closed in the circuit in Fig. P7.31, the voltage across the paralleled capacitors is 50 V and the voltage on the 0.25 μF capacitor is 40 V.

a) What percentage of the initial energy stored in the three capacitors is dissipated in the 24 kΩ resistor?

b) Repeat part (a) for the 0.4 and 16 kΩ resistors.

c) What percentage of the initial energy is trapped in the capacitors?

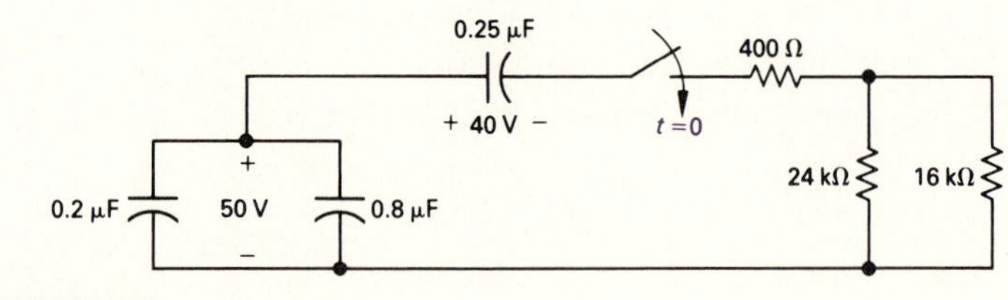

FIGURE P7.31

7.32 After the circuit in Fig. P7.32 has been in operation for a long time, a screwdriver was inadvertently connected across the terminals a,b. Assume the resistance of the screwdriver is negligible.

a) Find the current in the screwdriver at $t = 0^+$ and $t = \infty$.

b) Derive the expression for the current in the screwdriver for $t \geq 0^+$.

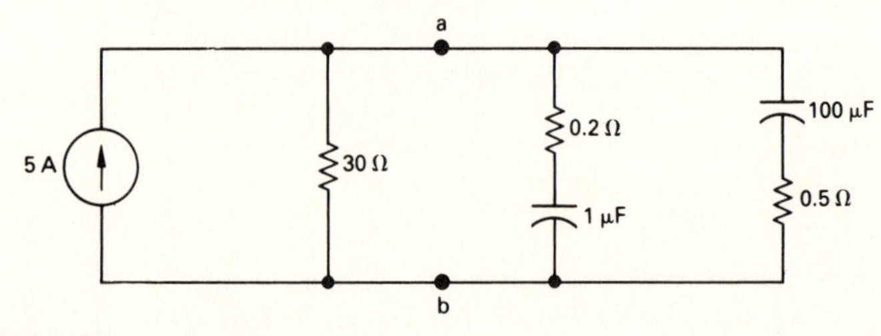

FIGURE P7.32

7.33 The current and voltage at the terminals of the inductor in the circuit in Fig. 7.16 are

$$i(t) = (10 - 10e^{-500t})\text{ A}, \quad t \geq 0;$$
$$v(t) = 200e^{-500t}\text{ V}, \quad t \geq 0^+.$$

a) Specify the numerical values of V_s, R, and L.

b) How many milliseconds after the switch has been closed does the energy stored in the inductor reach 25% of its final value?

7.34 The switch in the circuit shown in Fig. P7.34 has been in position a for a long time. At $t = 0$ the switch moves instantaneously to position b.

a) Find the numerical expression for $i_o(t)$ when $t \geq 0$.

b) Find the numerical expression for $v_o(t)$ for $t \geq 0^+$.

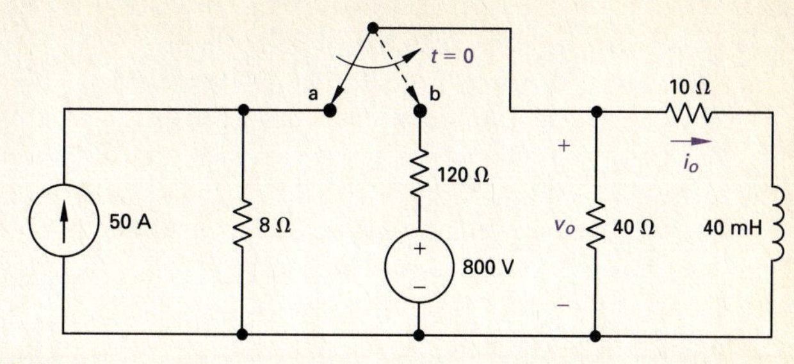

FIGURE P7.34

7.35 The switch in the circuit shown in Fig. P7.35 has been closed for a long time before opening at $t = 0$.

a) Find the numerical expressions for $i_L(t)$ and $v_o(t)$ for $t \geq 0$.

b) Find the numerical values of $v_L(0^+)$ and $v_o(0^+)$.

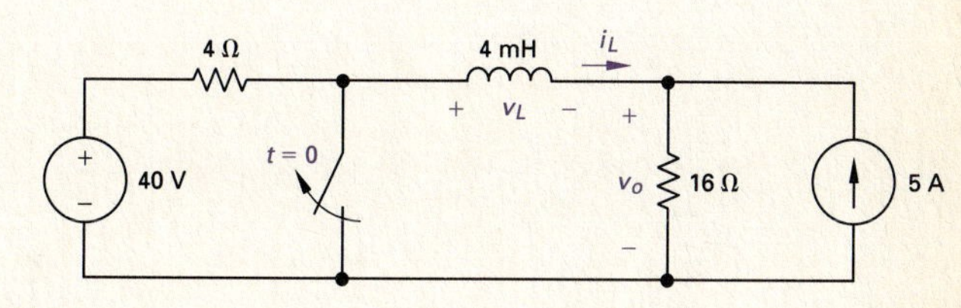

FIGURE P7.35

7.36 The switch in the circuit seen in Fig. P7.36 has been closed for a long time. The switch opens at $t = 0$. Find the numerical expressions for $i_o(t)$ and $v_o(t)$ when $t \geq 0^+$.

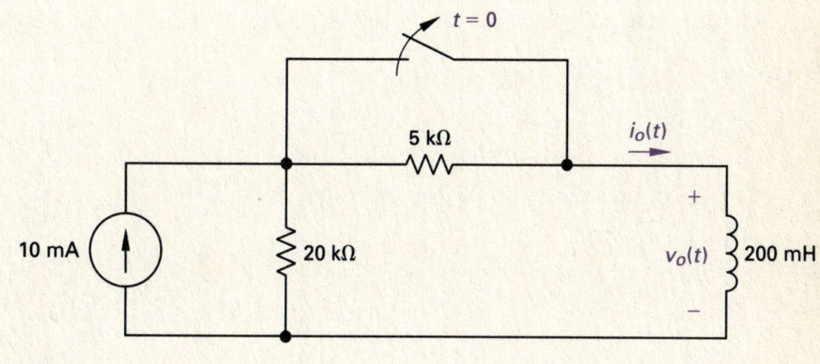

FIGURE P7.36

7.37 The switch in the circuit shown in Fig. P7.37 has been closed for a long time. The switch opens at $t = 0$. For $t \geq 0^+$,

a) find $v_o(t)$ as a function of I_g, R_1, R_2 and L;

b) verify your expression by using it to find $v_o(t)$ in the circuit of Fig. P7.36;

c) explain what happens to $v_o(t)$ as R_2 gets larger and larger;

d) find v_{sw} as a function of I_g, R_1, R_2, and L; and

e) explain what happens to v_{sw} as R_2 gets larger and larger.

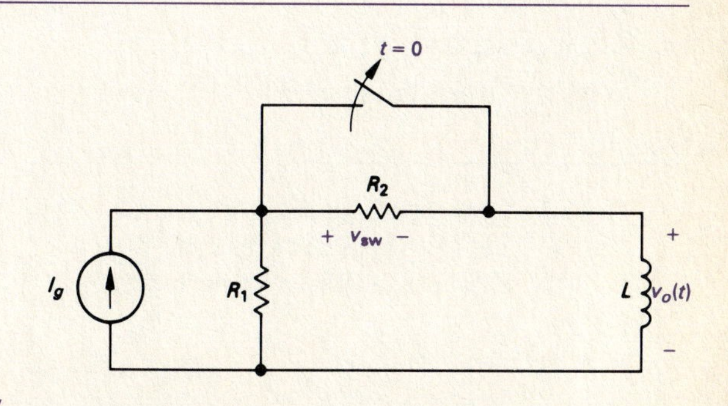

FIGURE P7.37

7.38 The switch in the circuit in Fig. P7.38 has been closed for a long time. A student abruptly opens the switch and reports to her instructor that when the switch opened, an electric arc with noticeable persistence was established across the switch, and at the same time the voltmeter placed across the coil was damaged. On the basis of your analysis of the circuit in Problem 7.37, can you explain to the student why this happened?

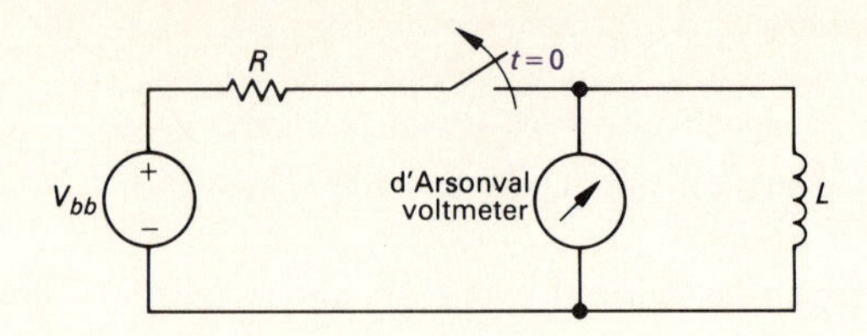

FIGURE P7.38

7.39 The switch in the circuit in Fig. P7.39 has been open a long time before closing at $t = 0$. Find $i_o(t)$ for $t \geq 0$.

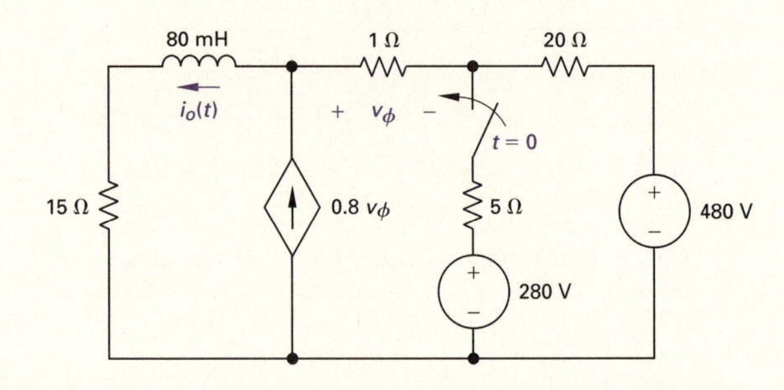

FIGURE P7.39

7.40 The switch in the circuit in Fig. P7.40 has been in position 1 for a long time. At $t = 0$ it moves instantaneously to position 2. How many milliseconds after the switch operates does v_o equal 100 V?

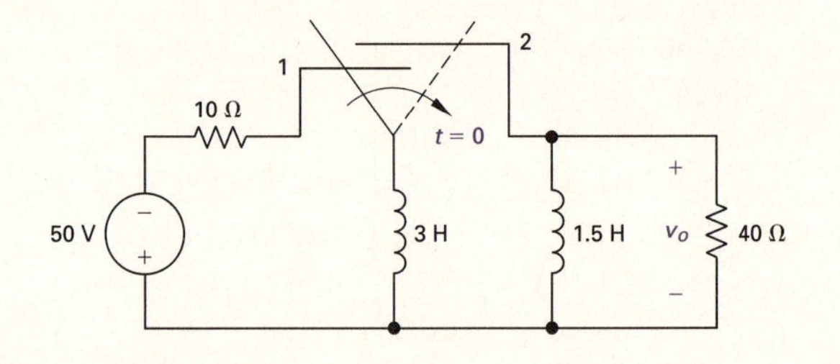

FIGURE P7.40

7.41 For the circuit in Fig. P7.40, find (in joules)

a) the total energy dissipated in the 40 Ω resistor;

b) the energy trapped in the inductors; and

c) the initial energy stored in the inductors.

7.42 The make-before-break switch in the circuit of Fig. P7.42 has been in position a for a long time. At $t = 0$ the switch moves instantaneously to position b. Find

a) $v_o(t)$, $t \geq 0^+$;

b) $i_1(t)$, $t \geq 0$; and

c) $i_2(t)$, $t \geq 0$.

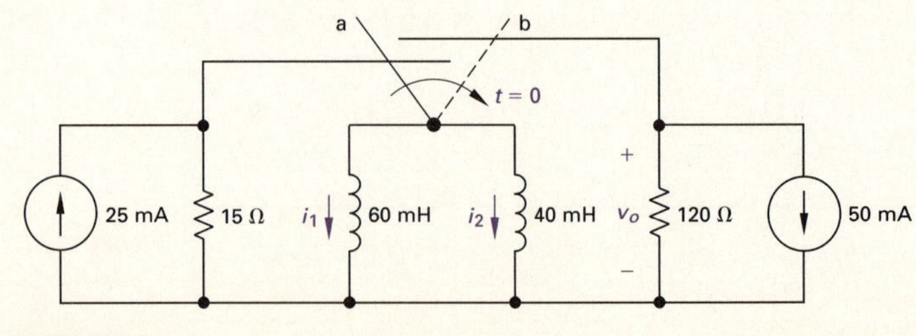

FIGURE P7.42

7.43 The switch in the circuit in Fig. P7.43 has been open a long time before closing at $t = 0$. Find $v_o(t)$ for $t \geq 0^+$.

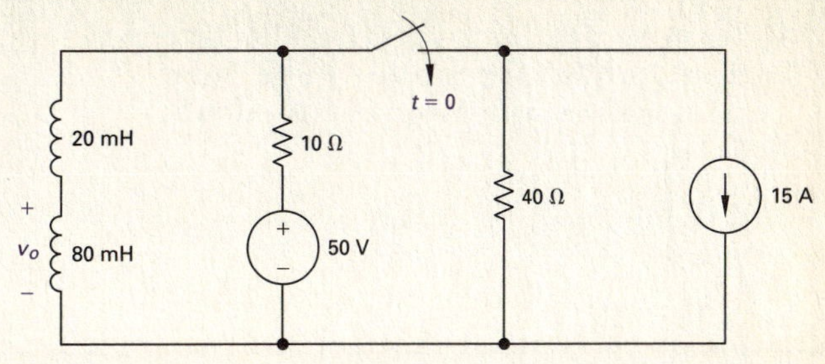

FIGURE P7.43

7.44 There is no energy stored in the inductors L_1 and L_2 at the time the switch is opened in the circuit shown in Fig. P7.44.

a) Derive the expressions for the currents $i_1(t)$ and $i_2(t)$ for $t \geq 0$.

b) Use the expressions derived in part (a) to find $i_1(\infty)$ and $i_2(\infty)$.

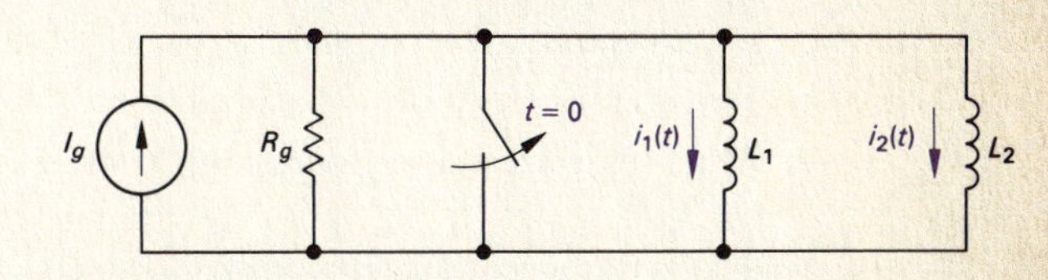

FIGURE P7.44

7.45 The current and voltage at the terminals of the capacitor in the circuit in Fig. 7.21 are

$$i(t) = 50e^{-2500t} \text{ mA}, \quad t \geq 0^+;$$
$$v(t) = (80 - 80e^{-2500t}) \text{ V}, \quad t \geq 0.$$

a) Specify the numerical values of I_s, R, C, and τ.

b) How many microseconds after the switch has been closed does the energy stored in the capacitor reach 64% of its final value?

7.46 The switch in the circuit shown in Fig. P7.46 has been closed a long time before opening at $t = 0$. For $t \geq 0^+$, find

a) $v_o(t)$;

b) $i_o(t)$;

c) $i_1(t)$;

d) $i_2(t)$; and

e) $i_1(0^+)$.

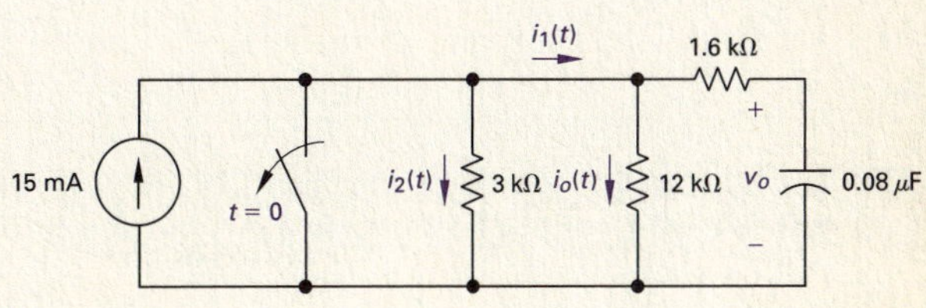

FIGURE P7.46

7.47 The switch in the circuit seen in Fig. P7.47 has been in position a for a long time. At $t = 0$ the switch moves instantaneously to position b. For $t \geq 0^+$, find

a) $v_o(t)$;

b) $i_o(t)$;

c) $v_g(t)$; and

d) $v_g(0^+)$.

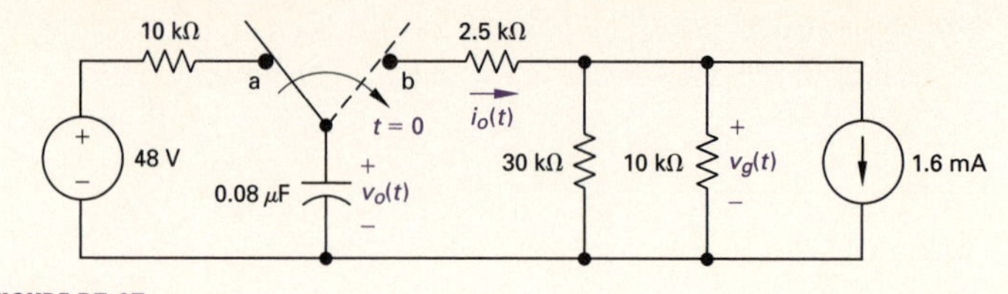

FIGURE P7.47

7.48 The switch in the circuit shown in Fig. P7.48 has been closed a long time before opening at $t = 0$.

a) What is the initial value of $i_o(t)$?

b) What is the final value of $i_o(t)$?

c) What is the time constant of the circuit for $t \geq 0$?

d) What is the numerical expression for $i_o(t)$ when $t \geq 0^+$?

e) What is the numerical expression for $v_o(t)$ when $t \geq 0^+$?

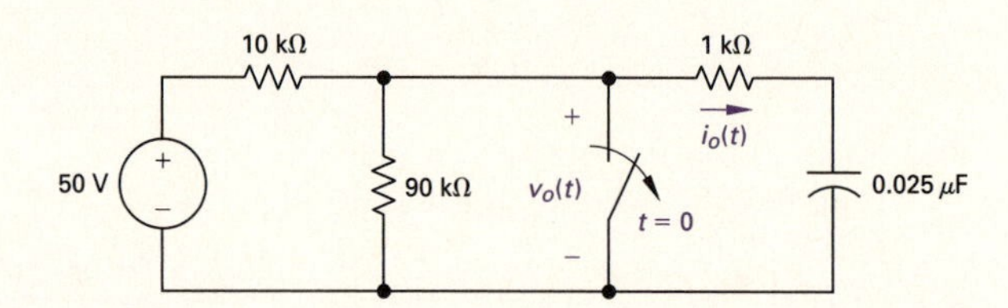

FIGURE P7.48

7.49 The switch in the circuit seen in Fig. P7.49 has been in position a for a long time. At $t = 0$ the switch moves instantaneously to position b. Find $v_o(t)$ and $i_o(t)$ for $t \geq 0^+$.

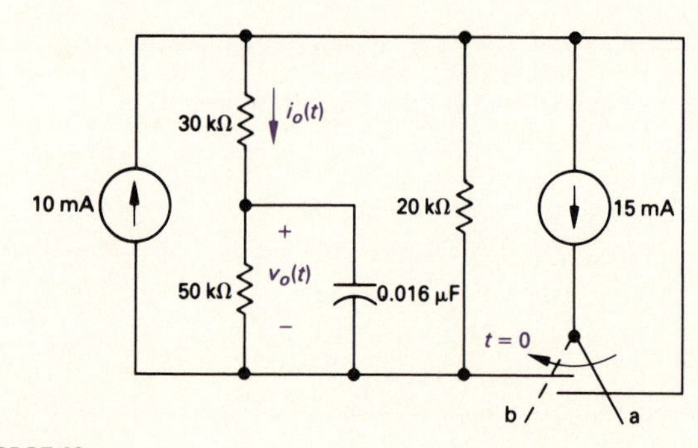

FIGURE P7.49

7.50 The switch in the circuit shown in Fig. P7.50 has been in the OFF position for a long time. At $t = 0$ the switch moves instantaneously to the ON position. Find $v_o(t)$ for $t \geq 0$.

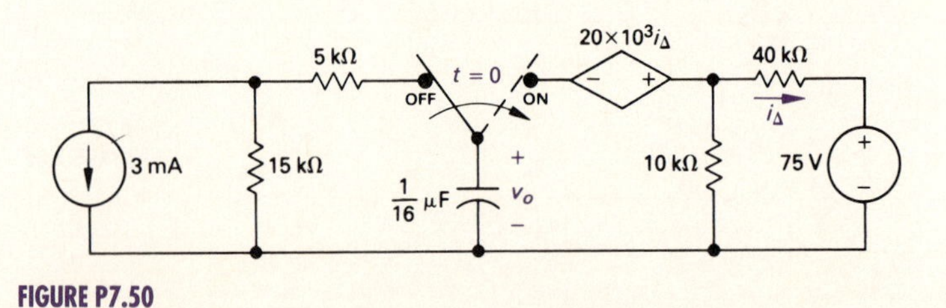

FIGURE P7.50

7.51 Assume that the switch in the circuit of Fig. P7.50 has been in the ON position for a long time before switching instantaneously to the OFF position at $t = 0$. Find $v_o(t)$ for $t \geq 0$.

7.52 The circuit in Fig. P7.52 has been in operation for a long time. At $t = 0$ the voltage source reverses polarity and the current source drops from 3 mA to 2 mA. Find $v_o(t)$ for $t \geq 0$.

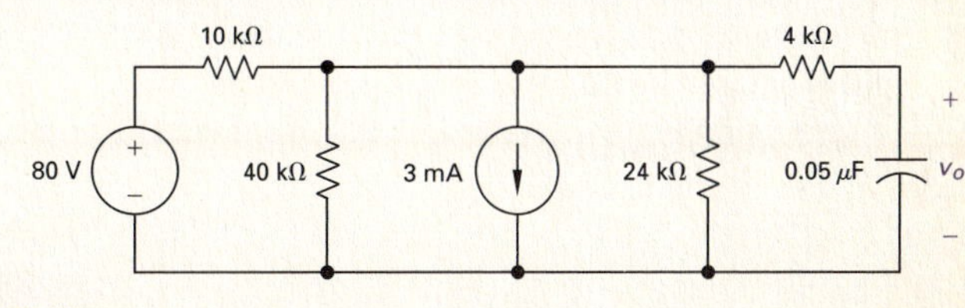

FIGURE P7.52

7.53 There is no energy stored in the capacitors C_1 and C_2 at the time the switch is closed in the circuit seen in Fig. P7.53.

a) Derive the expressions for $v_1(t)$ and $v_2(t)$ for $t \geq 0$.

b) Use the expressions derived in part (a) to find $v_1(\infty)$ and $v_2(\infty)$.

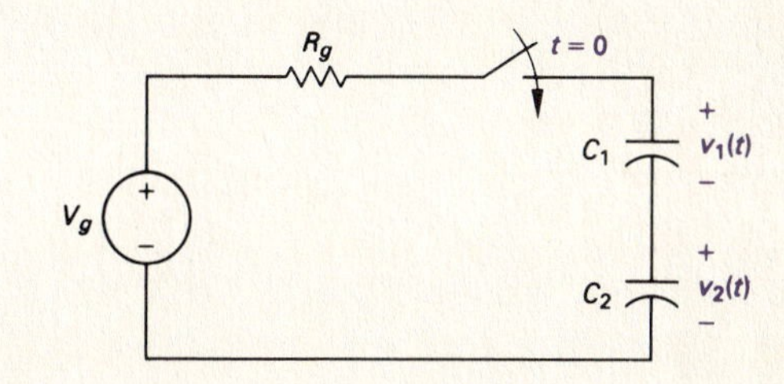

FIGURE P7.53

7.54 The switch in the circuit of Fig. P7.54 has been in position a for a long time. At $t = 0$ it moves instantaneously to position b. For $t \geq 0^+$, find

a) $v_o(t)$,

b) $i_o(t)$,

c) $v_1(t)$,

d) $v_2(t)$, and

e) the energy trapped in the capacitors as $t \to \infty$.

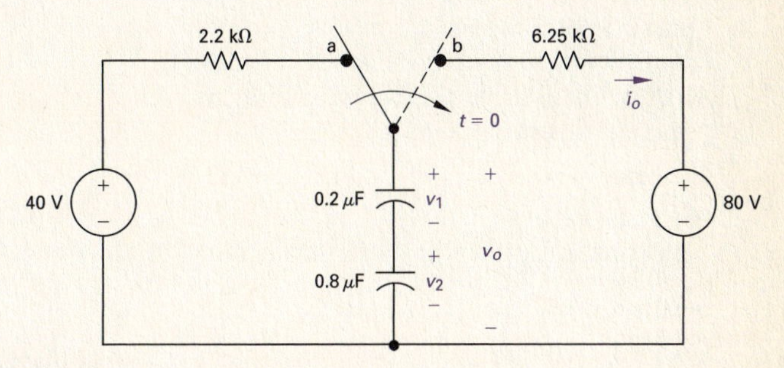

FIGURE P7.54

7.55 The switch in the circuit in Fig. P7.55 has been in position x for a long time. The initial charge on the 15 nF capacitor is zero. At $t = 0$ the switch moves instantaneously to position y.

a) Find $v_o(t)$ for $t \geq 0^+$.

b) Find $v_1(t)$ for $t \geq 0$.

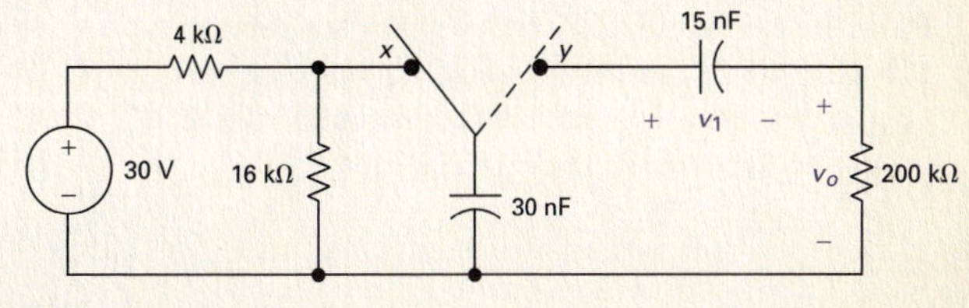

FIGURE P7.55

7.56 For the circuit in Fig. P7.55, find (in microjoules)

a) the energy delivered to the 200 kΩ resistor;

b) the energy trapped in the capacitors; and

c) the initial energy stored in the capacitors.

7.57 The switch in the circuit shown in Fig. P7.57 opens at $t = 0$ after being closed for a long time. How many milliseconds after the switch opens is the energy stored in the capacitor 25% of its final value?

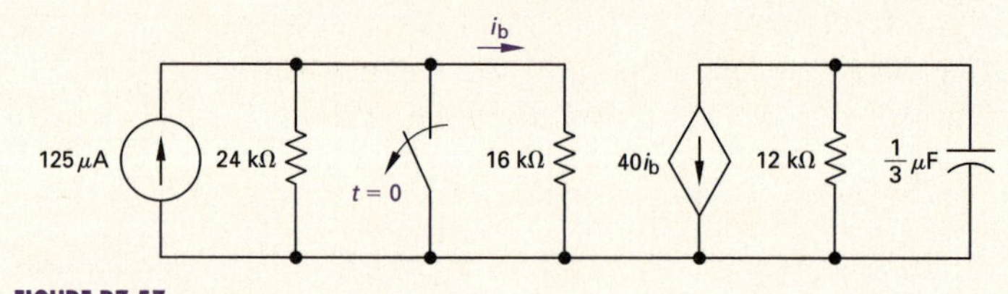

FIGURE P7.57

7.58 The switch in the circuit in Fig. P7.58 has been open a long time before closing at $t = 0$. Find $v_o(t)$ for $t \geq 0^+$.

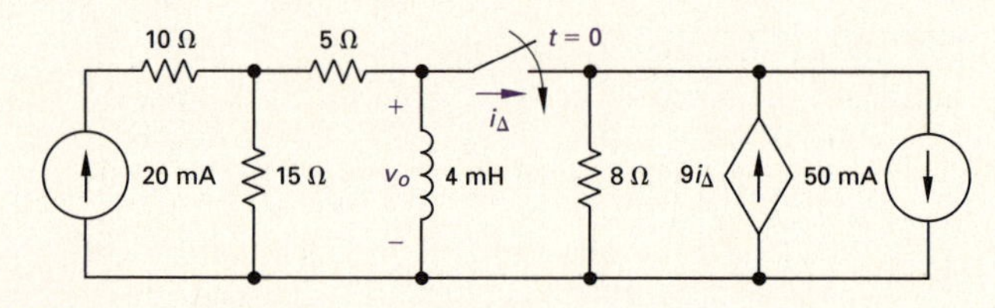

FIGURE P7.58

7.59 The switch in the circuit shown in Fig. P7.59 has been in position a for a long time. At $t = 0$ the switch is moved to position b, where it remains for 600 μs. The switch is then moved to position c, where it remains indefinitely.

a) Find $i(0^+)$.

b) Find i (200 μs).

c) Find i (1 ms).

d) Find v (600^- μs).

e) Find v (600^+ μs).

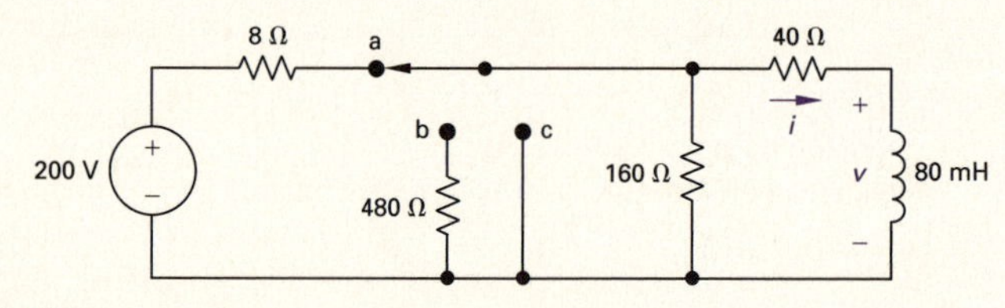

FIGURE P7.59

7.60 In the circuit in Fig. P7.60, switch A has been open and switch B has been closed for a long time. At $t = 0$, switch A closes. Five seconds after switch A closes, switch B opens. Find $i_L(t)$ for $t \geq 0$.

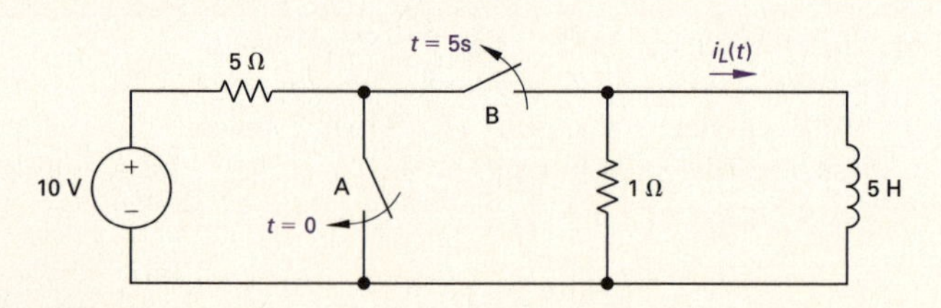

FIGURE P7.60

7.61 The action of the two switches in the circuit seen in Fig. P7.61 is as follows. For $t < 0$, switch 1 is in position a and switch 2 is open. This state has existed for a long time. At $t = 0$, switch 1 moves instantaneously from position a to position b, while switch 2 remains open. Two hundred microseconds after switch 1 operates, switch 2 closes, remains closed for 500 μs, and then opens. Find $v_o(t)$ 1.2 ms after switch 1 moves to position b.

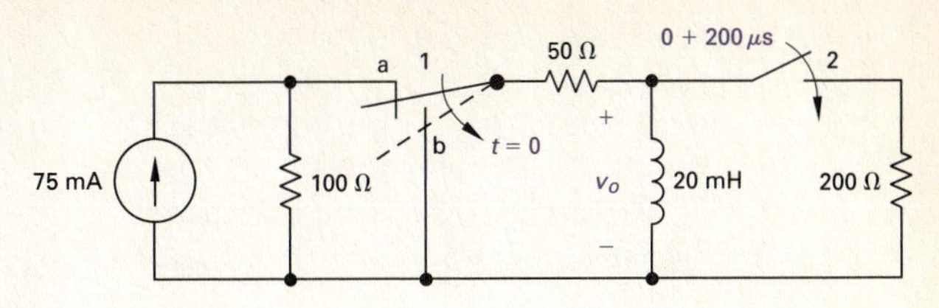

FIGURE P7.61

7.62 For the circuit in Fig. P7.61, how many microseconds after switch 1 moves to position b is the energy stored in the inductor 16% of its initial value?

7.63 The switch in the circuit in Fig. P7.63 has been in position a for a long time. At $t = 0$ it moves instantaneously to position b, where it remains for 50 ms before moving instantaneously to position c. Find v_o for $t \geq 0$.

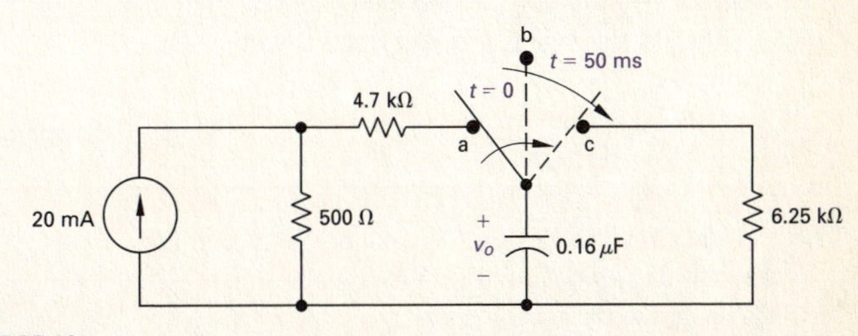

FIGURE P7.63

7.64 The switch in the circuit in Fig. P7.64 has been in position a for a long time. At $t = 0$, the switch moves instantaneously to position b. At the instant the switch makes contact with terminal b, switch 2 opens. Find $v_o(t)$ for $t \geq 0$.

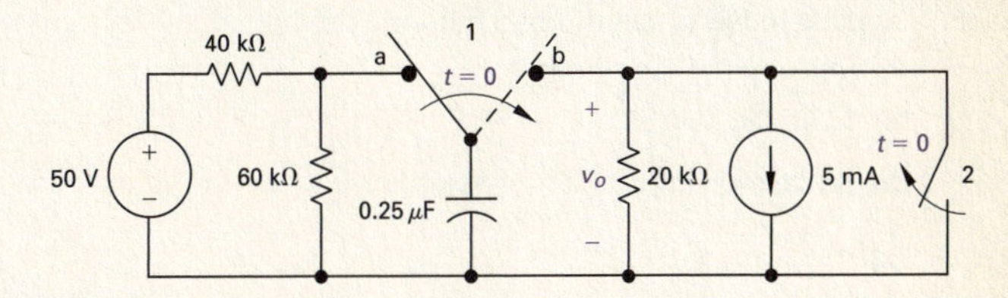

FIGURE P7.64

7.65 There is no energy stored in the capacitor in the circuit in Fig. P7.65 when switch 1 closes at $t = 0$. Ten microseconds later, switch 2 closes. Find $v_o(t)$ for $t \geq 0$.

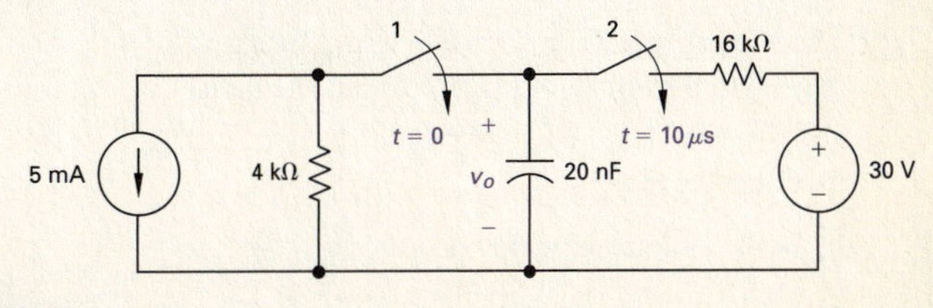

FIGURE P7.65

7.66 The capacitor in the circuit seen in Fig. P7.66 has been charged to 244.28 mV. At $t = 0$, switch 1 closes, causing the capacitor to discharge into the resistive network. Switch 2 closes 100 μs after switch 1 closes. Find the magnitude and direction of the current in the second switch 250 μs after switch 1 closes.

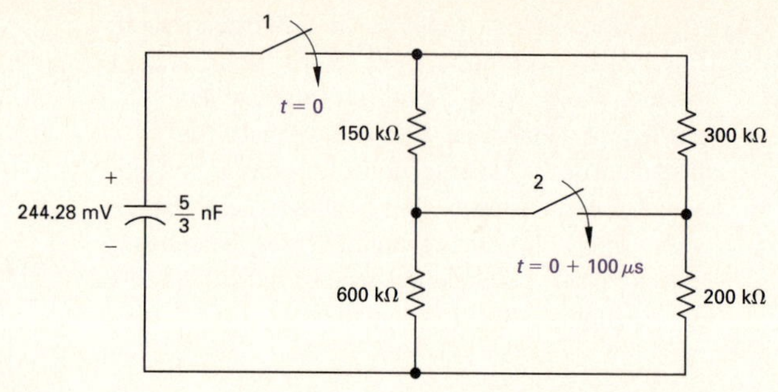

FIGURE P7.66

7.67 In the circuit in Fig. P7.67, switch 1 has been in position a, and switch 2 has been closed for a long time. At $t = 0$, switch 1 moves instantaneously to position b. Fifty microseconds later, switch 2 opens, remains open for 100 μs, and then recloses. Find v_o 200 μs after switch 1 makes contact with terminal b.

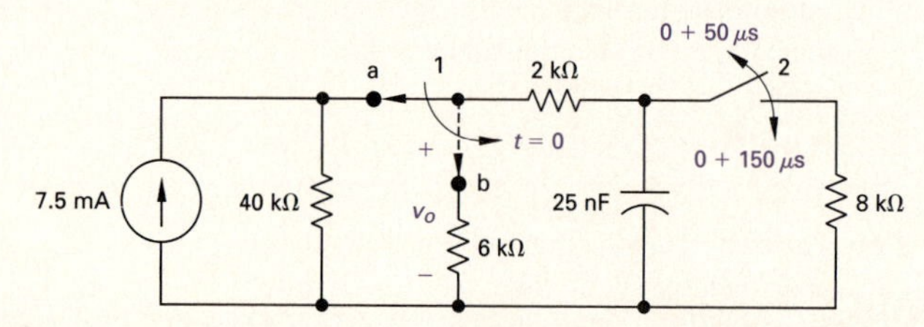

FIGURE P7.67

7.68 For the circuit in Fig. P7.67, what percentage of the initial energy stored in the 25 nF capacitor is dissipated in the 8 kΩ resistor?

7.69 The voltage waveform shown in Fig. P7.69(a) is applied to the circuit of Fig. P7.69(b). The initial voltage on the capacitor is zero.

a) Calculate $v_o(t)$.

b) Make a sketch of $v_o(t)$ versus t.

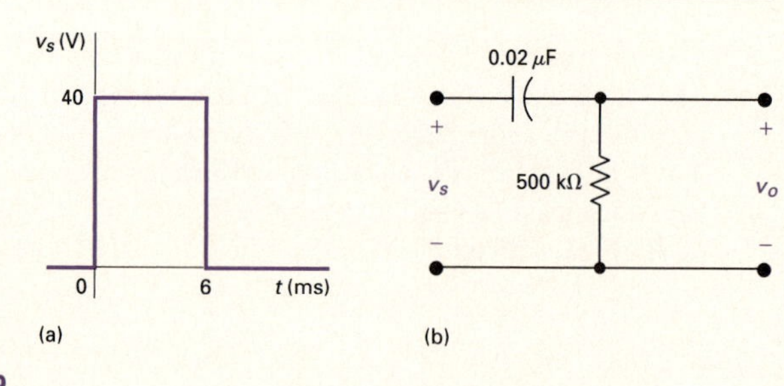

FIGURE P7.69

7.70 The voltage waveform shown in Fig. P7.70(a) is applied to the circuit of Fig. P7.70(b). The initial current in the inductor is zero.

a) Calculate $v_o(t)$.

b) Make a sketch of $v_o(t)$ versus t.

c) Find i_o at $t = 5$ ms.

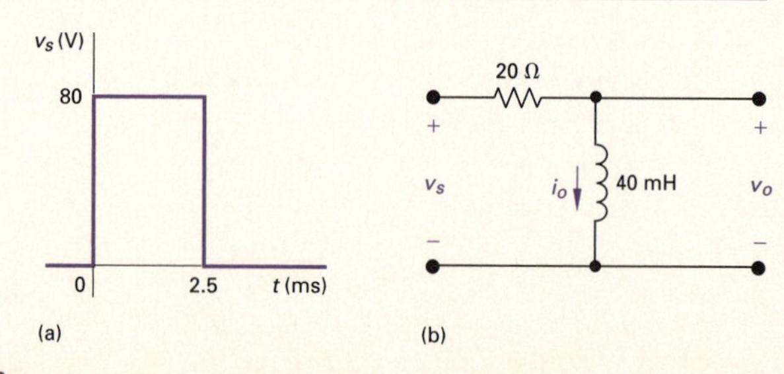

FIGURE P7.70

7.71 The current source in the circuit in Fig. P7.71(a) generates the current pulse shown in Fig. P7.71(b). There is no energy stored at $t = 0$.

a) Derive the numerical expressions for $v_o(t)$ for the time intervals $t < 0$, $0 < t < 80\ \mu s$, and $80\ \mu s < t < \infty$.

b) Calculate $v_o\ (80^-\ \mu s)$ and $v_o\ (80^+\ \mu s)$.

c) Calculate $i_o\ (80^-\ \mu s)$ and $i_o\ (80^+\ \mu s)$.

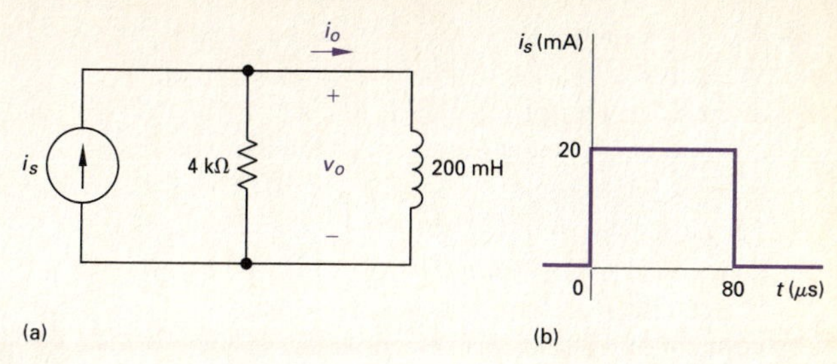

FIGURE P7.71

7.72 The current source in the circuit in Fig. P7.72(a) generates the current pulse shown in Fig. P7.72(b). There is no energy stored at $t = 0$.

a) Derive the expressions for $i_o(t)$ and $v_o(t)$ for the time intervals $t < 0$; $0 < t < 0.001$ s; and $0.001\ s < t < \infty$.

b) Calculate $i_o(0^-)$; $i_o(0^+)$; $i_o(0.001^-)$; and $i_o(0.001^+)$.

c) Calculate $v_o(0^-)$; $v_o(0^+)$; $v_o(0.001^-)$; and $v_o(0.001^+)$.

d) Sketch $i_o(t)$ versus t for the interval $-1\ ms < t < 3$ ms.

e) Sketch $v_o(t)$ versus t for the interval $-1\ ms < t < 3$ ms.

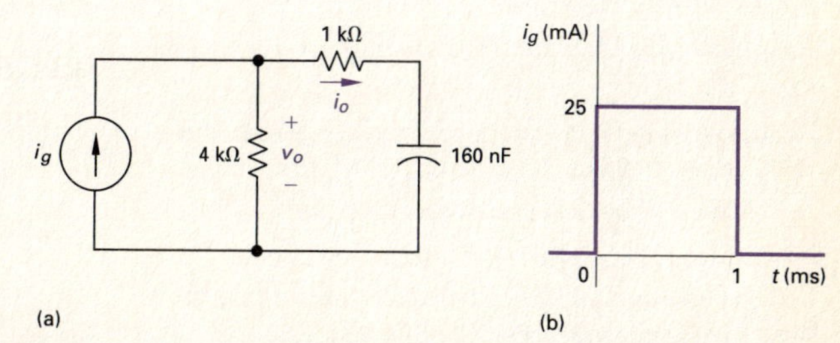

FIGURE P7.72

7.73 The voltage signal source in the circuit in Fig. P7.73(a) is generating the signal shown in Fig. P7.73(b). There is no stored energy at $t = 0$.

a) Derive the expressions for $v_o(t)$ that apply in the intervals $t < 0$; $0 \le t \le 4$ ms; $4\ ms \le t \le 8$ ms; and $8\ ms \le t \le \infty$.

b) Sketch v_o and v_s on the same coordinate axes.

c) Repeat parts (a) and (b) with R reduced to 50 kΩ.

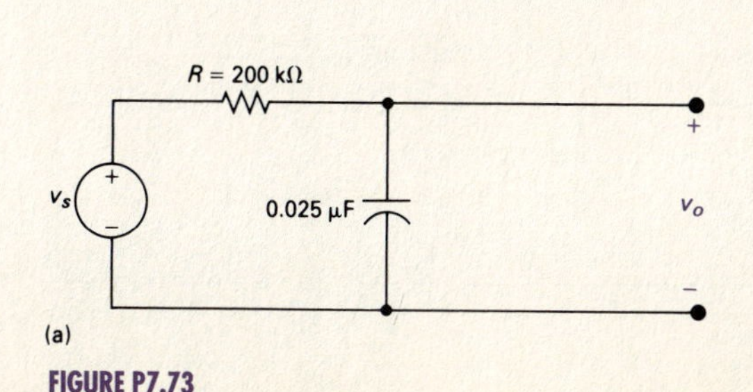

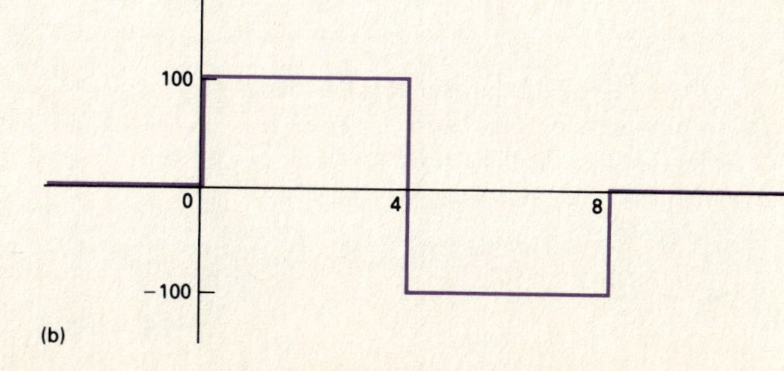

FIGURE P7.73

7.74 The circuit shown in Fig. P7.74 is used to close the switch between a and b for a predetermined length of time. The electric relay holds its contact arms down so long as the voltage across the relay coil exceeds 5 V. When the coil voltage equals 5 V, the relay contacts return to their initial position by a mechanical spring action. The switch between a and b is initially closed by momentarily pressing the push button. Assume that the capacitor is fully charged when the push button is first pushed down. The resistance of the relay coil is 25 kΩ, and the inductance of the coil is negligible.

a) How long will the switch between a and b remain closed?

b) Write the numerical expression for i from the time when the relay contacts first close to the time when the capacitor is completely charged.

c) How many milliseconds (after the circuit between a and b is interrupted) does it take the capacitor to reach 85% of its final value?

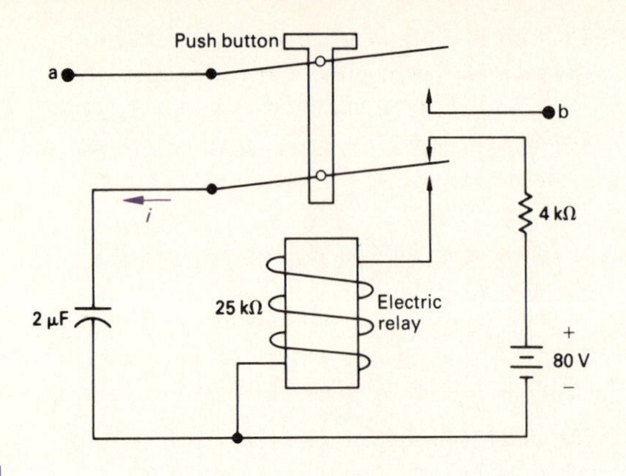

FIGURE P7.74

7.75 In the circuit of Fig. P7.75, the lamp starts to conduct whenever the lamp voltage reaches 15 V. During the time when the lamp conducts, it can be modeled as a 10 kΩ resistor. Once the lamp conducts, it will continue to conduct until the lamp voltage drops to 5 V. When the lamp is not conducting, it appears as an open circuit. Assume that the circuit has been in operation for a long time. Let $t = 0$ at the instant when the lamp stops conducting.

a) Derive the expression for the voltage across the lamp for one full cycle of operation.

b) How many times per minute will the lamp turn on?

c) The 800 kΩ resistor is replaced with a variable resistor R. The resistance is adjusted until the lamp flashes 12 times per minute. What is the value of R?

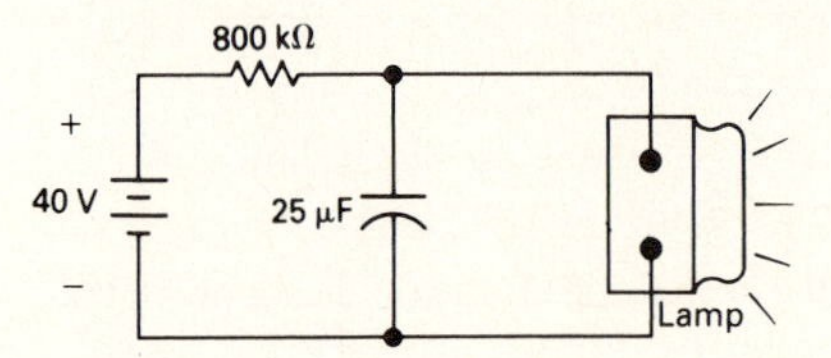

FIGURE P7.75

7.76 The capacitor in the circuit shown in Fig. P7.76 is charged to 50 V at the time the switch is closed. If the capacitor ruptures when its terminal voltage equals or exceeds 25 kV, how long does it take to rupture the capacitor?

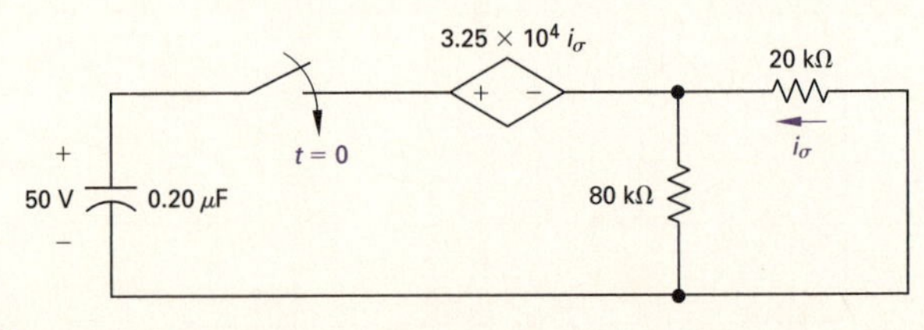

FIGURE P7.76

7.77 The inductor current in the circuit in Fig. P7.77 is 20 mA at the instant the switch is opened. The inductor will malfunction whenever the magnitude of the inductor current equals or exceeds 8 A. How long after the switch is opened does the inductor malfunction?

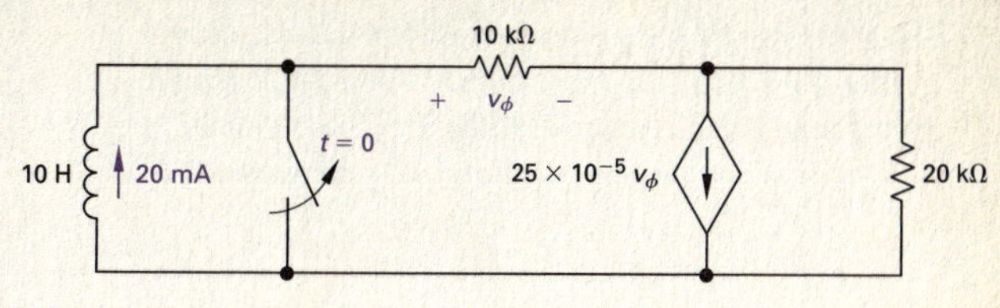

FIGURE P7.77

7.78 The gap in the circuit seen in Fig. P7.78 will arc over whenever the voltage across the gap reaches 45 kV. The initial current in the inductor is zero. The value of β is adjusted so the Thévenin resistance with respect to the terminals of the inductor is -5 kΩ.

a) What is the value of β?

b) How many microseconds after the switch has been closed will the gap arc over?

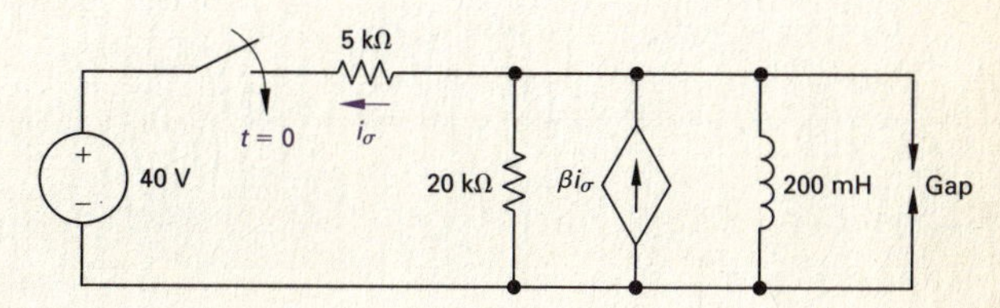

FIGURE P7.78

7.79 The switch in the circuit in Fig. P7.79 has been closed for a long time. The maximum voltage rating of the 12.5 nF capacitor is 600 V. How long after the switch is opened does the voltage across the capacitor reach the maximum voltage rating?

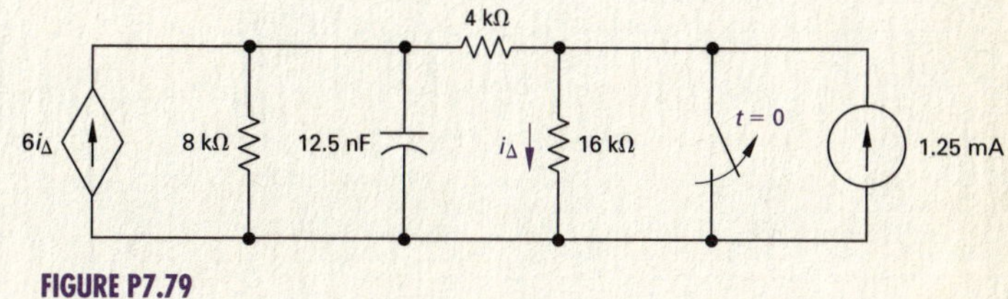

FIGURE P7.79

7.80 The energy stored in the capacitor in the circuit shown in Fig. P7.80 is zero at the instant the switch is closed. The ideal operational amplifier reaches saturation in 10 ms. What is the numerical value of R in kilo-ohms?

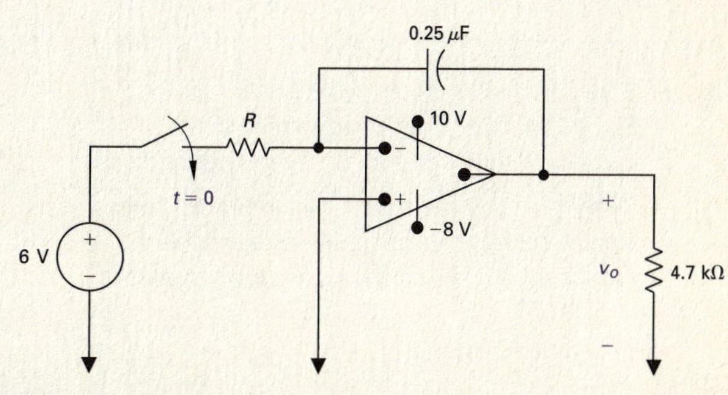

FIGURE P7.80

7.81 At the instant the switch is closed in the circuit of Fig. P7.80, the capacitor is charged to 3 V, positive at the left-hand terminal. If the ideal operational amplifier saturates in 15 ms, what is the value of R?

7.82 There is no energy stored in the capacitor at the time the switch in the circuit of Fig. P7.82 makes contact with terminal a. The switch remains at position a for 5 ms and then moves instantaneously to position b. How many milliseconds after making contact with terminal b does the op amp saturate?

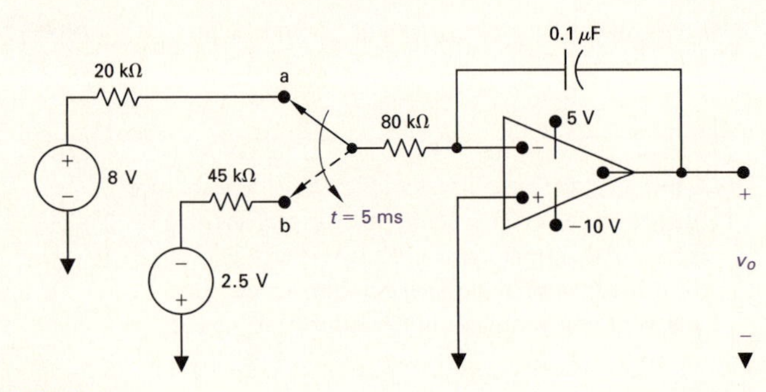

FIGURE P7.82

7.83 At the instant the switch of Fig. P7.83 is closed, the voltage on the capacitor is 16 V. Assume an ideal operational amplifier. How many milliseconds after the switch is closed will the output voltage v_o equal zero?

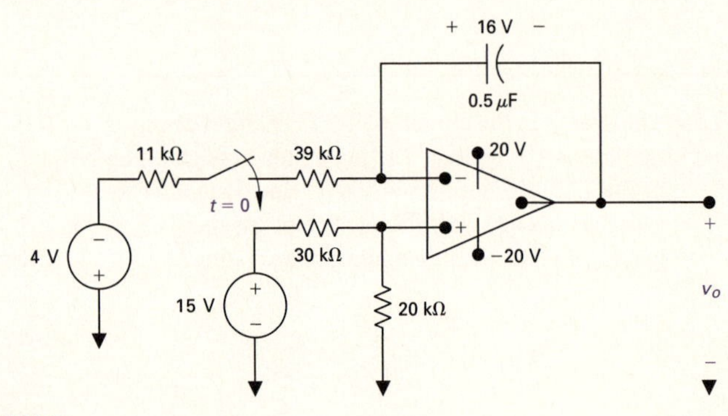

FIGURE P7.83

7.84 a) When the switch closes in the circuit seen in Fig. P7.84, there is no energy stored in the capacitor. How long does it take to saturate the op amp?

b) Repeat part (a) with an initial voltage on the capacitor of 1.5 V, positive at the lower terminal.

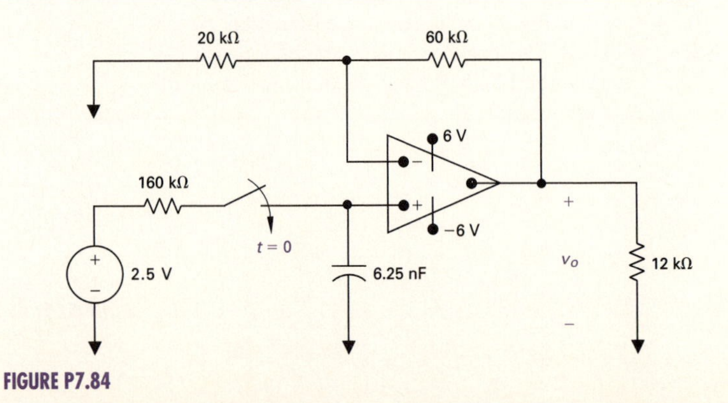

FIGURE P7.84

7.85 There is no energy stored in the capacitors in the circuit shown in Fig. P7.85 at the instant the two switches close.

a) Find v_o as a function of v_a, v_b, R, and C.

b) On the basis of the result obtained in part (a), describe the operation of the circuit.

c) How long will it take to saturate the amplifier if $v_a = 40$ mV; $v_b = 15$ mV; $R = 50$ kΩ; $C = 10$ nF; and $V_{CC} = 6$ V?

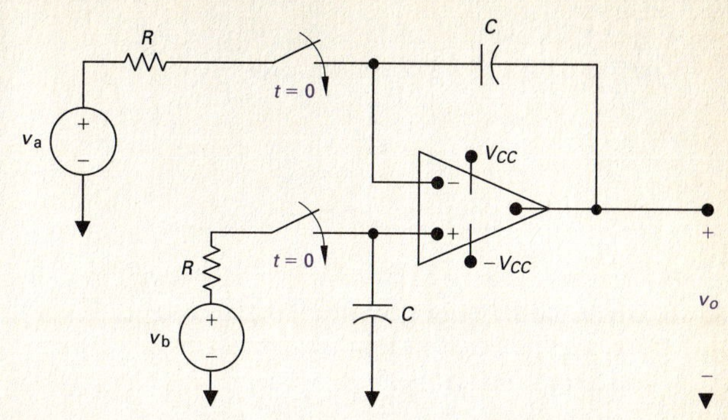

FIGURE P7.85

7.86 At the time the double-pole switch in the circuit shown in Fig. P7.86 is closed, the initial voltages on the capacitors are 15 mV and 5 mV, as shown. Find the numerical expressions for $v_o(t)$, $v_2(t)$, and $v_f(t)$ that are applicable as long as the ideal op amp operates in its linear range.

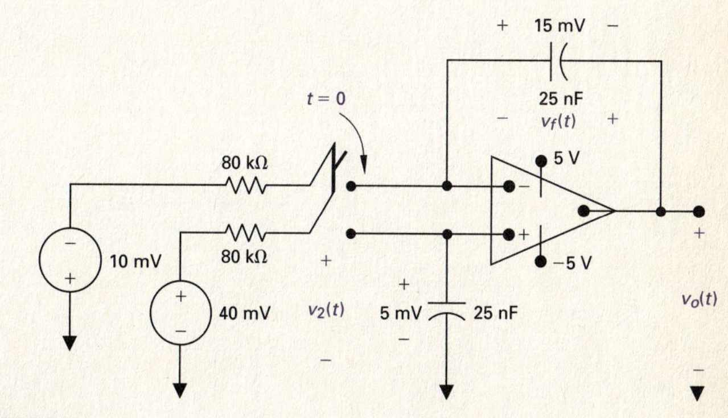

FIGURE P7.86

7.87 The voltage pulse shown in Fig. P7.87(a) is applied to the ideal integrating amplifier shown in Fig. P7.87(b). Derive the numerical expressions for $v_o(t)$ for the time intervals (a) $t < 0$; (b) $0 \leq t \leq 250$ ms; (c) 250 ms $\leq t \leq 500$ ms; and (d) 500 ms $\leq t \leq \infty$ when $v_o(0) = 0$.

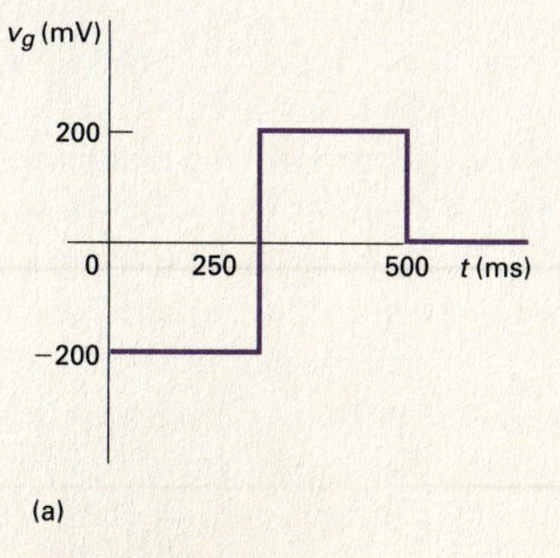

0.4 μF
25 kΩ
6 V
t = 0
–6 V
v_g
v_o

(b)

FIGURE P7.87

7.88 Repeat Problem 7.87 with a 5 MΩ resistor placed across the 0.4 μF feedback capacitor.

7.89 The voltage source in the circuit in Fig. P7.89(a) is generating the triangular waveform shown in Fig. P7.89(b). Assume the energy stored in the capacitor is zero at $t = 0$.

a) Derive the numerical expressions for $v_o(t)$ for the following time intervals: $0 \le t \le 1$ μs; 1 μs $\le t \le 3$ μs; and 3 μs $\le t \le 4$ μs.

b) Sketch the output waveform between 0 and 4 μs.

c) If the triangular input voltage continues to repeat itself for $t > 4$ μs, what would you expect the output voltage to be? Explain.

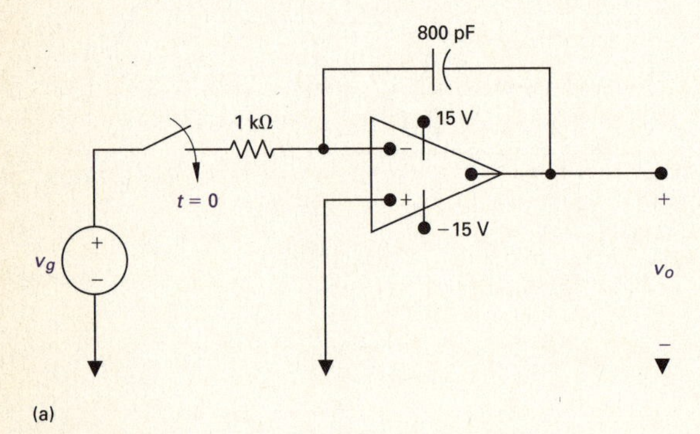

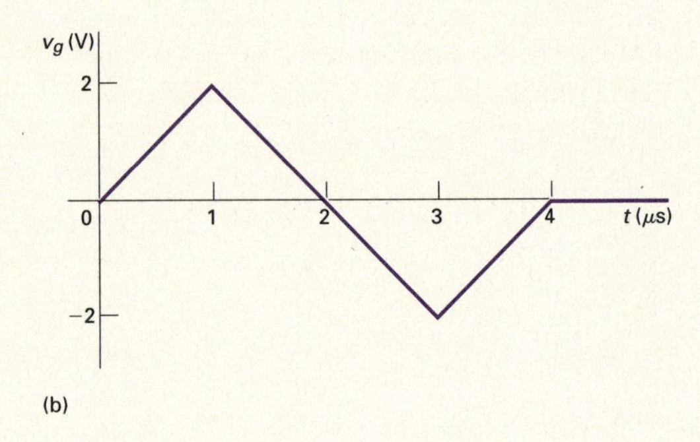

FIGURE P7.89

7.90 The circuit shown in Fig. P7.90 is known as an *astable multivibrator* and finds wide application in pulse circuits. The purpose of this problem is to relate the charging and discharging of the capacitors to the operation of the circuit. The key to analyzing the circuit is to understand the behavior of the ideal transistor switches T_1 and T_2. The circuit is designed so that the switches automatically alternate between ON and OFF. When T_1 is OFF, T_2 is ON, and vice versa. Thus in the analysis of this circuit, we assume a switch is either ON or OFF. We also assume that the ideal transistor switch can change its state instantaneously. In other words, it can snap from OFF to ON, and vice versa.

When a transistor switch is ON, (1) the base current i_b is greater than zero, (2) the terminal voltage v_{be} is zero, and (3) the terminal voltage v_{ce} is zero. Thus when a transistor switch is ON, it presents a short circuit between the terminals b,e and c,e.

When a transistor switch is OFF, (1) the terminal voltage v_{be} is negative, (2) the base current is zero, and (3) there is an open circuit between the terminals c,e. Thus when a transistor switch is OFF, it presents an open circuit between the terminal sets b,e and c,e.

Assume that T_2 has been ON and has just snapped OFF, while T_1 has been OFF and has just snapped ON. You may assume that at this instance, C_2 is charged to the supply voltage V_{CC}, and the charge on C_1 is zero. Also assume $C_1 = C_2$ and $R_1 = R_2 = 10R_L$.

a) Derive the expression for v_{be2} during the interval T_2 is OFF.
b) Derive the expression for v_{ce2} during the interval T_2 is OFF.
c) Find the length of time T_2 is OFF.
d) Find the value of v_{ce2} at the end of the interval T_2 is OFF.
e) Derive the expression for i_{b1} during the interval T_2 is OFF.
f) Find the value of i_{b1} at the end of the interval T_2 is OFF.
g) Sketch v_{ce2} versus t during the interval T_2 is OFF.
h) Sketch i_{b1} versus t during the interval T_2 is OFF.

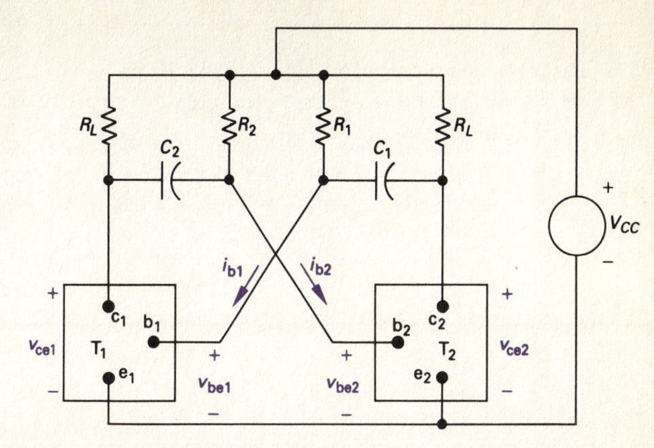

FIGURE P7.90

7.91 The component values in the circuit of Fig. P7.90 are $V_{CC} = 10$ V; $R_L = 1$ kΩ; $C_1 = C_2 = 1$ nF; and $R_1 = R_2 = 14.43$ kΩ.

a) How long is T_2 in the OFF state during one cycle of operation?
b) How long is T_2 in the ON state during one cycle of operation?
c) Repeat part (a) for T_1.
d) Repeat part (b) for T_1.
e) At the first instant after T_1 turns ON, what is the value of i_{b1}?
f) At the instant just before T_1 turns OFF, what is the value of i_{b1}?
g) What is the value of v_{ce2} at the instant just before T_2 turns ON?

7.92 Repeat Problem 7.91 with $C_1 = 1$ nF and $C_2 = 0.8$ nF. All other component values are unchanged.

7.93 An astable multivibrator circuit is to satisfy the following criteria: (1) One transistor switch is to be ON for 48 μs and OFF for 36 μs for each cycle; (2) $R_L = 2$ kΩ; (3) $V_{CC} = 5$ V; (4) $R_1 = R_2$; and (5) $6R_L \leq R_1 \leq 50R_L$. What are the limiting values for the capacitors C_1 and C_2?

7.94 The circuit shown in Fig. P7.94 is known as a *monostable multivibrator*. The adjective *monostable* is used to describe the fact that the circuit has one stable state. That is, if left alone, the electronic switch T_2 will be ON, and T_1 will be OFF. (The operation of the ideal transistor switch is described in Problem 7.90.) T_2 can be turned OFF by momentarily closing the switch S. After S returns to its open position, T_2 will return to its ON state.

a) Show that if T_2 is ON, T_1 is OFF and will stay OFF.
b) Explain why T_2 is turned OFF when S is momentarily closed.
c) Show that T_2 will stay OFF for $RC\ ln\ 2$ s.

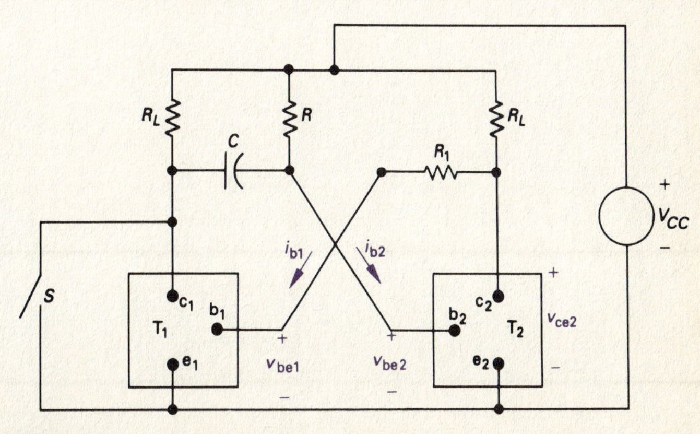

FIGURE P7.94

7.95 The parameter values in the circuit in Fig. P7.94 are $V_{CC} = 6$ V; $R_1 = 5.0$ kΩ; $R_L = 20$ kΩ; $C = 250$ pF; and $R = 23{,}083$ Ω.

a) Sketch v_{ce2} versus t, assuming that after S is momentarily closed, it remains open until the circuit has reached its stable state. Assume S is closed at $t = 0$. Make your sketch for the interval $-5 \leq t \leq 10$ μs.

b) Repeat part (a) for i_{b2} versus t.

8 NATURAL AND STEP RESPONSES OF *RLC* CIRCUITS

CHAPTER CONTENTS

In this chapter, discussion of the natural response and step response of circuits containing both inductors and capacitors is limited to two simple structures: the parallel *RLC* circuit and the series *RLC* circuit. Finding the natural response of a parallel *RLC* circuit consists of finding the voltage created across the parallel branches by the release of energy stored in the inductor or capacitor or both. The task is defined in terms of the circuit shown in Fig. 8.1. The initial voltage on the capacitor, V_0, represents the initial energy stored in the capacitor. The initial current through the inductor, I_0, represents the initial energy stored in the inductor. If the individual branch currents are of interest, you can find them after determining the terminal voltage.

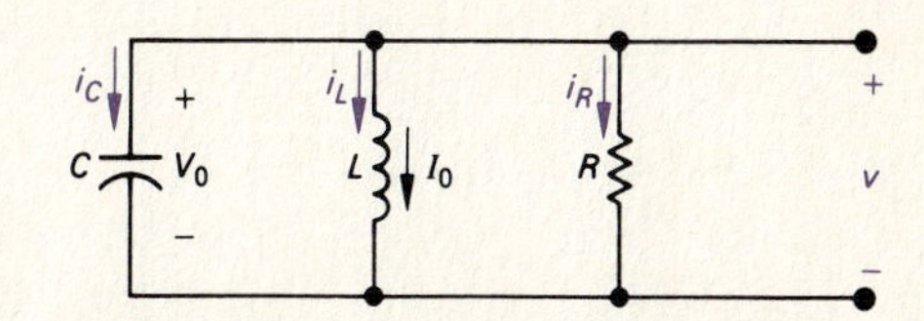

FIGURE 8.1 A circuit used to illustrate the natural response of a parallel *RLC* circuit.

We derive the step response of a parallel *RLC* circuit by using Fig. 8.2. We are interested in the voltage that appears across the parallel branches as a result of the sudden application of a dc current source. Energy may or may not be stored in the circuit when the current source is applied.

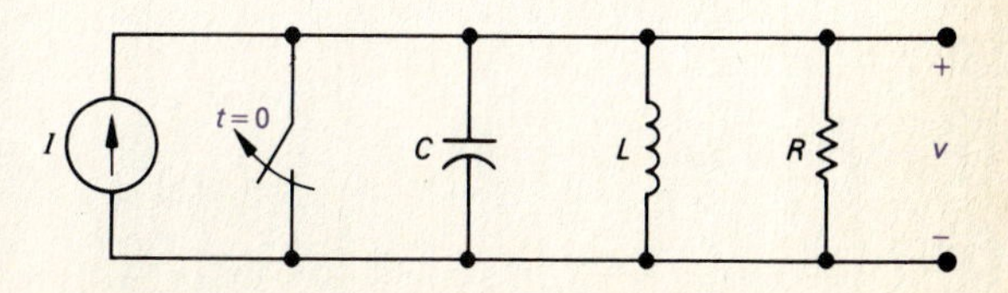

FIGURE 8.2 A circuit used to illustrate the step response of a parallel *RLC* circuit.

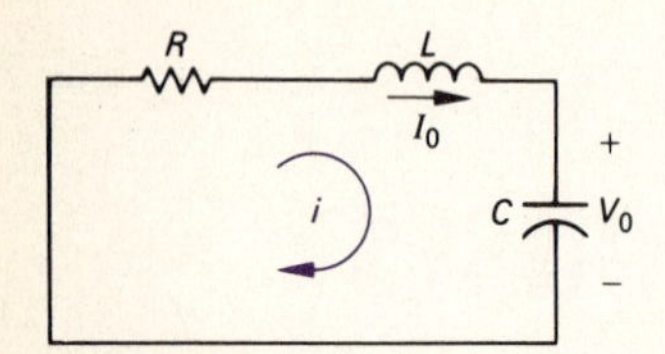

FIGURE 8.3 A circuit used to illustrate the natural response of a series *RLC* circuit.

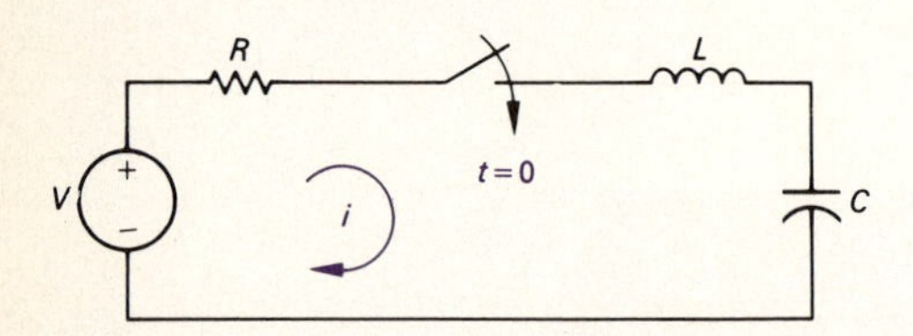

FIGURE 8.4 A circuit used to illustrate the step response of a series *RLC* circuit.

Finding the natural response of a series *RLC* circuit consists of finding the current generated in the series-connected elements by the release of initially stored energy in the inductor, capacitor, or both. The task is defined by the circuit shown in Fig. 8.3. As before, the initial inductor current, I_0, and the initial capacitor voltage, V_0, represent the initially stored energy. If any of the individual element voltages are of interest, you can find them after determining the current.

We describe the step response of a series *RLC* circuit in terms of the circuit shown in Fig. 8.4. We are interested in the current resulting from the sudden application of the dc voltage source. Energy may or may not be stored in the circuit when the switch is closed.

If you have not studied ordinary differential equations, derivation of the natural and step responses of parallel and series *RLC* circuits may be a bit difficult to follow. However, the results are important enough to warrant presentation at this time. We begin with the natural response of a parallel *RLC* circuit and cover this material over two sections: one to discuss the solution of the differential equation that describes the circuit, and one to present the three distinct forms that the solution can take. After introducing these three forms, we show that the same forms apply to the step response of a parallel *RLC* circuit, as well as to the natural and step responses of series *RLC* circuits.

8.1 Introduction to the Natural Response of a Parallel *RLC* Circuit

The first step in finding the natural response of the circuit shown in Fig. 8.1 is to derive the differential equation that the voltage v must satisfy. We choose to find the voltage first, because it is the same for each component. After that, a branch current can be found by using the current-voltage relationship for the branch component. We easily obtain the differential equation for the voltage by summing the currents away from the top node, where each current is expressed as a function of the unknown voltage v:

$$\frac{v}{R} + \frac{1}{L}\int_0^t v\,d\tau + I_0 + C\frac{dv}{dt} = 0. \tag{8.1}$$

We eliminate the integral in Eq. (8.1) by differentiating once with respect to t, and, because I_0 is a constant, we get

$$\frac{1}{R}\frac{dv}{dt} + \frac{v}{L} + C\frac{d^2v}{dt^2} = 0. \tag{8.2}$$

We now divide through Eq. (8.2) by the capacitance C and arrange the derivatives in descending order:

$$\frac{d^2v}{dt^2} + \frac{1}{RC}\frac{dv}{dt} + \frac{v}{LC} = 0. \tag{8.3}$$

Comparing Eq. (8.3) with the differential equations derived in Chapter 7 reveals that they differ by the presence of the term involving the second derivative. Equation (8.3) is an ordinary, second-order differential equation with constant coefficients. Circuits in this chapter contain both inductors and capacitors, so the differential equation describing these circuits is of the second order. Therefore, we sometimes call such circuits **second-order circuits**.

You can use a symbolic equation solver to find closed-form solutions for second-order differential equations.

THE GENERAL SOLUTION OF THE SECOND-ORDER DIFFERENTIAL EQUATION

We can't solve Eq. (8.3) by separating the variables and integrating as we were able to do with the first-order equations in Chapter 7. The classical approach to solving Eq. (8.3) is to assume that the solution is of exponential form, that is, to assume that the voltage is of the form

$$v = Ae^{st}, \tag{8.4}$$

where A and s are unknown constants.

Before showing how this assumption leads to the solution of Eq. (8.3), we need to show that it is rational. The strongest argument we can make in favor of Eq. (8.4) is to note from Eq. (8.3) that the second derivative of the solution, plus a constant times the first derivative, plus a constant times the solution itself, must sum to zero for all values of t. This can occur only if higher order derivatives of the solution have the same form as the solution. The exponential function satisfies this criterion. A second argument in favor of Eq. (8.4) is that the solutions of all the first-order equations we derived in Chapter 7 were exponential. It seems reasonable to assume that the solution of the second-order equation also involves the exponential function.

If Eq. (8.4) is a solution of Eq. (8.3), it must satisfy Eq. (8.3) for all values of t. Substituting Eq. (8.4) into Eq. (8.3) generates the expression

$$As^2e^{st} + \frac{As}{RC}e^{st} + \frac{Ae^{st}}{LC} = 0,$$

or

$$Ae^{st}\left(s^2 + \frac{s}{RC} + \frac{1}{LC}\right) = 0, \tag{8.5}$$

which can be satisfied for all values of t only if A is zero or the parenthetical term is zero, because $e^{st} \neq 0$ for any finite values of st. We cannot use $A = 0$ as a general solution because to do so implies that the voltage is zero for all time—a physical impossibility if energy is stored in either the inductor or capacitor. Therefore, in order for Eq. (8.4) to be a solution of Eq. (8.3), the parenthetical term in Eq. (8.5) must be zero, or

$$s^2 + \frac{s}{RC} + \frac{1}{LC} = 0. \tag{8.6}$$

Equation (8.6) is called the **characteristic equation** of the differential equation because the roots of this quadratic equation determine the mathematical character of $v(t)$.

The two roots of Eq. (8.6) are

$$s_1 = -\frac{1}{2RC} + \sqrt{\left(\frac{1}{2RC}\right)^2 - \frac{1}{LC}} \tag{8.7}$$

and

$$s_2 = -\frac{1}{2RC} - \sqrt{\left(\frac{1}{2RC}\right)^2 - \frac{1}{LC}}. \tag{8.8}$$

If either root is substituted into Eq. (8.4), the assumed solution satisfies the given differential equation, that is, Eq. (8.3). Note from Eq. (8.5) that this result holds regardless of the value of A. Therefore, both

$$v = A_1 e^{s_1 t} \quad \text{and} \quad v = A_2 e^{s_2 t}$$

satisfy Eq. (8.3). Denoting these two solutions v_1 and v_2, respectively, we can show that their sum also is a solution. Specifically, if we let

$$v = v_1 + v_2 = A_1 e^{s_1 t} + A_2 e^{s_2 t}, \tag{8.9}$$

then

$$\frac{dv}{dt} = A_1 s_1 e^{s_1 t} + A_2 s_2 e^{s_2 t} \tag{8.10}$$

and

$$\frac{d^2 v}{dt^2} = A_1 s_1^2 e^{s_1 t} + A_2 s_2^2 e^{s_2 t}. \tag{8.11}$$

Substituting Eqs. (8.9), (8.10), and (8.11) into Eq. (8.3) gives

$$A_1 e^{s_1 t}\left(s_1^2 + \frac{1}{RC}s_1 + \frac{1}{LC}\right) + A_2 e^{s_2 t}\left(s_2^2 + \frac{1}{RC}s_2 + \frac{1}{LC}\right) = 0. \tag{8.12}$$

But each parenthetical term is zero because by definition s_1 and s_2 are roots of the characteristic equation. Hence the natural response of the parallel *RLC* circuit shown in Fig. 8.1 is of the form

$$v = A_1 e^{s_1 t} + A_2 e^{s_2 t}. \tag{8.13}$$

Equation (8.13) is a repeat of the assumption made in Eq. (8.9). We have shown that v_1 is a solution, v_2 is a solution, and $v_1 + v_2$ is a solution. Therefore, the general solution of Eq. (8.3) has the form given in Eq. (8.13). The roots of the characteristic equation (s_1 and

s_2) are determined by the circuit parameters R, L, and C. The initial conditions determine the values of the constants A_1 and A_2.[1]

The behavior of $v(t)$ depends on the values of s_1 and s_2. Therefore the first step in finding the natural response is to determine the roots of the characteristic equation. We return to Eqs. (8.7) and (8.8) and rewrite them using a notation widely used in the literature:

$$s_1 = -\alpha + \sqrt{\alpha^2 - \omega_0^2} \quad \textbf{(8.14)}$$

$$s_2 = -\alpha - \sqrt{\alpha^2 - \omega_0^2}, \quad \textbf{(8.15)}$$

where

$$\alpha = \frac{1}{2RC} \quad \textbf{(8.16)}$$

and

$$\omega_0 = \frac{1}{\sqrt{LC}}. \quad \textbf{(8.17)}$$

These results are summarized in Table 8.1.

TABLE 8.1

NATURAL RESPONSE PARAMETERS OF THE PARALLEL *RLC* CIRCUIT

PARAMETER	TERMINOLOGY	VALUE IN NATURAL RESPONSE
s_1, s_2	Characteristic roots	$s_1 = -\alpha + \sqrt{\alpha^2 - \omega_0^2}$
		$s_2 = -\alpha - \sqrt{\alpha^2 - \omega_0^2}$
α	Neper frequency	$\alpha = \frac{1}{2RC}$
ω_0	Resonant radian frequency	$\omega_0 = \frac{1}{\sqrt{LC}}$

The exponent of e must be dimensionless, so both s_1 and s_2 (and hence α and ω_0) must have the dimension of the reciprocal of time, or frequency. To distinguish among the frequencies s_1, s_2, α, and ω_0, we use the following terminology: s_1 and s_2 are referred to as complex frequencies, α is called the neper frequency, and ω_0 is the resonant radian frequency. The full significance of this terminology unfolds as we move through the remaining chapters of this book. All these frequencies have the dimension of angular frequency per time. For complex frequencies, the neper frequency, and the resonant radian frequency, we specify values using the unit *radians per second (rad/s)*.

---

[1]The form of Eq. (8.13) must be modified if the two roots s_1 and s_2 are equal. We discuss this modification when we turn to the critically damped voltage response, in Section 8.2.

The nature of the roots s_1 and s_2 depends on the values of α and ω_0. There are three possible outcomes. First, if $\omega_0^2 < \alpha^2$, both roots will be real and distinct. For reasons to be discussed later, the voltage response is said to be **overdamped** in this case. Second, if $\omega_0^2 > \alpha^2$, both s_1 and s_2 will be complex and, in addition, will be conjugates of each other. In this situation, the voltage response is said to be **underdamped**. The third possible outcome is that $\omega_0^2 = \alpha^2$. In this case, s_1 and s_2 will be real and equal. Here the voltage response is said to be **critically damped**. As we shall see, damping affects the way the voltage response reaches its final (or steady-state) value. We discuss each case separately in Section 8.2.

Example 8.1 illustrates how the numerical values of s_1 and s_2 are determined by the values of R, L, and C.

EXAMPLE 8.1

a) Find the roots of the characteristic equation that governs the transient behavior of the voltage shown in Fig. 8.5 if $R = 200\ \Omega$, $L = 50$ mH, and $C = 0.2\ \mu$F.

b) Will the response be overdamped, underdamped, or critically damped?

c) Repeat (a) and (b) for $R = 312.5\ \Omega$.

d) What value of R causes the response to be critically damped?

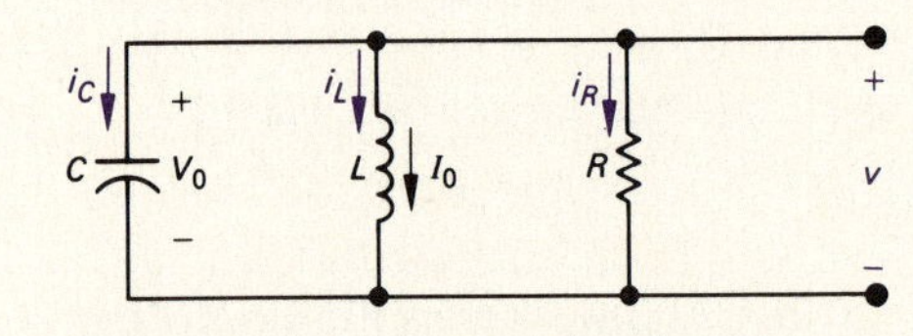

FIGURE 8.5 A circuit used to illustrate the natural response of a parallel *RLC* circuit.

SOLUTION

a) For the given values of R, L, and C,

$$\alpha = \frac{1}{2RC} = \frac{10^6}{(400)(0.2)} = 1.25 \times 10^4 \text{ rad/s},$$

$$\omega_0^2 = \frac{1}{LC} = \frac{(10^3)(10^6)}{(50)(0.2)} = 10^8 \text{ rad}^2/\text{s}^2.$$

From Eqs. (8.14) and (8.15),

$$s_1 = -1.25 \times 10^4 + \sqrt{1.5625 \times 10^8 - 10^8}$$
$$= -12{,}500 + 7500 = -5000 \text{ rad/s}$$

and

$$s_2 = -1.25 \times 10^4 - \sqrt{1.5625 \times 10^8 - 10^8}$$
$$= -12{,}500 - 7500 = -20{,}000 \text{ rad/s}.$$

b) The voltage response is overdamped because $\omega_0^2 < \alpha^2$.

c) For $R = 312.5\ \Omega$,

$$\alpha = \frac{10^6}{(625)(0.2)} = 8000 \text{ rad/s},$$

$$\alpha^2 = 64 \times 10^6 = 0.64 \times 10^8 \text{ rad}^2/\text{s}^2.$$

As ω_0^2 remains at $10^8 \text{ rad}^2/\text{s}^2$,

$$s_1 = -8000 + j6000 \text{ rad/s},$$

and

$$s_2 = -8000 - j6000 \text{ rad/s}.$$

(In electrical engineering, the imaginary number $\sqrt{-1}$ is represented by the letter j, because the letter i represents current.)
In this case, the voltage response is underdamped since $\omega_0^2 > \alpha^2$.

d) For critical damping, $\alpha^2 = \omega_0^2$, so

$$\left(\frac{1}{2RC}\right)^2 = \frac{1}{LC} = 10^8, \quad \text{or} \quad \frac{1}{2RC} = 10^4,$$

and

$$R = \frac{10^6}{(2 \times 10^4)(0.2)} = 250\ \Omega.$$

Use a spreadsheet to narrow in on this computed value of *R*.

DRILL EXERCISE

8.1 The resistance and inductance of the circuit in Fig. 8.5 are 200 Ω and 10 mH, respectively.

a) Find the value of C that makes the voltage response critically damped.

b) If C is adjusted to give a neper frequency of 10 krad/s, find the value of C and the roots of the characteristic equation.

c) If C is adjusted to give a resonant frequency of 50 krad/s, find the value of C and the roots of the characteristic equation.

ANSWER: (a) $C = 62.5$ nF; (b) $C = 0.25\ \mu$F, $s_1 = -10{,}000 + j17{,}320.51$ rad/s, $s_2 = -10{,}000 - j17{,}320.51$ rad/s; (c) $C = 40$ nF, $s_1 = -25{,}000$ rad/s, $s_2 = -100{,}000$ rad/s.

8.2 The Forms of the Natural Response of a Parallel *RLC* Circuit

So far we have seen that the behavior of a second-order *RLC* circuit depends on the values of s_1 and s_2, which in turn depend on the circuit parameters R, L, and C. Therefore, the first step in finding the natural response is to calculate these values and, relatedly, determine whether the response is over-, under-, or critically damped.

Completing the description of the natural response requires finding two unknown coefficients, such as A_1 and A_2 in Eq. (8.13). The method used to do this is based on matching the solution for the natural response to the initial conditions imposed by the circuit, which are the initial value of the current (or voltage) and the initial value of the first derivative of the current (or voltage). Note that these same initial conditions, plus the final value of the variable, will also be needed when finding the step response of a second-order circuit.

In this section we analyze the natural response form for each of the three types of damping, beginning with the overdamped response. As we will see, the response equations, as well as the equations for evaluating the unknown coefficients, are slightly different for each of the three damping configurations. This is why we want to determine at the outset of the problem whether the response is over-, under-, or critically damped.

The Overdamped Voltage Response

When the roots of the characteristic equation are real and distinct, the voltage response of a parallel *RLC* circuit is said to be overdamped. The solution for the voltage is of the form

$$v = A_1 e^{s_1 t} + A_2 e^{s_2 t}, \tag{8.18}$$

where s_1 and s_2 are the roots of the characteristic equation. The constants A_1 and A_2 are determined by the initial conditions, specifically from the values of $v(0^+)$ and $dv(0^+)/dt$, which are in turn determined from the initial voltage on the capacitor, V_0, and the initial current in the inductor, I_0.

Next, we show how to use the initial voltage on the capacitor and the initial current in the inductor to find A_1 and A_2. First we note from Eq. (8.18) that

$$v(0^+) = A_1 + A_2 \tag{8.19}$$

and

$$\frac{dv(0^+)}{dt} = s_1 A_1 + s_2 A_2. \tag{8.20}$$

With s_1 and s_2 known, the task of finding A_1 and A_2 reduces to finding $v(0^+)$ and $dv(0^+)/dt$. The value of $v(0^+)$ is the initial voltage on the capacitor V_0. We get the initial value of dv/dt by first finding the current in the capacitor branch at $t = 0^+$. Then,

$$\frac{dv(0^+)}{dt} = \frac{i_C(0^+)}{C}. \tag{8.21}$$

We use Kirchhoff's current law to find the initial current in the capacitor branch. We know that the sum of the three branch currents at $t = 0^+$ must be zero. The current in the resistive branch at $t = 0^+$ is the initial voltage V_0 divided by the resistance, and the current in the inductive branch is I_0. Using the reference system depicted in Fig. 8.5, we obtain

$$i_C(0^+) = \frac{-V_0}{R} - I_0. \tag{8.22}$$

After finding the numerical value of $i_C(0^+)$, we use Eq. (8.21) to find the initial value of dv/dt.

We can summarize the process for finding the overdamped response, $v(t)$, as follows:

1. Find the roots of the characteristic equation, s_1 and s_2, using the values of R, L, and C.
2. Find $v(0^+)$ and $dv(0^+)/dt$ using circuit analysis.
3. Find the values of A_1 and A_2 by solving Eqs. (8.23) and (8.24) simultaneously:

$$v(0^+) = A_1 + A_2, \tag{8.23}$$

and

$$\frac{dv(0^+)}{dt} = \frac{i_C(0^+)}{C} = s_1A_1 + s_2A_2. \tag{8.24}$$

4. Substitute the values for s_1, s_2, A_1, and A_2 into Eq. (8.18) to determine the expression for $v(t)$ for $t \geq 0$.

Examples 8.2 and 8.3 illustrate how to find the overdamped response of a parallel *RLC* circuit.

EXAMPLE 8.2

For the circuit in Fig. 8.6, $v(0^+) = 12$ V, and $i_L(0^+) = 30$ mA.

a) Find the initial current in each branch of the circuit.

b) Find the initial value of dv/dt.

c) Find the expression for $v(t)$.

d) Sketch $v(t)$ in the interval $0 \leq t \leq 250\ \mu$s.

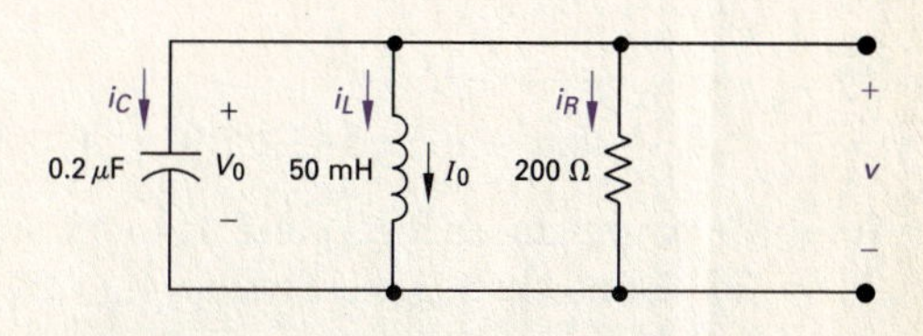

FIGURE 8.6 The circuit for Example 8.2.

SOLUTION

a) The inductor prevents an instantaneous change in its current, so the initial value of the inductor current is 30 mA:

$$i_L(0^-) = i_L(0) = i_L(0^+) = 30 \text{ mA}.$$

The capacitor holds the initial voltage across the parallel elements to 12 V. Thus the initial current in the resistive branch, $i_R(0^+)$, is 12/200, or 60 mA. Kirchhoff's current law requires the sum of the currents leaving the top node to equal zero at every instant. Hence

$$i_C(0^+) = -i_L(0^+) - i_R(0^+)$$
$$= -90 \text{ mA}.$$

Note that if we assumed the inductor current and capacitor voltage had reached their dc values at the instant energy begins to be released, $i_C(0^-) = 0$. In other words, there is an instantaneous change in the capacitor current at $t = 0$.

b) Because $i_C = C(dv/dt)$,

$$\frac{dv(0^+)}{dt} = \frac{-90 \times 10^{-3}}{0.2 \times 10^{-6}} = -450 \text{ kV/s}.$$

c) The roots of the characteristic equation come from the values of R, L, and C. For the values specified and from Eqs. (8.14) and (8.15),

$$s_1 = -1.25 \times 10^4 + \sqrt{1.5625 \times 10^8 - 10^8}$$
$$= -12{,}500 + 7500 = -5000 \text{ rad/s},$$

and

$$s_2 = -1.25 \times 10^4 - \sqrt{1.5625 \times 10^8 - 10^8}$$
$$= -12{,}500 - 7500 = -20{,}000 \text{ rad/s}.$$

Because the roots are real and distinct, we know that the response is overdamped and hence has the form of Eq. (8.18). We find the coefficients A_1 and A_2 from Eqs. (8.23) and (8.24). We've already determined s_1, s_2, $v(0^+)$, and $dv(0^+)/dt$, so

$$12 = A_1 + A_2,$$

and

$$-450 \times 10^3 = -5000A_1 - 20{,}000A_2.$$

We solve two equations for A_1 and A_2 to obtain $A_1 = -14$ V and $A_2 = 26$ V. Substituting these values into Eq. (8.18) yields the overdamped voltage response:

$$v(t) = (-14e^{-5000t} + 26e^{-20{,}000t}) \text{ V}, \quad t \geq 0.$$

As a check on these calculations, we note that the solution yields $v(0) = 12$ V and $dv(0^+)/dt = -450{,}000$ V/s.

d) Figure 8.7 shows a plot of $v(t)$ versus t over the interval $0 \leq t \leq 250$ μs.

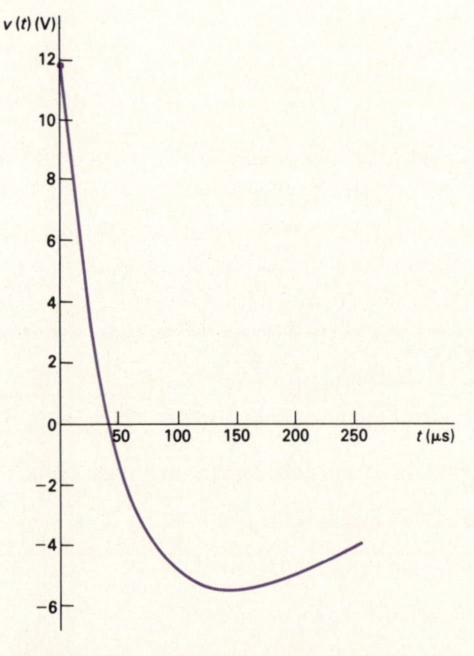

FIGURE 8.7 The voltage response for Example 8.2.

Use a computer to vary the initial conditions and plot the results.

EXAMPLE 8.3

Derive the expressions that describe the three branch currents i_R, i_L, and i_C in Example 8.2 (Fig. 8.6) during the time the stored energy is being released.

SOLUTION

We know the voltage across the three branches from the solution in Example 8.2, namely,

$$v(t) = (-14e^{-5000t} + 26e^{-20{,}000t}) \text{ V}, \quad t \geq 0.$$

The current in the resistive branch is then

$$i_R(t) = \frac{v(t)}{200} = (-70e^{-5000t} + 130e^{-20{,}000t}) \text{ mA}, \quad t \geq 0.$$

There are two ways to find the current in the inductive branch. One way is to use the integral relationship that exists between the current and the voltage at the terminals of an inductor:

$$i_L(t) = \frac{1}{L}\int_0^t v_L(x)dx + I_0.$$

A second approach is to find the current in the capacitive branch first and then use the fact that $i_R + i_L + i_C = 0$. Let's use this approach. The current in the capacitive branch is

$$\begin{aligned} i_C(t) &= C\frac{dv}{dt} \\ &= 0.2 \times 10^{-6}(70{,}000e^{-5000t} - 520{,}000e^{-20{,}000t}) \\ &= (14e^{-5000t} - 104e^{-20{,}000t}) \text{ mA}, \quad t \geq 0^+. \end{aligned}$$

Note that $i_C(0^+) = -90$ mA, which agrees with the result in Example 8.2.

Now we obtain the inductive branch current from the relationship

$$\begin{aligned} i_L(t) &= -i_R(t) - i_C(t) \\ &= (56e^{-5000t} - 26e^{-20{,}000t}) \text{ mA}, \quad t \geq 0. \end{aligned}$$

We leave it to you, in Drill Exercise 8.2, to show that the integral relationship alluded to leads to the same result. Note that the expression for i_L agrees with the initial inductor current, as it must.

DRILL EXERCISES

8.2 Use the integral relationship between i_L and v to find the expression for i_L in Fig. 8.6.

ANSWER: $i_L(t) = (56e^{-5000t} - 26e^{-20{,}000t})$ mA, $t \geq 0$.

8.3 The element values in the circuit shown are $R = 400\ \Omega$, $L = 50$ mH, and $C = 50$ nF. The initial current I_0 in the inductor is -4 A, and the initial voltage on the capacitor is 0 V. The output signal is the voltage v. Find (a) $i_R(0^+)$; (b) $i_C(0^+)$; (c) $dv(0^+)/dt$; (d) A_1; (e) A_2; and (f) $v(t)$ when $t \geq 0$.

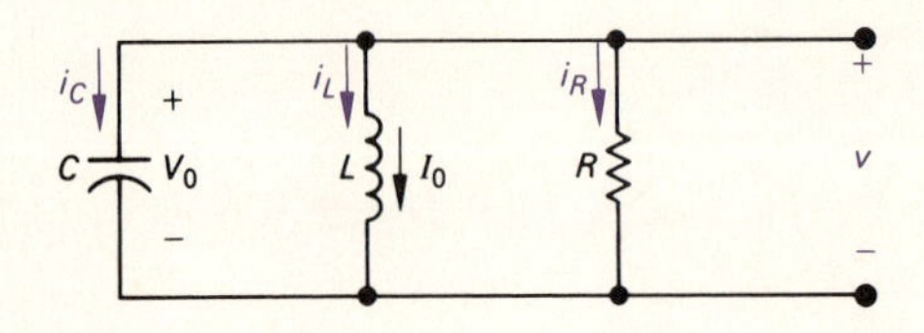

ANSWER: (a) 0; (b) 4 A; (c) 8×10^7 V/s; (d) 8000/3 V; (e) $-8000/3$ V; (f) $(8000/3)(e^{-10{,}000t} - e^{-40{,}000t})$ V when $t \geq 0$.

The Underdamped Voltage Response

When $\omega_0^2 > \alpha^2$, the roots of the characteristic equation are complex, and the response is underdamped. For convenience, we express the roots s_1 and s_2 as

$$s_1 = -\alpha + \sqrt{-(w_0^2 - \alpha^2)}$$
$$= \alpha + j\sqrt{\omega_0^2 - \alpha^2}$$
$$= -\alpha + j\omega_d \qquad \textbf{(8.25)}$$

and

$$s_2 = -\alpha - j\omega_d, \qquad \textbf{(8.26)}$$

where

$$\omega_d = \sqrt{\omega_0^2 - \alpha^2}. \qquad \textbf{(8.27)}$$

The term ω_d is called the **damped radian frequency**. We explain later the reason for this terminology.

The underdamped voltage response of a parallel RLC circuit is

$$v(t) = B_1 e^{-\alpha t} \cos \omega_d t + B_2 e^{-\alpha t} \sin \omega_d t, \qquad \textbf{(8.28)}$$

which follows from Eq. (8.18). In making the transition from Eq. (8.18) to Eq. (8.28), we use the Euler identity:

$$e^{\pm j\theta} = \cos\theta \pm j \sin\theta. \qquad \textbf{(8.29)}$$

Thus

$$\begin{aligned}
v(t) &= A_1 e^{(-\alpha+j\omega_d)t} + A_2 e^{-(\alpha+j\omega_d)t} \\
&= A_1 e^{-\alpha t} e^{j\omega_d t} + A_2 e^{-\alpha t} e^{-j\omega_d t} \\
&= e^{-\alpha t}(A_1 \cos\omega_d t + jA_1 \sin\omega_d t + A_2 \cos\omega_d t \\
&\quad - jA_2 \sin\omega_d t) \\
&= e^{-\alpha t}[(A_1 + A_2)\cos\omega_d t + j(A_1 - A_2)\sin\omega_d t].
\end{aligned}$$

At this point in the transition from Eq. (8.18) to (8.28), replace the arbitrary constants $A_1 + A_2$ and $j(A_1 - A_2)$ with new arbitrary constants denoted B_1 and B_2 to get

$$\begin{aligned}
v &= e^{-\alpha t}(B_1 \cos\omega_d t + B_2 \sin\omega_d t) \\
&= B_1 e^{-\alpha t} \cos\omega_d t + B_2 e^{-\alpha t} \sin\omega_d t.
\end{aligned}$$

The constants B_1 and B_2 are real, not complex, because the voltage is a real function. Don't be misled by the fact that $B_2 = j(A_1 - A_2)$. In this underdamped case, A_1 and A_2 are complex conjugates, and thus B_1 and B_2 are real. (See Problems 8.53 and 8.54.) The reason for defining the underdamped response in terms of the coefficients B_1 and B_2 is that it yields a simpler expression for the voltage, v. We determine B_1 and B_2 by the initial energy stored in the circuit, in the same way that we found A_1 and A_2 for the overdamped response: by evaluating v at $t = 0^+$, and its derivative at $t = 0^+$. As with s_1 and s_2, α and ω_d are fixed by the circuit parameters R, L, and C.

For the underdamped response, the two simultaneous equations which determine B_1 and B_2 are

$$v(0^+) = V_0 = B_1 \tag{8.30}$$

and

$$\frac{dv(0^+)}{dt} = \frac{i_c(0^+)}{C} = -\alpha B_1 + \omega_d B_2. \tag{8.31}$$

Let's look at the general nature of the underdamped response. First, the trigonometric functions indicate that this response is oscillatory; that is, the voltage alternates between positive and negative values. The rate at which the voltage oscillates is fixed by ω_d. Second, the amplitude of the oscillation decreases exponentially. The rate at which the amplitude falls off is determined by α. Because α determines how quickly the oscillations subside, it is also referred to as the **damping factor**, or **damping coefficient**. That explains why ω_d is called the damped radian frequency. If there is no damping, $\alpha = 0$ and the frequency of oscillation is ω_0. Whenever there is a dissipative element, R, in the circuit, α is not zero and the frequency of oscillation, ω_d, is less than ω_0. Thus when α is not zero, the frequency of oscillation is said to be damped.

The oscillatory behavior is possible because of the two types of energy-storage elements in the circuit: the inductor and the capacitor. (A mechanical analogy of this electric circuit is that of a mass suspended on a spring, where oscillation is possible because energy can be stored in both the spring and the moving mass.) We say more about the characteristics of the underdamped response following Example 8.4. In summary, note that the overall process for finding the underdamped response is the same as that for the overdamped response, although the response equations and the simultaneous equations used to find the constants are slightly different.

EXAMPLE 8.4

In the circuit shown in Fig. 8.8, $V_0 = 0$, and $I_0 = -12.25$ mA.

a) Calculate the roots of the characteristic equation.

b) Calculate v and dv/dt at $t = 0^+$.

c) Calculate the voltage response for $t \geq 0$.

d) Plot $v(t)$ versus t for the time interval $0 \leq t \leq 11$ ms.

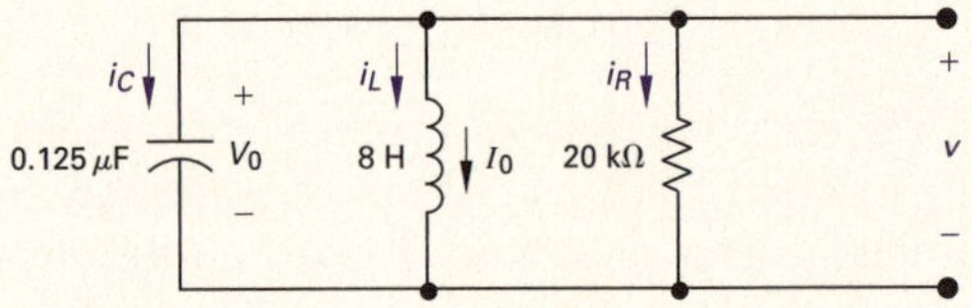

FIGURE 8.8 The circuit for Example 8.4.

SOLUTION

a) Because

$$\alpha = \frac{1}{2RC} = \frac{10^6}{2(20)10^3(0.125)} = 200 \text{ rad/s},$$

and

$$\omega_0 = \frac{1}{\sqrt{LC}} = \sqrt{\frac{10^6}{(8)(0.125)}} = 10^3 \text{ rad/s},$$

we have

$$\omega_0^2 > \alpha^2.$$

Therefore, the response is underdamped. Now,

$$\begin{aligned} \omega_d &= \sqrt{\omega_0^2 - \alpha^2} = \sqrt{10^6 - 4 \times 10^4} = 100\sqrt{96} \\ &= 979.80 \text{ rad/s}; \\ s_1 &= -\alpha + j\omega_d = -200 + j979.80 \text{ rad/s}; \\ s_2 &= -\alpha - j\omega_d = -200 - j979.80 \text{ rad/s}. \end{aligned}$$

For the underdamped case, we do not ordinarily solve for s_1 and s_2 because we do not use them explicitly. However, this example emphasizes why s_1 and s_2 are known as complex frequencies.

b) Because v is the voltage across the terminals of a capacitor, we have

$$v(0) = v(0^+) = V_0 = 0.$$

Because $v(0^+) = 0$, the current in the resistive branch is zero at $t = 0^+$. Hence the current in the capacitor at $t = 0^+$ is the negative of the inductor current:

$$i_C(0^+) = -(-12.25) = 12.25 \text{ mA}.$$

Therefore the initial value of the derivative is

$$\frac{dv(0^+)}{dt} = \frac{(12.25)(10^{-3})}{(0.125)(10^{-6})} = 98{,}000 \text{ V/s}.$$

c) From Eqs. (8.30) and (8.31), $B_1 = 0$ and

$$B_2 = \frac{98{,}000}{\omega_d} \approx 100 \text{ V}.$$

Substituting the numerical values of α, ω_d, B_1, and B_2 into the expression for $v(t)$ gives

$$v(t) = 100e^{-200t} \sin 979.80t \text{ V}, \quad t \geq 0.$$

d) Figure 8.9 shows the plot of $v(t)$ versus t for the first 11 ms after the stored energy is released. It clearly indicates the damped oscillatory nature of the underdamped response. The voltage $v(t)$ approaches its final value, alternating between values that are greater than and less than the final value. Furthermore, these swings about the final value decrease exponentially with time.

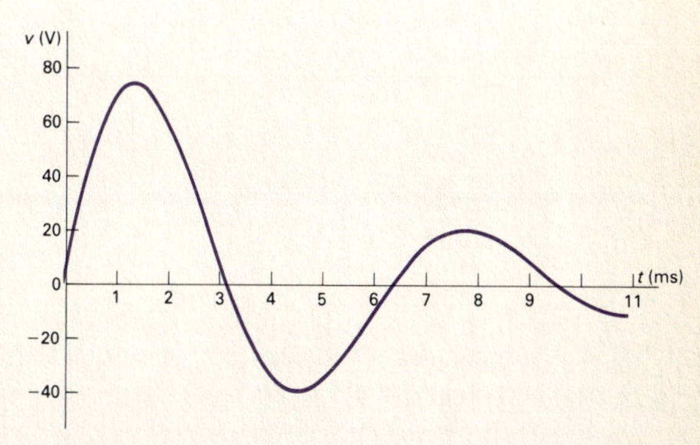

FIGURE 8.9 The voltage response for Example 8.4.

Use a computer to vary R and plot the effect on the output.

Characteristics of the Underdamped Response

The underdamped response has several important characteristics. First, as the dissipative losses in the circuit decrease, the persistence of the oscillations increases, and the frequency of the oscillations approaches ω_0. In other words, as $R \to \infty$, the dissipation in the circuit in Fig. 8.8 approaches zero because $p = v^2/R$. As $R \to \infty$, $\alpha \to 0$, which tells us that $\omega_d \to \omega_0$. When $\alpha = 0$, the maximum amplitude of the voltage remains constant; thus the oscillation at ω_0 is sustained. In Example 8.4, if R were increased to infinity, the solution for $v(t)$ would become

$$v(t) = 98 \sin 1000t \text{ V}, \quad t \geq 0.$$

Thus in this case the oscillation is sustained, the maximum amplitude of the voltage is 98 V, and the frequency of oscillation is 1000 rad/s.

We may now describe qualitatively the difference between an underdamped and an overdamped response. In an underdamped system, the response oscillates, or "bounces," about its final value. This oscillation is also referred to as *ringing*. In an overdamped system, the response approaches its final value without ringing, or in what is sometimes described as a "sluggish" manner. When specifying the desired response of a second-order system, you may want to reach the final value in the shortest time possible, and you may not be concerned with small oscillations about that final value. If so, you would design the system components to achieve an underdamped response. On the other hand, you may be concerned that the response not exceed its final value, perhaps to ensure that components are not damaged. In such a case, you would design the system components to achieve an overdamped response, and you would have to accept a relatively slow rise to the final value.

DRILL EXERCISE

8.4 A 10 mH inductor, a 1 μF capacitor, and a variable resistor are connected in parallel in the circuit shown. The resistor is adjusted so that the roots of the characteristic equation are $-8000 \pm j6000$ rad/s. The initial voltage on the capacitor is 10 V, and the initial current in the inductor is 80 mA. Find (a) R; (b) $dv(0^+)/dt$; (c) B_1 and B_2 in the solution for v; and (d) $i_L(t)$.

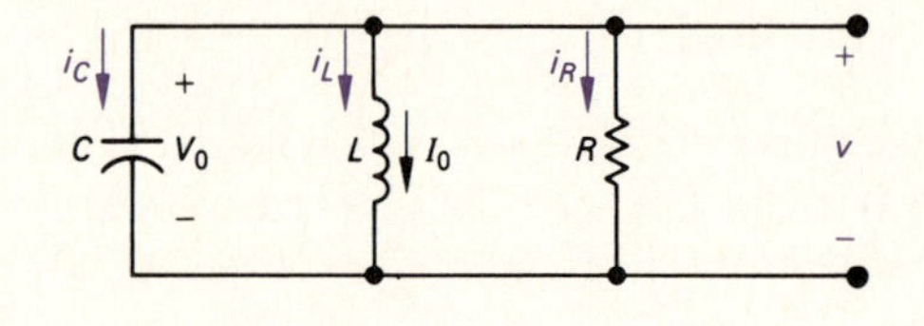

ANSWER: (a) 62.5 Ω; (b) −240,000 V/s; (c) $B_1 = 10$ V, $B_2 = -80/3$ V; (d) $i_L(t) = 10e^{-8000t}[8\cos 6000t + (82/3)\sin 6000t]$ mA when $t \geq 0$.

THE CRITICALLY DAMPED VOLTAGE RESPONSE

The second-order circuit in Fig. 8.8 is critically damped when $\omega_0^2 = \alpha^2$, or $\omega_0 = \alpha$. When a circuit is critically damped, the response is on the verge of oscillating. In addition, the two roots of the characteristic equation are real and equal; that is,

$$s_1 = s_2 = -\alpha = -\frac{1}{2RC}. \tag{8.32}$$

When this occurs, the solution for the voltage no longer takes the form of Eq. (8.18). This equation breaks down because if $s_1 = s_2 = -\alpha$, it predicts that

$$v = (A_1 + A_2)e^{-\alpha t} = A_0 e^{-\alpha t}, \tag{8.33}$$

where A_0 is an arbitrary constant. Equation (8.33) cannot satisfy two independent initial conditions (V_0, I_0) with only one arbitrary constant, A_0. Recall that the circuit parameters R and C fix α.

We can trace this dilemma back to the assumption that the solution takes the form of Eq. (8.18). When the roots of the characteristic equation are equal, the solution for the differential equation takes a different form, namely

$$v(t) = D_1te^{-\alpha t} + D_2e^{-\alpha t}. \tag{8.34}$$

Thus in the case of a repeated root, the solution involves a simple exponential term plus the product of a linear and an exponential term. The justification of Eq. (8.34) is left for an introductory course in differential equations. Finding the solution involves obtaining D_1 and D_2 by following the same pattern set in the overdamped and underdamped cases: We use the initial values of the voltage and the derivative of the voltage with respect to time to write two equations containing D_1 and/or D_2.

From Eq. (8.34), the two simultaneous equations needed to determine D_1 and D_2 are

$$v(0^+) = V_0 = D_2 \tag{8.35}$$

and

$$\frac{dv(0^+)}{dt} = \frac{i_C(0^+)}{C} = D_1 - \alpha D_2. \tag{8.36}$$

As we can see, in the case of a critically damped response, both the equation for $v(t)$ and the simultaneous equations for the constants D_1 and D_2 differ from those for over- and underdamped responses, but the general approach is the same. You will rarely encounter critically damped systems in practice, largely because ω_0 must equal α exactly. Both of these quantities depend on circuit parameters, and in a real circuit it is very difficult to choose component values which satisfy an exact equality relationship.

Example 8.5 illustrates the approach for finding the critically damped response of a parallel *RLC* circuit.

EXAMPLE 8.5

a) For the circuit in Example 8.4 (Fig. 8.8), find the value of R that results in a critically damped voltage response.

b) Calculate $v(t)$ for $t \geq 0$.

c) Plot $v(t)$ versus t for $0 \leq t \leq 7$ ms.

SOLUTION

a) From Example 8.4, we know that $\omega_0^2 = 10^6$. Therefore for critical damping

$$\alpha = 10^3 = \frac{1}{2RC}, \quad \text{or} \quad R = \frac{10^6}{(2000)(0.125)} = 4000\ \Omega.$$

b) From the solution of Example 8.4, we know that $v(0^+) = 0$ and $dv(0^+)/dt = 98{,}000$ V/s. From Eqs. (8.35) and (8.36), $D_2 = 0$ and $D_1 = 98{,}000$ V/s. Substituting these values for α, D_1, and D_2 into Eq. (8.34) gives

$$v(t) = 98{,}000te^{-1000t}\ \text{V}, \quad t \geq 0.$$

c) Figure 8.10 shows a plot of $v(t)$ versus t in the interval $0 \leq t \leq 7$ ms.

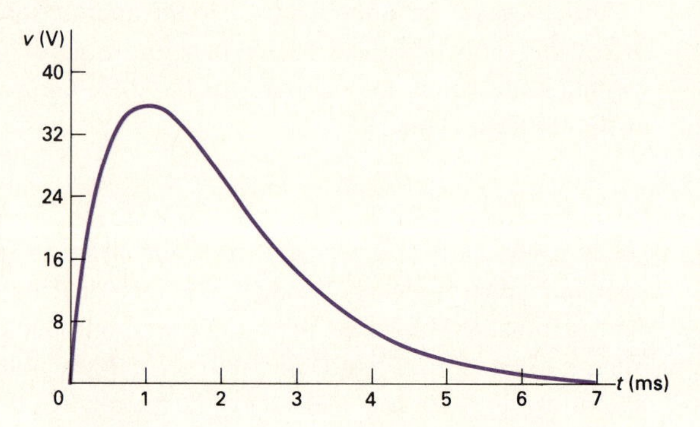

FIGURE 8.10 The voltage response for Example 8.5.

DRILL EXERCISE

8.5 The resistor in the circuit in Drill Exercise 8.4 is adjusted for critical damping. The inductance and capacitance values are 0.4 H and 10 μF, respectively. The initial energy stored in the circuit is 25 mJ and is distributed equally between the inductor and capacitor. Find (a) R; (b) V_0; (c) I_0; (d) D_1 and D_2 in the solution for v; and (e) i_R, $t \geq 0^+$.

ANSWER: (a) 100 Ω; (b) 50 V; (c) 250 mA; (d) −50,000 V/s, 50 V; (e) $i_R(t) = (-500te^{-500t} + 0.50e^{-500t})$ A, $t \geq 0^+$.

A SUMMARY OF THE RESULTS

We conclude our discussion of the parallel *RLC* circuit's natural response with a brief summary of the results. The first step in finding the natural response is to calculate the roots of the characteristic equation. You then know immediately whether the response is overdamped, underdamped, or critically damped.

If the roots are real and distinct ($\omega_0^2 < \alpha^2$), the response is overdamped and the voltage is

$$v(t) = A_1e^{s_1t} + A_2e^{s_2t},$$

where

$$s_1 = -\alpha + \sqrt{\alpha^2 - \omega_0^2}, \quad s_2 = -\alpha - \sqrt{\alpha^2 - \omega_0^2},$$

$$\alpha = \frac{1}{2RC}, \quad \text{and} \quad \omega_0^2 = \frac{1}{LC}.$$

The values of A_1 and A_2 are determined by solving the following simultaneous equations:

$$v(0^+) = A_1 + A_2;$$

$$\frac{dv(0^+)}{dt} = \frac{i_C(0^+)}{C} = s_1 A_1 + s_2 A_2.$$

If the roots are complex ($\omega_0^2 > \alpha^2$), the response is underdamped and the voltage is

$$v(t) = B_1 e^{-\alpha t} \cos \omega_d t + B_2 e^{-\alpha t} \sin \omega_d t,$$

where

$$\omega_d = \sqrt{\omega_0^2 - \alpha^2}.$$

The values of B_1 and B_2 are found by solving the following simultaneous equations:

$$v(0^+) = V_0 = B_1;$$

$$\frac{dv(0^+)}{dt} = \frac{i_C(0^+)}{C} = -\alpha B_1 + \omega_d B_2.$$

If the roots of the characteristic equation are real and equal ($\omega_0^2 = \alpha^2$), the voltage response is

$$v(t) = D_1 t e^{-\alpha t} + D_2 e^{-\alpha t},$$

where α is as in the other solution forms. To determine values for the constants D_1 and D_2, solve the following simultaneous equations:

$$v(0^+) = V_0 = D_2;$$

$$\frac{dv(0^+)}{dt} = \frac{i_C(0^+)}{C} = D_1 - \alpha D_2.$$

8.3 THE STEP RESPONSE OF A PARALLEL *RLC* CIRCUIT

Finding the step response of a parallel *RLC* circuit involves finding the voltage across the parallel branches or the current in the individual branches as a result of the sudden application of a dc current source. There may or may not be energy stored in the circuit when the current

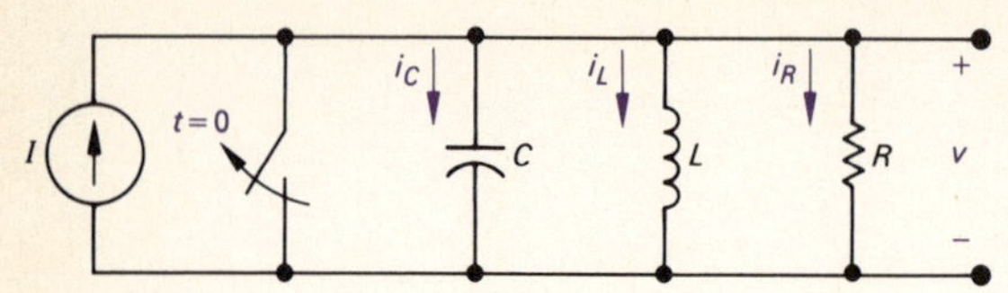

FIGURE 8.11 A circuit used to describe the step response of a parallel *RLC* circuit.

source is applied. The task is represented by the circuit shown in Fig. 8.11.

To develop a general approach to finding the step response of a second-order circuit, we focus on finding the current in the inductive branch (i_L). This current is of particular interest because it does not approach zero as t increases. Rather, after the switch has been open for a long time, the inductor current equals the dc source current I. Because we want to focus on the technique for finding the step response, we assume that the initial energy stored in the circuit is zero. This assumption simplifies the calculations and doesn't alter the basic process involved. In Example 8.10 we will see how the presence of initially stored energy enters into the general procedure.

To find the inductor current i_L, we must solve a second-order differential equation equated to the forcing function I, which we derive as follows. From Kirchhoff's current law, we have

$$i_L + i_R + i_C = I,$$

or

$$i_L + \frac{v}{R} + C\frac{dv}{dt} = I. \tag{8.37}$$

Because

$$v = L\frac{di_L}{dt}, \tag{8.38}$$

we get

$$\frac{dv}{dt} = L\frac{d^2 i_L}{dt^2}. \tag{8.39}$$

Substituting Eqs. (8.38) and (8.39) into Eq. (8.37) gives

$$i_L + \frac{L}{R}\frac{di_L}{dt} + LC\frac{d^2 i_L}{dt^2} = I. \tag{8.40}$$

For convenience, we divide through by LC and rearrange terms:

$$\frac{d^2 i_L}{dt^2} + \frac{1}{RC}\frac{di_L}{dt} + \frac{i_L}{LC} = \frac{I}{LC}. \tag{8.41}$$

A symbolic equation solver will generate a closed-form solution.

Comparing Eq. (8.41) with Eq. (8.3) reveals that the presence of a nonzero term on the right-hand side of the equation alters the task. Before showing how to solve Eq. (8.41) directly, we obtain the solution indirectly. When we know the solution of Eq. (8.41), explaining the direct approach is easier.

THE INDIRECT APPROACH

We can solve for i_L indirectly by first finding the voltage v. We do this with the techniques introduced in Section 8.2, because the differential equation that v must satisfy is identical to Eq. (8.3).

To see this, we simply return to Eq. (8.37) and express i_L as a function of v; thus

$$\frac{1}{L}\int_0^t v\,d\tau + \frac{v}{R} + C\frac{dv}{dt} = I. \tag{8.42}$$

Differentiating Eq. (8.42) once with respect to t reduces the right-hand side to zero because I is a constant. Thus

$$\frac{v}{L} + \frac{1}{R}\frac{dv}{dt} + C\frac{d^2v}{dt^2} = 0,$$

or

$$\frac{d^2v}{dt^2} + \frac{1}{RC}\frac{dv}{dt} + \frac{v}{LC} = 0. \tag{8.43}$$

As discussed in Section 8.2, the solution for v depends on the roots of the characteristic equation. Thus the three possible solutions are

$$v = A_1e^{s_1t} + A_2e^{s_2t}, \tag{8.44}$$

$$v = B_1e^{-\alpha t}\cos\omega_d t + B_2e^{-\alpha t}\sin\omega_d t, \tag{8.45}$$

and

$$v = D_1te^{-\alpha t} + D_2e^{-\alpha t}. \tag{8.46}$$

A word of caution: Because there is a source in the circuit for $t > 0$, you must take into account the value of the source current at $t = 0^+$ when you evaluate the coefficients in Eqs. (8.44)–(8.46).

To find the three possible solutions for i_L, we substitute Eqs. (8.44)–(8.46) into Eq. (8.37). You should be able to verify that, when this has been done, the three solutions for i_L will be

$$i_L = I + A_1'e^{s_1t} + A_2'e^{s_2t}, \tag{8.47}$$

$$i_L = I + B_1'e^{-\alpha t}\cos\omega_d t + B_2'e^{-\alpha t}\sin\omega_d t, \tag{8.48}$$

and

$$i_L = I + D_1'te^{-\alpha t} + D_2'e^{-\alpha t}, \tag{8.49}$$

where A_1', A_2', B_1', B_2', D_1', and D_2' are arbitrary constants.

In each case, the primed constants can be found indirectly in terms of the arbitrary constant associated with the voltage solution. However, this approach is cumbersome.

THE DIRECT APPROACH

It is much easier to find the primed constants directly in terms of the initial values of the response function. For the circuit being discussed, we would find the primed constants from $i_L(0)$ and $di_L(0)/dt$. We illustrate this approach in Examples 8.6 through 8.10.

The solution for a second-order differential equation with a constant forcing function equals the forced response plus a response function identical in form to the natural response. Thus we can always write the solution for the step response in the form

$$i = I_f + \left\{ \begin{array}{l} \text{function of the same form} \\ \text{as the natural response} \end{array} \right\} \qquad \textbf{(8.50)}$$

or

$$v = V_f + \left\{ \begin{array}{l} \text{function of the same form} \\ \text{as the natural response} \end{array} \right\}, \qquad \textbf{(8.51)}$$

where I_f and V_f represent the final value of the response function. The final value may be zero, as was, for example, the case with the voltage v in the circuit in Fig. 8.8.

Examples 8.6–8.10 illustrate the technique of finding the step response of a parallel *RLC* circuit.

EXAMPLE 8.6

The initial energy stored in the circuit in Fig. 8.12 is zero. At $t = 0$, a dc current source of 24 mA is applied to the circuit. The value of the resistor is 400 Ω.

a) What is the initial value of i_L?

b) What is the initial value of di_L/dt?

c) What are the roots of the characteristic equation?

d) What is the numerical expression for $i_L(t)$ when $t \geq 0$?

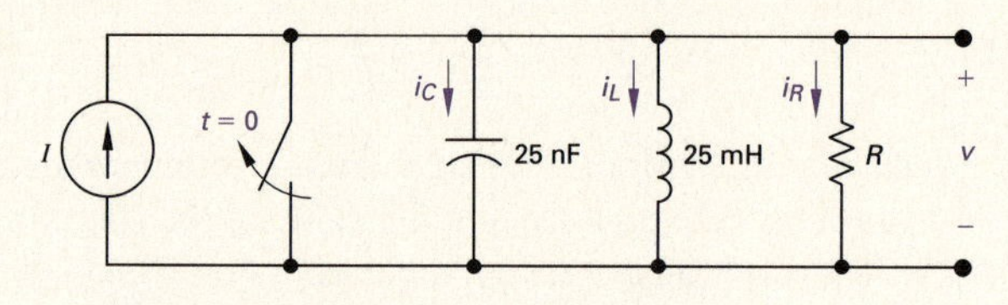

FIGURE 8.12 The circuit for Example 8.6.

SOLUTION

a) No energy is stored in the circuit prior to the application of the dc current source, so the initial current in the inductor is zero. The inductor prohibits an instantaneous change in inductor current; therefore $i_L(0) = 0$ immediately after the switch has been opened.

b) The initial voltage on the capacitor is zero before the switch has been opened; therefore it will be zero immediately after. Now, because $v = L di_L/dt$,

$$\frac{di_L}{dt}(0^+) = 0.$$

c) From the circuit elements, we obtain

$$\omega_0^2 = \frac{1}{LC} = \frac{10^{12}}{(25)(25)} = 16 \times 10^8$$

and

$$\alpha = \frac{1}{2RC} = \frac{10^9}{(2)(400)(25)} = 5 \times 10^4 \text{ rad/s},$$

or

$$\alpha^2 = 25 \times 10^8.$$

Because $\omega_0^2 < \alpha^2$, the roots of the characteristic equation are real and distinct. Thus

$$s_1 = -5 \times 10^4 + 3 \times 10^4 = -20{,}000 \text{ rad/s},$$

and

$$s_2 = -5 \times 10^4 - 3 \times 10^4 = -80{,}000 \text{ rad/s}.$$

d) Because the roots of the characteristic equation are real and distinct, the inductor current response will be overdamped. Thus $i_L(t)$ takes the form of Eq. (8.47), namely,

$$i_L = I_f + A_1' e^{s_1 t} + A_2' e^{s_2 t}.$$

Hence, from this solution, the two simultaneous equations which determine A_1' and A_2' are

$$i_L(0) = I_f + A_1' + A_2' = 0$$

and

$$\frac{di_L}{dt}(0) = s_1 A_1' + s_2 A_2' = 0.$$

Solving for A_1' and A_2' gives

$$A_1' = -32 \text{ mA} \quad \text{and} \quad A_2' = 8 \text{ mA}.$$

The numerical solution for $i_L(t)$ is

$$i_L(t) = (24 - 32e^{-20{,}000t} + 8e^{-80{,}000t}) \text{ mA}, \quad t \geq 0.$$

EXAMPLE 8.7

The resistor in the circuit in Example 8.6 (Fig. 8.12) is increased to 625 Ω. Find $i_L(t)$ for $t \geq 0$.

SOLUTION

Because L and C remain fixed, ω_0^2 has the same value as in Example 8.6; that is, $\omega_0^2 = 16 \times 10^8$. Increasing R to 625 Ω decreases α to

3.2×10^4 rad/s. With $\omega_0^2 > \alpha^2$, the roots of the characteristic equation are complex. Hence

$$s_1 = -3.2 \times 10^4 + j2.4 \times 10^4 \text{ rad/s},$$

and

$$s_2 = -3.2 \times 10^4 - j2.4 \times 10^4 \text{ rad/s}.$$

The current response is now underdamped and given by Eq. (8.48):

$$i_L(t) = I_f + B_1' e^{-\alpha t} \cos \omega_d t + B_2' e^{-\alpha t} \sin \omega_d t.$$

Here, α is 32,000 rad/s, ω_d is 24,000 rad/s, and I_f is 24 mA.

As in Example 8.6, B_1' and B_2' are determined from the initial conditions. Thus the two simultaneous equations are

$$i_L(0) = I_f + B_1' = 0$$

and

$$\frac{di_L}{dt}(0) = \omega_d B_2' - \alpha B_1' = 0.$$

Then,

$$B_1' = -24 \text{ mA} \quad \text{and} \quad B_2' = -32 \text{ mA}.$$

The numerical solution for $i_L(t)$ is

$$i_L(t) = (24 - 24e^{-32{,}000t} \cos 24{,}000t - 32e^{-32{,}000t} \sin 24{,}000t) \text{ mA}, \quad t \geq 0.$$

EXAMPLE 8.8

The resistor in the circuit in Example 8.6 (Fig. 8.12) is set at 500 Ω. Find i_L for $t \geq 0$.

SOLUTION

We know that ω_0^2 remains at 16×10^8. With R set at 500 Ω, α becomes 4×10^4 s^{-1}, which corresponds to critical damping. Therefore the solution for $i_L(t)$ takes the form of Eq. (8.49):

$$i_L(t) = I_f + D_1' t e^{-\alpha t} + D_2' e^{-\alpha t}.$$

Again, D_1' and D_2' are computed from initial conditions, or

$$i_L(0) = I_f + D_2' = 0$$

and

$$\frac{di_L}{dt}(0) = D_1' - \alpha D_2' = 0.$$

Thus

$$D_1' = -960{,}000 \text{ mA/s} \quad \text{and} \quad D_2' = -24 \text{ mA}.$$

The numerical expression for $i_L(t)$ is

$$i_L(t) = (24 - 960{,}000te^{-40{,}000t} - 24e^{-40{,}000t}) \text{ mA}, \quad t \geq 0.$$

EXAMPLE 8.9

a) Plot on a single graph, over a range from 0 to 220 μs, the overdamped, underdamped, and critically damped responses derived in Examples 8.6 through 8.8.

b) Use the plots of (a) to find the time required for i_L to reach 90% of its final value.

c) On the basis of the results obtained in (b), which response would you specify in a design that puts a premium on reaching the final value of the output in the shortest time?

d) Which response would you specify in a design that must ensure that the final value of the current is never exceeded?

SOLUTION

a) See Fig. 8.13.

b) The final value of i_L is 24 mA, so we can read the times off the plots corresponding to $i_L = 21.6$ mA. Thus $t_{od} = 130\ \mu$s, $t_{cd} = 97 \mu$s, and $t_{ud} = 74\ \mu$s.

c) The underdamped response reaches 90% of the final value in the fastest time, so it is the desired response type when speed is the most important design specification.

d) From the plot, you can see that the underdamped response overshoots the final value of current, while neither the critically damped nor the overdamped response produces currents in excess of 24 mA. While specifying either of the latter two responses would meet the design specification, it is best to use the overdamped response. It would be impractical to require a design to achieve the exact component values which ensure a critically damped response.

Use a computer to vary R over a wide range, and summarize the effect using a plot.

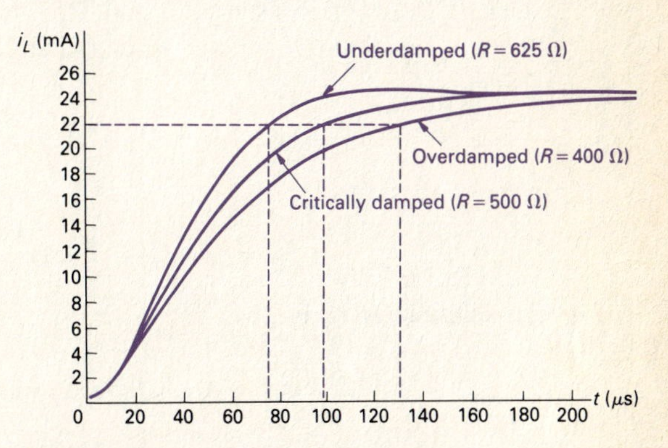

FIGURE 8.13 The current plots for Example 8.9.

EXAMPLE 8.10

Energy is stored in the circuit in Example 8.8 (Fig. 8.12, with $R = 500\ \Omega$) at the instant the dc current source is applied. The initial current in the inductor is 29 mA, and the initial voltage across the capacitor is 50 V. Find (a) $i_L(0)$; (b) $di_L(0)/dt$; (c) $i_L(t)$ for $t \geq 0$; (d) $v(t)$ for $t \geq 0^+$.

SOLUTION

a) There cannot be an instantaneous change of current in an inductor, so the initial value of i_L in the first instant after the dc current source has been applied must be 29 mA.

b) The capacitor holds the initial voltage across the inductor to 50 V. Therefore

$$L\frac{di_L}{dt}(0^+) = 50, \qquad \text{or} \qquad \frac{di_L}{dt}(0^+) = \frac{50}{25} \times 10^3 = 2000 \text{ A/s.}$$

c) From the solution of Example 8.8, we know that the current response is critically damped. Thus

$$i_L(t) = I_f + D_1'te^{-\alpha t} + D_2'e^{-\alpha t},$$

where

$$\alpha = \frac{1}{2RC} = 40{,}000 \text{ rad/s} \qquad \text{and} \qquad I_f = 24 \text{ mA.}$$

Notice that the effect of the nonzero initial stored energy is on the calculations for the constants D_1' and D_2', which we obtain from the initial conditions. First we use the initial value of the inductor current,

$$i_L(0) = I_f + D_2' = 29 \text{ mA,}$$

from which we get

$$D_2' = 29 - 24 = 5 \text{ mA.}$$

The solution for D_1' is

$$\frac{di_L}{dt}(0^+) = D_1' - \alpha D_2' = 2000,$$

or

$$\begin{aligned} D_1' &= 2000 + \alpha D_2' \\ &= 2000 + (40{,}000)(5 \times 10^{-3}) \\ &= 2200 \text{ A/s} = 2.2 \times 10^6 \text{ mA/s.} \end{aligned}$$

Thus the numerical expression for $i_L(t)$ is

$$i_L(t) = (24 + 2.2 \times 10^6 te^{-40,000t} + 5e^{-40,000t})\ \text{mA}, \quad t \geq 0.$$

d) We can get the expression for $v(t),\ t \geq 0^+$ by using the relationship between the voltage and current in an inductor:

$$\begin{aligned} v(t) &= L\frac{di_L}{dt} \\ &= (25 \times 10^{-3})[(2.2 \times 10^6)(-40{,}000)te^{-40,000t} \\ &\quad + 2.2 \times 10^6 e^{-40,000t} + (5)(-40{,}000)e^{-40,000t}] \times 10^{-3} \\ &= -2.2 \times 10^6 te^{-40,000t} + 50e^{-40,000t}\ \text{V}. \end{aligned}$$

To check this result, let's verify that the initial voltage across the inductor is 50 V:

$$v(0^+) = -2.2 \times 10^6(0)(1) + 50(1) = 50\ \text{V}.$$

DRILL EXERCISE

8.6 In the circuit shown, $R = 250\ \Omega$, $L = 0.32$ H, $C = 2\ \mu$F, $I_0 = 0.5$ A, $V_0 = 80$ V, and $I = -1.5$ A. Find (a) $i_R(0^+)$; (b) $i_C(0^+)$; (c) $di_L(0^+)/dt$; (d) s_1, s_2; (e) $i_L(t)$ for $t \geq 0$; and (f) $v(t)$ for $t \geq 0^+$.

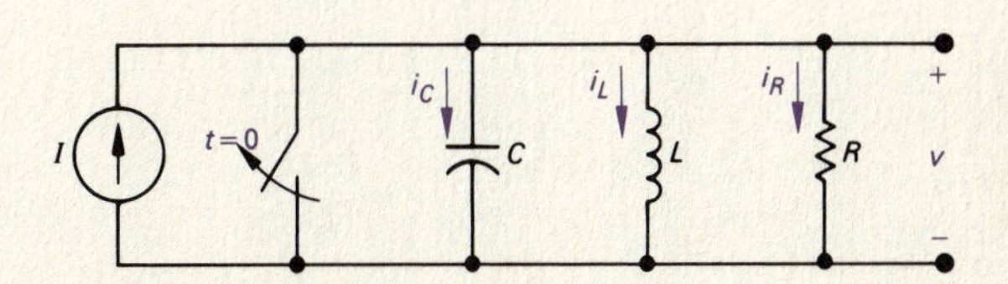

ANSWER: (a) 320 mA; (b) −2.32 A; (c) 250 A/s; (d) $(-1000 + j750)$ rad/s, $(-1000 - j750)$ rad/s; (e) $i_L(t) = [-1.5 + 2e^{-1000t}(\cos 750t + 1.5 \sin 750t]$ A, for $t \geq 0$; (f) $v(t) = 80e^{-1000t}(\cos 750t - 18 \sin 750t)$ V, for $t \geq 0^+$.

8.4 The Natural and Step Responses of a Series *RLC* Circuit

The procedures for finding the natural or step responses of a series *RLC* circuit are the same as those used to find the natural or step responses of a parallel *RLC* circuit, because both circuits are described by differential equations that have the same form. We begin by summing the

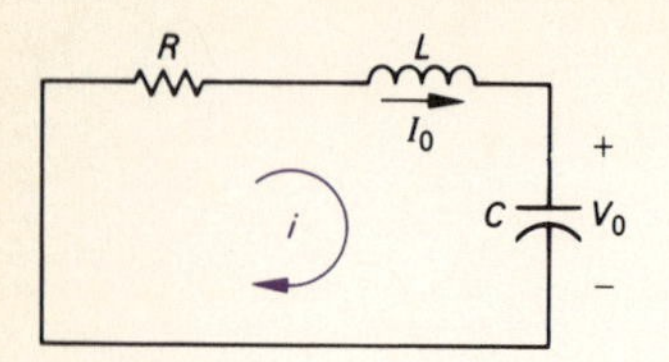

FIGURE 8.14 A circuit used to illustrate the natural response of a series *RLC* circuit.

voltages around the closed path in the circuit shown in Fig. 8.14. Thus

$$Ri + L\frac{di}{dt} + \frac{1}{C}\int_0^t i d\tau + V_o = 0. \quad \textbf{(8.52)}$$

We now differentiate Eq. (8.52) once with respect to t to get

$$R\frac{di}{dt} + L\frac{d^2i}{dt^2} + \frac{i}{C} = 0, \quad \textbf{(8.53)}$$

which we can rearrange as

$$\frac{d^2i}{dt^2} + \frac{R}{L}\frac{di}{dt} + \frac{i}{LC} = 0. \quad \textbf{(8.54)}$$

Comparing Eq. (8.54) with Eq. (8.3) reveals that they have the same form. Therefore, to find the solution of Eq. (8.54), we follow the same process that led us to the solution of Eq. (8.3).

From Eq. (8.54), the characteristic equation for the series *RLC* circuit is

$$s^2 + \frac{R}{L}s + \frac{1}{LC} = 0. \quad \textbf{(8.55)}$$

The roots of the characteristic equation are

$$s_{1,2} = -\frac{R}{2L} \pm \sqrt{\left(\frac{R}{2L}\right)^2 - \frac{1}{LC}}, \quad \textbf{(8.56)}$$

or

$$s_{1,2} = -\alpha \pm \sqrt{\alpha^2 - \omega_0^2}. \quad \textbf{(8.57)}$$

The neper frequency (α) for the series *RLC* circuit is

$$\alpha = \frac{R}{2L} \text{ rad/s}, \quad \textbf{(8.58)}$$

and the expression for the resonant radian frequency is

$$\omega_0 = \frac{1}{\sqrt{LC}} \text{ rad/s}. \quad \textbf{(8.59)}$$

Note that the neper frequency of the series *RLC* circuit differs from that of the parallel *RLC* circuit, but the resonant radian frequencies are the same.

The current response will be overdamped, underdamped, or critically damped according to whether $\omega_0^2 < \alpha^2$, $\omega_0^2 > \alpha^2$, or $\omega_0^2 = \alpha^2$, respectively. Thus the three possible solutions for the current are as follows:

$$i(t) = A_1e^{s_1t} + A_2e^{s_2t} \text{ (overdamped)}; \quad \textbf{(8.60)}$$

$$i(t) = B_1e^{\alpha t}\cos\omega_d t + B_2e^{-\alpha t}\sin\omega_d t \text{ (underdamped)}; \quad \textbf{(8.61)}$$

$$i(t) = D_1te^{-\alpha t} + D_2e^{-\alpha t} \text{ (critically damped)}. \quad \textbf{(8.62)}$$

When you have obtained the natural current response, you can find the natural voltage response across any circuit element.

To verify that the procedure for finding the step response of a series *RLC* circuit is the same as that for a parallel *RLC* circuit, we show that the differential equation that describes the capacitor voltage in Fig. 8.15 has the same form as the differential equation that describes the inductor current in Fig. 8.11. For convenience, we assume that zero energy is stored in the circuit at the instant the switch is closed.

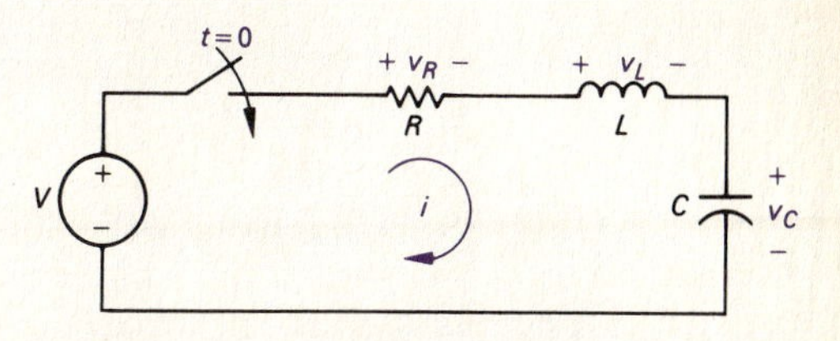

FIGURE 8.15 A circuit used to illustrate the step response of a series *RLC* circuit.

Applying Kirchhoff's voltage law to the circuit shown in Fig. 8.15 gives

$$V = Ri + L\frac{di}{dt} + v_C. \quad \textbf{(8.63)}$$

The current (i) is related to the capacitor voltage (v_C) by the expression

$$i = C\frac{dv_C}{dt}, \quad \textbf{(8.64)}$$

from which

$$\frac{di}{dt} = C\frac{d^2v_C}{dt^2}. \quad \textbf{(8.65)}$$

Substitute Eqs. (8.64) and (8.65) into Eq. (8.63) and write the resulting expression as

$$\frac{d^2v_C}{dt^2} + \frac{R}{L}\frac{dv_C}{dt} + \frac{v_C}{LC} = \frac{V}{LC}. \quad \textbf{(8.66)}$$

Equation (8.66) has the same form as Eq. (8.41); therefore the procedure for finding v_C parallels that for finding i_L. The three possible solutions for v_C are as follows:

$$v_C = V_f + A_1'e^{s_1t} + A_2'e^{s_2t} \text{ (overdamped);} \quad \textbf{(8.67)}$$

$$v_C = V_f + B_1'e^{-\alpha t}\cos\omega_d t + B_2'e^{-\alpha t}\sin\omega_d t \text{ (underdamped);} \quad \textbf{(8.68)}$$

$$v_C = V_f + D_1'te^{-\alpha t} + D_2'e^{-\alpha t} \text{ (critically damped),} \quad \textbf{(8.69)}$$

where V_f is the final value of v_C. Hence, from the circuit shown in Fig. 8.15, the final value of v_C is the dc source voltage V.

Examples 8.11 and 8.12 illustrate the mechanics of finding the natural and step responses of a series *RLC* circuit.

EXAMPLE 8.11

The 0.1 μF capacitor in the circuit shown in Fig. 8.16 is charged to 100 V. At $t = 0$ the capacitor is discharged through a series combination of a 100 mH inductor and a 560 Ω resistor.

a) Find $i(t)$ for $t \geq 0$.

b) Find $v_C(t)$ for $t \geq 0$.

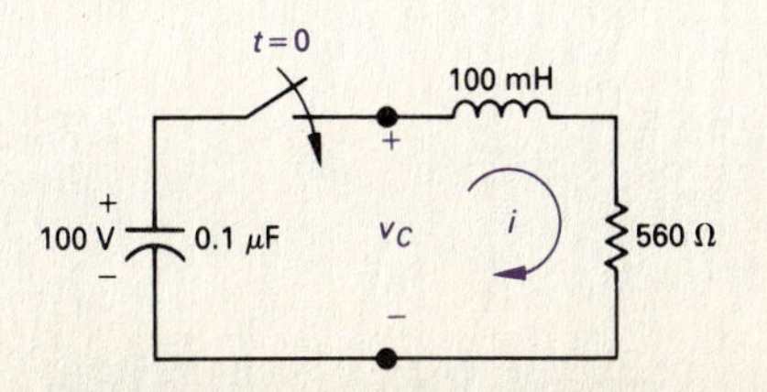

FIGURE 8.16 The circuit for Example 8.11

SOLUTION

a) The first step to finding $i(t)$ is to calculate the roots of the characteristic equation. For the given element values,

$$\omega_0^2 = \frac{1}{LC} = \frac{(10^3)(10^6)}{(100)(0.1)} = 10^8;$$

$$\alpha = \frac{R}{2L} = \frac{560}{2(100)} \times 10^3 = 2800 \text{ rad/s}.$$

Use a circuit simulator to confirm your results.

Next, we compare ω_0^2 to α^2 and note that $\omega_0^2 > \alpha^2$, because

$$\alpha^2 = 7.84 \times 10^6 = 0.0784 \times 10^8.$$

At this point, we know that the response is underdamped and that the solution for $i(t)$ is of the form

$$i(t) = B_1 e^{-\alpha t} \cos \omega_d t + B_2 e^{-\alpha t} \sin \omega_d t,$$

where $\alpha = 2800$ rad/s and $\omega_d = 9600$ rad/s.

The numerical values of B_1 and B_2 come from the initial conditions. The inductor current is zero before the switch has been closed, and hence it is zero immediately after. Therefore

$$i(0) = 0 = B_1.$$

To find B_2, we evaluate $di(0^+)/dt$. From the circuit, we note that, because $i(0) = 0$ immediately after the switch has been closed, there will be no voltage drop across the resistor. Thus the initial voltage on the capacitor appears across the terminals of the inductor, which leads to the expression

$$L\frac{di(0^+)}{dt} = V_0,$$

or

$$\frac{di(0^+)}{dt} = \frac{V_0}{L} = \frac{100}{100} \times 10^3 = 1000 \text{ A/s}.$$

Because $B_1 = 0$,

$$\frac{di}{dt} = 400B_2 e^{-2800t}(24\cos 9600t - 7\sin 9600t).$$

Thus

$$\frac{di(0^+)}{dt} = 9600B_2 \quad \text{and} \quad B_2 = \frac{1000}{9600} \approx 0.1042 \text{ A}.$$

The solution for $i(t)$ is

$$i(t) = 0.1042e^{-2800t} \sin 9600t \text{ A}, \quad t \geq 0.$$

b) To find $v_C(t)$, we can use either of the following relationships:

$$v_C = -\frac{1}{C}\int_0^t i\,d\tau + 100 \quad \text{or} \quad v_C = iR + L\frac{di}{dt}.$$

Whichever expression is used (the second is recommended), the result is

$$v_C(t) = (100\cos 9600t + 29.17\sin 9600t)e^{-2800t}\ \text{V}, \quad t \geq 0.$$

EXAMPLE 8.12

No energy is stored in the 100 mH inductor or the 0.4 μF capacitor when the switch in the circuit shown in Fig. 8.17 is closed. Find $v_C(t)$ for $t \geq 0$.

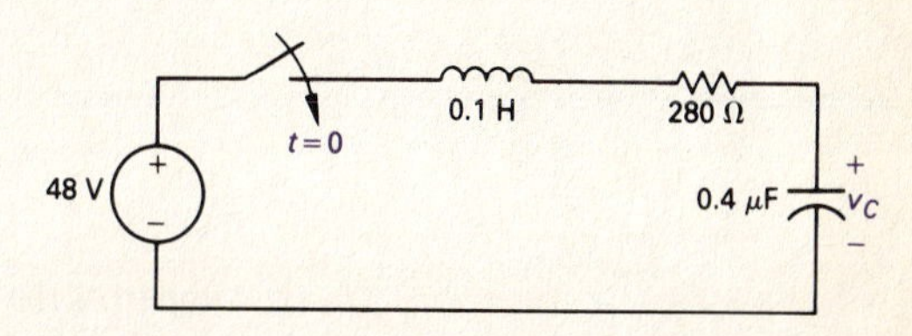

FIGURE 8.17 The circuit for Example 8.12.

SOLUTION

The roots of the characteristic equation are

$$s_1 = -\frac{280}{0.2} + \sqrt{\left(\frac{280}{0.2}\right)^2 - \frac{10^6}{(0.1)(0.4)}}$$
$$= (-1400 + j4800)\ \text{rad/s};$$
$$s_2 = (-1400 - j4800)\ \text{rad/s}.$$

The roots are complex, so the voltage response is underdamped. Thus

$$v_C(t) = 48 + B_1'e^{-1400t}\cos 4800t + B_2'e^{-1400t}\sin 4800t, \quad t \geq 0.$$

No energy is stored in the circuit initially, so both $v_C(0)$ and $dv_C(0^+)/dt$ are zero. Then,

$$v_C(0) = 0 = 48 + B_1'$$

and

$$\frac{dv_C(0^+)}{dt} = 0 = 4800B_2' - 1400B_1'.$$

Solving for B_1' and B_2' yields

$$B_1' = -48\ \text{V} \quad \text{and} \quad B_2' = -14\ \text{V}.$$

Therefore the solution for $v_C(t)$ is

$$v_C(t) = (48 - 48e^{-1400t}\cos 4800t - 14e^{-1400t}\sin 4800t)\ \text{V}, \quad t \geq 0.$$

DRILL EXERCISES

8.7 The switch in the circuit shown has been in position a for a long time. At $t = 0$ it moves to position b. Find (a) $i(0^+)$; (b) $v_C(0^+)$; (c) $di(0^+)/dt$; (d) s_1, s_2; and (e) $i(t)$ for $t \geq 0$.

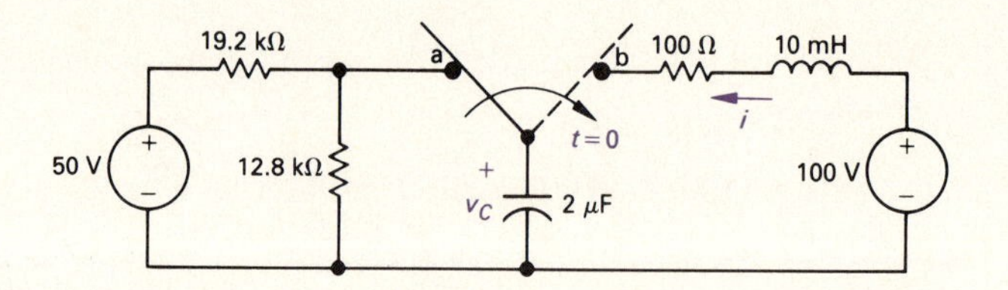

ANSWER: (a) 0; (b) 20 V; (c) 8000 A/s;
(d) $(-5000 + j5000)$ rad/s, $(-5000 - j5000)$ rad/s;
(e) $i(t) = (1.6e^{-5000t} \sin 5000t)$ A for $t \geq 0$.

8.8 Find $v_C(t)$ for $t \geq 0$ for the circuit in Drill Exercise 8.7.

ANSWER: $v_C = [100 - 80e^{-5000t}(\cos 5000t + \sin 5000t)]$ V for $t \geq 0$.

8.5 A Circuit with Two Integrating Amplifiers

A circuit containing two integrating amplifiers connected in cascade[2] is also a second-order circuit; that is, the output voltage of the second integrator is related to the input voltage of the first by a second-order differential equation. We begin our analysis of a circuit containing two cascaded amplifiers with the circuit shown in Fig. 8.18.

We assume that the operational amplifiers are ideal. The task is to derive the differential equation that establishes the relationship between v_o and v_g.

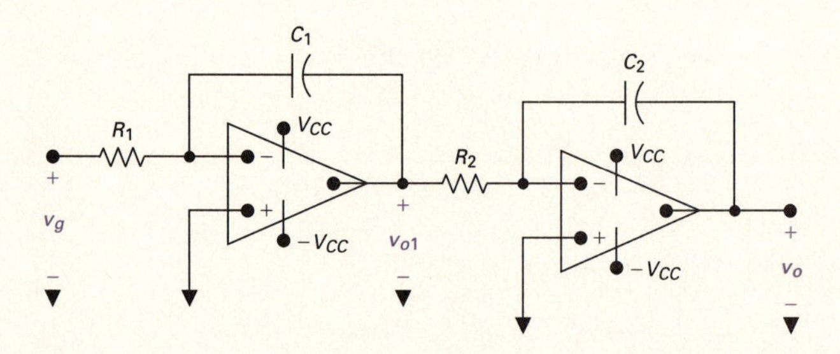

FIGURE 8.18 Two integrating amplifiers connected in cascade.

[2]In a cascade connection, the output signal of the first amplifier (v_{o1} in Fig. 8.18) is the input signal for the second amplifier.

We begin the derivation by summing the currents at the inverting input terminal of the first integrator. Because the operational amplifier is ideal,

$$\frac{0 - v_g}{R_1} + C_1 \frac{d}{dt}(0 - v_{o1}) = 0. \quad \textbf{(8.70)}$$

From Eq. (8.70),

$$\frac{dv_{o1}}{dt} = -\frac{1}{R_1 C_1} v_g. \quad \textbf{(8.71)}$$

Now we sum the currents away from the inverting input terminal of the second integrating amplifier:

$$\frac{0 - v_{o1}}{R_2} + C_2 \frac{d}{dt}(0 - v_o) = 0, \quad \textbf{(8.72)}$$

or

$$\frac{dv_o}{dt} = -\frac{1}{R_2 C_2} v_{o1}. \quad \textbf{(8.73)}$$

Differentiating Eq. (8.73) gives

$$\frac{d^2 v_o}{dt^2} = -\frac{1}{R_2 C_2} \frac{dv_{o1}}{dt}. \quad \textbf{(8.74)}$$

We find the differential equation that governs the relationship between v_o and v_g by substituting Eq. (8.71) into Eq. (8.74):

$$\frac{d^2 v_o}{dt^2} = \frac{1}{R_1 C_1} \frac{1}{R_2 C_2} v_g. \quad \textbf{(8.75)}$$

Example 8.13 illustrates the step response of a circuit containing two cascaded integrating amplifiers.

EXAMPLE 8.13

No energy is stored in the circuit shown in Fig. 8.19 when the input voltage v_g jumps instantaneously from 0 to 25 mV.

a) Derive the expression for $v_o(t)$ for $0 \le t \le t_{\text{sat}}$.

b) How long is it before the circuit saturates?

SOLUTION

a) Figure 8.19 indicates that the amplifier scaling factors are

$$\frac{1}{R_1 C_1} = \frac{1000}{(250)(0.1)} = 40$$

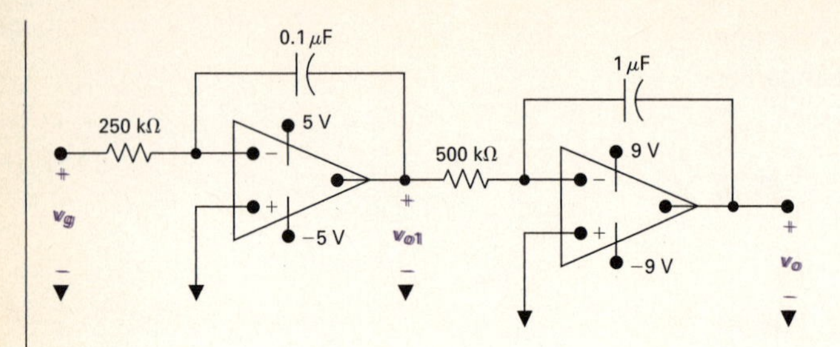

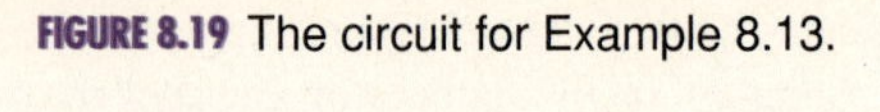
FIGURE 8.19 The circuit for Example 8.13.

and

$$\frac{1}{R_2C_2} = \frac{1000}{(500)(1)} = 2.$$

Now, because $v_g = 25$ mV for $t > 0$, Eq. (8.75) becomes

$$\frac{d^2v_o}{dt^2} = (40)(2)(25 \times 10^{-3}) = 2.$$

To solve for v_o, we let

$$g(t) = \frac{dv_o}{dt};$$

then,

$$\frac{dg(t)}{dt} = 2 \quad \text{and} \quad dg(t) = 2dt.$$

Hence

$$\int_{g(0)}^{g(t)} dy = 2\int_0^t dx,$$

from which

$$g(t) - g(0) = 2t.$$

However,

$$g(0) = \frac{dv_o(0)}{dt} = 0,$$

because the energy stored in the circuit initially is zero, and the operational amplifiers are ideal. (See Drill Exercise 8.9.) Then,

$$\frac{dv_o}{dt} = 2t \quad \text{and} \quad v_o = t^2 + v_o(0).$$

But $v_o(0) = 0$, so the expression for v_o becomes

$$v_o = t^2, \quad 0 \le t \le t_{\text{sat}}.$$

b) The second integrating amplifier saturates when v_o reaches 9 V or $t = 3$ s. But it is possible that the first integrating amplifier saturates before $t = 3$ s. To explore this possibility, use Eq. (8.71) to find dv_{o1}/dt:

$$\frac{dv_{o1}}{dt} = -40(25) \times 10^{-3} = -1.$$

Solving for v_{o1} yields

$$v_{o1} = -t.$$

Thus, at $t = 3$ s, $v_{o1} = -3$ V, and, because the power supply voltage on the first integrating amplifier is ± 5 V, the circuit reaches saturation when the second amplifier saturates. When one of the operational amplifiers saturates, we no longer can use the linear model to predict the behavior of the circuit.

DRILL EXERCISES

8.9 Show that, if no energy is stored in the circuit shown in Fig. 8.19 at the instant v_g jumps in value, then dv_o/dt equals zero at $t = 0$.

ANSWER: Derivation.

8.10 a) Find the equation for $v_o(t)$ for $0 \le t \le t_{sat}$ in the circuit shown in Fig. 8.19 if $v_{o1}(0) = 2$ V and $v_o(0) = 8$ V.

b) How long does the circuit take to reach saturation?

ANSWER: (a) $v_o = t^2 - 4t + 8$ V; (b) $t_{sat} = 4.24$ s.

TWO INTEGRATING AMPLIFIERS WITH FEEDBACK RESISTORS

Figure 8.20 depicts a variation of the circuit shown in Fig. 8.18. Recall from Section 7.7 that the reason the op amp in the integrating amplifier saturates is the feedback capacitor's accumulation of charge. Here, a resistor is placed in parallel with each feedback capacitor (C_1 and C_2) to overcome this problem. We rederive the equation for the output voltage, v_o, and determine the impact of these feedback resistors on the integrating amplifiers from Example 8.13.

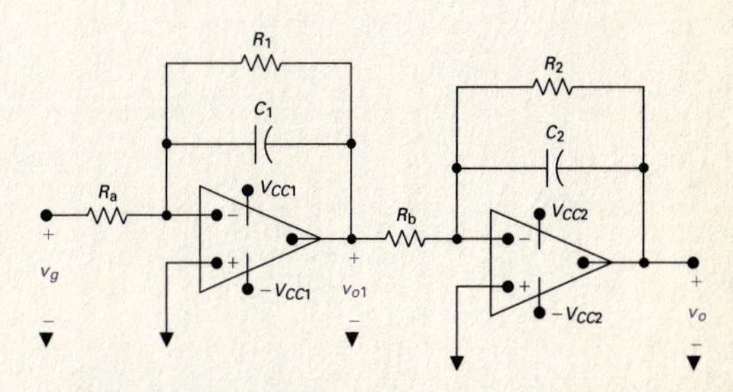

FIGURE 8.20 Cascaded integrating amplifiers with feedback resistors.

We begin the derivation of the second-order differential equation that relates v_o to v_g by summing the currents at the inverting input node of the first integrator:

$$\frac{0 - v_g}{R_a} + \frac{0 - v_{o1}}{R_1} + C_1\frac{d}{dt}(0 - v_{o1}) = 0. \qquad \textbf{(8.76)}$$

We simplify Eq. (8.76) to read

$$\frac{dv_{o1}}{dt} + \frac{1}{R_1C_1}v_{o1} = \frac{-v_g}{R_aC_1}. \tag{8.77}$$

For convenience, we let $\tau_1 = R_1C_1$ and write Eq. (8.77) as

$$\frac{dv_{o1}}{dt} + \frac{v_{o1}}{\tau_1} = \frac{-v_g}{R_aC_1}. \tag{8.78}$$

The next step is to sum the currents at the inverting input terminal of the second integrator:

$$\frac{0 - v_{o1}}{R_b} + \frac{0 - v_o}{R_2} + C_2\frac{d}{dt}(0 - v_o) = 0. \tag{8.79}$$

We rewrite Eq. (8.79) as

$$\frac{dv_o}{dt} + \frac{v_o}{\tau_2} = \frac{-v_{o1}}{R_bC_2}, \tag{8.80}$$

where $\tau_2 = R_2C_2$. Differentiating Eq. (8.80) yields

$$\frac{d^2v_o}{dt^2} + \frac{1}{\tau_2}\frac{dv_o}{dt} = -\frac{1}{R_bC_2}\frac{dv_{o1}}{dt}. \tag{8.81}$$

From Eq. (8.78),

$$\frac{dv_{o1}}{dt} = \frac{-v_{o1}}{\tau_1} - \frac{v_g}{R_aC_1}, \tag{8.82}$$

and from Eq. (8.80),

$$v_{o1} = -R_bC_2\frac{dv_o}{dt} - \frac{R_bC_2}{\tau_2}v_o. \tag{8.83}$$

We use Eqs. (8.82) and (8.83) to eliminate dv_{o1}/dt from Eq. (8.81) and obtain the desired relationship:

$$\frac{d^2v_o}{dt^2} + \left(\frac{1}{\tau_1} + \frac{1}{\tau_2}\right)\frac{dv_o}{dt} + \left(\frac{1}{\tau_1\tau_2}\right)v_o = \frac{v_g}{R_aC_1R_bC_2}. \tag{8.84}$$

From Eq. (8.84), the characteristic equation is

$$s^2 + \left(\frac{1}{\tau_1} + \frac{1}{\tau_2}\right)s + \frac{1}{\tau_1\tau_2} = 0. \tag{8.85}$$

The roots of the characteristic equation are real, namely,

$$s_1 = \frac{-1}{\tau_1} \tag{8.86}$$

and

$$s_2 = \frac{-1}{\tau_2}. \tag{8.87}$$

Example 8.14 illustrates the analysis of the step response of two cascaded integrating amplifiers when the feedback capacitors are shunted with feedback resistors.

EXAMPLE 8.14

The parameters for the circuit shown in Fig. 8.20 are $R_a = 100$ kΩ, $R_1 = 500$ kΩ, $C_1 = 0.1\ \mu$F, $R_b = 25$ kΩ, $R_2 = 100$ kΩ, and $C_2 = 1\ \mu$F. The power supply voltage for each operational amplifier is ±6 V. The signal voltage (v_g) for the cascaded integrating amplifiers jumps from 0 to 250 mV at $t = 0$. No energy is stored in the feedback capacitors at the instant the signal is applied.

a) Find the numerical expression of the differential equation for v_o.

b) Find $v_o(t)$ for $t \geq 0$.

c) Find the numerical expression of the differential equation for v_{o1}.

d) Find $v_{o1}(t)$ for $t \geq 0$.

SOLUTION

a) From the numerical values of the circuit parameters, we have $\tau_1 = R_1C_1 = 0.05$ s; $\tau_2 = R_2C_2 = 0.10$ s, and $v_g/R_aC_1R_bV_2 = 1000$ V/s^2. Substituting these values into Eq. (8.84) gives

$$\frac{d^2v_o}{dt^2} + 30\frac{dv_o}{dt} + 200v_o = 1000.$$

b) The roots of the characteristic equation are $s_1 = -20$ rad/s and $s_2 = -10$ rad/s. The final value of v_o is the input voltage times the gain of each stage, since the capacitors behave as open circuits as $t \to \infty$. Thus,

$$v_o(\infty) = (250 \times 10^{-3})\frac{(-500)}{100}\frac{(-100)}{25} = 5 \text{ V}.$$

The solution for v_o thus takes the form

$$v_o = 5 + A_1'e^{-10t} + A_2'e^{-20t}.$$

With $v_o(0) = 0$ and $dv_o(0)/dt = 0$, the numerical values of A_1' and A_2' are $A_1' = -10$ V and $A_2' = 5$ V. Therefore the solution for v_o is

$$v_o(t) = (5 - 10e^{-10t} + 5e^{-20t}) \text{ V}, \quad t \geq 0.$$

The solution assumes that neither operational amplifier saturates. We have already noted that the final value of v_o is 5 V, which is less than 6 V; hence the second op amp does not saturate. The final value of v_{o1} is $(250 \times 10^{-3})(-500/100)$, or -1.25 V. Therefore the first operational amplifier does not saturate, and our assumption and solution are correct.

c) Substituting the numerical values of the parameters into Eq. (8.78) generates the desired differential equation:

$$\frac{dv_{o1}}{dt} + 20v_{o1} = -25.$$

d) We have already noted the initial and final values of v_{o1}, along with the time constant τ_1. Thus we write the solution in accordance with the technique developed in Section 7.4:

$$\begin{aligned} v_{o1} &= -1.25 + [0 - (-1.25)]e^{-20t} \\ &= -1.25 + 1.25e^{-20t} \text{ V}, \quad t \geq 0. \end{aligned}$$

DRILL EXERCISES

8.11 Rework Example 8.14 with feedback resistors R_1 and R_2 removed.

ANSWER: (a) $d^2v_o/dt^2 = 1000$; (b) $v_o = 500t^2$ V, $0 \leq t \leq 0.1095$ s; (c) $dv_{o1}/dt = -25$; (d) $v_{o1} = -25t$ V, $0 \leq t \leq 0.1095$ s.

8.12 Rework Example 8.14 with $v_{o1}(0) = -2$ V and $v_o(0) = 4$ V.

ANSWER: (a) Same as in Example 8.14; (b) $v_o = (5 + 2e^{-10t} - 3e^{-20t})$ V, $t \geq 0$; (c) same as in Example 8.14; (d) $v_{o1} = -(1.25 + 0.75e^{-20t})$ V, $t \geq 0$.

SUMMARY

- The **characteristic equation** for both the parallel and series *RLC* circuits has the form

$$s^2 + 2\alpha s + \omega_0^2 = 0,$$

where $\alpha = 1/2RC$ for the parallel circuit, $\alpha = R/2L$ for the series circuit, and $\omega_0^2 = 1/LC$ for both the parallel and series circuits.

- The roots of the characteristic equation are

$$s_{1,2} = -\alpha \pm \sqrt{\alpha^2 - \omega_0^2}.$$

- The form of the natural and step responses of series and parallel *RLC* circuits depends on the values of α^2 and ω_0^2; such responses can be **overdamped**, **underdamped**, or **critically damped**. These terms describe the impact of the dissipative element (R) on the response. The **neper frequency**, α, reflects the effect of R.

- The response of a second-order circuit is overdamped, underdamped, or critically damped as follows:

The Circuit Is	When	Qualitative Nature of the Response
Overdamped	$\alpha^2 > \omega_o^2$	The voltage or current approaches its final value without oscillation.
Underdamped	$\alpha^2 < \omega_o^2$	The voltage or current oscillates about its final value.
Critically damped	$\alpha^2 = \omega_o^2$	The voltage or current is on the verge of oscillating about its final value.

- In determining the **natural response** of a second-order circuit, we first determine whether it is over-, under-, or critically damped, and then we solve the appropriate equations as follows:

Damping	Natural Response Equations	Coefficient Equations
Overdamped	$x(t) = A_1e^{s_1t} + A_2e^{s_2t}$	$x(0) = A_1 + A_2$; $dx/dt(0) = A_1s_1 + A_2s_2$
Underdamped	$x(t) = (B_1 \cos \omega_d t + B_2 \sin \omega_d t)e^{-\alpha t}$	$x(0) = B_1$; $dx/dt(0) = -\alpha B_1 + \omega_d B_2$, where $\omega_d = \sqrt{\omega_0^2 - \alpha^2}$
Critically damped	$x(t) = (D_1t + D_2)e^{-\alpha t}$	$x(0) = D_2$; $dx/dt(0) = D_1 - \alpha D_2$

- In determining the **step response** of a second-order circuit, we apply the appropriate equations depending on the damping, as follows:

Damping	Step Response Equations[3]	Coefficient Equations
Overdamped	$x(t) = X_f + A_1'e^{s_1t} + A_2'e^{s_2t}$	$x(0) = X_f + A_1' + A_2'$; $dx/dt(0) = A_1's_1 + A_2's_2$
Underdamped	$x(t) = X_f + (B_1' \cos \omega_d t + B_2' \sin \omega_d t)e^{-\alpha t}$	$x(0) = X_f + B_1'$; $dx/dt(0) = -\alpha B_1' + \omega_d B_2'$
Critically damped	$x(t) = X_f + D_1'te^{-\alpha t} + D_2'e^{-\alpha t}$	$x(0) = X_f + D_2'$; $dx/dt(0) = D_1' - \alpha D_2'$

[3]where X_f is the final value of $x(t)$.

- For each of the three forms of response, the unknown coefficients (i.e., the A's, B's, and D's) are obtained by evaluating the initial value of the response, $x(0)$, and the initial value of the first derivative of the response, $dx(0)/dt$.
- When two integrating amplifiers with ideal op amps are connected in cascade, the output voltage of the second integrator is related to the input voltage of the first by an ordinary, second-order differential equation. Therefore the techniques developed in this chapter may be used to analyze the behavior of a cascaded integrator.
- We can overcome the limitation of a simple integrating amplifier—the saturation of the op amp due to charge accumulating in the feedback capacitor—by placing a resistor in parallel with the capacitor in the feedback path.

PROBLEMS

8.1 The resistance, inductance, and capacitance in a parallel RLC circuit are 2000 Ω, 250 mH, and 10 nF, respectively.

a) Calculate the roots of the characteristic equation that describe the voltage response of the circuit.

b) Will the response be over-, under-, or critically damped?

c) What value of R will yield a damped frequency of 12 krad/s?

d) What are the roots of the characteristic equation for the value of R found in part (c)?

e) What value of R will result in a critically damped response?

8.2 The initial voltage on the 0.05 μF capacitor in the circuit shown in Fig. 8.1 is 15 V. The initial current in the inductor is zero. The voltage response for $t \geq 0$ is

$$v(t) = -5e^{-5000t} + 20e^{-20,000t} \text{ V}.$$

a) Determine the numerical values of R, L, α, and ω_0

b) Calculate $i_R(t)$, $i_L(t)$, and $i_C(t)$ for $t \geq 0^+$.

8.3 The circuit elements in the circuit in Fig. 8.1 are $R = 200$ Ω, $C = 0.2$ μF, and $L = 50$ mH. The initial inductor current is −45 mA, and the initial capacitor voltage is 15 V.

a) Calculate the initial current in each branch of the circuit.

b) Find $v(t)$ for $t \geq 0$.

c) Find $i_L(t)$ for $t \geq 0$.

8.4 The resistance in Problem 8.3 is increased to 312.5 Ω. Find the expression for $v(t)$ for $t \geq 0$.

8.5 The resistance in Problem 8.3 is increased to 250 Ω. Find the expression for $v(t)$ for $t \geq 0$.

8.6 The natural response for the circuit shown in Fig. 8.1 is known to be

$$v = -11e^{-100t} + 20e^{-400t}\ \text{V}, \quad t \geq 0.$$

If $L = 12.5$ H and $C = 2\ \mu$F, find $i_L(0^+)$ in milliamperes.

8.7 The natural voltage response of the circuit in Fig. 8.1 is

$$v = 75e^{-8000t}(\cos 6000t - 4\sin 6000t)\ \text{V}, \quad t \geq 0,$$

when the inductor is 400 mH. Find (a) C; (b) R; (c) V_0; (d) I_0; and (e) $i_L(t)$.

8.8 The initial value of the voltage v in the circuit in Fig. 8.1 is 15 V, and the initial value of the capacitor current, $i_c(0^+)$ is 45 mA. The expression for the capacitor current is known to be

$$i_c(t) = A_1e^{-200t} + A_2e^{-800t}, \quad t \geq 0^+,$$

when R is 250 Ω. Find the numerical

a) value of α, ω_0, L, C, A_1, and A_2;

b) expression for $v(t)$, $t \geq 0$;

c) expression for $i_R(t) \geq 0$; and

d) expression for $i_L(t) \geq 0$.

8.9 The voltage response for the circuit in Fig. 8.1 is known to be

$$v(t) = D_1t\,e^{-1000t} + D_2e^{-1000t}, \quad t \geq 0.$$

The initial current in the inductor (I_0) is 8 mA, and the initial voltage on the capacitor (V_0) is 5 V. The inductor has an inductance of 2.5 H.

a) Find the value of R, C, D_1, and D_2.

b) Find $i_C(t)$ for $t \geq 0^+$.

8.10 In the circuit in Fig. 8.1, $R = 12.5$ Ω, $L = 50/101$ H, $C = 0.08$ F, $V_0 = 0$ V, and $I_0 = -4$ A.

a) Find $v(t)$ for $t \geq 0$.

b) Find the first three values of t for which dv/dt is zero. Let these values of t be denoted t_1, t_2, and t_3.

c) Show that $t_3 - t_1 = T_d$.

d) Show that $t_2 = t_1 = T_d/2$.

e) Calculate $v(t_1)$, $v(t_2)$, and $v(t_3)$.

f) Sketch $v(t)$ versus t for $0 \leq t \leq t_2$.

8.11 a) Find $v(t)$ for $t \geq 0$ in the circuit in Problem 8.10 if the 12.5 Ω resistor is removed from the circuit.

b) Calculate the frequency of $v(t)$ in hertz.

c) Calculate the maximum amplitude of $v(t)$ in volts.

8.12 In the circuit shown in Fig. 8.1, a 2.5 H inductor is shunted by a 100 nF capacitor, the resistor R is adjusted for critical damping, $V_0 = -15$ V, and $I_0 = -5$ mA.

a) Calculate the numerical value of R.

b) Calculate $v(t)$ for $t \geq 0$.

c) Find $v(t)$ when $i_C(t) = 0$.

d) What percentage of the initially stored energy remains stored in the circuit at the instant $i_C(t)$ is 0?

8.13 The resistor in the circuit in Example 8.4 is changed to 3200 Ω.

a) Find the numerical expression for $v(t)$ when $t \geq 0$.

b) Plot $v(t)$ versus t for the time interval $0 \leq t \leq 7$ ms. Compare this response with the one in Example 8.4 ($R = 20$ kΩ) and Example 8.5 ($R = 4$ kΩ). In particular, compare peak values of $v(t)$ and the times when these peak values occur.

8.14 The two switches in the circuit seen in Fig. P8.14 operate synchronously. When switch 1 is in position a, switch 2 is in position d. When switch 1 moves to position b, switch 2 moves to position c, and vice versa. Switch 1 has been in position a for a long time. At $t = 0$ the switches move to their alternate positions. Find $v_o(t)$ for $t \geq 0$.

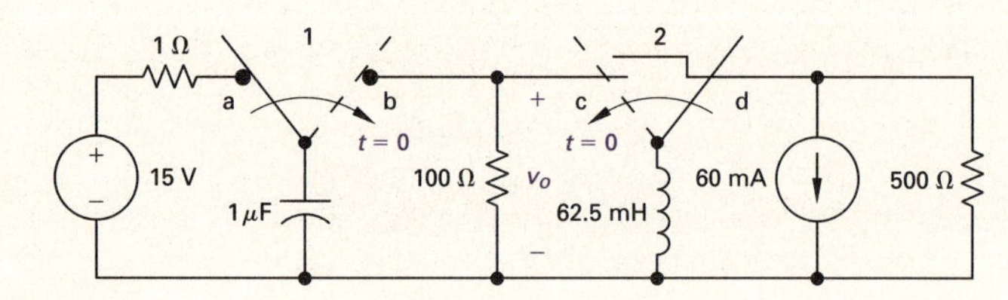

FIGURE P8.14

8.15 The resistor in the circuit of Fig. P8.14 is increased from 100 to 250 Ω. Find $v_o(t)$ for $t > 0$.

8.16 The resistor in the circuit of Fig. P8.14 is increased from 100 to 125 Ω. Find $v_o(t)$ for $t \geq 0$.

8.17 The switch in the circuit of Fig. P8.17 has been in position a for a long time. At $t = 0$ the switch moves instantaneously to position b. Find $v_o(t)$ for $t \geq 0$.

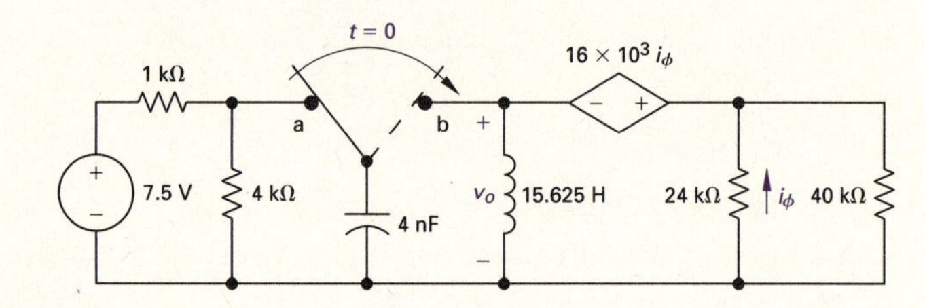

FIGURE P8.17

8.18 For the circuit in Example 8.6, find, for $t \geq 0$, (a) $v(t)$; (b) $i_R(t)$; and (c) $i_C(t)$.

8.19 For the circuit in Example 8.7, find, for $t \geq 0$, (a) $v(t)$ and (b) $i_C(t)$.

8.20 For the circuit in Example 8.8, find $v(t)$ for $t \geq 0$.

8.21 The switch in the circuit in Fig. P8.21 has been open a long time before closing at $t = 0$. Find $i_L(t)$ for $t \geq 0$.

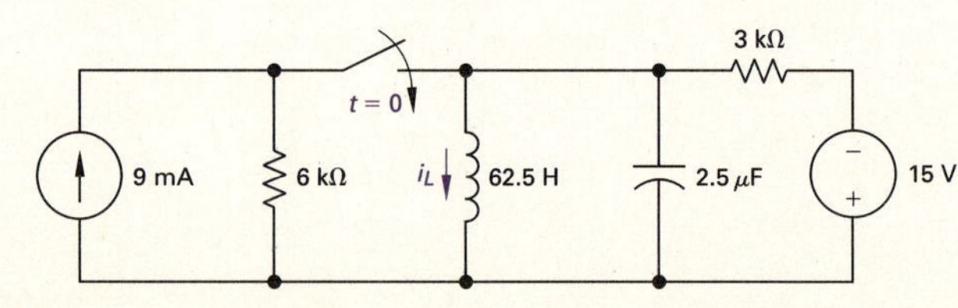

FIGURE P8.21

8.22 Assume that at the instant the 60 mA dc current source is applied to the circuit in Fig. P8.22, the initial current in the 50 mH inductor is −45 mA, and the initial voltage on the capacitor is 15 V (positive at the upper terminal).

Find the expression for $i_L(t)$ for $t \geq 0$ if R equals 200 Ω.

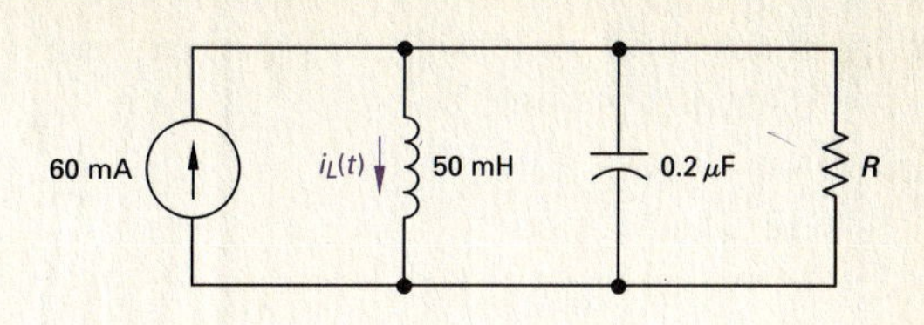

FIGURE P8.22

8.23 The resistance in the circuit in Fig. P8.22 is increased to 312.5 Ω. Find $i_L(t)$ for $t \geq 0$.

8.24 The resistance in the circuit in Fig. P8.22 is changed to 250 Ω. Find $i_L(t)$ for $t \geq 0$.

8.25 There is no energy stored in the circuit in Fig. P8.25 when the switch is closed at $t = 0$. Find $v_o(t)$ for $t \geq 0$.

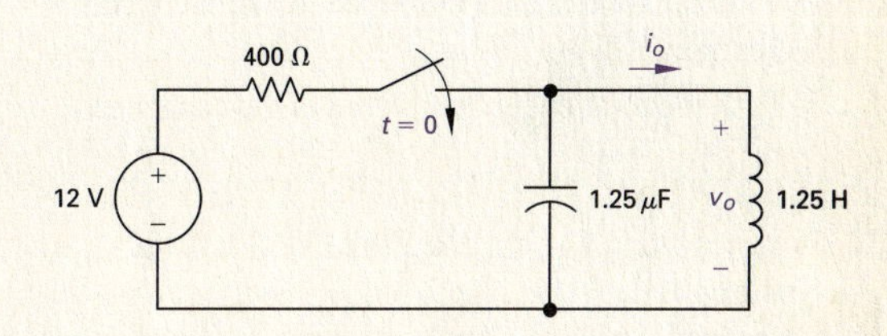

FIGURE P8.25

8.26 a) For the circuit in Fig. P8.25, find i_o for $t \geq 0$.

b) Show that your solution for i_o is consistent with the solution for v_o in Problem 8.25.

8.27 The switch in the circuit in Fig. P8.27 has been open for a long time before closing at $t = 0$. Find $v_o(t)$ for $t \geq 0$.

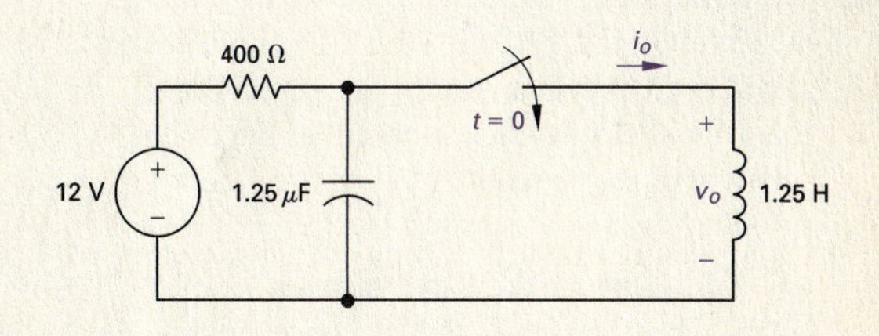

FIGURE P8.27

8.28 a) For the circuit in Fig. P8.27, find i_o for $t \geq 0$.

b) Show that your solution for i_o is consistent with the solution for v_o in Problem 8.27.

8.29 The switch in the circuit in Fig. P8.29 has been open a long time before closing at $t = 0$. Find v_o for $t \geq 0$.

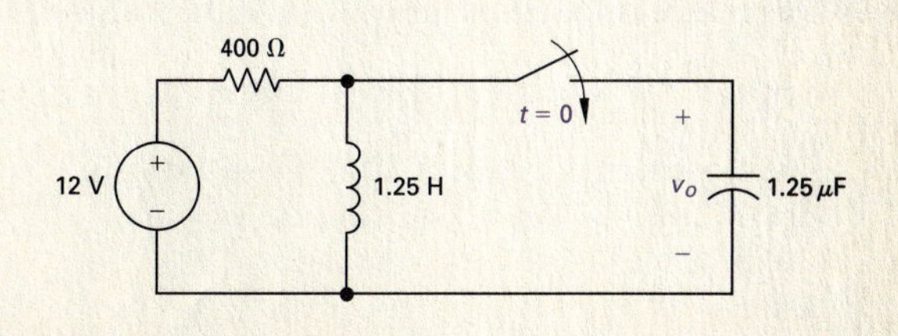

FIGURE P8.29

8.30 The switch in the circuit in Fig. P8.30 has been open a long time before closing at $t = 0$.

a) Find $v_o(t)$ for $t \geq 0^+$.

b) Find $i_L(t)$ for $t \geq 0$.

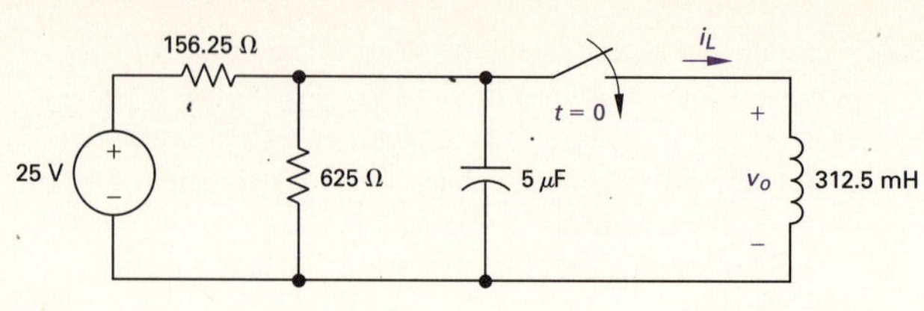

FIGURE P8.30

8.31 Use the circuit in Fig. P8.30.

a) Find the total energy delivered to the inductor.

b) Find the total energy delivered to the equivalent resistor.

c) Find the total energy delivered to the capacitor.

d) Find the total energy delivered by the equivalent current source.

e) Check the results of parts (a) through (d) against the conservation of energy principle.

8.32 Switches 1 and 2 in the circuit in Fig. P8.32 are synchronized. When switch 1 is opened, switch 2 closes, and vice versa. Switch 1 has been open a long time before closing at $t = 0$. Find $i_L(t)$ for $t \geq 0$.

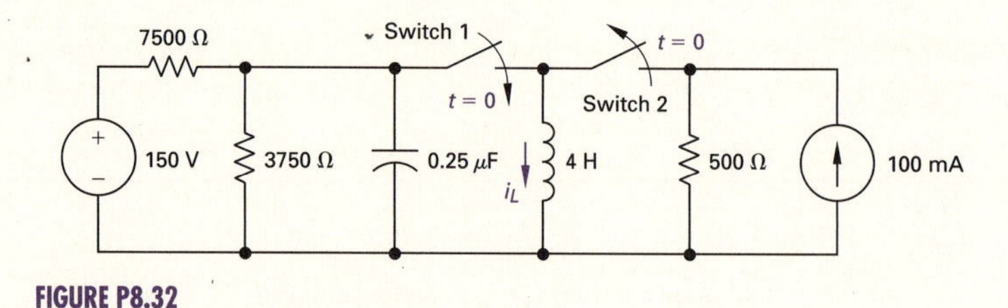

FIGURE P8.32

8.33 The initial energy stored in the 50 nF capacitor in the circuit in Fig. P8.33 is 90 μJ. The initial energy stored in the inductor is zero. The roots of the characteristic equation that describes the natural behavior of the current i are $-1000\ s^{-1}$ and $-4000\ s^{-1}$.

a) Find the numerical values of R and L.

b) Find the numerical values of $i(0)$ and $di(0)/dt$ immediately after the switch has been closed.

c) Find $i(t)$ for $t \geq 0$.

d) How many microseconds after the switch closes does the current reach its maximum value?

e) What is the maximum value of i in milliamperes?

f) Find $v_L(t)$ for $t \geq 0$.

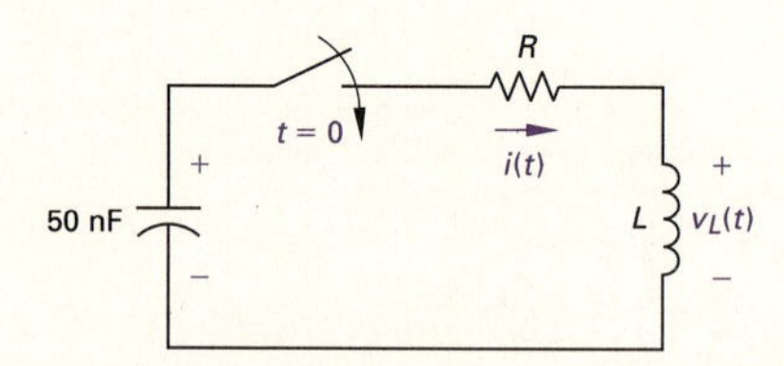

FIGURE P8.33

8.34 The current in the circuit in Fig. 8.3 is known to be

$$i = B_1 e^{-2000t} \cos 1500t + B_2 e^{-2000t} \sin 1500t, \quad t \geq 0.$$

The capacitor has a value of 80 nF; the initial value of the current is 7.5 mA; and the initial voltage on the capacitor is -30 V. Find the values of R, L, B_1, and B_2.

8.35 Find the voltage across the 80 nF capacitor for the circuit described in Problem 8.34. Assume the reference polarity for the capacitor voltage is positive at the upper terminal.

8.36 In the circuit in Fig. P8.36, the resistor is adjusted for critical damping. The initial capacitor voltage is 15 V, and the initial inductor current is 6 mA.

a) Find the numerical value of R

b) Find the numerical values of i and di/dt immediately after the switch is closed.

c) Find $v_C(t)$ for $t \geq 0$.

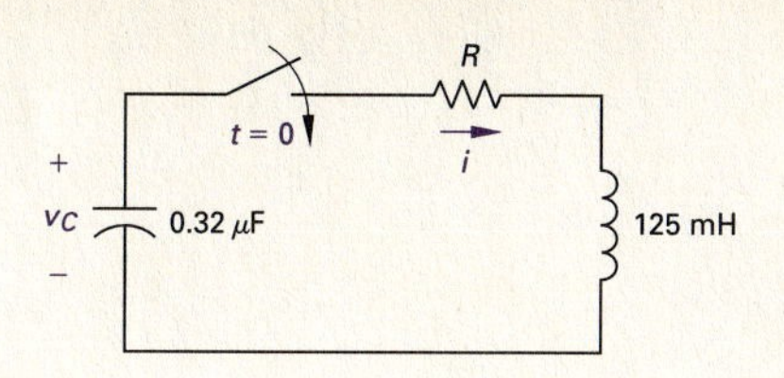

FIGURE P8.36

8.37 The switch in the circuit in Fig. P8.37 has been in position a for a long time. At $t = 0$, the switch moves instantaneously to position b.

a) What is the initial value of v_a?

b) What is the initial value of dv_a/dt?

c) What is the numerical expression for $v_a(t)$ for $t \geq 0$.

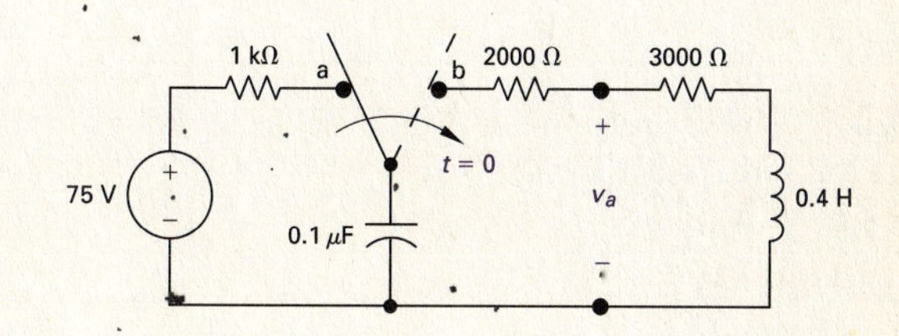

FIGURE P8.37

8.38 The make-before-break switch in the circuit shown in Fig. P8.38 has been in position a for a long time. At $t = 0$ the switch is moved instantaneously to position b. Find $i(t)$ for $t \geq 0$.

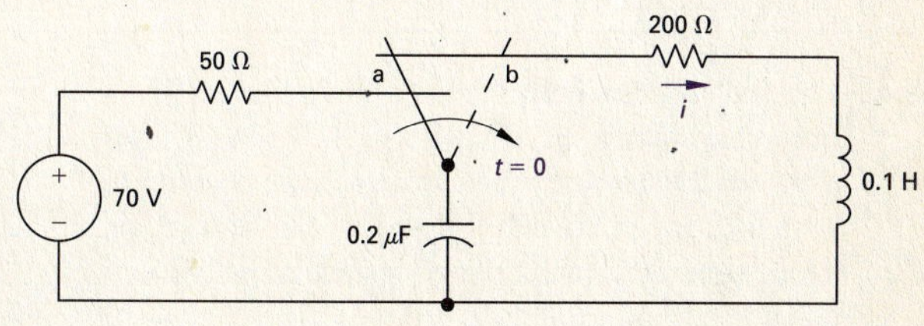

FIGURE P8.38

8.39 The switch in the circuit shown in Fig. P8.39 has been closed for a long time. The switch opens at $t = 0$. Find $v_o(t)$ for $t \geq 0$.

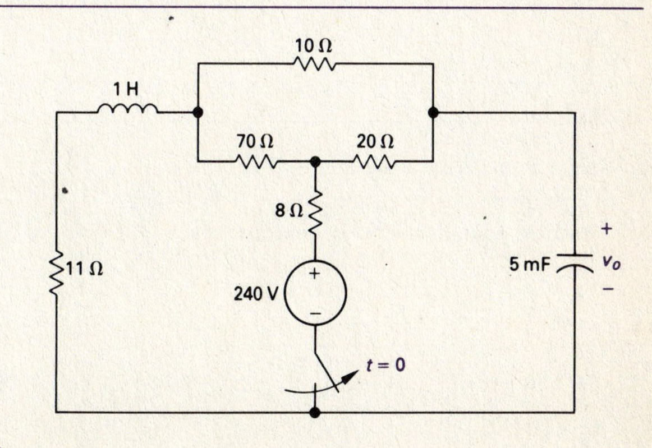

FIGURE P8.39

8.40 The switch in the circuit shown in Fig. P8.40 has been closed for a long time. The switch opens at $t = 0$.

a) Find $i_o(t)$ for $t \geq 0$.

b) Find $v_o(t)$ for $t \geq 0$.

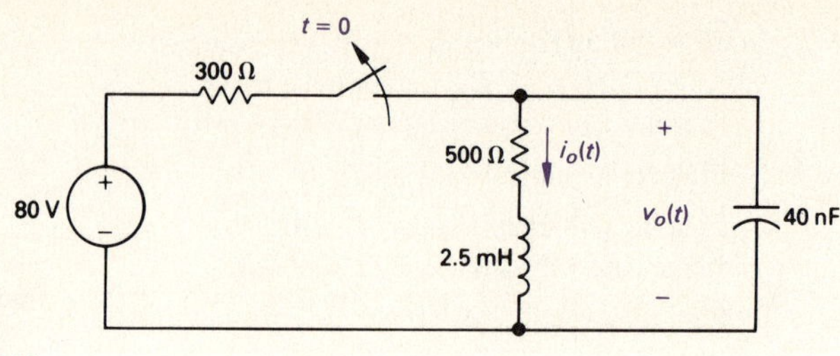

FIGURE P8.40

8.41 The initial energy stored in the circuit in Fig. P8.41 is zero. Find $v_o(t)$ for $t \geq 0$.

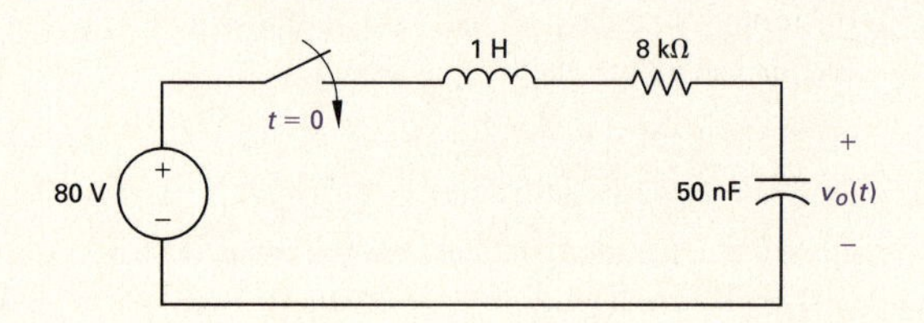

FIGURE P8.41

8.42 The two switches in the circuit seen in Fig. P8.42 operate synchronously. When switch 1 is in position a, switch 2 is closed. When switch 1 is in position b, switch 2 is open. Switch 1 has been in position a for a long time. At $t = 0$ it moves instantaneously to position b. Find $v_C(t)$ for $t \geq 0$.

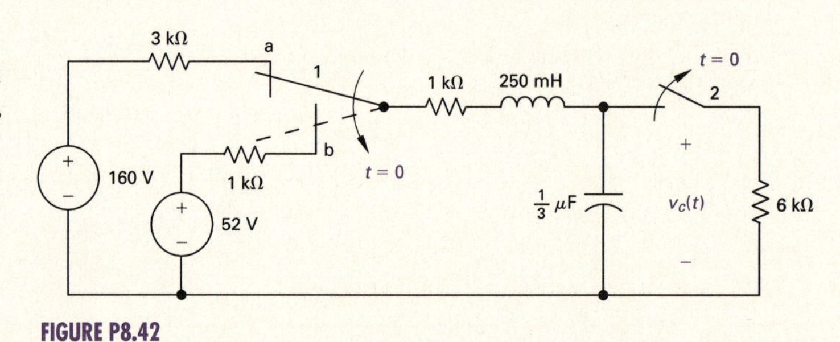

FIGURE P8.42

8.43 The circuit shown in Fig. P8.43 has been in operation for a long time. At $t = 0$ the voltage suddenly jumps to 250 V. Find $v_o(t)$ for $t \geq 0$.

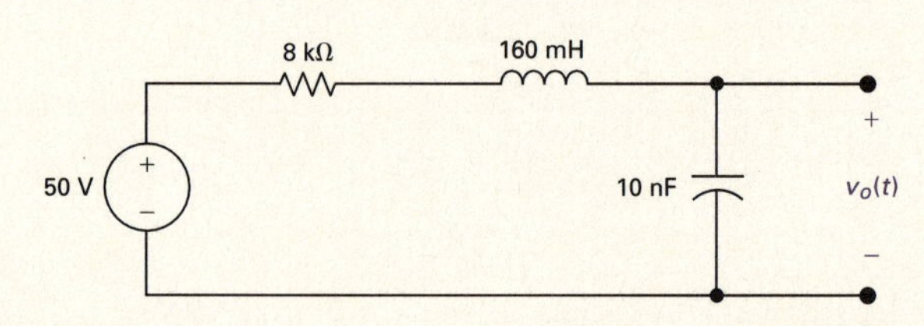

FIGURE P8.43

8.44 The switch in the circuit of Fig. P8.44 has been in position a for a long time. At $t = 0$ the switch moves instantaneously to position b.

a) Find $v_o(0^+)$.

b) Find $dv_o(0^+)/dt$.

c) Find $v_o(t)$ for $t \geq 0$.

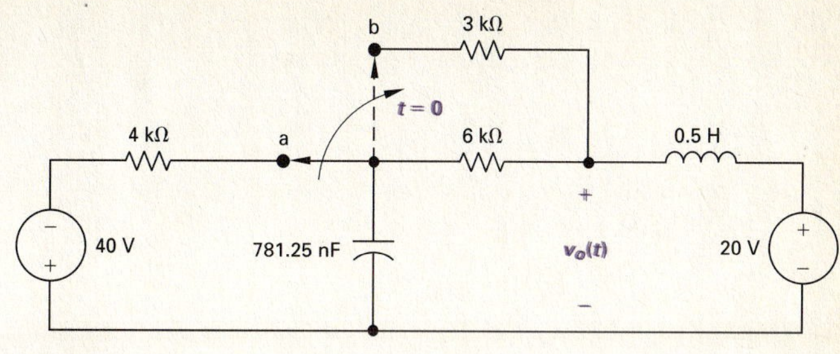

FIGURE P8.44

8.45 Assume that the capacitor voltage in the circuit of Fig. 8.11 is underdamped. Also assume that no energy is stored in the circuit elements when the switch is closed.

a) Show that $dv_C/dt = (\omega_0^2/\omega_d)Ve^{-\alpha t}\sin\omega_d t$.

b) Show that $dv_C/dt = 0$ when $t = n\pi/\omega_d$, where $n = 0, 1, 2, \ldots$.

c) Let $t_n = n\pi/\omega_d$, and show that $v_C(t_n) = V - V(-1)^n e^{-\alpha n\pi/\omega_d}$.

d) Show that

$$\alpha = \frac{1}{T_d}\ ln\ \frac{v_C(t_1) - V}{v_C(t_3) - V},$$

where $T_d = t_3 - t_1$.

8.46 The voltage across a 0.1 μF capacitor in the circuit of Fig. 8.11 is described as follows: After the switch has been closed for several seconds, the voltage is constant at 100 V. The first time the voltage exceeds 100 V, it reaches a peak of 163.84 V. This occurs $\pi/7$ ms after the switch has been closed. The second time the voltage exceeds 100 V, it reaches a peak of 126.02 V. This second peak occurs $3\pi/7$ ms after the switch has been closed. At the time when the switch is closed, there is no energy stored in either the capacitor or the inductor. Find the numerical values of R and L. (*Hint:* Work Problem 8.45 first.)

8.47 The switch in the circuit shown in Fig. P8.47 has been closed for a long time before it is opened at $t = 0$. Assume that the circuit parameters are such that the response is underdamped.

a) Derive the expression for $v_o(t)$ as a function of v_g, α, ω_d, C, and R for $t \geq 0$.

b) Derive the expression for the value of t when the magnitude of v_o is maximum.

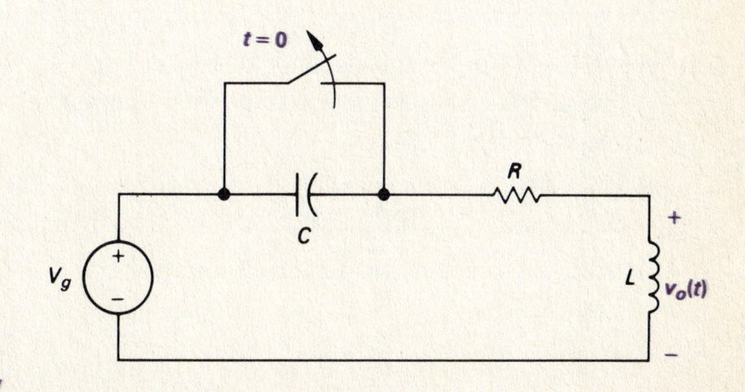

FIGURE P8.47

8.48 The circuit parameters in the circuit of Fig. P8.47 are $R = 4800\ \Omega$, $L = 64$ mH, $C = 4$ nF, and $v_g = -72$ V.

a) Express $v_o(t)$ numerically for $t \geq 0$.

b) How many microseconds after the switch opens is the inductor voltage maximum?

c) What is the maximum value of the inductor voltage?

d) Repeat parts (a), (b), and (c), with R reduced to $480\ \Omega$.

8.49 The voltage signal of Fig. P8.49(a) is applied to the cascaded integrating amplifiers shown in Fig. P8.49(b). There is no energy stored in the capacitors at the instant the signal is applied.

a) Derive the numerical expressions for $v_o(t)$ and $v_{o1}(t)$ for the time intervals $0 \leq t \leq 0.5$ s and $0.5\ \text{s} \leq t \leq t_{\text{sat}}$.

b) Compute the value of t_{sat}.

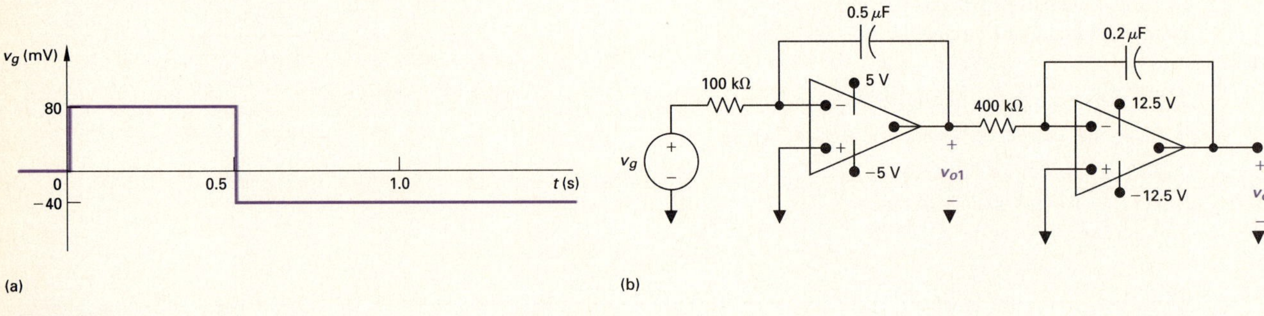

FIGURE P8.49

8.50 The circuit in Fig. P8.49(b) is modified by adding a $1\ \text{M}\Omega$ resistor in parallel with the $0.5\ \mu$F capacitor and a $5\ \text{M}\Omega$ resistor in parallel with the $0.2\ \mu$F capacitor. As in Problem 8.49, there is no energy stored in the capacitors at the time the signal is applied. Derive the numerical expressions for $v_o(t)$ and $v_{o1}(t)$ for the time intervals $0 \leq t \leq 0.5$ s and $0.5\ \text{s} \leq t \leq \infty$.

8.51 a) Derive the differential equation that relates the output voltage to the input voltage for the circuit shown in Fig. P8.51.

b) Compare the result with Eq. (8.75) when $R_1C_1 = R_2C_2 = RC$ in Fig. 8.18.

c) What is the advantage of the circuit shown in Fig. P8.51?

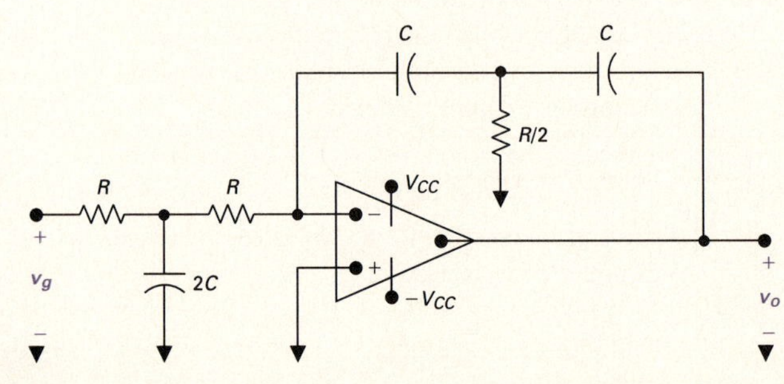

FIGURE P8.51

8.52 We now wish to illustrate how several op amp circuits can be interconnected to solve a differential equation.

a) Derive the differential equation for the spring-mass system shown in Fig. P8.52(a). Assume that the force exerted by the spring is directly proportional to the spring displacement, that the mass is constant, and that the frictional force is directly proportional to the velocity of the moving mass.

b) Rewrite the differential equation derived in part (a) so that the highest order derivative is expressed as a function of all the other terms in the equation. Now assume that a voltage equal to d^2x/dt^2 is available and by successive integrations generates dx/dt and x. We can synthesize the coefficients in the equations by scaling amplifiers, and we can combine the terms required to generate d^2x/dt^2 by using a summing amplifier. With these ideas in mind, analyze the interconnection shown in Fig. P8.52(b). In particular, describe the purpose of each shaded area in the circuit and describe the signal at the points labeled B, C, D, E, and F, assuming the signal at A represents d^2x/dt^2. Also discuss the parameters R; R_1, C_1; R_2, C_2; R_3, R_4; R_5, R_6; and R_7, R_8 in terms of the coefficients in the differential equation.

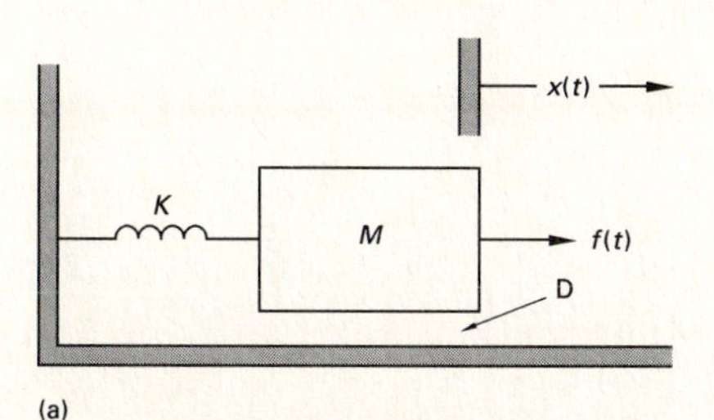

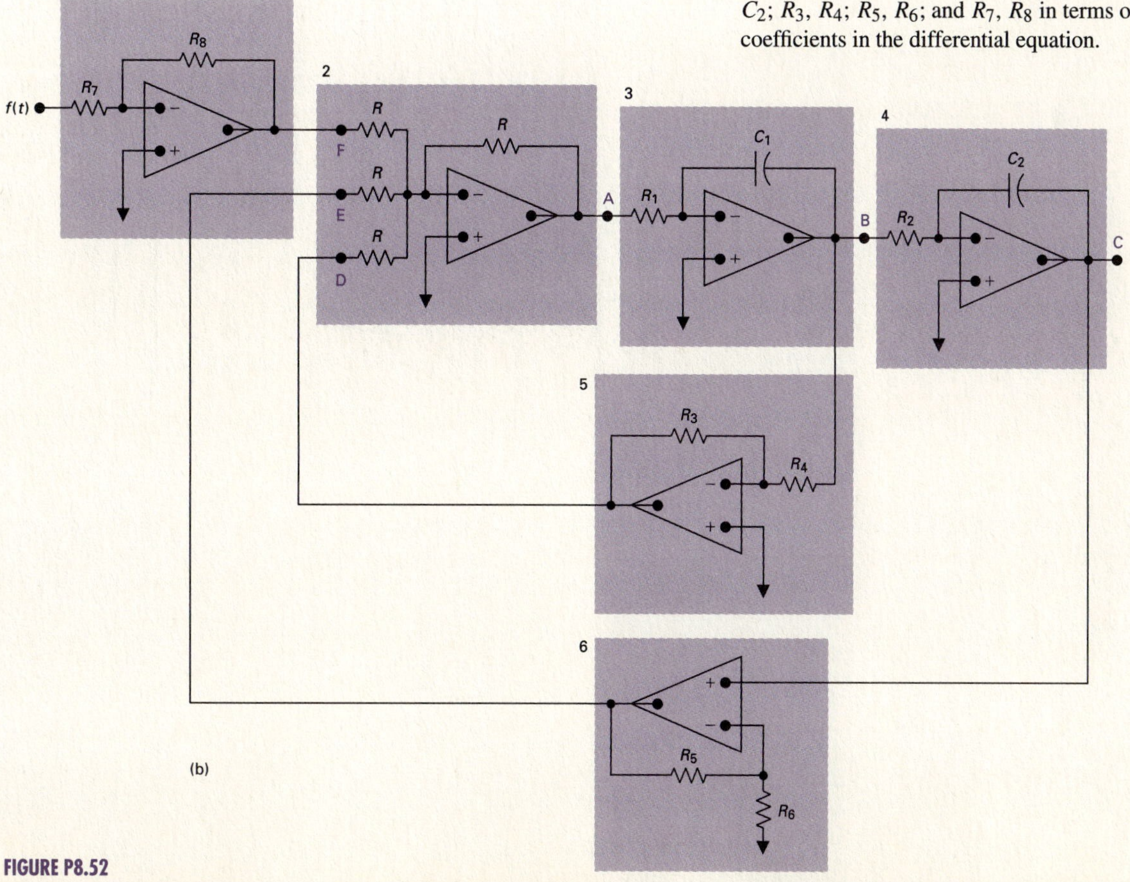

FIGURE P8.52

8.53 Assume the underdamped voltage response of the circuit in Fig. 8.1 is written as

$$v(t) = (A_1 + A_2)e^{-\alpha t} \cos \omega_d t$$

$$+j(A_1 - A_2)e^{-\alpha t} \sin \omega_d t$$

The initial value of the inductor current is I_0, and the initial value of the capacitor voltage is V_0. Show that A_2 is the conjugate of A_1. (*Hint:* Use the same process as outlined in the text to find A_1 and A_2.)

8.54 Show that the results obtained from Problem 8.53—that is, the expressions for A_1 and A_2—are consistent with Eqs. (8.30) and (8.31) in the text.

9 Sinusoidal Steady-State Analysis

Chapter Contents

Thus far, we have focused on circuits with constant sources; we are now ready to consider circuits energized by time-varying voltage or current sources. In particular, we are interested in sources in which the value of the voltage or current varies sinusoidally. **Sinusoidal sources** and their effect on circuit behavior form an important area of study for several reasons. First, the generation, transmission, distribution, and consumption of electric energy occur under essentially sinusoidal steady-state conditions. Second, an understanding of sinusoidal behavior makes it possible to predict the behavior of circuits with nonsinusoidal sources. Third, steady-state sinusoidal behavior often simplifies the design of electrical systems. Thus a designer can spell out specifications in terms of a desired steady-state sinusoidal response and design the circuit or system to meet those characteristics. If the device satisfies the specifications, the designer knows that the circuit will respond satisfactorily to nonsinusoidal inputs.

The subsequent chapters of this book are largely based on a thorough understanding of the techniques needed to analyze circuits driven by sinusoidal sources. Fortunately, the circuit analysis and simplification techniques first introduced in Chapters 1 through 4 work for circuits with sinusoidal as well as dc sources, so some of the material in this chapter will be very familiar to you. The challenges in first approaching sinusoidal analysis include developing the appropriate modeling equations and working in the mathematical realm of complex numbers.

9.1 THE SINUSOIDAL SOURCE

A **sinusoidal voltage source** (independent or dependent) produces a voltage that varies sinusoidally with time. A **sinusoidal current source** (independent or dependent) produces a current that varies sinusoidally with time. In reviewing the sinusoidal function, we use a voltage source, but our observations also apply to current sources.

We can express a sinusoidally varying function with either the sine function or the cosine function. Although either works equally well, we cannot use both functional forms simultaneously. We will use the cosine function throughout our discussion. Hence we write a sinusoidally varying voltage as

$$v = V_m \cos(\omega t + \phi). \tag{9.1}$$

To aid discussion of the parameters in Eq. (9.1), we show the voltage versus time plot in Fig. 9.1. Note that the sinusoidal function repeats at regular intervals. Such a function is called periodic. A **cycle** of the function is any portion beginning at a specific amplitude (say $-V_m$, as in Fig. 9.1) and returning to that same amplitude. One of the parameters of interest, therefore, is the length of time required for the sinusoidal function to pass through a cycle. This time is referred to as the **period** of the function and is denoted T. The reciprocal of T gives the number of cycles per second, or the frequency, of the sine function and is denoted f, or

$$f = \frac{1}{T}. \tag{9.2}$$

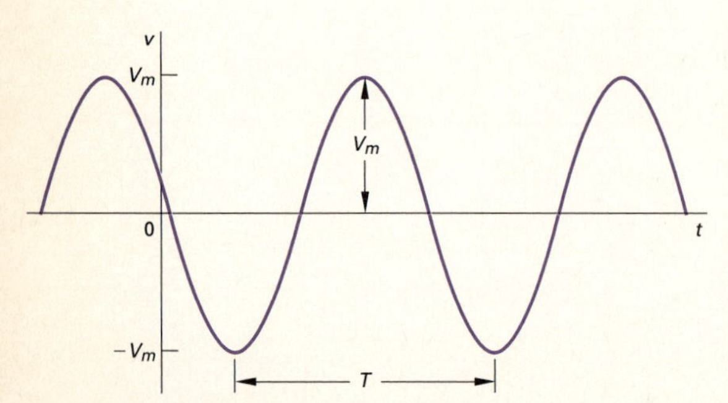

FIGURE 9.1 A sinusoidal voltage.

A cycle per second is referred to as a hertz, abbreviated Hz. (The term *cycles per second* rarely is used in contemporary technical literature.) The coefficient of t in Eq. (9.1) contains the numerical value of T or f. Omega (ω) represents the angular frequency of the sinusoidal function, or

$$\omega = 2\pi f = 2\pi/T \quad \text{(radians/second).} \tag{9.3}$$

Equation 9.3 is based on the fact that the cosine (or sine) function passes through a complete set of values each time its argument, ωt, passes through 2π rad (360°). From Eq. (9.3), note that, whenever t is an integral multiple of T, the argument ωt increases by an integral multiple of 2π rad.

The coefficient V_m gives the maximum amplitude of the sinusoidal voltage. Because ± 1 bounds the cosine function, $\pm V_m$ bounds the amplitude. Figure 9.1 shows these characteristics.

The angle ϕ in Eq. (9.1) is known as the **phase angle** of the sinusoidal voltage. It determines the value of the sinusoidal function at $t = 0$; therefore it fixes the point on the periodic wave at which we start measuring time. Changing the phase angle ϕ shifts the sinusoidal function along the time axis but has no effect on either the amplitude (V_m) or the angular frequency (ω). Note, for example, that reducing ϕ to zero shifts the sinusoidal function shown in Fig. 9.1 ϕ/ω time units to the right, as shown in Fig. 9.2. Note also that if ϕ is positive, the sinusoidal function shifts to the left, whereas if ϕ is negative, the function shifts to the right. (See Problem 9.2.)

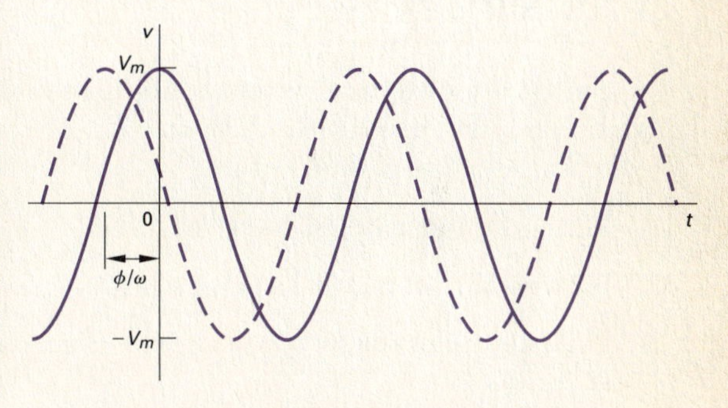

FIGURE 9.2 The sinusoidal voltage from Fig. 9.1 shifted to the right when $\phi = 0$.

A comment with regard to the phase angle is in order: ωt and ϕ must carry the same units because they are added together in the argument of the sinusoidal function. With ωt expressed in radians, you would expect ϕ to be also. However, ϕ normally is given in degrees, and ωt is converted from radians to degrees before the two quantities are added. We continue this bias toward degrees by expressing the phase angle in degrees. Recall from your studies of trigonometry that the conversion from radians to degrees is given by

$$\text{(number of degrees)} = \frac{180^\circ}{\pi}\text{(number of radians).} \tag{9.4}$$

Another important characteristic of the sinusoidal voltage (or current) is its rms value. The **rms value** of a periodic function is defined as the square root of the mean value of the squared function. Hence, if $v = V_m \cos(\omega t + \phi)$, the rms value of v is

$$V_{\text{rms}} = \sqrt{\frac{1}{T}\int_{t_0}^{t_0+T} V_m^2 \cos^2(\omega t + \phi)dt}. \tag{9.5}$$

Note from Eq. (9.5) that we obtain the mean value of the squared voltage by integrating v^2 over one period (that is, from t_0 to $t_0 + T$)

and then dividing by the range of integration, T. Note further that the starting point for the integration t_0 is arbitrary.

The quantity under the radical sign in Eq. (9.5) reduces to $V_m^2/2$. (See Problem 9.7.) Hence the rms value of v is

$$V_{\text{rms}} = \frac{V_m}{\sqrt{2}}. \tag{9.6}$$

The rms value of the sinusoidal voltage depends only on the maximum amplitude of v, namely, V_m. The rms value is not a function of either the frequency or the phase angle. We stress the importance of the rms value as it relates to power calculations in Chapter 10 (see Section 10.3).

Use a computer tool to plot sinusoids and explore their characteristics.

Thus, we can completely describe a specific sinusoidal signal if we know its frequency, phase angle, and amplitude (either the maximum or the rms value). Examples 9.1, 9.2, and 9.3 illustrate these basic properties of the sinusoidal function. In Example 9.4 we calculate the rms value of a periodic function, and in so doing we clarify the meaning of *root mean square*.

EXAMPLE 9.1

A sinusoidal current has a maximum amplitude of 20 A. The current passes through one complete cycle in 1 ms. The magnitude of the current at zero time is 10 A.

a) What is the frequency of the current in hertz?

b) What is the frequency in radians per second?

c) Write the expression for $i(t)$ using the cosine function. Express ϕ in degrees.

d) What is the rms value of the current?

SOLUTION

a) From the statement of the problem, $T = 1$ ms; hence $f = 1/T = 1000$ Hz.

b) $\omega = 2\pi f = 2000\pi$ rad/s.

c) We have $i(t) = I_m \cos(\omega t + \phi) = 20 \cos(2000\pi t + \phi)$, but $i(0) = 10$ A. Therefore $10 = 20 \cos \phi$ and $\phi = 60°$. Thus the expression for $i(t)$ becomes

$$i(t) = 20 \cos(2000\pi t + 60°).$$

d) From the derivation of Eq. (9.6), the rms value of a sinusoidal current is $I_m/\sqrt{2}$. Therefore the rms value is $20/\sqrt{2}$, or 14.14 A.

EXAMPLE 9.2

A sinusoidal voltage is given by the expression $v = 300 \cos(120\pi t + 30°)$.

a) What is the period of the voltage in milliseconds?

b) What is the frequency in hertz?

c) What is the magnitude of v at $t = 2.778$ ms?

d) What is the rms value of v?

SOLUTION

a) From the expression for v, $\omega = 120\pi$ rad/s. Because $\omega = 2\pi/T$, $T = 2\pi/\omega = \frac{1}{60}$ s, or 16.667 ms.

b) The frequency is $1/T$, or 60 Hz.

c) From part (a), $\omega = 2\pi/16.667$; thus, at $t = 2.778$ ms, ωt is nearly 1.047 rad, or 60°. Therefore $v(2.778 \text{ ms}) = 300 \cos(60° + 30°) = 0$ V.

d) $V_{\text{rms}} = 300/\sqrt{2} = 212.13$ V.

EXAMPLE 9.3

We can translate the sine function to the cosine function by subtracting 90° ($\pi/2$ rad) from the argument of the sine function.

a) Verify this translation by showing that

$$\sin(\omega t + \theta) = \cos(\omega t + \theta - 90°).$$

b) Use the result in (a) to express $\sin(\omega t + 30°)$ as a cosine function.

SOLUTION

a) Verification involves direct application of the trigonometric identity

$$\cos(\alpha - \beta) = \cos\alpha \cos\beta + \sin\alpha \sin\beta.$$

We let $\alpha = \omega t + \theta$ and $\beta = 90^\circ$. As $\cos 90^\circ = 0$ and $\sin 90^\circ = 1$, we have

$$\cos(\alpha - \beta) = \sin\alpha = \sin(\omega t + \theta) = \cos(\omega t + \theta - 90^\circ).$$

b) From (a) we have

$$\sin(\omega t + 30^\circ) = \cos(\omega t + 30^\circ - 90^\circ) = \cos(\omega t - 60^\circ).$$

EXAMPLE 9.4

Calculate the rms value of the periodic triangular current shown in Fig. 9.3. Express your answer in terms of the peak current I_p.

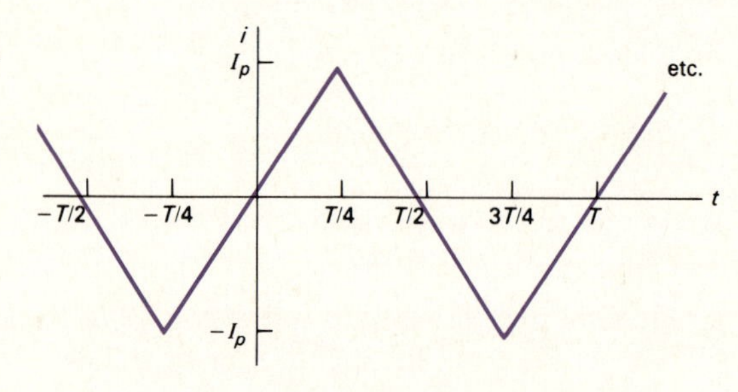

FIGURE 9.3 Periodic triangular current.

SOLUTION

From Eq. (9.5), the rms value of i is

$$I_{\text{rms}} = \sqrt{\frac{1}{T}\int_{t_0}^{t_0+T} i^2 dt}.$$

Interpreting the integral under the radical sign as the area under the squared function for an interval of one period is helpful for finding the rms value. The squared function with the area between 0 and T shaded is shown in Fig. 9.4, which also indicates that for this particular function, the area under the squared current for an interval of one period is equal to four times the area under the squared current for the interval 0 to $T/4$ seconds; that is,

$$\int_{t_0}^{t_0+T} i^2 dt = 4\int_0^{T/4} i^2 dt.$$

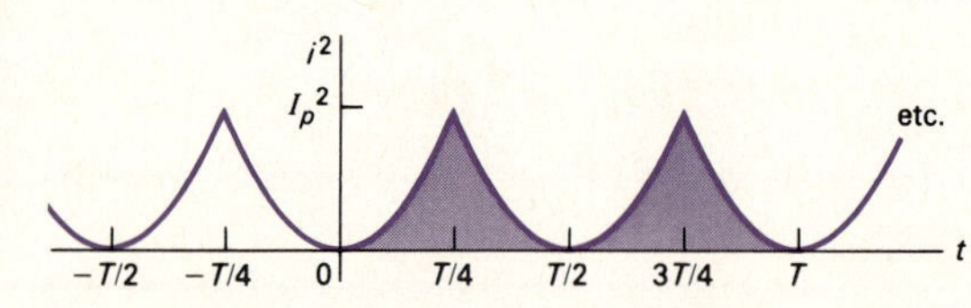

FIGURE 9.4 i^2 versus t.

The analytical expression for i in the interval 0 to $T/4$ is

$$i = \frac{4I_p}{T}t, \quad 0 < t < T/4.$$

The area under the squared function for one period is

$$\int_{t_0}^{t_0+T} i^2 dt = 4\int_0^{T/4} \frac{16 I_p^2}{T^2} t^2 dt = \frac{I_p^2 T}{3}.$$

The mean, or average, value of the function is simply the area for one period divided by the period. Thus

$$i_{\text{mean}} = \frac{1}{T}\frac{I_p^2 T}{3} = \frac{1}{3}I_p^2.$$

The rms value of the current is the square root of this mean value. Hence

$$I_{\text{rms}} = \frac{I_p}{\sqrt{3}}.$$

DRILL EXERCISES

9.1 A sinusoidal voltage is given by the expression

$$v = 40 \cos(2513.27t + 36.87°).$$

Find (a) f in hertz; (b) T in milliseconds; (c) V_m; (d) $v(0)$; (e) ϕ in degrees and radians; (f) the smallest positive value of t at which $v = 0$; and (g) the smallest positive value of t at which $dv/dt = 0$.

ANSWER: (a) 400 Hz; (b) 2.5 ms; (c) 40 V; (d) 32 V; (e) 36.87°, or 0.6435 rad; (f) 368.96 μs; (g) 993.96 μs.

9.2 Find the rms value of the half-wave rectified sinusoidal voltage shown.

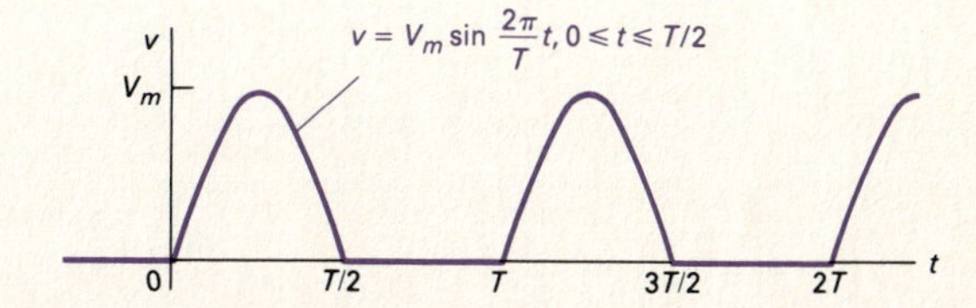

ANSWER: $V_{\text{rms}} = V_m/2$.

9.2 THE SINUSOIDAL RESPONSE

Before focusing on the steady-state response to sinusoidal sources, let's consider the problem in broader terms, that is, in terms of the total response. Such an overview will help you keep the steady-state solution in perspective. The circuit shown in Fig. 9.5 describes the general nature of the problem. There, v_s is a sinusoidal voltage, or

$$v_s = V_m \cos(\omega t + \phi). \tag{9.7}$$

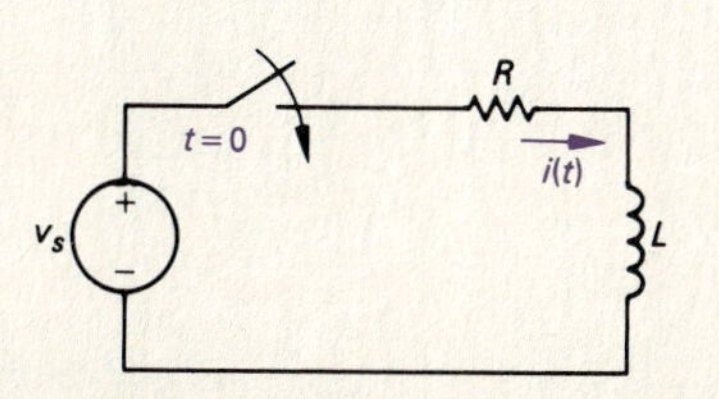

FIGURE 9.5 An RL circuit excited by a sinusoidal voltage source.

For convenience, we assume the initial current in the circuit to be zero and measure time from the moment the switch is closed. The task is to derive the expression for $i(t)$ when $t \geq 0$. It is similar to finding the step response of an RL circuit, as in Chapter 7. The only difference is that the voltage source is now a time-varying sinusoidal voltage rather than a constant, or dc, voltage. Direct application of Kirchhoff's voltage law to the circuit shown in Fig. 9.5 leads to the ordinary differential equation

$$L\frac{di}{dt} + Ri = V_m \cos(\omega t + \phi), \tag{9.8}$$

the formal solution of which is discussed in an introductory course in differential equations. We ask those of you who have not yet studied differential equations to accept that the solution for i is

$$\begin{aligned} i = & \frac{-V_m}{\sqrt{R^2 + \omega^2 L^2}} \cos(\phi - \theta) e^{-(R/L)t} \\ & + \frac{V_m}{\sqrt{R^2 + \omega^2 L^2}} \cos(\omega t + \phi - \theta), \end{aligned} \tag{9.9}$$

where θ is defined as the angle whose tangent is $\omega L/R$. Thus we can easily determine θ for a circuit driven by a sinusoidal source of known frequency.

We can check the validity of Eq. (9.9) by determining that it satisfies Eq. (9.8) for all values of $t \geq 0$, an exercise left for your exploration in Problem 9.5.

The first term on the right-hand side of Eq. (9.9) is referred to as the **transient component** of the current because it becomes infinitesimal as time elapses. The second term on the right-hand side is known as the **steady-state component** of the solution. It exists as long as the switch remains closed and the source continues to supply the sinusoidal voltage. In this chapter we develop a technique for calculating the steady-state response directly, thus avoiding the problem of solving the differential equation. However, in doing so, we forfeit obtaining either the transient component or the total response, which is the sum of the transient and steady-state components.

We now focus on the steady-state portion of Eq. (9.9). It is important to remember the following characteristics of the steady-state solution:

1. The steady-state solution is a sinusoidal function.
2. The frequency of the response signal is identical to the frequency of the source signal. This condition is always true in a linear circuit when the circuit parameters, R, L, and C, are constant. (If frequencies in the response signals are not present in the source signals, there is a nonlinear element in the circuit.)
3. The maximum amplitude of the steady-state response, in general, differs from the maximum amplitude of the source. For the circuit being discussed, the maximum amplitude of the response signal is $V_m/\sqrt{R^2 + \omega^2 L^2}$, and the maximum amplitude of the signal source is V_m.

4. The phase angle of the response signal, in general, differs from the phase angle of the source. For the circuit being discussed, the phase angle of the current is $\phi - \theta$, and that of the voltage source is ϕ.

These characteristics are worth remembering because they help you understand the motivation for the phasor method, which we begin developing in Section 9.3. In particular, note that once the decision has been made to find only the steady-state response, the task is reduced to finding the maximum amplitude and phase angle of the response signal. The waveform and frequency of the response are already known.

DRILL EXERCISE

9.3 The voltage applied to the circuit shown in Fig. 9.5 at $t = 0$ is $100 \cos(400t + 60^\circ)$. The circuit resistance is 40 Ω, and the initial current in the 75 mH inductor is zero.

a) Find $i(t)$ for $t \geq 0$.

b) Write the expressions for the transient and steady-state components of $i(t)$.

c) Find the numerical value of i after the switch has been closed for 1.875 ms.

d) What are the maximum amplitude, frequency (in radians per second), and phase angle of the steady-state current?

e) By how many degrees are the voltage and the steady-state current out of phase?

ANSWER: (a) $-1.84e^{-533.33t} + 2\cos(400t + 23.13^\circ)$ A; (b) $-1.84e^{-533.33t}$ A, $2\cos(400t + 23.13^\circ)$ A; (c) 133.61 mA; (d) 2 A, 400 rad/s, 23.13°; (e) 36.87°.

9.3 THE PHASOR

The **phasor** is a complex number that carries the amplitude and phase angle information of a sinusoidal function.[1] The phasor concept is rooted in Euler's identity, which relates the exponential function to the trigonometric function:

$$e^{\pm j\theta} = \cos\theta \pm j \sin\theta. \tag{9.10}$$

Equation (9.10) is important here because it gives us another way of expressing the cosine and sine functions. We can think of the cosine

[1]If you feel a bit uneasy about complex numbers, pause here and peruse Appendix B.

function as the real part of the exponential function, and the sine function as the imaginary part of the exponential function; that is,

$$\cos\theta = \Re\{e^{j\theta}\} \tag{9.11}$$

and

$$\sin\theta = \Im\{e^{j\theta}\}, \tag{9.12}$$

where $\Re$ means "the real part of" and $\Im$ means "the imaginary part of."

Because we have already chosen to use the cosine function in analyzing the sinusoidal steady state (see Section 9.1), we can apply Eq. (9.11) directly. In particular, we write the sinusoidal voltage function given by Eq. (9.1) in the form suggested by Eq. (9.11):

$$\begin{aligned} v &= V_m \cos(\omega t + \phi) \\ &= V_m \Re\{e^{j(\omega t+\phi)}\} \\ &= V_m \Re\{e^{j\omega t} e^{j\phi}\}. \end{aligned} \tag{9.13}$$

We can move the coefficient V_m inside the argument of the real part of the function without altering the result. We can also reverse the order of the two exponential functions inside the argument and write Eq. (9.13) as

$$v = \Re\{V_m e^{j\phi} e^{j\omega t}\}. \tag{9.14}$$

In Eq. (9.14) note that the quantity $V_m e^{j\phi}$ is a complex number that carries the amplitude and phase angle of the given sinusoidal function. This complex number is by definition the **phasor representation,** or **phasor transform,** of the given sinusoidal function. Thus

$$\mathbf{V} = V_m e^{j\phi} = \mathcal{P}\{V_m \cos(\omega t + \phi)\}, \tag{9.15}$$

where the notation $\mathcal{P}\{V_m \cos(\omega t + \phi)\}$ is read "the phasor transform of $V_m \cos(\omega t + \phi)$." Thus the phasor transform transfers the sinusoidal function from the time domain to the complex-number domain, which is also called the **frequency domain,** since the response depends, in general, on ω. As in Eq. (9.15), throughout this book we represent a phasor quantity by using a boldface letter.

Equation (9.15) is the polar form of a phasor, but we also can express a phasor in rectangular form. Thus we rewrite Eq. (9.15) as

$$\mathbf{V} = V_m \cos\phi + j V_m \sin\phi. \tag{9.16}$$

Both polar and rectangular forms are useful in circuit applications of the phasor concept.

One additional comment regarding Eq. (9.15) is in order. The frequent occurrence of the exponential function $e^{j\phi}$ has led to an abbreviation that lends itself to text material. This abbreviation is the angle notation

$$1\angle\phi \equiv 1e^{j\phi}.$$

We use this notation extensively in the material that follows.

INVERSE PHASOR TRANSFORM

So far we have emphasized moving from the sinusoidal function to its phasor transform. However, we may also reverse the process. That is, for a phasor we may write the expression for the sinusoidal function. Thus for $\mathbf{V} = 100\angle -26^\circ$, the expression for v is $100 \cos(\omega t - 26^\circ)$ because we have decided to use the cosine function for all sinusoids. Observe that we cannot deduce the value of ω from the phasor. The phasor carries only amplitude and phase information. The step of going from the phasor transform to the time-domain expression is referred to as *finding the inverse phasor transform* and is formalized by the equation

$$\mathcal{P}^{-1}\{V_m e^{j\phi}\} = \Re\{V_m e^{j\phi} e^{j\omega t}\}, \quad \textbf{(9.17)}$$

where the notation $\mathcal{P}^{-1}\{V_m e^{j\phi}\}$ is read as "the inverse phasor transform of $V_m e^{j\phi}$." Equation (9.17) indicates that to find the inverse phasor transform, we multiply the phasor by $e^{j\omega t}$ and then extract the real part of the product.

The phasor transform is useful in circuit analysis because it reduces the task of finding the maximum amplitude and phase angle of the steady-state sinusoidal response to the algebra of complex numbers. The following observations verify this conclusion:

1. The transient component vanishes as time elapses, so the steady-state component of the solution must also satisfy the differential equation. (See Problem 9.5[b].)

2. In a linear circuit driven by sinusoidal sources, the steady-state response also is sinusoidal, and the frequency of the sinusoidal response is the same as the frequency of the sinusoidal source.

3. Using the notation introduced in Eq. (9.11), we can postulate that the steady-state solution is of the form $\Re\{Ae^{j\beta}e^{j\omega t}\}$, where A is the maximum amplitude of the response and β is the phase angle of the response.

4. When we substitute the postulated steady-state solution into the differential equation, the exponential term $e^{j\omega t}$ cancels out, leaving the solution for A and β in the domain of complex numbers.

We illustrate these observations with the circuit shown in Fig. 9.5. We know that the steady-state solution for the current i is of the form

$$i_{ss}(t) = \Re\{I_m e^{j\beta} e^{j\omega t}\}, \quad \textbf{(9.18)}$$

where the subscript "ss" emphasizes that we are dealing with the steady-state solution. When we substitute Eq. (9.18) into Eq. (9.8), we generate the expression

$$\Re\{j\omega L I_m e^{j\beta} e^{j\omega t}\} + \Re\{R I_m e^{j\beta} e^{j\omega t}\} = \Re\{V_m e^{j\phi} e^{j\omega t}\}. \quad \textbf{(9.19)}$$

In deriving Eq. (9.19) we recognized that both differentiation and multiplication by a constant can be taken inside the real part of an operation. We also rewrote the right-hand side of Eq. (9.8), using the notation of Eq. (9.11). From the algebra of complex numbers, we know that the sum of the real parts is the same as the real part of the sum. Therefore we may reduce the left-hand side of Eq. (9.19) to a single term:

$$\Re\{(j\omega L + R) I_m e^{j\beta} e^{j\omega t}\} = \Re\{V_m e^{j\phi} e^{j\omega t}\}. \quad \textbf{(9.20)}$$

Recall that our decision to use the cosine function in analyzing the response of a circuit in the sinusoidal steady state results in the use of the $\Re$ operator in deriving Eq. (9.20). If instead we had chosen to use the sine function in our sinusoidal steady-state analysis, we would have applied Eq. (9.12) directly, in place of Eq. (9.11), and the result would be Eq. (9.21):

$$\Im\{(j\omega L + R) I_m e^{j\beta} e^{j\omega t}\} = \Im\{V_m e^{j\phi} e^{j\omega t}\}. \quad \textbf{(9.21)}$$

Note that the complex quantities on either side of Eq. (9.21) are identical to those on either side of Eq. (9.20). When both the real and imaginary parts of two complex quantities are equal, then the complex quantities are themselves equal. Therefore, from Eqs. (9.20) and (9.21),

$$(j\omega L + R) I_m e^{j\beta} = V_m e^{j\phi},$$

or

$$I_m e^{j\beta} = \frac{V_m e^{j\phi}}{R + j\omega L}. \quad \textbf{(9.22)}$$

Note that $e^{j\omega t}$ has been eliminated from the determination of the amplitude (I_m) and phase angle (β) of the response. Thus for this circuit the task of finding I_m and β involves the algebraic manipulation of the complex quantities $V_m e^{j\phi}$ and $R + j\omega L$. Note that we encountered both polar and rectangular forms.

An important warning is in order: the phasor transform, along with the inverse phasor transform, allows you to go back and forth between the time domain and the frequency domain. Therefore, when you obtain a solution, you are either in the time domain or the frequency domain. You cannot be in both domains simultaneously. Any solution which contains a mixture of time domain and phasor domain nomenclature is nonsensical.

Use the computer to transition between the time and frequency domains and make connections between time-domain and frequency-domain behavior.

The phasor transform is also useful in circuit analysis because it applies directly to the sum of sinusoidal functions. Circuit analysis involves summing currents and voltages, so the importance of this observation is obvious. We can formalize this property as follows: If

$$v = v_1 + v_2 + \cdots + v_n \quad \textbf{(9.23)}$$

where all the voltages on the right-hand side are sinusoidal voltages of the same frequency, then

$$\mathbf{V} = \mathbf{V}_1 + \mathbf{V}_2 + \cdots + \mathbf{V}_n. \quad \textbf{(9.24)}$$

Thus the phasor representation is the sum of the phasors of the individual terms. We discuss the development of Eq. (9.24) in Section 9.5.

Before applying the phasor transform to circuit analysis, we illustrate its usefulness in solving a problem with which you are already familiar: adding sinusoidal functions via trigonometric identities. Example 9.5 shows how the phasor transform greatly simplifies this type of problem.

EXAMPLE 9.5

If $y_1 = 20\cos(\omega t - 30°)$ and $y_2 = 40\cos(\omega t + 60°)$, express $y = y_1 + y_2$ as a single sinusoidal function.

a) Solve by using trigonometric identities.

b) Solve by using the phasor concept.

SOLUTION

a) First we expand both y_1 and y_2, using the cosine of the sum of two angles, to get

$$y_1 = 20\cos\omega t\cos 30° + 20\sin\omega t\sin 30°;$$
$$y_2 = 40\cos\omega t\cos 60° - 40\sin\omega t\sin 60°.$$

Adding y_1 and y_2, we obtain

$$y = (20\cos 30 + 40\cos 60)\cos\omega t + (20\sin 30 - 40\sin 60)\sin\omega t$$
$$= 37.32\cos\omega t - 24.64\sin\omega t.$$

To combine these two terms we treat the coefficients of the cosine and sine as sides of a right triangle (Fig. 9.6) and then multiply and

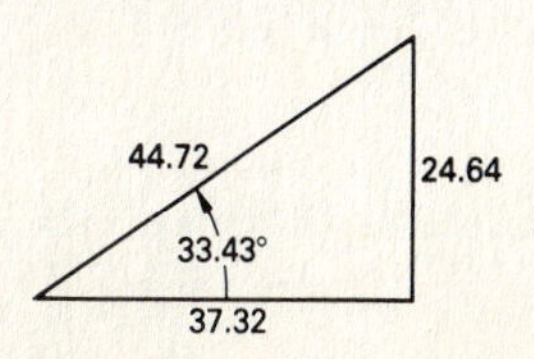

FIGURE 9.6 A right triangle used in the solution for y.

divide the right-hand side by the hypotenuse. Our expression for y becomes

$$
\begin{aligned}
y &= 44.72\left(\frac{37.32}{44.72}\cos\omega t - \frac{24.64}{44.72}\sin\omega t\right) \\
&= 44.72(\cos 33.43^\circ \cos\omega t - \sin 33.43^\circ \sin\omega t).
\end{aligned}
$$

Again, we invoke the identity involving the cosine of the sum of two angles and write

$$y = 44.72\cos(\omega t + 33.43^\circ).$$

b) We can solve the problem by using phasors as follows: Because

$$y = y_1 + y_2,$$

then, from Eq. (9.24),

$$
\begin{aligned}
\mathbf{Y} &= \mathbf{Y}_1 + \mathbf{Y}_2 \\
&= 20\underline{/-30^\circ} + 40\underline{/60^\circ} \\
&= (17.32 - j10) + (20 + j34.64) \\
&= 37.32 + j24.64 \\
&= 44.72\underline{/33.43^\circ}.
\end{aligned}
$$

Once we know the phasor $\mathbf{Y}$, we can write the corresponding trigonometric function for y by taking the inverse phasor transform:

$$
\begin{aligned}
y &= \mathcal{P}^{-1}\{44.72e^{j33.43}\} = \Re\{44.72e^{j33.43}e^{j\omega t}\} \\
&= 44.72\cos(\omega t + 33.43^\circ).
\end{aligned}
$$

The superiority of the phasor approach to adding sinusoidal functions should be apparent. Note that it requires the ability to move back and forth between the polar and rectangular forms of complex numbers.

Use a computer tool to transition between the two forms of a complex number.

DRILL EXERCISES

9.4 Find the phasor transform of each trigonometric function:

a) $v = 170\cos(377t - 40^\circ)$ V;

b) $i = 10\sin(1000t + 20^\circ)$ A;

c) $i = [5\cos(\omega t + 36.87^\circ) + 10\cos(\omega t - 53.13^\circ)]$ A;

d) $v = [300\cos(20{,}000\pi t + 45^\circ) - 100\sin(20{,}000\pi t + 30^\circ)]$ mV.

ANSWER: (a) $\mathbf{V} = 170\underline{/-40^\circ}$ V; (b) $\mathbf{I} = 10\underline{/-70^\circ}$ A; (c) $\mathbf{I} = 11.18\underline{/-26.57^\circ}$ A; (d) $\mathbf{V} = 339.90\underline{/61.51^\circ}$ mV.

9.5 Find the time-domain expression corresponding to each phasor:

a) $\mathbf{V} = 86.3\,\underline{/+26^\circ}$ V;

b) $\mathbf{I} = (10\,\underline{/30^\circ} + 25\,\underline{/60^\circ})$ mA;

c) $\mathbf{V} = (60 + j30 + 100\,\underline{/-28^\circ})$ V.

ANSWER: (a) $v = 86.3 \cos(\omega t + 26^\circ)$ V;
(b) $i = 34.03 \cos(\omega t + 51.55^\circ)$ mA;
(c) $v = 149.26 \cos(\omega t - 6.52^\circ)$ V.

9.4 THE PASSIVE CIRCUIT ELEMENTS IN THE FREQUENCY DOMAIN

The systematic application of the phasor transform in circuit analysis requires two steps. First, we must establish the relationship between the phasor current and the phasor voltage at the terminals of the passive circuit elements. Second, we must develop the phasor-domain version of Kirchhoff's laws, which we discuss in Section 9.5. In this section we establish the relationship between the phasor current and voltage at the terminals of the resistor, inductor, and capacitor. We begin with the resistor and use the passive sign convention in all the derivations.

THE V-I RELATIONSHIP FOR A RESISTOR

From Ohm's law, if the current in a resistor varies sinusoidally with time—that is, if $i = I_m \cos(\omega t + \theta_i)$—the voltage at the terminals of the resistor, as shown in Fig. 9.7, is

$$v = R[I_m \cos(\omega t + \theta_i)]$$
$$= RI_m[\cos(\omega t + \theta_i)], \qquad \textbf{(9.25)}$$

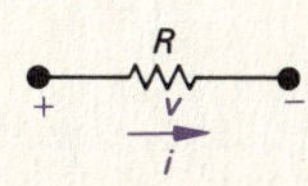

FIGURE 9.7 A resistive element carrying a sinusoidal current.

where I_m is the maximum amplitude of the current in amperes and θ_i is the phase angle of the current.

The phasor transform of this voltage is

$$\mathbf{V} = RI_m e^{j\theta_i} = RI_m \underline{/\theta_i}. \qquad \textbf{(9.26)}$$

But $I_m \underline{/\theta_i}$ is the phasor representation of the sinusoidal current, so we can write Eq. (9.26) as

$$\mathbf{V} = R\mathbf{I}, \qquad \textbf{(9.27)}$$

which states that the phasor voltage at the terminals of a resistor is simply the resistance times the phasor current. Figure 9.8 shows the circuit diagram for a resistor in the frequency domain.

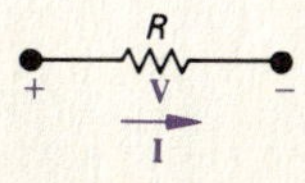

FIGURE 9.8 The frequency-domain equivalent circuit of a resistor.

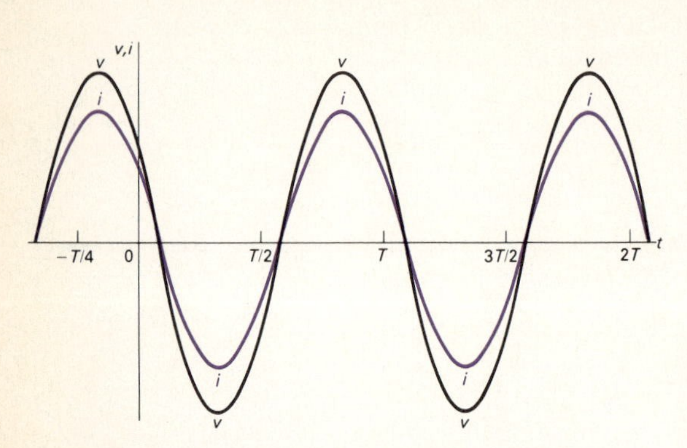

FIGURE 9.9 A plot showing that the voltage and current at the terminals of a resistor are in phase.

Equations (9.25) and (9.27) both contain another important piece of information—namely, that at the terminals of a resistor, there is no phase shift between the current and voltage. Figure 9.9 depicts this phase relationship, where the phase angle of both the voltage and the current waveforms is 60°. The signals are said to be **in phase** because they both reach corresponding values on their respective curves at the same time (for example, they are at their positive maxima at the same instant).

THE V-I RELATIONSHIP FOR AN INDUCTOR

We derive the relationship between the phasor current and phasor voltage at the terminals of an inductor by assuming a sinusoidal current and using $L di/dt$ to establish the corresponding voltage. Thus, for $i = I_m \cos(\omega t + \theta_i)$, the expression for the voltage is

$$v = L\frac{di}{dt} = -\omega L I_m \sin(\omega t + \theta_i). \tag{9.28}$$

We now rewrite Eq. (9.28) using the cosine function:

$$v = -\omega L I_m \cos(\omega t + \theta_i - 90^\circ). \tag{9.29}$$

The phasor representation of the voltage given by Eq. (9.29) is

$$\begin{aligned} \mathbf{V} &= -\omega L I_m e^{j(\theta_i - 90^\circ)} \\ &= -\omega L I_m e^{j\theta_i} e^{-j90^\circ} \\ &= j\omega L I_m e^{j\theta_i} \\ &= j\omega L \mathbf{I}. \end{aligned} \tag{9.30}$$

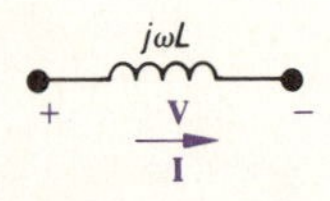

FIGURE 9.10 The frequency-domain equivalent circuit for an inductor.

Note that in deriving Eq. (9.30) we used the identity

$$e^{-j90^\circ} = \cos 90^\circ - j \sin 90^\circ = -j.$$

Equation (9.30) states that the phasor voltage at the terminals of an inductor equals $j\omega L$ times the phasor current. Figure 9.10 shows the frequency-domain equivalent circuit for the inductor. We can rewrite Eq. (9.30) as

$$\begin{aligned} \mathbf{V} &= (\omega L \angle 90^\circ) I_m \angle \theta_i \\ &= \omega L I_m \angle (\theta_i + 90^\circ), \end{aligned} \tag{9.31}$$

which indicates that the voltage and current are out of phase by exactly 90°. In particular, the voltage leads the current by 90°, or, equivalently, the current lags behind the voltage by 90°. Figure 9.11 illustrates this concept of *voltage leading current* or *current lagging voltage*. For example, the voltage reaches its negative peak exactly 90° before the current reaches its negative peak. The same observation can be made with respect to the zero-going-positive crossing or the positive peak.

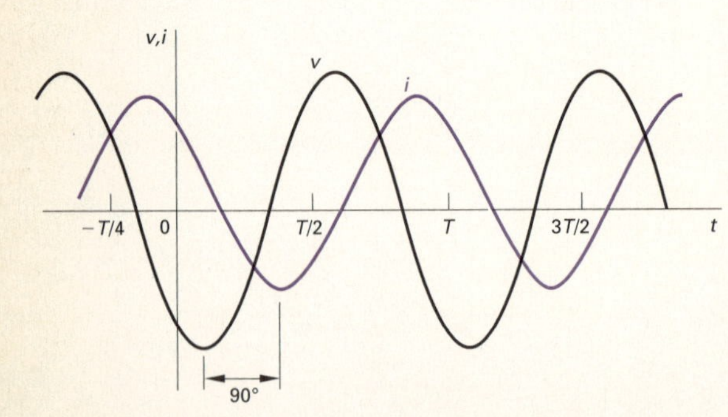

FIGURE 9.11 A plot showing the phase relationship between the current and voltage at the terminals of an inductor ($\theta_i = 60^\circ$).

We can also express the phase shift in seconds. A phase shift of 90° corresponds to one-fourth of a period; hence the voltage leads the current by $T/4$, or $\frac{1}{4f}$ second.

THE V-I RELATIONSHIP FOR A CAPACITOR

We obtain the relationship between the phasor current and phasor voltage at the terminals of a capacitor from the derivation of Eq. (9.30). In other words, if we note that for a capacitor

$$i = C\frac{dv}{dt},$$

and assume that

$$v = V_m \cos(\omega t + \theta_v),$$

then

$$\mathbf{I} = j\omega C\mathbf{V}. \tag{9.32}$$

Now if we solve Eq. (9.32) for the voltage as a function of the current, we get

$$\mathbf{V} = \frac{1}{j\omega C}\mathbf{I}. \tag{9.33}$$

Equation (9.33) demonstrates that the equivalent circuit for the capacitor in the phasor domain is as shown in Fig. 9.12.

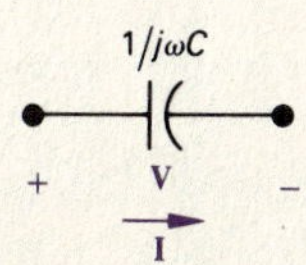

FIGURE 9.12 The frequency domain equivalent circuit for a capacitor.

The voltage across the terminals of a capacitor lags behind the current by exactly 90°. We can easily show this relationship by rewriting Eq. (9.33) as

$$\mathbf{V} = \frac{1}{\omega C}\underline{/-90^\circ}\, I_m \underline{/\theta_i}$$

$$= \frac{I_m}{\omega C}\underline{/(\theta_i - 90^\circ)}. \tag{9.34}$$

Use a computer tool to plot the *V-I* relationships.

The alternative way to express the phase relationship contained in Eq. (9.34) is to say that the current leads the voltage by 90°. Figure 9.13 shows the phase relationship between the current and voltage at the terminals of a capacitor.

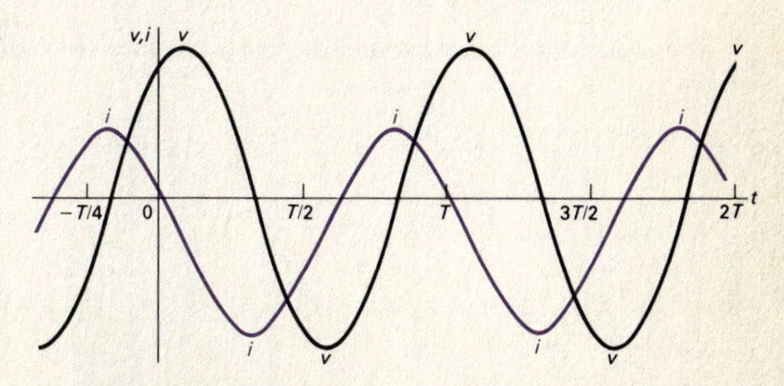

FIGURE 9.13 A plot showing the phase relationship between the current and voltage at the terminals of a capacitor ($\theta_i = 60^\circ$).

IMPEDANCE AND REACTANCE

We conclude this discussion of passive circuit elements in the frequency domain with an important observation. When we compare Eqs. (9.27), (9.30), and (9.33), we note that they are all of the form

$$\mathbf{V} = Z\mathbf{I}, \tag{9.35}$$

where Z represents the **impedance** of the circuit element. Thus the impedance of a resistor is R, the impedance of an inductor is $j\omega L$, and the impedance of a capacitor is $1/j\omega C$. In all cases, impedance is measured in ohms. Note that, although impedance is a complex number, it is not a phasor. Remember, a phasor is a complex number which shows up as the coefficient of $e^{j\omega t}$. Thus, although all phasors are complex numbers, not all complex numbers are phasors.

Impedance in the frequency domain is the quantity analogous to resistance, inductance, and capacitance in the time domain. The imaginary part of the impedance is called **reactance**. The values of impedance and reactance for each of the component values are summarized in Table 9.1.

And finally, a reminder. If the reference direction for the current in a passive circuit element is in the direction of the voltage rise across the element, you must insert a minus sign into the equation that relates the voltage to the current.

TABLE 9.1

IMPEDANCE AND REACTANCE VALUES

CIRCUIT ELEMENT	IMPEDANCE	REACTANCE
Resistor	R	—
Inductor	$j\omega L$	ωL
Capacitor	$1/j\omega C$	$-1/\omega C$

DRILL EXERCISES

75 mH
+ v
i

9.6 The current in the 75 mH inductor is $4 \cos (40{,}000t - 38°)$ mA. Calculate (a) the inductive reactance; (b) the impedance of the inductor; (c) the phasor voltage **V**; and (d) the steady-state expression for $v(t)$.

ANSWER: (a) 3000 Ω; (b) $j3000$ Ω; (c) $12\underline{/52°}$ V; (d) $12 \cos (40{,}000t + 52°)$ V.

0.2 μF
+ v −
i

9.7 The voltage across the terminals of the 0.2 μF capacitor is $40 \cos (10^5 t - 50°)$ V. Calculate (a) the capacitive reactance; (b) the impedance of the capacitor; (c) the phasor current **I**; and (d) the steady-state expression for $i(t)$.

ANSWER: (a) −50 Ω; (b) $-j50$ Ω; (c) $0.8\underline{/40°}$ A; (d) $0.8 \cos (10^5 t + 40°)$ A.

9.5 KIRCHHOFF'S LAWS IN THE FREQUENCY DOMAIN

We pointed out in Section 9.3—with reference to Eqs. (9.23) and (9.24)—that the phasor transform is useful in circuit analysis because it applies to the sum of sinusoidal functions. We illustrated this usefulness in Example 9.5. We now formalize this observation by developing Kirchhoff's laws in the frequency domain.

Kirchhoff's Voltage Law in the Frequency Domain

We begin by assuming that v_1–v_n represent voltages around a closed path in a circuit. We also assume that the circuit is operating in a sinusoidal steady state. Thus Kirchhoff's voltage law requires that

$$v_1 + v_2 + \cdots + v_n = 0, \tag{9.36}$$

which in the sinusoidal steady state becomes

$$V_{m_1} \cos(\omega t + \theta_1) + V_{m_2} \cos(\omega t + \theta_2) + \cdots + V_{m_n} \cos(\omega t + \theta_n) = 0. \tag{9.37}$$

We now use Euler's identity to write Eq. (9.37) as

$$\Re\{V_{m_1} e^{j\theta_1} e^{j\omega t}\} + \Re\{V_{m_2} e^{j\theta_2} e^{j\omega t}\} + \cdots + \Re\{V_{m_n} e^{j\theta_n} e^{j\omega t}\} = 0, \tag{9.38}$$

which we rewrite as

$$\Re\{V_{m_1} e^{j\theta_1} e^{j\omega t} + V_{m_2} e^{j\theta_2} e^{j\omega t} + \cdots + V_{m_n} e^{j\theta_n} e^{j\omega t}\} = 0. \tag{9.39}$$

Factoring the term $e^{j\omega t}$ from each term yields

$$\Re\{(V_{m_1} e^{j\theta_1} + V_{m_2} e^{j\theta_2} + \cdots + V_{m_n} e^{j\theta_n}) e^{j\omega t}\} = 0,$$

or

$$\Re\{(\mathbf{V}_1 + \mathbf{V}_2 + \cdots + \mathbf{V}_n) e^{j\omega t}\} = 0. \tag{9.40}$$

But $e^{j\omega t} \neq 0$, so

$$\mathbf{V}_1 + \mathbf{V}_2 + \cdots + \mathbf{V}_n = 0, \tag{9.41}$$

which is the statement of Kirchhoff's voltage law as it applies to phasor voltages. In other words, Eq. (9.36) applies to a set of sinusoidal voltages in the time domain, and Eq. (9.41) is the equivalent statement in the frequency domain.

Kirchhoff's Current Law in the Frequency Domain

A similar derivation applies to a set of sinusoidal currents. Thus if

$$i_1 + i_2 + \cdots + i_n = 0, \tag{9.42}$$

then

$$\mathbf{I}_1 + \mathbf{I}_2 + \cdots + \mathbf{I}_n = 0, \tag{9.43}$$

where $\mathbf{I}_1, \mathbf{I}_2, \ldots, \mathbf{I}_n$ are the phasor representations of the individual currents $i_1, i_2, \ldots, i_n$.

Equations (9.35), (9.41), and (9.43) form the basis for circuit analysis in the frequency domain. Note that Eq. (9.35) has the same algebraic form as Ohm's law and that Eqs. (9.41) and (9.43) state Kirchhoff's laws for phasor quantities. Therefore you may use all the techniques developed for analyzing resistive circuits to find phasor currents and voltages. You need learn no new analytic techniques; the basic circuit analysis and simplification tools covered in Chapters 2 through 4 can all be used to analyze circuits in the frequency domain. Phasor circuit analysis consists of two fundamental parts: (1) You must be able to construct the frequency-domain model of a circuit; and (2) you must be able to manipulate complex numbers and/or quantities algebraically. We illustrate these aspects of phasor analysis in the discussion that follows, beginning with series, parallel, and delta-to-wye simplifications.

DRILL EXERCISE

9.8 Four branches terminate at a common node. The reference direction of each branch current (i_1, i_2, i_3, and i_4) is toward the node. If $i_1 = 100\cos(\omega t + 25^\circ)$ A, $i_2 = 100\cos(\omega t + 145^\circ)$ A, and $i_3 = 100\cos(\omega t - 95^\circ)$ A, find i_4.

ANSWER: $i_4 = 0$.

9.6 Series, Parallel, and Delta-to-Wye Simplifications

The rules for combining impedances in series or parallel and for making delta-to-wye transformations are the same as those for resistors. The only difference is that combining impedances involves the algebraic manipulation of complex numbers.

Combining Impedances in Series and Parallel

Impedances in series can be combined into a single impedance by simply adding the individual impedances. The circuit shown in Fig. 9.14 defines the problem in general terms. The impedances $Z_1, Z_2, \ldots, Z_n$ are connected in series between terminals a,b. When impedances are in series, they carry the same phasor current **I**. From Eq. (9.35), the

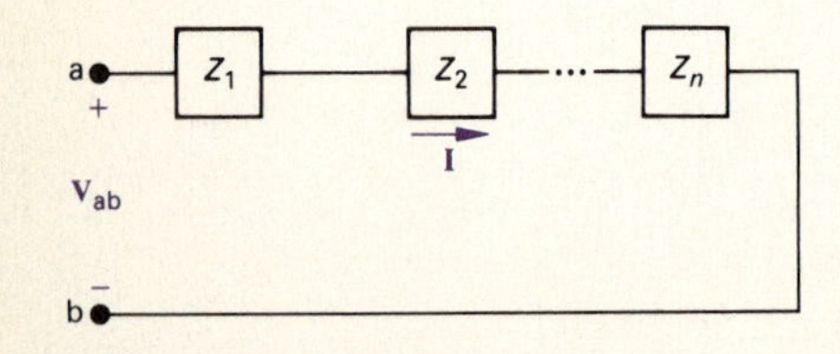

FIGURE 9.14 Impedances in series.

voltage drop across each impedance is $Z_1\mathbf{I}$, $Z_2\mathbf{I}$, ..., $Z_n\mathbf{I}$, and from Kirchhoff's voltage law,

$$\mathbf{V}_{\text{ab}} = Z_1\mathbf{I} + Z_2\mathbf{I} + \cdots + Z_n\mathbf{I}$$
$$= (Z_1 + Z_2 + \cdots + Z_n)\mathbf{I}. \quad (9.44)$$

The equivalent impedance between terminals a,b is

$$Z_{\text{ab}} = \frac{\mathbf{V}_{\text{ab}}}{\mathbf{I}} = Z_1 + Z_2 + \cdots + Z_n. \quad (9.45)$$

Example 9.6 illustrates a numerical application of Eq. (9.45).

EXAMPLE 9.6

A 90 Ω resistor, a 32 mH inductor, and a 5 μF capacitor are connected in series across the terminals of a sinusoidal voltage source, as shown in Fig. 9.15. The steady-state expression for the source voltage v_s is $750 \cos(5000t + 30^\circ)$.

a) Construct the frequency-domain equivalent circuit.

b) Calculate the steady-state current i by the phasor method.

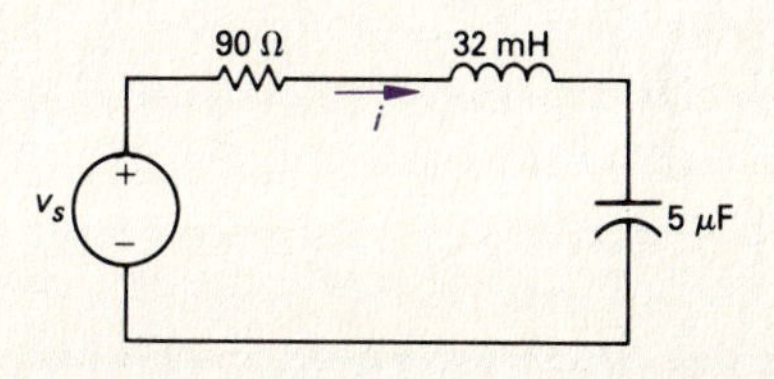

FIGURE 9.15 The circuit for Example 9.6.

SOLUTION

a) From the expression for v_s, we have $\omega = 5000$ rad/s. Therefore the impedance of the 32 mH inductor is

$$Z_L = j\omega L = j(5000)(32 \times 10^{-3}) = j160\ \Omega,$$

and the impedance of the capacitor is

$$Z_C = j\frac{-1}{\omega C} = -j\frac{10^6}{(5000)(5)} = -j40\ \Omega.$$

The phasor transform of v_s is

$$\mathbf{V}_s = 750\underline{/30^\circ}\ \text{V}.$$

Figure 9.16 illustrates the frequency-domain equivalent circuit of the circuit shown in Fig. 9.15.

b) We compute the phasor current simply by dividing the voltage of the voltage source by the equivalent impedance between the terminals a,b. From Eq. (9.45),

$$Z_{\text{ab}} = 90 + j160 - j40$$
$$= 90 + j120 = 150\underline{/53.13^\circ}\ \Omega.$$

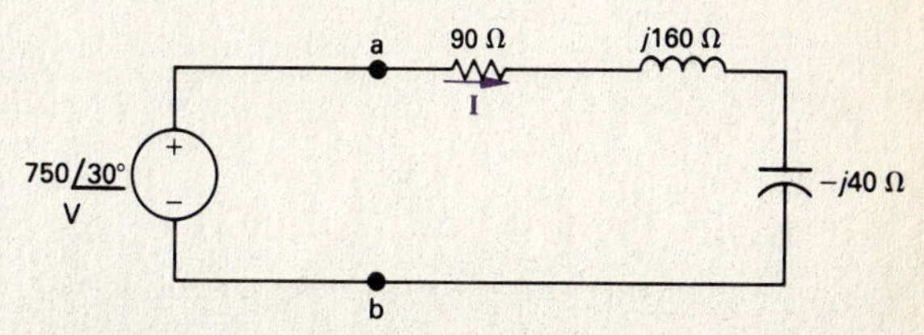

FIGURE 9.16 The frequency-domain equivalent circuit of the circuit shown in Fig. 9.15.

Thus

$$\mathbf{I} = \frac{750\angle 30^\circ}{150\angle 53.13^\circ} = 5\angle -23.13^\circ \text{ A}.$$

We may now write the steady-state expression for i directly:

$$i = 5\cos(5000t - 23.13^\circ) \text{ A}.$$

DRILL EXERCISE

9.9 For the circuit in Fig. 9.15, with $\mathbf{V}_s = 75\angle 30^\circ$ V

a) find the value of capacitance that yields a steady-state output current i with a phase angle of 75°; and

b) find the magnitude of the steady-state output current i.

ANSWER: (a) 0.8 μF; (b) 5.89 A.

Impedances connected in parallel may be reduced to a single equivalent impedance by the reciprocal relationship

$$\frac{1}{Z_{ab}} = \frac{1}{Z_1} + \frac{1}{Z_2} + \cdots + \frac{1}{Z_n}. \quad \textbf{(9.46)}$$

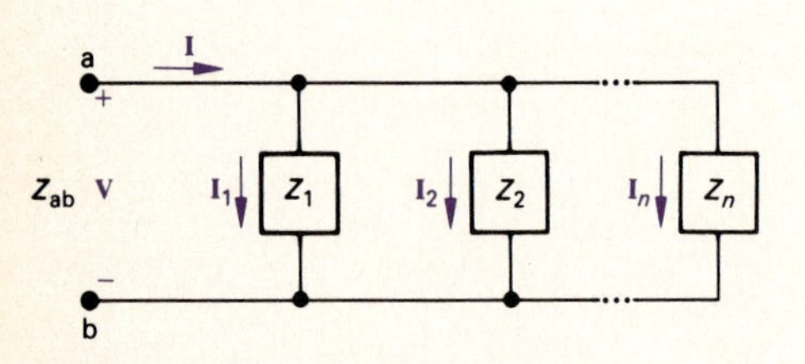

FIGURE 9.17 Impedances in parallel.

Figure 9.17 depicts the parallel connection of impedances. Note that when impedances are in parallel, they have the same voltage across their terminals. We derive Eq. (9.46) directly from Fig. 9.17 by simply combining Kirchhoff's current law with the phasor-domain version of Ohm's law, that is, Eq. (9.35). From Fig. 9.17,

$$\mathbf{I} = \mathbf{I}_1 + \mathbf{I}_2 + \cdots + \mathbf{I}_n$$

or

$$\frac{\mathbf{V}}{Z_{ab}} = \frac{\mathbf{V}}{Z_1} + \frac{\mathbf{V}}{Z_2} + \cdots + \frac{\mathbf{V}}{Z_n}. \quad \textbf{(9.47)}$$

Canceling the common voltage term out of Eq. (9.47) reveals Eq. (9.46).

From Eq. (9.46), for the special case of just two impedances in parallel,

$$Z_{ab} = \frac{Z_1 Z_2}{Z_1 + Z_2}. \quad \textbf{(9.48)}$$

We can also express Eq. (9.46) in terms of **admittance,** defined as the reciprocal of impedance and denoted Y. Thus

$$Y = \frac{1}{Z} = G + jB \quad \text{(siemens)}. \tag{9.49}$$

Admittance is, of course, a complex number, whose real part, G, is called **conductance** and whose imaginary part, B, is called **susceptance.** Like admittance, conductance and susceptance are measured in siemens. Using Eq. (9.49) in Eq. (9.46), we get

$$Y_{ab} = Y_1 + Y_2 + \cdots + Y_n. \tag{9.50}$$

The admittance of each of the ideal passive circuit elements also is worth noting and is summarized in Table 9.2.

Example 9.7 illustrates the application of Eqs. (9.49) and (9.50) to a specific circuit.

TABLE 9.2

ADMITTANCE AND SUSCEPTANCE VALUES

CIRCUIT ELEMENT	ADMITTANCE	SUSCEPTANCE
Resistor	G	—
Inductor	$1/j\omega L$	$-1/\omega L$
Capacitor	$j\omega C$	ωC

EXAMPLE 9.7

The sinusoidal current source in the circuit shown in Fig. 9.18 produces the current $i_s = 8 \cos 200{,}000t$ A.

a) Construct the frequency-domain equivalent circuit.

b) Find the steady-state expressions for v, i_1, i_2, and i_3.

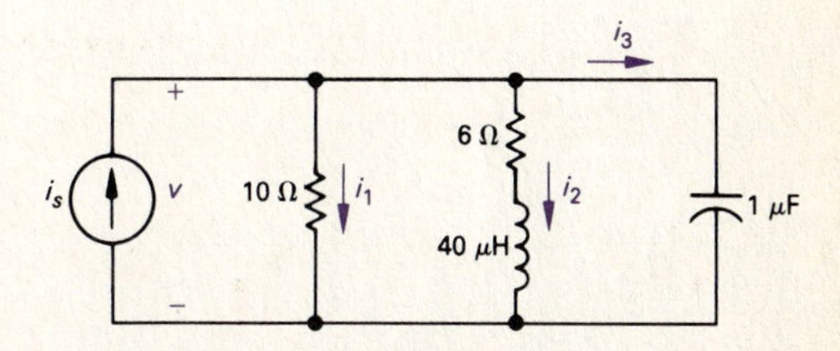

FIGURE 9.18 The circuit for Example 9.7.

SOLUTION

a) The phasor transform of the current source is $8\angle 0^\circ$; the resistors transform directly to the frequency domain as 10 and 6 Ω; the 40 μH inductor has an impedance of $j8$ Ω at the given frequency of 200,000 rad/s; and at this frequency the 1 μF capacitor has an impedance of $-j5$ Ω. Figure 9.19 shows the frequency-domain equivalent circuit and symbols representing the phasor transforms of the unknowns.

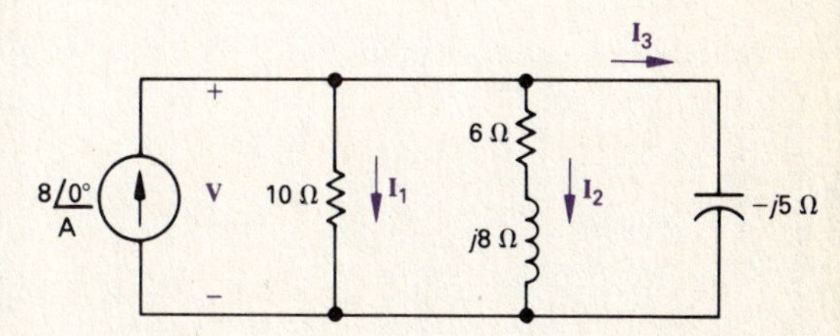

FIGURE 9.19 The frequency-domain equivalent circuit.

b) The circuit shown in Fig. 9.19 indicates that we can easily obtain the voltage across the current source once we know the equivalent impedance of the three parallel branches. Moreover, once we know $\mathbf{V}$, we can calculate the three phasor currents $\mathbf{I}_1$, $\mathbf{I}_2$, and $\mathbf{I}_3$ by using Eq. (9.35). To find the equivalent impedance of the three branches, we first find the equivalent admittance simply by adding the admittances of each branch. The admittance of the first branch is

$$Y_1 = \frac{1}{10} = 0.1 \text{ S};$$

the admittance of the second branch is

$$Y_2 = \frac{1}{6 + j8} = \frac{6 - j8}{100} = 0.06 - j0.08 \text{ S};$$

and the admittance of the third branch is

$$Y_3 = \frac{1}{-j5} = j0.2 \text{ S}.$$

The admittance of the three branches is

$$\begin{aligned} Y &= Y_1 + Y_2 + Y_3 \\ &= 0.16 + j0.12 \\ &= 0.2\underline{/36.87^\circ} \text{ S}. \end{aligned}$$

The impedance at the current source is

$$Z = \frac{1}{Y} = 5\underline{/-36.87^\circ}\ \Omega.$$

The voltage **V** is

$$\mathbf{V} = Z\mathbf{I} = 40\underline{/-36.87^\circ} \text{ V}.$$

Hence

$$\mathbf{I}_1 = \frac{40\underline{/-36.87^\circ}}{10} = 4\underline{/-36.87^\circ} = 3.2 - j2.4 \text{ A};$$

$$\mathbf{I}_2 = \frac{40\underline{/-36.87^\circ}}{6 + j8} = 4\underline{/-90^\circ} = -j4 \text{ A};$$

and

$$\mathbf{I}_3 = \frac{40\underline{/-36.87^\circ}}{5\underline{/-90^\circ}} = 8\underline{/53.13^\circ} = 4.8 + j6.4 \text{ A}.$$

We check the computations at this point by verifying that

$$\mathbf{I}_1 + \mathbf{I}_2 + \mathbf{I}_3 = \mathbf{I}.$$

Specifically,

$$3.2 - j2.4 - j4 + 4.8 + j6.4 = 8 + j0.$$

The corresponding steady-state time-domain expressions are

$$\begin{aligned} v &= 40 \cos (200{,}000t - 36.87^\circ) \text{ V}; \\ i_1 &= 4 \cos (200{,}000t - 36.87^\circ) \text{ A}; \\ i_2 &= 4 \cos (200{,}000t - 90^\circ) \text{ A}; \\ i_3 &= 8 \cos (200{,}000t + 53.13^\circ) \text{ A}. \end{aligned}$$

DRILL EXERCISES

9.10 A 100 Ω resistor is connected in parallel with a 50 mH inductor. This parallel combination is connected in series with a 10 Ω resistor and a 10 μF capacitor.

a) Calculate the impedance of this interconnection if the frequency is 1 krad/s.

b) Repeat (a) for a frequency of 4 krad/s.

c) At what finite frequency does the impedance of the interconnection become purely resistive?

d) What is the impedance at the frequency found in (c)?

ANSWER: (a) $30 - j60$ Ω; (b) $90 + j15$ Ω; (c) 2 krad/s; (d) 60 Ω.

9.11 The interconnection described in Drill Exercise 9.10 is connected across the terminals of a voltage source that is generating $v = 300 \cos 2000t$ V. What is the maximum amplitude of the current in the 50 mH inductor?

ANSWER: 3.54 A.

9.12 Three branches having impedances of $3 + j4$ Ω, $16 - j12$ Ω, and $-j4$ Ω, respectively, are connected in parallel. What are the equivalent (a) admittance, (b) conductance, and (c) susceptance of the parallel connection in millisiemens? (d) If the parallel branches are excited from a sinusoidal current source where $i = 8 \cos \omega t$ A, what is the maximum amplitude of the current in the purely capacitive branch?

ANSWER: (a) $200\underline{/36.87^\circ}$ mS; (b) 160 mS; (c) 120 mS; (d) 10 A.

9.13 Find the steady-state expression for v_o in the circuit shown if $i_g = 0.8 \cos 4000t$ A.

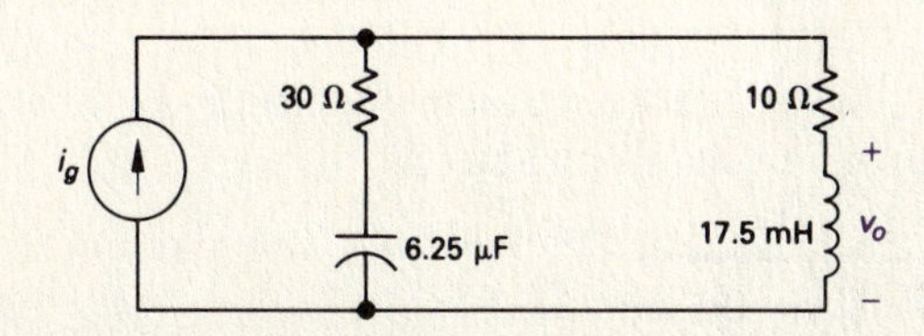

ANSWER: $v_o = 56 \cos 4000t$ V.

DELTA-TO-WYE TRANSFORMATIONS

The Δ-to-Y transformation that we discussed in Section 3.7 with regard to resistive circuits also applies to impedances. Figure 9.20 defines the Δ-connected impedances along with the Y-equivalent circuit. The Y impedances as functions of the Δ impedances are

$$Z_1 = \frac{Z_b Z_c}{Z_a + Z_b + Z_c}; \quad (9.51)$$

$$Z_2 = \frac{Z_c Z_a}{Z_a + Z_b + Z_c}; \quad (9.52)$$

$$Z_3 = \frac{Z_a Z_b}{Z_a + Z_b + Z_c}. \quad (9.53)$$

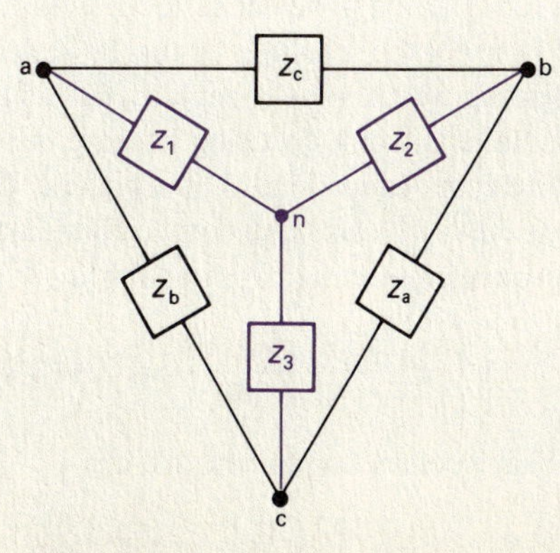

FIGURE 9.20 The delta-to-wye transformation.

The Δ-to-Y transformation also may be reversed; that is, we can start with the Y structure and replace it with an equivalent Δ structure. The Δ impedances as functions of the Y impedances are

$$Z_a = \frac{Z_1Z_2 + Z_2Z_3 + Z_3Z_1}{Z_1}; \tag{9.54}$$

$$Z_b = \frac{Z_1Z_2 + Z_2Z_3 + Z_3Z_1}{Z_2}; \tag{9.55}$$

$$Z_c = \frac{Z_1Z_2 + Z_2Z_3 + Z_3Z_1}{Z_3}. \tag{9.56}$$

Use a computer to calculate the equivalent impedances when transforming from Y to Δ, or vice versa.

The process used to derive Eqs. (9.51) through (9.53) or Eqs. (9.54) through (9.56) is the same as that used to derive the corresponding equations for pure resistive circuits. In fact, comparing Eqs. (3.37)–(3.39) with Eqs. (9.51)–(9.53), and Eqs. (3.40)–(3.42) with Eqs. (9.54)–(9.56), reveals that the symbol Z has replaced the symbol R. You may want to review Problem 3.54 concerning the derivation of the Δ-to-Y transformation.

Example 9.8 illustrates the usefulness of the Δ-to-Y transformation in phasor circuit analysis.

EXAMPLE 9.8

Use a Δ-to-Y impedance transformation to find $\mathbf{I}_0$, $\mathbf{I}_1$, $\mathbf{I}_2$, $\mathbf{I}_3$, $\mathbf{I}_4$, $\mathbf{I}_5$, $\mathbf{V}_1$, and $\mathbf{V}_2$ in the circuit in Fig. 9.21.

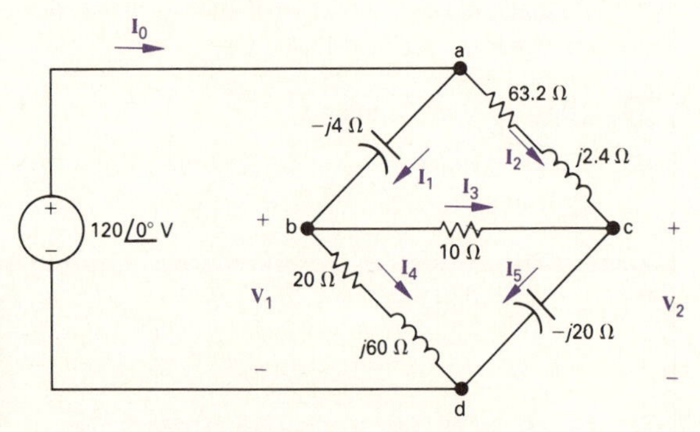

FIGURE 9.21 The circuit for Example 9.8.

SOLUTION

First note that the circuit is not amenable to series or parallel simplification as it now stands. A Δ-to-Y impedance transformation allows us to solve for all the branch currents without resorting to either the node-voltage or the mesh-current method. If we replace either the upper delta (abc) or the lower delta (bcd) with its Y equivalent, we can further simplify the resulting circuit by series-parallel combinations. In deciding which delta to replace, the sum of the impedances around each delta is worth checking because this quantity forms the denominator for the equivalent Y impedances. The sum around the lower delta is $30 + j40$, so we choose to eliminate it from the circuit. The Y impedance connecting to terminal b is

$$Z_1 = \frac{(20 + j60)(10)}{30 + j40} = 12 + j4\ \Omega;$$

the Y impedance connecting to terminal c is

$$Z_2 = \frac{10(-j20)}{30 + j40} = -3.2 - j2.4\ \Omega;$$

and the Y impedance connecting to terminal d is

$$Z_3 = \frac{(20 + j60)(-j20)}{30 + j40} = 8 - j24\ \Omega.$$

Inserting the Y-equivalent impedances into the circuit, we get the circuit shown in Fig. 9.22, which we can now simplify by series-parallel reductions. The impedance of the abn branch is

$$Z_{\text{abn}} = 12 + j4 - j4 = 12\ \Omega,$$

and the impedance of the acn branch is

$$Z_{\text{acn}} = 63.2 + j2.4 - j2.4 - 3.2 = 60\ \Omega.$$

Note that the abn branch is in parallel with the acn branch. Therefore we may replace these two branches with a single branch having an impedance of

$$Z_{\text{an}} = \frac{(60)(12)}{72} = 10\ \Omega.$$

Combining this 10 Ω resistor with the impedance between n and d reduces the circuit shown in Fig. 9.22 to the one shown in Fig. 9.23. From the latter circuit,

$$\mathbf{I}_0 = \frac{120\angle 0^\circ}{18 - j24} = 4\angle 53.13^\circ = 2.4 + j3.2\ \text{A}.$$

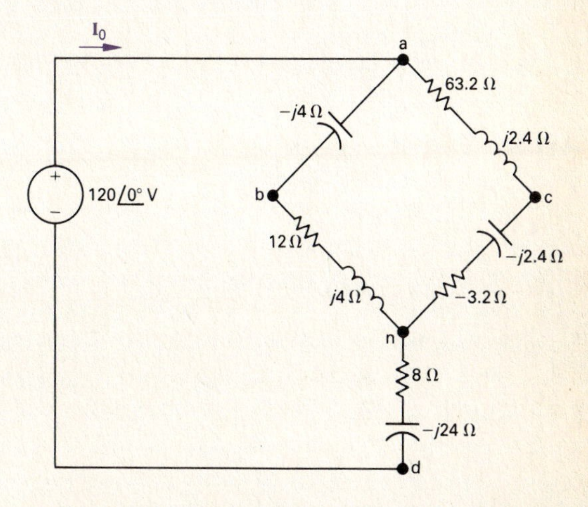

FIGURE 9.22 The circuit shown in Fig. 9.21, with the lower delta replaced by its equivalent wye.

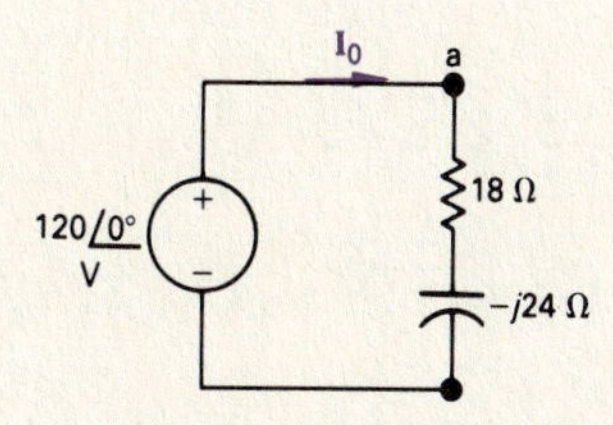

FIGURE 9.23 A simplified version of the circuit shown in Fig. 9.22.

Once we know $\mathbf{I}_0$, we can work back through the equivalent circuits to find the branch currents in the original circuit. We begin by noting that $\mathbf{I}_0$ is the current in the branch nd of Fig. 9.22. Therefore

$$\mathbf{V}_{\text{nd}} = (8 - j24)\mathbf{I}_0 = 96 - j32\ \text{V}.$$

We may now calculate the voltage $\mathbf{V}_{\text{an}}$ because

$$\mathbf{V} = \mathbf{V}_{\text{an}} + \mathbf{V}_{\text{nd}}$$

and both $\mathbf{V}$ and $\mathbf{V}_{\text{nd}}$ are known. Thus

$$\mathbf{V}_{\text{an}} = 120 - 96 + j32 = 24 + j32\ \text{V}.$$

We now compute the branch currents $\mathbf{I}_{\text{abn}}$ and $\mathbf{I}_{\text{acn}}$:

$$\mathbf{I}_{\text{abn}} = \frac{24 + j32}{12} = 2 + j\frac{8}{3}\ \text{A};$$

$$\mathbf{I}_{\text{acn}} = \frac{24 + j32}{60} = \frac{4}{10} + j\frac{8}{15}\ \text{A}.$$

In terms of the branch currents defined in Fig. 9.21,

$$\mathbf{I}_1 = \mathbf{I}_{\text{abn}} = 2 + j\frac{8}{3}\ \text{A};$$

$$\mathbf{I}_2 = \mathbf{I}_{\text{acn}} = \frac{4}{10} + j\frac{8}{15} \text{ A.}$$

We check the calculations of $\mathbf{I}_1$ and $\mathbf{I}_2$ by noting that

$$\mathbf{I}_1 + \mathbf{I}_2 = 2.4 + j3.2 = \mathbf{I}_0.$$

To find the branch currents $\mathbf{I}_3$, $\mathbf{I}_4$, and $\mathbf{I}_5$, we must first calculate the voltages $\mathbf{V}_1$ and $\mathbf{V}_2$. Referring to Fig. 9.21, we note that

$$\mathbf{V}_1 = 120\underline{/0^\circ} - (-j4)\mathbf{I}_1 = \frac{328}{3} + j8 \text{ V}$$

and

$$\mathbf{V}_2 = 120\underline{/0^\circ} - (63.2 + j2.4)\mathbf{I}_2 = 96 - j\frac{104}{3} \text{ V.}$$

We now calculate the branch currents $\mathbf{I}_3$, $\mathbf{I}_4$, and $\mathbf{I}_5$:

$$\mathbf{I}_3 = \frac{\mathbf{V}_1 - \mathbf{V}_2}{10} = \frac{4}{3} + j\frac{12.8}{3} \text{ A;}$$

$$\mathbf{I}_4 = \frac{\mathbf{V}_1}{20 + j60} = \frac{2}{3} - j1.6 \text{ A;}$$

$$\mathbf{I}_5 = \frac{\mathbf{V}_2}{-j20} = \frac{26}{15} + j4.8 \text{ A.}$$

We check the calculations by noting that

$$\mathbf{I}_4 + \mathbf{I}_5 = \frac{2}{3} + \frac{26}{15} - j1.6 + j4.8 = 2.4 + j3.2 = \mathbf{I}_0;$$

$$\mathbf{I}_3 + \mathbf{I}_4 = \frac{4}{3} + \frac{2}{3} + j\frac{12.8}{3} - j1.6 = 2 + j\frac{8}{3} = \mathbf{I}_1;$$

$$\mathbf{I}_3 + \mathbf{I}_2 = \frac{4}{3} + \frac{4}{10} + j\frac{12.8}{3} + j\frac{8}{15} = \frac{26}{15} + j4.8 = \mathbf{I}_5.$$

DRILL EXERCISE

9.14 Use a Δ-to-Y transformation to find the current **I** in the circuit shown.

ANSWER: $\mathbf{I} = 4\underline{/+28.07^\circ}$ A.

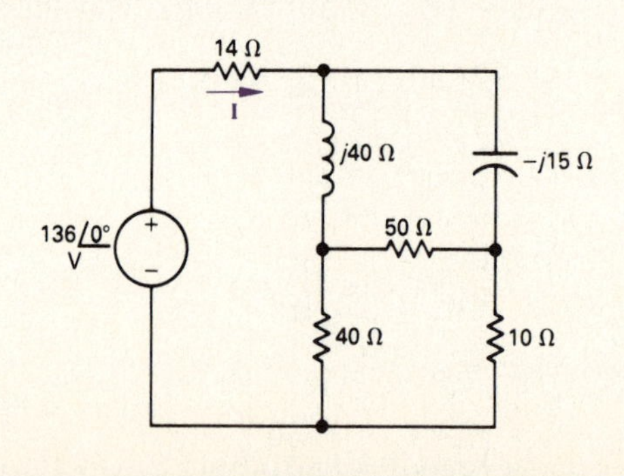

9.7 Source Transformations and Thévenin-Norton Equivalent Circuits

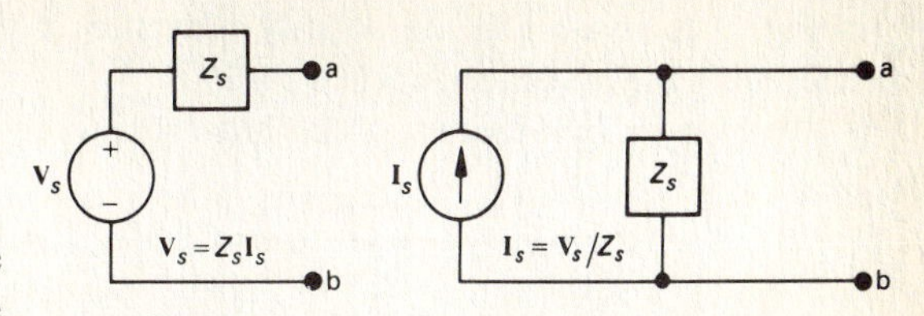

FIGURE 9.24 A source transformation in the frequency domain.

The source transformations introduced in Section 4.9 and the Thévenin-Norton equivalent circuits discussed in Section 4.10 are analytical techniques that also can be applied to frequency-domain circuits. We prove the validity of these techniques by following the same process used in Sections 4.9 and 4.10, except that we substitute impedance (Z) for resistance (R). Figure 9.24 shows a source-transformation equivalent circuit with the nomenclature of the frequency domain.

Figure 9.25 illustrates the frequency-domain version of a Thévenin equivalent circuit. Figure 9.26 shows a Norton equivalent circuit. The techniques for finding the Thévenin equivalent voltage and impedance are identical to those used for resistive circuits, except that the frequency-domain equivalent circuit involves the manipulation of complex quantities. The same holds for finding the Norton equivalent current and impedance.

Example 9.9 demonstrates the application of the source-transformation equivalent circuit to frequency-domain analysis. Example 9.10 illustrates the details of finding a Thévenin equivalent circuit in the frequency domain.

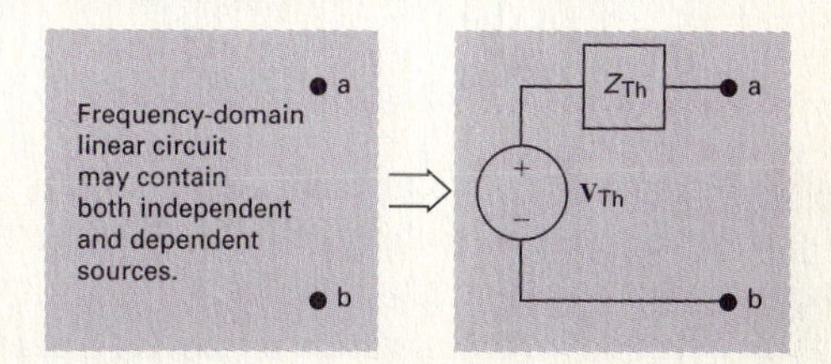

FIGURE 9.25 The frequency-domain version of a Thévenin equivalent circuit.

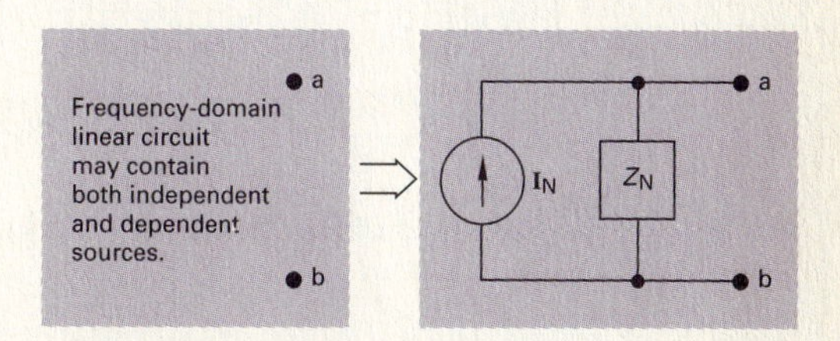

FIGURE 9.26 The frequency-domain version of a Norton equivalent circuit.

EXAMPLE 9.9

Use the concept of source transformation to find the phasor voltage V_0 in the circuit shown in Fig. 9.27.

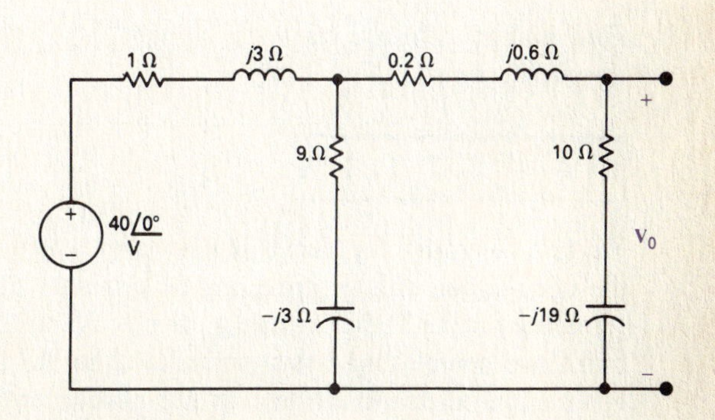

FIGURE 9.27 The circuit for Example 9.9.

SOLUTION

We can replace the series combination of the voltage source ($40\angle 0°$) and the impedance of $1 + j3\ \Omega$ with the parallel combination of a current source and the $1 + j3\ \Omega$ impedance. The source current is

$$\mathbf{I} = \frac{40}{1 + j3} = \frac{40}{10}(1 - j3) = 4 - j12 \text{ A}.$$

Thus we can modify the circuit shown in Fig. 9.27 to the one shown in Fig. 9.28. Note that the polarity reference of the 40 V source determines the reference direction for **I**.

Next, we combine the two parallel branches into a single impedance,

$$Z = \frac{(1 + j3)(9 - j3)}{10} = 1.8 + j2.4\ \Omega,$$

which is in parallel with the current source of $4 - j12$ A. Another source transformation converts this parallel combination to a series combination consisting of a voltage source in series with the impedance of $1.8 + j2.4\ \Omega$. The voltage of the voltage source is

$$\mathbf{V} = (4 - j12)(1.8 + j2.4) = 36 - j12\ \text{V}.$$

Using this source transformation, we redraw the circuit as Fig. 9.29. Note the polarity of the voltage source. We added the current $\mathbf{I}_0$ to the circuit to expedite the solution for $\mathbf{V}_0$.

Also note that we have reduced the circuit to a simple series circuit. We calculate the current $\mathbf{I}_0$ by dividing the voltage of the source by the total series impedance:

$$\mathbf{I}_0 = \frac{36 - j12}{12 - j16} = \frac{12(3 - j1)}{4(3 - j4)}$$

$$= \frac{39 + j27}{25} = 1.56 + j1.08\ \text{A}.$$

We now obtain the value of $\mathbf{V}_0$ by multiplying $\mathbf{I}_0$ by the impedance $10 - j19$:

$$\mathbf{V}_0 = (1.56 + j1.08)(10 - j19) = 36.12 - j18.84\ \text{V}.$$

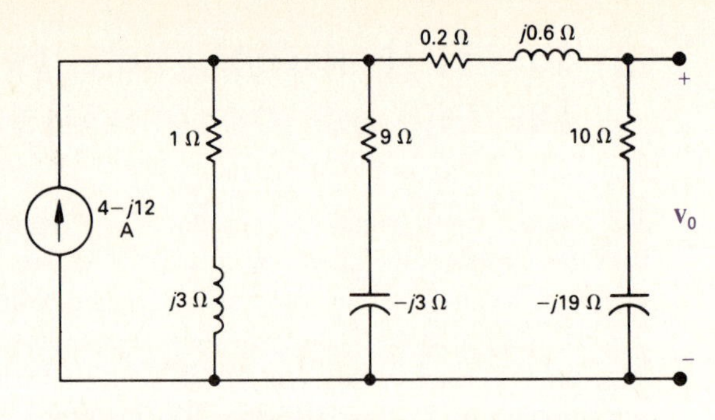

FIGURE 9.28 The first step in reducing the circuit shown in Fig. 9.27.

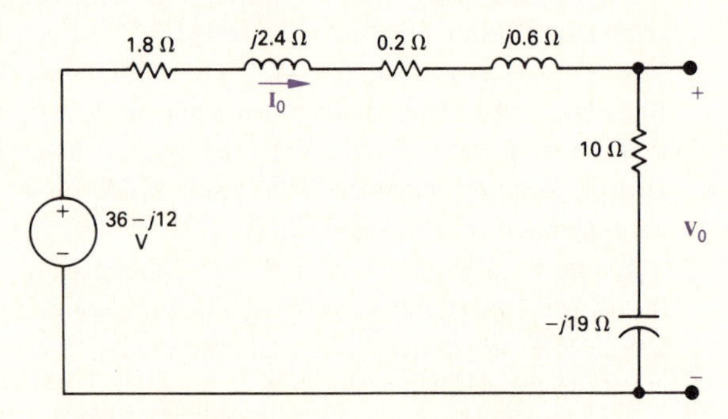

FIGURE 9.29 The second step in reducing the circuit shown in Fig. 9.27.

EXAMPLE 9.10

Find the Thévenin equivalent circuit with respect to terminals a,b for the circuit shown in Fig. 9.30.

SOLUTION

We first determine the Thévenin equivalent voltage. This voltage is the open-circuit voltage appearing at terminals a,b. We choose the reference for the Thévenin voltage as positive at terminal a. We can make two source transformations relative to the 120 V, 12 Ω, and 60 Ω circuit elements to simplify this portion of the circuit. At the same time, these transformations must preserve the identity of the controlling voltage $\mathbf{V}_x$ because of the dependent voltage source.

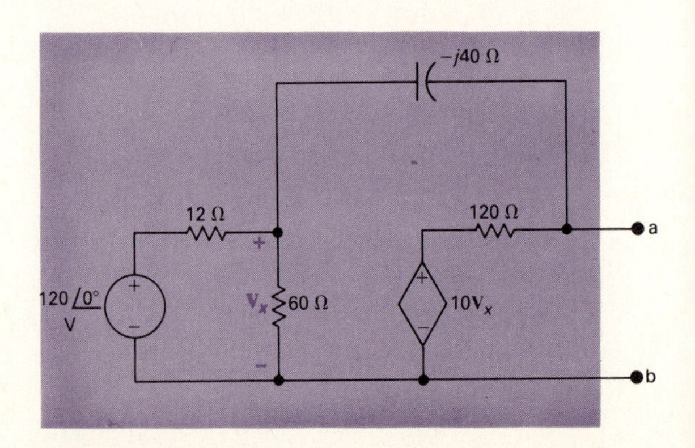

FIGURE 9.30 The circuit for Example 9.10.

We determine the two source transformations by first replacing the series combination of the 120 V source and 12 Ω resistor with a 10 A current source in parallel with 12 Ω. Next, we replace the parallel combination of the 12 and 60 Ω resistors with a single 10 Ω resistor. Finally, we replace the 10 A source in parallel with 10 Ω with a 100 V source in series with 10 Ω. Figure 9.31 shows the resulting circuit.

We added the current **I** to Fig. 9.31 to aid further discussion. Note that once we know the current **I** we can compute the Thévenin voltage. We find **I** by summing the voltages around the closed path in the circuit shown in Fig. 9.31. Hence

$$100 = 10\mathbf{I} - j40\mathbf{I} + 120\mathbf{I} + 10\mathbf{V}_x = (130 - j40)\mathbf{I} + 10\mathbf{V}_x.$$

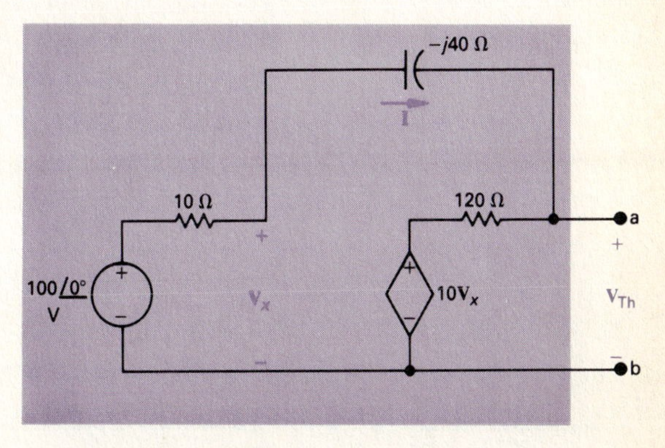

FIGURE 9.31 A simplified version of the circuit shown in Fig. 9.30.

We relate the controlling voltage $\mathbf{V}_x$ to the current **I** by noting from Fig. 9.31 that

$$\mathbf{V}_x = 100 - 10\mathbf{I}.$$

Then,

$$\mathbf{I} = \frac{-900}{30 - j40} = 18\underline{/-126.87^\circ} \text{ A}.$$

We now calculate $\mathbf{V}_x$:

$$\mathbf{V}_x = 100 - 180\underline{/-126.87^\circ} = 208 + j144 \text{ V}.$$

Finally, we note from Fig. 9.31 that

$$\begin{aligned}\mathbf{V}_{\text{Th}} &= 10\mathbf{V}_x + 120\mathbf{I}\\ &= 2080 + j1440 + 120(18)\underline{/-126.87^\circ}\\ &= 784 - j288 = 835.22\underline{/-20.17^\circ} \text{ V}.\end{aligned}$$

To obtain the Thévenin impedance, we may use any of the techniques previously used to find the Thévenin resistance. We illustrate the test-source method in this example. Recall that in using this method, we deactivate all independent sources from the circuit and then apply either a test voltage source or a test current source to the terminals of interest. The ratio of the voltage to the current at the source is the Thévenin impedance. Figure 9.32 shows the result of applying this technique to the circuit shown in Fig. 9.30. Note that we chose a test voltage source $\mathbf{V}_T$. Also note that we deactivated the independent voltage source with an appropriate short circuit and preserved the identity of $\mathbf{V}_x$. The branch currents $\mathbf{I}_a$ and $\mathbf{I}_b$ have been added to the circuit to simplify the calculation of $\mathbf{I}_T$. By straightforward applications of Kirchhoff's circuit laws, you should be able to verify the following relationships:

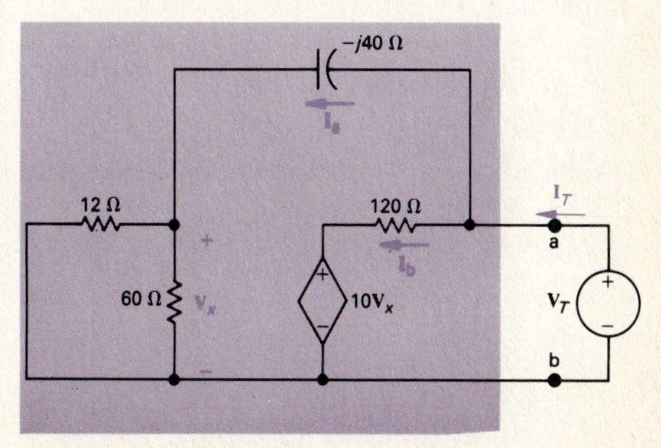

FIGURE 9.32 A circuit for calculating the Thévenin equivalent impedance.

$$\mathbf{I}_a = \frac{\mathbf{V}_T}{10 - j40}; \quad \mathbf{V}_x = 10\mathbf{I}_a;$$

$$\mathbf{I}_b = \frac{\mathbf{V}_T - 10\mathbf{V}_x}{120}$$

$$= \frac{-\mathbf{V}_T(9 + j4)}{120(1 - j4)};$$

$$\mathbf{I}_T = \mathbf{I}_a + \mathbf{I}_b$$

$$= \frac{\mathbf{V}_T}{10 - j40}\left(1 - \frac{9 + j4}{12}\right)$$

$$= \frac{\mathbf{V}_T(3 - j4)}{12(10 - j40)};$$

$$Z_{\text{Th}} = \frac{\mathbf{V}_T}{\mathbf{I}_T} = 91.2 - j38.4\ \Omega.$$

Figure 9.33 depicts the Thévenin equivalent circuit.

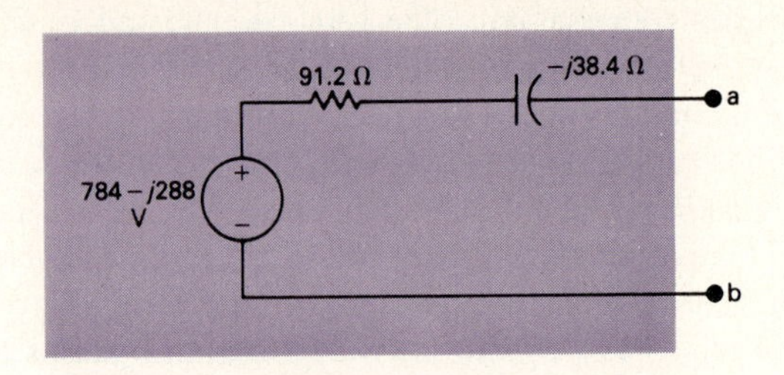

FIGURE 9.33 The Thévenin equivalent for the circuit shown in Fig. 9.30.

Use a circuit simulator to calculate the Thévenin equivalent.

DRILL EXERCISES

9.15 Find the Thévenin equivalent with respect to terminals a,b in the circuit shown.

ANSWER: $\mathbf{V}_{\text{Th}} = \mathbf{V}_{\text{ab}} = 20\underline{/-90^\circ}$ V; $Z_{\text{Th}} = 2.5 - j2.5\ \Omega$.

9.16 Find the Norton equivalent with respect to terminals a,b in the circuit shown.

ANSWER: $\mathbf{I}_N = 5\underline{/0^\circ}$ A; $Z_{\text{Th}} = 1.6 - j1.2\ \Omega$.

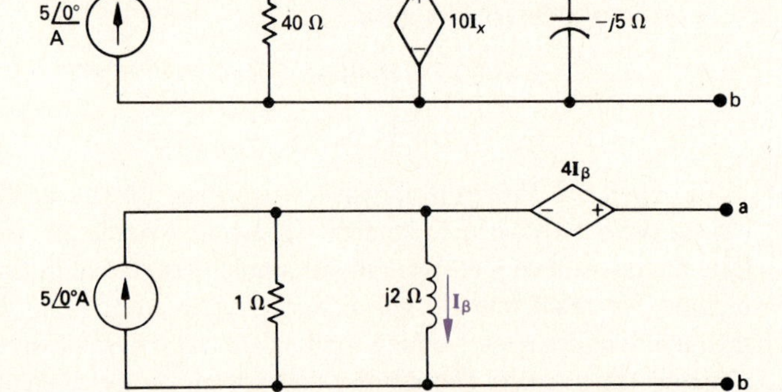

9.8 THE NODE-VOLTAGE METHOD

In Sections 4.2 through 4.4 we introduced the basic concepts of the node-voltage method of circuit analysis. The same concepts apply when we use the node-voltage method to analyze frequency-domain circuits. Example 9.11 illustrates the solution of such a circuit by the node-voltage technique. Drill Exercise 9.17 and Problems 9.36–9.41 give you an opportunity to use the node-voltage method to solve for steady-state sinusoidal responses.

EXAMPLE 9.11

Use the node-voltage method to find the branch currents $\mathbf{I}_a$, $\mathbf{I}_b$, and $\mathbf{I}_c$ in the circuit shown in Fig. 9.34.

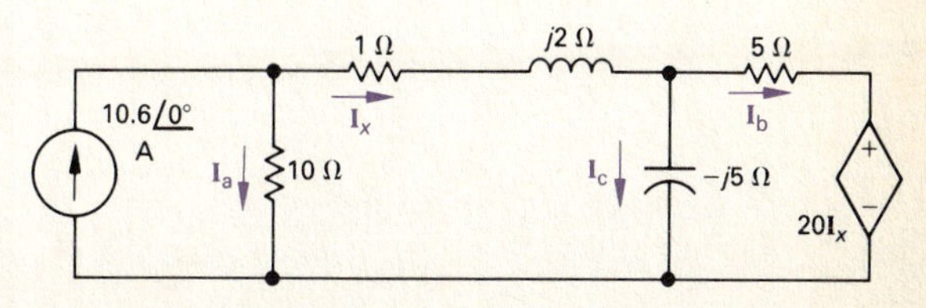

FIGURE 9.34 The circuit for Example 9.11.

SOLUTION

We can describe the circuit in terms of two node voltages because it contains three essential nodes. Four branches terminate at the essential node that stretches across the bottom of Fig. 9.34, so we use it as the reference node. The remaining two essential nodes are labeled 1 and 2, and the appropriate node voltages are designated $\mathbf{V}_1$ and $\mathbf{V}_2$. Figure 9.35 reflects the choice of reference node and the terminal labels. Summing the currents away from node 1 yields

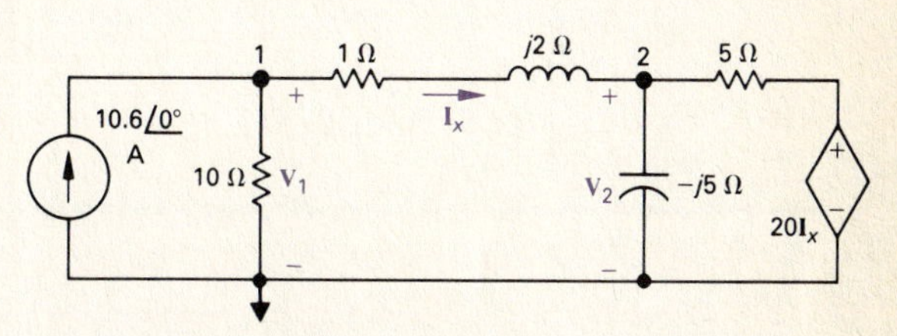

FIGURE 9.35 The circuit shown in Fig. 9.34, with the node voltages defined.

$$-10.6 + \frac{\mathbf{V}_1}{10} + \frac{\mathbf{V}_1 - \mathbf{V}_2}{1 + j2} = 0.$$

Multiplying by $1 + j2$ and collecting the coefficients of $\mathbf{V}_1$ and $\mathbf{V}_2$ generates the expression

$$\mathbf{V}_1(1.1 + j0.2) - \mathbf{V}_2 = 10.6 + j21.2.$$

Summing the currents away from node 2 gives

$$\frac{\mathbf{V}_2 - \mathbf{V}_1}{1 + j2} + \frac{\mathbf{V}_2}{-j5} + \frac{\mathbf{V}_2 - 20\mathbf{I}_x}{5} = 0.$$

The controlling current $\mathbf{I}_x$ is

$$\mathbf{I}_x = \frac{\mathbf{V}_1 - \mathbf{V}_2}{1 + j2}.$$

Substituting this expression for $\mathbf{I}_x$ into the node 2 equation, multiplying by $1 + j2$, and collecting coefficients of $\mathbf{V}_1$ and $\mathbf{V}_2$ produces the equation

$$-5\mathbf{V}_1 + (4.8 + j0.6)\mathbf{V}_2 = 0.$$

The solutions for $\mathbf{V}_1$ and $\mathbf{V}_2$ are

$$\mathbf{V}_1 = 68.40 - j16.80 \text{ V}$$

and

$$\mathbf{V}_2 = 68 - j26 \text{ V}.$$

Hence the branch currents are

$$\mathbf{I}_{\rm a} = \frac{\mathbf{V}_1}{10} = 6.84 - j1.68 \text{ A};$$

$$\mathbf{I}_x = \frac{\mathbf{V}_1 - \mathbf{V}_2}{1 + j2} = 3.76 + j1.68 \text{ A};$$

$$\mathbf{I}_{\rm b} = \frac{\mathbf{V}_2 - 20\mathbf{I}_x}{5} = -1.44 - j11.92 \text{ A};$$

$$\mathbf{I}_{\rm c} = \frac{\mathbf{V}_2}{-j5} = 5.2 + j13.6 \text{ A}.$$

To check our work, we note that

$$\mathbf{I}_{\rm a} + \mathbf{I}_x = 6.84 - j1.68 + 3.76 + j1.68$$
$$= 10.6 \text{ A}$$

and

$$\mathbf{I}_x = \mathbf{I}_{\rm b} + \mathbf{I}_{\rm c} = -1.44 - j11.92 + 5.2 + j13.6$$
$$= 3.76 + j1.68 \text{ A}.$$

DRILL EXERCISE

9.17 Use the node-voltage method to find the steady-state expression for $v(t)$ in the circuit shown. The sinusoidal sources are $i_s = 10 \cos \omega t$ A and $v_s = 100 \sin \omega t$ V, where $\omega = 50$ krad/s.

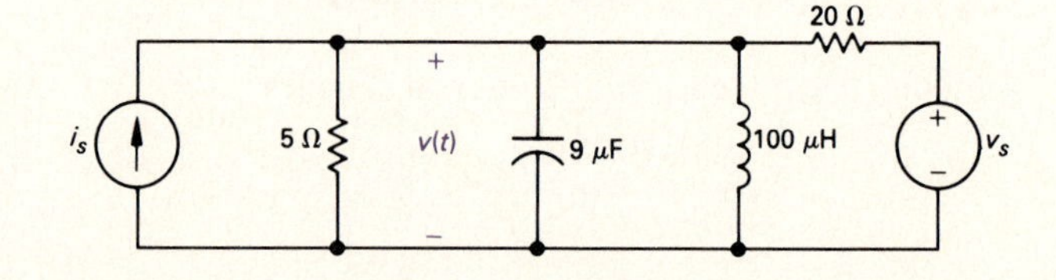

ANSWER: $v(t) = 31.62 \cos (50{,}000t - 71.57°)$ V.

9.9 THE MESH-CURRENT METHOD

We can also use the mesh-current method to analyze frequency-domain circuits. The procedures used in frequency-domain applications are the same as those used in analyzing resistive circuits. In Sections 4.5 through 4.7 we introduced the basic techniques of the mesh-current method; we demonstrate the extension of this method to frequency-domain circuits in Example 9.12.

EXAMPLE 9.12

Use the mesh-current method to find the voltages $\mathbf{V}_1$, $\mathbf{V}_2$, and $\mathbf{V}_3$ in the circuit shown in Fig. 9.36.

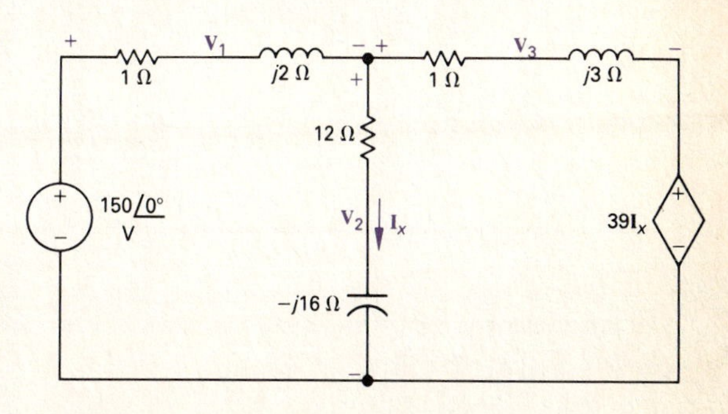

FIGURE 9.36 The circuit for Example 9.12.

SOLUTION

The circuit has two meshes and a dependent voltage source, so we must write two mesh-current equations and a constraint equation. The reference direction for the mesh currents $\mathbf{I}_1$ and $\mathbf{I}_2$ is clockwise, as shown in Fig. 9.37. Once we know $\mathbf{I}_1$ and $\mathbf{I}_2$, we can easily find the unknown voltages. Summing the voltages around mesh 1 gives

$$150 = (1 + j2)\mathbf{I}_1 + (12 - j16)(\mathbf{I}_1 - \mathbf{I}_2),$$

or

$$150 = (13 - j14)\mathbf{I}_1 - (12 - j16)\mathbf{I}_2.$$

Summing the voltages around mesh 2 generates the equation

$$0 = (12 - j16)(\mathbf{I}_2 - \mathbf{I}_1) + (1 + j3)\mathbf{I}_2 + 39\mathbf{I}_x.$$

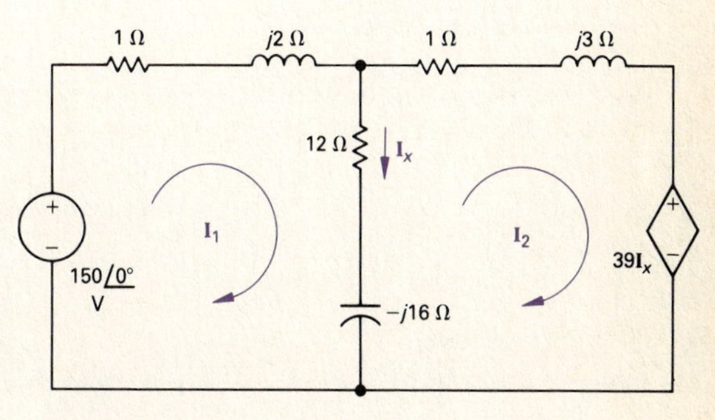

FIGURE 9.37 Mesh currents used to solve the circuit shown in Fig. 9.36.

Figure 9.37 reveals that the controlling current $\mathbf{I}_x$ is the difference between $\mathbf{I}_1$ and $\mathbf{I}_2$; that is, the constraint is

$$\mathbf{I}_x = \mathbf{I}_1 - \mathbf{I}_2.$$

Substituting this constraint into the mesh 2 equation and simplifying the resulting expression gives

$$0 = (27 + j16)\mathbf{I}_1 - (26 + j13)\mathbf{I}_2.$$

Solving for $\mathbf{I}_1$ and $\mathbf{I}_2$ yields

$$\mathbf{I}_1 = -26 - j52 \text{ A};$$
$$\mathbf{I}_2 = -24 - j58 \text{ A};$$
$$\mathbf{I}_x = -2 + j6 \text{ A}.$$

The three voltages are

$$\mathbf{V}_1 = (1 + j2)\mathbf{I}_1 = 78 - j104 \text{ V};$$
$$\mathbf{V}_2 = (12 - j16)\mathbf{I}_x = 72 + j104 \text{ V};$$
$$\mathbf{V}_3 = (1 + j3)\mathbf{I}_2 = 150 - j130 \text{ V};$$
$$39\mathbf{I}_x = -78 + j234 \text{ V}.$$

We check these calculations by summing the voltages around closed paths:

$$-150 + \mathbf{V}_1 + \mathbf{V}_2 = -150 + 78 - j104 + 72 + j104 = 0;$$

$$-\mathbf{V}_2 + \mathbf{V}_3 + 39\mathbf{I}_x = -72 - j104 + 150 - j130 - 78 + j234 = 0;$$

$$-150 + \mathbf{V}_1 + \mathbf{V}_3 + 39\mathbf{I}_x = -150 + 78 - j104 + 150 - j130 - 78 + j234 = 0.$$

DRILL EXERCISE

9.18 Use the mesh-current method to find the phasor current $\mathbf{I}$ in the circuit shown.

ANSWER: $\mathbf{I} = 29 + j2 = 29.07\angle 3.95^\circ$ A.

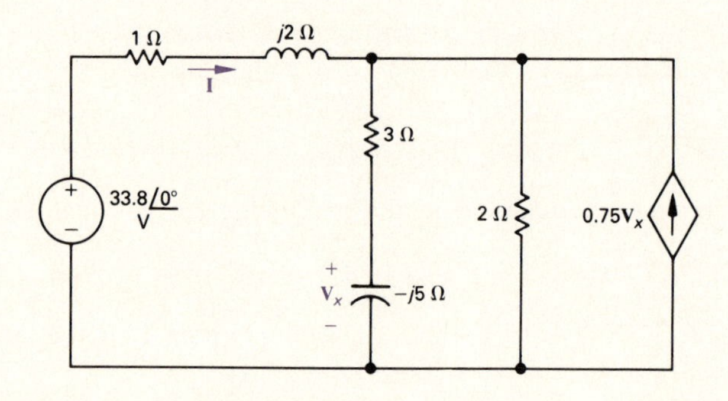

9.10 PHASOR DIAGRAMS

When we are using the phasor method to analyze the steady-state sinusoidal operation of a circuit, a diagram of the phasor currents and voltages may give further insight into the behavior of the circuit. A phasor diagram shows the magnitude and phase angle of each phasor quantity in the complex-number plane. Phase angles are measured counterclockwise from the positive real axis, and magnitudes are measured from the origin of the axes. For example, Fig. 9.38 shows the phasor quantities $10\angle 30^\circ$, $12\angle 150^\circ$, $5\angle -45^\circ$, and $8\angle -170^\circ$.

Constructing phasor diagrams of circuit quantities generally involves both currents and voltages. As a result, two different magnitude scales are necessary, one for currents and one for voltages.

FIGURE 9.38 A graphic representation of phasors.

The ability to visualize a phasor quantity on the complex-number plane can be useful when you are checking pocket calculator calculations. The typical pocket calculator doesn't offer a printout of the data entered. But when the calculated angle is displayed, you can compare it to your mental image as a check on whether you keyed in the appropriate values. For example, suppose that you are to compute the polar form of $-7 - j3$. Without making any calculations, you should anticipate a magnitude greater than 7 and an angle in the third quadrant that is more negative than $-135°$ or less positive than $225°$, as illustrated in Fig. 9.39.

Examples 9.13 and 9.14 illustrate the construction and use of phasor diagrams. We use such diagrams in subsequent chapters whenever they give additional insight into the steady-state sinusoidal operation of the circuit under investigation. Problem 9.62 shows how a phasor diagram can help explain the operation of a phase-shifting circuit.

Use a computer tool to aid in visualizing phasor diagrams.

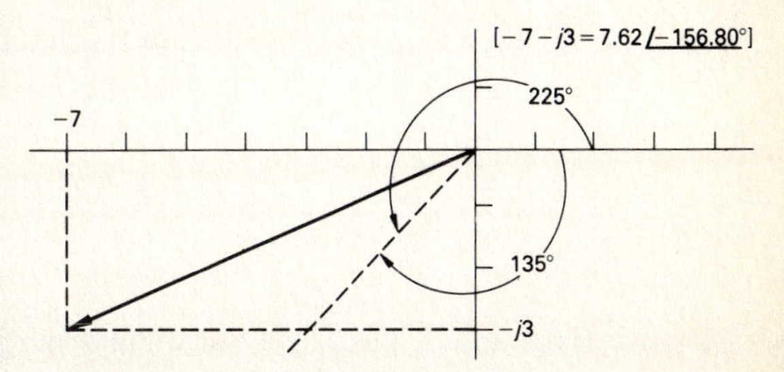

FIGURE 9.39 The complex number $-7 - j3$.

EXAMPLE 9.13

For the circuit in Fig. 9.40, use a phasor diagram to find the value of R that will cause the current through that resistor, $\mathbf{I}_R$, to lag the source current, $\mathbf{I}_s$, by 45°.

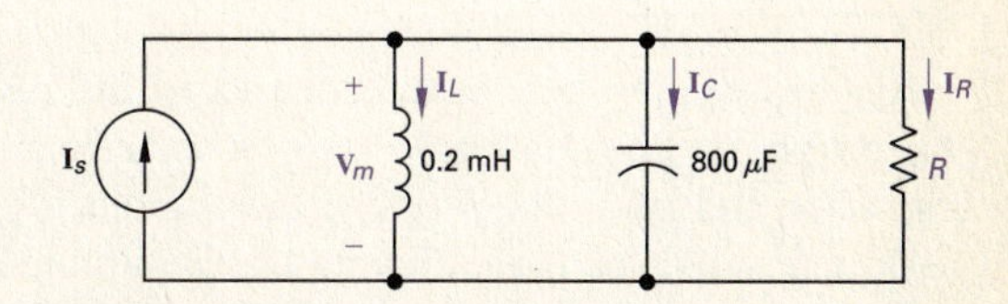

FIGURE 9.40 The circuit for Example 9.13.

SOLUTION

By Kirchhoff's current law, the sum of the currents $\mathbf{I}_R$, $\mathbf{I}_L$, and $\mathbf{I}_C$ must equal the source current $\mathbf{I}_s$. If we assume that the phase angle of the voltage $\mathbf{V}_m$ is zero, we can draw the current phasors for each of the components. The current phasor for the inductor is given by

$$\mathbf{I}_L = \frac{V_m \angle 0°}{j(5000)(0.2 \times 10^{-3})} = V_m \angle -90°,$$

while the current phasor for the capacitor is given by

$$\mathbf{I}_C = \frac{V_m \angle 0°}{-j/(5000)(50 \times 10^{-6})} = 4V_m \angle 90°,$$

and the current phasor for the resistor is given by

$$\mathbf{I}_R = \frac{V_m \angle 0°}{R} = \frac{V_m}{R} \angle 0°.$$

These phasors are shown in Fig. 9.41. The phasor diagram also shows the source current phasor, sketched as a dotted line, which must be the

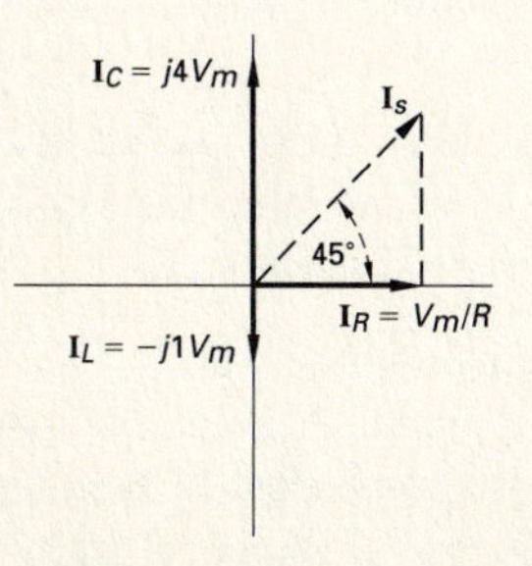

FIGURE 9.41 The phasor diagram for the currents in Fig. 9.40.

sum of the current phasors of the three circuit components and must be at an angle which is 45° more positive than the current phasor for the resistor. As you can see, summing the phasors makes an isosceles triangle, so the length of the current phasor for the resistor must equal $3V_m$. Therefore, the value of the resistor is $\frac{1}{3}\ \Omega$.

EXAMPLE 9.14

The circuit in Fig. 9.42 has a load consisting of the parallel combination of the resistor and inductor. Use phasor diagrams to explore the effect of adding a capacitor across the terminals of the load on the amplitude of $\mathbf{V}_s$ if we adjust $\mathbf{V}_s$ so that the amplitude of $\mathbf{V}_L$ remains constant. Utility companies use this technique to control the voltage drop on their lines.

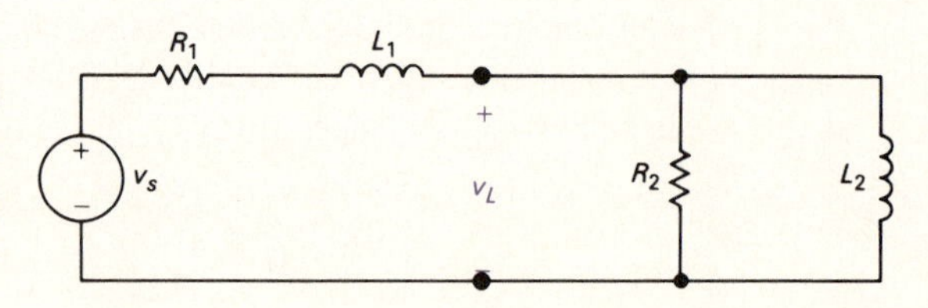

FIGURE 9.42 The circuit for Example 9.14.

SOLUTION

Begin by assuming zero capacitance across the load. After constructing the phasor diagram for the zero-capacitance case, we can add the capacitor and study its effect on the amplitude of $\mathbf{V}_s$, holding the amplitude of $\mathbf{V}_L$ constant. Figure 9.43 shows the frequency-domain equivalent of the circuit shown in Fig. 9.42. We added the phasor branch currents $\mathbf{I}$, $\mathbf{I}_a$, and $\mathbf{I}_b$ to Fig. 9.43 to aid discussion.

Figure 9.44 shows the stepwise evolution of the phasor diagram. Keep in mind that we are not interested in specific phasor values and positions in this example, but rather in the general effect of adding a capacitor across the terminals of the load. Thus, we want to develop the relative positions of the phasors before and after the capacitor has been added.

Relating the phasor diagram to the circuit shown in Fig. 9.43 reveals the following points:

1. Because we are holding the amplitude of the load voltage constant, we choose $\mathbf{V}_L$ as our reference. For convenience, we place this phasor on the positive real axis.

2. We know that $\mathbf{I}_a$ is in phase with $\mathbf{V}_L$ and that its magnitude is $|\mathbf{V}_L|/R_2$. (On the phasor diagram, the magnitude scale for the current phasors is independent of the magnitude scale for the voltage phasors.)

3. We know that $\mathbf{I}_b$ lags behind $\mathbf{V}_L$ by 90° and that its magnitude is $|\mathbf{V}_L|/\omega L_2$.

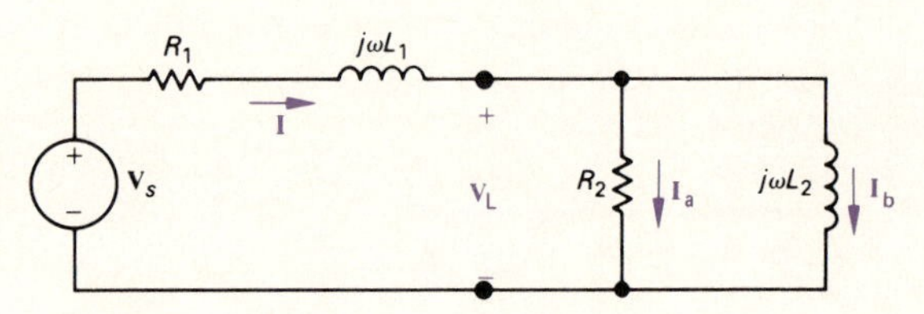

FIGURE 9.43 The frequency-domain equivalent of the circuit in Fig. 9.42.

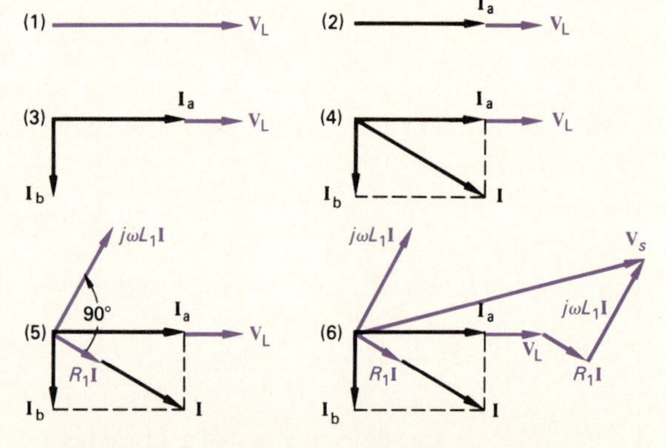

FIGURE 9.44 The step-by-step evolution of the phasor diagram for the circuit in Fig. 9.43.

4. The line current $\mathbf{I}$ is equal to the sum of $\mathbf{I}_a$ and $\mathbf{I}_b$.
5. The voltage drop across R_1 is in phase with the line current, and the voltage drop across $j\omega L_1$ leads the line current by 90°.
6. The source voltage is the sum of the load voltage and the drop along the line; that is, $\mathbf{V}_s = \mathbf{V}_L + (R_1 + j\omega L_1)\mathbf{I}$.

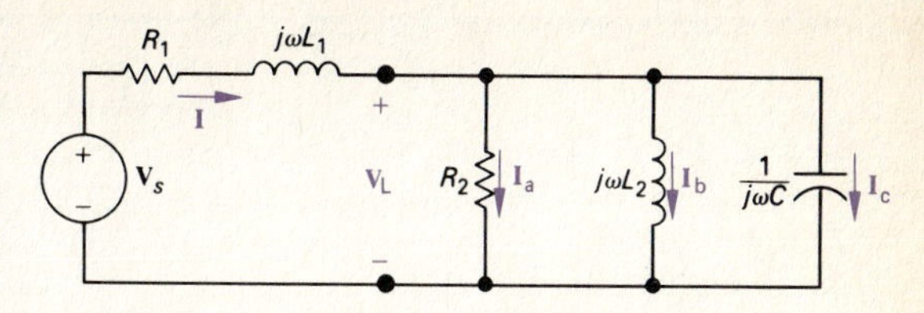

FIGURE 9.45 The addition of a capacitor to the circuit shown in Fig. 9.43.

Note that the completed phasor diagram shown in step 6 of Fig. 9.44 clearly shows the amplitude and phase angle relationships among all the currents and voltages in Fig. 9.43.

Now add the capacitor branch shown in Fig. 9.45. We are holding $\mathbf{V}_L$ constant, so we construct the phasor diagram for the circuit in Fig. 9.45 following the same steps as those in Fig. 9.44, except that, in step 4, we add the capacitor current $\mathbf{I}_c$ to the diagram. In so doing, $\mathbf{I}_c$ leads $\mathbf{V}_L$ by 90°, with its magnitude being $|\mathbf{V}_L\omega C|$. Figure 9.46 shows the effect of $\mathbf{I}_c$ on the line current: both the magnitude and phase angle of the line current $\mathbf{I}$ change with changes in the magnitude of $\mathbf{I}_c$. As $\mathbf{I}$ changes, so do the magnitude and phase angle of the voltage drop along the line. As the drop along the line changes, the magnitude and phase angle of $\mathbf{V}_s$ change. The phasor diagram shown in Fig. 9.47 depicts these observations. The dotted phasors represent the pertinent currents and voltages before the addition of the capacitor.

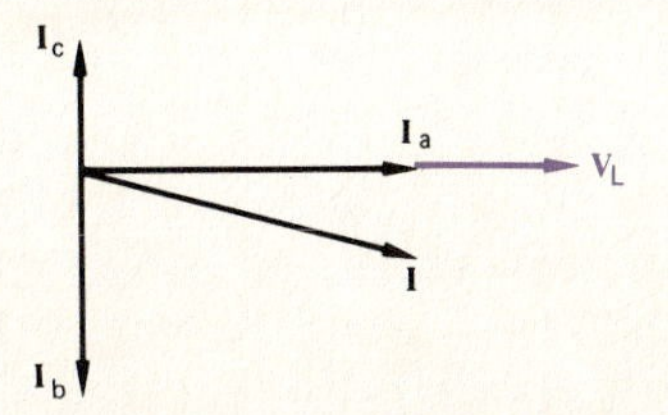

FIGURE 9.46 The effect of the capacitor current $\mathbf{I}_c$ on the line current $\mathbf{I}$.

Thus, comparing the dotted phasors of $\mathbf{I}$, $R_1\mathbf{I}$, $j\omega L_1\mathbf{I}$, and $\mathbf{V}_s$ with their solid counterparts clearly shows the effect of adding C to the circuit. In particular, note that this reduces the amplitude of the source voltage and still maintains the amplitude of the load voltage. Practically, this result means that, as the load increases (that is, as $\mathbf{I}_a$ and $\mathbf{I}_b$ increase), we can add capacitors to the system (that is, increase $\mathbf{I}_c$) so that under heavy load conditions we can maintain $\mathbf{V}_L$ without increasing the amplitude of the source voltage.

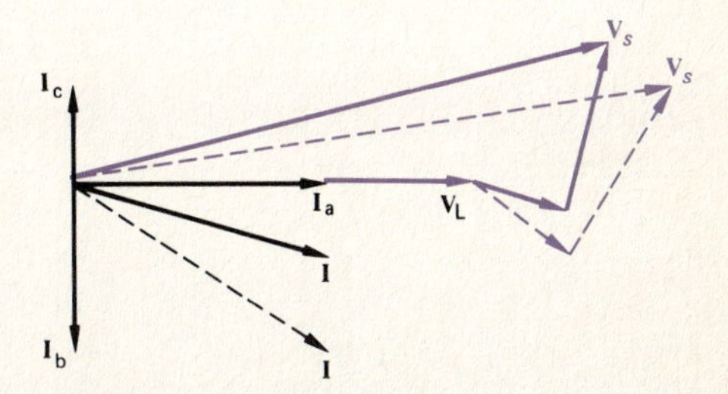

FIGURE 9.47 The effect of adding a load-shunting capacitor to the circuit shown in Fig. 9.43 if $\mathbf{V}_L$ is held constant.

DRILL EXERCISE

9.19 The parameters in the circuit shown in Fig. 9.43 are $R_1 = 0.1\ \Omega$, $\omega L_1 = 0.8\ \Omega$, $R_2 = 24\ \Omega$, $\omega L_2 = 32\ \Omega$, and $\mathbf{V}_L = 240 + j0$ V.

a) Calculate the phasor voltage $\mathbf{V}_s$.

b) Connect a capacitor in parallel with the inductor, hold $\mathbf{V}_L$ constant, and adjust the capacitor until the magnitude of $\mathbf{I}$ is a minimum. What is the capacitive reactance? What is the value of $\mathbf{V}_s$?

c) Find the value of the capacitive reactance that keeps the magnitude of $\mathbf{I}$ as small as possible and that at the same time makes

$$|\mathbf{V}_s| = |\mathbf{V}_L| = 240 \text{ V}.$$

ANSWER: (a) $247.11\underline{/1.68^\circ}$ V; (b) $-32\ \Omega$, $241.13\underline{/1.90^\circ}$ V; (c) $-26.90\ \Omega$.

SUMMARY

- The general equation for a **sinusoidal source** is

$$v = V_m \cos(\omega t + \phi) \quad \text{(voltage source)}$$

or

$$i = I_m \cos(\omega t + \phi) \quad \text{(current source)},$$

where V_m (or I_m) is the maximum amplitude, ω is the frequency, and ϕ is the phase angle.

- The frequency, ω, of a sinusoidal response is the same as the frequency of the sinusoidal source driving the circuit. The amplitude and phase angle of the response are usually different from those of the source.

- The best way to find the steady-state voltages and currents in a circuit driven by sinusoidal sources is to perform the analysis in the frequency domain. The following mathematical transforms allow us to move between the time and frequency domains.

  1. The phasor transform (from the time domain to the frequency domain):

  $$\mathbf{V} = V_m e^{j\phi} = \mathcal{P}\{V_m \cos(\omega t + \phi)\}$$

  2. The inverse phasor transform (from the frequency domain to the time domain):

  $$\mathcal{P}^{-1}\{V_m e^{j\phi}\} = \Re\{V_m e^{j\phi} e^{j\omega t}\}$$

- When working with sinusoidally varying signals, remember that voltage leads current by 90° at the terminals of an inductor, and current leads voltage by 90° at the terminals of a capacitor.

- **Impedance** (Z) plays the same role in the frequency domain as resistance, inductance, and capacitance play in the time domain. Specifically, the relationship between phasor current and phasor voltage for resistors, inductors, and capacitors is

$$\mathbf{V} = Z\mathbf{I},$$

where the reference direction for **I** obeys the passive sign convention. The reciprocal of impedance is **admittance** (Y), so another way to express the current-voltage relationship for resistors, inductors, and capacitors in the frequency domain is

$$\mathbf{V} = \mathbf{I}/Y.$$

Impedance and admittance are complex numbers. We call the imaginary part of impedance **reactance,** and we call the imaginary part of admittance **susceptance.** The impedance, reactance, admittance, and susceptance values for resistors, inductors, and capacitors are summarized in Table 9.3.

TABLE 9.3

IMPEDANCE AND RELATED VALUES

ELEMENT	IMPEDANCE (Z)	REACTANCE	ADMITTANCE (Y)	SUSCEPTANCE
Resistor	R (resistance)	—	G (conductance)	—
Capacitor	$1/j\omega C$	$-1/\omega C$	$j\omega C$	ωC
Inductor	$j\omega L$	ωL	$1/j\omega L$	$-1/\omega L$

- All the techniques used in dc circuit analysis (series-parallel and delta-to-wye simplifications, node-voltage equations, mesh-current equations, source transformations, Thévenin-Norton equivalents, and superposition) may be used to analyze a frequency-domain circuit.
- Phasor diagrams help in visualizing the relationships among voltage phasors and among current phasors in the frequency domain.

PROBLEMS

9.1 Consider the sinusoidal voltage

$$v = 80 \cos (1000\pi t - 30^\circ) \text{ V}.$$

a) What is the maximum amplitude of the voltage?

b) What is the frequency in hertz?

c) What is the frequency in radians per second?

d) What is the phase angle in radians?

e) What is the phase angle in degrees?

f) What is the period in milliseconds?

g) What is the first time after $t = 0$ that $v = 80$ V?

h) The sinusoidal function is shifted 2/3 ms to the left along the time axis. What is the expression for $v(t)$?

i) What is the minimum number of milliseconds that the function must be shifted to the right if the expression for $v(t)$ is $80 \cos 1000\pi t$ V?

j) What is the minimum number of milliseconds that the function must be shifted to the left if the expression for $v(t)$ is $80 \sin 1000\pi t$ V?

9.2 In a single graph, sketch $v = 60\cos(\omega t + \phi)$ versus ωt for $\phi = -60°, -30°, 0°, +30°$, and $60°$.

a) State whether the voltage function is shifting to the right or left as ϕ becomes more positive.

b) What is the direction of shift if ϕ changes from 0 to $-30°$?

9.3 At $t = -250/3\ \mu$s, a sinusoidal voltage is known to be zero and going positive. The voltage is next zero at $t = 1250/3\ \mu$s. It is also known that the voltage is 50 V at $t = 0$.

a) What is the frequency of v in hertz?

b) What is the expression for v?

9.4 A sinusoidal current is zero at $t = -625\ \mu$s and increasing at a rate of 8000π A/s. The maximum amplitude of the current is 20 A.

a) What is the frequency of i in radians per second?

b) What is the expression for i?

9.5 a) Verify that Eq. (9.9) is the solution of Eq. (9.8). This can be done by substituting Eq. (9.9) into the left-hand side of Eq. (9.8) and then noting that it equals the right-hand side for all values to $t > 0$. At $t = 0$, Eq. (9.9) should reduce to the initial value of the current.

b) Since the transient component vanishes as time elapses and since our solution must satisfy the differential equation for all values of t, the steady-state component, by itself, must also satisfy the differential equation. Verify this observation by showing that the steady-state component of Eq. (9.9) satisfies Eq. (9.8).

9.6 Use the concept of the phasor to combine the following sinusoidal functions into a single trigonometric expression:

a) $y = 50\cos(500t + 60°) + 100\cos(500t - 30°)$;

b) $y = 200\cos(377t + 50°) - 100\sin(377t + 150°)$;

c) $y = 80\cos(100t + 30°) - 100\sin(100t - 135°) + 50\cos(100t - 90°)$;

d) $y = 250\cos\omega t + 250\cos(\omega t + 120°) + 250\cos(\omega t - 120°)$.

9.7 Show that

$$\int_{t_o}^{t_o+T} V_m^2 \cos^2(\omega t + \phi)dt = \frac{V_m^2 T}{2}.$$

9.8 The rms value of the sinusoidal voltage supplied to the convenience outlet of a U.S. home is 120 V. What is the maximum value of the voltage at the outlet?

9.9 A 400 Hz sinusoidal voltage with a maximum amplitude of 100 V at $t = 0$ is applied across the terminals of an inductor. The maximum amplitude of the steady-state current in the inductor is 20 A.

a) What is the frequency of the inductor current?

b) What is the phase angle of the voltage?

c) What is the phase angle of the current?

d) What is the inductive reactance of the inductor?

e) What is the inductance of the inductor in millihenrys?

f) What is the impedance of the inductor?

9.10 An 80 kHz sinusoidal voltage has zero phase angle and a maximum amplitude of 25 mV. When this voltage is applied across the terminals of a capacitor, the resulting steady-state current has a maximum amplitude of 628.32 μA.

a) What is the frequency of the current in radians per second?

b) What is the phase angle of the current?

c) What is the capacitive reactance of the capacitor?

d) What is the capacitance of the capacitor in microfarads?

e) What is the impedance of the capacitor?

9.11 A 100 Ω resistor, a 6.25 mH inductor, and a 2.0 μF capacitor are connected in series. The series-connected elements are energized by a sinusoidal voltage source whose voltage is 300 cos $(4000t - 30°)$ V.

a) Draw the frequency-domain equivalent circuit.

b) Reference the current in the direction of the voltage rise across the source, and find the phasor current.

c) Find the steady-state expression for $i(t)$.

9.12 A 25 Ω resistor and a 0.4 μF capacitor are connected in parallel. This parallel combination is also in parallel with the series combination of a 6 Ω resistor and an 80 μH inductor. These three parallel branches are driven by a sinusoidal current source whose current is 10 cos $(10^5 t - 60°)$ A.

a) Draw the frequency-domain equivalent circuit.

b) Reference the voltage across the current source as a rise in the direction of the source current, and find the phasor voltage.

c) Find the steady-state expression for $v(t)$.

9.13 Find the steady-state expression for $i_o(t)$ in the circuit in Fig. P9.13 if $v_s = 500 \sin 4000t$ mV.

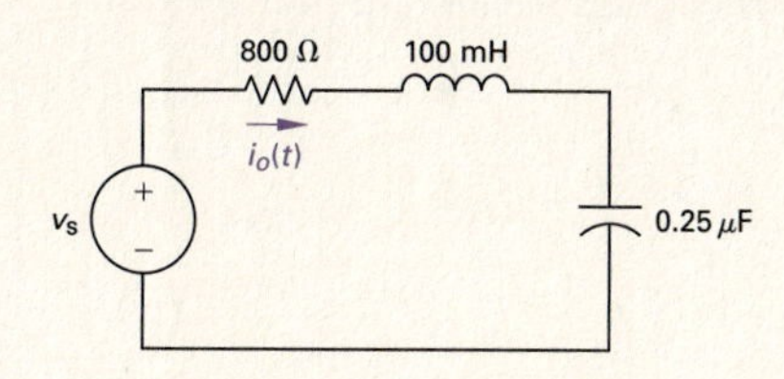

FIGURE P9.13

9.14 The circuit in Fig. P9.14 is operating in the sinusoidal steady state. Find the steady-state expression for $v_o(t)$ if $v_g = 40 \cos 50{,}000t$ V.

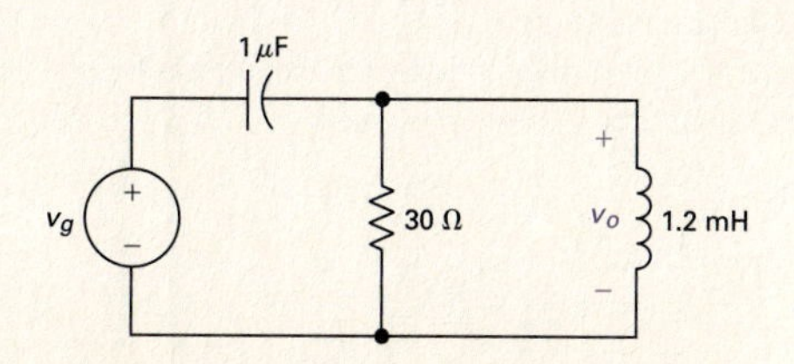

FIGURE P9.14

9.15 Find the impedance Z_{ab} in the circuit seen in Fig. P9.15. Express Z_{ab} in both polar and rectangular form.

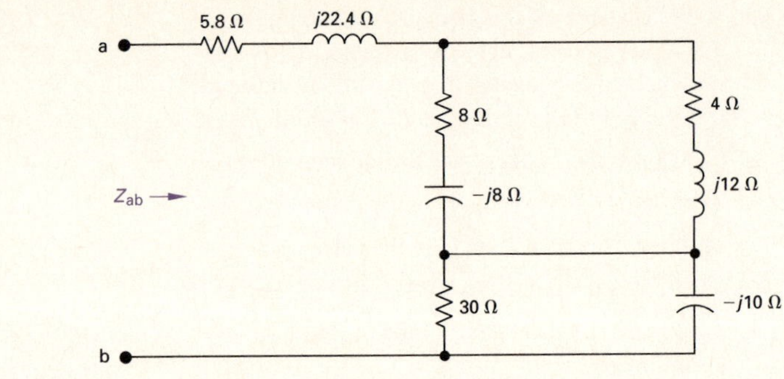

FIGURE P9.15

9.16 Find the admittance Y_{ab} in the circuit seen in Fig. P9.16. Express Y_{ab} in both polar and rectangular form. Give the value of Y_{ab} in millimhos.

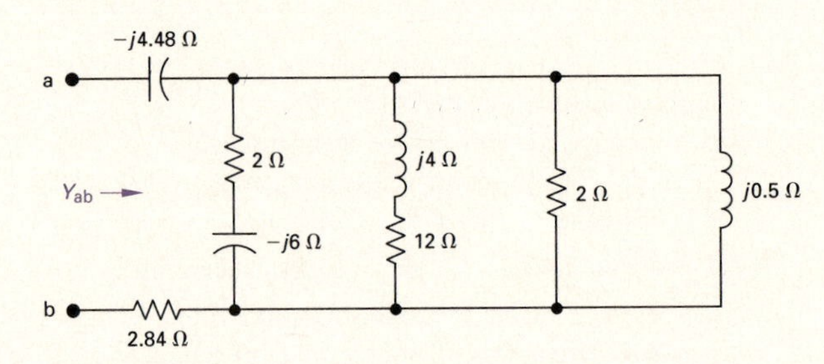

FIGURE P9.16

9.17 Find Z_{ab} in the circuit shown in Fig. P9.17 when the circuit is operating at a frequency of 100 krad/s.

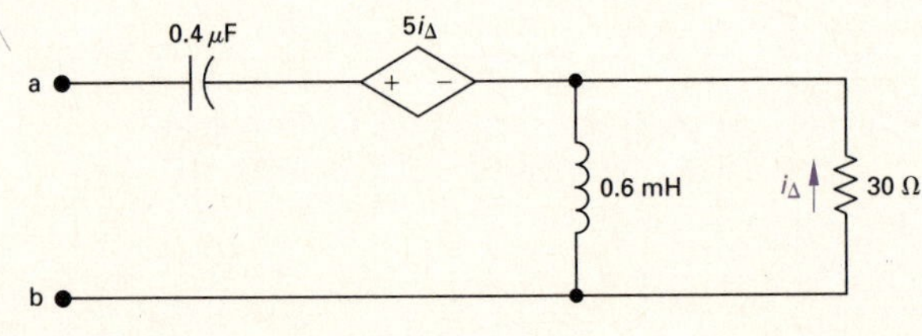

FIGURE P9.17

9.18 a) For the circuit shown in Fig. P9.18, find the frequency (in radians per second) at which the impedance Z_{ab} is purely resistive.

b) Find the value of Z_{ab} at the frequency of part (a).

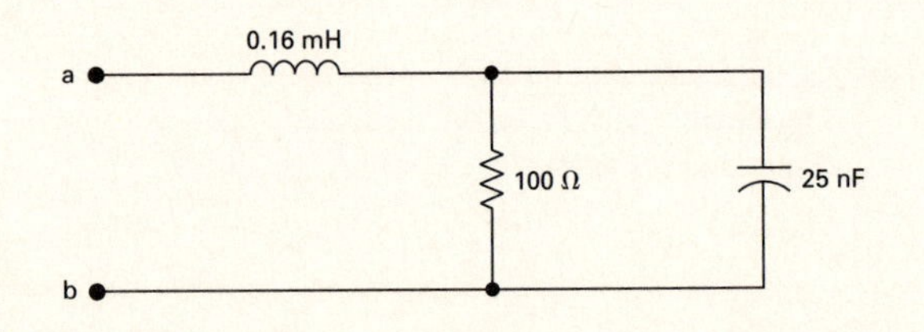

FIGURE P9.18

9.19 a) The source voltage in the circuit in Fig. P9.19 is $v_g = 50 \cos 50{,}000t$ V. Find the values of L such that i_g is in phase with v_g when the circuit is operating in the steady state.

b) For the values of L found in part (a), find the steady-state expressions for i_g.

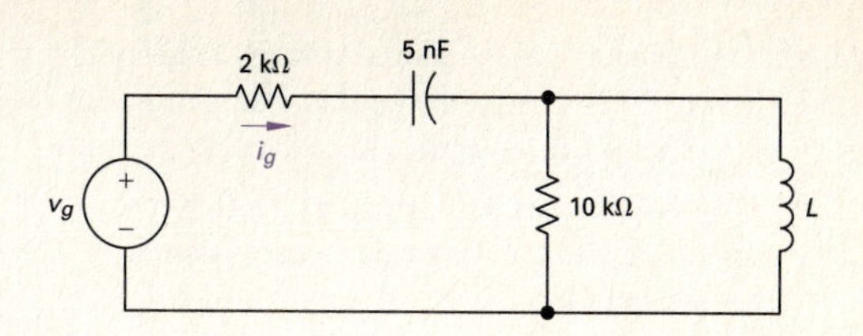

FIGURE P9.19

9.20 The circuit shown in Fig. P9.20 is operating in the sinusoidal steady state. The capacitor is adjusted until the current i_g is in phase with the sinusoidal voltage v_g.

a) Specify the values of capacitance in microfarads if $v_g = 80 \cos 5000t$ V.

b) Give the steady-state expressions for i_g when C has the values found in part (a).

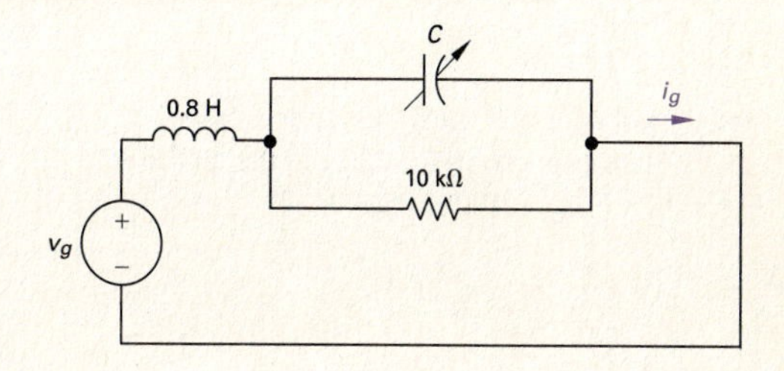

FIGURE P9.20

9.21 The frequency of the sinusoidal current source in the circuit in Fig. P9.21 is adjusted until v_o is in phase with i_g.

a) What is the value of ω in radians per second?

b) If $i_g = 0.75 \cos \omega t$ mA (where ω is the frequency found in part [a]), what is the steady-state expression for v_o?

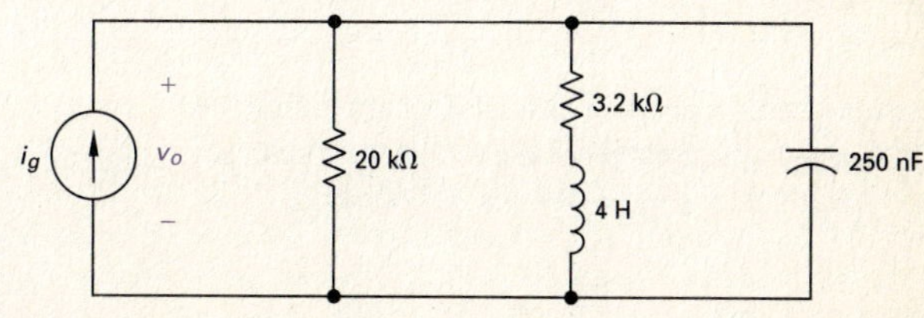

FIGURE P9.21

9.22 The frequency of the sinusoidal voltage source in the circuit in Fig. P9.22 is adjusted until the current i_o in phase with v_g.

a) Find the frequency in hertz.

b) Find the steady-state expression for i_o (at the frequency found in part [a]) if $v_g = 30 \cos \omega t$ V.

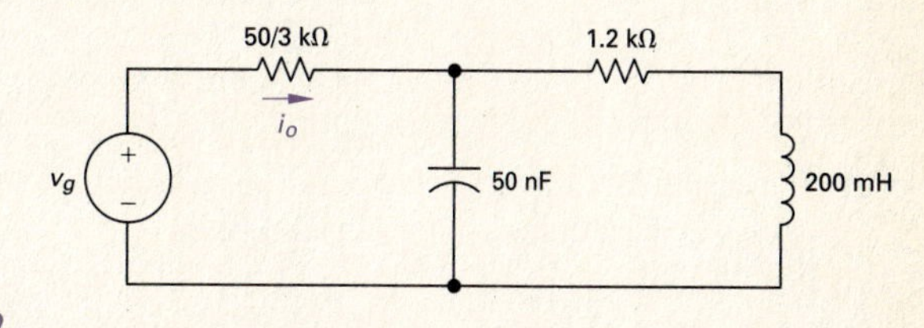

FIGURE P9.22

9.23 The circuit shown in Fig. P9.23 is operating in the sinusoidal steady state. Find the value of ω if

$$i_o = 40 \sin (\omega t + 21.87^\circ) \text{ mA}$$

and

$$v_g = 40 \cos (\omega t - 15^\circ) \text{ V}.$$

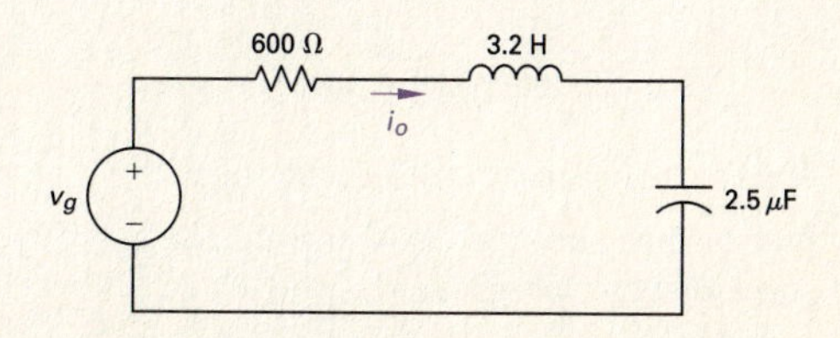

FIGURE P9.23

9.24 a) The frequency of the source voltage in the circuit in Fig. P9.24 is adjusted until i_g is in phase with v_g. What is the value of ω in radians per second?

b) If $v_g = 20 \cos \omega t$ V (where ω is the frequency found in part [a]), what is the steady-state expression for v_o?

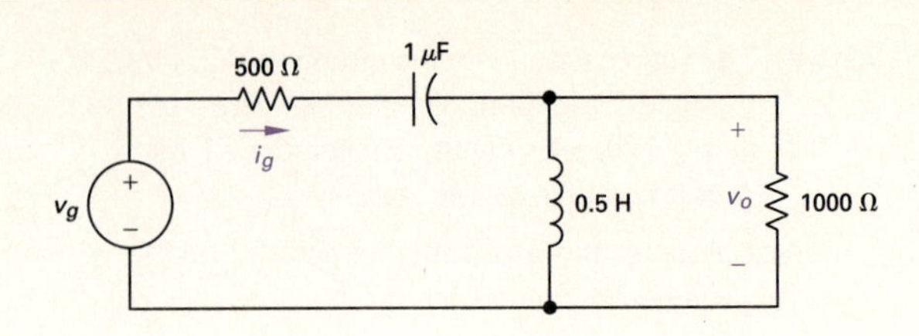

FIGURE P9.24

9.25 Use the concept of voltage division to find the steady-state expression for $v_o(t)$ in the circuit in Fig. P9.25 if $v_g = 100 \cos 8000t$ V.

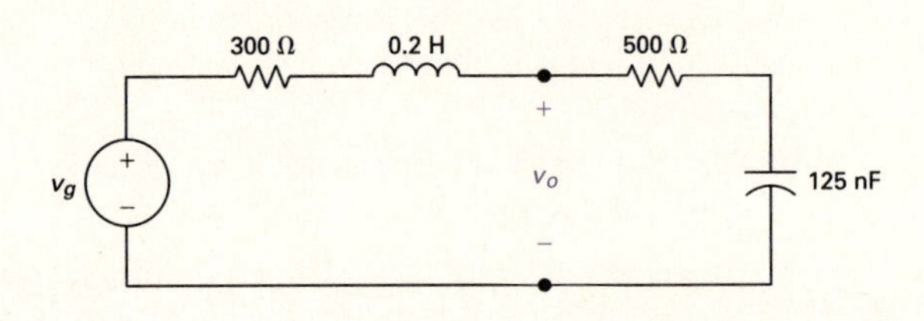

FIGURE P9.25

9.26 Use the concept of current division to find the steady-state expression for i_o in the circuit in Fig. P9.26 if $i_g = 400 \cos 20{,}000t$ mA.

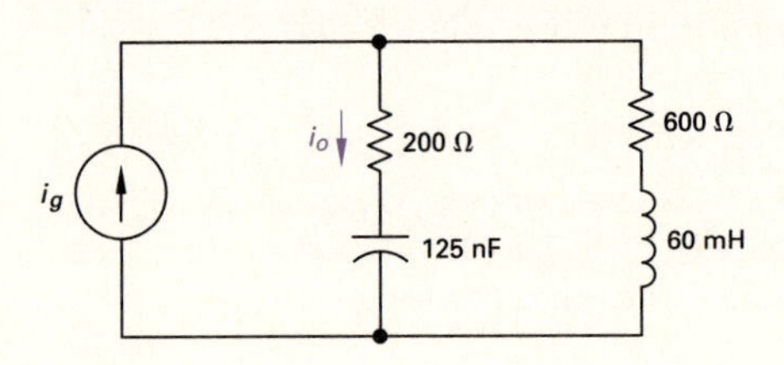

FIGURE P9.26

9.27 The expressions for the steady-state voltage and current at the terminals of the circuit seen in Fig. P9.27 are

$$v_g = 120 \cos (5000\pi t + 17.5^\circ) \text{ V}$$

and

$$i_g = 20 \sin (5000\pi t + 85^\circ) \text{ A}$$

a) What is the impedance seen by the source?

b) By how many microseconds is the current out of phase with the voltage?

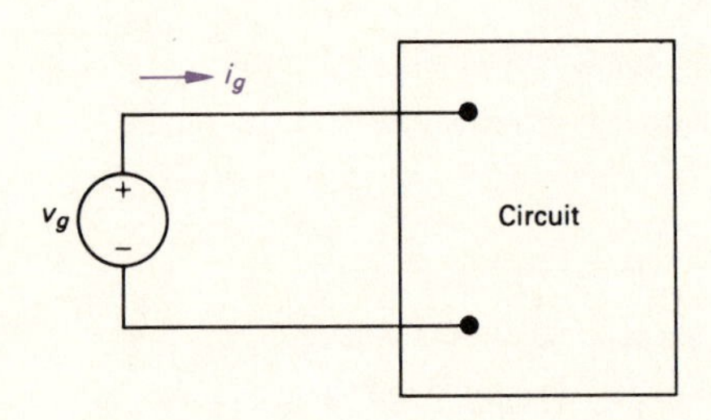

FIGURE P9.27

9.28 a) Show that at a given frequency ω, the circuits in Fig. P9.28(a) and (b) will have the same impedance between the terminals a,b if

$$R_1 = \frac{R_2}{1+\omega^2 R_2^2 C_2^2} \quad \text{and} \quad C_1 = \frac{1+\omega^2 R_2^2 C_2^2}{\omega^2 R_2^2 C_2}.$$

b) Find the values of resistance and capacitance that when connected in series will have the same impedance at 50 krad/s as that of a 400 Ω resistor connected in parallel with a 100 nF capacitor.

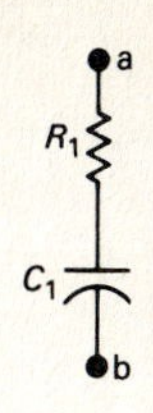

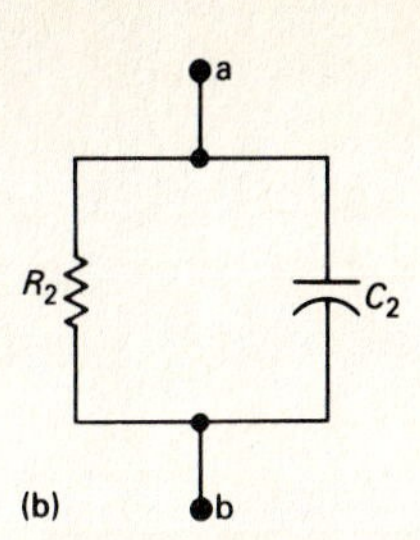

FIGURE P9.28

9.29 a) Show that at a given frequency ω, the circuits in Fig 9.28(a) and (b) will have the same impedance between the terminals a,b if

$$R_2 = \frac{1+\omega^2 R_1^2 C_1^2}{\omega^2 R_1 C_1^2} \quad \text{and} \quad C_2 = \frac{C_1}{1+\omega^2 R_1^2 C_1^2}.$$

(*Hint:* The two circuits will have the same impedance if they have the same admittance.)

b) Find the values of resistance and capacitance that when connected in parallel will give the same impedance at 40 krad/s as that of a 500 Ω resistor connected in series with a capacitance of 150 nF.

9.30 a) Show that at a given frequency ω, the circuits in Fig. P9.30(a) and (b) will have the same impedance between the terminals a,b if

$$R_1 = \frac{\omega^2 L_2^2 R_2}{R_2^2 + \omega^2 L_2^2} \quad \text{and} \quad L_1 = \frac{R_2^2 L_2}{R_2^2 + \omega^2 L_2^2}.$$

b) Find the values of resistance and inductance that when connected in series will have the same impedance at 10 krad/s as that of a 40 kΩ resistor connected in parallel with a 2.0 H inductor.

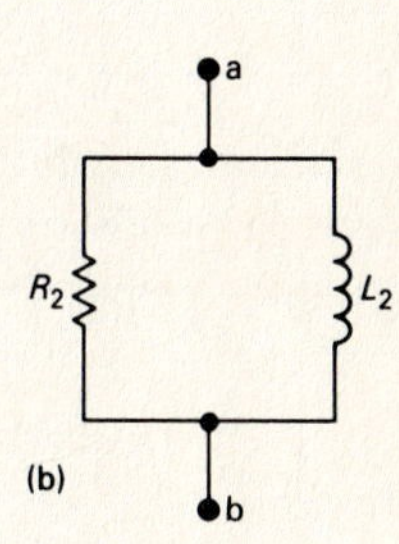

FIGURE P9.30

9.31 a) Show that at a given frequency ω, the circuits in Fig. P9.30(a) and (b) will have the same impedance between the terminals a,b if

$$R_2 = \frac{R_1^2 + \omega^2 L_1^2}{R_1} \quad \text{and} \quad L_2 = \frac{R_1^2 + \omega^2 L_1^2}{\omega^2 L_1}.$$

(*Hint:* The two circuits will have the same impedance if they have the same admittance.)

b) Find the values of resistance and inductance that when connected in parallel will have the same impedance at 5000 rad/s as a 1 kΩ resistor connected in series with a 0.4 H inductor.

9.32 The phasor current $\mathbf{I}_a$ in the circuit shown in Fig. P9.32 is $2\angle 0^\circ$ A.

a) Find $\mathbf{I}_b$, $\mathbf{I}_c$, and $\mathbf{V}_g$.

b) If $\omega = 800$ rad/s, write the expressions for $i_b(t)$, $i_c(t)$, and $v_g(t)$.

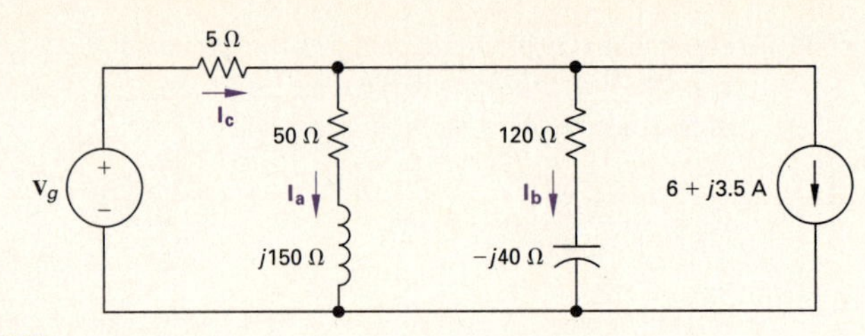

FIGURE P9.32

9.33 Find $\mathbf{I}_b$ and Z in the circuit shown in Fig. P9.33 if $\mathbf{V}_g = 25\angle 0^\circ$ V and $\mathbf{I}_a = 5\angle 90^\circ$ A.

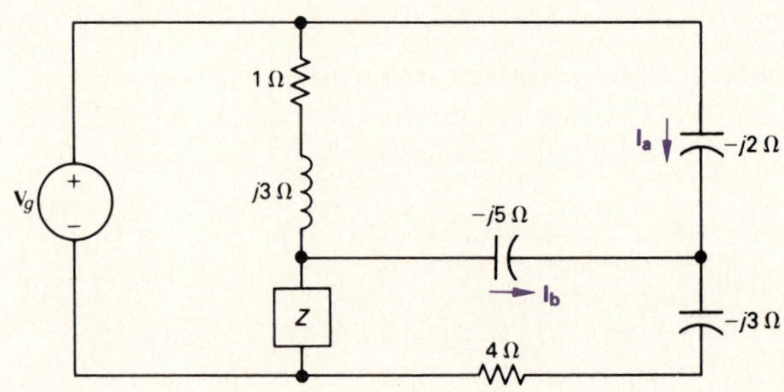

FIGURE P9.33

9.34 Find the steady-state expression for $v_o(t)$ in the circuit seen in Fig. P9.34 by using the technique of source transformations. The sinusoidal voltage sources are

$$v_1 = 500 \sin(8000t + 126.87^\circ) \text{ V}$$

and

$$v_2 = 1000 \cos(8000t - 90^\circ) \text{ V}$$

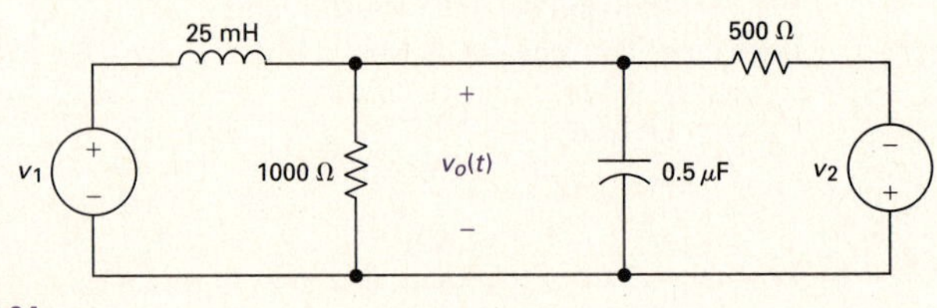

FIGURE P9.34

9.35 The circuit in Fig. P9.35 is operating in the sinusoidal steady state. Find $v_o(t)$ if $i_s(t) = 12.5 \cos 5000t$ mA.

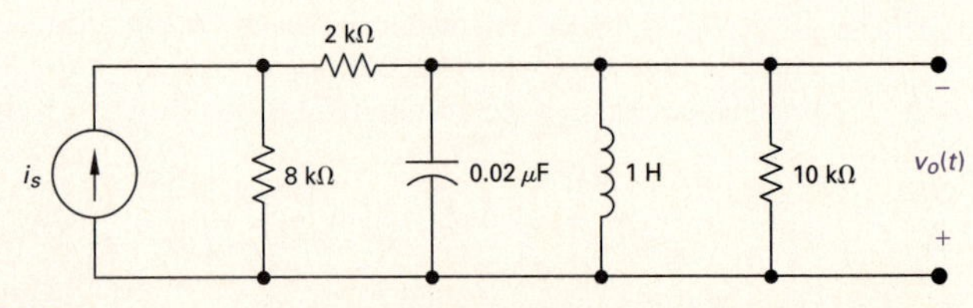

FIGURE P9.35

9.36 Use the node-voltage method to find the steady-state expression for $v_o(t)$ in the circuit in Fig. P9.36 if

$$v_{g1} = 20 \cos (2000t - 36.87^\circ) \text{ V}.$$

and

$$v_{g2} = 50 \sin (2000t - 16.26^\circ) \text{ V}.$$

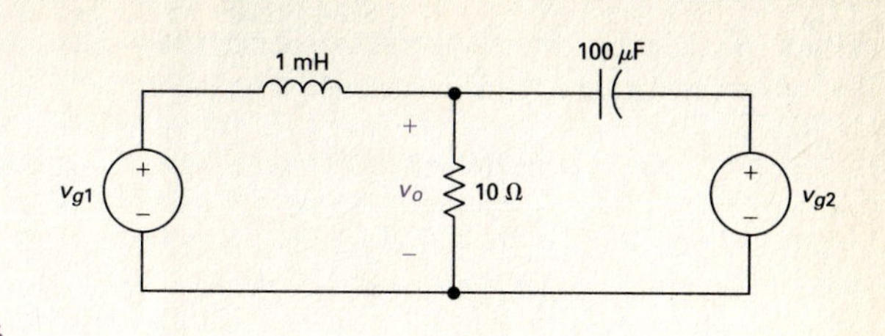

FIGURE P9.36

9.37 Use source transformations to find the steady-state expression for $v_o(t)$ in the circuit in Fig. P9.36.

9.38 Use the node-voltage method to find $\mathbf{V}_o$ in the circuit in Fig. P9.38.

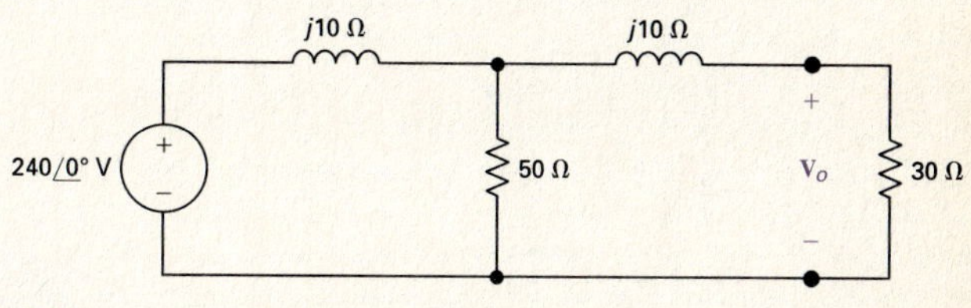

FIGURE P9.38

9.39 Use the node-voltage method to find the phasor voltage across the inductor in the circuit in Fig. P9.44. Assume the voltage is positive at the upper terminal of the inductor.

9.40 Use the node-voltage method to find the phasor voltage $\mathbf{V}_o$ in the circuit shown in Fig. P9.40. Express the voltage in both polar and rectangular form.

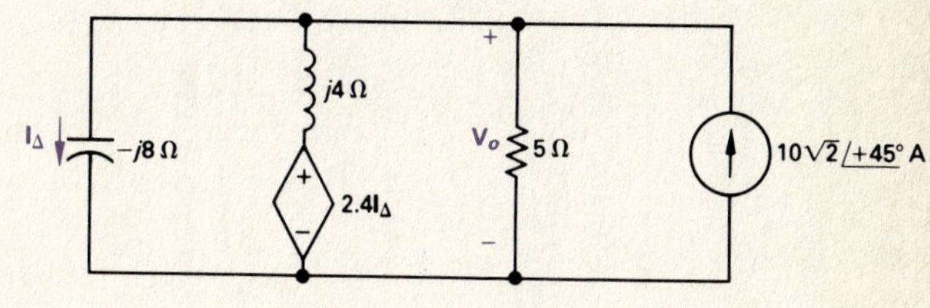

FIGURE P9.40

9.41 Use the node-voltage method to find $\mathbf{V}_o$ and $\mathbf{I}_o$ in the circuit seen in Fig. P9.41.

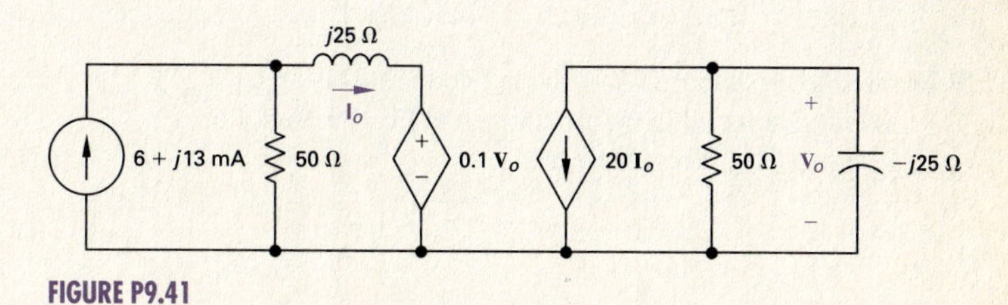

FIGURE P9.41

9.42 Use the mesh-current method to find the steady-state expression for $v_o(t)$ in the circuit in Fig. P9.36.

9.43 Use the mesh-current method to find the steady-state expression for $i_o(t)$ in the circuit in Fig. P9.43 if

$$v_a = 100\cos(50{,}000t)\text{ V}$$

and

$$v_b = 100\sin(50{,}000t + 180°)\text{ V}.$$

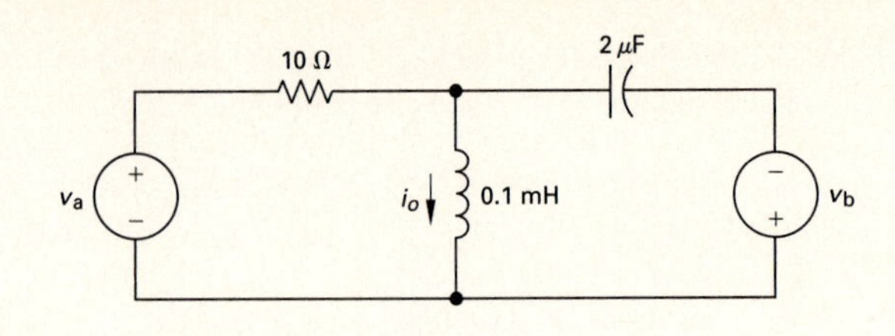

FIGURE P9.43

9.44 Use the mesh-current method to find the phasor current $\mathbf{I}_g$ in the circuit shown in Fig. P9.44.

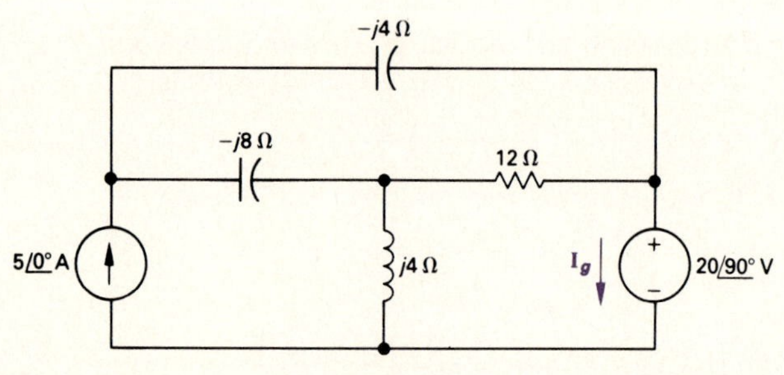

FIGURE P9.44

9.45 Use the mesh-current method to find the branch currents $\mathbf{I}_a$, $\mathbf{I}_b$, $\mathbf{I}_c$, and $\mathbf{I}_d$ in the circuit shown in Fig. P9.45.

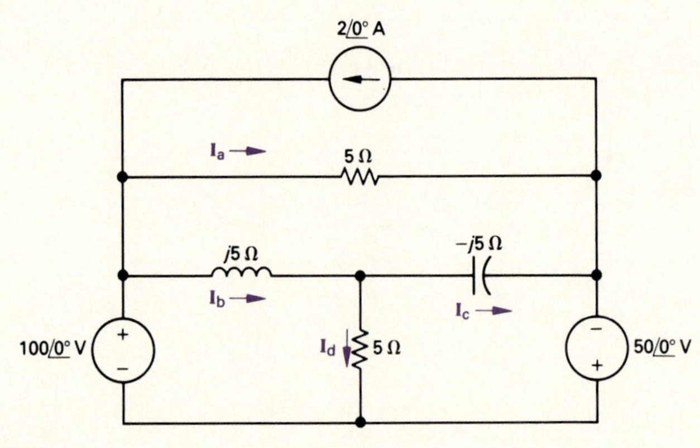

FIGURE P9.45

9.46 Use the mesh-current method to find the steady-state expression for v_o in the circuit seen in Fig. P9.46 if v_g equals 130 cos 10,000t V.

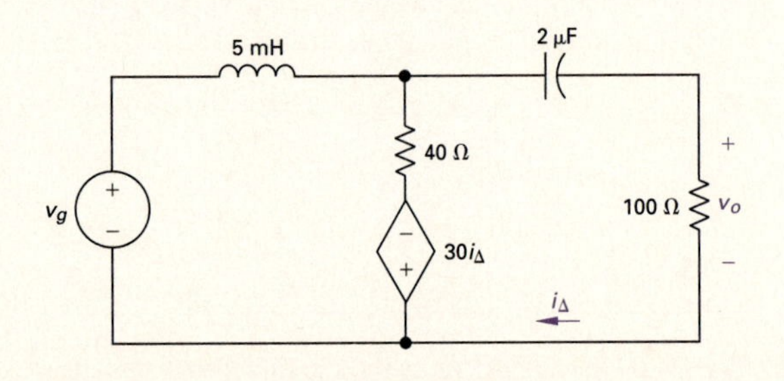

FIGURE P9.46

9.47 To introduce you to a circuit configuration widely used in residential wiring, we have shown a representational circuit in Fig. P9.47. In this simplified model, the resistor R_3 is used to model a 250 V appliance (such as an electric range), and the resistors R_1 and R_2 to model 125 V appliances (such as a lamp, toaster, and iron). The branches carrying $\mathbf{I}_1$ and $\mathbf{I}_2$ are modeling what electricians refer to as the hot conductors in the circuit, and the branch carrying $\mathbf{I}_n$ is modeling the neutral conductor. Our purpose in analyzing the circuit is to show the importance of the neutral conductor in the satisfactory operation of the circuit. You are to choose the method for analyzing the circuit.

a) Show that $\mathbf{I}_n$ is zero if $R_1 = R_2$.

b) Show that $\mathbf{V}_1 = \mathbf{V}_2$ if $R_1 = R_2$.

c) Open the neutral branch and calculate $\mathbf{V}_1$ and $\mathbf{V}_2$ if $R_1 = 40\ \Omega$, $R_2 = 400\ \Omega$, and $R_3 = 8\ \Omega$.

d) Close the neutral branch and repeat part (c).

e) On the basis of your calculations, explain why the neutral conductor is never fused in such a manner that it could open while the hot conductors are energized.

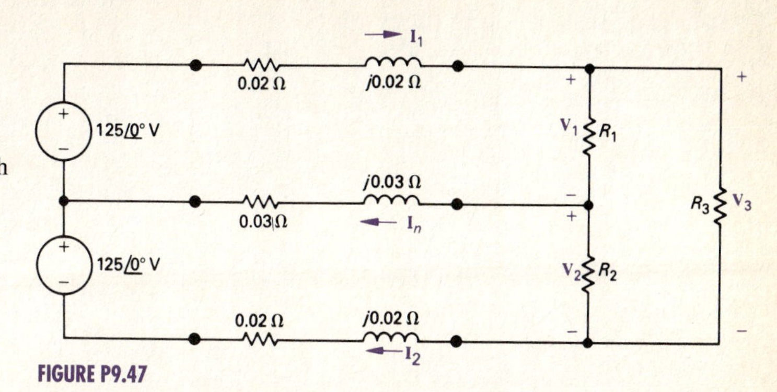

FIGURE P9.47

9.48 Find the steady-state expressions for the branch currents i_a, i_b, and i_c in the circuit seen in Fig. P9.48 if $v_a = 50 \sin 10^6 t$ V and $v_b = 25 \cos (10^6 t + 90°)$ V.

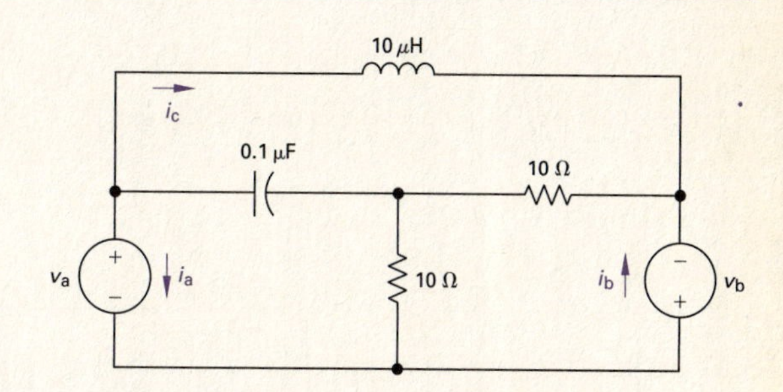

FIGURE P9.48

9.49 a) For the circuit shown in Fig. P9.49, find the steady-state expression for v_o if $i_g = 2 \cos (16 \times 10^5 t)$ A.

b) By how many nanoseconds does v_o lag i_g?

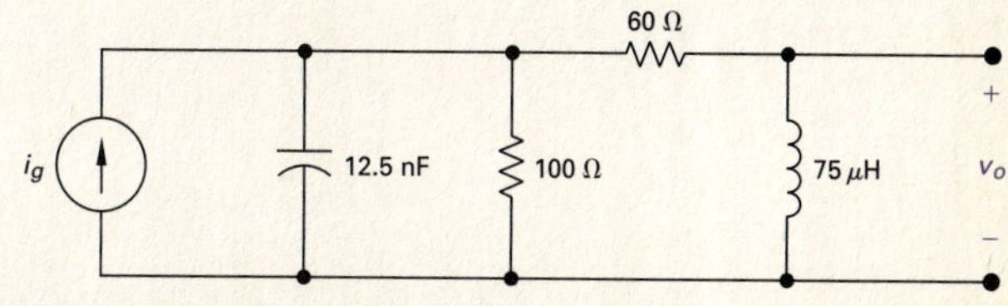

FIGURE P9.49

9.50 Find the value of Z in the circuit seen in Fig. P9.50 if $\mathbf{V}_g = 100 - j50$ V, $\mathbf{I}_g = 30 + j20$ A, and $\mathbf{V}_b = 140 + j30$ V.

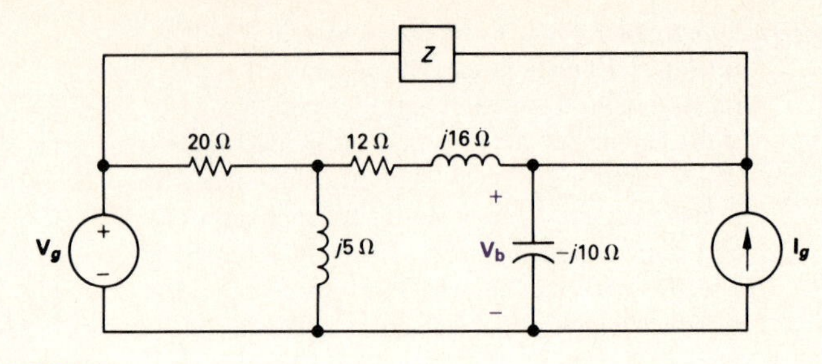

FIGURE P9.50

9.51 Find the Thévenin equivalent circuit with respect to the terminals a,b for the circuit shown in Fig. P9.51.

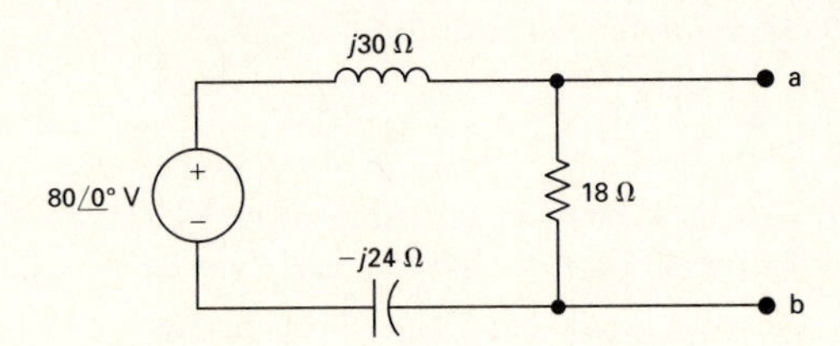

FIGURE P9.51

9.52 Find the Norton equivalent circuit with respect to the terminals a,b for the circuit shown in Fig. P9.52.

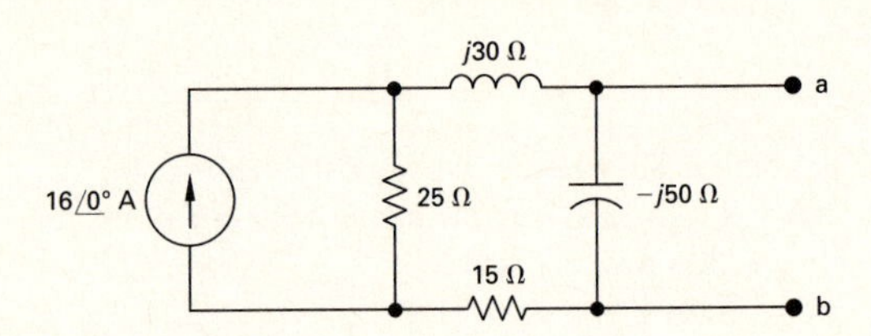

FIGURE P9.52

9.53 Find the Thévenin equivalent circuit with respect to the terminals a,b for the circuit shown in Fig. P9.53.

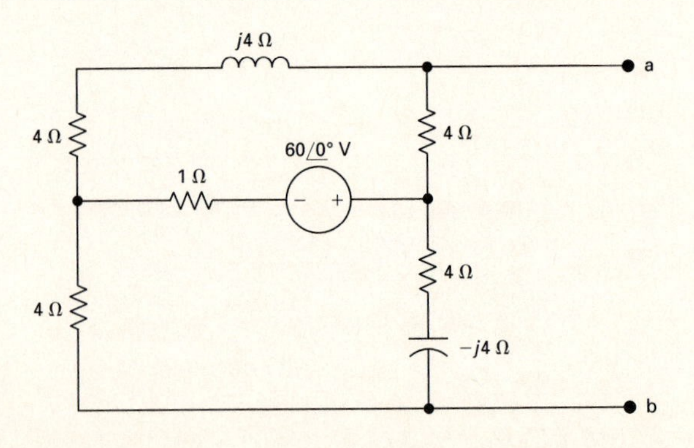

FIGURE P9.53

9.54 The sinusoidal voltage source in the circuit in Fig. P9.54 is developing a voltage equal to 247.49 cos (1000t + 45°) V.

a) Find the Thévenin voltage with respect to the terminals a,b.

b) Find the Thévenin impedance with respect to the terminals a,b.

c) Draw the Thévenin equivalent.

FIGURE P9.54

9.55 Find the Thévenin equivalent circuit with respect to the terminals a,b of the circuit shown in Fig. P9.55.

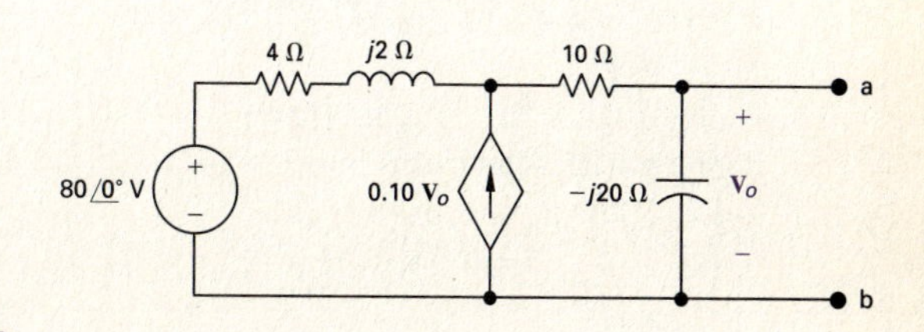

FIGURE P9.55

9.56 Find the Norton equivalent circuit with respect to the terminals a,b for the circuit shown in Fig. P9.56 when $\mathbf{V}_s = 100\underline{/0^\circ}$ mV.

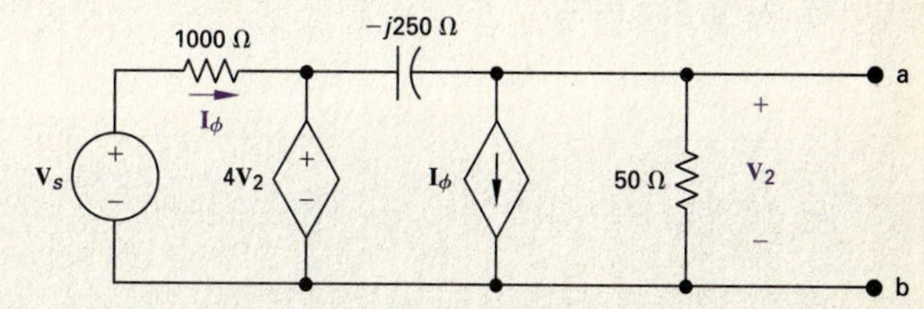

FIGURE P9.56

9.57 Find the Thévenin impedance seen looking into the terminals a,b of the circuit in Fig. P9.57 if the frequency of operation is $(25/\pi)$ kHz.

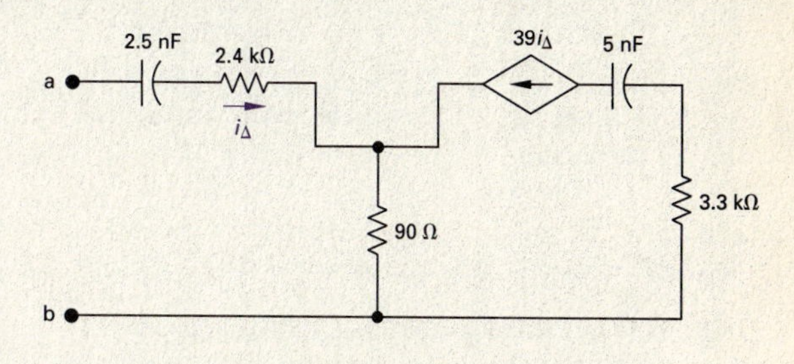

FIGURE P9.57

9.58 The device in Fig. P9.58 is represented in the frequency domain by a Norton equivalent. When a resistor having an impedance of 5 kΩ is connected across the device, the value of $\mathbf{V}_0$ is $5 - j15$ V. When a capacitor having an impedance of $-j3$ kΩ is connected across the device, the value of $\mathbf{I}_0$ is $4.5 - j6$ mA. Find the Norton current $\mathbf{I}_n$ and the Norton impedance Z_n.

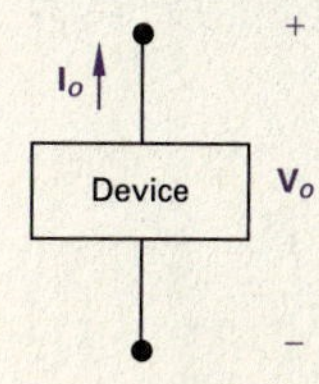

FIGURE P9.58

9.59 The circuit shown in Fig. P9.59 is operating at a frequency of 10 rad/s. Assume σ is real and lies between -10 and $+10$, that is, $-10 \leq \sigma \leq 10$.

a) Find the value of σ so that the Thévenin impedance looking into the terminals a,b is purely resistive.

b) What is the value of the Thévenin impedance for the σ found in part (a)?

c) Can σ be adjusted so that the Thévenin impedance equals $500 - j500\ \Omega$? If so, what is the value of σ?

d) For what values of σ will the Thévenin impedance be inductive?

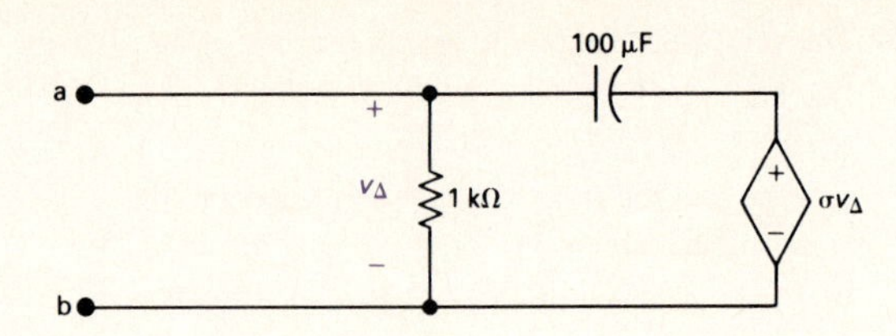

FIGURE P9.59

9.60 Use the principle of superposition to find the steady-state expression for the voltage $v_o(t)$ in the circuit in Fig. P9.36.

9.61 a) For the circuit shown in Fig. P9.61, compute $\mathbf{V}_s$ and $\mathbf{V}_l$.

b) Construct a phasor diagram showing the relationship between $\mathbf{V}_s$, $\mathbf{V}_l$, and the load voltage of $240\angle 0^\circ$ V.

c) Repeat parts (a) and (b), given that the load voltage remains at $240\angle 0^\circ$ V when a capacitive reactance of $-5\ \Omega$ is connected across the load terminals.

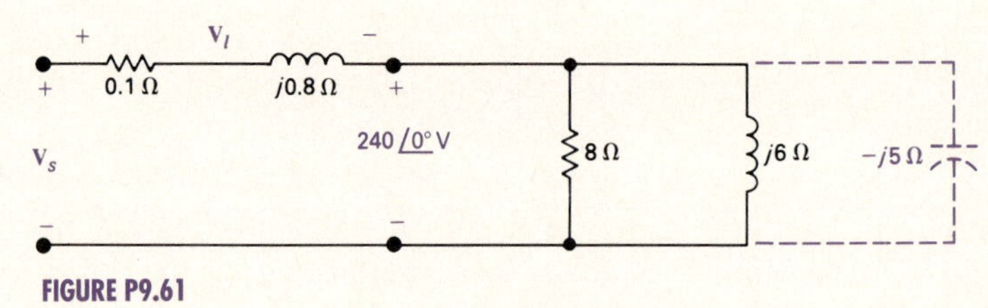

FIGURE P9.61

9.62 Show by using a phasor diagram what happens to the magnitude and phase angle of the voltage v_o in the circuit in Fig. P9.62 as R_x is varied from zero to infinity. The amplitude and phase angle of the source voltage are held constant as R_x varies.

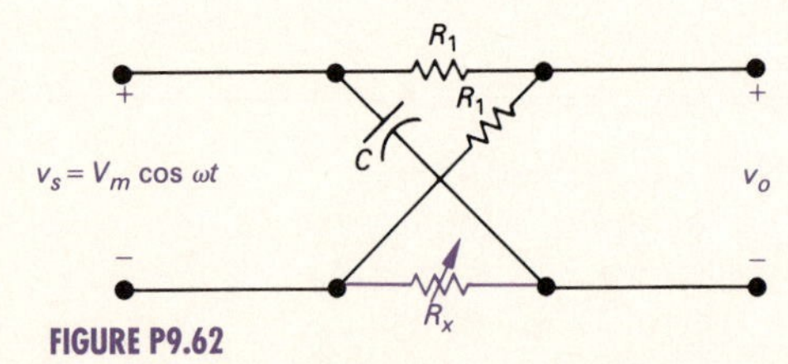

FIGURE P9.62

9.63 The operational amplifier in the circuit in Fig. P9.63 is ideal.

a) Find the steady-state expression for $v_o(t)$.

b) How large can the amplitude of v_g be before the amplifier saturates?

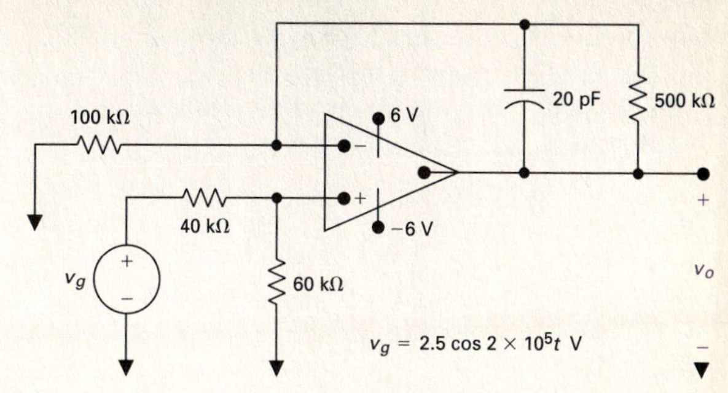

FIGURE P9.63

9.64 The operational amplifier in the circuit seen in Fig. P9.64 is ideal. Find the steady-state expression for $v_o(t)$ when $v_g = 2.5 \cos 10^7 t$ V.

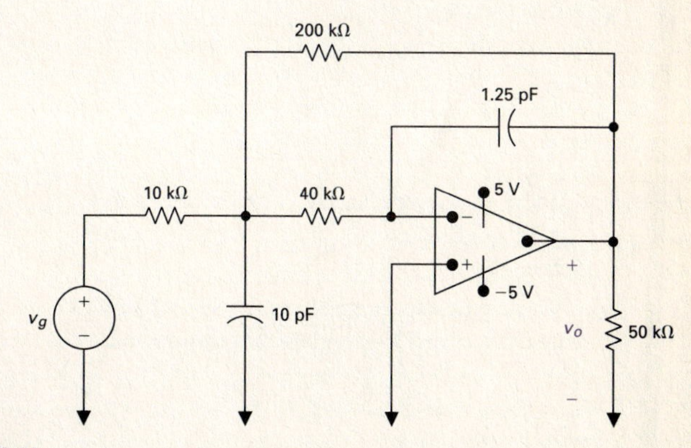

FIGURE P9.64

9.65 The operational amplifier in the circuit shown in Fig. P9.65 is ideal. The voltage of the ideal sinusoidal source is $v_g = 30 \cos 10^6 t$ V.

a) How small can C_o be before the steady-state output voltage no longer has a pure sinusoidal waveform?

b) For the value of C_o found in part (a), write the steady-state expression for v_o.

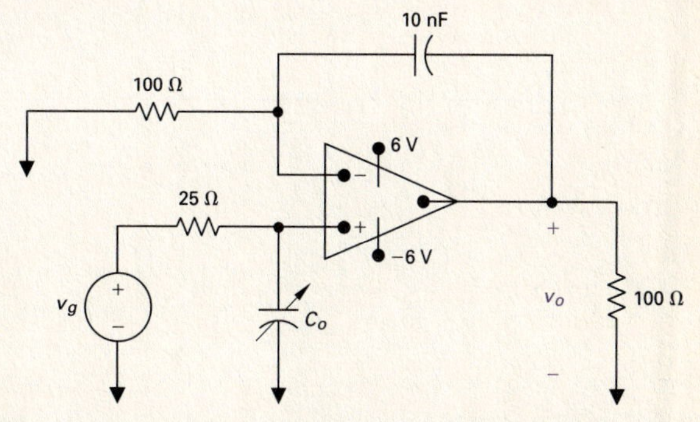

FIGURE P9.65

9.66 The sinusoidal voltage source in the circuit shown in Fig. P9.66 is generating the voltage $v_g = 3\cos 2500t$ V. If the op amp is ideal, what is the steady-state expression for $v_o(t)$?

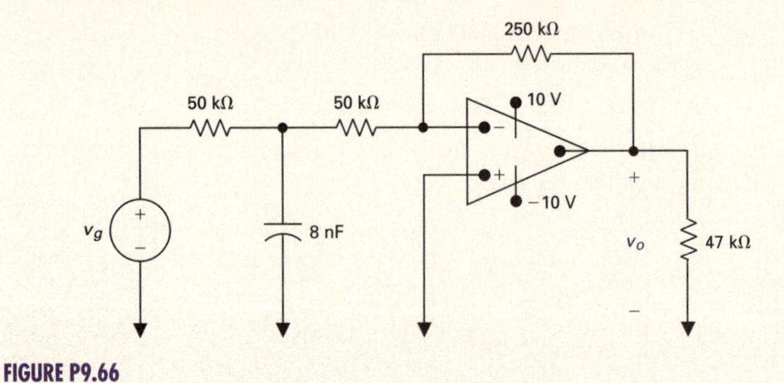

FIGURE P9.66

9.67 The 8 nF capacitor in the circuit seen in Fig. P9.66 is replaced with a variable capacitor. The capacitor is adjusted until the output voltage leads the input voltage by 120°.

a) Find the value of C in microfarads.

b) Write the steady-state expression for $v_o(t)$ when C has the value found in part (a).

9.68 a) Find the input impedance Z_{ab} for the circuit in Fig. P9.68. Express Z_{ab} as a function of Z and K where $K = (R_2/R_1)$.

b) If Z is a pure capacitive element, what is the capacitance seen looking into the terminals a,b?

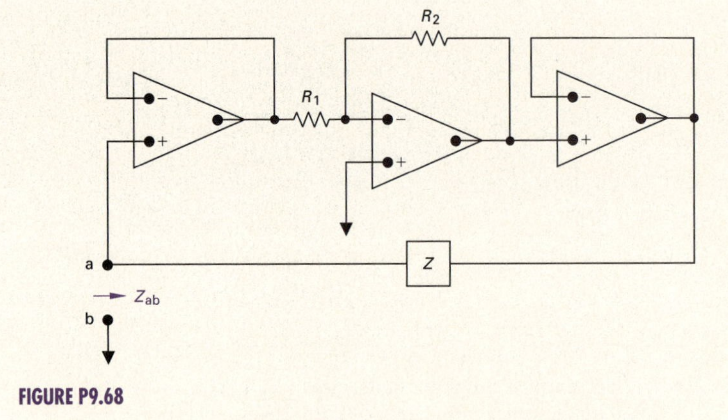

FIGURE P9.68

9.69 You may have the opportunity as an engineering graduate to serve as an expert witness in lawsuits involving either personal injury or property damage. As an example of the type of problem on which you may be asked to give an opinion, consider the following event.

At the end of a day of fieldwork, a farmer returns to his farmstead, checks his hog-confinement building, and finds to his dismay that the hogs are dead. The problem is traced to a blown fuse that caused a 240 V fan motor to stop. The loss of ventilation led to the suffocation of the livestock. The interrupted fuse is located in the main switch that connects the farmstead to the electrical service.

Before the insurance company settles the claim, it wants to know if the electric circuit supplying the farmstead functioned properly. The lawyers for the insurance company are puzzled because the farmer's wife, who was in the house on the day of the accident convalescing from minor surgery, was able to watch TV during the afternoon. Furthermore, when she went to the kitchen to start preparing the evening meal, the electric clock indicated the correct time.

The lawyers have hired you to explain (1) why the electric clock in the kitchen and the television set in the living room continued to operate after the fuse in the main switch blew and (2) why the second fuse in the main switch didn't blow after the fan motor stalled.

After ascertaining the loads on the three-wire distribution circuit prior to the interruption of fuse A, you are able to construct the circuit model shown in Fig. P9.69. The impedances of the line conductors and the neutral conductor are assumed negligible.

a) Calculate the branch currents $\mathbf{I}_1$, $\mathbf{I}_2$, $\mathbf{I}_3$, $\mathbf{I}_4$, $\mathbf{I}_5$, and $\mathbf{I}_6$ prior to the interruption of fuse A.

b) Calculate the branch currents after the interruption of fuse A. Assume the stalled fan motor behaves as a short circuit.

c) Explain why the clock and television set were not affected by the momentary short circuit that interrupted fuse A.

d) Assume the fan motor is equipped with a thermal cutout designed to interrupt the motor circuit if the motor current becomes excessive. Would you expect the thermal cutout to operate? Explain.

e) Explain why fuse B is not interrupted when the fan motor stalls.

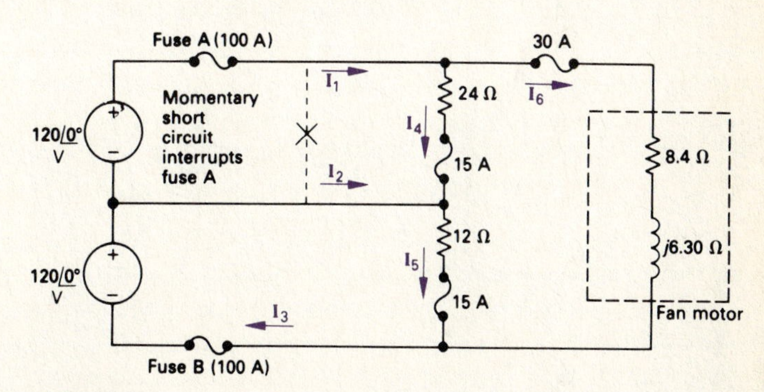

FIGURE P9.69

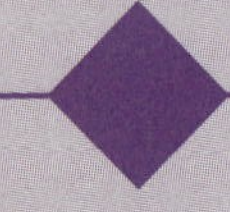

PRACTICAL PERSPECTIVE

Heating Appliances

In Chapter 9 we calculated the steady-state voltages and currents in electric circuits driven by sinusoidal sources. In this chapter we consider power in such circuits. The techniques we develop are useful for analyzing many of the electrical devices we encounter daily, because sinusoidal sources are the predominant means of providing electric power in our homes, schools, and businesses.

One common class of electrical devices is heaters, which transform electric energy into thermal energy. Examples include electric stoves and ovens, toasters, irons, electric water heaters, space heaters, electric clothes dryers, and hair dryers. One of the critical design concerns in a heater is power consumption. Power is important for two reasons: the more power a heater uses, the more it costs to operate, and the more heat it can produce.

Many electric heaters have different power settings corresponding to the amount of heat the device supplies. You may wonder just how these settings result in different amounts of heat output. The Practical Perspective example at the end of this chapter examines the design of a handheld hair dryer with three operating settings. You will see how the design provides for three different power levels which correspond to three different levels of heat output.

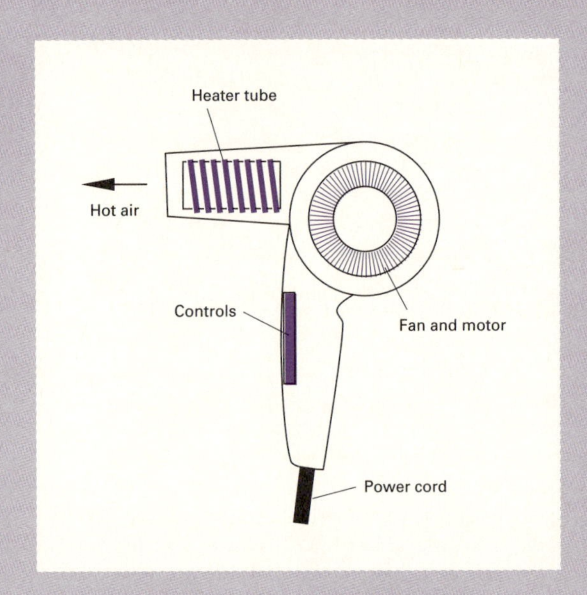

10 Sinusoidal Steady-State Power Calculations

CHAPTER CONTENTS

Power engineering has evolved into one of the important subdisciplines within electrical engineering. The range of problems dealing with the delivery of energy to do work is considerable, from determining the power rating within which an appliance operates safely and efficiently, to designing the vast array of generators, transformers, and wires which provide electric energy to household and industrial consumers.

Nearly all electric energy is supplied in the form of sinusoidal voltages and currents. Thus, after our Chapter 9 discussion of sinusoidal circuits, this is the logical place to consider sinusoidal steady-state power calculations. We are primarily interested in the average power delivered to or supplied from a pair of terminals due to sinusoidal voltages and currents. Other measures, such as reactive power, complex power, and apparent power, will also be presented. The concept of the rms value of a sinusoid, briefly introduced in Chapter 9, is particularly pertinent to power calculations.

We begin and end this chapter with two concepts which should be very familiar to you from previous chapters: the basic equation for power (Section 10.1) and maximum power transfer (Section 10.7). In between, we discuss the general processes for analyzing power, which will be familiar from your studies in Chapters 1 and 4, although some additional mathematical techniques are required here to deal with sinusoidal, rather than dc, signals.

10.1 INSTANTANEOUS POWER

We begin our investigation of sinusoidal power calculations with the familiar circuit in Fig. 10.1. Here, v and i are steady-state sinusoidal signals. Using the passive sign convention, the power at any instant of time is

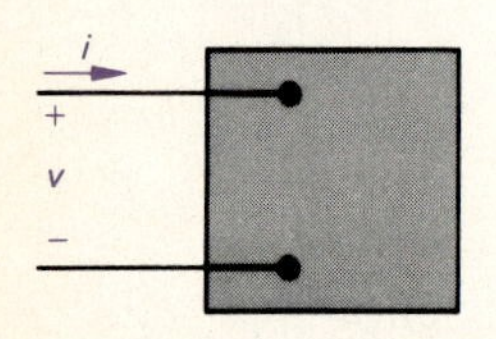

FIGURE 10.1 The black box representation of a circuit used for calculating power.

$$p = vi. \quad \textbf{(10.1)}$$

This is **instantaneous power**. Remember that if the reference direction of the current is in the direction of the voltage rise, Eq. (10.1) must be written with a minus sign. Instantaneous power is measured in watts when the voltage is in volts and the current is in amperes. First, we write expressions for v and i:

$$v = V_m \cos(\omega t + \theta_v) \quad \textbf{(10.2)}$$

and

$$i = I_m \cos(\omega t + \theta_i), \quad \textbf{(10.3)}$$

where θ_v is the voltage phase angle, and θ_i is the current phase angle.

We are operating in the sinusoidal steady state, so we may choose any convenient reference for zero time. Engineers designing systems which transfer large blocks of power have found it convenient to use a zero time corresponding to the instant when the current is passing through a positive maximum. This reference system requires a shift of both the voltage and current by θ_i. Thus Eqs. (10.2) and (10.3) become

$$v = V_m \cos(\omega t + \theta_v - \theta_i) \quad \textbf{(10.4)}$$

and

$$i = I_m \cos \omega t. \tag{10.5}$$

When we substitute Eqs. (10.4) and (10.5) into Eq. (10.1), the expression for the instantaneous power becomes

$$p = V_m I_m \cos (\omega t + \theta_v - \theta_i) \cos \omega t. \tag{10.6}$$

We could use Eq. (10.6) directly to find the average power; however, by simply applying a couple of trigonometric identities, we can put Eq. (10.6) into a much more informative form.

We begin with the trigonometric identity[1]

$$\cos \alpha \cos \beta = \frac{1}{2} \cos (\alpha - \beta) + \frac{1}{2} \cos (\alpha + \beta)$$

to expand Eq. (10.6); letting $\alpha = \omega t + \theta_v - \theta_i$ and $\beta = \omega t$ gives

$$p = \frac{V_m I_m}{2} \cos (\theta_v - \theta_i) + \frac{V_m I_m}{2} \cos (2\omega t + \theta_v - \theta_i). \tag{10.7}$$

Now use the trigonometric identity

$$\cos (\alpha + \beta) = \cos \alpha \cos \beta - \sin \alpha \sin \beta$$

to expand the second term on the right-hand side of Eq. (10.7), which gives

$$p = \frac{V_m I_m}{2} \cos (\theta_v - \theta_i) + \frac{V_m I_m}{2} \cos (\theta_v - \theta_i) \cos 2\omega t - \frac{V_m I_m}{2} \sin (\theta_v - \theta_i) \sin 2\omega t. \tag{10.8}$$

Figure 10.2 depicts a representative relationship among v, i, and p, based on the assumptions $\theta_v = 60°$ and $\theta_i = 0°$. You can see that the frequency of the instantaneous power is twice the frequency of the voltage or current. This observation also follows directly from the second two terms on the right-hand side of Eq. (10.8). Therefore, the instantaneous power goes through two complete cycles for every cycle of either the voltage or the current. Also note that the instantaneous power may be negative for a portion of each cycle, even if the network between the terminals is passive. In a completely passive network, negative power implies that energy stored in the inductors or capacitors is now being extracted. The fact that the instantaneous power varies with time in the sinusoidal steady-state operation of a circuit explains why some motor-driven appliances (such as refrigerators) experience vibration and require resilient motor mountings to prevent excessive vibration.

Use a computer to generate plots like those in the figure.

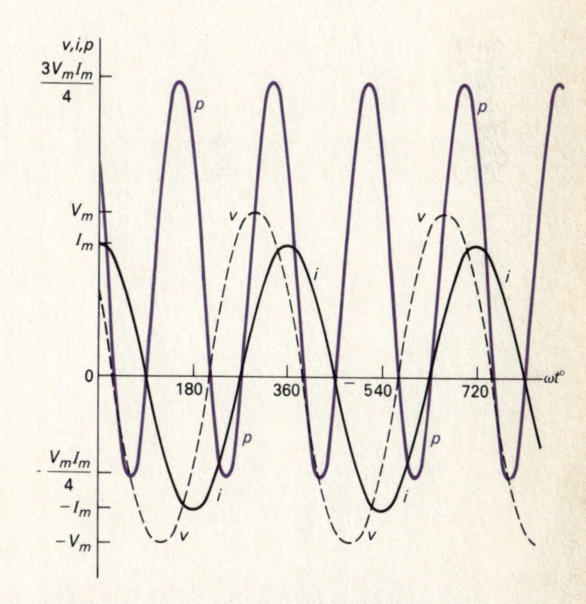

FIGURE 10.2 Instantaneous power, voltage, and current versus ωt for steady-state sinusoidal operation.

[1] See entry 8 in Appendix E.

We are now ready to use Eq. (10.8) to find the average power at the terminals of the circuit represented by Fig. 10.1 and, at the same time, introduce the concept of reactive power.

10.2 Average and Reactive Power

We begin by noting that Eq. (10.8) has three terms, which we can rewrite as follows:

$$p = P + P\cos 2\omega t - Q\sin 2\omega t, \tag{10.9}$$

where

$$P = \frac{V_m I_m}{2}\cos(\theta_v - \theta_i) \tag{10.10}$$

and

$$Q = \frac{V_m I_m}{2}\sin(\theta_v - \theta_i). \tag{10.11}$$

P is called the **average power**, and Q is called the **reactive power**. Average power is sometimes called **real power**, because it describes the power in a circuit which is transformed from electric to nonelectric energy. Although the two terms are interchangeable, we primarily use the term *average power* in this text.

It is easy to see why P is called the average power. The average power associated with sinusoidal signals is the average of the instantaneous power over one period, or, in equation form,

$$P = \frac{1}{T}\int_{t_0}^{t_0+T} p\,dt, \tag{10.12}$$

where T is the period of the sinusoidal function. The limits on Eq. (10.12) imply that we can initiate the integration process at any convenient time t_0 but that we must terminate the integration exactly one period later. (We could integrate over nT periods, where n is an integer, provided we multiply the integral by $1/nT$.)

We could find the average power by substituting Eq. (10.9) directly into Eq. (10.12) and then performing the integration. But note that the average value of p is given by the first term on the right-hand side of Eq. (10.9), because the integral of both $\cos 2\omega t$ and $\sin 2\omega t$ over one period is zero. Thus the average power is given in Eq. (10.10).

We can develop a better understanding of all the terms in Eq. (10.9) and the relationships among them by examining the power in circuits which are purely resistive, purely inductive, or purely capacitive.

Power for Purely Resistive Circuits

If the circuit between the terminals is purely resistive, the voltage and current are in phase, which means that $\theta_v = \theta_i$. Equation (10.9) then reduces to

$$p = P + P\cos 2\omega t. \tag{10.13}$$

The instantaneous power expressed in Eq. (10.13) is referred to as the **instantaneous real power**. Figure 10.3 shows a graph of Eq. (10.13) for a representative purely resistive circuit, assuming $\omega = 377$ rad/s. By definition, the average power, P, is the average of p over one period. Thus it is easy to see just by looking at the graph that $P = 1$ for this circuit. Note from Eq. (10.13) that the instantaneous real power can never be negative, which is also shown in Fig. 10.3. In other words, power cannot be extracted from a purely resistive network. Rather, all the electric energy is dissipated in the form of thermal energy.

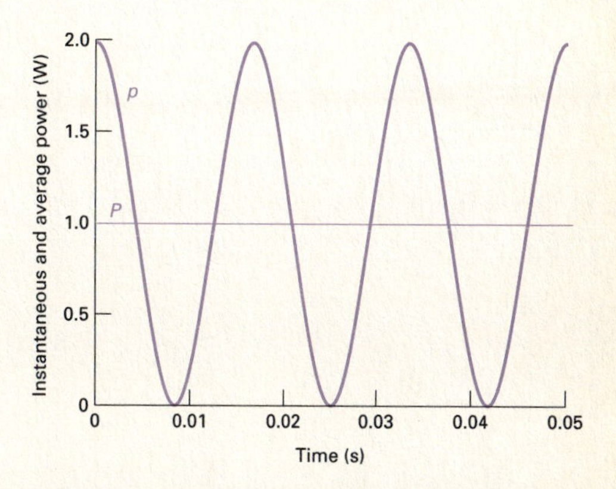

FIGURE 10.3 Instantaneous real power and average power for a purely resistive circuit.

Power for Purely Inductive Circuits

If the circuit between the terminals is purely inductive, the voltage and current are out of phase by precisely 90°. In particular, the current lags the voltage by 90° (that is, $\theta_i = \theta_v - 90°$); therefore $\theta_v - \theta_i = +90°$. The expression for the instantaneous power then reduces to

$$p = -Q\sin 2\omega t. \tag{10.14}$$

In a purely inductive circuit, the average power is zero. Therefore no transformation of energy from electric to nonelectric form takes place. The instantaneous power at the terminals in a purely inductive circuit is continually exchanged between the circuit and the source driving the circuit, at a frequency of 2ω. In other words, when p is positive, energy is being stored in the magnetic fields associated with the inductive elements, and when p is negative, energy is being extracted from the magnetic fields.

A measure of the power associated with purely inductive circuits is the reactive power Q. The name *reactive power* comes from the characterization of an inductor as a reactive element; its impedance is purely reactive. Note that average power P and reactive power Q carry the same dimension. To distinguish between average and reactive power, we use the units *watts* for average power and **var** (VAR) for reactive power (*var* is an acronym for the phrase *volt-amp reactive*). Figure 10.4 plots the instantaneous power for a representative purely inductive circuit, assuming $\omega = 377$ rad/s.

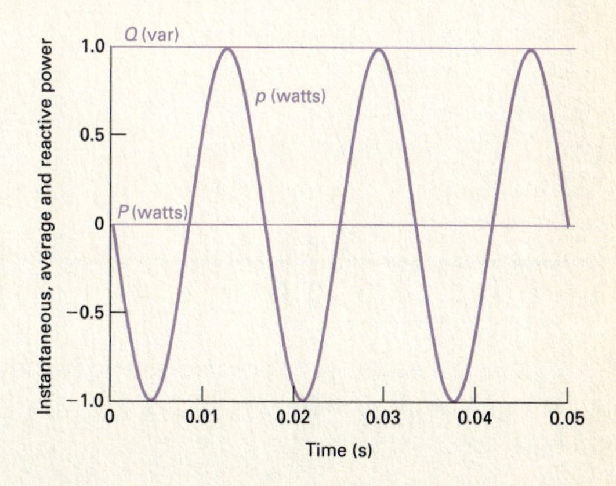

FIGURE 10.4 Instantaneous power, average power, and reactive power for a purely inductive circuit.

Power for Purely Capacitive Circuits

If the circuit between the terminals is purely capacitive, the voltage and current are precisely 90° out of phase. In this case, the current leads the voltage by 90° (that is, $\theta_i = \theta_v + 90°$); thus, $\theta_v - \theta_i = -90°$.

The expression for the instantaneous power then becomes

$$p = Q \sin 2\omega t. \tag{10.15}$$

Again, the average power is zero, so there is no transformation of energy from electric to nonelectric form. In a purely capacitive circuit, the power is continually exchanged between the source driving the circuit and the electric field associated with the capacitive elements. Figure 10.5 plots the instantaneous power for a representative purely capacitive circuit, assuming $\omega = 377$ rad/s.

Note that the decision to use the current as the reference leads to Q being positive for inductors (that is, $\theta_v - \theta_i = 90°$) and negative for capacitors (that is, $\theta_v - \theta_i = -90°$). Power engineers recognize this difference in the algebraic sign of Q by saying that inductors demand (or absorb) magnetizing vars, and capacitors furnish (or deliver) magnetizing vars. We say more about this convention later.

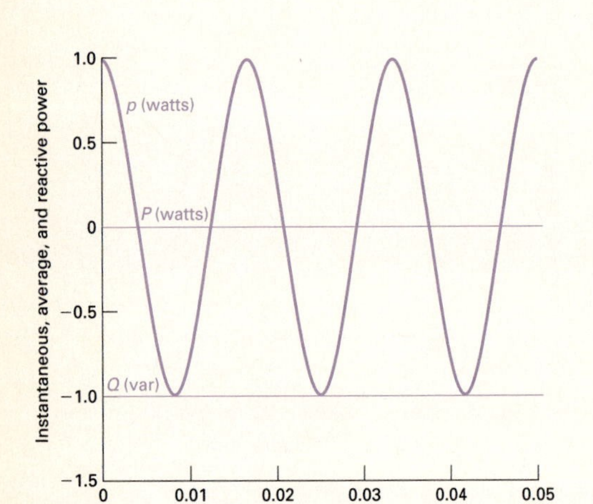

FIGURE 10.5 Instantaneous power, average power, and reactive power for a purely capacitive circuit.

THE POWER FACTOR

The angle $\theta_v - \theta_i$ plays a role in the computation of both average and reactive power and is referred to as the **power factor angle**. The cosine of this angle is called the **power factor**, abbreviated pf, and the sine of this angle is called the **reactive factor**, abbreviated rf. Thus

$$\text{pf} = \cos(\theta_v - \theta_i) \tag{10.16}$$

and

$$\text{rf} = \sin(\theta_v - \theta_i). \tag{10.17}$$

Knowing the value of the power factor does not tell you the value of the power factor angle, because $\cos(\theta_v - \theta_i) = \cos(\theta_i - \theta_v)$. To completely describe this angle, we use the descriptive phrases **lagging power factor** and **leading power factor**. Lagging power factor implies that current lags voltage—hence an inductive load. Leading power factor implies that current leads voltage—hence a capacitive load. Both the power factor and the reactive factor are convenient quantities to use in describing electrical loads.

Example 10.1 illustrates the interpretation of P and Q on the basis of a numerical calculation.

EXAMPLE 10.1

a) Calculate the average power and the reactive power at the terminals of the network shown in Fig. 10.6 if

$$v = 100 \cos(\omega t + 15°) \text{ V}$$

and

$$i = 4 \sin(\omega t - 15°) \text{ A}.$$

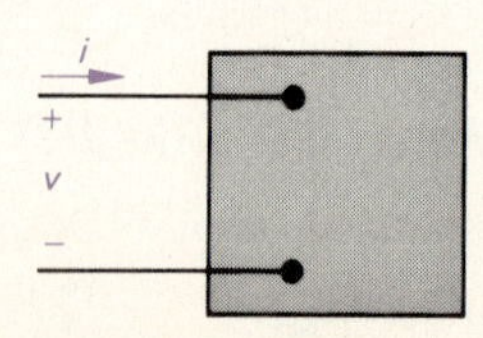

FIGURE 10.6 A pair of terminals used for calculating power.

b) State whether the network inside the box is absorbing or delivering average power.

c) State whether the network inside the box is absorbing or supplying magnetizing vars.

SOLUTION

a) Because i is expressed in terms of the sine function, the first step in the calculation for P and Q is to rewrite i as a cosine function:

$$i = 4\cos(\omega t - 105^\circ)\text{ A.}$$

We now calculate P and Q directly from Eqs. (10.10) and (10.11). Thus

$$P = \frac{1}{2}(100)(4)\cos[15 - (-105)] = -100\text{ W}$$

and

$$Q = \frac{1}{2}100(4)\sin[15 - (-105)] = 173.21\text{ VAR.}$$

b) Note from Fig. 10.6 the use of the passive sign convention. Because of this, the negative value of −100 W means that the network inside the box is delivering average power to the terminals.

c) The passive sign convention means that, because Q is positive, the network inside the box is absorbing magnetizing vars at its terminals.

DRILL EXERCISES

10.1 For each of the following sets of voltage and current, calculate the real and reactive power in the line between networks A and B in the circuit shown. In each case, state whether the power flow is from A to B or vice versa. Also state whether magnetizing vars are being transferred from A to B or vice versa.

a) $v = 250\cos(\omega t + 45^\circ)$ V; $i = 12\cos(\omega t - 15^\circ)$ A.

b) $v = 250\cos(\omega t + 45^\circ)$ V; $i = 12\cos(\omega t + 165^\circ)$ A.

c) $v = 250\cos(\omega t + 45^\circ)$ V; $i = 12\cos(\omega t + 105^\circ)$ A.

d) $v = 250\cos\omega t$ V; $i = 12\cos(\omega t - 120^\circ)$ A.

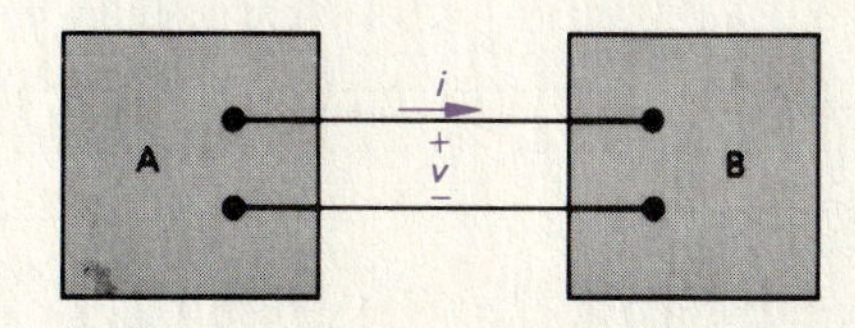

ANSWER: (a) $P = 750$ W (A to B), $Q = 1299.04$ VAR (A to B); (b) $P = -750$ W (B to A), $Q = -1299.04$ VAR (B to A); (c) $P = 750$ W (A to B), $Q = -1299.04$ VAR (B to A); (d) $P = -750$ W (B to A), $Q = 1299.04$ VAR (A to B).

10.2 Compute the power factor and the reactive factor for the network inside the box in Fig. 10.6, whose voltage and current are described in Example 10.1.

ANSWER: pf = 0.866 lagging; rf = 0.5.

10.3 The RMS Value and Power Calculations

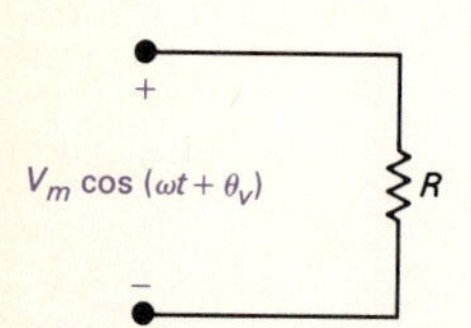

FIGURE 10.7 A sinusoidal voltage applied to the terminals of a resistor.

In introducing the rms value of a sinusoidal voltage (or current) in Section 9.1, we mentioned that it would play an important role in power calculations. We can now discuss this role.

Assume that a sinusoidal voltage is applied to the terminals of a resistor, as shown in Fig. 10.7, and that we want to determine the average power delivered to the resistor. From Eq. (10.12),

$$P = \frac{1}{T}\int_{t_0}^{t_0+T} \frac{V_m^2 \cos^2(\omega t + \phi_v)}{R} dt$$

$$= \frac{1}{R}\left[\frac{1}{T}\int_{t_0}^{t_0+T} V_m^2 \cos^2(\omega t + \phi_v) dt\right]. \quad \textbf{(10.18)}$$

Comparing Eq. (10.18) with Eq. (9.5) reveals that the average power delivered to R is simply the rms value of the voltage squared divided by R, or

$$P = \frac{V_{\text{rms}}^2}{R}. \quad \textbf{(10.19)}$$

If the resistor is carrying a sinusoidal current, say, $I_m \cos(\omega t + \phi_i)$, the average power delivered to the resistor is

$$P = I_{\text{rms}}^2 R. \quad \textbf{(10.20)}$$

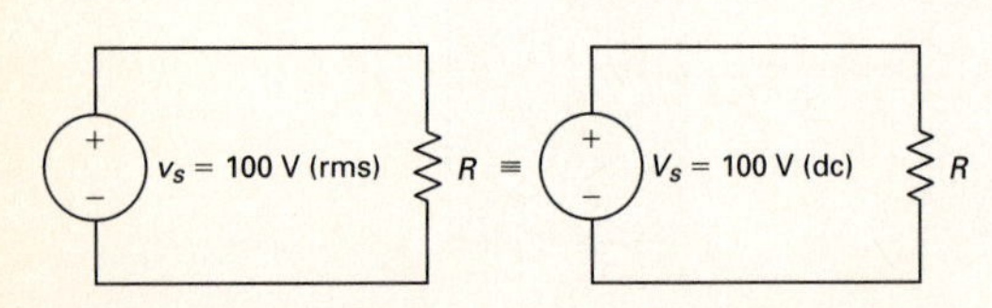

FIGURE 10.8 The effective value of v_s (100 V rms) delivers the same power to R as the dc voltage V_s (100 V dc).

The rms value is also referred to as the **effective value** of the sinusoidal voltage (or current). The rms value has an interesting property: given an equivalent resistive load, R, and an equivalent time period, T, the rms value of a sinusoidal source delivers the same energy to R as does a dc source of the same value. For example, a dc source of 100 V delivers the same energy in T seconds that a sinusoidal source of 100 V_{rms} delivers, assuming equivalent load resistances (see Problem 10.8). Figure 10.8 demonstrates this equivalence. Energywise, the effect of the two sources is identical. This has led to the term *effective value* being used interchangeably with *rms value.*

and $|S|$ as the sides of a right triangle, as shown in Fig. 10.9. It is easy to show that the angle θ in the power triangle is the power factor angle $\theta_v - \theta_i$. For the right triangle shown in Fig. 10.9,

$$\tan\theta = \frac{Q}{P}. \quad \textbf{(10.24)}$$

But from the definitions of P and Q (Eqs. [10.10] and [10.11], respectively),

$$\frac{Q}{P} = \frac{(V_m I_m/2)\sin(\theta_v - \theta_i)}{(V_m I_m/2)\cos(\theta_v - \theta_i)}$$

$$= \tan(\theta_v - \theta_i). \quad \textbf{(10.25)}$$

Therefore, $\theta = \theta_v - \theta_i$. The geometric relations for a right triangle mean also that the four power triangle dimensions (the three sides and the power factor angle) can be determined if any two of the four are known.

The magnitude of complex power is referred to as **apparent power**. Specifically,

$$|S| = \sqrt{P^2 + Q^2}. \quad \textbf{(10.26)}$$

Apparent power, like complex power, is measured in volt-amps. The apparent power, or volt-amp, requirement of a device designed to convert electric energy to a nonelectric form is more important than the average power requirement. Although the average power represents the useful output of the energy-converting device, the apparent power represents the volt-amp capacity required to supply the average power. As you can see from the power triangle in Fig. 10.9, unless the power factor angle is 0° (that is, the device is purely resistive, pf = 1, and $Q = 0$), the volt-amp capacity required by the device is larger than the average power used by the device. As we will see in Example 10.5, it makes sense to operate devices at a power factor close to 1.

Many useful appliances (such as refrigerators, fans, air conditioners, fluorescent lighting fixtures, and washing machines) and most industrial loads operate at a lagging power factor. The power factor of these loads sometimes is corrected either by adding a capacitor to the device itself or by connecting capacitors across the line feeding the load; the latter method is often used for large industrial loads. Drill Exercise 10.4 and Problems 10.37–10.39 give you a chance to make some calculations that correct a lagging power factor load and improve the operation of a circuit.

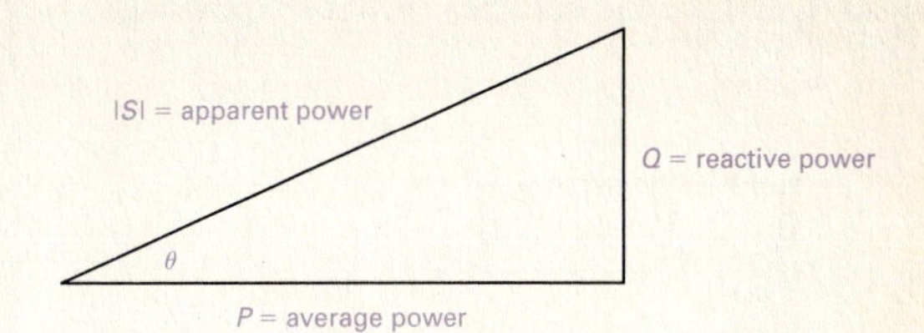

FIGURE 10.9 A power triangle.

EXAMPLE 10.3

An electrical load operates at 240 V rms. The load absorbs an average power of 8 kW at a lagging power factor of 0.8.

a) Calculate the complex power of the load.

b) Calculate the impedance of the load.

SOLUTION

a) The power factor is described as lagging, so we know that the load is inductive and that the algebraic sign of the reactive power is positive. From the power triangle shown in Fig. 10.10,

$$P = |S|\cos\theta \quad \text{and} \quad Q = |S|\sin\theta.$$

Now, because $\cos\theta = 0.8$, $\sin\theta = 0.6$. Therefore

$$|S| = \frac{P}{\cos\theta} = \frac{8\text{ kW}}{0.8} = 10\text{ kVA},$$

$$Q = 10\sin\theta = 6\text{ kVAR},$$

and

$$S = 8 + j6\text{ kVA}.$$

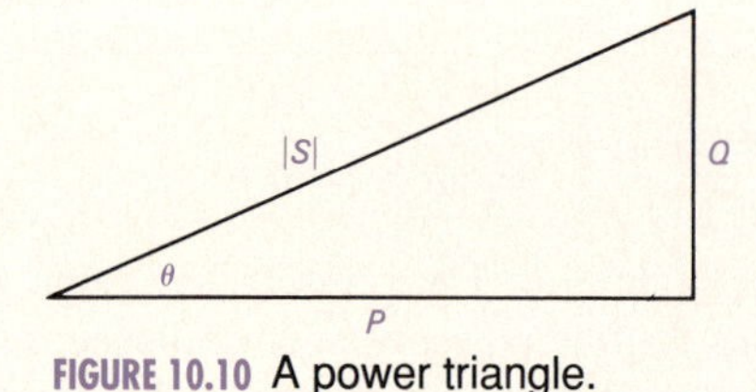

FIGURE 10.10 A power triangle.

b) From the computation of the complex power of the load, we see that $P = 8$ kW. Using Eq. (10.21),

$$\begin{aligned} P &= V_{\text{eff}} I_{\text{eff}} \cos(\theta_v - \theta_i) \\ &= (240) I_{\text{eff}} (0.8) \\ &= 8000. \end{aligned}$$

Solving for I_{eff},

$$I_{\text{eff}} = 41.67\text{ A}.$$

We already know the angle of the load impedance, since it is the power factor angle:

$$\theta = \cos^{-1}(0.8) = 36.87^\circ.$$

We also know that θ is positive because the power factor is lagging, indicating an inductive load. We compute the magnitude of the load impedance from its definition as the ratio of the magnitude of the voltage to the magnitude of the current:

$$|Z| = \frac{|V_{\text{eff}}|}{|I_{\text{eff}}|} = \frac{240}{41.67} = 5.76.$$

Hence,

$$Z = 5.76\underline{/36.87^\circ}\ \Omega = 4.608 + j3.456\ \Omega.$$

10.5 POWER CALCULATIONS

We are now ready to develop additional equations that can be used to calculate real, reactive, and complex power. We begin by combining Eqs. (10.10), (10.11), and (10.23) to get

$$\begin{aligned} S &= \frac{V_m I_m}{2} \cos(\theta_v - \theta_i) + j\frac{V_m I_m}{2} \sin(\theta_v - \theta_i) \\ &= \frac{V_m I_m}{2}[\cos(\theta_v - \theta_i) + j \sin(\theta_v - \theta_i)] \\ &= \frac{V_m I_m}{2} e^{j(\theta_v - \theta_i)} = \frac{1}{2} V_m I_m \underline{/ (\theta_v - \theta_i)}. \end{aligned} \quad \textbf{(10.27)}$$

If we use the effective values of the sinusoidal voltage and current, Eq. (10.27) becomes

$$S = V_{\text{eff}} I_{\text{eff}} \underline{/ (\theta_v - \theta_i)}. \quad \textbf{(10.28)}$$

Equations (10.27) and (10.28) are important relationships in power calculations because they show that if the phasor current and voltage are known at a pair of terminals, the complex power associated with that pair of terminals is either one-half the product of the voltage and the conjugate of the current, or the product of the rms phasor voltage and the conjugate of the rms phasor current. We can show this for the rms phasor voltage and current in Fig. 10.11 as follows:

$$\begin{aligned} S &= V_{\text{eff}} I_{\text{eff}} \underline{/ (\theta_v - \theta_i)} \\ &= V_{\text{eff}} I_{\text{eff}} e^{j(\theta_v - \theta_i)} \\ &= V_{\text{eff}} e^{j\theta_v} I_{\text{eff}} e^{-j\theta_i} \\ &= \mathbf{V}_{\text{eff}} \mathbf{I}^*_{\text{eff}}. \end{aligned} \quad \textbf{(10.29)}$$

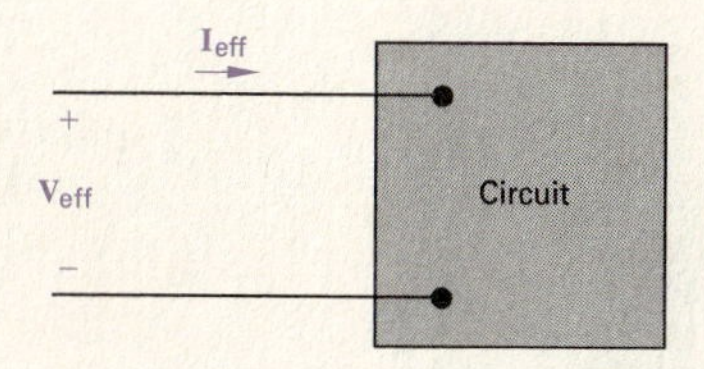

FIGURE 10.11 The phasor voltage and current associated with a pair of terminals.

Note that $\mathbf{I}^*_{\text{eff}} = I_{\text{eff}} e^{-j\theta_i}$ follows from Euler's identity and the trigonometric identities $\cos(-\theta) = \cos(\theta)$ and $\sin(-\theta) = -\sin(\theta)$:

$$\begin{aligned} I_{\text{eff}} e^{-j\theta_i} &= I_{\text{eff}} \cos(-\theta_i) + j I_{\text{eff}} \sin(-\theta_i) \\ &= I_{\text{eff}} \cos(\theta_i) - j I_{\text{eff}} \sin(\theta_i) \\ &= \mathbf{I}^*_{\text{eff}}. \end{aligned}$$

The same derivation technique could be applied to Eq. (10.27) to yield

$$S = \frac{1}{2}\mathbf{V}\mathbf{I}^*. \quad \textbf{(10.30)}$$

Both Eqs. (10.29) and (10.30) are based on the passive sign convention. If the current reference is in the direction of the voltage rise across the terminals, we insert a minus sign on the right-hand side of each equation.

To illustrate the use of Eq. (10.30) in a power calculation, let's use the same circuit that we used in Example 10.1. Expressed in terms of the phasor representation of the terminal voltage and current,

$$\mathbf{V} = 100\underline{/15^\circ}\text{ V} \quad \text{and} \quad \mathbf{I} = 4\underline{/-105^\circ}\text{ A}.$$

Therefore

$$\begin{aligned} S &= \frac{1}{2}(100\underline{/15^\circ})(4\underline{/+105^\circ}) = 200\underline{/120^\circ} \\ &= -100 + j173.21\text{ VA}. \end{aligned}$$

Once we calculate the complex power, we can read off both the real and reactive powers, since $S = P + jQ$. Thus

$$P = -100\text{ W} \quad \text{and} \quad Q = 173.21\text{ VAR}.$$

The interpretations of the algebraic signs on P and Q are identical to those given in the solution of Example 10.1.

ALTERNATE FORMS FOR COMPLEX POWER

Equations (10.29) and (10.30) have several useful variations. Here, we use the rms value form of the equations because rms values are the most common type of representation for voltages and currents in power computations.

The first variation of Eq. (10.29) is to replace the voltage with the product of the current times the impedance. In other words, we can always represent the circuit inside the box of Fig. 10.11 by an equivalent impedance, as shown in Fig. 10.12. Then,

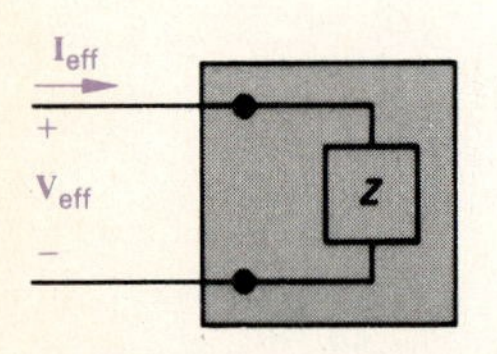

FIGURE 10.12 The general circuit of Fig. 10.11 replaced with an equivalent impedance.

$$\mathbf{V}_{\text{eff}} = Z\mathbf{I}_{\text{eff}}. \tag{10.31}$$

Substituting Eq. (10.31) into Eq. (10.29) yields

$$\begin{aligned} S &= Z\mathbf{I}_{\text{eff}}\mathbf{I}_{\text{eff}}^* \\ &= |\mathbf{I}_{\text{eff}}|^2 Z \\ &= |\mathbf{I}_{\text{eff}}|^2(R + jX) \\ &= |\mathbf{I}_{\text{eff}}|^2 R + j|\mathbf{I}_{\text{eff}}|^2 X = P + jQ, \end{aligned} \tag{10.32}$$

from which

$$P = |\mathbf{I}_{\text{eff}}|^2 R = \frac{1}{2}I_m^2 R \tag{10.33}$$

and

$$Q = |\mathbf{I}_{\text{eff}}|^2 X = \frac{1}{2}I_m^2 X. \tag{10.34}$$

In Eq. (10.34), X is the reactance of either the equivalent inductance or equivalent capacitance of the circuit. Recall from our earlier discussion of reactance that it is positive for inductive circuits and negative for capacitive circuits.

A second useful variation of Eq. (10.29) comes from replacing the current with the voltage divided by the impedance:

$$S = \mathbf{V}_{\text{eff}} \left(\frac{\mathbf{V}_{\text{eff}}}{Z} \right)^* = \frac{|\mathbf{V}_{\text{eff}}|^2}{Z^*} = P + jQ. \quad \textbf{(10.35)}$$

Note that if Z is a pure resistance element,

$$P = \frac{|\mathbf{V}_{\text{eff}}|^2}{R}, \quad \textbf{(10.36)}$$

and if Z is a pure reactive element,

$$Q = \frac{|\mathbf{V}_{\text{eff}}|^2}{X}. \quad \textbf{(10.37)}$$

In Eq. (10.37), X is positive for an inductor and negative for a capacitor.

The following examples demonstrate various power calculations in circuits operating in the sinusoidal steady state.

EXAMPLE 10.4

In the circuit shown in Fig. 10.13, a load having an impedance of $39 + j26\ \Omega$ is fed from a voltage source through a line having an impedance of $1 + j4\ \Omega$. The effective, or rms, value of the source voltage is 250 V.

a) Calculate the load current $\mathbf{I}_L$ and voltage $\mathbf{V}_L$.

b) Calculate the average and reactive power delivered to the load.

c) Calculate the average and reactive power delivered to the line.

d) Calculate the average and reactive power supplied by the source.

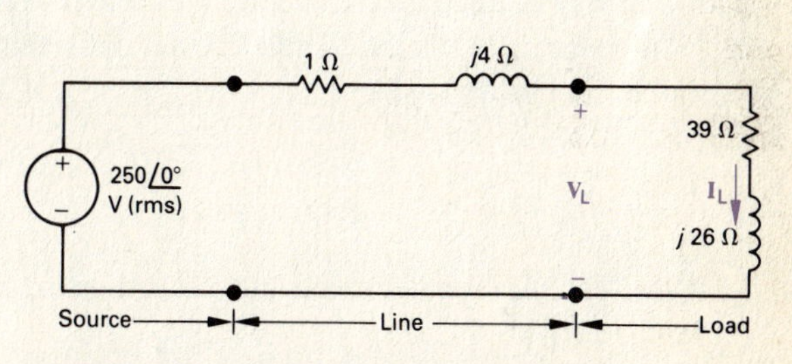

FIGURE 10.13 The circuit for Example 10.4.

SOLUTION

a) The line and load impedances are in series across the voltage source, so the load current equals the voltage divided by the total impedance, or

$$\mathbf{I}_L = \frac{250\underline{/0^\circ}}{40 + j30} = 4 - j3 = 5\underline{/-36.87^\circ}\ \text{A (rms)}.$$

Because the voltage is given in terms of its rms value, the current also is rms. The load voltage is the product of the load current and load impedance:

$$\mathbf{V}_\text{L} = (39 + j26)\mathbf{I}_\text{L} = 234 - j13$$
$$= 234.36\angle{-3.18^\circ} \text{ V (rms)}.$$

b) The average and reactive power delivered to the load can be computed using Eq. (10.29). Therefore

$$S = \mathbf{V}_\text{L}\mathbf{I}_\text{L}^* = (234 - j13)(4 + j3)$$
$$= 975 + j650 \text{ VA}.$$

Thus the load is absorbing an average power of 975 W and a reactive power of 650 VAR.

c) The average and reactive power delivered to the line are most easily calculated from Eqs. (10.33) and (10.34) because the line current is known. Thus

$$P = (5)^2(1) = 25 \text{ W}$$

and

$$Q = (5)^2(4) = 100 \text{ VAR}.$$

Note that the reactive power associated with the line is positive because the line reactance is inductive.

d) One way to calculate the average and reactive power delivered by the source is to add the complex power delivered to the line to that delivered to the load, or

$$S = 25 + j100 + 974 + j650$$
$$= 1000 + j750 \text{ VA}.$$

The complex power at the source can also be calculated from Eq. (10.29):

$$S_s = -250\mathbf{I}_\text{L}^*.$$

The minus sign is inserted in Eq. (10.29) whenever the current reference is in the direction of a voltage rise. Thus

$$S_s = -250(4 + j3) = -(1000 + j750) \text{ VA}.$$

The minus sign implies that both average power and magnetizing reactive power are being delivered by the source. Note that this result agrees with the previous calculation of S, as it must, because the source must furnish all the average and reactive power absorbed by the line and load.

EXAMPLE 10.5

The two loads in the circuit shown in Fig. 10.14 can be described as follows: Load 1 absorbs an average power of 8 kW at a leading power factor of 0.8. Load 2 absorbs 20 kVA at a lagging power factor of 0.6.

a) Determine the power factor of the two loads in parallel.

b) Determine the apparent power required to supply the loads, the magnitude of the current, I_s, and the average power loss in the transmission line.

c) Given that the frequency of the source is 60 Hz, compute the value of the capacitor which, if placed in parallel with the two loads, would correct the power factor to 1. Recompute the values in part (b) for the load with the corrected power factor.

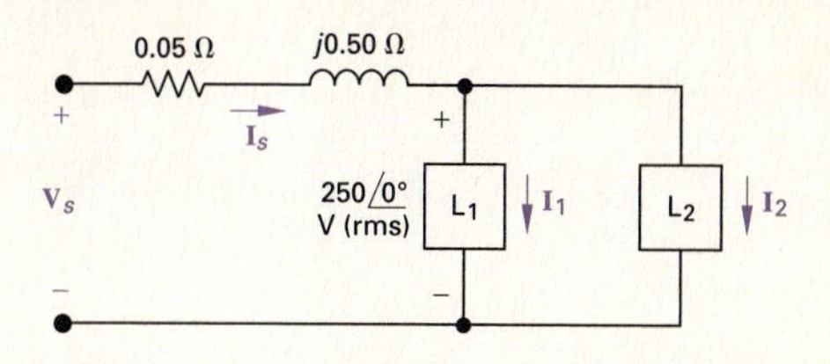

FIGURE 10.14 The circuit for Example 10.5.

SOLUTION

a) All voltage and current phasors in this problem are assumed to represent effective values. Note from the circuit diagram in Fig. 10.14 that $\mathbf{I}_s = \mathbf{I}_1 + \mathbf{I}_2$. The total complex power absorbed by the two loads is

$$\begin{aligned} S &= (250)\mathbf{I}_s^* \\ &= (250)(\mathbf{I}_1 + \mathbf{I}_2)^* \\ &= (250)\mathbf{I}_1^* + (250)\mathbf{I}_2^* \\ &= S_1 + S_2. \end{aligned}$$

We can sum the complex powers geometrically, using the power triangles for each load, as shown in Fig. 10.15. By hypothesis,

$$\begin{aligned} S_1 &= 8000 - j\frac{8000(.6)}{(.8)} \\ &= 8000 - j6000 \text{ VA}, \end{aligned}$$

and

$$\begin{aligned} S_2 &= 20{,}000(.6) + j20{,}000(.8) \\ &= 12{,}000 + j16{,}000 \text{ VA}. \end{aligned}$$

It follows that

$$S = 20{,}000 + j10{,}000 \text{ VA}$$

and

$$I_s^* = \frac{20{,}000 + j10{,}000}{250} = 80 + j40 \text{ A}.$$

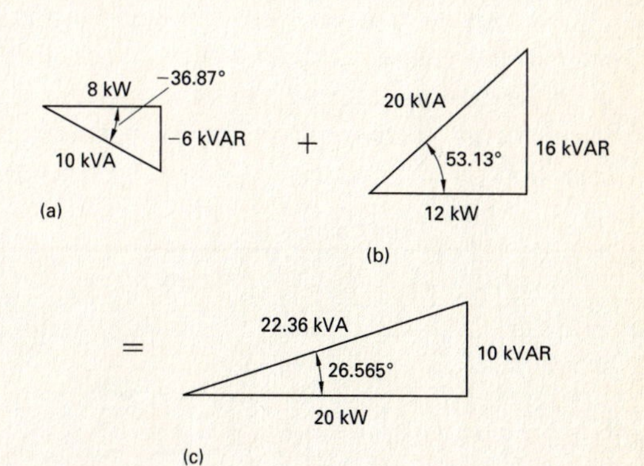

FIGURE 10.15 (a) The power triangle for load 1; (b) the power triangle for load 2; and (c) the sum of the power triangles.

Therefore

$$I_s = 80 - j40 = 89.44\underline{/-26.57^\circ} \text{ A}.$$

Thus the power factor of the combined load is

$$\begin{aligned}\text{pf} &= \cos(0 + 26.57^\circ)\\ &= .8944 \text{ lagging}.\end{aligned}$$

The power factor of the two loads in parallel is lagging because the net reactive power is positive.

b) The apparent power which must be supplied to these loads is

$$\begin{aligned}|S| &= |20 + j10|\\ &= 22.36 \text{ kVA}.\end{aligned}$$

The magnitude of the current which supplies this apparent power is

$$\begin{aligned}I_s &= |80 - j40|\\ &= 89.89 \text{ A}.\end{aligned}$$

The average power lost in the line is due to the current flowing through the line resistance:

$$P_{\text{line}} = I_s^2 R = (89.89)^2(0.05) = 404 \text{ W}.$$

Note that the power company must supply a total of 20,000 + 404 = 20,404 W, even though the loads require a total of only 20,000 W.

c) As we can see from the power triangle in Fig. 10.15(c), we can correct the power factor to 1 if we place a capacitor in parallel with the existing loads such that the capacitor supplies 10 kVAR of magnetizing reactive power. The value of the capacitor is calculated as follows. First, find the capacitive reactance from Eq. (10.37):

$$\begin{aligned}X &= \frac{|V_{\text{eff}}|^2}{Q}\\ &= \frac{(250)^2}{-10{,}000}\\ &= -6.25\ \Omega.\end{aligned}$$

Recall that the reactive impedance of a capacitor is $-1/\omega C$, and $\omega = 2\pi(60) = 376.99$ rad/s if the source frequency is 60 Hz. Thus,

$$C = \frac{-1}{\omega X} = \frac{-1}{(376.99)(-6.25)} = 424.4\ \mu\text{F}.$$

The addition of the capacitor as the third load is represented in geometric form as the sum of the three power triangles shown in Fig. 10.16.

When the power factor is 1, the apparent power and the average power are the same, as seen from the power triangle in Fig. 10.16(c). Therefore, the apparent power once the power factor has been corrected is

$$|S| = P = 20\text{ kVA}.$$

The magnitude of the current which supplies this apparent power is

$$I_s = \frac{20{,}000}{250} = 80\text{ A}$$

The average power lost in the line is thus reduced to

$$P_{\text{line}} = I_s^2 R = (80)^2(0.05) = 320\text{ W}.$$

Therefore, the power company must supply a total of 20,000 + 320 = 20,320 W. Note that the addition of the capacitor has reduced the line loss from 404 W to 320 W.

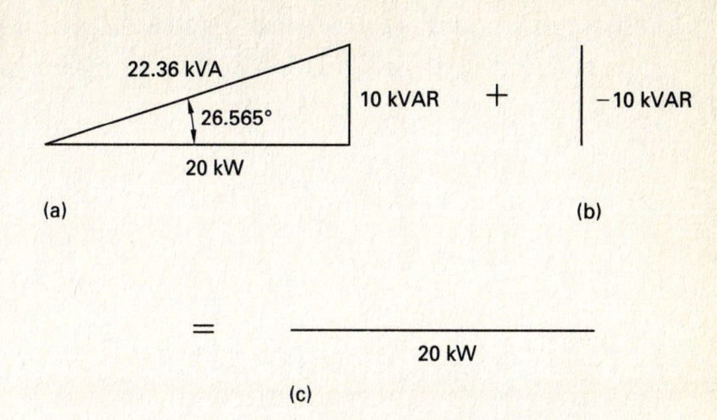

FIGURE 10.16 (a) The sum of the power triangles for loads 1 and 2; (b) the power triangle for a 424.4 μF capacitor at 60 Hz; and (c) the sum of the power triangles in (a) and (b).

EXAMPLE 10.6

a) Calculate the total average and reactive power delivered to each impedance in the circuit shown in Fig. 10.17.

b) Calculate the average and reactive powers associated with each source in the circuit.

c) Verify that the average power delivered equals the average power absorbed, and that the magnetizing reactive power delivered equals the magnetizing reactive power absorbed.

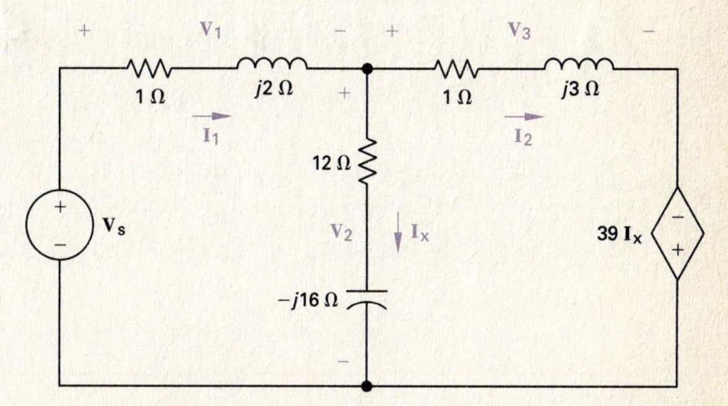

$\mathbf{V}_s = 150\angle 0^\circ$ V

$\mathbf{V}_1 = (78 - j104)$ V $\mathbf{I}_1 = (-26 - j52)$ A

$\mathbf{V}_2 = (72 + j104)$ V $\mathbf{I}_x = (-2 + j6)$ A

$\mathbf{V}_3 = (150 - j130)$ V $\mathbf{I}_3 = (-24 - j58)$ A

FIGURE 10.17 The circuit, with solution, for Example 10.6.

SOLUTION

a) The complex power delivered to the $(1 + j2)\ \Omega$ impedance is

$$\begin{aligned} S_1 &= \frac{1}{2}\mathbf{V}_1\mathbf{I}_1^* = P_1 + jQ_1 \\ &= \frac{1}{2}(78 - j104)(-26 + j52) \\ &= \frac{1}{2}(3380 + j6760) \\ &= 1690 + j3380\text{ VA}. \end{aligned}$$

Thus this impedance is absorbing an average power of 1690 W and a reactive power of 3380 VAR. The complex power delivered to the $(12 - j16)$ Ω impedance is

$$S_2 = \frac{1}{2}\mathbf{V}_2\mathbf{I}_x^* = P_2 + jQ_2$$
$$= \frac{1}{2}(72 + j104)(-2 - j6)$$
$$= 240 - j320 \text{ VA}.$$

Therefore the impedance in the vertical branch is absorbing 240 W and delivering 320 VAR. The complex power delivered to the $(1 + j3)$ Ω impedance is

$$S_3 = \frac{1}{2}\mathbf{V}_3\mathbf{I}_2^* = P_3 + jQ_3$$
$$= \frac{1}{2}(150 - j130)(-24 + j58)$$
$$= 1970 + j5910 \text{ VA}.$$

This impedance is absorbing 1970 W and 5910 VAR.

b) The complex power associated with the independent voltage source is

$$S_s = -\frac{1}{2}\mathbf{V}_s\mathbf{I}_1^* = P_s + jQ_s$$
$$= -\frac{1}{2}(150)(-26 + j52)$$
$$= 1950 - j3900 \text{ VA}.$$

Note that the independent voltage source is absorbing an average power of 1950 W and delivering 3900 VAR. The complex power associated with the current-controlled voltage source is

$$S_x = \frac{1}{2}(39\mathbf{I}_x)(\mathbf{I}_2^*) = P_x + jQ_x$$
$$= \frac{1}{2}(-78 + j234)(-24 + j58)$$
$$= -5850 - j5070 \text{ VA}.$$

Both average power and magnetizing reactive power are being delivered by the dependent source.

c) The total power absorbed by the passive impedances and the independent voltage source is

$$P_{\text{absorbed}} = P_1 + P_2 + P_3 + P_s = 5850 \text{ W}.$$

The dependent voltage source is the only circuit element delivering average power. Thus

$$P_{\text{delivered}} = 5850 \text{ W}.$$

Magnetizing reactive power is being absorbed by the two horizontal branches. Thus

$$Q_{\text{absorbed}} = Q_1 + Q_3 = 9290 \text{ VAR}.$$

Magnetizing reactive power is being delivered by the independent voltage source, the capacitor in the vertical impedance branch, and the dependent voltage source. Therefore

$$Q_{\text{delivered}} = 9290 \text{ VAR}.$$

DRILL EXERCISES

10.4 The load impedance in the circuit shown is shunted by a capacitor having a capacitive reactance of $-52\ \Omega$. Calculate:

a) the rms phasors $\mathbf{V}_L$ and $\mathbf{I}_L$;

b) the average power and magnetizing reactive power absorbed by the $(39 + j26)\ \Omega$ load impedance;

c) the average power and magnetizing reactive power absorbed by the $(1 + j4)\ \Omega$ line impedance;

d) the average power and magnetizing reactive power delivered by the source; and

e) the magnetizing reactive power delivered by the shunting capacitor.

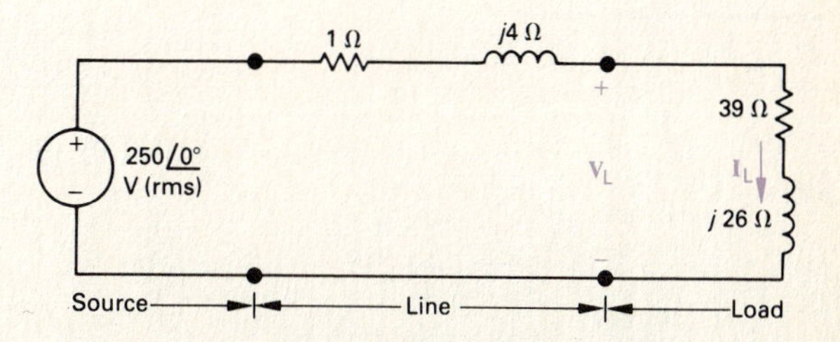

ANSWER: (a) $252.20\angle{-4.54^\circ}$ V (rms), $5.38\angle{-38.23^\circ}$ A (rms); (b) 1129.09 W, 752.73 VAR; (c) 23.52 W, 94.09 VAR; (d) 1152.62 W, −376.36 VAR; (e) 1223.18 VAR.

10.5 The rms voltage at the terminals of a load is 440 V. The load is absorbing an average power of 20 kW and a magnetizing reactive power of 10 kVAR. Derive two equivalent impedance models of the load.

ANSWER: 7.744 Ω in series with 3.872 Ω of inductive reactance; 9.68 Ω in parallel with 19.36 Ω of inductive reactance.

10.6 Find the phasor voltage $\mathbf{V}_s$ (rms) in the circuit shown if loads L_1 and L_2 are 13 kVA at 0.8 pf lag and 10 kVA at 0.96 pf lead, respectively. Express $\mathbf{V}_s$ in polar form.

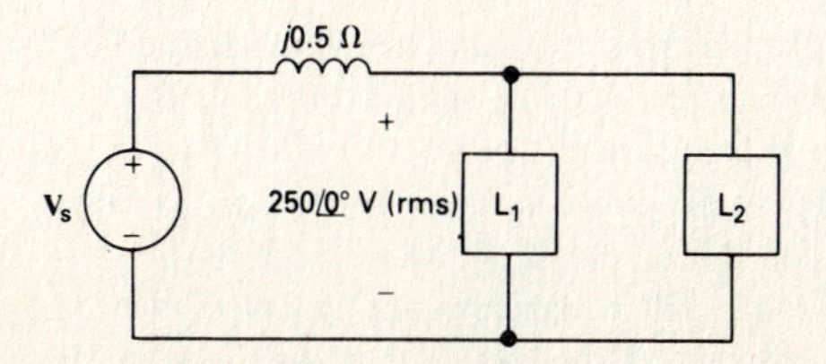

ANSWER: $263.06\angle{8.75^\circ}$ V.

10.6 APPLIANCE RATINGS

The average power rating and estimated annual kilowatt-hour consumption of some common appliances are presented in Table 10.2. The energy consumption values are obtained by estimating the number of hours annually that the appliances are in use. For example, a coffeemaker has an estimated annual consumption of 140 kWh and an average power consumption during operation of 1.2 kW. Therefore a coffeemaker is assumed to be in operation 140/1.2, or 116.67, hours per year, or approximately 19 minutes per day.

EXAMPLE 10.7

The branch circuit supplying the outlets in a typical home kitchen is wired with #12 conductor and is protected by either a 20 A fuse or a 20 A circuit breaker. Assume that the following 120 V appliances are in operation at the same time: a coffeemaker, egg cooker, frying pan, and toaster. Will the circuit be interrupted by the protective device?

SOLUTION

From Table 10.2, the total average power demanded by the four appliances is

$$P = 1200 + 516 + 1196 + 1146 = 4058 \text{ W}.$$

The total current in the protective device is

$$I_{\text{eff}} = \frac{4058}{120} \approx 33.82 \text{ A}.$$

Yes, the protective device will interrupt the circuit.

DRILL EXERCISES

10.7 a) A university student is drying her hair with a hair dryer while sitting under a sunlamp and watching a basketball game on a color tube-type television set. At the same time, her roommate is vacuuming the rug in their air-conditioned bedroom. If all these appliances are supplied from a 120 V branch circuit protected by a 15 A circuit breaker, will the breaker interrupt the game?

b) Will the student be able to watch television if she turns off the sunlamp and her roommate turns off the vacuum cleaner?

ANSWER: (a) Yes, the breaker current is approximately 22 A. (b) Yes, if she can locate the distribution panel and reclose the breaker. The current will be approximately 14 A.

TABLE 10.2

ANNUAL ENERGY REQUIREMENTS OF ELECTRIC HOUSEHOLD APPLIANCES

APPLIANCE	AVERAGE WATTAGE	EST. kWh CONSUMED ANNUALLY*
Food preparation		
Coffeemaker	1,200	140
Dishwasher	1,201	165
Egg cooker	516	14
Frying pan	1,196	100
Mixer	127	2
Oven, microwave (only)	1,450	190
Range, with oven	12,200	596
Toaster	1,146	39
Laundry		
Clothes dryer	4,856	993
Washing machine, automatic	512	103
Water heater	2,475	4,219
Quick recovery type	4,474	4,811
Comfort conditioning		
Air conditioner (room)	860	860†
Dehumidifier	257	377
Fan (circulating)	88	43
Heater (portable)	1,322	176
Health and beauty		
Hair dryer	600	25
Shaver	15	0.5
Sunlamp	279	16
Home entertainment		
Radio	71	86
Television, color, tube type	240	528
Solid-state type	145	320
Housewares		
Clock	2	17
Vacuum cleaner	630	46

*Based on normal usage. When using these figures for projections, such factors as the size of the specific appliance, the geographical area of use, and individual usage should be taken into consideration. Note that the wattages are not additive, since all units are normally not in operation at the same time.

†Based on 1000 hours of operation per year. This figure will vary widely depending on the area and the specific size of the unit. See EEI-Pub #76-2, "Air Conditioning Usage Study," for an estimate for your location.

Source: Edison Electric Institute, 1111 19th Street N.W., Washington, D.C.

10.8 a) A personal computer with a built-in monitor and keyboard requires 40 W at 115 V (rms). Calculate the rms value of the current carried by its power cord.

b) A floppy disk drive for the personal computer in part (a) is rated at 90 W at 115 V (rms). If the disk drive is plugged into the same wall outlet as the computer, what is the rms value of the current drawn from the outlet?

ANSWER: (a) 0.35 A; (b) 1.13 A.

10.7 Maximum Power Transfer

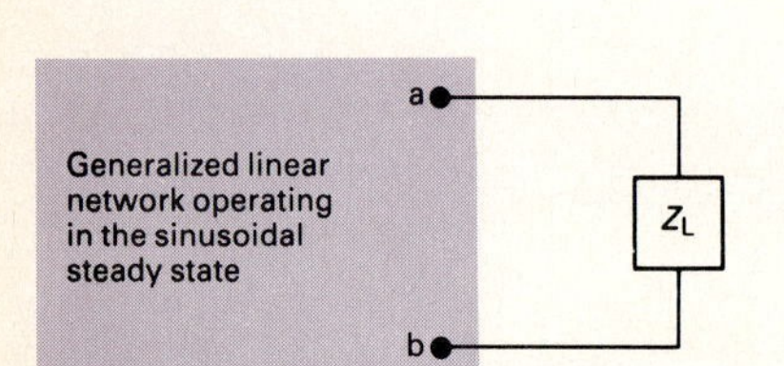

FIGURE 10.18 A circuit describing maximum power transfer.

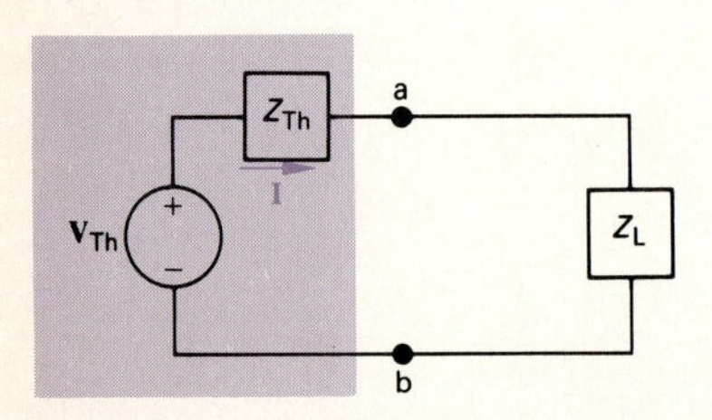

FIGURE 10.19 The circuit shown in Fig. 10.18, with the network replaced by its Thévenin equivalent.

Recall from Chapter 4 that certain systems—for example, those that transmit information via electric signals—depend on being able to transfer a maximum amount of power from the source to the load. We now reexamine maximum power transfer in the context of a sinusoidal steady-state network, beginning with Fig. 10.18. We must determine the load impedance Z_L that results in the delivery of maximum average power to terminals a and b. Any linear network may be viewed from the terminals of the load in terms of a Thévenin equivalent circuit. Thus the task reduces to finding the value of Z_L that results in maximum average power delivered to Z_L in the circuit shown in Fig. 10.19.

For maximum average power transfer, Z_L must equal the conjugate of the Thévenin impedance; that is,

$$Z_{\mathrm{L}} = Z_{\mathrm{Th}}^{*}. \tag{10.38}$$

We derive Eq. (10.38) by a straightforward application of elementary calculus. We begin by expressing Z_{Th} and Z_L in rectangular form:

$$Z_{\mathrm{Th}} = R_{\mathrm{Th}} + jX_{\mathrm{Th}} \tag{10.39}$$

and

$$Z_{\mathrm{L}} = R_{\mathrm{L}} + jX_{\mathrm{L}}. \tag{10.40}$$

In both Eqs. (10.39) and (10.40), the reactance term carries its own algebraic sign—positive for inductance and negative for capacitance. Because we are making an average-power calculation, we assume that the amplitude of the Thévenin voltage is expressed in terms of its rms value. We also use the Thévenin voltage as the reference phasor. Then, from Fig. 10.19, the rms value of the load current $\mathbf{I}$ is

$$\mathbf{I} = \frac{\mathbf{V}_{\mathrm{Th}}\angle 0^{\circ}}{(R_{\mathrm{Th}} + R_{\mathrm{L}}) + j(X_{\mathrm{Th}} + X_{\mathrm{L}})}. \tag{10.41}$$

The average power delivered to the load is

$$P = |\mathbf{I}|^2 R_{\mathrm{L}}. \tag{10.42}$$

Substituting Eq. (10.41) into Eq. (10.42) yields

$$P = \frac{|\mathbf{V}_{\mathrm{Th}}|^2 R_{\mathrm{L}}}{(R_{\mathrm{Th}} + R_{\mathrm{L}})^2 + (X_{\mathrm{Th}} + X_{\mathrm{L}})^2}. \quad (10.43)$$

When working with Eq. (10.43), always remember that V_{Th}, R_{Th}, and X_{Th} are fixed quantities, whereas R_{L} and X_{L} are independent variables. Therefore, to maximize P, we must find the values of R_{L} and X_{L} where $\partial P/\partial R_{\mathrm{L}}$ and $\partial P/\partial X_{\mathrm{L}}$ both are zero. From Eq. (10.43),

Use a symbolic equation solver to maximize P.

$$\frac{\partial P}{\partial X_{\mathrm{L}}} = \frac{-|\mathbf{V}_{\mathrm{Th}}|^2 2R_{\mathrm{L}}(X_{\mathrm{L}} + X_{\mathrm{Th}})}{[(R_{\mathrm{L}} + R_{\mathrm{Th}})^2 + (X_{\mathrm{L}} + X_{\mathrm{Th}})^2]^2}; \quad (10.44)$$

$$\frac{\partial P}{\partial R_{\mathrm{L}}} = \frac{-|\mathbf{V}_{\mathrm{Th}}|^2[(R_{\mathrm{L}} + R_{\mathrm{Th}})^2 + (X_{\mathrm{L}} + X_{\mathrm{Th}})^2 - 2R_{\mathrm{L}}(R_{\mathrm{L}} + R_{\mathrm{Th}})]}{[(R_{\mathrm{L}} + R_{\mathrm{Th}})^2 + (X_{\mathrm{L}} + X_{\mathrm{Th}})^2]^2}. \quad (10.45)$$

From Eq. (10.44), $\partial P/\partial X_{\mathrm{L}}$ is zero when

$$X_{\mathrm{L}} = -X_{\mathrm{Th}}. \quad (10.46)$$

From Eq. (10.45), $\partial P/\partial R_{\mathrm{L}}$ is zero when

$$R_{\mathrm{L}} = \sqrt{R_{\mathrm{Th}}^2 + (X_{\mathrm{L}} + X_{\mathrm{Th}})^2}. \quad (10.47)$$

Note that when we combine Eq. (10.46) with Eq. (10.47), both derivatives are zero when $Z_{\mathrm{L}} = Z_{\mathrm{Th}}^*$.

The Maximum Value of Average Power Absorbed

The maximum average power that can be delivered to Z_{L} when it is set equal to the conjugate of Z_{Th} is calculated directly from the circuit in Fig. 10.19. When $Z_{\mathrm{L}} = Z_{\mathrm{Th}}^*$, the rms load current is $\mathbf{V}_{\mathrm{Th}}/2R_{\mathrm{L}}$ and the maximum average power delivered to the load is

$$P_{\max} = \frac{|\mathbf{V}_{\mathrm{Th}}|^2 R_{\mathrm{L}}}{4R_{\mathrm{L}}^2} = \frac{1}{4}\frac{|\mathbf{V}_{\mathrm{Th}}|^2}{R_{\mathrm{L}}}. \quad (10.48)$$

If the Thévenin voltage is expressed in terms of its maximum amplitude rather than its rms amplitude, Eq. (10.48) becomes

$$P_{\max} = \frac{1}{8}\frac{V_m^2}{R_{\mathrm{L}}}. \quad (10.49)$$

Maximum Power Transfer When Z is Restricted

Maximum average power can be delivered to Z_{L} only if Z_{L} can be set equal to the conjugate of Z_{Th}. There are situations in which this

Use a computer to vary R_L and X_L within a range and to plot the power transferred.

is not possible. First, R_L and X_L may be restricted to a limited range of values. In this situation, the optimum condition for R_L and X_L is to adjust X_L as near to $-X_{Th}$ as possible and then adjust R_L as close to $\sqrt{R_{Th}^2 + (X_L + X_{Th})^2}$ as possible (see Example 10.9).

A second type of restriction occurs when the magnitude of Z_L can be varied but its phase angle cannot. Under this restriction, the greatest amount of power is transferred to the load when the magnitude of Z_L is set equal to the magnitude of Z_{Th}, that is, when

$$|Z_L| = |Z_{Th}|. \quad \textbf{(10.50)}$$

The proof of Eq. (10.50) is left to you as Problem 10.33.

For purely resistive networks, maximum power transfer occurs when the load resistance equals the Thévenin resistance. Note that we first derived this result in the introduction to maximum power transfer in Chapter 4.

Examples 10.8, 10.9, and 10.10 illustrate the problem of obtaining maximum power transfer in the situations just discussed.

EXAMPLE 10.8

a) For the circuit shown in Fig. 10.20, determine the impedance Z_L that results in maximum average power transferred to Z_L.

b) What is the maximum average power transferred to the load impedance determined in (a)?

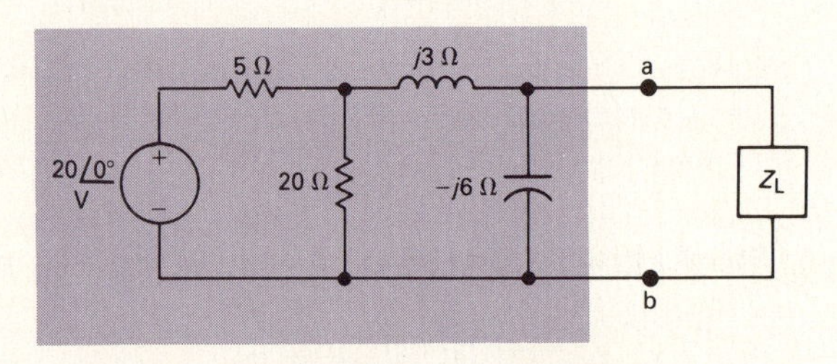

FIGURE 10.20 The circuit for Example 10.8.

SOLUTION

a) We begin by determining the Thévenin equivalent with respect to the load terminals a,b. After two source transformations involving the 20 V source, the 5 Ω resistor, and the 20 Ω resistor, we simplify the circuit shown in Fig. 10.20 to the one shown in Fig. 10.21. Then,

$$\mathbf{V}_{Th} = \frac{16\angle 0^\circ}{4 + j3 - j6}(-j6)$$

$$= 19.2\angle -53.13^\circ = 11.52 - j15.36 \text{ V}.$$

We find the Thévenin impedance by deactivating the independent source and calculating the impedance seen looking into the terminals a and b. Thus,

$$Z_{Th} = \frac{(-j6)(4 + j3)}{4 + j3 - j6} = 5.76 - j1.68\ \Omega.$$

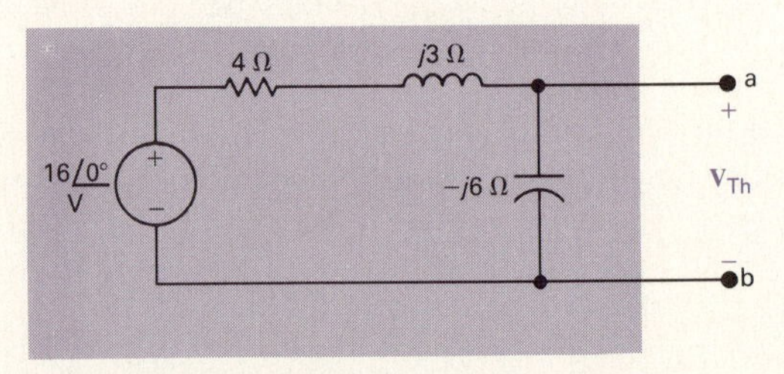

FIGURE 10.21 A simplification of Fig. 10.20 by source transformations.

For maximum average-power transfer, the load impedance must be the conjugate of Z_{Th}, so

$$Z_L = 5.76 + j1.68\ \Omega.$$

b) We calculate the maximum average power delivered to Z_L from the circuit shown in Fig. 10.22, in which we replaced the original network with its Thévenin equivalent. From Fig. 10.22, the rms magnitude of the load current **I** is

$$I_{\text{eff}} = \frac{19.2/\sqrt{2}}{2(5.76)} = 1.1785\ \text{A}.$$

The average power delivered to the load is

$$P = I_{\text{eff}}^2(5.76) = 8\ \text{W}.$$

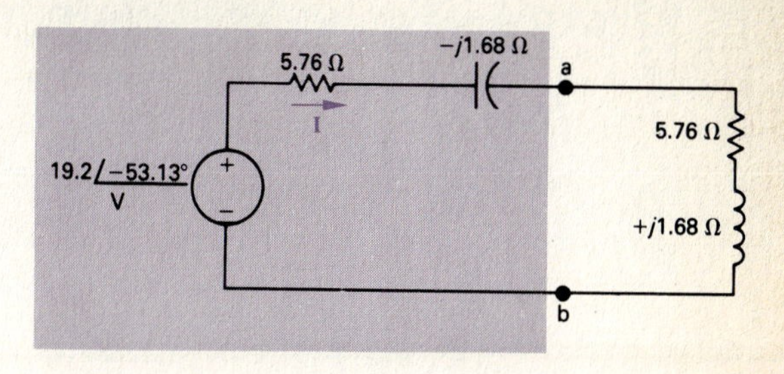

FIGURE 10.22 The circuit shown in Fig. 10.20, with the original network replaced by its Thévenin equivalent.

EXAMPLE 10.9

a) For the circuit shown in Fig. 10.23, what value of Z_L results in maximum average-power transfer to Z_L? What is the maximum power in milliwatts?

b) Assume that the load resistance can be varied between 0 and 4000 Ω and that the capacitive reactance of the load can be varied between 0 and −2000 Ω. What settings of R_L and X_L transfer the most average power to the load? What is the maximum average power that can be transferred under these restrictions?

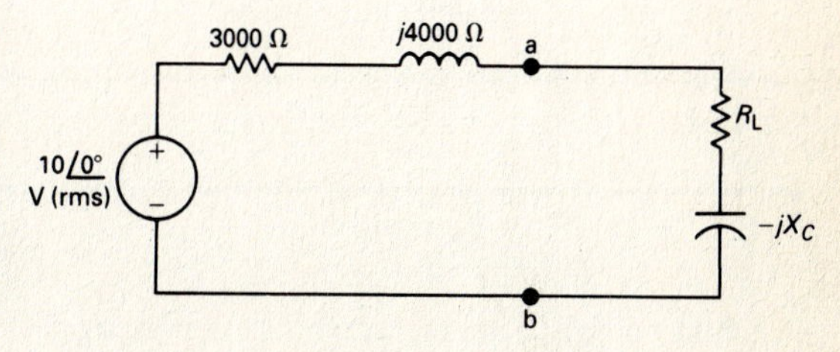

FIGURE 10.23 The circuit for Examples 10.9 and 10.10.

SOLUTION

a) If there are no restrictions on R_L and X_L, the load impedance is set equal to the conjugate of the output or the Thévenin impedance. Therefore we set

$$R_L = 3000\ \Omega \quad \text{and} \quad X_C = -4000\ \Omega,$$

or

$$Z_L = 3000 - j4000\ \Omega.$$

Because the source voltage is given in terms of its rms value, the average power delivered to Z_L is

$$P = \frac{1}{4}\frac{10^2}{3000} = \frac{25}{3}\ \text{mW} = 8.33\ \text{mW}.$$

b) Because R_L and X_L are restricted, we first set X_C as close to $-4000\ \Omega$ as possible; thus $X_C = -2000\ \Omega$. Next, we set R_L as close to $\sqrt{R_{Th}^2 + (X_L + X_{Th})^2}$ as possible. Thus

$$R_L = \sqrt{3000^2 + (-2000 + 4000)^2} = 3605.55\ \Omega.$$

Now, because R_L can be varied from 0 to 4000 Ω, we can set R_L to 3605.55 Ω. Therefore the load impedance is adjusted to a value of

$$Z_L = 3605.55 - j2000\ \Omega.$$

With Z_L set at this value, the value of the load current is

$$\mathbf{I}_{eff} = \frac{10\underline{/0^\circ}}{6605.55 + j2000} = 1.4489\underline{/-16.85^\circ}\ \text{mA}.$$

The average power delivered to the load is

$$P = (1.4489 \times 10^{-3})^2(3605.55) = 7.567\ \text{mW}.$$

This quantity is the maximum power that we can deliver to a load, given the restrictions on R_L and X_L. Note that this is less than the power that can be delivered if there are no restrictions; in (a) we found that we can deliver 8.33 mW.

EXAMPLE 10.10

A load impedance having a constant phase angle of -36.87° is connected across the load terminals a,b in the circuit shown in Fig. 10.23. The magnitude of Z_L is varied until the average power delivered is the most possible under the given restriction.

a) Specify Z_L in rectangular form.

b) Calculate the average power delivered to Z_L.

SOLUTION

a) From Eq. (10.50), we know that the magnitude of Z_L must equal the magnitude of Z_{Th}. Therefore

$$|Z_L| = |Z_{Th}| = |3000 + j4000| = 5000\ \Omega.$$

Now, as we know that the phase angle of Z_L is -36.87°, we have

$$Z_L = 5000\underline{/-36.87^\circ} = 4000 - j3000\ \Omega.$$

b) With Z_L set equal to $4000 - j3000\ \Omega$, the load current is

$$\mathbf{I}_{\text{eff}} = \frac{10}{7000 + j1000} = 1.4142\underline{/-8.13^\circ}\ \text{mA},$$

and the average power delivered to the load is

$$P = 1.4142^2(4) = 8\ \text{mW}.$$

This quantity is the maximum power that can be delivered by this circuit to a load impedance whose angle is constant at -36.87°. Again, this quantity is less than the maximum power that can be delivered if there are no restrictions on Z_L.

DRILL EXERCISE

10.9 The source current in the circuit shown is $5 \cos 8000t$ A.

a) What impedance should be connected across terminals a,b for maximum average-power transfer?

b) What is the average power transferred to the impedance in (a)?

c) Assume that the load is restricted to pure resistance. What size resistor connected across a,b will result in the maximum average power transferred?

d) What is the average power transferred to the resistor in (c)?

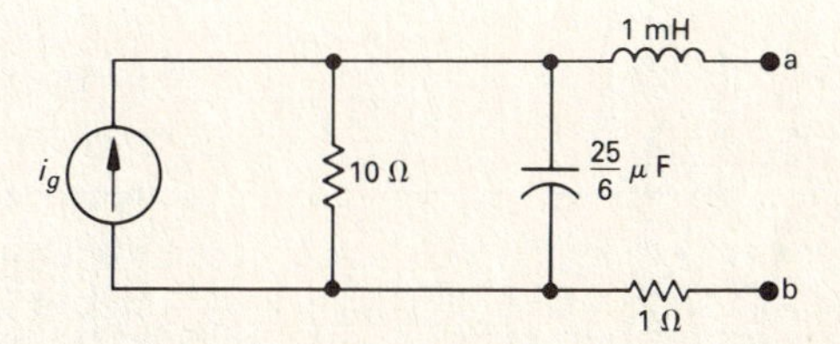

ANSWER: (a) $10 - j5\ \Omega$; (b) 28.125 W; (c) 11.18 Ω; (d) 26.56 W.

PRACTICAL PERSPECTIVE

Heating Appliances

A handheld hair dryer contains a heating element, which is just a resistor heated by the sinusoidal current passing through it, and a fan that blows the warm air surrounding the resistor out the front of the

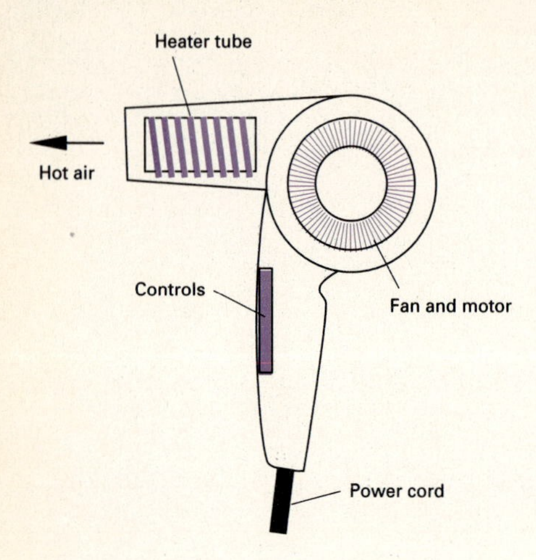

FIGURE 10.24 Schematic representation of a handheld hair dryer.

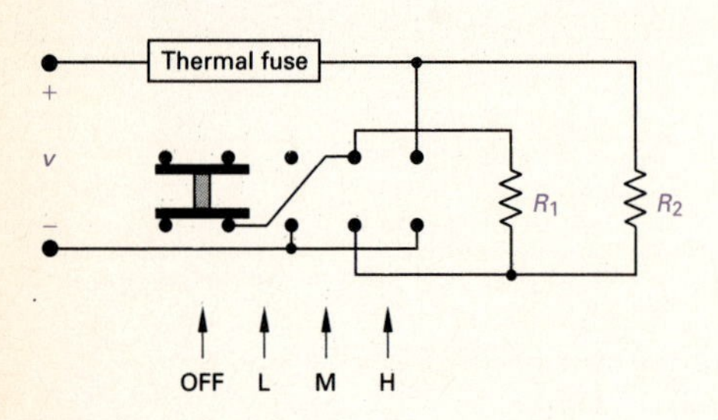

FIGURE 10.25 A circuit diagram for the hair dryer controls.

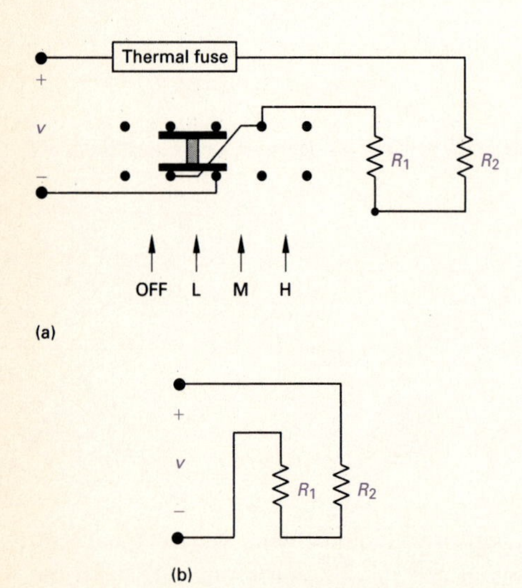

FIGURE 10.26 (a) The circuit in Fig. 10.25 redrawn for the LOW switch setting; (b) a simplified equivalent circuit for part (a).

unit. This is shown schematically in Fig. 10.24. The heater tube in this figure is a resistor made of coiled nichrome wire. Nichrome is an alloy of iron, chromium, and nickel. Two properties make it ideal for use in heaters. First, it is more resistive than most other metals, so less material is required to achieve the needed resistance. This allows the heater to be very compact. Second, unlike many other metals, nichrome does not oxidize when heated red hot in air. Thus the heater element lasts a long time.

A circuit diagram for the hair dryer controls is shown in Fig. 10.25. This is only the part of the hair dryer circuit which gives you control over the heat setting. The rest of the circuit provides power to the fan motor and is not of interest here. The coiled wire which comprises the heater tube has a connection partway along the coil, dividing the coil into two pieces. We model this in Fig. 10.25 with two series resistors, R_1 and R_2. The controls to turn the dryer on and select the heat setting use a four-position switch in which two pairs of terminals in the circuit will be shorted together by a pair of sliding metal bars. The position of the switch determines which pairs of terminals are shorted together. The metal bars are connected by an insulator, so there is no conduction path between the pairs of shorted terminals.

The circuit in Fig. 10.25 contains a thermal fuse. This is a protective device which normally acts like a short circuit. But if the temperature near the heater becomes dangerously high, the thermal fuse becomes an open circuit, discontinuing the flow of current and reducing the risk of fire or injury. The thermal fuse provides protection in case the motor fails or the airflow becomes blocked. While the design of the protection system is not part of this example, it is important to point out that safety analysis is an essential part of an electrical engineer's work.

EXAMPLE

Now that we have modeled the controls for the hair dryer, let's design the circuit component values:

a) Redraw the circuit in Fig. 10.25 as three separate circuit diagrams corresponding to the LOW, MEDIUM, and HIGH switch settings.

b) The hair dryer uses a 60 Hz sinusoidal input voltage of 120 V (rms). The specifications require the heater element to dissipate 250 W at the LOW setting, 500 W at the MEDIUM setting, and 1000 W at the HIGH setting. What values should be used for resistors R_1 and R_2?

SOLUTION

a) The circuit in Fig. 10.25 is redrawn in Fig. 10.26(a) for the LOW switch setting. The open-circuited wires have been removed for clarity. A simplified equivalent circuit is shown in Fig. 10.26(b). A similar pair of figures is shown for the MEDIUM setting

(Fig. 10.27) and the HIGH setting (Fig. 10.28). Note from these figures that at the LOW setting, the voltage source sees the resistors R_1 and R_2 in series; at the MEDIUM setting, the voltage source sees only the R_2 resistor; and at the HIGH setting, the voltage source sees the resistors in parallel.

b) From part (a) we see that there are three different circuit configurations for the three different power requirements. We begin with the MEDIUM setting, shown in Fig. 10.27, because it involves only the resistor R_2. This resistor is connected across the 120 V (rms) source. We can use Eq. (10.19) to find the average power delivered to the resistor:

$$P = \frac{V^2}{R_2} = \frac{120^2}{R_2} = 500 \text{ W}.$$

Therefore,

$$R_2 = \frac{120^2}{500} = 28.8 \ \Omega.$$

When the switch is in the LOW position, R_1 and R_2 are in series, and the average power delivered to them by the sinusoidal source is

$$P = \frac{V^2}{R_1 + R_2} = \frac{120^2}{R_1 + 28.8} = 250 \text{ W}.$$

Therefore,

$$R_1 = \frac{120^2}{250} - 28.8 = 28.8 \ \Omega.$$

Note that we have satisfied two of the three design conditions by specifying the two resistor values. How can we satisfy the third design condition, for the HIGH setting, when we have already specified all of the component values for the circuit? The answer is, we can't! In general, only two of the three power specifications can be met with a two-resistor design (see Problems 10.42, 10.43, and 10.44). But let's check the average power for the HIGH setting, using the values we have already calculated for the LOW and MEDIUM settings:

$$P = \frac{V^2}{R_1 \| R_2}$$
$$= \frac{120^2}{28.8 \| 28.8}$$
$$= \frac{120^2}{14.4} = 1000 \text{ W}.$$

This is the specified value, so the design is complete. The power specifications for this circuit were carefully constructed with this two-resistor design in mind!

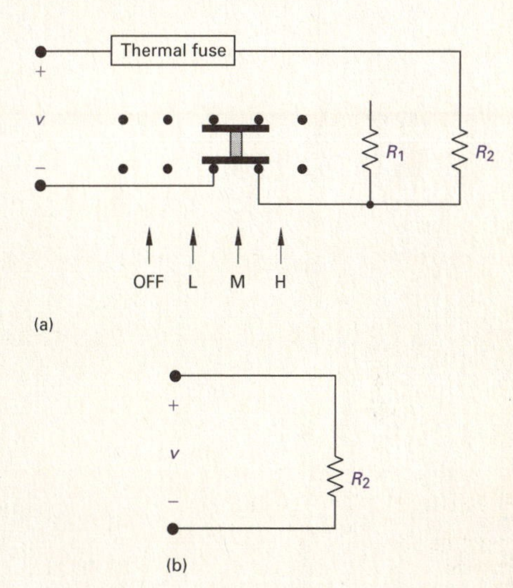

FIGURE 10.27 (a) The circuit in Fig. 10.25 redrawn for the MEDIUM switch setting; (b) a simplified equivalent circuit for part (a).

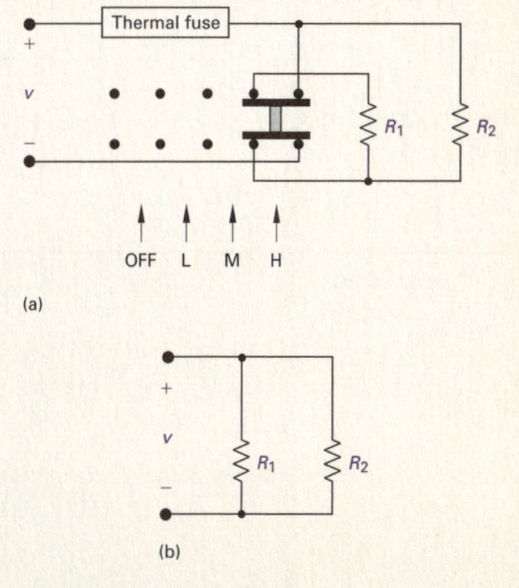

FIGURE 10.28 (a) The circuit in Fig. 10.25 redrawn for the HIGH switch setting; (b) a simplified equivalent circuit for part (a).

SUMMARY

- **Instantaneous power** is the product of the instantaneous terminal voltage and current, or $p = \pm vi$. The positive sign is used when the reference direction for the current is from the positive to the negative reference polarity of the voltage. The frequency of the instantaneous power is twice the frequency of the voltage (or current).

- **Average power** is the average value of the instantaneous power over one period. It is the power converted from electric to nonelectric form, and vice versa. This conversion is the reason that average power is also referred to as real power. Average power, with the passive sign convention, is expressed as

$$P = \frac{1}{2} V_m I_m \cos(\theta_v - \theta_i)$$
$$= V_{\text{eff}} I_{\text{eff}} \cos(\theta_v - \theta_i).$$

- **Reactive power** is the electric power exchanged between the magnetic field of an inductor and the source that drives it, or between the electric field of a capacitor and the source that drives it. Reactive power is never converted to nonelectric power. Reactive power, with the passive sign convention, is expressed as

$$Q = \frac{1}{2} V_m I_m \sin(\theta_v - \theta_i)$$
$$= V_{\text{eff}} I_{\text{eff}} \sin(\theta_v - \theta_i).$$

  Both average power and reactive power can be expressed in terms of either peak $(V_m, \; I_m)$ or effective $(V_{\text{eff}}, \; I_{\text{eff}})$ current and voltage. Effective values are widely used in both household and industrial applications. *Effective value* and *rms value* are interchangeable terms for the same value.

- The **power factor** is the cosine of the phase angle between the voltage and the current:

$$\text{pf} = \cos(\theta_v - \theta_i).$$

  The terms *lagging* and *leading* added to the description of the power factor indicate whether the current is lagging or leading the voltage and thus whether the load is inductive or capacitive.

- The **reactive factor** is the sine of the phase angle between the voltage and the current:

$$\text{rf} = \sin(\theta_v - \theta_i).$$

- **Complex power** is the complex sum of the real and reactive powers, or

$$S = P + jQ$$
$$= \frac{1}{2}\mathbf{V}\mathbf{I}^* = \mathbf{V}_{\text{eff}}\mathbf{I}^*_{\text{eff}}$$
$$= I^2_{\text{eff}}Z = \frac{V^2_{\text{eff}}}{Z^*}.$$

- **Apparent power** is the magnitude of the complex power:

$$|S| = \sqrt{P^2 + Q^2}.$$

- The **watt** is used as the unit for both instantaneous and real power.
- The **var** (volt amp reactive, or VAR) is used as the unit for reactive power.
- The **volt-amp** (VA) is used as the unit for complex and apparent power.
- **Maximum power transfer** occurs in circuits operating in the sinusoidal steady state when the load impedance is the conjugate of the Thévenin impedance as viewed from the terminals of the load impedance.

PROBLEMS

10.1 The following sets of values for v and i pertain to the circuit seen in Fig. 10.1. For each set of values, calculate P and Q and state whether the circuit inside the box is absorbing or delivering (1) average power and (2) magnetizing vars.

a) $v = 170 \cos(\omega t + 55^\circ)$ V
$i = 15 \cos(\omega t + 20^\circ)$ A

b) $v = 50 \cos(\omega t - 20^\circ)$ V
$i = 10 \cos(\omega t + 50^\circ)$ A

c) $v = 250 \cos(\omega t + 35^\circ)$ V
$i = 8 \sin(\omega t + 250^\circ)$ A

d) $v = 180 \sin(\omega t + 240^\circ)$ V
$i = 5 \cos(\omega t + 30^\circ)$ A

10.2 Show that the maximum value of the instantaneous power given by Eq. (10.9) is $P + \sqrt{P^2 + Q^2}$ and that the minimum value is $P - \sqrt{P^2 + Q^2}$.

10.3 A load consisting of a 1.35 kΩ resistor in parallel with a 405 mH inductor is connected across the terminals of a sinusoidal voltage source v_g, where $v_g = 90 \cos 2500t$ V.

a) What is the peak value of the instantaneous power delivered by the source?

b) What is the peak value of the instantaneous power absorbed by the source?

c) What is the average power delivered to the load?

d) What is the reactive power?

e) Does the load absorb or generate magnetizing vars?

f) What is the power factor of the load?

g) What is the reactive factor of the load?

10.4 The three loads in the circuit in Fig. P10.4 can be described as follows: Load 1 is a 12 Ω resistor in series with a 15 mH inductor; load 2 is a 16 μF capacitor in series with an 80 Ω resistor; and load 3 is a 400 Ω resistor in series with the parallel combination of a 20 H inductor and a 5 μF capacitor. Give the power factor and reactive factor of each load if the frequency of the voltage source is 60 Hz.

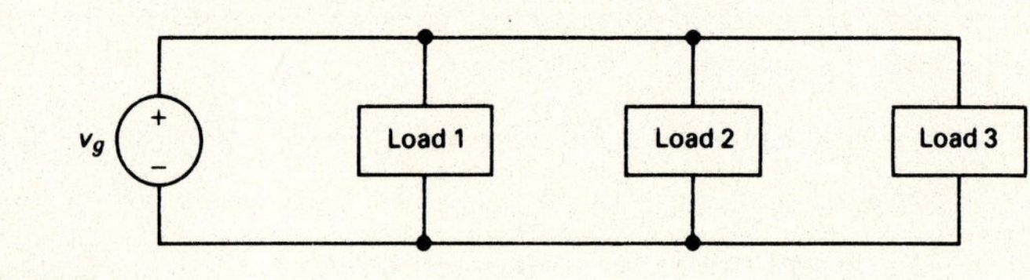

FIGURE P10.4

10.5 Find the rms value of the periodic current shown in Fig. P10.5.

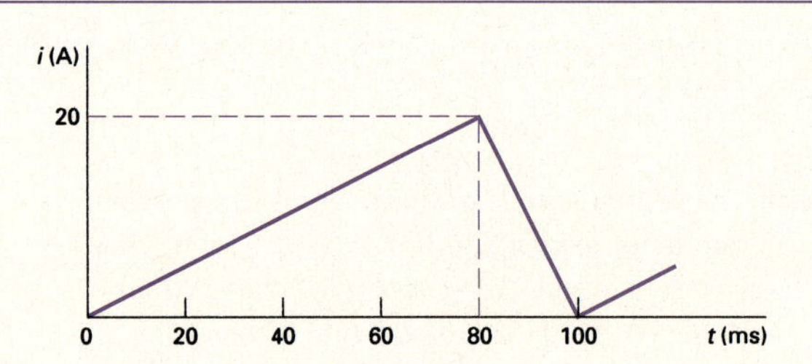

FIGURE P10.5

10.6 The periodic current shown in Fig. P10.5 dissipates an average power of 1280 W in a resistor. What is the value of the resistor?

10.7 a) Find the rms value of the periodic voltage shown in Fig. P10.7.

b) If this voltage is applied to the terminals of a 4 Ω resistor, what is the average power dissipated in the resistor?

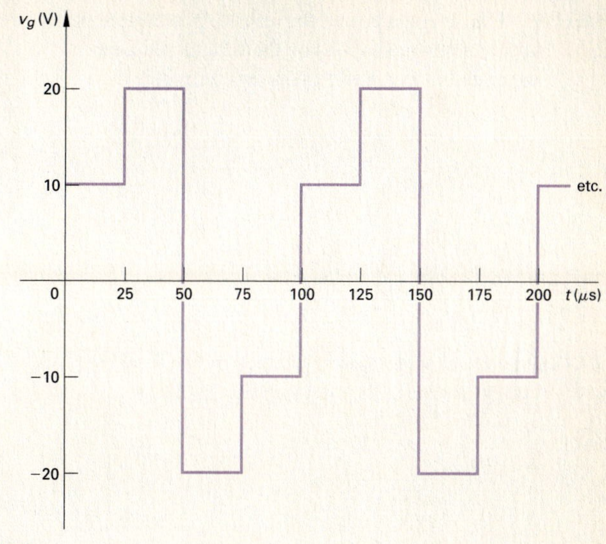

FIGURE P10.7

10.8 A dc voltage equal to V_{dc} V is applied to a resistor of R Ω. A sinusoidal voltage equal to v_s V is also applied to a resistor of R Ω. Show that the dc voltage will deliver the same amount of energy in T seconds (where T is the period of the sinusoidal voltage) as the sinusoidal voltage provided V_{dc} equals the rms value of v_s. (*Hint:* Equate the two expressions for the energy delivered to the resistor.)

10.9 The voltage $\mathbf{V}_g$ in the frequency-domain circuit shown in Fig. P10.9 is $240\angle 0^\circ$ V (rms).

a) Find the average and reactive powers at the terminals of the voltage source.

b) Is the voltage source absorbing or delivering average power?

c) Is the voltage source absorbing or delivering magnetizing vars?

d) Find the average and reactive powers associated with each impedance branch in the circuit.

e) Check the balance between delivered and absorbed average power.

f) Check the balance between delivered and absorbed magnetizing vars.

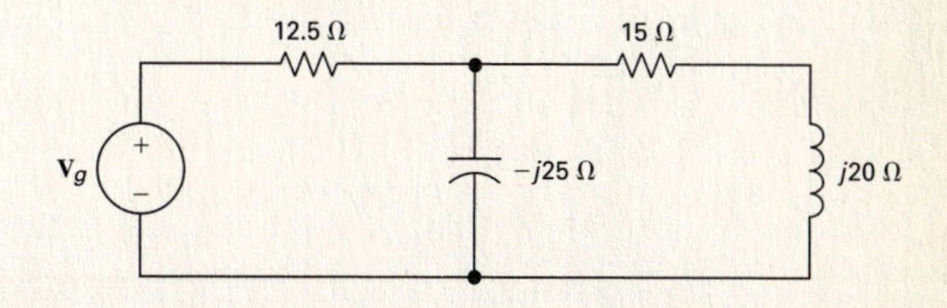

FIGURE P10.9

10.10 Find the average power, the reactive power, and the apparent power absorbed by the load in the circuit in Fig. P10.10 if i_g equals $40 \cos 1250t$ mA.

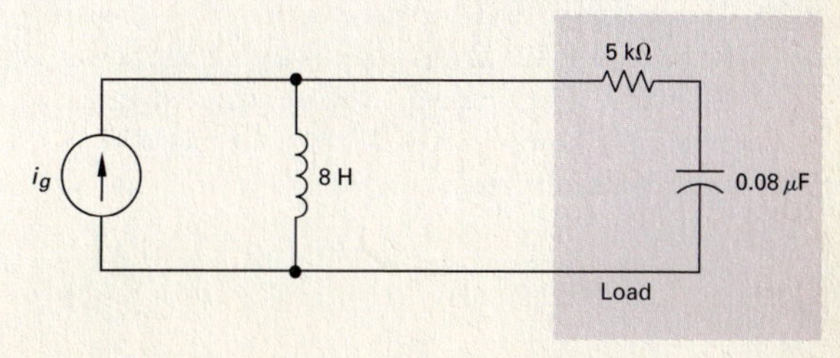

FIGURE P10.10

10.11 Find the average power, the reactive power, and the apparent power supplied by the voltage source in the circuit in Fig. P10.11 if $v_g = 40 \cos 10^6 t$ V.

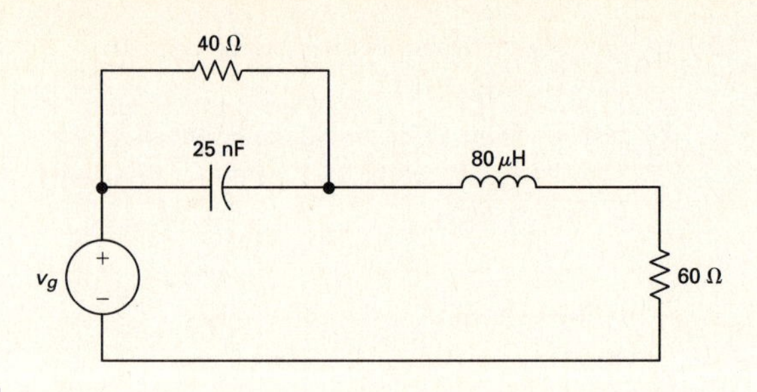

FIGURE P10.11

10.12 Find the average power delivered by the ideal current source in the circuit in Fig. P10.12 if $i_g = 4 \cos 5000t$ mA.

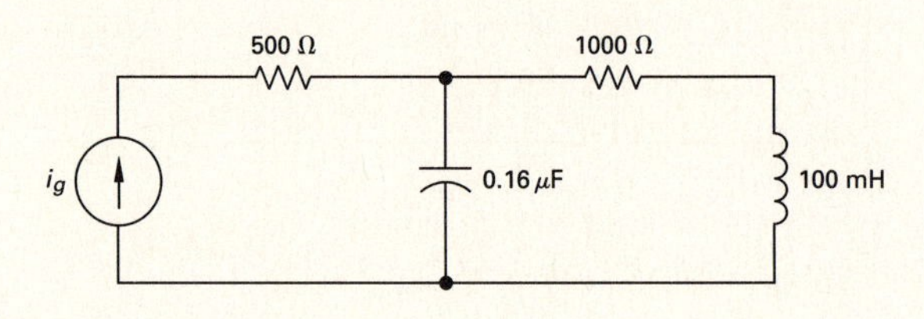

FIGURE P10.12

10.13 a) Calculate the real and reactive power associated with each circuit element in the circuit in Fig. P9.44.

b) Verify that the average power generated equals the average power absorbed.

c) Verify that the magnetizing vars generated equal the magnetizing vars absorbed.

10.14 Repeat Problem 10.13 for the circuit shown in Fig. P9.46.

10.15 Find the average power dissipated in the 30 Ω resistor in the circuit seen in Fig. P10.15 if $i_g = 6 \cos 20{,}000t$ A.

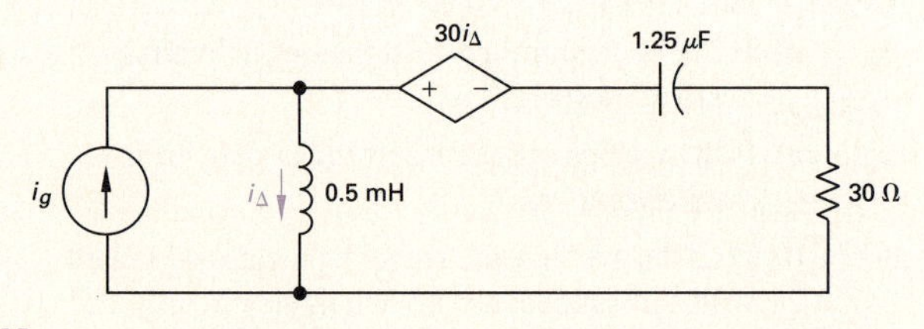

FIGURE P10.15

10.16 The capacitive load impedance in Fig. P10.16 has a magnitude of 125 Ω and absorbs an average power of 7500 W. The sinusoidal voltage source supplies an average power of 8000 W. Find the inductive reactance of the line.

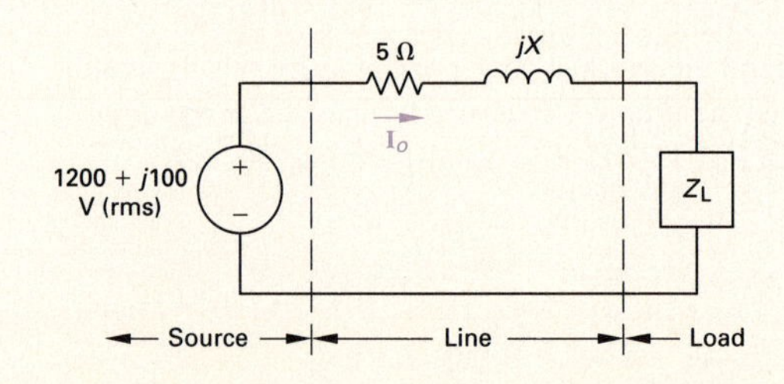

FIGURE P10.16

10.17 Three loads are connected in parallel across a 250 V (rms) line, as shown in Fig. P10.17. Load 1 absorbs 3 kW and 4 kVAR. Load 2 absorbs 2.5 kVA at 0.6 pf lead. Load 3 absorbs 1.5 kW at unity power factor.

a) Find the impedance that is equivalent to the three parallel loads.

b) Find the power factor of the equivalent load as seen from the line's input terminals.

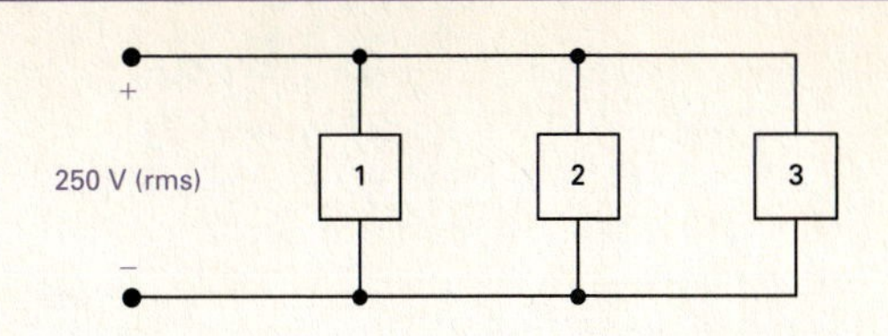

FIGURE P10.17

10.18 Two 480 V (rms) loads are connected in parallel. The two loads draw a total average power of 40,800 W at a power factor of 0.80 lagging. One of the loads draws 20 kVA at a power factor of 0.96 leading. What is the reactive factor of the other load?

10.19 The two loads shown in Fig. P10.19 can be described as follows: Load 1 absorbs an average power of 218.24 kW and 56.32 kVAR magnetizing reactive power; load 2 has an impedance of $40 - j30\ \Omega$. The voltage at the terminals of the loads is $4400\sqrt{2}\cos 120\pi t$ V.

a) Find the rms value of the source voltage.

b) By how many microseconds is the load voltage out of phase with the source voltage?

c) Does the load voltage lead or lag the source voltage?

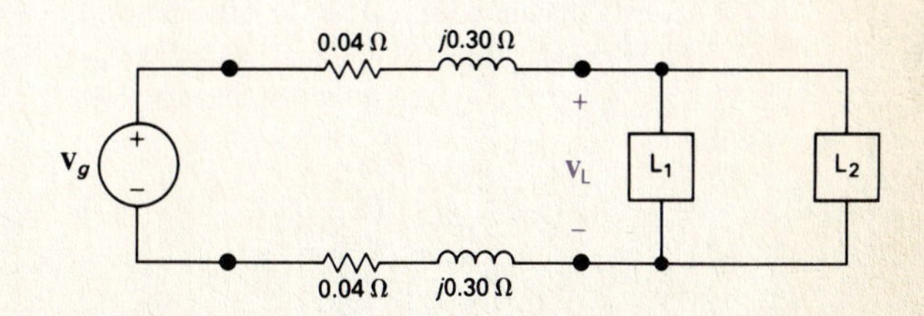

FIGURE P10.19

10.20 The three parallel loads in the circuit shown in Fig. P10.20 can be described as follows: Load 1 is absorbing an average power of 7.5 kW and 9 kVAR of magnetizing vars; load 2 is absorbing an average power of 2.1 kW and generating 1.8 kVAR of magnetizing reactive power; load 3 consists of a 48 Ω resistor in parallel with an inductive reactance of 19.2 Ω. Find the rms magnitude and the phase angle of $\mathbf{V}_g$ if $\mathbf{V}_o = 480\angle 0^\circ$ V (rms).

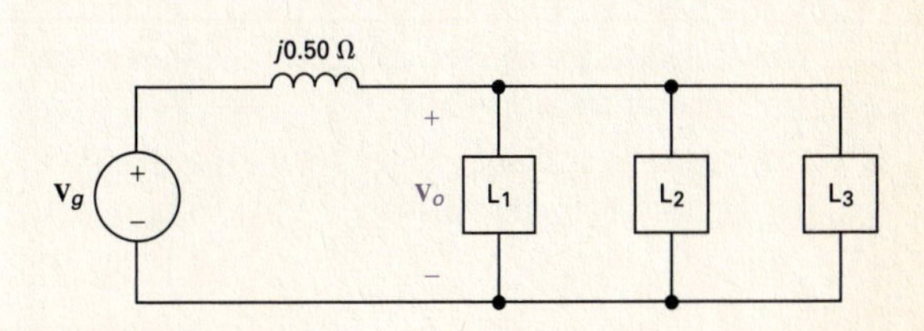

FIGURE P10.20

10.21 The three loads in the circuit seen in Fig. P10.21 are described as follows: Load 1 is absorbing 1.2 kW and 240 VAR; load 2 is 1 kVA at a 0.96 pf lead; load 3 is a 6.25 Ω resistor in parallel with an inductor that has a reactance of 25 Ω. Calculate the average power and the magnetizing reactive power delivered by each source if $\mathbf{V}_{g1} = \mathbf{V}_{g2} = 125\angle 0^\circ$ V (rms).

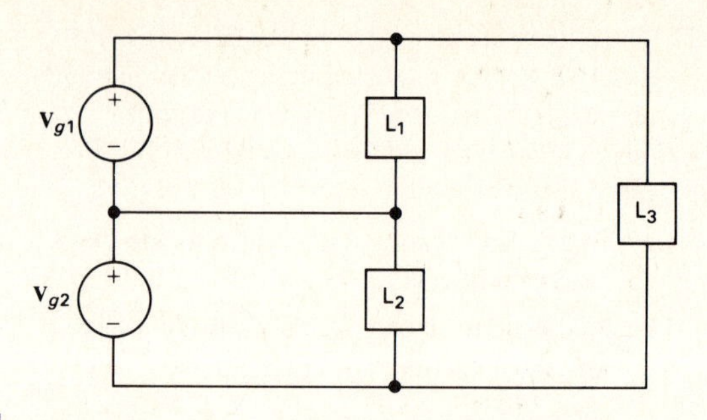

FIGURE P10.21

10.22 The three loads in the circuit shown in Fig. P10.22 are $S_1 = 5 + j1.25$ kVA, $S_2 = 6.25 + j2.5$ kVA, and $S_3 = 8 + j0$ kVA.

a) Calculate the complex power associated with each voltage source, $\mathbf{V}_{g1}$ and $\mathbf{V}_{g2}$.

b) Verify that the total real and reactive power delivered by the sources equals the total real and reactive power absorbed by the network.

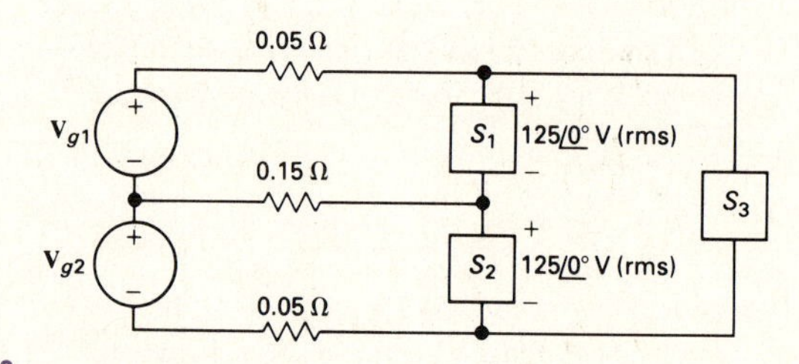

FIGURE P10.22

10.23 The three loads in Problem 10.17 are fed from a line having a series impedance $0.04 + j0.32$ Ω, as shown in Fig. P10.23.

a) Calculate the rms value of the voltage ($\mathbf{V}_s$) at the sending end of the line.

b) Calculate the average and reactive powers associated with the line impedance.

c) Calculate the average and reactive powers at the sending end of the line.

d) Calculate the efficiency (η) of the line if the efficiency is defined as

$$\eta = (P_{\text{load}}/P_{\text{sending end}}) \times 100.$$

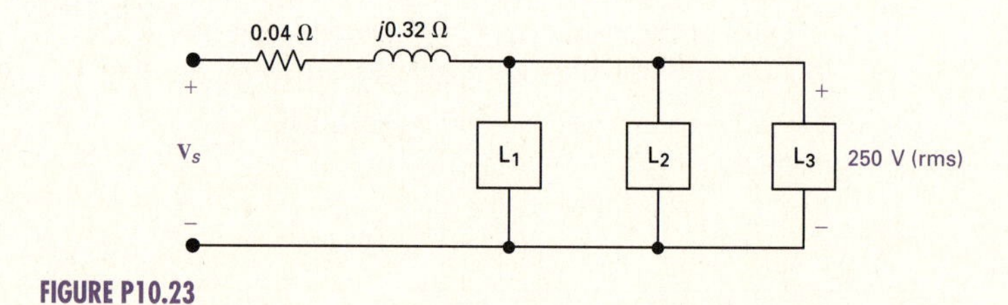

FIGURE P10.23

10.24 A group of small appliances on a 60 Hz system requires 20 kVA at 0.85 pf lagging when operated at 125 V (rms). The impedance of the feeder supplying the appliances is $0.01 + j0.08\ \Omega$. The voltage at the load end of the feeder is 125 V.

a) What is the rms magnitude of the voltage at the source end of the feeder?

b) What is the average power loss in the feeder?

c) What size capacitor (in microfarads) at the load end of the feeder is needed to improve the load power factor to unity?

d) After the capacitor is installed, what is the rms magnitude of the voltage at the source end of the feeder if the load voltage is maintained at 125 V?

e) What is the average power loss in the feeder for part (d)?

10.25 The circuit shown in Fig. P10.21 represents a residential distribution circuit in which the impedances of the service conductors are negligible and $\mathbf{V}_{g1} = \mathbf{V}_{g2} = 125\angle 0^\circ$ V (rms). The three loads in the circuit are L_1 (a dishwasher, a mixer, and a microwave oven); L_2 (a portable heater and a sunlamp); and L_3 (a clothes dryer, a water heater, and a range with an oven). Assume that all these appliances are in operation at the same time. The service conductors are protected with 100 A circuit breakers. Will the service to this residence be interrupted? Explain.

10.26 a) Find V (rms) and θ for the circuit in Fig. P10.26 if the load absorbs 2500 VA at a lagging power factor of 0.8.

b) Construct a phasor diagram of each solution obtained in part (a).

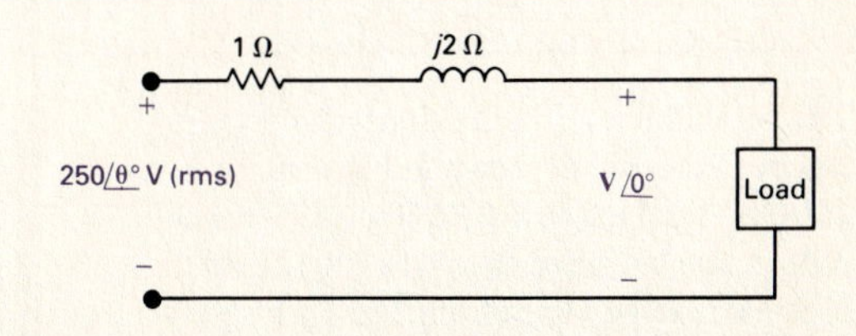

FIGURE P10.26

10.27 a) Find the average power dissipated in the line in Fig. P10.27.

b) Find the capacitive reactance that when connected in parallel with the load will make the load look purely resistive.

c) What is the equivalent impedance of the load in part (b)?

d) Find the average power dissipated in the line when the capacitive reactance is connected across the load.

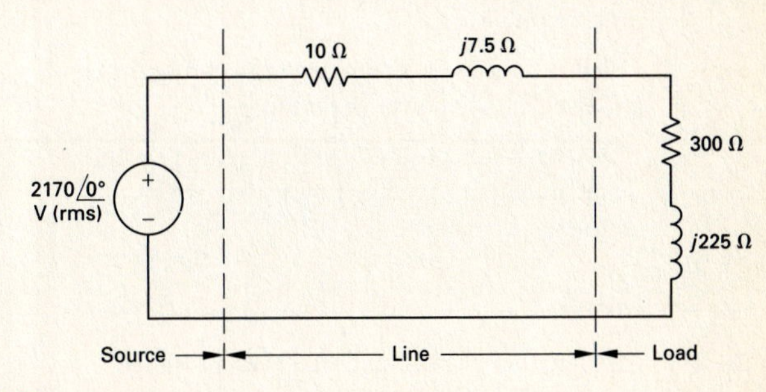

FIGURE P10.27

10.28 The steady-state voltage drop between the load and the sending end of the line seen in Fig. P10.28 is excessive. A capacitor is placed in parallel with the 200 kVA load and is adjusted until the steady-state voltage at the sending end of the line has the same magnitude as the voltage at the load end, that is, 6400 V (rms). The 200 kVA load is operating at a power factor of 0.6 lag. Calculate the size of the capacitor in microfarads if the circuit is operating at 60 Hz. In selecting the capacitor, keep in mind the need to keep the power loss in the line at a reasonable level.

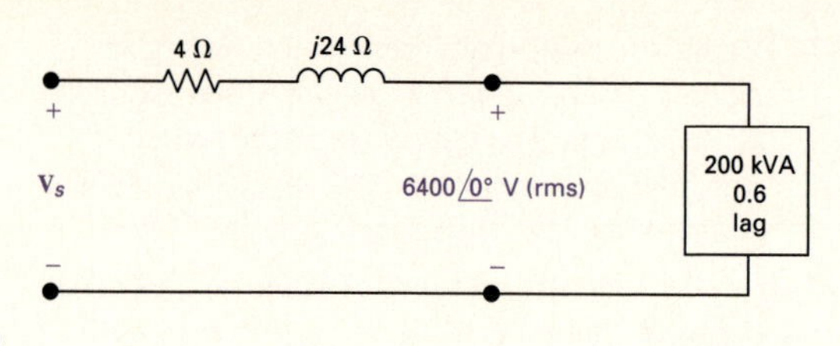

FIGURE P10.28

10.29 a) Determine the load impedance for the circuit shown in Fig. P10.29 that will result in maximum average power being transferred to the load if $\omega = 5$ krad/s.

b) Determine the maximum average power if $v_g = 80 \cos 5000t$ V.

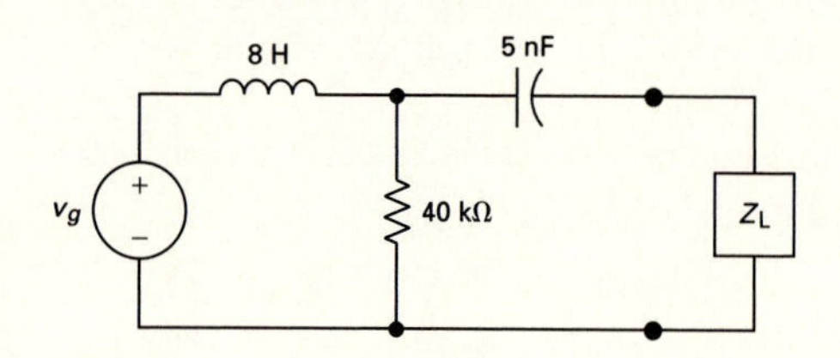

FIGURE P10.29

10.30 The peak amplitude of the sinusoidal voltage source in the circuit shown in Fig. P10.30 is $100\sqrt{2}$ V, and its period is 250π μs. The load resistor can be varied from 0 to 200 Ω, and the load capacitor can be varied from 1 to 4 μF.

a) Calculate the average power delivered to the load when $R_L = 100$ Ω and $C_L = 4$ μF.

b) Determine the settings of R_L and C_L that will result in the most average power being transferred to R_L.

c) What is the most average power in part (b)? Is it greater than the power in part (a)?

d) If there are no constraints on R_L and C_L, what is the maximum average power that can be delivered to a load?

e) What are the values of R_L and C_L for the condition of part (d)?

f) Is the average power calculated in part (d) larger than that calculated in part (c)?

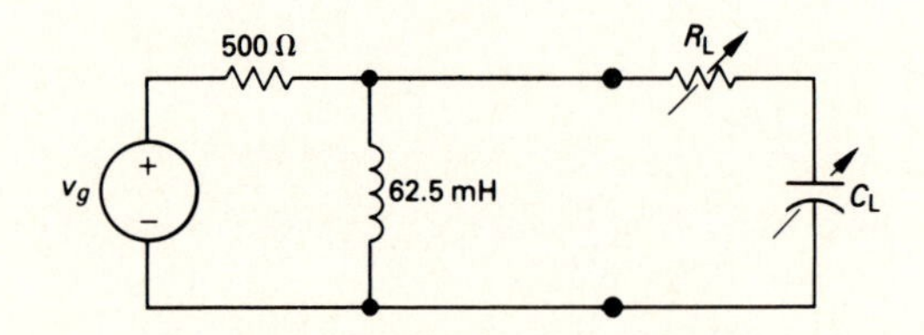

FIGURE P10.30

10.31 a) Assume that R_L in Fig. P10.30 can be varied between 0 and 500 Ω. Repeat parts (b) and (c) of Problem 10.30.

b) Is the new average power calculated in part (a) greater than that found in Problem 10.30?

c) Is the new average power calculated in part (a) less than that found in 10.30(d)?

10.32 The phasor voltage $\mathbf{V}_{ab}$ in the circuit shown in Fig. P10.32 is $240\angle 0^\circ$ V (rms) when no external load is connected to the terminals a,b. When a load is having an impedance of $90 - j30\ \Omega$ is connected across a,b, the value of $\mathbf{V}_{ab}$ is $115.2 - j86.4$ V(rms).

a) Find the impedance that should be connected across a,b for maximum average power transfer.

b) Find the maximum average power transferred to the load of part (a).

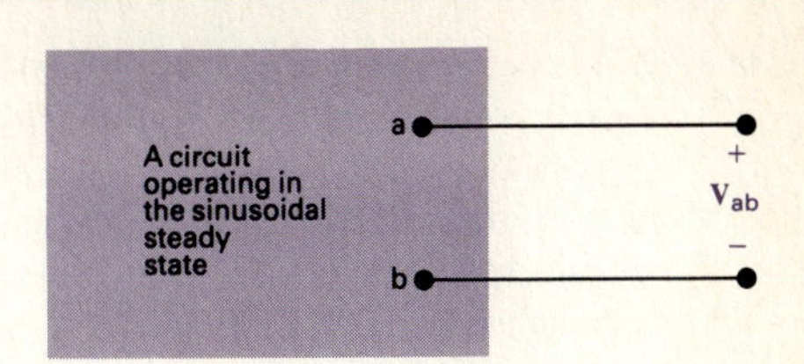

FIGURE P10.32

10.33 Prove that if only the magnitude of the load impedance can be varied, most average power is transferred to the load when $|Z_L| = |Z_{Th}|$. (*Hint:* In deriving the expression for the average load power, write the load impedance (Z_L) in the form $Z_L = |Z_L|\cos\theta + j|Z_L|\sin\theta$, and note that only $|Z_L|$ is variable.)

10.34 The variable resistor in the circuit shown in Fig. P10.34 is adjusted until the average power it absorbs is maximum.

a) Find R.

b) Find the maximum average power.

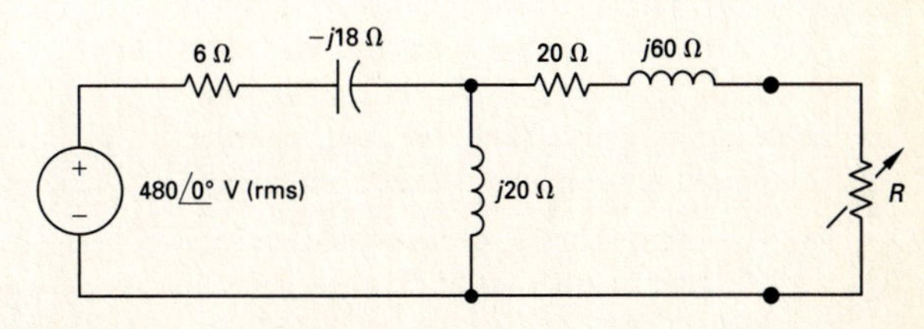

FIGURE P10.34

10.35 The load impedance Z_L for the circuit shown in Fig. P10.35 is adjusted until maximum average power is delivered to Z_L.

a) Find the maximum average power delivered to Z_L.

b) What percentage of the total power developed in the circuit is delivered to Z_L?

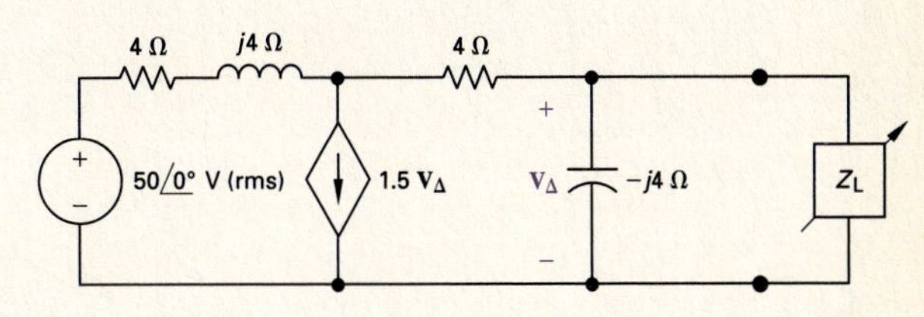

FIGURE P10.35

10.36 The variable resistor R_σ in the circuit shown in Fig. P10.36 is adjusted until maximum average power is delivered to R_σ.

a) What is the value of R_σ in ohms?

b) Calculate the average power delivered to R_σ.

c) If R_σ is replaced with a variable impedance Z_σ, what is the maximum average power that can be delivered to Z_σ?

d) In part (c), what percentage of the circuit's developed power is delivered to the load Z_σ?

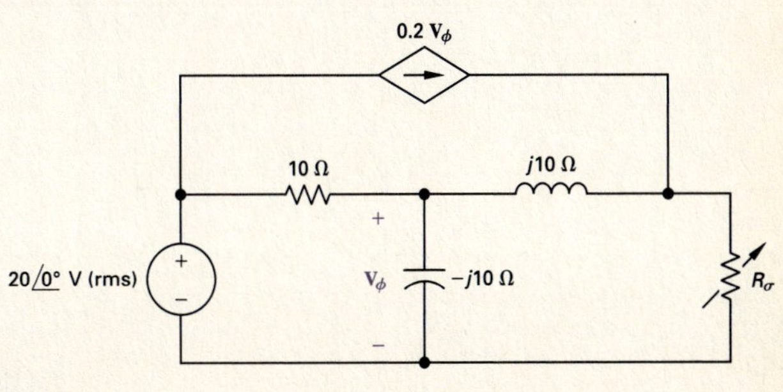

FIGURE P10.36

10.37 A factory has an electrical load of 1600 kW at a lagging power factor of 0.8. An additional variable power factor load is to be added to the factory. The new load will add 320 kW to the real power load of the factory. The power factor of the added load is to be adjusted so that the overall power factor of the factory is 0.96 leading.

a) Specify the reactive power associated with the added load.

b) Does the added load absorb or deliver magnetizing vars?

c) What is the power factor of the additional load?

d) Assume that the rms voltage at the input to the factory is 2400 V. What is the rms magnitude of the current into the factory before the variable power factor load is added?

e) What is the rms magnitude of the current into the factory after the variable power factor load has been added?

10.38 Assume the factory described in Problem 10.37 is fed from a line having an impedance of $0.05 + j0.40\ \Omega$. If the voltage at the factory is maintained at 2400 V (rms), find the magnitude of the voltage at the sending end of the line before and after the new load is added.

10.39 The sending-end voltage in the circuit seen in Fig. P10.39 is adjusted so that the rms value of the load voltage is always 4000 V. The variable capacitor is adjusted until the average power dissipated in the line resistance is minimum.

a) If the frequency of the sinusoidal source is 60 Hz, what is the value of the capacitance in microfarads?

b) If the capacitor is removed from the circuit, what percentage increase in the magnitude of $\mathbf{V}_s$ is necessary to maintain 4000 V at the load?

c) If the capacitor is removed from the circuit, what is the percentage increase in line loss?

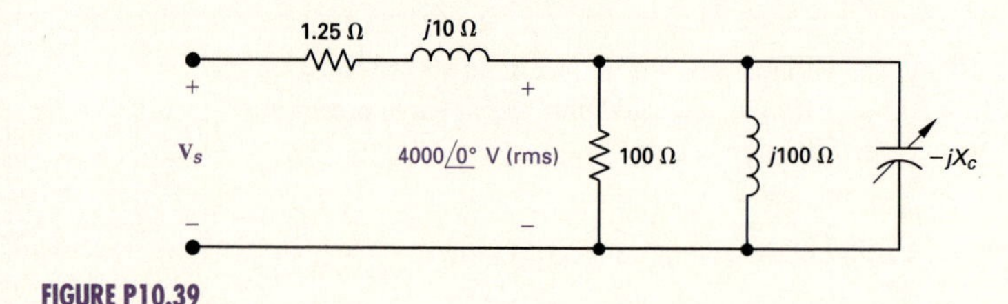

FIGURE P10.39

10.40 The operational amplifier in the circuit shown in Fig. P10.40 is ideal. Calculate the average power delivered to the 800 Ω resistor when $v_g = 2\cos(8000t - 90°)$ V.

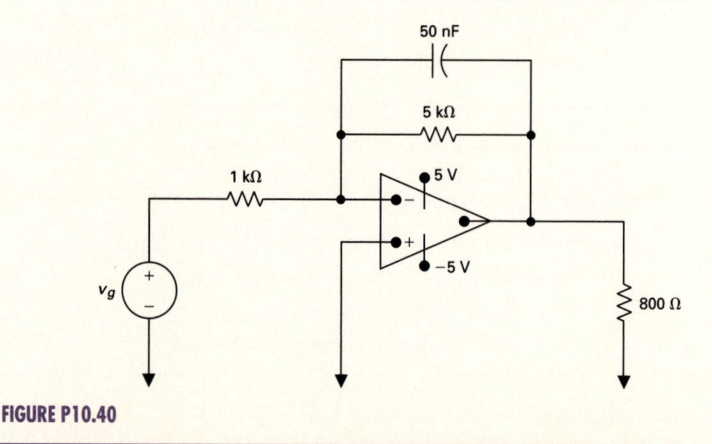

FIGURE P10.40

◆ **10.41** You have been given the job of redesigning the hair dryer discussed in the Practical Perspective Example for use in England. The standard supply voltage in England is 220 V (rms). What resistor values will you use in your design to meet the same power specifications?

◆ **10.42** As mentioned in the Practical Perspectives example, only two independent power specifications can be made when two resistors make up the heating element of the hair dryer.

a) Show that the expression for the high power rating (P_H) is

$$P_H = \frac{P_M^2}{P_M - P_L},$$

where P_M = the medium power rating and P_L = the lower power rating.

b) If $P_L = 250$ W and $P_M = 750$ W, what must the high power rating be?

◆ **10.43** Specify the values of R_1 and R_2 in the hair dryer circuit in Fig. 10.25 if the low power rating is 240 W and the high power rating is 1000 W. Assume the supply voltage is 120 V (rms). (*Hint:* Work Problem 10.42 first.)

◆ **10.44** If a third resistor is added to the hair dryer circuit in Fig. 10.25, it is possible to design to three independent power specifications. If the resistor R_3 is added in series with the Thermal fuse, then the corresponding LOW, MEDIUM, and HIGH power circuit diagrams are as shown in Fig. P10.44. If the three power settings are 600 W, 900 W, and 1200 W, respectively, when connected to a 120 V (rms) supply, what resistor values should be used?

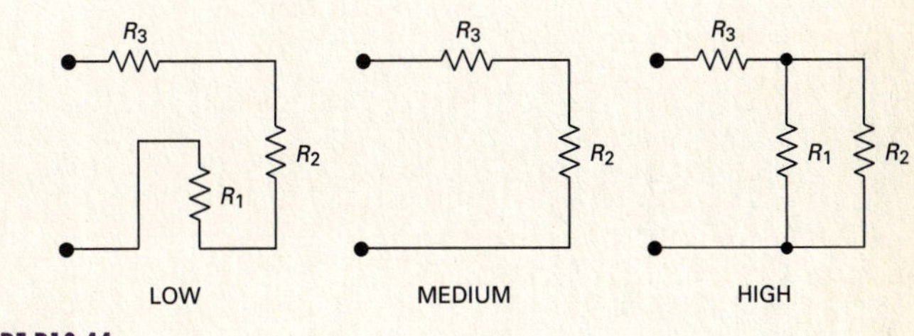

FIGURE P10.44

11 Balanced Three-Phase Circuits

CHAPTER CONTENTS

Generating, transmitting, distributing, and using large blocks of electric power is accomplished with three-phase circuits. The comprehensive analysis of such systems is a field of study in its own right; we cannot hope to cover it in a single chapter. Fortunately, an understanding of only the steady-state sinusoidal behavior of balanced three-phase circuits is sufficient for engineers who do not specialize in power systems. We define what we mean by a balanced circuit later in the discussion. The same circuit analysis techniques discussed in earlier chapters can be applied to either unbalanced or balanced three-phase circuits. Here, we use these familiar techniques to develop several shortcuts to the analysis of balanced three-phase circuits.

For economic reasons, three-phase systems are usually designed to operate in the balanced state. Thus, in this introductory treatment, we can justify considering only balanced circuits. The analysis of unbalanced three-phase circuits, which you will encounter if you study electric power in later courses, relies heavily on an understanding of balanced circuits.

The basic structure of a three-phase system consists of voltage sources connected to loads by means of transformers and transmission lines.[1] To analyze such a circuit, we can reduce it to a voltage source connected to a load via a line. The omission of the transformer simplifies the discussion without jeopardizing a basic understanding of the calculations involved. Figure 11.1 shows a basic circuit. A defining characteristic of a balanced three-phase circuit is that it contains a set of balanced three-phase voltages at its source. We begin by considering these voltages, and then we move to the voltage and current relationships for the Y-Y and Y-Δ circuits. After considering voltage and current in such circuits, we conclude with sections on power and power measurement.

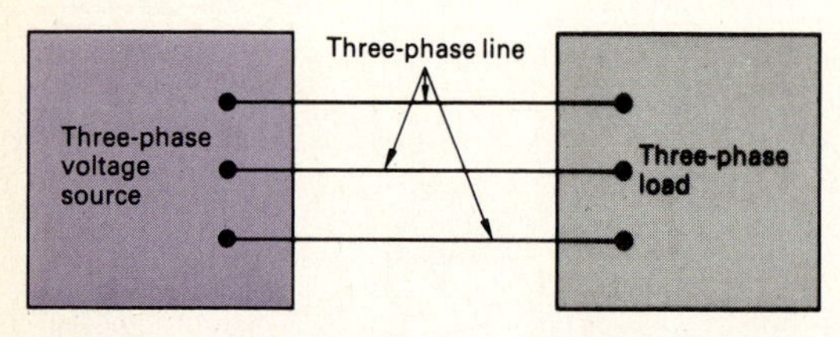

FIGURE 11.1 A basic three-phase circuit.

11.1 BALANCED THREE-PHASE VOLTAGES

A set of balanced three-phase voltages consists of three sinusoidal voltages that have identical amplitudes and frequencies but are out of phase with each other by exactly 120°. Standard practice is to refer to the three phases as a, b, and c and to use the a-phase as the reference phase. The three voltages are referred to as the **a-phase voltage**, the **b-phase voltage**, and the **c-phase voltage**.

Only two possible phase relationships can exist between the a-phase voltage and the b- and c-phase voltages. One possibility is for the b-phase voltage to lag the a-phase voltage by 120°, in which case the c-phase voltage must lead the a-phase voltage by 120°. This phase relationship is known as the **abc**, or **positive**, **phase sequence**. The only other possibility is for the b-phase voltage to lead the a-phase voltage by 120°, in which case the c-phase voltage must lag the a-phase voltage by 120°. This phase relationship is known as the **acb**, or **negative**, **phase sequence**. In phasor notation, the two possible sets of balanced phase voltages are

$$\mathbf{V}_{\rm a} = V_m \underline{/0^\circ}; \quad \mathbf{V}_{\rm b} = V_m \underline{/-120^\circ}; \quad \mathbf{V}_{\rm c} = V_m \underline{/+120^\circ}; \qquad (11.1)$$

[1] We introduce transformers in Chapter 12.

and

$$\mathbf{V}_a = V_m\angle 0^\circ;$$
$$\mathbf{V}_b = V_m\angle +120^\circ; \quad \mathbf{(11.2)}$$
$$\mathbf{V}_c = V_m\angle -120^\circ.$$

Equations (11.1) are for the abc, or positive, sequence. Equations (11.2) are for the acb, or negative, sequence. Figure 11.2 shows the phasor diagrams of the voltage sets in Eqs. (11.1) and (11.2). The phase sequence is the clockwise order of the subscripts around the diagram from $\mathbf{V}_a$. The fact that a three-phase circuit can have one of two phase sequences must be taken into account whenever two such circuits are operated in parallel. The circuits can operate in parallel only if they have the same phase sequence.

Another important characteristic of a set of balanced three-phase voltages is that the sum of the voltages is zero. Thus, from either Eqs. (11.1) or Eqs. (11.2),

$$\mathbf{V}_a + \mathbf{V}_b + \mathbf{V}_c = 0. \quad \mathbf{(11.3)}$$

Because the sum of the phasor voltages is zero, the sum of the instantaneous voltages also is zero; that is,

$$v_a + v_b + v_c = 0. \quad \mathbf{(11.4)}$$

Now that we know the nature of a balanced set of three-phase voltages, we can state the first of the analytical shortcuts alluded to in the introduction to this chapter: if we know the phase sequence and one voltage in the set, we know the entire set. Thus for a balanced three-phase system, we can focus on determining the voltage (or current) in one phase, because once we know one phase quantity, we know the others.

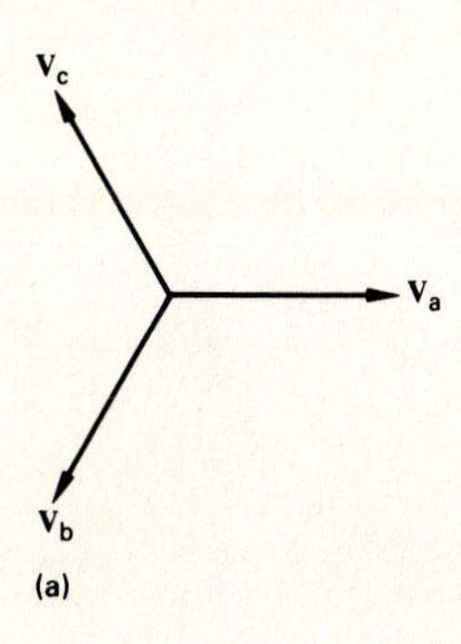

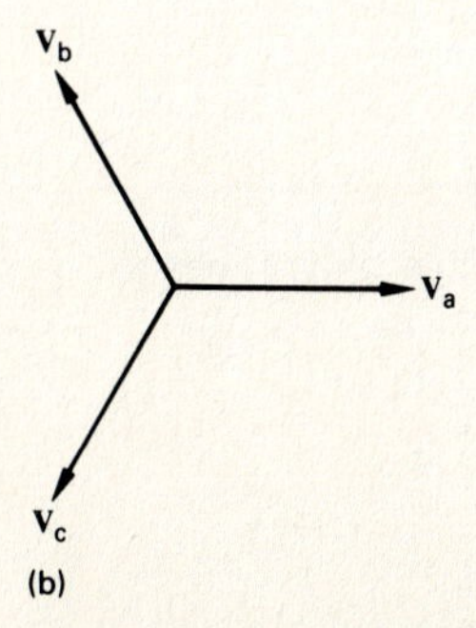

FIGURE 11.2 Phasor diagrams of a balanced set of three-phase voltages: (a) the abc (positive) sequence; and (b) the acb (negative) sequence.

DRILL EXERCISE

11.1 What is the phase sequence of each of the following sets of voltages?

a) $v_a = 208\cos(\omega t + 76^\circ)$ V,
$v_b = 208\cos(\omega t + 316^\circ)$ V,
$v_c = 208\cos(\omega t - 164^\circ)$ V.

b) $v_a = 4160\cos(\omega t - 49^\circ)$ V,
$v_b = 4160\cos(\omega t - 289^\circ)$ V,
$v_c = 4160\cos(\omega t + 191^\circ)$ V.

ANSWER: (a) abc; (b) acb.

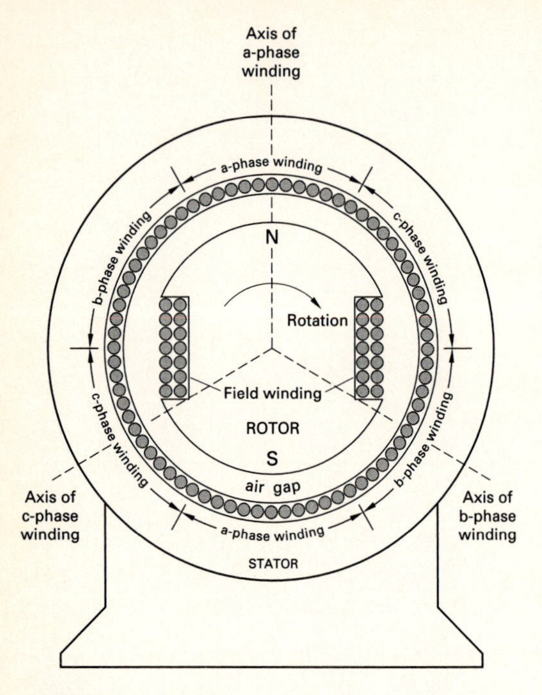

FIGURE 11.3 A sketch of a three-phase voltage source.

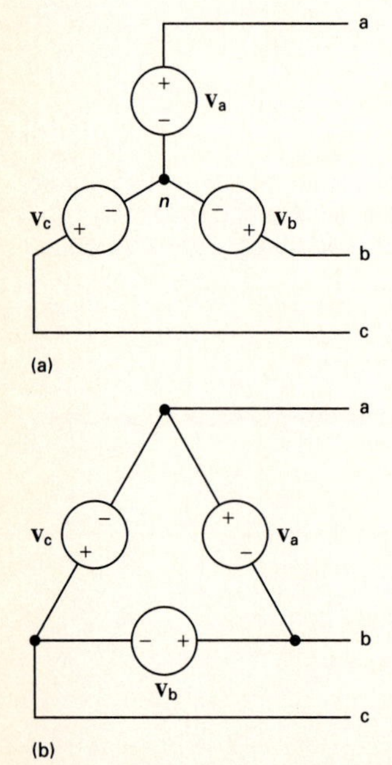

FIGURE 11.4 The two basic connections of an ideal three-phase source: (a) a Y-connected source; and (b) a Δ-connected source.

11.2 THREE-PHASE VOLTAGE SOURCES

A three-phase voltage source is a generator with three separate windings distributed around the periphery of the stator. Each winding comprises one phase of the generator. The rotor of the generator is an electromagnet driven at synchronous speed by a prime mover, such as a steam or gas turbine. Rotation of the electromagnet induces a sinusoidal voltage in each winding. The phase windings are designed so that the sinusoidal voltages induced in them are equal in amplitude and out of phase with each other by 120°. The phase windings are stationary with respect to the rotating electromagnet, so the frequency of the voltage induced in each winding is the same. Figure 11.3 shows a sketch of a two-pole three-phase source.

There are two ways of interconnecting the separate phase windings to form a three-phase source: in either a wye (Y) or a delta (Δ) configuration. Figure 11.4 shows both, with ideal voltage sources used to model the phase windings of the three-phase generator. The common terminal in the Y-connected source, labeled n in Fig. 11.4(a), is called the **neutral terminal** of the source. The neutral terminal may or may not be available for external connections.

Sometimes, the impedance of each phase winding is so small (compared to other impedances in the circuit) that we need not account for it in modeling the generator; the model consists solely of ideal voltage sources, as in Fig. 11.4. However, if the impedance of each phase winding is not negligible, we place the winding impedance in series with an ideal sinusoidal voltage source. All windings on the machine are of the same construction, so we assume the winding impedances to be identical. The winding impedance of a three-phase generator is inductive. Figure 11.5 shows a model of such a machine, in which R_w is the winding resistance, and X_w is the inductive reactance of the winding.

Because three-phase sources and loads can be either Y connected or Δ connected, the basic circuit in Fig. 11.1 represents four different configurations:

Source	Load
Y	Y
Y	Δ
Δ	Y
Δ	Δ

We begin by analyzing the Y-Y circuit. The remaining three arrangements can be reduced to a Y-Y equivalent circuit, so analysis of the Y-Y circuit is the key to solving all balanced three-phase arrangements. We then illustrate the reduction of the Y-Δ arrangement and leave the analysis of the Δ-Y and Δ-Δ arrangements to you in the Problems.

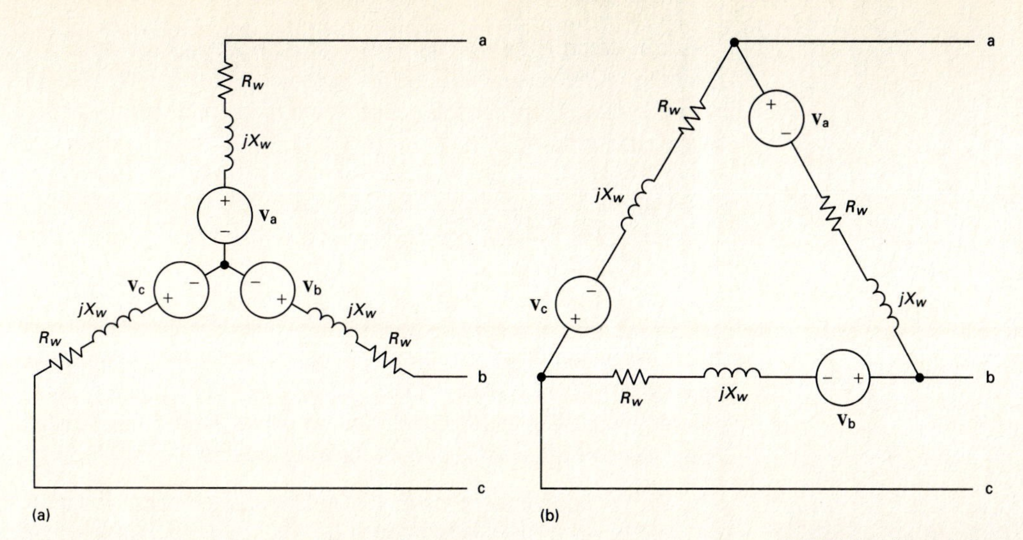

FIGURE 11.5 A model of a three-phase source with winding impedance: (a) a Y-connected source; and (b) a Δ-connected source.

11.3 ANALYSIS OF THE WYE-WYE CIRCUIT

Figure 11.6 illustrates a general Y-Y circuit, in which we included a fourth conductor that connects the source neutral to the load neutral. A fourth conductor is possible only in the Y-Y arrangement. (More about this later.) For convenience, we transformed the Y connections into "tipped-over tees." In Fig. 11.6, Z_{ga}, Z_{gb}, and Z_{gc} represent the internal impedance associated with each phase winding of the voltage generator; Z_{1a}, Z_{1b}, and Z_{1c} represent the impedance of the lines connecting a phase of the source to a phase of the load; Z_o is the impedance of the neutral conductor connecting the source neutral to the load neutral; and Z_A, Z_B, and Z_C represent the impedance of each phase of the load.

Use computer tools to analyze a complex circuit like this one.

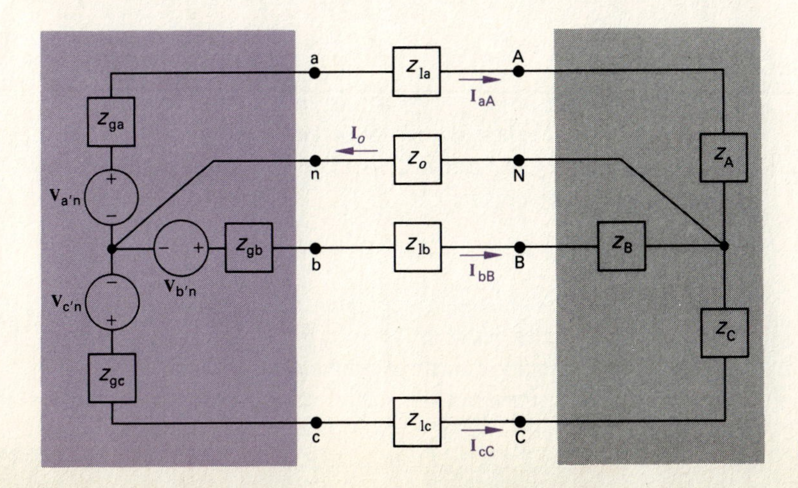

FIGURE 11.6 A three-phase Y-Y system.

We can describe this circuit with a single node-voltage equation. Using the source neutral as the reference node and letting $\mathbf{V}_{\mathrm{N}}$ denote the node voltage between the nodes N and n, we find that the node-voltage equation is

$$\frac{\mathbf{V}_{\mathrm{N}}}{Z_o} + \frac{\mathbf{V}_{\mathrm{N}} - \mathbf{V}_{\mathrm{a'n}}}{Z_{\mathrm{A}} + Z_{1\mathrm{a}} + Z_{\mathrm{ga}}} + \frac{\mathbf{V}_{\mathrm{N}} - \mathbf{V}_{\mathrm{b'n}}}{Z_{\mathrm{B}} + Z_{1\mathrm{b}} + Z_{\mathrm{gb}}} + \frac{\mathbf{V}_{\mathrm{N}} - \mathbf{V}_{\mathrm{c'n}}}{Z_{\mathrm{C}} + Z_{1\mathrm{c}} + Z_{\mathrm{gc}}} = 0. \quad \textbf{(11.5)}$$

This is the general equation for any circuit of the Y-Y configuration depicted in Fig. 11.6. But we can simplify Eq. (11.5) significantly if we now consider the formal definition of a balanced three-phase circuit. Such a circuit satisfies the following criteria:

1. The voltage sources form a set of balanced three-phase voltages. In Fig. 11.6, this means that $\mathbf{V}_{\mathrm{a'n}}$, $\mathbf{V}_{\mathrm{b'n}}$, and $\mathbf{V}_{\mathrm{c'n}}$ are a set of balanced three-phase voltages.

2. The impedance of each phase of the voltage source is the same. In Fig. 11.6, this means that $Z_{\mathrm{ga}} = Z_{\mathrm{gb}} = Z_{\mathrm{gc}}$.

3. The impedance of each line (or phase) conductor is the same. In Fig. 11.6, this means that $Z_{1\mathrm{a}} = Z_{1\mathrm{b}} = Z_{1\mathrm{c}}$.

4. The impedance of each phase of the load is the same. In Fig. 11.6, this means that $Z_{\mathrm{A}} = Z_{\mathrm{B}} = Z_{\mathrm{C}}$.

There is no restriction on the impedance of a neutral conductor; its value has no effect on whether the system is balanced.

If the circuit in Fig. 11.6 is balanced, we may rewrite Eq. (11.5) as

$$\mathbf{V}_{\mathrm{N}}\left(\frac{1}{Z_o} + \frac{3}{Z_\phi}\right) = \frac{\mathbf{V}_{\mathrm{a'n}} + \mathbf{V}_{\mathrm{b'n}} + \mathbf{V}_{\mathrm{c'n}}}{Z_\phi}, \quad \textbf{(11.6)}$$

where

$$Z_\phi = Z_{\mathrm{A}} + Z_{1\mathrm{a}} + Z_{\mathrm{ga}} = Z_{\mathrm{B}} + Z_{1\mathrm{b}} + Z_{\mathrm{gb}} = Z_{\mathrm{C}} + Z_{1\mathrm{c}} + Z_{\mathrm{gc}}.$$

The right-hand side of Eq. (11.6) is zero because by hypothesis the numerator is a set of balanced three-phase voltages and Z_ϕ is not zero. The only value of $\mathbf{V}_{\mathrm{N}}$ that satisfies Eq. (11.6) is zero. Therefore, for a balanced three-phase circuit,

$$\mathbf{V}_{\mathrm{N}} = 0. \quad \textbf{(11.7)}$$

Equation (11.7) is extremely important. If $\mathbf{V}_{\mathrm{N}}$ is zero, there is no difference in potential between the source neutral, n, and the load neutral, N; consequently, the current in the neutral conductor is zero. Hence we may either remove the neutral conductor from a balanced Y-Y configuration ($I_0 = 0$) or replace it with a perfect short circuit between the nodes n and N ($\mathbf{V}_N = 0$). Both equivalents are convenient to use when modeling balanced three-phase circuits.

We now turn to the effect that balanced conditions have on the three line currents. With reference to Fig. 11.6, when the system is balanced the three line currents are

$$\mathbf{I}_{aA} = \frac{\mathbf{V}_{a'n} - \mathbf{V}_N}{Z_A + Z_{1a} + Z_{ga}} = \frac{\mathbf{V}_{a'n}}{Z_\phi}; \quad (11.8)$$

$$\mathbf{I}_{bB} = \frac{\mathbf{V}_{b'n} - \mathbf{V}_N}{Z_B + Z_{1b} + Z_{gb}} = \frac{\mathbf{V}_{b'n}}{Z_\phi}; \quad (11.9)$$

$$\mathbf{I}_{cC} = \frac{\mathbf{V}_{c'n} - \mathbf{V}_N}{Z_C + Z_{1c} + Z_{gc}} = \frac{\mathbf{V}_{c'n}}{Z_\phi}. \quad (11.10)$$

We see that the three line currents form a balanced set of three-phase currents; that is, the current in each line is equal in amplitude and frequency and is 120° out of phase with the other two line currents. Thus, if we calculate the current $\mathbf{I}_{aA}$ and we know the phase sequence, we have a shortcut for finding $\mathbf{I}_{bB}$ and $\mathbf{I}_{cC}$. This procedure parallels the shortcut used to find the b- and c-phase source voltages from the a-phase source voltage.

We can use Eq. (11.8) to construct an equivalent circuit for the a-phase of the balanced Y-Y circuit. From this equation, the current in the a-phase conductor line is simply the voltage generated in the a-phase winding of the generator divided by the total impedance in the a-phase of the circuit. Thus Eq. (11.8) describes the simple circuit shown in Fig. 11.7, in which the neutral conductor has been replaced by a perfect short circuit. The circuit in Fig. 11.7 is referred to as the **single-phase equivalent circuit** of a balanced three-phase circuit. Because of the established relationships between phases, once we solve this circuit, we can easily write down the voltages and currents in the other two phases. Thus, drawing a single-phase equivalent circuit is an important first step in analyzing a three-phase circuit.

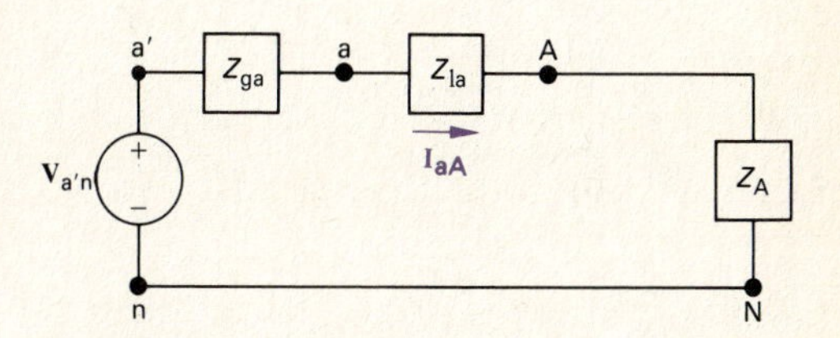

FIGURE 11.7 A single-phase equivalent circuit.

A word of caution here. The current in the neutral conductor in Fig. 11.7 is $\mathbf{I}_{aA}$, which is not the same as the current in the neutral conductor of the balanced three-phase circuit, which is

$$\mathbf{I}_o = \mathbf{I}_{aA} + \mathbf{I}_{bB} + \mathbf{I}_{cC}. \quad (11.11)$$

Thus the circuit shown in Fig. 11.7 gives the correct value of the line current but only the a-phase component of the neutral current. Whenever this single-phase equivalent circuit is applicable, the line currents form a balanced three-phase set, and the right-hand side of Eq. (11.11) sums to zero.

Once we know the line current in Fig. 11.7, calculating any voltages of interest is relatively simple. Of particular interest is the relationship between the line-to-line voltages and the line-to-neutral voltages. We establish this relationship at the load terminals, but our observations

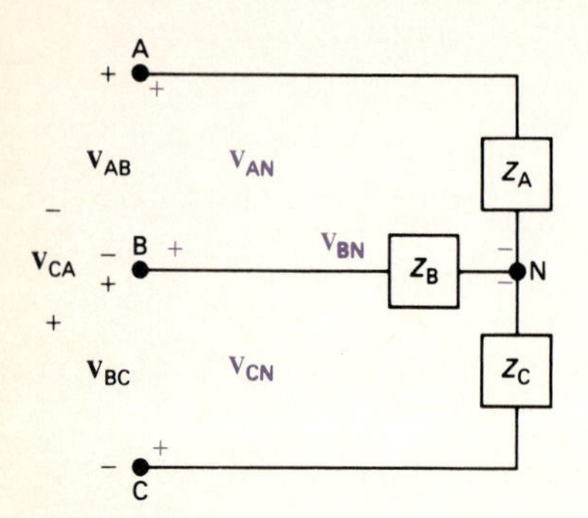

FIGURE 11.8 Line-to-line and line-to-neutral voltages.

also apply at the source terminals. The line-to-line voltages at the load terminals can be seen in Fig. 11.8. They are $\mathbf{V}_{AB}$, $\mathbf{V}_{BC}$, and $\mathbf{V}_{CA}$, where the double subscript notation indicates a voltage drop from the first-named node to the second. (Because we are interested in the balanced state, we have omitted the neutral conductor from Fig. 11.8.)

The line-to-neutral voltages are $\mathbf{V}_{AN}$, $\mathbf{V}_{BN}$, and $\mathbf{V}_{CN}$. We can now describe the line-to-line voltages in terms of the line-to-neutral voltages, using Kirchhoff's voltage law:

$$\mathbf{V}_{AB} = \mathbf{V}_{AN} - \mathbf{V}_{BN}, \tag{11.12}$$

$$\mathbf{V}_{BC} = \mathbf{V}_{BN} - \mathbf{V}_{CN}, \tag{11.13}$$

and

$$\mathbf{V}_{CA} = \mathbf{V}_{CN} - \mathbf{V}_{AN}. \tag{11.14}$$

Use a computer tool to assist in analyzing the negative sequence.

To show the relationship between the line-to-line voltages and the line-to-neutral voltages, we assume a positive, or abc, sequence. Using the line-to-neutral voltage of the a-phase as the reference,

$$\mathbf{V}_{AN} = V_\phi \angle 0^\circ, \tag{11.15}$$

$$\mathbf{V}_{BN} = V_\phi \angle -120^\circ, \tag{11.16}$$

and

$$\mathbf{V}_{CN} = V_\phi \angle +120^\circ, \tag{11.17}$$

where V_ϕ represents the magnitude of the line-to-neutral voltage. Substituting Eqs. (11.15)–(11.17) into Eqs. (11.12)–(11.14), respectively, yields

$$\mathbf{V}_{AB} = V_\phi \angle 0^\circ - V_\phi \angle -120^\circ = \sqrt{3} V_\phi \angle 30^\circ, \tag{11.18}$$

$$\mathbf{V}_{BC} = V_\phi \angle -120^\circ - V_\phi \angle 120^\circ = \sqrt{3} V_\phi \angle -90^\circ, \tag{11.19}$$

and

$$\mathbf{V}_{CA} = V_\phi \angle 120^\circ - V_\phi \angle 0^\circ = \sqrt{3} V_\phi \angle 150^\circ. \tag{11.20}$$

Equations (11.18)–(11.20) reveal that (1) the magnitude of the line-to-line voltage is $\sqrt{3}$ times the magnitude of the line-to-neutral voltage, (2) the line-to-line voltages form a balanced three-phase set of voltages, and (3) the set of line-to-line voltages leads the set of line-to-neutral voltages by 30°. We leave to you the demonstration that for a negative sequence, the only change is that the set of line-to-line voltages lags the set of line-to-neutral voltages by 30°. The phasor diagrams shown in Fig. 11.9 summarize these observations. Here, again, is a shortcut in the analysis of a balanced system: if you know the line-to-neutral voltage at some point in the circuit, you can easily determine the line-to-line voltage at the same point, and vice versa.

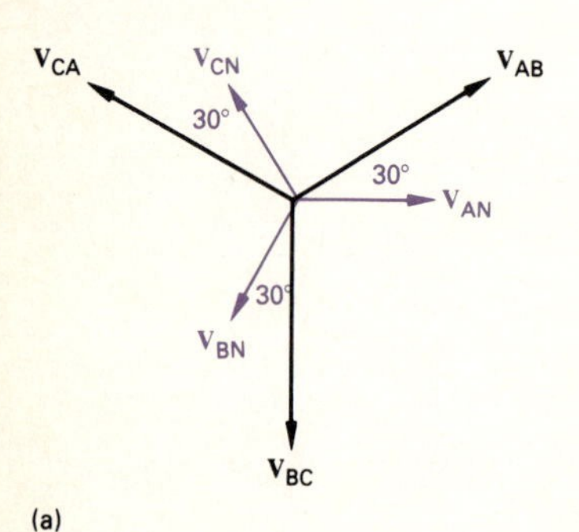

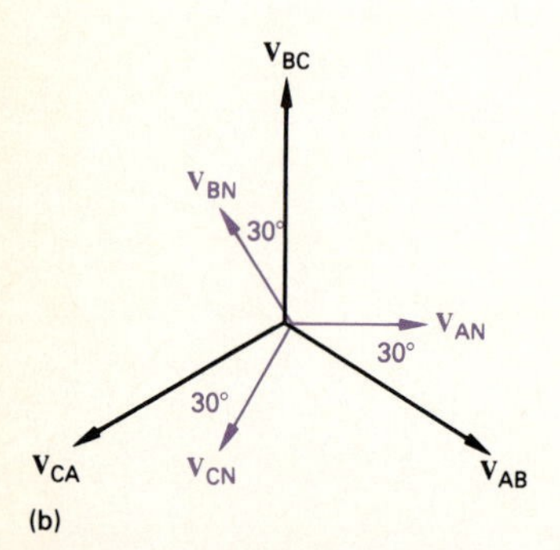

FIGURE 11.9 Phasor diagrams showing the relationship between line-to-line and line-to-neutral voltages in a balanced system: (a) the abc sequence; and (b) the acb sequence.

DRILL EXERCISES

11.2 The voltage from B to N in a balanced three-phase circuit is $120\angle 60^\circ$ V. If the phase sequence is positive, what is the value of $\mathbf{V}_{BC}$?

ANSWER: $207.85\angle +90^\circ$ V.

11.3 The c-phase voltage of a balanced three-phase Y-connected system is $660\angle 160^\circ$ V. If the phase sequence is negative, what is the value of $\mathbf{V}_{AB}$?

ANSWER: $1143.15\angle -110^\circ$ V.

We now pause to elaborate on terminology. **Line voltage** refers to the voltage across any pair of lines; **phase voltage** refers to the voltage across a single phase. **Line current** refers to the current in a single line; **phase current** refers to current in a single phase. Observe that in a Δ connection, line voltage and phase voltage are identical, and in a Y connection, line current and phase current are identical.

Because three-phase systems are designed to handle large blocks of electric power, all voltage and current specifications are given as rms values. When voltage ratings are given, they refer specifically to the rating of the line voltage. Thus when a three-phase transmission line is rated at 345 kV, the nominal value of the rms line-to-line voltage is 345,000 V. In this chapter we express all voltages and currents as rms values.

Finally, the Greek letter phi (ϕ) is widely used in the literature to denote a per-phase quantity. Thus $\mathbf{V}_\phi$, $\mathbf{I}_\phi$, Z_ϕ, P_ϕ, and Q_ϕ are interpreted as voltage/phase, current/phase, impedance/phase, power/phase, and reactive power/phase, respectively.

Example 11.1 shows how to use the observations made so far to solve a balanced three-phase Y-Y circuit.

EXAMPLE 11.1

A balanced three-phase Y-connected generator with positive sequence has an impedance of $0.2 + j0.5\ \Omega/\phi$ and an internal voltage of $120\ \text{V}/\phi$. The generator feeds a balanced three-phase Y-connected load having an impedance of $39 + j28\ \Omega/\phi$. The impedance of the line connecting the generator to the load is $0.8 + j1.5\ \Omega/\phi$. The a-phase internal voltage of the generator is specified as the reference phasor.

a) Construct the a-phase equivalent circuit of the system.

b) Calculate the three line currents $\mathbf{I}_{aA}$, $\mathbf{I}_{bB}$, and $\mathbf{I}_{cC}$.

c) Calculate the three phase voltages at the load, $\mathbf{V}_{AN}$, $\mathbf{V}_{BN}$, and $\mathbf{V}_{CN}$.

d) Calculate the line voltages $\mathbf{V}_{AB}$, $\mathbf{V}_{BC}$, and $\mathbf{V}_{CA}$ at the terminals of the load.

e) Calculate the phase voltages at the terminals of the generator, $\mathbf{V}_{an}$, $\mathbf{V}_{bn}$, and $\mathbf{V}_{cn}$.

f) Calculate the line voltages $\mathbf{V}_{ab}$, $\mathbf{V}_{bc}$, and $\mathbf{V}_{ca}$ at the terminals of the generator.

g) Repeat (a)–(f) for a negative phase sequence.

SOLUTION

a) Figure 11.10 shows the single-phase equivalent circuit.

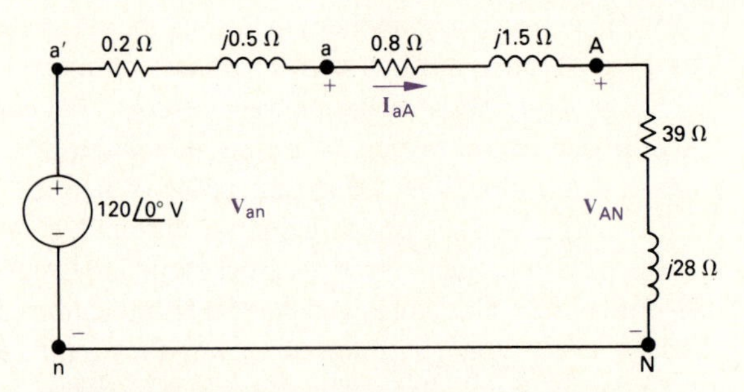

FIGURE 11.10 The single-phase equivalent circuit for Example 11.1.

b) The a-phase line current is

$$\mathbf{I}_{aA} = \frac{120\angle 0^\circ}{(0.2 + 0.8 + 39) + j(0.5 + 1.5 + 28)}$$

$$= \frac{120\angle 0^\circ}{40 + j30}$$

$$= 2.4\angle -36.87^\circ \text{ A}.$$

For a positive phase sequence,

$$\mathbf{I}_{bB} = 2.4\angle -156.87^\circ \text{ A};$$

$$\mathbf{I}_{cC} = 2.4\angle 83.13^\circ \text{ A}.$$

c) The phase voltage at the A terminal of the load is

$$\mathbf{V}_{AN} = (39 + j28)(2.4\angle -36.87^\circ)$$

$$= 115.22\angle -1.19^\circ \text{ V}.$$

For a positive phase sequence,

$$\mathbf{V}_{BN} = 115.22\angle -121.19^\circ \text{ V};$$

$$\mathbf{V}_{CN} = 115.22\angle +118.81^\circ \text{ V}.$$

d) For a positive phase sequence, the line voltages lead the phase voltages by 30°; thus

$$\mathbf{V}_{AB} = (\sqrt{3}\angle 30^\circ)\mathbf{V}_{AN}$$

$$= 199.58\angle 28.81^\circ \text{ V};$$

$$\mathbf{V}_{BC} = 199.58\angle -91.19^\circ \text{ V};$$

$$\mathbf{V}_{CA} = 199.58\angle 148.81^\circ \text{ V}.$$

e) The phase voltage at the a terminal of the source is

$$\begin{aligned}\mathbf{V}_{an} &= 120 - (0.2 + j0.5)(2.4\angle{-36.87^\circ}) \\ &= 120 - 1.29\angle{31.33^\circ} \\ &= 118.90 - j0.67 \\ &= 118.90\angle{-0.32^\circ}\text{ V.}\end{aligned}$$

For a positive phase sequence,

$$\mathbf{V}_{bn} = 118.90\angle{-120.32^\circ}\text{ V;}$$
$$\mathbf{V}_{cn} = 118.90\angle{119.68^\circ}\text{ V.}$$

f) The line voltages at the source terminals are

$$\begin{aligned}\mathbf{V}_{ab} &= (\sqrt{3}\angle{30^\circ})\mathbf{V}_{an} \\ &= 205.94\angle{29.68^\circ}\text{ V;}\end{aligned}$$
$$\mathbf{V}_{bc} = 205.94\angle{-90.32^\circ}\text{ V;}$$
$$\mathbf{V}_{ca} = 205.94\angle{149.68^\circ}\text{ V.}$$

g) Changing the phase sequence has no effect on the single-phase equivalent circuit. The three line currents are

$$\mathbf{I}_{aA} = 2.4\angle{-36.87^\circ}\text{ A;}$$
$$\mathbf{I}_{bB} = 2.4\angle{83.13^\circ}\text{ A;}$$
$$\mathbf{I}_{cC} = 2.4\angle{-156.87^\circ}\text{ A.}$$

The phase voltages at the load are

$$\mathbf{V}_{AN} = 115.22\angle{-1.19^\circ}\text{ V;}$$
$$\mathbf{V}_{BN} = 115.22\angle{118.81^\circ}\text{ V;}$$
$$\mathbf{V}_{CN} = 115.22\angle{-121.19^\circ}\text{ V.}$$

For a negative phase sequence, the line voltages lag the phase voltages by 30°:

$$\begin{aligned}\mathbf{V}_{AB} &= (\sqrt{3}\angle{-30^\circ})\mathbf{V}_{AN} \\ &= 199.58\angle{-31.19^\circ}\text{ V;}\end{aligned}$$
$$\mathbf{V}_{BC} = 199.58\angle{88.81^\circ}\text{ V;}$$
$$\mathbf{V}_{CA} = 199.58\angle{-151.19^\circ}\text{ V.}$$

The phase voltages at the terminals of the generator are

$$\mathbf{V}_{an} = 118.90\angle{-0.32^\circ}\text{ V;}$$
$$\mathbf{V}_{bn} = 118.90\angle{119.68^\circ}\text{ V;}$$
$$\mathbf{V}_{cn} = 118.90\angle{-120.32^\circ}\text{ V.}$$

The line voltages at the terminals of the generator are

$$\begin{aligned}\mathbf{V}_{ab} &= (\sqrt{3}\angle{-30^\circ})\mathbf{V}_{an} \\ &= 205.94\angle{-30.32^\circ}\ \text{V}; \\ \mathbf{V}_{bc} &= 205.94\angle{89.68^\circ}\ \text{V}; \\ \mathbf{V}_{ca} &= 205.94\angle{-150.32^\circ}\ \text{V}.\end{aligned}$$

DRILL EXERCISE

11.4 The phase voltage at the terminals of a balanced three-phase Y-connected load is 2400 V. The load has an impedance of $16 + j12\ \Omega/\phi$ and is fed from a line having an impedance of $0.10 + j0.80\ \Omega/\phi$. The Y-connected source at the sending end of the line has a phase sequence of acb and an internal impedance of $0.02 + j0.16\ \Omega/\phi$. Use the a-phase voltage at the load as the reference and calculate (a) the line currents $\mathbf{I}_{aA}$, $\mathbf{I}_{bB}$, and $\mathbf{I}_{cC}$; (b) the line voltages at the source, $\mathbf{V}_{ab}$, $\mathbf{V}_{bc}$, and $\mathbf{V}_{ca}$; and (c) the internal phase-to-neutral voltages at the source, $\mathbf{V}_{a'n}$, $\mathbf{V}_{b'n}$, and $\mathbf{V}_{c'n}$.

ANSWER: (a) $\mathbf{I}_{aA} = 120\angle{-36.87^\circ}$ A, $\mathbf{I}_{bB} = 120\angle{83.13^\circ}$ A, and $\mathbf{I}_{cC} = 120\angle{-156.87^\circ}$ A; (b) $\mathbf{V}_{ab} = 4275.02\angle{-28.38^\circ}$ V, $\mathbf{V}_{bc} = 4275.02\angle{91.62^\circ}$ V, and $\mathbf{V}_{ca} = 4275.02\angle{-148.38^\circ}$ V; (c) $\mathbf{V}_{a'n} = 2482.05\angle{1.93^\circ}$ V, $\mathbf{V}_{b'n} = 2482.05\angle{121.93^\circ}$ V, and $\mathbf{V}_{c'n} = 2482.05\angle{-118.07^\circ}$ V.

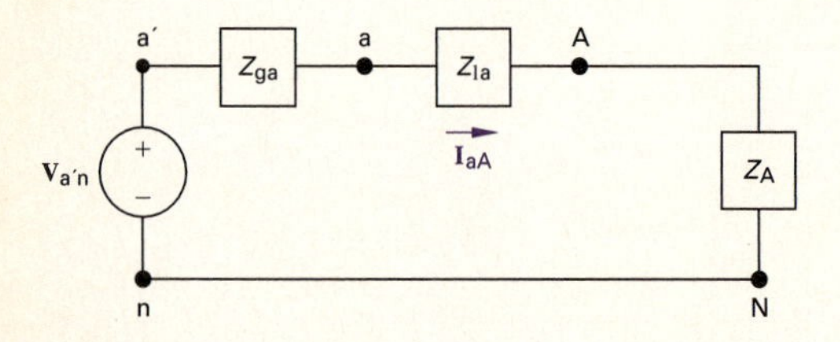

FIGURE 11.11 A single-phase equivalent circuit.

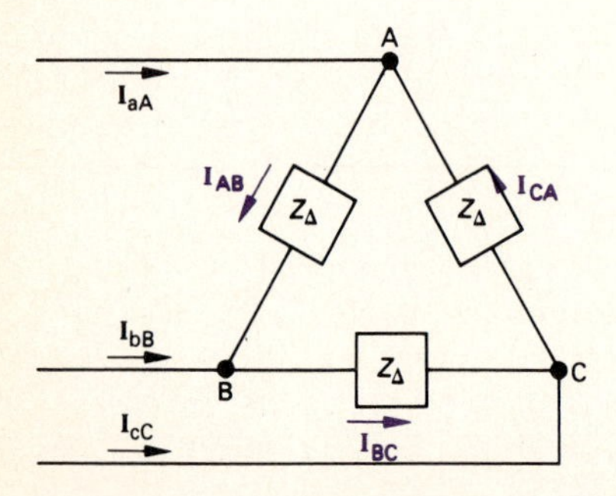

FIGURE 11.12 A circuit used to establish the relationship between line currents and phase currents in a balanced Δ load.

11.4 ANALYSIS OF THE WYE-DELTA CIRCUIT

If the load in a three-phase circuit is connected in a delta, it can be transformed into a wye by using the delta-to-wye transformation discussed in Section 9.6. When the load is balanced, the impedance of each leg of the wye is one-third the impedance of each leg of the delta, or

$$Z_Y = \frac{Z_\Delta}{3}, \tag{11.21}$$

which follows directly from Eqs. (9.49) through (9.51). After the Δ load has been replaced by its Y equivalent, the a-phase can be modeled by the single-phase equivalent circuit shown in Fig. 11.11.

We use this circuit to calculate the line currents, and we then use the line currents to find the currents in each leg of the original Δ load. The relationship between the line currents and the currents in each leg of the delta can be derived using the circuit shown in Fig. 11.12.

When a load (or source) is connected in a delta, the current in each leg of the delta is the phase current, and the voltage across each leg is the phase voltage. Figure 11.12 shows that, in the Δ configuration, the phase voltage is identical to the line voltage.

To demonstrate the relationship between the phase currents and line currents, we assume a positive phase sequence and let I_ϕ represent the magnitude of the phase current. Then

$$\mathbf{I}_{AB} = I_\phi \angle 0^\circ, \qquad (11.22)$$

$$\mathbf{I}_{BC} = I_\phi \angle -120^\circ, \qquad (11.23)$$

and

$$\mathbf{I}_{CA} = I_\phi \angle 120^\circ. \qquad (11.24)$$

In writing these equations, we arbitrarily selected $\mathbf{I}_{AB}$ as the reference phasor.

We can write the line currents in terms of the phase currents by direct application of Kirchhoff's current law:

$$\mathbf{I}_{aA} = \mathbf{I}_{AB} - \mathbf{I}_{CA} = I_\phi \angle 0^\circ - I_\phi \angle 120^\circ$$
$$= \sqrt{3} I_\phi \angle -30^\circ; \qquad (11.25)$$

$$\mathbf{I}_{bB} = \mathbf{I}_{BC} - \mathbf{I}_{AB} = I_\phi \angle -120^\circ - I_\phi \angle 0^\circ$$
$$= \sqrt{3} I_\phi \angle -150^\circ; \qquad (11.26)$$

$$\mathbf{I}_{cC} = \mathbf{I}_{CA} - \mathbf{I}_{BC} = I_\phi \angle 120^\circ - I_\phi \angle -120^\circ$$
$$= \sqrt{3} I_\phi \angle 90^\circ. \qquad (11.27)$$

Comparing Eqs. (11.25)–(11.27) with Eqs. (11.22)–(11.24) reveals that the magnitude of the line currents is $\sqrt{3}$ times the magnitude of the phase currents and that the set of line currents lags the set of phase currents by 30°.

We leave to you to verify that, for a negative phase sequence, the line currents are $\sqrt{3}$ times larger than the phase currents and lead the phase currents by 30°. Thus, we have a shortcut for calculating line currents from phase currents (or vice versa) for a balanced three-phase Δ-connected load. Figure 11.13 summarizes this shortcut graphically.

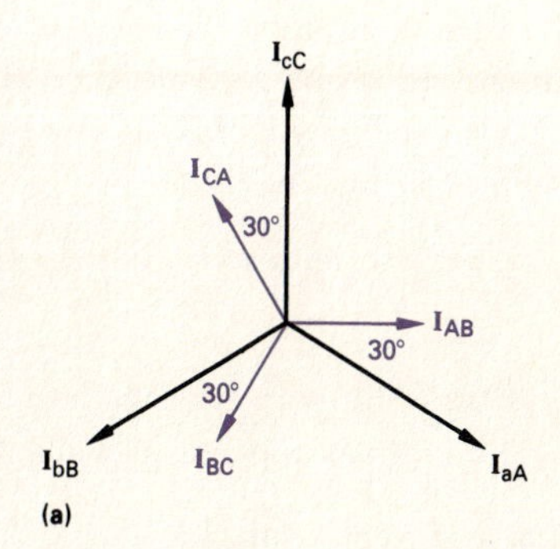

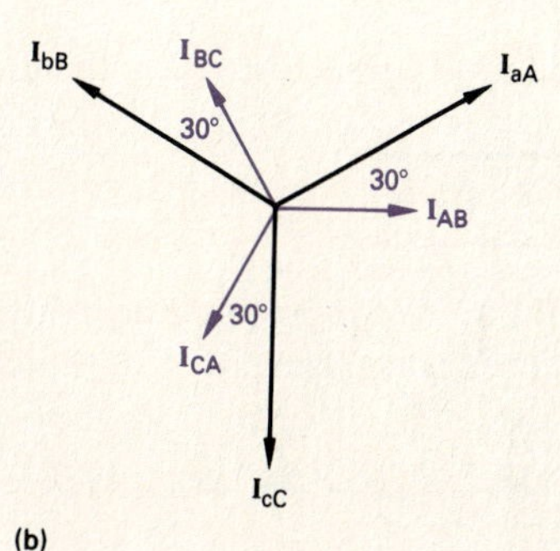

FIGURE 11.13 Phasor diagrams showing the relationship between line currents and phase currents in a Δ-connected load: (a) the positive sequence; and (b) the negative sequence.

DRILL EXERCISES

11.5 The current $\mathbf{I}_{CA}$ in a balanced three-phase Δ-connected load is $15 \angle 38^\circ$ A. If the phase sequence is positive, what is the value of $\mathbf{I}_{cC}$?

ANSWER: $25.98 \angle 8^\circ$ A.

11.6 A balanced three-phase Δ-connected load is fed from a balanced three-phase circuit. The reference for the b-phase line current is toward the load. The value of the current in the b-phase is $26 \angle -50^\circ$ A. If the phase sequence is negative, what is the value of $\mathbf{I}_{AB}$?

ANSWER: $15.01 \angle 160^\circ$ A.

Example 11.2 illustrates the calculations involved in analyzing a balanced three-phase circuit having a Y-connected source and a Δ-connected load.

EXAMPLE 11.2

The Y-connected source in Example 11.1 feeds a Δ-connected load through a distribution line having an impedance of $0.3 + j0.9\ \Omega/\phi$. The load impedance is $118.5 + j85.8\ \Omega/\phi$. Use the a-phase internal voltage of the generator as the reference.

a) Construct a single-phase equivalent circuit of the three-phase system.

b) Calculate the line currents $\mathbf{I}_{aA}$, $\mathbf{I}_{bB}$, and $\mathbf{I}_{cC}$.

c) Calculate the phase voltages at the load terminals.

d) Calculate the phase currents of the load.

e) Calculate the line voltages at the source terminals.

SOLUTION

a) Figure 11.14 shows the single-phase equivalent circuit. The load impedance of the Y equivalent is

$$\left(\frac{1}{3}\right)(118.5 + j85.8), \quad \text{or} \quad 39.5 + j28.6\ \Omega/\phi.$$

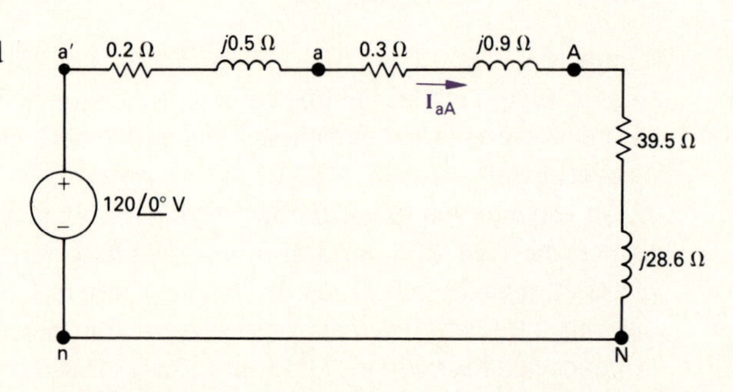

FIGURE 11.14 The single-phase equivalent circuit for Example 11.2.

b) The a-phase line current is

$$\mathbf{I}_{aA} = \frac{120\underline{/0^\circ}}{(0.2 + 0.3 + 39.5) + j(0.5 + 0.9 + 28.6)}$$

$$= \frac{120\underline{/0^\circ}}{40 + j30} = 2.4\underline{/-36.87^\circ}\text{ A}.$$

Hence

$$\mathbf{I}_{bB} = 2.4\underline{/-156.87^\circ}\text{ A};$$

$$\mathbf{I}_{cC} = 2.4\underline{/83.13^\circ}\text{ A}.$$

c) Because the load is Δ connected, the phase voltages are the same as the line voltages. To calculate the line voltages, we first calculate $\mathbf{V}_{AN}$:

$$\mathbf{V}_{AN} = (39.5 + j28.6)(2.4\underline{/-36.87^\circ})$$

$$= 117.04\underline{/-0.96^\circ}\text{ V}.$$

Because the phase sequence is positive, the line voltage $\mathbf{V}_{AB}$ is

$$\mathbf{V}_{AB} = \sqrt{3}\angle 30^\circ \, \mathbf{V}_{AN} = 202.72\angle 29.04^\circ \text{ V}.$$

Therefore

$$\mathbf{V}_{BC} = 202.72\angle -90.96^\circ \text{ V};$$

$$\mathbf{V}_{CA} = 202.72\angle 149.04^\circ \text{ V}.$$

d) The phase currents of the load may be calculated directly from the line currents:

$$\mathbf{I}_{AB} = \frac{1}{\sqrt{3}}\angle 30^\circ \, \mathbf{I}_{aA} = 1.39\angle -6.87^\circ \text{ A}.$$

Once we know $\mathbf{I}_{AB}$, we also know the other load phase currents:

$$\mathbf{I}_{BC} = 1.39\angle -126.87^\circ \text{ A};$$

$$\mathbf{I}_{CA} = 1.39\angle 113.13^\circ \text{ A}.$$

Note that we can check the calculation of $\mathbf{I}_{AB}$ by using the previously calculated $\mathbf{V}_{AB}$ and the impedance of the Δ-connected load; that is,

$$\mathbf{I}_{AB} = \frac{\mathbf{V}_{AB}}{Z_\phi} = \frac{202.72\angle 29.04^\circ}{118.5 + j85.8} = 1.39\angle -6.87^\circ \text{ A}.$$

e) To calculate the line voltage at the terminals of the source, we first calculate $\mathbf{V}_{an}$. Figure 11.14 shows that $\mathbf{V}_{an}$ is the voltage drop across the line impedance plus the load impedance, so

$$\mathbf{V}_{an} = (39.8 + j29.5)(2.4\angle -36.87^\circ) = 118.90\angle -0.32^\circ \text{ V}.$$

The line voltage $\mathbf{V}_{ab}$ is

$$\mathbf{V}_{ab} = \sqrt{3}\angle 30^\circ \, \mathbf{V}_{an},$$

or

$$\mathbf{V}_{ab} = 205.94\angle 29.68^\circ \text{ V}.$$

Therefore

$$\mathbf{V}_{bc} = 205.94\angle -90.32^\circ \text{ V};$$

$$\mathbf{V}_{ca} = 205.94\angle 149.68^\circ \text{ V}.$$

DRILL EXERCISES

11.7 The line voltage $\mathbf{V}_{AB}$ at the terminals of a balanced three-phase Δ-connected load is $4160\angle 0^\circ$ V. The line current $\mathbf{I}_{aA}$ is $69.28\angle -10^\circ$ A.

a) Calculate the per-phase impedance of the load if the phase sequence is positive.

b) Repeat (a) for a negative phase sequence.

ANSWER: (a) $104\angle -20^\circ\ \Omega$; (b) $104\angle +40^\circ\ \Omega$.

11.8 The line voltage at the terminals of a balanced Δ-connected load is 208 V. Each phase of the load consists of a 5.2 Ω resistor in parallel with a 6.933 Ω inductor. What is the magnitude of the current in the line feeding the load?

ANSWER: 86.60 A.

11.5 POWER CALCULATIONS IN BALANCED THREE-PHASE CIRCUITS

So far, we have limited our analysis of balanced three-phase circuits to determining currents and voltages. We now discuss three-phase power calculations. We begin by considering the average power delivered to a balanced Y-connected load.

AVERAGE POWER IN A BALANCED Y-LOAD

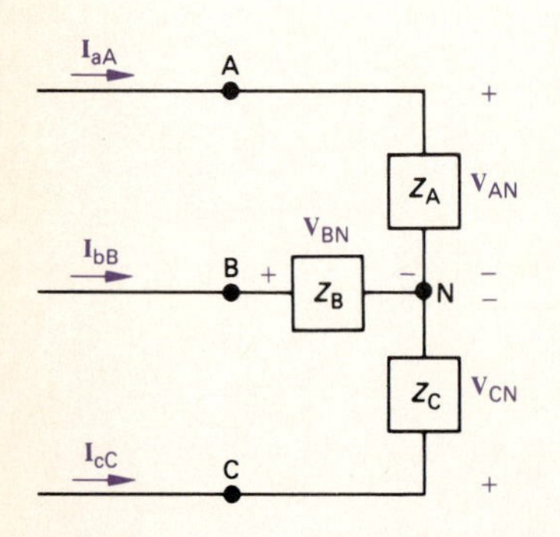

FIGURE 11.15 A balanced Y load used to introduce average power calculations in three-phase circuits.

Figure 11.15 shows a Y-connected load, along with its pertinent currents and voltages. We calculate the average power associated with any one phase by using the techniques introduced in Chapter 10. With Eq. (10.21) as a starting point, we express the average power associated with the a-phase as

$$P_A = |\mathbf{V}_{AN}||\mathbf{I}_{aA}|\cos(\theta_{vA} - \theta_{iA}), \tag{11.28}$$

where θ_{vA} and θ_{iA} denote the phase angles of $\mathbf{V}_{AN}$ and $\mathbf{I}_{aA}$, respectively. Using the notation introduced in Eq. (11.28), we can find the power associated with the b- and c-phases:

$$P_B = |\mathbf{V}_{BN}||\mathbf{I}_{bB}|\cos(\theta_{vB} - \theta_{iB}); \tag{11.29}$$

$$P_C = |\mathbf{V}_{CN}||\mathbf{I}_{cC}|\cos(\theta_{vC} - \theta_{iC}). \tag{11.30}$$

In Eqs. (11.28) through (11.30), all phasor currents and voltages are written in terms of the rms value of the sinusoidal function they represent.

In a balanced three-phase system, the magnitude of each line-to-neutral voltage is the same, as is the magnitude of each phase current. The argument of the cosine functions is also the same for all three phases. We emphasize these observations by introducing the following notation:

$$V_\phi = |\mathbf{V}_{\rm AN}| = |\mathbf{V}_{\rm BN}| = |\mathbf{V}_{\rm CN}|, \quad (11.31)$$

$$I_\phi = |\mathbf{I}_{\rm aA}| = |\mathbf{I}_{\rm bB}| = |\mathbf{I}_{\rm cC}|, \quad (11.32)$$

and

$$\theta_\phi = \theta_{v\rm A} - \theta_{i\rm A} = \theta_{v\rm B} - \theta_{i\rm B} = \theta_{v\rm C} - \theta_{i\rm C}. \quad (11.33)$$

Moreover, for a balanced system, the power delivered to each phase of the load is the same, so

$$P_{\rm A} = P_{\rm B} = P_{\rm C} = P_\phi = V_\phi I_\phi \cos\theta_\phi, \quad (11.34)$$

where P_ϕ represents the average power per phase.

The total average power delivered to the balanced Y-connected load is simply three times the power per phase, or

$$P_T = 3P_\phi = 3V_\phi I_\phi \cos\theta_\phi. \quad (11.35)$$

Expressing the total power in terms of the rms magnitudes of the line voltage and current is also desirable. If we let $V_{\rm L}$ and $I_{\rm L}$ represent the rms magnitudes of the line voltage and current, respectively, we can modify Eq. (11.35) as follows:

$$P_T = 3\left(\frac{V_{\rm L}}{\sqrt{3}}\right) I_{\rm L} \cos\theta_\phi$$

$$= \sqrt{3} V_{\rm L} I_{\rm L} \cos\theta_\phi. \quad (11.36)$$

In deriving Eq. (11.36), we recognized that, for a balanced Y-connected load, the magnitude of the phase voltage is the magnitude of the line voltage divided by $\sqrt{3}$, and that the magnitude of the line current is equal to the magnitude of the phase current. When using Eq. (11.36) to calculate the total power delivered to the load, remember that θ_ϕ is the phase angle between the phase voltage and current.

COMPLEX POWER IN A BALANCED Y LOAD

We can also calculate the reactive power and complex power associated with any one phase of a Y-connected load by using the techniques introduced in Chapter 10. For a balanced load, the expressions for the reactive power are

$$Q_\phi = V_\phi I_\phi \sin\theta_\phi; \quad (11.37)$$

$$Q_T = 3Q_\phi = \sqrt{3} V_{\rm L} I_{\rm L} \sin\theta_\phi. \quad (11.38)$$

Equation (10.30) is the basis for expressing the complex power associated with any phase. For a balanced load,

$$S_\phi = \mathbf{V}_{\text{AN}}\mathbf{I}^*_{\text{aA}} = \mathbf{V}_{\text{BN}}\mathbf{I}^*_{\text{bB}} = \mathbf{V}_{\text{CN}}\mathbf{I}^*_{\text{cC}} = \mathbf{V}_\phi\mathbf{I}^*_\phi, \tag{11.39}$$

where $\mathbf{V}_\phi$ and $\mathbf{I}_\phi$ represent a phase voltage and current taken from the same phase. Thus, in general,

$$S_\phi = P_\phi + jQ_\phi = \mathbf{V}_\phi\mathbf{I}^*_\phi; \tag{11.40}$$

$$S_T = 3S_\phi = \sqrt{3}V_{\text{L}}I_{\text{L}}\underline{/\theta_\phi}. \tag{11.41}$$

Power Calculations in a Balanced Δ Load

If the load is Δ connected, the calculation of power—reactive or complex—is basically the same as that for a Y-connected load. Figure 11.16 shows a Δ-connected load, along with its pertinent currents and voltages. The power associated with each phase is

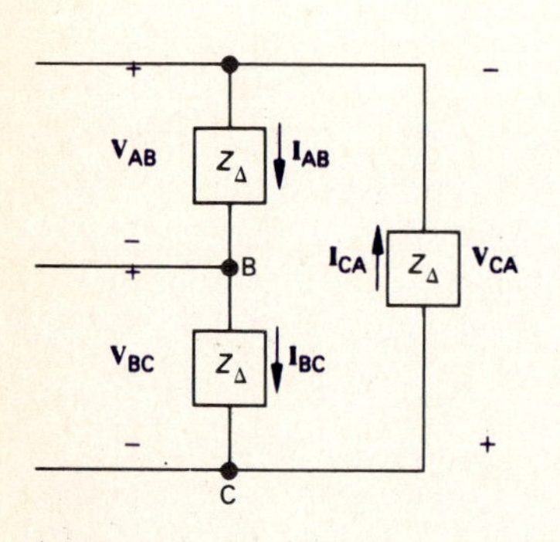

FIGURE 11.16 A Δ-connected load used to discuss power calculations.

$$P_{\text{A}} = |\mathbf{V}_{\text{AB}}||\mathbf{I}_{\text{AB}}|\cos(\theta_{v\text{AB}} - \theta_{i\text{AB}}); \tag{11.42}$$

$$P_{\text{B}} = |\mathbf{V}_{\text{BC}}||\mathbf{I}_{\text{BC}}|\cos(\theta_{v\text{BC}} - \theta_{i\text{BC}}); \tag{11.43}$$

$$P_{\text{C}} = |\mathbf{V}_{\text{CA}}||\mathbf{I}_{\text{CA}}|\cos(\theta_{v\text{CA}} - \theta_{i\text{CA}}). \tag{11.44}$$

For a balanced load,

$$|\mathbf{V}_{\text{AB}}| = |\mathbf{V}_{\text{BC}}| = |\mathbf{V}_{\text{CA}}| = V_\phi, \tag{11.45}$$

$$|\mathbf{I}_{\text{AB}}| = |\mathbf{I}_{\text{BC}}| = |\mathbf{I}_{\text{CA}}| = I_\phi, \tag{11.46}$$

$$\theta_{v\text{AB}} - \theta_{i\text{AB}} = \theta_{v\text{BC}} - \theta_{i\text{BC}} = \theta_{v\text{CA}} - \theta_{i\text{CA}} = \theta_\phi, \tag{11.47}$$

and

$$P_{\text{A}} = P_{\text{B}} = P_{\text{C}} = P_\phi = V_\phi I_\phi \cos\theta_\phi. \tag{11.48}$$

Note that Eq. (11.48) is the same as Eq. (11.34). Thus, in a balanced load, regardless of whether it is Y or Δ connected, the average power per phase is equal to the product of the rms magnitude of the phase voltage, the rms magnitude of the phase current, and the cosine of the angle between the phase voltage and current.

The total power delivered to a balanced Δ-connected load is

$$\begin{aligned} P_T &= 3P_\phi = 3V_\phi I_\phi \cos\theta_\phi \\ &= 3V_{\text{L}}\left(\frac{I_{\text{L}}}{\sqrt{3}}\right)\cos\theta_\phi \\ &= \sqrt{3}V_{\text{L}}I_{\text{L}}\cos\theta_\phi. \end{aligned} \tag{11.49}$$

Note that Eq. (11.49) is the same as Eq. (11.36).

The expressions for reactive power and complex power also have the same form as those developed for the Y load:

$$Q_\phi = V_\phi I_\phi \sin\theta_\phi; \tag{11.50}$$

$$Q_T = 3Q_\phi = 3V_\phi I_\phi \sin\theta_\phi; \tag{11.51}$$

$$S_\phi = P_\phi + jQ_\phi = \mathbf{V}_\phi \mathbf{I}_\phi^*; \tag{11.52}$$

$$S_T = 3S_\phi = \sqrt{3}V_L I_L \underline{/\theta_\phi}. \tag{11.53}$$

Examples 11.3–11.5 illustrate power calculations in balanced three-phase circuits.

EXAMPLE 11.3

a) Calculate the average power per phase delivered to the Y-connected load of Example 11.1.

b) Calculate the total average power delivered to the load.

c) Calculate the total average power lost in the line.

d) Calculate the total average power lost in the generator.

e) Calculate the total number of magnetizing vars absorbed by the load.

f) Calculate the total complex power delivered by the source.

SOLUTION

a) From Example 11.1, $V_\phi = 115.22$ V, $I_\phi = 2.4$ A, and $\theta_\phi = -1.19 - (-36.87) = 35.68°$. Therefore

$$P_\phi = (115.22)(2.4)\cos 35.68°$$
$$= 224.64 \text{ W}.$$

The power per phase may also be calculated from $I_\phi^2 R_\phi$, or

$$P_\phi = (2.4)^2(39) = 224.64 \text{ W}.$$

b) The total average power delivered to the load is $P_T = 3P_\phi = 673.92$ W. We calculated the line voltage in Example 11.1, so we may also use Eq. (11.36):

$$P_T = \sqrt{3}(199.58)(2.4)\cos 35.68°$$
$$= 673.92 \text{ W}.$$

c) The total power lost in the line is

$$P_{\text{line}} = 3(2.4)^2(0.8) = 13.824 \text{ W}.$$

d) The total internal power lost in the generator is

$$P_{\text{gen}} = 3(2.4)^2(0.2) = 3.456 \text{ W}.$$

e) The total number of magnetizing vars absorbed by the load is

$$Q_T = \sqrt{3}(199.58)(2.4)\sin 35.68^\circ$$
$$= 483.84 \text{ VAR}.$$

f) The total complex power associated with the source is

$$S_T = 3S_\phi = -3(120)(2.4)\underline{/36.87^\circ}$$
$$= -691.20 - j518.40 \text{ VA}.$$

The minus sign indicates that the internal power and magnetizing reactive power are being delivered to the circuit. We check this result by calculating the total and reactive power absorbed by the circuit:

$$P = 673.92 + 13.824 + 3.456$$
$$= 691.20 \text{ W} \quad \text{(check)};$$
$$Q = 483.84 + 3(2.4)^2(1.5) + 3(2.4)^2(0.5)$$
$$= 483.84 + 25.92 + 8.64$$
$$= 518.40 \text{ VAR} \quad \text{(check)}.$$

A computer tool can help to verify the results of these calculations.

EXAMPLE 11.4

a) Calculate the total complex power delivered to the Δ-connected load of Example 11.2.

b) What percentage of the average power at the sending end of the line is delivered to the load?

SOLUTION

a) Using the a-phase values from the solution of Example 11.2, we obtain

$$\mathbf{V}_\phi = \mathbf{V}_{\text{AB}} = 202.72\underline{/29.04^\circ} \text{ V};$$
$$\mathbf{I}_\phi = \mathbf{I}_{\text{AB}} = 1.39\underline{/-6.87^\circ} \text{ A}.$$

Using Eqs. (11.52) and (11.53), we have

$$S_T = 3(202.72\underline{/29.04^\circ})(1.39\underline{/6.87^\circ})$$
$$= 682.56 + j494.21 \text{ VA}.$$

b) The total power at the sending end of the distribution line equals the total power delivered to the load plus the total power lost in the line; therefore

$$P_{\text{input}} = 682.56 + 3(2.4)^2(0.3)$$
$$= 687.74 \text{ W}.$$

The percentage of the average power reaching the load is 682.56/687.744, or 99.25%. Nearly 100% of the average power at the input is delivered to the load because the impedance of the line is quite small compared to the load impedance.

EXAMPLE 11.5

A balanced three-phase load requires 480 kW at a lagging power factor of 0.8. The load is fed from a line having an impedance of $0.005 + j0.025\ \Omega/\phi$. The line voltage at the terminals of the load is 600 V.

a) Construct a single-phase equivalent circuit of the system.

b) Calculate the magnitude of the line current.

c) Calculate the magnitude of the line voltage at the sending end of the line.

d) Calculate the power factor at the sending end of the line.

SOLUTION

a) Figure 11.17 shows the single-phase equivalent circuit. We arbitrarily selected the line-to-neutral voltage at the load as the reference.

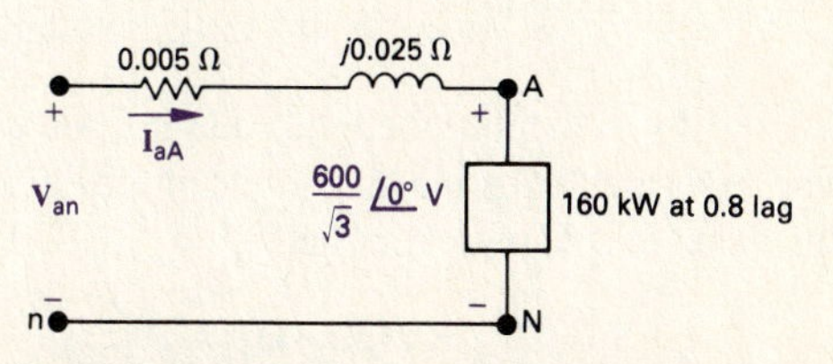

FIGURE 11.17 The single-phase equivalent circuit for Example 11.5.

b) The line current $\mathbf{I}_{aA}^*$ is given by

$$\left(\frac{600}{\sqrt{3}}\right)\mathbf{I}_{aA}^* = (160 + j120)10^3,$$

or

$$\mathbf{I}_{aA}^* = 577.35\underline{/36.87^\circ} \text{ A}.$$

Therefore $\mathbf{I}_{aA} = 577.35\underline{/-36.87^\circ}$ A. The magnitude of the line current is the magnitude of $\mathbf{I}_{aA}$:

$$I_L = 577.35 \text{ A}.$$

We obtain an alternative solution for I_L from the expression

$$\begin{aligned} P_T &= \sqrt{3}V_L I_L \cos\theta_p \\ &= \sqrt{3}(600) I_L (0.8) = 480{,}000 \text{ W}; \end{aligned}$$

$$I_L = \frac{480{,}000}{\sqrt{3}(600)(0.8)} = \frac{1000}{\sqrt{3}} = 577.35 \text{ A}.$$

c) To calculate the magnitude of the line voltage at the sending end, we first calculate $\mathbf{V}_{an}$. From Fig. 11.17,

$$\begin{aligned} \mathbf{V}_{an} &= \mathbf{V}_{AN} + Z_L\mathbf{I}_{aA} \\ &= \frac{600}{\sqrt{3}} + (0.005 + j0.025)(577.35\underline{/-36.87^\circ}) \\ &= 357.51\underline{/1.57^\circ} \text{ V}. \end{aligned}$$

Thus

$$V_L = \sqrt{3}|\mathbf{V}_{an}| = 619.23 \text{ V}.$$

d) The power factor at the sending end of the line is the cosine of the phase angle between $\mathbf{V}_{an}$ and $\mathbf{I}_{aA}$:

$$\begin{aligned} \text{pf} &= \cos\,[1.57^\circ - (-36.87^\circ)] \\ &= \cos 38.44^\circ = 0.783 \text{ lagging}. \end{aligned}$$

An alternative method for calculating the power factor is to first calculate the complex power at the sending end of the line:

$$\begin{aligned} S_\phi &= (160 + j120)10^3 + (577.35)^2(0.005 + j0.025) \\ &= 161.67 + j128.33 \text{ kVA} \\ &= 206.41\underline{/38.44^\circ} \text{ kVA}. \end{aligned}$$

The power factor is

$$\text{pf} = \cos 38.44^\circ = 0.783 \text{ lagging}.$$

Finally, if we calculate the total complex power at the sending end, after first calculating the magnitude of the line current, we may use this value to calculate V_L. That is,

$$\sqrt{3}V_L I_L = 3(206.41) \times 10^3;$$

$$V_L = \frac{3(206.41) \times 10^3}{\sqrt{3}(577.35)} = 619.23 \text{ V}.$$

Instantaneous Power in Three-Phase Circuits

Although we are primarily interested in average, reactive, and complex power calculations, the computation of the total instantaneous power is also important. In a balanced three-phase circuit, this power has an interesting property: it is invariant with time! Thus the torque developed at the shaft of a three-phase motor is constant, which in turn means less vibration in machinery powered by three-phase motors.

Let the instantaneous line-to-neutral voltage v_{AN} be the reference, and, as before, θ_ϕ is the phase angle $\theta_{vA} - \theta_{iA}$. Then, for a positive phase sequence, the instantaneous power in each phase is

$$p_A = v_{AN} i_{aA} = V_\phi I_\phi \cos \omega t \cos (\omega t - \theta_\phi),$$

$$p_B = v_{BN} i_{bB} = V_\phi I_\phi \cos (\omega t - 120^\circ) \cos (\omega t - \theta_\phi - 120^\circ),$$

and

$$p_C = v_{CN} i_{cC} = V_\phi I_\phi \cos (\omega t + 120^\circ) \cos (\omega t - \theta_\phi + 120^\circ),$$

where V_ϕ and I_ϕ represent the rms values of the phase voltage and line current, respectively. The instantaneous total power is the sum of the instantaneous phase powers, which reduces to $1.5 V_\phi I_\phi \cos \theta_\phi$; that is,

$$p_T = p_A + p_B + p_C = 1.5 V_\phi I_\phi \cos \theta_\phi.$$

We leave this reduction to you (see Problem 11.28).

DRILL EXERCISES

11.9 The three-phase average power rating of the central processing unit (CPU) on a mainframe digital computer is 22,659 W. The three-phase line supplying the computer has a line voltage rating of 208 V (rms). The line current is 73.8 A (rms).

a) Calculate the total magnetizing reactive power absorbed by the CPU.

b) Calculate the power factor.

ANSWER: (a) 13,909.50 VAR; (b) 0.852 lagging.

11.10 The complex power associated with each phase of a balanced load is $384 + j288$ kVA. The line voltage at the terminals of the load is 4160 V.

a) What is the magnitude of the line current feeding the load?

b) The load is delta connected, and the impedance of each phase consists of a resistance in parallel with a reactance. Calculate R and X.

c) The load is wye connected, and the impedance of each phase consists of a resistance in series with a reactance. Calculate R and X.

ANSWER: (a) 199.85 A; (b) $R = 45.07\ \Omega$, $X = 60.09\ \Omega$; (c) $R = 9.61\ \Omega$, $X = 7.21\ \Omega$.

11.11 A balanced bank of delta-connected capacitors is connected in parallel with the load described in Drill Exercise 11.10. The effect is to place a capacitor in parallel with the load in each phase. The line voltage at the terminals of the load thus remains at 4160 V. The circuit is operating at a frequency of 60 Hz. The capacitors are adjusted so that the magnitude of the line current feeding the parallel combination of the load and capacitor bank is at its minimum.

a) What is the size of each capacitor in microfarads?

b) Repeat (a) for wye-connected capacitors.

c) What is the magnitude of the line current?

ANSWER: (a) 44.14 μF; (b) 132.42 μF; (c) 159.88 A.

11.6 Measuring Average Power in Three-Phase Circuits

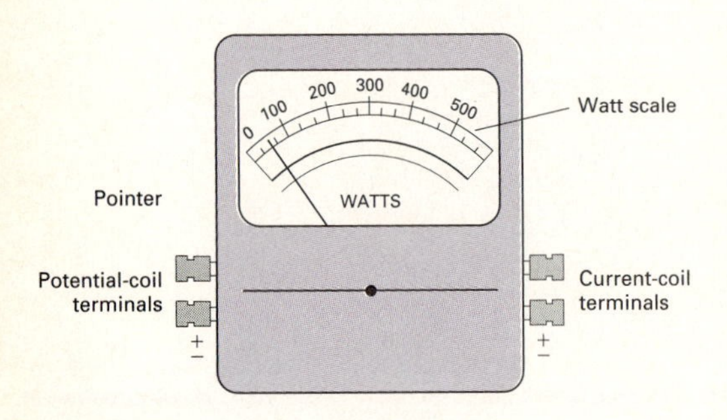

FIGURE 11.18 The key features of the electrodynamometer wattmeter.

The basic instrument used to measure power in three-phase circuits is the electrodynamometer wattmeter. It contains two coils. One coil, called the **current coil**, is stationary and is designed to carry a current proportional to the load current. The second coil, called the **potential coil**, is movable and carries a current proportional to the load voltage. The important features of the wattmeter are shown in Fig. 11.18.

The average deflection of the pointer attached to the movable coil is proportional to the product of the effective value of the current in the current coil, the effective value of the voltage impressed on the potential coil, and the cosine of the phase angle between the voltage and current. The direction in which the pointer deflects depends on the instantaneous polarity of the current-coil current and the potential-coil voltage. Therefore each coil has one terminal with a polarity mark—usually a plus sign—but sometimes the double polarity mark $\pm$ is used. The wattmeter deflects upscale when (1) the polarity-marked terminal of the current coil is toward the source and (2) the polarity-marked terminal of the potential coil is connected to the same line in which the current coil has been inserted.

The Two-Wattmeter Method

Consider a general network inside a box to which power is supplied by n conducting lines. Such a system is shown in Fig. 11.19. If we wish to measure the total power at the terminals of the box, we need to know $n - 1$ currents and voltages. This follows because if we choose one terminal as a reference, there are only $n - 1$ independent voltages. Likewise, only $n - 1$ independent currents can exist in the n conductors entering the box. Thus the total power is the sum of $n - 1$ product terms; that is, $p = v_1 i_1 + v_2 i_2 + \cdots + v_{n-1} i_{n-1}$.

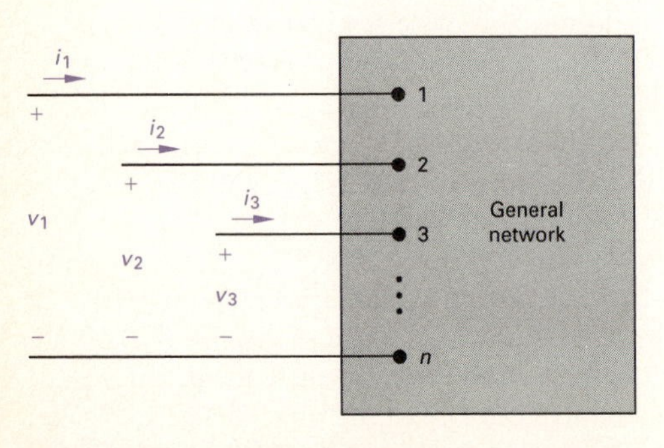

FIGURE 11.19 A general circuit whose power is supplied by n conductors.

Applying this general observation, we can see that for a three-conductor circuit, whether balanced or not, we need only two wattmeters to measure the total power. For a four-conductor circuit we need three wattmeters if the three-phase circuit is unbalanced, but only two wattmeters if it is balanced, since in the latter case there is no current in the neutral line. Thus, only two wattmeters are needed to measure the total average power in any balanced three-phase system.

The two-wattmeter method reduces to determining the magnitude and algebraic sign of the average power indicated by each wattmeter. We can describe the basic problem in terms of the circuit shown in Fig. 11.20, where the two wattmeters are indicated by the shaded boxes and labeled W_1 and W_2. The coil notations cc and pc stand for current coil and potential coil, respectively. We have elected to insert the current coils of the wattmeters in lines aA and cC. Thus, line bB is the reference line for the two potential coils. The load is connected as a wye, and the per-phase load impedance is designated as $Z_\phi = |Z|\underline{/\theta}$. This is a general representation, since any Δ-connected load can be represented by its Y equivalent; furthermore, for the balanced case, the impedance angle θ is unaffected by the Δ-to-Y transformation.

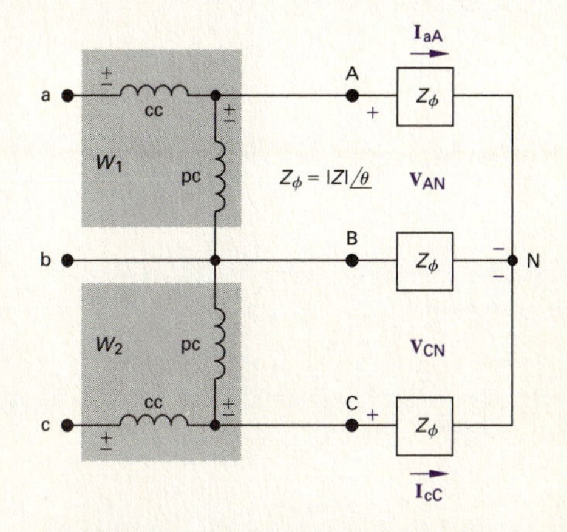

FIGURE 11.20 A circuit used to analyze the two-wattmeter method of measuring average power delivered to a balanced load.

We now develop general equations for the readings of the two wattmeters. We assume that the current drawn by the potential coil of the wattmeter is negligible compared with the line current measured by the current coil. We further assume that the loads can be modeled by passive circuit elements so that the phase angle of the load impedance (θ in Fig. 11.20) lies between $-90°$ (pure capacitance) and $+90°$ (pure inductance). Finally, we assume a positive phase sequence.

From our introductory discussion of the average deflection of the wattmeter, we can see that wattmeter 1 will respond to the product of $|\mathbf{V}_{AB}|$, $|\mathbf{I}_{aA}|$, and the cosine of the angle between $\mathbf{V}_{AB}$ and $\mathbf{I}_{aA}$. If we denote this wattmeter reading as W_1, we can write

$$\begin{aligned} W_1 &= |\mathbf{V}_{AB}||\mathbf{I}_{aA}|\cos\theta_1 \\ &= V_L I_L \cos\theta_1. \end{aligned} \tag{11.54}$$

It follows that

$$\begin{aligned} W_2 &= |\mathbf{V}_{CB}||\mathbf{I}_{cC}|\cos\theta_2 \\ &= V_L I_L \cos\theta_2. \end{aligned} \tag{11.55}$$

In Eq. (11.54), θ_1 is the phase angle between $\mathbf{V}_{AB}$ and $\mathbf{I}_{aA}$, and θ_2 is the phase angle between $\mathbf{V}_{CB}$ and $\mathbf{I}_{cC}$.

To calculate W_1 and W_2, we express θ_1 and θ_2 in terms of the impedance angle θ, which is also the same as the phase angle between the phase voltage and current. For a positive phase sequence,

$$\theta_1 = \theta + 30° = \theta_\phi + 30° \tag{11.56}$$

and

$$\theta_2 = \theta - 30° = \theta_\phi - 30°. \tag{11.57}$$

The derivation of Eqs. (11.56) and (11.57) is left as an exercise (see Problem 11.30). When we substitute Eqs. (11.56) and (11.57) into

Eqs. (11.54) and (11.55), respectively, we get

$$W_1 = V_L I_L \cos(\theta + 30°) \tag{11.58}$$

and

$$W_2 = V_L I_L \cos(\theta - 30°). \tag{11.59}$$

To find the total power, we add W_1 and W_2; thus

$$P_T = W_1 + W_2 = 2V_L I_L \cos\theta_\phi \cos 30°$$
$$= \sqrt{3} V_L I_L \cos\theta_\phi, \tag{11.60}$$

which is the expression for the total power in a three-phase circuit. Therefore we have confirmed that the sum of the two wattmeter readings yields the total average power.

Use a circuit simulator to create a model of a wattmeter.

A closer look at Eqs. (11.58) and (11.59) reveals the following about the readings of the two wattmeters:

1. If the power factor is greater than 0.5, both wattmeters read positive.
2. If the power factor equals 0.5, one wattmeter reads zero.
3. If the power factor is less than 0.5, one wattmeter reads negative.
4. Reversing the phase sequence will interchange the readings on the two wattmeters.

These observations are illustrated in the following example and in Problems 11.31 through 11.38.

EXAMPLE 11.6

Calculate the reading of each wattmeter in the circuit in Fig. 11.20 if the phase voltage at the load is 120 V and (a) $Z_\phi = 8 + j6\ \Omega$; (b) $Z_\phi = 8 - j6\ \Omega$; (c) $Z_\phi = 5 + j5\sqrt{3}\ \Omega$; and (d) $Z_\phi = 10\angle{-75°}\ \Omega$. (e) Verify for parts (a) through (d) that the sum of the wattmeter readings equals the total power delivered to the load.

SOLUTION

a) $Z_\phi = 10\angle{36.87°}\ \Omega$, $V_L = 120\sqrt{3}$ V, and $I_L = 120/10 = 12$ A.

$$W_1 = (120\sqrt{3})(12)\cos(36.87° + 30°)$$
$$= 979.75 \text{ W};$$
$$W_2 = (120\sqrt{3})(12)\cos(36.87° - 30°)$$
$$= 2476.25 \text{ W}.$$

b) $Z_\phi = 10\underline{/-36.87^\circ}\ \Omega$, $V_L = 120\sqrt{3}$ V, and $I_L = 120/10 = 12$ A.

$$\begin{aligned} W_1 &= (120\sqrt{3})(12)\cos(-36.87^\circ + 30^\circ) \\ &= 2476.25 \text{ W}; \\ W_2 &= (120\sqrt{3})(12)\cos(-36.87^\circ - 30^\circ) \\ &= 979.75 \text{ W}. \end{aligned}$$

c) $Z_\phi = 5(1 + j\sqrt{3}) = 10\underline{/60^\circ}\ \Omega$, $V_L = 120\sqrt{3}$ V, and $I_L = 12$ A.

$$\begin{aligned} W_1 &= (120\sqrt{3})(12)\cos(60^\circ + 30^\circ) = 0; \\ W_2 &= (120\sqrt{3})(12)\cos(60^\circ - 30^\circ) \\ &= 2160 \text{ W}. \end{aligned}$$

d) $Z_\phi = 10\underline{/-75^\circ}\ \Omega$, $V_L = 120\sqrt{3}$ V, and $I_L = 12$ A.

$$\begin{aligned} W_1 &= (120\sqrt{3})(12)\cos(-75^\circ + 30^\circ) = 1763.63 \text{ W}; \\ W_2 &= (120\sqrt{3})(12)\cos(-75^\circ - 30^\circ) = -645.53 \text{ W}. \end{aligned}$$

e) $P_T(\text{a}) = 3(12)^2(8) = 3456$ W,

$$\begin{aligned} W_1 + W_2 &= 979.75 + 2476.25 \\ &= 3456 \text{ W}; \end{aligned}$$

$P_T(\text{b}) = P_T(\text{a}) = 3456$ W,

$$\begin{aligned} W_1 + W_2 &= 2476.25 + 979.75 \\ &= 3456 \text{ W}; \end{aligned}$$

$P_T(\text{c}) = 3(12)^2(5) = 2160$ W,

$$\begin{aligned} W_1 + W_2 &= 0 + 2160 \\ &= 2160 \text{ W}; \end{aligned}$$

$P_T(\text{d}) = 3(12)^2(2.5882) = 1118.10$ W,

$$\begin{aligned} W_1 + W_2 &= 1763.63 - 645.53 \\ &= 1118.10 \text{ W}. \end{aligned}$$

DRILL EXERCISES

11.12 The two-wattmeter method is used to measure the power at the load end of the line in Example 11.1. Calculate the reading of each wattmeter.

ANSWER: 197.29 W; 476.63 W.

11.13 The two wattmeters in Fig. 11.20 can be used to compute the total reactive power of the load.

a) Prove this statement by showing that $\sqrt{3}(W_2 - W_1) = \sqrt{3}V_L I_L \sin\theta_\phi$.

b) Compute the total reactive power from the wattmeter readings for each of the loads in Example 11.6.

Check your computations by calculating the total reactive power directly from the given voltage and impedance.

ANSWER: (a) derivation; (b) 2592 VAR, −2592 VAR, 3741.23 VAR, −4172.80 VAR.

SUMMARY

- When analyzing balanced three-phase circuits, the first step is to transform any Δ connections into Y connections, so that the overall circuit is of the Y-Y configuration.
- A **single-phase equivalent circuit** is used to calculate the line current and the phase voltage in one phase of the Y-Y structure. The a-phase is normally chosen for this purpose.
- Once we know the line current and phase voltage in the a-phase equivalent circuit, we can take analytical shortcuts to find any current or voltage in a balanced three-phase circuit, based on the following facts:

1. The b- and c-phase currents and voltages are identical to the a-phase current and voltage except for a 120° shift in phase. In a positive-sequence circuit, the b-phase quantity lags the a-phase quantity by 120°, and the c-phase quantity leads the a-phase quantity by 120°. For a negative sequence circuit, phases b and c are interchanged with respect to phase a.
2. The set of line voltages is out of phase with the set of phase voltages by ±30°. The plus or minus sign corresponds to positive and negative sequence, respectively.
3. In a Y-Y circuit the magnitude of a line voltage is $\sqrt{3}$ times the magnitude of a phase voltage.
4. The set of line currents is out of phase with the set of phase currents in Δ-connected sources and loads by ∓30°. The minus or plus sign corresponds to positive and negative sequence, respectively.
5. The magnitude of a line current is $\sqrt{3}$ times the magnitude of a phase current in a Δ-connected source or load.

- The techniques for calculating per-phase average power, reactive power, and complex power are identical to those introduced in Chapter 10.
- The total real, reactive, and complex power can be determined either by multiplying the corresponding per phase quantity by 3 or by using the expressions based on line current and line voltage, as given by Eqs. (11.36), (11.38), and (11.41).
- The total instantaneous power in a balanced three-phase circuit is constant and equals 1.5 times the average power per phase.
- A wattmeter measures the average power delivered to a load by using a current coil connected in series with the load and a potential coil connected in parallel with the load.
- The total average power in a balanced three-phase circuit can be measured by summing the readings of two wattmeters connected in two different phases of the circuit.

PROBLEMS

All phasor voltages are stated in terms of the rms value.

11.1 For each set of voltages, state whether or not the voltages form a balanced three-phase set. If the set is balanced, state whether the phase sequence is positive or negative. If the set is not balanced, explain why.

a) $v_a = 139 \cos 377t$ V
$v_b = 139 \cos (377t + 120°)$ V
$v_c = 139 \cos (377t - 120°)$ V

b) $v_a = 381 \cos 377t$ V
$v_b = 381 \cos (377t + 240°)$ V
$v_c = 381 \cos (377t + 120°)$ V

c) $v_a = 2771 \sin (377t - 30°)$ V
$v_b = 2771 \cos 377t$ V
$v_c = 2771 \sin (377t + 210°)$ V

d) $v_a = 170 \sin (\omega t + 30°)$ V
$v_b = -170 \cos \omega t$ V
$v_c = 170 \cos (\omega t + 60°)$ V

e) $v_a = 339 \cos \omega t$ V
$v_b = 339 \cos (\omega t + 120°)$ V
$v_c = 393 \cos (\omega t - 120°)$ V

f) $v_a = 3983 \sin (\omega t + 50°)$ V
$v_b = 3983 \cos (\omega t - 160°)$ V
$v_c = 3983 \cos (\omega t + 70°)$ V

11.2 Verify that Eq. (11.3) is true for either Eq. (11.1) or Eq. (11.2).

11.3 The time-domain expressions for three line-to-neutral voltages at the terminals of a Y-connected load are

$v_{AN} = 7620 \cos(\omega t + 30^\circ)$ V,

$v_{BN} = 7620 \cos(\omega t + 150^\circ)$ V,

$v_{CN} = 7620 \cos(\omega t - 90^\circ)$ V.

What are the time-domain expressions for the three line-to-line voltages v_{AB}, v_{BC}, and v_{CA}?

11.4 Refer to the circuit in Fig. 11.4(b). Assume that there are no external connections to the terminals a,b,c. Assume further that the three windings are from a balanced three-phase generator. How much current will circulate in the Δ-connected generator?

11.5 a) Is the circuit in Fig. P11.5 a balanced or unbalanced three-phase system? Explain.

b) Find $\mathbf{I}_o$.

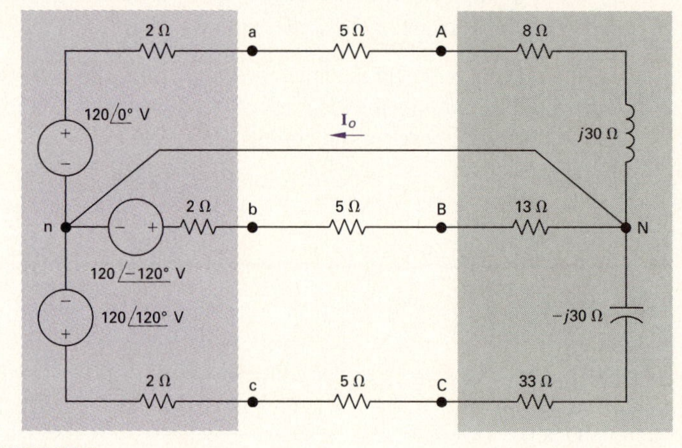

FIGURE P11.5

11.6 a) Find $\mathbf{I}_o$ in the circuit in Fig. P11.6.

b) Find $\mathbf{V}_{BN}$.

c) Find $\mathbf{V}_{BC}$.

d) Is the circuit a balanced or unbalanced three-phase system?

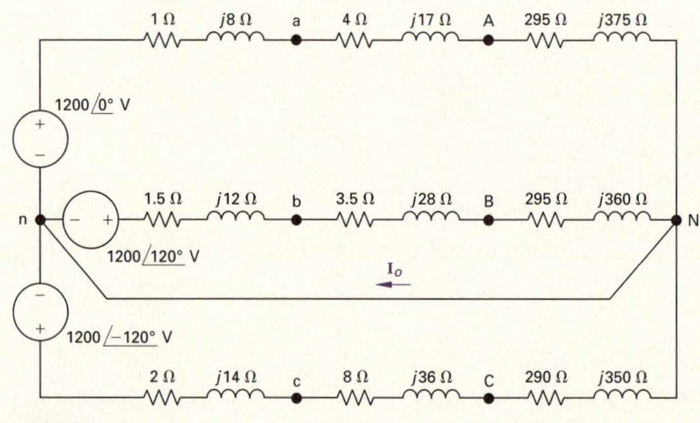

FIGURE P11.6

11.7 Find the rms value of $\mathbf{I}_0$ in the unbalanced three-phase circuit seen in Fig. P11.7.

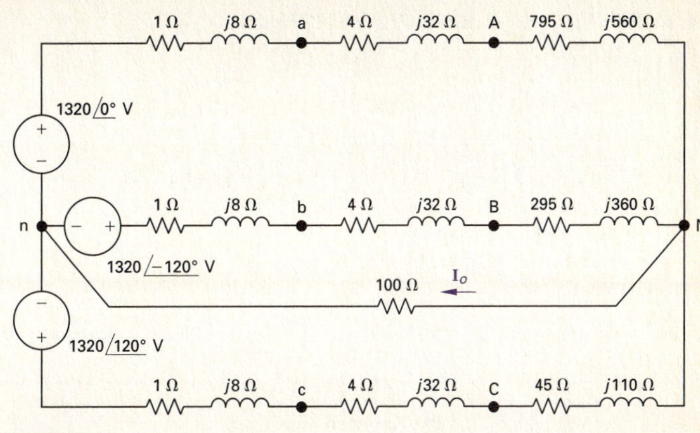

FIGURE P11.7

11.8 The magnitude of the phase voltage of an ideal balanced three-phase Y-connected source is 4000 V. The source is connected to a balanced Y-connected load by a distribution line that has an impedance of $1 + j8\ \Omega/\phi$. The load impedance is $119 + j27\ \Omega/\phi$. The phase sequence of the source is abc. Use the a-phase voltage of the source as the reference. Specify the magnitude and phase angle of the following quantities: (a) the three line currents, (b) the three line voltages at the source, (c) the three phase voltages at the load, and (d) the three line voltages at the load.

11.9 The magnitude of the line voltage at the terminals of a balanced Y-connected load is 6600 V. The load impedance is $240 - j70\ \Omega/\phi$. The load is fed from a line that has an impedance of $0.5 + j4\ \Omega/\phi$.

a) What is the magnitude of the line current?

b) What is the magnitude of the line voltage at the source?

11.10 A balanced Δ-connected load has an impedance of $864 - j252\ \Omega/\phi$. The load is fed through a line having an impedance of $0.5 + j4\ \Omega/\phi$. The phase voltage at the terminals of the load is 69 kV. The phase sequence is positive. Use $\mathbf{V}_{AB}$ as the reference.

a) Calculate the three phase currents of the load.

b) Calculate the three line currents.

c) Calculate the three line voltages at the sending end of the line.

11.11 A balanced Y-connected load having an impedance of $72 + j21\ \Omega/\phi$ is connected in parallel with a balanced Δ-connected load having an impedance of $150\angle 0^\circ\ \Omega/\phi$. The paralleled loads are fed from a line having an impedance of $j1\ \Omega/\phi$. The magnitude of the line-to-neutral voltage of the Y-load is 7650 V.

a) Calculate the magnitude of the current in the line feeding the loads.

b) Calculate the magnitude of the phase current in the Δ-connected load.

c) Calculate the magnitude of the phase current in the Y-connected load.

d) Calculate the magnitude of the line voltage at the sending end of the line.

11.12 For the circuit shown in Fig. P11.12, find

a) the phase currents $\mathbf{I}_{AB}$, $\mathbf{I}_{BC}$, and $\mathbf{I}_{CA}$;

b) the line currents $\mathbf{I}_{aA}$, $\mathbf{I}_{bB}$, and $\mathbf{I}_{cC}$;

c) the phase currents $\mathbf{I}_{ba}$, $\mathbf{I}_{ac}$, and $\mathbf{I}_{cb}$,

when $Z_1 = 3.6 - j1.05\ \Omega$, $Z_2 = 12 + j9\ \Omega$, and $Z_3 = 30 + j0\ \Omega$.

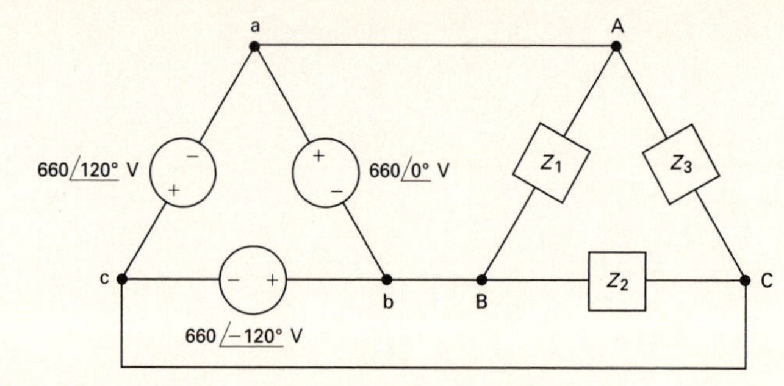

FIGURE P11.12

11.13 The impedance Z in the balanced three-phase circuit in Fig. P11.13 is $100 - j75\ \Omega$. Find

a) $\mathbf{I}_{AB}$, $\mathbf{I}_{BC}$, and $\mathbf{I}_{CA}$;

b) $\mathbf{I}_{aA}$, $\mathbf{I}_{bB}$, and $\mathbf{I}_{cC}$; and

c) $\mathbf{I}_{ba}$, $\mathbf{I}_{cb}$, and $\mathbf{I}_{ac}$.

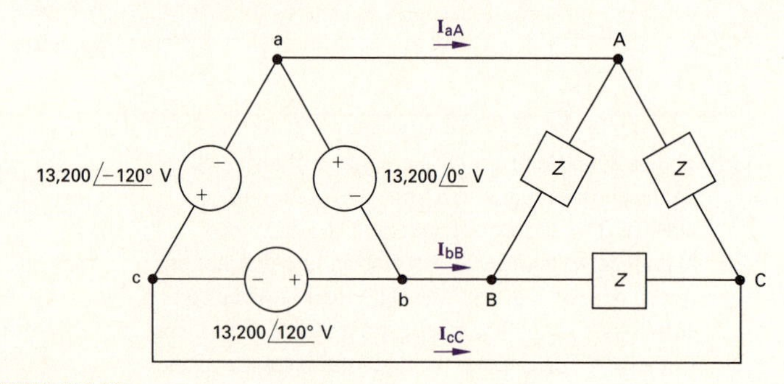

FIGURE P11.13

11.14 A balanced three-phase Δ-connected source is shown in Fig. P11.14.

a) Find the Y-connected equivalent circuit.

b) Show that the Y-connected equivalent circuit delivers the same open-circuit voltage as the original Δ-connected source.

c) Apply an external short circuit to the terminals A, B, and C. Use the Δ-connected source to find the three line currents I_{aA}, I_{bB}, and I_{cC}.

d) Repeat part (c) but use the Y-equivalent source to find the three line currents.

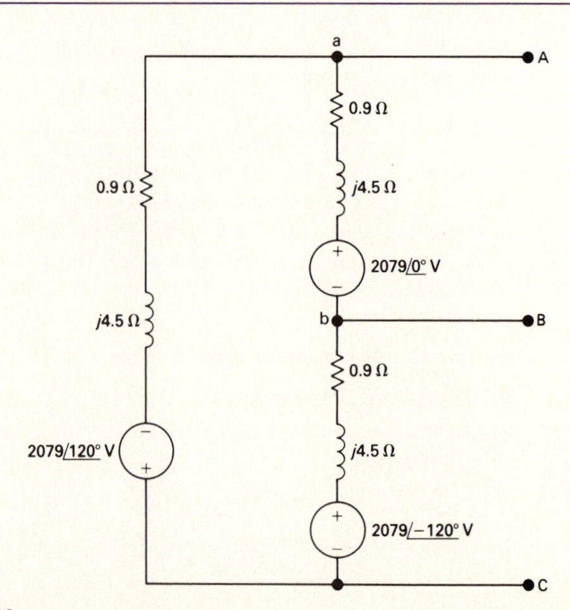

FIGURE P11.14

11.15 The Δ-connected source of Problem 11.14 is connected to a Y-connected load by means of a balanced three-phase distribution line. The load impedance is $318 - j255\ \Omega/\phi$, and the line impedance is $1.7 + j13.5\ \Omega/\phi$.

a) Construct a single-phase equivalent circuit of the system.

b) Determine the magnitude of the line voltage at the terminals of the load.

c) Determine the magnitude of the phase current in the Δ-source.

d) Determine the magnitude of the line voltage at the terminals of the source.

11.16 A three-phase Δ-connected generator has an internal impedance of $0.9 + j9.0\ \Omega/\phi$. When the load is removed from the generator, the magnitude of the terminal voltage is 13,200 V. The generator feeds a Δ-connected load through a transmission line with an impedance of $0.7 + j3.0\ \Omega/\phi$. The per-phase impedance of the load is $645 + j171\ \Omega$.

a) Construct a single-phase equivalent circuit.

b) Calculate the magnitude of the line current.

c) Calculate the magnitude of the line voltage at the terminals of the load.

d) Calculate the magnitude of the line voltage at the terminals of the source.

e) Calculate the magnitude of the phase current in the load.

f) Calculate the magnitude of the phase current in the source.

11.17 a) Find the rms magnitude and the phase angle of $\mathbf{I}_{BC}$ in the circuit shown in Fig. P11.17.

b) What percentage of the average power delivered by the three-phase source is dissipated in the three-phase load?

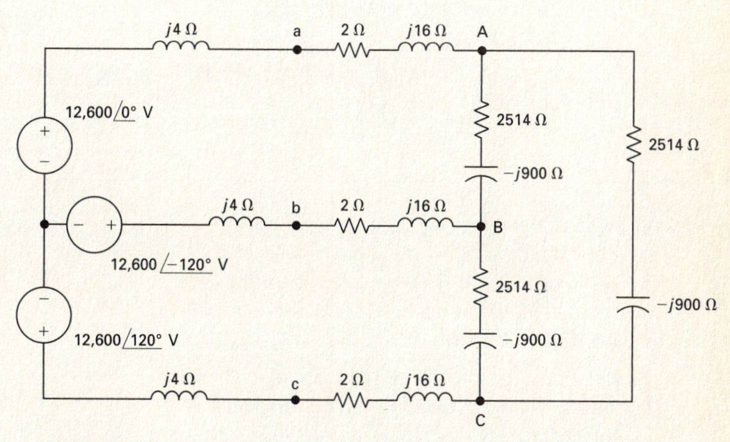

FIGURE P11.17

11.18 A three-phase line has an impedance of $0.5 + j4\ \Omega/\phi$. The line feeds two balanced three-phase loads connected in parallel. The first load is absorbing a total of 691.20 kW and delivering 201.6 kVAR magnetizing vars. The second load is Δ connected and has an impedance of $622.08 + j181.44\ \Omega/\phi$. The line-to-neutral voltage at the load end of the line is 7200 V. What is the magnitude of the line voltage at the source end of the line?

11.19 Calculate the complex power in each phase of the unbalanced load in Problem 11.12.

11.20 Three balanced three-phase loads are connected in parallel. Load 1 is Y connected with an impedance of $300 + j150\ \Omega/\phi$; load 2 is Δ connected with an impedance of $3600 - j2700\ \Omega/\phi$; and load 3 is 450 kVA at 0.8 pf lagging. The loads are fed from a distribution line with an impedance of $1 + j8\ \Omega/\phi$. The magnitude of the line-to-neutral voltage at the load end of the line is 7.5 kV.

a) Calculate the total complex power at the sending end of the line.

b) What percentage of the average power at the sending end of the line is delivered to the loads?

11.21 The line-to-neutral voltage at the terminals of the balanced three-phase load in the circuit shown in Fig. P11.21 is 600 V. At this voltage, the load is absorbing 135 kVA at 0.96 pf lag.

a) Use $\mathbf{V}_{AN}$ as the reference and express $\mathbf{I}_{na}$ in polar form.

b) Calculate the complex power associated with the ideal three-phase source.

c) Check that the total average power delivered equals the total average power absorbed.

d) Check that the total magnetizing reactive power delivered equals the total magnetizing reactive power absorbed.

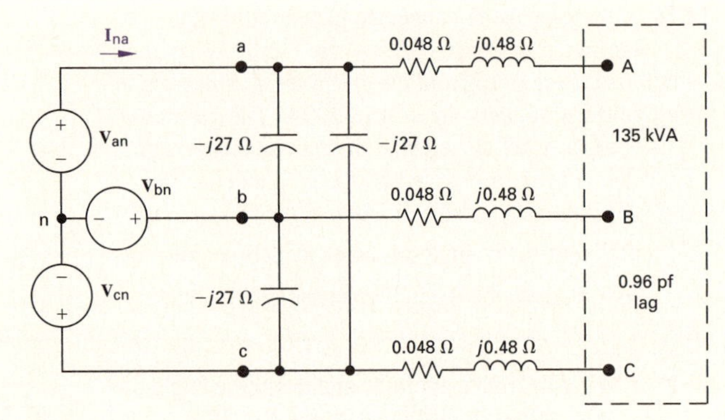

FIGURE P11.21

11.22 A balanced three-phase source is supplying 150 kVA at 0.8 pf lead to two balanced Y-connected parallel loads. The distribution line connecting the source to the load has negligible impedance. Load 1 equals $30 + j30$ kVA.

a) Determine the impedance per phase of load 2 if the line voltage is $240\sqrt{3}$ V and the impedance components are in series.

b) Repeat part (a), with the impedance components in parallel.

11.23 The three pieces of computer equipment described below are installed as part of a computation center. Each piece of equipment is a balanced three-phase load rated at 208 V. Calculate (a) the magnitude of the line current supplying these three devices and (b) the power factor of the combined load.

Disk: 4.864 kW at 0.79 pf lag

Drum: 17.636 kVA at 0.96 pf lag

CPU: line current 73.8 A, 13.853 kVAR

11.24 A balanced three-phase distribution line has an impedance of $0.5 + j5\ \Omega/\phi$. This line is used to supply three balanced three-phase loads that are connected in parallel. The three loads are

$L_1 = 125$ kVA at 0.96 pf lag

$L_2 = 120$ kW at 0.60 pf lead, and

$L_3 = 120$ kW and 20 kVAR (magnetizing).

The magnitude of the line voltage at the terminals of the loads is $2500\sqrt{3}$ V.

a) What is the magnitude of the line voltage at the sending end of the line?

b) What is the percentage of efficiency of the distribution line with respect to average power?

11.25 The output of the balanced positive-sequence three-phase source in Fig. P11.25 is 75 kVA at a leading power factor of 0.96. The line voltage at the source is 720 V.

a) Find the line voltage at the load.

b) Find the total complex power at the terminals of the load.

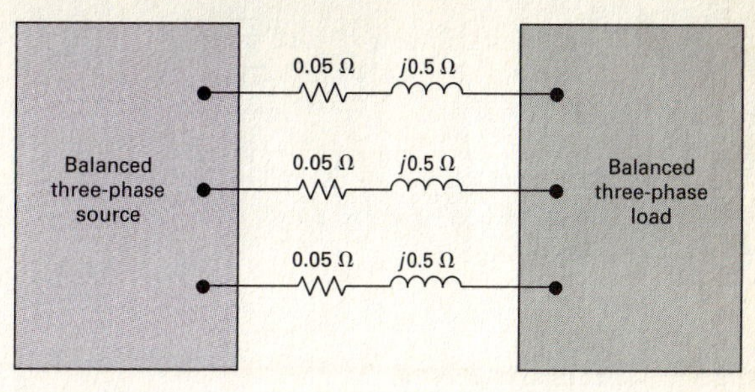

FIGURE P11.25

11.26 The total power delivered to a balanced three-phase load when operating at a line voltage of $4800\sqrt{3}$ V is 900 kW at a lagging power factor of 0.60. The impedance of the distribution line supplying the load is $0.6 + j4.8\ \Omega/\phi$. Under these operating conditions, the drop in the magnitude of the line voltage between the sending end and the load end of the line is excessive. To compensate, a bank of Δ-connected capacitors is placed in parallel with the load. The capacitor bank is designed to furnish 1200 kVAR of magnetizing reactive power when operated at a line voltage of $4800\sqrt{3}$ V.

a) What is the magnitude of the voltage at the sending end of the line when the load is operating at a line voltage of $4800\sqrt{3}$ V and the capacitor bank is disconnected?

b) Repeat part (a), with the capacitor bank connected.

c) What is the average power efficiency of the line in part (a)?

d) What is the average power efficiency in part (b)?

e) If the system is operating at a frequency of 60 Hz, what is the size of each capacitor in microfarads?

11.27 A balanced three-phase load absorbs 1500 kVA at a lagging power factor of 0.8 when the line voltage at the terminals of the load is 13,200 V. Find four equivalent circuits that can be used to model this load.

11.28 Show that the total instantaneous power in a balanced three-phase circuit is constant and equal to $1.5 V_\phi I_\phi \cos\theta_\phi$, where V_ϕ and I_ϕ represent the maximum amplitudes of the phase voltage and phase current, respectively.

11.29 At full load, a commercially available 100 hp, three-phase induction motor operates at an efficiency of 97% and a power factor of 0.88 lag. The motor is supplied from a three-phase outlet with a line-voltage rating of 208 V.

a) What is the magnitude of the line current drawn from the 208 V outlet? (1 hp = 746 W.)

b) Calculate the reactive power supplied to the motor.

11.30 a) Calculate the reading of each wattmeter in the circuit shown in Fig. P11.30. The value of Z_ϕ is $40\angle{-30^\circ}\ \Omega$.

b) Verify that the sum of the wattmeter readings equals the total average power delivered to the Δ-connected load.

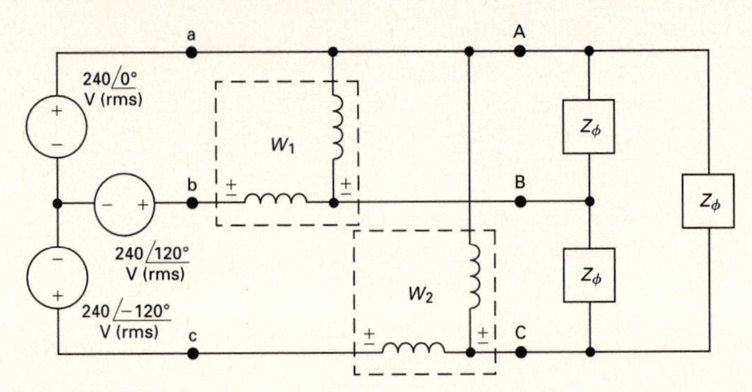

FIGURE P11.30

11.31 The balanced three-phase load shown in Fig. P11.31 is fed from a balanced, negative-sequence, three-phase Y-connected source. The impedance of the line connecting the source to the load is negligible. The line-to-neutral voltage of the source is 7200 V.

a) Find the reading of the wattmeter in watts.

b) Explain how you would connect a second wattmeter in the circuit so that the two wattmeters would measure the total power.

c) Calculate the reading of the second wattmeter.

d) Verify that the sum of the two wattmeter readings equals the total average power delivered to the load.

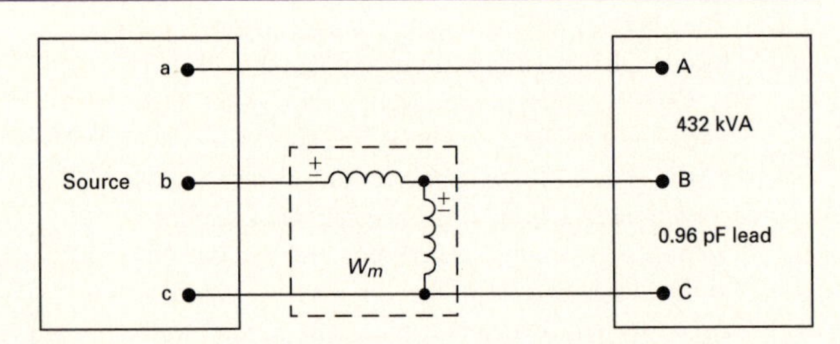

FIGURE P11.31

11.32 a) Find the reading of each wattmeter in the circuit shown in Fig. P11.32 if $Z_A = 20\angle{30^\circ}\ \Omega$, $Z_B = 60\angle{0^\circ}\ \Omega$, and $Z_C = 40\angle{-30^\circ}\ \Omega$.

b) Show that the sum of the wattmeter readings equals the total average power delivered to the unbalanced three-phase load.

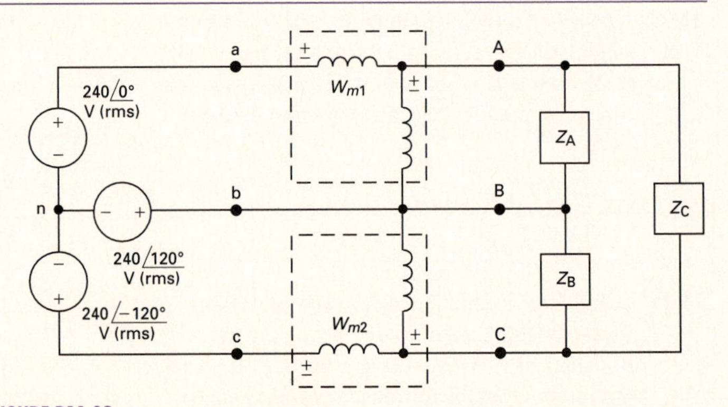

FIGURE P11.32

11.33 The wattmeters in the circuit in Fig. 11.20 read as follows: $W_1 = 37{,}297.54$ W, and $W_2 = 139{,}196.31$ W. The magnitude of the line voltage is 4160 V. The phase sequence is positive. Find Z_ϕ.

11.34 a) Calculate the complex power associated with each phase of the balanced load in Problem 11.13.

b) If the two-wattmeter method is used to measure the average power delivered to the load, specify the reading of each meter.

11.35 The two-wattmeter method is used to measure the power delivered to the unbalanced load in Problem 11.12. The current coil of wattmeter 1 is placed in line aA, and that of wattmeter 2 is placed in line bB.

a) Calculate the reading of wattmeter 1.

b) Calculate the reading of wattmeter 2.

c) Show that the sum of the two wattmeter readings equals the total power delivered to the unbalanced load.

11.36 In the balanced three-phase circuit shown in Fig. P11.36, the current coil of the wattmeter is connected in line aA, and the potential coil of the wattmeter is connected across lines b and c. Show that the wattmeter reading multiplied by $\sqrt{3}$ equals the total reactive power associated with the load. The phase sequence is positive.

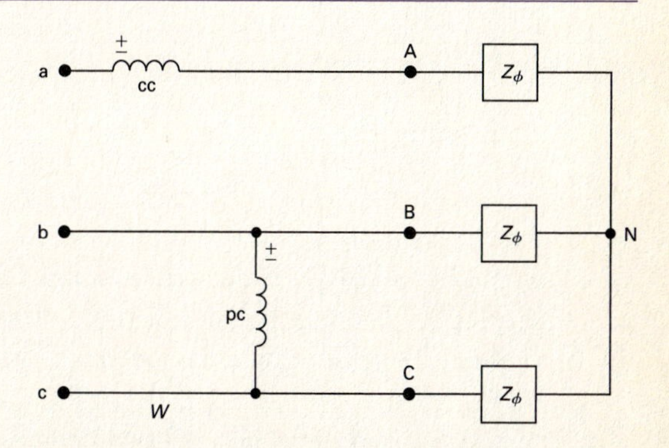

FIGURE P11.36

11.37 The line voltage in the circuit in Fig. P11.36 is 680 V, the phase sequence is positive, and the load impedance is $16 - j12\ \Omega/\phi$.

a) Calculate the wattmeter reading.

b) Calculate the total reactive power associated with the load.

11.38 a) Calculate the reading of each wattmeter in the circuit shown in Fig. P11.38 when $Z = 13.44 + j46.08\ \Omega$.

b) Check that the sum of the two wattmeter readings equals the total power delivered to the load.

c) Check that $\sqrt{3}(W_1 - W_2)$ equals the total magnetizing vars delivered to the load.

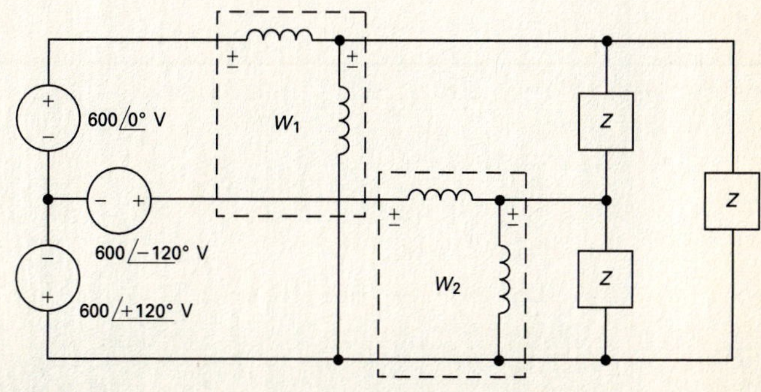

FIGURE P11.38

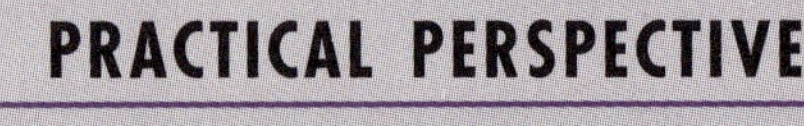

PRACTICAL PERSPECTIVE

An Ignition Circuit

In this chapter we introduce the concept of mutual inductance. An automobile ignition circuit makes use of this concept. In such a circuit, mutual inductance is the basis of a two-winding device known as an autotransformer. The autotransformer can be designed to either step up or step down a time-varying voltage. The ignition circuit uses a switch, either mechanical or electronic, to cause a rapid change in the current in one winding (the primary winding) of the autotransformer. The high voltage induced in the primary winding is stepped up to an even higher value in the secondary winding. This high voltage, which peaks at from 20 to 40 kV, is used to generate a spark across the gap of the spark plug. The spark ignites the fuel-air mixture in the cylinder.

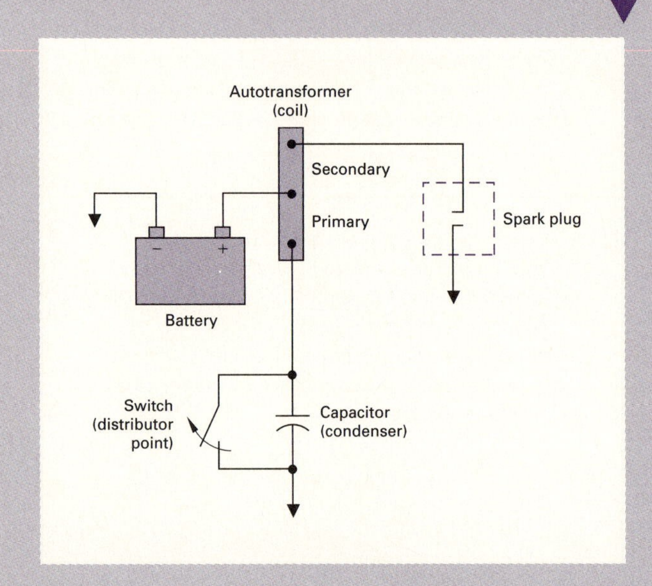

A schematic diagram showing the basic components of an ignition system is shown in the accompanying figure. In today's automobile, electronic (as opposed to mechanical) switching is used to cause the rapid change in the primary winding current. An understanding of the electronic switching circuit requires a knowledge of electronic components that is beyond the scope of this text. However, an analysis of the older, conventional ignition circuit will serve as an introduction to the types of problems encountered in the design of a useful circuit.

12 Mutual Inductance

CHAPTER CONTENTS

The magnetic field we considered in our study of inductors in Chapter 6 was restricted to a single circuit. We said that inductance is the parameter that relates a voltage to a time-varying current in the same circuit; thus, inductance is more precisely referred to as self-inductance.

We now consider the situation in which two circuits are linked by a magnetic field. In this case, the voltage induced in the second circuit can be related to the time-varying current in the first circuit by a parameter known as mutual inductance. By first offering, in Section 12.1, a quantitative review of self-inductance, we pave the way for a complete description of mutual inductance in the sections that follow.

Two circuits linked by a magnetic field are said to be magnetically coupled. A transformer is a device designed to take advantage of magnetic coupling. The practical significance of magnetic coupling, as embodied in transformers, unfolds as we study the relationships between current, voltage, power, and several new parameters specific to mutual inductance. Transformers allow us to increase or decrease voltages from a source to a load. This phenomenon is used effectively in power applications. A familiar example is in the electric utility industry, where voltage levels are decreased, or stepped down, to safe levels from a power line to a residence using transformers. We can also use transformers for impedance matching. This application is used in communications systems.

Although we will use differential equations in this chapter to describe circuits containing mutual inductance, we will not discuss the solution of these equations. Such analysis (of the transient response) will wait until we introduce, in Chapter 13, the Laplace transform method of solving differential equations. Here, we will limit our analysis to the sinusoidal steady-state response of magnetically coupled circuits.

12.1 A Review of Self-Inductance

The concept of inductance can be traced to Michael Faraday, who did pioneering work in this area early in the 1800s. Faraday postulated that a magnetic field consists of lines of force surrounding the current-carrying conductor. Visualize these lines of force as energy-storing elastic bands that close upon themselves. As the current increases and decreases, the elastic bands spread and collapse about the conductor. The voltage induced in the conductor is proportional to the number of lines that collapse into, or cut, the conductor. This image of induced voltage is expressed by what is called Faraday's law; that is,

$$v = \frac{d\lambda}{dt}, \tag{12.1}$$

where λ is referred to as the flux linkage and is measured in weber-turns.

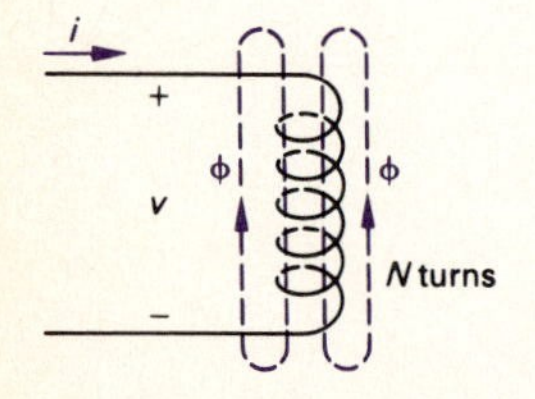

FIGURE 12.1 Representation of a magnetic field linking an N-turn coil.

How do we get from Faraday's law to the definition of self-inductance presented in Chapter 6? We can begin to draw this connection using Fig. 12.1 as a reference. The lines threading the N turns and labeled ϕ represent the magnetic lines of force which make up the magnetic field. The strength of the magnetic field depends on the strength of the current, and the spatial orientation of the field depends on the direction of the current. The right-hand rule relates the orientation of the field to the direction of the current: When the fingers of the right hand are wrapped around the coil so that the fingers point in the direction of the current, the thumb points in the direction of that portion of the magnetic field inside the coil. The flux linkage is the product of the magnetic field (ϕ), measured in webers (Wb), and the number of turns linked by the field (N):

$$\lambda = N\phi. \quad \textbf{(12.2)}$$

The magnitude of the flux, ϕ, is related to the magnitude of the coil current by the relationship

$$\phi = \mathcal{P}Ni, \quad \textbf{(12.3)}$$

where N is the number of turns on the coil, and $\mathcal{P}$ is the permeance of the space occupied by the flux. Permeance is a quantity that describes the magnetic properties of this space, and as such, a detailed discussion of permeance is outside the scope of this text. Here, we need only observe that, when the space containing the flux is made up of magnetic materials (such as iron, nickel, and cobalt), the permeance varies with the flux, giving a nonlinear relationship between ϕ and i. But when the space containing the flux is comprised of nonmagnetic materials, the permeance is constant, giving a linear relationship between ϕ and i. Note from Eq. (12.3) that the flux is also proportional to the number of turns on the coil.

Here, we assume that the core material—the space containing the flux—is nonmagnetic. Then, substituting Eqs. (12.2) and (12.3) into Eq. (12.1) yields

$$\begin{aligned} v &= \frac{d\lambda}{dt} = \frac{d(N\phi)}{dt} \\ &= N\frac{d\phi}{dt} = N\frac{d}{dt}(\mathcal{P}Ni) \\ &= N^2\mathcal{P}\frac{di}{dt} = L\frac{di}{dt}, \end{aligned} \quad \textbf{(12.4)}$$

which shows that self-inductance is proportional to the square of the number of turns on the coil. We make use of this observation later.

The polarity of the induced voltage in the circuit in Fig. 12.1 reflects the reaction of the field to the current creating the field. For example, when i is increasing, di/dt is positive and v is positive. Thus energy is required to establish the magnetic field. The product vi gives the rate at which energy is stored in the field. When the field collapses, di/dt is negative, and again the polarity of the induced voltage is in opposition to the change. As the field collapses about the coil, energy is returned to the circuit.

Keeping in mind this further insight into the concept of self-inductance, we now turn to mutual inductance.

12.2 THE CONCEPT OF MUTUAL INDUCTANCE

Mutual inductance is the circuit parameter that relates the voltage induced in one circuit to a time-varying current in another circuit. This situation arises whenever a common magnetic field links two or more circuits. We restrict our discussion to the magnetic coupling of only two circuits.

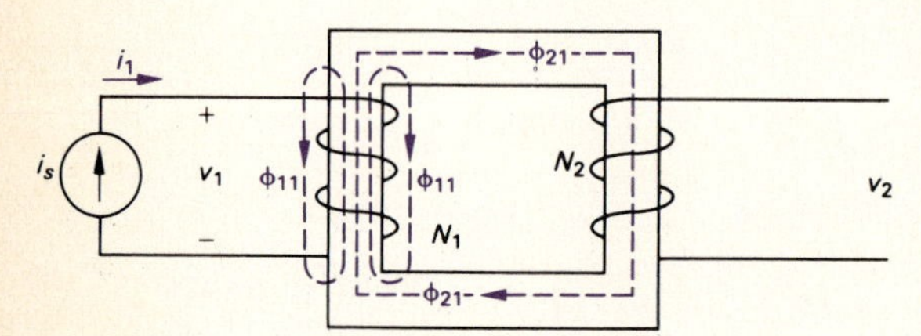

FIGURE 12.2 Two magnetically coupled coils.

Figure 12.2 shows two magnetically coupled coils. The number of turns on each coil are N_1 and N_2, respectively. Coil 1 is energized by a time-varying current source that establishes the current i_1 in the N_1 turns. Coil 2 is not energized and is open. The coils are wound on a nonmagnetic core. The flux produced by the current i_1 can be divided into two components, labeled ϕ_{11} and ϕ_{21} in Fig. 12.2. The flux component ϕ_{11} is the flux produced by i_1 that links only the N_1 turns. The component ϕ_{21} is the flux produced by i_1 that links the N_2 turns and the N_1 turns. The first digit in the subscript to the flux gives the coil number, and the second digit refers to the coil current. Thus ϕ_{11} is a flux linking coil 1 and produced by a current in coil 1, whereas ϕ_{21} is a flux linking coil 2 and produced by a current in coil 1.

The total flux linking coil 1 is ϕ_1, the sum of ϕ_{11} and ϕ_{21}:

$$\phi_1 = \phi_{11} + \phi_{21}. \quad \textbf{(12.5)}$$

The flux ϕ_1 and its components ϕ_{11} and ϕ_{21} are related to the coil current i_1 as follows:

$$\phi_1 = \mathcal{P}_1 N_1 i_1, \quad \textbf{(12.6)}$$

$$\phi_{11} = \mathcal{P}_{11} N_1 i_1, \quad \textbf{(12.7)}$$

and

$$\phi_{21} = \mathcal{P}_{21} N_1 i_1, \quad \textbf{(12.8)}$$

where $\mathcal{P}_1$ is the permeance of the space occupied by the flux ϕ_1, $\mathcal{P}_{11}$ is the permeance of the space occupied by the flux ϕ_{11}, and $\mathcal{P}_{21}$ is the permeance of the space occupied by the flux ϕ_{21}. Substituting Eqs. (12.6), (12.7), and (12.8) into Eq. (12.5) yields the relationship between the permeance of the space occupied by the total flux ϕ_1 and the permeances of the spaces occupied by its components ϕ_{11} and ϕ_{21}:

$$\mathcal{P}_1 = \mathcal{P}_{11} + \mathcal{P}_{21}. \quad \textbf{(12.9)}$$

We use Faraday's law to derive expressions for v_1 and v_2,

$$v_1 = \frac{d\lambda_1}{dt} = \frac{d(N_1\phi_1)}{dt} = N_1 \frac{d}{dt}(\phi_{11} + \phi_{21})$$

$$= N_1^2(\mathcal{P}_{11} + \mathcal{P}_{21})\frac{di_1}{dt} = N_1^2\mathcal{P}_1\frac{di_1}{dt} = L_1\frac{di_1}{dt} \quad \textbf{(12.10)}$$

and

$$v_2 = \frac{d\lambda_2}{dt} = \frac{d(N_2\phi_{21})}{dt} = N_2\frac{d}{dt}(\mathcal{P}_{21}N_1i_1)$$

$$= N_2N_1\mathcal{P}_{21}\frac{di_1}{dt}. \quad \textbf{(12.11)}$$

The coefficient of di_1/dt in Eq. (12.10) is the self-inductance of coil 1. The coefficient of di_1/dt in Eq. (12.11) is the mutual inductance between coils 1 and 2. Thus

$$M_{21} = N_2N_1\mathcal{P}_{21}. \quad \textbf{(12.12)}$$

The subscript on M specifies an inductance that relates the voltage induced in coil 2 to the current in coil 1.

The coefficient of mutual inductance gives

$$v_2 = M_{21}\frac{di_1}{dt}. \quad \textbf{(12.13)}$$

Note that no polarity references are assigned to v_2 in Fig. 12.2. We discuss determining the polarity of mutually induced voltages in Section 12.3.

For the coupled coils in Fig. 12.2, exciting coil 2 from a time-varying current source (i_2) and leaving coil 1 open produces the circuit arrangement shown in Fig. 12.3. In this case, no polarity references are assigned to v_1.

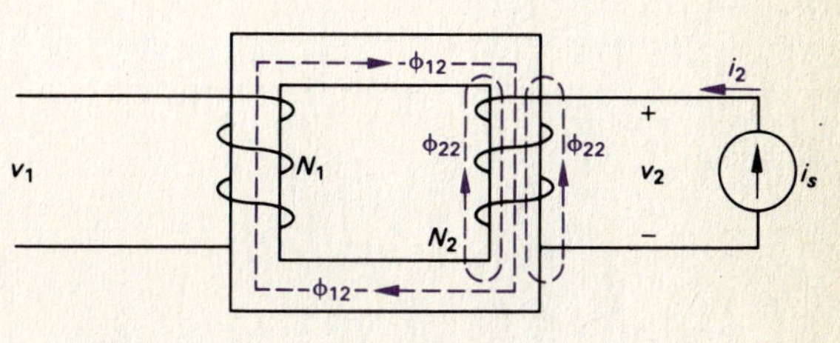

FIGURE 12.3 The magnetically coupled coils of Fig. 12.2, with coil 2 excited and coil 1 open.

The total flux linking coil 2 is

$$\phi_2 = \phi_{22} + \phi_{12}. \quad \textbf{(12.14)}$$

The flux ϕ_2 and its components ϕ_{22} and ϕ_{12} are related to the coil current i_2 as follows:

$$\phi_2 = \mathcal{P}_2N_2i_2, \quad \textbf{(12.15)}$$

$$\phi_{22} = \mathcal{P}_{22}N_2i_2, \quad \textbf{(12.16)}$$

and

$$\phi_{12} = \mathcal{P}_{12}N_2i_2. \quad \textbf{(12.17)}$$

The voltages v_2 and v_1 are

$$v_2 = \frac{d\lambda_2}{dt} = \mathcal{P}_2N_2^2\frac{di_2}{dt} = L_2\frac{di_2}{dt} \quad \textbf{(12.18)}$$

and

$$v_1 = \frac{d\lambda_1}{dt} = \frac{d}{dt}(N_1\phi_{12}) = N_1N_2\mathcal{P}_{12}\frac{di_2}{dt}. \quad \textbf{(12.19)}$$

The coefficient of mutual inductance that relates the voltage induced in coil 1 to the time-varying current in coil 2 is the coefficient of di_2/dt in Eq. (12.19):

$$M_{12} = N_1 N_2 \mathcal{P}_{12}. \tag{12.20}$$

For nonmagnetic materials, the permeances $\mathcal{P}_{12}$ and $\mathcal{P}_{21}$ are equal, and therefore

$$M_{12} = M_{21} = M. \tag{12.21}$$

Hence for linear circuits with just two magnetically coupled coils, attaching subscripts to the coefficient of mutual inductance is not necessary.

Mutual Inductance in Terms of Self-Inductance

The value of mutual inductance is a function of the self-inductances. We derive this relationship as follows. From Eqs. (12.10) and (12.18),

$$L_1 = N_1^2 \mathcal{P}_1 \tag{12.22}$$

and

$$L_2 = N_2^2 \mathcal{P}_2, \tag{12.23}$$

respectively. From Eqs. (12.22) and (12.23),

$$L_1 L_2 = N_1^2 N_2^2 \mathcal{P}_1 \mathcal{P}_2. \tag{12.24}$$

We now use Eq. (12.9) and the corresponding expression for $\mathcal{P}_2$ to write

$$L_1 L_2 = N_1^2 N_2^2 (\mathcal{P}_{11} + \mathcal{P}_{21})(\mathcal{P}_{22} + \mathcal{P}_{12}). \tag{12.25}$$

But for a linear system, $\mathcal{P}_{21} = \mathcal{P}_{12}$, so Eq. (12.25) becomes

$$\begin{aligned} L_1 L_2 &= (N_1 N_2 \mathcal{P}_{12})^2 \left(1 + \frac{\mathcal{P}_{11}}{\mathcal{P}_{12}}\right)\left(1 + \frac{\mathcal{P}_{22}}{\mathcal{P}_{12}}\right) \\ &= M^2 \left(1 + \frac{\mathcal{P}_{11}}{\mathcal{P}_{12}}\right)\left(1 + \frac{\mathcal{P}_{22}}{\mathcal{P}_{12}}\right). \end{aligned} \tag{12.26}$$

Replacing the two terms involving permeances by a single constant expresses Eq. (12.26) in a more meaningful form:

$$\frac{1}{k^2} = \left(1 + \frac{\mathcal{P}_{11}}{\mathcal{P}_{12}}\right)\left(1 + \frac{\mathcal{P}_{22}}{\mathcal{P}_{12}}\right). \tag{12.27}$$

Substituting Eq. (12.27) into Eq. (12.26) yields

$$M^2 = k^2 L_1 L_2$$

or

$$M = k\sqrt{L_1 L_2}, \tag{12.28}$$

where the constant k is called the **coefficient of coupling**. According to Eq. (12.27), $1/k^2$ must be greater than 1, which means that k must be less than 1. In fact, the coefficient of coupling must lie between 0 and 1, or

$$0 \leq k \leq 1. \tag{12.29}$$

The coefficient of coupling is 0 when the two coils have no common flux, that is, when $\phi_{12} = \phi_{21} = 0$. This condition implies that $\mathcal{P}_{12} = 0$, and Eq. (12.27) indicates that $1/k^2 = \infty$, or $k = 0$. If there is no flux linkage between the coils, obviously M is 0.

The coefficient of coupling is equal to 1 when ϕ_{11} and ϕ_{22} are 0. This condition implies that all the flux that links coil 1 also links coil 2. In terms of Eq. (12.27), $\mathcal{P}_{11} = \mathcal{P}_{22} = 0$, which obviously represents an ideal state; in reality, winding two coils so that they share precisely the same flux is physically impossible. Magnetic materials (such as alloys of iron, cobalt, and nickel) create a space with high permeance and are used to establish coefficients of coupling that approach unity. (We say more about this important quality of magnetic materials later.)

To review, in this section we showed how to extend the concept of inductance to relate the voltage induced in one circuit to a time-varying current in another circuit. We also showed how inductance relates to the number of turns on the magnetically coupled coils and a constant, known as permeance, which characterizes the magnetic properties of the space occupied by the flux. For linearly coupled coils, Eqs. (12.21), (12.28), and (12.29) contain the important results for circuit analysis. Next we discuss the importance of determining the polarity of the mutually induced voltages.

DRILL EXERCISES

12.1 Two magnetically coupled coils are wound on a nonmagnetic core. The self-inductance of coil 1 is 6 H, the mutual inductance is 9.6 H, the coefficient of coupling is 0.8, and the physical structure of the coils is such that $\mathcal{P}_{11} = \mathcal{P}_{22}$.

a) Find L_2 and the turns ratio N_1/N_2.

b) If $N_1 = 800$, what is the value of $\mathcal{P}_1$ and $\mathcal{P}_2$?

ANSWER: (a) 24 H, 0.5; (b) 9.375×10^{-6} Wb/A.

12.2 The self-inductances of two magnetically coupled coils are $L_1 = 50$ mH and $L_2 = 72$ mH. The coupling medium is nonmagnetic. If coil 1 has 500 turns and coil 2 has 600 turns, find $\mathcal{P}_{11}$ and $\mathcal{P}_{21}$ (in microwebers per ampere) when the coefficient of coupling is 0.85.

ANSWER: 0.03 μWb/A, 0.17 μWb/A.

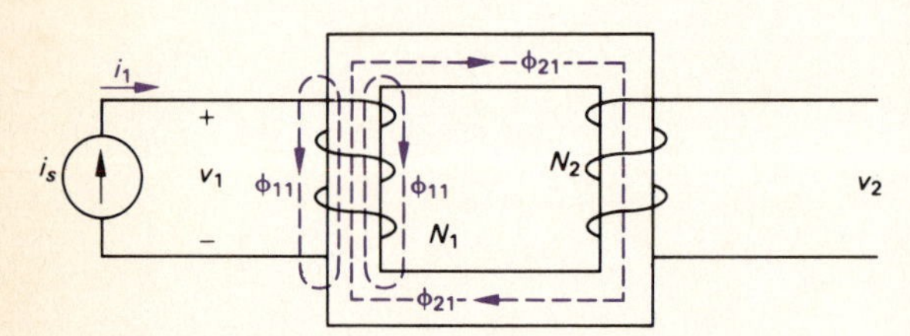

FIGURE 12.4 Two magnetically coupled coils.

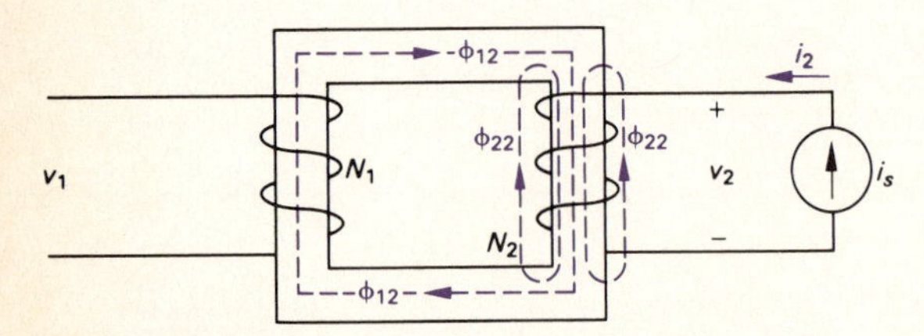

FIGURE 12.5 The magnetically coupled coils of Fig. 12.4, with coil 2 excited and coil 1 open.

12.3 THE POLARITY OF MUTUALLY INDUCED VOLTAGES (THE DOT CONVENTION)

The polarity of a mutually induced voltage also reflects a reaction against the time-varying flux that creates the voltage. For example, when i_1 in the circuit in Fig. 12.4 is increasing in the reference direction, the polarity of v_2 is positive at the upper terminal of coil 2. The reason is that the polarity of v_2 would establish a current out of the upper terminal of the N_2 coil to produce a flux in opposition to ϕ_{21}. Thus when v_1 is positive at the upper terminal of coil 1, v_2 is positive at the upper terminal of coil 2, and vice versa. (Note that exciting coil 2 as in Fig. 12.5 yields the same conclusion concerning the relative polarities of v_1 and v_2.)

Having deduced the polarity of v_2 in relation to the current i_1, we now express v_2 as a function of i_1, with the proper algebraic sign. If the reference for v_2 is positive at the upper terminal of coil 2,

$$v_2 = M\frac{di_1}{dt}. \qquad \textbf{(12.30)}$$

But if the reference for v_2 is positive at the lower terminal of coil 2,

$$v_2 = -M\frac{di_1}{dt}. \qquad \textbf{(12.31)}$$

Similarly, for the circuit in Fig. 12.5, where coil 2 is energized by i_2 and coil 1 is open,

$$v_1 = \pm M\frac{di_2}{dt}, \qquad \textbf{(12.32)}$$

where the positive sign is a positive reference for v_1 at the upper terminal of coil 1, and the minus sign is a positive reference at the lower terminal.

In general, showing the details of mutually coupled windings is too cumbersome, so we do not use the reaction method for determining polarity signs in writing circuit equations. Instead, we keep track of the polarities by a method known as the **dot convention**, in which a dot is placed on one terminal of each winding. These dots carry the sign information. The rule for determining the polarity of mutually induced voltage by the dot convention can be summarized as follows:

> When the reference direction for a current enters the dotted terminal of a coil, the reference polarity of the voltage that it induces in the other coil is positive at its dotted terminal.

Or, stated alternatively,

> When the reference direction for a current leaves the dotted terminal of a coil, the reference polarity of the voltage that it induces in the other coil is negative at its dotted terminal.

The dot markings in Fig. 12.6 reflect the dot convention for the circuit in Fig. 12.4. Note that the dot markings allow us to draw the coils schematically rather than showing how they wrap around a core structure.

For the most part, dot markings will be provided for you in the circuit diagrams in this text. The important skill is to be able to write the appropriate circuit equations given your understanding of mutual inductance and the dot convention. Figuring out where to place the polarity dots if they are not given may be possible by examining the physical configuration of an actual circuit, or by testing it in the laboratory. We will discuss these procedures briefly, but we turn first to the use of dot markings.

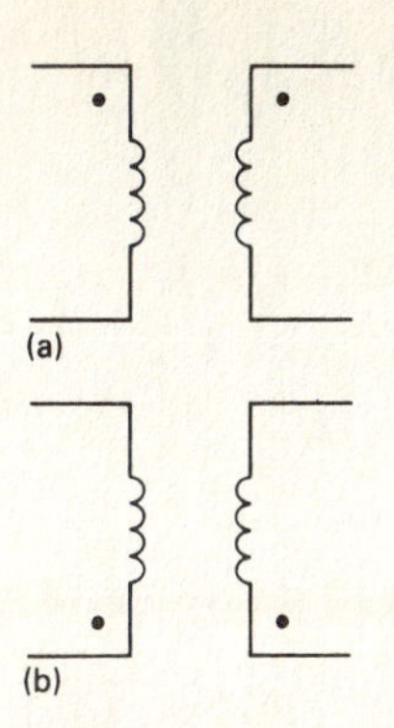

FIGURE 12.6 Dot markings for the coils of Fig. 12.4.

The Use of Dot Markings in Circuit Analysis

The easiest way to analyze circuits containing mutual inductance is with mesh currents. Another possibility is the node-voltage method, but it is somewhat clumsy to use. The following analysis shows the mesh-current approach, with Fig. 12.7 illustrating the procedure.

A circuit simulator can be used to model mutually coupled coils.

The circuit in Fig. 12.7 represents two magnetically coupled coils with dot markings as shown. The self-inductances of the two coils are labeled L_1 and L_2, and the mutual inductance is labeled M. The double-headed arrow adjacent to M indicates the pair of coils with this value of mutual inductance. This notation is needed particularly in circuits containing more than one pair of magnetically coupled coils.

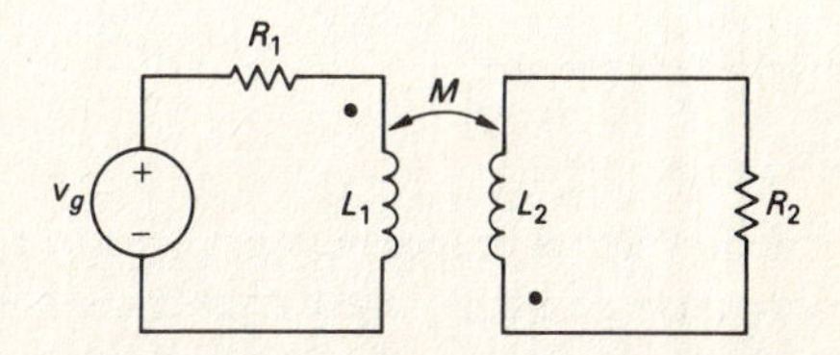

FIGURE 12.7 A circuit illustrating how dot markings are used in writing circuit equations.

The problem is to write the circuit equations that describe the circuit in terms of the coil currents. First, choose the reference direction for each coil current. Figure 12.8 shows arbitrarily selected reference currents. After choosing the reference directions for i_1 and i_2, sum the voltages around each closed path. Because of the mutual inductance M, there will be two voltages across each coil, namely, a self-induced voltage and a mutually induced voltage. The self-induced voltage is a voltage drop in the direction of the current producing the voltage. Then determine the polarity of the mutually induced voltage, using the dot convention rule stated earlier.

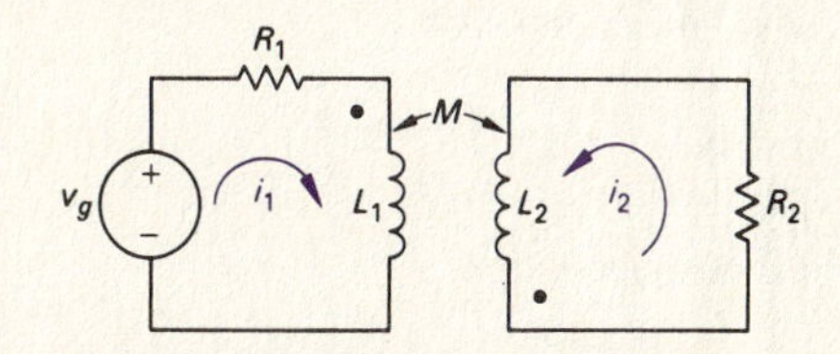

FIGURE 12.8 Coil currents i_1 and i_2 used to describe the circuit shown in Fig. 12.7.

The rule indicates that the reference polarity for the voltage induced in coil 1 by the current i_2 is negative at the dotted terminal of coil 1. This voltage (Mdi_2/dt) is a voltage rise with respect to i_1. The voltage induced in coil 2 by the current i_1 is Mdi_1/dt, and its reference polarity is positive at the dotted terminal of coil 2. This voltage is a voltage rise in the direction of i_2. Figure 12.9 shows the self- and mutually induced voltages across coils 1 and 2, along with their polarity marks.

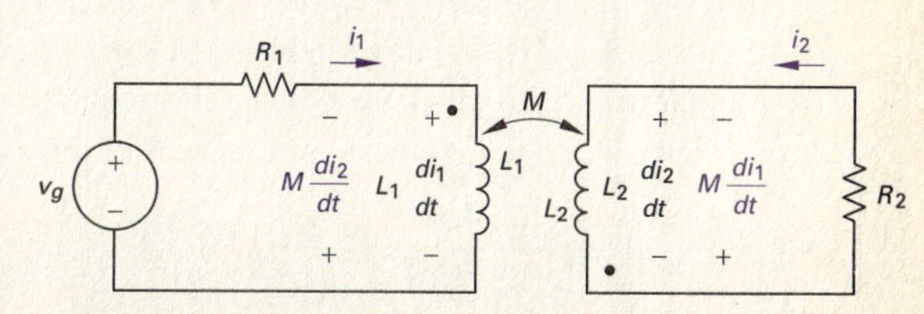

FIGURE 12.9 The self- and mutually-induced voltages appearing across the coils shown in Fig. 12.8.

Now sum the voltages around each closed loop. In Eqs. (12.33) and (12.34), voltage rises in the reference direction of a current are negative:

$$-v_g + i_1R_1 + L_1\frac{di_1}{dt} - M\frac{di_2}{dt} = 0; \quad (12.33)$$

$$i_2R_2 + L_2\frac{di_2}{dt} - M\frac{di_1}{dt} = 0. \quad (12.34)$$

Example 12.1 shows how to use the dot markings to formulate a set of circuit equations.

EXAMPLE 12.1

a) Write a set of mesh-current equations that describe the circuit in Fig. 12.10 in terms of the currents i_1, i_2, and i_3.

b) What is the coefficient of coupling of the magnetically coupled coils?

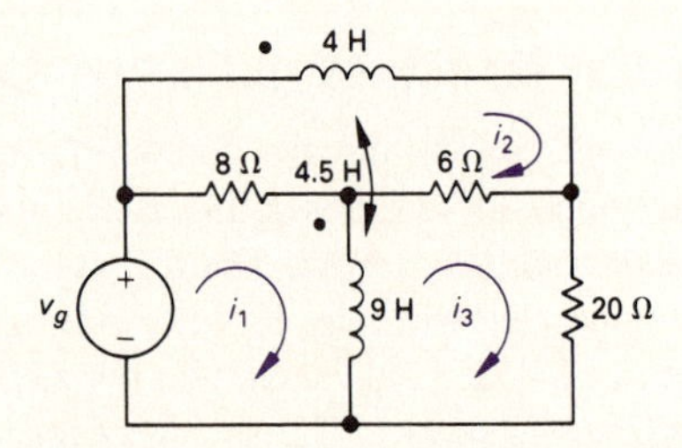

FIGURE 12.10 The circuit for Example 12.1.

SOLUTION

a) (In the following set of mesh-current equations, voltage drops appear as positive quantities on the right-hand side of each equation.) Summing the voltages around the first mesh yields

$$v_g = 8(i_1 - i_2) + 9\frac{d}{dt}(i_1 - i_3) + 4.5\frac{di_2}{dt}.$$

The second mesh equation is

$$0 = 4\frac{di_2}{dt} + 4.5\frac{d}{dt}(i_1 - i_3) + 6(i_2 - i_3) + 8(i_2 - i_1).$$

The third mesh equation is

$$0 = 9\frac{d}{dt}(i_3 - i_1) - 4.5\frac{di_2}{dt} + 6(i_3 - i_2) + 20i_3.$$

Use a computer tool to help solve these equations.

Note that the voltage induced across the 4 H coil by the current $(i_1 - i_3)$—that is, $4.5d(i_1 - i_3)/dt$—is a voltage drop in the reference direction of i_2. The voltage induced in the 9 H coil by the current i_2—that is, $4.5di_2/dt$—is a voltage drop in the reference direction of i_1 but a voltage rise in the reference direction of i_3.

b) $k = \dfrac{M}{\sqrt{L_1L_2}} = \dfrac{4.5}{\sqrt{(9)(4)}} = \dfrac{4.5}{6} = 0.75.$

DRILL EXERCISE

12.3 Write a set of mesh-current equations for the circuit in Example 12.1, with the dot on the 9 H inductor at the lower terminal and the reference direction of i_3 reversed.

ANSWER:

$$v_g = 8(i_1 - i_2) + 9\frac{d}{dt}(i_1 + i_3) - 4.5\frac{di_2}{dt},$$

$$0 = 4\frac{di_2}{dt} - 4.5\frac{d}{dt}(i_1 + i_3) + 6(i_2 + i_3) + 8(i_2 - i_1),$$

$$0 = 20i_3 + 6(i_3 + i_2) + 9\frac{d}{dt}(i_3 + i_1) - 4.5\frac{di_2}{dt}.$$

THE PROCEDURE FOR DETERMINING DOT MARKINGS

We shift now to two methods of determining dot markings. The first assumes that we know the physical arrangement of the two coils and the mode of each winding in a magnetically coupled circuit. The following six steps, applied here to Fig. 12.11, determine a set of dot markings:

1. Arbitrarily select one terminal—say, the D terminal—of one coil and give it a dot.
2. Assign a current into the dotted terminal and label it i_D.
3. Use the right-hand rule to determine the direction of the magnetic field established by i_D inside the coupled coils and label this field ϕ_D.
4. Arbitrarily pick one terminal of the second coil—say, terminal A—and assign a current into this terminal, showing the current as i_A.
5. Use the right-hand rule to determine the direction of the flux established by i_A inside the coupled coils and label this flux ϕ_A.
6. Compare the directions of the two fluxes ϕ_D and ϕ_A. If the fluxes have the same reference direction, place a dot on the terminal of the second coil where the test current (i_A) enters. (In Fig. 12.11, the fluxes ϕ_D and ϕ_A have the same reference direction, and therefore a dot goes on terminal A.) If the fluxes have different reference directions, place a dot on the terminal of the second coil where the test current leaves.

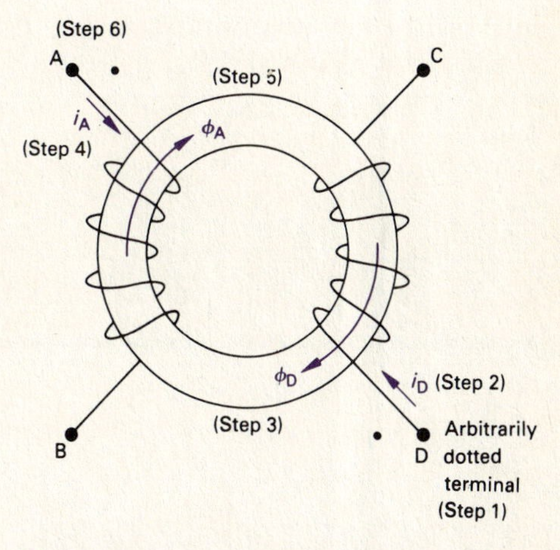

FIGURE 12.11 A set of coils showing a method for determining a set of dot markings.

The relative polarities of magnetically coupled coils can also be determined experimentally. This capability is important because in

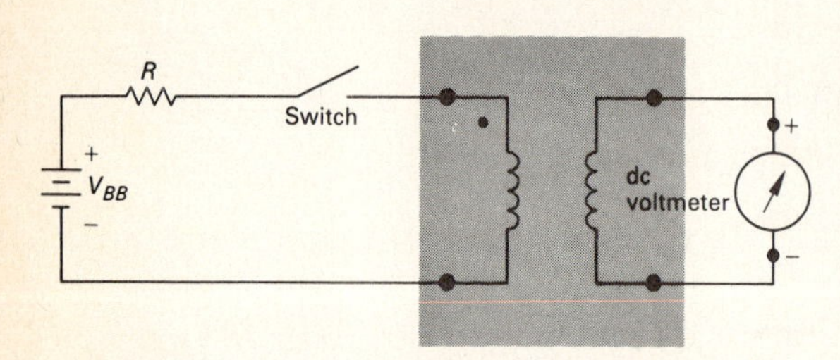

FIGURE 12.12 An experimental setup for determining polarity marks.

some situations, determining how the coils are wound on the core is impossible. One experimental method is to connect a dc voltage source, a resistor, a switch, and a dc voltmeter to the pair of coils, as shown in Fig. 12.12. The shaded box covering the coils implies that physical inspection of the coils is not possible. The resistor R limits the magnitude of the current supplied by the dc voltage source.

The coil terminal connected to the positive terminal of the dc source via the switch and limiting resistor receives a polarity mark, as shown in Fig. 12.12. When the switch is closed, the voltmeter deflection is observed. If the momentary deflection is upscale, the coil terminal connected to the positive terminal of the voltmeter receives the polarity mark. If the deflection is downscale, the coil terminal connected to the negative terminal of the voltmeter receives the polarity mark.

DRILL EXERCISE

12.4 Assume that the magnetic flux is confined to the core material in the structure shown. Which terminal of coil 2 should be given a dot marking?

ANSWER: b.

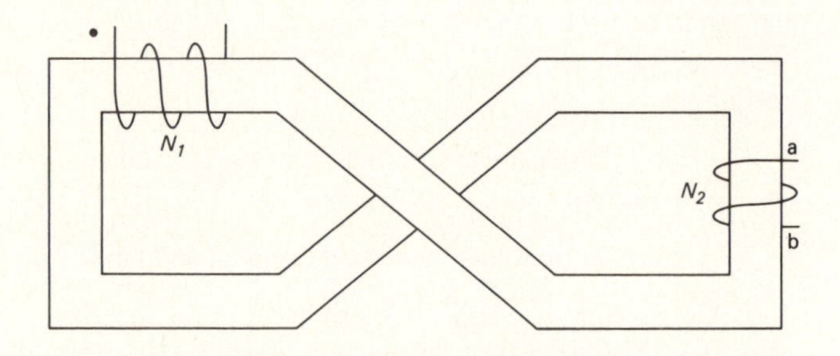

12.4 ENERGY CALCULATIONS

We turn now to a discussion of the total energy stored in magnetically coupled coils. In doing so, we will confirm two observations made earlier: for linear magnetic coupling, (1) $M_{12} = M_{21} = M$, and (2) $M = k\sqrt{L_1 L_2}$, where $0 \leq k \leq 1$.

We use the circuit shown in Fig. 12.13 to derive the expression for the total energy stored in the magnetic fields associated with a pair of linearly coupled coils. We begin by assuming that the currents i_1 and i_2 are zero and that this zero-current state corresponds to zero energy stored in the coils. Then we let i_1 increase from zero to some arbitrary value I_1 and compute the energy stored when $i_1 = I_1$. Because $i_2 = 0$,

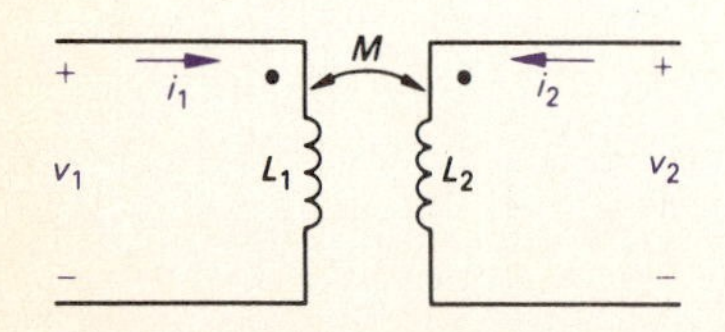

FIGURE 12.13 The circuit used to derive the basic energy relationships.

the total power input into the pair of coils is $v_1 i_1$, and the energy stored is

$$\int_0^{W_1} dw = L_1 \int_0^{I_1} i_1 di_1;$$

$$W_1 = \tfrac{1}{2} L_1 I_1^2. \tag{12.35}$$

Now we hold i_1 constant at I_1 and increase i_2 from zero to some arbitrary value I_2. During this time interval, the voltage induced in coil 2 by i_1 is zero because I_1 is constant. The voltage induced in coil 1 by i_2 is $M_{12} di_2/dt$. Therefore the power input to the pair of coils is

$$p = I_1 M_{12} \frac{di_2}{dt} + i_2 v_2.$$

The total energy stored in the pair of coils when $i_2 = I_2$ is

$$\int_{W_1}^{W} dw = \int_0^{I_2} I_1 M_{12} di_2 + \int_0^{I_2} L_2 i_2 di_2$$

or

$$\begin{aligned} W &= W_1 + I_1 I_2 M_{12} + \tfrac{1}{2} L_2 I_2^2 \\ &= \tfrac{1}{2} L_1 I_1^2 + \tfrac{1}{2} L_2 I_2^2 + I_1 I_2 M_{12}. \end{aligned} \tag{12.36}$$

If we reverse the procedure—that is, if we first increase i_2 from zero to I_2 and then increase i_1 from zero to I_1—the total energy stored is

$$W = \tfrac{1}{2} L_1 I_1^2 + \tfrac{1}{2} L_2 I_2^2 + I_1 I_2 M_{21}. \tag{12.37}$$

Equations (12.36) and (12.37) express the total energy stored in a pair of linearly coupled coils as a function of the coil currents, the self-inductances, and the mutual inductance. Note that the only difference between these equations is the coefficient of the current product $I_1 I_2$. We use Eq. (12.36) if i_1 is established first, and Eq. (12.37) if i_2 is established first.

When the coupling medium is linear, the total energy stored is the same regardless of the order used to establish I_1 and I_2. The reason is that in a linear coupling, the resultant magnetic flux depends on only the final values of i_1 and i_2—not on how the currents reached their final values. If the resultant flux is the same, the stored energy is the same. Therefore for linear coupling, $M_{12} = M_{21}$. Also, because I_1 and I_2 are arbitrary values of i_1 and i_2, respectively, we represent the coil currents by their instantaneous values i_1 and i_2. Thus at any instant of time, the total energy stored in the coupled coils is

$$w(t) = \tfrac{1}{2} L_1 i_1^2 + \tfrac{1}{2} L_2 i_2^2 + M i_1 i_2. \tag{12.38}$$

We derived Eq. (12.38) by assuming that both coil currents entered polarity-marked terminals. We leave it to you to verify that, if one current enters a polarity-marked terminal while the other leaves such a terminal, the algebraic sign of the term Mi_1i_2 reverses. Thus, in general,

$$w(t) = \tfrac{1}{2}L_1i_1^2 + \tfrac{1}{2}L_2i_2^2 \pm Mi_1i_2. \tag{12.39}$$

We use Eq. (12.39) to show that M cannot exceed $\sqrt{L_1L_2}$. The magnetically coupled coils are passive elements, so the total energy stored can never be negative. If $w(t)$ can never be negative, Eq. (12.39) indicates that the quantity

$$\tfrac{1}{2}L_1i_1^2 + \tfrac{1}{2}L_2i_2^2 - Mi_1i_2$$

must be greater than or equal to zero when i_1 and i_2 are either both positive or both negative. The limiting value of M corresponds to setting the quantity equal to zero:

$$\tfrac{1}{2}L_1i_1^2 + \tfrac{1}{2}L_2i_2^2 - Mi_1i_2 = 0. \tag{12.40}$$

To find the limiting value of M we add and subtract the term $i_1i_2\sqrt{L_1L_2}$ to the left-hand side of Eq. (12.40). Doing so generates a term that is a perfect square:

$$\left(\sqrt{\frac{L_1}{2}}i_1 - \sqrt{\frac{L_2}{2}}i_2\right)^2 + i_1i_2(\sqrt{L_1L_2} - M) = 0. \tag{12.41}$$

The squared term in Eq. (12.41) can never be negative, but it can be zero. Therefore $w(t) \geq 0$ only if

$$\sqrt{L_1L_2} \geq M, \tag{12.42}$$

which is another way of saying that

$$M = k\sqrt{L_1L_2} \quad (0 \leq k \leq 1).$$

We derived Eq. (12.42) by assuming that i_1 and i_2 are either both positive or both negative. However, we get the same result if i_1 and i_2 are of opposite sign, because in this case we obtain the limiting value of M by selecting the plus sign in Eq. (12.39).

DRILL EXERCISES

12.5 The self-inductances of the coils in Fig. 12.13 are $L_1 =$ 5 mH and $L_2 = 33.8$ mH. If the coefficient of coupling is 0.96, calculate the energy stored in the system in millijoules when (a) $i_1 = 10$ A, $i_2 = 5$ A; (b) $i_1 = -10$ A, $i_2 = -5$ A; (c) $i_1 = -10$ A, $i_2 = 5$ A; and (d) $i_1 = 10$ A, $i_2 = -5$ A.

ANSWER: (a) 1296.50 mJ; (b) 1296.50 mJ; (c) 48.50 mJ; (d) 48.50 mJ.

12.6 The coefficient of coupling in Drill Exercise 12.5 is increased to 1.0.

a) If i_1 equals 10 A, what value of i_2 results in zero stored energy?

b) Is there any physically realizable value of i_2 that can make the stored energy negative?

ANSWER: (a) −3.846 A; (b) no.

12.5 THE LINEAR TRANSFORMER

Two coils form a simple **transformer** when they are wound on a single core to ensure magnetic coupling. We now analyze the sinusoidal steady-state response of a linear transformer.

Figure 12.14 shows the basic circuit using a transformer as a coupling device. The problem is to determine how the transformer affects the relationship between the load and the source. We show how the transformer affects steady-state sinusoidal currents and voltages using a frequency-domain circuit model of the source, transformer, and load.

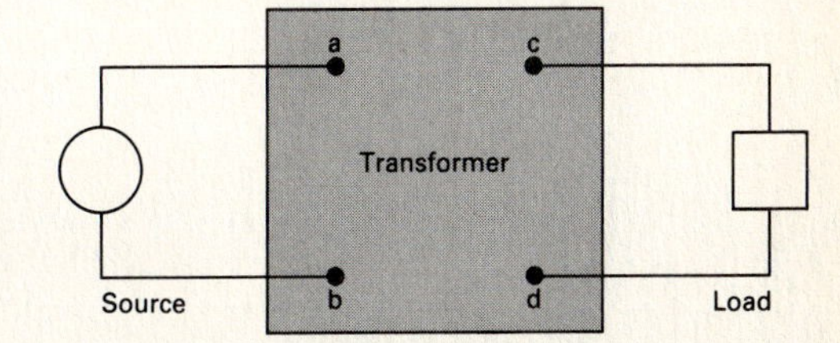

FIGURE 12.14 The basic arrangement of a transformer used to connect a load to a source.

THE ANALYSIS OF A LINEAR TRANSFORMER CIRCUIT

Figure 12.15 shows the frequency-domain circuit model of the system in Fig. 12.14. In discussing this circuit, we refer to the transformer winding connected to the source as the **primary winding** and the winding connected to the load as the **secondary winding.** Based on this terminology, the transformer circuit parameters are

$R_1 =$ the resistance of the primary winding,

$R_2 =$ the resistance of the secondary winding,

$L_1 =$ the self-inductance of the primary winding,

$L_2 =$ the self-inductance of the secondary winding, and

$M =$ the mutual inductance.

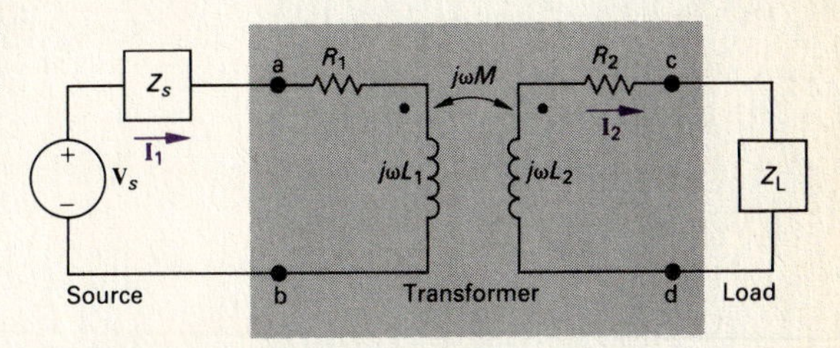

FIGURE 12.15 The frequency-domain circuit model for the system of Fig. 12.14.

The internal voltage of the sinusoidal source is $\mathbf{V}_s$, and the internal impedance of the source is Z_s. The impedance Z_L represents the load connected to the secondary winding of the transformer. The phasor currents $\mathbf{I}_1$ and $\mathbf{I}_2$ represent the primary and secondary currents of the transformer, respectively.

The analysis of the circuit in Fig. 12.15 consists of finding $\mathbf{I}_1$ and $\mathbf{I}_2$ as functions of the circuit parameters $\mathbf{V}_s$, Z_s, R_1, L_1, L_2, R_2, M, Z_L, and ω. We are also interested in finding the impedance seen looking into the transformer from the terminals a,b. To find $\mathbf{I}_1$ and $\mathbf{I}_2$, we first write the two mesh-current equations that describe the circuit:

$$\mathbf{V}_s = (Z_s + R_1 + j\omega L_1)\mathbf{I}_1 - j\omega M\mathbf{I}_2 \tag{12.43}$$

and

$$0 = -j\omega M\mathbf{I}_1 + (R_2 + j\omega L_2 + Z_\text{L})\mathbf{I}_2. \tag{12.44}$$

To facilitate the algebraic manipulation of Eqs. (12.43) and (12.44), we let

$$Z_{11} = Z_s + R_1 + j\omega L_1 \tag{12.45}$$

and

$$Z_{22} = R_2 + j\omega L_2 + Z_\text{L}, \tag{12.46}$$

where Z_{11} is the total self-impedance of the mesh containing the primary winding of the transformer, and Z_{22} is the total self-impedance of the mesh containing the secondary winding. Based on the notation introduced in Eqs. (12.45) and (12.46), the solutions for $\mathbf{I}_1$ and $\mathbf{I}_2$ from Eqs. (12.43) and (12.44) are

$$\mathbf{I}_1 = \frac{Z_{22}}{Z_{11}Z_{22} + \omega^2 M^2}\mathbf{V}_s \tag{12.47}$$

and

$$\mathbf{I}_2 = \frac{j\omega M}{Z_{11}Z_{22} + \omega^2 M^2}\mathbf{V}_s = \frac{j\omega M}{Z_{22}}\mathbf{I}_1. \tag{12.48}$$

To the internal source voltage $\mathbf{V}_s$, the impedance appears as $\mathbf{V}_s/\mathbf{I}_1$, or

$$\frac{\mathbf{V}_s}{\mathbf{I}_1} = Z_\text{int} = \frac{Z_{11}Z_{22} + \omega^2 M^2}{Z_{22}} = Z_{11} + \frac{\omega^2 M^2}{Z_{22}}. \tag{12.49}$$

The impedance at the terminals of the source is $Z_\text{int} - Z_s$, so

$$\begin{aligned} Z_\text{ab} &= Z_{11} + \frac{\omega^2 M^2}{Z_{22}} - Z_s \\ &= R_1 + j\omega L_1 + \frac{\omega^2 M^2}{(R_2 + j\omega L_2 + Z_\text{L})}. \end{aligned} \tag{12.50}$$

Note that the impedance Z_ab is independent of the magnetic polarity of the transformer. The reason is that the mutual inductance appears in Eq. (12.50) as a squared quantity. This impedance is of particular interest because it shows how the transformer affects the impedance of the load as seen from the source. Without the transformer, the load would be connected directly to the source, and the source would see a load impedance of Z_L; with the transformer, the load is connected to the source through the transformer, and the source sees a load impedance which is a modified version of Z_L, as seen in the third term of Eq. (12.50).

REFLECTED IMPEDANCE

The third term in Eq. (12.50) is called the **reflected impedance** (Z_r), because it is the equivalent impedance of the secondary coil and load impedance transmitted, or reflected, to the primary side of the transformer. Note that the reflected impedance is due solely to the existence of mutual inductance; that is, if the two coils are decoupled, M becomes zero, Z_r becomes zero, and Z_{ab} reduces to the self-impedance of the primary coil.

To consider reflected impedance in more detail, we first express the load impedance in rectangular form:

$$Z_L = R_L + jX_L, \tag{12.51}$$

where the load reactance X_L carries its own algebraic sign. In other words, X_L is a positive number if the load is inductive, and a negative number if the load is capacitive. We now use Eq. (12.51) to write the reflected impedance in rectangular form:

$$\begin{aligned} Z_r &= \frac{\omega^2 M^2}{R_2 + R_L + j(\omega L_2 + X_L)} \\ &= \frac{\omega^2 M^2[(R_2 + R_L) - j(\omega L_2 + X_L)]}{(R_2 + R_L)^2 + (\omega L_2 + X_L)^2} \\ &= \frac{\omega^2 M^2}{|Z_{22}|^2}[(R_2 + R_L) - j(\omega L_2 + X_L)]. \end{aligned} \tag{12.52}$$

The derivation of Eq. (12.52) takes advantage of the fact that, when Z_L is written in rectangular form, the self-impedance of the mesh containing the secondary winding is

$$Z_{22} = R_2 + R_L + j(\omega L_2 + X_L). \tag{12.53}$$

Now observe from Eq. (12.52) that the self-impedance of the secondary circuit is reflected into the primary circuit by a scaling factor of $(\omega M/|Z_{22}|)^2$ and that the sign of the reactive component $(\omega L_2 + X_L)$ is reversed. Thus the linear transformer reflects the conjugate of the self-impedance of the secondary circuit (Z_{22}^*) into the primary winding by a scalar multiplier. Before describing Eq. (12.52) further, we use Example 12.2 to illustrate the results discussed so far.

EXAMPLE 12.2

The parameters of a certain linear transformer are $R_1 = 200\ \Omega$, $R_2 = 100\ \Omega$, $L_1 = 9$ H, $L_2 = 4$ H, and $k = 0.5$. The transformer couples an impedance consisting of an 800 Ω resistor in series with a 1 μF capacitor to a sinusoidal voltage source. The 300 V (rms) source has an internal impedance of $500 + j100\ \Omega$ and a frequency of 400 rad/s.

a) Construct a frequency-domain equivalent circuit of the system.

b) Calculate the self-impedance of the primary circuit.

c) Calculate the self-impedance of the secondary circuit.

d) Calculate the impedance reflected into the primary winding.

e) Calculate the scaling factor for the reflected impedance.

f) Calculate the impedance seen looking into the primary terminals of the transformer.

g) Calculate the rms value of the primary and secondary current.

h) Calculate the rms value of the voltage at the terminals of the load and source.

i) Calculate the average power delivered to the 800 Ω resistor.

j) What percentage of the average power delivered to the transformer is delivered to the load?

SOLUTION

a) Figure 12.16 shows the frequency-domain equivalent circuit. Note that the internal voltage of the source serves as the reference phasor and that $\mathbf{V}_1$ and $\mathbf{V}_2$ represent the terminal voltages of the transformer. In constructing the circuit in Fig. 12.16, we made the following calculations:

$$j\omega L_1 = j(400)(9) = j3600\ \Omega;$$
$$j\omega L_2 = j(400)(4) = j1600\ \Omega;$$
$$M = 0.5\sqrt{(9)(4)} = 3\ \text{H};$$
$$j\omega M = j(400)(3) = j1200\ \Omega;$$
$$\frac{1}{j\omega C} = \frac{10^6}{j400} = -j2500\ \Omega.$$

b) The self-impedance of the primary circuit is

$$Z_{11} = 500 + j100 + 200 + j3600 = 700 + j3700\ \Omega.$$

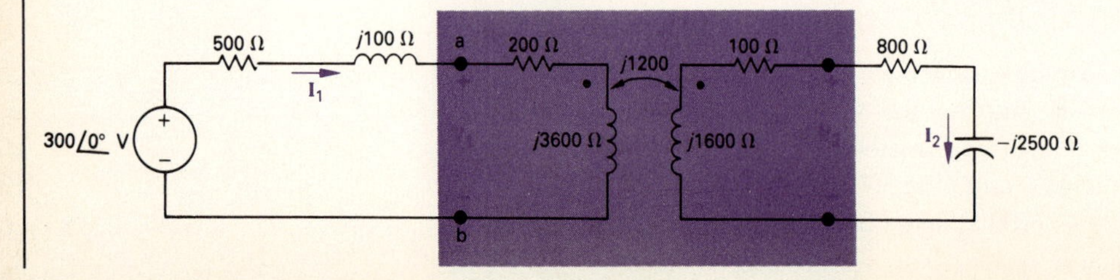

FIGURE 12.16 The frequency-domain equivalent circuit for Example 12.2.

c) The self-impedance of the secondary circuit is

$$Z_{22} = 100 + j1600 + 800 - j2500 = 900 - j900\ \Omega.$$

d) The impedance reflected into the primary winding is

$$Z_r = \left(\frac{1200}{|900 - j900|}\right)^2 (900 + j900)$$
$$= \frac{8}{9}(900 + j900) = 800 + j800\ \Omega.$$

e) The scaling factor by which Z_{22}^* is reflected is 8/9.

f) The impedance seen looking into the primary terminals of the transformer is the impedance of the primary winding plus the reflected impedance; thus

$$Z_{ab} = 200 + j3600 + 800 + j800 = 1000 + j4400\ \Omega.$$

g) Knowing the input impedance to the transformer, we can easily calculate $\mathbf{I}_1$:

$$\mathbf{I}_1 = \frac{\mathbf{V}_s}{Z_s + Z_{ab}} = \frac{300\angle 0^\circ}{1500 + j4500}$$
$$= 20 - j60 = 63.25\angle -71.57^\circ \text{ mA (rms)}.$$

We calculate $\mathbf{I}_2$ from Eq. (12.48):

$$\mathbf{I}_2 = \frac{j(1200)}{900 - j900}\mathbf{I}_1 = 59.63\angle 63.43^\circ \text{ mA (rms)}.$$

h) The voltages at the terminals of the transformer are

$$\mathbf{V}_2 = (800 - j2500)\mathbf{I}_2 = 156.52\angle -8.82^\circ \text{ V (rms)}$$

and

$$\mathbf{V}_1 = Z_{ab}\mathbf{I}_1$$
$$= (1000 + j4400)\mathbf{I}_1$$
$$= 285.38\angle 5.63^\circ \text{ V (rms)}.$$

i) The average power delivered to the load is

$$P = (800)|\mathbf{I}_2|^2 = 2.84 \text{ W}.$$

j) The average power delivered to the transformer is

$$P_{ab} = (1000)|\mathbf{I}_1|^2 = 4.00 \text{ W};$$

therefore

$$\eta = \frac{2.84}{4.00} \times 100 = 71.11\%.$$

Use a computer tool to verify results like these.

Exploring Limiting Values

A useful relationship between the input impedance and load impedance emerges as L_1 and L_2 each become infinitely large and, at the same time, the coefficient of coupling approaches unity. Transformers wound on ferromagnetic cores can approach this condition. Even though such transformers are nonlinear, we can obtain some useful information by constructing an ideal model that ignores the nonlinearities.

A computer tool can realize the resulting ideal model.

To show how Z_{ab} changes when $k = 1$ and L_1 and L_2 approach infinity, we first introduce the notation

$$Z_{22} = R_2 + R_{\text{L}} + j(\omega L_2 + X_{\text{L}}) = R_{22} + jX_{22}$$

and then rearrange Eq. (12.50):

$$\begin{aligned} Z_{\text{ab}} &= R_1 + \frac{\omega^2 M^2 R_{22}}{R_{22}^2 + X_{22}^2} + j\left(\omega L_1 - \frac{\omega^2 M^2 X_{22}}{R_{22}^2 + X_{22}^2}\right) \\ &= R_{\text{ab}} + jX_{\text{ab}}. \end{aligned} \quad \textbf{(12.54)}$$

At this point, we must be careful with the coefficient of j in Eq. (12.54) because, as L_1 and L_2 approach infinity, this coefficient is the difference between two large quantities. Thus before letting L_1 and L_2 increase, we write the coefficient as

$$X_{\text{ab}} = \omega L_1 - \frac{(\omega L_1)(\omega L_2)X_{22}}{R_{22}^2 + X_{22}^2} = \omega L_1\left(1 - \frac{\omega L_2 X_{22}}{R_{22}^2 + X_{22}^2}\right), \quad \textbf{(12.55)}$$

where we recognize that, when $k = 1$, $M^2 = L_1 L_2$. Putting the term multiplying ωL_1 over a common denominator gives

$$X_{\text{ab}} = \omega L_1\left(\frac{R_{22}^2 + \omega L_2 X_{\text{L}} + X_{\text{L}}^2}{R_{22}^2 + X_{22}^2}\right). \quad \textbf{(12.56)}$$

Factoring ωL_2 out of the numerator and denominator of Eq. (12.56) yields

$$X_{\text{ab}} = \frac{L_1}{L_2}\frac{X_{\text{L}} + (R_{22}^2 + X_{\text{L}}^2)/\omega L_2}{(R_{22}/\omega L_2)^2 + [1 + (X_{\text{L}}/\omega L_2)]^2}. \quad \textbf{(12.57)}$$

As k approaches 1.0, the ratio L_1/L_2 approaches the constant value of $(N_1/N_2)^2$, which follows from Eqs. (12.22) and (12.23). The reason is that, as the coupling becomes extremely tight, the two permeances $\mathcal{P}_1$ and $\mathcal{P}_2$ become equal. Equation (12.57) then reduces to

$$X_{\text{ab}} = \left(\frac{N_1}{N_2}\right)^2 X_{\text{L}}, \quad \textbf{(12.58)}$$

as $L_1 \to \infty$, $L_2 \to \infty$, and $k \to 1.0$.

The same reasoning leads to simplification of the reflected resistance in Eq. (12.54):

$$\frac{\omega^2 M^2 R_{22}}{R_{22}^2 + X_{22}^2} = \frac{L_1}{L_2} R_{22} = \left(\frac{N_1}{N_2}\right)^2 R_{22}. \quad \textbf{(12.59)}$$

Applying the results given by Eqs. (12.58) and (12.59) to Eq. (12.54) yields

$$Z_{\text{ab}} = R_1 + \left(\frac{N_1}{N_2}\right)^2 R_2 + \left(\frac{N_1}{N_2}\right)^2 (R_{\text{L}} + jX_{\text{L}}). \quad \textbf{(12.60)}$$

Compare this result to the result in Eq. (12.50). Here we see that when the coefficient of coupling approaches unity and the self-inductances of the coupled coils approach infinity, the transformer reflects the secondary winding resistance and the load impedance to the primary side by a scaling factor equal to the turns ratio (N_1/N_2) squared. By lumping the winding resistance R_2 with the load, we can use this special type of transformer to raise or lower a load's impedance level. In Section 12.6 we expand on this desirable feature by defining the ideal transformer.

DRILL EXERCISE

12.7 A linear transformer couples a load consisting of a 360 Ω resistor in series with a 0.25 H inductor to a sinusoidal voltage source, as shown. The voltage source has an internal impedance of $184 + j0\ \Omega$ and a maximum voltage of 245.20 V, and it is operating at 800 rad/s. The transformer parameters are $R_1 = 100\ \Omega$, $L_1 = 0.5$ H, $R_2 = 40\ \Omega$, $L_2 = 0.125$ H, and $k = 0.4$. Calculate (a) the reflected impedance; (b) the primary current; (c) the secondary current; and (d) the average power delivered to the primary terminals of the transformer.

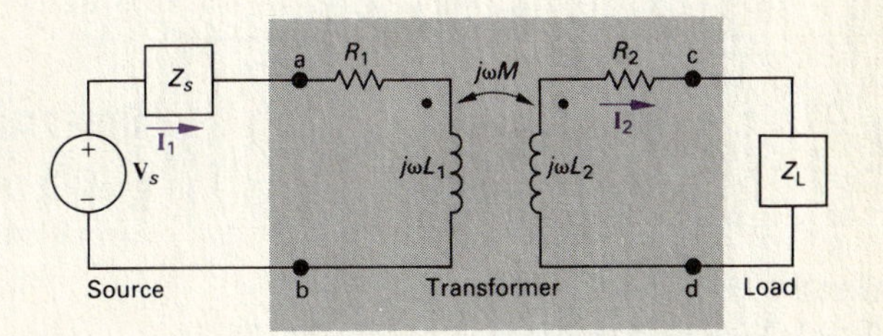

ANSWER: (a) $10.24 - j7.68\ \Omega$; (b) $0.5 \cos(800t - 53.13°)$ A; (c) $0.08 \cos 800t$ A; (d) 13.78 W.

12.6 THE IDEAL TRANSFORMER

An **ideal transformer** consists of two magnetically coupled coils having N_1 and N_2 turns, respectively, and exhibiting these three properties:

1. The coefficient of coupling is unity ($k = 1$).
2. The self-inductance of each coil is infinite ($L_1 = L_2 = \infty$).
3. The coil losses, due to parasitic resistance, are negligible.

Hence we may describe the terminal behavior of the ideal transformer in terms of two characteristics. First, the magnitude of the volts per turn is the same for each coil, or

$$\left|\frac{v_1}{N_1}\right| = \left|\frac{v_2}{N_2}\right|. \tag{12.61}$$

Second, the magnitude of the ampere-turns is the same for each coil, or

$$|i_1 N_1| = |i_2 N_2|. \tag{12.62}$$

We are forced to use magnitude signs in Eqs. (12.61) and (12.62) because we have not yet established reference polarities for the currents and voltages; we discuss the removal of the magnitude signs shortly.

Figure 12.17 shows two lossless ($R_1 = R_2 = 0$) magnetically coupled coils. We use Fig. 12.17 to validate Eqs. (12.61) and (12.62). In Fig. 12.17(a), coil 2 is open; in Fig. 12.17(b), coil 2 is shorted. Although we carry out the following analysis in terms of sinusoidal steady-state operation, the results also apply to instantaneous values of v and i.

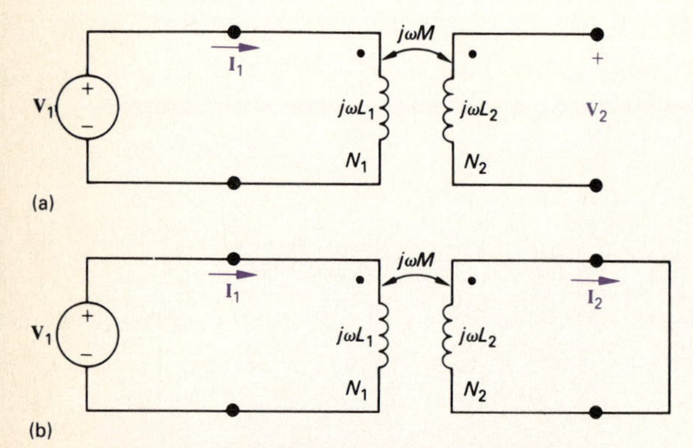

FIGURE 12.17 The circuits used to verify the volts-per-turn and ampere-turn relationships for an ideal transformer.

DETERMINING THE VOLTAGE AND CURRENT RATIOS

Note in Fig. 12.17(a) that the voltage at the terminals of the open-circuit coil is entirely the result of the current in coil 1; therefore

$$\mathbf{V}_2 = j\omega M \mathbf{I}_1. \tag{12.63}$$

The current in coil 1 is

$$\mathbf{I}_1 = \frac{\mathbf{V}_1}{j\omega L_1}. \tag{12.64}$$

From Eqs. (12.63) and (12.64),

$$\mathbf{V}_2 = \frac{M}{L_1}\mathbf{V}_1. \tag{12.65}$$

For unity coupling, the mutual inductance equals $\sqrt{L_1 L_2}$, so Eq. (12.65) becomes

$$\mathbf{V}_2 = \sqrt{\frac{L_2}{L_1}}\mathbf{V}_1. \tag{12.66}$$

For unity coupling, the flux linking winding 1 is the same as the flux linking winding 2, so we need only one permeance to describe the self-inductance of each winding. Thus Eq. (12.66) becomes

$$\mathbf{V}_2 = \sqrt{\frac{N_2^2\mathcal{P}}{N_1^2\mathcal{P}}}\mathbf{V}_1 = \frac{N_2}{N_1}\mathbf{V}_1 \tag{12.67}$$

or

$$\frac{\mathbf{V}_1}{N_1} = \frac{\mathbf{V}_2}{N_2}. \tag{12.68}$$

Summing the voltages around the shorted coil of Fig. 12.17(b) yields

$$0 = -j\omega M\mathbf{I}_1 + j\omega L_2\mathbf{I}_2, \tag{12.69}$$

from which, for $k = 1$,

$$\frac{\mathbf{I}_1}{\mathbf{I}_2} = \frac{L_2}{M} = \frac{L_2}{\sqrt{L_1L_2}} = \sqrt{\frac{L_2}{L_1}} = \frac{N_2}{N_1}. \tag{12.70}$$

Equation (12.70) is equivalent to

$$\mathbf{I}_1N_1 = \mathbf{I}_2N_2. \tag{12.71}$$

Figure 12.18 shows the graphic symbol for an ideal transformer. The vertical lines in the symbol represent the layers of magnetic material from which ferromagnetic cores are often made. Thus, the symbol reminds us that coils wound on a ferromagnetic core behave very much like an ideal transformer.

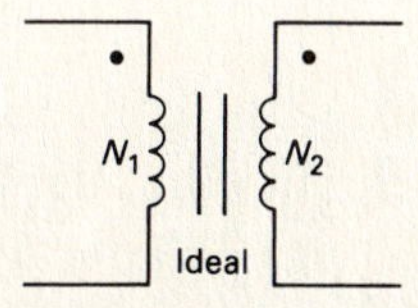

FIGURE 12.18 The graphic symbol for an ideal transformer.

There are several reasons for this. The ferromagnetic material creates a space with high permeance. Thus most of the magnetic flux is trapped inside the core material, establishing tight magnetic coupling between coils that share the same core. High permeance also means high self-inductance, because $L = N^2\mathcal{P}$. Finally, ferromagnetically coupled coils efficiently transfer power from one coil to the other. Efficiencies in excess of 95% are common, so neglecting losses is not a crippling approximation for many applications.

Determining the Polarity of the Voltage and Current Ratios

We now turn to the removal of the magnitude signs from Eqs. (12.61) and (12.62). Note that magnitude signs did not show up in the derivations of Eqs. (12.68) and (12.71). We did not need them there because we had established reference polarities for voltages and reference directions for currents. In addition, we knew the magnetic polarity dots of the two coupled coils.

The rules for assigning the proper algebraic sign to Eqs. (12.61) and (12.62) are as follows:

1. If the coil voltages v_1 and v_2 are both positive or negative at the dot-marked terminal, use a plus sign in Eq. (12.61). Otherwise, use a negative sign.

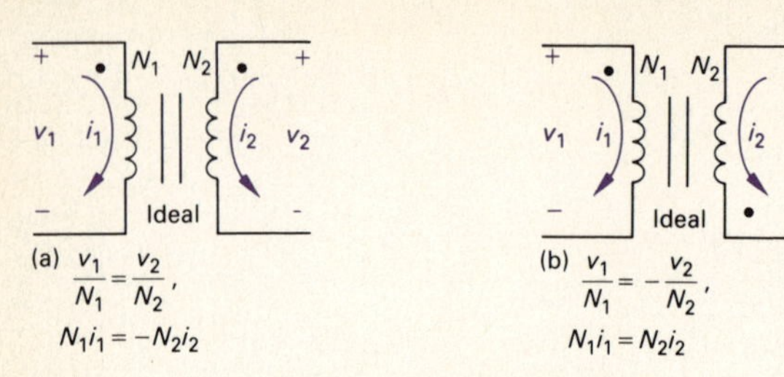

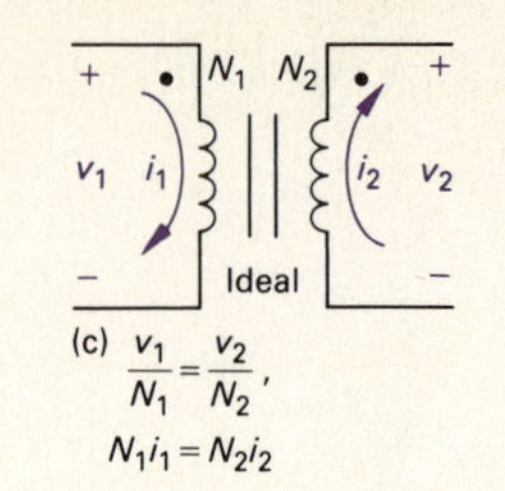

(d) $\frac{v_1}{N_1} = -\frac{v_2}{N_2}$, $N_1 i_1 = -N_2 i_2$

FIGURE 12.19 Circuits that show the proper algebraic signs for relating the terminal voltages and currents of an ideal transformer.

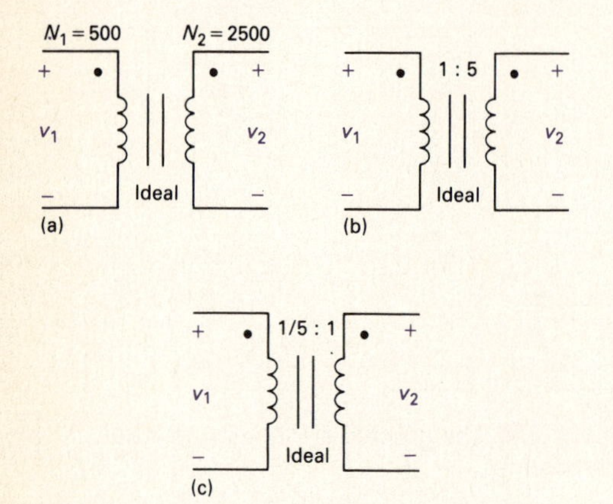

FIGURE 12.20 Three ways to show that the turns ratio of an ideal transformer is 5.

2. If the coil currents i_1 and i_2 are both directed into or out of the dot-marked terminal, use a minus sign in Eq. (12.62). Otherwise, use a plus sign.

The four circuits shown in Fig. 12.19 illustrate these rules.

The ratio of the turns on the two windings is an important parameter of the ideal transformer. The turns ratio is defined as either N_1/N_2 or N_2/N_1; both ratios appear in various writings. In this text, we use a to denote the ratio N_2/N_1, or

$$a = \frac{N_2}{N_1}. \qquad \textbf{(12.72)}$$

Figure 12.20 shows three ways to represent the turns ratio of an ideal transformer. Figure 12.20(a) shows the number of turns in each coil explicitly. Figure 12.20(b) shows that the ratio N_2/N_1 is 5 to 1, and Fig. 12.20(c) shows that the ratio N_2/N_1 is 1 to $\frac{1}{5}$.

Example 12.3 illustrates the analysis of a circuit containing an ideal transformer.

EXAMPLE 12.3

a) Find the average power delivered by the sinusoidal current source in the circuit shown in Fig. 12.21.

b) Find the average power delivered to the 20 Ω resistor.

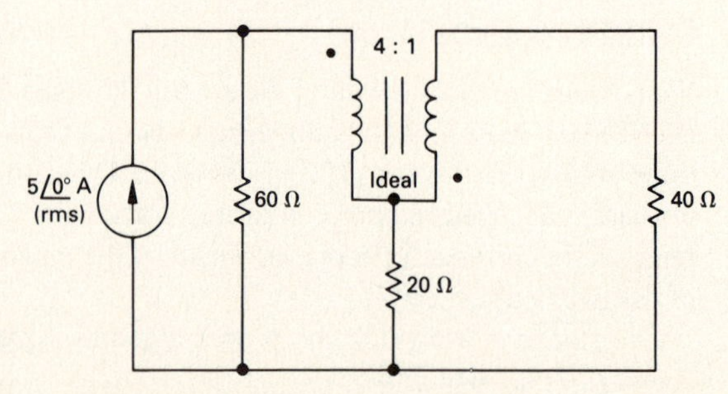

FIGURE 12.21 The circuit for Example 12.3.

SOLUTION

a) The ideal transformer encourages use of the mesh-current method of analysis. Therefore we begin by transforming the current source to an equivalent voltage source. Figure 12.22 shows the transformation, along with the phasor mesh currents and transformer terminal voltages used in the solution. Summing the voltages around meshes 1 and 2 generates the equations

$$300 = 60\mathbf{I}_1 + \mathbf{V}_1 + 20(\mathbf{I}_1 - \mathbf{I}_2)$$

and

$$0 = 20(\mathbf{I}_2 - \mathbf{I}_1) + \mathbf{V}_2 + 40\mathbf{I}_2.$$

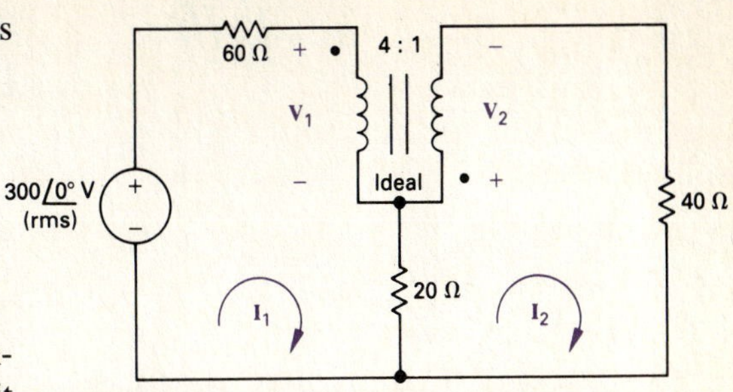

FIGURE 12.22 The circuit for the solution of Example 12.3.

In addition to these two mesh-current equations, we need the constraint equations imposed by the ideal transformer. From the circuit diagram, $N_2/N_1 = 1/4$. In terms of the currents and voltages defined at the terminals of the ideal transformer,

$$\mathbf{V}_2 = \tfrac{1}{4}\mathbf{V}_1 \quad \text{and} \quad \mathbf{I}_2 = -4\mathbf{I}_1.$$

We now have four equations and four unknowns. The solutions for $\mathbf{V}_1$, $\mathbf{V}_2$, $\mathbf{I}_1$, and $\mathbf{I}_2$ are

$$\mathbf{V}_1 = 260 \text{ V (rms)};$$
$$\mathbf{V}_2 = 65 \text{ V (rms)};$$
$$\mathbf{I}_1 = 0.25 \text{ A (rms)};$$
$$\mathbf{I}_2 = -1.0 \text{ A (rms)}.$$

Use a computer tool to verify these solutions.

The voltage across the 5 A current source in the circuit in Fig. 12.21 is

$$\begin{aligned}\mathbf{V}_{5A} &= \mathbf{V}_1 + 20(\mathbf{I}_1 - \mathbf{I}_2)\\ &= 260 + 20[0.25 - (-1)]\\ &= 285 \text{ V (rms)}.\end{aligned}$$

Note that $\mathbf{V}_{5A}$ is a voltage rise in the direction of the source current. The average power associated with the current source is

$$P = -(285)(5) = -1425 \text{ W}.$$

The minus sign means that the source is delivering power to the circuit.

b) To find the average power delivered to the 20 Ω resistor, we first calculate the current in the resistor. From the circuit shown in Fig. 12.22, the current oriented down through the 20 Ω resistor is

$$\begin{aligned}\mathbf{I}_{20\Omega} &= \mathbf{I}_1 - \mathbf{I}_2 = 0.25 - (-1)\\ &= 1.25 \text{ A (rms)}.\end{aligned}$$

Therefore the average power dissipated in the resistor is

$$P_{20\Omega} = (1.25)^2(20) = 31.25 \text{ W}.$$

THE USE OF AN IDEAL TRANSFORMER FOR IMPEDANCE MATCHING

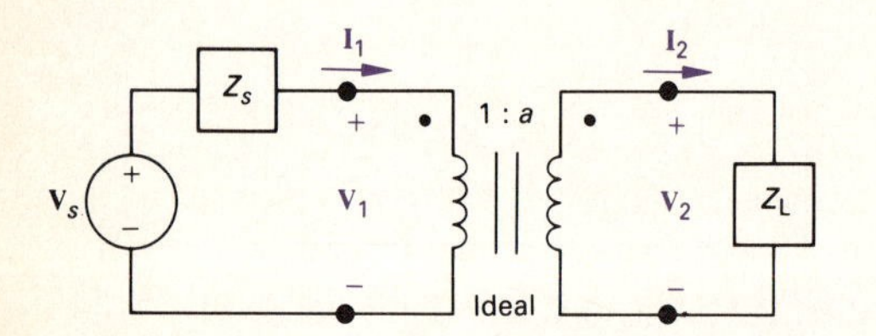

FIGURE 12.23 Using an ideal transformer to couple a load to a source.

At the end of Section 12.5, we implied that an ideal transformer can be used to raise or lower the impedance level of a load. The circuit shown in Fig. 12.23 confirms this fact. The impedance seen by the practical voltage source ($\mathbf{V}_s$ in series with $\mathbf{Z}_s$) is $\mathbf{V}_1/\mathbf{I}_1$. The voltage and current at the terminals of the load impedance ($\mathbf{V}_2$ and $\mathbf{I}_2$) are related to $\mathbf{V}_1$ and $\mathbf{I}_1$ by the transformer turns ratio; thus

$$\mathbf{V}_1 = \frac{\mathbf{V}_2}{a} \quad (12.73)$$

and

$$\mathbf{I}_1 = a\mathbf{I}_2. \quad (12.74)$$

Therefore the impedance seen by the practical source is

$$Z_{\text{IN}} = \frac{\mathbf{V}_1}{\mathbf{I}_1} = \frac{1}{a^2}\frac{\mathbf{V}_2}{\mathbf{I}_2}, \quad (12.75)$$

but the ratio $\mathbf{V}_2/\mathbf{I}_2$ is the load impedance Z_L, so Eq. (12.75) becomes

$$Z_{\text{IN}} = \frac{1}{a^2}Z_L. \quad (12.76)$$

Thus, the ideal transformer's secondary coil reflects the load impedance back to the primary coil, with the scaling factor $1/a^2$.

Note that the ideal transformer changes the magnitude of Z_L but does not affect its phase angle. Whether Z_{IN} is greater or less than Z_L depends on the turns ratio a.

The ideal transformer—or its practical counterpart, the ferromagnetic core transformer—can be used to match the magnitude of Z_L to the magnitude of Z_s, thus maximizing the amount of average power transferred from the source to the load.

DRILL EXERCISES

12.8 Make the following changes in the ideal transformer in the circuit shown in Fig. 12.21: (1) Place the dot on the right-hand side of the transformer at the upper terminal, and (2) change the turns ratio from 4:1 to 3:1. Calculate the average power delivered to the 20 Ω resistor.

ANSWER: 28.8 W.

12.9 Find the average power delivered to the 4 kΩ resistor in the circuit shown.

ANSWER: 160 W.

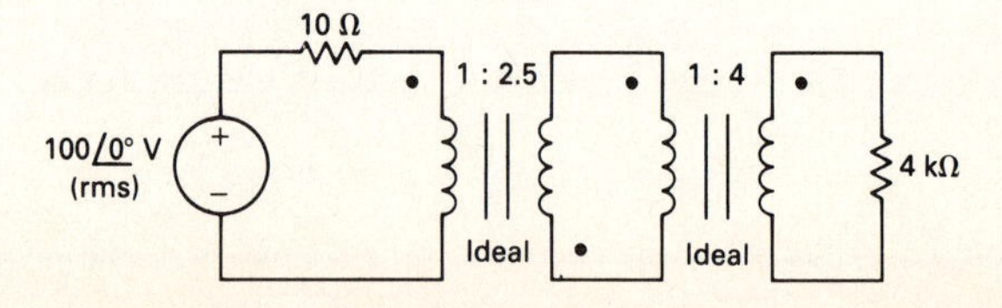

12.10 The ideal transformer connected to the 4 kΩ load in Drill Exercise 12.9 is replaced with an ideal transformer that has a turns ratio of $1{:}a$.

a) What value of a results in maximum average power being delivered to the 4 kΩ resistor?

b) What is the maximum average power?

ANSWER: (a) 8; (b) 250 W.

12.7 EQUIVALENT CIRCUITS FOR MAGNETICALLY COUPLED COILS

At times, it is convenient to model magnetically coupled coils with an equivalent circuit that does not involve magnetic coupling. Consider the two magnetically coupled coils shown in Fig. 12.24. The resistances R_1 and R_2 represent the winding resistance of each coil. The goal is to replace the magnetically coupled coils inside the shaded area with a set of inductors that are not magnetically coupled. Before deriving the equivalent circuits, we must point out an important restriction: The voltage between terminals b and d must be zero. In other words, if terminals b and d can be shorted together without disturbing the voltages and currents in the original circuit, the equivalent circuits derived in the material that follows can be used to model the coils (see Problem 12.43). This restriction is imposed because, while the equivalent circuits we develop both have four terminals, two of those four terminals are shorted together. Thus, the same requirement is placed on the original circuits.

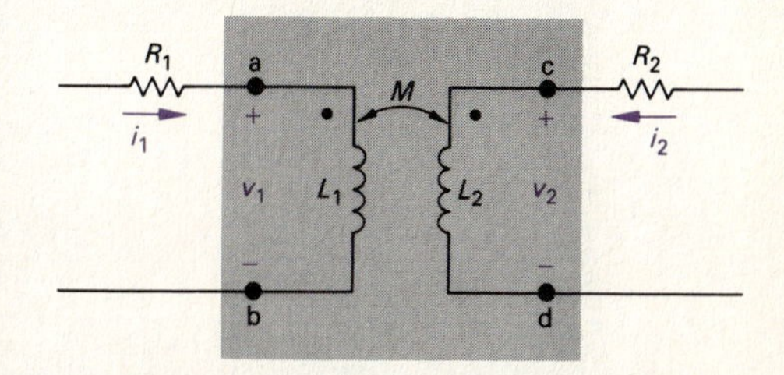

FIGURE 12.24 The circuit used to develop an equivalent circuit for magnetically coupled coils.

We begin developing the circuit models by writing the two equations that relate the terminal voltages v_1 and v_2 to the terminal currents i_1 and i_2. For the given references and polarity dots,

$$v_1 = L_1\frac{di_1}{dt} + M\frac{di_2}{dt} \tag{12.77}$$

and

$$v_2 = M\frac{di_1}{dt} + L_2\frac{di_2}{dt}. \tag{12.78}$$

THE T-EQUIVALENT CIRCUIT

To arrive at an equivalent circuit for these two magnetically coupled coils, we seek an arrangement of inductors that can be described by a set of equations equivalent to Eqs. (12.77) and (12.78). The key to finding the arrangement is to regard Eqs. (12.77) and (12.78) as mesh-current equations with i_1 and i_2 as the mesh variables. Then we

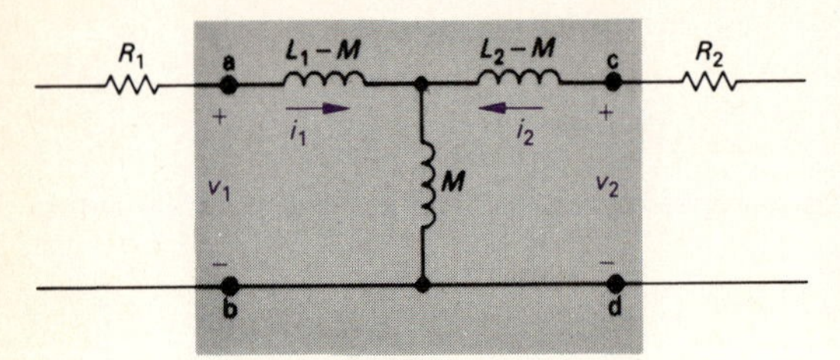

FIGURE 12.25 The T-equivalent circuit for the magnetically coupled coils of Fig. 12.24.

need one mesh with a total inductance of L_1 H and a second mesh with a total inductance of L_2 H. Furthermore, the two meshes must have a common inductance of M H. The T-arrangement of coils shown in Fig. 12.25 satisfies these requirements. You should verify that the equations relating v_1 and v_2 to i_1 and i_2 reduce to Eqs. (12.77) and (12.78). Note the absence of magnetic coupling between the inductors and the zero voltage between b and d.

The π-Equivalent Circuit

We can derive a π-equivalent circuit for the magnetically coupled coils shown in Fig. 12.24. This derivation is based on solving Eqs. (12.77) and (12.78) for the derivatives di_1/dt and di_2/dt and then regarding the resulting expressions as a pair of node-voltage equations. Using Cramer's method for solving simultaneous equations, we obtain expressions for di_1/dt and di_2/dt:

$$\frac{di_1}{dt} = \frac{\begin{vmatrix} v_1 & M \\ v_2 & L_2 \end{vmatrix}}{\begin{vmatrix} L_1 & M \\ M & L_2 \end{vmatrix}} = \frac{L_2}{L_1L_2 - M^2}v_1 - \frac{M}{L_1L_2 - M^2}v_2; \quad \textbf{(12.79)}$$

$$\frac{di_2}{dt} = \frac{\begin{vmatrix} L_1 & v_1 \\ M & v_2 \end{vmatrix}}{L_1L_2 - M^2} = \frac{-M}{L_1L_2 - M^2}v_1 + \frac{L_1}{L_1L_2 - M^2}v_2. \quad \textbf{(12.80)}$$

Now we solve for i_1 and i_2 by multiplying both sides of Eqs. (12.79) and (12.80) by dt and then integrating:

$$i_1 = i_1(0) + \frac{L_2}{L_1L_2 - M^2}\int_0^t v_1 d\tau - \frac{M}{L_1L_2 - M^2}\int_0^t v_2 d\tau \quad \textbf{(12.81)}$$

and

$$i_2 = i_2(0) - \frac{M}{L_1L_2 - M^2}\int_0^t v_1 d\tau + \frac{L_1}{L_1L_2 - M^2}\int_0^t v_2 d\tau \quad \textbf{(12.82)}$$

If we regard v_1 and v_2 as node voltages, Eqs. (12.81) and (12.82) describe a circuit of the form shown in Fig. 12.26.

All that remains to be done in deriving the π-equivalent circuit is to find L_A, L_B, and L_C as functions of L_1, L_2, and M. We easily do so by writing the equations for i_1 and i_2 in Fig. 12.26 and then comparing them with Eqs. (12.81) and (12.82). Thus

$$i_1 = i_1(0) + \frac{1}{L_A}\int_0^t v_1 d\tau + \frac{1}{L_B}\int_0^t (v_1 - v_2)d\tau$$

$$= i_1(0) + \left(\frac{1}{L_A} + \frac{1}{L_B}\right)\int_0^t v_1 d\tau - \frac{1}{L_B}\int_0^t v_2 d\tau \quad \textbf{(12.83)}$$

and

$$
\begin{aligned}
i_2 &= i_2(0) + \frac{1}{L_C}\int_0^t v_2 d\tau + \frac{1}{L_B}\int_0^t (v_2 - v_1) d\tau \\
&= i_2(0) - \frac{1}{L_B}\int_0^t v_1 d\tau + \left(\frac{1}{L_B} + \frac{1}{L_C}\right)\int_0^t v_2 d\tau. \quad \textbf{(12.84)}
\end{aligned}
$$

Then

$$\frac{1}{L_B} = \frac{M}{L_1 L_2 - M^2}, \quad \textbf{(12.85)}$$

$$\frac{1}{L_A} = \frac{L_2 - M}{L_1 L_2 - M^2}, \quad \textbf{(12.86)}$$

$$\frac{1}{L_C} = \frac{L_1 - M}{L_1 L_2 - M^2}. \quad \textbf{(12.87)}$$

When we incorporate Eqs. (12.85)–(12.87) into the circuit shown in Fig. 12.26, the π-equivalent circuit for the magnetically coupled coils shown in Fig. 12.24 is as shown in Fig. 12.27.

Note that the initial values of i_1 and i_2 are explicit in the π-equivalent circuit but implicit in the T-equivalent circuit. In this chapter we are focusing on the sinusoidal steady-state behavior of circuits containing mutual inductance, so we can assume that the initial values of i_1 and i_2 are zero. We can thus eliminate the current sources in the π-equivalent circuit, and the circuit shown in Fig. 12.27 simplifies to the one shown in Fig. 12.28.

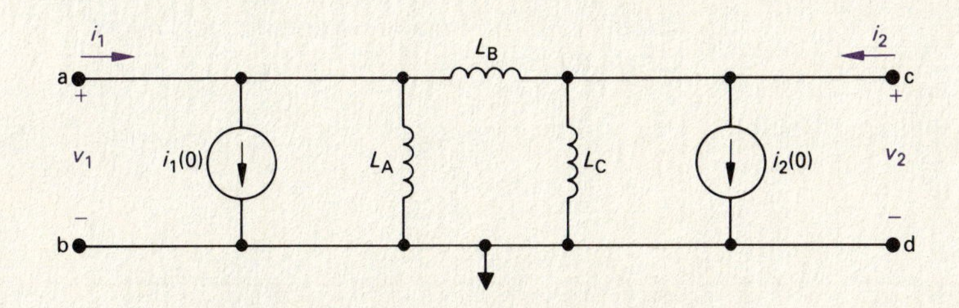

FIGURE 12.26 The circuit used to derive the π-equivalent circuit for magnetically coupled coils.

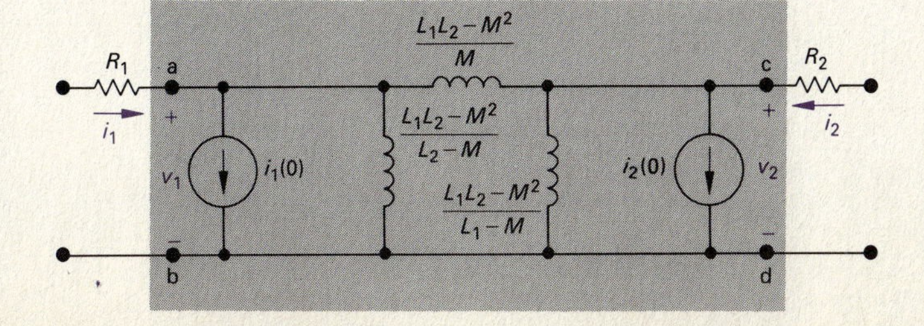

FIGURE 12.27 The π-equivalent circuit for the magnetically coupled coils of Fig. 12.24.

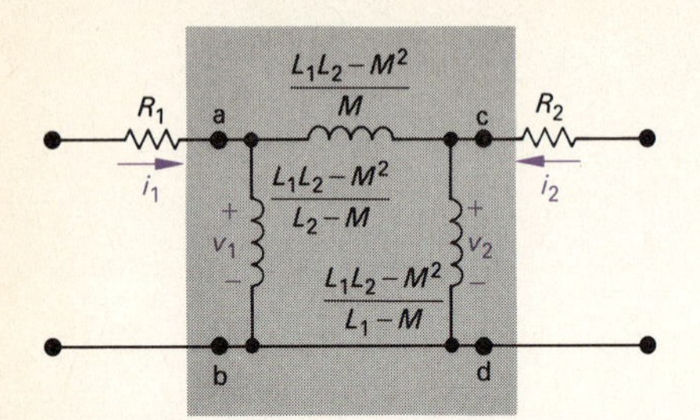

FIGURE 12.28 The π-equivalent circuit used for sinusoidal steady-state analysis.

The mutual inductance carries its own algebraic sign in the T- and π-equivalent circuits. In other words, if the magnetic polarity of the coupled coils is reversed from that given in Fig. 12.24, the algebraic sign of M reverses. A reversal in magnetic polarity requires moving one polarity dot without changing the reference polarities of the terminal currents and voltages.

Example 12.4 illustrates the application of the T-equivalent circuit.

EXAMPLE 12.4

a) Use the T-equivalent circuit for the magnetically coupled coils described in Example 12.2 to find the phasor currents $\mathbf{I}_1$ and $\mathbf{I}_2$. These currents are defined in the circuit shown in Fig. 12.29.

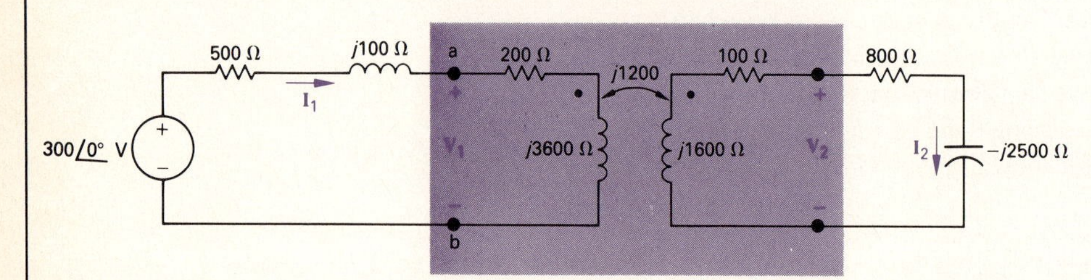

FIGURE 12.29 The frequency-domain equivalent circuit for Example 12.4.

b) Repeat (a), but with the polarity dot on the secondary winding moved to the lower terminal.

SOLUTION

a) For the polarity dots shown in Fig. 12.29, M carries a value of $+3$ H in the T-equivalent circuit. Therefore the three inductances in the equivalent circuit are

$$L_1 - M = 9 - 3 = 6 \text{ H};$$

$$L_2 - M = 4 - 3 = 1 \text{ H};$$

$$M = 3 \text{ H}.$$

Figure 12.30 shows the T-equivalent circuit, and Fig. 12.31 shows the frequency-domain equivalent circuit at a frequency of 400 rad/s.

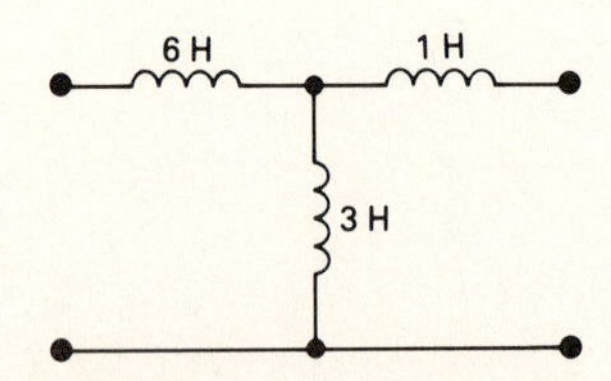

FIGURE 12.30 The T-equivalent circuit for the magnetically coupled coils in Example 12.2.

Figure 12.32 shows the frequency-domain circuit for the original system. Here the magnetically coupled coils are modeled by the circuit shown in Fig. 12.31. To find the phasor currents $\mathbf{I}_1$ and $\mathbf{I}_2$, we first find the node voltage across the 1200 Ω inductive reactance. If we use the lower node as the reference, the single node-voltage equation is

$$\frac{\mathbf{V}-300}{700+j2500}+\frac{\mathbf{V}}{j1200}+\frac{\mathbf{V}}{900-j2100}=0.$$

Solving for $\mathbf{V}$ yields

$$\mathbf{V}=136-j8=136.24\angle{-3.37^\circ}\text{ V (rms)}.$$

Then

$$\mathbf{I}_1=\frac{300-(136-j8)}{700+j2500}=63.25\angle{-71.57^\circ}\text{ mA (rms)}$$

and

$$\mathbf{I}_2=\frac{136-j8}{900-j2100}=59.63\angle{63.43^\circ}\text{ mA (rms)}.$$

Checking these results against those obtained in Example 12.2 shows that they are identical.

b) When the polarity dot is moved to the lower terminal of the secondary coil, M carries a value of -3 H in the T-equivalent circuit. Before carrying out the solution with the new T-equivalent circuit, we note from Example 12.2 that reversing the algebraic sign of M has no effect on the solution for $\mathbf{I}_1$ and shifts $\mathbf{I}_2$ by 180°. Therefore we anticipate that

$$\mathbf{I}_1=63.25\angle{-71.57^\circ}\text{ mA (rms)}$$

and

$$\mathbf{I}_2=59.63\angle{-116.57^\circ}\text{ mA (rms)}.$$

We now proceed to find these solutions by using the new T-equivalent circuit. With $M=-3$ H, the three inductances in the equivalent circuit are

$$L_1-M=9-(-3)=12\text{ H};$$
$$L_2-M=4-(-3)=7\text{ H};$$
$$M=-3\text{ H}.$$

At an operating frequency of 400 rad/s, the frequency-domain equivalent circuit requires two inductors and a capacitor, as shown in Fig. 12.33. The resulting frequency-domain circuit for the original system appears in Fig. 12.34. As before, we first find the node

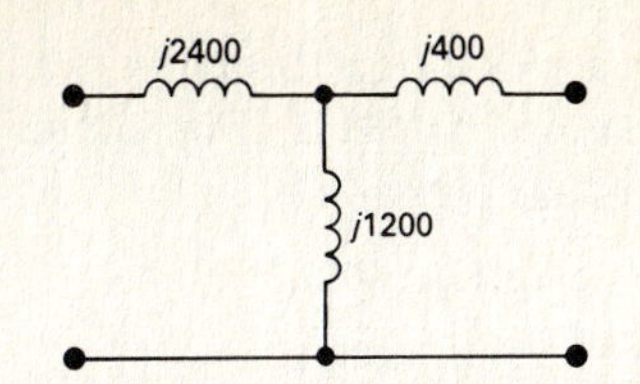

FIGURE 12.31 The frequency-domain model of the equivalent circuit at 400 rad/s.

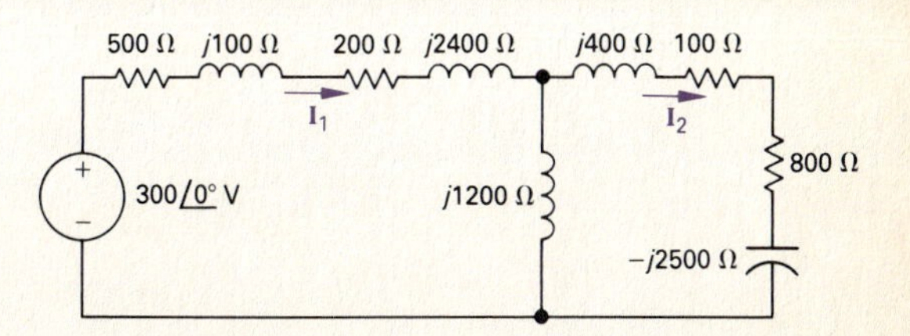

FIGURE 12.32 The circuit of Fig. 12.29, with the magnetically coupled coils replaced by their T-equivalent circuit.

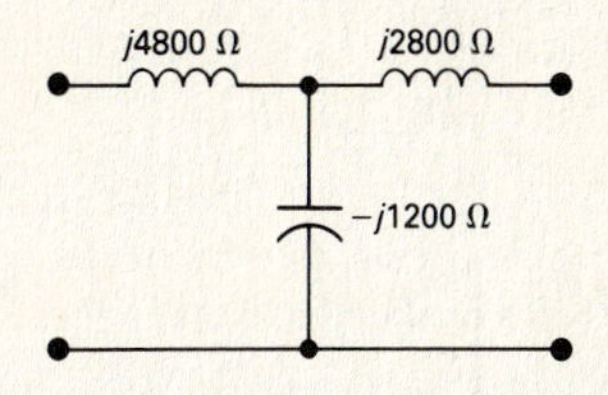

FIGURE 12.33 The frequency-domain equivalent circuit for $M=-3$ H and $\omega=400$ rad/s.

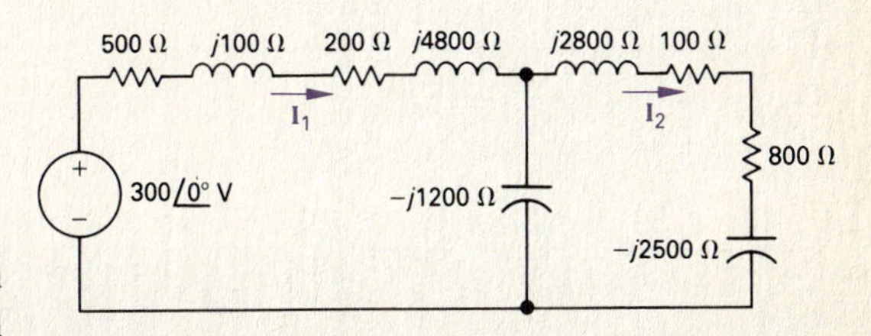

FIGURE 12.34 The frequency-domain equivalent circuit for Example 12.4(b).

voltage across the center branch, which in this case is a capacitive reactance of $-j1200\ \Omega$. If we use the lower node as reference, the node-voltage equation is

$$\frac{\mathbf{V}-300}{700+j4900}+\frac{\mathbf{V}}{-j1200}+\frac{\mathbf{V}}{900+j300}=0.$$

Solving for **V** gives

$$\mathbf{V}=-8-j56$$
$$=56.57\underline{/-98.13^\circ}\text{ V (rms).}$$

Then

$$\mathbf{I}_1=\frac{300-(-8-j56)}{700+j4900}$$
$$=63.25\underline{/-71.57^\circ}\text{ mA (rms)}$$

and

$$\mathbf{I}_2=\frac{-8-j56}{900+j300}$$
$$=59.63\underline{/-116.57^\circ}\text{ mA (rms).}$$

DRILL EXERCISES

12.11 a) Show that, if the reference direction for i_2 is reversed in both circuits shown here, the T-equivalent circuit shown on the right is still valid.

b) Show that if the dot on the L_2 coil in the circuit shown on the left is moved to the lower terminal, the three inductors in the circuit shown on the right are L_1+M, L_2+M, and $-M$.

c) Will the circuit derived in (b) be valid if the reference direction for i_2 is reversed in both circuits?

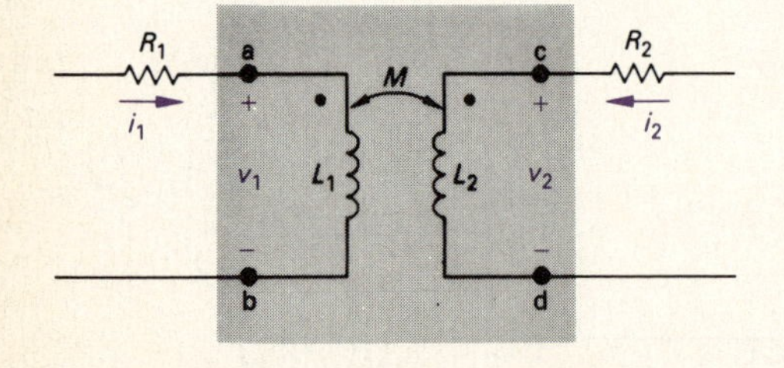

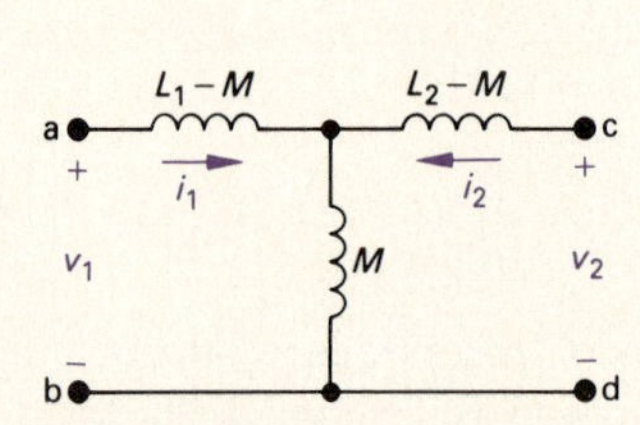

ANSWER: (a) Show; (b) show; (c) yes.

12.12 Use the T-equivalent circuit for the linear transformer in Drill Exercise 12.7 to find $\mathbf{I}_1$ and $\mathbf{I}_2$. The phasor currents $\mathbf{I}_1$ and $\mathbf{I}_2$ are defined as shown.

ANSWER: $\mathbf{I}_1 = 500\underline{/-53.13^\circ}$ mA; $\mathbf{I}_2 = 80\underline{/0^\circ}$ mA.

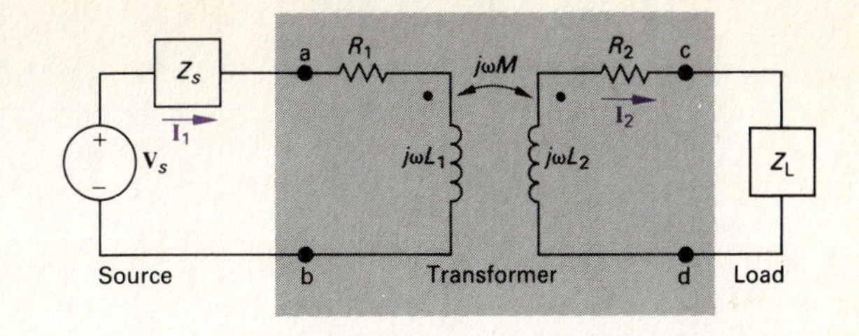

12.8 THE NEED FOR IDEAL TRANSFORMERS IN THE EQUIVALENT CIRCUITS

The inductors in the T- and π-equivalent circuits of magnetically coupled coils can have negative values. For example, if $L_1 = 3$ mH, $L_2 = 12$ mH, and $M = 5$ mH, the T-equivalent circuit requires an inductor of -2 mH, and the π-equivalent circuit requires an inductor of -5.5 mH. These negative inductance values are not troublesome when you are using the equivalent circuits in computations. However, if you are to build the equivalent circuits with circuit components, the negative inductors can be bothersome. The reason is that whenever the frequency of the sinusoidal source changes, you must change the capacitor used to simulate the negative reactance. For example, at a frequency of 50 krad/s, a -2 mH inductor has an impedance of $-j100\ \Omega$. This impedance can be modeled with a capacitor having a capacitance of 0.2 μF. If the frequency changes to 25 krad/s, the -2 mH inductor impedance changes to $-j50\ \Omega$. At 25 krad/s, this requires a capacitor with a capacitance of 0.8 μF. Obviously, in a situation where the frequency is varied continuously, the use of a capacitor to simulate negative inductance is practically worthless.

You can circumvent the problem of dealing with negative inductances by introducing an ideal transformer into the equivalent circuit. This doesn't completely solve the modeling problem, because ideal transformers can only be approximated. However, in some situations the approximation is good enough to warrant a discussion of using an ideal transformer in the T- and π-equivalent circuits of magnetically coupled coils.

An ideal transformer can be used in two different ways in either the T-equivalent or the π-equivalent circuit. Figure 12.35 shows the two arrangements for each type of equivalent circuit.

Verifying any of the equivalent circuits in Fig. 12.35 requires showing only that, for any circuit, the equations relating v_1 and v_2 to di_1/dt and di_2/dt are identical to Eqs. (12.77) and (12.78). Here, we validate

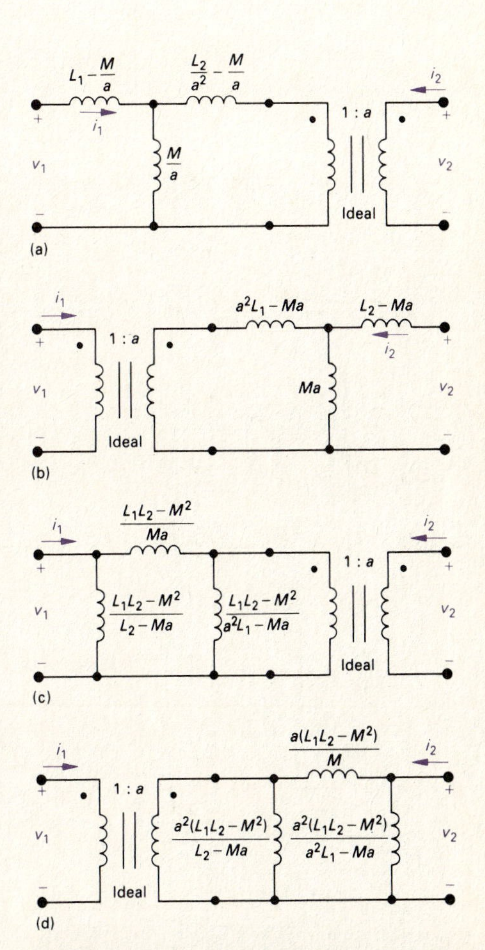

FIGURE 12.35 The four ways of using an ideal transformer in the T- and π-equivalent circuit for magnetically coupled coils.

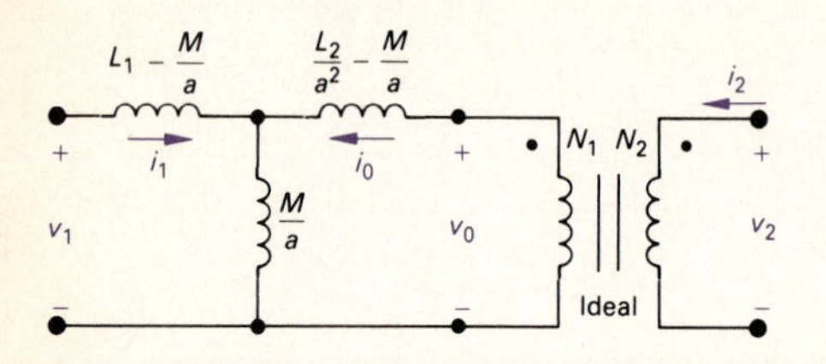

FIGURE 12.36 The circuit of Fig. 12.35(a) with i_0 and v_0 defined.

the circuit shown in Fig. 12.35(a); we leave it to you to verify the circuits in Figs. 12.35(b), (c), and (d). To aid the discussion, we redrew the circuit shown in Fig. 12.35(a) as Fig. 12.36, adding the variables i_0 and v_0. From this circuit,

$$v_1 = \left(L_1 - \frac{M}{a}\right)\frac{di_1}{dt} + \frac{M}{a}\frac{d}{dt}(i_1 + i_0) \qquad \textbf{(12.88)}$$

and

$$v_0 = \left(\frac{L_2}{a^2} - \frac{M}{a}\right)\frac{di_0}{dt} + \frac{M}{a}\frac{d}{dt}(i_0 + i_1). \qquad \textbf{(12.89)}$$

The ideal transformer imposes constraints on v_0 and i_0:

$$v_0 = \frac{v_2}{a}; \qquad \textbf{(12.90)}$$

$$i_0 = ai_2. \qquad \textbf{(12.91)}$$

Substituting Eqs. (12.90) and (12.91) into Eqs. (12.88) and (12.89) gives

$$v_1 = L_1\frac{di_1}{dt} + \frac{M}{a}\frac{d}{dt}(ai_2) \qquad \textbf{(12.92)}$$

and

$$\frac{v_2}{a} = \frac{L_2}{a^2}\frac{d}{dt}(ai_2) + \frac{M}{a}\frac{di_1}{dt}. \qquad \textbf{(12.93)}$$

From Eqs. (12.92) and (12.93),

$$v_1 = L_1\frac{di_1}{dt} + M\frac{di_2}{dt} \qquad \textbf{(12.94)}$$

and

$$v_2 = M\frac{di_1}{dt} + L_2\frac{di_2}{dt}. \qquad \textbf{(12.95)}$$

Equations (12.95) and (12.96) are identical to Eqs. (12.77) and (12.78); thus, insofar as terminal behavior is concerned, the circuit shown in Fig. 12.36 is equivalent to the magnetically coupled coils shown inside the box in Fig. 12.24.

In showing that the circuit in Fig. 12.36 is equivalent to the magnetically coupled coils in Fig. 12.24, we placed no restrictions on the turns ratio a. Therefore, an infinite number of equivalent circuits are possible. Furthermore, we can always find a turns ratio to make all the inductances positive. Three values of a are of particular interest:

$$a = \frac{M}{L_1}, \qquad \textbf{(12.96)}$$

$$a = \frac{L_2}{M}, \qquad \textbf{(12.97)}$$

and

$$a = \sqrt{\frac{L_2}{L_1}}. \tag{12.98}$$

The value of a given by Eq. (12.96) eliminates the inductances $L_1 - M/a$ and $a^2L_1 - aM$ from the T-equivalent circuits and the inductances $(L_1L_2 - M^2)/(a^2L_1 - aM)$ and $a^2(L_1L_2 - M^2)/(a^2L_1 - aM)$ from the π-equivalent circuits. The value of a given by Eq. (12.97) eliminates the inductances $(L_2/a^2) - (M/a)$ and $L_2 - aM$ from the T-equivalent circuits and the inductances $(L_1L_2 - M^2)/(L_2 - aM)$ and $a^2(L_1L_2 - M^2)/(L_2 - aM)$ from the π-equivalent circuits.

Also note that when $a = M/L_1$, the circuits in Figs. 12.35(a) and (c) become identical, and when $a = L_2/M$, the circuits in Figs. 12.35(b) and (d) become identical. Figures 12.37 and 12.38 summarize these observations. In deriving the expressions for the inductances there, we used the relationship $M = k\sqrt{L_1L_2}$. Expressing the inductances as functions of the self-inductances L_1 and L_2 and the coefficient of coupling k allows the values of a given by Eqs. (12.96) and (12.97) not only to reduce the number of inductances needed in the equivalent circuit, but also to guarantee that all the inductances will be positive. We leave to you to investigate the consequences of choosing the value of a given by Eq. (12.98).

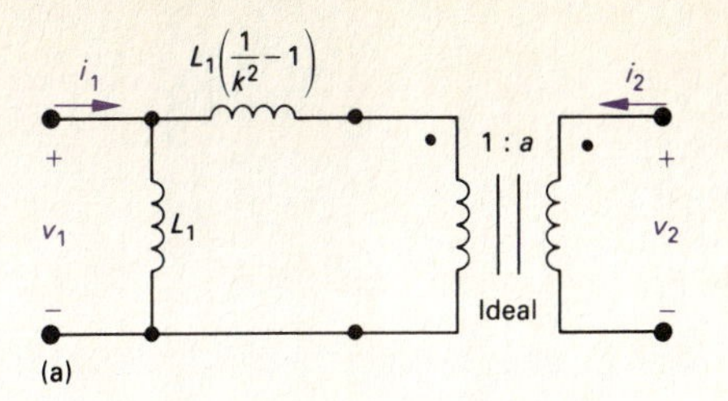

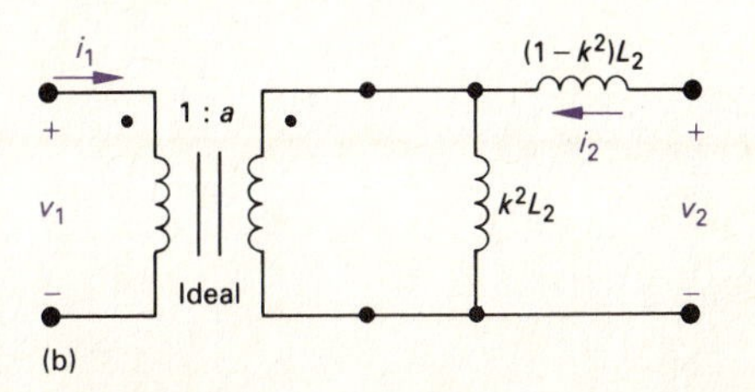

FIGURE 12.37 Two equivalent circuits when $a = M/L_1$.

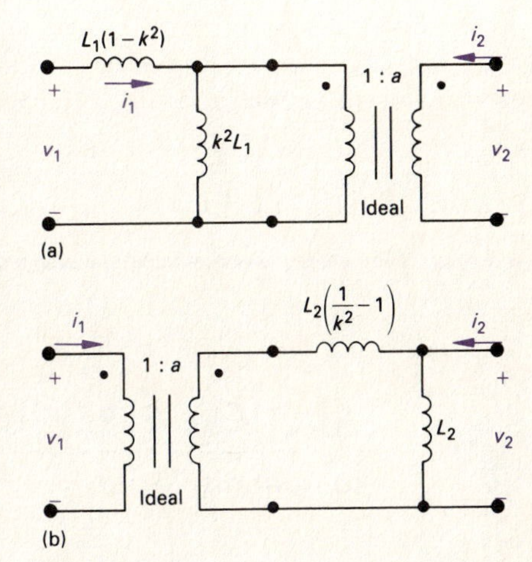

FIGURE 12.38 Two equivalent circuits when $a = L_2/M$.

The values of a given by Eqs. (12.96)–(12.98) can be determined experimentally. The ratio M/L_1 is obtained by driving the coil designated as having N_1 turns by a sinusoidal voltage source. The source frequency is set high enough that $\omega L_1 \gg R_1$, and the N_2 coil is left open. Figure 12.39 shows this arrangement.

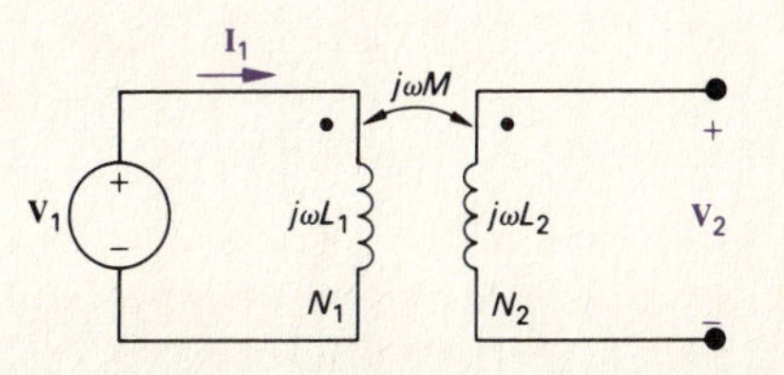

FIGURE 12.39 Experimental determination of the ratio M/L_1.

With the N_2 coil open,

$$\mathbf{V}_2 = j\omega M\mathbf{I}_1. \tag{12.99}$$

Now, as $j\omega L_1 \gg R_1$, the current $\mathbf{I}_1$ is

$$\mathbf{I}_1 = \frac{\mathbf{V}_1}{j\omega L_1}. \tag{12.100}$$

Substituting Eq. (12.100) into Eq. (12.99) yields

$$\left(\frac{\mathbf{V}_2}{\mathbf{V}_1}\right)_{I_2=0} = \frac{M}{L_1}, \tag{12.101}$$

in which the ratio M/L_1 is the terminal voltage ratio corresponding to coil 2 being open; that is, $\mathbf{I}_2 = 0$.

We obtain the ratio L_2/M by reversing the procedure; that is, coil 2 is energized and coil 1 is left open. Then

$$\frac{L_2}{M} = \left(\frac{\mathbf{V}_2}{\mathbf{V}_1}\right)_{I_1=0}. \tag{12.102}$$

Finally, we observe that the value of a given by Eq. (12.98) is the geometric mean of these two voltage ratios; thus

$$\sqrt{\left(\frac{\mathbf{V}_2}{\mathbf{V}_1}\right)_{I_2=0}\left(\frac{\mathbf{V}_2}{\mathbf{V}_1}\right)_{I_1=0}} = \sqrt{\frac{M}{L_1}\frac{L_2}{M}} = \sqrt{\frac{L_2}{L_1}}. \quad (12.103)$$

For coils wound on nonmagnetic cores, the voltage ratio is not the same as the turns ratio, as it very nearly is for coils wound on ferromagnetic cores. Because the self-inductances vary as the square of the number of turns, Eq. (12.103) reveals that the turns ratio is approximately equal to the geometric mean of the two voltage ratios, or

$$\sqrt{\frac{L_2}{L_1}} = \frac{N_2}{N_1} = \sqrt{\left(\frac{\mathbf{V}_2}{\mathbf{V}_1}\right)_{I_2=0}\left(\frac{\mathbf{V}_2}{\mathbf{V}_1}\right)_{I_1=0}}. \quad (12.104)$$

DRILL EXERCISE

12.13 The circuit shown is the equivalent circuit of a lossless linear transformer. Compute (a) L_1, (b) L_2, (c) M, and (d) k.

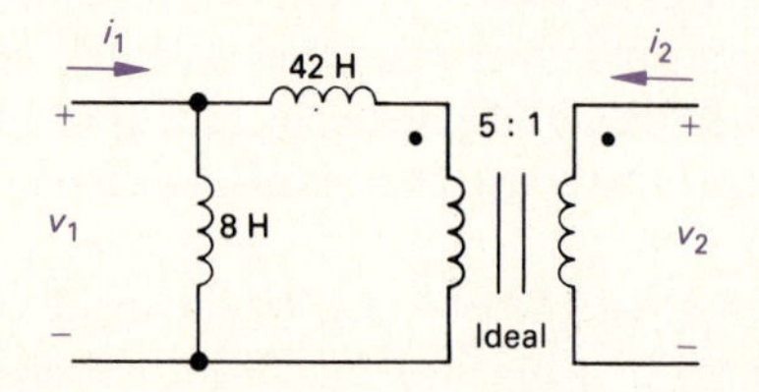

ANSWER: (a) 8 H; (b) 2 H; (c) 1.6 H; (d) 0.4.

PRACTICAL PERSPECTIVE

An Ignition Circuit

Now let us return to the conventional ignition system introduced at the beginning of the chapter. A circuit diagram of the system is shown in Fig. 12.40. Consider the circuit characteristics that provide the energy to ignite the fuel-air mixture in the cylinder. First, the maximum voltage available at the spark plug, v_{sp}, must be high enough to ignite the fuel. Second, the voltage across the capacitor must be limited to prevent arcing across the switch or distributor points. Third, the current in the primary winding of the autotransformer must cause sufficient energy to be stored in the system to ignite the fuel-air mixture in the cylinder. Remember that the energy stored in the circuit at the instant of switching is proportional to the primary current squared, that is, $\mathrm{w}_0 = \frac{1}{2}\mathrm{L}i^2(0)$.

EXAMPLE

a) Find the maximum voltage at the spark plug, assuming the following values in the circuit of Fig. 12.40: $V_{dc} = 12$ V, $R = 4\ \Omega$, $L = 3$ mH, $C = 0.4\ \mu$F, and $a = 100$.

b) What distance must separate the switch contacts to prevent arcing at the time the voltage at the spark plug is maximum?

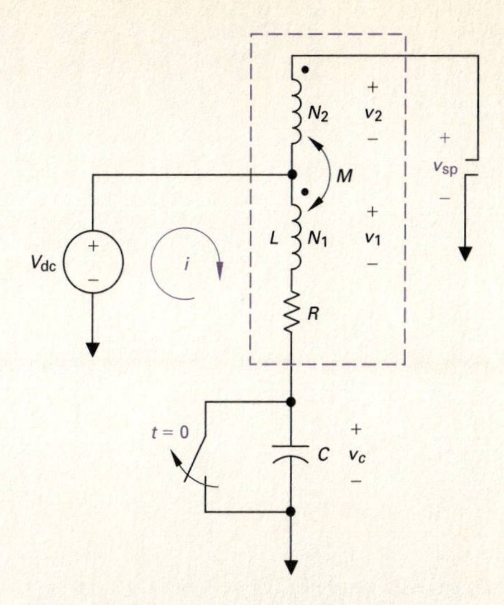

FIGURE 12.40 The circuit diagram of the conventional automobile ignition system.

SOLUTION

a) We analyze the circuit in Fig. 12.40 to find an expression for the spark plug voltage v_{sp}. We limit our analysis to a study of the voltages in the circuit prior to the firing of the spark plug. We assume that the current in the primary winding at the time of switching has its maximum possible value V_{dc}/R, where R is the total resistance in the primary circuit. We also assume that the ratio of the secondary voltage (v_2) to the primary voltage (v_1) is the same as the turns ratio N_2/N_1. We can justify this assumption as follows. With the secondary circuit open, the voltage induced in the secondary winding is

$$v_2 = M\frac{di}{dt}, \tag{12.105}$$

and the voltage induced in the primary winding is

$$v_1 = L\frac{di}{dt}. \tag{12.106}$$

It follows from Eqs. (12.105) and (12.106) that

$$\frac{v_2}{v_1} = \frac{M}{L}. \tag{12.107}$$

It is reasonable to assume that the permeance is the same for the fluxes ϕ_{11} and ϕ_{21} in the iron-core autotransformer; hence Eq. (12.107) reduces to

$$\frac{v_2}{v_1} = \frac{N_1 N_2 \mathcal{P}}{N_1^2 \mathcal{P}} = \frac{N_2}{N_1} = a. \tag{12.108}$$

We are now ready to analyze the voltages in the ignition circuit. The values of R, L, and C are such that when the switch is opened, the primary coil current response is underdamped. Using the techniques developed in Section 8.4 and assuming $t = 0$ at the instant the switch is opened, the expression for the primary coil current is found to be

$$i = \frac{V_{dc}}{\mathrm{R}} e^{-\alpha t}[\cos \omega_d t + \left(\frac{\alpha}{\omega_d}\right) \sin \omega_d t], \tag{12.109}$$

where

$$\alpha = \frac{R}{2L} \quad \text{and} \quad \omega_d = \sqrt{\frac{1}{\text{LC}} - \alpha^2}.$$

(See Problem 12.45[a].) The voltage induced in the primary winding of the autotransformer is

$$v_1 = L\frac{di}{dt} = \frac{-V_{\text{dc}}}{\omega_d RC}e^{-\alpha t}\sin\omega_d t. \quad \textbf{(12.110)}$$

(See Problem 12.45[b].) It follows from Eq. (12.108) that

$$v_2 = \frac{-aV_{\text{dc}}}{\omega_d RC}e^{-\alpha t}\sin\omega_d t. \quad \textbf{(12.111)}$$

The voltage across the capacitor can be derived either by using the relationship

$$v_c = \frac{1}{C}\int_0^t i\,dx + v_c(0) \quad \textbf{(12.112)}$$

or by summing the voltages around the mesh containing the primary winding:

$$v_c = V_{\text{dc}} - iR - L\frac{di}{dt}. \quad \textbf{(12.113)}$$

In either case we find

$$v_c = V_{\text{dc}}[1 - e^{-\alpha t}\cos\omega_d t + Ke^{-\alpha t}\sin\omega_d t], \quad \textbf{(12.114)}$$

where

$$K = \frac{1}{\omega_d}\left(\frac{1}{RC} - \alpha\right).$$

(See Problem 12.45[c].) As can be seen from Fig. 12.40, the voltage across the spark plug is

$$\begin{aligned} v_{\text{sp}} &= V_{\text{dc}} + v_2 \\ &= V_{\text{dc}} - \frac{aV_{\text{dc}}}{\omega_d RC}e^{-\alpha t}\sin\omega_d t \\ &= V_{\text{dc}}\left[1 - \frac{a}{\omega_d RC}e^{-\alpha t}\sin\omega_d t\right]. \end{aligned} \quad \textbf{(12.115)}$$

To find the maximum value of v_{sp}, we find the smallest positive value of time where dv_{sp}/dt is zero and then evaluate v_{sp} at this instant. The expression for t_{max} is

$$t_{\text{max}} = \frac{1}{\omega_d}\tan^{-1}\left(\frac{\omega_d}{\alpha}\right). \quad \textbf{(12.116)}$$

(See Problem 12.46.) For the component values in the problem statement, we have

$$\alpha = \frac{R}{2L} = \frac{4 \times 10^3}{6} = 666.67 \text{ rad/s}$$

and

$$\omega_d = \sqrt{\frac{10^9}{1.2} - (666.67)^2} = 28{,}859.81 \text{ rad/s}.$$

Substituting these values into Eq. (12.116) gives

$$t_{\max} = 53.63 \ \mu\text{s}.$$

Now use Eq. (12.115) to find the maximum spark plug voltage, $v_{sp}(t_{\max})$:

$$v_{sp}(t_{\max}) = -25{,}975.69 \text{ V}.$$

b) The voltage across the capacitor at $t_{\max}$ is obtained from Eq. (12.114) as

$$v_c(t_{\max}) = 262.15 \text{ V}.$$

The dielectric strength of air is approximately 3×10^6 V/m, so this result tells us that the switch contacts must be separated by $262.15/3 \times 10^6$, or 87.38, microns to prevent arcing at the points at $t_{\max}$. You can explore this circuit further by working Problems 12.45 through 12.48 at the end of the chapter.

In the design and testing of ignition systems, consideration must be given to nonuniform fuel-air mixtures; the widening of the spark plug gap over time due to the erosion of the plug electrodes; the relationship between available spark plug voltage and engine speed; the time it takes the primary current to build up to its initial value after the switch is closed; and the amount of maintenance required to ensure reliable operation.

We can use the preceding analysis of a conventional ignition system to explain why electronic switching has replaced mechanical switching in today's automobiles. First, the current emphasis on fuel economy and exhaust emissions requires a spark plug with a wider gap. This, in turn, requires a higher available spark plug voltage. These higher voltages (up to 40 kV) cannot be achieved with mechanical switching. Electronic switching also permits higher initial currents in the primary winding of the autotransformer. This means the initial stored energy in the system is larger, and hence a wider range of fuel-air mixtures and running conditions can be accommodated. Finally, the electronic switching circuit eliminates the need for the point contacts. This means the deleterious effects of point contact arcing can be removed from the system.

SUMMARY

- **Mutual inductance**, M, is the circuit parameter relating the voltage induced in one circuit to a time-varying current in another circuit. Specifically,

$$v_1 = L_1 \frac{di_1}{dt} + M_{12} \frac{di_2}{dt} \quad \text{and} \quad v_2 = M_{21} \frac{di_1}{dt} + L_2 \frac{di_2}{dt},$$

  where v_1 and i_1 are the voltage and current in circuit 1, and v_2 and i_2 are the voltage and current in circuit 2.

- For coils wound on nonmagnetic cores, $M_{12} = M_{21} = M$.

- The **coefficient of coupling**, k, is the measure of the degree of magnetic coupling. By definition, $0 \leq k \leq 1$.

- The relationship between the self-inductance of each winding and the mutual inductance between the windings is

$$M = k\sqrt{L_1 L_2}.$$

- The **dot convention** establishes the polarity of mutually induced voltages:

  > When the reference direction for a current enters the dotted terminal of a coil, the reference polarity of the voltage that it induces in the other coil is positive at its dotted terminal.

  Or, alternatively,

  > When the reference direction for a current leaves the dotted terminal of a coil, the reference polarity of the voltage that it induces in the other coil is negative at its dotted terminal.

- The energy stored in magnetically coupled coils in a linear medium is related to the coil currents and inductances by the relationship

$$w = \frac{1}{2}L_1 i_1^2 + \frac{1}{2}L_2 i_2^2 \pm M i_1 i_2.$$

- **Reflected impedance** is the impedance of the secondary circuit as seen from the terminals of the primary circuit, or vice versa.

- The two-winding **linear transformer** is a coupling device made up of two coils wound on the same nonmagnetic core. The reflected impedance of a linear transformer seen from the primary side is the conjugate of the self-impedance of the secondary circuit scaled by the factor $(\omega M/|Z_{22}|)^2$.

- The two-winding **ideal transformer** is a linear transformer with the following special properties: perfect coupling ($k = 1$); infinite self-inductance in each coil ($L_1 = L_2 = \infty$); and lossless coils ($R_1 = R_2 = 0$). The circuit behavior is governed by the turns ratio $a = N_2/N_1$. In particular, the volts per turn is the same for each winding, or

$$\frac{v_1}{N_1} = \pm\frac{v_2}{N_2},$$

and the ampere turns are the same for each winding, or

$$N_1 i_1 = \pm N_2 i_2.$$

- An ideal transformer can be used to match the magnitude of the load impedance, Z_L, to the magnitude of the source impedance, Z_s, thus maximizing the amount of average power transferred.
- The **reflected impedance** of an ideal transformer is defined quantitatively as follows:

  Looking into coil 1, reflected impedance is the impedance connected across coil 2 multiplied by $1/a^2$.

  or

  Looking into coil 2, reflected impedance is the impedance connected across coil 1 multiplied by a^2.

- Magnetically coupled coils can be replaced by three uncoupled coils arranged in either a T or π configuration. (See pg. 521 for restriction.) Ideal transformers are used in T- or π-equivalent circuits to eliminate the need for negative inductors.

PROBLEMS

12.1 Two magnetically coupled coils have self-inductances of 52 and 13 mH, respectively. The mutual inductance between the coils is 19.5 mH.

a) What is the coefficient of coupling?

b) For these two coils, what is the largest value that M can have?

c) Assume that the physical structure of these coupled coils is such that $\mathcal{P}_1 = \mathcal{P}_2$. What is the turns ratio N_1/N_2 if N_1 is the number of turns on the 52 mH coil?

12.2 The self-inductances of two magnetically coupled coils are 288 and 162 mH, respectively. The 288 mH coil has 1000 turns, and the coefficient of coupling between the coils is 1/3. The coupling medium is nonmagnetic. When coil 1 is excited with coil 2 open, the flux linking coil 1 is only 0.5 as large as the flux linking coil 2.

a) How many turns does coil 2 have?

b) What is the value of $\mathscr{P}_2$ in nanowebers per ampere?

c) What is the value of $\mathscr{P}_{11}$ in nanowebers per ampere?

d) What is the ratio (ϕ_{22}/ϕ_{12})?

12.3 The physical construction of four pairs of magnetically coupled coils is shown in Fig. P12.3. Assume that the magnetic flux is confined to the core material in each structure. Show two possible locations for the dot markings on each pair of coils.

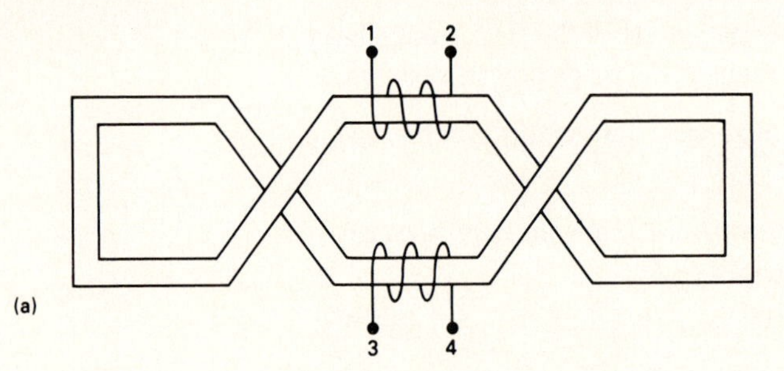

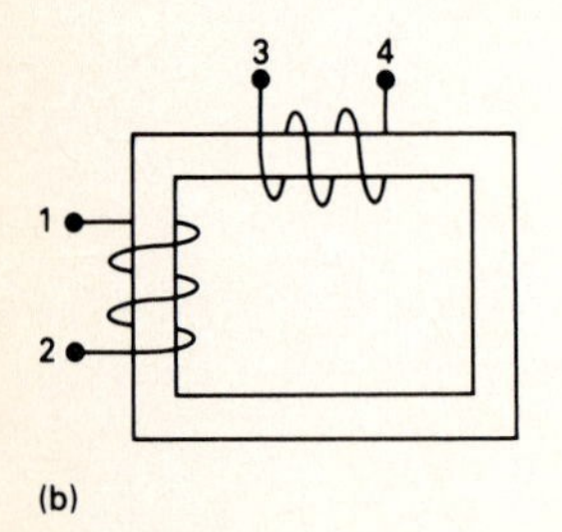

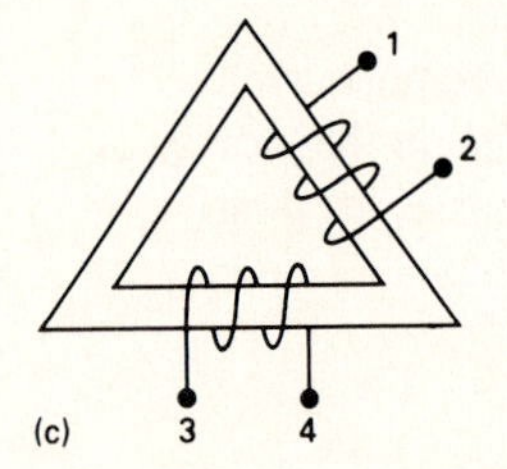

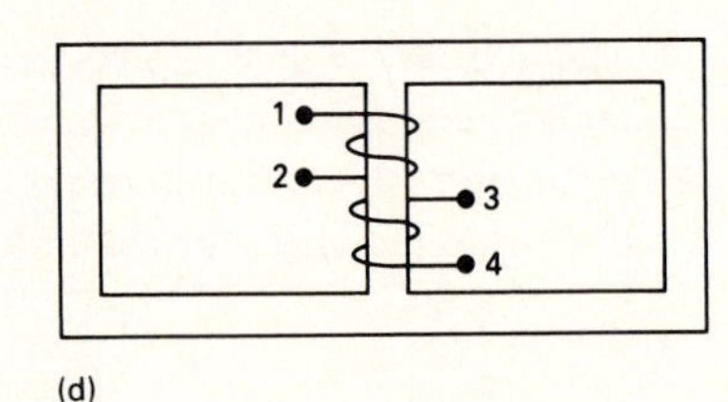

FIGURE P12.3

12.4 The polarity markings on two coils are to be determined experimentally. The experimental setup is shown in Fig. 12.4. Assume that the terminal connected to the negative terminal of the battery has been given a polarity mark as shown. When the switch is *closed,* the dc voltmeter kicks upscale. Where should the polarity mark be placed on the coil connected to the voltmeter?

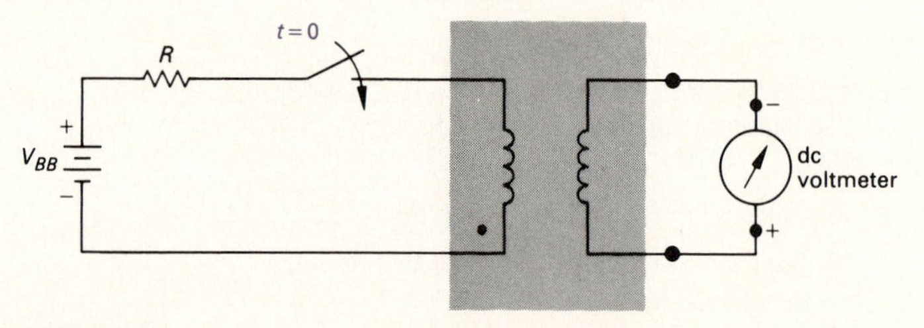

FIGURE P12.4

12.5 a) Starting with Eq. (12.27), show that the coefficient of coupling can also be expressed as

$$k = \sqrt{\left(\frac{\phi_{21}}{\phi_1}\right)\left(\frac{\phi_{12}}{\phi_2}\right)}.$$

b) On the basis of the fractions ϕ_{21}/ϕ_1 and ϕ_{12}/ϕ_2, explain why k is less than 1.0.

12.6 a) Show that the two coupled coils in Fig. P12.6 can be replaced by a single coil having an inductance of $L_{ab} = L_1 + L_2 + 2M$. (*Hint:* Express v_{ab} as a function of i_{ab}.)

b) Show that if the connections to the terminals of the coil labeled L_2 are reversed, $L_{ab} = L_1 + L_2 - 2M$.

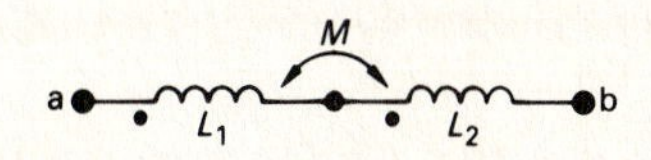

FIGURE P12.6

12.7 a) Show that the two magnetically coupled coils in Fig. P12.7 can be replaced by a single coil having an inductance of

$$L_{ab} = \frac{L_1L_2 - M^2}{L_1 + L_2 - 2M}.$$

(*Hint:* Let i_1 and i_2 be clockwise mesh currents in the left and right "windows" of Fig. P12.7, respectively. Sum the voltages around the two meshes. In mesh 1 let v_{ab} be the unspecified applied voltage. Solve for di_1/dt as a function of v_{ab}.)

b) Show that if the magnetic polarity of coil 2 is reversed, then

$$L_{ab} = \frac{L_1L_2 - M^2}{L_1 + L_2 + 2M}.$$

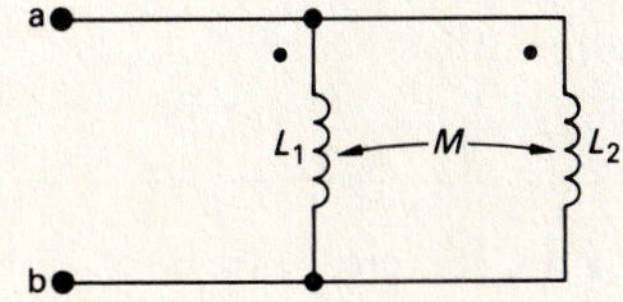

FIGURE P12.7

12.8 A series combination of a 60 Ω resistor and a 15.625 nF capacitor is connected to a sinusoidal voltage source by a linear transformer. The source is operating at a frequency of 800 krad/s. At this frequency the internal impedance of the source is $10 + j27.2$ Ω. The rms voltage at the terminals of the source is 80 V when is is not loaded. The parameters of the linear transformer are $R_1 = 15.6$ Ω, $L_1 = 90$ μH, $R_2 = 30$ Ω, $L_2 = 250$ μH, and $M = 75$ μH.

a) What is the value of the impedance reflected into the primary?

b) What is the value of the impedance seen from the terminals of the practical source?

c) What is the rms magnitude of the voltage across the load impedance?

d) What percentage of the average power developed by the practical source is delivered to the load impedance?

12.9 a) Find the steady-state expressions for the currents i_g and i_L in the circuit in Fig. P12.9 when $v_g = 200 \cos 10{,}000t$ V.

b) Find the coefficient of coupling.

c) Find the energy stored in the magnetically coupled coils at $t = 50\pi$ μs and $t = 100\pi$ μs.

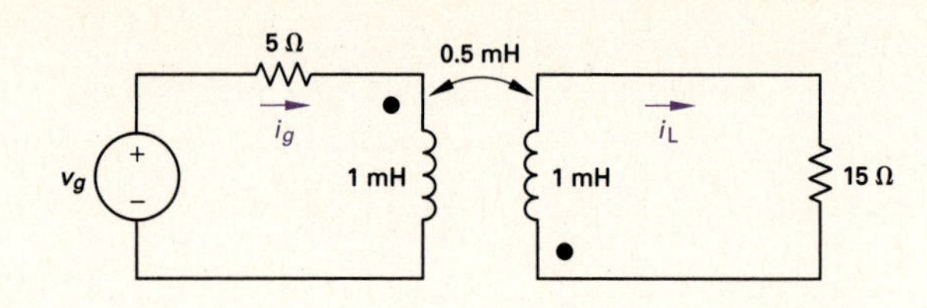

FIGURE P12.9

12.10 The value of k in the circuit in Fig. P12.10 is adjusted so that Z_{ab} is purely resistive when $\omega = 5$ krad/s. Find Z_{ab}.

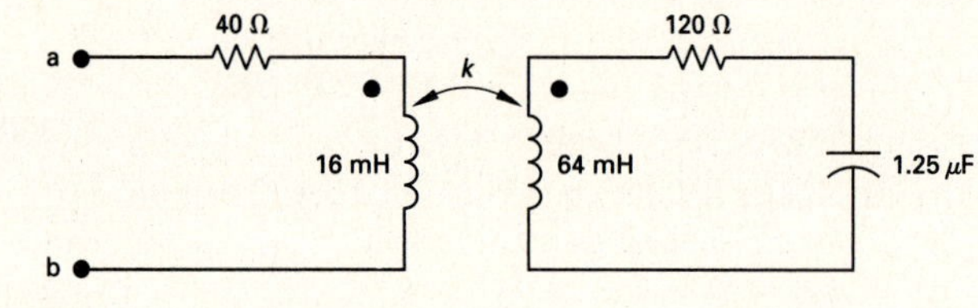

FIGURE P12.10

12.11 The sinusoidal voltage source in the circuit seen in Fig. P12.11 is operating at a frequency of 40 krad/s. The coefficient of coupling is adjusted until the peak amplitude of i_1 is maximum.

a) What is the value of k?

b) What is the peak amplitude of i_1 if $v_g = 500 \cos (4 \times 10^4 t)$ V?

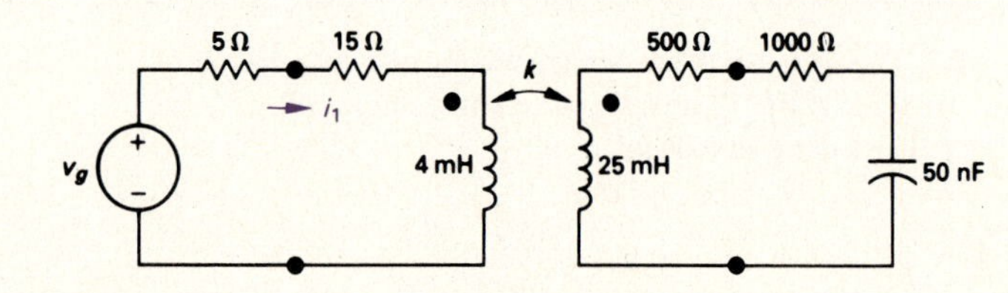

FIGURE P12.11

12.12 Find the average power delivered to the 60 Ω resistor in the circuit shown in Fig. P12.12 if $v_g = 150 \cos 4000t$ V.

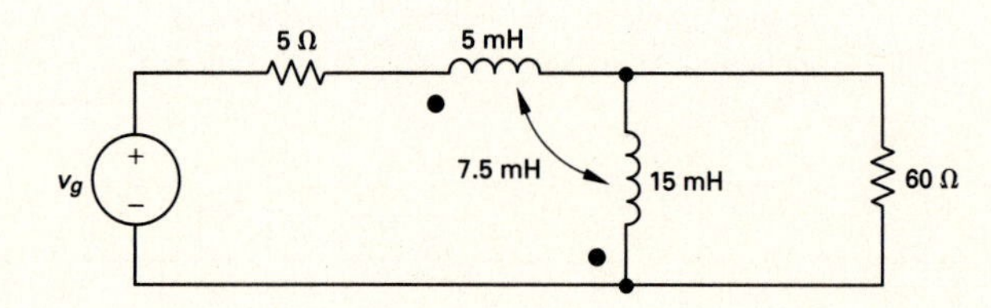

FIGURE P12.12

12.13 Find the average power delivered to the 200 Ω resistor in the circuit shown in Fig P12.13 if $v_g = 424 \cos 8000t$ V.

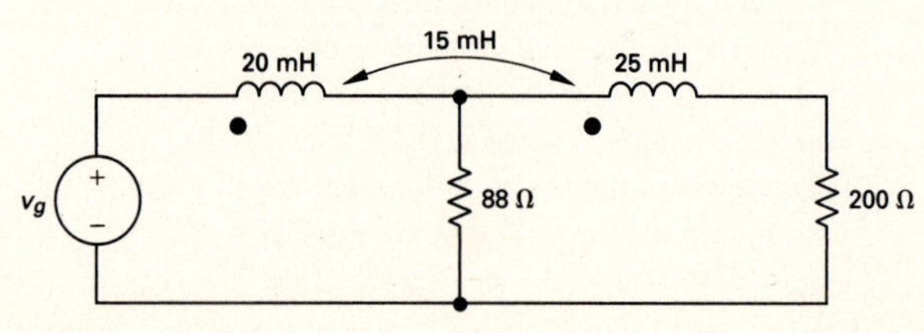

FIGURE P12.13

12.14 a) For the circuit in Fig. P12.14, find the Thévenin equivalent with respect to the terminals c,d.

b) Find the average power developed by the sinusoidal voltage source if an impedance equal to the conjugate of the Thévenin impedance is connected to the terminals c,d.

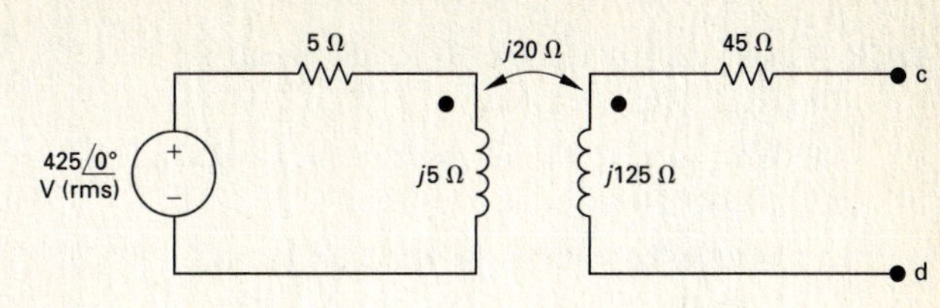

FIGURE P12.14

12.15 The impedance Z_L in the circuit in Fig. P12.15 is adjusted for maximum average power transfer to Z_L. The internal impedance of the sinusoidal voltage source is $20 + j35\ \Omega$.

a) What is the maximum average power delivered to Z_L?

b) What percentage of the average power delivered to the linear transformer is delivered to Z_L?

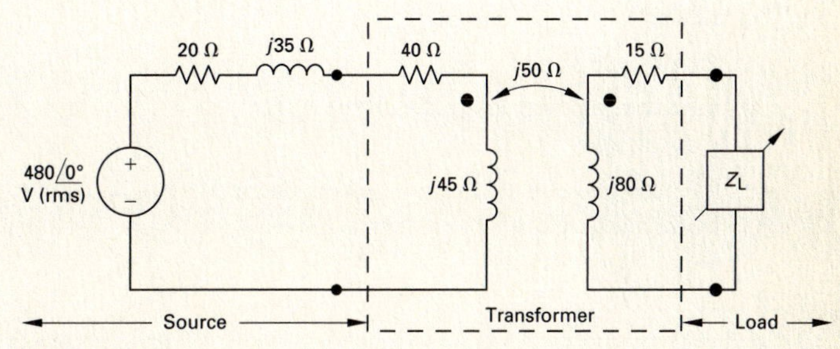

FIGURE P12.15

12.16 a) Find the power delivered to the 15 Ω resistor in the circuit in Fig. P12.9.

b) If the 15 Ω resistor is replaced by a variable resistor R_L, what value of R_L will yield maximum average power transfer to R_L?

c) What is the maximum average power in part (b)?

d) Assume the 15 Ω resistor is replaced by a variable impedance Z_L. What value of Z_L will result in maximum average power transfer to Z_L?

e) What is the maximum average power in part (d)?

12.17 The values of the parameters in the circuit shown in Fig. P12.17 are $L_1 = 20$ mH; $L_2 = 5$ mH; $k = 0.8$; $R_g = 10\ \Omega$; and $R_L = 90\ \Omega$. If $v_g = 86\sqrt{2}\cos 5000t$ V, find

a) the rms magnitude of v_o;

b) the average power delivered to R_L; and

c) the percentage of the average power generated by the ideal voltage source that is delivered to R_L.

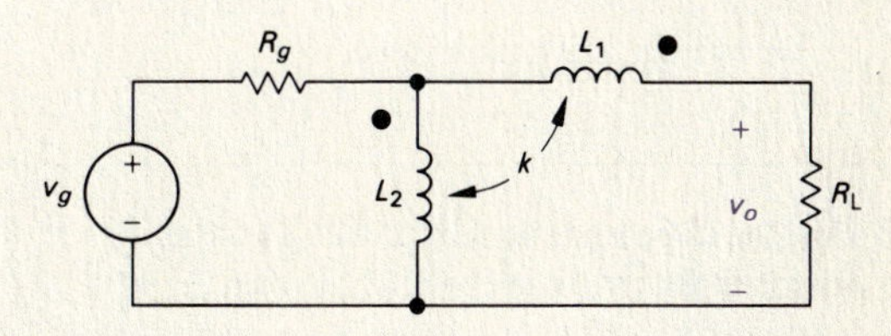

FIGURE P12.17

12.18 Assume the load resistor (R_L) in the circuit in Fig. P12.17 is adjustable.

a) What value of R_L will result in the maximum average power being transferred to R_L?

b) What is the value of the maximum power transferred?

12.19 The polarity dot on L_1 in the circuit in Fig. P12.17 is reversed.

a) Find the value of k that makes v_o equal to zero.

b) Find the power developed by the source when k has the value found in part (a).

12.20 For the frequency-domain circuit in Fig. P12.20, calculate:

a) the rms magnitude of $\mathbf{V}_o$;

b) the average power dissipated in the 80 Ω resistor; and

c) the percentage of the average power generated by the ideal voltage source that is delivered to the 80 Ω load resistor.

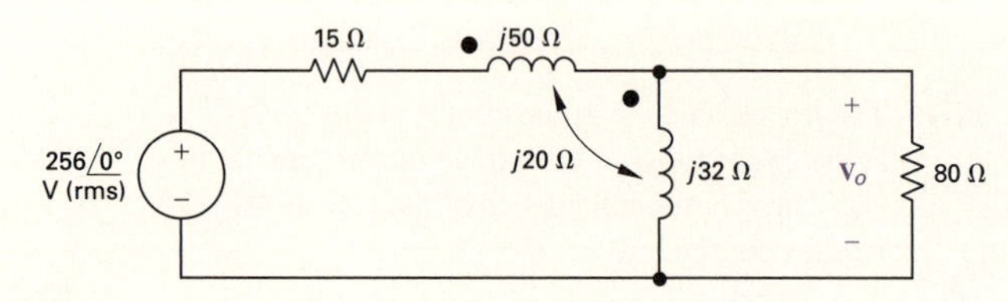

FIGURE P12.20

12.21 The 80 Ω resistor in the circuit in Fig. P12.20 is replaced with a variable impedance Z_o. Assume Z_o is adjusted for maximum average power transfer to Z_o.

a) What is the maximum average power that can be delivered to Z_o?

b) What is the average power developed by the ideal voltage source when maximum average power is delivered to Z_o?

12.22 a) Find the average power delivered to the 8 Ω resistor in the circuit in Fig. P12.22.

b) Find the average power developed by the ideal sinusoidal voltage source.

c) Find Z_{ab}.

d) Show that the average power developed equals the average power dissipated.

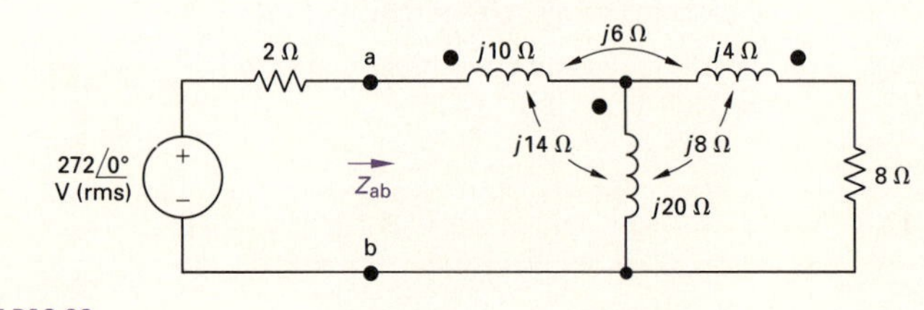

FIGURE P12.22

12.23 Find the impedance seen by the ideal voltage source in the circuit in Fig. P12.23 when Z_o is adjusted for maximum average power transfer to Z_o.

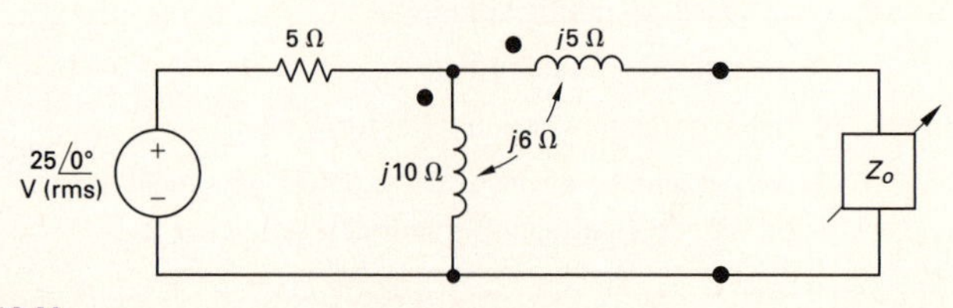

FIGURE P12.23

12.24 a) Find the six branch currents $\mathbf{I}_a$ through $\mathbf{I}_f$ in the circuit in Fig. P12.24.

b) Find the complex power in each branch of the circuit.

c) Check your calculations by verifying that the average power developed equals the average power dissipated.

d) Check your calculations by verifying that the magnetizing vars generated equal the magnetizing vars absorbed.

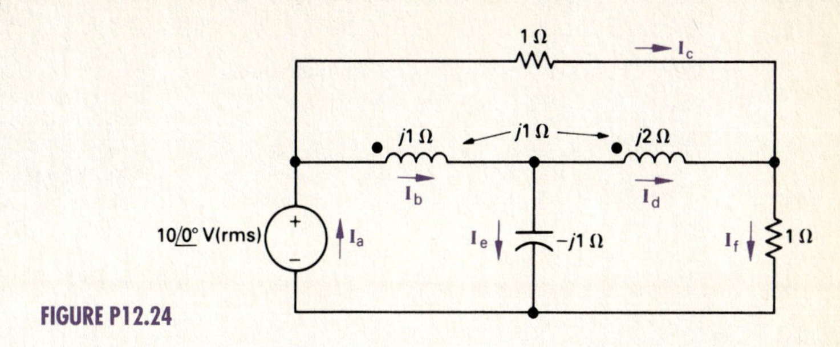

FIGURE P12.24

12.25 At first glance, it may appear from Eq. (12.54) that an inductive load could make the reactance seen looking into the primary terminals (i.e., X_{ab}) look capacitive. Intuitively, we know this is impossible. Show that X_{ab} can never be negative if X_L is an inductive reactance.

12.26 a) Show that the impedance seen looking into the terminals a,b in the circuit in Fig. P12.26 is given by the expression

$$Z_{ab} = \left(1 + \frac{N_1}{N_2}\right)^2 Z_L.$$

b) Show that if the polarity terminals of either one of the coils is reversed,

$$Z_{ab} = \left(1 - \frac{N_1}{N_2}\right)^2 Z_L.$$

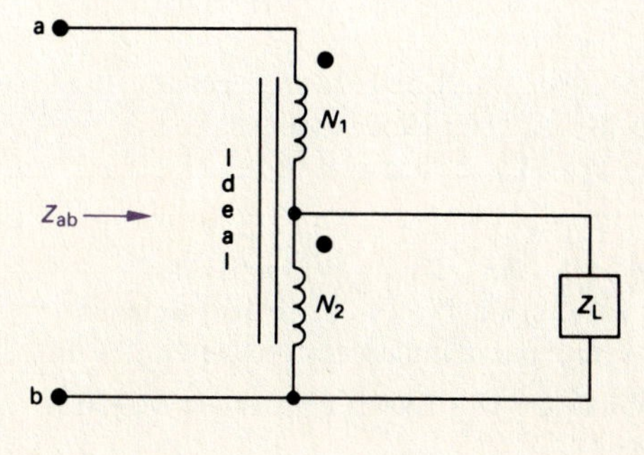

FIGURE P12.26

12.27 a) Show that the impedance seen looking into the terminals a,b in the circuit in Fig. P12.27 is given by the expression

$$Z_{ab} = \frac{Z_L}{\left(1 + \frac{N_1}{N_2}\right)^2}.$$

b) Show that if the polarity terminal of either one of the coils is reversed that

$$Z_{ab} = \frac{Z_L}{\left(1 - \frac{N_1}{N_2}\right)^2}.$$

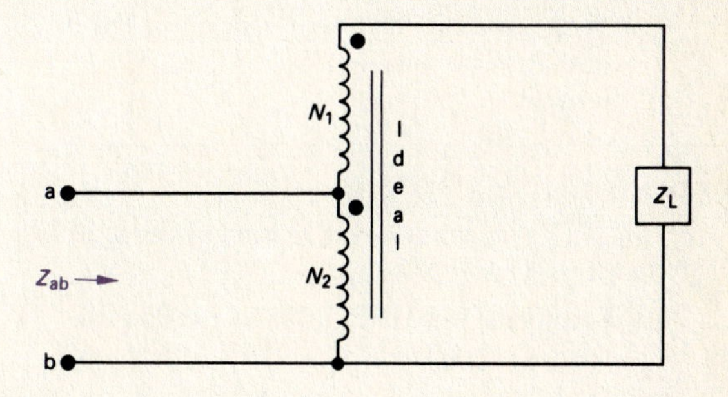

FIGURE P12.27

12.28 a) Find the average power dissipated in the 1 Ω resistor in the circuit in Fig. P12.28.

b) Check your answer by showing that the total power developed equals the total power absorbed.

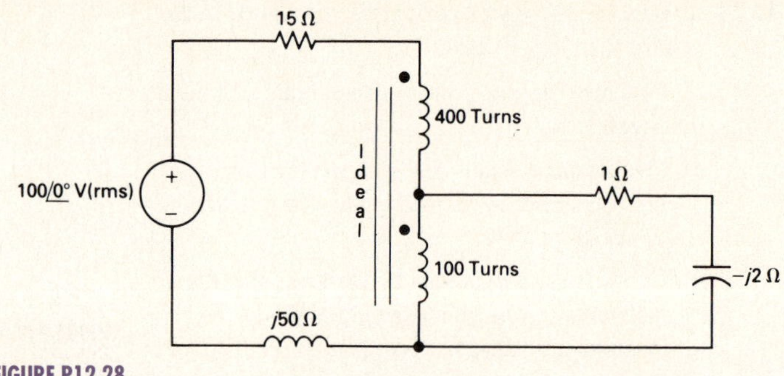

FIGURE P12.28

12.29 a) If N_1 equals 1000 turns, how many turns should be placed on the N_2 winding of the ideal transformer in the circuit seen in Fig. P12.29 so that maximum average power is delivered to the 8500 Ω load?

b) Find the average power delivered to the 8500 Ω resistor.

c) What percentage of the average power delivered by the ideal voltage source is dissipated in the linear transformer?

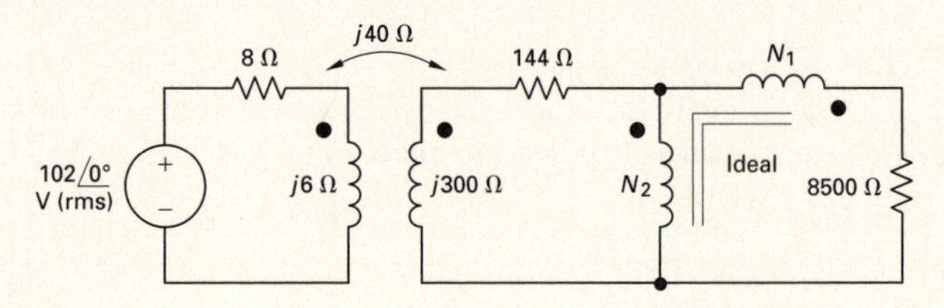

FIGURE P12.29

12.30 a) Find the turns ratio N_1/N_2 for the ideal transformer in Fig. P12.30 so that maximum average power is transferred to the 400 Ω load.

b) Find the average power delivered to the 400 Ω load.

c) Find the rms magnitude of the voltage at the input of the ideal transformer, that is, $\mathbf{V}_i$.

d) What percentage of the power developed by the ideal current source is delivered to the 400 Ω resistor?

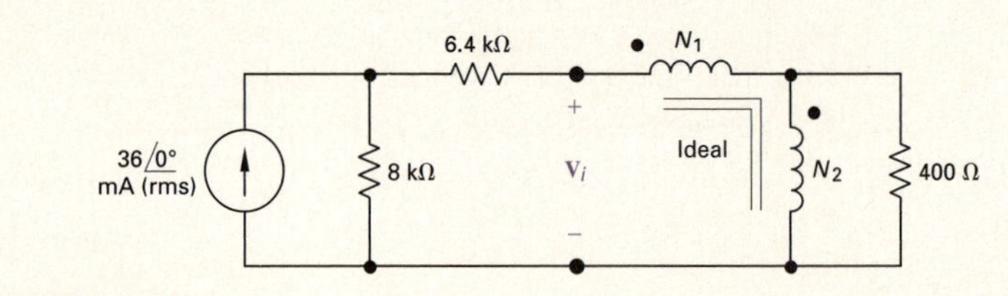

FIGURE P12.30

12.31 The load impedance Z_L in the circuit in Fig. P12.31 is adjusted until maximum average power is transferred to Z_L.

a) Specify the value of Z_L if $N_1 = 10{,}000$ turns and $N_2 = 2000$ turns.

b) Specify the values of $\mathbf{I}_L$ and $\mathbf{V}_L$ when Z_L is absorbing maximum average power.

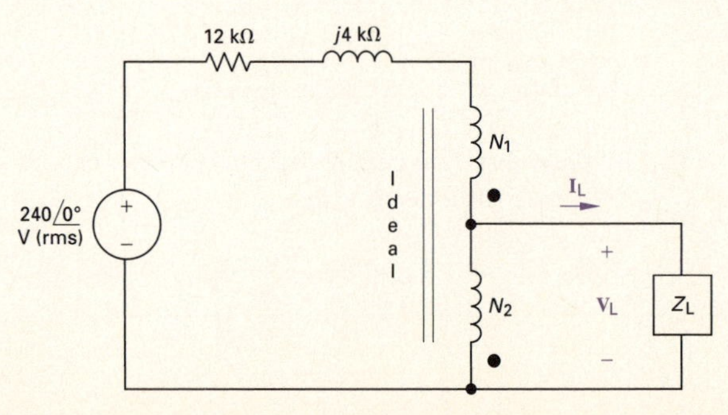

FIGURE P12.31

12.32 The variable load resistor R_L in the circuit shown in Fig. 12.32 is adjusted for maximum average power transfer to R_L.

a) Find the maximum average power.

b) What percentage of the average power developed by the ideal voltage source is delivered to R_L when R_L is absorbing maximum average power?

c) Test your solution by showing that the power developed by the ideal voltage source equals the power dissipated in the circuit.

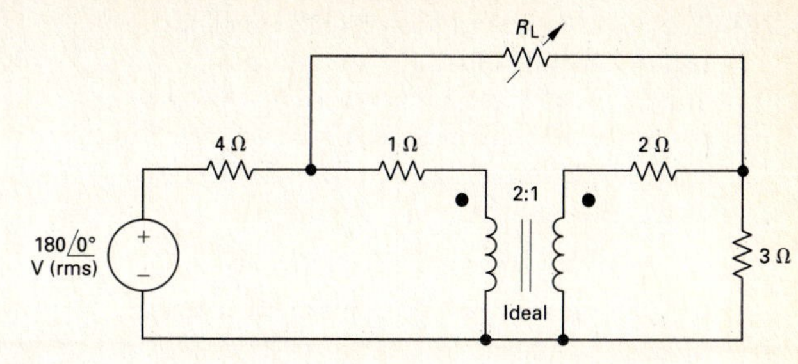

FIGURE P12.32

12.33 Repeat Problem 12.32 when the turns ratio is 1:2 instead of 2:1.

12.34 The sinusoidal current source in the circuit in Fig. P12.34 is operating at a frequency of 20 krad/s. The variable capacitor in the circuit is adjusted until the average power delivered to the 100 Ω resistor is as large as is possible.

a) Find the value of C in microfarads.

b) When C has the value found in part (a), what is the average power delivered to the 100 Ω resistor?

c) Replace the 100 Ω resistor with a variable resistor R_o. Specify the value of R_o so that maximum average power is delivered to R_o.

d) What is the maximum average power that can be delivered to R_o?

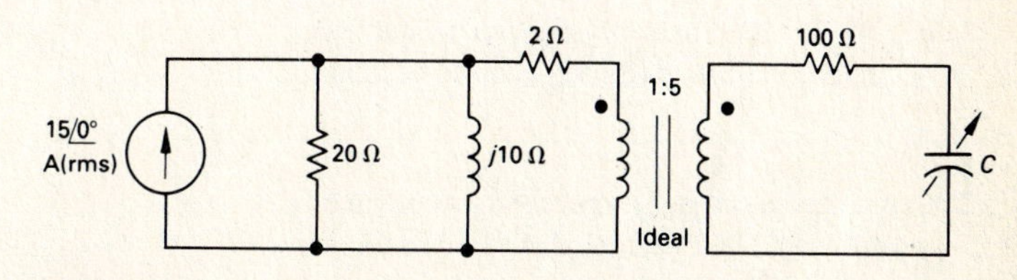

FIGURE P12.34

12.35 Find the impedance Z_{ab} in the circuit in Fig. P12.35 if $Z_L = 120\angle 60°$ Ω.

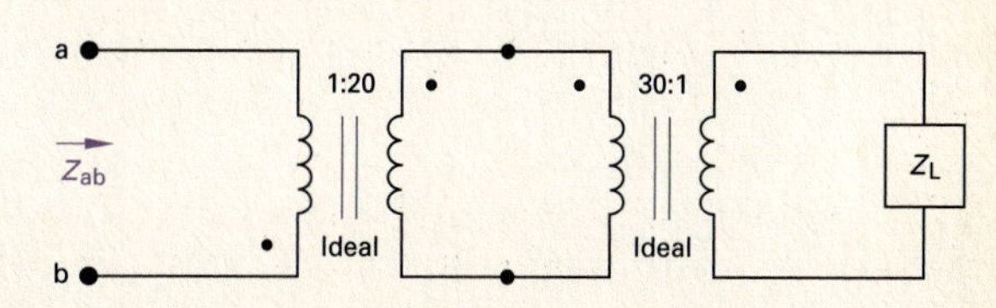

FIGURE P12.35

12.36 The sinusoidal voltage source in the circuit in Fig. P12.36 is developing an rms voltage of 208 V. The 16 Ω load in the circuit is absorbing 25 times as much average power as the 64 Ω load. The two loads are matched to the sinusoidal source that has an internal impedance of $41.6\underline{/0^\circ}$ kΩ.

a) Specify the numerical values of a_1 and a_2.

b) Calculate the power delivered to the 16 Ω load.

c) Calculate the rms value of the voltage across the 64 Ω resistor.

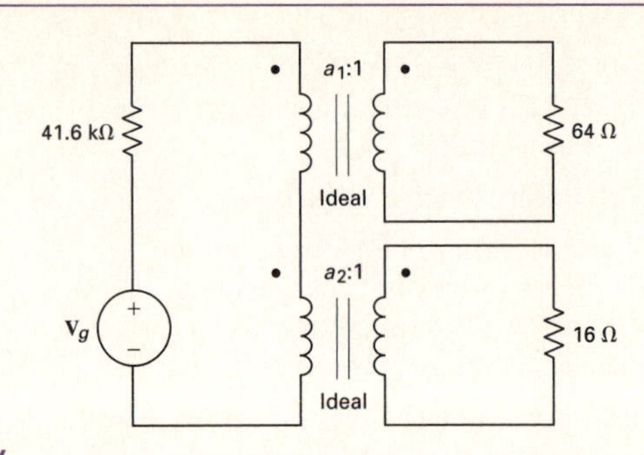

FIGURE P12.36

12.37 a) Find the T-equivalent circuit for the magnetically coupled coils shown in Fig. P12.37.

b) Find the π-equivalent circuit for the same set of coils. Assume that $i_1 0 = i_2(0) = 0$.

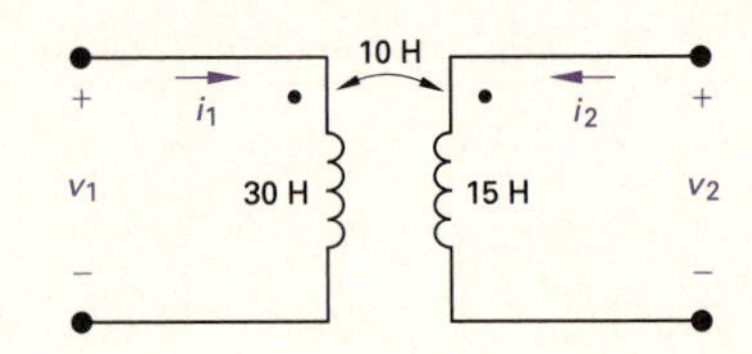

FIGURE P12.37

12.38 Show that the frequency-domain π-equivalent circuit for the magnetically coupled coils in Fig. 12.22 can be derived from the T-equivalent circuit by simply making a Y-to-Δ transformation.

12.39 A sinusoidal voltage source with an internal resistance of 28 Ω is coupled to a load by means of a lossless linear transformer, as shown in Fig. P12.39. The transformer inductances are $L_1 = 20$ mH and $L_2 = 10$ mH. The transformer coefficient of coupling is $0.5\sqrt{2}$.

a) Specify the frequency-domain T-equivalent circuit for the lossless transformer when the source frequency is 4 krad/s.

b) Use the T-equivalent circuit of part (a) to find the steady-state expression for v_L when $v_g = 240 \cos 4000t$ V.

c) Repeat parts (a) and (b), given that the polarity dot on the secondary side of the transformer is shifted to the lower terminal.

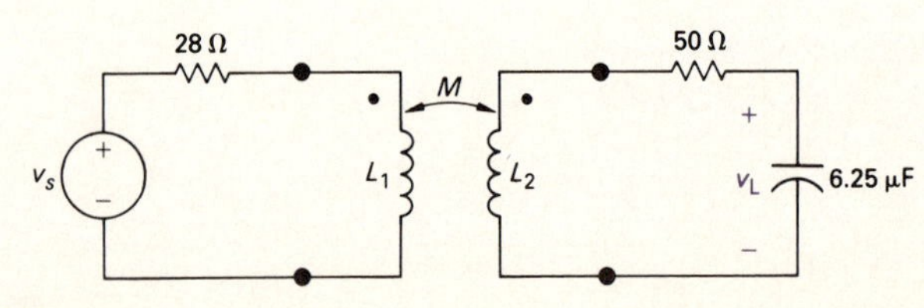

FIGURE P12.39

12.40 The equivalent circuit of a lossless linear transformer is shown in Fig. P12.40.

a) Find L_1, L_2, and M.

b) Find the coefficient of coupling k.

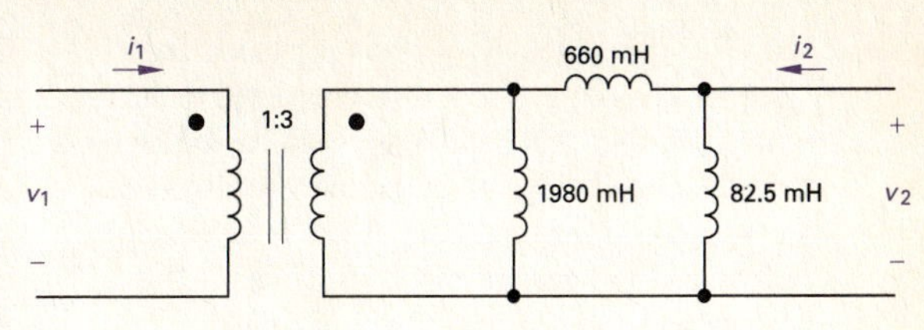

FIGURE P12.40

12.41 The following measurements were made on a lossless linear transformer. With the high-voltage side open, the inductance measured looking into the low-voltage side is 10 mH. With a 15 V (rms) sinusoidal voltage applied to the low-voltage winding, the open-circuit voltage measured on the high-voltage winding is 30 V (rms). With a 40 V (rms) sinusoidal voltage applied to the high-voltage side, the open-circuit voltage measured on the low-voltage side is 5 V.

a) Specify the numerical values of L_1, L_2, M, and k for the transformer.

b) Calculate the turns ratio of the transformer.

c) Construct two possible equivalent circuits for the linear transformer if a is chosen to equal M/L_1.

d) Repeat part (c), given that a is chosen to equal L_2/M.

12.42 a) Use one (your choice) of the equivalent circuits derived in Problem 12.41 to calculate the rms voltage at the terminals of the high-voltage winding when the transformer is used in the circuit in Fig. P12.42. The 50 V (rms) sinusoidal voltage source is operating at a frequency of 25 krad/s.

b) Verify your calculation in part(a) by finding the same voltage without using an equivalent circuit for the linear transformer.

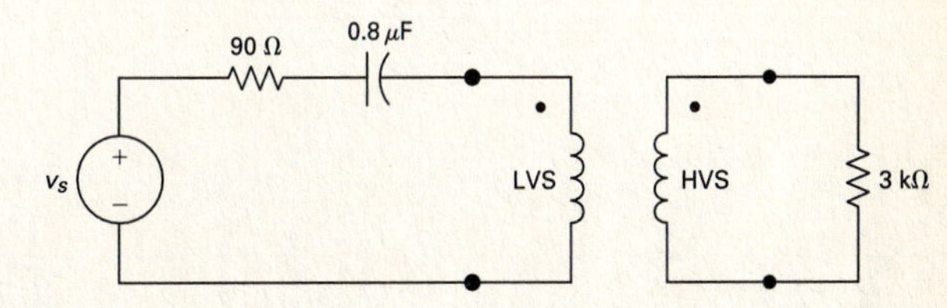

FIGURE P12.42

12.43 The purpose of this problem is to illustrate a circuit structure in which the T- or π-equivalent circuits derived in Section 12.7 cannot be used because the voltage $\mathbf{V}_{bd}$ in Fig. 12.24 is not zero. With this in mind, calculate the voltage $\mathbf{V}_{bd}$ in the circuit in Fig. P12.43.

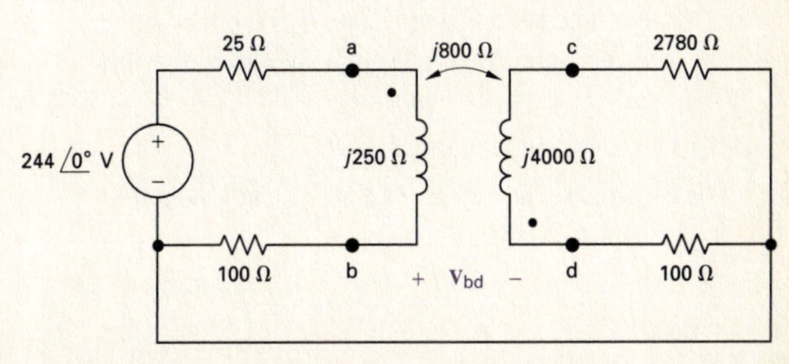

FIGURE P12.43

12.44 The circuit shown in Fig. P12.44 is a balanced three-phase system. The load is connected to the source through ideal transformers. As can be seen from the circuit diagram, the primary windings of the three transformers are connected in wye and the secondary windings are connected in delta. The primary and secondary windings of each transformer are identified by a circled number. For example, the primary winding of transformer 1 is connected between terminals A and N, and the secondary winding of transformer 1 is connected between terminals A' and B'. Each transformer reduces the load side voltage by a factor of 2, that is $a = 1/2$.

a) Calculate $\mathbf{V}_{ab}$

b) Calculate $\mathbf{V}_{A'B'}$.

c) Calculate $\mathbf{I}_{A'B'}$.

d) Calculate the total average power delivered to the balanced Δ-connected 15 Ω load.

e) Calculate the total average power delivered by the balanced three-phase source.

f) What percentage of the average power delivered by the source reaches the load?

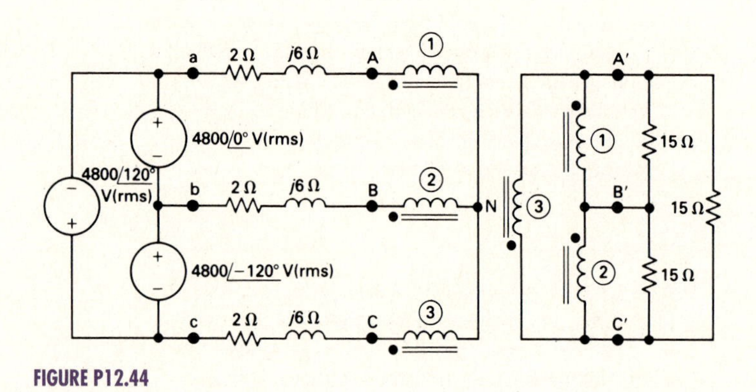

FIGURE P12.44

12.45 a) Derive Eq. 12.109.

b) Derive Eq. 12.110.

c) Derive Eq. 12.114.

12.46 Derive Eq. 12.116.

12.47 a) Using the same numerical values used in the Practical Perspective example in the text, find the instant of time when the voltage across the capacitor is maximum.

b) Find the maximum value of v_c.

c) Compare the values obtained in parts (a) and (b) with t_{max} and $v_c(t_{max})$.

12.48 The values of the parameters in the circuit in Fig. 12.40 are $R = 3\ \Omega$; $L = 5$ mH; $C = 0.25\ \mu$F; $V_{dc} = 12$ V; and $a = 50$. Assume the switch opens when the primary winding current is 4 A.

a) How much energy is stored in the circuit at $t = 0^+$?

b) Assume the spark plug does not fire. What is the maximum voltage available at the spark plug?

c) What is the voltage across the capacitor when the voltage across the spark plug is at its maximum value?

13 Introduction to the Laplace Transform

Chapter Contents

We now introduce a powerful analytical technique widely used to study the behavior of linear, lumped-parameter circuits. The method is based on the Laplace transform, which we define mathematically in Section 13.1. Before doing so, we need to explain why another analytical technique is needed. First, we wish to consider the transient behavior of circuits whose describing equations consist of more than a single node-voltage or mesh-current differential equation. In other words, we want to consider multiple-node and multiple-mesh circuits that are described by sets of linear differential equations.

Second, we wish to determine the transient response of circuits whose signal sources vary in ways more complicated than the simple dc level jumps considered in Chapters 7 and 8. Third, we can use the Laplace transform to introduce the concept of the transfer function as a tool for analyzing the steady-state sinusoidal response of a circuit when the frequency of the sinusoidal source is varied. We discuss the transfer function in Chapter 14. Finally, we wish to relate, in a systematic fashion, the time-domain behavior of a circuit to its frequency-domain behavior. Using the Laplace transform will provide a broader understanding of circuit functions.

In this chapter, we introduce the Laplace transform, discuss its pertinent characteristics, and present a systematic method for transforming from the frequency domain to the time domain.

13.1 DEFINITION OF THE LAPLACE TRANSFORM

The **Laplace transform** of a function is given by the expression

$$\mathcal{L}\{f(t)\} = \int_0^\infty f(t)e^{-st}dt, \tag{13.1}$$

where the symbol $\mathcal{L}\{f(t)\}$ is read "the Laplace transform of $f(t)$."

The Laplace transform of $f(t)$ is also denoted $F(s)$; that is,

$$F(s) = \mathcal{L}\{f(t)\}. \tag{13.2}$$

This notation emphasizes that when the integral in Eq. (13.1) has been evaluated, the resulting expression is a function of s. In our applications, t represents the time domain, and, because the exponent of e in the integral of Eq. (13.1) must be dimensionless, s must have the dimension of reciprocal time, or frequency. The Laplace transform transforms the problem from the time domain to the frequency domain. More specifically, it transforms a set of integrodifferential equations in the time domain to a set of algebraic equations in the frequency domain. It therefore reduces the solution for an unknown quantity to the manipulation of algebraic equations. After obtaining the frequency-domain expression for the unknown, we inverse-transform it back to the time domain.

If the idea behind the Laplace transform seems foreign, consider another familiar mathematical transform. Logarithms are used to change a multiplication or division problem, such as A = BC, into a simpler addition or subtraction problem: log A = log BC = log B + log C. Antilogs are used to carry out the inverse process. The phasor is another transform; as we know from Chapter 9, it converts a sinusoidal signal into a complex number for easier, algebraic computation of circuit values. After determining the phasor value of a signal, we transform it back to its time-domain expression. Both of these examples point out the essential feature of mathematical transforms: They are designed to create a new domain to make the mathematical manipulations easier. After finding the unknown in the new domain, we inverse-transform it back to the original domain. In circuit analysis, we use the Laplace transform to transform a set of integrodifferential equations from the time domain to a set of algebraic equations in the frequency domain. We therefore simplify the solution for an unknown quantity to the manipulation of a set of algebraic equations.

Before we illustrate some of the important properties of the Laplace transform, some general comments are in order. First, note that the integral in Eq. (13.1) is improper because the upper limit is infinite. Thus we are confronted immediately with the question of whether the integral converges. In other words, does a given $f(t)$ have a Laplace transform? Obviously, the functions of primary interest in engineering analysis have Laplace transforms; otherwise we would not be interested in the transform. In linear circuit analysis, we excite circuits with sources that have Laplace transforms. Excitation functions such as t^t or e^{t^2}, which do not have Laplace transforms, are of no interest here.

Second, because the lower limit on the integral is zero, the Laplace transform ignores $f(t)$ for negative values of t. Put another way, $F(s)$ is determined by the behavior of $f(t)$ only for positive values of t. To emphasize that the lower limit is zero, Eq. (13.1) is frequently referred to as the **one-sided**, or **unilateral, Laplace transform**. In the two-sided, or bilateral, Laplace transform, the lower limit is $-\infty$. We do not use the bilateral form here; hence $F(s)$ is understood to be the one-sided transform.

Another point regarding the lower limit concerns the situation when $f(t)$ has a discontinuity at the origin. If $f(t)$ is continuous at the origin—as, for example, in Fig. 13.1(a)—$f(0)$ is not ambiguous. However, if $f(t)$ has a finite discontinuity at the origin—as, for example, in Fig. 13.1(b)—the question arises as to whether the Laplace transform integral should include or exclude the discontinuity. In other words, should we make the lower limit 0^- and include the discontinuity, or should we exclude the discontinuity by making the lower limit 0^+? (We use the notation 0^- and 0^+ to denote values of t just to the left and right of the origin, respectively.) Actually, we may choose either so long as we are consistent. For reasons explained later, we choose 0^- as the lower limit.

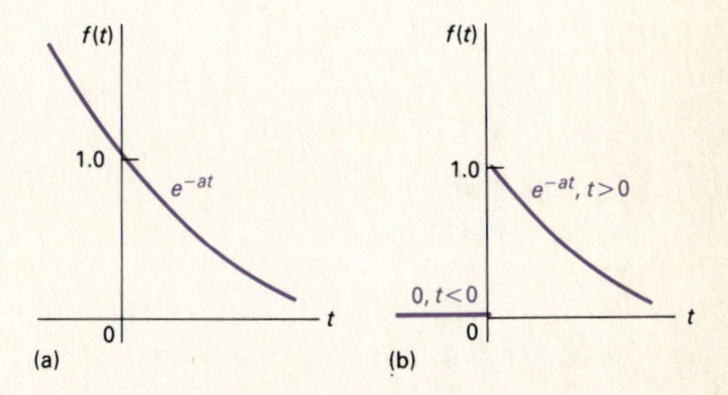

FIGURE 13.1 A continuous and discontinuous function at the origin; (a) $f(t)$ is continuous at the origin; and (b) $f(t)$ is discontinuous at the origin.

Because we are using 0^- as the lower limit, we note immediately that the integration from 0^- to 0^+ is zero. The only exception is when the discontinuity at the origin is an impulse function, a situation we discuss in Section 13.3. The important point now is that the two functions shown in Fig. 13.1 have the same unilateral Laplace transform because there is no impulse function at the origin.

The one-sided Laplace transform ignores $f(t)$ for $t < 0^-$. What happens prior to 0^- is accounted for by the initial conditions. Thus we use the Laplace transform to predict the response to a disturbance that occurs after initial conditions have been established.

In the discussion that follows, we divide the Laplace transforms into two types: functional transforms and operational transforms. A **functional transform** is the Laplace transform of a specific function, such as $\sin \omega t$, t, e^{-at}, and so on. An **operational transform** defines a general mathematical property of the Laplace transform, such as finding the transform of the derivative of $f(t)$. Before considering functional and operational transforms, however, we need to introduce the step and impulse functions.

13.2 THE STEP FUNCTION

We may encounter functions that have a discontinuity, or jump, at the origin. For example, we know from earlier discussions of transient behavior that switching operations create abrupt changes in currents and voltages. We accommodate these discontinuities mathematically by introducing the step and impulse functions.

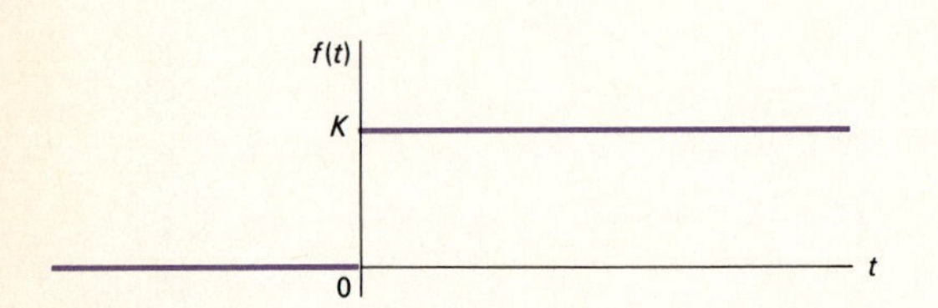

FIGURE 13.2 The step function.

Figure 13.2 illustrates the step function. It is zero for $t < 0$. The symbol for the step function is $Ku(t)$. Thus, the mathematical definition of the **step function** is

$$Ku(t) = 0, \quad t < 0;$$
$$Ku(t) = K, \quad t > 0. \tag{13.3}$$

If K is 1, the function defined by Eq. (13.3) is the **unit step**.

The step function is not defined at $t = 0$. In situations where we need to define the transition between 0^- and 0^+, we assume that it is linear and that

$$Ku(0) = 0.5K. \tag{13.4}$$

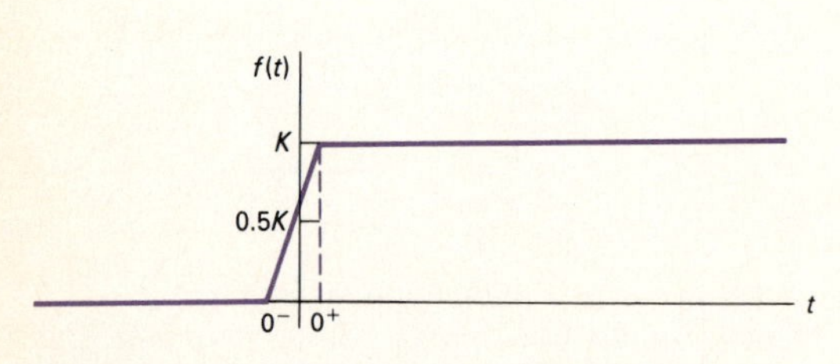

FIGURE 13.3 The linear approximation to the step function.

As before, 0^- and 0^+ represent symmetric points arbitrarily close to the left and right of the origin. Figure 13.3 illustrates the linear transition from 0^- and 0^+.

A discontinuity may occur at some time other than $t = 0$—for example, in sequential switching. A step that occurs at $t = a$ is expressed as $Ku(t - a)$. Thus

$$Ku(t-a) = 0, \quad t < a;$$
$$Ku(t-a) = K, \quad t > a. \tag{13.5}$$

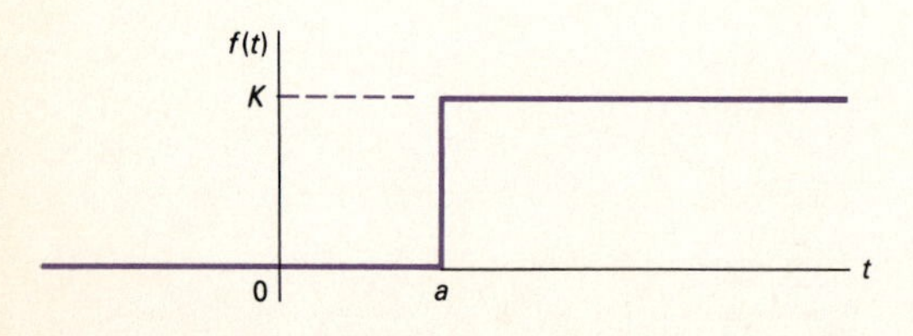

FIGURE 13.4 A step function occurring at $t = a$ when $a > 0$.

If $a > 0$, the step occurs to the right of the origin, and if $a < 0$, the step occurs to the left of the origin. Figure 13.4 illustrates Eq. (13.5). Note that the step function is 0 when the argument $t - a$ is negative, and it is K when the argument is positive.

A step function equal to K for $t < a$ is written as $Ku(a - t)$. Thus

$$Ku(a-t) = K, \quad t < a;$$
$$Ku(a-t) = 0, \quad t > a. \tag{13.6}$$

The discontinuity is to the left of the origin when $a < 0$. Equation (13.6) is shown in Fig. 13.5.

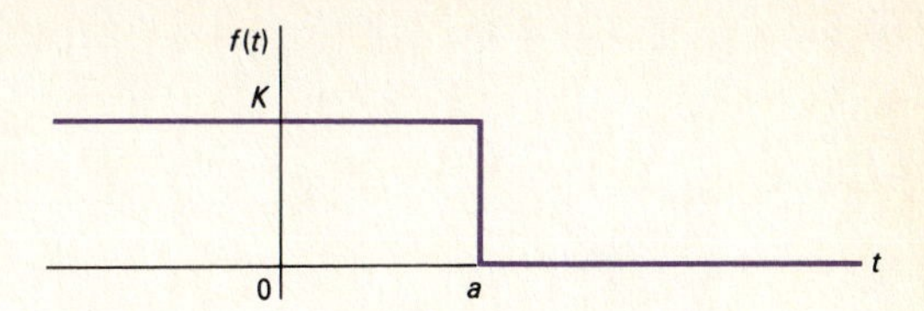

FIGURE 13.5 A step function $Ku(a - t)$ for $a > 0$.

One application of the step function is using it to write the mathematical expression for a function that is nonzero for a finite duration but is defined for all positive time. One example useful in circuit analysis is a finite-width pulse, which we can create by adding two step functions. The function $K[u(t-1) - u(t-3)]$ has the value K for $1 < t < 3$ and the value 0 everywhere else, so it is a finite-width pulse of height K initiated at $t = 1$ and terminated at $t = 3$. In defining this pulse using step functions, it is helpful to think of the step function $u(t-1)$ as "turning on" the constant value K at $t = 1$, and the step function $-u(t-3)$ as "turning off" the constant value K at $t = 3$. We use step functions to turn on and turn off linear functions at desired times in Example 13.1.

Use a computer tool to plot this and other pulse functions.

EXAMPLE 13.1

Use step functions to write an expression for the function illustrated in Fig. 13.6.

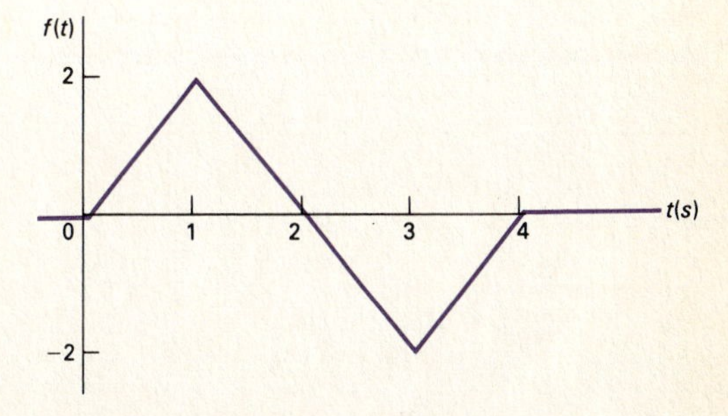

FIGURE 13.6 The function for Example 13.1.

SOLUTION

The function shown in Fig. 13.6 is made up of linear segments with break points at 0, 1, 3, and 4 s. To construct this function, we must add and subtract linear functions of the proper slope. We use the step function to initiate and terminate these linear segments at the proper times. In other words, we use the step function to turn on and turn off a straight line with the following equations: $+2t$, on at $t = 0$, off at $t = 1$; $-2t + 4$, on at $t = 1$, off at $t = 3$; and $+2t - 8$, on at $t = 3$, off at $t = 4$. These straight line segments and their equations are shown in Fig. 13.7. The expression for $f(t)$ is

$$f(t) = 2t[u(t) - u(t-1)] + (-2t+4)[u(t-1) - u(t-3)] + (2t-8)[u(t-3) - u(t-4)].$$

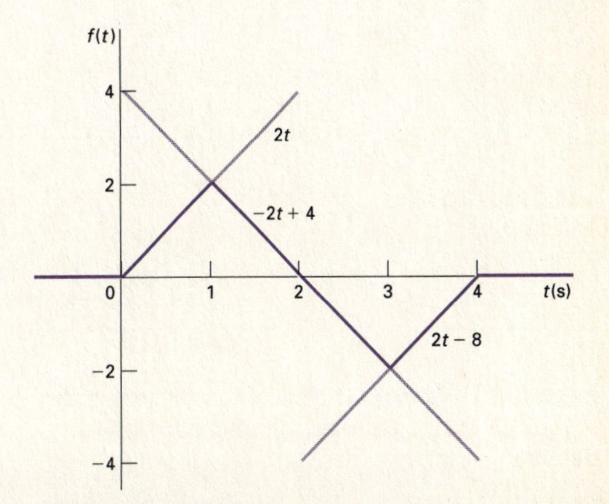

FIGURE 13.7 Definition of the three line segments turned on and off with step functions to form the function shown in Fig. 13.6.

DRILL EXERCISE

13.1 Use step functions to write the expression for each function shown.

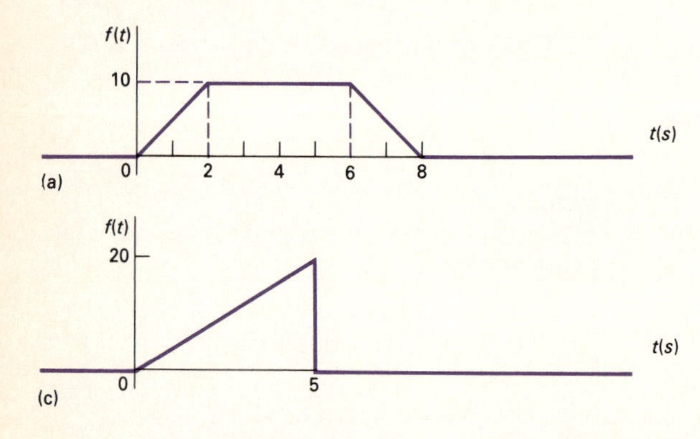

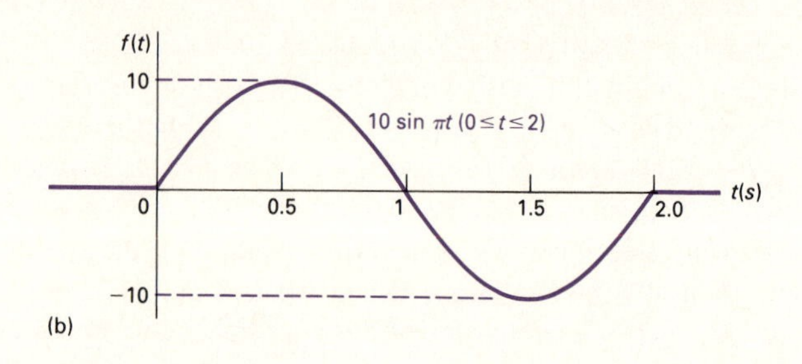

ANSWER: (a) $f(t) = 5t[u(t) - u(t-2)] + 10[u(t-2) - u(t-6)] + (-5t+40)[u(t-6) - u(t-8)]$;
(b) $f(t) = 10\sin(\pi t)[u(t) - u(t-2)]$;
(c) $f(t) = 4t[u(t) - u(t-5)]$.

13.3 The Impulse Function

When we have a finite discontinuity in a function, such as that illustrated in Fig. 13.1(b), the derivative of the function is not defined at the point of the discontinuity. The concept of an impulse function[1] enables us to define the derivative at a discontinuity, and thus to define the Laplace transform of that derivative. An impulse is a signal of infinite amplitude and zero duration. Such signals don't exist in nature, but some circuit signals come very close to approximating this definition, so we find a mathematical model of an impulse useful. Impulsive voltages and currents occur in circuit analysis either because of a switching operation or because the circuit is excited by an impulsive source. We will analyze these situations in Chapter 14, but here we focus on defining the impulse function generally.

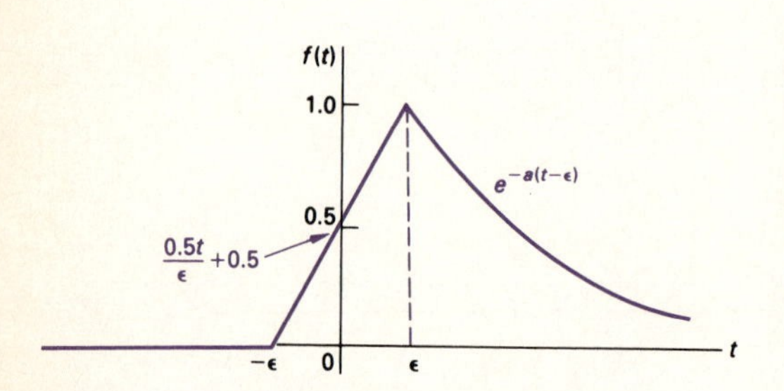

FIGURE 13.8 A magnified view of the discontinuity in Fig. 13.1(b), assuming a linear transition between $-\epsilon$ and $+\epsilon$.

To define the derivative of a function at a discontinuity, we first assume that the function varies linearly across the discontinuity, as shown in Fig. 13.8, where we observe that as $\epsilon \to 0$, an abrupt dis-

[1] The impulse function is also known as the Dirac delta function.

continuity occurs at the origin. When we differentiate the function, the derivative between $-\epsilon$ and $+\epsilon$ is constant at a value of $1/2\epsilon$. For $t > \epsilon$, the derivative is $-ae^{-a(t-\epsilon)}$. Figure 13.9 shows these observations graphically. As ϵ approaches zero, the value of $f'(t)$ between $\pm\epsilon$ approaches infinity. At the same time, the duration of this large value is approaching zero. Furthermore, the area under $f'(t)$ between $\pm\epsilon$ remains constant as $\epsilon \to 0$. In this example, the area is unity. As ϵ approaches zero, we say that the function between $\pm\epsilon$ approaches a **unit impulse function**, denoted $\delta(t)$. Thus the derivative of $f(t)$ at the origin approaches a unit impulse function as ϵ approaches zero, or

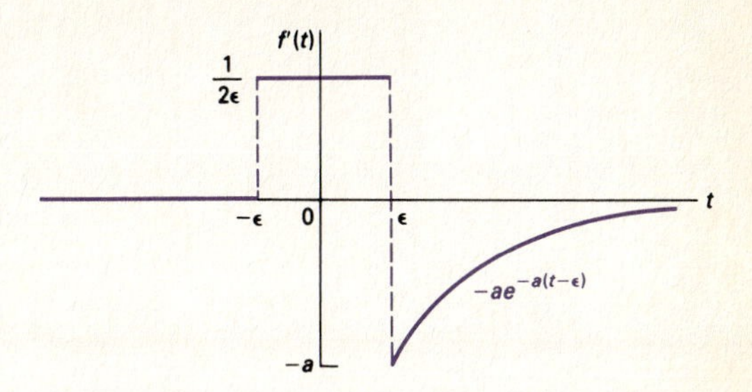

FIGURE 13.9 The derivative of the function shown in Fig. 13.8.

$$f'(0) \to \delta(t) \quad \text{as} \quad \epsilon \to 0.$$

If the area under the impulse function curve is other than unity, the impulse function is denoted $K\delta(t)$, where K is the area. K is often referred to as the **strength** of the impulse function.

To summarize, an impulse function is created from a variable-parameter function whose parameter approaches zero. The variable-parameter function must exhibit the following three characteristics as the parameter approaches zero:

Use a computer tool to graph such functions as their parameters change.

1. The amplitude approaches infinity.
2. The duration of the function approaches zero.
3. The area under the variable-parameter function is constant as the parameter changes.

Many different variable-parameter functions have the aforementioned characteristics. In Fig. 13.8, we used a linear function $f(t) = 0.5t/\epsilon + 0.5$. Another example of a variable-parameter function is the exponential function:

$$f(t) = \frac{K}{2\epsilon}e^{-|t|/\epsilon}. \tag{13.7}$$

As ϵ approaches zero, the function becomes infinite at the origin and at the same time decays to zero in an infinitesimal length of time. Figure 13.10 illustrates the character of $f(t)$ as $\epsilon \to 0$. To show that an impulse function is created as $\epsilon \to 0$, we must also show that the area under the function is independent of ϵ. Thus,

$$\begin{aligned}\text{Area} &= \int_{-\infty}^{0} \frac{K}{2\epsilon}e^{t/\epsilon}dt + \int_{0}^{\infty} \frac{K}{2\epsilon}e^{-t/\epsilon}dt \\ &= \frac{K}{2\epsilon} \cdot \frac{e^{t/\epsilon}}{1/\epsilon}\Big|_{-\infty}^{0} + \frac{K}{2\epsilon} \cdot \frac{e^{-t/\epsilon}}{-1/\epsilon}\Big|_{0}^{\infty} \\ &= \frac{K}{2} + \frac{K}{2} = K,\end{aligned} \tag{13.8}$$

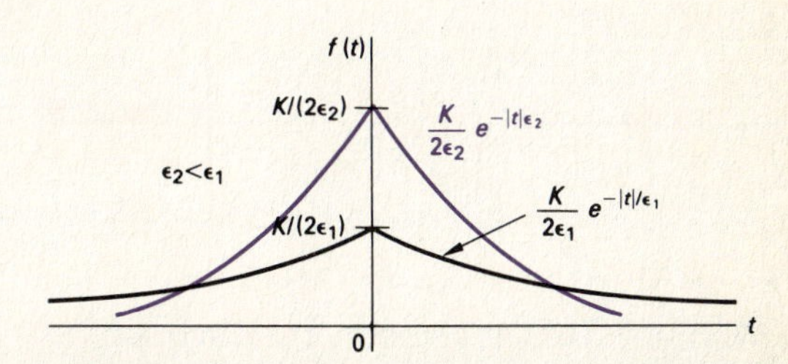

FIGURE 13.10 A variable-parameter function used to generate an impulse function.

which tells us that the area under the curve is constant and equal to K units. Therefore as $\epsilon \to 0$, $f(t) \to K\delta(t)$.

Mathematically, the **impulse function** is defined

$$\int_{-\infty}^{\infty} K\delta(t)dt = K; \tag{13.9}$$

$$\delta(t) = 0, \quad t \neq 0. \tag{13.10}$$

Equation (13.9) states that the area under the impulse function is constant. This area represents the strength of the impulse. Equation (13.10) states that the impulse is zero everywhere except at $t = 0$. An impulse that occurs at $t = a$ is denoted $K\delta(t - a)$.

The graphic symbol for the impulse function is an arrow. The strength of the impulse is given parenthetically next to the head of the arrow. Figure 13.11 shows the impulses $K\delta(t)$ and $K\delta(t - a)$.

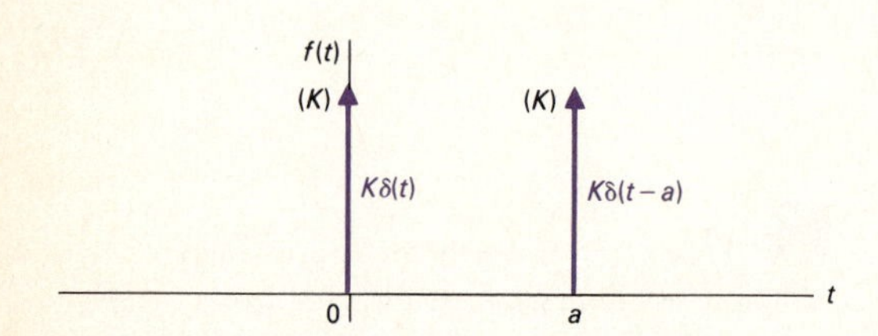

FIGURE 13.11 A graphic representation of the impulse $K\delta(t)$ and $K\delta(t-a)$.

An important property of the impulse function is the **sifting property**, which is expressed as

$$\int_{-\infty}^{\infty} f(t)\delta(t - a)dt = f(a), \tag{13.11}$$

where the function $f(t)$ is assumed to be continuous at $t = a$, that is, at the location of the impulse. Equation (13.11) shows that the impulse function sifts out everything except the value of $f(t)$ at $t = a$. The validity of Eq. (13.11) follows from noting that $\delta(t - a)$ is zero everywhere except at $t = a$, and hence the integral can be written

$$I = \int_{-\infty}^{\infty} f(t)\delta(t - a)dt = \int_{a-\epsilon}^{a+\epsilon} f(t)\delta(t - a)dt. \tag{13.12}$$

But because $f(t)$ is continuous at a, it takes on the value $f(a)$ as $t \to a$, so

$$I = \int_{a-\epsilon}^{a+\epsilon} f(a)\delta(t-a)dt = f(a)\int_{a-\epsilon}^{a+\epsilon} \delta(t-a)dt = f(a). \tag{13.13}$$

We use the sifting property of the impulse function to find its Laplace transform:

$$\mathscr{L}\{\delta(t)\} = \int_{0^-}^{\infty} \delta(t)e^{-st}dt = \int_{0^-}^{\infty} \delta(t)dt = 1, \tag{13.14}$$

which is an important Laplace transform pair that we make good use of in circuit analysis.

We can also define the derivatives of the impulse function and the Laplace transform of these derivatives. We discuss the first derivative, along with its transform, and then state the result for the higher-order derivatives.

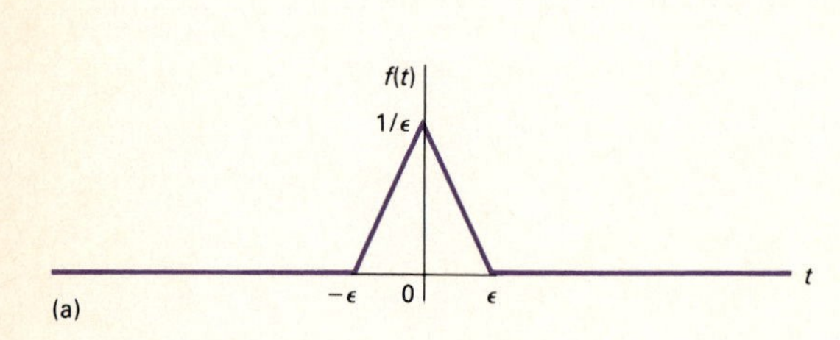

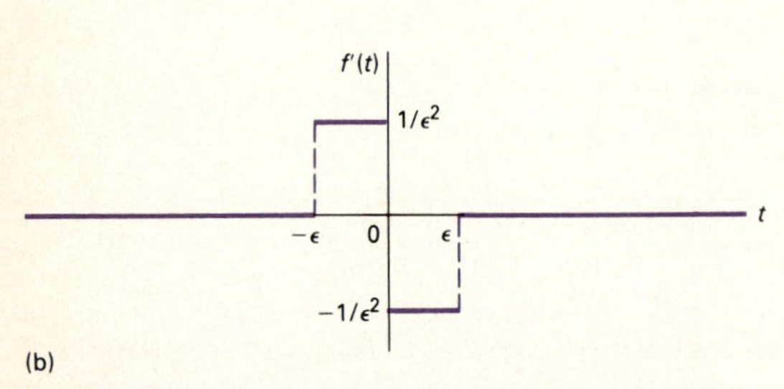

FIGURE 13.12 The first derivative of the impulse function: (a) The impulse-generating function used to define the first derivative of the impulse; and (b) the first derivative of the impulse-generating function that approaches $\delta'(t)$ as $\epsilon \to 0$.

The function illustrated in Fig. 13.12(a) generates an impulse function as $\epsilon \to 0$. Figure 13.12(b) shows the derivative of this impulse-generating function, which is defined as the derivative of the impulse

$[\delta'(t)]$ as $\epsilon \to 0$. The derivative of the impulse function sometimes is referred to as a moment function, or unit doublet.

To find the Laplace transform of $\delta'(t)$, we simply apply the defining integral to the function shown in Fig. 13.12(b) and, after integrating, let $\epsilon \to 0$. Then

$$\begin{aligned}
\mathcal{L}\{\delta'(t)\} &= \lim_{\epsilon\to 0}\left[\int_{-\epsilon}^{0^-} \frac{1}{\epsilon^2} e^{-st}dt + \int_{0^+}^{\epsilon}\left(-\frac{1}{\epsilon^2}\right)e^{-st}dt\right] \\
&= \lim_{\epsilon\to 0} \frac{e^{s\epsilon} + e^{-s\epsilon} - 2}{s\epsilon^2} \\
&= \lim_{\epsilon\to 0} \frac{se^{s\epsilon} - se^{-s\epsilon}}{2\epsilon s} \\
&= \lim_{\epsilon\to 0} \frac{s^2e^{s\epsilon} + s^2e^{-s\epsilon}}{2s} \\
&= s. \qquad \textbf{(13.15)}
\end{aligned}$$

In deriving Eq. (13.15), we had to use l'Hôpital's rule twice to evaluate the indeterminate form 0/0.

Higher-order derivatives may be generated in a manner similar to that used to generate the first derivative (see Problem 13.6), and the defining integral may then be used to find its Laplace transform. For the nth derivative of the impulse function, we find that its Laplace transform simply is s^n; that is,

$$\mathcal{L}\{\delta^{(n)}(t)\} = s^n. \qquad \textbf{(13.16)}$$

Finally, an impulse function can be thought of as a derivative of a step function; that is,

$$\delta(t) = \frac{du(t)}{dt}. \qquad \textbf{(13.17)}$$

Figure 13.13 presents the graphic interpretation of Eq. (13.17). The function shown in Fig. 13.13(a) approaches a unit step function as $\epsilon \to 0$. The function shown in Fig. 13.13(b)—the derivative of the function in Fig. 13.13(a)—approaches a unit impulse as $\epsilon \to 0$.

The impulse function is an extremely useful concept in circuit analysis, and we say more about it in the following chapters. We introduced the concept here so that we can include discontinuities at the origin in our definition of the Laplace transform.

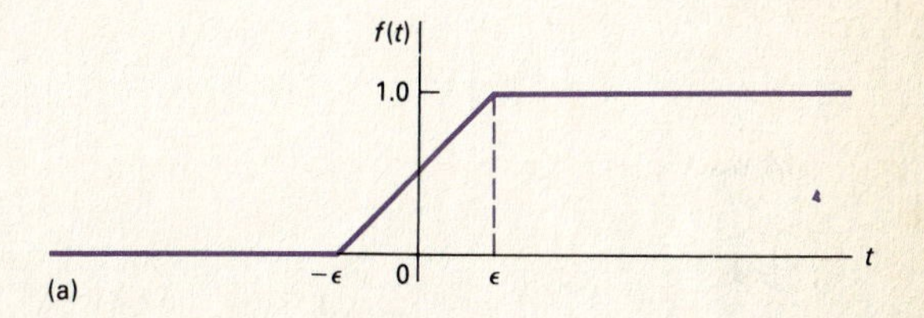

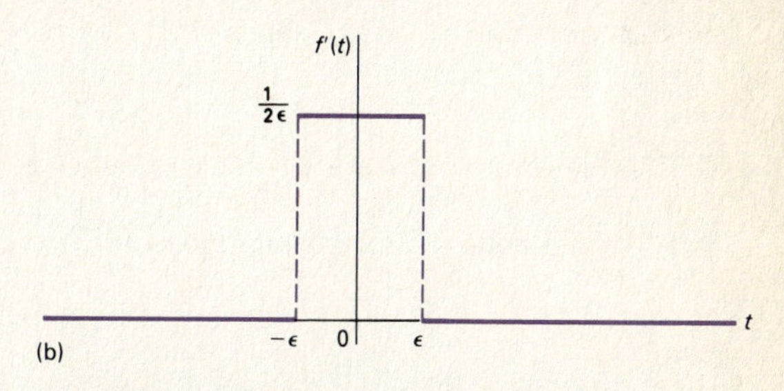

FIGURE 13.13 The impulse function as the derivative of the step function: (a) $f(t) \to u(t)$ as $\epsilon \to 0$; and (b) $f'(t) \to \delta(t)$ as $\epsilon \to 0$.

DRILL EXERCISES

13.2 a) Find the area under the function shown in Fig. 13.12(a).

b) What is the duration of the function when $\epsilon = 0$?

c) What is the magnitude of $f(t)$ when $\epsilon = 0$?

ANSWER: (a) 1; (b) 0; (c) ∞.

13.3 Evaluate the following integrals:

a) $I = \int_{-2}^{4} (t^3 + 4)[\delta(t) + 4\delta(t-2)]dt;$

b) $I = \int_{-3}^{4} t^2[\delta(t) + \delta(t+2.5) + \delta(t-5)]dt.$

ANSWER: (a) 52; (b) 6.25.

13.4 Find $f(t)$ if

$$f(t) = \frac{1}{2\pi}\int_{-\infty}^{\infty} F(\omega)e^{jt\omega}d\omega$$

and

$$F(\omega) = \frac{3 + j\omega}{4 + j\omega}\pi\delta(\omega).$$

ANSWER: 3/8.

13.4 Functional Transforms

A functional transform is simply the Laplace transform of a specified function of t. Because we are limiting our introduction to the unilateral, or one-sided, Laplace transform, we define all functions to be zero for $t < 0^-$.

We derived one functional transform pair in Section 13.3, where we showed that the Laplace transform of the unit impulse function equals 1—see Eq. (13.14). A second illustration is the unit step function of Fig. 13.13(a), where

$$\mathcal{L}\{u(t)\} = \int_{0^-}^{\infty} f(t)e^{-st}dt = \int_{0^+}^{\infty} 1e^{-st}dt$$

$$= \frac{e^{-st}}{-s}\Bigg|_{0^+}^{\infty} = \frac{1}{s}. \tag{13.18}$$

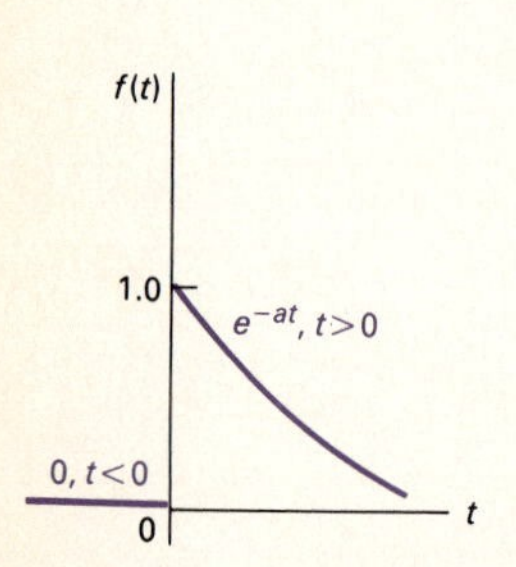

FIGURE 13.14 A decaying exponential function.

Equation (13.18) shows that the Laplace transform of the unit step function is $1/s$.

The Laplace transform of the decaying exponential function shown in Fig. 13.14 is

$$\mathcal{L}\{e^{-at}\} = \int_{0^+}^{\infty} e^{-at}e^{-st}dt = \int_{0^+}^{\infty} e^{-(a+s)t}dt = \frac{1}{s+a}. \tag{13.19}$$

In deriving Eqs. (13.18) and (13.19), we used the fact that integration across the discontinuity at the origin is zero.

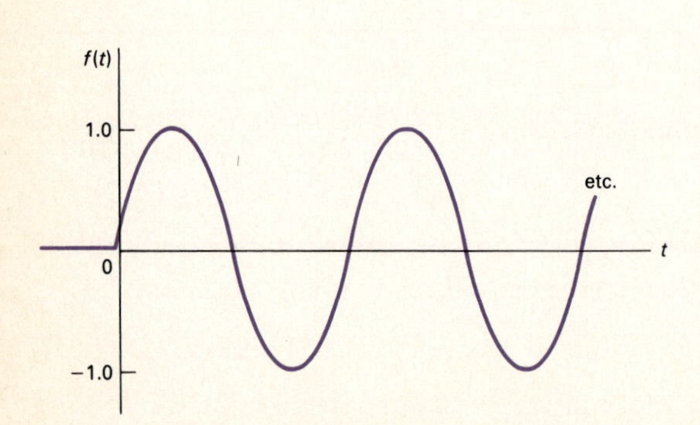

FIGURE 13.15 A sinusoidal function for $t > 0$.

A third illustration of finding a functional transform is the sinusoidal function shown in Fig. 13.15. The expression for $f(t)$ for $t > 0^-$ is $\sin \omega t$; hence the Laplace transform is

$$\mathcal{L}\{\sin \omega t\} = \int_{0^-}^{\infty} (\sin \omega t)e^{-st}dt$$

$$= \int_{0^-}^{\infty} \left(\frac{e^{j\omega t} - e^{-j\omega t}}{2j}\right)e^{-st}dt$$

$$= \int_{0^-}^{\infty} \frac{e^{-(s-j\omega)t} - e^{-(s+j\omega)t}}{2j} dt$$

$$= \frac{1}{2j}\left(\frac{1}{s-j\omega} - \frac{1}{s+j\omega}\right)$$

$$= \frac{\omega}{s^2+\omega^2}. \quad \textbf{(13.20)}$$

Table 13.1 gives an abbreviated list of Laplace transform pairs. It includes the functions of most interest in an introductory course on circuit applications.

TABLE 13.1

AN ABBREVIATED LIST OF LAPLACE TRANSFORM PAIRS

$f(t)(t > 0^-)$	TYPE	$F(s)$
$\delta(t)$	(impulse)	1
$u(t)$	(step)	$\frac{1}{s}$
t	(ramp)	$\frac{1}{s^2}$
e^{-at}	(exponential)	$\frac{1}{s+a}$
$\sin \omega t$	(sine)	$\frac{\omega}{s^2+\omega^2}$
$\cos \omega t$	(cosine)	$\frac{s}{s^2+\omega^2}$
te^{-at}	(damped ramp)	$\frac{1}{(s+a)^2}$
$e^{-at} \sin \omega t$	(damped sine)	$\frac{\omega}{(s+a)^2+\omega^2}$
$e^{-at} \cos \omega t$	(damped cosine)	$\frac{s+a}{(s+a)^2+\omega^2}$

DRILL EXERCISE

13.5 Use the defining integral to

a) prove that the Laplace transform of t is $1/s^2$; and

b) find the Laplace transform of $\cosh \beta t$.

ANSWER: (a) derivation; (b) $s/(s^2 - \beta^2)$.

13.5 Operational Transforms

Operational transforms indicate how mathematical operations performed on either $f(t)$ or $F(s)$ are converted into the opposite domain. The operations of primary interest are (1) multiplication by a constant; (2) addition (subtraction); (3) differentiation; (4) integration; (5) translation in the time domain; (6) translation in the frequency domain; and (7) scale changing.

Multiplication by a Constant

From the defining integral, if

$$\mathcal{L}\{f(t)\} = F(s),$$

then

$$\mathcal{L}\{Kf(t)\} = KF(s). \tag{13.21}$$

Thus, multiplication of $f(t)$ by a constant corresponds to multiplying $F(s)$ by the same constant.

Addition (Subtraction)

Addition (subtraction) in the time domain translates into addition (subtraction) in the frequency domain. Thus if

$$\mathcal{L}\{f_1(t)\} = F_1(s),$$

$$\mathcal{L}\{f_2(t)\} = F_2(s),$$

and

$$\mathcal{L}\{f_3(t)\} = F_3(s),$$

then

$$\mathcal{L}\{f_1(t) + f_2(t) - f_3(t)\} = F_1(s) + F_2(s) - F_3(s), \tag{13.22}$$

which is derived by simply substituting the algebraic sum of time-domain functions into the defining integral.

Differentiation

Differentiation in the time domain corresponds to multiplying $F(s)$ by s and then subtracting the initial value of $f(t)$—that is, $f(0^-)$—from this product:

$$\mathcal{L}\left\{\frac{df(t)}{dt}\right\} = sF(s) - f(0^-), \tag{13.23}$$

which is obtained directly from the definition of the Laplace transform, or

$$\mathcal{L}\left\{\frac{df(t)}{dt}\right\} = \int_{0^-}^{\infty}\left[\frac{df(t)}{dt}\right]e^{-st}dt. \quad \textbf{(13.24)}$$

We evaluate the integral in Eq. (13.24) by integrating by parts. Letting $u = e^{-st}$ and $dv = [df(t)/dt]dt$ yields

$$\mathcal{L}\left\{\frac{df(t)}{dt}\right\} = e^{-st}f(t)\Big|_{0^-}^{\infty} - \int_{0^-}^{\infty} f(t)(-se^{-st}dt). \quad \textbf{(13.25)}$$

Because we are assuming that $f(t)$ is Laplace transformable, the evaluation of $e^{-st}f(t)$ at $t = \infty$ is zero. Therefore the right-hand side of Eq. (13.25) reduces to

$$-f(0^-) + s\int_{0^-}^{\infty} f(t)e^{-st}dt = sF(s) - f(0^-).$$

This observation completes the derivation of Eq. (13.23). It is an important result because it states that differentiation in the time domain reduces to an algebraic operation in the s domain.

We determine the Laplace transform of higher-order derivatives by using Eq. (13.23) as the starting point. For example, to find the Laplace transform of the second derivative of $f(t)$, we first let

$$g(t) = \frac{df(t)}{dt}. \quad \textbf{(13.26)}$$

Now we use Eq. (13.23) to write

$$G(s) = sF(s) - f(0^-). \quad \textbf{(13.27)}$$

But because

$$\frac{dg(t)}{dt} = \frac{d^2f(t)}{dt^2},$$

we write

$$\mathcal{L}\left\{\frac{dg(t)}{dt}\right\} = \mathcal{L}\left\{\frac{d^2f(t)}{dt^2}\right\} = sG(s) - g(0^-). \quad \textbf{(13.28)}$$

Combining Eqs. (13.26), (13.27), and (13.28) gives

$$\mathcal{L}\left\{\frac{d^2f(t)}{dt^2}\right\} = s^2F(s) - sf(0^-) - \frac{df(0^-)}{dt}. \quad \textbf{(13.29)}$$

We find the Laplace transform of the nth derivative by successively applying the preceding process, which leads to the general result

$$\mathcal{L}\left\{\frac{d^nf(t)}{dt^n}\right\} = s^nF(s) - s^{n-1}f(0^-) - s^{n-2}\frac{df(0^-)}{dt} - s^{n-3}\frac{d^2f(0^-)}{dt^2} - \cdots - \frac{d^{n-1}f(0^-)}{dt^{n-1}}. \quad \textbf{(13.30)}$$

INTEGRATION

Integration in the time domain corresponds to dividing by s in the s domain. As before, we establish the relationship by the defining integral:

$$\mathcal{L}\left\{\int_{0^-}^{t} f(x)dx\right\} = \int_{0^-}^{\infty}\left[\int_{0^-}^{t} f(x)dx\right]e^{-st}dt. \tag{13.31}$$

We evaluate the integral on the right-hand side of Eq. (13.31) by integrating by parts, first letting

$$u = \int_{0^-}^{t} f(x)dx \quad \text{and} \quad dv = e^{-st}dt.$$

Then

$$du = f(t)dt \quad \text{and} \quad v = -\frac{e^{-st}}{s}.$$

The integration-by-parts formula yields

$$\mathcal{L}\left\{\int_{0^-}^{t} f(x)dx\right\} = -\frac{e^{-st}}{s}\int_{0^-}^{t} f(x)dx\Bigg|_{0^-}^{\infty} + \int_{0^-}^{\infty}\frac{e^{-st}}{s}f(t)dt. \tag{13.32}$$

The first term on the right-hand side of Eq. (13.32) is zero at both the upper and lower limits. The evaluation at the lower limit obviously is zero, whereas the evaluation at the upper limit is zero because we are assuming that $f(t)$ has a Laplace transform. The second term on the right-hand side of Eq. (13.32) is $F(s)/s$; therefore

$$\mathcal{L}\left\{\int_{0^-}^{t} f(x)dx\right\} = \frac{F(s)}{s}, \tag{13.33}$$

which reveals that the operation of integration in the time domain is transformed to the algebraic operation of multiplying by $1/s$ in the s domain. Equation (13.33) and Eq. (13.30) form the basis of the earlier statement that the Laplace transform translates a set of integrodifferential equations into a set of algebraic equations.

TRANSLATION IN THE TIME DOMAIN

If we start with any function $f(t)u(t)$, we can represent the same function, translated in time by the constant a, as $f(t-a)u(t-a)$.[2] Translation in the time domain corresponds to multiplication by an exponential in the frequency domain. Thus

$$\mathcal{L}\{f(t-a)u(t-a)\} = e^{-as}F(s), \quad a > 0. \tag{13.34}$$

[2]Note that throughout, we multiply any arbitrary function $f(t)$ by the unit step function $u(t)$ to ensure that the resulting function is defined for all positive time.

For example, knowing that

$$\mathcal{L}\{tu(t)\} = \frac{1}{s^2},$$

Eq. (13.34) permits writing the Laplace transform of $(t-a)u(t-a)$ directly:

$$\mathcal{L}\{(t-a)u(t-a)\} = \frac{e^{-as}}{s^2}.$$

The proof of Eq. (13.34) follows from the defining integral:

$$\mathcal{L}\{f(t-a)u(t-a)\} = \int_{0^-}^{\infty} u(t-a)f(t-a)e^{-st}dt$$

$$= \int_{a}^{\infty} f(t-a)e^{-st}dt. \qquad \textbf{(13.35)}$$

In writing Eq. (13.35), we took advantage of $u(t-a) = 1$ for $t > a$. Now we change the variable of integration. Specifically, we let $x = t - a$. Then $x = 0$ when $t = a$, $x = \infty$ when $t = \infty$, and $dx = dt$. Thus we write the integral in Eq. (13.35) as

$$\mathcal{L}\{f(t-a)u(t-a)\} = \int_{0}^{\infty} f(x)e^{-s(x+a)}dx$$

$$= e^{-sa}\int_{0}^{\infty} f(x)e^{-sx}dx$$

$$= e^{-as}F(s),$$

which is what we set out to prove.

Translation in the Frequency Domain

Translation in the frequency domain corresponds to multiplication by an exponential in the time domain:

$$\mathcal{L}\{e^{-at}f(t)\} = F(s+a), \qquad \textbf{(13.36)}$$

which follows from the defining integral. The derivation of Eq. (13.36) is left to Problem 13.15.

We may use the relationship in Eq. (13.36) to derive new transform pairs. Thus, knowing that

$$\mathcal{L}\{\cos\omega t\} = \frac{s}{s^2+\omega^2},$$

we use Eq. (13.36) to deduce that

$$\mathcal{L}\{e^{-at}\cos\omega t\} = \frac{s+a}{(s+a)^2+\omega^2}.$$

SCALE CHANGING

The scale-change property gives the relationship between $f(t)$ and $F(s)$ when the time variable is multiplied by a positive constant:

$$\mathscr{L}\{f(at)\} = \frac{1}{a}F\left(\frac{s}{a}\right), \quad a > 0, \tag{13.37}$$

the derivation of which is left to Problem 13.19. The scale-change property is particularly useful in experimental work, especially where time-scale changes are made to facilitate building a model of a system.

We use Eq. (13.37) to formulate new transform pairs. Thus, knowing that

$$\mathscr{L}\{\cos t\} = \frac{s}{s^2+1},$$

we deduce from Eq. (13.37) that

$$\mathscr{L}\{\cos \omega t\} = \frac{1}{\omega}\frac{s/\omega}{(s/\omega)^2+1} = \frac{s}{s^2+\omega^2}.$$

Table 13.2 gives an abbreviated list of operational transforms. Some entries were not discussed in this section, but you will become more familiar with them by working Problems 13.20 and 13.21.

TABLE 13.2

AN ABBREVIATED LIST OF OPERATIONAL TRANSFORMS

OPERATION	$f(t)$	$F(s)$
Multiplication by a constant	$Kf(t)$	$KF(s)$
Addition/subtraction	$f_1(t)+f_2(t)-f_3(t)+\cdots$	$F_1(s)+F_2(s)-F_3(s)+\cdots$
First derivative (time)	$\frac{df(t)}{dt}$	$sF(s)-f(0^-)$
Second derivative (time)	$\frac{d^2f(t)}{dt^2}$	$s^2F(s)-sf(0^-)-\frac{df(0^-)}{dt}$
nth derivative (time)	$\frac{d^nf(t)}{dt^n}$	$s^nF(s)-s^{n-1}f(0^-)-s^{n-2}\frac{df(0^-)}{dt} - s^{n-3}\frac{df^2(0^-)}{dt^2}-\cdots-\frac{d^{n-1}f(0^-)}{dt^{n-1}}$
Time integral	$\int_0^t f(x)dx$	$\frac{F(s)}{s}$
Translation in time	$f(t-a)u(t-a), a>0$	$e^{-as}F(s)$
Translation in frequency	$e^{-at}f(t)$	$F(s+a)$
Scale changing	$f(at), a>0$	$\frac{1}{a}F\left(\frac{s}{a}\right)$
First derivative (s)	$tf(t)$	$-\frac{dF(s)}{ds}$
nth derivative (s)	$t^nf(t)$	$(-1)^n\frac{d^nF(s)}{ds^n}$
s integral	$\frac{f(t)}{t}$	$\int_s^\infty F(u)du$

DRILL EXERCISE

13.6 Use the appropriate operational transform from Table 13.2 to find the Laplace transform of each function: (a) t^2e^{-at}; (b) $\frac{d}{dt}(e^{-at}\cosh\beta t)$; (c) $t\cos\omega t$.

ANSWER: (a) $\frac{2}{(s+a)^3}$; (b) $\frac{\beta^2 - a(s+a)}{(s+a)^2 - \beta^2}$; (c) $\frac{s^2 - \omega^2}{(s^2+\omega^2)^2}$.

13.6 AN EXAMPLE

We now illustrate how to use the Laplace transform to solve the ordinary integrodifferential equations that describe the behavior of lumped-parameter circuits. Consider the circuit shown in Fig. 13.16. We assume that no initial energy is stored in the circuit at the instant when the switch, which is shorting the dc current source, is opened. The problem is to find the time-domain expression for $v(t)$ when $t \geq 0$.

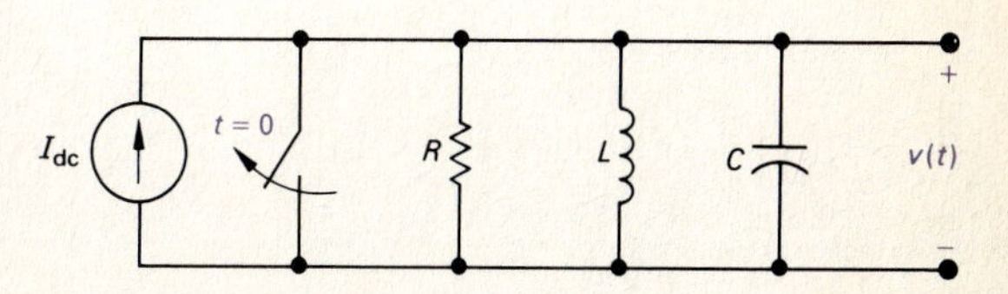

FIGURE 13.16 A parallel *RLC* circuit.

We begin by writing the integrodifferential equation that $v(t)$ must satisfy. We need only a single node-voltage equation to describe the circuit. Summing the currents away from the top node in the circuit generates the equation:

$$\frac{v(t)}{R} + \frac{1}{L}\int_0^t v(x)dx + C\frac{dv(t)}{dt} = I_{dc}u(t). \tag{13.38}$$

Note that in writing Eq. (13.38), we indicated the opening of the switch in the step jump of the source current from zero to I_{dc}.

After deriving the integrodifferential equations (in this example, just one), we transform the equations to the s domain. We will not go through the steps of the transformation in detail, because in Chapter 14 we will discover how to bypass them and generate the s-domain equations directly. Briefly though, we use three operational transforms and one functional transform on Eq. (13.38) to obtain

$$\frac{V(s)}{R} + \frac{1}{L}\frac{V(s)}{s} + C[sV(s) - v(0^-)] = I_{dc}\left(\frac{1}{s}\right), \tag{13.39}$$

an algebraic equation in which $V(s)$ is the unknown variable. We are assuming that the circuit parameters R, L, and C, as well as the source current I_{dc}, are known; the initial voltage on the capacitor $v(0^-)$ is zero because the initial energy stored in the circuit is zero. Thus we have reduced the problem to solving an algebraic equation.

Next we solve the algebraic equations (again, just one in this case) for the unknowns. Solving Eq. (13.39) for $V(s)$ gives

$$V(s)\left(\frac{1}{R}+\frac{1}{sL}+sC\right)=\frac{I_{\text{dc}}}{s}$$

$$V(s)=\frac{I_{\text{dc}}/C}{s^2+(1/RC)s+(1/LC)}. \quad \textbf{(13.40)}$$

To find $v(t)$ we must inverse-transform the expression for $V(s)$. We denote this inverse operation

$$v(t)=\mathcal{L}^{-1}\{V(s)\}. \quad \textbf{(13.41)}$$

The next step in the analysis is to find the inverse transform of the s-domain expression; this is the subject of Section 13.7. In that section we also present a final, critical step: checking the validity of the resulting time-domain expression. The need for such checking is not unique to the Laplace transform; conscientious and prudent engineers always test any derived solution to be sure it makes sense in terms of known system behavior.

Simplifying the notation now is advantageous. We do so by dropping the parenthetical t in time-domain expressions and the parenthetical s in frequency-domain expressions. We use lowercase letters for all time-domain variables, and we represent the corresponding s-domain variables with uppercase letters. Thus

$$\mathcal{L}\{v\}=V \quad \text{or} \quad v=\mathcal{L}^{-1}\{V\},$$

$$\mathcal{L}\{i\}=I \quad \text{or} \quad i=\mathcal{L}^{-1}\{I\},$$

$$\mathcal{L}\{f\}=F \quad \text{or} \quad f=\mathcal{L}^{-1}\{F\},$$

and so on.

DRILL EXERCISE

13.7 In the circuit shown in Fig. 13.16, the dc current source is replaced with a sinusoidal source that delivers a current of $1.2\cos t$ A. The circuit components are $R = 1\ \Omega$, $C = 0.625$ F, and $L = 1.6$ H. Find the numerical expression for V.

ANSWER: $V=\dfrac{1.92s^2}{(s^2+1.6s+1)(s^2+1)}$.

13.7 INVERSE TRANSFORMS

The expression for $V(s)$ in Eq. (13.40) is a **rational** function of s—that is, one that can be expressed in the form of a ratio of two polynomials in s such that no nonintegral powers of s appear in the polynomials. In fact, for linear, lumped-parameter circuits whose component values are constant, the s-domain expressions for the unknown voltages and currents are always rational functions of s. (You may verify this observation by working Problems 13.22, 13.29, and 13.31.) If we can inverse-transform rational functions of s, we can solve for the time-domain expressions for the voltages and currents. The purpose of this section is to present a straightforward and systematic technique for finding the inverse transform of a rational function.

In general, we need to find the inverse transform of a function that has the form

$$F(s) = \frac{N(s)}{D(s)} = \frac{a_n s^n + a_{n-1}s^{n-1} + \cdots + a_1 s + a_0}{b_m s^m + b_{m-1}s^{m-1} + \cdots + b_1 s + b_0}. \quad \textbf{(13.42)}$$

The coefficients a and b are real constants, and the exponents m and n are positive integers. The ratio $N(s)/D(s)$ is called a **proper rational function** if $m > n$, and an **improper rational function** if $m \leq n$. Only a proper rational function can be expanded as a sum of partial fractions. This restriction poses no problem, as we show at the end of this section.

PARTIAL FRACTION EXPANSION: PROPER RATIONAL FUNCTIONS

A proper rational function is expanded into a sum of partial fractions by writing a term or a series of terms for each root of $D(s)$. Thus $D(s)$ must be in factored form before we can make a partial fraction expansion. For each distinct root of $D(s)$, a single term appears in the sum of partial fractions. For each multiple root of $D(s)$ of multiplicity r, the expansion contains r terms. For example, in the rational function

$$\frac{s+6}{s(s+3)(s+1)^2},$$

the denominator has four roots. Two of these roots are distinct—namely, at $s = 0$ and $s = -3$. A multiple root of multiplicity 2 occurs at $s = -1$. Thus the partial fraction expansion of this function takes the form

$$\frac{s+6}{s(s+3)(s+1)^2} \equiv \frac{K_1}{s} + \frac{K_2}{s+3} + \frac{K_3}{(s+1)^2} + \frac{K_4}{s+1}. \quad \textbf{(13.43)}$$

The key to the partial fraction technique for finding inverse transforms lies in recognizing the $f(t)$ corresponding to each term in the sum of partial fractions. From Table 13.1 you should be able to verify that

$$\mathscr{L}^{-1}\left\{\frac{s+6}{s(s+3)(s+1)^2}\right\} = (K_1 + K_2e^{-3t} + K_3te^{-t} + K_4e^{-t})u(t). \tag{13.44}$$

All that remains is to establish a technique for determining the coefficients (K_1, K_2, K_3, ...) generated by making a partial fraction expansion. There are four general forms this problem can take. Specifically, the roots of $D(s)$ are either (1) real and distinct; (2) complex and distinct; (3) real and repeated; or (4) complex and repeated. Before we consider each situation in turn, a few general comments are in order.

We used the identity sign $\equiv$ in Eq. (13.43) to emphasize that expanding a rational function into a sum of partial fractions establishes an identical equation. Thus both sides of the equation must be the same for all values of the variable s. Also, the identity relationship must hold when both sides are subjected to the same mathematical operation. These characteristics are pertinent to determining the coefficients, as we will see.

Be sure to verify that the rational function is proper. This check is important because nothing in the procedure for finding the various K's will alert you to nonsense results if the rational function is improper. We present a procedure for checking the K's, but you can avoid wasted effort by forming the habit of asking yourself, "Is $F(s)$ a proper rational function?"

Use a computer tool to check your coefficient values.

Partial Fraction Expansion: Distinct Real Roots of $D(s)$

We first consider determining the coefficients in a partial fraction expansion when all the roots of $D(s)$ are real and distinct. To find a K associated with a term that arises because of a distinct root of $D(s)$, we multiply both sides of the identity by a factor equal to the denominator beneath the desired K. Then when we evaluate both sides of the identity at the root corresponding to the multiplying factor, the right-hand side is always the desired K, and the left-hand side is always its numerical value. For example,

$$F(s) = \frac{96(s+5)(s+12)}{s(s+8)(s+6)} \equiv \frac{K_1}{s} + \frac{K_2}{s+8} + \frac{K_3}{s+6}. \tag{13.45}$$

To find the value of K_1, we multiply both sides by s and then evaluate both sides at $s = 0$:

$$\left.\frac{96(s+5)(s+12)}{(s+8)(s+6)}\right|_{s=0} \equiv K_1 + \left.\frac{K_2 s}{s+8}\right|_{s=0} + \left.\frac{K_3 s}{s+6}\right|_{s=0}$$

or

$$\frac{96(5)(12)}{8(6)} \equiv K_1 = 120. \tag{13.46}$$

To find the value of K_2, we multiply both sides by $s + 8$ and then evaluate both sides at $s = -8$:

$$\left.\frac{96(s+5)(s+12)}{s(s+6)}\right|_{s=-8} \equiv \left.\frac{K_1(s+8)}{s}\right|_{s=-8} + K_2 + \left.\frac{K_3(s+8)}{(s+6)}\right|_{s=-8}$$

or

$$\frac{96(-3)(4)}{(-8)(-2)} = K_2 = -72. \tag{13.47}$$

Then K_3 is

$$\left.\frac{96(s+5)(s+12)}{s(s+8)}\right|_{s=-6} = K_3 = 48. \tag{13.48}$$

From Eq. (13.45) and the K values obtained,

$$\frac{96(s+5)(s+12)}{s(s+8)(s+6)} \equiv \frac{120}{s} + \frac{48}{s+6} - \frac{72}{s+8}. \tag{13.49}$$

At this point, testing the result to protect against computational errors is a good idea. As we already mentioned, a partial fraction expansion creates an identity; thus both sides of Eq. (13.49) must be the same for all s values. The choice of test values is completely open; hence we choose values that are easy to verify. For example, in Eq. (13.49), testing at either -5 or -12 is attractive because in both cases the left-hand side reduces to zero. Choosing -5 yields

$$\frac{120}{-5} + \frac{48}{1} - \frac{72}{3} = -24 + 48 - 24 = 0,$$

whereas testing -12 gives

$$\frac{120}{-12} + \frac{48}{-6} - \frac{72}{-4} = -10 - 8 + 18 = 0.$$

Now confident that the numerical values of the various K's are correct, we proceed to find the inverse transform:

$$\mathscr{L}^{-1}\left\{\frac{96(s+5)(s+12)}{s(s+8)(s+6)}\right\} = (120 + 48e^{-6t} - 72e^{-8t})u(t). \tag{13.50}$$

DRILL EXERCISES

13.8 Find $f(t)$ if

$$F(s) = \frac{16(s^2+3s+9)}{(s+2)(s+4)(s+6)}.$$

ANSWER: $f(t) = (14e^{-2t} - 52e^{-4t} + 54e^{-6t})u(t)$.

13.9 Find $f(t)$ if

$$F(s) = \frac{2s+12}{(s+1)(s^2+5s+6)}.$$

ANSWER: $f(t) = (5e^{-t} - 8e^{-2t} + 3e^{-3t})u(t)$.

PARTIAL FRACTION EXPANSION: DISTINCT COMPLEX ROOTS OF $D(s)$

The only difference between finding the coefficients associated with distinct complex roots and finding those associated with distinct real roots is that the algebra in the former involves complex numbers. We illustrate by expanding the rational function:

$$F(s) = \frac{100(s+3)}{(s+6)(s^2+6s+25)}. \tag{13.51}$$

We begin by noting that $F(s)$ is a proper rational function. Next we must find the roots of the quadratic term $s^2 + 6s + 25$:

$$s^2 + 6s + 25 = (s+3-j4)(s+3+j4). \tag{13.52}$$

With the denominator in factored form, we proceed as before:

$$\frac{100(s+3)}{(s+6)(s^2+6s+25)} \equiv \frac{K_1}{s+6} + \frac{K_2}{s+3-j4} + \frac{K_3}{s+3+j4}. \tag{13.53}$$

To find K_1, K_2, and K_3, we use the same process as before:

$$K_1 = \left.\frac{100(s+3)}{s^2+6s+25}\right|_{s=-6} = \frac{100(-3)}{25} = -12; \tag{13.54}$$

$$K_2 = \left.\frac{100(s+3)}{(s+6)(s+3+j4)}\right|_{s=-3+j4} = \frac{100(j4)}{(3+j4)(j8)}$$

$$= 6 - j8 = 10e^{-j53.13^\circ}; \tag{13.55}$$

$$K_3 = \left.\frac{100(s+3)}{(s+6)(s+3-j4)}\right|_{s=-3-j4} = \frac{100(-j4)}{(3-j4)(-j8)}$$

$$= 6 + j8 = 10e^{j53.13^\circ}. \tag{13.56}$$

Then

$$\frac{100(s+3)}{(s+6)(s^2+6s+25)} = \frac{-12}{s+6} + \frac{10\angle{-53.13^\circ}}{s+3-j4} + \frac{10\angle{53.13^\circ}}{s+3+j4}. \quad \textbf{(13.57)}$$

Again, we need to make some observations. First, in physically realizable circuits, complex roots always appear in conjugate pairs. Second, the coefficients associated with these conjugate pairs are themselves conjugates. Note, for example, that K_3 (Eq. [13.56]) is the conjugate of K_2 (Eq. [13.55]). Thus for complex conjugate roots, you actually need to calculate only half the coefficients.

Before inverse-transforming Eq. (13.57), we check the partial fraction expansion numerically. Testing at -3 is attractive because the left-hand side reduces to zero at this value:

$$\begin{aligned} F(s) &= \frac{-12}{3} + \frac{10\angle{-53.13^\circ}}{-j4} + \frac{10\angle{53.13^\circ}}{j4} \\ &= -4 + 2.5\angle{36.87^\circ} + 2.5\angle{-36.87^\circ} \\ &= -4 + 2.0 + j1.5 + 2.0 - j1.5 = 0. \end{aligned}$$

We now proceed to inverse-transform Eq. (13.57):

$$\mathscr{L}^{-1}\left\{\frac{100(s+3)}{(s+6)(s^2+6s+25)}\right\} = (-12e^{-6t} + 10e^{-j53.13^\circ}e^{-(3-j4)t} + 10e^{j53.13^\circ}e^{-(3+j4)t})u(t). \quad \textbf{(13.58)}$$

In general, having the function in the time domain contain imaginary components is undesirable. Fortunately, because the terms involving imaginary components always come in conjugate pairs, we can eliminate the imaginary components simply by adding the pairs:

$$\begin{aligned} &10e^{-j53.13^\circ}e^{-(3-j4)t} + 10e^{j53.13^\circ}e^{-(3+j4)t} \\ &= 10e^{-3t}\left(e^{j(4t-53.13^\circ)} + e^{-j(4t-53.13^\circ)}\right) \\ &= 20e^{-3t}\cos(4t - 53.13^\circ), \end{aligned} \quad \textbf{(13.59)}$$

which enables us to simplify Eq. (13.58):

$$\mathscr{L}^{-1}\left\{\frac{100(s+3)}{(s+6)(s^2+6s+25)}\right\} = [-12e^{-6t} + 20e^{-3t}\cos(4t - 53.13^\circ)]u(t). \quad \textbf{(13.60)}$$

Because distinct complex roots appear frequently in lumped-parameter linear circuit analysis, we need to summarize these results with a new transform pair. Whenever $D(s)$ contains distinct complex roots—that is, factors of the form $(s+\alpha-j\beta)(s+\alpha+j\beta)$—a pair of terms of the form

$$\frac{K}{s+\alpha-j\beta}+\frac{K^*}{s+\alpha+j\beta} \tag{13.61}$$

appears in the partial fraction expansion, where the partial fraction coefficient is, in general, a complex number. In polar form,

$$K = |K|e^{j\theta} = |K|\underline{/\theta}, \tag{13.62}$$

where $|K|$ denotes the magnitude of the complex coefficient. Then

$$K^* = |K|e^{-j\theta} = |K|\underline{/-\theta}. \tag{13.63}$$

The complex conjugate pair in Eq. (13.61) always inverse-transforms as

$$\mathcal{L}^{-1}\left\{\frac{K}{s+\alpha-j\beta}+\frac{K^*}{s+\alpha+j\beta}\right\} = 2|K|e^{-\alpha t}\cos(\beta t+\theta). \tag{13.64}$$

In applying Eq. (13.64) it is important to note that K is defined as the coefficient associated with the denominator term $s+\alpha-j\beta$, while K^* is defined as the coefficient associated with the denominator $s+\alpha+j\beta$.

DRILL EXERCISES

13.10 Find $f(t)$ if

$$F(s) = 10(s^2+119)/[(s+5)(s^2+10s+169)].$$

ANSWER: $f(t) = (10e^{-5t} - 8.33e^{-5t}\sin 12t)u(t)$.

13.11 Find $v(t)$ in Drill Exercise 13.7.

ANSWER:
$v(t) = [2e^{-0.8t}\cos(0.6t + 233.13°) + 1.2\cos t]u(t)$.

PARTIAL FRACTION EXPANSION: REPEATED REAL ROOTS OF $D(s)$

To find the coefficients associated with the terms generated by a multiple root of multiplicity r, we multiply both sides of the identity by the multiple root raised to its rth power. We find the K appearing over the factor raised to the rth power by evaluating both sides of the identity at the multiple root. To find the remaining $(r-1)$ coefficients, we differentiate both sides of the identity $(r-1)$ times. At the end of each differentiation, we evaluate both sides of the identity at the multiple root. The right-hand side is always the desired K, and the left-hand side is always its numerical value. For example,

$$\frac{180(s+30)}{s(s+5)(s+3)^2} = \frac{K_1}{s} + \frac{K_2}{s+5} + \frac{K_3}{(s+3)^2} + \frac{K_4}{s+3}. \quad \textbf{(13.65)}$$

We find K_1 and K_2 as previously described; that is,

$$K_1 = \left.\frac{180(s+30)}{(s+5)(s+3)^2}\right|_{s=0} = \frac{180(30)}{(5)(9)} = 120 \quad \textbf{(13.66)}$$

and

$$K_2 = \left.\frac{180(s+30)}{s(s+3)^2}\right|_{s=-5} = \frac{180(25)}{(-5)(4)} = -225. \quad \textbf{(13.67)}$$

To find K_3, we multiply both sides by $(s+3)^2$ and then evaluate both sides at -3:

$$\left.\frac{180(s+30)}{s(s+5)}\right|_{s=-3} = \left.\frac{K_1(s+3)^2}{s}\right|_{s=-3} + \left.\frac{K_2(s+3)^2}{s+5}\right|_{s=-3}$$
$$+ K_3 + K_4(s+3)|_{s=-3}; \quad \textbf{(13.68)}$$

$$\frac{180(27)}{(-3)(2)} = K_1 \times 0 + K_2 \times 0 + K_3 + K_4 \times 0$$
$$= K_3 = -810. \quad \textbf{(13.69)}$$

To find K_4 we first must multiply both sides of Eq. (13.65) by $(s+3)^2$. Next we differentiate both sides once with respect to s and then evaluate at $s=-3$:

$$\frac{d}{ds}\left[\frac{180(s+30)}{s(s+5)}\right]_{s=-3} = \frac{d}{ds}\left[\frac{K_1(s+3)^2}{s}\right]_{s=-3}$$
$$+ \frac{d}{ds}\left[\frac{K_2(s+3)^2}{s+5}\right]_{s=-3}$$
$$+ \left.\frac{d}{ds}(K_3)\right|_{s=-3}$$
$$+ \left.\frac{d}{ds}[K_4(s+3)]\right|_{s=-3}, \quad \textbf{(13.70)}$$

$$180\left[\frac{s(s+5)-(s+30)(2s+5)}{s^2(s+5)^2}\right]_{s=-3} = K_4, \quad \textbf{(13.71)}$$

or

$$180\left[\frac{(-3)(2)-(27)(-1)}{(9)(4)}\right] = K_4 = 105. \quad \textbf{(13.72)}$$

Then

$$\frac{180(s+30)}{s(s+5)(s+3)^2} = \frac{120}{s} - \frac{225}{s+5} - \frac{810}{(s+3)^2} + \frac{105}{s+3}. \quad (13.73)$$

At this point, check your coefficient values by testing both sides of Eq. (13.73) at $s = 30$. If the right-hand side yields zero, you can then inverse-transform each partial fraction to yield

$$\mathscr{L}^{-1}\left\{\frac{180(s+30)}{s(s+5)(s+3)^2}\right\} = (120{-}225e^{-5t}{-}810te^{-3t}{+}105e^{-3t})u(t). \quad (13.74)$$

DRILL EXERCISE

13.12 Find $f(t)$ if $F(s) = (4s^2 + 7s + 1)/[s(s+1)^2]$.

ANSWER: $f(t) = (1 + 2te^{-t} + 3e^{-t})u(t)$.

PARTIAL FRACTION EXPANSION: REPEATED COMPLEX ROOTS OF *D*(*s*)

We handle repeated complex roots in the same way that we did repeated real roots; the only difference is that the algebra involves complex numbers. Recall that complex roots always appear in conjugate pairs and that the coefficients associated with a conjugate pair are also conjugates, so that only half the K's need be evaluated. For example,

$$F(s) = \frac{768}{(s^2 + 6s + 25)^2}. \quad (13.75)$$

After factoring the denominator polynomial, we write

$$\begin{aligned} F(s) &= \frac{768}{(s+3-j4)^2(s+3+j4)^2} \\ &= \frac{K_1}{(s+3-j4)^2} + \frac{K_2}{s+3-j4} \\ &\quad + \frac{K_1^*}{(s+3+j4)^2} + \frac{K_2^*}{s+3+j4}. \end{aligned} \quad (13.76)$$

Now we need to evaluate only K_1 and K_2, because K_1^* and K_2^* are conjugate values. The value of K_1 is

$$K_1 = \left.\frac{768}{(s+3+j4)^2}\right|_{s=-3+j4}$$

$$= \frac{768}{(j8)^2} = -12. \tag{13.77}$$

The value of K_2 is

$$K_2 = \frac{d}{ds}\left[\left.\frac{768}{(s+3+j4)^2}\right|_{s=-3+j4}\right]$$

$$= -\left.\frac{2(768)}{(s+3+j4)^3}\right|_{s=-3+j4}$$

$$= -\frac{2(768)}{(j8)^3}$$

$$= -j3 = 3\angle{-90^\circ}. \tag{13.78}$$

From Eqs. (13.77) and (13.78),

$$K_1^* = -12 \tag{13.79}$$

and

$$K_2^* = j3 = 3\angle{90^\circ}. \tag{13.80}$$

We now group the partial fraction expansion by conjugate terms to obtain

$$F(s) = \left[\frac{-12}{(s+3-j4)^2} + \frac{-12}{(s+3+j4)^2}\right] + \left(\frac{3\angle{-90^\circ}}{s+3-j4} + \frac{3\angle{90^\circ}}{s+3+j4}\right). \tag{13.81}$$

We now write the inverse transform of $F(s)$:

$$f(t) = [-24te^{-3t}\cos 4t + 6e^{-3t}\cos(4t - 90^\circ)]u(t). \tag{13.82}$$

Note that if $F(s)$ has a real root a of multiplicity r in its denominator, the term in a partial fraction expansion is of the form

$$\frac{K}{(s+a)^r}.$$

The inverse transform of this term is

$$\mathcal{L}^{-1}\left\{\frac{K}{(s+a)^r}\right\} = \frac{Kt^{r-1}e^{-at}}{(r-1)!}u(t). \tag{13.83}$$

If $F(s)$ has a complex root of $\alpha + j\beta$ of multiplicity r in its denominator, the term in partial fraction expansion is the conjugate pair

$$\frac{K}{(s+\alpha-j\beta)^r} + \frac{K^*}{(s+\alpha+j\beta)^r}.$$

The inverse transform of this pair is

$$\mathscr{L}^{-1}\left\{\frac{K}{(s+\alpha-j\beta)^r} + \frac{K^*}{(s+\alpha+j\beta)^r}\right\}$$

$$= \left[\frac{2|K|t^{r-1}}{(r-1)!}e^{-\alpha t}\cos(\beta t+\theta)\right]u(t). \tag{13.84}$$

Equations (13.83) and (13.84) are the key to being able to inverse-transform any partial fraction expansion by inspection. One further note regarding these two equations: In most circuit analysis problems, r is seldom greater than 2. Therefore the inverse transform of a rational function can be handled with four transform pairs. Table 13.3 lists these pairs.

TABLE 13.3

FOUR USEFUL TRANSFORM PAIRS

PAIR NUMBER	NATURE OF ROOTS	$F(S)$	$f(t)$
1	Distinct real	$\frac{K}{s+a}$	$Ke^{-at}u(t)$
2	Repeated real	$\frac{K}{(s+a)^2}$	$Kte^{-at}u(t)$
3	Distinct complex	$\frac{K}{s+\alpha-j\beta} + \frac{K^*}{s+\alpha+j\beta}$	$2\|K\|e^{-\alpha t}\cos(\beta t+\theta)u(t)$
4	Repeated complex	$\frac{K}{(s+\alpha-j\beta)^2} + \frac{K^*}{(s+\alpha+j\beta)^2}$	$2t\|K\|e^{-\alpha t}\cos(\beta t+\theta)u(t)$

Note: In pairs 1 and 2, K is a real quantity, whereas in pairs 3 and 4, K is the complex quantity $|K|\underline{/\theta}$.

DRILL EXERCISE

13.13 Find $f(t)$ if $F(s) = 40/(s^2+4s+5)^2$.

ANSWER: $f(t) = (-20te^{-2t}\cos t + 20e^{-2t}\sin t)u(t)$.

Partial Fraction Expansion: Improper Rational Functions

We conclude the discussion of partial fraction expansions by returning to an observation made at the beginning of this section, namely, that improper rational functions pose no serious problem in finding inverse transforms. An improper rational function can always be expanded into a polynomial plus a proper rational function. The polynomial is then inverse-transformed into impulse functions and derivatives of impulse functions. The proper rational function is inverse-transformed by the techniques outlined in this section. To illustrate the procedure, we use the function

$$F(s) = \frac{s^4 + 13s^3 + 66s^2 + 200s + 300}{s^2 + 9s + 20}. \quad \textbf{(13.85)}$$

Dividing the denominator into the numerator until the remainder is a proper rational function gives

$$F(s) = s^2 + 4s + 10 + \frac{30s + 100}{s^2 + 9s + 20}, \quad \textbf{(13.86)}$$

where the term $(30s + 100)/(s^2 + 9s + 20)$ is the remainder.

Next we expand the proper rational function into a sum of partial fractions:

$$\frac{30s + 100}{s^2 + 9s + 20} = \frac{30s + 100}{(s + 4)(s + 5)} = \frac{-20}{s + 4} + \frac{50}{s + 5}. \quad \textbf{(13.87)}$$

Substituting Eq. (13.87) into Eq. (13.86) yields

$$F(s) = s^2 + 4s + 10 - \frac{20}{s + 4} + \frac{50}{s + 5}. \quad \textbf{(13.88)}$$

Now we can inverse-transform Eq. (13.88) by inspection. Hence

$$f(t) = \frac{d^2\delta(t)}{dt^2} + 4\frac{d\delta(t)}{dt} + 10\delta(t) - (20e^{-4t} - 50e^{-5t})u(t). \quad \textbf{(13.89)}$$

DRILL EXERCISES

13.14 Find $f(t)$ if $F(s) = s^2/[(s + 1)(s + 2)]$.

ANSWER: $f(t) = \delta(t) + (e^{-t} - 4e^{-2t})u(t)$.

13.15 Find $f(t)$ if

$$F(s) = (2s^3 + 8s^2 + 2s - 4)/(s^2 + 5s + 4).$$

ANSWER: $f(t) = 2\frac{d\delta}{dt} - 2\delta + 4e^{-4t}u(t)$.

13.8 POLES AND ZEROS OF F(s)

The rational function of Eq. (13.42) also may be expressed as the ratio of two factored polynomials. In other words, we may write $F(s)$ as

$$F(s) = \frac{K(s+z_1)(s+z_2)\cdots(s+z_n)}{(s+p_1)(s+p_2)\cdots(s+p_m)}, \tag{13.90}$$

where K is the constant a_n/b_m. For example, we may also write the function

$$F(s) = \frac{8s^2+120s+400}{2s^4+20s^3+70s^2+100s+48}$$

as

$$F(s) = \frac{8(s^2+15s+50)}{2(s^4+10s^3+35s^2+50s+24)}$$

$$= \frac{4(s+5)(s+10)}{(s+1)(s+2)(s+3)(s+4)}. \tag{13.91}$$

Use a computer tool to factor the numerator and denominator polynomials.

The roots of the denominator polynomial—that is, $-p_1, -p_2, -p_3, \ldots, -p_m$—are called the **poles of F(s)**; they are the values of s at which $F(s)$ becomes infinitely large. In the function described by Eq. (13.91), the poles of $F(s)$ are -1, -2, -3, and -4.

The roots of the numerator polynomial—that is, $-z_1, -z_2, -z_3, \ldots, -z_n$—are called the **zeros of F(s)**; they are the values of s at which $F(s)$ becomes zero. In the function described by Eq. (13.91), the zeros of $F(s)$ are -5 and -10.

In what follows you may find that being able to visualize the poles and zeros of $F(s)$ as points on a complex s plane is helpful. A complex plane is needed because the roots of the polynomials may be complex. In the complex s plane we use the horizontal axis to plot the real values of s, and the vertical axis to plot the imaginary values of s.

As an example of plotting the poles and zeros of $F(s)$, consider the function

$$F(s) = \frac{10(s+5)(s+3-j4)(s+3+j4)}{s(s+10)(s+6-j8)(s+6+j8)}. \tag{13.92}$$

The poles of $F(s)$ are at 0, -10, $-6+j8$, and $-6-j8$. The zeros are at -5, $-3+j4$, and $-3-j4$. Figure 13.17 shows the poles and zeros plotted on the s plane, where X's represent poles and O's represent zeros.

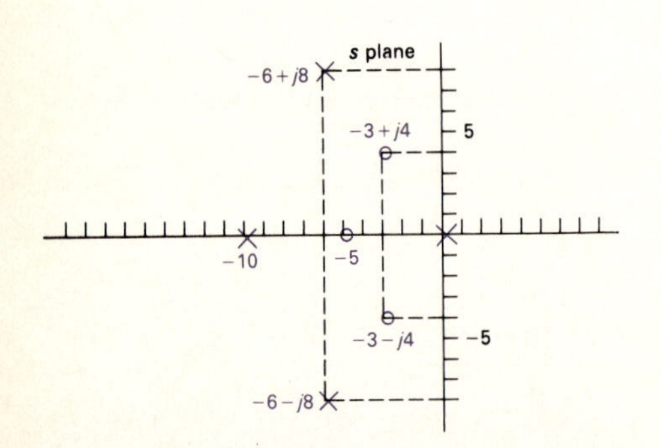

FIGURE 13.17 Plotting poles and zeros on the *s* plane.

Note that the poles and zeros for Eq. (13.90) are located in the finite s plane. $F(s)$ can also have either an rth-order pole or an rth-order zero at infinity. For example, the function described by Eq. (13.91) has a second-order zero at infinity, because for large values of s the function reduces to $4/s^2$, and $F(s) = 0$ when $s = \infty$. In this text, we are interested in the poles and zeros located in the finite s plane. Therefore, when we refer to the poles and zeros of a rational function of s, we are referring to the finite poles and zeros.

13.9 Initial- and Final-Value Theorems

The initial- and final-value theorems are useful because they enable us to determine from $F(s)$ the behavior of $f(t)$ at 0 and ∞. Hence we can check the initial and final values of $f(t)$ to see if they conform with known circuit behavior, before actually finding the inverse transform of $F(s)$.

The initial-value theorem states that

$$\lim_{t\to 0^+} f(t) = \lim_{s\to\infty} sF(s), \tag{13.93}$$

and the final-value theorem states that

$$\lim_{t\to\infty} f(t) = \lim_{s\to 0} sF(s). \tag{13.94}$$

The initial-value theorem is based on the assumption that $f(t)$ contains no impulse functions. In Eq. (13.94), we must add the restriction that the theorem is valid only if the poles of $F(s)$—except for a first-order pole at the origin—lie in the left half of the s plane.

To prove Eq. (13.93), we start with the operational transform of the first derivative:

$$\mathcal{L}\left\{\frac{df}{dt}\right\} = sF(s) - f(0^-) = \int_{0^-}^{\infty} \frac{df}{dt} e^{-st}\,dt. \tag{13.95}$$

Now we take the limit as $s \to \infty$:

$$\lim_{s\to\infty} [sF(s) - f(0^-)] = \lim_{s\to\infty} \int_{0^-}^{\infty} \frac{df}{dt} e^{-st}\,dt. \tag{13.96}$$

Observe that the right-hand side of Eq. (13.96) may be written as

$$\lim_{s\to\infty}\left(\int_{0^-}^{0^+} \frac{df}{dt} e^{0}\,dt + \int_{0^+}^{\infty} \frac{df}{dt} e^{-st}\,dt\right).$$

As $s \to \infty$, $(df/dt)e^{-st} \to 0$; hence the second integral vanishes in the limit. The first integral reduces to $f(0^+) - f(0^-)$, which is independent of s. Thus the right-hand side of Eq. (13.96) becomes

$$\lim_{s\to\infty} \int_{0^-}^{\infty} \frac{df}{dt} e^{-st}\,dt = f(0^+) - f(0^-). \tag{13.97}$$

Because $f(0^-)$ is independent of s, the left-hand side of Eq. (13.96) may be written

$$\lim_{s\to\infty}[sF(s) - f(0^-)] = \lim_{s\to\infty}[sF(s)] - f(0^-). \tag{13.98}$$

From Eqs. (13.97) and (13.98),

$$\lim_{s\to\infty} sF(s) = f(0^+) = \lim_{t\to 0^+} f(t),$$

which completes the proof of the initial-value theorem.

The proof of the final-value theorem also starts with Eq. (13.95). Here we take the limit as $s \to 0$:

$$\lim_{s\to 0}[sF(s) - f(0^-)] = \lim_{s\to 0}\left(\int_{0^-}^{\infty} \frac{df}{dt}e^{-st}dt\right). \tag{13.99}$$

The integration is with respect to t and the limit operation is with respect to s, so the right-hand side of Eq. (13.99) reduces to

$$\lim_{s\to 0}\left(\int_{0^-}^{\infty} \frac{df}{dt}e^{-st}dt\right) = \int_{0^-}^{\infty} \frac{df}{dt}dt. \tag{13.100}$$

Because the upper limit on the integral is infinite, this integral may also be written as a limit process:

$$\int_{0^-}^{\infty} \frac{df}{dt}dt = \lim_{t\to\infty}\int_{0^-}^{t} \frac{df}{dy}dy, \tag{13.101}$$

where we use y as the symbol of integration to avoid confusion with the upper limit on the integral. Carrying out the integration process yields

$$\lim_{t\to\infty}[f(t) - f(0^-)] = \lim_{t\to\infty}[f(t)] - f(0^-). \tag{13.102}$$

Substituting Eq. (13.102) into Eq. (13.99) gives

$$\lim_{s\to 0}[sF(s)] - f(0^-) = \lim_{t\to\infty}[f(t)] - f(0^-). \tag{13.103}$$

Because $f(0^-)$ cancels, Eq. (13.103) reduces to the final-value theorem, namely,

$$\lim_{s\to 0} sF(s) = \lim_{t\to\infty} f(t).$$

The final-value theorem is useful only if $f(\infty)$ exists. This condition is true only if all the poles of $F(s)$—except for a simple pole at the origin—lie in the left half of the s plane.

THE APPLICATION OF INITIAL- AND FINAL-VALUE THEOREMS

To illustrate the application of the initial- and final-value theorems, we apply them to a function we used to illustrate partial fraction expansions. Consider the transform pair given by Eq. (13.60). The initial-value theorem gives

$$\lim_{s\to\infty} sF(s) = \lim_{s\to\infty} \frac{100s^2[1+(3/s)]}{s^3[1+(6/s)][1+(6/s)+(25/s^2)]} = 0;$$

$$\lim_{t\to 0^+} f(t) = [-12 + 20\cos(-53.13^\circ)](1) = -12 + 12 = 0.$$

The final-value theorem gives

$$\lim_{s\to 0} sF(s) = \lim_{s\to 0} \frac{100s(s+3)}{(s+6)(s^2+6s+25)} = 0;$$

$$\lim_{t\to\infty} f(t) = \lim_{t\to\infty} [-12e^{-6t} + 20e^{-3t}\cos(4t - 53.13^\circ)]u(t) = 0.$$

In applying the theorems to Eq. (13.60), we already had the time-domain expression and were merely testing our understanding. But the real value of the initial- and final-value theorems lies in being able to test the s-domain expressions before working out the inverse transform. For example, consider the expression for $V(s)$ given by Eq. (13.40). Although we cannot calculate $v(t)$ until the circuit parameters are specified, we can check to see if $V(s)$ predicts the correct values of $v(0^+)$ and $v(\infty)$. We know from the statement of the problem that generated $V(s)$ that $v(0^+)$ is zero. We also know that $v(\infty)$ must be zero because the ideal inductor is a perfect short circuit across the dc current source. Finally, we know that the poles of $V(s)$ must lie in the left half of the s plane because R, L, and C are positive constants. Hence the poles of $sV(s)$ also lie in the left half of the s plane.

Applying the initial-value theorem yields

$$\lim_{s\to\infty} sV(s) = \lim_{s\to\infty} \frac{s(I_{dc}/C)}{s^2[1+1/(RCs)+1/(LCs^2)]} = 0.$$

Applying the final-value theorem gives

$$\lim_{s\to 0} sV(s) = \lim_{s\to 0} \frac{s(I_{dc}/C)}{s^2+(s/RC)+(1/LC)} = 0.$$

The derived expression for $V(s)$ correctly predicts the initial and final values of $v(t)$.

DRILL EXERCISES

13.16 Use the initial- and final-value theorems to find the initial and final values of $f(t)$ in Drill Exercises 13.9, 13.12, and 13.13.

ANSWER: 0, 0; 4, 1; and 0, 0.

13.17 a) Use the initial-value theorem to find the initial value of v in Drill Exercise 13.7.

b) Can the final-value theorem be used to find the steady-state value of v? Why?

ANSWER: (a) 0; (b) no, because V has a pair of poles on the imaginary axis.

SUMMARY

- The **Laplace transform** is a tool for converting time-domain equations into frequency-domain equations, according to the following general definition:

$$\mathcal{L}\{f(t)\} = \int_0^\infty f(t)e^{-st}dt = F(s),$$

where $f(t)$ is the time-domain expression, and $F(s)$ is the frequency-domain expression.

- A **functional transform** is the Laplace transform of a specific function. Important functional transform pairs are summarized in Table 13.1.

- **Operational transforms** define the general mathematical properties of the Laplace transform. Important operational transform pairs are summarized in Table 13.2.

- The **step function** $Ku(t)$ describes a function that experiences a discontinuity from one constant level to another at some point in time. K is the magnitude of the jump; if $K = 1$, $Ku(t)$ is the **unit step function**.

- The **impulse function** $K\delta(t)$ is defined

$$\int_{-\infty}^{\infty} K\delta(t)dt = K;$$

$$\delta(t) = 0, \quad t \neq 0.$$

K is the strength of the impulse; if $K = 1$, $K\delta(t)$ is the **unit impulse function**.

- In linear lumped-parameter circuits, $F(s)$ is a rational function of s.

- If $F(s)$ is a proper rational function, the inverse transform is found by a partial fraction expansion.

- If $F(s)$ is an improper rational function, it can be inverse-transformed by first expanding it into a sum of a polynomial and a proper rational function.
- $F(s)$ can be expressed as the ratio of two factored polynomials. The roots of the denominator are called **poles** and are plotted as X's on the complex s plane. The roots of the numerator are called **zeros** and are plotted as O's on the complex s plane.
- The initial-value theorem states that

$$\lim_{t\to 0^+} f(t) = \lim_{s\to\infty} sF(s).$$

  The theorem assumes that $f(t)$ contains no impulse function.
- The final-value theorem states that

$$\lim_{t\to\infty} f(t) = \lim_{s\to 0^+} sF(s).$$

  The theorem is valid only if the poles of $F(s)$—except for a first-order pole at the origin—lie in the left half of the s plane.
- The initial- and final-value theorems allow us to predict the initial and final values of $f(t)$ from an s-domain expression.

PROBLEMS

13.1 Use step functions to write the expression for each of the functions shown in Fig. P13.1.

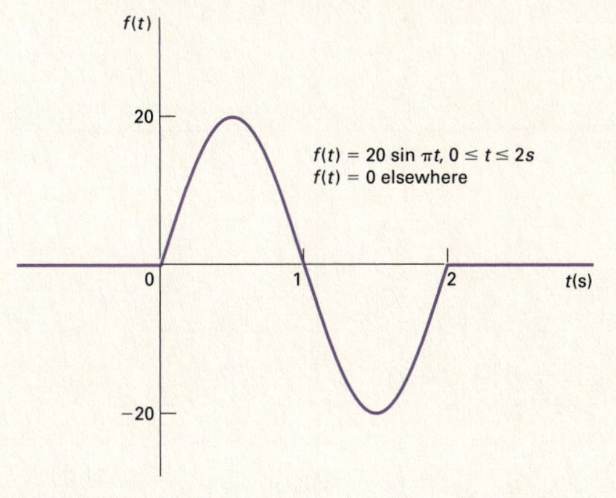

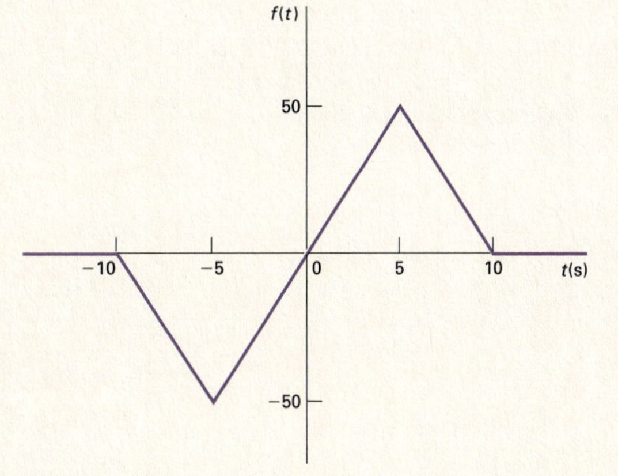

FIGURE P13.1

13.2 Step functions can be used to define a *window* function. Thus $u(t-1) - u(t-4)$ defines a window 1 unit high and 3 units wide located on the time axis between 1 and 4.

A function $f(t)$ is defined as follows:

$$f(t) = 0, \quad t \le 0;$$

$$f(t) = -20t, \quad 0 \le t \le 1s;$$

$$f(t) = -20, \quad 1s \le t \le 2s;$$

$$f(t) = 20 \cos \frac{\pi}{2}t, \quad 2s \le t \le 4s;$$

$$f(t) = 100 - 20t, \quad 4s \le t \le 5s;$$

$$f(t) = 0, \quad 5s \le t \le \infty.$$

a) Sketch $f(t)$ over the interval $-1s \le t \le 6s$

b) Use the concept of the window function to write an expression for $f(t)$.

13.3 Make a sketch of $f(t)$ for $-15s \le t \le 35s$ when $f(t)$ is given by the following expression:
$f(t) = (100 + 10t)u(t+10) - (50 + 10t)u(t+5)$
$+ (50 - 10t)u(t-5) - (150 - 10t)u(t-15)$
$+ (10t - 250)u(t-25) - (10t - 300)u(t-30)$.

13.4 Explain why the following function generates an impulse function as $\epsilon \to 0$:

$$f(t) = \frac{\epsilon/\pi}{\epsilon^2 + t^2}, \quad -\infty \le t \le \infty.$$

13.5 In Section 13.3, we used the sifting property of the impulse function to show that $\mathcal{L}\{\delta(t)\} = 1$. Show that we can obtain the same result by finding the Laplace transform of the rectangular pulse that exists between $\pm\epsilon$ in Fig. 13.9 and then finding the limit of this transform as $\epsilon \to 0$.

13.6 The triangular pulses shown in Fig. P13.6 are equivalent to the rectangular pulses in Fig. 13.12(b) because they both enclose the same area ($1/\epsilon$) and they both approach infinity proportional to $1/\epsilon^2$ as $\epsilon \to 0$. Use this triangular-pulse representation for $\delta'(t)$ to find the Laplace transform of $\delta''(t)$.

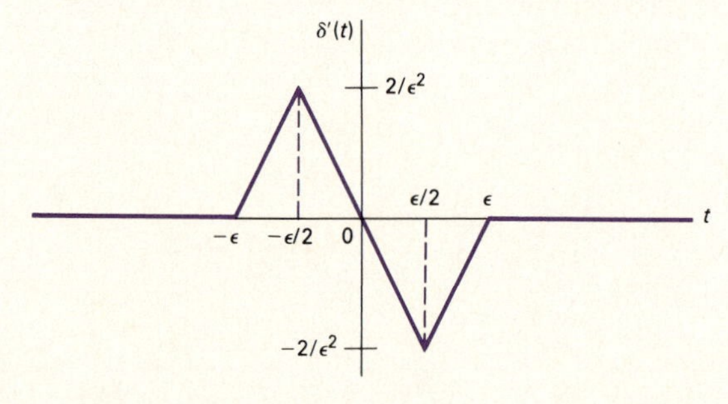

FIGURE P13.6

13.7 a) Show that

$$\int_{-\infty}^{\infty} f(t)\delta'(t-a)dt = -f'(a).$$

(*Hint:* Integrate by parts.)

b) Use the formula in part (a) to show that

$$\mathcal{L}\{\delta'(t)\} = s.$$

13.8 Find the Laplace transform of each of the following functions:

a) $f(t) = te^{-at}$;

b) $f(t) = \cos\ \omega t$;

c) $f(t) = \cos\ (\omega t + \theta)$;

d) $f(t) = \sinh\ t$;

e) $f(t) = \sinh\ (t + \theta)$.

13.9 Find the Laplace transform (when $\epsilon \to 0$) of the derivative of the exponential function illustrated in Fig. 13.8, using each of the following two methods:

a) First differentiate the function and then find the transform of the resulting function.

b) Use the operational transform given by Eq. (13.23).

13.10 Show that

$$\mathcal{L}\{\delta^{(n)}(t)\} = s^n.$$

13.11 a) Find the Laplace transform of te^{-at}.

b) Use the operational transform given by Eq. (13.23) to find the Laplace transform of $\frac{d}{dt}(te^{-at})$.

c) Check your result in part (b) by first differentiating and then transforming the resulting expression.

13.12 a) Find $\mathcal{L}\left\{\frac{d}{dt}\sin\omega t\right\}$.

b) Find $\mathcal{L}\left\{\frac{d}{dt}\cos\omega t\right\}$.

c) Find $\mathcal{L}\left\{\frac{d^3}{dt^3}t^2\right\}$.

d) Check the results of parts (a), (b), and (c) by first differentiating and then transforming.

13.13 a) Find the Laplace transform of $\int_{0^-}^{t} x\,dx$ by first integrating and then transforming.

b) Check the result obtained in part (a) by using the operational transform given by Eq. (13.33).

13.14 a) Find $\mathcal{L}\left\{\int_{0^-}^{t} e^{-ax}\,dx\right\}$.

b) Find $\mathcal{L}\left\{\int_{0^-}^{t} y\,dy\right\}$.

c) Check the results of parts (a) and (b) by first integrating and then transforming.

13.15 Show that

$$\mathcal{L}\{e^{-at}f(t)\} = F(s+a).$$

13.16 Find the Laplace transform of each of the following functions:

a) $f(t) = 40e^{-8(t-3)}u(t-3)$;

b) $f(t) = (5t-10)[u(t-2)-u(t-4)] + (30-5t)[u(t-4)-u(t-8)] - (5t-50)[u(t-8)-u(t-10)]$.

13.17 a) Find the Laplace transform of the function illustrated in Fig. P13.17.

b) Find the Laplace transform of the first derivative of the function illustrated in Fig. P13.17.

c) Find the Laplace transform of the second derivative of the function illustrated in Fig. P13.17.

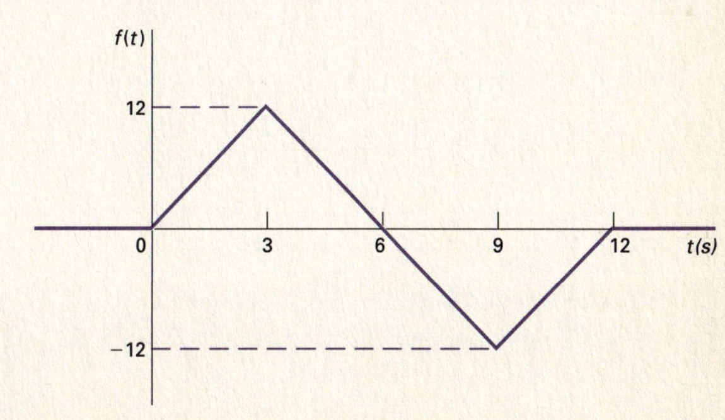

FIGURE P13.17

13.18 Find the Laplace transform for (a) and (b).

a) $f(t) = \frac{d}{dt}(e^{-at} \sin \omega t)$.

b) $f(t) = \int_{0^-}^{t} e^{-ax} \cos \omega x \, dx$.

c) Verify the results obtained in parts (a) and (b) by first carrying out the indicated mathematical operation and then finding the Laplace transform.

13.19 a) Show that

$$\mathcal{L}\{f(at)\} = \frac{1}{a} F\left(\frac{s}{a}\right).$$

b) Use the result of part (a) along with the answer derived in Problem 13.8(d) to find

$$\mathcal{L}\{\sinh \beta t\}.$$

13.20 a) Given that $F(s) = \mathcal{L}\{f(t)\}$, show that

$$-\frac{dF(s)}{ds} = \mathcal{L}\{tf(t)\}.$$

b) Show that

$$(-1)^n \frac{d^n F(s)}{ds^n} = \mathcal{L}\{t^n f(t)\}.$$

c) Use the result of part (b) to find $\mathcal{L}\{t^5\}$, $\mathcal{L}\{t \sin \beta t\}$, and $\mathcal{L}\{te^{-t} \cosh t\}$.

13.21 a) Show that if $F(s) = \mathcal{L}\{f(t)\}$ and $\{f(t)/t\}$ is Laplace-transformable, then

$$\int_s^\infty F(u) du = \mathcal{L}\left\{\frac{f(t)}{t}\right\}.$$

(*Hint:* Use the defining integral to write

$$\int_s^\infty F(u) du = \int_s^\infty \left(\int_{0-}^\infty f(t) e^{-ut} dt \right) du$$

and then reverse the order of integration.)

b) Start with the result obtained in Problem 13.20(c) for $\mathcal{L}\{t \sin \beta t\}$ and use the operational transform given in part (a) of this problem to find $\mathcal{L}\{\sin \beta t\}$.

13.22 There is no energy stored in the circuit shown in Fig. P13.22 at the time the switch is closed.

a) Derive the integrodifferential equation that governs the behavior of the current i_o.

b) Show that

$$I_o(s) = \frac{V_{dc}/L}{s^2 + (R/L)s + (1/LC)}.$$

c) Show that

$$V_o(s) = \frac{V_{dc}/LC}{s[s^2 + (R/L)s + (1/LC)]}.$$

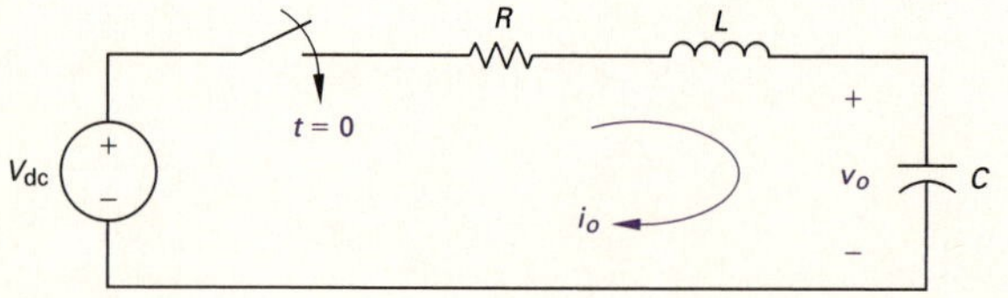

FIGURE P13.22

13.23 The circuit parameters in the circuit seen in Fig. P13.22 have the following values: $R = 250\ \Omega$, $L = 50$ mH, $C = 5\ \mu$F, and $V_{dc} = 48$ V.

a) Find $i_o(t)$ for $t \geq 0$.

b) Find $v_o(t)$ for $t \geq 0$.

13.24 The switch in the circuit in Fig. P13.24 has been open for a long time. At $t = 0$, the switch closes.

a) Derive the integrodifferential equation that governs the behavior of the voltage v_o for $t \geq 0$.

b) Show that

$$V_o(s) = \frac{(1/RC)V_{dc}}{s^2 + (1/RC)s + (1/LC)}.$$

c) Show that

$$I_o(s) = \frac{(1/RLC)V_{dc}}{s[s^2 + (1/RC)s + (1/LC)]}.$$

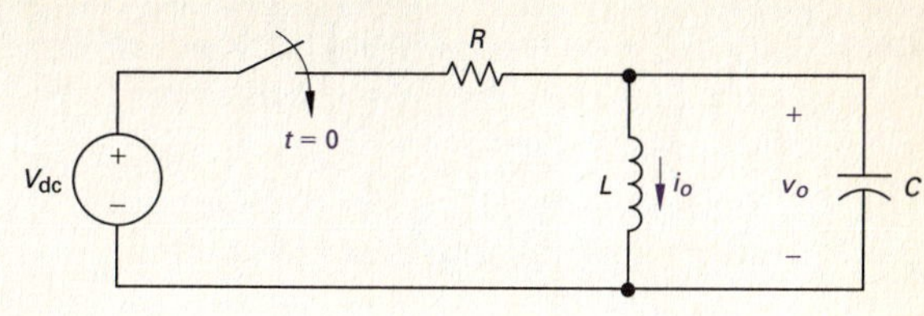

FIGURE P13.24

13.25 The circuit parameters in the circuit in Fig. P13.24 are $R = 5$ kΩ; $L = 200$ mH; and $C = 0.1$ μF. If V_{dc} is 35 V, find

a) $v_o(t)$ for $t \geq 0$; and

b) $i_o(t)$ for $t \geq 0$.

13.26 Find $f(t)$ for each of the following functions:

a) $F(s) = \dfrac{6s^2 + 26s + 26}{(s + 1)(s + 2)(s + 3)}$.

b) $F(s) = \dfrac{13s^3 + 134s^2 + 392s + 288}{s(s + 2)(s + 4)(s + 6)}$.

c) $F(s) = \dfrac{20s^2 + 16s + 12}{(s + 1)(s^2 + 2s + 5)}$.

d) $F(s) = \dfrac{16s^2 - 30s + 500}{s(s^2 + 2s + 50)}$.

13.27 Find $f(t)$ for each of the following functions:

a) $F(s) = \dfrac{100}{s^2(s + 5)}$.

b) $F(s) = \dfrac{50(s + 5)}{s(s + 1)^2}$.

c) $F(s) = \dfrac{100(s + 3)}{s^2(s^2 + 6s + 10)}$.

d) $F(s) = \dfrac{5(s + 2)^2}{s(s + 1)^3}$.

e) $F(s) = \dfrac{400}{s(s^2 + 4s + 5)^2}$.

13.28 Find $f(t)$ for each of the following functions:

a) $F(s) = \dfrac{s^2 + 6s + 7}{(s + 1)(s + 2)}$.

b) $F(s) = \dfrac{6s^2 + 100s + 4150}{s^2 + 14s + 625}$.

c) $F(s) = \dfrac{10s^3 + 105s^2 + 216s + 104}{s^2 + 10s + 16}$.

13.29 The switch in the circuit in Fig. P13.29 has been in position a for a long time. At $t = 0$, the switch moves instantaneously to position b.

a) Derive the integrodifferential equation that governs the behavior of the current i_o for $t \geq 0^+$.

b) Show that

$$I_o(s) = \frac{I_{dc}[s + (1/RC)]}{[s^2 + (1/RC)s + (1/LC)]}.$$

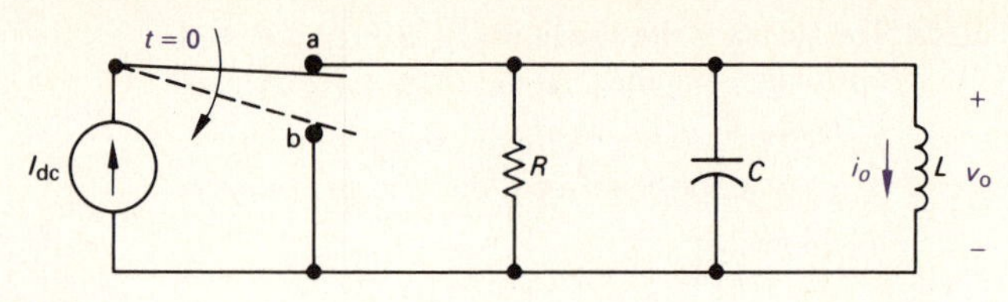

FIGURE P13.29

13.30 The circuit parameters in the circuit in Fig. P13.29 are $R = 500\ \Omega$, $L = 250$ mH, and $C = 0.25\ \mu$F. If $I_{dc} = 5$ mA, find $i_o(t)$ for $t \geq 0$.

13.31 There is no energy stored in the circuit shown in Fig. P13.31 at the time the switch is opened.

a) Derive the integrodifferential equations that govern the behavior of the node voltages v_1 and v_2.

b) Show that

$$V_2(s) = \frac{sI_g}{C[s^2 + (R/L)s + (1/LC)]}.$$

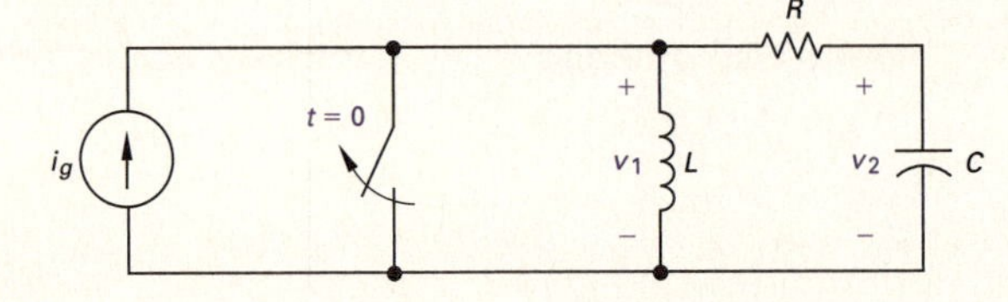

FIGURE P13.31

13.32 The circuit parameters in the circuit in Fig. P13.31 are $R = 2500\ \Omega$; $L = 0.5$ H; and $C = 0.5\ \mu$F. If $i_g(t) = 15u(t)$ mA, find $v_2(t)$.

13.33 Derive the transform pair given by Eq. (13.64).

13.34 a) Derive the transform pair given by Eq. (13.83).

b) Derive the transform pair given by Eq. (13.84).

13.35 Apply the initial- and final-value theorems to each transform pair in Problem 13.26.

13.36 Apply the initial- and final-value theorems to each transform pair in Problem 13.27.

13.37 Use the initial- and final-value theorems to check the initial and final values of the current and voltage in Problem 13.22.

13.38 Use the initial- and final-value theorems to check the initial and final values of the voltage and current in Problem 13.24.

13.39 Use the initial- and final-value theorems to check the initial and final values of the current in Problem 13.29.

13.40 a) Write the two simultaneous differential equations that describe the circuit shown in Fig. P13.40 in terms of the mesh currents i_1 and i_2.

b) Laplace-transform the equations derived in part (a). Assume that the initial energy stored in the circuit is zero.

c) Solve the equations in part (b) for $I_1(s)$ and $I_2(s)$.

d) Find $i_1(t)$ and $i_2(t)$.

e) Find $i_1(\infty)$ and $i_2(\infty)$.

f) Do the solutions for i_1 and i_2 make sense? Explain.

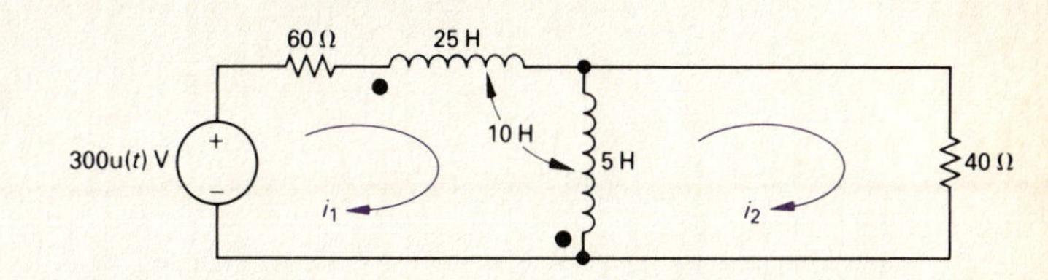

FIGURE P13.40

14 THE LAPLACE TRANSFORM IN CIRCUIT ANALYSIS

CHAPTER CONTENTS

The Laplace transform has two characteristics that make it an attractive tool in circuit analysis. First, it transforms a set of linear constant-coefficient differential equations into a set of linear polynomial equations, which are easier to manipulate. Second, it automatically introduces into the polynomial equations the initial values of the current and voltage variables. Thus, initial conditions are an inherent part of the transform process. (This contrasts with the classical approach to the solution of differential equations, in which initial conditions are considered when the unknown coefficients are evaluated.)

We begin this chapter by showing how we can skip the step of writing time-domain integrodifferential equations and transforming them into the s domain. In Section 14.1, we'll develop the s-domain circuit models for resistors, inductors, and capacitors so that we can write s-domain equations for all circuits directly. Section 14.2 reviews Ohm's and Kirchhoff's laws in the context of the s domain. After establishing these fundamentals, we apply the Laplace transform method to a variety of circuit problems in Section 14.3.

Analytical and simplification techniques first introduced with resistive circuits—such as mesh-current and node-voltage methods, and source transformations—can be used in the s-domain as well. After solving for the circuit response in the s domain, we inverse-transform back to the time domain, using partial fraction expansion (as demonstrated in the preceding chapter). As before, checking the final time-domain equations in terms of the initial conditions and final values is an important step in the solution process.

The s-domain descriptions of circuit input and output lead us, in Section 14.4, to the concept of the transfer function. The transfer function for a particular circuit is the ratio of the Laplace transform of its output to the Laplace transform of its input. In future chapters we'll examine the design uses of the transfer function, but here we focus on its use as an analytical tool. We continue this chapter with a look at the role of partial fraction expansion (Section 14.5) and the convolution integral (Section 14.6) in employing the transfer function in circuit analysis. We conclude with a look at the impulse function in circuit analysis.

14.1 CIRCUIT ELEMENTS IN THE *s* DOMAIN

The procedure for developing an s-domain equivalent circuit for each circuit element is simple. First, we write the time-domain equation that relates the terminal voltage to the terminal current. Next, we take the Laplace transform of the time-domain equation. This step generates an algebraic relationship between the s-domain current and voltage. Note that the dimension of a transformed voltage is volt-seconds, and the dimension of a transformed current is ampere-seconds. A voltage-to-current ratio in the s domain carries the dimension of volts per ampere. An impedance in the s domain is measured in ohms, and an admittance is measured in siemens. Finally, we construct a circuit model that satisfies the relationship between the s-domain current and voltage. We use the passive sign convention in all the derivations.

A Resistor in the s Domain

We begin with the resistance element. From Ohm's law,

$$v = Ri. \tag{14.1}$$

Because R is a constant, the Laplace transform of Eq. (14.1) is

$$V = RI, \tag{14.2}$$

where

$$V = \mathcal{L}\{v\} \quad \text{and} \quad I = \mathcal{L}\{i\}.$$

Equation (14.3) states that the s-domain equivalent circuit of a resistor is simply a resistance of R ohms that carries a current of I ampere-seconds and has a terminal voltage of V volt-seconds.

Figure 14.1 shows the time- and frequency-domain circuits of the resistor. Note that going from the time domain to the frequency domain does not change the resistance element.

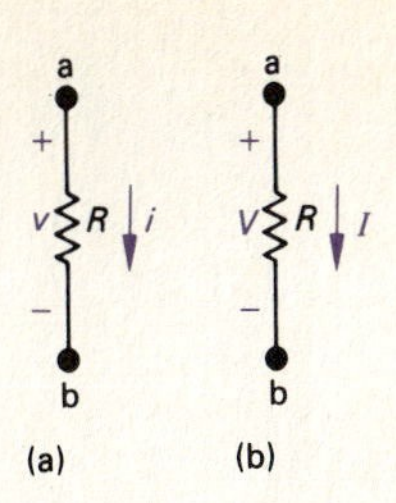

FIGURE 14.1 The resistance element: (a) time domain and (b) frequency domain.

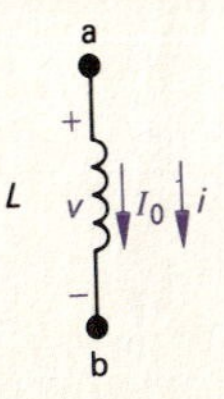

FIGURE 14.2 An inductor of L henrys carrying an initial current of I_0 amperes.

An Inductor in the s Domain

Figure 14.2 shows an inductor carrying an initial current of I_0 amperes. The time-domain equation that relates the terminal voltage to the terminal current is

$$v = L\frac{di}{dt}. \tag{14.3}$$

The Laplace transform of Eq. (14.3) gives

$$V = L[sI - i(0^-)] = sLI - LI_0. \tag{14.4}$$

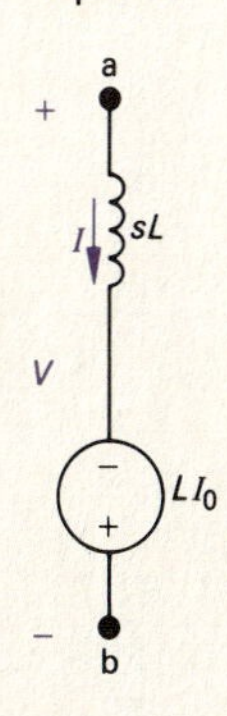

FIGURE 14.3 The series equivalent circuit for an inductor of L henrys carrying an initial current of I_0 amperes.

Two different circuit configurations satisfy Eq. (14.4). The first consists of an impedance of sL ohms in series with an independent voltage source of LI_0 volt-seconds, as shown in Fig. 14.3. Note that the polarity marks on the voltage source LI_0 agree with the minus sign in Eq. (14.4). Note also that LI_0 carries its own algebraic sign; that is, if the initial value of i is opposite to the reference direction for i, then I_0 has a negative value.

The second s-domain equivalent circuit that satisfies Eq. (14.4) consists of an impedance of sL ohms in parallel with an independent current source of I_0/s ampere-seconds, as shown in Fig. 14.4. We can derive the alternative equivalent circuit shown in Fig. 14.4 in several ways. One way is simply to solve Eq. (14.4) for the current I and then construct the circuit to satisfy the resulting equation. Thus

$$I = \frac{V + LI_0}{sL} = \frac{V}{sL} + \frac{I_0}{s}. \tag{14.5}$$

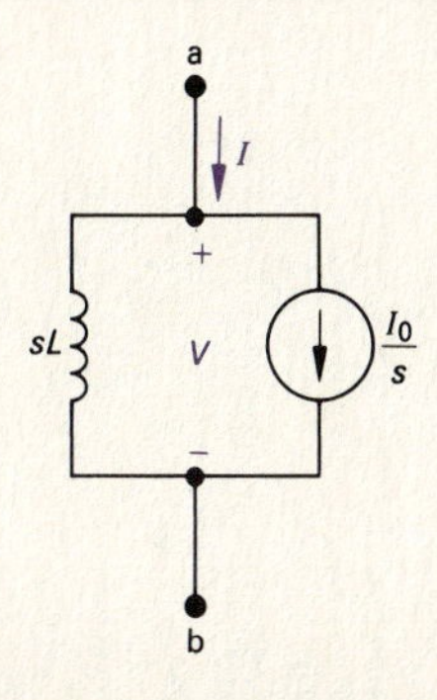

FIGURE 14.4 The parallel equivalent circuit for an inductor of L henrys carrying an initial current of I_0 amperes.

Two other ways are (1) to find the Norton equivalent of the circuit shown in Fig. 14.3 and (2) to start with the inductor current as a function of the inductor voltage and then find the Laplace transform of the resulting integral equation. We leave these two approaches to Problems 14.1 and 14.2.

If the initial energy stored in the inductor is zero—that is, if $I_0 = 0$—the s-domain equivalent circuit of the inductor reduces to an inductor with an impedance of sL ohms. Figure 14.5 shows this circuit.

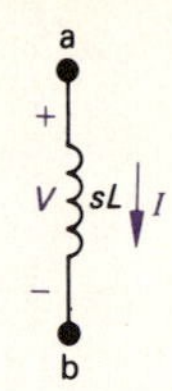

FIGURE 14.5 The s-domain circuit for an inductor when the initial current is zero.

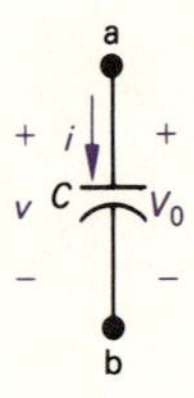

FIGURE 14.6 A capacitor of C farads initially charged to V_0 volts.

A CAPACITOR IN THE s DOMAIN

An initially charged capacitor also has two s-domain equivalent circuits. Figure 14.6 shows a capacitor initially charged to V_0 volts. The terminal current is

$$i = C\frac{dv}{dt}. \tag{14.6}$$

Transforming Eq. (14.6) yields

$$I = C[sV - v(0^-)]$$

or

$$I = sCV - CV_0, \tag{14.7}$$

which indicates that the s-domain current I is the sum of two branch currents. One branch consists of an admittance of sC siemens, and the second branch consists of an independent current source of CV_0 ampere-seconds. Figure 14.7 shows this parallel equivalent circuit.

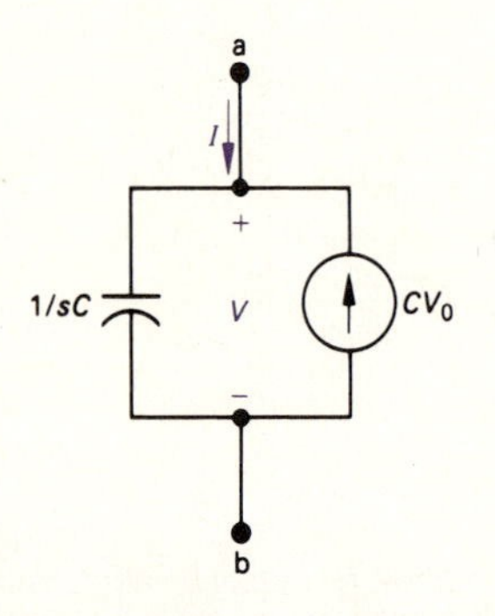

FIGURE 14.7 The parallel equivalent circuit for a capacitor initially charged to V_0 volts.

We derive the series equivalent circuit for the charged capacitor by solving Eq. (14.7) for V:

$$V = \left(\frac{1}{sC}\right)I + \frac{V_0}{s}. \tag{14.8}$$

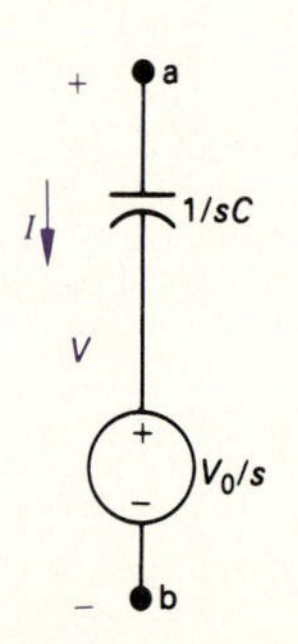

FIGURE 14.8 The series equivalent circuit for a capacitor initially charged to V_0 volts.

Figure 14.8 shows the circuit that satisfies Eq. (14.8).

In the equivalent circuits shown in Figs. 14.7 and 14.8, V_0 carries its own algebraic sign. In other words, if the polarity of V_0 is opposite to the reference polarity for v, V_0 is a negative quantity. If the initial voltage on the capacitor is zero, both equivalent circuits reduce to an impedance of $1/sC$ ohms, as shown in Fig. 14.9.

In this chapter, an important first problem-solving step will be to choose between the parallel or series equivalents when inductors and capacitors are present. With a little forethought and some experience, the correct choice will often be quite evident. The equivalent circuits are summarized in Table 14.1.

TABLE 14.1

SUMMARY OF THE s-DOMAIN EQUIVALENT CIRCUITS

TIME DOMAIN	FREQUENCY DOMAIN	
$v = Ri$	$V = RI$	
$v = L\,di/dt$, $i = \frac{1}{L}\int_{0^-}^{t} v\,dx + I_0$	$V = sLI - LI_0$	$I = \frac{V}{sL} + \frac{I_0}{s}$
$i = C\,dv/dt$, $v = \frac{1}{C}\int_{0^-}^{t} i\,dx + V_0$	$V = \frac{I}{sC} + \frac{V_0}{s}$	$I = sCV - CV_0$

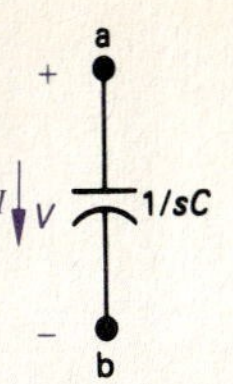

FIGURE 14.9 The s-domain circuit for a capacitor when the initial voltage is zero.

14.2 CIRCUIT ANALYSIS IN THE s DOMAIN

Before illustrating how to use the s-domain equivalent circuits in analysis, we need to lay some groundwork.

First, we know that if no energy is stored in the inductor or capacitor, the relationship between the terminal voltage and current for each passive element takes the form

$$V = ZI, \tag{14.9}$$

where Z refers to the s-domain impedance of the element. Thus a resistor has an impedance of R ohms, an inductor has an impedance of sL ohms, and a capacitor has an impedance of $1/sC$ ohms. The relationship contained in Eq. (14.9) is also contained in Figs. 14.1(b), 14.5, and 14.9. Equation (14.9) is sometimes referred to as Ohm's law for the s domain.

The reciprocal of the impedance is admittance. Therefore the s-domain admittance of a resistor is $1/R$ siemens, an inductor has an admittance of $1/sL$ siemens, and a capacitor has an admittance of sC siemens.

The rules for combining impedances and admittances in the s domain are the same as those for frequency-domain circuits. Thus series-parallel simplifications and Δ-to-Y conversions also are applicable to s-domain analysis.

In addition, Kirchhoff's laws apply to s-domain currents and voltages. Their applicability stems from the operational transform stating that the Laplace transform of a sum of time-domain functions is the sum of the transforms of the individual functions (see Table 13.2). Because the algebraic sum of the currents at a node is zero in the time domain, the algebraic sum of the transformed currents is also zero. A similar statement holds for the algebraic sum of the transformed voltages around a closed path. The s-domain version of Kirchhoff's laws is

$$\text{alg} \sum I\text{'s} = 0; \quad \textbf{(14.10)}$$

$$\text{alg} \sum V\text{'s} = 0. \quad \textbf{(14.11)}$$

Because the voltage and current at the terminals of a passive element are related by an algebraic equation and because Kirchhoff's laws still hold, all the techniques of circuit analysis developed for pure resistive networks may be used in s-domain analysis. Thus node voltages, mesh currents, source transformations, and Thévenin-Norton equivalents all are valid techniques, even when energy is stored initially in the inductors and capacitors. Initially stored energy requires that we modify Eq. (14.9) by simply adding independent sources either in series or parallel with the element impedances. The addition of these sources is governed by Kirchhoff's laws.

DRILL EXERCISES

14.1 A 125 Ω resistor, a 4 mH inductor, and a 0.1 μF capacitor are connected in parallel.

a) Express the admittance of this parallel combination of elements as a rational function of s.

b) Compute the numerical values of the zeros and poles.

ANSWER: (a) $10^{-7}(s^2 + 80{,}000s + 25 \times 10^8)/s$; (b) $-z_1 = -40{,}000 - j30{,}000$; $-z_2 = -40{,}000 + j30{,}000$; $p_1 = 0$.

14.2 The parallel circuit in Drill Exercise 14.1 is placed in series with a 500 Ω resistor.

a) Express the impedance of this series combination as a rational function of s.

b) Compute the numerical values of the zeros and poles.

ANSWER: (a) $500(s + 50{,}000)^2/(s^2 + 80{,}000s + 25 \times 10^8)$; (b) $-z_1 = -z_2 = -50{,}000$; $-p_1 = -40{,}000 - j30{,}000$, $-p_2 = -40{,}000 + j30{,}000$.

14.3 EXAMPLES

We now illustrate how to use the Laplace transform to determine the transient behavior of several linear lumped-parameter circuits. We start by analyzing familiar circuits from Chapters 7 and 8 because they represent a simple starting place and because they show that the Laplace transform approach yields the same results. In all the examples, the ease of manipulating algebraic equations instead of differential equations should be apparent.

THE NATURAL RESPONSE OF AN *RC* CIRCUIT

We first revisit the natural response of an RC circuit (Fig. 14.10) via Laplace transform techniques. (You may want to review the classical analysis of this same circuit in Section 7.2).

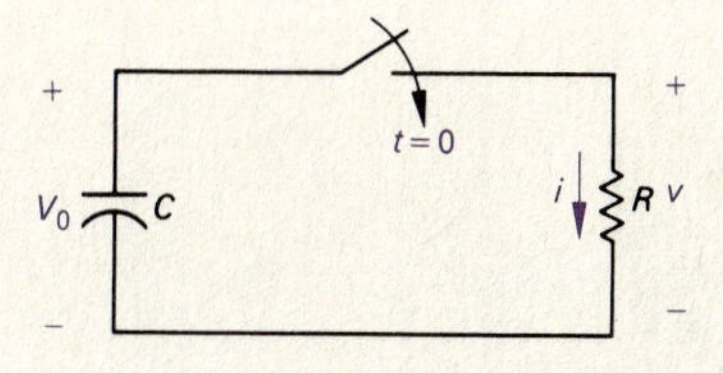

FIGURE 14.10 The capacitor discharge circuit.

The capacitor is initially charged to V_0 volts, and we are interested in the time-domain expressions for i and v. We start by finding i. In transferring the circuit in Fig. 14.10 to the s domain, we have a choice of two equivalent circuits for the charged capacitor. Because we are interested in the current, the series-equivalent circuit is more attractive; it results in a single-mesh circuit in the frequency domain. Thus we construct the s-domain circuit shown in Fig. 14.11.

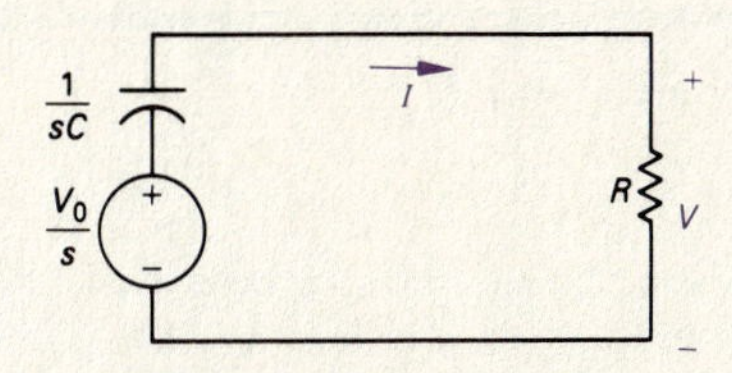

FIGURE 14.11 An s-domain equivalent circuit for the circuit shown in Fig. 14.10.

Summing the voltages around the mesh generates the expression

$$\frac{V_0}{s} = \frac{1}{sC}I + RI. \tag{14.12}$$

Solving Eq. (14.12) for I yields

$$I = \frac{CV_0}{RCs + 1} = \frac{V_0/R}{s + (1/RC)}. \tag{14.13}$$

Note that the expression for I is a proper rational function of s and can be inverse-transformed by inspection:

$$i = \frac{V_0}{R}e^{-t/RC}u(t), \tag{14.14}$$

which is equivalent to the expression for the current derived by the classical methods discussed in Chapter 7. In that chapter, the current is given by Eq. (7.26), where τ is used in place of RC.

After we have found i, the easiest way to determine v is simply to apply Ohm's law; that is, from the circuit,

$$v = Ri = V_0 e^{-t/RC} u(t). \tag{14.15}$$

We now illustrate a way to find v from the circuit without first finding i. In this alternative approach, we return to the original circuit of Fig. 14.10 and transfer it to the s domain using the parallel equivalent circuit for the charged capacitor. Using the parallel equivalent circuit is attractive now because we can describe the resulting circuit in terms of a single node voltage. Figure 14.12 shows the new s-domain equivalent circuit.

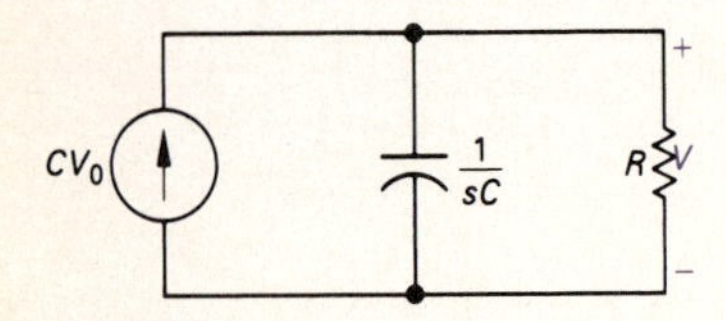

FIGURE 14.12 An s-domain equivalent circuit for the circuit shown in Fig. 14.10.

The node-voltage equation that describes the new circuit is

$$\frac{V}{R} + sCV = CV_0. \tag{14.16}$$

Solving Eq. (14.16) for V gives

$$V = \frac{V_0}{s + (1/RC)}. \tag{14.17}$$

Inverse-transforming Eq. (14.17) leads to the same expression for v given by Eq. (14.15), namely,

$$v = V_0 e^{-t/RC} = V_0 e^{-t/\tau} u(t). \tag{14.18}$$

Our purpose in deriving v by direct use of the transform method is to show that the choice of which s-domain equivalent circuit to use is influenced by which response signal is of interest.

DRILL EXERCISE

14.3 The switch in the circuit shown has been in position a for a long time. At $t = 0$, the switch is thrown to position b.

a) Find I, V_1, and V_2 as rational functions of s.

b) Find the time-domain expressions for i, v_1, and v_2.

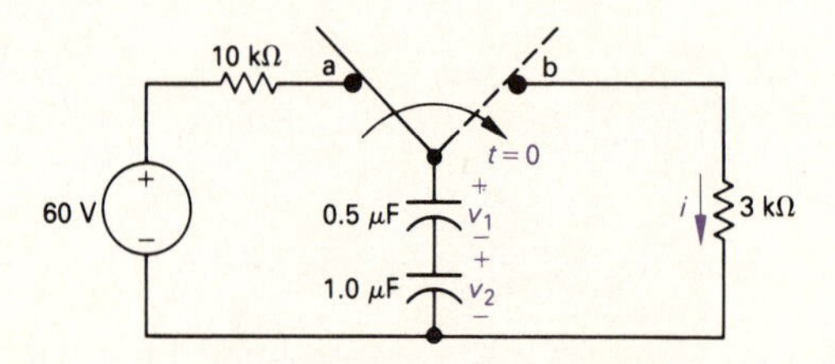

ANSWER: (a) $I = 0.02/(s + 1000)$, $V_1 = 40/(s + 1000)$, $V_2 = 20/(s + 1000)$; (b) $i = 20e^{-1000t}u(t)$ mA, $v_1 = 40e^{-1000t}u(t)$ V, $v_2 = 20e^{-1000t}u(t)$ V.

THE STEP RESPONSE OF A PARALLEL *RLC* CIRCUIT

Next we analyze the parallel RLC circuit, shown in Fig. 14.13, that we first analyzed in Example 8.7. The problem is to find the expression for i_L after the constant current source is switched across the parallel elements. The initial energy stored in the circuit is zero.

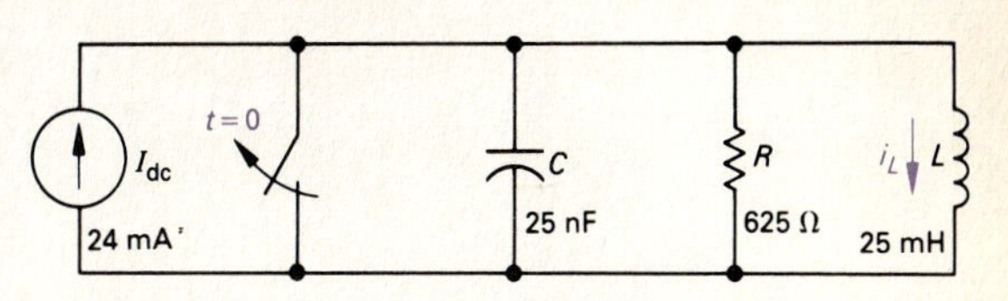

FIGURE 14.13 The step response of a parallel *RLC* circuit.

As before, we begin by constructing the s-domain equivalent circuit shown in Fig. 14.14. Note how easily an independent source can be transformed from the time domain to the frequency domain. We transform the source to the s domain simply by determining the Laplace transform of its time-domain function. Here, opening the switch results in a step change in the current applied to the circuit. Therefore the s-domain current source is $\mathcal{L}\{I_{dc}u(t)\}$, or I_{dc}/s. To find I_L, we first solve for V and then use

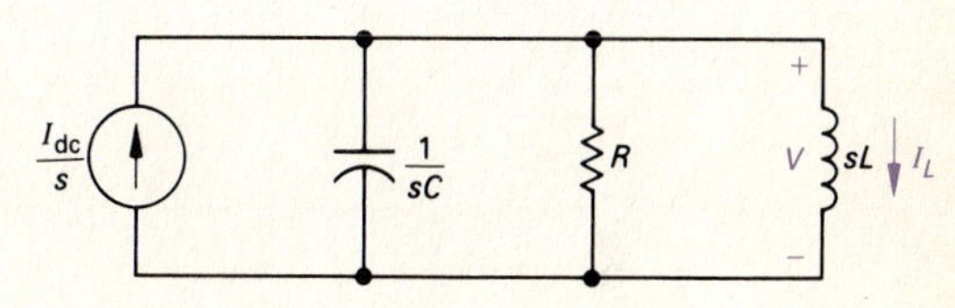

FIGURE 14.14 The s-domain equivalent circuit for the circuit shown in Fig. 14.13.

$$I_L = \frac{V}{sL} \tag{14.19}$$

to establish the s-domain expression for I_L. Summing the currents away from the top node generates the expression

$$sCV + \frac{V}{R} + \frac{V}{sL} = \frac{I_{dc}}{s}. \tag{14.20}$$

Solving Eq. (14.20) for V gives

$$V = \frac{I_{dc}/C}{s^2 + (1/RC)s + (1/LC)}. \tag{14.21}$$

Substituting Eq. (14.21) into Eq. (14.19) gives

$$I_L = \frac{I_{dc}/LC}{s[s^2 + (1/RC)s + (1/LC)]}. \tag{14.22}$$

Substituting the numerical values of R, L, C, and I_{dc} into Eq. (14.22) yields

$$I_L = \frac{384 \times 10^5}{s(s^2 + 64{,}000s + 16 \times 10^8)}. \tag{14.23}$$

Before expanding Eq. (14.23) into a sum of partial fractions, we factor the quadratic term in the denominator:

$$I_L = \frac{384 \times 10^5}{s(s + 32{,}000 - j24{,}000)(s + 32{,}000 + j24{,}000)}. \tag{14.24}$$

Now, we can test the s-domain expression for I_L by checking to see whether the final-value theorem predicts the correct value for i_L at $t = \infty$. All the poles of I_L, except for the first-order pole at the origin, lie in the left half of the s plane, so the theorem is applicable. We know from the behavior of the circuit that after the switch has been open for a long time, the inductor will short circuit the current source. Therefore the final value of i_L must be 24 mA. The limit of

sI_L as $s \rightarrow 0$ is

$$\lim_{s \to 0} sI_L = \frac{384 \times 10^5}{16 \times 10^8} = 24 \text{ mA}. \quad \textbf{(14.25)}$$

(Currents in the s domain carry the dimension of ampere-seconds, so the dimension of sI_L will be amperes.) Thus, our s-domain expression checks out.

We now proceed with the partial fraction expansion of Eq. (14.24):

$$I_L = \frac{K_1}{s} + \frac{K_2}{s + 32{,}000 - j24{,}000} + \frac{K_2^*}{s + 32{,}000 + j24{,}000}. \quad \textbf{(14.26)}$$

The partial fraction coefficients are

$$K_1 = \frac{384 \times 10^5}{16 \times 10^8} = 24 \times 10^{-3}; \quad \textbf{(14.27)}$$

$$K_2 = \frac{384 \times 10^5}{(-32{,}000 + j24{,}000)(j48{,}000)} = 20 \times 10^{-3} \angle 126.87^\circ. \quad \textbf{(14.28)}$$

Substituting the numerical values of K_1 and K_2 into Eq. (14.26) and inverse-transforming the resulting expression yields

$$i_L = [24 + 40e^{-32{,}000t} \cos{(24{,}000t + 126.87^\circ)}]u(t) \text{ mA}. \quad \textbf{(14.29)}$$

The answer given by Eq. (14.29) is equivalent to the answer given for Example 8.7 because

$$40 \cos{(24{,}000t + 126.87^\circ)} = -24 \cos 24{,}000t - 32 \sin 24{,}000t.$$

If we weren't using a previous solution as a check, we would test Eq. (14.29) to make sure that $i_L(0)$ satisfied the given initial conditions and $i_L(\infty)$ satisfied the known behavior of the circuit.

DRILL EXERCISE

14.4 The energy stored in the circuit shown is zero at the time when the switch is closed.

a) Find the s-domain expression for I.

b) Find the time-domain expression for i when $t > 0$.

c) Find the s-domain expression for V.

d) Find the time-domain expression for v when $t > 0$.

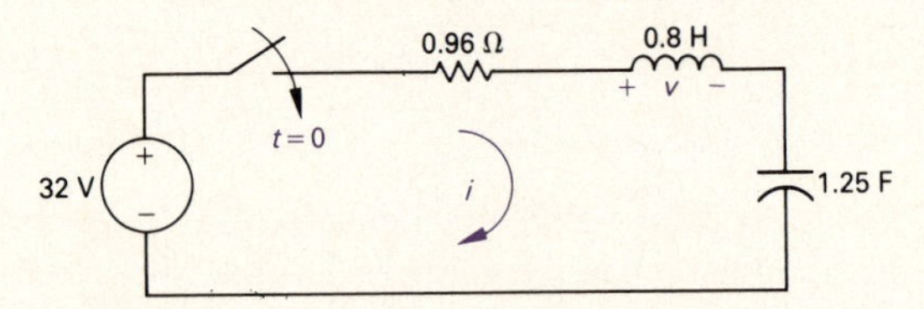

ANSWER: (a) $I = 40/(s^2 + 1.2s + 1)$; (b) $i = (50e^{-0.6t} \sin 0.8t)u(t)$ A; (c) $V = 32s/(s^2 + 1.2s + 1)$; (d) $v = [40e^{-0.6t} \cos{(0.8t + 36.87^\circ)}]u(t)$ V.

The Transient Response of a Parallel *RLC* Circuit

Another example of using the Laplace transform to find the transient behavior of a circuit arises from replacing the dc current source in the circuit shown in Fig. 14.13 with a sinusoidal current source. The new current source is

$$i_g = I_m \cos \omega t \text{ A}, \tag{14.30}$$

where $I_m = 24$ mA and $\omega = 40{,}000$ rad/s. As before, we assume that the initial energy stored in the circuit is zero.

The s-domain expression for the source current is

$$I_g = \frac{sI_m}{s^2 + \omega^2}. \tag{14.31}$$

The voltage across the parallel elements is

$$V = \frac{(I_g/C)s}{s^2 + (1/RC)s + (1/LC)}. \tag{14.32}$$

Substituting Eq. (14.31) into Eq. (14.32) results in

$$V = \frac{(I_m/C)s^2}{(s^2 + \omega^2)[s^2 + (1/RC)s + (1/LC)]}, \tag{14.33}$$

from which

$$I_L = \frac{V}{sL} = \frac{(I_m/LC)s}{(s^2 + \omega^2)[s^2 + (1/RC)s + (1/LC)]}. \tag{14.34}$$

Substituting the numerical values of I_m, ω, R, L, and C into Eq. (14.34) gives

$$I_L = \frac{384 \times 10^5 s}{(s^2 + 16 \times 10^8)(s^2 + 64{,}000s + 16 \times 10^8)}. \tag{14.35}$$

We now write the denominator in factored form:

$$I_L = \frac{384 \times 10^5 s}{(s - j\omega)(s + j\omega)(s + \alpha - j\beta)(s + \alpha + j\beta)}, \tag{14.36}$$

where $\omega = 40{,}000$, $\alpha = 32{,}000$, and $\beta = 24{,}000$.

We can't test the final value of i_L with the final-value theorem because I_L has a pair of poles on the imaginary axis, that is, poles at $\pm j4 \times 10^4$. Thus we must first find i_L and then check the validity of the expression from known circuit behavior.

When we expand Eq. (14.36) into a sum of partial fractions, we generate the equation

$$I_L = \frac{K_1}{s - j40{,}000} + \frac{K_1^*}{s + j40{,}000} + \frac{K_2}{s + 32{,}000 - j24{,}000} + \frac{K_2^*}{s + 32{,}000 + j24{,}000}. \tag{14.37}$$

The numerical values of the coefficients K_1 and K_2 are

$$K_1 = \frac{384 \times 10^5(j40{,}000)}{(j80{,}000)(32{,}000 + j16{,}000)(32{,}000 + j64{,}000)}$$
$$= 7.5 \times 10^{-3}\underline{/-90^\circ}; \quad (14.38)$$

$$K_2 = \frac{384 \times 10^5(-32{,}000 + j24{,}000)}{(-32{,}000 - j16{,}000)(-32{,}000 + j64{,}000)(j48{,}000)}$$
$$= 12.5 \times 10^{-3}\underline{/90^\circ}. \quad (14.39)$$

Substituting the numerical values from Eqs. (14.38) and (14.39) into Eq. (14.37) and inverse-transforming the resulting expression yields

$$i_L = [15\cos(40{,}000t - 90^\circ) + 25e^{-32{,}000t}\cos(24{,}000t + 90^\circ)]\text{ mA}$$
$$= (15\sin 40{,}000t - 25e^{-32{,}000t}\sin 24{,}000t)u(t)\text{ mA}. \quad (14.40)$$

We now test Eq. (14.40) to see whether it makes sense in terms of the given initial conditions and the known circuit behavior after the switch has been open for a long time. For $t = 0$, Eq. (14.40) predicts zero initial current, which agrees with the initial energy of zero in the circuit. Equation (14.40) also predicts a steady-state current of

$$i_{L_{ss}} = 15\sin 40{,}000t\text{ mA}, \quad (14.41)$$

which can be verified by the phasor method (Chapter 9).

The Step Response of a Multiple-Mesh Circuit

Use a computer tool to solve sets of simultaneous differential equations.

Until now, we avoided circuits which required two or more node-voltage or mesh-current equations, because the techniques for solving simultaneous differential equations are beyond the scope of this text. However, using Laplace techniques, we can solve a problem like the one posed by the multiple-mesh circuit in Fig. 14.15.

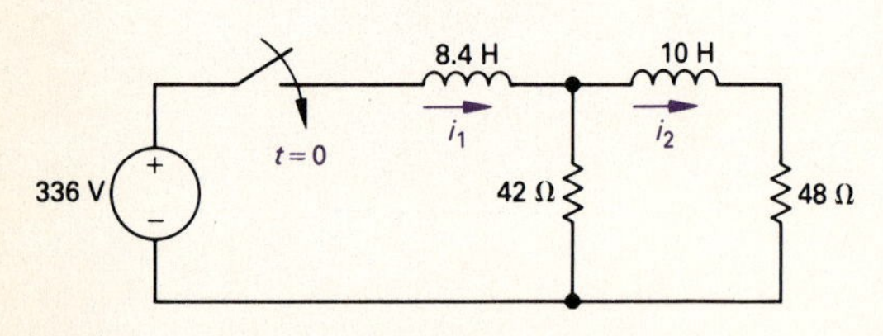

FIGURE 14.15 A multiple-mesh RL circuit.

Here we want to find the branch currents i_1 and i_2 that arise when the 336 V dc voltage source is applied suddenly to the circuit. The initial energy stored in the circuit is zero. Figure 14.16 shows the s-domain equivalent circuit of Fig. 14.15. The two mesh-current equations are

$$\frac{336}{s} = (42 + 8.4s)I_1 - 42I_2; \quad (14.42)$$
$$0 = -42I_1 + (90 + 10s)I_2. \quad (14.43)$$

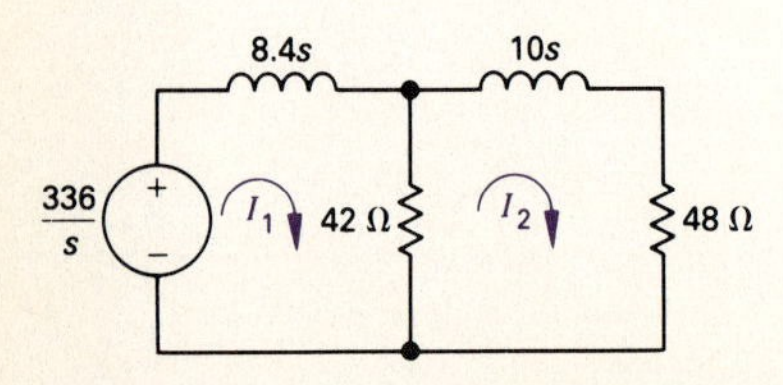

FIGURE 14.16 The s-domain equivalent circuit for the circuit shown in Fig. 14.15.

Using Cramer's method to solve for I_1 and I_2, we obtain

$$\Delta = \begin{vmatrix} 42 + 8.4s & -42 \\ -42 & 90 + 10s \end{vmatrix}$$
$$= 84(s^2 + 14s + 24)$$
$$= 84(s + 2)(s + 12); \quad (14.44)$$

$$N_1 = \begin{vmatrix} 336/s & -42 \\ 0 & 90 + 10s \end{vmatrix}$$

$$= \frac{3360(s+9)}{s}; \quad \textbf{(14.45)}$$

$$N_2 = \begin{vmatrix} 42 + 8.4s & 336/s \\ -42 & 0 \end{vmatrix}$$

$$= \frac{14{,}112}{s}. \quad \textbf{(14.46)}$$

Based on Eqs. (14.44) through (14.46),

$$I_1 = \frac{N_1}{\Delta} = \frac{40(s+9)}{s(s+2)(s+12)}; \quad \textbf{(14.47)}$$

$$I_2 = \frac{N_2}{\Delta} = \frac{168}{s(s+2)(s+12)}. \quad \textbf{(14.48)}$$

Expanding I_1 and I_2 into a sum of partial fractions gives

$$I_1 = \frac{15}{s} - \frac{14}{s+2} - \frac{1}{s+12}; \quad \textbf{(14.49)}$$

$$I_2 = \frac{7}{s} - \frac{8.4}{s+2} + \frac{1.4}{s+12}. \quad \textbf{(14.50)}$$

We obtain the expressions for i_1 and i_2 by inverse-transforming Eqs. (14.49) and (14.50), respectively:

$$i_1 = (15 - 14e^{-2t} - e^{-12t})u(t) \text{ A}; \quad \textbf{(14.51)}$$

$$i_2 = (7 - 8.4e^{-2t} + 1.4e^{-12t})u(t) \text{ A}. \quad \textbf{(14.52)}$$

Next we test the solutions to see whether they make sense in terms of the circuit. Because no energy is stored in the circuit at the instant the switch is closed, both $i_1(0^-)$ and $i_2(0^-)$ must be zero. The solutions agree with these initial values. After the switch has been closed for a long time, the two inductors appear as short circuits. Therefore the final values of i_1 and i_2 are

$$i_1(\infty) = \frac{336(90)}{42(48)} = 15 \text{ A}, \quad \textbf{(14.53)}$$

and

$$i_2(\infty) = \frac{15(42)}{90} = 7 \text{ A}. \quad \textbf{(14.54)}$$

One final test involves the numerical values of the exponents and calculating the voltage drop across the 42 Ω resistor by three different methods. From the circuit, the voltage across the 42 Ω resistor (positive at the top) is

$$v = 42(i_1 - i_2) = 336 - 8.4\frac{di_1}{dt} = 48i_2 + 10\frac{di_2}{dt}. \quad \textbf{(14.55)}$$

Use a computer tool to help in verifying this voltage.

You should verify that regardless of which form of Eq. (14.55) is used, the voltage is

$$v = (336 - 235.2e^{-2t} - 100.80e^{-12t})u(t)\text{ V}.$$

We are thus confident that the solutions for i_1 and i_2 are correct.

DRILL EXERCISE

14.5 The dc current and voltage sources are applied simultaneously to the circuit shown. No energy is stored in the circuit at the instant of application.

a) Derive the s-domain expressions for V_1 and V_2.

b) For $t > 0$, derive the time-domain expressions for v_1 and v_2.

c) Calculate $v_1(0^+)$ and $v_2(0^+)$.

d) Compute the steady-state values of v_1 and v_2.

ANSWER: (a) $V_1 = [5(s+3)]/[s(s+0.5)(s+2)]$, $V_2 = [2.5(s^2+6)]/[s(s+0.5)(s+2)]$; (b) $v_1 = (15 - \frac{50}{3}e^{-0.5t} + \frac{5}{3}e^{-2t})u(t)$ V, $v_2 = (15 - \frac{125}{6}e^{-0.5t} + \frac{25}{3}e^{-2t})u(t)$ V; (c) $v_1(0^+) = 0,\ v_2(0^+) = 2.5$ V; (d) $v_1 = v_2 = 15$ V.

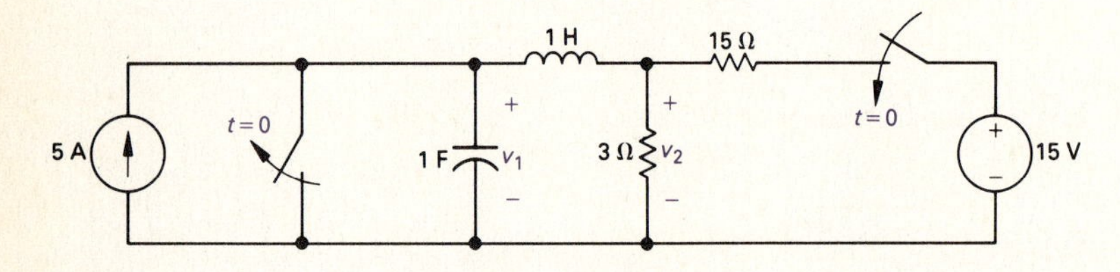

THE USE OF THÉVENIN'S EQUIVALENT

In this section we show how to use Thévenin's equivalent in the s domain. Figure 14.17 shows the circuit to be analyzed. The problem is to find the capacitor current that results from closing the switch. The energy stored in the circuit prior to closing is zero.

To find i_C we first construct the s-domain equivalent circuit and then find the Thévenin equivalent of this circuit with respect to the terminals of the capacitor. Figure 14.18 shows the s-domain circuit.

The Thévenin voltage is the open-circuit voltage across terminals a and b. Under open-circuit conditions, there is no voltage across the 60 Ω resistor. Hence

$$V_{\text{Th}} = \frac{(480/s)(0.002s)}{20 + 0.002s} = \frac{480}{s + 10^4}. \tag{14.56}$$

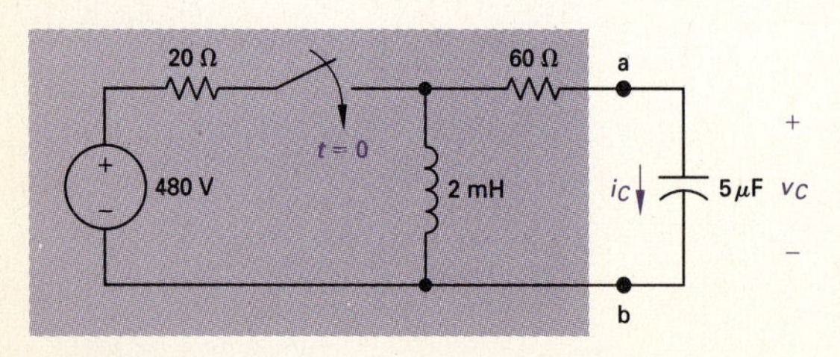

FIGURE 14.17 A circuit showing the use of Thévenin's equivalent in the s domain.

The Thévenin impedance seen from terminals a and b equals the 60 Ω resistor in series with the parallel combination of the 20 Ω resistor and the 2 mH inductor. Thus

$$Z_{\text{Th}} = 60 + \frac{0.002s(20)}{20 + 0.002s} = \frac{80(s + 7500)}{s + 10^4}. \tag{14.57}$$

Using the Thévenin equivalent, we reduce the circuit shown in Fig. 14.18 to the one shown in Fig. 14.19. It indicates that the capacitor current I_C equals the Thévenin voltage divided by the total series impedance. Thus

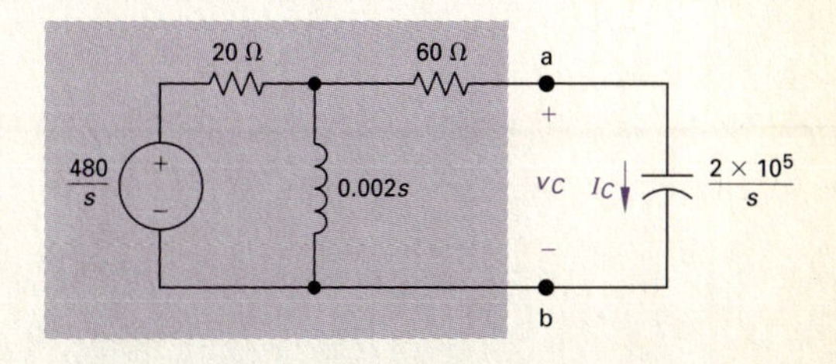

FIGURE 14.18 The s-domain model of the circuit shown in Fig.14.17.

$$I_C = \frac{480/(s + 10^4)}{[80(s + 7500)/(s + 10^4)] + [(2 \times 10^5)/s]}. \tag{14.58}$$

We simplify Eq. (14.58) to

$$\begin{aligned} I_C &= \frac{6s}{s^2 + 10{,}000s + 25 \times 10^6} \\ &= \frac{6s}{(s + 5000)^2}. \end{aligned} \tag{14.59}$$

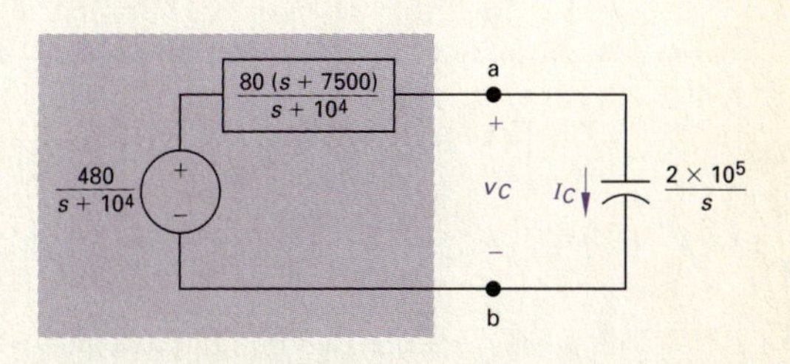

FIGURE 14.19 A simplified version of the circuit shown in Fig. 14.18, using a Thévenin equivalent.

A partial fraction expansion of Eq. (14.59) generates

$$I_C = \frac{-30{,}000}{(s + 5000)^2} + \frac{6}{(s + 5000)}, \tag{14.60}$$

the inverse transform of which is

$$i_C = (-30{,}000te^{-5000t} + 6e^{-5000t})u(t) \text{ A}. \tag{14.61}$$

We now test Eq. (14.61) to see whether it makes sense in terms of known circuit behavior. From Eq. (14.61),

$$i_C(0) = 6 \text{ A}. \tag{14.62}$$

This result agrees with the initial current in the capacitor, as calculated from the circuit in Fig. 14.17. The initial inductor current is zero and the initial capacitor voltage is zero, so the initial capacitor current is 480/80, or 6 A. The final value of the current is zero, which also agrees with Eq. (14.61). Note also from this equation that the current reverses sign when t exceeds 6/30,000, or 200 μs. The fact that i_C reverses sign makes sense because, when the switch first closes, the capacitor begins to charge. Eventually this charge is reduced to zero because the inductor is a short circuit at $t = \infty$. The sign reversal of i_C reflects the charging and discharging of the capacitor.

Let's assume that the voltage drop across the capacitor v_C is also of interest. Once we know i_C, we find v_C by integration in the time domain; that is,

$$v_C = 2 \times 10^5 \int_{0^-}^{t} (6 - 30{,}000x)e^{-5000x}\,dx. \tag{14.63}$$

Although the integration called for in Eq. (14.63) is not difficult, we may avoid it altogether by first finding the s-domain expression for V_C and then finding v_C by an inverse transform. Thus

$$V_C = \frac{1}{sC} I_C = \frac{2 \times 10^5}{s} \frac{6s}{(s+5000)^2}$$

$$= \frac{12 \times 10^5}{(s+5000)^2}, \quad (14.64)$$

from which

$$v_C = 12 \times 10^5 t e^{-5000t} u(t). \quad (14.65)$$

You should verify that Eq. (14.65) is consistent with Eq. (14.61) and that it also supports the observations made with regard to the behavior of i_C (see Problem 14.35).

DRILL EXERCISE

14.6 The initial charge on the capacitor in the circuit shown is zero.

a) Find the s-domain Thévenin equivalent circuit with respect to terminals a and b.

b) Find the s-domain expression for the current that the circuit delivers to a load consisting of a 0.4 H inductor in series with a 1 Ω resistor.

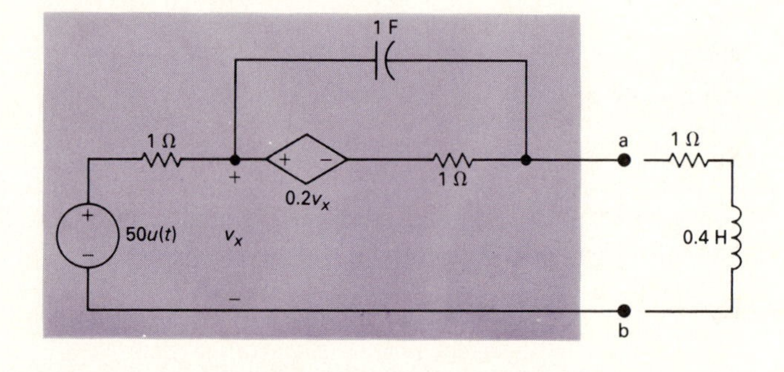

ANSWER: (a) $V_{Th} = V_{ab} = [50(s+0.8)]/[s(s+1)]$, $Z_{Th} = (s+1.8)/(s+1)$;
(b) $I_{ab} = [125(s+0.8)]/[s(s^2+6s+7)]$.

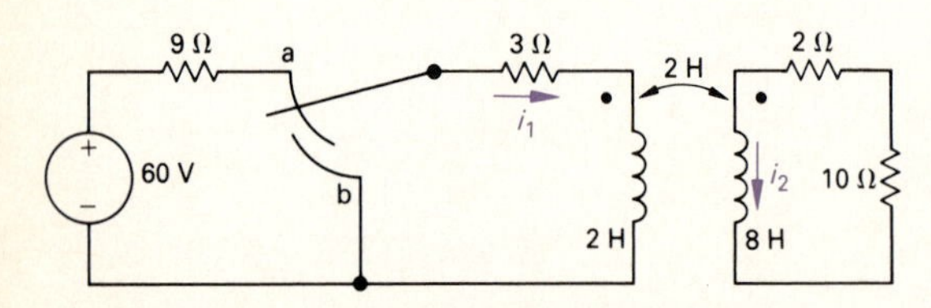

FIGURE 14.20 A circuit containing magnetically coupled coils.

A Circuit With Mutual Inductance

The next example illustrates how to use the Laplace transform to analyze the transient response of a circuit that contains mutual inductance. Figure 14.20 shows the circuit. The make-before-break switch has been in position a for a long time. At $t = 0$, the switch moves instantaneously to position b. The problem is to derive the time-domain expression for i_2.

We begin by redrawing the circuit in Fig. 14.20, with the switch in position b and the magnetically coupled coils replaced with a T-equivalent circuit. Figure 14.21 shows the new circuit.

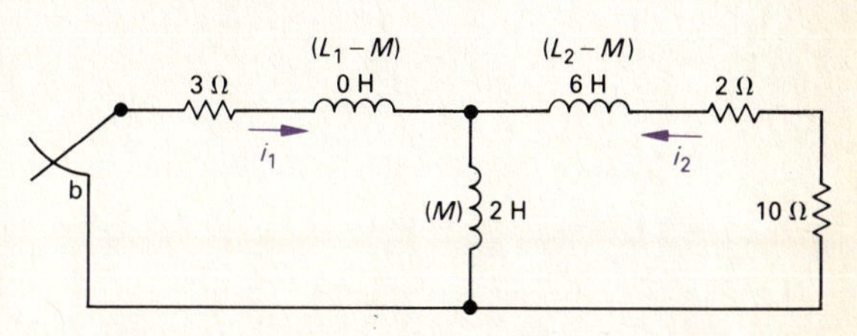

FIGURE 14.21 The circuit shown in Fig. 14.20, with the magnetically coupled coils replaced by a T-equivalent circuit.

We now transform this circuit to the s domain. In so doing, we note that

$$i_1(0^-) = \frac{60}{12} = 5 \text{ A}; \qquad \textbf{(14.66)}$$

$$i_2(0^-) = 0. \qquad \textbf{(14.67)}$$

Because we plan to use mesh analysis in the s domain, we use the series equivalent circuit for an inductor carrying an initial current. Figure 14.22 shows the s-domain circuit. Note that there is only one independent voltage source. This source appears in the vertical leg of the tee to account for the initial value of the current in the 2 H inductor of $i_1(0^-) + i_2(0^-)$, or 5 A. The branch carrying i_1 has no voltage source because $L_1 - M = 0$.

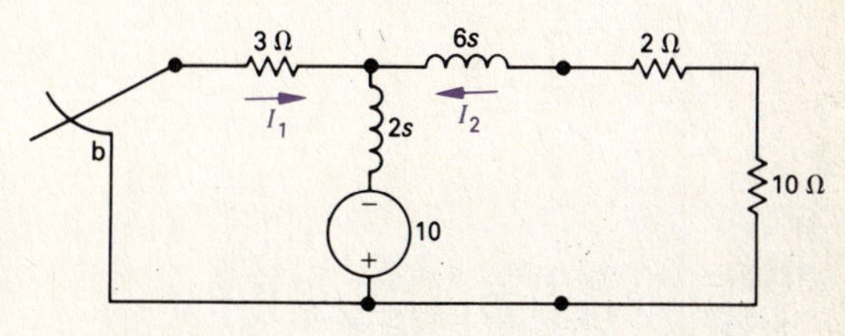

FIGURE 14.22 The s-domain equivalent circuit for the circuit shown in Fig. 14.21.

The two s-domain mesh equations that describe the circuit in Fig. 14.22 are

$$(3 + 2s)I_1 + 2sI_2 = 10 \qquad \textbf{(14.68)}$$

and

$$2sI_1 + (12 + 8s)I_2 = 10. \qquad \textbf{(14.69)}$$

Solving for I_2 yields

$$I_2 = \frac{2.5}{(s+1)(s+3)}. \qquad \textbf{(14.70)}$$

Expanding Eq. (14.70) into a sum of partial fractions generates

$$I_2 = \frac{1.25}{s+1} - \frac{1.25}{s+3}. \qquad \textbf{(14.71)}$$

Then,

$$i_2 = (1.25e^{-t} - 1.25e^{-3t})u(t) \text{ A}. \qquad \textbf{(14.72)}$$

Equation (14.72) reveals that i_2 increases from zero to a peak value of 481.13 mA in 549.31 ms after the switch is moved to position b. Thereafter, i_2 decreases exponentially toward zero. Figure 14.23 shows a plot of i_2 versus t. This response makes sense in terms of the known physical behavior of the magnetically coupled coils. A current can exist in the L_2 inductor only if there is a time-varying current in the L_1 inductor. As i_1 decreases from its initial value of 5 A, i_2 increases from zero and then approaches zero as i_1 approaches zero.

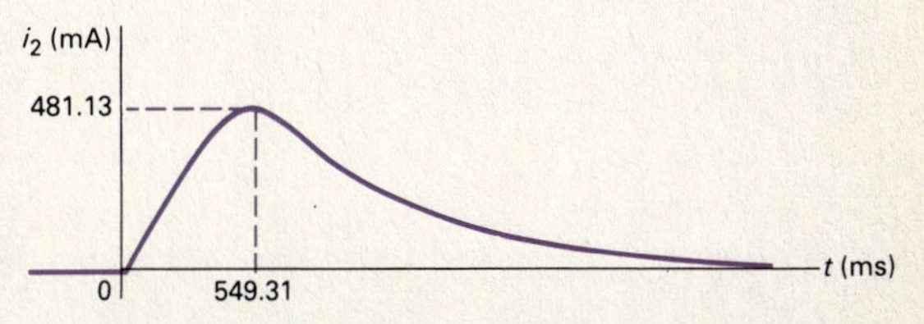

FIGURE 14.23 The plot of i_2 versus t for the circuit shown in Fig. 14.20.

DRILL EXERCISE

14.7 a) Verify from Eq. (14.72) that i_2 reaches a peak value of 481.13 mA at $t = 549.31$ ms.

b) Find i_1, for $t > 0$, for the circuit shown in Fig. 14.20.

c) Compute di_1/dt when i_2 is at its peak value.

d) Express i_2 as a function of di_1/dt when i_2 is at its peak value.

e) Use the results obtained in (c) and (d) to calculate the peak value of i_2.

ANSWER: (a) $di_2/dt = 0$ when $t = \frac{1}{2}\ ln\ 3$ (s); (b) $i_1 = 2.5(e^{-t} + e^{-3t})u(t)$ A; (c) -2.89 A/s; (d) $i_2 = -(M di_1/dt)/12$; (e) 481.13 mA.

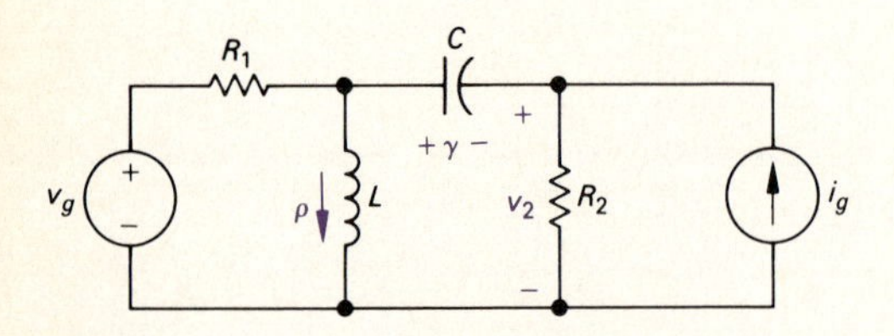

FIGURE 14.24 A circuit showing the use of superposition in s-domain analysis.

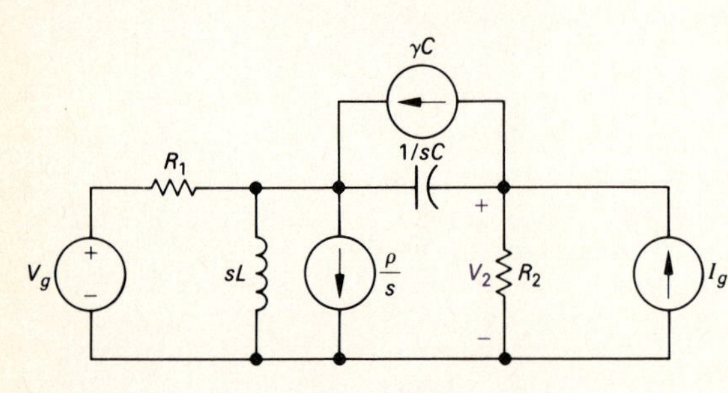

FIGURE 14.25 The s-domain equivalent for the circuit of Fig. 14.24.

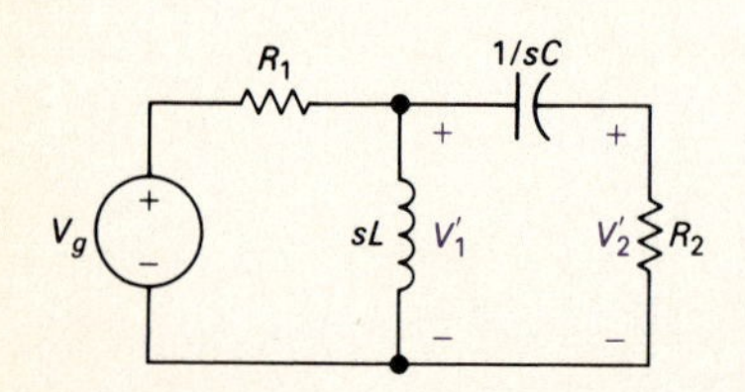

FIGURE 14.26 The circuit shown in Fig. 14.25 with V_g acting alone.

THE USE OF SUPERPOSITION

Because we are analyzing linear lumped-parameter circuits, we can use superposition to divide the response into components that can be identified with particular sources and initial conditions. Distinguishing these components is critical to being able to use the transfer function, which we introduce in the next section.

Figure 14.24 shows our illustrative circuit. We assume that at the instant when the two sources are applied to the circuit, the inductor is carrying an initial current of ρ amperes and that the capacitor is carrying an initial voltage of γ volts. The desired response of the circuit is the voltage across the resistor R_2, labeled v_2.

Figure 14.25 shows the s-domain equivalent circuit. We opted for the parallel equivalents for L and C because we anticipated solving for V_2 using the node-voltage method.

To find V_2 by superposition, we calculate the component of V_2 resulting from each source acting alone, and then we sum the components. We begin with V_g acting alone. Opening each of the three current sources deactivates them. Figure 14.26 shows the resulting circuit. We added the node voltage V_1' to aid the analysis. The primes on V_1 and V_2 indicate that they are the components of V_1 and V_2 attributable to V_g acting alone. The two equations that describe the circuit in Fig. 14.26 are

$$\left(\frac{1}{R_1} + \frac{1}{sL} + sC\right) V_1' - sCV_2' = \frac{V_g}{R_1}; \quad \textbf{(14.73)}$$

$$-sCV_1' + \left(\frac{1}{R_2} + sC\right) V_2' = 0. \quad \textbf{(14.74)}$$

For convenience, we introduce the notation

$$Y_{11} = \frac{1}{R_1} + \frac{1}{sL} + sC; \tag{14.75}$$

$$Y_{12} = -sC; \tag{14.76}$$

$$Y_{22} = \frac{1}{R_2} + sC. \tag{14.77}$$

Substituting Eqs. (14.75), (14.76), and (14.77) into Eqs. (14.73) and (14.74) gives

$$Y_{11}V_1' + Y_{12}V_2' = V_g/R_1; \tag{14.78}$$

$$Y_{12}V_1' + Y_{22}V_2' = 0. \tag{14.79}$$

Solving Eqs. (14.78) and (14.79) for V_2' gives

$$V_2' = \frac{-Y_{12}/R_1}{Y_{11}Y_{22} - Y_{12}^2}V_g. \tag{14.80}$$

With the current source I_g acting alone, the circuit shown in Fig. 14.25 reduces to the one shown in Fig. 14.27. Here, V_1'' and V_2'' are the components of V_1 and V_2 resulting from I_g. If we use the notation introduced in Eqs. (14.75) through (14.77), the two node-voltage equations that describe the circuit in Fig. 14.27 are

$$Y_{11}V_1'' + Y_{12}V_2'' = 0; \tag{14.81}$$

$$Y_{12}V_1'' + Y_{22}V_2'' = I_g. \tag{14.82}$$

Solving Eqs. (14.81) and (14.82) for V_2'' yields

$$V_2'' = \frac{Y_{11}}{Y_{11}Y_{22} - Y_{12}^2}I_g. \tag{14.83}$$

To find the component of V_2 resulting from the initial energy stored in the inductor (V_2'''), we must solve the circuit shown in Fig. 14.28, where

$$Y_{11}V_1''' + Y_{12}V_2''' = -\rho/s; \tag{14.84}$$

$$Y_{12}V_1''' + Y_{22}V_2''' = 0. \tag{14.85}$$

Thus

$$V_2''' = \frac{Y_{12}/s}{Y_{11}Y_{22} - Y_{12}^2}\rho. \tag{14.86}$$

From the circuit shown in Fig. 14.29, we find the component of V_2 (V_2'''') resulting from the initial energy stored in the capacitor. The

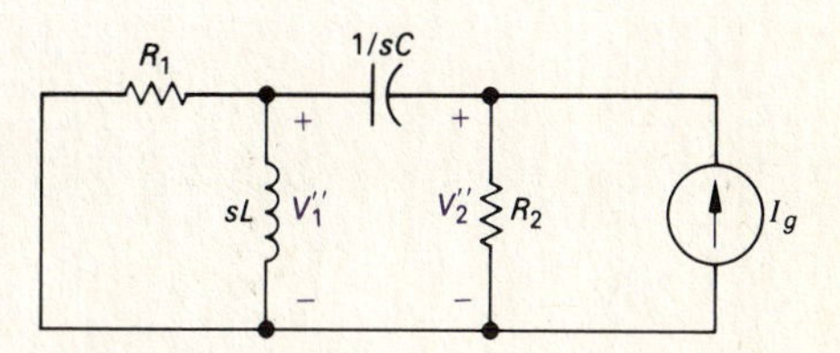

FIGURE 14.27 The circuit shown in Fig. 14.25, with I_g acting alone.

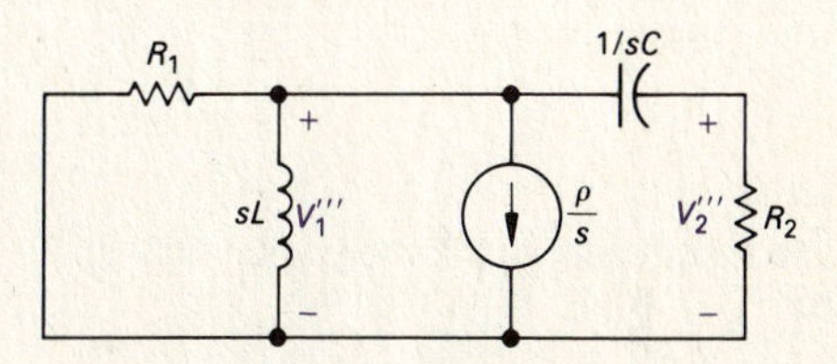

FIGURE 14.28 The circuit shown in Fig. 14.25, with the energized inductor acting alone.

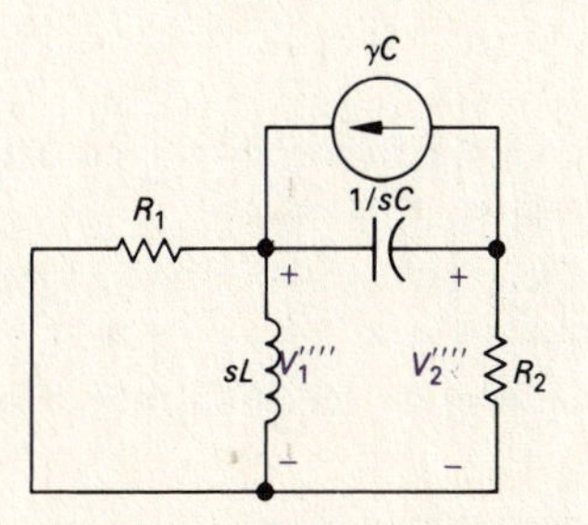

FIGURE 14.29 The circuit shown in Fig. 14.25, with the energized capacitor acting alone.

node-voltage equations describing this circuit are

$$Y_{11}V_1'''' + Y_{12}V_2'''' = \gamma C; \quad \textbf{(14.87)}$$

$$Y_{12}V_1'''' + Y_{22}V_2'''' = -\gamma C. \quad \textbf{(14.88)}$$

Solving for V_2'''' yields

$$V_2'''' = \frac{-(Y_{11} + Y_{12})C}{Y_{11}Y_{22} - Y_{12}^2}\gamma. \quad \textbf{(14.89)}$$

The expression for V_2 is

$$\begin{aligned} V_2 &= V_2' + V_2'' + V_2''' + V_2'''' \\ &= \frac{-(Y_{12}/R_1)}{Y_{11}Y_{22} - Y_{12}^2}V_g + \frac{Y_{11}}{Y_{11}Y_{22} - Y_{12}^2}I_g \\ &\quad + \frac{Y_{12}/s}{Y_{11}Y_{22} - Y_{12}^2}\rho + \frac{-C(Y_{11} + Y_{12})}{Y_{11}Y_{22} - Y_{12}^2}\gamma. \end{aligned} \quad \textbf{(14.90)}$$

We can find V_2 without using superposition by solving the two node-voltage equations that describe the circuit shown in Fig. 14.25. Thus

$$Y_{11}V_1 + Y_{12}V_2 = \frac{V_g}{R_1} + \gamma C - \frac{\rho}{s}; \quad \textbf{(14.91)}$$

$$Y_{12}V_1 + Y_{22}V_2 = I_g - \gamma C. \quad \textbf{(14.92)}$$

Use a computer tool to help you verify this solution.

You should verify (in Problem 14.48) that the solution of Eqs. (14.91) and (14.92) for V_2 gives the same result as Eq. (14.90).

DRILL EXERCISE

14.8 The energy stored in the circuit shown is zero at the instant the two sources are turned on.

a) Find the component of v for $t > 0$ owing to the voltage source.

b) Find the component of v for $t > 0$ owing to the current source.

c) Find the expression for v when $t > 0$.

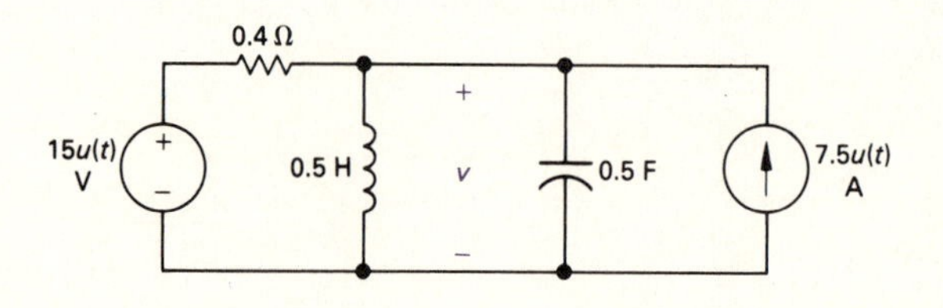

ANSWER: (a) $25e^{-t} - 25e^{-4t}$ V; (b) $5e^{-t} - 5e^{-4t}$ V; (c) $30e^{-t} - 30e^{-4t}$ V.

14.4 THE TRANSFER FUNCTION

The **transfer function** is defined as the s-domain ratio of the Laplace transform of the output (response) to the Laplace transform of the input (source). In computing the transfer function, we restrict our attention to circuits where all initial conditions are zero. If a circuit has multiple independent sources, we can find the transfer function for each source and use superposition to find the response to all sources.

A circuit simulator can compute the transfer function of a circuit.

The transfer function is

$$H(s) = \frac{Y(s)}{X(s)}, \tag{14.93}$$

where $Y(s)$ is the Laplace transform of the output signal, and $X(s)$ is the Laplace transform of the input signal. Note that the transfer function depends on what is defined as the output signal. Consider, for example, the series circuit shown in Fig. 14.30. If the current is defined as the response signal of the circuit,

$$H(s) = \frac{I}{V_g} = \frac{1}{R + sL + 1/sC} = \frac{sC}{s^2LC + RCs + 1}. \tag{14.94}$$

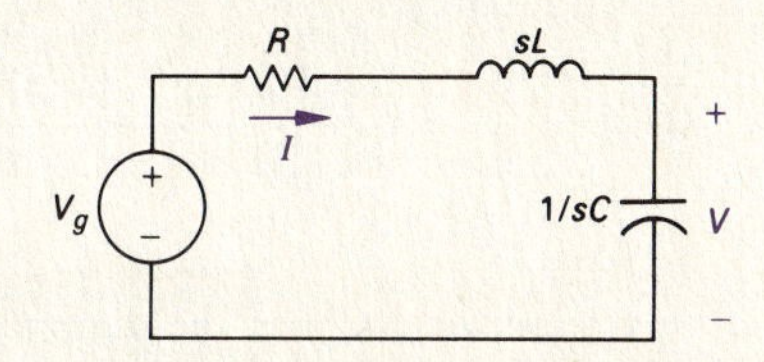

FIGURE 14.30 A series RLC circuit.

In deriving Eq. (14.94), we recognized that I corresponds to the output $Y(s)$ and V_g corresponds to the input $X(s)$.

If the voltage across the capacitor is defined as the output signal of the circuit shown in Fig. 14.30, the transfer function is

$$H(s) = \frac{V}{V_g} = \frac{1/sC}{R + sL + 1/sC} = \frac{1}{s^2LC + RCs + 1}. \tag{14.95}$$

Thus, because circuits may have multiple sources and because the definition of the output signal of interest can vary, a single circuit can generate many transfer functions. Remember that when multiple sources are involved, no single transfer function can represent the total output—transfer functions associated with each source must be combined using superposition to yield the total response. Example 14.1 illustrates the computation of a transfer function for known numerical values of R, L, and C.

EXAMPLE 14.1

The voltage source v_g drives the circuit shown in Fig. 14.31. The response signal is the voltage across the capacitor, v_o.

a) Calculate the numerical expression for the transfer function.

b) Calculate the numerical values for the poles and zeros of the transfer function.

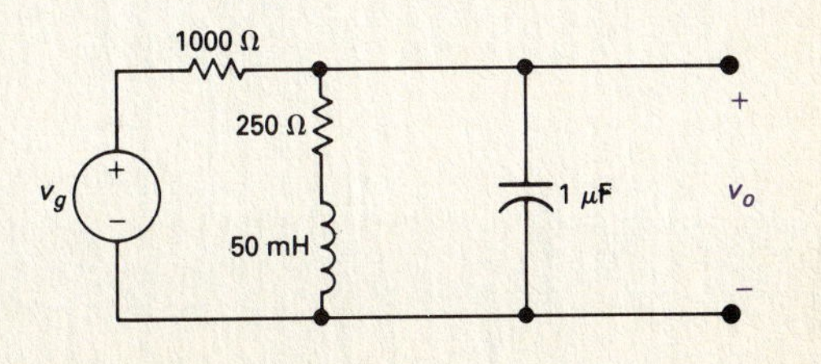

FIGURE 14.31 The circuit for Example 14.1.

SOLUTION

a) The first step in finding the transfer function is to construct the s-domain equivalent circuit, as shown in Fig. 14.32. By definition, the transfer function is the ratio of V_o/V_g, which can be computed from a single node-voltage equation. Summing the currents away from the upper node generates

$$\frac{V_o - V_g}{1000} + \frac{V_o}{250 + 0.05s} + \frac{V_o s}{10^6} = 0.$$

Solving for V_o yields

$$V_o = \frac{1000(s + 5000)\,V_g}{s^2 + 6000s + 25 \times 10^6}.$$

Hence the transfer function is

$$H(s) = \frac{V_o}{V_g} = \frac{1000(s + 5000)}{s^2 + 6000s + 25 \times 10^6}.$$

b) The poles of $H(s)$ are the roots of the denominator polynomial. Therefore

$$-p_1 = -3000 - j4000;$$
$$-p_2 = -3000 + j4000.$$

The zeros of $H(s)$ are the roots of the numerator polynomial; thus $H(s)$ has a zero at

$$-z_1 = -5000.$$

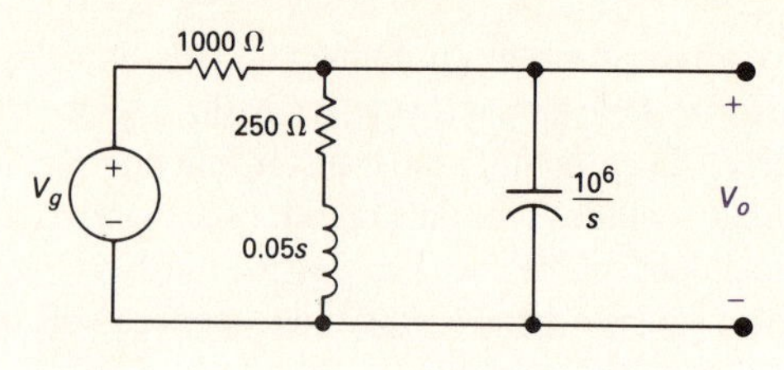

FIGURE 14.32 The s-domain equivalent circuit for the circuit shown in Fig. 14.31.

THE LOCATION OF POLES AND ZEROS OF $H(s)$

For linear lumped-parameter circuits, $H(s)$ is always a rational function of s. Complex poles and zeros always appear in conjugate pairs. The poles of $H(s)$ must lie in the left half of the s plane if the response to a bounded source (one whose values lie within some finite bounds) is to be bounded. The zeros of $H(s)$ may lie in either the right half or the left half of the s plane.

With these general characteristics in mind, we next discuss the role that $H(s)$ plays in determining the response function. We begin with the partial fraction expansion technique for finding $y(t)$.

DRILL EXERCISE

14.9 a) Derive the numerical expression for the transfer function V_o/I_g for the circuit shown.

b) Give the numerical value of each pole and zero of $H(s)$.

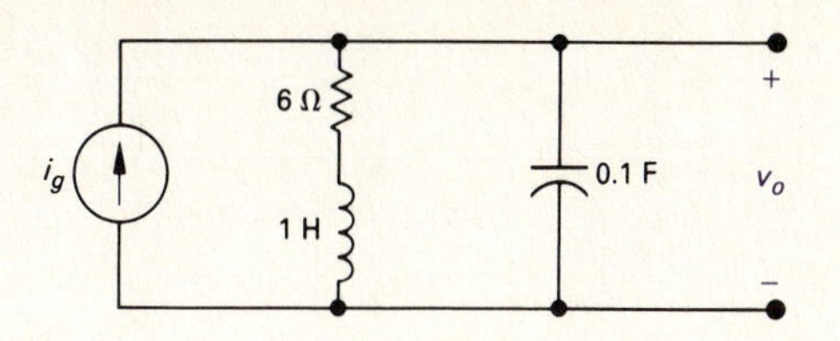

ANSWER: (a) $H(s) = 10(s+6)/(s^2+6s+10)$;
(b) $-p_1 = -3+j1, -p_2 = -3-j1, -z = -6$.

14.5 THE TRANSFER FUNCTION IN PARTIAL FRACTION EXPANSIONS

From Eq. 14.93 we can write the circuit output as the product of the transfer function and the driving function:

$$Y(s) = H(s)X(s). \tag{14.96}$$

We have already noted that $H(s)$ is a rational function of s. Reference to Table 13.1 shows that $X(s)$ also is a rational function of s for the excitation functions of most interest in circuit analysis.

Expanding the right-hand side of Eq. (14.96) into a sum of partial fractions produces a term for each pole of $H(s)$ and $X(s)$. Remember from Chapter 13 that poles are the roots of the denominator polynomial; zeros are the roots of the numerator polynomial. The terms generated by the poles of $H(s)$ give rise to the transient component of the total response, whereas the terms generated by the poles of $X(s)$ give rise to the steady-state component of the response. By steady-state response, we mean the response that exists after the transient components have become negligible. Example 14.2 illustrates these general observations.

EXAMPLE 14.2

The circuit in Example 14.1 (Fig. 14.31) is driven by a voltage source whose voltage increases linearly with time, namely, $v_g = 50tu(t)$.

a) Use the transfer function to find v_o.

b) Identify the transient component of the response.

c) Identify the steady-state component of the response.

d) Sketch v_o versus t for $0 \leq t \leq 1.5$ ms.

SOLUTION

a) From Example 14.1,

$$H(s) = \frac{1000(s + 5000)}{s^2 + 6000s + 25 \times 10^6}.$$

The transform of the driving voltage is $50/s^2$; therefore the s-domain expression for the output voltage is

$$V_o = \frac{1000(s + 5000)}{(s^2 + 6000s + 25 \times 10^6)} \frac{50}{s^2}.$$

The partial fraction expansion of V_o is

$$V_o = \frac{K_1}{s + 3000 - j4000} + \frac{K_1^*}{s + 3000 + j4000} + \frac{K_2}{s^2} + \frac{K_3}{s}.$$

We evaluate the coefficients K_1, K_2, and K_3 by using the techniques described in Section 13.7:

$$K_1 = 5\sqrt{5} \times 10^{-4} \angle 79.70^\circ;$$
$$K_1^* = 5\sqrt{5} \times 10^{-4} \angle -79.70^\circ;$$
$$K_2 = 10;$$
$$K_3 = -4 \times 10^{-4}.$$

The time-domain expression for v_o is

$$v_o = [10\sqrt{5} \times 10^{-4} e^{-3000t} \cos (4000t + 79.70^\circ) + 10t - 4 \times 10^{-4}]u(t) \text{ V}.$$

b) The transient component of v_o is

$$10\sqrt{5} \times 10^{-4} e^{-3000t} \cos (4000t + 79.70^\circ).$$

Note that this term is generated by the poles $(-3000 + j4000)$ and $(-3000 - j4000)$ of the transfer function.

c) The steady-state component of the response is

$$(10t - 4 \times 10^{-4})u(t).$$

These two terms are generated by the second-order pole (K/s^2) of the driving voltage.

d) Figure 14.33 shows a sketch of v_o versus t. Note that the deviation from the steady-state solution $10{,}000t - 0.4$ mV is imperceptible after approximately 1 ms.

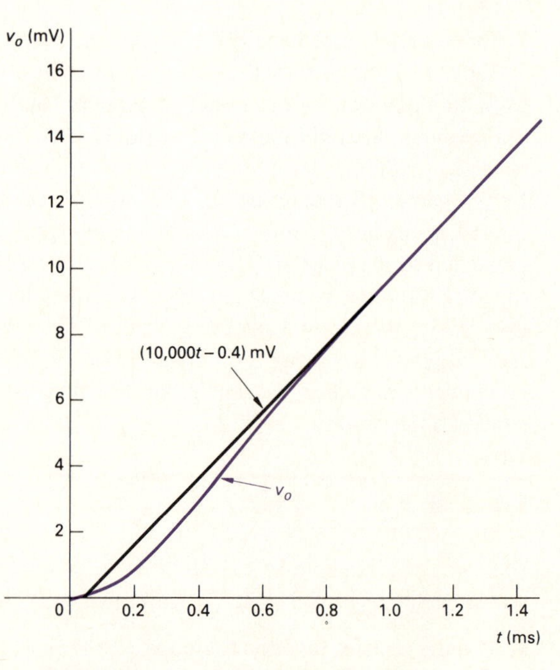

FIGURE 14.33 The graph of v_o versus t for Example 14.2.

OBSERVATIONS ON THE USE OF *H(S)* IN CIRCUIT ANALYSIS

Example 14.2 clearly shows how the transfer function $H(s)$ relates to the response of a circuit through a partial fraction expansion. However, the example raises questions about the practicality of driving a circuit with an increasing ramp voltage that generates an increasing ramp response. Eventually the circuit components will fail under the stress of excessive voltage, and when that happens our linear model is no longer valid. The ramp response is of interest in practical applications where the ramp function increases to a maximum value over a finite time interval. If the time taken to reach this maximum value is long compared to the time constants of the circuit, the solution assuming an unbounded ramp is valid for this finite time interval.

We make two additional observations regarding Eq. (14.96). First, let's look at the response of the circuit due to a delayed input. If the input is delayed by a seconds,

$$\mathcal{L}\{x(t-a)u(t-a)\} = e^{-as}X(s),$$

and, from Eq. (14.96), the response becomes

$$Y(s) = H(s)X(s)e^{-as}. \tag{14.97}$$

If $y(t) = \mathcal{L}^{-1}\{H(s)X(s)\}$, then, from Eq. (14.97),

$$y(t-a)u(t-a) = \mathcal{L}^{-1}\{H(s)X(s)e^{-as}\}. \tag{14.98}$$

Therefore delaying the input by a seconds simply delays the response function by a seconds. A circuit that exhibits this characteristic is said to be **time-invariant**.

Second, if a unit impulse source drives the circuit, the response of the circuit equals the inverse transform of the transfer function. Thus if

$$x(t) = \delta(t), \qquad \text{then} \qquad X(s) = 1$$

and

$$Y(s) = H(s). \tag{14.99}$$

Hence, from Eq. (14.99),

$$y(t) = h(t), \tag{14.100}$$

where the inverse transform of the transfer function equals the unit impulse response of the circuit. Note that this is also the natural response of the circuit because the application of an impulsive source is equivalent to instantaneously storing energy in the circuit (see Section 14.8). The subsequent release of this stored energy gives rise to the natural response. (See Problem 14.83.)

Actually, the unit impulse response of a circuit, $h(t)$, contains enough information to compute the response to any source which drives the circuit. The convolution integral is used to extract the response of a circuit to an arbitrary source, as demonstrated in the next section.

DRILL EXERCISES

14.10 Find (a) the unit step and (b) the unit impulse response of the circuit shown in Drill Exercise 14.9.

ANSWER: (a) $[6 + 10e^{-3t} \cos (t + 126.87°)]u(t)$ V; (b) $31.62e^{-3t} \cos (t - 71.57°)u(t)$ V.

14.11 The unit impulse response of a circuit is

$$v_o(t) = 10{,}000e^{-70t} \cos (240t + \theta)V,$$

where $\tan \theta = \frac{7}{24}$.

a) Find the transfer function of the circuit.

b) Find the unit step response of the circuit.

ANSWER: (a) $9600s/(s^2 + 140s + 62{,}500)$; (b) $40e^{-70t} \sin 240t$ V.

14.6 THE TRANSFER FUNCTION AND THE CONVOLUTION INTEGRAL

The convolution integral relates the output $y(t)$ of a linear time-invariant circuit to the input $x(t)$ of the circuit and the circuit's impulse response $h(t)$. The integral relationship can be expressed in two ways:

$$y(t) = \int_{-\infty}^{\infty} h(\lambda)x(t - \lambda)d\lambda = \int_{-\infty}^{\infty} h(t - \lambda)x(\lambda)d\lambda. \qquad (14.101)$$

We are interested in the convolution integral for several reasons. First, it allows us to work entirely in the time domain. Doing so may be beneficial in situations where $x(t)$ and $h(t)$ are known only through experimental data. In such cases, the transform method may be awkward or even impossible, as it would require us to compute the Laplace transform of experimental data. Second, the convolution integral introduces the concepts of memory and the weighting function into analysis. We will show how the concept of memory enables us to look at the impulse response (or the weighting function) $h(t)$ and predict, to some degree, how closely the output waveform replicates the input waveform. Finally, the convolution integral provides a formal procedure for finding the inverse transform of products of Laplace transforms.

We based the derivation of Eq. (14.101) on the assumption that the circuit is linear and time-invariant. Because the circuit is linear, the principle of superposition is valid, and because it is time-invariant, the amount of the response delay is exactly the same as that of the input delay. Now consider Fig. 14.34, in which the block containing $h(t)$ represents any linear time-invariant circuit whose impulse response is known, $x(t)$ represents the excitation signal, and $y(t)$ represents the desired output signal.

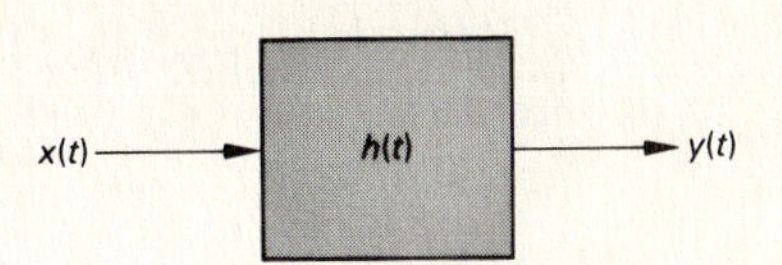

FIGURE 14.34 A block diagram of a general circuit.

We assume that $x(t)$ is the general excitation signal shown in Fig. 14.35(a). For convenience we also assume that $x(t) = 0$ for $t < 0^-$. Once you see the derivation of the convolution integral assuming $x(t) = 0$ for $t < 0^-$, the extension of the integral to include excitation functions that exist over all time becomes apparent. Note also that we permit a discontinuity in $x(t)$ at the origin, that is, a jump between 0^- and 0^+.

Now we approximate $x(t)$ by a series of rectangular pulses of uniform width $\Delta\lambda$, as shown in Fig. 14.35(b). Thus

$$x(t) = x_0(t) + x_1(t) + \cdots + x_i(t) + \cdots, \qquad \textbf{(14.102)}$$

where $x_i(t)$ is a rectangular pulse that equals $x(\lambda_i)$ between λ_i and λ_{i+1} and is zero elsewhere. Note that the ith pulse can be expressed in terms of step functions; that is,

$$x_i(t) = x(\lambda_i)\{u(t - \lambda_i) - u[t - (\lambda_i + \Delta\lambda)]\}.$$

The next step in the approximation of $x(t)$ is to make $\Delta\lambda$ small enough that the ith component can be approximated by an impulse function of strength $x(\lambda_i)\Delta\lambda$. Figure 14.35(c) shows the impulse representation, with the strength of each impulse shown in brackets beside each arrow. The impulse representation of $x(t)$ is

$$x(t) = x(\lambda_0)\Delta\lambda\delta(t - \lambda_0) + x(\lambda_1)\Delta\lambda\delta(t - \lambda_1) + \cdots + x(\lambda_i)\Delta\lambda\delta(t - \lambda_i) + \cdots. \qquad \textbf{(14.103)}$$

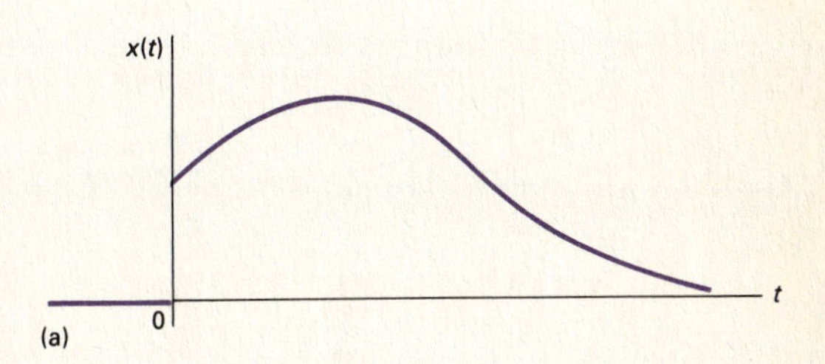

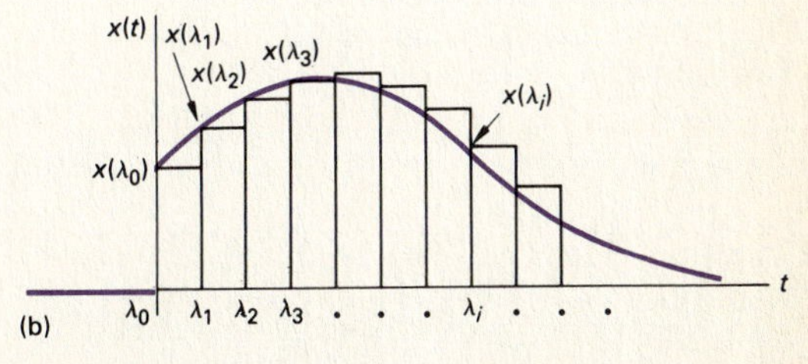

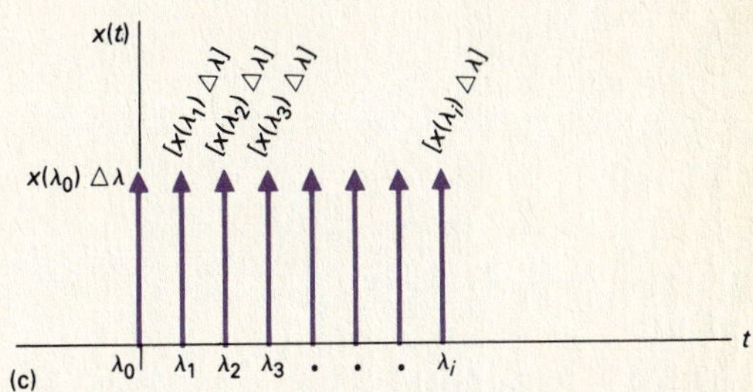

FIGURE 14.35 The excitation signal of $x(t)$: (a) a general excitation signal; (b) approximating $x(t)$ with a series of pulses; and (c) approximating $x(t)$ with a series of impulses.

Now when $x(t)$ is represented by a series of impulse functions (which occur at equally spaced intervals of time, that is, at λ_0, λ_1, $\lambda_2, \ldots$), the response function $y(t)$ consists of the sum of a series of uniformly delayed impulse responses. The strength of each response depends on the strength of the impulse driving the circuit. For example, let's assume that the unit impulse response of the circuit contained in the box in Fig. 14.34 is the exponential decay function shown in Fig. 14.36(a). Then the approximation of $y(t)$ is the sum of the impulse responses shown in Fig. 14.36(b).

Analytically, the expression for $y(t)$ is

$$y(t) = x(\lambda_0)\Delta\lambda h(t - \lambda_0) + x(\lambda_1)\Delta\lambda h(t - \lambda_1) + x(\lambda_2)\Delta\lambda h(t - \lambda_2) + \cdots + x(\lambda_i)\Delta\lambda h(t - \lambda_i) + \cdots. \qquad \textbf{(14.104)}$$

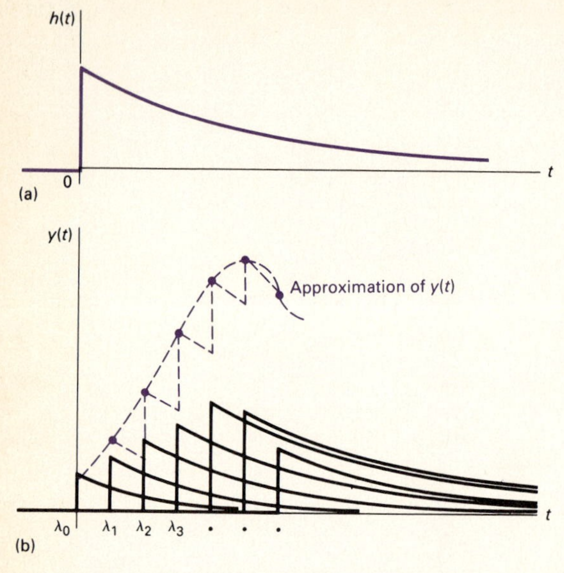

FIGURE 14.36 The approximation of $y(t)$: (a) the impulse response of the box shown in Fig. 14.34; and (b) summing the impulse responses.

As $\Delta\lambda \to 0$, the summation in Eq. (14.104) approaches a continuous integration, or

$$\sum_{i=0}^{\infty} x(\lambda_i)h(t-\lambda_i)\Delta\lambda \to \int_0^{\infty} x(\lambda)h(t-\lambda)d\lambda. \quad \textbf{(14.105)}$$

Therefore

$$y(t) = \int_0^{\infty} x(\lambda)h(t-\lambda)d\lambda. \quad \textbf{(14.106)}$$

If $x(t)$ exists over all time, then the lower limit on Eq. (14.106) becomes $-\infty$; thus, in general,

$$y(t) = \int_{-\infty}^{\infty} x(\lambda)h(t-\lambda)d\lambda, \quad \textbf{(14.107)}$$

which is the second form of the convolution integral given in Eq. (14.101). We derive the first form of the integral from Eq. (14.107) by making a change in the variable of integration. We let $u = t - \lambda$, and then we note that $du = -d\lambda$, $u = -\infty$ when $\lambda = \infty$, and $u = +\infty$ when $\lambda = -\infty$. Now we can write Eq. (14.107) as

$$y(t) = \int_{\infty}^{-\infty} x(t-u)h(u)(-du)$$

or

$$y(t) = \int_{-\infty}^{\infty} x(t-u)h(u)(du). \quad \textbf{(14.108)}$$

But because u is just a symbol of integration, Eq. (14.108) is equivalent to the first form of the convolution integral, Eq. (14.101).

The integral relationship between $y(t)$, $h(t)$, and $x(t)$, expressed in Eq. (14.101), often is written in a shorthand notation:

$$y(t) = h(t) * x(t) = x(t) * h(t), \quad \textbf{(14.109)}$$

where the asterisk signifies the integral relationship between $h(t)$ and $x(t)$. Thus $h(t) * x(t)$ is read as "$h(t)$ is convolved with $x(t)$" and implies that

$$h(t) * x(t) = \int_{-\infty}^{\infty} h(\lambda)x(t-\lambda)d\lambda,$$

whereas $x(t) * h(t)$ is read as "$x(t)$ is convolved with $h(t)$" and implies that

$$x(t) * h(t) = \int_{-\infty}^{\infty} x(\lambda)h(t-\lambda)d\lambda.$$

The integrals in Eq. (14.101) give the most general relationship for the convolution of two functions. However, in our applications of the convolution integral, we can change the lower limit to zero and the upper limit to t. Then we can write Eq. (14.101) as

$$y(t) = \int_0^t h(\lambda)x(t-\lambda)d\lambda = \int_0^t x(\lambda)h(t-\lambda)d\lambda. \qquad \textbf{(14.110)}$$

We change the limits for two reasons. First, for physically realizable circuits, $h(t)$ is zero for $t < 0$. In other words, there can be no impulse response before an impulse is applied. Second, we start measuring time at the instant the excitation $x(t)$ is turned on; therefore $x(t) = 0$ for $t < 0^-$.

A graphic interpretation of the convolution integrals contained in Eq. (14.110) is important in the use of the integral as a computational tool. We begin with the first integral. For purposes of discussion, we assume that the impulse response of our circuit is the exponential decay function shown in Fig. 14.37(a) and that the excitation function has the waveform shown in Fig. 14.37(b). In each of these plots, we replaced t with λ, the symbol of integration. Figures 14.37(c) and (d) illustrate $x(t-\lambda)$. Replacing λ with $-\lambda$ simply folds the excitation function over the vertical axis, and replacing $-\lambda$ with $t-\lambda$ slides the folded function to the right. This folding operation gives rise to the term *convolution.* At any specified value of t, the response function $y(t)$ is the area under the product function $h(\lambda)x(t-\lambda)$, as shown in Fig. 14.37(e). It should be apparent from this plot why the lower limit

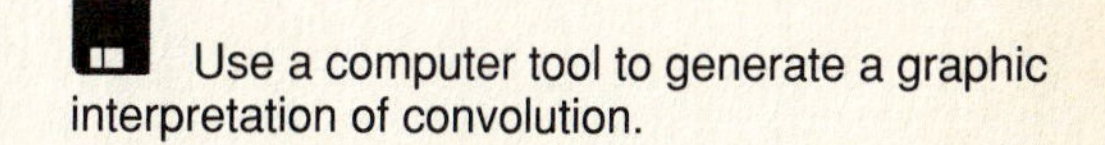
Use a computer tool to generate a graphic interpretation of convolution.

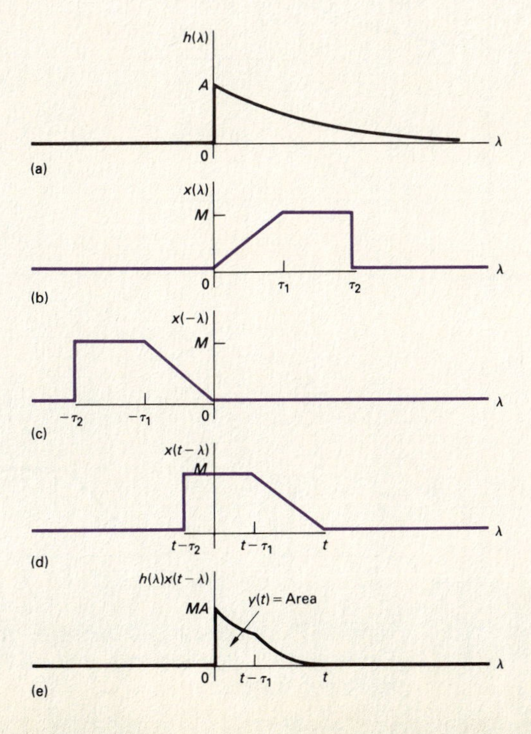

FIGURE 14.37 A graphic interpretation of the convolution integral $\int_0^t h(\lambda)\, x(t-\lambda)d\lambda$: (a) the impulse response; (b) the excitation function; (c) the folded excitation function; (d) the folded excitation function displaced t units; and (e) the product $h(\lambda)x(t-\lambda)$.

on the convolution integral is zero and the upper limit is t. For $\lambda < 0$, the product $h(\lambda)x(t - \lambda)$ is zero because $h(\lambda)$ is zero. For $\lambda > t$, the product $h(\lambda)x(t - \lambda)$ is zero because $x(t - \lambda)$ is zero.

Figure 14.38 shows the second form of the convolution integral. Note that the product function in Fig. 14.38(e) confirms the use of zero for the lower limit and t for the upper limit.

Example 14.3 illustrates how to use the convolution integral, in conjunction with the unit impulse response, to find the response of a circuit.

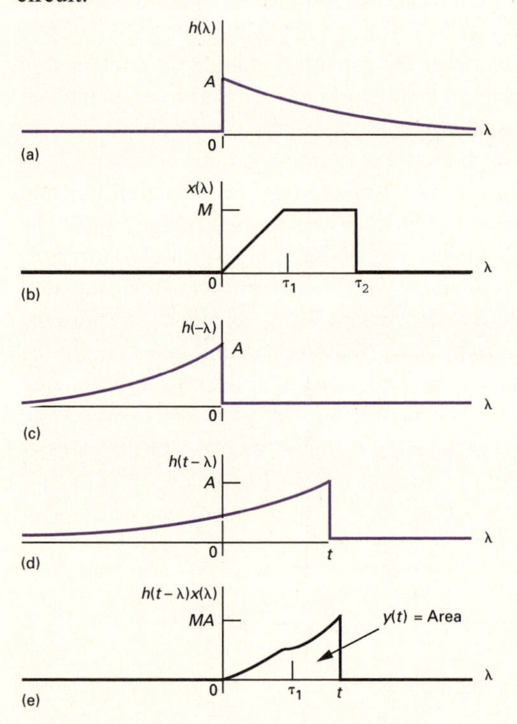

FIGURE 14.38 A graphic interpretation of the convolution integral $\int_0^t h(t-\lambda)\, x(\lambda)\, d\lambda$: (a) the impulse response; (b) the excitation function; (c) the folded excitation function; (d) the folded excitation function displaced t units; and (e) the product $h(t-\lambda)\, x(\lambda)$.

EXAMPLE 14.3

The excitation voltage v_i for the circuit shown in Fig. 14.39(a) is shown in Fig. 14.39(b).

a) Use the convolution integral to find v_o.

b) Plot v_o over the range of $0 \le t \le 15$ s.

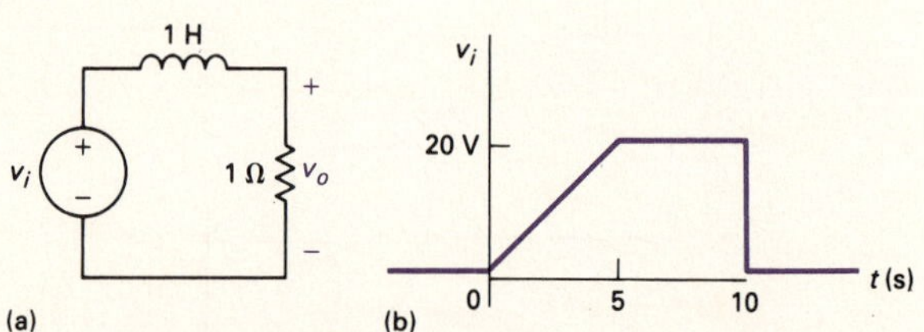

FIGURE 14.39 The circuit and excitation voltage for Example 14.3: (a) the circuit; and (b) the excitation voltage.

SOLUTION

a) The first step in using the convolution integral is to find the unit impulse response of the circuit. We obtain the expression for V_o from the s-domain equivalent of the circuit in Fig. 14.39(a):

$$V_o = \frac{V_i}{s+1}(1).$$

When v_i is a unit impulse function $\delta(t)$,

$$v_o = h(t) = e^{-t}u(t),$$

from which

$$h(\lambda) = e^{-\lambda}u(\lambda).$$

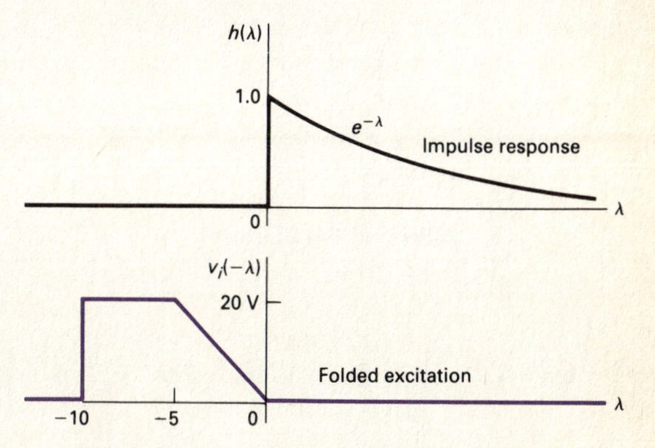

FIGURE 14.40 The impulse response and the folded excitation function for Example 14.3.

Using the first form of the convolution integral in Eq. (14.110), we construct the impulse response and folded excitation function shown in Fig. 14.40, which are helpful in selecting the limits on the convolution integral. Sliding the folded excitation function to the right requires breaking the integration into three intervals: $0 \le t \le 5$; $5 \le t \le 10$; and $10 \le t \le \infty$. The breaks in the excitation function at 0, 5, and 10 s dictate these break points. Figure 14.41 shows the positioning of the folded excitation for each of these intervals. The analytical expression for v_i in the time interval $0 \le t \le 5$ is

$$v_i = 4t \quad (0 \le t \le 5 \text{ s}).$$

Hence the analytical expression for the folded excitation function in the interval $t - 5 \le \lambda \le t$ is

$$v_i(t-\lambda) = 4(t-\lambda) \quad (t-5 \le \lambda \le t).$$

We can now set up the three integral expressions for v_o. For $0 \le t \le 5$ s:

$$v_o = \int_0^t 4(t-\lambda)e^{-\lambda}d\lambda = 4(e^{-t} + t - 1) \text{ V}.$$

For $5 \le t \le 10$ s,

$$v_o = \int_0^{t-5} 20e^{-\lambda}d\lambda + \int_{t-5}^{t} 4(t-\lambda)e^{-\lambda}d\lambda = 4(5 + e^{-t} - e^{-(t-5)}) \text{ V}.$$

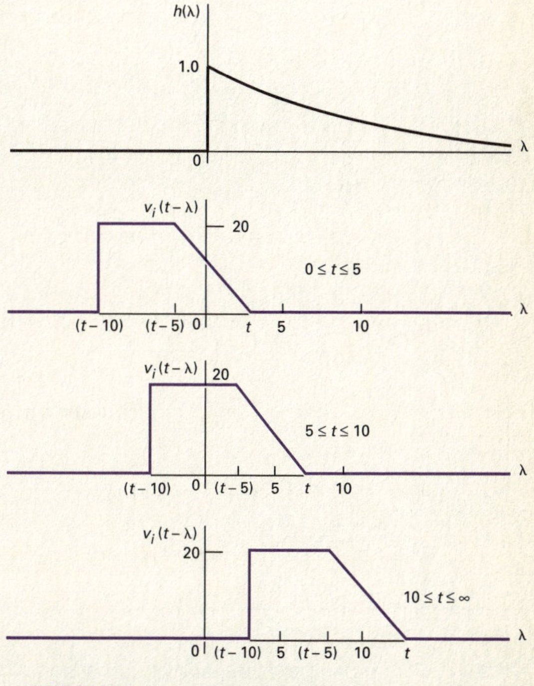

FIGURE 14.41 The displacement of $v_i(t-\lambda)$ for three different time intervals.

And for $10 \leq t \leq \infty$ s,

$$v_o = \int_{t-10}^{t-5} 20e^{-\lambda}d\lambda + \int_{t-5}^{t} 4(t-\lambda)e^{-\lambda}d\lambda$$
$$= 4(e^{-t} - e^{-(t-5)} + 5e^{-(t-10)}) \text{ V}.$$

b) We have computed v_o for 1 s intervals of time, using the appropriate equation. The results are tabulated in Table 14.2 and shown graphically in Fig. 14.42.

TABLE 14.2

NUMERICAL VALUES OF $v_o(t)$

t	v_o	t	v_o	t	v_o
1	1.47	6	18.54	11	7.35
2	4.54	7	19.56	12	2.70
3	8.20	8	19.80	13	0.99
4	12.07	9	19.93	14	0.37
5	16.03	10	19.97	15	0.13

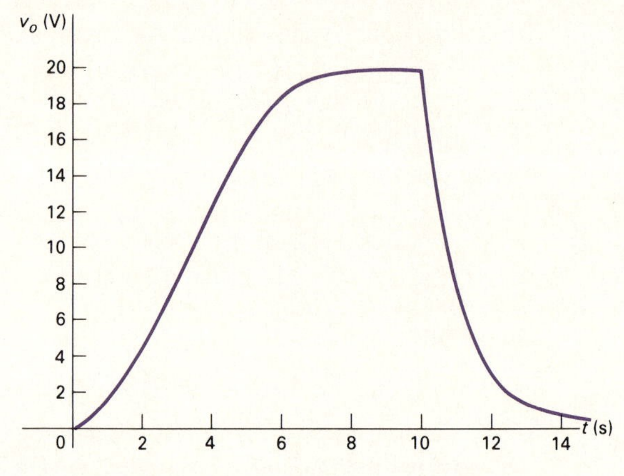

FIGURE 14.42 The voltage response versus time for Example 14.3.

THE CONCEPTS OF MEMORY AND THE WEIGHTING FUNCTION

We mentioned at the beginning of this section that the convolution integral introduces the concepts of memory and the weighting function into circuit analysis. The graphic interpretation of the convolution integral is the easiest way to begin to grasp these concepts. We can view the folding and sliding of the excitation function on a timescale characterized as past, present, and future. The vertical axis, over which the excitation function $x(t)$ is folded, represents the present value; past values of $x(t)$ lie to the right of the vertical axis, and future values lie to the left. Figure 14.43 shows this description of $x(t)$. For illustrative purposes, we used the excitation function from Example 14.3.

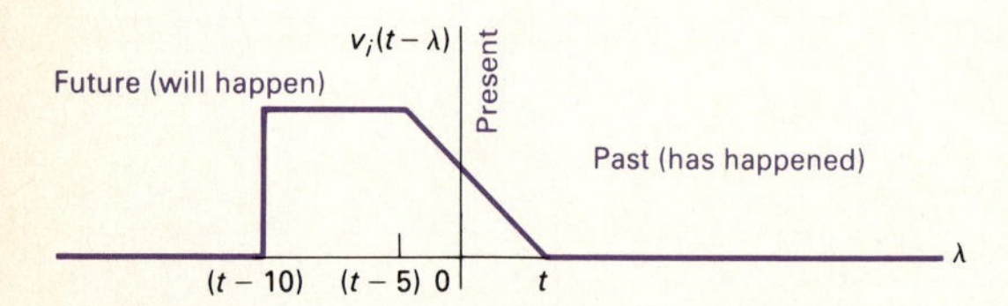

FIGURE 14.43 The past, present, and future values of the excitation function.

When we combine the past, present, and future views of $x(t-\tau)$ with the impulse response of the circuit, we see that the impulse response weights $x(t)$ according to present and past values. For example, Fig. 14.41 shows that the impulse response in Example 14.3 gives less weight to past values of $x(t)$ than to the present value of $x(t)$. In other words, the circuit retains less and less about past input values. Therefore, in Fig. 14.42, v_o quickly approaches zero when the present value of the input is zero (that is, when $t > 10$ s). In other words, because the present value of the input receives more weight than the past values, the output quickly approaches the present value of the input.

The multiplication of $x(t - \lambda)$ by $h(\lambda)$ gives rise to the practice of referring to the impulse response as the circuit **weighting function**. The weighting function, in turn, determines how much memory the circuit has. **Memory** is the extent to which the circuit's response matches its input. For example, if the impulse response, or weighting function, is flat, as shown in Fig. 14.44(a), it gives equal weight to all values of $x(t)$, past and present. Such a circuit has a perfect memory. However, if the impulse response is an impulse function, as shown in Fig. 14.44(b), it gives no weight to past values of $x(t)$. Such a circuit has no memory. Thus the more memory a circuit has, the more distortion there is between the waveform of the excitation function and the waveform of the response function. We can show this relationship by assuming that the circuit has no memory—that is, $h(t) = A\delta(t)$—and then noting from the convolution integral that

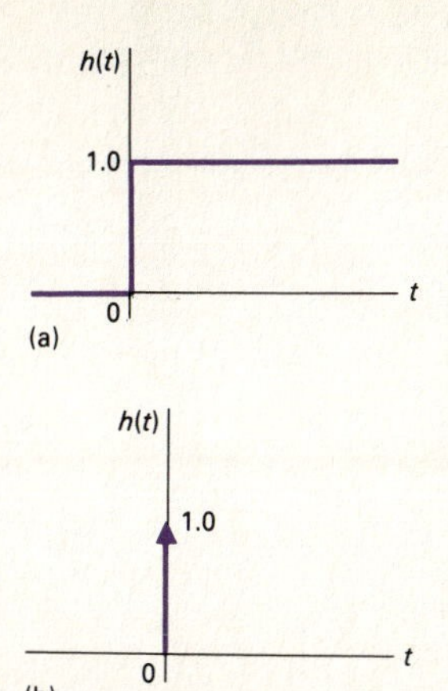

FIGURE 14.44 Weighting functions: (a) perfect memory; (b) no memory.

$$\begin{aligned} y(t) &= \int_0^t h(\lambda)x(t-\lambda)d\lambda \\ &= \int_0^t A\delta(\lambda)x(t-\lambda)d\lambda \\ &= Ax(t). \end{aligned} \quad \textbf{(14.111)}$$

Equation (14.111) shows that, if the circuit has no memory, the output is a scaled replica of the input.

The circuit shown in Example 14.3 illustrates the distortion between input and output for a circuit that has some memory. This distortion is clear when we plot the input and output waveforms on the same graph, as in Fig. 14.45.

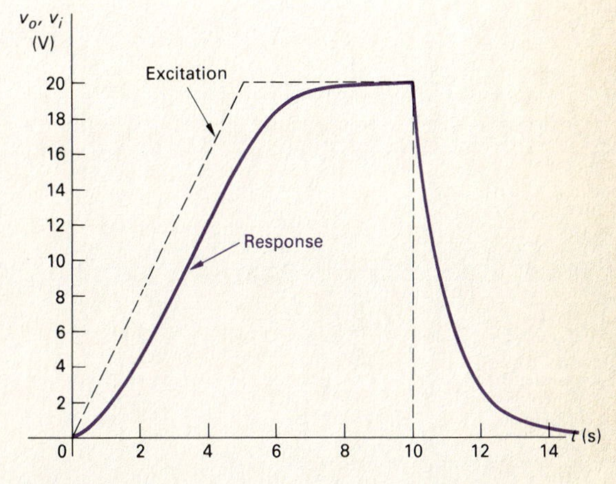

FIGURE 14.45 The input and output waveforms for Example 14.3.

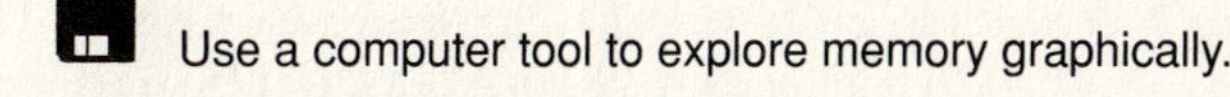

Use a computer tool to explore memory graphically.

DRILL EXERCISES

14.12 A rectangular voltage pulse $v_i = [u(t) - u(t-1)]$ V is applied to the circuit shown. Use the convolution integral to find v_o.

ANSWER: $v_o = 1 - e^{-t}$ V, $0 \leq t \leq 1$; $v_o = (e-1)e^{-t}$ V, $1 \leq t \leq \infty$.

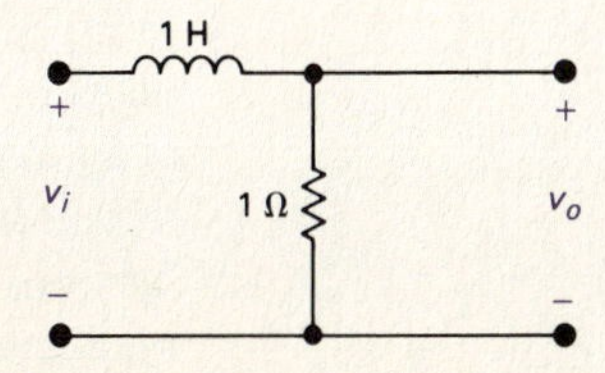

14.13 Interchange the inductor and resistor in Drill Exercise 14.12 and again use the convolution integral to find v_o.

ANSWER: $v_o = e^{-t}$ V, $0 < t < 1$; $v_o = (1 - e)e^{-t}$ V, $1 < t \leq \infty$.

14.7 The Transfer Function and the Steady-State Sinusoidal Response

Once we have computed a circuit's transfer function, we no longer need to perform a separate phasor analysis of the circuit to determine its steady-state response. Instead, we use the transfer function to relate the steady-state response to the excitation source. First we assume that

$$x(t) = A\cos(\omega t + \phi), \tag{14.112}$$

and then we use Eq. (14.96) to find the steady-state solution of $y(t)$. To find the Laplace transform of $x(t)$, we first write $x(t)$ as

$$x(t) = A\cos\omega t\cos\phi - A\sin\omega t\sin\phi, \tag{14.113}$$

from which

$$X(s) = \frac{(A\cos\phi)s}{s^2+\omega^2} - \frac{(A\sin\phi)\omega}{s^2+\omega^2} = \frac{A(s\cos\phi - \omega\sin\phi)}{s^2+\omega^2}. \tag{14.114}$$

Substituting Eq. (14.114) into Eq. (14.96) gives the s-domain expression for the response:

$$Y(s) = H(s)\frac{A(s\cos\phi - \omega\sin\phi)}{s^2+\omega^2}. \tag{14.115}$$

We now visualize the partial fraction expansion of Eq. (14.115). The number of terms in the expansion depends on the number of poles of $H(s)$. Because $H(s)$ is not specified beyond being the transfer function of a physically realizable circuit, the expansion of Eq. (14.115) is

$$Y(s) = \frac{K_1}{s - j\omega} + \frac{K_1^*}{s + j\omega} + \sum \text{terms generated by the poles of } H(s). \tag{14.116}$$

In Eq. (14.116) the first two terms result from the complex conjugate poles of the driving source; that is, $s^2 + \omega^2 = (s - j\omega)(s + j\omega)$. However, the terms generated by the poles of $H(s)$ do not contribute to the steady-state response of $y(t)$, because of all these poles lie in the left half of the s plane; consequently, the corresponding time-domain terms approach zero as t increases. Thus the first two terms on the right-hand side of Eq. (14.115) determine the steady-state response. The problem is reduced to finding the partial fraction coefficient K_1:

$$K_1 = \left.\frac{H(s)A(s\cos\phi - \omega\sin\phi)}{s + j\omega}\right|_{s=j\omega}$$
$$= \frac{H(j\omega)A(j\omega\cos\phi - \omega\sin\phi)}{2j\omega}$$
$$= \frac{H(j\omega)A(\cos\phi + j\sin\phi)}{2} = \frac{1}{2}H(j\omega)Ae^{j\phi}. \quad \textbf{(14.117)}$$

In general, $H(j\omega)$ is a complex quantity, which we recognize by writing it in polar form; thus

$$H(j\omega) = |H(j\omega)|e^{j\theta(\omega)}. \quad \textbf{(14.118)}$$

Note from Eq. (14.118) that both the magnitude, $|H(j\omega)|$, and phase angle, $\theta(\omega)$, of the transfer function vary with the frequency ω. When we substitute Eq. (14.118) into Eq. (14.117), the expression for K_1 becomes

$$K_1 = \frac{A}{2}|H(j\omega)|e^{j[\theta(\omega)+\phi]}. \quad \textbf{(14.119)}$$

We obtain the steady-state solution for $y(t)$ by inverse-transforming Eq. (14.116) and, in the process, ignoring the terms generated by the poles of $H(s)$. Thus

$$y_{ss}(t) = A|H(j\omega)|\cos[\omega t + \phi + \theta(\omega)], \quad \textbf{(14.120)}$$

which indicates how to use the transfer function to find the steady-state sinusoidal response of a circuit. The amplitude of the response equals the amplitude of the source, A, times the magnitude of the transfer function, $|H(j\omega)|$. The phase angle of the response, $\phi + \theta(\omega)$, equals the phase angle of the source, ϕ, plus the phase angle of the transfer function, $\theta(\omega)$. We evaluate both $|H(j\omega)|$ and $\theta(\omega)$ at the frequency of the source, ω.

Example 14.4 illustrates how to use the transfer function to find the steady-state sinusoidal response of a circuit.

EXAMPLE 14.4

The circuit from Example 14.1 is shown in Fig. 14.46. The sinusoidal source voltage is $120\cos(5000t + 30°)$ V. Find the steady-state expression for v_o.

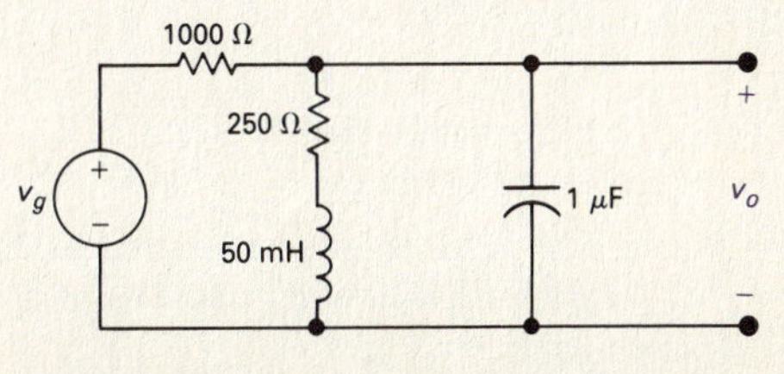

FIGURE 14.46 The circuit for Example 14.4.

SOLUTION

From Example 14.1,

$$H(s) = \frac{1000(s + 5000)}{s^2 + 6000s + 25 \times 10^6}.$$

The frequency of the voltage source is 5000 rad/s; hence we evaluate $H(s)$ at $H(j5000)$:

$$H(j5000) = \frac{1000(5000 + j5000)}{-25 \times 10^6 + j5000(6000) + 25 \times 10^6}$$

$$= \frac{1 + j1}{j6} = \frac{1 - j1}{6} = \frac{\sqrt{2}}{6}\angle{-45^\circ}.$$

Then, from Eq. (14.120),

$$v_{o_{ss}} = \frac{(120)\sqrt{2}}{6}\cos(5000t + 30^\circ - 45^\circ)$$

$$= 20\sqrt{2}\cos(5000t - 15^\circ)\text{ V}.$$

The ability to use the transfer function to calculate the steady-state sinusoidal response of a circuit is important. Note that if we know $H(j\omega)$, we also know $H(s)$, at least theoretically. In other words, we can reverse the process; instead of using $H(s)$ to find $H(j\omega)$, we use $H(j\omega)$ to find $H(s)$. Once we know $H(s)$, we can find the response to other excitation sources. In this application, we determine $H(j\omega)$ experimentally and then construct $H(s)$ from the data. Practically, this experimental approach is not always possible; however, in some cases it does provide a useful method for deriving $H(s)$. In theory, the relationship between $H(s)$ and $H(j\omega)$ provides a link between the time domain and the frequency domain. The transfer function is also a very useful tool in problems concerning the frequency response of a circuit, a concept we introduce in the next chapter.

DRILL EXERCISES

14.14 The current source in the circuit shown is delivering $3\sqrt{2}\cos 2t$ A. Use the transfer function to compute the steady-state expression for v_o.

ANSWER: $20\cos(2t - 45^\circ)$ V.

14.15 a) For the circuit shown, find the steady-state expression for v_o when $v_g = 10\cos 50{,}000t$ V.

b) Replace the 50 kΩ resistor with a variable resistor and compute the value of resistance necessary to cause v_o to lead v_g by 120°.

ANSWER: (a) $10\cos(50{,}000t + 90^\circ)$ V; (b) 28,867.51 Ω.

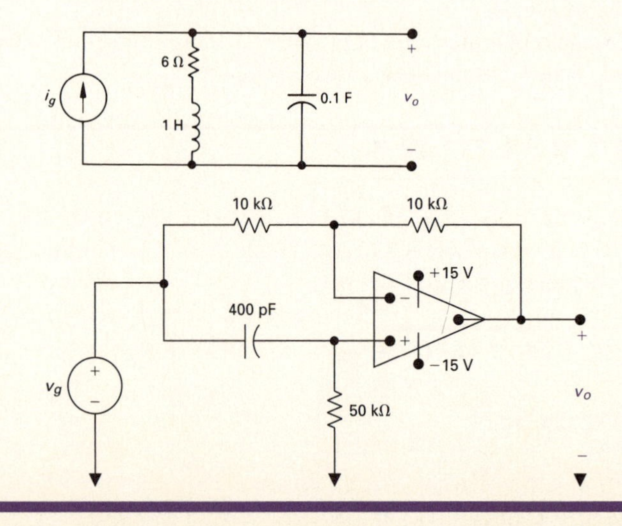

14.8 The Impulse Function in Circuit Analysis

Impulse functions occur in circuit analysis either because of a switching operation or because a circuit is excited by an impulsive source. The Laplace transform can be used to predict the impulsive currents and voltages created during switching and the response of a circuit to an impulsive source. We begin our discussion by showing how to create an impulse function with a switching operation.

Switching Operations

Capacitor Circuit In the circuit shown in Fig. 14.47, the capacitor C_1 is charged to an initial voltage of V_0 at the time the switch is closed. The initial charge on C_2 is zero. The problem is to find the expression for $i(t)$ as $R \to 0$. Figure 14.48 shows the s-domain equivalent circuit.

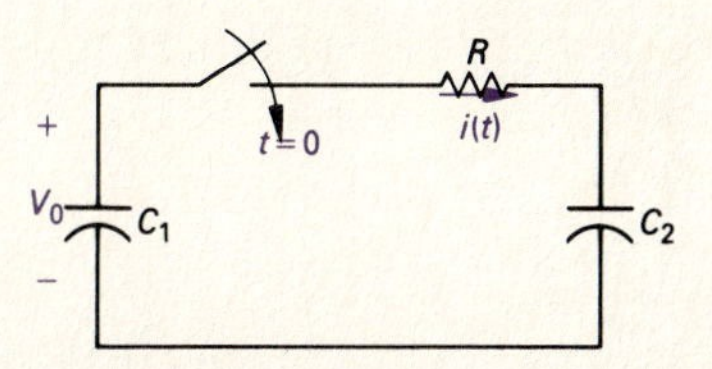

FIGURE 14.47 A circuit showing the creation of an impulsive current.

From Fig. 14.48,

$$I = \frac{V_0/s}{R + (1/sC_1) + (1/sC_2)}$$

$$= \frac{V_0/R}{s + (1/RC_e)}, \tag{14.121}$$

where the equivalent capacitance $C_1C_2/(C_1 + C_2)$ is replaced by C_e.

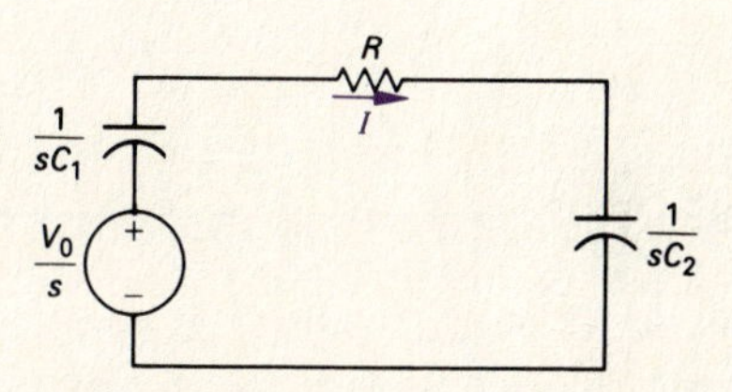

FIGURE 14.48 The s-domain equivalent circuit for the circuit shown in Fig. 14.47.

We inverse-transform Eq. (14.121) by inspection to obtain

$$i = \left(\frac{V_0}{R}e^{-t/RC_e}\right)u(t), \tag{14.122}$$

which indicates that as R decreases, the initial current (V_0/R) increases and the time constant (RC_e) decreases. Thus as R gets smaller, the current starts from a larger initial value and then drops off more rapidly. Figure 14.49 shows these characteristics of i.

Apparently i is approaching an impulse function as R approaches zero because the initial value of i is approaching infinity and the duration of i is approaching zero. We still have to determine whether the area under the current function is independent of R. Physically, the total area under the i versus t curve represents the total charge transferred to C_2 after the switch is closed. Thus

$$\text{Area} = q = \int_{0^-}^{\infty} \frac{V_0}{R}e^{-t/RC_e}\,dt = V_0C_e, \tag{14.123}$$

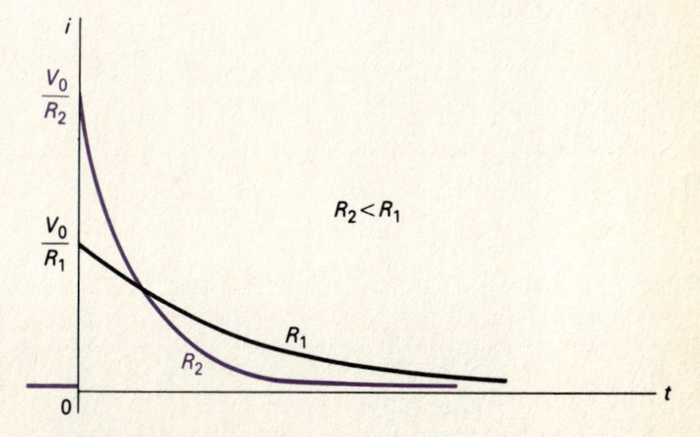

FIGURE 14.49 The plot of $i(t)$ versus t for two different values of R.

which says that the total charge transferred to C_2 is independent of R and equals V_0C_e coulombs. Thus, as R approaches zero, the current approaches an impulse strength $V_0 C_e$; that is,

$$i \to V_0 C_e \delta(t). \tag{14.124}$$

The physical interpretation of Eq. (14.124) is that when $R = 0$, a finite amount of charge is transferred to C_2 instantaneously. Making R zero in the circuit shown in Fig. 14.47 shows why we get an instantaneous transfer of charge. With $R = 0$, we create a contradiction when we close the switch; that is, we apply a voltage across a capacitor that has a zero initial voltage. The only way to have an instantaneous change in capacitor voltage is to have an instantaneous transfer of charge. When the switch is closed, the voltage across C_2 does not jump to V_0 but to its final value of

$$v_2 = \frac{C_1 V_0}{C_1 + C_2}. \tag{14.125}$$

We leave the derivation of Eq. (14.125) to you (see Problem 14.74).

If we set R equal to zero at the outset, the Laplace transform analysis will predict the impulsive current response. Thus

$$I = \frac{V_0/s}{(1/sC_1) + (1/sC_2)} = \frac{C_1 C_2 V_0}{C_1 + C_2} = C_e V_0. \tag{14.126}$$

In writing Eq. (14.126), we use the capacitor voltages at $t = 0^-$. The inverse transform of a constant is the constant times the impulse function; therefore, from Eq. (14.126),

$$i = C_e V_0 \delta(t). \tag{14.127}$$

The ability of the Laplace transform to predict correctly the occurrence of an impulsive response is one reason why the transform is widely used to analyze the transient behavior of linear lumped-parameter time-invariant circuits.

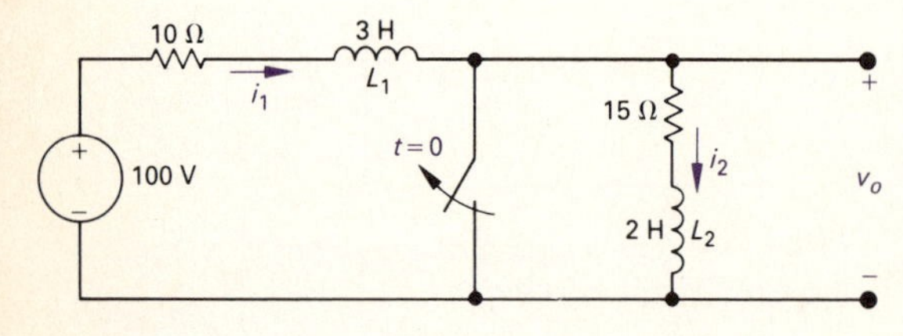

FIGURE 14.50 A circuit showing the creation of an impulsive voltage.

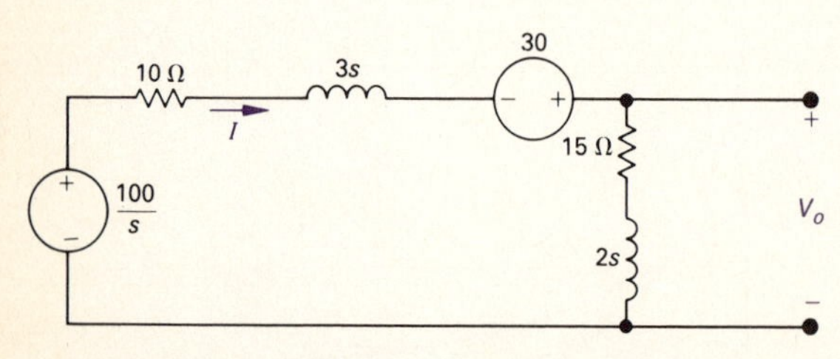

FIGURE 14.51 The s-domain equivalent circuit for the circuit shown in Fig. 14.50.

SERIES INDUCTOR CIRCUIT The circuit shown in Fig. 14.50 illustrates a second switching operation that produces an impulsive response. The problem is to find the time-domain expression for v_o after the switch has been opened. Note that opening the switch forces an instantaneous change in the current of L_2, which causes v_o to contain an impulsive component.

Figure 14.51 shows the s-domain equivalent with the switch open. In deriving this circuit we recognized that the current in the 3 H inductor at $t = 0^-$ is 10 A and the current in the 2 H inductor at $t = 0^-$ is zero. Using the initial conditions at $t = 0^-$ is a direct consequence of our using 0^- as the lower limit on the defining integral of the Laplace transform.

We derive the expression for V_0 from a single node-voltage equation. Summing the currents away from the node between the 15 Ω resistor and the 30 V source gives

$$\frac{V_0}{2s+15} + \frac{V_0 - [(100/s)+30]}{3s+10} = 0. \quad (14.128)$$

Solving for V_0 yields

$$V_0 = \frac{40(s+7.5)}{s(s+5)} + \frac{12(s+7.5)}{s+5}. \quad (14.129)$$

We anticipate that v_o will contain an impulse term because the second term on the right-hand side of Eq. (14.129) is an improper rational function. We can express this improper fraction as a constant plus a rational function by simply dividing the denominator into the numerator; that is,

$$\frac{12(s+7.5)}{s+5} = 12 + \frac{30}{s+5}. \quad (14.130)$$

Combining Eq. (14.130) with the partial fraction expansion of the first term on the right-hand side of Eq. (14.129) gives

$$\begin{aligned} V_0 &= \frac{60}{s} - \frac{20}{s+5} + 12 + \frac{30}{s+5} \\ &= 12 + \frac{60}{s} + \frac{10}{s+5}, \end{aligned} \quad (14.131)$$

from which

$$v_o = 12\delta(t) + (60 + 10e^{-5t})u(t) \text{ V}. \quad (14.132)$$

Does this solution make sense? Before answering that question, let's first derive the expression for the current when $t > 0^-$. After the switch has been opened, the current in L_1 is the same as the current in L_2. If we reference the current clockwise around the mesh, the s-domain expression is

$$\begin{aligned} I &= \frac{(100/s)+30}{5s+25} = \frac{20}{s(s+5)} + \frac{6}{s+5} \\ &= \frac{4}{s} - \frac{4}{s+5} + \frac{6}{s+5} \\ &= \frac{4}{s} + \frac{2}{s+5}. \end{aligned} \quad (14.133)$$

Inverse-transforming Eq. (14.133) gives

$$i = (4 + 2e^{-5t})u(t) \text{ A}. \quad (14.134)$$

Before the switch is opened, the current in L_1 is 10 A and the current in L_2 is 0 A; from Eq. (14.134) we know that at $t = 0^+$, the current in L_1 and in L_2 is 6 A. Then, the current in L_1 changes instantaneously from 10 to 6 A, while the current in L_2 changes instantaneously

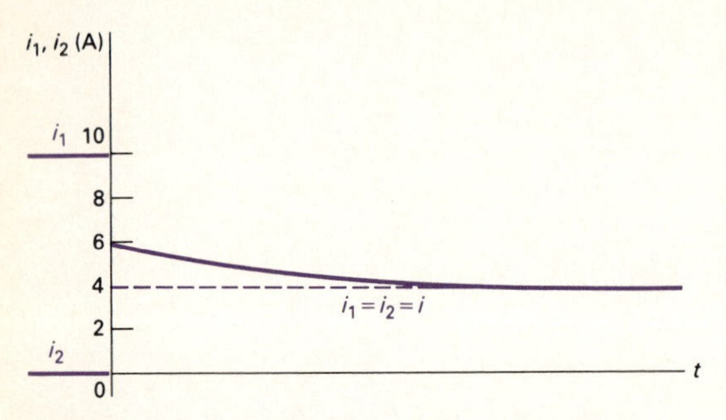

FIGURE 14.52 The inductor currents versus t for the circuit shown in Fig. 14.50.

from 0 to 6 A. From this value of 6 A, the current decreases exponentially to a final value of 4 A. This final value is easily verified from the circuit; that is, it should equal 100/25, or 4 A. Figure 14.52 shows these characteristics of i_1 and i_2.

How can we verify that these instantaneous jumps in the inductor current make sense in terms of the physical behavior of the circuit? First, we note that the switching operation places the two inductors in series. Any impulsive voltage appearing across the 3 H inductor must be exactly balanced by an impulsive voltage across the 2 H inductor, because the sum of the impulsive voltages around a closed path must equal zero. Faraday's law states that the induced voltage is proportional to the change in flux linkage ($v = d\lambda/dt$). Therefore the change in flux linkage must sum to zero. In other words, the total flux linkage immediately after switching is the same as that before switching. For the circuit here, the flux linkage before switching is

$$\lambda = L_1 i_1 + L_2 i_2 = 3(10) + 2(0) = 30 \text{ Wb-turns.} \quad \textbf{(14.135)}$$

Immediately after switching, it is

$$\lambda = (L_1 + L_2)i(0^+) = 5i(0^+). \quad \textbf{(14.136)}$$

Combining Eqs. (14.135) and (14.136) gives

$$i(0^+) = 30/5 = 6 \text{ A.} \quad \textbf{(14.137)}$$

Thus the solution for i (Eq. [14.134]) agrees with the principle of the conservation of flux linkage.

We now test the validity of Eq. (14.132). First we check the impulsive term $12\delta(t)$. The instantaneous jump of i_2 from 0 to 6 A at $t = 0$ gives rise to an impulse of strength $6\delta(t)$ in the derivative of i_2. This impulse gives rise to the $12\delta(t)$ in the voltage across the 2 H inductor. For $t > 0^+$, di_2/dt is $-10e^{-5t}$ A/s; therefore the voltage v_o is

$$\begin{aligned} v_o &= 15(4 + 2e^{-5t}) + 2(-10e^{-5t}) \\ &= (60 + 10e^{-5t})u(t) \text{ V.} \end{aligned} \quad \textbf{(14.138)}$$

Equation (14.138) agrees with the last two terms on the right-hand side of Eq. (14.132); thus we have confirmed that Eq. (14.132) does make sense in terms of known circuit behavior.

We can also check the instantaneous drop from 10 to 6 A in the current i_1. This drop gives rise to an impulse of $-4\delta(t)$ in the derivative of i_1. Therefore the voltage across L_1 contains an impulse of $-12\delta(t)$ at the origin. This impulse exactly balances the impulse across L_2; that is, the sum of the impulsive voltages around a closed path equals zero.

Impulsive Sources

Impulse functions can occur in sources as well as responses; such sources are called impulsive sources. An impulsive source driving a circuit imparts a finite amount of energy into the system instantaneously. A mechanical analogy is striking a bell with an impulsive clapper blow. After the energy has been transferred to the bell, the natural response of the bell determines the tone emitted (that is, the frequency of the resulting sound waves) and the tone's duration.

In the circuit shown in Fig. 14.53, an impulsive voltage source having a strength of V_0 volt-seconds is applied to a series connection of a resistor and an inductor. When the voltage source is applied, the initial energy in the inductor is zero; therefore the initial current is zero. There is no voltage drop across R, so the impulsive voltage source appears directly across L. An impulsive voltage at the terminals of an inductor establishes an instantaneous current. The current is

Use a computer tool to model an impulsive source.

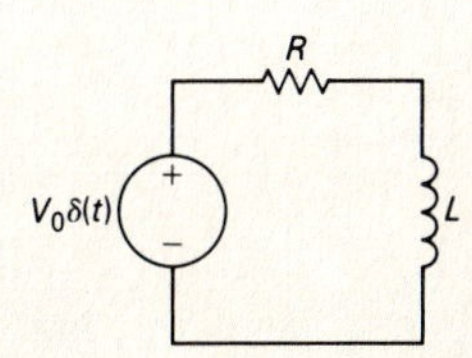

FIGURE 14.53 An *RL* circuit excited by an impulsive voltage source.

$$i = \frac{1}{L}\int_{0^-}^{t} V_0\delta(x)dx. \quad \textbf{(14.139)}$$

Given that the integral of $\delta(t)$ over any interval which includes zero is 1, we find that Eq. (14.139) yields

$$i(0^+) = \frac{V_0}{L}\ \text{A}. \quad \textbf{(14.140)}$$

Thus, in an infinitesimal moment, the impulsive voltage source has stored

$$w = \frac{1}{2}L\left(\frac{V_0}{L}\right)^2 = \frac{1}{2}\frac{V_0^2}{L}\ \text{J} \quad \textbf{(14.141)}$$

in the inductor.

The current V_0/L now decays to zero in accordance with the natural response of the circuit; that is,

$$i = \frac{V_0}{L}e^{-t/\tau}u(t), \quad \textbf{(14.142)}$$

where $\tau = L/R$. Remember from Chapter 7 that the natural response is attributable only to passive elements releasing or storing energy, and not to the effects of sources. When a circuit is driven by only an impulsive source, the total response is completely defined by the natural response; the duration of the impulsive source is so infinitesimal that it does not contribute to any forced response.

We may also obtain Eq. (14.142) by direct application of the Laplace transform method. Figure 14.54 shows the s-domain equivalent of the circuit in Fig. 14.53. Hence

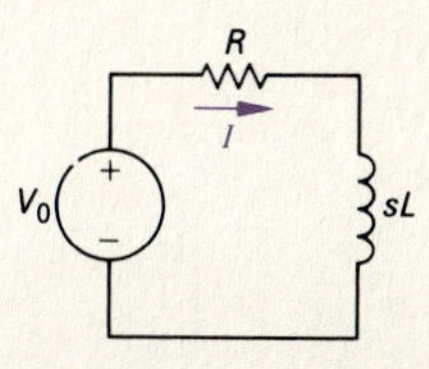

$$I = \frac{V_0}{R + sL} = \frac{V_0/L}{s + (R/L)} \quad \textbf{(14.143)}$$

FIGURE 14.54 The s-domain equivalent circuit for the circuit shown in Fig. 14.53.

and

$$i = \frac{V_0}{L}e^{-(R/L)t} = \frac{V_0}{L}e^{-t/\tau}u(t). \quad \textbf{(14.144)}$$

Thus the Laplace transform method gives the correct solution for $i \geq 0^+$.

Finally, we consider the case in which internally generated impulses and externally applied impulses occur simultaneously. The Laplace transform approach automatically ensures the correct solution for $t > 0^+$ if inductor currents and capacitor voltages at $t = 0^-$ are used in constructing the s-domain equivalent circuit and if externally applied impulses are represented by their transforms. To illustrate, we add an impulsive voltage source of $50\delta(t)$ in series with the 100 V source to the circuit shown in Fig. 14.50. Figure 14.55 shows the new arrangement.

FIGURE 14.55 The circuit shown in Fig. 14.50 with an impulsive voltage source added in series with the 100 V source.

At $t = 0^-$, $i_1(0^-) = 10$ A and $i_2(0^-) = 0$ A. The Laplace transform of $50\delta(t) = 50$. If we use these values, the s-domain equivalent circuit is as shown in Fig. 14.56.

The expression for I is

$$\begin{aligned} I &= \frac{50 + (100/s) + 30}{25 + 5s} = \frac{16}{s+5} + \frac{20}{s(s+5)} \\ &= \frac{16}{s+5} + \frac{4}{s} - \frac{4}{s+5} \\ &= \frac{12}{s+5} + \frac{4}{s}, \end{aligned} \quad \textbf{(14.145)}$$

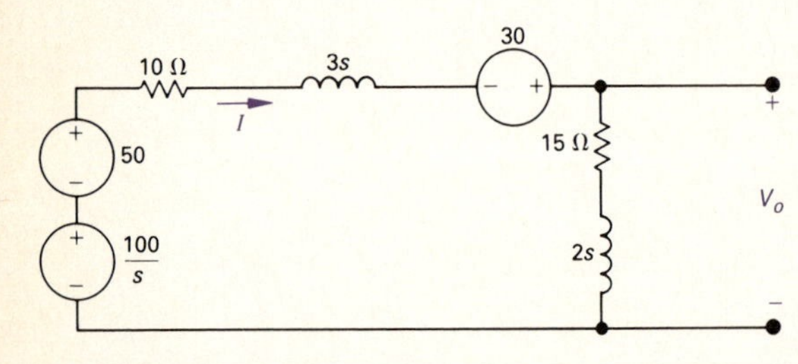

FIGURE 14.56 The s-domain equivalent circuit for the circuit shown in Fig. 14.55.

from which

$$i(t) = (12e^{-5t} + 4)u(t) \text{ A}. \quad \textbf{(14.146)}$$

The expression for V_0 is

$$\begin{aligned} V_0 = (15 + 2s)I &= \frac{32(s + 7.5)}{s+5} + \frac{40(s+7.5)}{s(s+5)} \\ &= 32\left(1 + \frac{2.5}{s+5}\right) + \frac{60}{s} - \frac{20}{s+5} \\ &= 32 + \frac{60}{s+5} + \frac{60}{s}, \end{aligned} \quad \textbf{(14.147)}$$

from which

$$v_0 = 32\delta(t) + (60e^{-5t} + 60)u(t) \text{ V}. \quad \textbf{(14.148)}$$

Now we test the results to see whether they make sense. From Eq. (14.146), we see that the current in L_1 and L_2 is 16 A at $t = 0^+$. As in the previous case, the switch operation causes i_1 to decrease instantaneously from 10 to 6 A and, at the same time, causes i_2 to increase from 0 to 6 A. Superimposed on these changes is the establishment of 10 A in L_1 and L_2 by the impulsive voltage source; that is,

$$i = \frac{1}{3+2}\int_{0^-}^{t} 50\delta(x)dx = 10 \text{ A}. \quad \textbf{(14.149)}$$

Therefore i_1 increases suddenly from 10 to 16 A, while i_2 increases suddenly from 0 to 16 A. The final value of i is 4 A. Figure 14.57 shows i_1, i_2, and i graphically.

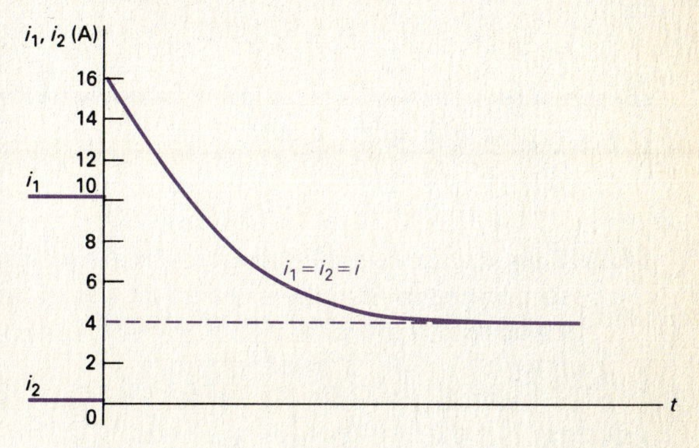

FIGURE 14.57 The inductor currents versus t for the circuit shown in Fig. 14.55.

We may also find the abrupt changes in i_1 and i_2 without using superposition. The sum of the impulsive voltages across L_1 (3 H) and L_2 (2 H) equals $50\delta(t)$. Thus the change in flux linkage must sum to 50; that is,

$$\Delta\lambda_1 + \Delta\lambda_2 = 50. \quad \textbf{(14.150)}$$

Because $\lambda = Li$, we express Eq. (14.150) as

$$3\Delta i_1 + 2\Delta i_2 = 50. \quad \textbf{(14.151)}$$

But because i_1 and i_2 must be equal after the switching takes place,

$$i_1(0^-) + \Delta i_1 = i_2(0^-) + \Delta i_2. \quad \textbf{(14.152)}$$

Then,

$$10 + \Delta i_1 = 0 + \Delta i_2. \quad \textbf{(14.153)}$$

Solving Eqs. (14.151) and (14.153) for Δi_1 and Δi_2 yields

$$\Delta i_1 = 6 \text{ A}; \quad \textbf{(14.154)}$$

$$\Delta i_2 = 16 \text{ A}. \quad \textbf{(14.155)}$$

These expressions agree with the previous check.

Figure 14.57 also indicates that the derivatives of i_1 and i_2 will contain an impulse at $t = 0$. Specifically, the derivative of i_1 will have an impulse of $6\delta(t)$, and the derivative of i_2 will have an impulse of $16\delta(t)$. Figures 14.58(a) and (b), respectively, illustrate the derivatives of i_1 and i_2.

Now let's turn to Eq. (14.148). The impulsive component $32\delta(t)$ agrees with the impulse of $16\delta(t)$ of di_2/dt at the origin. The terms $60e^{-5t} + 60$ agree with the fact that for $t > 0^+$,

$$v_o = 15i + 2\frac{di}{dt}.$$

We test the impulsive component of di_1/dt by noting that it produces an impulsive voltage of $(3)6\delta(t)$, or $18\delta(t)$, across L_1. This voltage, along with $32\delta(t)$ across L_2, adds to $50\delta(t)$. Thus the algebraic sum of the impulsive voltages around the mesh adds to zero.

To summarize, the Laplace transform will correctly predict the creation of impulsive currents and voltages that arise from switching.

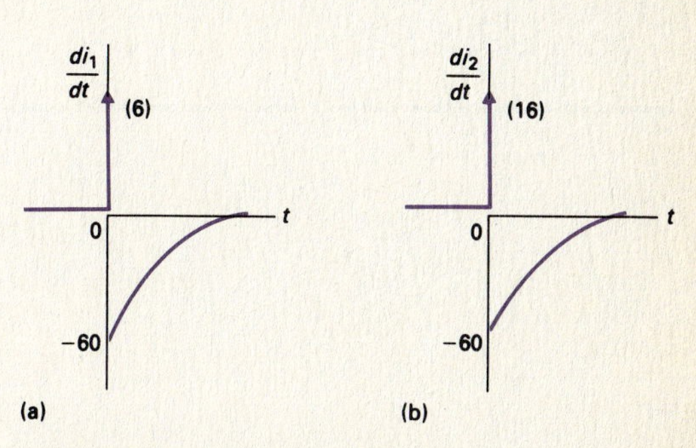

FIGURE 14.58 The derivative of i_1 and i_2.

However, the s-domain equivalent circuits must be based on initial conditions at $t = 0^-$, that is, on the initial conditions that exist prior to the disturbance caused by the switching. The Laplace transform will correctly predict the response to impulsive driving sources by simply representing these sources in the s domain by their correct transforms.

DRILL EXERCISES

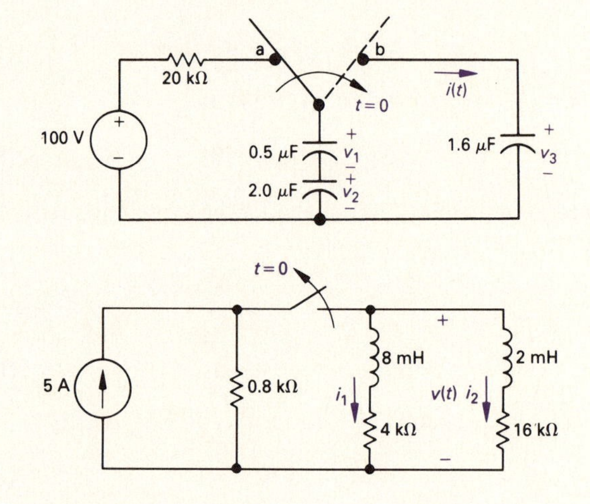

14.16 The switch in the circuit shown has been in position a for a long time. At $t = 0$, the switch moves to position b. Compute (a) $v_1(0^-)$; (b) $v_2(0^-)$; (c) $v_3(0^-)$; (d) $i(t)$; (e) $v_1(0^+)$; (f) $v_2(0^+)$; and (g) $v_3(0^+)$.

ANSWER: (a) 80 V; (b) 20 V; (c) 0 V; (d) $32\delta(t)$ μA; (e) 16 V; (f) 4 V; (g) 20 V.

14.17 The switch in the circuit shown has been closed for a long time. The switch opens at $t = 0$. Compute (a) $i_1(0^-)$;
(b) $i_1(0^+)$; (c) $i_2(0^-)$; (d) $i_2(0^+)$; (e) $i_1(t)$; (f) $i_2(t)$; and (g) $v(t)$.

ANSWER: (a) 0.8 A; (b) 0.6 A; (c) 0.2 A; (d) -0.6 A; (e) $0.6e^{-2\times10^6 t}u(t)$ A; (f) $-0.6e^{-2\times10^6 t}u(t)$ A; (g) $-1.6 \times 10^{-3}\delta(t) - 7200e^{-2\times10^6 t}u(t)$ V.

SUMMARY

- We can represent each of the circuit elements as an s-domain equivalent circuit by Laplace-transforming the voltage-current equation for each element:

  Resistor: $V = RI$

  Inductor: $V = sLI - LI_0$

  Capacitor: $V = (1/sC)I + V_0/s$

  In these equations, $V = \mathcal{L}\{v\}$, $I = \mathcal{L}\{i\}$, I_0 is the initial current through the inductor, and V_0 is the initial voltage across the capacitor.

- We can perform circuit analysis in the s domain by replacing each circuit element with its s-domain equivalent circuit. The resulting equivalent circuit is solved by writing algebraic equations using the circuit analysis techniques from resistive circuits. Table 14.1 summarizes the equivalent circuits for resistors, inductors, and capacitors in the s domain.
- Circuit analysis in the s domain is particularly advantageous for solving transient response problems in linear lumped parameter circuits when initial conditions are known. It is also useful for problems involving multiple simultaneous mesh-current or node-voltage equations, because it reduces problems to algebraic rather than differential equations.
- The **transfer function** is the s-domain ratio of a circuit's output to its input. It is represented as

$$H(s) = \frac{Y(s)}{X(s)},$$

  where $Y(s)$ is the Laplace transform of the output signal, and $X(s)$ is the Laplace transform of the input signal.
- The partial fraction expansion of the product $H(s)X(s)$ yields a term for each pole of $H(s)$ and $X(s)$. The $H(s)$ terms correspond to the transient component of the total response; the $X(s)$ terms correspond to the steady-state component.
- If a circuit is driven by a unit impulse, $x(t) = \delta(t)$, then the response of the circuit equals the inverse Laplace transform of the transfer function, $y(t) = \mathcal{L}^{-1}\{H(s)\} = h(t)$.
- A **time-invariant** circuit is one for which, if the input is delayed by a seconds, the response function is also delayed by a seconds.
- The output of a circuit, $y(t)$, can be computed by convolving the input, $x(t)$, with the impulse response of the circuit, $h(t)$:

$$\begin{aligned} y(t) = h(t) * x(t) &= \int_0^t h(\lambda)x(t-\lambda)d\lambda \\ = x(t) * h(t) &= \int_0^t x(\lambda)h(t-\lambda)d\lambda. \end{aligned}$$

  A graphical interpretation of the convolution integral often provides an easier computational method to generate $y(t)$.
- We can use the transfer function of a circuit to compute its steady-state response to a sinusoidal source. To do so, make the substitution $s = j\omega$ in $H(s)$ and represent the resulting complex number as a magnitude and phase angle. If

$$x(t) = A\cos(\omega t + \phi)$$

and

$$H(j\omega) = |H(j\omega)|e^{j\theta(\omega)},$$

then

$$y_{ss}(t) = A|H(j\omega)|\cos\left[\omega t + \phi + \theta(\omega)\right].$$

- Laplace transform analysis correctly predicts impulsive currents and voltages arising from switching and impulsive sources. You must ensure that the s-domain equivalent circuits are based on initial conditions at $t = 0^-$, that is, prior to the switching.

PROBLEMS

14.1 Find the Norton equivalent of the circuit shown in Fig. 14.3.

14.2 Derive the s-domain equivalent circuit shown in Fig. 14.4 by expressing the inductor current i as a function of the terminal voltage v and then finding the Laplace transform of this time-domain integral equation.

14.3 Find the Thévenin equivalent of the circuit shown in Fig. 14.7.

14.4 An 8 kΩ resistor, a 1 H inductor, and a 40 nF capacitor are in series.

a) Express the s-domain impedance of this series combination as a rational function.

b) Give the numerical value of the poles and zeros of the impedance.

14.5 A 2 kΩ resistor, a 312.5 mH inductor, and a 12.5 nF capacitor are in parallel.

a) Express the s-domain impedance of this parallel combination as a rational function.

b) Give the numerical values of the poles and zeros of the impedance.

14.6 A 1 kΩ resistor is in series with a 625 nF capacitor. This series combination is in parallel with a 100 mH inductor.

a) Express the equivalent s-domain impedance of these parallel branches as a rational function.

b) Determine the numerical values of the poles and zeros.

14.7 Find the poles and zeros of the impedance seen looking into the terminals a,b of the circuit shown in Fig. P14.7.

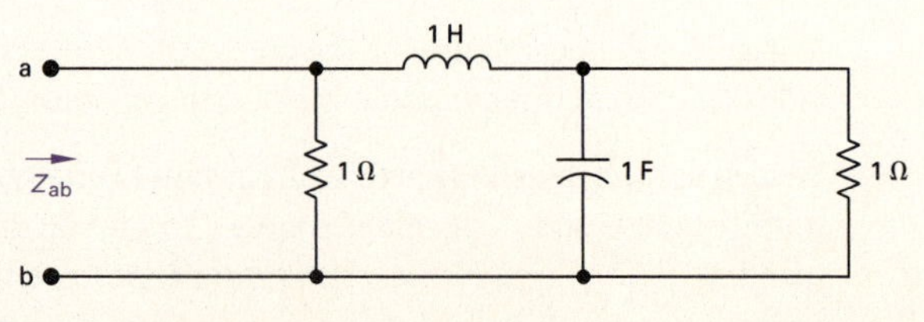

FIGURE P14.7

14.8 Find the poles and zeros of the impedance seen looking into the terminals a,b of the circuit shown in Fig. P14.8.

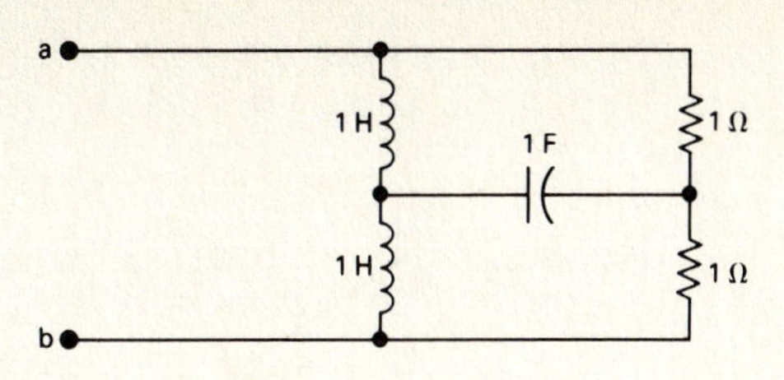

FIGURE P14.8

14.9 The switch in the circuit shown in Fig. P14.9 has been in position x for a long time. At $t = 0$, the switch moves instantaneously to position y.

a) Construct an s-domain circuit for $t > 0$.

b) Find V_a.

c) Find v_a.

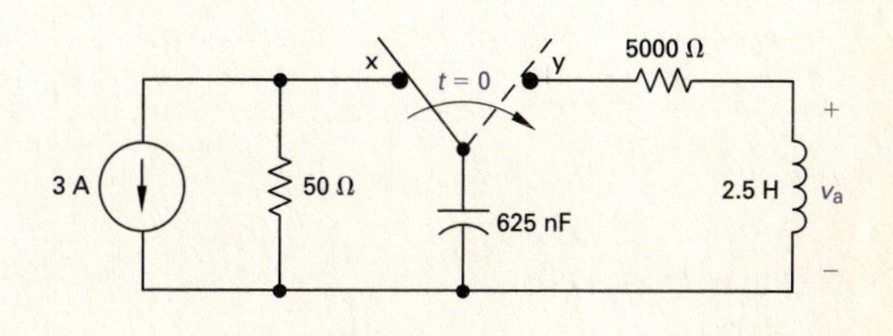

FIGURE P14.9

14.10 The switch in the circuit in Fig. P14.10 has been in position a for a long time. At $t = 0$, the switch moves instantaneously to position b.

a) Construct the s-domain circuit for $t > 0$.

b) Find V_o.

c) Find I_L.

d) Find v_o for $t > 0$.

e) Find i_L for $t > 0$.

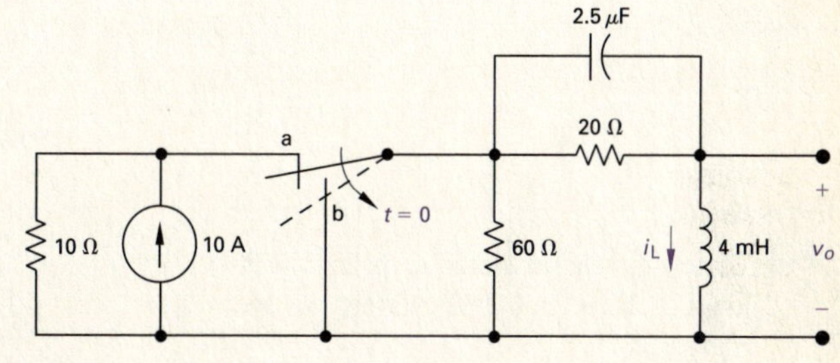

FIGURE P14.10

14.11 a) Find the s-domain expression for V_o in the circuit in Fig. P14.11.

b) Use the s-domain expression derived in part (a) to predict the initial and final values of v_o.

c) Find the time-domain expression for v_o.

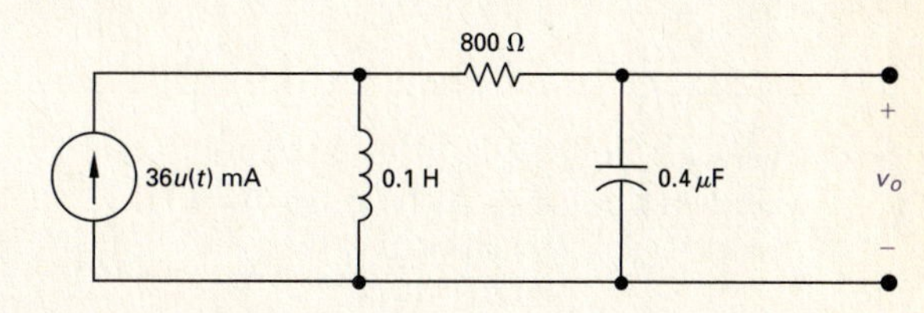

FIGURE P14.11

14.12 Find the time-domain expression for the current in the inductor in Fig. P14.11. Assume the reference direction for i_L is down.

14.13 Find V_o and v_o in the circuit shown in Fig. P14.13 if the initial energy is zero and the switch is closed at $t = 0$.

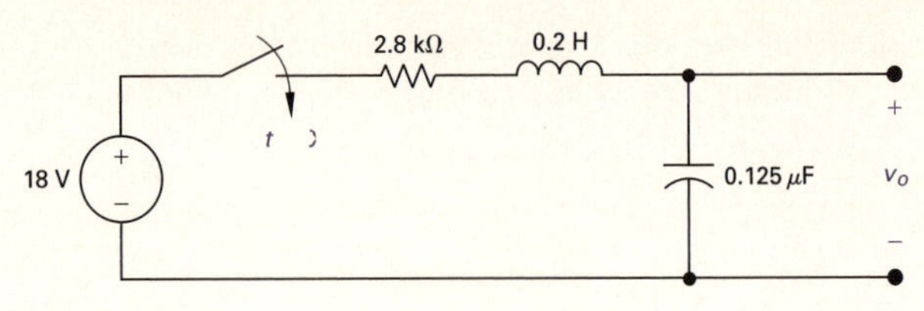

FIGURE P14.13

14.14 Repeat Problem 14.13 if the initial voltage on the capacitor is 30 V positive at the upper terminal.

14.15 The switch in the circuit in Fig. P14.15 has been in position a for a long time. At $t = 0$, it moves instantaneously from a to b.

a) Construct the s-domain circuit for $t > 0$.

b) Find $I_o(s)$.

c) Find $i_o(t)$ for $t \geq 0$.

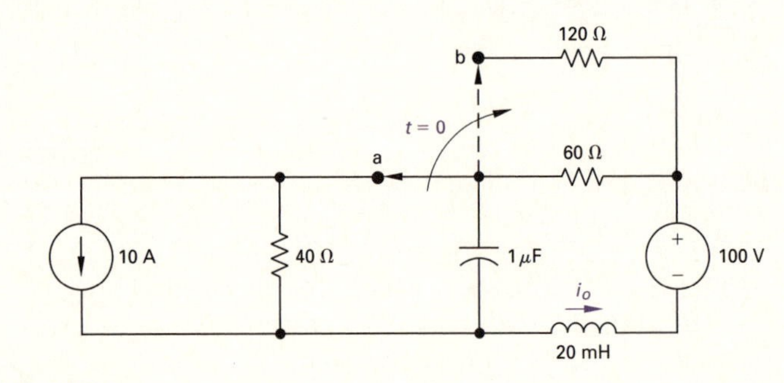

FIGURE P14.15

14.16 There is no energy stored in the circuit in Fig. P14.16 at the time the voltage source is energized.

a) Find I_o and V_o.

b) Find i_o and v_o for $t \geq 0$.

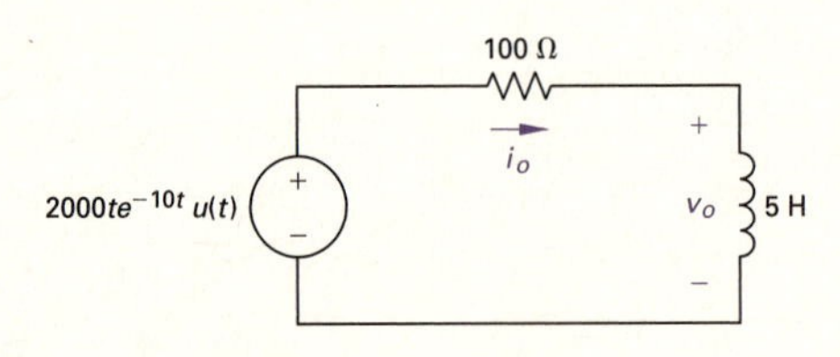

FIGURE P14.16

14.17 The switch in the circuit in Fig. P14.17 has been closed for a long time. At $t = 0$, the switch is opened. Find $v_o(t)$ for $t \geq 0$.

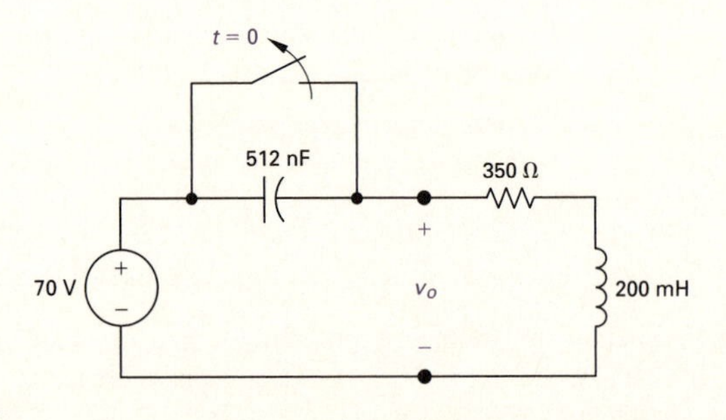

FIGURE P14.17

14.18 The switch in the circuit in Fig. P14.18 has been closed for a long time before opening at $t = 0$.

a) Construct the s-domain equivalent circuit for $t > 0$.

b) Find V_o.

c) Find v_o for $t \geq 0$.

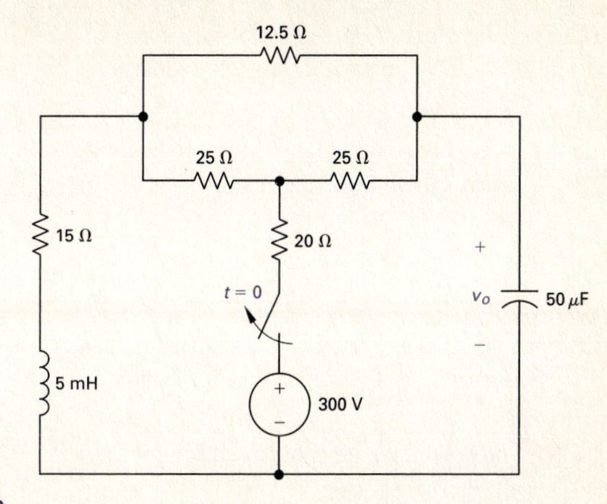

FIGURE P14.18

14.19 The switch in the circuit seen in Fig. P14.19 has been in position a for a long time. At $t = 0$, it moves instantaneously to position b.

a) Find V_o.

b) Find v_o.

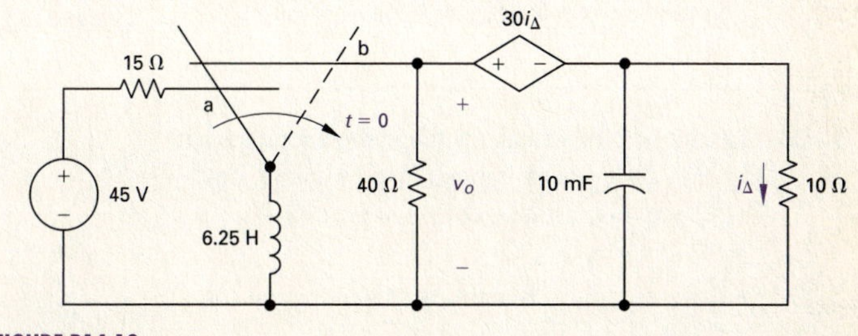

FIGURE P14.19

14.20 Find v_o in the circuit shown in Fig. P14.20 if $i_g = 5u(t)$ mA. There is no energy stored in the circuit at $t = 0$.

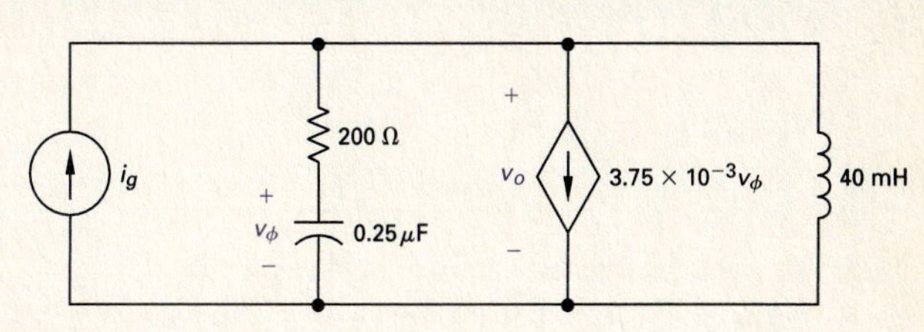

FIGURE P14.20

14.21 The switch in the circuit in Fig. P14.21 has been closed for a long time before opening at $t = 0$. Find v_o for $t \geq 0$.

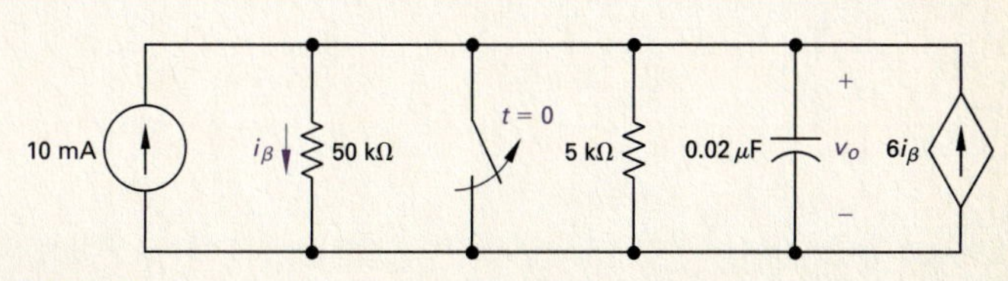

FIGURE P14.21

14.22 There is no energy stored in the circuit in Fig. P14.22 at the time the switch is closed.

a) Find v_o for $t \geq 0$.

b) Does your solution make sense in terms of known circuit behavior? Explain.

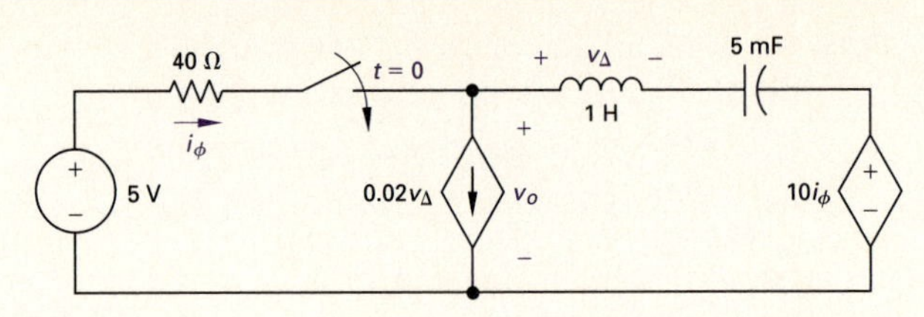

FIGURE P14.22

14.23 There is no energy stored in the capacitors in the circuit in Fig. P14.23 at the time the switch is closed.

a) Construct the s-domain circuit for $t > 0$.

b) Find I_1, V_1, and V_2.

c) Find i_1, v_1, and v_2.

d) Do your answers for i_1, v_1, and v_2 make sense in terms of known circuit behavior? Explain.

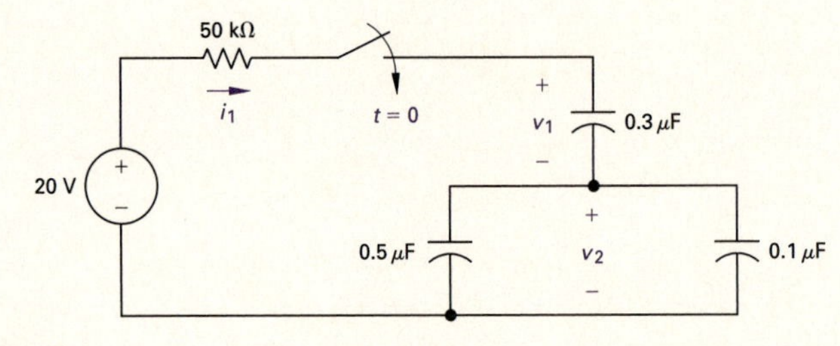

FIGURE P14.23

14.24 The make-before-break switch in the circuit seen in Fig. P14.24 has been in position a for a long time before moving instantaneously to position b at $t = 0$.

a) Construct the s-domain equivalent circuit for $t > 0$.

b) Find V_L and v_L.

c) Find V_C and v_C.

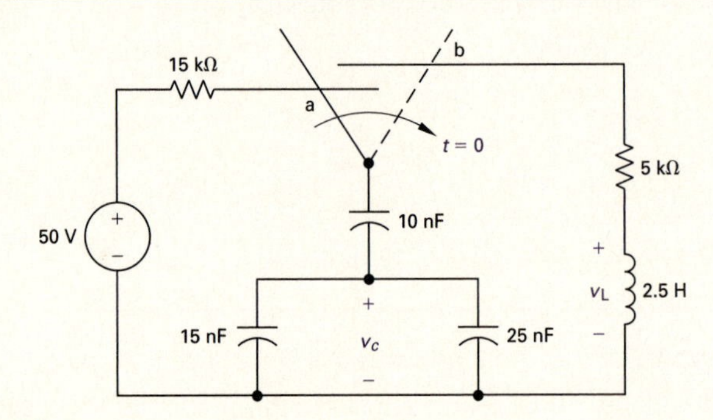

FIGURE P14.24

14.25 The make-before-break switch in the circuit in Fig. P14.25 has been in position a for a long time. At $t = 0$, it moves instantaneously to position b. Find i_o for $t \geq 0$.

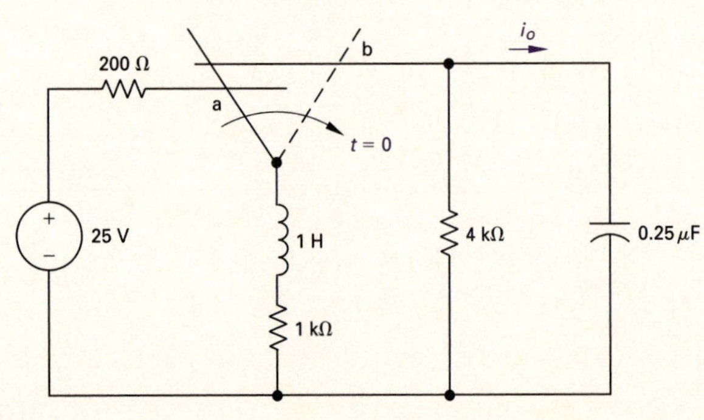

FIGURE P14.25

14.26 There is no energy stored in the circuit in Fig. P14.26 at the time the sources are energized.

a) Find $I_1(s)$ and $I_2(s)$.

b) Use the initial- and final-value theorems to check the initial and final values of $i_1(t)$ and $i_2(t)$.

c) Find $i_1(t)$ and $i_2(t)$ for $t \geq 0$.

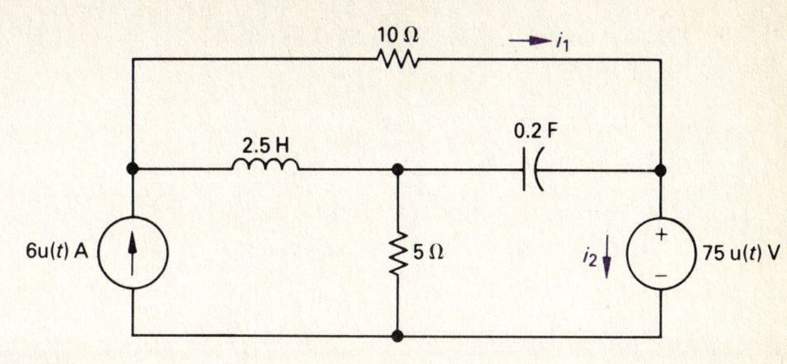

FIGURE P14.26

14.27 There is no energy stored in the circuit in Fig. P14.27 at the time the voltage source is turned on, and $v_g = 54u(t)$ V.

a) Find V_o and I_o.

b) Find v_o and i_o.

c) Do the solutions for v_o and i_o make sense in terms of known circuit behavior? Explain.

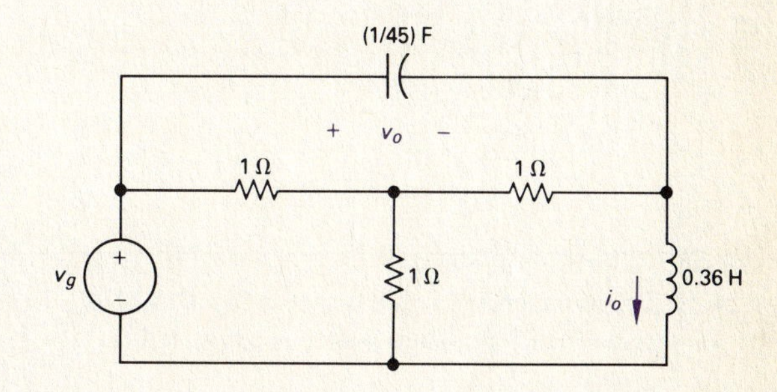

FIGURE P14.27

14.28 There is no energy stored in the circuit in Fig. P14.28 at the time the current source is energized.

a) Find I_a and I_b.

b) Find i_a and i_b.

c) Find V_a, V_b, and V_c.

d) Find v_a, v_b, and v_c.

e) Assume a capacitor will break down whenever its terminal voltage is 1000 V. How long after the current source turns on will one of the capacitors break down?

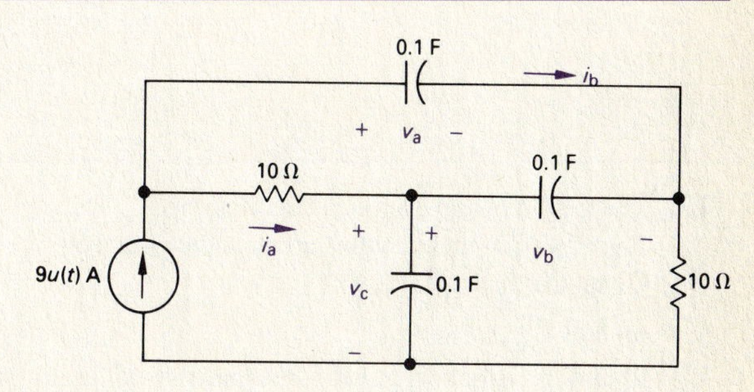

FIGURE P14.28

14.29 There is no energy stored in the circuit in Fig. P14.29 at $t = 0^-$.

a) Find V_o.

b) Find v_o.

c) Does your solution for v_o make sense in terms of known circuit behavior? Explain.

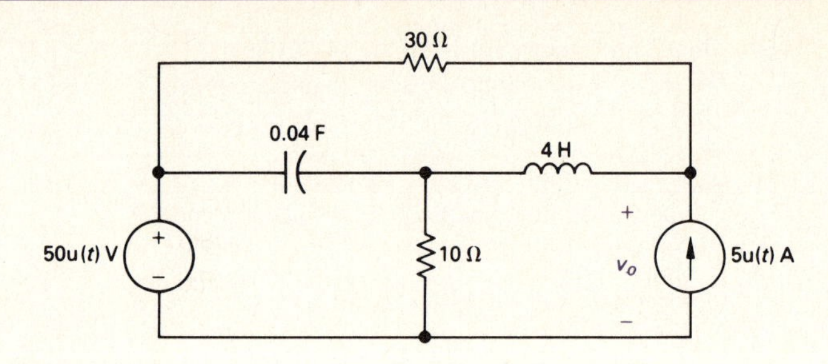

FIGURE P14.29

14.30 There is no energy stored in the circuit in Fig. P14.30 at $t = 0^-$. Use the mesh-current method to find i_o.

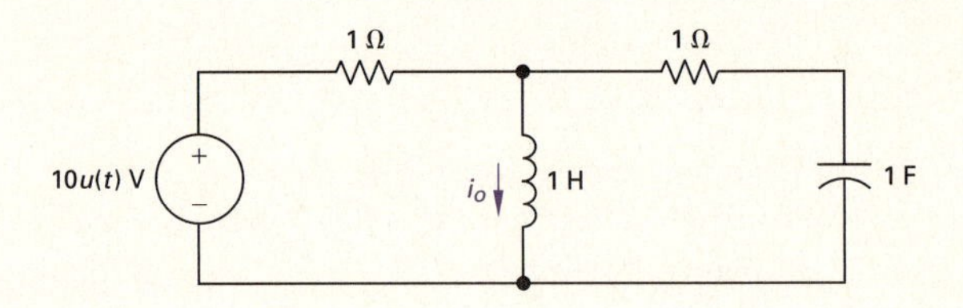

FIGURE P14.30

14.31 The initial energy in the circuit in Fig. P14.31 is zero. The ideal voltage source is $120u(t)$ V.

a) Find $I_o(s)$.

b) Use the initial- and final-value theorems to find $i_o(0^+)$ and $i_o(\infty)$.

c) Do the values obtained in part (b) agree with known circuit behavior? Explain.

d) Find $i_o(t)$.

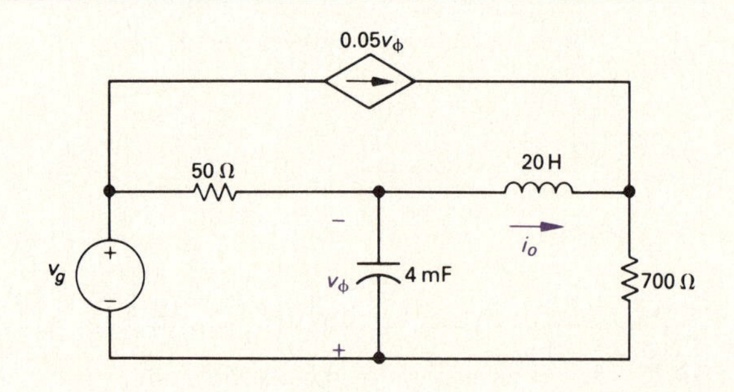

FIGURE P14.31

14.32 There is no energy stored in the circuit in Fig. P14.32 at the time the current source turns on. Given that $i_g = 50u(t)$ A,

a) find $V_o(s)$;

b) use the initial- and final-value theorems to find $v_o(0^+)$ and $v_o(\infty)$.

c) determine if the results obtained in part (b) agree with known circuit behavior; and

d) find $v_o(t)$.

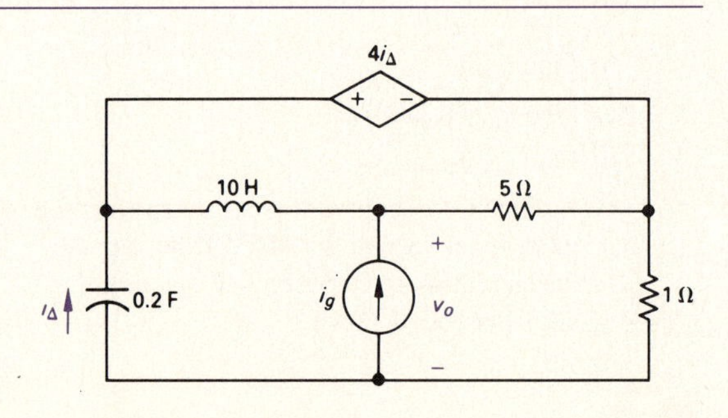

FIGURE P14.32

14.33 The switch in the circuit shown in Fig. P14.33 has been open for a long time. The voltage of the sinusoidal source is $v_g = V_m \sin(\omega t + \phi)$. The switch closes at $t = 0$. Note that the angle ϕ in the voltage expression determines the value of the voltage at the moment when the switch closes, that is, $v_g(0) = V_m \sin\phi$.

a) Use the Laplace transform method to find i for $t > 0$.

b) Using the expression derived in part (a), write the expression for the current after the switch has been closed for a long time.

c) Using the expression derived in part (a), write the expression for the transient component of i.

d) Find the steady-state expression for i using the phasor method. Verify that your expression is equivalent to that obtained in part (b).

e) Specify the value of ϕ so that the circuit passes directly into steady-state operation when the switch is closed.

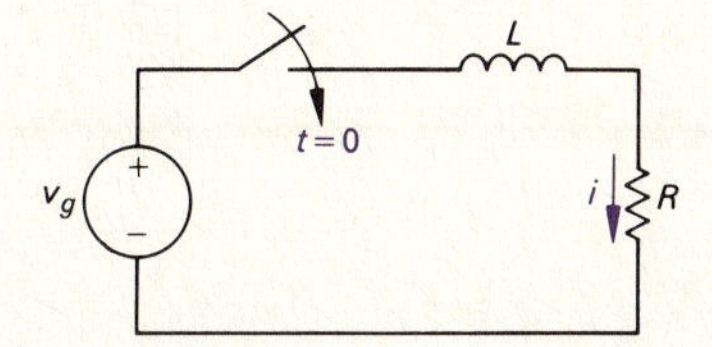

FIGURE P14.33

14.34 The two switches in the circuit shown in Fig. P14.34 operate simultaneously. There is no energy stored in the circuit at the instant the switches close. Find $i(t)$ for $t \geq 0^+$ by first finding the s-domain Thévenin equivalent of the circuit to the left of the terminals a,b.

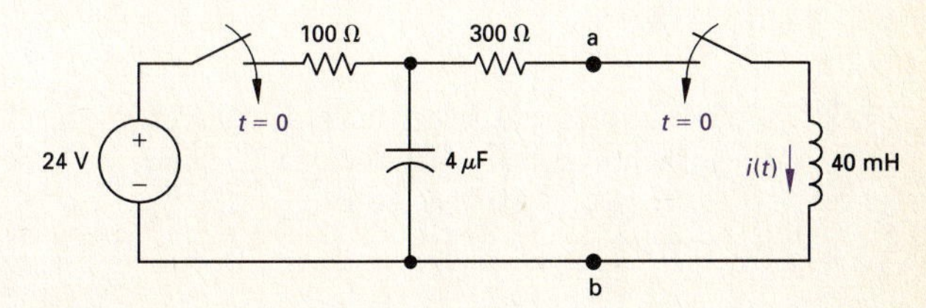

FIGURE P14.34

14.35 Beginning with Eq. (14.65), show that the capacitor current in the circuit in Fig. 14.19 is positive for $0 < t < 200\ \mu s$ and negative for $t > 200\ \mu s$. Also show that at $200\ \mu s$, the current is zero and that this corresponds to when dv_C/dt is zero.

14.36 The switch in the circuit seen in Fig. P14.36 has been closed for a long time before opening at $t = 0$. Use the Laplace transform method of analysis to find v_o.

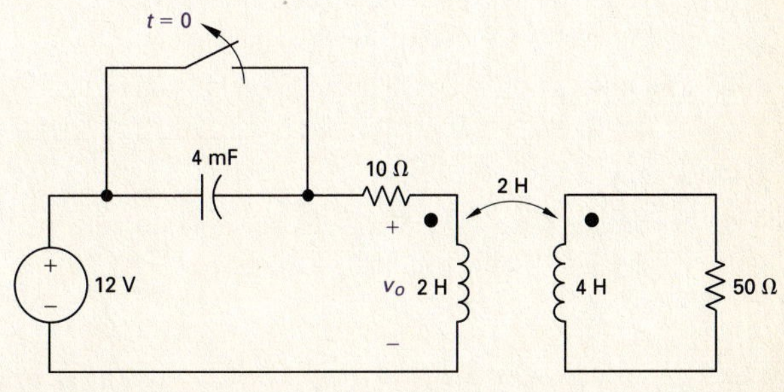

FIGURE P14.36

14.37 There is no energy stored in the circuit in Fig. P14.37 at the time the switch is closed.

a) Find I_1.

b) Use the initial- and final-value theorems to find $i_1(0^+)$ and $i_1(\infty)$.

c) Find i_1.

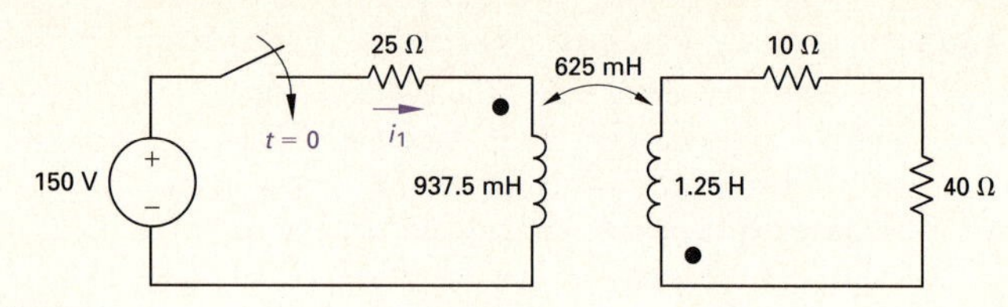

FIGURE P14.37

14.38 a) Find the current in the 40 Ω resistor in the circuit in Fig. P14.37. The reference direction for the current is down through the resistor.

b) Repeat part (a) if the dot on the 1.25 H coil is reversed.

14.39 The make-before-break switch in the circuit seen in Fig. P14.39 has been in position a for a long time. At $t = 0$, it moves instantaneously to position b. Find i_o for $t \geq 0$.

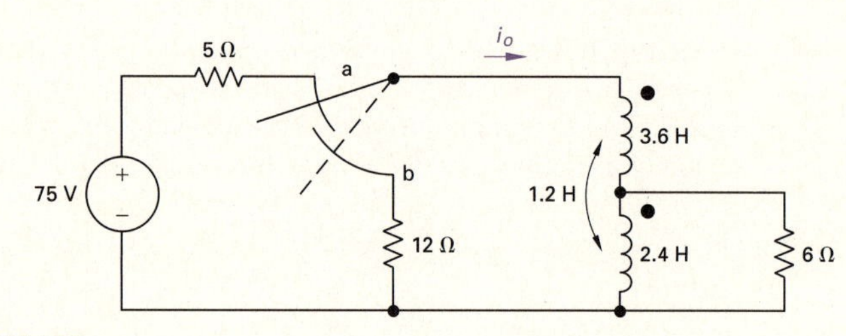

FIGURE P14.39

14.40 In the circuit in Fig. P14.40, switch 1 closes at $t = 0$ and the make-before-break switch moves instantaneously from position a to position b.

a) Construct the s-domain equivalent circuit for $t > 0$.

b) Find I_1.

c) Use the initial- and final-value theorems to check the initial and final values of i_1.

d) Find i_1 for $t \geq 0^+$.

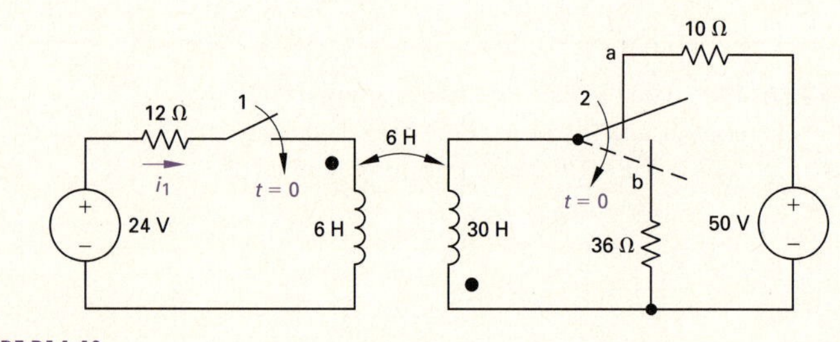

FIGURE P14.40

14.41 The operational amplifier in the circuit shown in Fig. P14.41 is ideal. There is no energy stored in the circuit at the time it is energized. If $v_g = 16{,}000tu(t)$ V, find (a) V_o, (b) v_o, (c) how long it takes to saturate the operational amplifier, and (d) how small the rate of increase in v_g must be to prevent saturation.

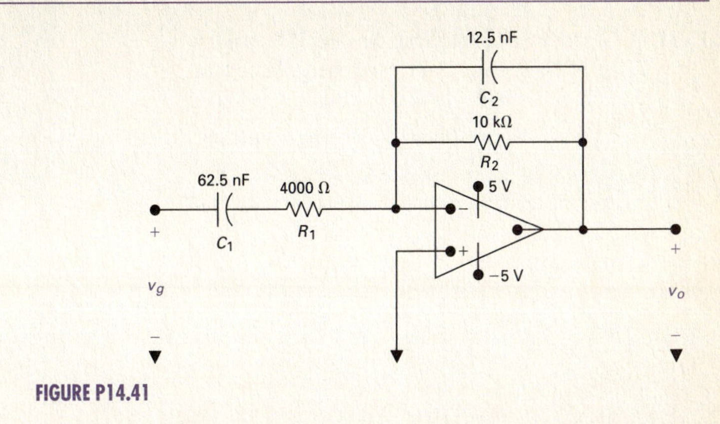

FIGURE P14.41

14.42 The operational amplifier in the circuit seen in Fig. P14.42 is ideal. There is no energy stored in the capacitors at the time the circuit is energized. Determine (a) V_o, (b) v_o, and (c) how long it takes to saturate the operational amplifier.

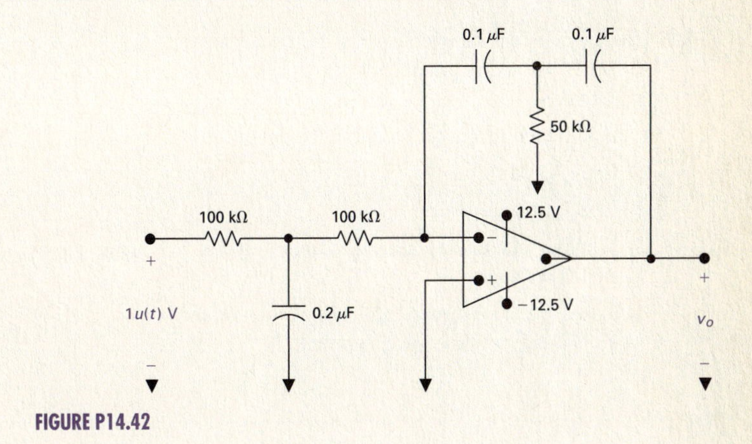

FIGURE P14.42

14.43 Find $v_o(t)$ in the circuit shown in Fig. P14.43 if the ideal op amp operates within its linear range and $v_g = 16u(t)$ mV.

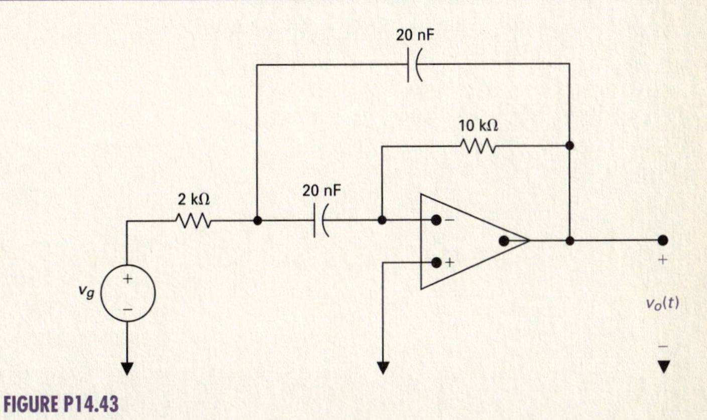

FIGURE P14.43

14.44 The operational amplifier in the circuit shown in Fig. P14.44 is ideal. There is no energy stored in the capacitors at the instant the circuit is energized.

a) Find v_o if $v_{g1} = 16u(t)$ V and $v_{g2} = 8u(t)$ V.

b) How many milliseconds after the two voltage sources are turned on does the op amp saturate?

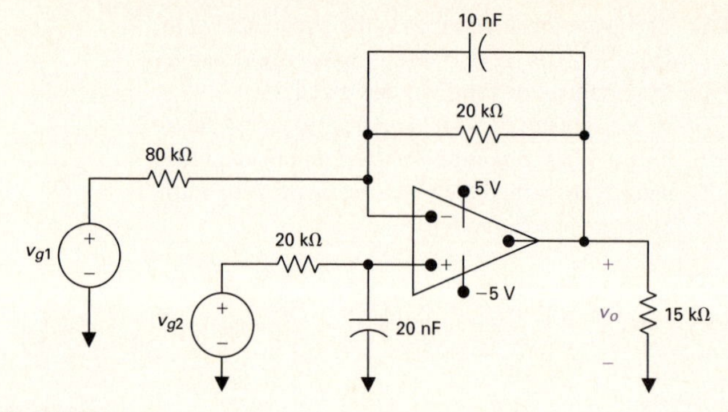

FIGURE P14.44

14.45 The magnetically coupled coils in the circuit seen in Fig. P14.45 carry initial currents of 15 and 10 A, as shown.

a) Find the initial energy stored in the circuit.

b) Find I_1 and I_2.

c) Find i_1 and i_2.

d) Find the total energy dissipated in the 120 and 270 Ω resistors.

e) Repeat parts (a) through (d), with the dot on the 18 H inductor at the lower terminal.

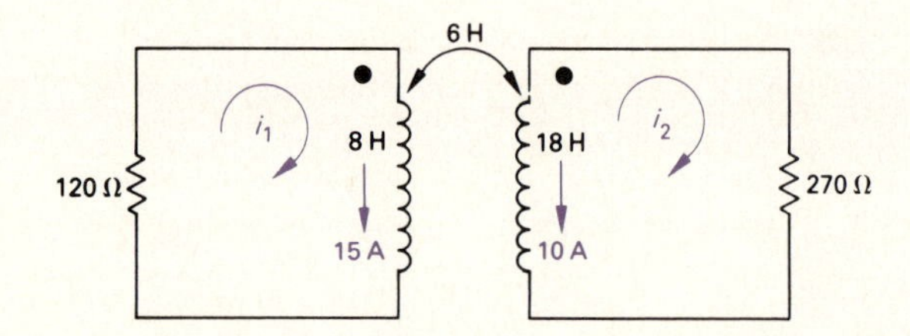

FIGURE P14.45

14.46 The operational amplifiers in the circuit shown in Fig. P14.46 are ideal. There is no energy stored in the capacitors at $t = 0^-$. If $v_g = 16u(t)$ mV, how many milliseconds elapse before an operational amplifier saturates?

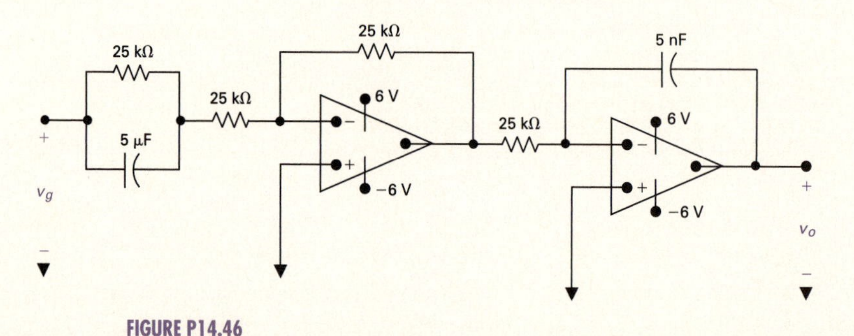

FIGURE P14.46

14.47 There is no energy stored in the circuit seen in Fig. P14.47 at the time the two sources are energized.

a) Use the principle of superposition to find V_o.

b) Find v_o for $t > 0$.

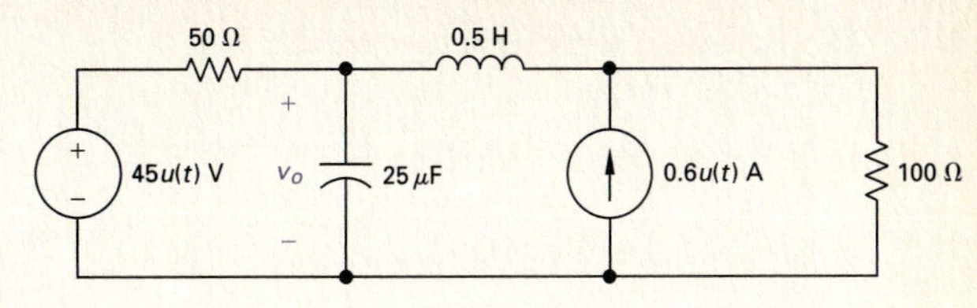

FIGURE P14.47

14.48 Verify that the solution of Eqs. (14.91) and (14.92) for V_2 yields the same expression as that given by Eq. (14.90).

14.49 Find the numerical expression for the transfer function (V_o/V_i) of each circuit in Fig. P14.49 and give the numerical value of the poles and zeros of each transfer function.

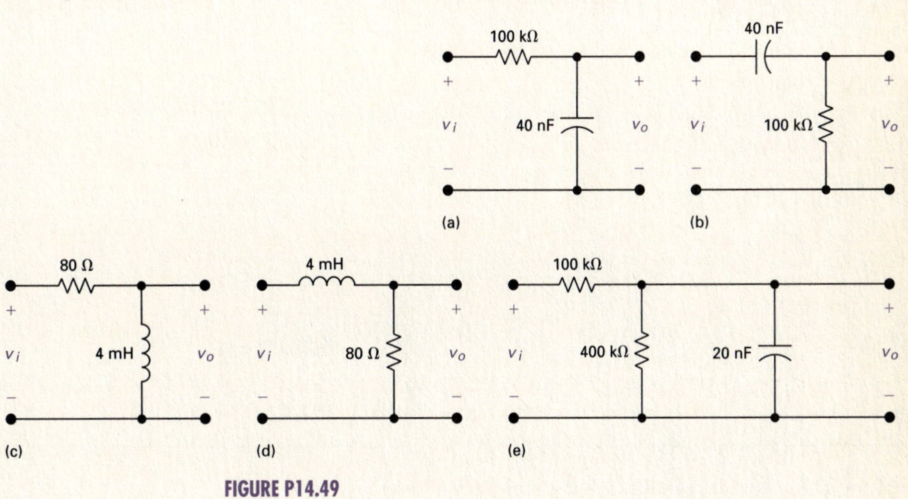

FIGURE P14.49

14.50 a) Find the numerical expression for the transfer function $H(s) = V_o/V_i$ for the circuit in Fig. P14.50.

b) Give the numerical value of each pole and zero of $H(s)$.

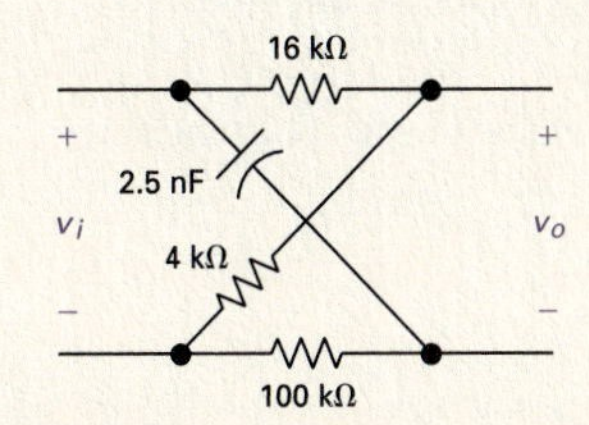

FIGURE P14.50

14.51 The operational amplifier in the circuit in Fig. P14.51 is ideal.

a) Find the numerical expression for the transfer function $H(s) = V_o/V_g$.

b) Give the numerical value of each zero and pole of $H(s)$.

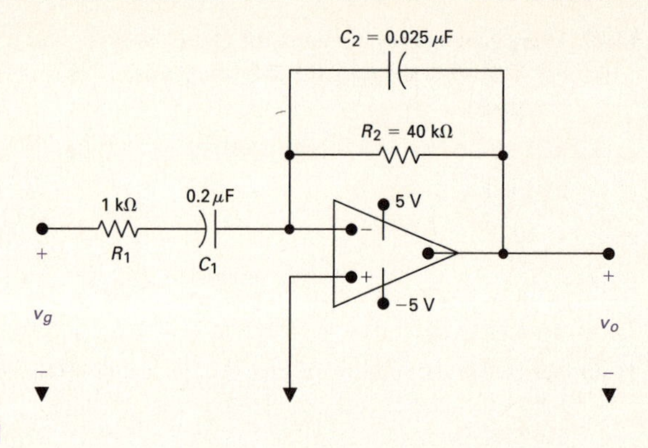

FIGURE P14.51

14.52 The operational amplifier in the circuit in Fig. P14.52 is ideal.

a) Find the numerical expression for the transfer function $H(s) = V_o/V_g$.

b) Give the numerical value of each zero and pole of $H(s)$.

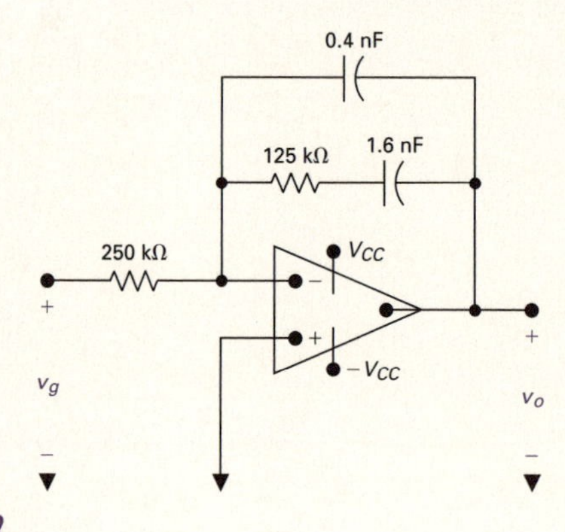

FIGURE P14.52

14.53 The operation amplifier in the circuit in Fig. P14.53 is ideal.

a) Derive the numerical expression of the transfer function $H(s) = V_o/V_g$ for the circuit in Fig. P14.53.

b) Give the numerical value of each pole and zero of $H(s)$.

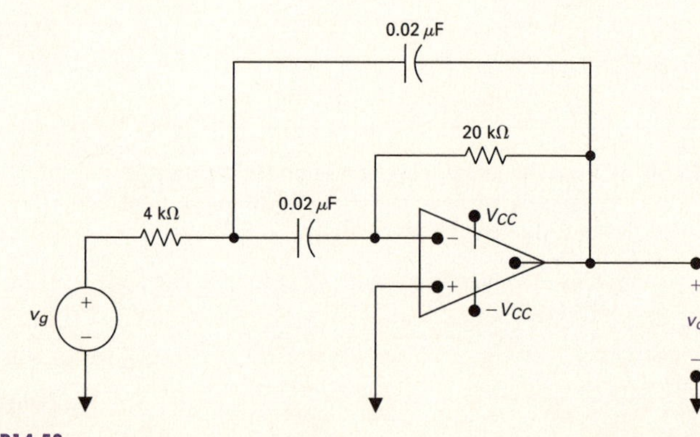

FIGURE P14.53

14.54 There is no energy stored in the circuit in Fig. P14.54 at the time the switch is opened. The sinusoidal current source is generating the signal 28.8 cos 5t mA. The response signal is the current i_o.

a) Find the transfer function I_o/I_g.

b) Find $I_o(s)$.

c) Describe the nature of the transient component of $i_o(t)$ without solving for $i_o(t)$.

d) Describe the nature of the steady-state component of $i_o(t)$ without solving for $i_o(t)$.

e) Verify the observations made in parts (c) and (d) by finding $i_o(t)$.

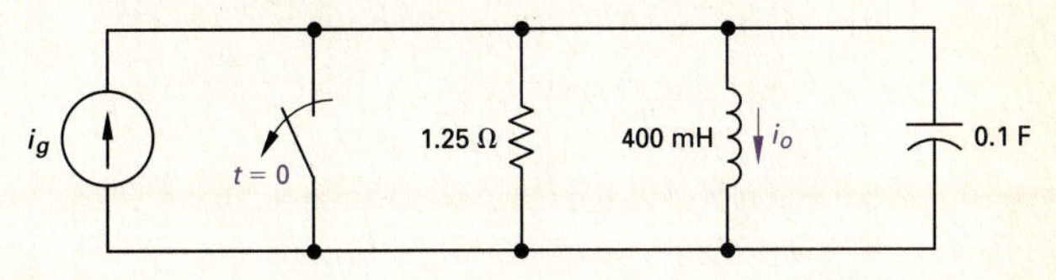

FIGURE P14.54

14.55 a) Find the transfer function I_o/I_g as a function of μ for the circuit seen in Fig. P14.55.

b) Find the largest value of μ that will produce a bounded output signal for a bounded input signal.

c) Find i_o for $\mu = -3$, 0, 4, 5, and 6 if $i_g = 5u(t)$ A.

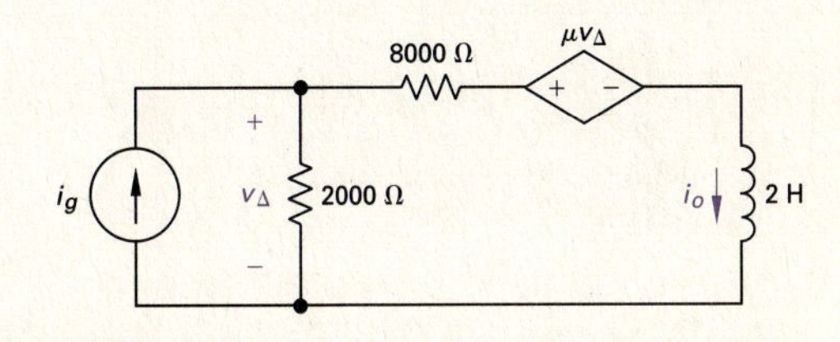

FIGURE P14.55

14.56 In the circuit of Fig. P14.56, v_o is the output signal and v_g is the input signal. Find the poles and zeros of the transfer function.

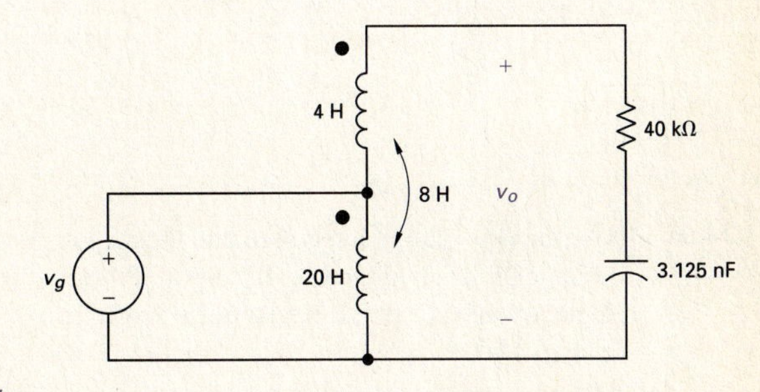

FIGURE P14.56

14.57 a) Find $h(t) * x(t)$ when $h(t)$ and $x(t)$ are the rectangular pulses shown in Fig. P14.57(a).

b) Repeat part (a) when $x(t)$ changes to the rectangular pulse shown in Fig. P14.57(b).

c) Repeat part (a) when $h(t)$ changes to the rectangular pulse shown in Fig. P14.57(c).

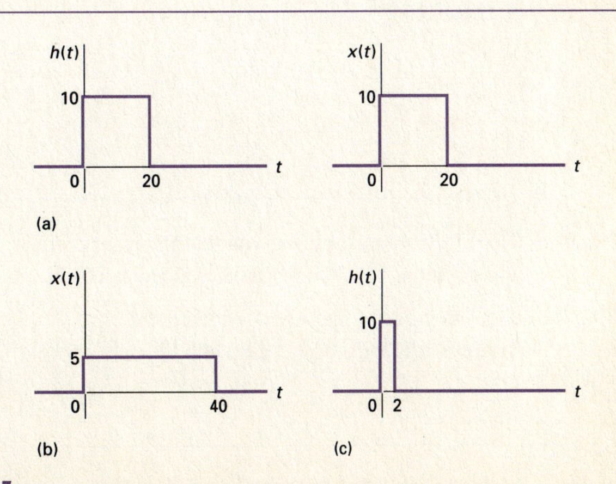

FIGURE P14.57

14.58 Assume the voltage impulse response of a circuit can be modeled by the triangular waveform shown in Fig. P14.58. The voltage input signal to this circuit is the step function $10u(t)$ V.

a) Use the convolution integral to derive the expressions for the output voltage.

b) Sketch the output voltage over the interval 0 to 15 s.

c) Repeat parts (a) and (b) if the area under the voltage impulse response stays the same but the width of the impulse response narrows to 4 s.

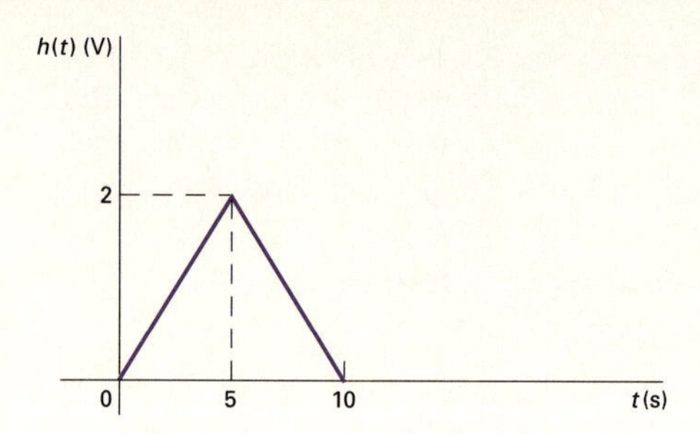

FIGURE P14.58

14.59 The voltage impulse response of a circuit is shown in Fig. P14.59(a). The input signal to the circuit is the rectangular voltage pulse shown in Fig. P14.59(b).

a) Derive the equations for the output voltage. Note the range of time for which each equation is applicable.

b) Sketch v_o for $-2 \leq t \leq 35$ s.

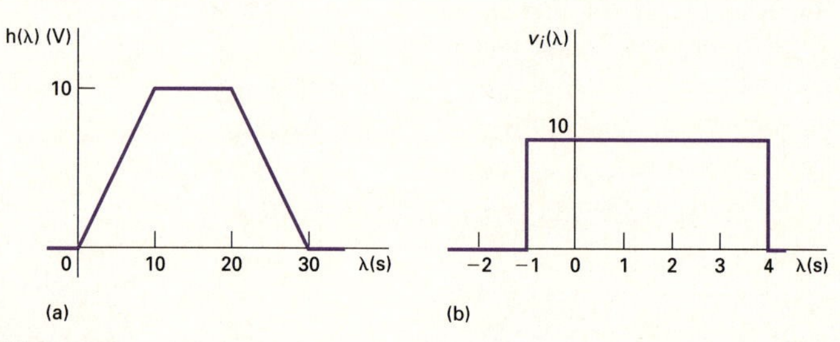

FIGURE P14.59

14.60 a) Use the convolution integral to find the output voltage of the circuit in Fig. P14.49(a) if the input voltage is the rectangular pulse shown in Fig. P14.60.

b) Sketch $v_o(t)$ versus t for the time interval $0 \leq t \leq 16$ ms.

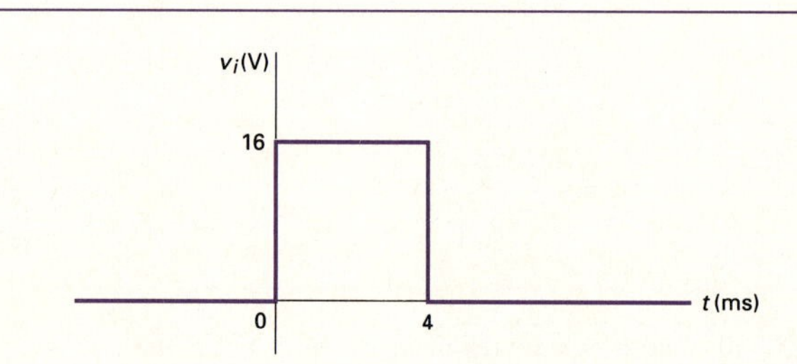

FIGURE P14.60

14.61 a) Repeat Problem 14.60, given that the resistor in the circuit in Fig. P14.49(a) is reduced to 10 kΩ.

b) Does decreasing the resistor increase or decrease the memory of the circuit?

c) Which circuit comes closest to transmitting a replica of the input voltage?

14.62 The input voltage in the circuit seen in Fig. P14.62 is

$$v_i = 8[u(t) - u(t - 0.1)] \text{ V}.$$

a) Use the convolution integral to find v_o.

b) Sketch v_o for $0 \leq t \leq 1$ s.

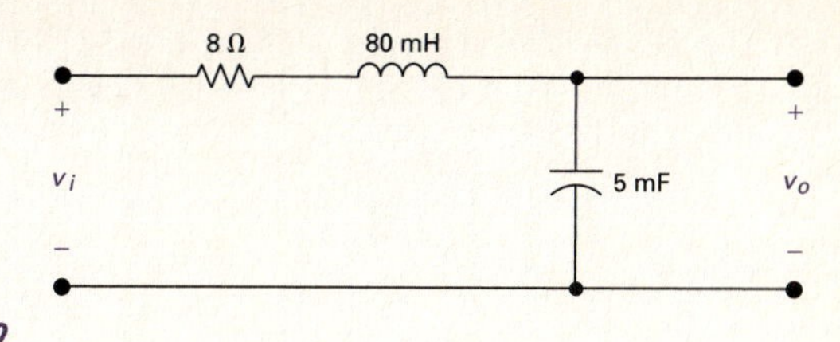

FIGURE P14.62

14.63 Use the convolution integral to find v_o in the circuit seen in Fig. P14.63 if $v_i = 25u(t)$ V.

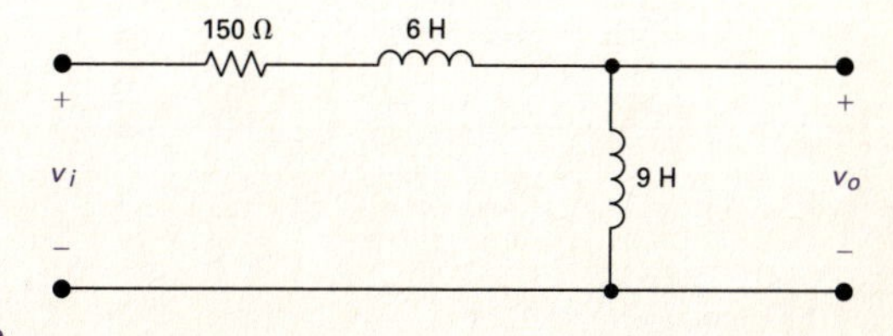

FIGURE P14.63

14.64 The sinusoidal voltage pulse shown in Fig. P14.64(a) is applied to the circuit shown in Fig. P14.64(b). Use the convolution integral to find the value of v_o at $t = 75$ ms.

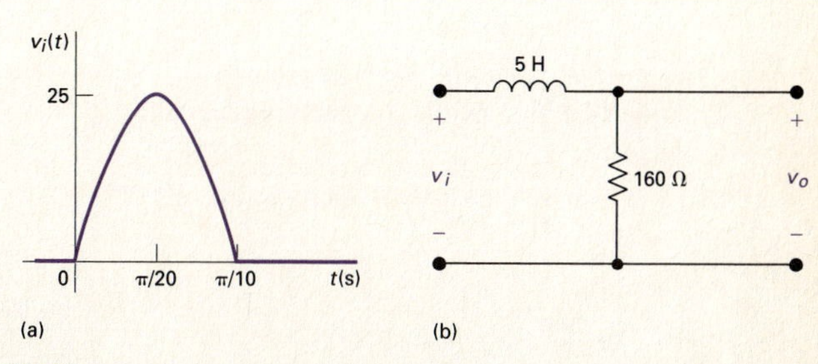

FIGURE P14.64

14.65 The current source in the circuit shown in Fig. P14.65(a) is generating the waveform shown in Fig. P14.65(b). Use the convolution integral to find v_o at $t = 5$ ms.

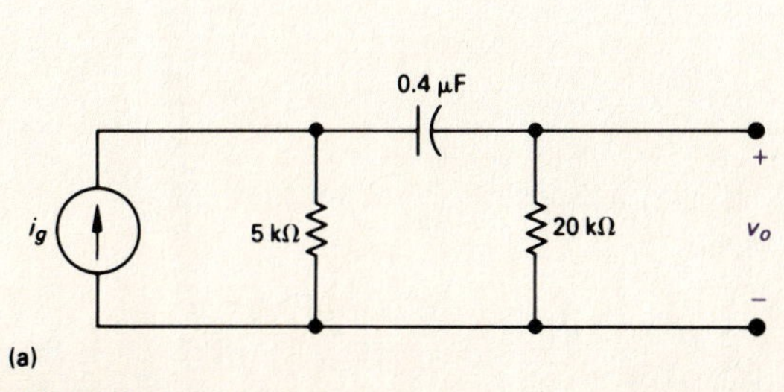

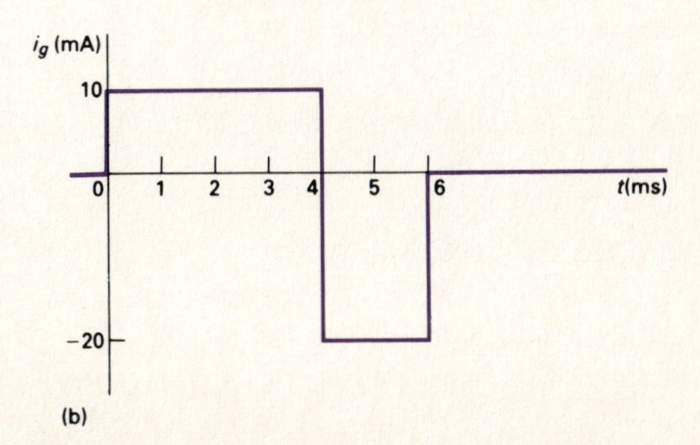

FIGURE P14.65

14.66 a) Use the convolution integral to find v_o in the circuit in Fig. P14.66(a) if i_g is the pulse shown in Fig. P14.66(b).

b) Use the convolution integral to find i_o.

c) Show that your solutions for v_o and i_o are consistent by calculating i_o at 100^- ms, 100^+ms, 200^- ms, and 200^+ ms.

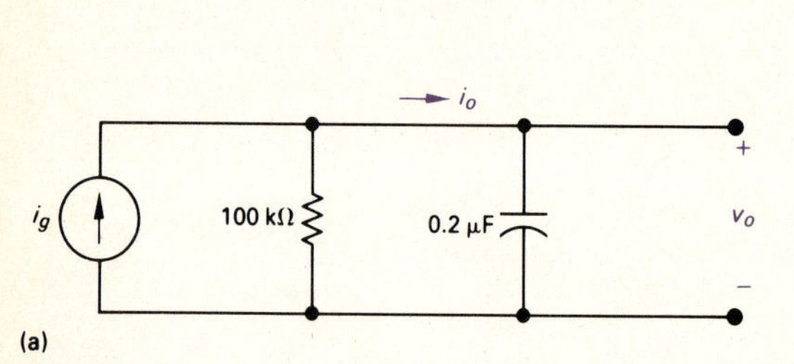

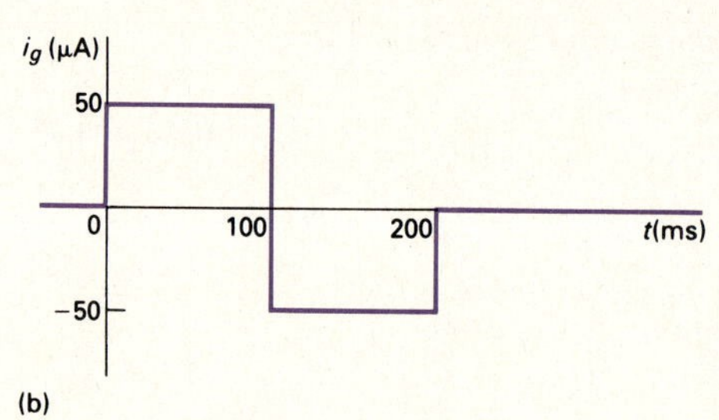

FIGURE P14.66

14.67 a) Find the impulse response of the circuit shown in Fig. P14.67(a) if v_g is the input signal and v_o is the output signal.

b) Given that v_g has the waveform shown in Fig. P14.67(b), use the convolution integral to find v_o.

c) Does v_o have the same waveform as v_g? Why?

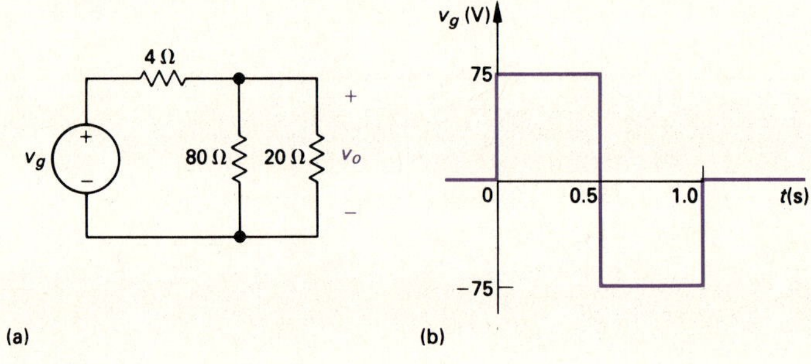

FIGURE P14.67

14.68 a) Find the impulse response of the circuit seen in Fig. P14.68 if v_g is the input signal and v_o is the output signal.

b) Assume that the voltage source has the waveform shown in Fig. P14.67(b). Use the convolution integral to find v_o.

c) Sketch v_o for $0 \leq t \leq 2$ s.

d) Does v_o have the same waveform as v_g? Why?

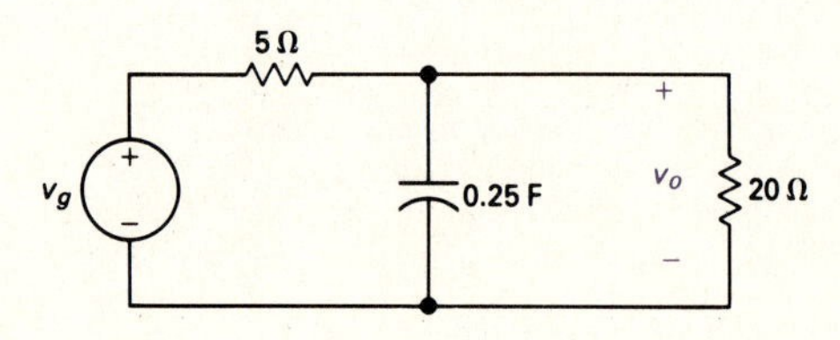

FIGURE P14.68

14.69 a) Show that if $y(t) = h(t) * x(t)$, then $Y(s) = H(s)X(s)$.

b) Use the result given in part (a) to find $f(t)$ if

$$F(s) = \frac{a}{s(s+a)^2}.$$

14.70 The transfer function for a linear time-invariant circuit is

$$H(s) = \frac{V_o}{V_g} = \frac{10^4(s + 5000)}{s^2 + 2000s + 10^8}$$

If $v_g = 2.2\cos(10{,}000t)$ V, what is the steady-state expression for v_o?

14.71 The operational amplifier in the circuit seen in Fig. P14.71 is ideal and is operating within its linear region.

a) Calculate the transfer function V_o/V_g.

b) If $v_g = 601 \cos 300t$ mV, what is the steady-state expression for v_o?

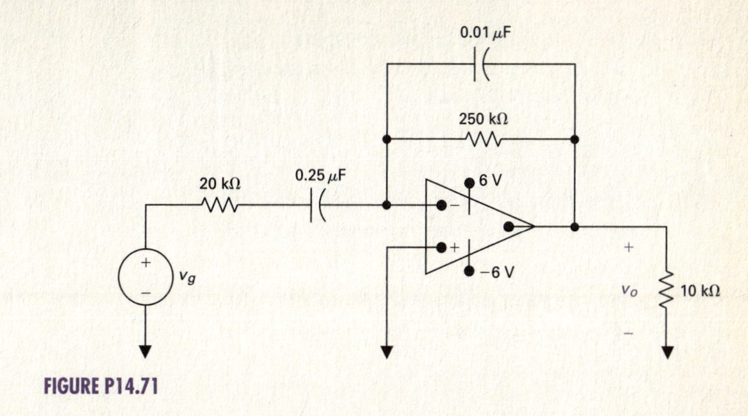

FIGURE P14.71

14.72 The operational amplifier in the circuit seen in Fig. P14.72 is ideal.

a) Find the transfer function V_o/V_g.

b) Find v_o if $v_g = u(t)$ V.

c) Find the steady-state expression for v_o if $v_g = 2 \cos 10{,}000t$ V.

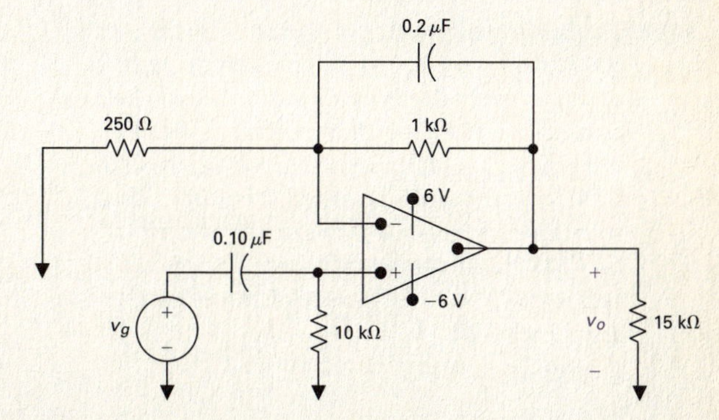

FIGURE P14.72

14.73 When an input voltage of $30u(t)$ V is applied to a circuit, the response is known to be

$$v_o = (50e^{-8000t} - 20e^{-5000t})u(t) \text{ V}.$$

What will the steady-state response be if $v_g = 120 \cos 6000t$ V?

14.74 Show that after $V_0 C_e$ coulombs are transferred from C_1 to C_2 in the circuit shown in Fig. 14.47, the voltage across each capacitor is $C_1 V_0/(C_1 + C_2)$. (*Hint:* Use the conservation-of-charge principle.)

14.75 The inductor L_1 in the circuit shown in Fig. P14.75 is carrying an initial current of ρ A at the instant the switch opens. Find (a) $v(t)$; (b) $i_1(t)$; (c) $i_2(t)$; and (d) $\lambda(t)$, where $\lambda(t)$ is the total flux linkage in the circuit.

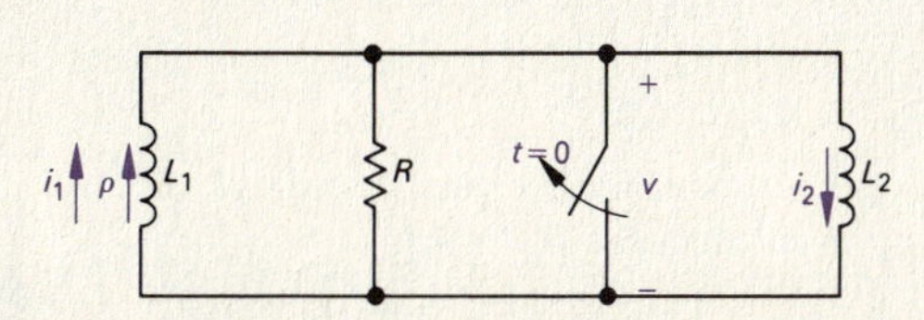

FIGURE P14.75

14.76 a) Let $R \to \infty$ in the circuit shown in Fig. P14.75, and use the solutions derived in Problem 14.75 to find $v(t)$, $i_1(t)$, and $i_2(t)$.

b) Let $R = \infty$ in the circuit shown in Fig. P14.75 and use the Laplace transform method to find $v(t)$, $i_1(t)$, and $i_2(t)$.

14.77 The switch in the circuit shown in Fig. P14.77 has been open for a long time before closing at $t = 0$.

a) Find v_o and i_o for $t \geq 0$.

b) Test your solutions and make sure they are in agreement with known circuit behavior.

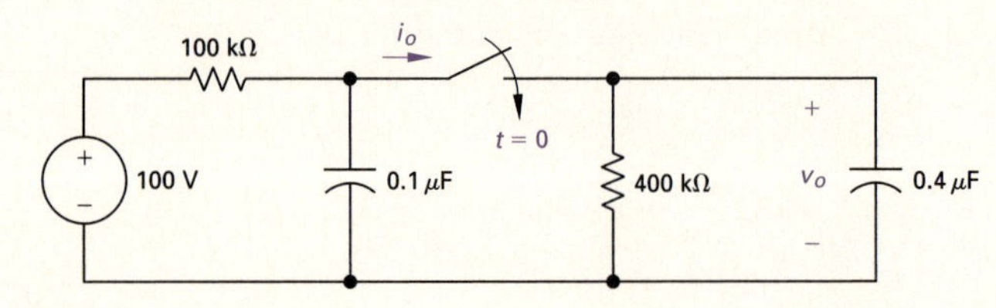

FIGURE P14.77

14.78 The parallel combination of R_2 and C_2 in the circuit shown in Fig. P14.78 represents the input circuit to a cathode-ray oscilloscope. The parallel combination of R_1 and C_1 is a circuit model of a compensating lead that is used to connect the CRO to the source. There is no energy stored in C_1 or C_2 at the time when the 10 V source is connected to the CRO via the compensating lead. The circuit values are $C_1 = 5$ pF, $C_2 = 20$ pF, $R_1 = 1$ MΩ, and $R_2 = 4$ MΩ.

a) Find v_o.

b) Find i_o.

c) Repeat parts (a) and (b) given C_1 is changed to 80 pF.

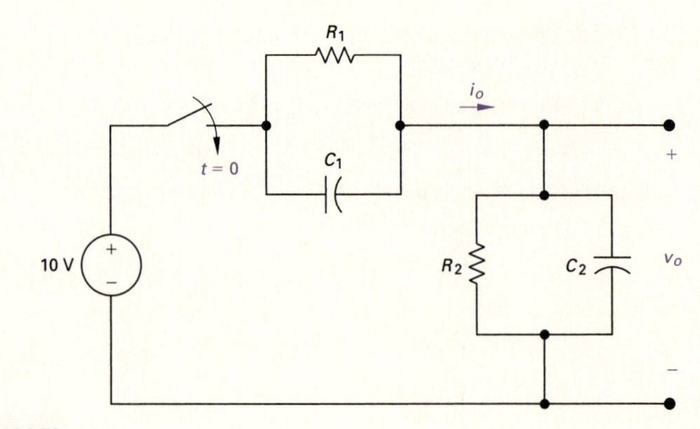

FIGURE P14.78

14.79 Show that if $R_1C_1 = R_2C_2$ in the circuit shown in Fig. P14.78, v_o will be a scaled replica of the source voltage.

14.80 There is no energy stored in the circuit in Fig. P14.80 at the time the impulsive current is applied.

a) Find v_o for $t \geq 0^+$.

b) Does your solution make sense in terms of known circuit behavior? Explain.

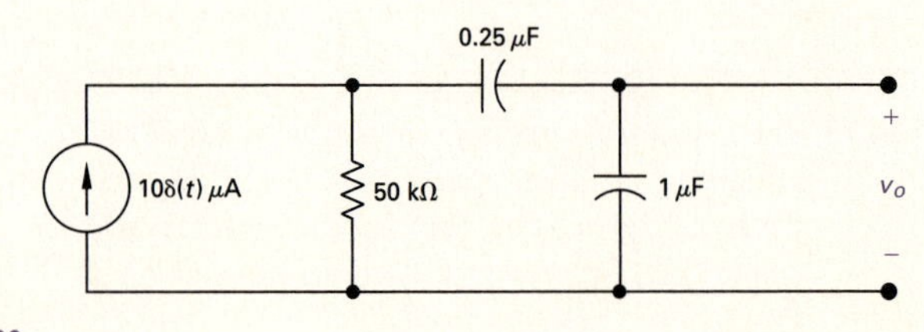

FIGURE P14.80

14.81 There is no energy stored in the circuit in Fig. P14.81 at the time the impulsive voltage is applied.

a) Find $v_o(t)$ for $t \geq 0$.

b) Does your solution make sense in terms of known circuit behavior? Explain.

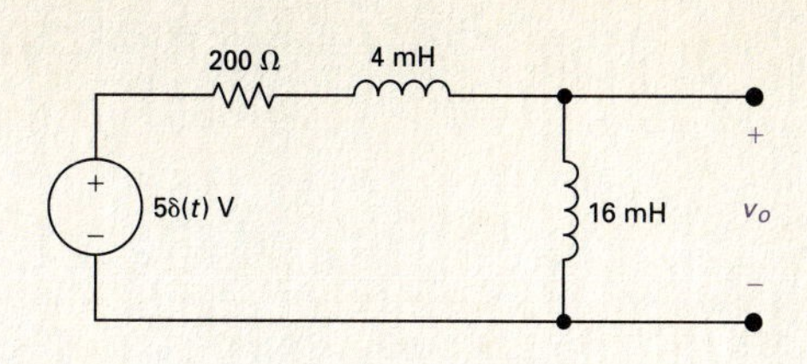

FIGURE P14.81

14.82 There is no energy stored in the circuit in Fig. P14.82 at the time the impulse voltage is applied.

a) Find i_1 for $t \geq 0^+$.

b) Find i_2 for $t \geq 0^+$.

c) Find v_o for $t \geq 0^+$.

d) Do your solutions for i_1, i_2, and v_o make sense in terms of known circuit behavior? Explain.

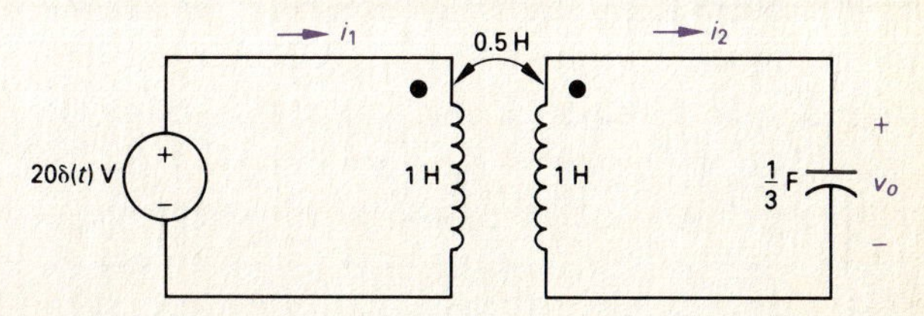

FIGURE P14.82

14.83 The voltage source in the circuit in Example 14.1 is changed to a unit impulse; that is, $v_g = \delta(t)$.

a) How much energy does the impulsive voltage source store in the capacitor?

b) How much energy does it store in the inductor?

c) Use the transfer function to find $v_o(t)$.

d) Show that the response found in part (c) is identical with the response generated by first charging the capacitor to 1000 V and then releasing the charge to the circuit, as shown in Fig. P14.83.

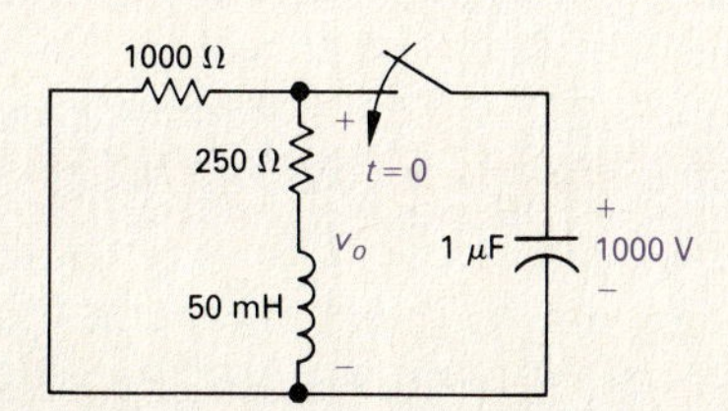

FIGURE P14.83

PRACTICAL PERSPECTIVE

Pushbutton Telephone Circuits

In this chapter we examine circuits in which the source frequency varies. The behavior of these circuits varies as the source frequency varies, because the impedance of the reactive components is a function of the source frequency. These frequency-dependent circuits are called *filters* and are used in many common electrical devices. In radios, filters are used to select one radio station's signal while rejecting the signals from others transmitting at different frequencies. In stereo systems, filters are used to adjust the relative strengths of the low- and high-frequency components of the audio signal. Filters are also used throughout telephone systems.

A pushbutton telephone produces tones which you hear when you press a button. You may have wondered about these tones. How are they used to tell the telephone system which button was pushed? Why are tones used at all? Why do the tones sound musical? How does the phone system tell the difference between button tones and the normal sounds of people talking or singing?

The telephone system was designed to handle audio signals—those with frequencies between 300 Hz and 3 kHz. Thus, all signals from the system to the user have to be audible—including the dial tone and the busy signal. Similarly, all signals from the user to the system have to be audible, including the signal that the user has pressed a button. It is important to distinguish button signals from the normal audio signal, so a dual-tone-multiple-frequency (DTMF) design is employed. When a number button is pressed, a unique pair of sinusoidal tones with very precise frequencies is sent by the phone to the telephone system. The DTMF frequency and timing specifications make it unlikely that a human voice could produce the exact tone pairs, even if the person were trying. In the central telephone facility, electric circuits monitor the audio signal, listening for the tone pairs that signal a number. In the Practical Perspective example at the end of the chapter, we will examine the design of the DTMF filters used to determine which button has been pushed.

15 INTRODUCTION TO FREQUENCY-SELECTIVE CIRCUITS

CHAPTER CONTENTS

Up to this point in our analysis of circuits with sinusoidal sources, the source frequency was held constant. In this chapter, we analyze the effect of varying source frequency on circuit voltages and currents. The result of this analysis is the **frequency response** of a circuit.

We've seen in previous chapters that a circuit's response depends on the types of elements in the circuit, the way the elements are connected, and the impedance of the elements. While varying the frequency of a sinusoidal source does not change the element types or their connections, it does alter the impedance of capacitors and inductors, because the impedance of these elements is a function of frequency. As we will see, the careful choice of circuit elements, their values, and their connections to other elements enables us to construct circuits which pass to the output only those input signals which reside in a desired range of frequencies. Such circuits are called **frequency-selective circuits**. Many devices which communicate via electric signals, such as telephones, radios, televisions, and satellites, employ frequency-selective circuits.

Frequency-selective circuits are also called **filters** because of their ability to filter out certain input signals on the basis of frequency. Figure 15.1 represents this ability in a simplistic way. To be more accurate, we should note that no practical frequency-selective circuit can perfectly or completely filter out selected frequencies. Rather, filters **attenuate**—that is, weaken or lessen the effect of—any input signals with frequencies outside a particular frequency band. Your home stereo system may have a graphic equalizer, which is an excellent example of a collection of filter circuits. Each band in the graphic equalizer is a filter which amplifies sounds (audible frequencies) in the frequency range of the band and attenuates frequencies outside of that band. Thus the graphic equalizer enables you to change the sound volume in each frequency band.

FIGURE 15.1 The action of a filter on an input signal results in an output signal.

We begin this chapter by analyzing circuits from each of the four major categories of filters: low pass, high pass, band pass, and band reject. The transfer function of a circuit is the starting point for the frequency response analysis. Pay close attention to the similarities among the transfer functions of circuits which perform the same filtering function. We will employ these similarities when designing filter circuits in Chapter 16.

We conclude this chapter with an introduction to Bode plots. Bode plot methods permit you to quickly sketch graphs of the magnitude and phase angle of a circuit's transfer function versus frequency. These plots approximate the actual frequency response of the circuit, providing a powerful visual representation of this important circuit characteristic.

15.1 SOME PRELIMINARIES

Recall from Section 14.7 that the transfer function of a circuit provides an easy way to compute the steady-state response to a sinusoidal input. There, we considered only fixed-frequency sources. To study the frequency response of a circuit, we replace a fixed-frequency sinusoidal source with a varying-frequency sinusoidal source. The transfer function is still an immensely useful tool because the magnitude and phase of the output signal depend only on the magnitude and phase of the transfer function $H(j\omega)$.

Note that the approach just outlined assumes that we can vary the frequency of a sinusoidal source without changing its magnitude or phase angle. Therefore, the amplitude and phase of the output will vary only if those of the transfer function vary as the frequency of the sinusoidal source is changed.

To further simplify this first look at frequency-selective circuits, we will also restrict our attention to cases where both the input and output signals are sinusoidal voltages, as illustrated in Fig. 15.2. Thus, the transfer function of interest to us will be the ratio of the Laplace transform of the output voltage to the Laplace transform of the input voltage, or $H(s) = V_o(s)/V_i(s)$. We should keep in mind, however, that for a particular application, a current may be either the input signal or output signal of interest.

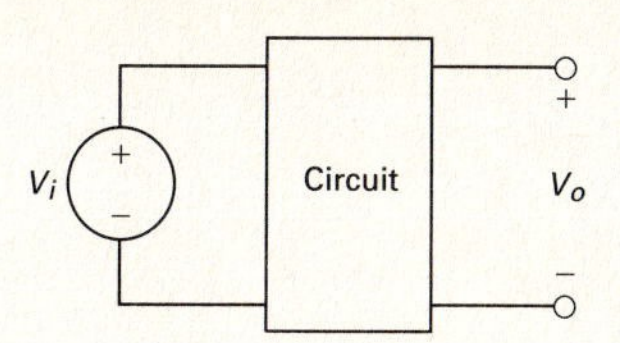

FIGURE 15.2 A circuit with voltage input and output.

The signals which are passed from the input to the output fall within a band of frequencies called the **passband**. Input voltages outside this band have their magnitudes attenuated by the circuit and are thus effectively prevented from reaching the output terminals of the circuit. Frequencies not in a circuit's passband are in its **stopband**. Frequency-selective circuits are categorized by the location of the passband.

One way of identifying the type of frequency-selective circuit is to examine a **frequency response plot**. A frequency response plot shows how a circuit's transfer function (both amplitude and phase) changes as the source frequency changes. A frequency response plot has two parts. One is a graph of $|H(j\omega)|$ versus frequency ω. This part of the plot is called the **magnitude plot**. The other part is a graph of $\theta(j\omega)$ versus frequency ω. This part is called the **phase angle plot**.

The ideal frequency response plots for the four major categories of filters are shown in Fig. 15.3. Parts (a) and (b) of the figure illustrate the ideal plots for a low-pass and a high-pass filter, respectively. Both

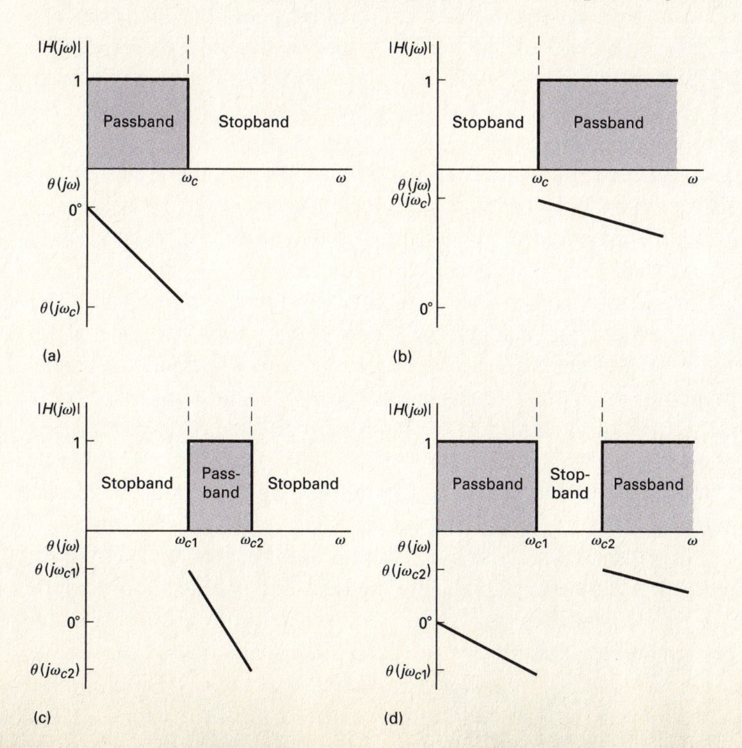

FIGURE 15.3 Ideal frequency response plots of the four types of filter circuits: (a) an ideal low-pass filter; (b) an ideal high-pass filter; (c) an ideal bandpass filter; and (d) an ideal bandreject filter.

filters have one passband and one stopband, which are defined by the **cutoff frequency** that separates them. The names *low pass* and *high pass* are derived from the magnitude plots: a **low-pass filter** passes signals at frequencies lower than the cutoff frequency from the input to the output, while a **high-pass filter** passes signals at frequencies higher than the cutoff frequency. The terms *low* and *high* as used here thus do not refer to any absolute values of frequency, but rather to relative values with respect to the cutoff frequency.

Note from the graphs for both these filters (as well as those for the bandpass and bandreject filters) that the phase angle plot for an ideal filter varies linearly in the passband. It is of no interest outside the passband because there the magnitude is zero. Linear phase variation is necessary to avoid phase distortion (see Chapter 17).

The two remaining categories of filters each have two cutoff frequencies. Figure 15.3(c) illustrates the ideal frequency response plot of a **bandpass filter**, which passes a source voltage to the output only when the source frequency is within the band defined by the two cutoff frequencies. Figure 15.3(d) shows the ideal plot of a **bandreject filter**, which passes a source voltage to the output only when the source frequency is outside the band defined by the two cutoff frequencies. The bandreject filter thus rejects, or stops, the source voltage from reaching the output when its frequency is within the band defined by the cutoff frequencies.

In specifying a realizable filter using any of the circuits from this chapter, it is important to note that the magnitude and phase angle characteristics are not independent. In other words, the characteristics of a circuit that result in a particular magnitude plot will also dictate the form of the phase angle plot, and vice versa. For example, once we select a desired form for the magnitude response of a circuit, the phase angle response is also determined. Alternatively, if we select a desired form for the phase angle response, the magnitude response is also determined. While there are some frequency-selective circuits for which the magnitude and phase angle behavior can be independently specified, these circuits are not presented here.

The next sections present examples of circuits from each of the four filter categories. They are a few of the many circuits that act as filters. You should focus your attention on trying to identify what properties of a circuit determine its behavior as a filter. Look closely at the form of the transfer function for circuits that perform the same filtering functions. Identifying the form of a filter's transfer function will ultimately help you in designing filtering circuits for particular applications.

All of the filters we will consider in this chapter are **passive filters**, so called because their filtering capabilities depend only on the passive elements: resistors, capacitors, and inductors. The largest output amplitude such filters can achieve is usually 1, and placing an impedance in series with the source or in parallel with the load will decrease this amplitude. Because many practical filter applications require increasing the amplitude of thc output, passive filters have some significant disadvantages. The only passive filter described in this chapter which can amplify its output is the series RLC resonant filter. A much greater selection of amplifying filters is found among the active filter circuits, the subject of Chapter 16.

15.2 LOW-PASS FILTERS

Here, we examine two circuits which behave as low-pass filters—the series RL circuit and the series RC circuit—and discover what characteristics of these circuits determine the cutoff frequency.

THE SERIES *RL* CIRCUIT—QUALITATIVE ANALYSIS

A series RL circuit is shown in Fig. 15.4(a). The circuit's input is a sinusoidal voltage source with varying frequency. The circuit's output is defined as the voltage across the resistor. Suppose the frequency of the source starts very low and increases gradually. We know that the behavior of the ideal resistor will not change, since its impedance is independent of frequency. But consider how the behavior of the inductor changes.

Recall that the impedance of an inductor is $j\omega L$. At low frequencies, the inductor's impedance is very small compared to the resistor's impedance, and the inductor effectively functions as a short circuit. The term *low frequencies* thus refers to any frequencies for which $\omega L \ll R$. The equivalent circuit for $\omega = 0$ is shown in Fig. 15.4(b). In this equivalent circuit, the output voltage and the input voltage are equal both in magnitude and in phase angle.

As the frequency increases, the impedance of the inductor increases relative to that of the resistor. Increasing the inductor's impedance causes a corresponding increase in the magnitude of the voltage drop across the inductor, and a corresponding decrease in the output voltage magnitude. Increasing the inductor's impedance also introduces a shift in phase angle between the inductor's voltage and current. This results in a phase angle difference between the input and output voltage. The output voltage lags the input voltage, and as the frequency increases, this phase lag approaches 90°.

At high frequencies, the inductor's impedance is very large compared to the resistor's impedance, and the inductor thus functions as an open circuit, effectively blocking the flow of current in the circuit. The term *high frequencies* thus refers to any frequencies for which

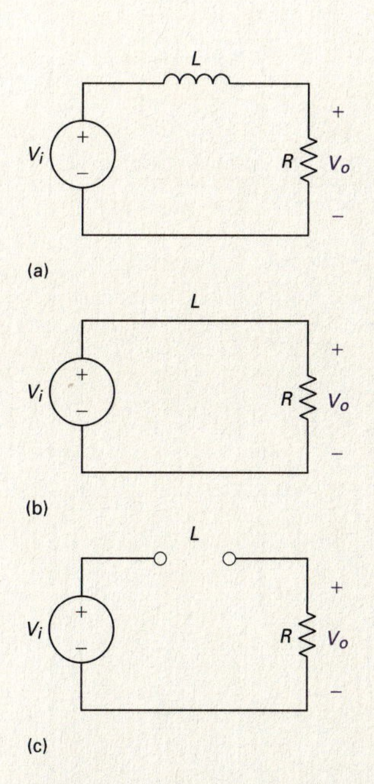

FIGURE 15.4 (a) A series RL low-pass filter; (b) the equivalent circuit at $\omega = 0$; and (c) the equivalent circuit at $\omega = \infty$.

$\omega L \gg R$. The equivalent circuit for $\omega = \infty$ is shown in Fig. 15.4(c), where the output voltage magnitude is zero. The phase angle of the output voltage is 90° more negative than that of the input voltage.

Based on the behavior of the output voltage magnitude, this series RL circuit selectively passes low-frequency inputs to the output, and it blocks high-frequency inputs from reaching the output. This circuit's response to varying input frequency thus has the shape shown in Fig. 15.5. These two plots comprise the frequency response plots of the series RL circuit in Fig. 15.4(a). The upper plot shows how $|H(j\omega)|$ varies with frequency. The lower plot shows how $\theta(j\omega)$ varies as a function of frequency. We will present a more formal method for constructing these plots at the end of this chapter.

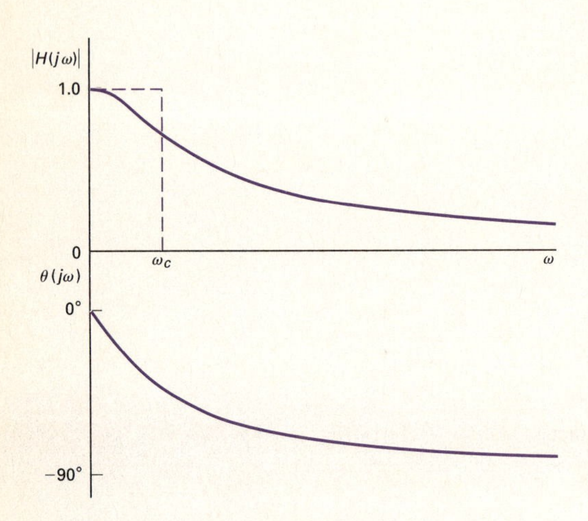

FIGURE 15.5 The frequency response plot for the series RL circuit in Fig. 15.4(a).

We have also superimposed the ideal magnitude plot for a low-pass filter from Fig. 15.3(a) on the magnitude plot of the RL filter in Fig. 15.5. There is obviously a difference between the magnitude plots of an ideal filter and the frequency response of an actual RL filter. The ideal filter exhibits a discontinuity in magnitude at the cutoff frequency, ω_c, which creates an abrupt transition into and out of the passband. While this is, ideally, how we would like our filters to perform, it is not possible to use real components to construct a circuit which has this abrupt transition in magnitude. Circuits acting as low-pass filters have a magnitude response which changes gradually from the passband to the stopband. Hence the magnitude plot of a real circuit requires us to define what we mean by the cutoff frequency, ω_c.

DEFINING THE CUTOFF FREQUENCY

We need to define the cutoff frequency, ω_c, for realistic filter circuits when the magnitude plot does not allow us to identify a single frequency that divides the passband and the stopband. The definition for cutoff frequency widely used by electrical engineers is the frequency for which the transfer function magnitude is decreased by the factor $1/\sqrt{2}$ from its maximum value:

$$|H(j\omega_c)| = \frac{1}{\sqrt{2}} H_{\text{max}}, \tag{15.1}$$

where H_{max} is the maximum magnitude of the transfer function. It follows from Eq. (15.1) that the passband of a realizable filter is defined as the range of frequencies in which the amplitude of the output voltage is at least 70.7% of the maximum possible amplitude.

The constant $1/\sqrt{2}$ used in defining the cutoff frequency may seem like an arbitrary choice. Examining another consequence of the cutoff frequency will make this choice seem more reasonable. Recall from

Section 10.5 that the average power delivered by any circuit to a load is proportional to V_L^2, where V_L is the amplitude of the voltage drop across the load:

$$P = \frac{1}{2}\frac{V_L^2}{R}. \tag{15.2}$$

If the circuit has a sinusoidal voltage source, $V_i(j\omega)$, then the load voltage is also a sinusoid, and its amplitude is a function of the frequency ω. Define P_{max} as the value of the average power delivered to a load when the magnitude of the load voltage is maximum:

$$P_{\text{max}} = \frac{1}{2}\frac{V_{L\text{max}}^2}{R}. \tag{15.3}$$

If we vary the frequency of the sinusoidal voltage source, $V_i(j\omega)$, the load voltage is a maximum when the magnitude of the circuit's transfer function is also a maximum:

$$V_{L\text{max}} = H_{\text{max}}|V_i|. \tag{15.4}$$

Now consider what happens to the average power when the frequency of the voltage source is ω_c. Using Eq. (15.1), we determine the magnitude of the load voltage at ω_c to be

$$\begin{aligned} |V_L(j\omega_c)| &= |H(j\omega_c)||V_i| \\ &= \frac{1}{\sqrt{2}}H_{\text{max}}|V_i| \\ &= \frac{1}{\sqrt{2}}V_{L\text{max}}. \end{aligned} \tag{15.5}$$

Substituting Eq. (15.5) into Eq. (15.2),

$$\begin{aligned} P(j\omega_c) &= \frac{1}{2}\frac{|V_L^2(j\omega_c)|}{R} \\ &= \frac{1}{2}\frac{\left(\frac{1}{\sqrt{2}}V_{L\text{max}}\right)^2}{R} \\ &= \frac{1}{2}\frac{V_{L\text{max}}^2/2}{R} \\ &= \frac{P_{\text{max}}}{2}. \end{aligned} \tag{15.6}$$

Equation (15.6) shows that at the cutoff frequency ω_c, the average power delivered by the circuit is one-half the maximum average power. Thus, ω_c is also called the **half-power frequency**. Therefore, in the passband, the average power delivered to a load is at least 50% of the maximum average power.

The Series *RL* Circuit—Quantitative Analysis

Now that we have defined the cutoff frequency for real filter circuits, we can analyze the series RL circuit to discover the relationship between the component values and the cutoff frequency for this low-pass filter. We begin by constructing the s-domain equivalent of the circuit in Fig. 15.4(a), assuming initial conditions of zero. The resulting equivalent circuit is shown in Fig. 15.6.

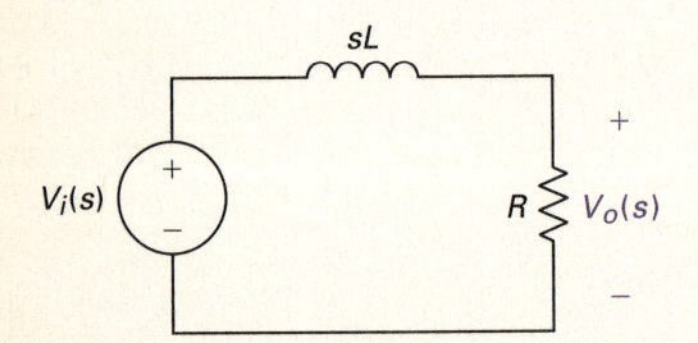

FIGURE 15.6 The s-domain equivalent for the circuit in Fig. 15.4(a).

The voltage transfer function for this circuit is

$$H(s) = \frac{\frac{R}{L}}{s + \frac{R}{L}}. \qquad (15.7)$$

To study the frequency response, we make the substitution $s = j\omega$ in Eq. (15.7):

$$H(j\omega) = \frac{\frac{R}{L}}{j\omega + \frac{R}{L}}. \qquad (15.8)$$

We can now separate Eq. (15.8) into two equations. The first defines the transfer function magnitude as a function of frequency; the second defines the transfer function phase angle as a function of frequency:

$$|H(j\omega)| = \frac{\frac{R}{L}}{\sqrt{\omega^2 + \left(\frac{R}{L}\right)^2}}; \qquad (15.9)$$

$$\theta(j\omega) = -\tan^{-1}\left(\frac{\omega L}{R}\right). \qquad (15.10)$$

Close examination of Eq. (15.9) provides the quantitative support for the magnitude plot shown in Fig. 15.5. When $\omega = 0$, the denominator and the numerator are equal and $|H(j0)| = 1$. This means that at $\omega = 0$, the input voltage is passed to the output terminals without a change in the voltage magnitude.

As the frequency increases, the numerator of Eq. (15.9) is unchanged, but the denominator gets larger. Thus $|H(j\omega)|$ decreases as the frequency increases, as shown in the plot in Fig. 15.5. Likewise, as the frequency increases, the phase angle changes from its dc value of 0°, becoming more negative, as seen from Eq. (15.10).

When $\omega = \infty$, the denominator of Eq. (15.9) is infinite and $|H(j\infty)| = 0$, as seen in Fig. 15.5. At $\omega = \infty$, the phase angle reaches a limit of −90°, as seen from Eq. (15.10) and the phase angle plot in Fig. 15.5.

Using Eq. (15.9), we can compute the cutoff frequency, ω_c. Remember that ω_c is defined as the frequency at which $|H(j\omega_c)| = (1/\sqrt{2})H_{\max}$. For the low-pass filter, $H_{\max} = |H(j0)|$, as seen in

Fig. 15.5. Thus, for the circuit in Fig. 15.4(a),

$$|H(j\omega_c)| = \frac{1}{\sqrt{2}}|1| = \frac{\frac{R}{L}}{\sqrt{\omega_c^2 + \left(\frac{R}{L}\right)^2}}. \quad \textbf{(15.11)}$$

Solving Eq. (15.11) for ω_c, we get

$$\omega_c = \frac{R}{L}. \quad \textbf{(15.12)}$$

Equation (15.12) provides an important result. The cutoff frequency, ω_c, can be set to any desired value by appropriately selecting values for R and L. We can therefore design a low-pass filter with whatever cutoff frequency is needed. Example 15.1 demonstrates the design potential of Eq. (15.12).

EXAMPLE 15.1

Electrocardiology is the study of the electric signals produced by the heart. These signals maintain the heart's rhythmic beat, and they are measured by an instrument called an electrocardiograph. This instrument must be capable of detecting periodic signals whose frequency is about 1 Hz (since the normal heart rate is 72 beats per minute). The instrument must operate in the presence of sinusoidal noise consisting of signals from the surrounding electrical environment, whose fundamental frequency is 60 Hz—the frequency at which electric power is supplied.

Choose values for R and L in the circuit of Fig. 15.4(a) such that the resulting circuit could be used in an electrocardiograph to filter out any noise above 10 Hz and pass the electric signals from the heart at or near 1 Hz. Then compute the magnitude of V_o at 1 Hz, 10 Hz, and 60 Hz to see how well the filter performs.

SOLUTION

The problem is to select values for R and L which yield a low-pass filter with a cutoff frequency of 10 Hz. From Eq. (15.12), we see that R and L cannot be specified independently to generate a value for ω_c. Therefore, let's choose a commonly available value of L, 100 mH. Before we use Eq. (15.12) to compute the value of R needed to obtain the desired cutoff frequency, we need to convert the cutoff frequency from hertz to radians per second:

$$\omega_c = 2\pi(10) = 62.83 \text{ rad/s}.$$

Now, solve for the value of R which, together with $L = 100$ mH, will yield a low-pass filter with a cutoff frequency of 10 Hz:

$$\begin{aligned} R &= \omega_c L \\ &= (62.8)(100 \times 10^{-3}) \\ &= 6.28\ \Omega. \end{aligned}$$

We can compute the magnitude of V_o using the equation $|V_o| = |H(j\omega)| \cdot |V_i|$:

$$\begin{aligned} |V_o(\omega)| &= \frac{\frac{R}{L}}{\sqrt{\omega^2 + \left(\frac{R}{L}\right)^2}} |V_i| \\ &= \frac{20\pi}{\sqrt{\omega^2 + 400\pi^2}}. \end{aligned}$$

Table 15.1 summarizes the computed magnitude values for the frequencies 1 Hz, 10 Hz, and 60 Hz. As expected, the input and output voltages have the same magnitudes at the low frequency, since the circuit is a low-pass filter. At the cutoff frequency, the output voltage magnitude has been reduced by approximately $1/\sqrt{2}$ from the unity passband magnitude. At 60 Hz, the output voltage magnitude has been reduced by a factor of about 6, achieving the desired attenuation of the noise which could corrupt the signal the electrocardiograph is designed to measure.

TABLE 15.1

INPUT AND OUTPUT VOLTAGE MAGNITUDES FOR SEVERAL FREQUENCIES

f (Hz)	$\|V_i\|$ (V)	$\|V_o\|$ (V)
1	1.0	0.99
10	1.0	0.69
60	1.0	0.16

Use a computer tool to confirm these voltage magnitudes or calculate others.

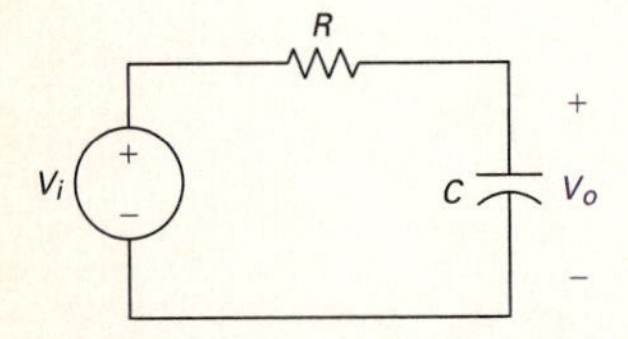

FIGURE 15.7 A series RC low-pass filter.

A SERIES *RC* CIRCUIT

The series RC circuit shown in Fig. 15.7 also behaves as a low-pass filter. We can verify this via the same qualitative analysis we used previously. In fact, such a qualitative examination is an important problem-solving step which you should get in the habit of performing when analyzing filters. Doing so will enable you to predict the filtering characteristics (low pass, high pass, and so forth) and thus also predict the general form of the transfer function. If the calculated transfer function matches the qualitatively predicted form, you have an important accuracy check.

Note that the circuit's output is defined as the output across the capacitor. As we did in the previous qualitative analysis, we use three frequency regions to develop the behavior of the series RC circuit in Fig. 15.7:

1. *Zero frequency* ($\omega = 0$): The impedance of the capacitor is infinite, and the capacitor acts as an open circuit. The input and output voltages are thus the same.

2. *Frequencies increasing from zero:* The impedance of the capacitor decreases relative to the impedance of the resistor, and the source voltage divides between the resistive impedance and the capacitive impedance. The output voltage is thus smaller than the source voltage.

3. *Infinite frequency* ($\omega = \infty$): The impedance of the capacitor is zero, and the capacitor acts as a short circuit. The output voltage is thus zero.

Based on this analysis of how the output voltage changes as a function of frequency, the series RC circuit functions as a low-pass filter. Example 15.2 explores this circuit quantitatively.

EXAMPLE 15.2

For the series RC circuit in Fig. 15.7,

a) find the transfer function between the source voltage and the output voltage;

b) determine an equation for the cutoff frequency in the series RC circuit; and

c) choose values for R and C that will yield a low-pass filter with a cutoff frequency of 3 kHz.

SOLUTION

a) To derive an expression for the transfer function, we first construct the s-domain equivalent of the circuit in Fig. 15.7, as shown in Fig. 15.8.

Using s-domain voltage division on the equivalent circuit, we find

$$H(s) = \frac{\frac{1}{RC}}{s + \frac{1}{RC}}.$$

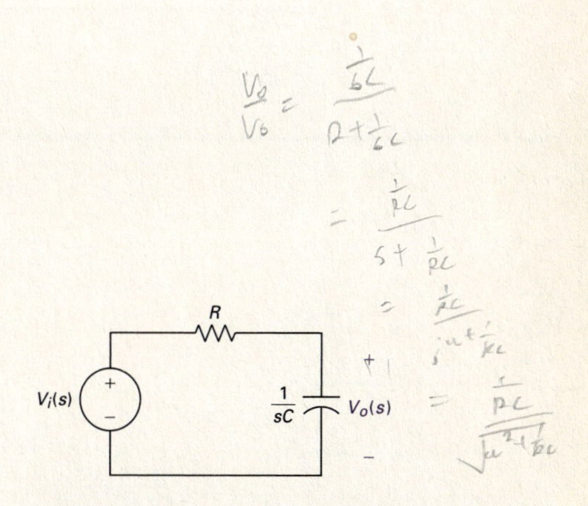

FIGURE 15.8 The s-domain equivalent for the circuit in Fig. 15.7.

Now, substitute $s = j\omega$ and compute the magnitude of the resulting complex expression:

$$|H(j\omega)| = \frac{\frac{1}{RC}}{\sqrt{\omega^2 + \left(\frac{1}{RC}\right)^2}}.$$

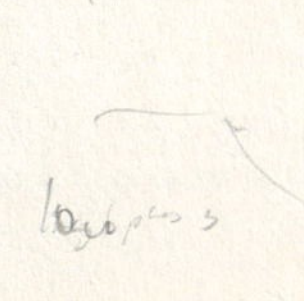

b) At the cutoff frequency ω_c, $|H(j\omega)|$ is equal to $(1/\sqrt{2})H_{max}$. For a low-pass filter, $H_{max} = H(j0)$, and for the circuit in Fig. 15.8,

$H(j0) = 1$. We can then describe the relationship among the quantities R, C, and ω_c:

$$|H(j\omega_c)| = \frac{1}{\sqrt{2}}(1) = \frac{\frac{1}{RC}}{\sqrt{\omega_c^2 + \left(\frac{1}{RC}\right)^2}}.$$

Solving this equation for ω_c, we get

$$\omega_c = \frac{1}{RC}.$$

c) From the results in part (b), we see that the cutoff frequency is determined by the values of R and C. Since R and C cannot be computed independently, let's choose $C = 1\ \mu$F. Given a choice, we will usually specify a value for C first, rather than for R or L, because the number of available capacitor values is much smaller than the number of resistor or inductor values. Remember that we have to convert the specified cutoff frequency from 3 kHz to $(2\pi)(3)$ krad/s:

$$R = \frac{1}{\omega_c C}$$
$$= \frac{1}{(2\pi)(3 \times 10^3)(1 \times 10^{-6})}$$
$$= 53.05\ \Omega.$$

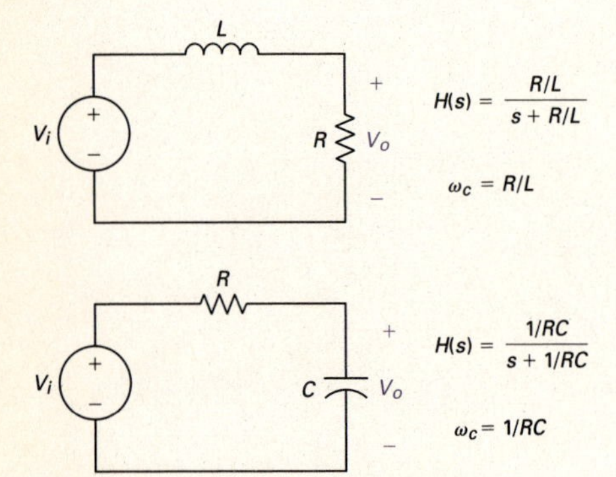

FIGURE 15.9 Two low-pass filters, the series *RL* and the series *RC*, together with their transfer functions and cutoff frequencies.

Figure 15.9 summarizes the two low-pass filter circuits we have examined. Look carefully at the transfer functions. Notice how similar in form they are—they differ only in the terms which specify the cutoff frequency. In fact, we can state a general form for the transfer functions of these two low-pass filters:

$$H(s) = \frac{\omega_c}{s + \omega_c}. \qquad \textbf{(15.13)}$$

Any circuit with the voltage ratio in Eq. (15.13) would behave as a low-pass filter with a cutoff frequency of ω_c. The problems at the end of the chapter give you other examples of circuits with this voltage ratio.

Relating the Frequency Domain to the Time Domain

Finally, you might have noticed one other important relationship. Remember our discussion of the natural responses of the first-order RL and RC circuits in Chapter 6. An important parameter for these circuits is the time constant, τ, which characterizes the shape of the time response. For the RL circuit, the time constant has the value L/R (Eq. [7.14]); for the RC circuit, the time constant is RC (Eq. [7.24]). Compare the time constants to the cutoff frequencies for these circuits and notice that

$$\tau = 1/\omega_c. \tag{15.14}$$

This result is a direct consequence of the relationship between the time response of a circuit and its frequency response, as revealed by the Laplace transform. The discussion of memory and weighting as represented in the convolution integral of Section 14.6 shows that as $\omega_c \to \infty$, the filter has no memory and the output approaches a scaled replica of the input; that is, no filtering has occurred. As $\omega_c \to 0$, the filter has increased memory and the output voltage is a distortion of the input, because filtering has occurred.

Use a computer tool to examine both the time and frequency response of an *RL* or *RC* circuit.

15.3 HIGH-PASS FILTERS

We next examine two circuits which function as high-pass filters. Once again, they are the series RL circuit and the series RC circuit. We will see that the same series circuit can act as either a low-pass or a high-pass filter, depending on where the output voltage is defined. We will also determine the relationship between the component values and the cutoff frequency of these filters.

THE SERIES *RC* CIRCUIT—QUALITATIVE ANALYSIS

A series RC circuit is shown in Fig. 15.10(a). In contrast to its low-pass counterpart in Fig. 15.7, the output voltage here is defined across the resistor, not the capacitor. Because of this, the effect of the changing capacitive impedance is different than it was in the low-pass configuration.

At $\omega = 0$, the capacitor behaves like an open circuit, so there is no current flowing in the resistor. This is illustrated in the equivalent circuit in Fig. 15.10(b). In this circuit, there is no voltage across the resistor, and the circuit filters out the low-frequency source voltage before it reaches the circuit's output.

As the frequency of the voltage source increases, the impedance of the capacitor decreases relative to the impedance of the resistor, and the source voltage is now divided between the capacitor and the resistor. The output voltage magnitude thus begins to increase.

When the frequency of the source is infinite ($\omega = \infty$), the capacitor behaves as a short circuit, and thus there is no voltage across the

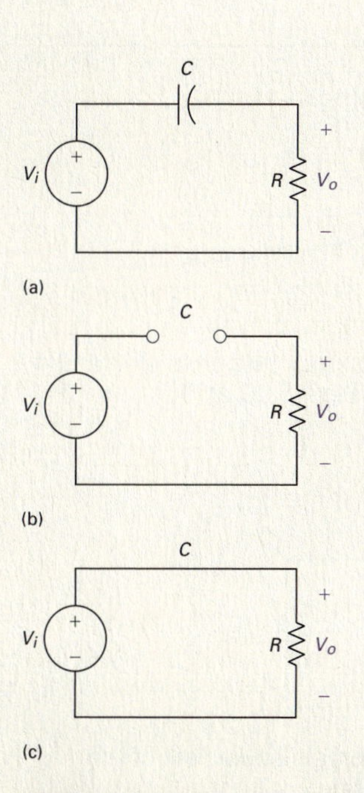

FIGURE 15.10 (a) A series *RC* high-pass filter; (b) the equivalent circuit at $\omega = 0$; and (c) the equivalent circuit at $\omega = \infty$.

capacitor. This is illustrated in the equivalent circuit in Fig. 15.10(c). In this circuit, the input voltage and output voltage are the same.

The phase angle difference between the source and output voltages also varies as the frequency of the source changes. For $\omega = \infty$, the output voltage is the same as the input voltage, so the phase angle difference is zero. As the frequency of the source decreases and the impedance of the capacitor increases, a phase shift is introduced between the voltage and the current in the capacitor. This creates a phase difference between the source and output voltages. The phase angle of the output voltage leads that of the source voltage. When $\omega = 0$, this phase angle difference reaches its maximum of $+90°$.

Based on our qualitative analysis, we see that when the output is defined as the voltage across the resistor, the series RC circuit behaves as a high-pass filter. The components and connections are identical to the low-pass series RC circuit, but the choice of output is different. Thus, we have confirmed the earlier observation that the filtering characteristics of a circuit depend on the definition of the output as well as on circuit components, values, and connections.

Figure 15.11 shows the frequency response plot for the series RC high-pass filter. For reference, the dashed lines indicate the magnitude plot for an ideal high-pass filter. We now turn to a quantitative analysis of this same circuit.

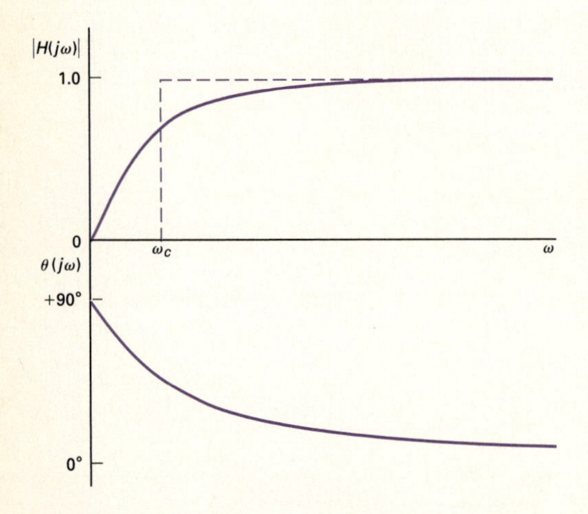

FIGURE 15.11 The frequency response plot for the series *RC* circuit in Fig. 15.10(a).

THE SERIES *RC* CIRCUIT—QUANTITATIVE ANALYSIS

To begin, we construct the s-domain equivalent of the circuit in Fig. 15.10(a). This equivalent is shown in Fig. 15.12. Applying s-domain voltage division to the circuit, we write the transfer function:

$$H(s) = \frac{s}{s + \dfrac{1}{RC}}.$$

Making the substitution $s = j\omega$ results in

$$H(j\omega) = \frac{j\omega}{j\omega + \dfrac{1}{RC}}. \quad (15.15)$$

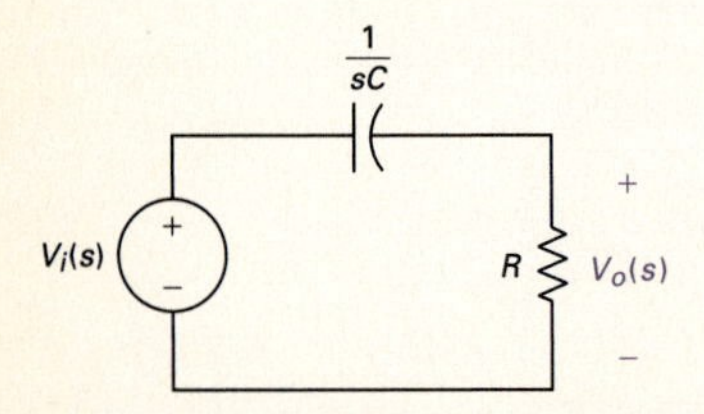

FIGURE 15.12 The s-domain equivalent of the circuit in Fig. 15.10(a).

Next, we separate Eq. (15.15) into two equations. The first is the equation describing the magnitude of the transfer function; the second is the equation describing the phase angle of the transfer function:

$$|H(j\omega)| = \frac{\omega}{\sqrt{\omega^2 + \left(\dfrac{1}{RC}\right)^2}}; \quad (15.16)$$

$$\theta(j\omega) = 90° - \tan^{-1}\omega RC. \quad (15.17)$$

A close look at Eqs. (15.16) and (15.17) confirms the shape of the frequency response plot in Fig. 15.11. Using Eq. (15.16), we can calculate the cutoff frequency for the series RC high-pass filter. Recall that at the cutoff frequency, the magnitude of the transfer function is $(1/\sqrt{2})H_{max}$. For a high-pass filter, $H_{max} = |H(j\omega)|_{\omega=\infty} = |H(j\infty)|$, as seen from Fig. 15.11. We can construct an equation for ω_c by setting the left-hand side of Eq. (15.16) to $(1/\sqrt{2})|H(j\infty)|$, noting that for this series RC circuit, $|H(j\infty)| = 1$:

$$\frac{1}{\sqrt{2}} = \frac{\omega_c}{\sqrt{\omega_c^2 + \left(\frac{1}{RC}\right)^2}}. \tag{15.18}$$

Solving Eq. (15.18) for ω_c, we get

$$\omega_c = \frac{1}{RC}. \tag{15.19}$$

Equation (15.19) presents a familiar result. The cutoff frequency for the series RC circuit has the value $1/RC$, whether the circuit is configured as a low-pass filter in Fig. 15.4(a) or as a high-pass filter in Fig. 15.10(a). This is perhaps not a surprising result, since we have already discovered a connection between the cutoff frequency, ω_c, and the time constant, τ, of a circuit.

Example 15.3 analyzes a series RL circuit, this time configured as a high-pass filter. Example 15.4 examines the effect of adding a load resistor in parallel with the inductor.

EXAMPLE 15.3

Show that the series RL circuit in Fig. 15.13 also acts like a high-pass filter:

a) Derive an expression for the circuit's transfer function;

b) Use the result from (a) to determine an equation for the cutoff frequency in the series RL circuit; and

c) Choose values for R and L that will yield a high-pass filter with a cutoff frequency of 15 kHz.

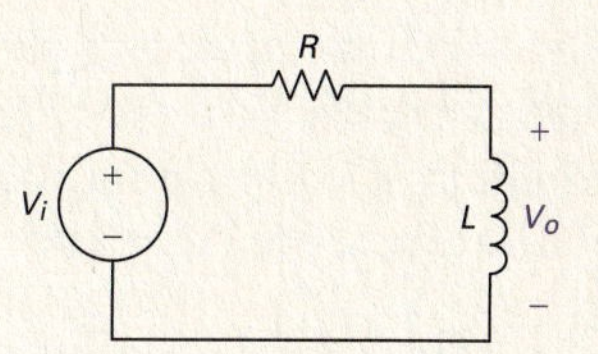

FIGURE 15.13 The circuit for Example 15.3.

SOLUTION

a) Begin by constructing the s-domain equivalent of the series RL circuit, as shown in Fig. 15.14. Then use s-domain voltage division on the equivalent circuit to construct the transfer function:

$$H(s) = \frac{s}{s + \frac{R}{L}}.$$

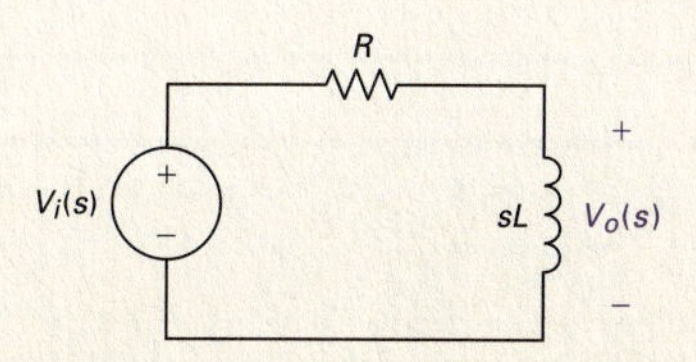

FIGURE 15.14 The s-domain equivalent of the circuit in Fig. 15.13.

Making the substitution $s = j\omega$, we get

$$H(j\omega) = \frac{j\omega}{j\omega + \frac{R}{L}}.$$

Notice that this equation has the same form as Eq. (15.15) for the series RC high-pass filter.

b) To find an equation for the cutoff frequency, first compute the magnitude of $H(j\omega)$:

$$|H(j\omega)| = \frac{\omega}{\sqrt{\omega^2 + \left(\frac{R}{L}\right)^2}}.$$

Then, as before, we set the left-hand side of this equation to $(1/\sqrt{2})H_{\text{max}}$, based on the definition of the cutoff frequency ω_c. Remember that $H_{\text{max}} = |H(j\infty)|$ for a high-pass filter, and for the series RL circuit, $|H(j\infty)| = 1$. We solve the resulting equation for the cutoff frequency:

$$\frac{1}{\sqrt{2}} = \frac{\omega_c}{\sqrt{\omega_c^2 + \left(\frac{R}{L}\right)^2}};$$

$$\omega_c = \frac{R}{L}.$$

This is the same cutoff frequency we computed for the series RL low-pass filter.

c) Using the equation for ω_c computed in part (b), we recognize that it is not possible to specify values for R and L independently. Therefore, let's arbitrarily select a value of 500 Ω for R. Remember to convert the cutoff frequency to radians per second:

$$L = \frac{R}{\omega_c}$$

$$= \frac{500}{(2\pi)(15{,}000)}$$

$$= 5.31 \text{ mH}.$$

EXAMPLE 15.4

Examine the effect of placing a load resistor in parallel with the inductor in the RL high-pass filter shown in Fig. 15.15:

a) Determine the transfer function for the circuit in Fig. 15.15.

b) Sketch the magnitude plot for the loaded RL high-pass filter, using the values for R and L from the circuit in Example 15.3(c) and letting $R_L = R$. On the same graph, sketch the magnitude plot for the unloaded RL high-pass filter of Example 15.3(c).

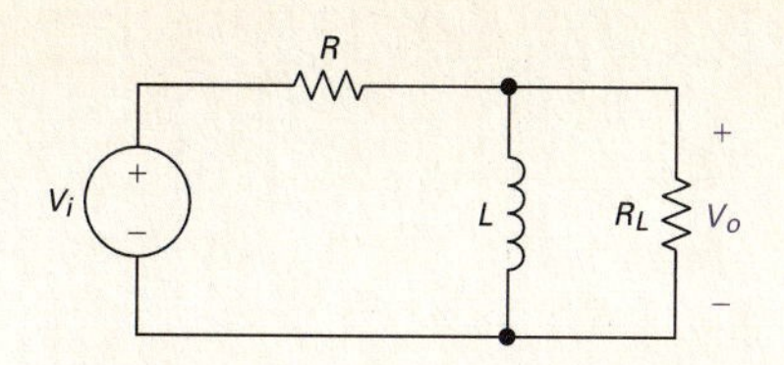

FIGURE 15.15 The circuit for Example 15.4.

SOLUTION

a) Begin by transforming the circuit in Fig. 15.15 to the s-domain, as shown in Fig. 15.16. Use voltage division across the parallel combination of inductor and load resistor to compute the transfer function:

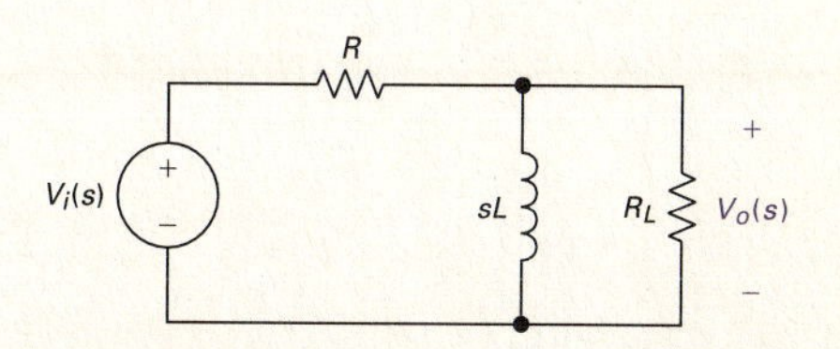

FIGURE 15.16 The s-domain equivalent of the circuit in Fig. 15.15.

$$H(s) = \frac{\dfrac{R_L sL}{R_L + sL}}{R + \dfrac{R_L sL}{R_L + sL}}$$

$$= \frac{\left(\dfrac{R_L}{R + R_L}\right) s}{s + \left(\dfrac{R_L}{R + R_L}\right) \dfrac{R}{L}}$$

$$= \frac{Ks}{s + \omega_c},$$

where

$$K = \frac{R_L}{R + R_L} \quad \text{and} \quad \omega_c = KR/L.$$

Note that ω_c is the cutoff frequency of the loaded filter.

b) For the unloaded RL high-pass filter from Example 15.3(c), the passband magnitude is 1, and the cutoff frequency is 15 kHz. For the loaded RL high-pass filter, $R = R_L = 500\ \Omega$, so $K = 1/2$. Thus, for the loaded filter, the passband magnitude is $(1)(1/2) = 1/2$, and the cutoff frequency is $(15{,}000)(1/2) = 7.5$ kHz. A sketch of the magnitude plots of the loaded and unloaded circuits is shown in Fig. 15.17.

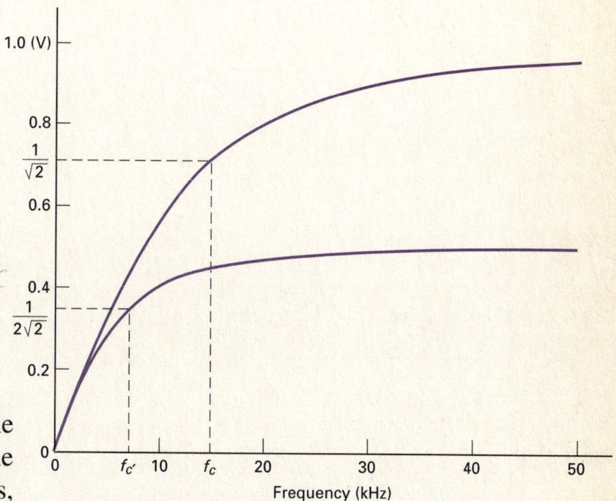

FIGURE 15.17 The magnitude plots for the unloaded RL high-pass filter of Fig 15.13 and the loaded RL high-pass filter of Fig. 15.15.

Comparing the transfer functions of the unloaded filter in Example 15.3 and the loaded filter in Example 15.4 is useful at this point. Both transfer functions are in the form

$$H(s) = \frac{Ks}{s + K\left(\dfrac{R}{L}\right)},$$

with $K = 1$ for the unloaded filter and $K = R_L/(R + R_L)$ for the loaded filter. Note that the value of K for the loaded circuit reduces to the value of K for the unloaded circuit when $R_L = \infty$, that is, when there is no load resistor. The cutoff frequencies for both filters can be seen directly from their transfer functions. In both cases, $\omega_c = K(R/L)$, where $K = 1$ for the unloaded circuit, and $K = R_L/(R + R_L)$ for the loaded circuit. Again, the cutoff frequency for the loaded circuit reduces to that of the unloaded circuit when $R_L = \infty$. Since $R_L/(R + R_L) < 1$, the effect of the load resistor is to reduce the passband magnitude by the factor K and to lower the cutoff frequency by the same factor. We predicted these results at the beginning of this chapter. The largest output amplitude a passive high-pass filter can achieve is 1, and placing a load across the filter, as we did in Example 15.4, has served to decrease the amplitude. When we need to amplify signals in the passband, we must turn to active filters, such as those discussed in Chapter 16.

The effect of a load on a filter's transfer function poses another dilemma in circuit design. We typically begin with a transfer function specification and then design a filter to produce that function. We may or may not know what the load on the filter will be, but in any event, we usually want the filter's transfer function to remain the same regardless of the load upon it. This desired behavior cannot be achieved with the passive filters presented in this chapter.

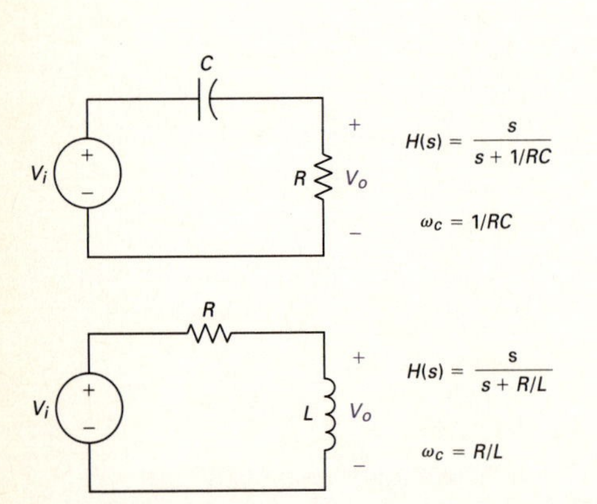

FIGURE 15.18 Two high-pass filters, the series *RC* and the series *RL*, together with their transfer functions and cutoff frequencies.

Figure 15.18 summarizes the high-pass filter circuits we have examined. Look carefully at the expressions for $H(s)$. Notice how similar in form these expressions are—they differ only in the denominator, which includes the cutoff frequency. As we did with the low-pass filters in Eq. (15.13), we state a general form for the transfer function of these two high-pass filters:

$$H(s) = \frac{s}{s + \omega_c}. \quad (15.20)$$

Any circuit with the transfer function in Eq. (15.20) would behave as a high-pass filter with a cutoff frequency of ω_c. The problems at the end of the chapter give you other examples of circuits with this voltage ratio.

We have drawn attention to another important relationship. We have discovered that a series RC circuit has the same cutoff frequency whether it is configured as a low-pass filter or as a high-pass filter. The same is true of a series RL circuit. Having previously noted the connection between the cutoff frequency of a filter circuit and the time constant of that same circuit, we should expect the cutoff frequency to be a characteristic parameter of the circuit whose value depends only on the circuit components, their values and the way they are connected.

DRILL EXERCISES

15.1 A series RC low-pass filter requires a cutoff frequency of 8 kHz. Use $R = 10\text{ k}\Omega$ and compute the value of C required.

ANSWER: 1.98 nF.

15.2 A series RL high-pass filter has $R = 5\text{ k}\Omega$ and $L = 3.5$ mH. What is ω_c for this filter?

ANSWER: 1.43 Mrad/s.

15.3 A series RL low-pass filter with a cutoff frequency of 2 kHz is needed. Using $R = 5\text{ k}\Omega$, compute (a) L; (b) $|H(j\omega)|$ at 50 kHz; and (c) $\theta(j\omega)$ at 50 kHz.

ANSWER: (a) 0.40 H; (b) 0.04; (c) -87.71°.

15.4 A series RC high-pass filter has $C = 1\ \mu$F. Compute the cutoff frequency for the following values of R: (a) $100\ \Omega$; (b) $5\text{ k}\Omega$; and (c) $30\text{ k}\Omega$.

ANSWER: (a) 10 krad/s; (b) 200 rad/s; (c) 33.33 rad/s.

15.5 Compute the transfer function of a series RC low-pass filter that has a load resistor R_L in parallel with its capacitor.

ANSWER: $H(s) = \dfrac{\dfrac{1}{RC}}{s + \dfrac{1}{KRC}}$, where $K = \dfrac{R_L}{R + R_L}$.

15.4 BANDPASS FILTERS

The next filters we examine are those which pass voltages within a band of frequencies to the output while filtering out voltages at frequencies outside this band. These filters are somewhat more complicated than the low-pass and high-pass filters of the previous sections. As we have already seen in Fig. 15.3(c), ideal bandpass filters have two cutoff frequencies, ω_{c1} and ω_{c2}, which identify the passband. For realistic bandpass filters, these cutoff frequencies are again defined as the frequencies for which the magnitude of the transfer function equals $(1/\sqrt{2})H_{\text{max}}$.

CENTER FREQUENCY, BANDWIDTH, AND QUALITY FACTOR

There are three other important parameters which characterize a bandpass filter. The first is the **center frequency**, ω_o, defined as the frequency for which a circuit's transfer function is purely real. Another name for the center frequency is the **resonant frequency**. This is the same name given to the frequency which characterizes the natural response of the second-order circuits in Chapter 7, because they are the same frequencies! When a circuit is driven at the resonant frequency, we say that the circuit is *in resonance,* since the frequency of the forcing function is the same as the natural frequency of the circuit. The center frequency is the geometric center of the passband, that is, $\omega_o = \sqrt{\omega_{c1}\omega_{c2}}$. For bandpass filters, the magnitude of the transfer function is a maximum at the center frequency ($H_{\text{max}} = |H(j\omega_o)|$).

The second parameter is the **bandwidth**, β, which is the width of the passband. The final parameter is the **quality factor**, which is the ratio of the center frequency to the bandwidth. The quality factor gives a measure of the width of the passband, independent of its location on the frequency axis. It also describes the shape of the magnitude plot, independent of frequency.

Although there are five different parameters which characterize the bandpass filter—ω_{c1}, ω_{c2}, ω_o, β, and Q—only two of the five can be specified independently. In other words, once we are able to solve for any two of these parameters, the other three can be calculated from the dependent relationships among them. We will define these quantities more specifically once we have analyzed a bandpass filter. In the next section, we examine two RLC circuits which act as bandpass filters, and then we derive expressions for all of their characteristic parameters.

The Series *RLC* Circuit—Qualitative Analysis

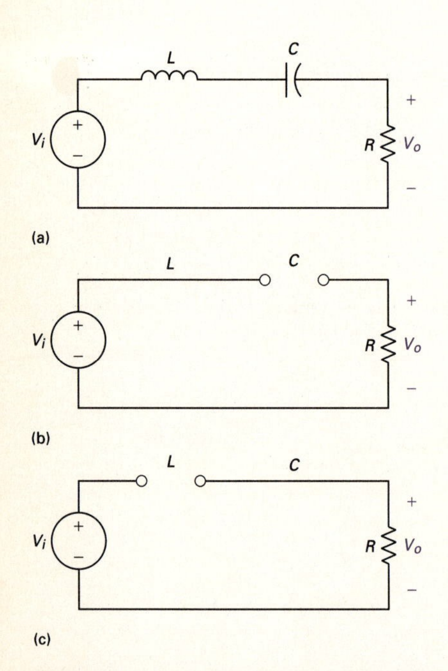

FIGURE 15.19 (a) A series *RLC* bandpass filter; (b) the equivalent circuit for $\omega = 0$; and (c) the equivalent circuit for $\omega = \infty$.

Figure 15.19(a) depicts a series RLC circuit. We want to consider the effect of changing the source frequency on the magnitude of the output voltage. As before, changes to the source frequency result in changes to the impedance of the capacitor and the inductor. This time, the qualitative analysis is somewhat more complicated, because the circuit has both an inductor and a capacitor.

At $\omega = 0$, the capacitor behaves like an open circuit, and the inductor behaves like a short circuit. The equivalent circuit is shown in Fig. 15.19(b). The open circuit representing the impedance of the capacitor prevents current from reaching the resistor, and the resulting output voltage is zero.

At $\omega = \infty$, the capacitor behaves like a short circuit, and the inductor behaves like an open circuit. The equivalent circuit is shown in Fig. 15.19(c). The inductor now prevents current from reaching the resistor, and again the output voltage is zero.

But what happens in the frequency region between $\omega = 0$ and $\omega = \infty$? Between these two extremes, both the capacitor and the inductor have finite impedances. In this region, voltage supplied by the source will drop across both the inductor and the capacitor, but some voltage will reach the resistor. Remember that the impedance of the capacitor is negative, while the impedance of the inductor is positive. Thus, at some frequency, the impedance of the capacitor and the impedance of the inductor have equal magnitudes and opposite signs—the two impedances cancel out, causing the output voltage to equal the source voltage. This special frequency is the center frequency, ω_o. On either side of ω_o, the output voltage is less than the source voltage. Note that at ω_o, the series combination of the inductor and capacitor appears as a short circuit.

The plot of the voltage magnitude ratio is shown in Fig. 15.20. Note that the ideal bandpass filter magnitude plot is overlaid on the plot of the series RLC transfer function magnitude.

Now consider what happens to the phase angle of the output voltage. At the frequency where the source and output voltage are the same, the phase angles are the same. As the frequency decreases, the phase angle contribution from the inductor is larger than that from the capacitor. Since the inductor contributes positive phase shift, the net phase angle at the output is positive. At very low frequencies, the phase angle at the output maximizes at $+90°$.

Conversely, if the frequency increases from the frequency at which the source and the output voltage are in phase, the phase angle contribution from the capacitor is larger than that from the inductor. The capacitor contributes negative phase shift, so the net phase angle at the output is negative. At very high frequencies, the phase angle at the output reaches its negative maximum of $-90°$. The plot of the phase angle difference thus has the shape shown in Fig. 15.20.

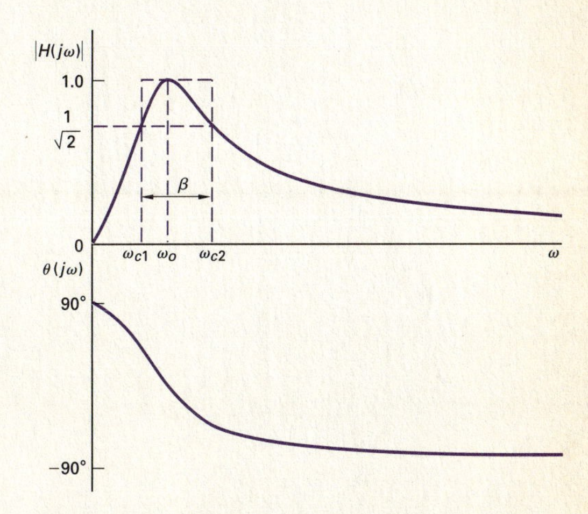

FIGURE 15.20 The frequency response plot for the series *RLC* bandpass filter circuit in Fig. 15.19.

The Series *RLC* Circuit—Quantitative Analysis

We begin by drawing the s-domain equivalent for the series RLC circuit, as shown in Fig. 15.21. Use s-domain voltage division to write an equation for the transfer function:

$$H(s) = \frac{\left(\frac{R}{L}\right)s}{s^2 + \left(\frac{R}{L}\right)s + \frac{1}{LC}}. \tag{15.21}$$

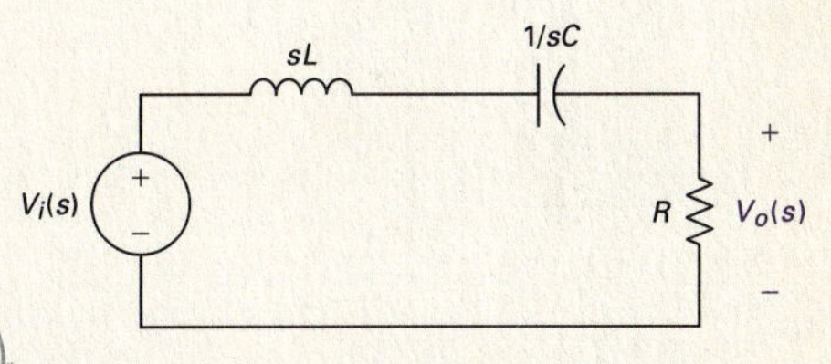

FIGURE 15.21 The s-domain equivalent for the circuit in Fig. 15.19.

As before, we substitute $s = j\omega$ into Eq. (15.21) and produce the equations for the magnitude and the phase angle of the transfer function:

$$|H(j\omega)| = \frac{\omega \frac{R}{L}}{\sqrt{\left(\frac{1}{LC} - \omega^2\right)^2 + \left(\omega \frac{R}{L}\right)^2}}; \tag{15.22}$$

$$\theta(j\omega) = 90° - \tan^{-1}\left(\frac{\omega \frac{R}{L}}{\frac{1}{LC} - \omega^2}\right). \tag{15.23}$$

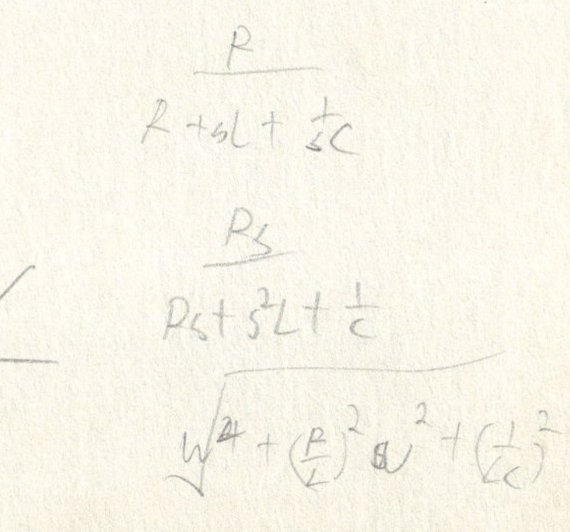

We now calculate the five parameters that characterize this RLC bandpass filter. Recall that the center frequency, ω_o, is defined as the frequency for which the circuit's transfer function is purely real. The transfer function for the RLC circuit in Fig. 15.19 will be real when the frequency of the voltage source makes the sum of the capacitor and inductor impedances zero:

$$j\omega_o L + \frac{1}{j\omega_o C} = 0. \tag{15.24}$$

Solving Eq. (15.24) for ω_o, we get

$$\omega_o = \sqrt{\frac{1}{LC}} \tag{15.25}$$

Next, calculate the cutoff frequencies, ω_{c1} and ω_{c2}. Remember that at the cutoff frequencies, the magnitude of the transfer function is $(1/\sqrt{2})H_{\max}$. Since $H_{\max} = |H(j\omega_o)|$, we can calculate $H_{\max}$ by substituting Eq. (15.25) into Eq. (15.22):

$$\begin{aligned} H_{\max} &= |H(j\omega_o)| \\ &= \frac{\omega_o \frac{R}{L}}{\sqrt{\left(\frac{1}{LC} - \omega_o^2\right)^2 + \left(\frac{\omega_o R}{L}\right)^2}} \\ &= \frac{\sqrt{\frac{1}{LC}}\frac{R}{L}}{\sqrt{\left(\frac{1}{LC} - \frac{1}{LC}\right)^2 + \left(\sqrt{\frac{1}{LC}}\frac{R}{L}\right)^2}} = 1. \end{aligned}$$

Now set the left-hand side of Eq. (15.22) to $(1/\sqrt{2})H_{\max}$ (which equals $1/\sqrt{2}$) and prepare to solve for ω_c:

$$\begin{aligned} \frac{1}{\sqrt{2}} &= \frac{\omega_c \frac{R}{L}}{\sqrt{\left(\frac{1}{LC} - \omega_c^2\right)^2 + \left(\omega_c \frac{R}{L}\right)^2}} \\ &= \frac{1}{\sqrt{\left(\omega_c \frac{L}{R} - \frac{1}{\omega_c RC}\right)^2 + 1}}. \end{aligned} \tag{15.26}$$

We can equate the denominators of the two sides of Eq. (15.26) to get

$$\pm 1 = \omega_c \frac{L}{R} - \frac{1}{\omega_c RC}. \tag{15.27}$$

Rearranging Eq. (15.27) results in the following quadratic equation:

$$\omega_c^2 L \pm \omega_c R - 1/C = 0. \tag{15.28}$$

The solution of Eq. (15.28) yields four values for the cutoff frequency. Only two of these values are positive and have physical significance; they identify the passband of this filter:

$$\omega_{c1} = -\frac{R}{2L} + \sqrt{\left(\frac{R}{2L}\right)^2 + \left(\frac{1}{LC}\right)}; \tag{15.29}$$

$$\omega_{c2} = \frac{R}{2L} + \sqrt{\left(\frac{R}{2L}\right)^2 + \left(\frac{1}{LC}\right)}. \tag{15.30}$$

We can use Eqs. (15.29) and (15.30) to confirm that the center frequency, ω_o, is the geometric mean of the two cutoff frequencies:

$$\begin{aligned} \omega_o &= \sqrt{\omega_{c1} \cdot \omega_{c2}} \\ &= \sqrt{\left[-\frac{R}{2L} + \sqrt{\left(\frac{R}{2L}\right)^2 + \left(\frac{1}{LC}\right)}\right]\left[\frac{R}{2L} + \sqrt{\left(\frac{R}{2L}\right)^2 + \left(\frac{1}{LC}\right)}\right]} \\ &= \sqrt{\frac{1}{LC}}. \end{aligned} \tag{15.31}$$

Recall that the bandwidth of a bandpass filter is defined as the difference between the two cutoff frequencies. Since $\omega_{c2} > \omega_{c1}$, we can compute the bandwidth by subtracting Eq. (15.29) from Eq. (15.30):

$$\begin{aligned} \beta &= \omega_{c2} - \omega_{c1} \\ &= \left[\frac{R}{2L} + \sqrt{\left(\frac{R}{2L}\right)^2 + \left(\frac{1}{LC}\right)}\right] - \left[-\frac{R}{2L} + \sqrt{\left(\frac{R}{2L}\right)^2 + \left(\frac{1}{LC}\right)}\right] \\ &= \frac{R}{L}. \end{aligned} \tag{15.32}$$

The quality factor, the last of the five characteristic parameters, is defined as the ratio of center frequency to bandwidth. Using Eqs. (15.25) and (15.32);

$$\begin{aligned} Q &= \omega_o/\beta \\ &= \frac{\sqrt{\frac{1}{LC}}}{\frac{R}{L}} \\ &= \sqrt{\frac{L}{CR^2}}. \end{aligned} \tag{15.33}$$

We now have five parameters which characterize the series RLC bandpass filter: two cutoff frequencies, ω_{c1} and ω_{c2}, which delimit the passband; the center frequency, ω_o, at which the magnitude of the transfer function is maximum; the bandwidth, β, a measure of the width of the passband; and the quality factor, Q, a second measure of passband width. As previously noted, only two of these parameters can be specified independently in a design. We have already observed that the quality factor is specified in terms of the center frequency and the bandwidth. We can also rewrite the equations for the cutoff frequencies in terms of the center frequency and the bandwidth:

$$\omega_{c1} = -\frac{\beta}{2} + \sqrt{\left(\frac{\beta}{2}\right)^2 + \omega_o^2}; \tag{15.34}$$

$$\omega_{c2} = \frac{\beta}{2} + \sqrt{\left(\frac{\beta}{2}\right)^2 + \omega_o^2}. \tag{15.35}$$

Alternative forms for these equations express the cutoff frequencies in terms of the quality factor and the center frequency:

$$\omega_{c1} = \omega_o \cdot \left[-\frac{1}{2Q} + \sqrt{1 + \left(\frac{1}{2Q}\right)^2}\right]; \tag{15.36}$$

$$\omega_{c2} = \omega_o \cdot \left[\frac{1}{2Q} + \sqrt{1 + \left(\frac{1}{2Q}\right)^2}\right]. \tag{15.37}$$

Also see Problem 15.11 at the end of the chapter.

The examples which follow illustrate the design of bandpass filters, introduce another RLC circuit that behaves as a bandpass filter, and examine the effects of source resistance on the characteristic parameters of a series RLC bandpass filter.

EXAMPLE 15.5

A graphic equalizer is an audio amplifier which allows you to select different levels of amplification within different frequency regions. Using the series RLC circuit in Fig. 15.19(a), choose values for R, L, and C that yield a bandpass circuit able to select inputs within the 1 kHz to 10 kHz frequency band. Such a circuit might be used in a graphic equalizer to select this frequency band from the larger audio band (generally 0 to 20 kHz) prior to amplification.

SOLUTION

We need to compute values for R, L, and C which produce a bandpass filter with cutoff frequencies of 1 kHz and 10 kHz. There are many possible approaches to a solution. For instance, we could use Eqs. (15.29) and (15.30), which specify ω_{c1} and ω_{c2} in terms of R, L, and C. Because of the form of these equations, the algebraic manipulations might get complicated. Instead, we will use the fact that the center frequency is the geometric mean of the cutoff frequencies to compute ω_o, and we will then use Eq. (15.31) to compute L and C from ω_o. Next we will use the definition of *quality factor* to compute Q, and last we will use Eq. (15.33) to compute R. While this approach involves more individual computational steps, each calculation is fairly simple.

Any approach we choose will provide only two equations—insufficient to solve for the three unknowns—because of the dependencies among the bandpass filter characteristics. Thus, we need to select a value for either R, L, or C and use the two equations we've chosen to calculate the remaining component values. Here, we choose 1 μF as the capacitor value, because there are stricter limitations on commercially available capacitors than on inductors or resistors.

We compute the center frequency as the geometric mean of the cutoff frequencies:

$$\begin{aligned} f_o &= \sqrt{f_{c1} f_{c2}} \\ &= \sqrt{(1000)(10{,}000)} \\ &= 3162.28 \text{ Hz}. \end{aligned}$$

Next, compute the value of L using the computed center frequency and the selected value for C. We must remember to convert the center frequency to radians per second before we can use Eq. (15.31).

$$\begin{aligned} L &= \frac{1}{\omega_o^2 C} \\ &= \frac{1}{[2\pi(3162.28)]^2(10^{-6})} \\ &= 2.53 \text{ mH}. \end{aligned}$$

The quality factor, Q, is defined as the ratio of the center frequency to the bandwidth. The bandwidth is the difference between the two cutoff frequency values. Thus,

$$\begin{aligned} Q &= \frac{f_o}{f_{c2} - f_{c1}} \\ &= \frac{3162.28}{10{,}000 - 1000} \\ &= 0.3514. \end{aligned}$$

Now use Eq. (15.33) to calculate R:

$$R = \sqrt{\frac{L}{CQ^2}} = \sqrt{\frac{0.0025}{(10^{-6})(0.3514)^2}} = 143.24\ \Omega.$$

To check whether these component values produce the bandpass filter we want, substitute them into Eqs. (15.29) and (15.30). We find that

$$\omega_{c1} = 6283.19 \text{ rad/s} = 1000 \text{ Hz}$$

and that

$$\omega_{c2} = 62{,}831.85 \text{ rad/s} = 10{,}000 \text{ Hz},$$

which are the cutoff frequencies specified for the filter.

This example reminds us that only two of the five bandpass filter parameters can be specified independently. The other three parameters can always be computed from the two that are specified. In turn, these five parameter values depend on the three component values, R, L, and C, of which only two can be specified independently.

EXAMPLE 15.6

a) Show that the RLC circuit in Fig. 15.22 is also a bandpass filter by deriving an expression for the transfer function $H(s)$.

b) Compute the center frequency, ω_o.

c) Calculate the cutoff frequencies, ω_{c1} and ω_{c2}, the bandwidth, β, and the quality factor, Q.

d) Compute values for R and L to yield a bandpass filter with a center frequency of 5 kHz and a bandwidth of 200 Hz, using a 5 μF capacitor.

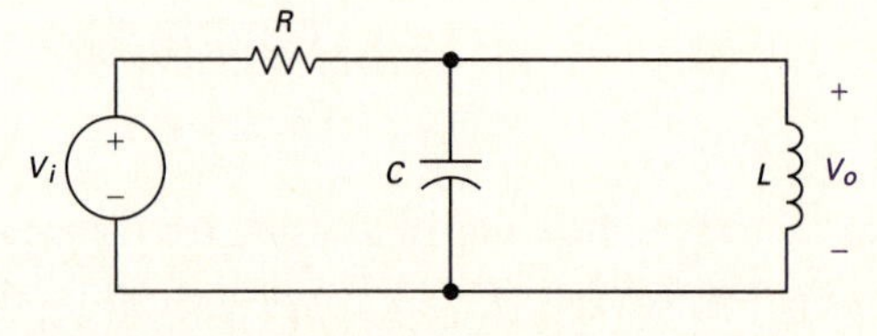

FIGURE 15.22 The circuit for Example 15.6.

SOLUTION

a) Begin by drawing the s-domain equivalent of the circuit in Fig. 15.22, as shown in Fig. 15.23. Using voltage division, we can compute the transfer function for the equivalent circuit if we first compute the equivalent impedance of the parallel combination of L and C, identified as $Z_{eq}(s)$ in Fig. 15.23:

$$Z_{eq}(s) = \frac{\frac{L}{C}}{sL + \frac{1}{sC}}.$$

Now,

$$H(s) = \frac{\frac{s}{RC}}{s^2 + \frac{s}{RC} + \frac{1}{LC}}.$$

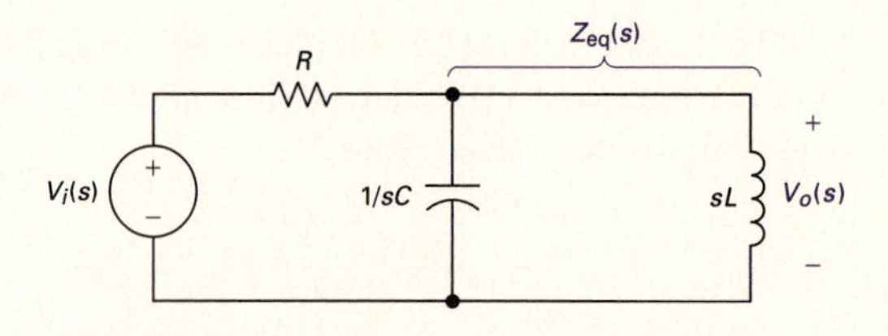

FIGURE 15.23 The s-domain equivalent of the circuit in Fig. 15.22.

b) To find the center frequency, ω_o, we need to calculate where the transfer function magnitude is maximum. Substituting $s = j\omega$ in $H(s)$,

$$|H(j\omega)| = \frac{\frac{\omega}{RC}}{\sqrt{\left(\frac{1}{LC} - \omega^2\right)^2 + \left(\frac{\omega}{RC}\right)^2}}$$

$$= \frac{1}{\sqrt{1 + \left(\omega RC - \frac{1}{\omega \frac{L}{R}}\right)^2}}.$$

The magnitude of this transfer function is maximum when the term

$$\left(\frac{1}{LC} - \omega^2\right)^2$$

is zero. Thus,

$$\omega_o = \sqrt{\frac{1}{LC}}$$

and

$$H_{\max} = |H(j\omega_o)| = 1.$$

c) At the cutoff frequencies, the magnitude of the transfer function is $(1/\sqrt{2})H_{\max} = 1/\sqrt{2}$. Substituting this constant on the left-hand side of the magnitude equation and then simplifying, we get

$$\left[\omega_c RC - \frac{1}{\omega_c \frac{L}{R}}\right] = \pm 1.$$

Squaring the left-hand side of this equation once again produces two quadratic equations for the cutoff frequencies, with four solutions. Only two of them are positive and therefore have physical significance:

$$\omega_{c1} = -\frac{1}{2RC} + \sqrt{\left(\frac{1}{2RC}\right)^2 + \frac{1}{LC}};$$

$$\omega_{c2} = \frac{1}{2RC} + \sqrt{\left(\frac{1}{2RC}\right)^2 + \frac{1}{LC}}.$$

We compute the bandwidth from the cutoff frequencies:

$$\beta = \omega_{c2} - \omega_{c1}$$
$$= \frac{1}{RC}.$$

Finally, use the definition of *quality factor* to calculate Q:

$$Q = \omega_o/\beta$$

$$= \sqrt{\frac{R^2 C}{L}}.$$

Notice that once again we can specify the cutoff frequencies for this bandpass filter in terms of its center frequency and bandwidth:

$$\omega_{c1} = -\frac{\beta}{2} + \sqrt{\left(\frac{\beta}{2}\right)^2 + \omega_o^2};$$

$$\omega_{c2} = \frac{\beta}{2} + \sqrt{\left(\frac{\beta}{2}\right)^2 + \omega_o^2}.$$

d) Use the equation for bandwidth in part (c) to compute a value for R, given a capacitance of 5 μF. Remember to convert the bandwidth to the appropriate units:

$$R = \frac{1}{\beta C}$$

$$= \frac{1}{(2\pi)(200)(5 \times 10^{-6})}$$

$$= 159.15\ \Omega.$$

Using the value of capacitance and the equation for center frequency in part (c), compute the inductor value:

$$L = \frac{1}{\omega_o^2 C}$$

$$= \frac{1}{[2\pi(5000)]^2(5 \times 10^{-6})}$$

$$= 202.64\ \mu\text{H}.$$

EXAMPLE 15.7

For each of the bandpass filters we have constructed, we have always assumed an ideal voltage source, that is, a voltage source with no series resistance. While this assumption is often valid, sometimes it is not, as in the case where the filter design can be achieved only with values of R, L, and C whose equivalent impedance has a magnitude close to the actual impedance of the voltage source. Examine the effect of assuming a nonzero source resistance, R_i, on the characteristics of a series RLC bandpass filter.

a) Determine the transfer function for the circuit in Fig. 15.24.

b) Sketch the magnitude plot for the circuit in Fig. 15.24, using the values for R, L, and C from Example 15.5 and setting $R_i = R$. On the same graph, sketch the magnitude plot for the circuit in Example 15.5, where $R_i = 0$.

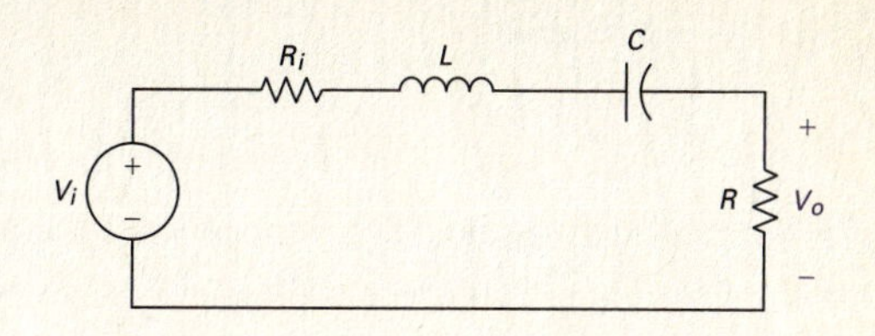

FIGURE 15.24 The circuit for Example 15.7.

SOLUTION

a) Begin by transforming the circuit in Fig. 15.24 to its s-domain equivalent, as shown in Fig. 15.25. Now use voltage division to construct the transfer function:

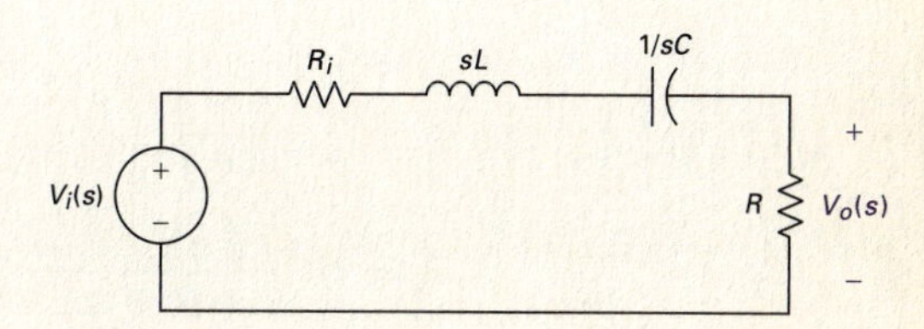

FIGURE 15.25 The s-domain equivalent of the circuit in Fig. 15.24.

$$H(s) = \frac{\frac{R}{L}s}{s^2 + \left(\frac{R + R_i}{L}\right)s + \frac{1}{LC}}.$$

Substitute $s = j\omega$ and calculate the transfer function magnitude:

$$|H(j\omega)| = \frac{\frac{R}{L}\omega}{\sqrt{\left(\frac{1}{LC} - \omega^2\right)^2 + \left(\omega\frac{R + R_i}{L}\right)^2}}.$$

The center frequency, ω_o, is the frequency at which this transfer function magnitude is maximum, which is

$$\omega_o = \sqrt{\frac{1}{LC}}.$$

At the center frequency, the maximum magnitude is

$$H_{\max} = |H(j\omega_o)| = \frac{R}{R_i + R}.$$

The cutoff frequencies can be computed by setting the transfer function magnitude equal to $(1/\sqrt{2})H_{\max}$:

$$\omega_{c1} = -\frac{R + R_i}{2L} + \sqrt{\left(\frac{R + R_i}{2L}\right)^2 + \frac{1}{LC}};$$

$$\omega_{c2} = \frac{R + R_i}{2L} + \sqrt{\left(\frac{R + R_i}{2L}\right)^2 + \frac{1}{LC}}.$$

The bandwidth is calculated from the cutoff frequencies:

$$\beta = \frac{R + R_i}{L}.$$

Finally, the quality factor is computed from the center frequency and the bandwidth:

$$Q = \frac{\sqrt{L/C}}{R + R_i}.$$

From this analysis, note that we can write the transfer function of the series RLC bandpass filter with nonzero source resistance as

$$H(s) = \frac{K\beta s}{s^2 + \beta s + \omega_o^2},$$

where

$$K = \frac{R}{R + R_i}.$$

Note that when $R_i = 0$, $K = 1$ and the transfer function is

$$H(s) = \frac{\beta s}{s^2 + \beta s + \omega_o^2}$$

b) A series RLC circuit has a center frequency of 1 kHz and a bandwidth of 20 Hz, and $H_{\text{max}} = 1$. If we use the same values for R, L, and C in the circuit in Fig. 15.24 and let $R_i = R$, then the center frequency remains at 1 kHz, but $\beta = (R + R_i)/L = 40$ Hz, and $H_{\text{max}} = R/(R + R_i) = 1/2$. The transfer function magnitudes for these two bandpass filters are plotted on the same graph in Fig. 15.26.

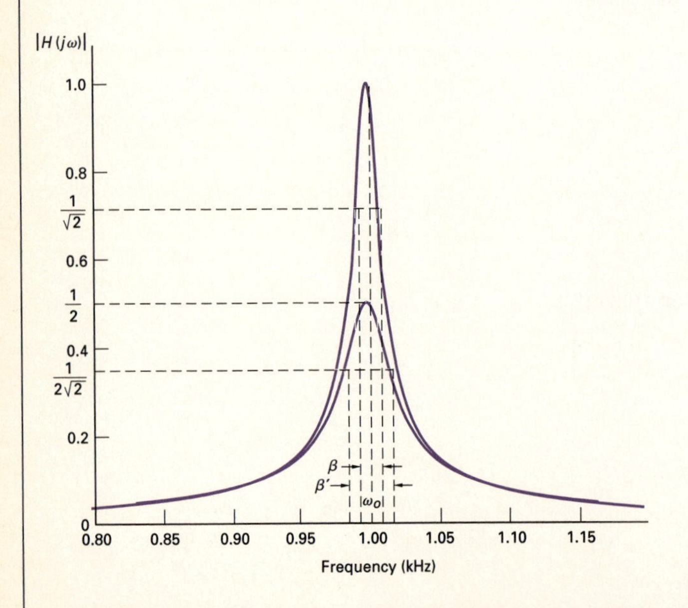

FIGURE 15.26 The magnitude plots for a series *RLC* bandpass filter with a zero source resistance and a nonzero source resistance.

If we compare the characteristic parameter values for the filter with $R_i = 0$ to the values for the filter with $R_i \neq 0$, we see the following:

- The center frequencies are the same.
- The maximum transfer function magnitude for the filter with $R_i \neq 0$ is smaller than for the filter with $R_i = 0$.
- The bandwidth for the filter with $R_i \neq 0$ is larger than that for the filter with $R_i = 0$. Thus, the cutoff frequencies and the quality factors for the two circuits are also different.

The addition of a nonzero source resistance to a series RLC bandpass filter leaves the center frequency unchanged but widens the passband and reduces the passband magnitude.

Here we see the same design challenge we saw with the addition of a load resistor to the high-pass filter—that is, we would like to design a bandpass filter which will have the same filtering properties regardless of any internal resistance associated with the voltage source. Unfortunately, filters constructed from passive elements have their filtering action altered with the addition of source resistance. In Chapter 16, we will discover that active filters are insensitive to changes in source resistance and thus are better suited to designs in which this is an important issue.

Figure 15.27 summarizes the two RLC bandpass filters we have studied. Note that the expressions for the circuit transfer functions have the same form. As we have done previously, we can create a general form for the transfer functions of these two bandpass filters:

$$H(s) = \frac{\beta s}{s^2 + \beta s + \omega_o^2}. \tag{15.38}$$

Any circuit with the transfer function in Eq. (15.38) acts as a bandpass filter with a center frequency ω_o and a bandwidth β.

In Example 15.7, we saw that the transfer function can also be written in the form

$$H(s) = \frac{K\beta s}{s^2 + \beta s + \omega_o^2}, \tag{15.39}$$

where the values for K and β depend on whether the series resistance of the voltage source is zero or nonzero.

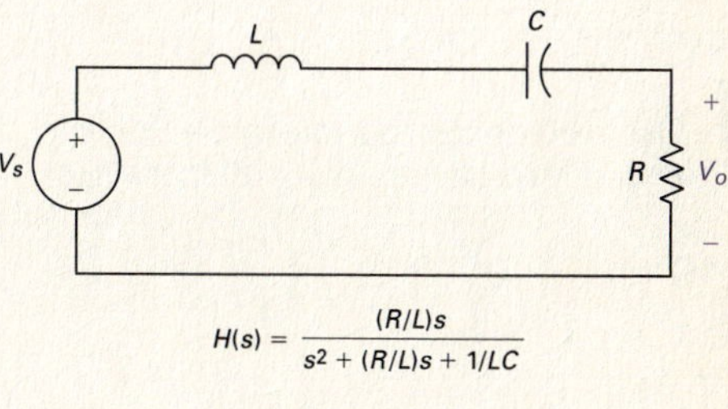

$H(s) = \dfrac{(R/L)s}{s^2 + (R/L)s + 1/LC}$

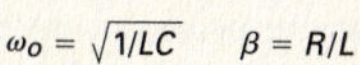

$\omega_o = \sqrt{1/LC} \qquad \beta = R/L$

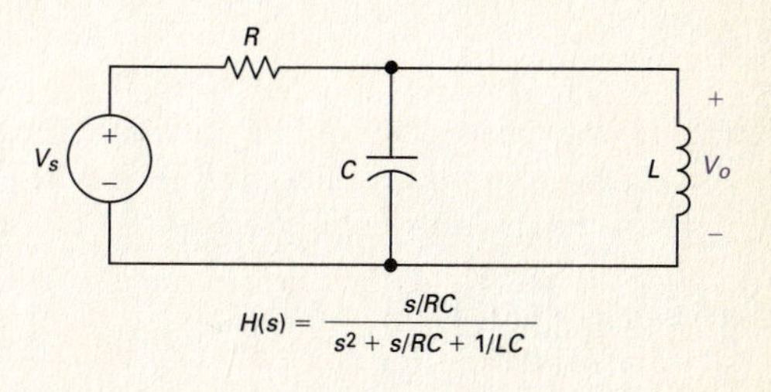

$H(s) = \dfrac{s/RC}{s^2 + s/RC + 1/LC}$

$\omega_o = \sqrt{1/LC} \qquad \beta = 1/RC$

FIGURE 15.27 Two RLC bandpass filters, together with equations for the transfer function, center frequency, and bandwidth of each.

Relating the Frequency Domain to the Time Domain

We can identify a relationship between the parameters which characterize the frequency response of RLC bandpass filters and the parameters which characterize the time response of RLC circuits. Consider the series RLC circuit in Fig. 15.19(a). In Chapter 8 we discovered that the natural response of this circuit is characterized by the neper frequency (α) and the resonant frequency (ω_o). These parameters were expressed in terms of the circuit components in Eqs. (8.58) and (8.59), which are repeated here for convenience:

$$\alpha = \frac{R}{2L} \text{ rad/s;} \quad (15.40)$$

$$\omega_o = \sqrt{\frac{1}{LC}} \text{ rad/s.} \quad (15.41)$$

We see that the same parameter ω_o is used to characterize both the time response and the frequency response. That's why the center frequency is also called the resonant frequency. The bandwidth and the neper frequency are related by the equation

$$\beta = 2\alpha. \quad (15.42)$$

Recall that the natural response of a series RLC circuit may be underdamped, overdamped, or critically damped. The transition from overdamped to underdamped occurs when $\omega_o^2 = \alpha^2$. Consider the relationship between α and β from Eq. (15.42) and the definition of the quality factor Q. The transition from an overdamped to an underdamped response occurs when $Q = 1/2$. Thus, a circuit whose frequency response contains a sharp peak at ω_o, indicating a high Q and a narrow bandwidth, will have an underdamped natural response. Conversely, a circuit whose frequency response has a broad bandwidth and a low Q will have an overdamped natural response.

Use a computer tool to explore the time and frequency responses of a bandpass filter.

DRILL EXERCISES

15.6 Using the circuit in Fig. 15.19(a), compute the values of R and L to give a bandpass filter with a center frequency of 12 kHz and a quality factor of 6. Use a 0.1 μF capacitor.

ANSWER: $L = 1.76$ mH, $R = 22.10\ \Omega$.

15.7 Using the circuit in Fig. 15.22, compute the values of L and C to give a bandpass filter with a center frequency of 2 kHz and a bandwidth of 500 Hz. Use a 250 Ω resistor.

ANSWER: $L = 4.97$ mH, $C = 1.27\ \mu$F.

15.8 Recalculate the component values for the circuit in Example 15.6(d) so that the frequency response of the resulting circuit is unchanged using a 0.2 μF capacitor.

ANSWER: $L = 5.07$ mH, $R = 3.98$ kΩ.

15.9 Recalculate the component values for the circuit in Example 15.6(d) so that the quality factor of the resulting circuit is unchanged but the center frequency has been moved to 2 kHz. Use a 0.2 μF capacitor.

ANSWER: $R = 9.95$ kΩ, $L = 31.66$ mH.

15.5 Bandreject Filters

We turn now to the last of the four filter categories—the bandreject filter. This filter passes source voltages outside the band between the two cutoff frequencies to the output (which is the passband), and attenuates source voltages before they reach the output at frequencies between the two cutoff frequencies (which is the stopband). Bandpass filters and bandreject filters thus perform complementary functions in the frequency domain.

Bandreject filters are characterized by the same parameters as bandpass filters: the two cutoff frequencies, the center frequency, the bandwidth, and the quality factor. Again, only two of these five parameters can be specified independently.

In the next sections, we examine two circuits which function as bandreject filters and then compute equations which relate the circuit component values to the characteristic parameters for each circuit.

The Series *RLC* Circuit—Qualitative Analysis

Figure 15.28(a) shows a series RLC circuit. While the circuit components and connections are identical to those in the series RLC bandpass filter in Fig. 15.19(a), the circuit in Fig. 15.28(a) has an important difference—the output voltage is now defined across the inductor-capacitor pair. As we saw in the case of low- and high-pass filters, the same circuit may perform two different filtering functions, depending on the definition of the output voltage.

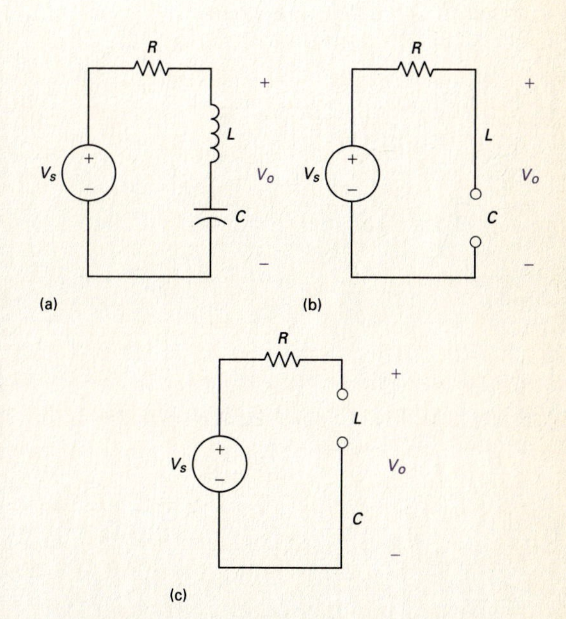

FIGURE 15.28 (a) A series RLC bandreject filter; (b) the equivalent circuit for $\omega = 0$; and (c) the equivalent circuit for $\omega = \infty$.

We have already noted that at $\omega = 0$, the inductor behaves like a short circuit and the capacitor behaves like an open circuit, while at $\omega = \infty$, these roles switch. Figure 15.28(b) presents the equivalent circuit for $\omega = 0$; Fig. 15.28(c) presents the equivalent circuit for $\omega = \infty$. In both equivalent circuits, the output voltage is defined over an effective open circuit, and thus the output and input voltages have the same magnitude. This series RLC bandreject filter circuit then has two passbands—one below a lower cutoff frequency, and the other above an upper cutoff frequency.

In between these two passbands, both the inductor and the capacitor have finite impedances of opposite signs. As the frequency is increased above the low-frequency passband, the impedance of the inductor increases and that of the capacitor decreases. The phase shift from the input to the output must therefore be increasing from its value of -90° in the low-frequency passband. As the frequency is decreased below the high-frequency passband, the impedance of the inductor decreases and that of the capacitor increases. The phase shift from the input to the output must therefore be decreasing from its value of $+90^\circ$ in the high-frequency passband.

At some frequency between the two passbands, the impedances of the inductor and capacitor are equal but of opposite sign. At this frequency, the series combination of the inductor and capacitor is that of a short circuit, so the magnitude of the output voltage must be zero. This is the center frequency of this series RLC bandreject filter.

Figure 15.29 presents a sketch of the frequency response of the series RLC bandreject filter from Fig. 15.28(a). Note that the magnitude plot is overlaid with that of the ideal bandreject filter from Fig. 15.3(d). Our qualitative analysis has confirmed the shape of the magnitude and phase angle plots. We now turn to a quantitative analysis of the circuit to confirm this frequency response and to compute values for the parameters which characterize this response.

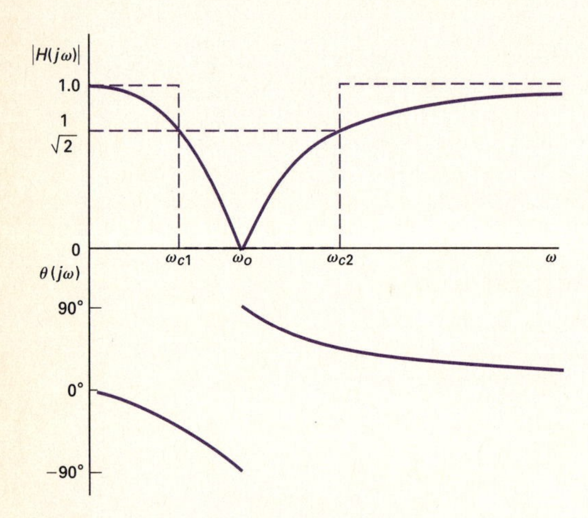

FIGURE 15.29 The frequency response plot for the series RLC bandreject filter circuit in Fig. 15.28(a).

THE SERIES RLC CIRCUIT—QUANTITATIVE ANALYSIS

After transforming to the s-domain, as shown in Fig. 15.30, we use voltage division to construct an equation for the transfer function:

$$H(s) = \frac{sL + \frac{1}{sC}}{R + sL + \frac{1}{sC}} = \frac{s^2 + \frac{1}{LC}}{s^2 + \frac{R}{L}s + \frac{1}{LC}}. \tag{15.43}$$

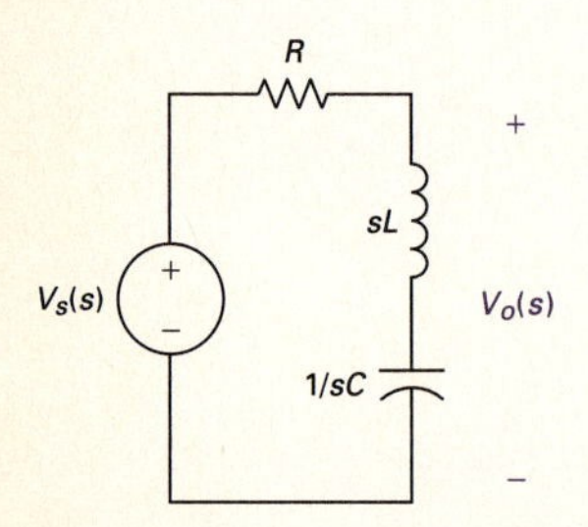

FIGURE 15.30 The s-domain equivalent of the circuit in Fig. 15.28(a).

Substitute $j\omega$ for s in Eq. (15.43) and generate equations for the transfer function magnitude and the phase angle:

$$|H(j\omega)| = \frac{\frac{1}{LC} - \omega^2}{\sqrt{\left(\frac{1}{LC} - \omega^2\right)^2 + \left(\frac{\omega R}{L}\right)^2}}; \tag{15.44}$$

$$\theta(j\omega) = -\tan^{-1}\left(\frac{\frac{\omega R}{L}}{\frac{1}{LC} - \omega^2}\right). \tag{15.45}$$

Note that Eqs. (15.44) and (15.45) confirm the frequency response shape pictured in Fig. 15.29, which we developed based on the qualitative analysis.

We use the circuit in Fig. 15.30 to calculate the center frequency. For the bandreject filter, the center frequency is still defined as the frequency for which the sum of the impedances of the capacitor and inductor is zero. In the bandpass filter, the magnitude at the center frequency was a maximum, but in the bandreject filter, this magnitude is a minimum. This is because in the bandreject filter, the center frequency is not in the passband; rather, it is in the stopband. It is easy to show that the center frequency is given by

$$\omega_o = \sqrt{\frac{1}{LC}}. \tag{15.46}$$

Substituting Eq. (15.46) into Eq. (15.44) shows that $|H(j\omega_o)| = 0$.

The cutoff frequencies, the bandwidth, and the quality factor are defined for the bandreject filter in exactly the way they were for the bandpass filters. Compute the cutoff frequencies by substituting the constant $(1/\sqrt{2})H_{max}$ for the left-hand side of Eq. (15.44) and then solving for ω_{c1} and ω_{c2}. Note that for the bandreject filter, $H_{max} = |H(j0)| = |H(j\infty)|$, and for the series RLC bandreject filter in Fig. 15.28(a), $H_{max} = 1$. Thus,

$$\omega_{c1} = -\frac{R}{2L} + \sqrt{\left(\frac{R}{2L}\right)^2 + \frac{1}{LC}}; \quad \textbf{(15.47)}$$

$$\omega_{c2} = \frac{R}{2L} + \sqrt{\left(\frac{R}{2L}\right)^2 + \frac{1}{LC}}. \quad \textbf{(15.48)}$$

Use the cutoff frequencies to generate an expression for the bandwidth, β:

$$\beta = R/L. \quad \textbf{(15.49)}$$

Finally, the center frequency and the bandwidth produce an equation for the quality factor, Q:

$$Q = \sqrt{\frac{L}{R^2C}}. \quad \textbf{(15.50)}$$

Again, we can represent the expressions for the two cutoff frequencies in terms of the bandwidth and center frequency, as we did for the bandpass filter:

$$\omega_{c1} = -\frac{\beta}{2} + \sqrt{\left(\frac{\beta}{2}\right)^2 + \omega_o^2}; \quad \textbf{(15.51)}$$

$$\omega_{c2} = \frac{\beta}{2} + \sqrt{\left(\frac{\beta}{2}\right)^2 + \omega_o^2}. \quad \textbf{(15.52)}$$

Alternative forms for these equations express the cutoff frequencies in terms of the quality factor and the center frequency:

Use a computer tool to verify the alternative forms for ω_{c1} and ω_{c2}.

$$\omega_{c1} = \omega_o \cdot \left[-\frac{1}{2Q} + \sqrt{1 + \left(\frac{1}{2Q}\right)^2}\right]; \quad \textbf{(15.53)}$$

$$\omega_{c2} = \omega_o \cdot \left[\frac{1}{2Q} + \sqrt{1 + \left(\frac{1}{2Q}\right)^2}\right]. \quad \textbf{(15.54)}$$

The example which follows presents the design of a series RLC bandreject filter.

EXAMPLE 15.8

Using the series RLC circuit in Fig. 15.28(a), compute the component values which yield a bandreject filter with a bandwidth of 250 Hz and a center frequency of 750 Hz. Use a 100 nF capacitor. Compute values for R, L, ω_{c1}, ω_{c2}, and Q.

SOLUTION

We begin by using the definition of quality factor to compute its value for this filter:

$$Q = \omega_o/\beta = 3.$$

Use Eq. (15.46) to compute L, remembering to convert ω_o to radians per second:

$$\begin{aligned} L &= \frac{1}{\omega_o^2 C} \\ &= \frac{1}{[2\pi(750)]^2(100 \times 10^{-9})} \\ &= 450 \text{ mH}. \end{aligned}$$

Use Eq. (15.49) to calculate R:

$$\begin{aligned} R &= \beta L \\ &= 2\pi(250)(450 \times 10^{-3}) \\ &= 707\ \Omega. \end{aligned}$$

The values for the center frequency and bandwidth can be used in Eqs. (15.51) and (15.52) to compute the two cutoff frequencies:

$$\begin{aligned} \omega_{c1} &= -\frac{\beta}{2} + \sqrt{\left(\frac{\beta}{2}\right)^2 + \omega_o^2} \\ &= 3992.0 \text{ rad/s;} \end{aligned}$$

$$\begin{aligned} \omega_{c2} &= \frac{\beta}{2} + \sqrt{\left(\frac{\beta}{2}\right)^2 + \omega_o^2} \\ &= 5562.8 \text{ rad/s.} \end{aligned}$$

The cutoff frequencies are at 635.3 Hz and 885.3 Hz. Their difference is $885.3 - 635.3 = 250$ Hz, confirming the specified bandwidth. The geometric mean is $\sqrt{(635.3)(885.3)} = 750$ Hz, confirming the specified center frequency.

As you might suspect by now, another configuration which produces a bandreject filter is a parallel LC circuit. While the analysis details of the parallel LC circuit are left to Problem 15.23, the results are summarized in Fig. 15.31, along with the series RLC bandreject filter. As we did for other categories of filters, we can state a general form for the transfer functions of bandreject filters, replacing the constant terms with β and ω_o:

$$H(s) = \frac{s^2 + \omega_o^2}{s^2 + \beta s + \omega_o^2}. \quad \textbf{(15.55)}$$

Equation (15.55) is useful in filter design, because any circuit with a transfer function in this form can be used as a bandreject filter.

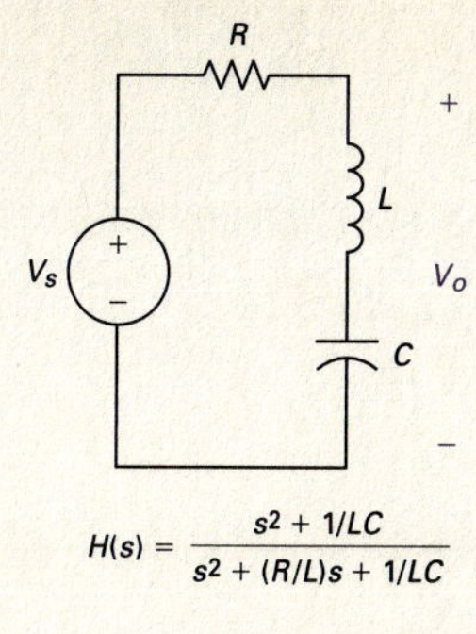

$$H(s) = \frac{s^2 + 1/LC}{s^2 + (R/L)s + 1/LC}$$

$\omega_o = \sqrt{1/LC} \quad \beta = R/L$

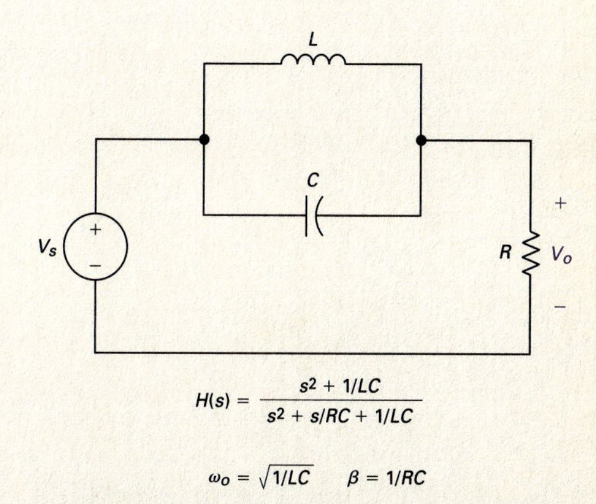

$$H(s) = \frac{s^2 + 1/LC}{s^2 + s/RC + 1/LC}$$

$\omega_o = \sqrt{1/LC} \quad \beta = 1/RC$

FIGURE 15.31 Two RLC bandreject filters, together with equations for the transfer function, center frequency, and bandwidth of each.

DRILL EXERCISES

15.10 Design the component values for the series RLC bandreject filter shown in Fig. 15.28(a) so that the center frequency is 4 kHz and the quality factor is 5. Use a 500 nF capacitor.

ANSWER: $L = 3.17$ mH, $R = 15.92\ \Omega$.

15.11 Recompute the component values for Drill Exercise 15.10 to achieve a bandreject filter with a center frequency of 20 kHz. The filter has a 100 Ω resistor. The quality factor remains at 5.

ANSWER: $L = 3.98$ mH, $C = 15.92$ nF.

15.6 BODE DIAGRAMS

As we have seen, the frequency response plot is a very important tool for analyzing a circuit's behavior. Up to this point, however, we

Computer tools can generate frequency response plots.

have shown qualitative sketches of the frequency response without discussing how to create such diagrams. The most efficient method for generating and plotting the amplitude and phase data is to use a digital computer; we can rely on it to give us accurate numerical plots of $|H(j\omega)|$ and $\theta(j\omega)$ versus ω. However, in some situations, preliminary sketches using Bode diagrams can help ensure the intelligent use of the computer.

A Bode diagram, or plot, is a graphical technique that gives a feel for the frequency response of a circuit. These diagrams are named in recognition of the pioneering work done by H. W. Bode.[1] They are most useful for circuits in which the poles and zeros of $H(s)$ are reasonably well separated.

Like the qualitative frequency response plots seen thus far, a Bode diagram consists of two separate plots: One shows how the amplitude of $H(j\omega)$ varies with frequency, and the other shows how the phase angle of $H(j\omega)$ varies with frequency. In Bode diagrams, the plots are made on semilog graph paper for greater accuracy in representing the wide range of frequency values. In both the amplitude and phase plots, the frequency is plotted on the horizontal log scale, and the amplitude and phase angle are plotted on the linear vertical scale.

Real, First-Order Poles and Zeros

To simplify the development of Bode diagrams, we begin by considering only cases where all the poles and zeros of $H(s)$ are real and first order. Later we will present cases with complex and repeated poles and zeros. For our purposes, having a specific expression for $H(s)$ is helpful. Hence we base the discussion on

$$H(s) = \frac{K(s + z_1)}{s(s + p_1)}, \tag{15.56}$$

from which

$$H(j\omega) = \frac{K(j\omega + z_1)}{j\omega(j\omega + p_1)}. \tag{15.57}$$

The first step in making Bode diagrams is to put the expression for $H(j\omega)$ in a **standard form**, which we derive simply by dividing out the poles and zeros:

$$H(j\omega) = \frac{Kz_1(1 + j\omega/z_1)}{p_1(j\omega)(1 + j\omega/p_1)}. \tag{15.58}$$

Next we let K_o represent the constant quantity Kz_1/p_1, and at

[1] See H. W. Bode, *Network Analysis and Feedback Design* (New York: Van Nostrand, 1945).

the same time we express $H(j\omega)$ in polar form:

$$H(j\omega) = \frac{K_o|1 + j\omega/z_1|\angle\psi_1}{|\omega|\angle 90^\circ |1 + j\omega/p_1|\angle\beta_1}$$

$$= \frac{K_o|1 + j\omega/z_1|}{|\omega||1 + j\omega/p_1|}\angle(\psi_1 - 90^\circ - \beta_1). \quad \textbf{(15.59)}$$

From Eq. (15.59),

$$|H(j\omega)| = \frac{K_o|1 + j\omega/z_1|}{\omega|1 + j\omega/p_1|}; \quad \textbf{(15.60)}$$

$$\theta(\omega) = \psi_1 - 90^\circ - \beta_1. \quad \textbf{(15.61)}$$

By definition, the phase angles ψ_1 and β_1 are

$$\psi_1 = \tan^{-1}\omega/z_1; \quad \textbf{(15.62)}$$

$$\beta_1 = \tan^{-1}\omega/p_1. \quad \textbf{(15.63)}$$

The Bode diagrams consist of plotting Eq. (15.60) (amplitude) and Eq. (15.61) (phase) as functions of ω.

Straight-Line Amplitude Plots

The amplitude plot involves the multiplication and division of factors associated with the poles and zeros of $H(s)$. We reduce this multiplication and division to addition and subtraction by expressing the amplitude of $H(j\omega)$ in terms of a logarithmic value: the decibel (dB).[2] The amplitude of $H(j\omega)$ in decibels is

$$A_{dB} = 20\log_{10}|H(j\omega)|. \quad \textbf{(15.64)}$$

To give you a feel for the unit of decibels, Table 15.2 provides a translation between the actual value of several amplitudes and their values in decibels. Expressing Eq. (15.64) in terms of decibels gives

$$A_{dB} = 20\log_{10}\frac{K_o|1 + j\omega/z_1|}{\omega|1 + j\omega/p_1|}$$

$$= 20\log_{10}K_o + 20\log_{10}|1 + j\omega/z_1|$$

$$- 20\log_{10}\omega - 20\log_{10}|1 + j\omega/p_1|. \quad \textbf{(15.65)}$$

The key to plotting Eq. (15.65) is to plot each term in the equation separately and then combine the separate plots graphically. The individual factors are easy to plot because they can be approximated in all cases by straight lines.

TABLE 15.2

Actual Amplitudes and Their Decibel Values

A_{dB}	A	A_{dB}	A
0	1.00	30	31.62
3	1.41	40	100.00
6	2.00	60	10^3
10	3.16	80	10^4
15	5.62	100	10^5
20	10.00	120	10^6

[2]See Appendix D for more information regarding the decibel.

The plot of $20\log_{10} K_o$ is a horizontal straight line because K_o is not a function of frequency. The value of this term is positive for $K_o > 1$, zero for $K_o = 1$, and negative for $K_o < 1$.

Two straight lines approximate the plot of $20\log_{10}|1 + j\omega/z_1|$. For small values of ω, the magnitude $|1 + j\omega/z_1|$ is approximately 1, and therefore

$$20\log_{10}|1 + j\omega/z_1| \to 0 \quad \text{as} \quad \omega \to 0. \tag{15.66}$$

For large values of ω, the magnitude $|1 + j\omega/z_1|$ is approximately ω/z_1, and therefore

$$20\log_{10}|1 + j\omega/z_1| \to 20\log_1(\omega/z_1) \quad \text{as} \quad \omega \to \infty. \tag{15.67}$$

On a log frequency scale, $20\log_{10}(\omega/z_1)$ is a straight line with a slope of 20 dB/decade (a decade is a 10-to-1 change in frequency). This straight line intersects the 0 dB axis at $\omega = z_1$. This value of ω is called the **corner frequency**. Thus, on the basis of Eqs. (15.66) and (15.67), two straight lines can approximate the amplitude plot of a first-order zero, as shown in Fig. 15.32.

The plot of $-20\log_{10}\omega$ is a straight line having a slope of -20 dB/decade that intersects the 0 dB axis at $\omega = 1$. Two straight lines approximate the plot of $-20\log_{10}|1 + j\omega/p_1|$. Here the two straight lines intersect on the 0 dB axis at $\omega = p_1$. For large values of ω, the straight line $20\log_{10}(\omega/p_1)$ has a slope of -20 dB/decade. Figure 15.33 shows the straight-line approximation of the amplitude plot of a first-order pole.

Figure 15.34 shows a plot of Eq. (15.65) for $K_o = \sqrt{10}$, $z_1 = 0.1$ rad/s, and $p_1 = 5$ rad/s. Each term in Eq. (15.65) is labeled on Fig. 15.34, so you can verify that the individual terms sum to create the resultant plot, labeled $20\log_{10}|H(j\omega)|$.

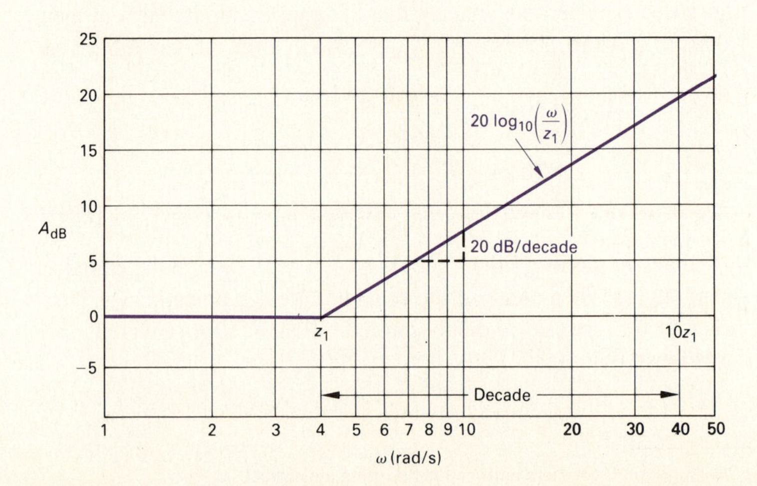

FIGURE 15.32 A straight-line approximation of the amplitude plot of a first-order zero.

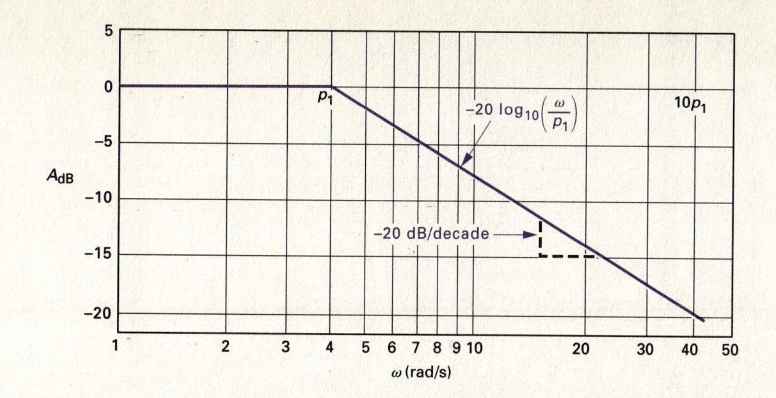

FIGURE 15.33 A straight-line approximation of the amplitude plot of a first-order pole.

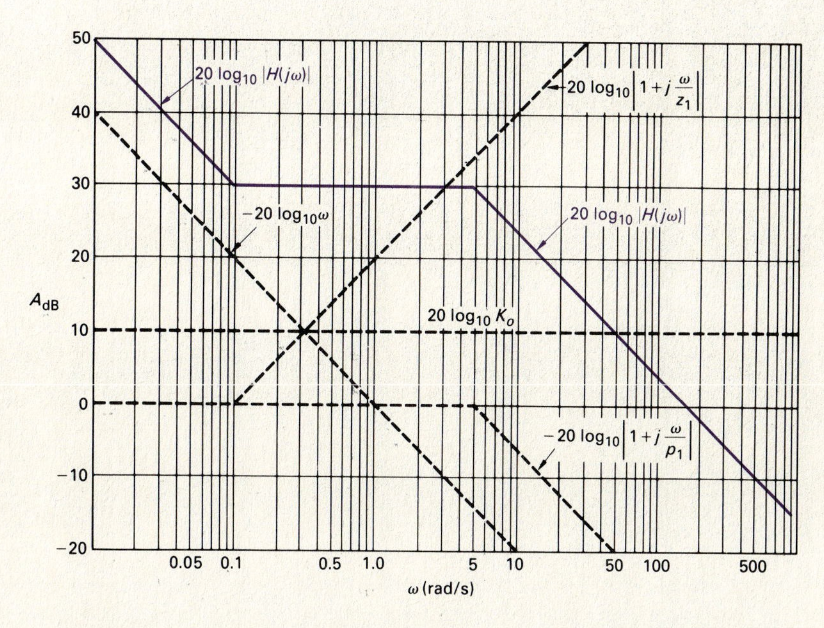

FIGURE 15.34 A straight-line approximation of the amplitude plot for Eq. (15.65).

Example 15.9 illustrates the construction of a straight-line amplitude plot for a transfer function characterized by first-order poles and zeros.

EXAMPLE 15.9

For the circuit in Fig. 15.35,

a) compute the transfer function, $H(s)$;

b) construct a straight-line approximation of the Bode amplitude plot;

c) calculate $20\log_{10}|H(j\omega)|$ at $\omega = 50$ rad/s and $\omega = 1000$ rad/s;

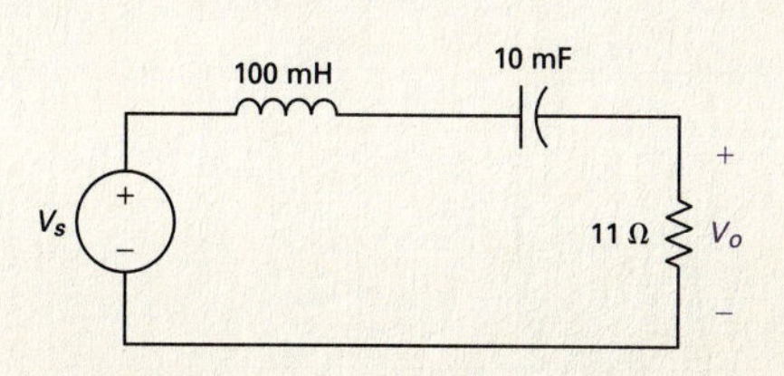

FIGURE 15.35 The circuit for Example 15.9.

d) plot the values computed in (c) on the straight-line graph; and

e) suppose $v_i(t) = 5\cos(500t + 15°)$ V, and then use the Bode plot you constructed to predict the amplitude of $v_o(t)$ in the steady state.

SOLUTION

a) Transforming the circuit in Fig. 15.35 into the s-domain and then using s-domain voltage division gives

$$H(s) = \frac{(R/L)s}{s^2 + (R/L)s + \dfrac{1}{LC}}.$$

Substituting the numerical values from the circuit, we get

$$H(s) = \frac{110s}{s^2 + 110s + 1000} = \frac{110s}{(s+10)(s+100)}.$$

b) We begin by writing $H(j\omega)$ in standard form:

$$H(j\omega) = \frac{0.11j\omega}{[1 + j(\omega/10)][1 + j(\omega/100)]}.$$

The expression for the amplitude of $H(j\omega)$ in decibels is

$$\begin{aligned} A_{\text{dB}} &= 20\log_{10}|H(j\omega)| \\ &= 20\log_{10}0.11 + 20\log_{10}|j\omega| \\ &\quad - 20\log_{10}\left|1 + j\frac{\omega}{10}\right| - 20\log_{10}\left|1 + j\frac{\omega}{100}\right|. \end{aligned}$$

Figure 15.36 shows the straight-line plot. Each term contributing to the overall amplitude is identified.

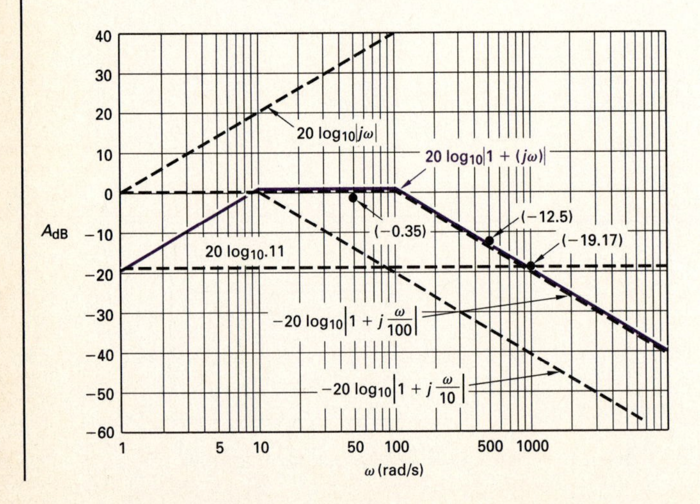

FIGURE 15.36 The straight-line amplitude plot for the transfer function of the circuit in Fig. 15.35.

c) We have

$$H(j50) = \frac{0.11(j50)}{(1 + j5)(1 + j0.5)}$$
$$= 0.9648\underline{/-15.25^\circ};$$

$$20\log_{10}|H(j50)| = 20\log_{10} 0.96$$
$$= -0.3116\text{ dB};$$

$$H(j1000) = \frac{0.11(j1000)}{(1 + j100)(1 + j10)}$$
$$= 0.1094\underline{/-83.72^\circ};$$

$$20\log_{10} 0.11 = -19.22\text{ dB}.$$

d) See Fig. 15.36.

e) As we can see from the Bode plot in Fig. 15.36, the value of A_{dB} at $\omega = 500$ rad/s is approximately -12.5 dB. Therefore,

$$|A| = 10^{(-12.5/20)} = 0.24$$

and

$$V_{mo} = |A|V_{mi} = (0.24)(5) = 1.19\text{ V}.$$

We can compute the actual value of $|H(j\omega)|$ by substituting $\omega = 500$ into the equation for $|H(j\omega)|$:

$$H(j500) = \frac{0.11(j500)}{(1 + j50)(1 + j5)} = 0.22\underline{/-77.54^\circ}.$$

Thus, the actual output voltage magnitude for the specified signal source at a frequency of 500 rad/s is

$$V_{mo} = |A|V_{mi} = (0.22)(5) = 1.1\text{ V}.$$

More Accurate Amplitude Plots

We can make the straight-line plots for first-order poles and zeros more accurate by correcting the amplitude values at the corner frequency, one-half the corner frequency, and twice the corner frequency. At the corner frequency, the actual value in decibels is

$$A_{\text{dB}_c} = \pm 20\log_{10}|1 + j1|$$
$$= \pm 20\log_{10}\sqrt{2}$$
$$\approx \pm 3\text{ dB}. \quad \textbf{(15.68)}$$

The actual value at one-half of the corner frequency is

$$\begin{aligned} A_{\text{dB}_{c/2}} &= \pm 20\log_{10}\left|1 + j\frac{1}{2}\right| \\ &= \pm 20\log_{10}\sqrt{5/4} \\ &\approx \pm 1 \text{ dB}. \end{aligned} \quad \textbf{(15.69)}$$

At twice the corner frequency, the actual value in decibels is

$$\begin{aligned} A_{\text{dB}_{2c}} &= \pm 20\log_{10}|1 + j2| \\ &= \pm 20\log_{10}\sqrt{5} \\ &\approx \pm 7 \text{ dB}. \end{aligned} \quad \textbf{(15.70)}$$

In Eqs. (15.68), (15.69), and (15.70), the plus sign applies to a first-order zero, and the minus sign applies to a first-order pole. The straight-line approximation of the amplitude plot gives 0 dB at the corner and one-half the corner frequencies, and ±6 dB at twice the corner frequency. Hence the corrections are ±3 dB at the corner frequency and ±1 dB at both one-half the corner frequency and twice the corner frequency. Figure 15.37 summarizes these corrections.

A 2-to-1 change in frequency is called an **octave.** A slope of 20 dB/decade is equivalent to 6.02 dB/octave, which for graphical purposes is equivalent to 6 dB/octave. Thus the corrections enumerated correspond to one octave below and one octave above the corner frequency.

If the poles and zeros of $H(s)$ are well separated, inserting these corrections into the overall amplitude plot and achieving a reasonably accurate curve is relatively easy. However, if the poles and zeros are close together, the overlapping corrections are difficult to evaluate, and you're better off using the straight-line plot as a first estimate of the amplitude characteristic. Then use a computer to refine the calculations in the frequency range of interest.

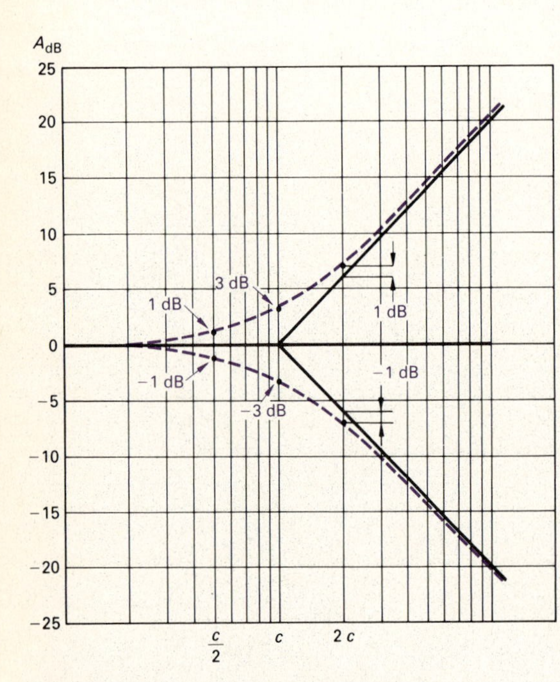

FIGURE 15.37 Corrected amplitude plots for a first-order zero and pole.

STRAIGHT-LINE PHASE ANGLE PLOTS

We can also make phase angle plots by using straight-line approximations. The phase angle associated with the constant K_o is zero, and the phase angle associated with a first-order zero or pole at the origin is a constant ±90°. For a first-order zero or pole not at the origin, the straight-line approximations are as follows. For frequencies less than one-tenth the corner frequency, the phase angle is assumed to be zero. For frequencies greater than 10 times the corner frequency, the phase angle is assumed to be ±90°. Between one-tenth the corner frequency and 10 times the corner frequency, the phase angle plot is a straight line that goes through 0° at one-tenth the corner frequency, ±45° at the corner frequency, and ±90° at 10 times the corner frequency. In all these cases, the plus sign applies to the first-order zero, and the minus sign to the first-order pole. Figure 15.38 depicts the straight-line approximation for a first-order zero and pole. The dashed curves show the exact variation of the phase angle as the frequency varies. Note how closely the straight-line plot approximates the actual variation in phase angle. The maximum deviation between the straight-line plot and the actual plot is approximately 6°.

Figure 15.39 depicts the straight-line approximation of the phase angle of the transfer function given by Eq. (15.56). Equation (15.61)

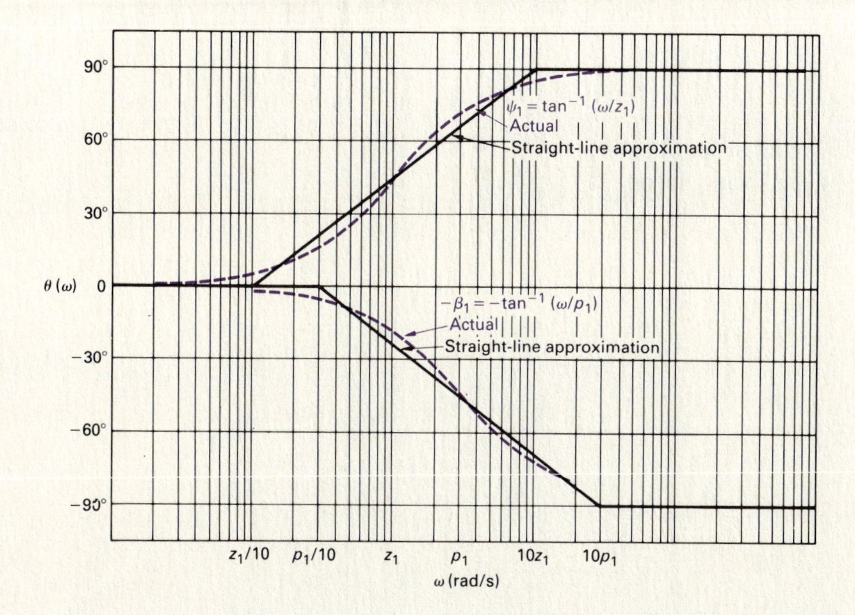

FIGURE 15.38 Phase angle plots for a first-order zero and pole.

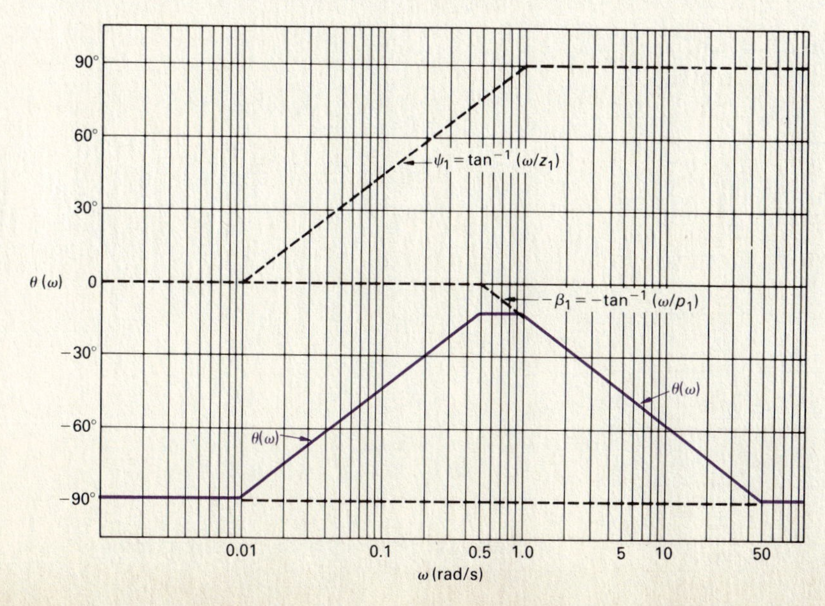

FIGURE 15.39 A straight-line approximation of the phase angle plot for Eq. (15.56).

gives the equation for the phase angle; the plot corresponds to $z_1 = 0.1$ rad/s, and $p_1 = 5$ rad/s.

An illustration of a phase angle plot using a straight-line approximation is given in Example 15.10.

EXAMPLE 15.10

a) Make a straight-line phase angle plot for the transfer function in Example 15.9.

b) Compute the phase angle $\theta(\omega)$ at $\omega = 50,\ 500,$ and 1000 rad/s.

c) Plot the values of (b) on the diagram of (a).

d) Using the results from Example 15.9(e) and part (b) of this example, compute the steady-state output voltage if the source voltage is given by $v_i(t) = 10\cos(500t - 25°)$ V.

SOLUTION

a) From Example 15.9,

$$H(j\omega) = \frac{0.11(j\omega)}{[1 + j(\omega/10)][1 + j(\omega/100)]}$$

$$= \frac{0.11|j\omega|}{|1 + j(\omega/10)||1 + j(\omega/100)|}\underline{/(\psi_1 - \beta_1 - \beta_2)}.$$

Therefore

$$\theta(\omega) = \psi_1 - \beta_1 - \beta_2,$$

where $\psi_1 = 90°$, $\beta_1 = \tan^{-1}(\omega/10)$, and $\beta_2 = \tan^{-1}(\omega/100)$. Figure 15.40 depicts the straight-line approximation of $\theta(\omega)$.

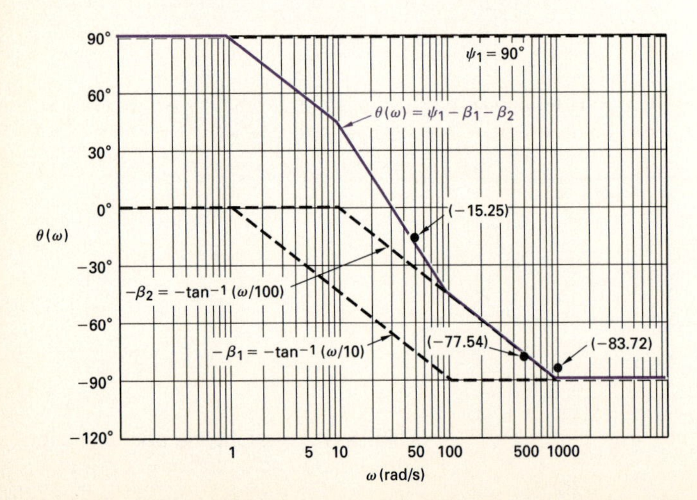

FIGURE 15.40 A straight-line approximation of $\theta(\omega)$ for Example 15.10.

b) We have

$$H(j50) = 0.96\underline{/-15.25^\circ};$$
$$H(j500) = 0.22\underline{/-77.54^\circ};$$
$$H(j1000) = 0.11\underline{/-83.72^\circ}.$$

Thus, $\theta(j50) = -15.25^\circ$, $\theta(j500) = -77.54^\circ$, and $\theta(j1000) = -83.72^\circ$.

c) See Fig. 15.40.

d) We have

$$V_{mo} = |H(j500)|V_{mi} = (0.22)(10) = 2.2 \text{ V};$$

$$\theta_o = \theta(j\omega) + \theta_i = -77.54^\circ - 25^\circ = -102.54^\circ.$$

Thus,

$$v_o(t) = 2.2 \cos(500t - 102.54^\circ) \text{ V}.$$

DRILL EXERCISES

15.12 The numerical expression for a transfer function is

$$H(s) = \frac{10^5(s+5)}{(s+100)(s+5000)}.$$

On the basis of a straight-line approximation of $|H(j\omega)|$ versus ω, estimate (a) the maximum amplitude of $H(j\omega)$ in decibels and (b) the value of $\omega > 0$, where the amplitude of $H(j\omega)$ equals unity.

ANSWER: (a) 26 dB; (b) 98 krad/s.

15.13 Approximate the phase angle of the transfer function in Drill Exercise 15.12 by means of a straight-line plot.

a) Use the straight-line plot to predict the phase angle of $H(s)$ at frequencies of 30, 50, 100, and 5000 rad/s.

b) Calculate the actual value of the phase angle at 30, 50, 100, and 5000 rad/s.

ANSWER: (a) 58.5°, 58.5°, 45°, –45°; (b) 63.49°, 57.15°, 40.99°, –43.91°.

15.7 BODE DIAGRAMS: COMPLEX POLES AND ZEROS

Complex poles and zeros in the expression for $H(s)$ require special attention when you make amplitude and phase angle plots. Let's focus on the contribution that a pair of complex poles makes to the amplitude and phase angle plots. Once you understand the rules for handling complex poles, their application to a pair of complex zeros becomes apparent.

The complex poles and zeros of $H(s)$ always appear in conjugate pairs. The first step in making either an amplitude or a phase angle plot of a transfer function that contains complex poles is to combine the conjugate pair into a single quadratic term. Thus for

$$H(s) = \frac{K}{(s+\alpha-j\beta)(s+\alpha+j\beta)}, \tag{15.71}$$

we first rewrite the product $(s+\alpha-j\beta)(s+\alpha+j\beta)$ as

$$(s+\alpha)^2+\beta^2 = s^2+2\alpha s+\alpha^2+\beta^2. \tag{15.72}$$

When making Bode diagrams, we write the quadratic term in a more convenient form:

$$s^2+2\alpha s+\alpha^2+\beta^2 = s^2+2\zeta\omega_n s+\omega_n^2. \tag{15.73}$$

A direct comparison of the two forms shows that

$$\omega_n^2 = \alpha^2+\beta^2 \tag{15.74}$$

and

$$\zeta\omega_n = \alpha. \tag{15.75}$$

The term ω_n is the corner frequency of the quadratic factor, and ζ is the damping coefficient of the quadratic term. The critical value of ζ is 1. If $\zeta < 1$, the roots of the quadratic factor are complex, and we use Eq. (15.73) to represent the complex poles. If $\zeta \geq 1$, we factor the quadratic factor into $(s+p_1)(s+p_2)$ and then plot amplitude and phase in accordance with the discussion in Section 15.6. Assuming that $\zeta < 1$, we rewrite Eq. (15.71) as

$$H(s) = \frac{K}{s^2+2\zeta\omega_n s+\omega_n^2}. \tag{15.76}$$

We then write Eq. (15.76) in standard form by dividing through by the poles and zeros. For the quadratic term, we divide through by ω_n, so

$$H(s) = \frac{K}{\omega_n^2}\frac{1}{1+(s/\omega_n)^2+2\zeta(s/\omega_n)}, \tag{15.77}$$

from which

$$H(j\omega) = \frac{K_o}{1-(\omega^2/\omega_n^2)+j(2\zeta\omega/\omega_n)}, \tag{15.78}$$

where

$$K_o = \frac{K}{\omega_n^2}.$$

Before discussing the amplitude and phase angle diagrams associated with Eq. (15.78), for convenience we replace the ratio ω/ω_n by a new variable, u. Then

$$H(j\omega) = \frac{K_o}{1 - u^2 + j2\zeta u}. \qquad (15.79)$$

Now we write $H(j\omega)$ in polar form:

$$H(j\omega) = \frac{K_o}{|(1-u^2) + j2\zeta u| \underline{/\beta_1}}, \qquad (15.80)$$

from which

$$\begin{aligned} A_{\text{dB}} &= 20\log_{10}|H(j\omega)| \\ &= 20\log_{10} K_o - 20\log_{10}|(1-u^2) + j2\zeta u| \end{aligned} \qquad (15.81)$$

and

$$\theta(\omega) = -\beta_1 = -\tan^{-1}\frac{2\zeta u}{1-u^2}. \qquad (15.82)$$

FIGURE 15.41 The amplitude plot for a pair of complex poles.

Amplitude Plots

The quadratic factor contributes to the amplitude of $H(j\omega)$ by means of the term $-20\log_{10}|1 - u^2 + j2\zeta u|$. Because $u = \omega/\omega_n$, $u \to 0$ as $\omega \to 0$, and $u \to \infty$ as $\omega \to \infty$. To see how the term behaves as ω ranges from 0 to ∞, we note that

$$\begin{aligned} -20\log_{10}|(1-u^2) + j2\zeta u| &= -20\log_{10}\sqrt{(1-u^2)^2 + 4\zeta^2u^2} \\ &= -10\log_{10}[u^4 + 2u^2(2\zeta^2 - 1) + 1]; \end{aligned} \qquad (15.83)$$

as $u \to 0$,

$$-10\log_{10}[u^4 + 2u^2(2\zeta^2 - 1) + 1] \to 0; \qquad (15.84)$$

and as $u \to \infty$,

$$-10\log_{10}[u^4 + 2u^2(2\zeta^2 - 1) + 1] \to -40\log_{10} u. \qquad (15.85)$$

From Eqs. (15.84) and (15.85), we conclude that the approximate amplitude plot consists of two straight lines. For $\omega < \omega_n$, the straight line lies along the 0 dB axis, and for $\omega > \omega_n$, the straight line has a slope of -40 dB/decade. These two straight lines join on the 0 dB axis at $u = 1$ or $\omega = \omega_n$. Figure 15.41 shows the straight-line approximation for a quadratic factor with $\zeta < 1$.

Correcting Straight-Line Amplitude Plots

Correcting the straight-line amplitude plot for a pair of complex poles is not as easy as correcting a first-order real pole, because the corrections depend on the damping coefficient ζ. Figure 15.42 shows the

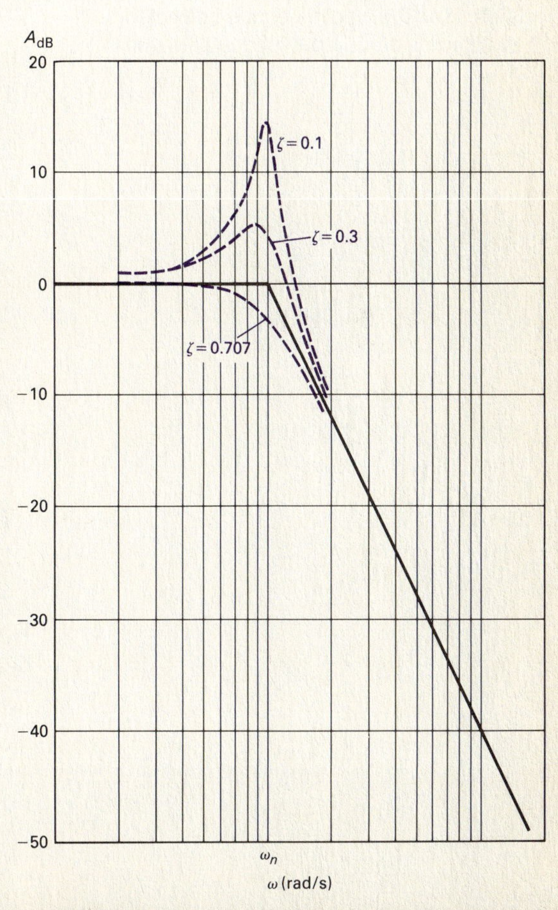

FIGURE 15.42 The effect of ζ on the amplitude plot.

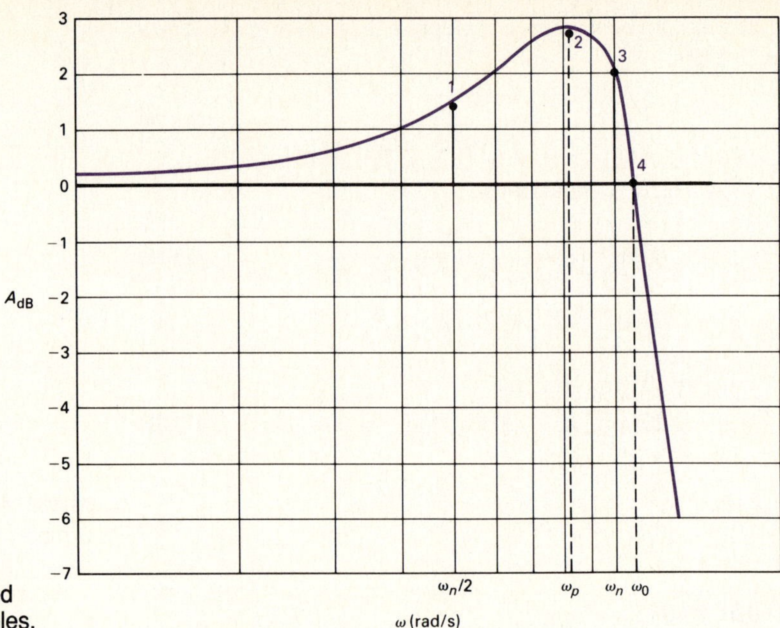

FIGURE 15.43 Four points on the corrected amplitude plot for a pair of complex poles.

effect of ζ on the amplitude plot. Note that as ζ becomes very small, a large peak in the amplitude occurs in the neighborhood of the corner frequency $\omega_n(u = 1)$. When $\zeta \geq 1/\sqrt{2}$, the corrected amplitude plot lies entirely below the straight-line approximation. For sketching purposes, the straight-line amplitude plot can be corrected by locating four points on the actual curve. These four points correspond to (1) one-half the corner frequency, (2) the frequency at which the amplitude reaches its peak value, (3) the corner frequency, and (4) the frequency at which the amplitude is zero. Figure 15.43 shows these four points.

At one-half the corner frequency (point 1), the actual amplitude is

$$A_{dB}(\omega_n/2) = -10\log_{10}(\zeta^2 + 0.5625). \tag{15.86}$$

The amplitude peaks (point 2) at a frequency of

$$\omega_p = \omega_n\sqrt{1 - 2\zeta^2}, \tag{15.87}$$

and it has a peak amplitude of

$$A_{dB}(\omega_p) = -10\log_{10}[4\zeta^2(1 - \zeta^2)]. \tag{15.88}$$

At the corner frequency (point 3), the actual amplitude is

$$A_{dB}(\omega_n) = -20\log_{10} 2\zeta. \tag{15.89}$$

The corrected amplitude plot crosses the 0 dB axis (point 4) at

$$\omega_o = \omega_n\sqrt{2(1 - 2\zeta^2)} = \sqrt{2}\omega_p. \tag{15.90}$$

The derivations of Eqs. (15.86), (15.89), and (15.90) follow from Eq. (15.83). Evaluating Eq. (15.83) at $u = 0.5$ and $u = 1.0$, respectively, yields Eqs. (15.86) and (15.89). Equation (15.90) corresponds to finding the value of u that makes $u^4 + 2u^2(2\zeta^2 - 1) + 1 = 1$. The derivation of Eq. (15.87) requires differentiating Eq. (15.83) with respect to u, and then finding the value of u where the derivative is zero. Equation (15.88) is the evaluation of Eq. (15.83) at the value of u found in Eq. (15.87).

Use a computer tool to show this.

Example 15.11 illustrates the amplitude plot for a transfer function with a pair of complex poles.

EXAMPLE 15.11

Compute the transfer function for the circuit shown in Fig. 15.44.

a) What is the value of the corner frequency in radians per second?

b) What is the value of K_o?

c) What is the value of the damping coefficient?

d) Make a straight-line amplitude plot ranging from 10 to 500 rad/s.

e) Calculate and sketch the actual amplitude in decibels at $\omega_n/2$, ω_p, ω_n, and ω_o.

f) From the straight-line amplitude plot, describe the type of filter represented by the circuit in Fig. 15.44 and estimate its cutoff frequency, ω_c.

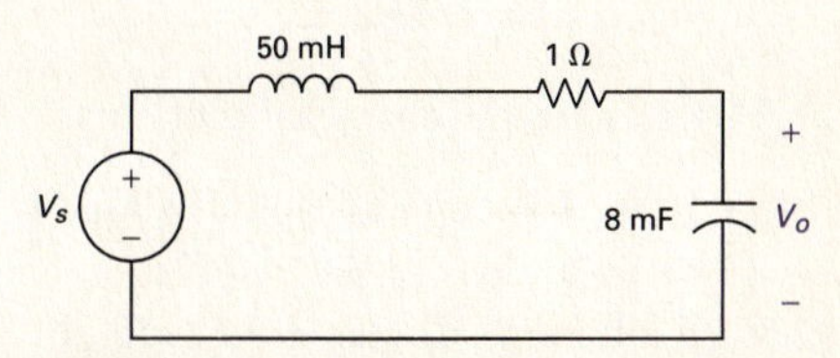

FIGURE 15.44 The circuit for Example 15.11.

SOLUTION

Transform the circuit in Fig. 15.44 to the s-domain and then use s-domain voltage division to get

$$H(s) = \frac{\dfrac{1}{LC}}{s^2 + \left(\dfrac{R}{L}\right)s + \dfrac{1}{LC}}.$$

Substituting the component values,

$$H(s) = \frac{2500}{s^2 + 20s + 2500}.$$

a) From the expression for $H(s)$, $\omega_n^2 = 2500$; therefore, $\omega_n = 50$ rad/s.

b) By definition, K_o is $2500/\omega_n^2$, or 1.

c) The coefficient of s equals $2\zeta\omega_n$; therefore

$$\zeta = \frac{20}{2\omega_n} = 0.20.$$

d) See Fig. 15.45.

e) The actual amplitudes are

$$A_{\text{dB}}(\omega_n/2) = -10\log_{10}(0.6025) = 2.2 \text{ dB};$$
$$\omega_p = 50\sqrt{0.92} = 47.96 \text{ rad/s};$$
$$A_{\text{dB}}(\omega_p) = -10\log_{10}(0.16)(0.96) = 8.14 \text{ dB};$$
$$A_{\text{dB}}(\omega_n) = -20\log_{10}(0.4) = 7.96 \text{ dB};$$
$$\omega_o = \sqrt{2}\omega_p = 67.82 \text{ rad/s};$$
$$A_{\text{dB}}(\omega_o) = 0 \text{ dB}.$$

Figure 15.45 shows the corrected plot.

f) It is clear from the amplitude plot in Fig. 15.45 that this circuit acts as a low-pass filter. At the cutoff frequency, the magnitude of the transfer function, $|H(j\omega_c)|$, is 3 dB less than the maximum magnitude. From the corrected plot, the cutoff frequency appears to be about 55 rad/s, almost the same as that predicted by the straight-line Bode diagram.

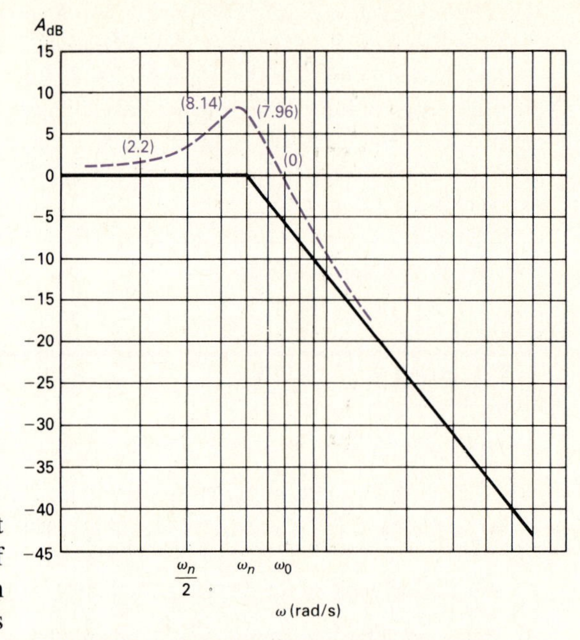

FIGURE 15.45 The amplitude plot for Example 15.11.

PHASE ANGLE PLOTS

The phase angle plot for a pair of complex poles is a plot of Eq. (15.82). The phase angle is zero at zero frequency and is $-90°$ at the corner frequency. It approaches $-180°$ as $\omega(u)$ becomes large. As in the case of the amplitude plot, ζ is important in determining the exact shape of the phase angle plot. For small values of ζ, the phase angle changes rapidly in the vicinity of the corner frequency. Figure 15.46 shows the effect of ζ on the phase angle plot.

We can also make a straight-line approximation of the phase angle plot for a pair of complex poles. We do so by drawing a line tangent to the phase angle curve at the corner frequency and extending this line until it intersects with the 0° and $-180°$ lines. The line tangent to the phase angle curve at $-90°$ has a slope of $-2.3/\zeta$ rad/decade ($-132/\zeta$ degrees/decade), and it intersects the 0° and $-180°$ lines at $u_1 = 4.81^{-\zeta}$ and $u_2 = 4.81^{\zeta}$, respectively. Figure 15.47 depicts the straight-line approximation for $\zeta = 0.3$ and shows the actual phase angle plot. Comparing the straight-line approximation to the actual curve indicates that the approximation is reasonable in the vicinity of the corner frequency. However, in the neighborhood of u_1 and u_2, the error is quite large.

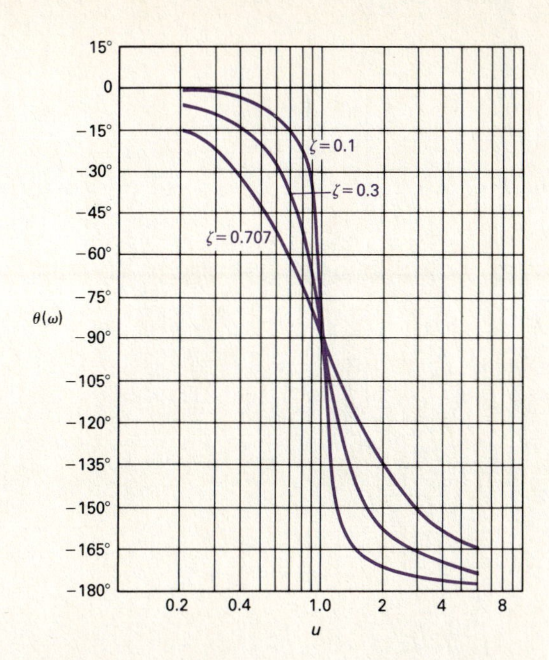

FIGURE 15.46 The effect of ζ on the phase angle plot.

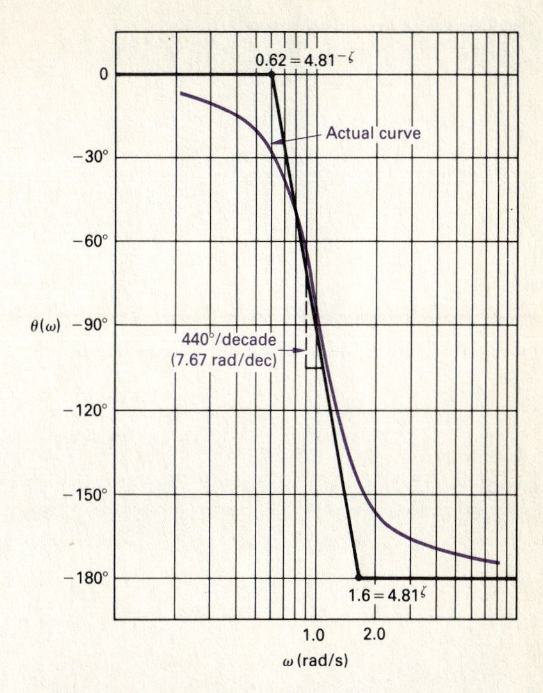

FIGURE 15.47 A straight-line approximation of the phase angle for a pair of complex poles.

In Example 15.12 we summarize our discussion of Bode diagrams.

EXAMPLE 15.12

a) Compute the transfer function for the circuit shown in Fig. 15.48.

b) Make a straight-line amplitude plot of $20 \log_{10} |H(j\omega)|$.

c) Use the straight-line amplitude plot to determine the type of filter represented by this circuit, and then estimate its cutoff frequency.

d) What is the actual cutoff frequency?

e) Make a straight-line phase angle plot of $H(j\omega)$.

f) What is the value of $\theta(\omega)$ at the cutoff frequency from part (c)?

g) What is the actual value of $\theta(\omega)$ at the cutoff frequency?

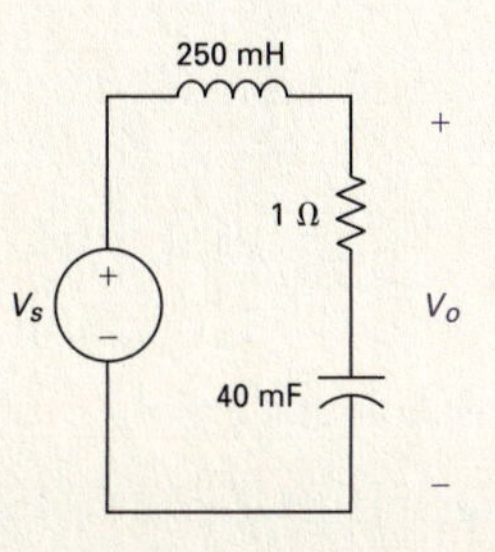

FIGURE 15.48 The circuit for Example 15.12.

SOLUTION

a) Transform the circuit in Fig. 15.48 to the s-domain and then perform s-domain voltage division to get

$$H(s) = \frac{\frac{R}{L}s + \frac{1}{LC}}{s^2 + \frac{R}{L}s + \frac{1}{LC}}.$$

Substituting the component values from the circuit gives

$$H(s) = \frac{4(s+25)}{s^2 + 4s + 100}.$$

b) The first step in making Bode diagrams is to put $H(j\omega)$ in standard form. Because $H(s)$ contains a quadratic factor, we first check the value of ζ. We find that $\zeta = 0.2$ and $\omega_n = 10$, so

$$H(s) = \frac{s/25 + 1}{1 + (s/10)^2 + 0.4(s/10)},$$

from which

$$H(j\omega) = \frac{|1 + j\omega/25| \underline{/\psi_1}}{|1 - (\omega/10)^2 + j0.4(\omega/10)| \underline{/\beta_1}}.$$

Note that for the quadratic factor, $u = \omega/10$. The amplitude of $H(j\omega)$ in decibels is

$$A_{\text{dB}} = 20\log_{10}|1 + j\omega/25| - 20\log_{10}\left[\left|1 - \left(\frac{\omega}{10}\right)^2 + j0.4\left(\frac{\omega}{10}\right)\right|\right],$$

and the phase angle is

$$\theta(\omega) = \psi_1 - \beta_1,$$

where

$$\psi_1 = \tan^{-1}(\omega/25)$$

and

$$\beta_1 = \tan^{-1}\frac{0.4(\omega/10)}{1 - (\omega/10)^2}.$$

Figure 15.49 shows the amplitude plot.

c) From the straight-line amplitude plot in Fig. 15.49, this circuit acts as a low-pass filter. At the cutoff frequency, the amplitude of $H(j\omega)$ is 3 dB less than the amplitude in the passband. From the plot, we predict that the cutoff frequency is approximately 13 rad/s.

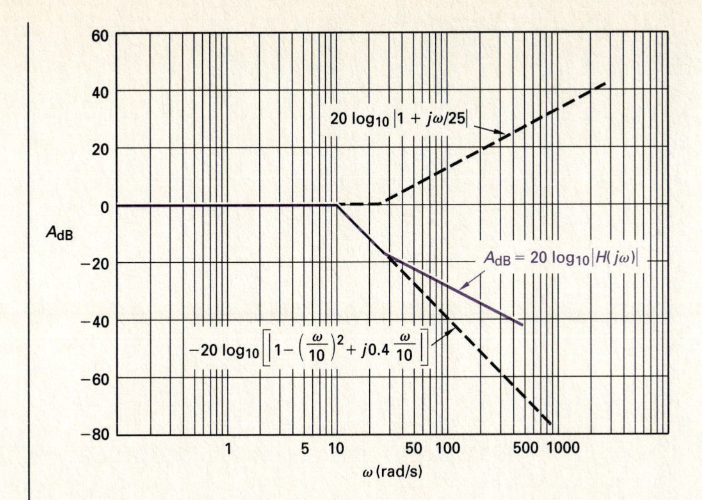

FIGURE 15.49 The amplitude plot for Example 15.12.

d) To solve for the actual cutoff frequency, replace s with $j\omega$ in $H(s)$, compute the expression for $|H(j\omega)|$, set $|H(j\omega_c)| = (1/\sqrt{2})H_{\max} = 1/\sqrt{2}$, and solve for ω_c. First,

$$H(j\omega) = \frac{4(j\omega) + 100}{(j\omega)^2 + 4(j\omega) + 100}.$$

Then,

$$|H(j\omega_c)| = \frac{\sqrt{(4\omega_c)^2 + 100^2}}{\sqrt{(100 - \omega_c^2)^2 + (4\omega_c)^2}} = \frac{1}{\sqrt{2}}.$$

Solving for ω_c gives us

$$\omega_c = 16 \text{ rad/s}.$$

e) Figure 15.50 shows the phase angle plot. Note that the straight-line segment of $\theta(\omega)$ between 1.0 and 2.5 rad/s does not have the same slope as the segment between 2.5 and 100 rad/s.

f) From the phase angle plot in Fig. 15.50, we estimate the phase angle at the cutoff frequency of 16 rad/s to be $-65°$.

g) We can compute the exact phase angle at the cutoff frequency by substituting $s = j16$ into the transfer function $H(s)$:

$$H(j16) = \frac{4(j16 + 25)}{(j16)^2 + 4(j16) + 100}.$$

Computing the phase angle, we see

$$\theta(\omega_c) = \theta(j16) = -125.0°.$$

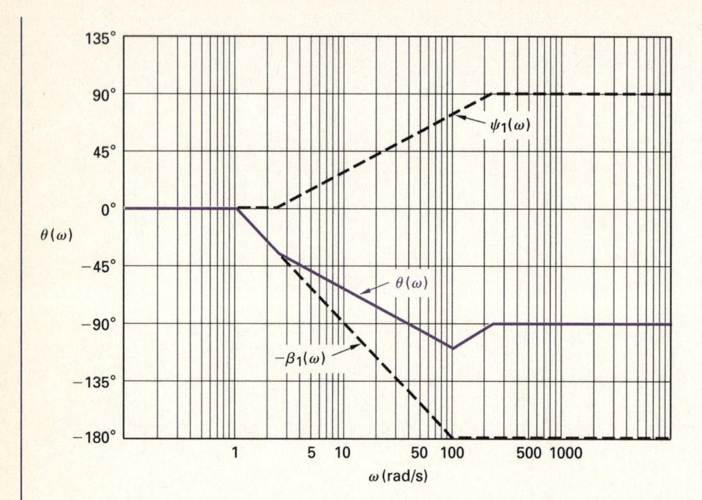

FIGURE 15.50 The phase angle plot for Example 15.12.

Note the large error in the predicted angle. In general, straight-line phase angle plots do not give satisfactory results in the frequency band where the phase angle is changing. The straight-line phase angle plot is useful only in predicting the general behavior of the phase angle, not in estimating actual phase angle values at particular frequencies.

DRILL EXERCISES

15.14 The numerical expression for a current transfer function is

$$H(s) = \frac{I_o}{I_i} = \frac{25 \times 10^8}{s^2 + 20{,}000s + 25 \times 10^8}.$$

Compute the

a) corner frequency;

b) damping coefficient;

c) frequencies when $H(j\omega)$ is unity;

d) peak amplitude of $H(j\omega)$ in decibels;

e) frequency at which the peak occurs; and

f) amplitude of $H(j\omega)$ at one-half the corner frequency.

ANSWER: (a) 50 krad/s; (b) 0.2; (c) 0, 67.82 krad/s; (d) 8.14 dB; (e) 47.96 krad/s; (f) 2.20 dB.

15.15 The numerical expression for a voltage transfer function is

$$H(s) = \frac{V_o}{V_i} = \frac{32 \times 10^5}{s^2 + 400s + 64 \times 10^4}.$$

a) Use a straight-line amplitude plot to find the frequency when the amplitude of $H(j\omega)$ equals unity.

b) What is the actual amplitude at the frequency found in (a)?

c) What is the peak amplitude of $H(j\omega)$ in decibels?

d) At what frequency does the output voltage reach its peak amplitude?

e) If the amplitude of the source voltage is 10 V, what is the peak amplitude of the output voltage in volts?

ANSWER: (a) 1800 rad/s; (b) 1.19; (c) 20.28 dB; (d) 748.33 rad/s; (e) 103.28 V.

PRACTICAL PERSPECTIVE

Pushbutton Telephone Circuits

In the Practical Perspective at the start of this chapter, we described the dual-tone-multiple-frequency (DTMF) system used to signal that a button has been pushed on a pushbutton telephone. In the following example, we design a DTMF receiver—a circuit that decodes the tones produced by pushing a button and determines which button was pushed.

Before we begin the design, we need a better understanding of the DTMF system. As you can see from Fig. 15.51, the buttons on the telephone are organized into rows and columns. The pair of tones generated by pushing a button depends on the button's row and column. The button's row determines its low-frequency tone, while the button's column determines its high-frequency tone.[3] For example, pressing the 6 button produces sinusoidal tones with the frequencies 770 Hz and 1477 Hz.

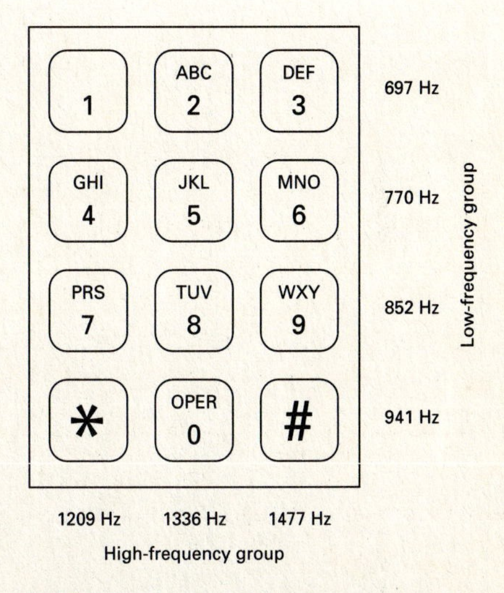

FIGURE 15.51 Tones generated by the rows and columns of telephone pushbuttons.

At the telephone switching facility, bandpass filters in the DTMF receiver first detect whether tones from both the low-frequency and high-frequency groups are simultaneously present. This test rejects many extraneous audio signals that are not DTMF. If tones are present in both bands, other filters are used to select among the possible tones in each band so that the frequencies can be decoded into a unique button signal. Additional tests are performed to prevent false button detection. For example, only one tone per frequency band is allowed; the high- and low-band frequencies must start and stop within a few milliseconds of one another to be considered valid; and the high- and low-band signal amplitudes must be sufficiently close to one another.

You may wonder why bandpass filters are used instead of a high-pass filter for the high-frequency group of DTMF tones and a low-pass filter for the low-frequency group of DTMF tones. The reason is that the telephone system uses frequencies outside of the 300 Hz to 3 kHz band for other signaling purposes, such as ringing the phone's bell. Bandpass filters prevent the DTMF receiver from erroneously detecting these other signals.

EXAMPLE

Now we are ready to design a series RLC bandpass filter (see Fig. 15.27) for detecting the low frequency tone.

a) Calculate the values of L and C that place the cutoff frequencies at the edges of the DTMF low-frequency band. Note that the resistance in standard telephone circuits is always $R = 600\ \Omega$.

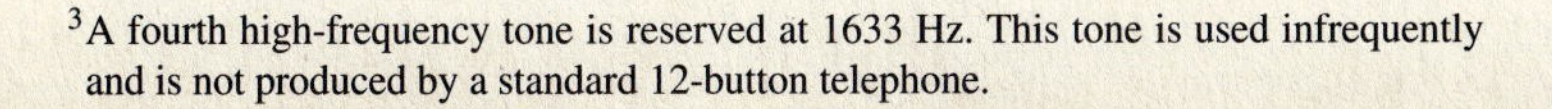

[3] A fourth high-frequency tone is reserved at 1633 Hz. This tone is used infrequently and is not produced by a standard 12-button telephone.

b) What is the output amplitude of this circuit at each of the low-band frequencies, relative to the peak amplitude of the bandpass filter?

c) What is the output amplitude of this circuit at the lowest of the high-band frequencies?

SOLUTION

a) Use the cutoff frequencies

$$\omega_{c1} = 2\pi(697) = 4379.38 \text{ rad/s}$$

and

$$\omega_{c2} = 2\pi(941) = 5912.48 \text{ rad/s}$$

to calculate the filter bandwidth as

$$\beta = \omega_{c2} - \omega_{c1} = 1533.10 \text{ rad/s}.$$

From Eq. (15.32), we calculate the inductance as

$$L = \frac{R}{\beta} = \frac{600}{1533.10} = 0.39 \text{ H}.$$

The capacitance can be calculated using Eq. (15.31) as

$$C = \frac{1}{L\omega_{c1}\omega_{c2}} = \frac{1}{(0.39)(4379.38)(5912.48)} = 0.10\,\mu\text{F}.$$

b) At the outermost two frequencies in the low-frequency group (697 Hz and 941 Hz), the amplitudes are

$$|V_{697\text{Hz}}| = |V_{941\text{Hz}}| = \frac{|V_{\text{peak}}|}{\sqrt{2}} = 0.707|V_{\text{peak}}|$$

because those are the cutoff frequencies. At the other two low-band frequencies, the amplitudes can be calculated using Eq. (15.22), with the substitution of the bandwidth from Eq. (15.32):

$$|V| = (|V_{\text{peak}}|)(|H(j\omega)|) = |V_{\text{peak}}|\frac{\omega\beta}{\sqrt{(\omega_o^2 - \omega^2)^2 + (\omega\beta)^2}}.$$

Therefore,

$$|V_{770\text{Hz}}| = |V_{\text{peak}}| =$$

$$\frac{(4838.05)(1533.10)}{\sqrt{(5088.52^2 - 4838.05^2)^2 + [(4838.05)(1533.10)]^2}} = 0.948|V_{\text{peak}}|$$

and

$$|V_{852\text{Hz magnitude line}}| = |V_{\text{peak}}| =$$

$$\frac{(5353.27)(1533.10)}{\sqrt{(5088.52^2 - 5353.27^2)^2 + [(5353.27)(1533.10)]^2}} = 0.948|V_{\text{peak}}|.$$

The fact that these two magnitudes are the same is not a coincidence. The frequencies in both bands of the DTMF system were carefully chosen to produce this type of predictable behavior with linear filters. In other words, the frequencies were chosen to be equally far apart with respect to the response produced by a linear filter. Most musical scales consist of tones designed with this same property—note intervals are selected to place the notes equally far apart. That is why the DTMF tones remind us of musical notes and allow us to play simple melodies using the telephone buttons.

Unlike musical scales, however, the DTMF frequencies were selected to be harmonically unrelated. This means that none of the frequencies can be formed as a linear combination of any of the others. This lowers the risk of misidentifying a tone's frequency if the circuit elements are not perfectly linear.

c) The high-band frequency closest to the low-frequency band is 1209 Hz. The amplitude of a tone with this frequency is

$$|V_{1209\text{Hz}}| = |V_{\text{peak}}| = \frac{(7596.37)(1533.10)}{\sqrt{(5088.52^2 - 7596.37^2)^2 + [(7596.37)(1533.10)]^2}} = 0.344|V_{\text{peak}}|.$$

This is less than one-half the amplitude of the signals with the low-band cutoff frequencies, ensuring adequate separation of the bands. In practice, such results are somewhat unrealistic. One fact we overlooked in our calculations is that the amplitudes of the high-band tones are deliberately made slightly larger than the amplitudes of the low-band tones to compensate for some other filtering that occurs in the telephone system.

SUMMARY

- A **frequency selective circuit**, or **filter**, enables signals at certain frequencies to reach the output, and it attenuates signals at other frequencies to prevent them from reaching the output. The **passband** contains the frequencies of those signals which are passed; the **stopband** contains the frequencies of those signals which are attenuated.

- The **cutoff frequency**, ω_c, identifies the location on the frequency axis which separates the stopband from the passband. At the cutoff frequency, the magnitude of the transfer function equals $(1/\sqrt{2})H_{\text{max}}$.

  A **low-pass filter** passes voltages at frequencies below ω_c and attenuates frequencies above ω_c. Any circuit with the transfer function

$$H(s) = \frac{\omega_c}{s + \omega_c}$$

functions as a low-pass filter.

- A **high-pass filter** passes voltages at frequencies above ω_c and attenuates voltages at frequencies below ω_c. Any circuit with the transfer function

$$H(s) = \frac{s}{s + \omega_c}$$

functions as a high-pass filter.

- Bandpass filters and bandreject filters each have two cutoff frequencies, ω_{c1} and ω_{c2}. These filters are further characterized by their **center frequency** (ω_o), **bandwidth** (β), and **quality factor** (Q). These quantities are defined as

$$\omega_o = \sqrt{\omega_{c1} \cdot \omega_{c2}};$$
$$\beta = \omega_{c2} - \omega_{c1};$$
$$Q = \omega_o / \beta.$$

- A **bandpass filter** passes voltages at frequencies within the passband, which is between ω_{c1} and ω_{c2}. It attenuates frequencies outside of the passband. Any circuit with the transfer function

$$H(s) = \frac{\beta s}{s^2 + \beta s + \omega_o^2}$$

functions as a bandpass filter.

- A **bandreject filter** attenuates voltages at frequencies within the stopband, which is between ω_{c1} and ω_{c2}. It passes frequencies outside of the stopband. Any circuit with the transfer function

$$H(s) = \frac{s^2 + \omega_o^2}{s^2 + \beta s + \omega_o^2}$$

functions as a bandreject filter.

- Adding a load to the output of a passive filter changes its filtering properties by altering the location and magnitude of the passband. Replacing an ideal voltage source with one whose source resistance is nonzero also changes the filtering properties of the rest of the circuit, again by altering the location and magnitude of the passband.

- A **Bode diagram** is a frequency response plot (on semilog graph paper) in which the amplitude is plotted in decibels, the phase angle is plotted in degrees, and the frequency is plotted by decades. Techniques for constructing straight-line approximations to Bode diagrams help you to sketch the frequency response of a transfer function very quickly.

PROBLEMS

15.1 a) Find the cutoff frequency in hertz for the RL filter shown in Fig. P15.1.

b) Calculate $H(j\omega)$ at ω_c, $0.2\omega_c$, and $5\omega_c$.

c) If $v_i = 10\cos\omega t$ V, write the steady-state expression for v_o when $\omega = \omega_c$, $\omega = 0.2\omega_c$, and $\omega = 5\omega_c$.

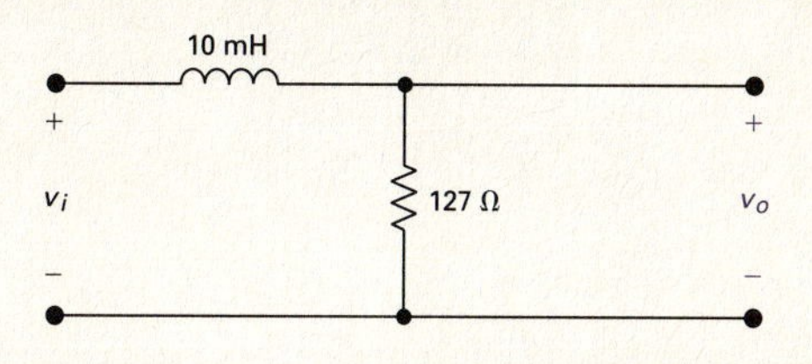

FIGURE P15.1

❖ **15.2** Use a 5 mH inductor to design a low-pass, RL, passive filter with a cutoff frequency of 1000 Hz.

a) Specify the value of the resistor.

b) A load having a resistance of 270 Ω is connected across the output terminals of the filter. What is the corner, or cutoff, frequency of the loaded filter in hertz?

15.3 A resistor, denoted as R_l, is added in series with the inductor in the circuit in Fig. 15.4(a). The new low-pass filter circuit is shown in Fig. P15.3.

a) Derive the expression for $H(s)$ where $H(s) = V_o/V_i$.

b) At what frequency will the magnitude of $H(j\omega)$ be maximum?

c) What is the maximum value of the magnitude of $H(j\omega)$?

d) At what frequency will the magnitude of $H(j\omega)$ equal its maximum value divided by $\sqrt{2}$?

e) Assume a resistance of 6.7 Ω is added in series with the 10 mH inductor in the circuit in Fig. P15.1. Find ω_c, $H(j0)$, $H(j\omega_c)$, $H(j0.2\omega_c)$, and $H(j5\omega_c)$.

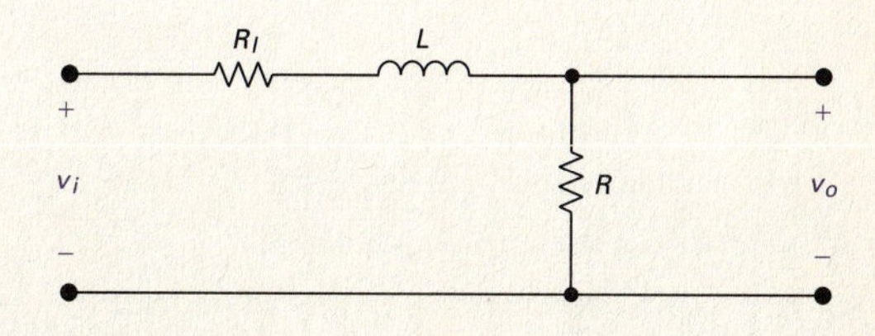

FIGURE P15.3

15.4 a) Find the cutoff frequency (in hertz) of the low-pass filter shown in Fig. P15.4.

b) Calculate $H(j\omega)$ at ω_c, $0.1\omega_c$, and $10\omega_c$.

c) If $v_i = 200\cos\omega t$ mV, write the steady-state expression for v_o when $\omega = \omega_c$, $0.1\omega_c$, and $10\omega_c$.

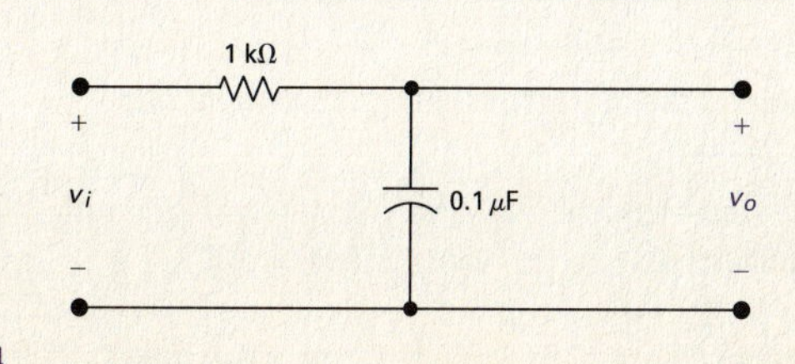

FIGURE P15.4

15.5 A resistor denoted as R_L is connected in parallel with the capacitor in the circuit in Fig. 15.7. The loaded low-pass filter circuit is shown in Fig. P15.5.

a) Derive the expression for the voltage transfer function V_o/V_i.

b) At what frequency will the magnitude of $H(j\omega)$ be maximum?

c) What is the maximum value of the magnitude of $H(j\omega)$?

d) At what frequency will the magnitude of $H(j\omega)$ equal its maximum value divided by $\sqrt{2}$?

e) Assume a resistance of 10 $k\Omega$ is added in parallel with the 0.1 μF capacitor in the circuit in Fig. P15.4. Find ω_c, $H(j0)$, $H(j\omega_c)$, $H(j0.1\omega_c)$, and $H(j10\omega_c)$.

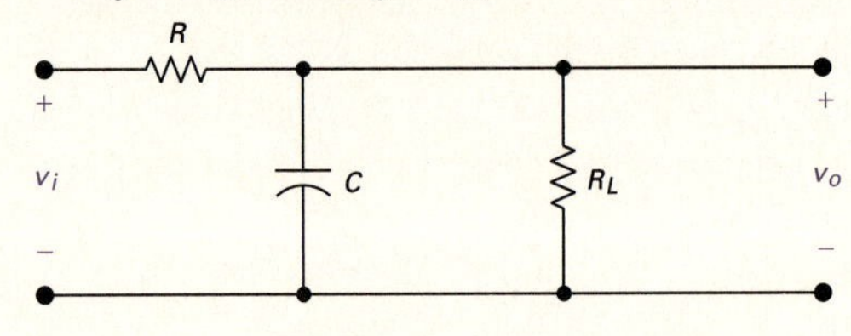

FIGURE P15.5

❖ **15.6** Use a 0.5 μF capacitor to design a low-pass passive filter with a cutoff frequency of 50 krad/s.

a) Specify the cutoff frequency in hertz.

b) Specify the value of the filter resistor.

c) Assume the cutoff frequency cannot increase by more than 5%. What is the smallest value of load resistance that can be connected across the output terminals of the filter?

d) If the resistor found in part (c) is connected across the output terminals, what is the magnitude of $H(j\omega)$ when $\omega = 0$?

15.7 a) Find the cutoff frequency (in hertz) for the high-pass filter shown in Fig. P15.7.

b) Find $H(j\omega)$ at ω_c, $0.2\omega_c$, and $5\omega_c$.

c) If $v_i = 500 \cos \omega t$ mV, write the steady-state expression for v_o when $\omega = \omega_c$, $\omega = 0.2\omega_c$, and $\omega = 5\omega_c$.

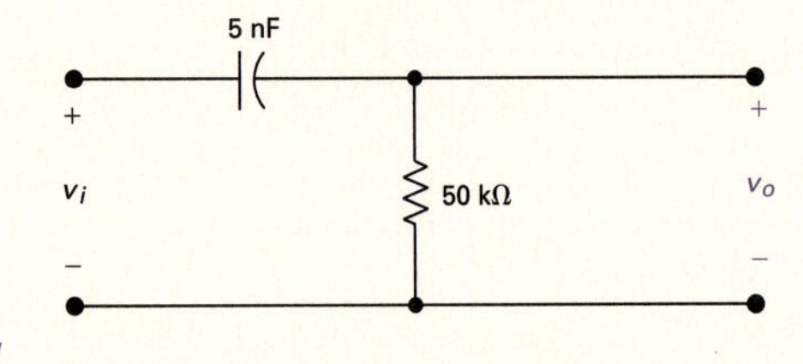

FIGURE P15.7

15.8 A resistor, denoted as R_c, is connected in series with the capacitor in the circuit in Fig. 15.10(a). The new high-pass filter circuit is shown in Fig. P15.8.

a) Derive the expression for $H(s)$ where $H(s) = V_o/V_i$.

b) At what frequency will the magnitude of $H(j\omega)$ be maximum?

c) What is the maximum value of the magnitude of $H(j\omega)$?

d) At what frequency will the magnitude of $H(j\omega)$ equal its maximum value divided by $\sqrt{2}$?

e) Assume a resistance of 12.5 kΩ is connected in series with the 5 nF capacitor in the circuit in Fig. P15.7. Calculate ω_c, $H(j\omega_c)$, $H(j0.2\omega_c)$, and $H(j5\omega_c)$.

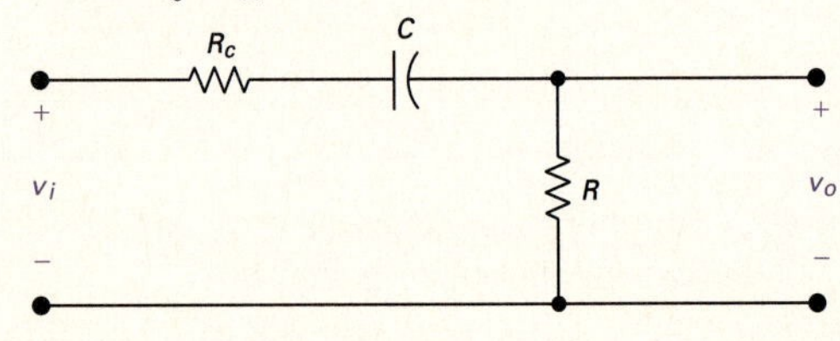

FIGURE P15.8

❖ **15.9** Using a 100 nF capacitor, design a high-pass passive filter with a cutoff frequency of 300 Hz.

a) Specify the value of R in kilohms.

b) A 47 kΩ resistor is connected across the output terminals of the filter. What is the cutoff frequency, in hertz, of the loaded filter?

❖ **15.10** Using a 5 mH inductor, design a high-pass, RL, passive filter with a cutoff frequency of 25 krad/s.

a) Specify the value of the resistance.

b) Assume the filter is connected to a pure resistive load. The cutoff frequency is not to drop below 24 krad/s. What is the smallest load resistor that can be connected across the output terminals of the filter?

15.11 Show that the alternative forms for the cutoff frequencies of a bandpass filter, given in Eqs. (15.36) and (15.37), can be derived from Eqs. (15.34) and (15.35).

15.12 Calculate the center frequency, the bandwidth, and the quality factor of a bandpass filter that has an upper cutoff frequency of 121 krad/s and a lower cutoff frequency of 100 krad/s.

15.13 A bandpass filter has a center, or resonant, frequency of 50 krad/s and a quality factor of 4. Find the bandwidth, the upper cutoff frequency, and the lower cutoff frequency. Express all answers in kilohertz.

❖ **15.14** Use a 5 nF capacitor to design a series RLC bandpass filter, as shown in Fig. 15.19(a). The center frequency of the filter is 8 kHz, and the quality factor is 2.

a) Specify the values of R and L.

b) What is the lower cutoff frequency in kilohertz?

c) What is the upper cutoff frequency in kilohertz?

d) What is the bandwidth of the filter in kilohertz?

15.15 For the bandpass filter shown in Fig. P15.15, find (a) ω_o, (b) f_o, (c) Q, (d) ω_{c1}, (e) f_{c1}, (f) ω_{c2}, (g) f_{c2}, and (h) β.

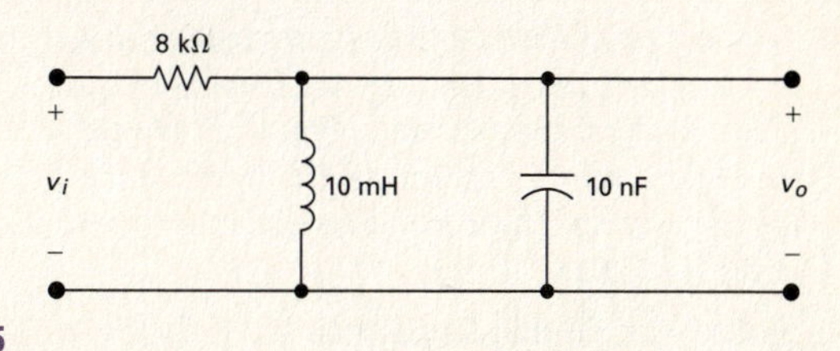

FIGURE P15.15

❖ **15.16** Using a 0.05 μF capacitor in the bandpass circuit shown in Fig. 15.22, design a filter with a quality factor of 5 and a center frequency of 20 krad/s.

a) Specify the numerical values of R and L.

b) Calculate the upper and lower cutoff frequencies in kilohertz.

c) Calculate the bandwidth in hertz.

15.17 For the bandpass filter shown in Fig. P15.17, calculate the following: (a) f_o; (b) Q; (c) f_{c1}; (d) f_{c2}; and (e) β.

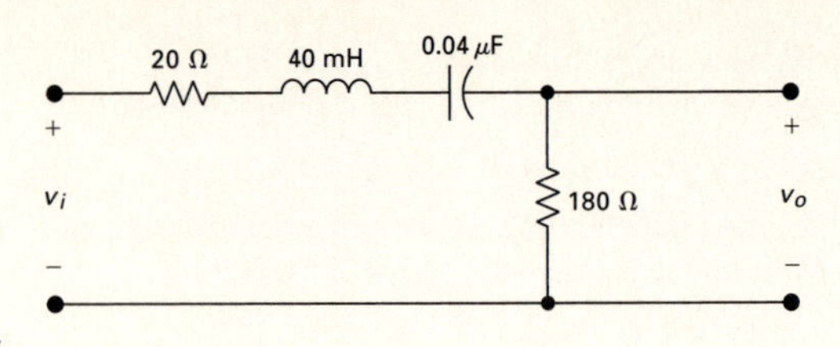

FIGURE P15.17

15.18 The input voltage in the circuit in Fig. P15.17 is $500 \cos \omega t$ mV. Calculate the output voltage when (a) $\omega = \omega_o$; (b) $\omega = \omega_{c1}$; and (c) $\omega = \omega_{c2}$.

15.19 A block diagram of a system consisting of a sinusoidal voltage source, an RLC series bandpass filter, and a load is shown in Fig. P15.19. The internal impedance of the sinusoidal source is $80 + j0\ \Omega$, and the impedance of the load is $480 + j0\ \Omega$.

The RLC series bandpass filter has a 0.02 μF capacitor, a center frequency of 50 krad/s, and a quality factor of 6.25.

a) Draw a circuit diagram of the system.

b) Specify the numerical values of L and R for the filter section of the system.

c) What is the quality factor of the interconnected system?

d) What is the bandwidth (in hertz) of the interconnected system?

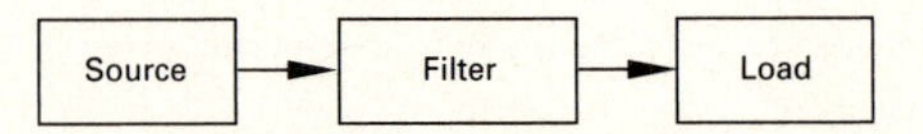

FIGURE P15.19

15.20 The purpose of this problem is to investigate how a resistive load connected across the output terminals of the bandpass filter shown in Fig. 15.22 affects the quality factor and hence the bandwidth of the filtering system. The loaded filter circuit is shown in Fig. P15.20.

a) Calculate the transfer function V_o / V_i for the circuit shown in Fig. P15.20.

b) What is the expression for the bandwidth of the system?

c) What is the expression for the loaded bandwidth (β_L) as a function of the unloaded bandwidth (β_U)?

d) What is the expression for the quality factor of the system?

e) What is the expression for the loaded quality factor (Q_L) as a function of the unloaded quality factor (Q_U)?

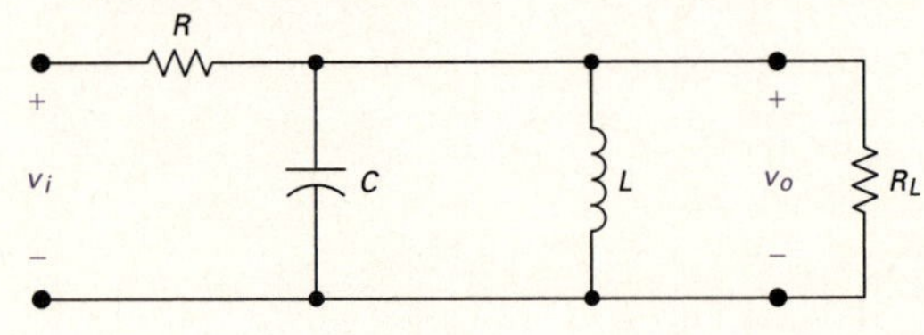

FIGURE P15.20

15.21 For the circuit shown in Fig. P15.21, find (a) ω_o; (b) β; (c) Q; and (d) the steady-state expression for v_o when $v_i = 250 \cos \omega_o t$ mV.

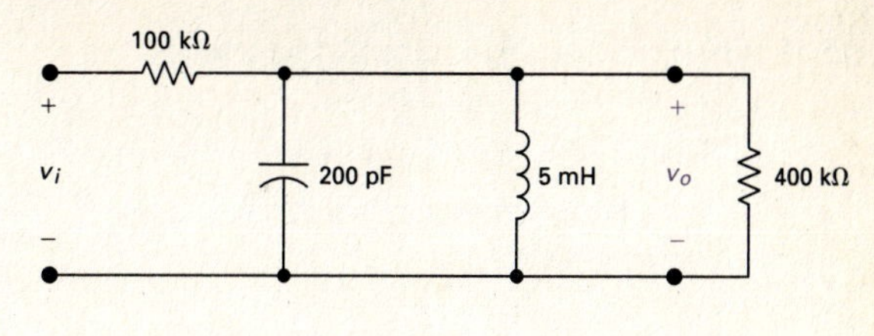

FIGURE P15.21

15.22 The parameters in the circuit in Fig. P15.20 are $R = 2400\ \Omega$, $C = 50$ pF, and $L = 2\ \mu$H. The quality factor of the circuit is not to drop below 7.5. What is the smallest permissible value of the load resistor R_L?

15.23 a) Show (via a qualitative analysis) that the circuit in Fig. P15.23 is a bandreject filter.

b) Support the qualitative analysis of part (a) by finding the voltage transfer function of the filter.

c) Derive the expression for the center frequency of the filter.

d) Derive the expressions for the cutoff frequencies ω_{c1} and ω_{c2}.

e) What is the expression for the bandwidth of the filter?

f) What is the expression for the quality factor of the circuit?

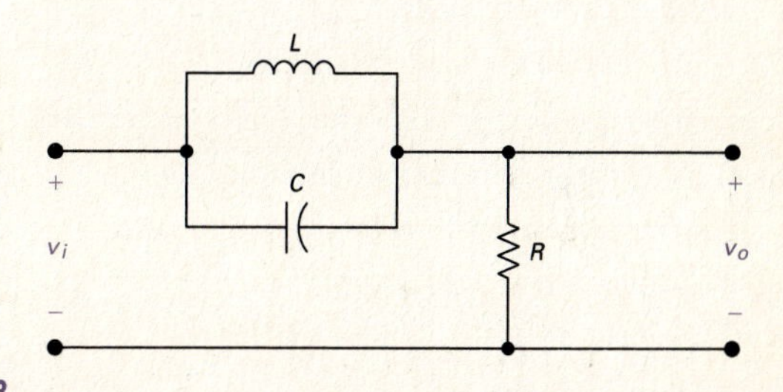

FIGURE P15.23

15.24 For the bandreject filter in Fig. P15.24, calculate (a) ω_o; (b) f_o; (c) Q; (d) ω_{c1}; (e) f_{c1}; (f) ω_{c2}; (g) f_{c2}; and (h) β in kilohertz.

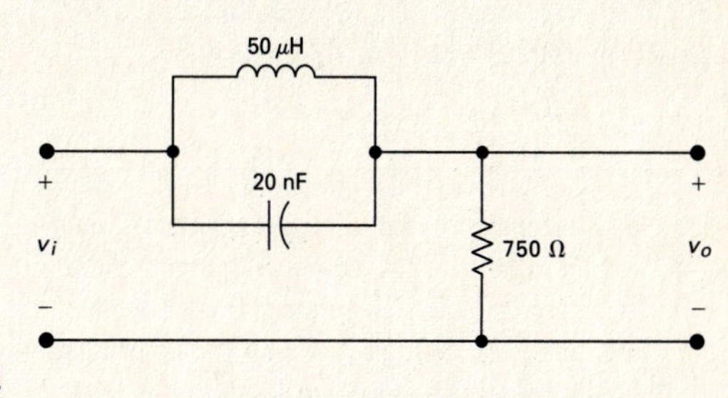

FIGURE P15.24

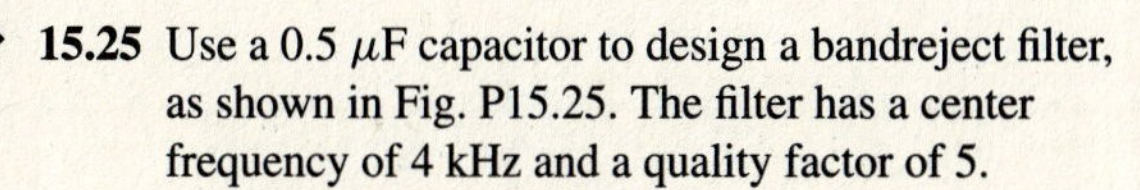

❖ **15.25** Use a 0.5 μF capacitor to design a bandreject filter, as shown in Fig. P15.25. The filter has a center frequency of 4 kHz and a quality factor of 5.

a) Specify the numerical values of R and L.

b) Calculate the upper and lower corner, or cutoff, frequencies in kilohertz.

c) Calculate the filter bandwidth in hertz.

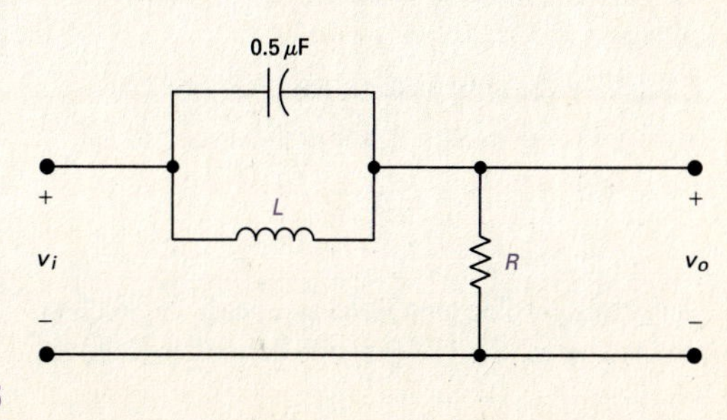

FIGURE P15.25

15.26 Assume the bandreject filter in Problem 15.25 is loaded with a 1 kΩ resistor.

a) What is the quality factor of the loaded circuit?

b) What is the bandwidth (in kilohertz) of the loaded circuit?

c) What is the upper cutoff frequency in kilohertz?

d) What is the lower cutoff frequency in kilohertz?

15.27 The purpose of this problem is to investigate how a resistive load connected across the output terminals of the bandreject filter shown in Fig. 15.28(a) affects the behavior of the filter. The loaded filter circuit is shown in Fig. P15.27.

a) Find the voltage transfer function V_o/V_i.

b) What is the expression for the center frequency?

c) What is the expression for the bandwidth?

d) What is the expression for the quality factor?

e) Evaluate $H(j\omega_o)$.

f) Evaluate $H(j0)$.

g) Evaluate $H(j\infty)$.

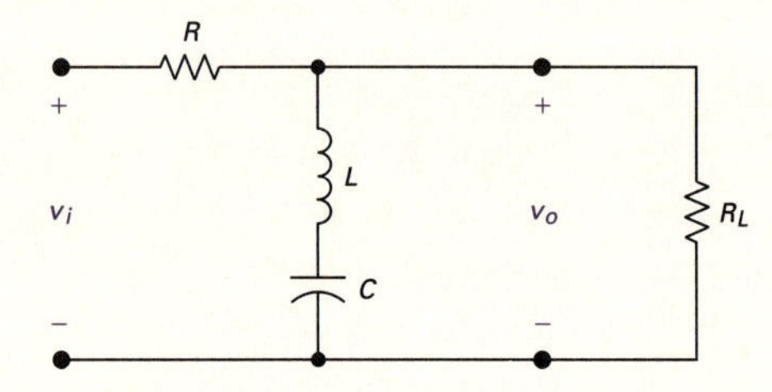

FIGURE P15.27

15.28 The parameters in the circuit in Fig. P15.27 are $R = 5$ kΩ, $L = 400$ mH, $C = 250$ pF, and $R_L = 20$ kΩ. Find (a) ω_o; (b) β in kilohertz; (c) Q; (d) $H(j0)$; (e) $H(j\infty)$; (f) f_{c2}; and (g) f_{c1}.

15.29 The load in the bandreject filter circuit shown in Fig. P15.27 is 36 kΩ. The center frequency of the filter is 1 Mrad/s, and the capacitor is 400 pF. At very low and very high frequencies, the amplitude of the sinusoidal output voltage should be at least 96% of the amplitude of the sinusoidal input voltage.

a) Specify the numerical values of R and L.

b) What is the quality factor of the circuit?

15.30 Make straight-line (uncorrected) amplitude and phase angle plots for each of the transfer functions derived in Problem 14.49.

15.31 Make straight-line amplitude and phase angle plots for the voltage transfer function derived in Problem 14.50.

15.32 a) Derive the numerical expression of the transfer function V_o/V_g for the circuit in Fig. P15.32.

b) Make a corrected amplitude plot for the transfer function derived in part (a).

c) At what frequency is the amplitude maximum?

d) What is the maximum amplitude in decibels?

e) At what frequencies is the amplitude down 3 dB from the maximum?

f) What is the bandwidth of the circuit?

g) Check your graphical results by calculating the actual amplitude in decibels at the frequencies read from the plot.

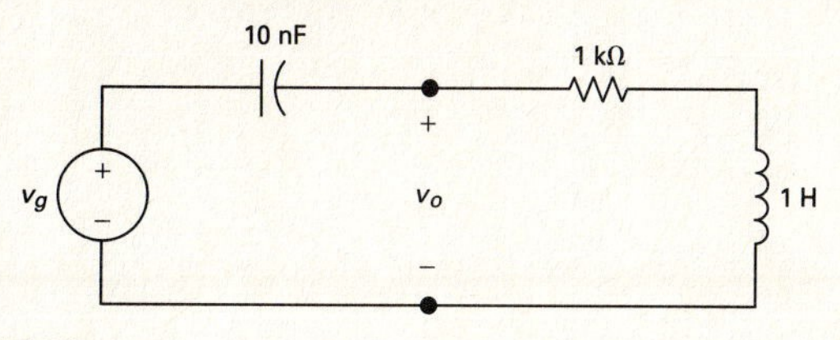

FIGURE P15.32

15.33 Use Bode diagrams to describe the behavior of the circuit in Problem 9.62 as R_x is varied from zero to infinity.

15.34 Given the following voltage transfer function:

$$H(s) = \frac{V_o}{V_i} = \frac{10^{10}}{s^2 + 50{,}000 + 10^{10}}.$$

a) At what frequencies (in radians per second) is the ratio of V_o/V_i equal to unity?

b) At what frequency is the ratio maximum?

c) What is the maximum value of the ratio?

15.35 The circuit shown in Fig. P15.35 resembles the interstage coupling network of an amplifier.

a) Show that

$$H(s) = \frac{V_o}{V_i}$$

$$= \frac{\left(\frac{1}{R_1C_2}\right)s}{s^2 + \left[\left(\frac{1}{R_1C_1}\right) + \left(\frac{1}{R_2C_2}\right) + \left(\frac{1}{R_1C_2}\right)\right]s + \left(\frac{1}{R_1C_1R_2C_2}\right)}.$$

b) Find the numerical expression for $H(s)$ if $R_1 = 40$ kΩ, $C_1 = 0.1\ \mu$F, $R_2 = 10$ kΩ, and $C_2 = 250$ pF.

c) Give the numerical values of the poles and zeros of $H(s)$.

d) Give an approximate numerical expression for $H(s)$ for values of s much less than the highest corner frequency, that is, for values of s close to the lowest corner frequency.

e) Give an approximate numerical expression for $H(s)$ for values of s much larger than the lowest corner frequency.

f) Show that the expression derived in part (d) is equivalent to the transfer function of the circuit in Fig. P15.35 when C_2 is neglected at low frequencies.

g) Show that the expression derived in part (e) is equivalent to the transfer function of the circuit in Fig. P15.35 when C_1 is neglected at high frequencies.

This problem illustrates that when the numerical values of the circuit parameters are known, it is sometimes possible to use different circuit models in different frequency ranges. Quite often, equivalent circuits for electronic amplifiers can be divided into models that apply to low-, mid-, and high-frequency ranges.

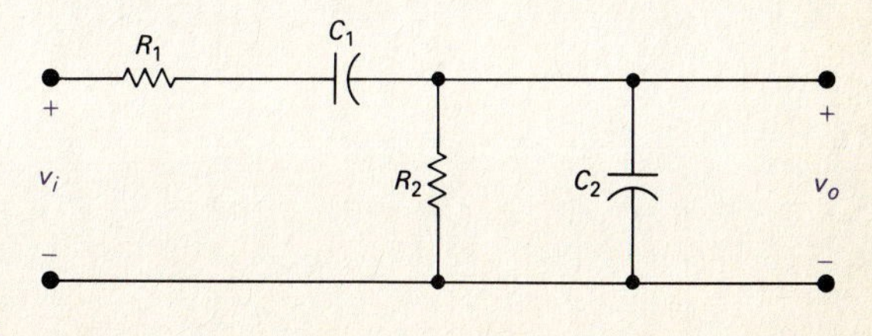

FIGURE P15.35

15.36 a) Find the resistance seen looking into the terminals a,b of the circuit in Fig. P15.36.

b) Find the power loss through the network, in decibels, when the output power is the power delivered to the 50 Ω resistor.

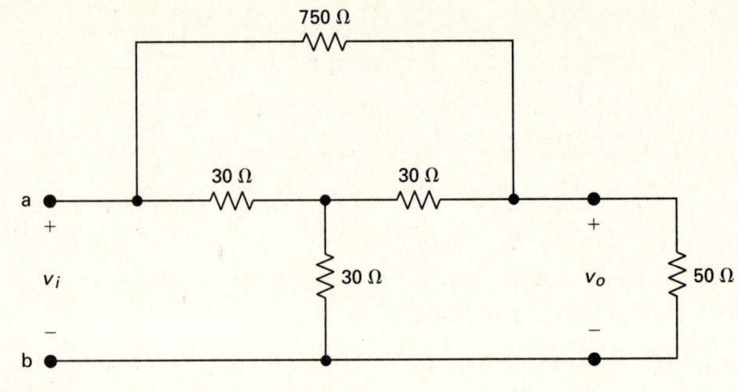

FIGURE P15.36

15.37 The amplitude plot of a transfer function is shown in Fig. P15.37. What is the unit step response of the system?

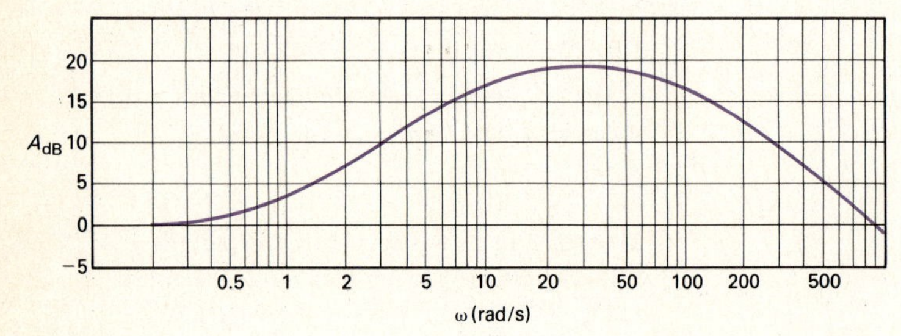

FIGURE P15.37

15.38 Design a DTMF high-band bandpass filter similar to the low-band filter design in the Practical Perspective example. Be sure to include the fourth high-frequency tone, 1633 Hz, in your design. What is the response amplitude of your filter to the highest of the low-frequency DTMF tones?

15.39 The 20 Hz signal that rings a telephone's bell has to have a very large amplitude to produce a loud enough bell signal. How much larger can the ringing signal amplitude be, relative to the low-bank DTMF signal, so that the response of the filter in the Practical Perspective example is no more than half as large as the largest of the DTMF tones?

16 ACTIVE FILTER CIRCUITS

CHAPTER CONTENTS

Up to this point, we have considered only passive filter circuits, that is, filter circuits consisting of resistors, inductors, and capacitors. There are areas of application, however, where active circuits—those which employ op amps—have certain advantages over passive filters. For instance, active circuits can produce bandpass and bandreject filters without using inductors. This is desirable because inductors are usually large, heavy, and costly, and they may introduce electromagnetic field effects which compromise the desired frequency response characteristics.

Examine the transfer functions of all the filter circuits from Chapter 15 and you will notice that the maximum magnitude does not exceed 1. While passive resonant filters can achieve voltage and current amplification at the resonant frequency, passive filters in general are incapable of amplification, because the output magnitude does not exceed the input magnitude. This is not a surprising observation, as many of the transfer functions in Chapter 15 were derived using voltage or current division. Active filters provide a control over amplification not available in passive filter circuits.

Finally, recall that both the cutoff frequency and the passband magnitude of passive filters were altered with the addition of a resistive load at the output of the filter. This is not the case with active filters, due to the properties of op amps. Thus, we use active circuits to implement filter designs when gain, load variation, and physical size are important parameters in the design specifications.

In this chapter we examine a few of the many filter circuits which employ op amps. As you will see, these op amp circuits overcome the disadvantages of passive filter circuits. Also, we will show how the basic op amp filter circuits can be combined to achieve specific frequency responses and to attain a more nearly ideal filter response. Note that throughout this chapter we assume that every op amp behaves as an ideal op amp.

16.1 FIRST-ORDER LOW-PASS AND HIGH-PASS FILTERS

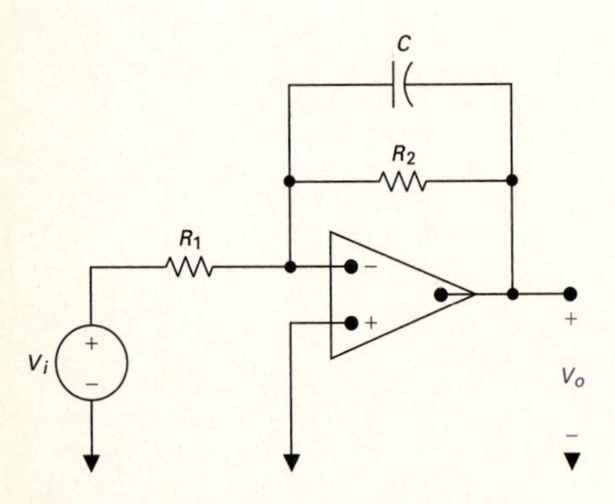

FIGURE 16.1 A first-order low-pass filter.

Consider the circuit in Fig. 16.1. Qualitatively, when the frequency of the source is varied, only the impedance of the capacitor is affected. At very low frequencies, the capacitor acts like an open circuit, and the op amp circuit acts like an amplifier with a gain of $-R_2/R_1$. At very high frequencies, the capacitor acts like a short circuit, thereby connecting the output of the op amp circuit to ground. The op amp circuit in Fig. 16.1 thus functions as a low-pass filter with a passband gain of $-R_2/R_1$.

To confirm this qualitative assessment, we can compute the transfer function $H(s) = V_o(s)/V_i(s)$. Note that the circuit in Fig. 16.1 has the general form of the circuit shown in Fig. 16.2, where the impedance in the input path (Z_i) is the resistor R_1, and the impedance in the feedback path (Z_f) is the parallel combination of the resistor R_2 and the capacitor C. The circuit in Fig. 16.2 is analogous to the inverting amplifier circuit from Chapter 5, so its transfer function is $-Z_f/Z_i$. Therefore, the transfer function for the circuit in Fig. 16.1 is

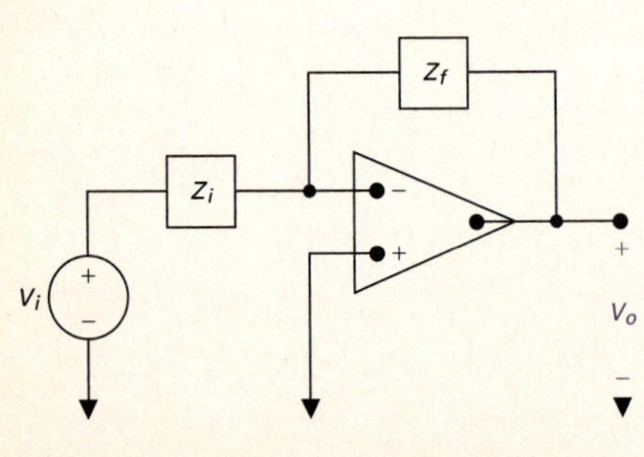

FIGURE 16.2 A general op amp circuit.

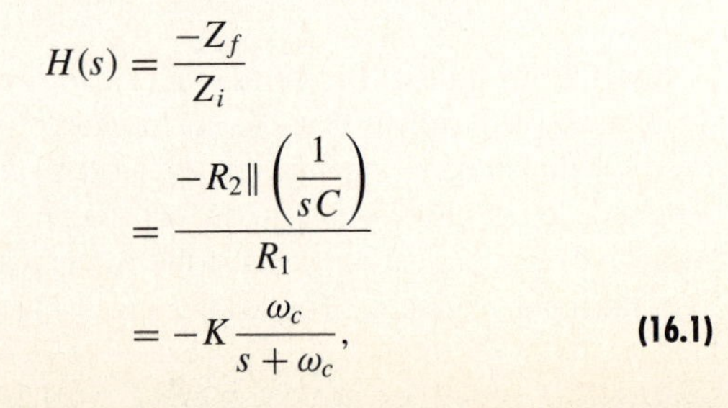

$$
\begin{aligned}
H(s) &= \frac{-Z_f}{Z_i} \\
&= \frac{-R_2 \| \left(\frac{1}{sC}\right)}{R_1} \\
&= -K \frac{\omega_c}{s + \omega_c}, \qquad (16.1)
\end{aligned}
$$

where

$$K = \frac{R_2}{R_1} \tag{16.2}$$

and

$$\omega_c = \frac{1}{R_2 C}. \tag{16.3}$$

Note that Eq. (16.1) has the same form as the general equation for low-pass filters given in Chapter 15, with an important exception: the gain in the passband, K, is set by the ratio R_2/R_1. The op amp low-pass filter thus permits the passband gain and the cutoff frequency to be specified independently, as Example 16.1 illustrates.

EXAMPLE 16.1

Using the circuit shown in Fig. 16.1, calculate values for C and R_2 which, together with $R_1 = 1\ \Omega$, produce a low-pass filter having a gain of 1 in the passband and a cutoff frequency of 1 rad/s. Construct the transfer function for this filter and use it to sketch a Bode magnitude plot of the filter's frequency response.

SOLUTION

Equation (16.2) gives the passband gain in terms of R_1 and R_2, so it allows us to calculate the required value of R_2:

$$\begin{aligned} R_2 &= KR_1 \\ &= (1)(1) \\ &= 1\ \Omega. \end{aligned}$$

Equation (16.3) then permits us to calculate C to meet the specified cutoff frequency:

$$\begin{aligned} C &= \frac{1}{R_2\omega_c} \\ &= \frac{1}{(1)(1)} \\ &= 1\text{ F}. \end{aligned}$$

The transfer function for the low-pass filter is given by Eq. (16.1):

$$\begin{aligned} H(s) &= -K\frac{\omega_c}{s+\omega_c} \\ &= \frac{-1}{s+1}. \end{aligned}$$

The Bode plot of $|H(j\omega)|$ is shown in Fig. 16.3. This is the so-called **prototype** low-pass op amp filter, because it uses a resistor value of 1 Ω and a capacitor value of 1 F, and it provides a cutoff frequency of 1 rad/s. As we shall see in the next section, prototype filters provide a useful starting point for the design of filters using more realistic component values to achieve a desired frequency response.

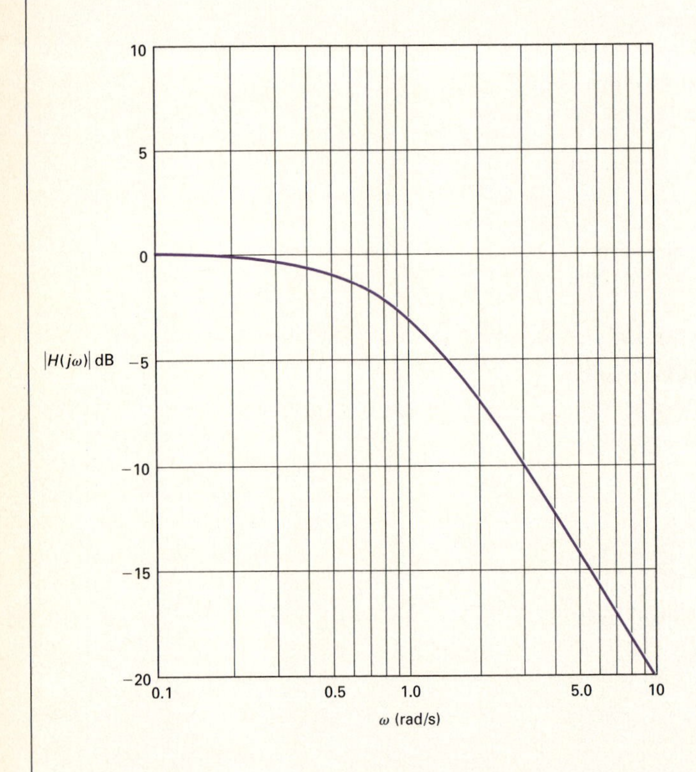

FIGURE 16.3 The Bode magnitude plot of the low-pass filter from Example 16.1.

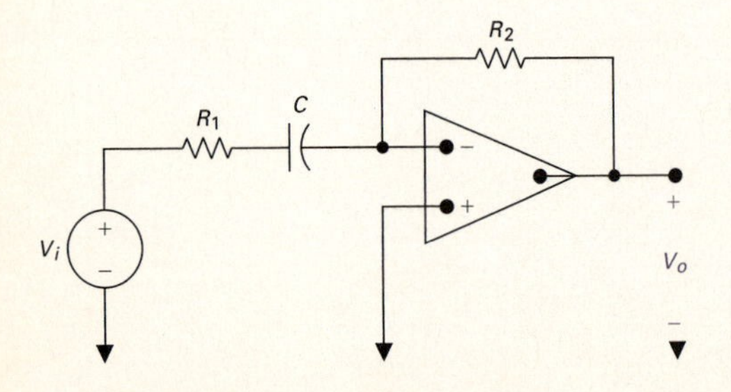

FIGURE 16.4 A first-order high-pass filter.

You may have recognized the circuit in Fig. 16.1 as the integrating amplifier circuit introduced in Chapter 7. They are indeed the same circuit, so integration in the time domain corresponds to low-pass filtering in the frequency domain. This relationship between integration and low-pass filtering is further confirmed by the operational Laplace transform for integration derived in Chapter 13.

The circuit in Fig. 16.4 is a first-order high-pass filter. This circuit also has the general form of the circuit in Fig. 16.2, only now the impedance in the input path is the series combination of R_1 and C, and the impedance in the feedback path is the resistor R_2. The transfer function for the circuit in Fig. 16.4 is thus

$$H(s) = \frac{-Z_f}{Z_i}$$

$$= \frac{-R_2}{R_1 + \frac{1}{sC}}$$

$$= -K\frac{s}{s+\omega_c}, \tag{16.4}$$

where

$$K = \frac{R_2}{R_1} \tag{16.5}$$

and

$$\omega_c = \frac{1}{R_1 C}. \tag{16.6}$$

Again, the form of the transfer function given in Eq. (16.4) is the same as that given in Eq. (15.20), the equation for passive high-pass filters. And again, the active filter permits the design of a passband gain greater than 1.

Example 16.2 considers the design of an active high-pass filter which must meet frequency response specifications from a Bode plot.

EXAMPLE 16.2

Figure 16.5 shows the Bode magnitude plot of a high-pass filter. Using the active high-pass filter circuit in Fig. 16.4, calculate values of R_1 and R_2 which produce the desired magnitude response. Use a 0.1 μF capacitor. If a 10 kΩ load resistor is added to this filter, how will the magnitude response change?

SOLUTION

Begin by writing a transfer function which has the magnitude plot shown in Fig. 16.5. To do this, note that the gain in the passband is 20 dB; therefore, $K = 10$. Also note that the 3 dB point is 500 rad/s. Equation (16.4) is the transfer function for a high-pass filter, so the transfer function which has the magnitude response shown in Fig. 16.5 is given by

$$H(s) = \frac{-10s}{s+500}.$$

We can compute the values of R_1 and R_2 needed to yield this transfer

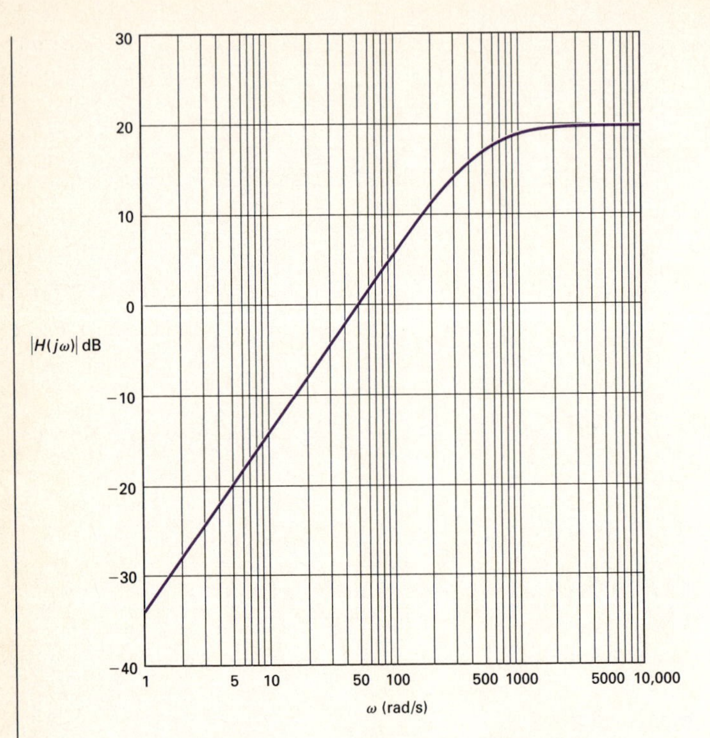

FIGURE 16.5 The Bode magnitude plot of the high-pass filter for Example 16.2.

function by equating the transfer function with Eq. (16.4):

$$H(s) = \frac{-10s}{s+500} = \frac{\frac{-R_2}{R_1}s}{s + \frac{1}{R_1 C}}.$$

Use a computer tool to confirm this result.

Equating the numerators and denominators and then simplifying, we get two equations:

$$10 = \frac{R_2}{R_1};$$

$$500 = \frac{1}{R_1 C}.$$

Using the specified value of C (0.1 μF), we find

$$R_1 = 20\text{ k}\Omega; \qquad R_2 = 200\text{ k}\Omega.$$

The circuit is shown in Fig. 16.6.

Since we have made the assumption that the op amp in this high-pass filter circuit is ideal, the addition of any load resistor, regardless of its resistance, has no effect on the behavior of the op amp. Thus, the magnitude response of a high-pass filter with a load resistor is the same as that of a high-pass filter with no load resistor, which is depicted in Fig. 16.5.

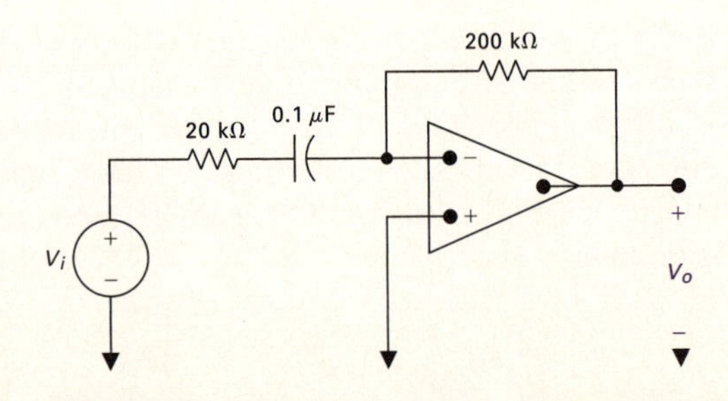

FIGURE 16.6 The high-pass filter for Example 16.2.

DRILL EXERCISES

16.1 Compute the values for R_1, R_2, and C that yield a high-pass filter with a passband gain of 1 and a cutoff frequency of 1 rad/s. (*Note:* This is the prototype high-pass filter.)

ANSWER: $R_1 = 1\ \Omega$, $R_2 = 1\ \Omega$, $C = 1$ F.

16.2 Compute the resistor values needed for the low-pass filter circuit in Fig. 16.1 to have the transfer function

$$H(s) = \frac{-20{,}000}{s + 5000}.$$

Use a 5 μF capacitor.

ANSWER: $R_1 = 10\ \Omega$, $R_2 = 40\ \Omega$.

16.2 SCALING

In the design and analysis of both passive and active filter circuits, working with element values such as 1 Ω, 1 H, and 1 F is convenient. Although these values are unrealistic for specifying practical components, they greatly simplify computations. After making computations using convenient values of R, L, and C, the designer can transform the convenient values into realistic values using the process known as **scaling**.

There are two types of scaling: **magnitude** and **frequency**. We scale a circuit in magnitude by multiplying the impedance at a given frequency by the scale factor k_m. Thus we multiply all resistors and inductors by k_m, and all capacitors by $1/k_m$. If we let unprimed variables represent the initial values of the parameters, and we let primed variables represent the scaled values of the variables, we have

$$R' = k_m R, \quad L' = k_m L, \quad \text{and} \quad C' = C/k_m. \qquad \textbf{(16.7)}$$

Note that k_m is by definition a positive real number which can be either less than or greater than 1.

In frequency scaling, we change the circuit parameters so that at the new frequency, the impedance of each element is the same as it was at the original frequency. Because resistance values are assumed to be independent of frequency, resistors are unaffected by frequency scaling. If we let k_f denote the frequency scale factor, both inductors and capacitors are multiplied by $1/k_f$. Thus for frequency scaling,

$$R' = R, \quad L' = L/k_f, \quad \text{and} \quad C' = C/k_f. \qquad \textbf{(16.8)}$$

The frequency scale factor k_f is also a positive real number that can be less than or greater than unity.

A circuit can be scaled simultaneously in both magnitude and frequency. The scaled values (primed) in terms of the original values (unprimed) are

$$R' = k_m R, \quad L' = \frac{k_m}{k_f} L, \quad \text{and} \quad C' = \frac{1}{k_m k_f} C. \tag{16.9}$$

Example 16.3 illustrates the scaling process.

EXAMPLE 16.3

The series RLC circuit shown in Fig. 16.7 has a center frequency of $\sqrt{1/LC} = 1$ rad/s, a bandwidth of $R/L = 1$ rad/s, and thus a quality factor of 1. Use scaling to compute new values of R and L which yield a circuit with the same quality factor but a center frequency of 500 Hz. Use a 2 μF capacitor.

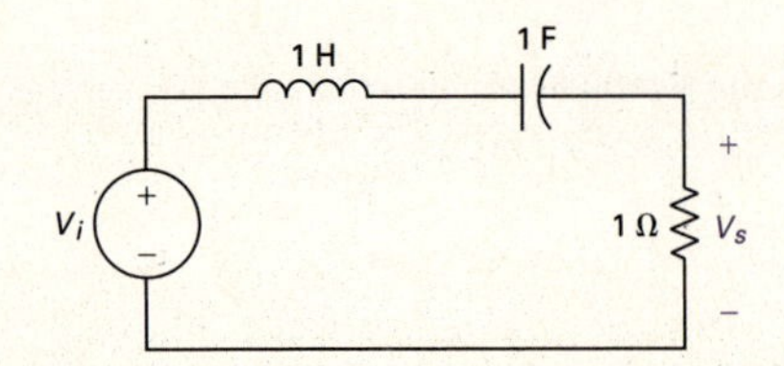

FIGURE 16.7 The series *RLC* circuit for Example 16.3.

SOLUTION

Begin by computing the frequency scale factor which will shift the center frequency from 1 rad/s to 500 Hz. The unprimed variables represent values before scaling, while the primed variables represent values after scaling.

$$k_f = \frac{\omega_o'}{\omega_o} = \frac{2\pi(500)}{1} = 3141.59.$$

Now, use Eq. (16.9) to compute the magnitude scale factor which, together with the frequency scale factor, will yield a capacitor value of 2 μF:

$$k_m = \frac{1}{k_f}\frac{C}{C'} = \frac{1}{(3141.59)(2 \times 10^{-6})} = 159.155.$$

Use Eq. (16.9) again to compute the magnitude and frequency scaled values of R and L:

$$R' = k_m R = 159.155\ \Omega;$$

$$L' = \frac{k_m}{k_f} L = 50.66 \text{ mH}.$$

With these component values, the center frequency of the series RLC circuit is $\sqrt{1/LC} = 3141.61$ rad/s or 500 Hz, and the bandwidth is $R/L = 3141.61$ rad/s or 500 Hz; thus the quality factor is still 1.

The Use of Scaling in the Design of Op Amp Filters

To use the concept of scaling in the design of op amp filters, first select the cutoff frequency, ω_c, to be 1 rad/s (if you are designing low- or high-pass filters), or select the center frequency, ω_o, to be 1 rad/s (if you are designing bandpass or bandreject filters). Then select a 1 F capacitor and calculate the values of the resistors needed to give the desired passband gain and the 1 rad/s cutoff or center frequency. Finally, use scaling to compute more realistic component values which give the desired cutoff or center frequency.

Example 16.4 illustrates the use of scaling in the design of a low-pass filter.

Example 16.4

Use the prototype low-pass op amp filter from Example 16.1 along with magnitude and frequency scaling to compute the resistor values for a low-pass filter with a gain of 5, a cutoff frequency of 1000 Hz, and a feedback capacitor of 0.01 μF. Construct a Bode plot of the resulting transfer function's magnitude.

Solution

To begin, use frequency scaling to place the cutoff frequency at 1000 Hz:

$$k_f = \omega_c'/\omega_c = 2\pi(1000)/1 = 6283.185,$$

where the primed variable has the new value and the unprimed variable has the old value of the cutoff frequency. Then compute the magnitude scale factor which, together with $k_f = 6283.185$, will scale the capacitor to 0.01 μF:

$$k_m = \frac{1}{k_f}\frac{C}{C'} = \frac{1}{(6283.185)(10^{-8})} = 15{,}915.5.$$

Since resistors are scaled only by using magnitude scaling,

$$R_1' = R_2' = k_m R = (15{,}915.5)(1) = 15{,}915.5\ \Omega.$$

Finally, we need to meet the passband gain specification. We can adjust the scaled values of either R_1 or R_2, since $K = R_2/R_1$. If we adjust R_2, we will change the cutoff frequency, because $\omega_c = 1/R_2C$. Therefore, we can adjust the value of R_1 to alter only the passband gain:

$$R_1 = R_2/K = (15{,}915.5)/(5) = 3183.1\ \Omega.$$

The final component values are

$$R_1 = 3183.1\ \Omega;$$
$$R_2 = 15{,}915.5\ \Omega;$$
$$C = 0.01\ \mu\text{F}.$$

The transfer function of the filter is given by

$$H(s) = \frac{-31{,}415.93}{s + 6283.185}.$$

The Bode plot of the magnitude of this transfer function is shown in Fig. 16.8.

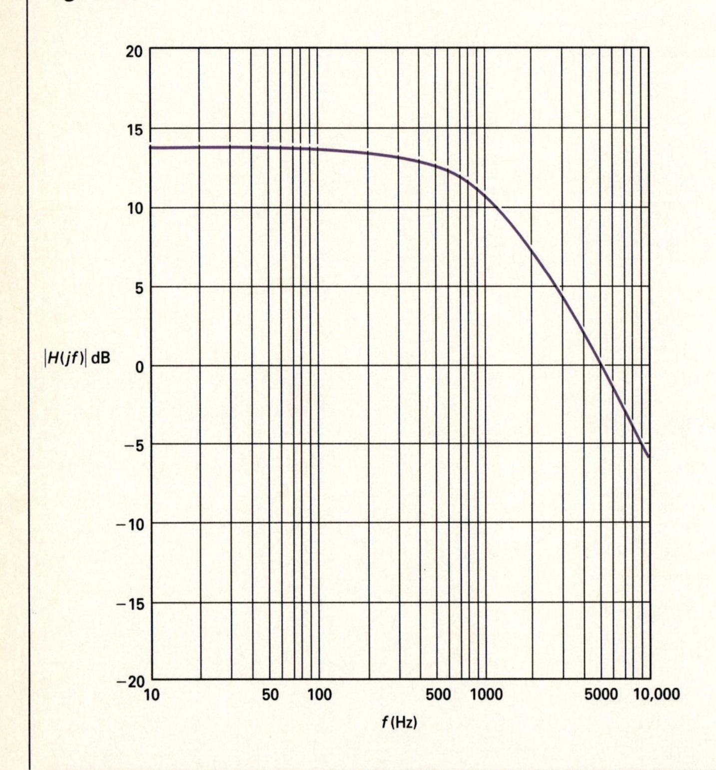

FIGURE 16.8 The Bode magnitude plot of the low-pass filter from Example 16.4.

DRILL EXERCISE

16.3 What magnitude and frequency scale factors will transform the prototype high-pass filter into a high-pass filter with a 0.5 μF capacitor and a cutoff frequency of 10 kHz?

ANSWER: $k_f = 62{,}831.85$, $k_m = 31.31$.

16.3 OP AMP BANDPASS AND BANDREJECT FILTERS

We now turn to the analysis and design of op amp circuits which act as bandpass and bandreject filters. While there is a wide variety of such op amp circuits, our approach is motivated by the Bode plot construction shown in Fig. 16.9. We can see from the plot that the bandpass filter consists of three separate components:

1. a unity-gain low-pass filter whose cutoff frequency is ω_{c2}, the larger of the two cutoff frequencies;
2. a unity-gain high-pass filter whose cutoff frequency is ω_{c1}, the smaller of the two cutoff frequencies; and
3. a gain component to provide the desired level of gain in the passband.

These three components are cascaded in series. They combine additively in the Bode plot construction, and so will combine multiplicatively in the s domain.

It is important to note that this method of constructing a bandpass magnitude response assumes that the lower cutoff frequency (ω_{c1})

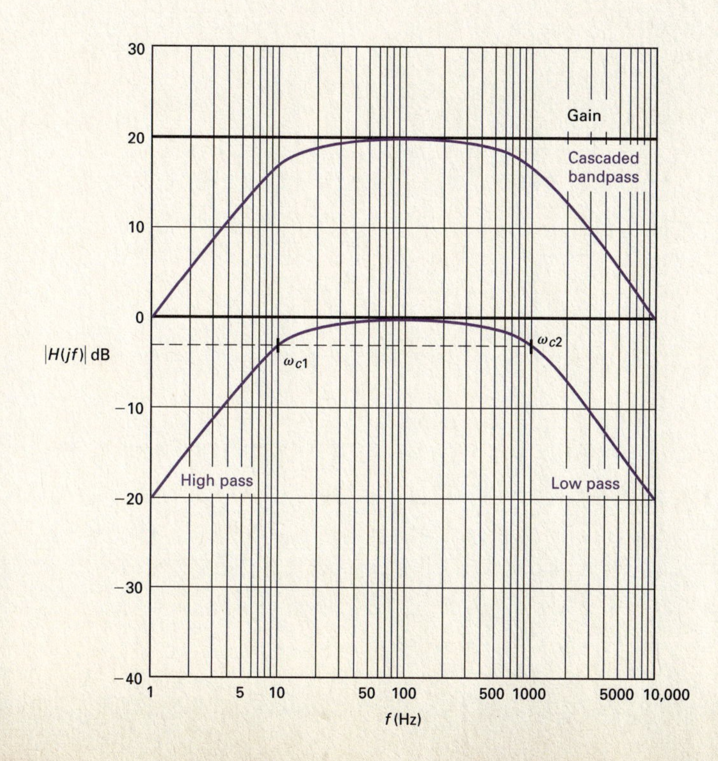

FIGURE 16.9 Constructing the Bode magnitude plot of a bandpass filter.

is smaller than the upper cutoff frequency (ω_{c2}). The resulting filter is called a **broadband** bandpass filter, since the band of frequencies passed is wide. The formal definition of a broadband filter requires the two cutoff frequencies to satisfy the equation

$$\frac{\omega_{c2}}{\omega_{c1}} \geq 2. \quad \textbf{(16.10)}$$

When Eq. (16.10) is satisfied, the magnitude of the high-pass filter is unity at the cutoff frequency of the low-pass filter, and the magnitude of the low-pass filter is unity at the cutoff frequency of the high-pass filter. Thus the bandpass filter will have the cutoff frequencies specified by the low-pass and high-pass filters. If the cutoff frequencies do not satisfy Eq. (16.10), the bandpass filter will not have the same cutoff frequencies as the low-pass and high-pass components of the cascade. Thus, the cascade approach to the design of op amp bandpass filters is useful only for broadband filters. We will have more to say about this in Section 16.5.

We can construct a circuit which provides each of the three components by cascading a low-pass op amp filter, a high-pass op amp filter, and an inverting amplifier (see Section 5.3), as shown in Fig. 16.10(a). Figure 16.10(a) is a form of illustration called a **block diagram**. Each block represents a component or subcircuit, and the output of one block is the input to the next, in the direction indicated. Because each circuit of the cascade is an op amp,

FIGURE 16.10 A cascaded op amp bandpass filter: (a) the block diagram; (b) the circuit.

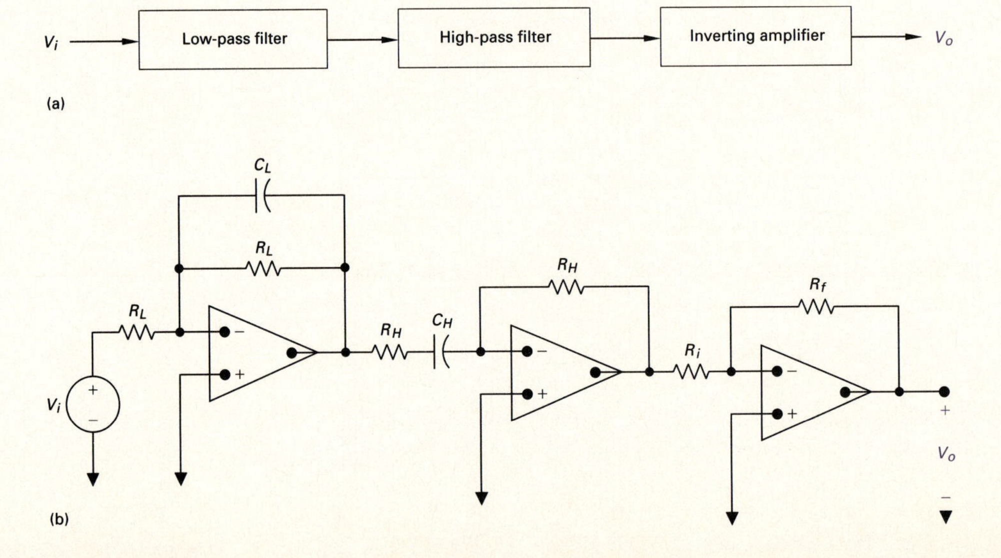

and because all of the op amps are assumed to be ideal, each circuit can be designed independently, without concern for the other circuits in the cascade. This reduces the design of the bandpass filter to the design of a unity-gain first-order low-pass filter, a unity-gain first-order high-pass filter, and an inverting amplifier, each of which is a simple circuit.

The transfer function of the cascaded bandpass filter is the product of the transfer functions of the three cascaded components:

$$\begin{aligned} H(s) &= \frac{V_o}{V_i} \\ &= \left(\frac{-\omega_{c2}}{s+\omega_{c2}}\right)\left(\frac{-s}{s+\omega_{c1}}\right)\left(\frac{-R_f}{R_i}\right) \\ &= \frac{-K\omega_{c2}s}{(s+\omega_{c1})(s+\omega_{c2})}. \end{aligned} \tag{16.11}$$

We compute the values of R_L and C_L in the low-pass filter to give us the desired upper cutoff frequency, ω_{c2}:

$$\omega_{c2} = \frac{1}{R_L C_L}. \tag{16.12}$$

We compute the values of R_H and C_H in the high-pass filter to give us the desired lower cutoff frequency, ω_{c1}:

$$\omega_{c1} = \frac{1}{R_H C_H}. \tag{16.13}$$

Now we compute the values of R_i and R_f in the inverting amplifier to provide the desired passband gain. To do this, we consider the magnitude of the bandpass filter's transfer function, evaluated at the center frequency, ω_o:

$$\begin{aligned} |H(j\omega_o)| &= \left|\frac{-K\omega_{c2}(j\omega_o)}{(j\omega_o)^2+(\omega_{c1}+\omega_{c2})(j\omega_o)+\omega_{c1}\omega_{c2}}\right| \\ &= \frac{K\omega_{c2}}{\omega_{c1}+\omega_{c2}}. \end{aligned} \tag{16.14}$$

Recall from Chapter 5 that the gain of the inverting amplifier is R_f/R_i. Therefore,

$$|H(j\omega_o)| = \frac{R_f}{R_i}\frac{\omega_{c2}}{(\omega_{c1}+\omega_{c2})}. \tag{16.15}$$

Any choice of resistors which satisfies Eq. (16.15) will produce the desired passband gain.

Example 16.5 illustrates the design process for the cascaded bandpass filter.

EXAMPLE 16.5

Design a bandpass filter for a graphic equalizer to provide an amplification of 2 within the band of frequencies between 500 and 5000 Hz. Use 0.2 μF capacitors.

SOLUTION

Begin with the low-pass stage. From Eq. (16.12),

$$\omega_{c2} = \frac{1}{R_L C_L} = 2\pi(5000);$$

$$R_L = \frac{1}{[2\pi(5000)](0.2 \times 10^{-6})}$$

$$= 159\ \Omega.$$

Next, we turn to the high-pass stage. From Eq. (16.13),

$$\omega_{c1} = \frac{1}{R_H C_H} = 2\pi(500);$$

$$R_H = \frac{1}{[2\pi(500)](0.2 \times 10^{-6})}$$

$$= 1591\ \Omega.$$

Finally, we need the gain stage. From Eq. (16.15), we see there are two unknowns, so one of the resistors can be selected arbitrarily. Let's select a 1 kΩ resistor for R_i. Then, from Eq. (16.15),

$$R_f = 2\left[\frac{\omega_{c1} + \omega_{c2}}{\omega_{c2}}\right]1000$$

$$= 2200\ \Omega.$$

The resulting circuit is shown in Fig. 16.11.

Use a computer tool to confirm the frequency response of this three-stage circuit.

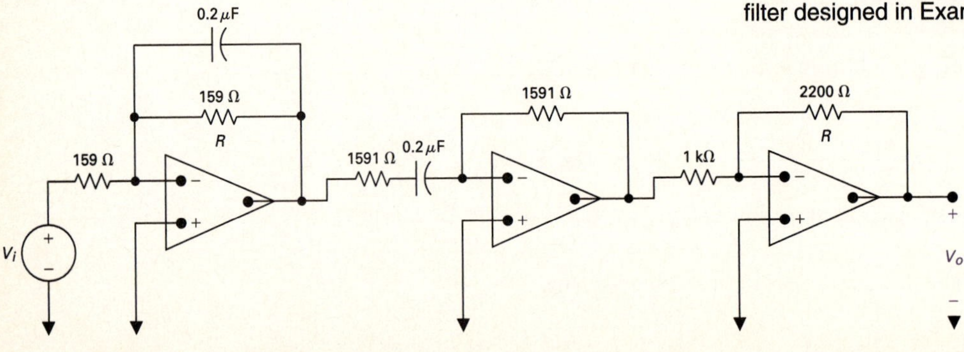

FIGURE 16.11 The cascaded op amp bandpass filter designed in Example 16.5.

We can use a component approach to the design of op amp bandreject filters too, as illustrated in Fig. 16.12. Like the bandpass filter, the bandreject filter consists of three separate components. There are important differences, however:

1. The unity-gain low-pass filter has a cutoff frequency of ω_{c1}, which is the smaller of the two cutoff frequencies.

2. The unity-gain high-pass filter has a cutoff frequency of ω_{c2}, which is the larger of the two cutoff frequencies.

3. The gain component provides the desired level of gain in the passbands.

The most important difference is that these three components cannot be cascaded in series, since they do not combine additively on the Bode plot. Instead, we use a parallel connection and a summing amplifier, as shown both in block diagram form and as a circuit in Fig. 16.13. Again it is assumed that the two cutoff frequencies are widely separated, so that the resulting design is a broadband bandreject filter, and $\omega_{c2}/\omega_{c1} \geq 2$. Then each component of the parallel design can be created independently. The transfer function of the resulting circuit is the sum of the low-pass and high-pass filter transfer

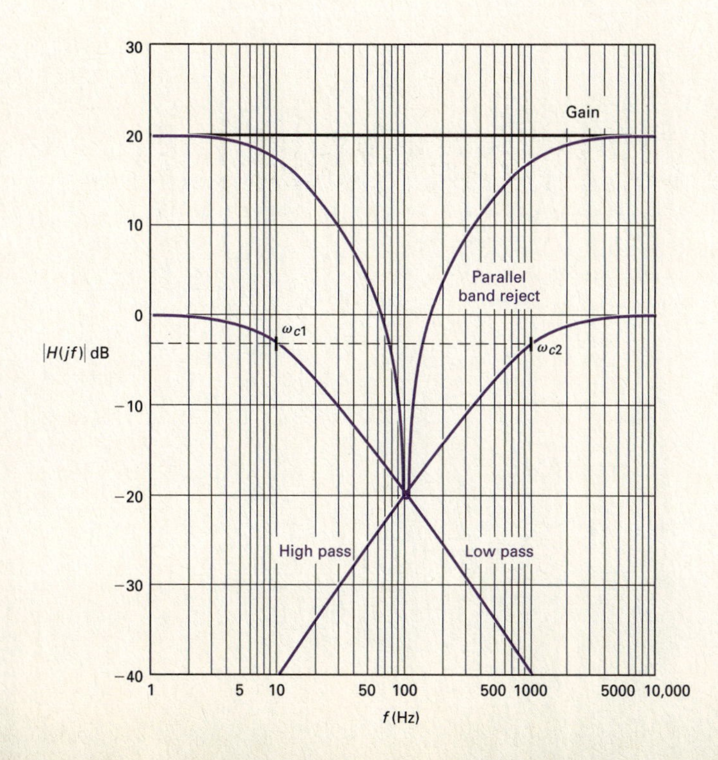

FIGURE 16.12 Constructing the Bode magnitude plot of a bandreject filter.

functions. From Fig. 16.13(b),

$$H(s) = \left(-\frac{R_f}{R_i}\right)\left[\frac{-\omega_{c1}}{s+\omega_{c1}} + \frac{-s}{s+\omega_{c2}}\right]$$

$$= \frac{R_f}{R_i}\left(\frac{\omega_{c1}(s+\omega_{c2}) + s(s+\omega_{c1})}{(s+\omega_{c1})(s+\omega_{c2})}\right)$$

$$= \frac{R_f}{R_i}\left(\frac{s^2 + 2\omega_{c1}s + \omega_{c1}\omega_{c2}}{(s+\omega_{c1})(s+\omega_{c2})}\right). \quad \textbf{(16.16)}$$

From the transfer function in Eq. (16.16), it is clear that the two cutoff frequencies are ω_{c1} and ω_{c2}. They are given by the equations

$$\omega_{c1} = \frac{1}{R_L C_L} \quad \textbf{(16.17)}$$

and

$$\omega_{c2} = \frac{1}{R_H C_H}. \quad \textbf{(16.18)}$$

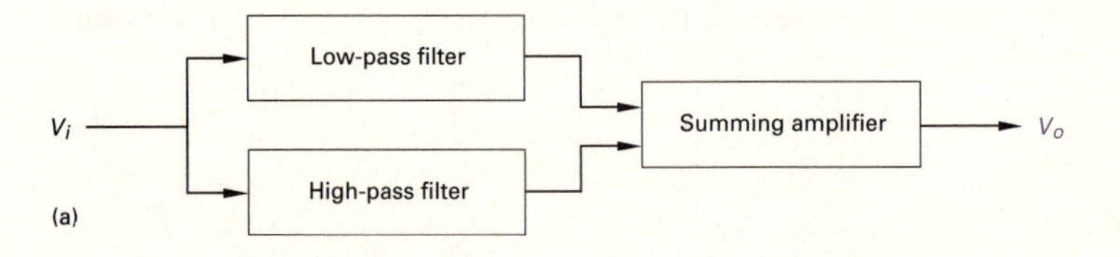

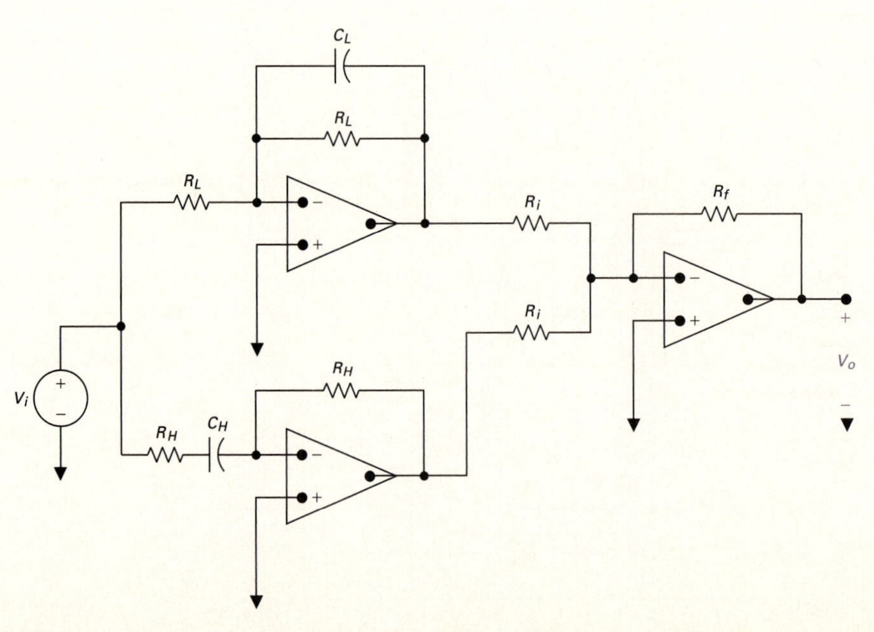

FIGURE 16.13 A parallel op amp bandreject filter: (a) the block diagram; (b) the circuit.

In the two passbands (as $s \to 0$ and $s \to \infty$), the gain of the transfer function is R_f/R_i. Therefore,

$$K = \frac{R_f}{R_i}. \tag{16.19}$$

As with the design of the cascaded bandpass filter, we have six unknowns and three equations. Typically we choose a commercially available capacitor value for C_L and C_H. Then Eqs. (16.17) and (16.18) permit us to calculate R_L and R_H to meet the specified cutoff frequencies. Finally, we choose a value for either R_f or R_i and then use Eq. (16.19) to compute the other resistance.

Note the magnitude of the transfer function in Eq. (16.16) at the center frequency, ω_o:

$$\begin{aligned}|H(j\omega_o)| &= \left|\frac{R_f}{R_i}\left(\frac{(j\omega_o)^2 + 2\omega_{c1}(j\omega_o) + \omega_{c1}\omega_{c2}}{(j\omega_o)^2 + (\omega_{c1} + \omega_{c2})(j\omega_o) + \omega_{c1}\omega_{c2}}\right)\right| \\ &= \frac{R_f}{R_i}\frac{2\omega_{c1}}{\omega_{c1} + \omega_{c2}} \\ &= \frac{R_f}{R_i}\frac{2}{1 + \omega_{c2}/\omega_{c1}}. \end{aligned} \tag{16.20}$$

If $\omega_{c2}/\omega_{c1} \geq 2$, then $|H(j\omega_o)| \leq \frac{2}{3}R_f/R_i$, so the magnitude at the center frequency is less than two-thirds of the passband magnitude. Thus the bandreject filter successfully rejects frequencies near the center frequency, again confirming our assumption that the parallel implementation is meant for broadband bandreject designs.

Example 16.6 illustrates the design process for the parallel bandreject filter.

EXAMPLE 16.6

Design a circuit based on the parallel bandreject op amp filter in Fig. 16.13(b). The Bode magnitude response of this filter is shown in Fig. 16.14. Use 0.5 μF capacitors in your design.

SOLUTION

From the Bode magnitude plot in Fig. 16.14, we see that the bandreject filter has cutoff frequencies of 100 rad/s and 2000 rad/s, and a gain of 3 in the passbands. Begin with the prototype low-pass filter and use scaling to meet the specifications for cutoff frequency and capacitor value. The frequency scale factor k_f is 100, which shifts the cutoff frequency from 1 rad/s to 100 rad/s. The magnitude scale factor

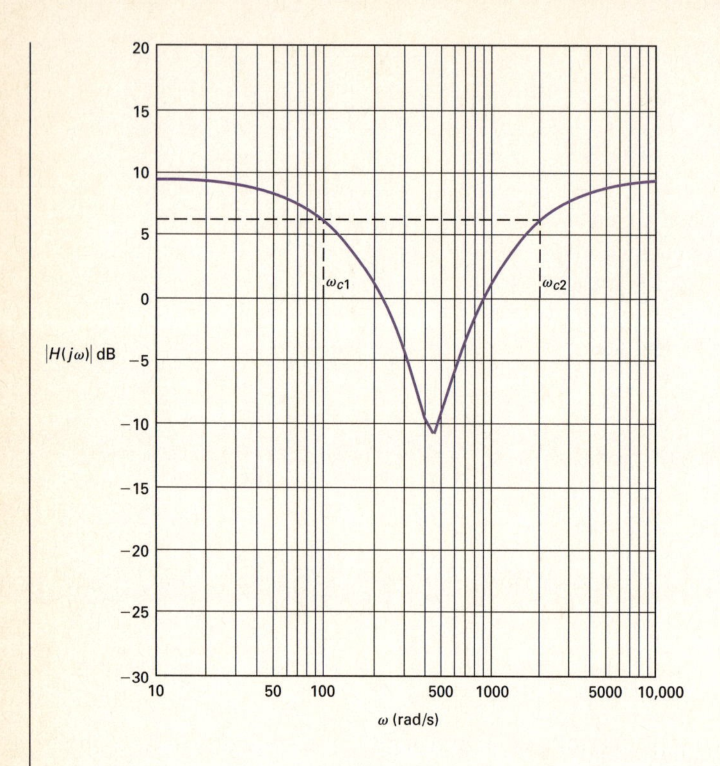

FIGURE 16.14 The Bode magnitude plot for the circuit to be designed in Example 16.6.

k_m is 20,000, which permits the use of a 0.5 μF capacitor. Using these scale factors results in the following scaled component values:

$$R_L = 20 \text{ k}\Omega;$$
$$C_L = 0.5\ \mu\text{F}.$$

The resulting cutoff frequency of the low-pass filter component is

$$\omega_{c1} = \frac{1}{R_L C_L} = \frac{1}{(20 \times 10^3)(0.5 \times 10^{-6})} = 100 \text{ rad/s}.$$

We use the same approach to design the high-pass filter, starting with the prototype high-pass op amp filter. Here, the frequency scale factor is $k_f = 2000$, and the magnitude scale factor is $k_m = 1000$, resulting in the following scaled component values:

$$R_H = 1 \text{ k}\Omega;$$
$$C_H = 0.5\ \mu\text{F}.$$

Finally, since the cutoff frequencies are widely separated, we can use the ratio R_f/R_i to establish the desired passband gain of 3. Let's choose $R_i = 1$ kΩ, since we are already using that resistance for R_H. Then $R_f = 3$ kΩ, and $K = R_f/R_i = 3000/1000 = 3$. The resulting parallel op amp bandreject filter circuit is shown in Fig. 16.15.

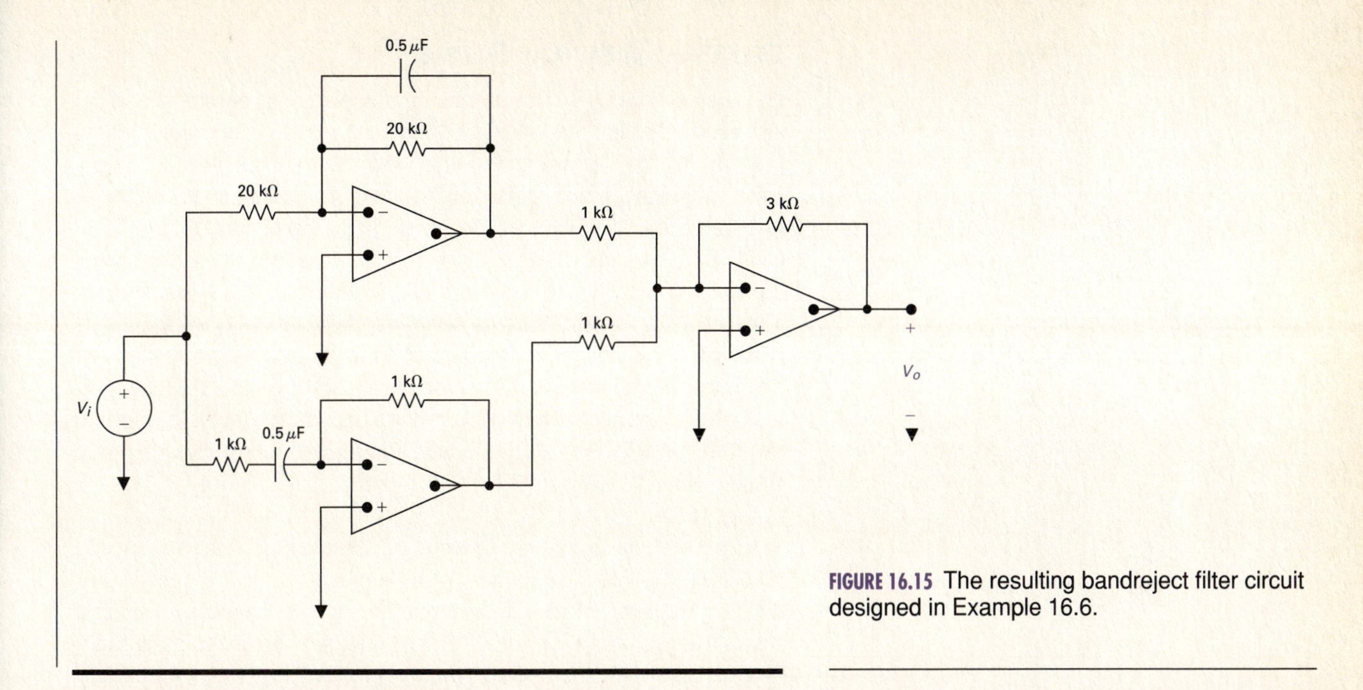

FIGURE 16.15 The resulting bandreject filter circuit designed in Example 16.6.

DRILL EXERCISES

16.4 Design a unity-gain bandpass filter, using a cascade connection, to give a center frequency of 200 Hz and a bandwidth of 1000 Hz. Use 5 μF capacitors. Specify ω_{c1}, ω_{c2}, R_L, and R_H.

ANSWER: $f_{c1} = 38.52$ Hz, $f_{c2} = 1038.52$ Hz, $R_L = 30.65\ \Omega$, and $R_H = 826.35\ \Omega$.

16.5 Design a parallel bandreject filter with a center frequency of 1000 rad/s, a bandwidth of 4000 rad/s, and a passband gain of 6. Use 0.2 μF capacitors, and specify all resistor values.

ANSWER: $R_L = 21.18$ kΩ, $R_H = 1.18$ kΩ, and $R_f/R_i = 6$.

16.4 HIGHER ORDER OP AMP FILTERS

You have probably noticed that all of the filter circuits we have examined so far, both passive and active, are nonideal. Remember from Chapter 15 that an ideal filter has a discontinuity at the point of cutoff, which sharply divides the passband and the stopband. Although we cannot hope to construct a circuit with a discontinuous frequency response, we can construct circuits with a sharper, yet still continuous, transition at the cutoff frequency.

Cascading Identical Filters

How can we obtain a sharper transition between the passband and the stopband? One approach is suggested by the Bode magnitude plots in Fig. 16.16. This figure shows the Bode magnitude plots of a cascade of identical prototype low-pass filters and includes plots of just one filter, two in cascade, three in cascade, and four in cascade. It is obvious that as more filters are added to the cascade, the transition from the passband to the stopband becomes sharper. The rules for constructing Bode plots (from Section 15.6) tell us that with one filter, the transition occurs with an asymptotic slope of 20 dB/decade. Since circuits in cascade are additive on a Bode magnitude plot, a cascade with two filters has a transition with an asymptotic slope of $20 + 20 = 40$ dB/decade; for three filters, the asymptotic slope is 60 dB/decade, and for four filters, it is 80 dB/decade, as seen in Fig. 16.16.

In general, an n-element cascade of identical low-pass filters will transition from the passband to the stopband with a slope of $20n$ dB/dec. Both the block diagram and the circuit diagram for such a cascade are shown in Fig. 16.17. It is easy to compute the transfer function for a cascade of n prototype low-pass filters—we just multiply

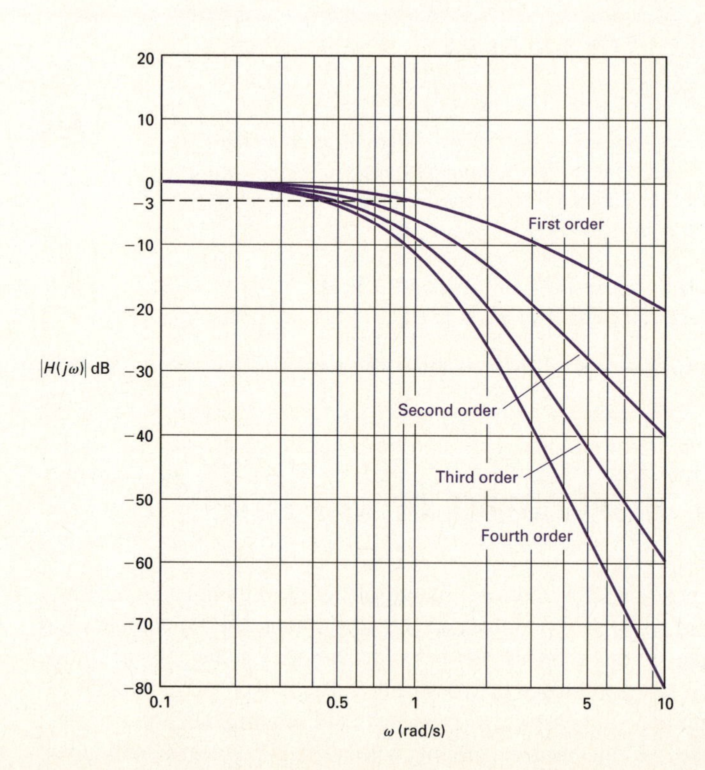

FIGURE 16.16 The Bode magnitude plot of a cascade of identical prototype low-pass filters.

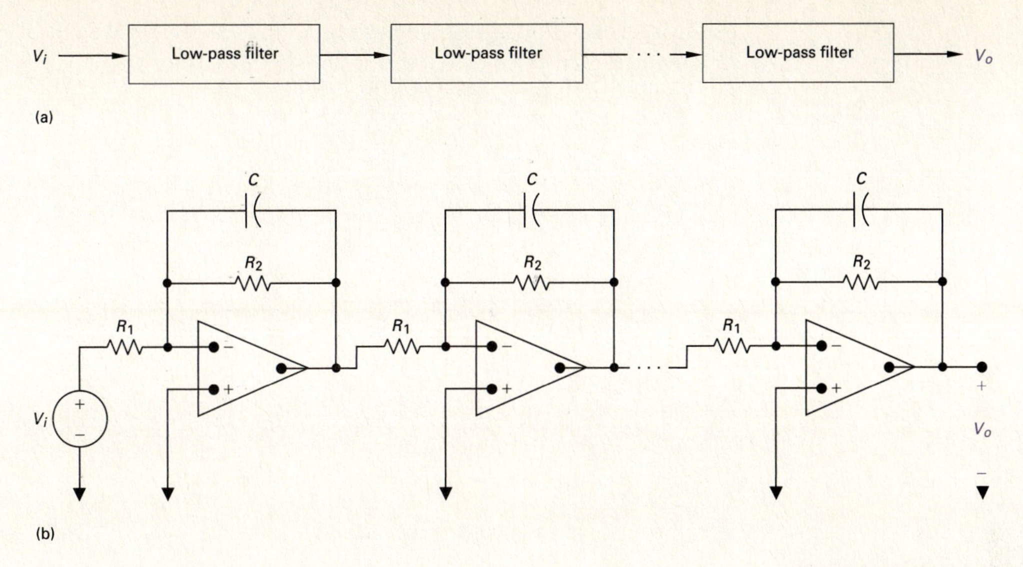

FIGURE 16.17 A cascade of identical unity-gain low-pass filters: (a) the block diagram; (b) the circuit.

the individual transfer functions:

$$H(s) = \left(\frac{-1}{s+1}\right)\left(\frac{-1}{s+1}\right)\cdots\left(\frac{-1}{s+1}\right)$$

$$= \frac{(-1)^n}{(s+1)^n}. \qquad \textbf{(16.21)}$$

The order of a filter is determined by the number of poles in its transfer function. From Eq. (16.21) we see that a cascade of first-order low-pass filters yields a higher order filter. In fact, a cascade of n first-order filters produces an nth-order filter, having n poles in its transfer function and a final slope of $20n$ dB/dec in the transition band.

There is an important issue yet to be resolved, as you will see if you look closely at Fig. 16.16. As the order of the low-pass filter is increased by adding prototype low-pass filters to the cascade, the cutoff frequency also changes. For example, in a cascade of two first-order low-pass filters, the magnitude of the resulting second-order filter at ω_c is -6 dB, so the cutoff frequency of the second-order filter is not ω_c. In fact, the cutoff frequency is less than ω_c.

As long as we are able to calculate the cutoff frequency of the higher order filters formed in the cascade of first-order filters, we can use frequency scaling to calculate component values which move the cutoff frequency to its specified location. If we start with a cascade of n prototype low-pass filters, we can compute the cutoff frequency for the resulting nth-order low-pass filter. We do so by solving for the value of ω_{cn} that results in $|H(j\omega)| = 1/\sqrt{2}$:

$$H(s) = \frac{(-1)^n}{(s+1)^n};$$

$$|H(j\omega_{cn})| = \left|\frac{1}{(j\omega_{cn}+1)^n}\right| = \frac{1}{\sqrt{2}};$$

$$\frac{1}{(\sqrt{\omega_{cn}^2+1})^n} = \frac{1}{\sqrt{2}};$$

$$\frac{1}{\omega_{cn}^2+1} = \left(\frac{1}{\sqrt{2}}\right)^{2/n};$$

$$\sqrt[n]{2} = \omega_{cn}^2 + 1;$$

$$\omega_{cn} = \sqrt{\sqrt[n]{2} - 1}. \qquad \textbf{(16.22)}$$

To demonstate the use of Eq. (16.22), let's compute the cutoff frequency of a fourth-order unity-gain low-pass filter constructed from a cascade of four prototype low-pass filters:

Use a computer tool to calculate ω_{cn} for other values of n.

$$\omega_{c4} = \sqrt{\sqrt[4]{2} - 1} = 0.435 \text{ rad/s}. \qquad \textbf{(16.23)}$$

Thus, we can design a fourth-order low-pass filter with any arbitrary cutoff frequency by starting with a fourth-order cascade consisting of prototype low-pass filters and then scaling the components by $k_f = \omega_c/0.435$ to place the cutoff frequency at any value of ω_c desired.

Note that we can build a higher order low-pass filter with a nonunity gain by adding an inverting amplifier circuit to the cascade. Example 16.7 illustrates the design of a fourth-order low-pass filter with nonunity gain.

EXAMPLE 16.7

Design a fourth-order low-pass filter with a cutoff frequency of 500 Hz and a passband gain of 10. Use 1 μF capacitors. Sketch the Bode magnitude plot for this filter.

SOLUTION

We begin our design with a cascade of four prototype low-pass filters. We have already used Eq. (16.23) to calculate the cutoff frequency for the resulting fourth-order low-pass filter as 0.435 rad/s. A frequency scale factor of $k_f = 7222.39$ will scale the component values to give a 500 Hz cutoff frequency. A magnitude scale factor of $k_m = 138.46$ permits the use of 1 μF capacitors. The scaled component values are thus

$$R = 138.46\ \Omega;$$
$$C = 1\ \mu\text{F}.$$

Finally, add an inverting amplifier stage with a gain of $R_f/R_i = 10$. As usual, we can arbitrarily select one of the two resistor values. Since we are already using 138.46 Ω resistors, let $R_i = 138.46\ \Omega$; then,

$$R_f = 10R_i = 1384.6\ \Omega.$$

The circuit for this cascaded the fourth-order low-pass filter is shown in Fig. 16.18. It has the transfer function

$$H(s) = -10\left[\frac{7222.39}{s + 7222.39}\right]^4.$$

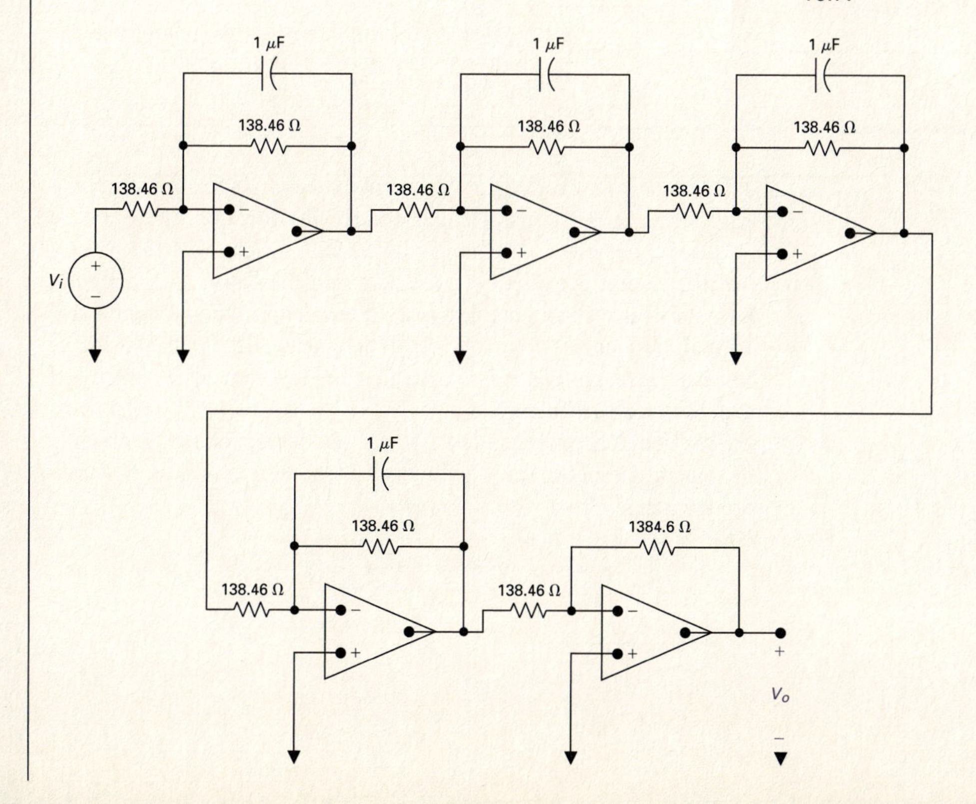

FIGURE 16.18 The cascade circuit for the fourth-order low-pass filter designed in Example 16.7.

The Bode magnitude plot for this transfer function is sketched in Fig. 16.19.

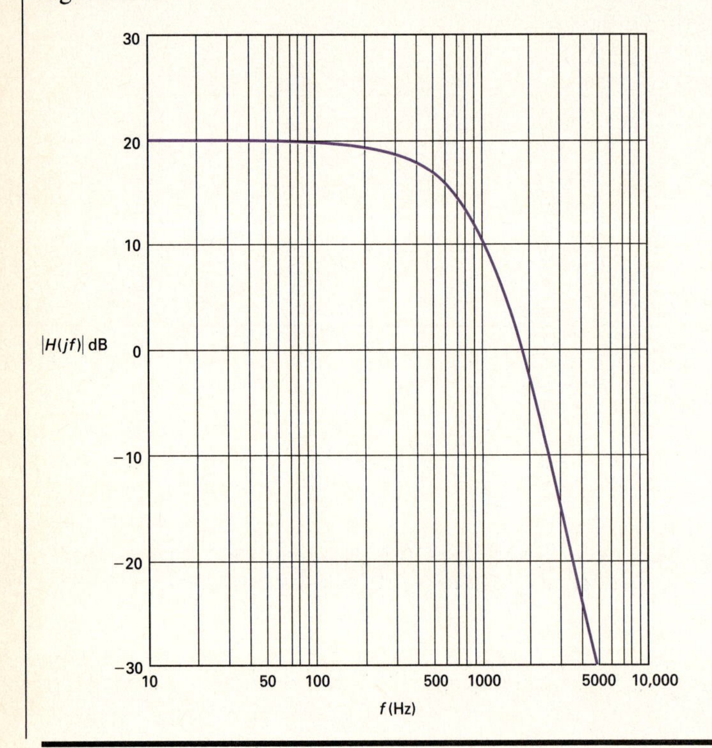

FIGURE 16.19 The Bode magnitude plot for the fourth-order low-pass filter designed in Example 16.7.

By cascading identical low-pass filters, we can increase the asymptotic slope in the transition and control the location of the cutoff frequency, but our approach has a serious shortcoming: the gain of the filter is not constant between zero and the cutoff frequency ω_c. Remember that in an ideal low-pass filter, the passband magnitude is 1 for all frequencies below the cutoff frequency. But in Fig. 16.16, we see that the magnitude is less than 1 (0 dB) for frequencies much less than the cutoff frequency.

This nonideal passband behavior is best understood by looking at the magnitude of the transfer function for a unity-gain low-pass nth-order cascade. Since

$$H(s) = \frac{\omega_{cn}^n}{(s + \omega_{cn})^n},$$

the magnitude is given by

$$|H(j\omega)| = \frac{\omega_{cn}^n}{\left(\sqrt{\omega^2 + \omega_{cn}^2}\right)^n}$$

$$= \frac{1}{\left(\sqrt{(\omega/\omega_{cn})^2 + 1}\right)^n}. \quad \textbf{(16.24)}$$

As we can see from Eq. (16.25), when $\omega \ll \omega_{cn}$, the denominator is approximately 1, and the magnitude of the transfer function is also nearly 1. But as $\omega \rightarrow \omega_{cn}$, the denominator becomes larger than 1, so the magnitude becomes smaller than 1. Because the cascade of low-pass filters results in this nonideal behavior in the passband, other approaches are taken in the design of higher order filters. One such approach is examined next.

BUTTERWORTH FILTERS

A unity-gain **Butterworth low-pass filter** has a transfer function whose magnitude is given by

$$|H(j\omega)| = \frac{1}{\sqrt{1 + (\omega/\omega_c)^{2n}}}, \quad \textbf{(16.25)}$$

where n is an integer that denotes the order of the filter.[1] In studying Eq. (16.25), note the following:

1. The cutoff frequency is ω_c rad/s for all values of n.
2. If n is large enough, the denominator is always close to unity when $\omega < \omega_c$.
3. In the expression for $|H(j\omega)|$, the exponent of ω/ω_c is always even.

This last observation is important because an even exponent is required for a physically realizable circuit (see Problem 16.24).

Given an equation for the magnitude of the transfer function, how do we find $H(s)$? The derivation for $H(s)$ is greatly simplified by using a prototype filter. Therefore, we set ω_c equal to 1 rad/s in Eq. (16.25). As before, we will use scaling to transform the prototype filter to a filter that meets the given filtering specifications.

To find $H(s)$, first note that if N is a complex quantity, then $|N|^2 = NN^*$, where N^* is the conjugate of N. It follows that

$$|H(j\omega)|^2 = H(j\omega)H(-j\omega). \quad \textbf{(16.26)}$$

But since $s = j\omega$, we can write

$$|H(j\omega)|^2 = H(s)H(-s). \quad \textbf{(16.27)}$$

---

[1] This filter was developed by the British engineer S. Butterworth and reported in *Wireless Engineering* 7 (1930):536–41.

Now observe that $s^2 = -\omega^2$. Thus,

$$|H(j\omega)|^2 = \frac{1}{1+\omega^{2n}} = \frac{1}{1+(\omega^2)^n}$$
$$= \frac{1}{1+(-s^2)^n} = \frac{1}{1+(-1)^n s^{2n}},$$

or

$$H(s)H(-s) = \frac{1}{1+(-1)^n s^{2n}}. \quad \textbf{(16.28)}$$

The procedure for finding $H(s)$ for a given value of n is as follows:

1. Find the roots of the polynomial
$$1+(-1)^n s^{2n} = 0.$$
2. Assign the left-half plane roots to $H(s)$ and the right-half plane roots to $H(-s)$.
3. Combine terms in the denominator of $H(s)$ to form first- and second-order factors.

Example 16.8 illustrates this process.

EXAMPLE 16.8

Find the Butterworth transfer functions for $n = 2$ and $n = 3$.

SOLUTION

Use a computer tool to find the roots.

For $n = 2$, we find the roots of the polynomial

$$1+(-1)^2 s^4 = 0.$$

Rearranging terms, we find

$$s^4 = -1 = 1\underline{/180^\circ}.$$

Therefore, the four roots are

$$s_1 = 1\underline{/45^\circ} = 1/\sqrt{2} + j/\sqrt{2};$$
$$s_2 = 1\underline{/135^\circ} = -1/\sqrt{2} + j/\sqrt{2};$$
$$s_3 = 1\underline{/225^\circ} = -1/\sqrt{2} + -j/\sqrt{2};$$
$$s_4 = 1\underline{/315^\circ} = 1/\sqrt{2} + -j/\sqrt{2}.$$

Roots s_2 and s_3 are in the left-half plane. Thus,

$$H(s) = \frac{1}{(s + 1/\sqrt{2} - j/\sqrt{2})(s + 1/\sqrt{2} + j/\sqrt{2})}$$
$$= \frac{1}{(s^2 + \sqrt{2}s + 1)}.$$

For $n = 3$, we find the roots of the polynomial

$$1 + (-1)^3 s^6 = 0.$$

Rearranging terms,

$$s^6 = 1\underline{/0^\circ} = 1\underline{/360^\circ}.$$

Therefore, the six roots are

$$s_1 = 1\underline{/0^\circ} = 1;$$
$$s_2 = 1\underline{/60^\circ} = 1/2 + j\sqrt{3}/2;$$
$$s_3 = 1\underline{/120^\circ} = -1/2 + j\sqrt{3}/2;$$
$$s_4 = 1\underline{/180^\circ} = -1 + j0;$$
$$s_5 = 1\underline{/240^\circ} = -1/2 + -j\sqrt{3}/2;$$
$$s_6 = 1\underline{/300^\circ} = 1/2 + -j\sqrt{3}/2.$$

Roots s_3, s_4, and s_5 are in the left-half plane. Thus,

$$H(s) = \frac{1}{(s + 1)(s + 1/2 - j\sqrt{3}/2)(s + 1/2 + j\sqrt{3}/2)}$$
$$= \frac{1}{(s + 1)(s^2 + s + 1)}.$$

We note in passing that the roots of the Butterworth polynomial are always equally spaced around the unit circle in the s plane. To assist in the design of Butterworth filters, Table 16.1 lists the Butterworth polynomials up to $n = 8$.

BUTTERWORTH FILTER CIRCUITS

Now that we know how to specify the transfer function for a Butterworth filter circuit (either by calculating the poles of the transfer function directly or by using Table 16.1), we turn to the problem of designing a circuit with such a transfer function. Notice the form of the

TABLE 16.1

NORMALIZED (SO THAT $\omega_c = 1$ RAD/S) BUTTERWORTH POLYNOMIALS UP TO THE EIGHTH ORDER

n	nth-Order Butterworth Polynomial
1	$(s+1)$
2	$(s^2+\sqrt{2}s+1)$
3	$(s+1)(s^2+s+1)$
4	$(s^2+0.765s+1)(s^2+1.848s+1)$
5	$(s+1)(s^2+0.618s+1)(s^2+1.618s+1)$
6	$(s^2+0.518s+1)(s^2+\sqrt{2}s+1)(s^2+1.932s+1)$
7	$(s+1)(s^2+0.445s+1)(s^2+1.247s+1)(s^2+1.802s+1)$
8	$(s^2+0.390s+1)(s^2+1.111s+1)(s^2+1.663s+1)(s^2+1.962s+1)$

Butterworth polynomials in Table 16.1. They are the product of first- and second-order factors; therefore, we can construct a circuit whose transfer function has a Butterworth polynomial in its denominator by cascading op amp circuits, each of which provides one of the needed factors. A block diagram of such a cascade is shown in Fig. 16.20, using a fifth-order Butterworth polynomial as an example.

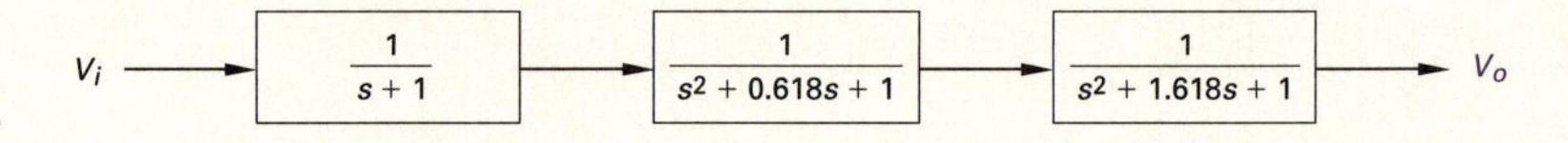

FIGURE 16.20 A cascade of first- and second-order circuits with the indicated transfer functions yielding a fifth-order low-pass Butterworth filter with $\omega_c = 1$ rad/s.

All odd-order Butterworth polynomials include the factor $(s+1)$, so all odd-order Butterworth filter circuits must have a subcircuit which provides the transfer function $H(s) = 1/(s+1)$. This is the transfer function of the prototype low-pass op amp filter from Fig. 16.1! So what remains is to find a circuit which provides a transfer function of the form $H(s) = 1/(s^2 + b_1 s + 1)$.

Such a circuit is shown in Fig. 16.21. The analysis of this circuit begins by writing the s-domain nodal equations at the noninverting terminal of the op amp and at the node labeled v_a:

$$\frac{V_a - V_i}{R} + (V_a - V_o)sC_1 + \frac{V_a - V_o}{R} = 0; \tag{16.29}$$

$$V_o s C_2 + \frac{V_o - V_a}{R} = 0. \tag{16.30}$$

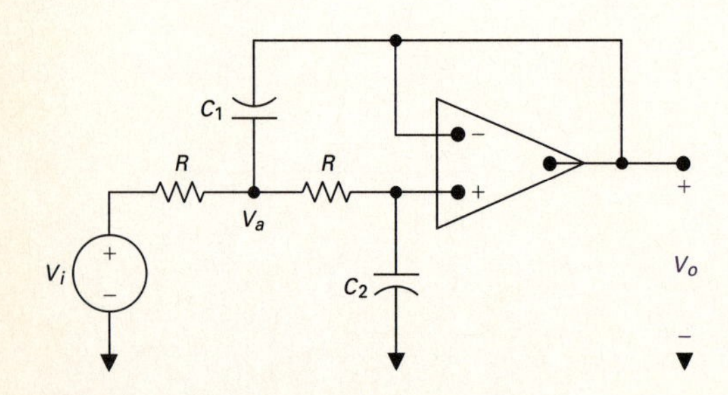

FIGURE 16.21 A circuit which provides the second-order transfer function for the Butterworth filter cascade.

Simplifying Eqs. (16.29) and (16.30) yields

$$(2 + RC_1 s)V_a - (1 + RC_1 s)V_o = V_i; \tag{16.31}$$

$$-V_a + (1 + RC_2 s)V_o = 0. \tag{16.32}$$

Using Cramer's rule with Eqs. (16.31) and (16.32), we solve for V_o:

$$V_o = \frac{\begin{vmatrix} 2+RC_1s & V_i \\ -1 & 0 \end{vmatrix}}{\begin{vmatrix} (2+RC_1s) & -(1+RC_1s) \\ -1 & (1+RC_2s) \end{vmatrix}}$$

$$= \frac{V_i}{R^2C_1C_2s^2 + 2RC_2s + 1}. \quad \textbf{(16.33)}$$

Then, rearrange Eq. (16.33) to write the transfer function for the circuit in Fig. 16.21:

$$H(s) = \frac{V_o}{V_i} = \frac{\frac{1}{R^2C_1C_2}}{s^2 + \frac{2}{RC_1}s + \frac{1}{R^2C_1C_2}} \quad \textbf{(16.34)}$$

Finally, set $R = 1\ \Omega$ in Eq. (16.34); then

$$H(s) = \frac{\frac{1}{C_1C_2}}{s^2 + \frac{2}{C_1}s + \frac{1}{C_1C_2}}. \quad \textbf{(16.35)}$$

Note that Eq. (16.35) has the form required for the second-order circuit in the Butterworth cascade. In other words, to get a transfer function of the form

$$H(s) = \frac{1}{s^2 + b_1s + 1},$$

we use the circuit in Fig. 16.21 and choose capacitor values so that

$$b_1 = \frac{2}{C_1} \quad \text{and} \quad 1 = \frac{1}{C_1C_2}. \quad \textbf{(16.36)}$$

We have thus outlined the procedure for designing an nth-order Butterworth low-pass filter circuit with a cutoff frequency of $\omega_c = 1$ rad/s and a gain of 1 in the passband. We can use frequency scaling to calculate revised capacitor values which yield any other cutoff frequency, and we can use magnitude scaling to provide more realistic or practical component values in our design. We can cascade an inverting amplifier circuit to provide a gain other than 1 in the passband.

Example 16.9 illustrates this design process.

EXAMPLE 16.9

Design a fourth-order Butterworth low-pass filter with a cutoff frequency of 500 Hz and a passband gain of 10. Use as many 1 kΩ resistors as possible. Compare the Bode magnitude plot for this Butterworth filter with that of the identical cascade filter in Example 16.7.

SOLUTION

From Table 16.1, we find that the fourth-order Butterworth polynomial is

$$(s^2 + 0.765s + 1)(s^2 + 1.848s + 1).$$

We will thus need a cascade of two second-order filters to yield the fourth-order transfer function, plus an inverting amplifier circuit for the passband gain of 10. The circuit is shown in Fig. 16.22.

Let the first stage of the cascade implement the transfer function for the polynomial $(s^2 + 0.765s + 1)$. From Eq. (16.36),

$$C_1 = 2.61 \text{ F};$$
$$C_2 = 0.38 \text{ F}.$$

Let the second stage of the cascade implement the transfer function for the polynomial $(s^2 + 1.848s + 1)$. From Eq. (16.36),

$$C_3 = 1.08 \text{ F};$$
$$C_4 = 0.924 \text{ F}.$$

The preceding values for C_1, C_2, C_3, and C_4 yield a fourth-order Butterworth filter with a cutoff frequency of 1 rad/s. A frequency scale factor of $k_f = 3141.6$ will move the cutoff frequency to 500 Hz. A magnitude scale factor of $k_m = 1000$ will permit the use of 1 kΩ resistors in place of 1 Ω resistors. The resulting scaled component values are

$$R = 1 \text{ k}\Omega;$$
$$C_1 = 831 \text{ nF};$$
$$C_2 = 121 \text{ nF};$$
$$C_3 = 344 \text{ nF};$$
$$C_4 = 294 \text{ nF}.$$

FIGURE 16.22 A fourth-order Butterworth filter with nonunity gain.

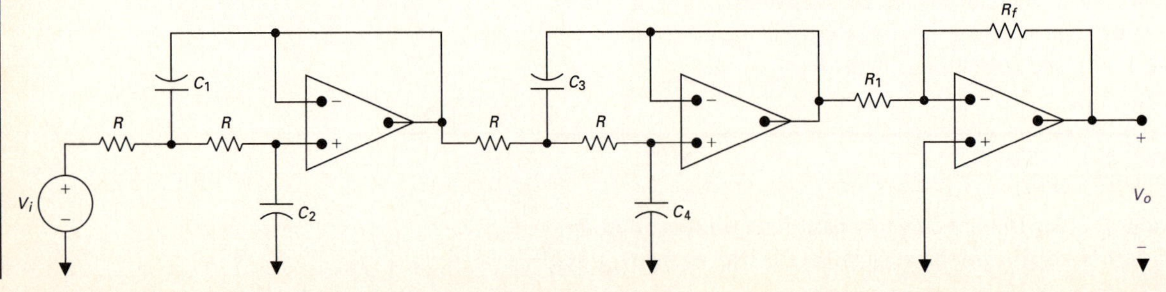

Finally, we need to specify the resistor values in the inverting amplifier stage to yield a passband gain of 10. Let $R_1 = 1\ \text{k}\Omega$; then

$$R_f = 10R_1 = 10\ \text{k}\Omega.$$

Figure 16.23 compares the magnitude responses of the fourth-order identical cascade filter from Example 16.7 and the Butterworth filter we just designed. Note that while both filters provide a passband gain of 10 (20 dB) and a cutoff frequency of 500 Hz, the Butterworth filter is closer to an ideal low-pass filter due to its flatter passband and steeper rolloff at the cutoff frequency. Thus, the Butterworth design is preferred over the identical cascade design.

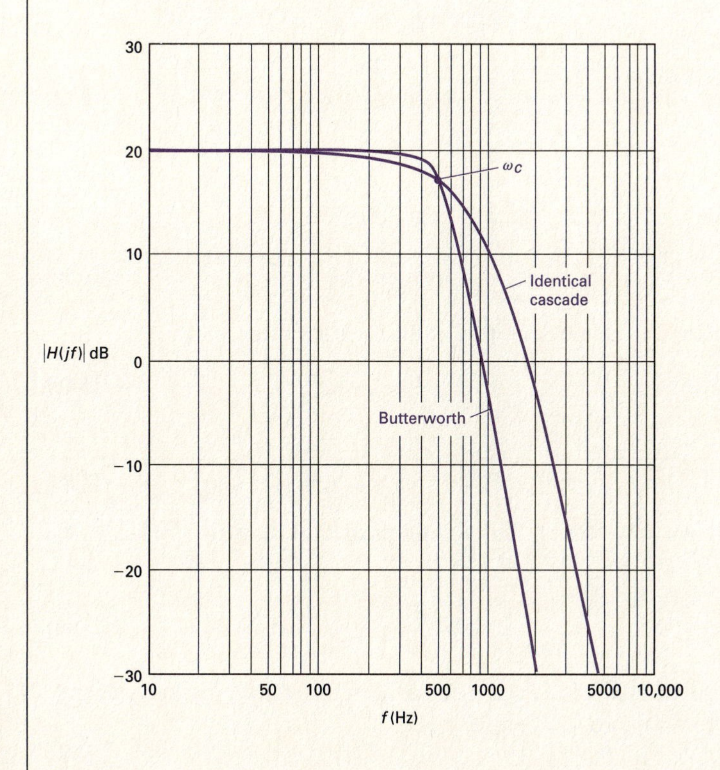

FIGURE 16.23 A comparison of the magnitude responses for a fourth-order low-pass filter using the identical cascade and Butterworth designs.

THE ORDER OF A BUTTERWORTH FILTER

It should be apparent at this point that the higher the order of the Butterworth filter, the closer the magnitude characteristic comes to that of an ideal low-pass filter. In other words, as n increases, the magnitude stays close to unity in the passband, the transition band narrows, and the magnitude stays close to zero in the stopband. At the same time, as the order increases, the number of circuit components increases. It follows then that a fundamental problem in the design of a filter is to determine the smallest value of n that will meet the filtering specifications.

In the design of a low-pass filter, the filtering specifications are usually given in terms of the abruptness of the transition region, as shown in Fig. 16.24. Once A_p, ω_p, A_s, and ω_s are specified, the order of the Butterworth filter can be determined.

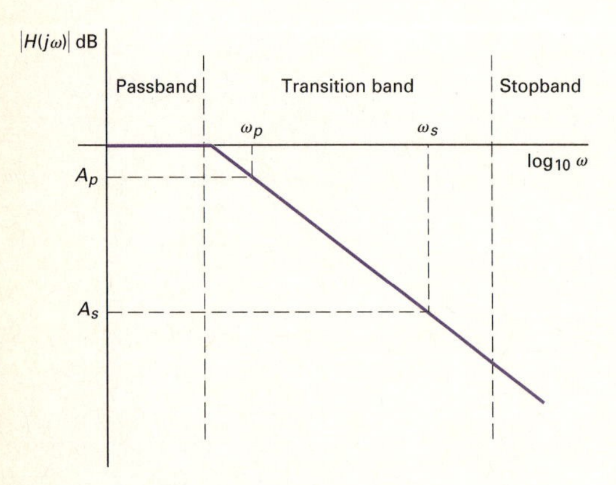

FIGURE 16.24 Defining the transition region for a low-pass filter.

For the Butterworth filter,

$$A_p = 20\log_{10}\frac{1}{\sqrt{1+\omega_p^{2n}}} = -10\log_{10}\,(1+\omega_p^{2n}) \tag{16.37}$$

and

$$A_s = 20\log_{10}\frac{1}{\sqrt{1+\omega_s^{2n}}} = -10\log_{10}\,(1+\omega_s^{2n}). \tag{16.38}$$

It follows from the definition of the logarithm that

$$10^{-0.1A_p} = 1+\omega_p^{2n} \tag{16.39}$$

and

$$10^{-0.1A_s} = 1+\omega_s^{2n}. \tag{16.40}$$

Now we solve for ω_p^n and ω_s^n and then form the ratio $(\omega_s/\omega_p)^n$. We get

$$\left(\frac{\omega_s}{\omega_p}\right)^n = \frac{\sqrt{10^{-0.1A_s}-1}}{\sqrt{10^{-0.1A_p}-1}} = \frac{\sigma_s}{\sigma_p}, \tag{16.41}$$

where the symbols σ_s and σ_p have been introduced for convenience.

From Eq. (16.41) we can write

$$n\log_{10}(\omega_s/\omega_p) = \log_{10}\,(\sigma_s/\sigma_p)$$

or

$$n = \frac{\log_{10}\,(\sigma_s/\sigma_p)}{\log_{10}\,(\omega_s/\omega_p)}. \tag{16.42}$$

We can simplify Eq. (16.42) if ω_p is the cutoff frequency because then A_p equals $-20\log_{10}\sqrt{2}$, and $\sigma_p = 1$. Hence

$$n = \frac{\log_{10}\sigma_s}{\log_{10}\,(\omega_s/\omega_p)}. \tag{16.43}$$

One further simplification is possible. We are using a Butterworth filter to achieve a steep transition region. Therefore, the filtering specification will make $10^{-0.1A_s} \gg 1$. Thus

$$\sigma_s \approx 10^{-0.05A_s} \quad \textbf{(16.44)}$$

and

$$\log_{10} \sigma_s \approx -0.05A_s. \quad \textbf{(16.45)}$$

Therefore, a good approximation for the calculation of n is

$$n = \frac{-0.05A_s}{\log_{10}(\omega_s/\omega_p)}. \quad \textbf{(16.46)}$$

Note that $\omega_s/\omega_p = f_s/f_p$, so we can work with either radians per second or hertz to calculate n.

The order of the filter must be an integer, hence in using either Eq. (16.42) or Eq. (16.46) we must select the nearest integer value greater than the result given by the equation. The following examples illustrate the usefulness of Eqs. (16.42) and (16.46).

EXAMPLE 16.10

a) Determine the order of a Butterworth filter that has a cutoff frequency of 1000 Hz and a gain of at least −50 dB at 6000 Hz.

b) What is the actual gain in dB at 6000 Hz?

SOLUTION

a) Since the cutoff frequency is given, we know $\sigma_p = 1$. We also note from the specification that $10^{-0.1(-50)}$ is much greater than 1. Hence, we can use Eq. (16.46) with confidence:

$$n = \frac{(-0.05)(-50)}{\log_{10}(6000/1000)} = 3.21.$$

Therefore, we need a fourth-order Butterworth filter.

b) We can use Eq. (16.25) to calculate the actual gain at 6000 Hz. The gain in decibels will be

$$K = 20\log_{10}\left(\frac{1}{\sqrt{1+6^8}}\right) = -62.25 \text{ dB}.$$

EXAMPLE 16.11

a) Determine the order of a Butterworth filter whose magnitude is 10 dB less than the passband magnitude at 500 Hz and at least 60 dB less than the passband magnitude at 5000 Hz.

b) Determine the cutoff frequency of the filter, in hertz.

c) What is the actual gain of the filter (in decibels) at 5000 Hz?

SOLUTION

a) Since the cutoff frequency is not given, we use Eq. (16.42) to determine the order of the filter:

$$\sigma_p = \sqrt{10^{-0.1(-10)} - 1} = 3;$$

$$\sigma_s = \sqrt{10^{-0.1(-60)} - 1} \approx 1000;$$

$$\omega_s/\omega_p = f_s/f_p = 5000/500 = 10;$$

$$n = \frac{\log_{10}(1000/3)}{\log_{10}(10)} = 2.52.$$

Therefore we need a third-order Butterworth filter to meet the specifications.

b) Knowing that the gain at 500 Hz is -10 dB, we can determine the cutoff frequency. From Eq. (16.25) we can write

$$-10\log_{10}[1 + (\omega/\omega_c)^6] = -10,$$

where $\omega = 1000\pi$ rad/s. Therefore

$$1 + (\omega/\omega_c)^6 = 10$$

and

$$\omega_c = \frac{\omega}{\sqrt[6]{9}}$$

$$= 2178.26 \text{ rad/s}.$$

It follows that

$$f_c = 346.68 \text{ Hz}.$$

c) The actual gain of the filter at 5000 Hz is

$$K = -10\log_{10}[1 + (5000/346.68)^6]$$

$$= -69.54 \text{ dB}.$$

Butterworth High-Pass, Bandpass, and Bandreject Filters

An nth-order Butterworth high-pass filter has a transfer function with the nth-order Butterworth polynomial in the denominator, just like the nth-order Butterworth low-pass filter. But in the high-pass filter, the numerator of the transfer function is s^n, whereas in the low-pass filter, the numerator is 1. Again, we use a cascade approach in designing the Butterworth high-pass filter. The first-order factor is achieved by including a prototype high-pass filter (Fig. 16.4, with $R_1 = R_2 = 1\ \Omega$, and $C = 1$ F) in the cascade.

To produce the second-order factors in the Butterworth polynomial, we need a circuit with a transfer function of the form

$$H(s) = \frac{s^2}{s^2 + b_1 s + 1}.$$

Such a circuit is shown in Fig. 16.25. This circuit has the transfer function

$$H(s) = \frac{V_o}{V_i} = \frac{s^2}{s^2 + \frac{2}{R_2 C}s + \frac{1}{R_1 R_2 C^2}}. \quad \textbf{(16.47)}$$

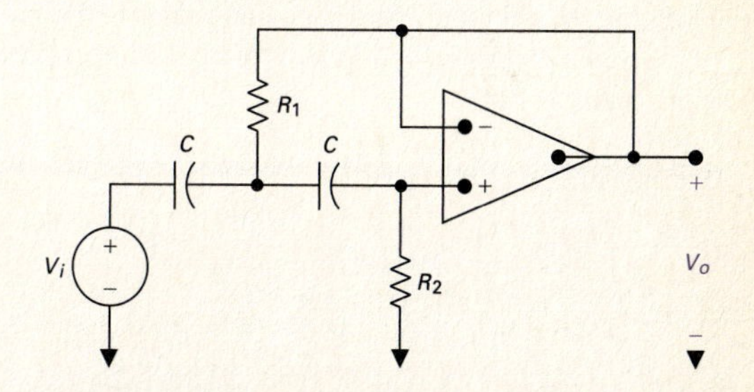

FIGURE 16.25 A second-order Butterworth high-pass filter circuit.

Setting $C = 1$ F yields

$$H(s) = \frac{s^2}{s^2 + \frac{2}{R_2}s + \frac{1}{R_1 R_2}}. \quad \textbf{(16.48)}$$

Thus, we can realize any second-order factor in a Butterworth polynomial of the form $(s^2 + b_1 s + 1)$ by including in the cascade the second-order circuit in Fig. 16.25 with resistor values which satisfy Eq. (16.49):

$$b_1 = \frac{2}{R_2} \quad \text{and} \quad 1 = \frac{1}{R_1 R_2}. \quad \textbf{(16.49)}$$

At this point, we pause to make a couple of observations relative to Figs. 16.21 and 16.25 and their prototype transfer functions $1/(s^2 + b_1 s + 1)$ and $s^2/(s^2 + b_1 s + 1)$. These observations are important because they are true in general. First, the high-pass circuit in Fig. 16.25 was obtained from the low-pass circuit in Fig. 16.21 by interchanging resistors and capacitors. Second, the prototype transfer function of a high-pass filter can be obtained from that of a low-pass filter by replacing s in the low-pass expression with $1/s$ (see Problem 16.45).

We can use frequency and magnitude scaling to design a Butterworth high-pass filter with practical component values and a cutoff frequency other than 1 rad/s. Adding an inverting amplifier to the cascade will accommodate designs with nonunity passband gains.

The problems at the end of the chapter include several Butterworth high-pass filter designs.

Now that we can design both nth-order low-pass and high-pass Butterworth filters with arbitrary cutoff frequencies and passband gains, we can combine these filters in cascade (as we did in Section 16.3) to produce nth-order Butterworth bandpass filters. We can combine these filters in parallel with a summing amplifier (again, as we did in Section 16.3) to produce nth-order Butterworth bandreject filters. This chapter's problems also include Butterworth bandpass and bandreject filter designs.

DRILL EXERCISE

16.6 For the circuit in Fig. 16.25, find values of R_1 and R_2 which yield a second-order prototype Butterworth high-pass filter.

ANSWER: $R_1 = 0.707\ \Omega$; $R_2 = 1.41\ \Omega$.

16.5 NARROW-BAND BANDPASS AND BANDREJECT FILTERS

The cascade and parallel component designs for synthesizing bandpass and bandreject filters from simpler low-pass and high-pass filters have the restriction that only broadband, or low-Q, filters will result. (The Q, of course, stands for *quality factor*.) This limitation is due principally to the fact that the transfer functions for cascaded bandpass and parallel bandreject filters have discrete real poles. The synthesis techniques work best for cutoff frequencies which are widely separated and therefore yield the lowest quality factors. But the largest quality factor we can achieve with discrete real poles arises when the cutoff frequencies, and thus the pole locations, are the same. Consider the transfer function which results:

$$\begin{aligned} H(s) &= \left(\frac{-\omega_c}{s+\omega_c}\right)\left(\frac{-s}{s+\omega_c}\right) \\ &= \frac{s\omega_c}{s^2+2\omega_c s+\omega_c^2} \\ &= \frac{0.5\beta s}{s^2+\beta s+\omega_c^2}. \end{aligned} \quad \text{(16.50)}$$

Eq. (16.50) is in the standard form of the transfer function of a bandpass filter, and thus we can determine the bandwidth and center frequency directly:

$$\beta = 2\omega_c; \tag{16.51}$$

$$\omega_o^2 = \omega_c^2. \tag{16.52}$$

From Eqs. (16.51) and (16.52) and the definition of Q, we see that

$$Q = \frac{\omega_o}{\beta} = \frac{\omega_c}{2\omega_c} = \frac{1}{2}. \tag{16.53}$$

Thus with discrete real poles, the highest quality bandpass filter (or bandreject filter) we can achieve has $Q = 1/2$.

To build active filters with high quality factor values, we need an op amp circuit which can produce a transfer function with complex conjugate poles. Figure 16.26 depicts one such circuit for us to analyze. At the inverting input of the op amp, we sum the currents to get

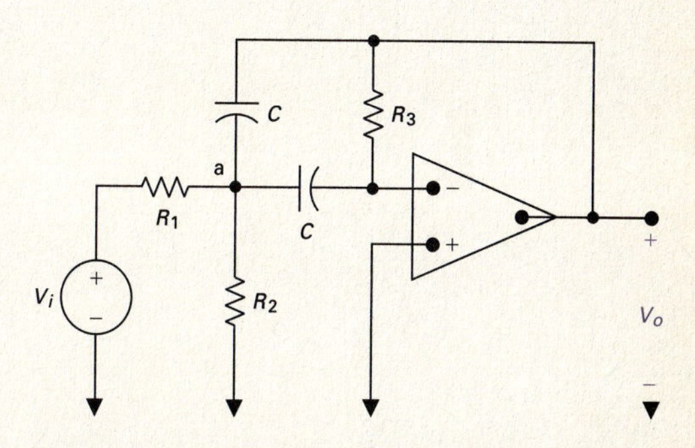

FIGURE 16.26 An active high-Q bandpass filter.

$$\frac{V_a}{\frac{1}{sC}} = \frac{-V_o}{R_3}.$$

Solving for V_a,

$$V_a = \frac{-V_o}{sR_3C}. \tag{16.54}$$

At the node labeled a, we sum the currents to get

$$\frac{V_i - V_a}{R_1} = \frac{V_a - V_o}{\frac{1}{sC}} + \frac{V_a}{\frac{1}{sC}} + \frac{V_a}{R_2}.$$

Solving for V_i,

$$V_i = (1 + 2sR_1C + R_1/R_2)V_a - sR_1CV_o. \tag{16.55}$$

Substituting Eq. (16.54) into Eq. (16.55) and then rearranging, we get an expression for the transfer function V_o/V_i:

$$H(s) = \frac{\frac{-s}{R_1C}}{s^2 + \frac{2}{R_3C}s + \frac{1}{R_{eq}R_3C^2}}, \tag{16.56}$$

where

$$R_{eq} = R_1||R_2 = \frac{R_1R_2}{R_1 + R_2}.$$

Since Eq. (16.56) is in the standard form of the transfer function for a bandpass filter, that is,

$$H(s) = \frac{K\beta s}{s^2 + \beta s + \omega_o^2},$$

we can equate terms and solve for the values of the resistors which will achieve a specified center frequency (ω_o), quality factor (Q), and passband gain (K):

$$\beta = \frac{2}{R_3 C}; \quad \textbf{(16.57)}$$

$$K\beta = \frac{1}{R_1 C}; \quad \textbf{(16.58)}$$

$$\omega_o^2 = \frac{1}{R_{eq} R_3 C^2}. \quad \textbf{(16.59)}$$

At this point, it is convenient to define the prototype version of the circuit in Fig. 16.25 as a circuit in which $\omega_o = 1$ rad/s and $C = 1$ F. Then the expressions for R_1, R_2, and R_3 can be given in terms of the desired quality factor and passband gain. We leave you to show (in Problem 16.36) that for the prototype circuit, the expressions for R_1, R_2, and R_3 are

$$R_1 = Q/K$$
$$R_2 = Q/(2Q^2 - K);$$
$$R_3 = 2Q.$$

Scaling is used to specify practical values for the circuit components. This design process is illustrated in Example 16.12.

EXAMPLE 16.12

Design a bandpass filter, using the circuit in Fig. 16.26, which has a center frequency of 3000 Hz, a quality factor of 10, and a passband gain of 2. Use 0.01 μF capacitors in your design. Compute the transfer function of your circuit, and sketch a Bode plot of its magnitude response.

SOLUTION

Since $Q = 10$ and $K = 2$, the values for R_1, R_2, and R_3 in the prototype circuit are

$$R_1 = 10/2 = 5;$$
$$R_2 = 10/(200 - 2) = 10/198;$$
$$R_3 = 2(10) = 20.$$

The scaling factors are $k_f = 6000\pi$ and $k_m = 10^8/k_f$. After scaling,

$$R_1 = 26.5\ \text{k}\Omega;$$
$$R_2 = 268.0\ \Omega;$$
$$R_3 = 106.1\ \text{k}\Omega.$$

The circuit is shown in Fig. 16.27.

Substituting the values of resistance and capacitance in Eq. (16.56) gives the transfer function for this circuit:

$$H(s) = \frac{-3770s}{s^2 + 1885.0s + 355 \times 10^6}.$$

It is easy to see that this transfer function meets the specification of the bandpass filter defined in the example. A Bode plot of its magnitude response is sketched in Fig. 16.28.

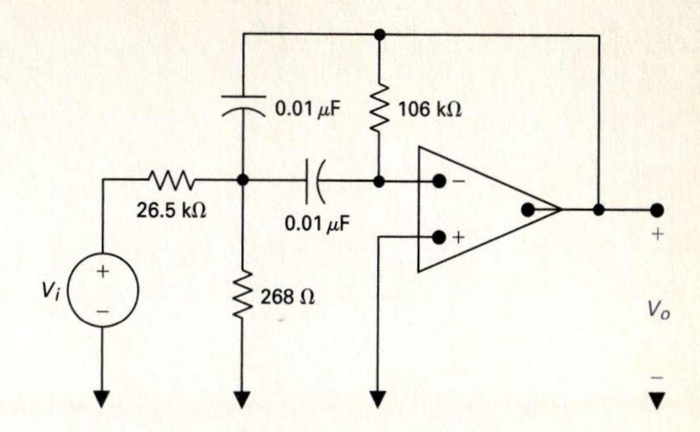

FIGURE 16.27 The high-Q bandpass filter designed in Example 16.12.

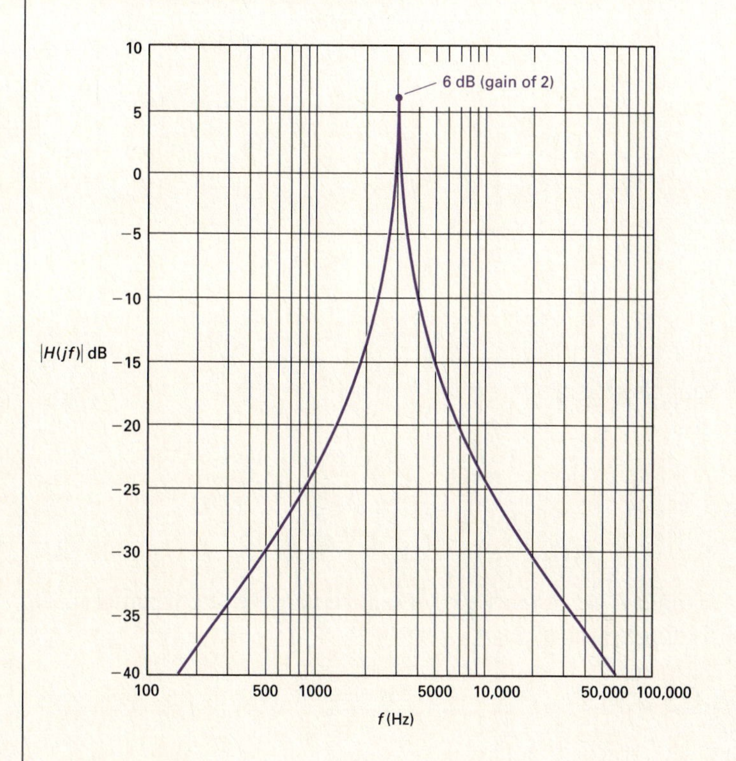

FIGURE 16.28 The Bode magnitude plot for the high-Q bandpass filter designed in Example 16.12.

The parallel implementation of a bandreject filter which combines low-pass and high-pass filter components with a summing amplifier has the same low-Q restriction as the cascaded bandpass filter. The circuit in Fig. 16.29 is an active high-Q bandreject filter known as the *twin-T notch filter* because of the two T-shaped parts of the circuit at the nodes labeled a and b.

We begin the analysis of this circuit by summing the currents away from node a:

$$(V_a - V_i)sC + (V_a - V_o)sC + \frac{2(V_a - \sigma V_o)}{R} = 0$$

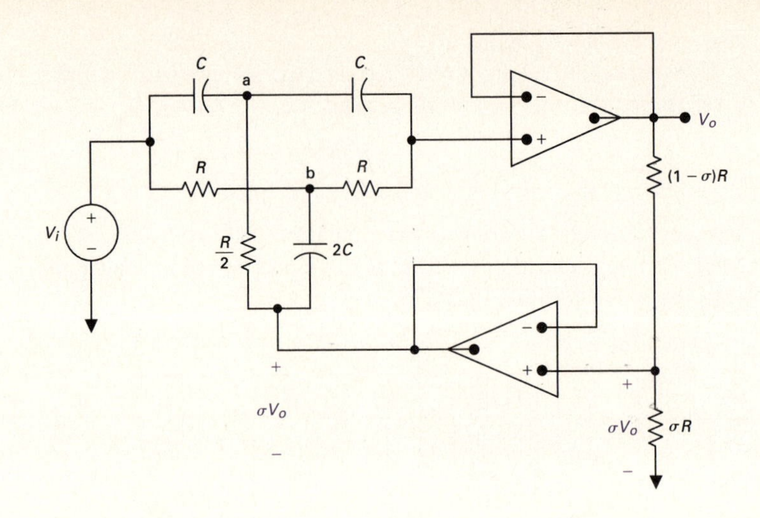

FIGURE 16.29 A high-Q active bandreject filter.

or

$$V_a[2sCR+2]-V_o[sCR+2\sigma]=sCRV_i. \quad \textbf{(16.60)}$$

Summing the currents away from node b yields

$$\frac{V_b-V_i}{R}+\frac{V_b-V_o}{R}+(V_b-\sigma V_o)2sC=0$$

or

$$V_b[2+2RCs]-V_o[1+2\sigma RCs]=V_i. \quad \textbf{(16.61)}$$

Summing the currents away from the noninverting input terminal of the top op amp gives

$$(V_o-V_a)sC+\frac{V_o-V_b}{R}=0$$

or

$$-sRCV_a-V_b+(sRC+1)V_o=0. \quad \textbf{(16.62)}$$

From Eqs. (16.60), (16.61), and (16.62), we can use Cramer's rule to solve for V_o:

$$V_o=\frac{\begin{vmatrix} 2(RCs+1) & 0 & sCRV_i \\ 0 & 2(RCs+1) & V_i \\ -RCs & -1 & 0 \end{vmatrix}}{\begin{vmatrix} 2(RCs+1) & 0 & -(RCs+2\sigma) \\ 0 & 2(RCs+1) & -(2\sigma RCs+1) \\ -RCs & -1 & (RCs+1) \end{vmatrix}}$$

$$=\frac{(R^2C^2s^2+1)V_i}{R^2C^2s^2+4RC(1-\sigma)s+1}. \quad \textbf{(16.63)}$$

Rearranging Eq. (16.63), we can solve for the transfer function:

$$H(s) = \frac{V_o}{V_i} = \frac{\left(s^2 + \frac{1}{R^2C^2}\right)}{\left[s^2 + \frac{4(1-\sigma)}{RC}s + \frac{1}{R^2C^2}\right]}, \tag{16.64}$$

which is in the standard form for the transfer function of a bandreject filter:

$$H(s) = \frac{s^2 + \omega_0^2}{s^2 + \beta s + \omega_0^2} \tag{16.65}$$

Equating Eqs. (16.64) and (16.65) gives

$$\omega_o^2 = \frac{1}{R^2C^2} \tag{16.66}$$

and

$$\beta = \frac{4(1-\sigma)}{RC}. \tag{16.67}$$

In this circuit, we have three parameters (R, C, and σ) and two design constraints (ω_o and β). Thus one parameter is chosen arbitrarily; it is usually the capacitor value because this value typically provides the fewest commercially available options. Once C is chosen,

$$R = \frac{1}{\omega_o C} \tag{16.68}$$

and

$$\sigma = 1 - \frac{\beta}{4\omega_o} = 1 - \frac{1}{4Q}. \tag{16.69}$$

Example 16.13 illustrates the design of a high-Q active bandreject filter.

EXAMPLE 16.13

Design a high-Q active bandreject filter (based on the circuit in Fig. 16.29) with a center frequency of 5000 rad/s and a bandwidth of 1000 rad/s. Use 1 μF capacitors in your design.

SOLUTION

In the bandreject prototype filter, $\omega_o = 1$ rad/s, $R = 1\ \Omega$, and $C = 1$ F. As just discussed, once ω_o and Q are given, C can be chosen arbitrarily, and R and σ can be found from Eqs. (16.68) and (16.69). From the specifications, $Q = 5$. Using Eqs. (16.68) and (16.69), we see that

$$R = 200\ \Omega$$

and

$$\sigma = 0.95.$$

Therefore we need resistors with the values 200 Ω (R), 100 Ω ($R/2$), 190 Ω (σR), and 10 Ω [$(1 - \sigma)R$]. The final design is depicted in Fig. 16.30, and the Bode magnitude plot is shown in Fig. 16.31.

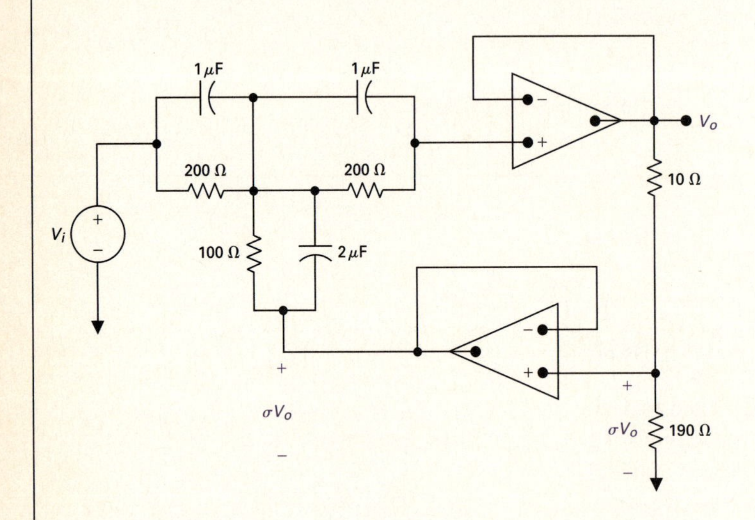

FIGURE 16.30 The high-Q active bandreject filter designed in Example 16.13.

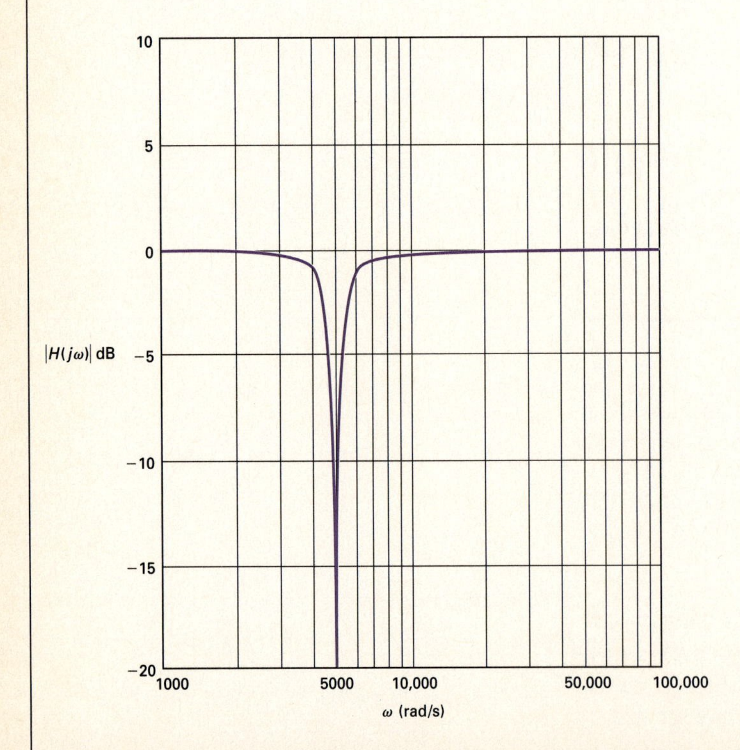

FIGURE 16.31 The Bode magnitude plot for the high-Q active bandreject filter designed in Example 16.13.

DRILL EXERCISES

16.7 Design an active bandpass filter with $Q = 8$, $K = 5$, and $\omega_o = 1000$ rad/s. Use 1 μF capacitors, and specify the values of all resistors.

ANSWER: $R_1 = 1.6$ kΩ, $R_2 = 65.04$ Ω, $R_3 = 16$ kΩ.

16.8 Design an active unity-gain bandreject filter with $\omega_o = 1000$ rad/s and $Q = 4$. Use 2 μF capacitors in your design, and specify the values of R and σ.

ANSWER: $R = 500$ Ω, $\sigma = 0.9375$.

SUMMARY

- Active filters consist of op amps, resistors, and capacitors. They can be configured as low-pass, high-pass, bandpass, and band-reject filters. They overcome many of the disadvantages associated with passive filters.
- A **prototype low-pass filter** has component values of $R_1 = R_2 = 1$ Ω and $C = 1$ F, and it produces a unity passband gain and a cutoff frequency of 1 rad/s. The **prototype high-pass filter** has the same component values and also produces a unity passband gain and a cutoff frequency of 1 rad/s.
- **Magnitude scaling** can be used to alter component values without changing the frequency response of a circuit. For a magnitude scale factor of k_m, the scaled (primed) values of resistance, capacitance, and inductance are

$$R' = k_m R, \quad L' = k_m L, \quad \text{and} \quad C' = C/k_m.$$

- **Frequency scaling** can be used to shift the frequency response of a circuit to another frequency region without changing the overall shape of the frequency response. For a frequency scale factor of k_f, the scaled (primed) values of resistance, capacitance, and inductance are

$$R' = R, \quad L' = L/k_f, \quad \text{and} \quad C' = C/k_f.$$

- Components can be scaled in both magnitude and frequency, with the scaled (primed) component values given by

$$R' = k_m R, \quad L' = (k_m/k_f)L, \quad \text{and} \quad C' = C/(k_m k_f).$$

- The design of active low-pass and high-pass filters can begin with a prototype filter circuit. Scaling can then be applied to shift the frequency response to the desired cutoff frequency, using component values which are commercially available.

- In a broadband filter, the upper cutoff frequency is at least twice the lower cutoff frequency, or

$$\omega_{c2}/\omega_{c1} \geq 2.$$

- An active broadband bandpass filter can be constructed using a cascade of a low-pass filter with the bandpass filter's upper cutoff frequency, a high-pass filter with the bandpass filter's lower cutoff frequency, and (optionally) an inverting amplifier gain stage to achieve nonunity gain in the passband. Bandpass filters implemented in this fashion must be broadband filters, so that the elements of the cascade can be specified independently of one another.

- An active broadband bandreject filter can be constructed using a parallel combination of a low-pass filter with the bandreject filter's lower cutoff frequency, and a high-pass filter with the bandreject filter's upper cutoff frequency. The outputs are then fed into a summing amplifier which can produce nonunity gain in the passband. Bandreject filters implemented in this way must be broadband filters, so that the low-pass and high-pass filter circuits can be designed independently of one another.

- Higher order active filters have multiple poles in their transfer functions, resulting in a sharper transition from the passband to the stopband and thus a more nearly ideal frequency response.

- The transfer function of an nth-order Butterworth low-pass filter with a cutoff frequency of 1 rad/s can be determined from the equation

$$H(s)H(-s) = \frac{1}{1 + (-1)^n s^{2n}}$$

  by

  1. finding the roots of the denominator polynomial;

  2. assigning the left-half plane roots to $H(s)$; and

  3. writing the denominator of $H(s)$ as a product of first- and second-order factors.

- The fundamental problem in the design of a Butterworth filter is to determine the order of the filter. The filter specification usually defines the sharpness of the transition band in terms of the quantities A_p, ω_p, A_s, and ω_s. From these quantities, we calculate the smallest integer larger than the solution to either Eq. (16.42) or Eq. (16.46).

- A cascade of second-order low-pass op amp filters (Fig. 16.21) with 1 Ω resistors and capacitor values chosen to produce each factor in the Butterworth polynomial will produce an even-order Butterworth low-pass filter. Adding a prototype low-pass op amp filter will produce an odd-order Butterworth low-pass filter.
- A cascade of second-order high-pass op amp filters (Fig. 16.25) with 1 F capacitors and resistor values chosen to produce each factor in the Butterworth polynomial will produce an even-order Butterworth high-pass filter. Adding a prototype high-pass op amp filter will produce an odd-order Butterworth high-pass filter.
- For both high- and low-pass Butterworth filters, frequency and magnitude scaling can be used to shift the cutoff frequency from 1 rad/s and to include realistic component values in the design. Cascading an inverting amplifier will produce a nonunity passband gain.
- Butterworth low-pass and high-pass filters can be cascaded to produce Butterworth bandpass filters of any order n. Butterworth low-pass and high-pass filters can be combined in parallel with a summing amplifier to produce a Butterworth bandreject filter of any order n.
- If a high-Q, or narrow band, bandpass or bandreject filter is needed, the cascade or parallel combination will not work. Instead, the circuits shown in Figs. 16.26 and 16.29 are used with the appropriate design equations. Typically, capacitor values are chosen from those commercially available, and the design equations are used to specify the resistor values.

PROBLEMS

16.1 Find the transfer function V_o/V_i for the circuit shown in Fig. P16.1 if Z_f is the equivalent impedance of the feedback circuit, Z_i is the equivalent impedance of the input circuit, and the operational amplifier is ideal.

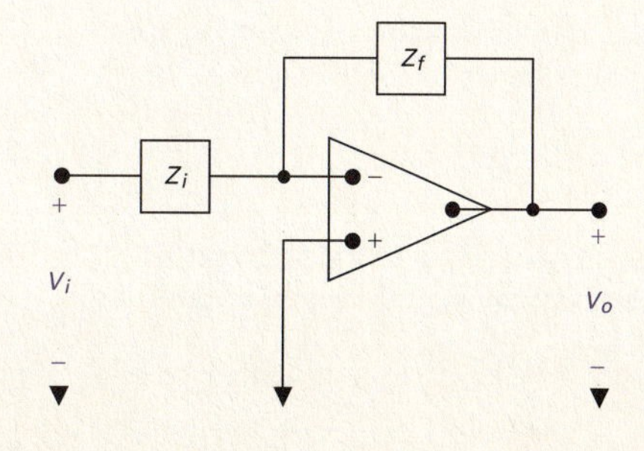

FIGURE P16.1

16.2 a) Use the results of Problem 16.1 to find the transfer function of the circuit shown in Fig. P16.2.

b) What is the gain of the circuit as $\omega \to 0$?

c) What is the gain of the circuit as $\omega \to \infty$?

d) Do your answers to (b) and (c) make sense in terms of known circuit behavior?

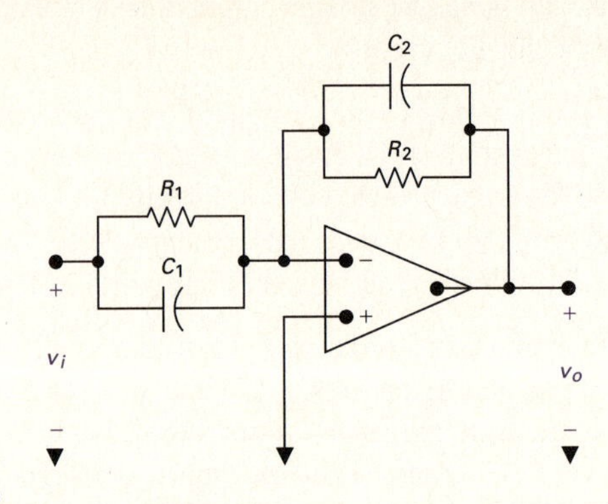

FIGURE P16.2

16.3 Repeat Problem 16.2, using the circuit shown in Fig. P16.3.

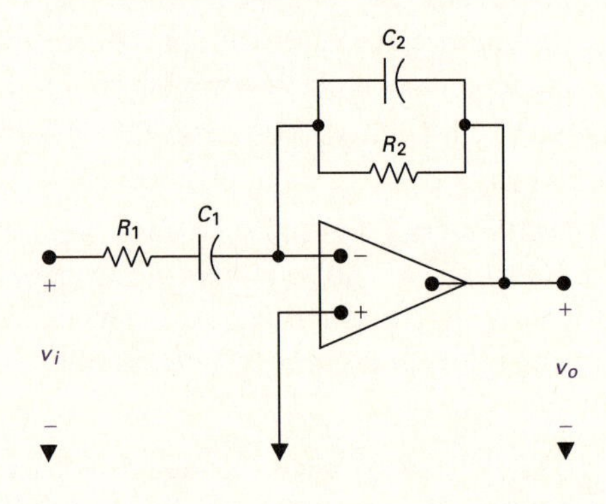

FIGURE P16.3

❖ **16.4** a) Using the circuit in Fig. 16.1, design a low-pass filter with a passband gain of 10 dB and a cutoff frequency of 1 kHz. Assume a 750 nF capacitor is available.

b) Draw the circuit diagram and label all components.

❖ **16.5** a) Use the circuit in Fig. 16.4 to design a high-pass filter with a cutoff frequency of 8 kHz and a passband gain of 14 dB. Use a 3.9 nF capacitor in the design.

b) Draw the circuit diagram of the filter and label all the components.

16.6 The voltage transfer function of either low-pass prototype filter shown in Fig. P16.6 is

$$H(s) = \frac{1}{s+1}.$$

Show that if either circuit is scaled in both magnitude and frequency, the scaled transfer function is

$$H'(s) = \frac{1}{(s/k_f)+1}.$$

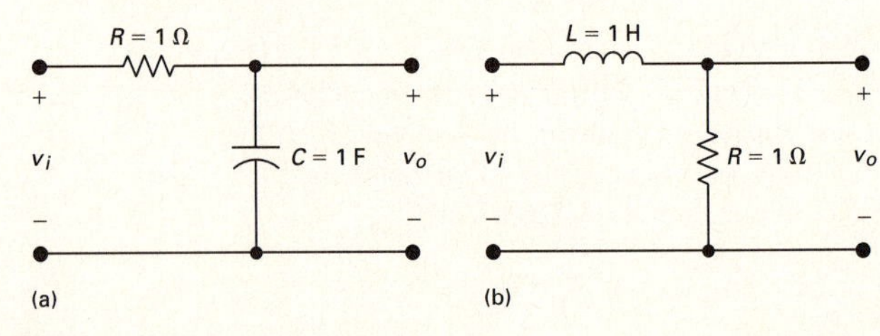

FIGURE P16.6

16.7 The voltage transfer function for either high-pass prototype filter shown in Fig. P16.7 is

$$H(s) = \frac{s}{s+1}.$$

Show that if either circuit is scaled in both magnitude and frequency, the scaled transfer function is

$$H'(s) = \frac{(s/k_f)}{(s/k_f)+1}.$$

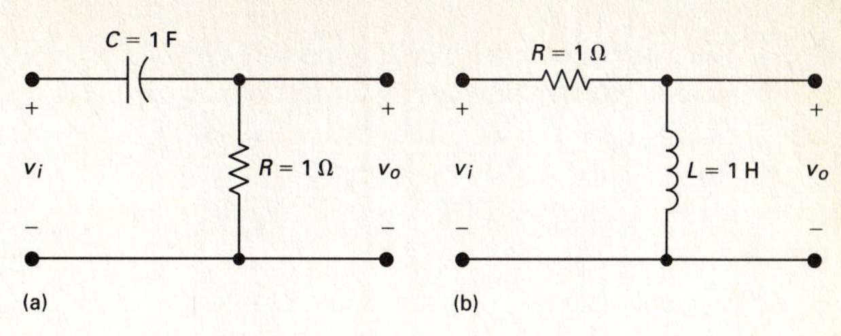

FIGURE P16.7

16.8 The voltage transfer function of the prototype bandpass filter shown in Fig. P16.8 is

$$H(s) = \frac{\left(\frac{1}{Q}\right)s}{s^2 + \left(\frac{1}{Q}\right)s + 1}.$$

Show that if the circuit is scaled in both magnitude and frequency, the scaled transfer function is

$$H'(s) = \frac{\left(\frac{1}{Q}\right)\left(\frac{s}{k_f}\right)}{\left(\frac{s}{k_f}\right)^2 + \left(\frac{1}{Q}\right)\left(\frac{s}{k_f}\right) + 1}.$$

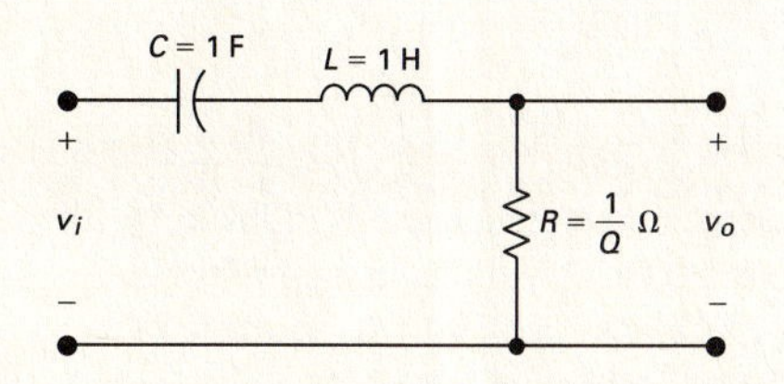

FIGURE P16.8

16.9 a) Specify the component values for the prototype passive bandpass filter described in Problem 16.8 if the quality factor of the filter is 20.

b) Specify the component values for the bandpass filter described in Problem 16.8 if the quality factor is 20; the center, or resonant, frequency is 40 krad/s; and the impedance at resonance is 5 kΩ.

c) Draw a circuit diagram of the scaled filter and label all the components.

16.10 An alternative to the prototype bandpass filter illustrated in Fig. P16.8 is to make $\omega_o = 1$ rad/s, $R = 1\ \Omega$, and $L = Q$ henrys.

a) What is the value of C in the prototype filter circuit?

b) What is the transfer function of the prototype filter?

c) Use the alternative prototype circuit just described to design a passive bandpass filter that has a quality factor of 16, a center frequency of 25 krad/s, and an impedance of 10 kΩ at resonance.

d) Draw a diagram of the scaled filter and label all the components.

e) Use the results obtained in Problem 16.8 to write the transfer function of the scaled circuit.

❖ **16.11** The passive bandpass filter illustrated in Fig. 15.22 has two prototype circuits. In the first prototype circuit, $\omega_o = 1$ rad/s, $C = 1$ F, $L = 1$ H, and $R = Q$ ohms. In the second prototype circuit, $\omega_o = 1$ rad/s, $R = 1\ \Omega$, $C = Q$ farads, and $L = (1/Q)$ henrys.

a) Use one of these prototype circuits (your choice) to design a passive bandpass filter which has a quality factor of 25 and a center frequency of 50 krad/s. The resistor R is 40 kΩ.

b) Draw a circuit diagram of the scaled filter and label all components.

16.12 The transfer function for the bandreject filter shown in Fig. 15.28(a) is

$$H(s) = \frac{s^2 + \left(\frac{1}{LC}\right)}{s^2 + \left(\frac{R}{L}\right)s + \left(\frac{1}{LC}\right)}.$$

Show that if the circuit is scaled in both magnitude and frequency, the transfer function of the scaled circuit is equal to the transfer function of the unscaled circuit with s replaced by (s/k_f), where k_f is the frequency scale factor.

16.13 Show that the observation made in Problem 16.12 with respect to the transfer function for the circuit in Fig. 15.28(a) also applies to the bandreject filter circuit (lower one) in Fig. 15.31.

16.14 The passive bandreject filter illustrated in Fig. 15.28(a) has the two prototype circuits shown in Fig. P16.14.

a) Show that for both circuits, the transfer function is

$$H(s) = \frac{s^2 + 1}{s^2 + \left(\frac{1}{Q}\right)s + 1}.$$

b) Write the transfer function for a bandreject filter that has a center frequency of 10 krad/s and a quality factor of 8.

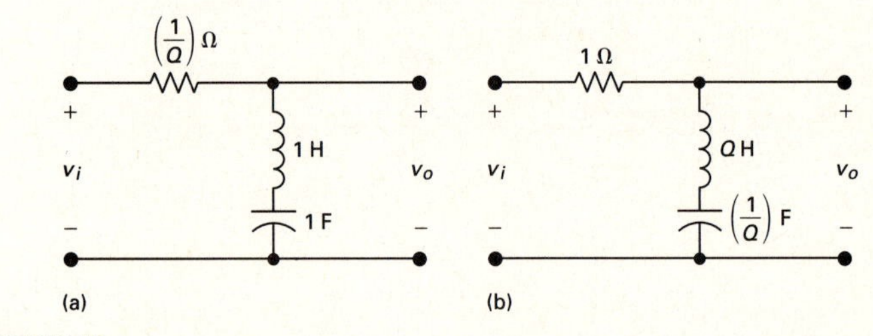

FIGURE P16.14

16.15 The two prototype versions of the passive bandreject filter shown in Fig. 15.31 (lower circuit) are shown in Figs. P16.15(a) and (b).

Show that the transfer function for either version is

$$H(s) = \frac{s^2 + 1}{s^2 + \left(\frac{1}{Q}\right)s + 1}.$$

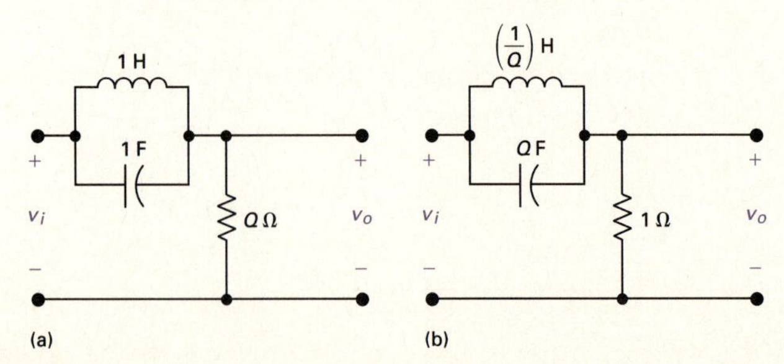

FIGURE P16.15

16.16 The circuit in Fig. P14.30 is scaled so that the 1 Ω resistors are replaced by 1 kΩ resistors and the 1 F capacitor is replaced by a 0.2 μF capacitor.

a) What is the scaled value of L?

b) What is the expression for i_o in the scaled circuit?

16.17 Scale the circuit in Problem 14.31 so that the 50 Ω resistor is increased to 5 kΩ and the frequency of the current response is increased by a factor of 5000. Find $i_o(t)$.

16.18 Scale the circuit in Problem 14.32 so that the inductor is reduced to 250 mH and the capacitor is reduced to 8 μF.

a) Specify the scaling factors k_m and k_f.

b) Find $v_o(t)$ for the scaled circuit.

16.19 The circuit in Fig. P14.36 is scaled so that the capacitor changes to 0.10 μF and the 4 H inductance changes to 40 mH. Write the expression for v_o.

16.20 a) Show that if the low-pass filter circuit illustrated in Fig. 16.1 is scaled in both magnitude and frequency, the transfer function of the scaled circuit is the same as Eq. (16.1) with s replaced by s/k_f, where k_f is the frequency scale factor.

b) In the prototype version of the low-pass filter circuit in Fig. 16.1, $\omega_c = 1$ rad/s, $C = 1$ F, $R_2 = 1\ \Omega$, and $R_1 = 1/K$ ohms. What is the transfer function of the prototype circuit?

c) Using the result obtained in part (a), derive the transfer function of the scaled filter.

16.21 a) Show that if the high-pass filter illustrated in Fig. 16.4 is scaled in both magnitude and frequency, the transfer function is the same as Eq. (16.4) with s replaced by s/k_f, where k_f is the frequency scale factor.

b) In the prototype version of the high-pass filter circuit in Fig. 16.4, $\omega_c = 1$ rad/s, $R_1 = 1\ \Omega$, $C = 1$ F, and $R_2 = K$ ohms. What is the transfer function of the prototype circuit?

c) Using the result in part (a), derive the transfer function of the scaled filter.

❖ **16.22** a) Using 0.1 μF capacitors, design an active broadband first-order passband filter that has a lower cutoff frequency of 1000 Hz, an upper cutoff frequency of 5000 Hz, and a passband gain of 0 dB. Use prototype versions of the low-pass and high-pass filters in the design process. (See Problems 16.20 and 16.21.)

b) Write the transfer function for the scaled filter.

c) Use the transfer function derived in part (b) to find $H(j\omega_o)$, where ω_o is the center frequency of the filter.

d) What is the passband gain (in decibels) of the filter at ω_o?

e) Make a Bode magnitude plot of the filter.

❖ **16.23** a) Using 0.01 μF capacitors, design an active broadband first-order bandreject filter with a lower cutoff frequency of 400 Hz, an upper cutoff frequency of 4000 Hz, and a passband gain of 0 dB. Use the prototype filter circuits introduced in Problems 16.20 and 16.21 in the design process.

b) Draw the circuit diagram of the filter and label all the components.

c) What is the transfer function of the scaled filter?

d) Evaluate the transfer function derived in part (c) at the center frequency of the filter.

e) What is the gain (in decibels) at the center frequency?

f) Make a Bode magnitude plot of the filter transfer function.

16.24 For circuits consisting of resistors, capacitors, inductors, and op amps, $|H(j\omega)|^2$ involves only even powers of ω. To illustrate this, compute $|H(j\omega)|^2$ for the three circuits in Fig. P16.24 when

$$H(s) = \frac{V_o}{V_i}.$$

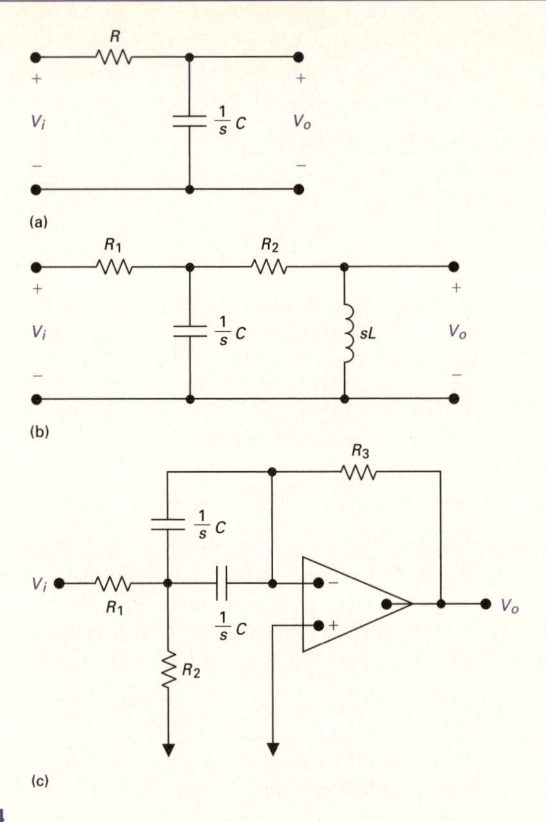

FIGURE P16.24

16.25 The purpose of this problem is to illustrate the advantage of an nth-order low-pass Butterworth filter over the cascade of n identical low-pass sections by calculating the slope (in decibels per decade) of each magnitude plot at the corner frequency ω_c. To facilitate the calculation, let y represent the magnitude of the plot (in decibels), and let $x = \log_{10}\omega$. Then calculate dy/dx at ω_c for each plot.

a) Show that at the corner frequency ($\omega_c = 1$ rad/s) of an nth-order low-pass prototype Butterworth filter,

$$\frac{dy}{dx} = -10n \text{ dB/dec.}$$

b) Show that for a cascade of n identical low-pass prototype sections, the slope at ω_c is

$$\frac{dy}{dx} = \frac{-20n(2^{1/n} - 1)}{2^{1/n}} \text{ dB/dec.}$$

c) Compute dy/dx for each type of filter for $n = 1$, 2, 3, 4, and ∞.

d) Discuss the significance of the results obtained in part (c).

16.26 a) Determine the order of a low-pass Butterworth filter that has a cutoff frequency of 2000 Hz and a gain of at least −30 dB at 7000 Hz.

b) What is the actual gain, in decibels, at 7000 Hz?

16.27 The circuit in Fig. 16.21 has the transfer function given by Eq. (16.34). Show that if the circuit in Fig. 16.21 is scaled in both magnitude and frequency, the transfer function of the scaled circuit is

$$H'(s) = \frac{\frac{1}{R^2C_1C_2}}{\left(\frac{s}{k_f}\right)^2 + \frac{2}{RC_1}\left(\frac{s}{k_f}\right) + \frac{1}{R^2C_1C_2}}.$$

16.28 a) Write the transfer function for the prototype low-pass Butterworth filter obtained in Problem 16.26(a).

b) Write the transfer function for the scaled filter in part (a). (See Problem 16.26.)

c) Check the expression derived in part (b) by using it to calculate the gain (in decibels) at 7000 Hz. Compare your result with that found in Problem 16.26(b).

❖ **16.29** a) Using 1 kΩ resistors and ideal op amps, design a circuit that will implement the low-pass Butterworth filter specified in Problem 16.26.

b) Construct the circuit diagram and label all component values.

❖ **16.30** a) Using 10 nF capacitors and ideal op amps, design a high-pass unity-gain Butterworth filter with a cutoff frequency of 2 kHz and a gain of at least −48 dB at 500 Hz.

b) Draw a circuit diagram of the filter and label all component values.

16.31 Verify the entries in Table 16.1 for $n = 5$ and $n = 6$.

16.32 The circuit in Fig. 16.25 has the transfer function given by Eq. (16.47). Show that if the circuit is scaled in both magnitude and frequency, the transfer function of the scaled circuit is

$$H'(s) = \frac{\left(\frac{s}{k_f}\right)^2}{\left(\frac{s}{k_f}\right)^2 + \frac{2}{R_2C}\left(\frac{s}{k_f}\right) + \frac{1}{R_1R_2C^2}}.$$

Hence the transfer function of a scaled circuit is obtained from the transfer function of an unscaled circuit by simply replacing s in the unscaled transfer function by s/k_f, where k_f is the frequency scaling factor.

❖ **16.33** a) Using 1 kΩ resistors and ideal op amps, design a low-pass unity-gain Butterworth filter that has a cutoff frequency of 8 kHz and is down at least 48 dB at 32 kHz.

b) Draw a circuit diagram of the filter and label all the components.

16.34 The high-pass filter designed in Problem 16.30 is cascaded with the low-pass filter designed in Problem 16.33.

a) Describe the type of filter formed by this interconnection.

b) Specify the cutoff frequencies, the midfrequency, and the quality factor of the filter.

c) Use the results of Problems 16.26 and 16.31 to derive the scaled transfer function of the filter.

d) Check the derivation of part (c) by using it to calculate $H(j\omega_o)$, where ω_o is the midfrequency of the filter.

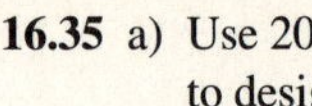

16.35 a) Use 20 nF capacitors in the circuit in Fig. 16.26 to design a bandpass filter with a quality factor of 16, a center frequency of 6.4 kHz, and a passband gain of 20 dB.

b) Draw the circuit diagram of the filter and label all the components.

16.36 Show that if $\omega_o = 1$ rad/s and $C = 1$ F in the circuit in Fig. 16.26, the prototype values of R_1, R_2, and R_3 are

$$R_1 = \frac{Q}{K},$$

$$R_2 = \frac{Q}{2Q^2 - K},$$

and

$$R_3 = 2Q.$$

❖ **16.37** a) Design a broadband Butterworth bandpass filter with a lower cutoff frequency of 500 Hz and an upper cutoff frequency of 4500 Hz. The passband gain of the filter is 20 dB. The gain should be down at least 20 dB at 200 Hz and 11.25 kHz. Use 15 nF capacitors in the high-pass circuit and 10 kΩ resistors in the low-pass circuit.

b) Draw a circuit diagram of the filter and label all the components.

16.38 a) Derive the expression for the scaled transfer function for the filter designed in Problem 16.37.

b) Using the expression derived in part (a), find the gain (in decibels) at 200 Hz and 1500 Hz.

c) Do the values obtained in part (b) satisfy the filtering specifications given in Problem 16.37?

16.39 a) Show that the transfer function for a prototype bandreject filter is

$$H(s) = \frac{s^2 + 1}{s^2 + \frac{1}{Q}s + 1}.$$

b) Use the result found in part (a) to find the transfer function of the filter designed in Example 16.13.

❖ **16.40** a) Using the circuit shown in Fig. 16.29, design a narrow-band bandreject filter having a center frequency of 1000 Hz and a quality factor of 20. Base the design on $C = 15$ nF.

b) Draw the circuit diagram of the filter and label all component values on the diagram.

c) What is the scaled transfer function of the filter?

❖ **16.41** The purpose of this problem is to guide you through the analysis necessary to establish a design procedure for determining the circuit components in a filter circuit. The circuit to be analyzed is shown in Fig. P16.41.

a) Analyze the circuit qualitatively and convince yourself that the circuit is a low-pass filter with a passband gain of R_2/R_1.

b) Support your qualitative analysis by deriving the transfer function V_o/V_i. (*Hint:* In deriving the transfer function, represent the resistors with their equivalent conductances, that is, $G_1 = 1/R_1$, and so forth.) To make the transfer function useful in terms of the entries in Table 16.1, put it in the form

$$H(s) = \frac{-Kb_o}{s^2 + b_1 s + b_o}.$$

c) Now observe that we have five circuit components—R_1, R_2, R_3, C_1, and C_2—and three transfer function constraints—K, b_1, and b_o. At first glance, it appears we have two free choices among the five components. However, when we investigate the relationships between the circuit components and the transfer function constraints, we see that if C_2 is chosen, there is an upper limit on C_1 in order for $R_2(G_2)$ to be realizable. With this in mind, show that if $C_2 = 1$ F, the three conductances are given by the expressions

$$G_1 = KG_2;$$
$$G_3 = \left(\frac{b_o}{G_2}\right) C_1;$$

and

$$G_2 = \frac{b_1 \pm \sqrt{b_1^2 - 4b_o(1+K)C_1}}{2(1+K)}.$$

For G_2 to be realizable,

$$C_1 \leq \frac{b_1^2}{4b_o(1+K)}.$$

d) Based on the results obtained in part (c), outline the design procedure for selecting the circuit components once K, b_o, and b_1 are known.

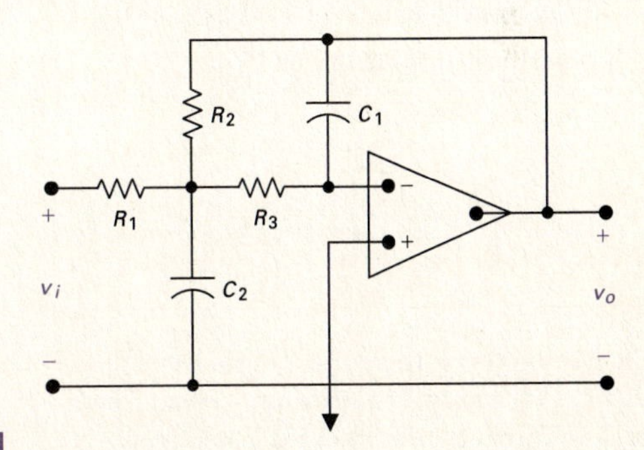

FIGURE P16.41

❖ **16.42** Assume the circuit analyzed in Problem 16.41 is part of a third-order low-pass Butterworth filter having a passband gain of 4.

a) If $C_2 = 1$ F in the prototype second-order section, what is the upper limit on C_1?

b) If the limiting value of C_1 is chosen, what are the prototype values of R_1, R_2, and R_3?

c) If the corner frequency of the filter is 2.5 kHz and C_2 is chosen to be 10 nF, calculate the scaled values of C_1, R_1, R_2, and R_3.

d) Specify the scaled values of the resistors and the capacitor in the first-order section of the filter.

e) Construct a circuit diagram of the filter and label all the component values on the diagram.

❖ **16.43** The purpose of this problem is to develop the design equations for the circuit in Fig. P16.43. (See Problem 16.41 for suggestions on the development of design equations.)

a) Based on a qualitative analysis, describe the type of filter implemented by the circuit.

b) Verify the conclusion reached in part (a) by deriving the transfer function V_o/V_i. Write the transfer function in a form that makes it compatible with the entries in Table 16.1.

c) How many free choices are there in the selection of the circuit components?

d) Derive the expressions for the conductances $G_1 = 1/R_1$ and $G_2 = 1/R_2$ in terms of C_1, C_2, and the coefficients b_o and b_1. (See Problem 16.41 for the definition of b_o and b_1.)

e) Are there any restrictions on C_1 or C_2?

f) Assume the circuit in Fig. P16.43 is used to design a fourth-order low-pass unity-gain Butterworth filter. Specify the prototype values of R_1 and R_2 in each second-order section if 1 F capacitors are used in the prototype circuit.

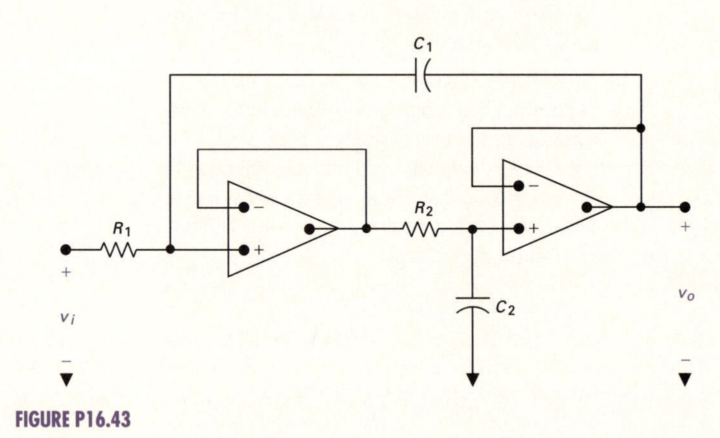

FIGURE P16.43

❖ **16.44** The fourth-order low-pass unity-gain Butterworth filter in Problem 16.43 is used in a system where the cutoff frequency is 3000 Hz. The filter has 4.7 nF capacitors.

a) Specify the numerical values of R_1 and R_2 in each section of the filter.

b) Draw a circuit diagram of the filter and label all the components.

16.45 Derive the prototype transfer function for a sixth-order high-pass Butterworth filter by first writing the transfer function for a sixth-order prototype low-pass Butterworth filter and then replacing s by $1/s$ in the low-pass expression.

16.46 The sixth-order Butterworth filter in Problem 16.45 is used in a system where the cutoff frequency is 25 krad/s.

a) What is the scaled transfer function for the filter?

b) Test your expression by finding the gain (in decibels) at the cutoff frequency.

❖ **16.47** Interchange the R's and C's in the circuit in Fig. P16.43, that is, replace R_1 with C_1, R_2 with C_2, and vice versa.

a) Analyze the circuit qualitatively and predict the type of filter implemented by the circuit.

b) Verify the conclusion reached in part (a) by deriving the transfer function V_o/V_i. Write the transfer function in a form that makes it compatible with the entries in Table 16.1.

c) How many free choices are there in the selection of the circuit components?

d) Find R_1 and R_2 as functions of b_o, b_1, C_1, and C_2.

e) Are there any restrictions on C_1 and C_2?

f) Assume the circuit is used in a third-order Butterworth filter of the type found in part (a). Specify the prototype values of R_1 and R_2 in the second-order section of the filter if $C_1 = C_2 = 1$ F.

❖ **16.48** a) The circuit in Problem 16.47 is used in a third-order high-pass Butterworth filter that has a cutoff frequency of 5 kHz. Specify the values of R_1 and R_2 if 75 nF capacitors are available to construct the filter.

b) Specify the values of resistance and capacitance in the first-order section of the filter.

c) Draw the circuit diagram and label all the components.

d) Give the numerical expression for the scaled transfer function of the filter.

❖ **16.49** Interchange the R's and C's in the circuit in Fig. P16.41; that is, replace R_1 with C_1, R_2 with C_2, R_3 with C_3, C_1 with R_1, and C_2 with R_2.

a) Describe the type of filter implemented as a result of the interchange.

b) Confirm the filter type described in part (a) by deriving the transfer function V_o/V_i. Write the transfer function in a form that makes it compatible with Table 16.1.

c) Set $C_2 = C_3 = 1$ F and derive the expressions for C_1, R_1, and R_2 in terms of K, b_1, and b_o. (See Problem 16.41 for the definition of b_1 and b_o.)

d) Assume the filter described in part (a) is used in the same type of third-order Butterworth filter which has a passband gain of 8. With $C_2 = C_3 = 1$ F, calculate the prototype values of C_1, R_1, and R_2 in the second-order section of the filter.

❖ **16.50** a) Use the circuits analyzed in Problems 16.41 and 16.49 to implement a broadband bandreject filter having a passband gain of 0 dB, a lower corner frequency of 400 Hz, an upper corner frequency of 6400 Hz, and an attenuation of at least 30 dB at both 1000 Hz and 2560 Hz. Use 10 nF capacitors whenever possible.

b) Draw a circuit diagram of the filter and label all the components.

16.51 a) Derive the transfer function for the bandreject filter described in Problem 16.50.

b) Use the transfer function derived in part (a) to find the attenuation (in decibels) at the center frequency of the filter.

17 FOURIER SERIES

CHAPTER CONTENTS

In the preceding chapters, we devoted a considerable amount of discussion to steady-state sinusoidal analysis. One reason for this interest in the sinusoidal excitation function is that it allows us to find the steady-state response to nonsinusoidal, but periodic, excitations. A **periodic function** is a function that repeats itself every T seconds. For example, the triangular wave illustrated in Fig. 17.1 is a nonsinusoidal, but periodic, waveform.

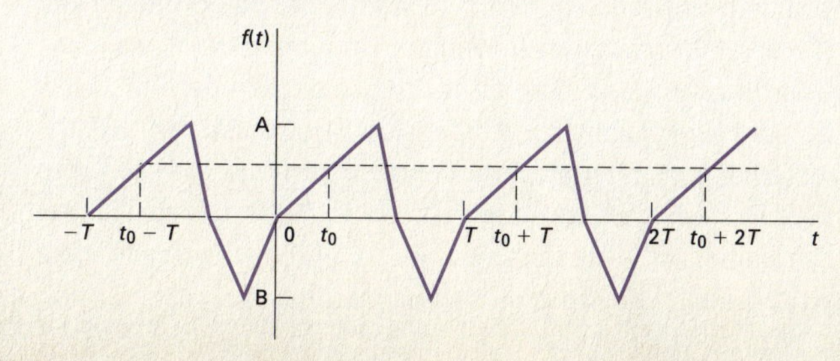

FIGURE 17.1 A periodic waveform.

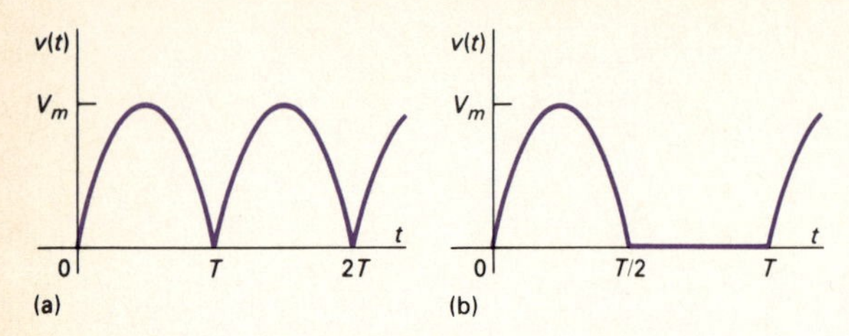

FIGURE 17.2 Output waveforms of a nonfiltered sinusoidal rectifier: (a) full-wave rectification; (b) half-wave rectification.

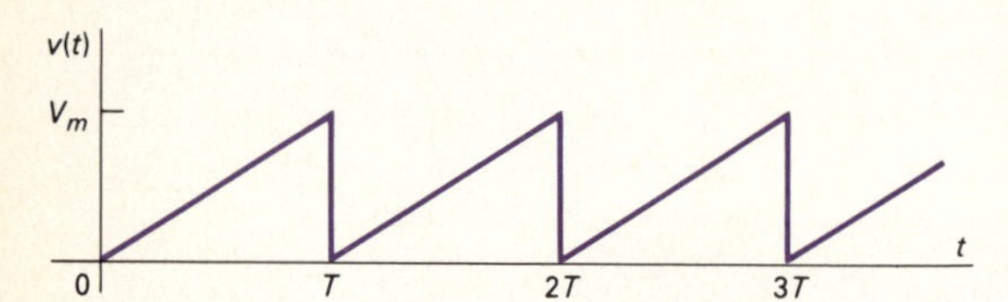

FIGURE 17.3 The triangular waveform of a cathode-ray oscilloscope sweep generator.

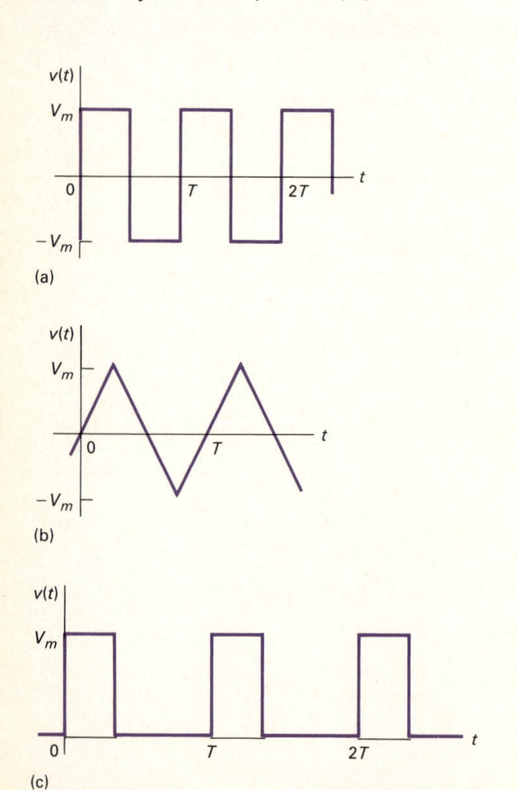

FIGURE 17.4 Waveforms produced by function generators used in laboratory testing: (a) square wave; (b) triangular wave; (c) rectangular pulse.

A periodic function is one that satisfies the relationship

$$f(t) = f(t \pm nT), \tag{17.1}$$

where n is an integer (1, 2, 3, ...) and T is the period. The function shown in Fig. 17.1 is periodic because

$$f(t_0) = f(t_0 - T) = f(t_0 + T) = f(t_0 + 2T) = \cdots$$

for any arbitrarily chosen value of t_0. Note that T is the smallest time interval that a periodic function may be shifted (in either direction) to produce a function that is identical with itself.

Why the interest in periodic functions? One reason is that many electrical sources of practical value generate periodic waveforms. For example, nonfiltered electronic rectifiers driven from a sinusoidal source produce rectified sine waves that are nonsinusoidal, but periodic. Figures 17.2(a) and (b) show the waveforms of the full-wave and half-wave sinusoidal rectifiers, respectively.

The sweep generator used to control the electron beam of a cathode-ray oscilloscope produces a periodic triangular wave like the one shown in Fig. 17.3.

Electronic oscillators, which are useful in laboratory testing of equipment, are designed to produce nonsinusoidal periodic waveforms. Function generators, which are capable of producing square-wave, triangular-wave, and rectangular-pulse waveforms, are found in most testing laboratories. Figure 17.4 illustrates typical waveforms.

Another practical problem that stimulates interest in periodic functions is that power generators, although designed to produce a sinusoidal waveform, cannot in practice be made to produce a pure sine wave. The distorted sinusoidal wave, however, is periodic. Engineers naturally are interested in ascertaining the consequences of exciting power systems with a slightly distorted sinusoidal voltage.

Interest in periodic functions also stems from the general observation that any nonlinearity in an otherwise linear circuit creates a nonsinusoidal periodic function. The rectifier circuit alluded to earlier is one example of this phenomenon. Magnetic saturation, which occurs in both machines and transformers, is another example of a nonlinearity that generates a nonsinusoidal periodic function. An electronic clipping circuit, which uses transistor saturation, is yet another example.

Moreover, nonsinusoidal periodic functions are important in the analysis of nonelectrical systems. Problems involving mechanical vibration, fluid flow, and heat flow all make use of periodic functions. In fact, the study and analysis of heat flow in a metal rod led the French mathematician Jean Baptiste Joseph Fourier (1768–1830) to the trigonometric series representation of a periodic function. This series bears his name and is the starting point for finding the steady-state response to periodic excitations of electric circuits.

17.1 Fourier Series Analysis: An Overview

What Fourier discovered in investigating heat-flow problems is that a periodic function can be represented by an infinite sum of sine or cosine functions that are harmonically related. In other words, the period of any trigonometric term in the infinite series is an integral multiple, or harmonic, of the fundamental period T of the periodic function. Thus for periodic $f(t)$, Fourier showed that $f(t)$ can be expressed as

$$f(t) = a_v + \sum_{n=1}^{\infty} a_n \cos n\omega_0 t + b_n \sin n\omega_0 t, \qquad \textbf{(17.2)}$$

where n is the integer sequence 1, 2, 3, ...

In Eq. (17.2), a_v, a_n, and b_n are known as the **Fourier coefficients** and are calculated from $f(t)$. The term ω_0 (which equals $2\pi/T$) represents the **fundamental frequency** of the periodic function $f(t)$. The integral multiples of ω_0—that is, $2\omega_0$, $3\omega_0$, $4\omega_0$, and so on—are known as the **harmonic frequencies of *f*(*t*)**. Thus $2\omega_0$ is the second harmonic, $3\omega_0$ is the third harmonic, and $n\omega_0$ is the nth harmonic of $f(t)$.

We discuss the determination of the Fourier coefficients in Section 17.2. Before pursuing the details of using a Fourier series in circuit analysis, we first need to look at the process in general terms. From an applications point of view, we can express all the periodic functions of interest in terms of a Fourier series. Mathematically, the conditions on a periodic function $f(t)$ that ensure expressing $f(t)$ as a convergent Fourier series (known as **Dirichlet's conditions**) are that

1. $f(t)$ be single-valued,
2. $f(t)$ have a finite number of discontinuities in the periodic interval,
3. $f(t)$ have a finite number of maxima and minima in the periodic interval, and
4. the integral

$$\int_{t_0}^{t_0+T} |f(t)|dt$$

exists.

Any periodic function generated by a physically realizable source satisfies Dirichlet's conditions. These are **sufficient** conditions, not **necessary** conditions. Thus if $f(t)$ meets these requirements, we know that we can express it as a Fourier series. However, if $f(t)$ does not meet these requirements, we still may be able to express it as a Fourier series. The necessary conditions on $f(t)$ are not known.

After we have determined $f(t)$ and calculated the Fourier coefficients (a_v, a_n, and b_n), we resolve the periodic source into a dc source (a_v) plus a sum of sinusoidal sources (a_n and b_n). Because the periodic source is driving a linear circuit, we may use the principle of superposition to find the steady-state response. In particular, we first calculate the response to each source generated by the Fourier series representation of $f(t)$ and then add the individual responses to obtain the total response. The steady-state response owing to a specific sinusoidal source is most easily found with the phasor method of analysis.

The procedure is straightforward and involves no new techniques of circuit analysis. It produces the Fourier series representation of the steady-state response; consequently, the actual shape of the response is unknown. Furthermore, the response waveform can be estimated only by adding a sufficient number of terms together. Even though the Fourier series approach to finding the steady-state response does have some drawbacks, it introduces a way of thinking about a problem that is as important as getting quantitative results. In fact, the conceptual picture is even more important in some respects than the quantitative one.

17.2 THE FOURIER COEFFICIENTS

After defining a periodic function over its fundamental period, we determine the Fourier coefficients from the relationships

$$a_v = \frac{1}{T}\int_{t_0}^{t_0+T} f(t)\,dt, \tag{17.3}$$

$$a_k = \frac{2}{T}\int_{t_0}^{t_0+T} f(t)\cos k\omega_0 t\,dt, \tag{17.4}$$

and

$$b_k = \frac{2}{T}\int_{t_0}^{t_0+T} f(t)\sin k\omega_0 t\,dt. \tag{17.5}$$

In Eqs. (17.4) and (17.5), the subscript k indicates the kth coefficient in the integer sequence 1, 2, 3, ... Note that a_v is the average value of $f(t)$, a_k is twice the average value of $f(t)\cos k\omega_0 t$, and b_k is twice the average value of $f(t)\sin k\omega_0 t$.

We easily derive Eqs. (17.3), (17.4), and (17.5) from Eq. (17.2) by recalling the following integral relationships, which hold when m and n are integers:

$$\int_{t_0}^{t_0+T} \sin m\omega_0 t\,dt = 0 \quad \text{for all } m; \tag{17.6}$$

$$\int_{t_0}^{t_0+T} \cos m\omega_0 t \, dt = 0 \quad \text{for all } m; \tag{17.7}$$

$$\int_{t_0}^{t_0+T} \cos m\omega_0 t \sin n\omega_0 t \, dt = 0 \quad \text{for all } m \text{ and } n; \tag{17.8}$$

$$\begin{aligned} \int_{t_0}^{t_0+T} \sin m\omega_0 t \sin n\omega_0 t \, dt &= 0 \quad \text{for all } m \neq n; \\ &= \frac{T}{2} \quad \text{for } m = n; \end{aligned} \tag{17.9}$$

and

$$\begin{aligned} \int_{t_0}^{t_0+T} \cos m\omega_0 t \cos n\omega_0 t \, dt &= 0 \quad \text{for all } m \neq n; \\ &= \frac{T}{2} \quad \text{for } m = n. \end{aligned} \tag{17.10}$$

We leave you to verify Eqs. (17.6)–(17.10) in Problem 17.1.

To derive Eq. (17.3), we simply integrate both sides of Eq. (17.2) over one period:

$$\begin{aligned} \int_{t_0}^{t_0+T} f(t)\, dt &= \int_{t_0}^{t_0+T} \left(a_v + \sum_{n=1}^{\infty} a_n \cos n\omega_0 t + b_n \sin n\omega_0 t \right) dt \\ &= \int_{t_0}^{t_0+T} a_v dt + \sum_{n=1}^{\infty} \int_{t_0}^{t_0+T} (a_n \cos n\omega_0 t + b_n \sin n\omega_0 t)\, dt \\ &= a_v T + 0. \end{aligned} \tag{17.11}$$

Equation (17.3) follows directly from Eq. (17.11).

To derive the expression for the kth value of a_n, we first multiply Eq. (17.2) by $\cos k\omega_0 t$ and then integrate both sides over one period of $f(t)$:

$$\begin{aligned} \int_{t_0}^{t_0+T} f(t) \cos k\omega_0 t \, dt &= \int_{t_0}^{t_0+T} a_v \cos k\omega_0 t \, dt \\ &\quad + \sum_{n=1}^{\infty} \int_{t_0}^{t_0+T} (a_n \cos n\omega_0 t \cos k\omega_0 t \\ &\quad + b_n \sin n\omega_0 t \cos k\omega_0 t)\, dt \\ &= 0 + a_k \left(\frac{T}{2} \right) + 0. \end{aligned} \tag{17.12}$$

Solving Eq. (17.12) for a_k yields the expression in Eq. (17.4).

We obtain the expression for the kth value of b_n by first multiplying both sides of Eq. (17.2) by $\sin k\omega_0 t$ and then integrating each side over one period of $f(t)$.

Example 17.1 shows how to use Eqs. (17.3)–(17.5) to find the Fourier coefficients for a specific periodic function.

EXAMPLE 17.1

Find the Fourier series for the periodic voltage shown in Fig. 17.5.

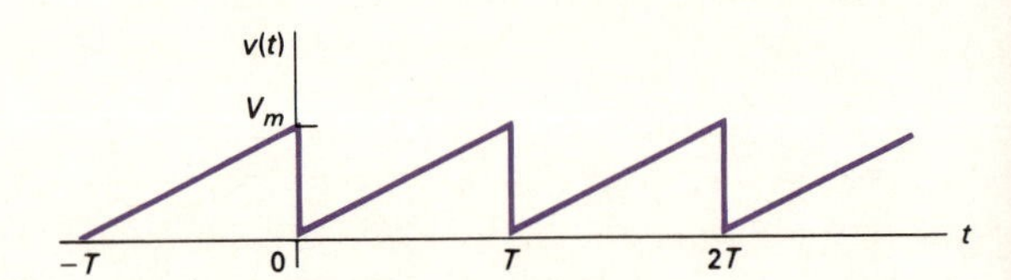

FIGURE 17.5 The periodic voltage for Example 17.1.

SOLUTION

When using Eqs. (17.3), (17.4), and (17.5) to find a_v, a_k, and b_k, we may choose the value of t_0. For the periodic voltage of Fig. 17.5, the best choice for t_0 is zero. Any other choice makes the required integrations more cumbersome. The expression for $v(t)$ between 0 and T is

$$v(t) = \left(\frac{V_m}{T}\right)t.$$

The equation for a_v is

$$a_v = \frac{1}{T}\int_0^T \left(\frac{V_m}{T}\right)t\,dt = \frac{1}{2}V_m.$$

This is clearly the average value of the waveform in Fig. 17.5.

The equation for the kth value of a_n is

$$\begin{aligned}
a_k &= \frac{2}{T}\int_0^T \left(\frac{V_m}{T}\right)t\cos k\omega_0 t\,dt \\
&= \frac{2V_m}{T^2}\left(\frac{1}{k^2\omega_0^2}\cos k\omega_0 t + \frac{t}{k\omega_0}\sin k\omega_0 t\right)\Bigg|_0^T \\
&= \frac{2V_m}{T^2}\left[\frac{1}{k^2\omega_0^2}(\cos 2\pi k - 1)\right] = 0 \quad \text{for all } k.
\end{aligned}$$

The equation for the kth value of b_n is

$$\begin{aligned}
b_k &= \frac{2}{T}\int_0^T \left(\frac{V_m}{T}\right)t\sin k\omega_0 t\,dt \\
&= \frac{2V_m}{T^2}\left(\frac{1}{k^2\omega_0^2}\sin k\omega_0 t - \frac{t}{k\omega_0}\cos k\omega_0 t\right)\Bigg|_0^T \\
&= \frac{2V_m}{T^2}\left(0 - \frac{T}{k\omega_0}\cos 2\pi k\right) \\
&= \frac{-V_m}{\pi k}.
\end{aligned}$$

The Fourier series for $v(t)$ is

$$v(t) = \frac{V_m}{2} - \frac{V_m}{\pi}\sum_{n=1}^{\infty}\frac{1}{n}\sin n\omega_0 t$$
$$= \frac{V_m}{2} - \frac{V_m}{\pi}\sin\omega_0 t - \frac{V_m}{2\pi}\sin 2\omega_0 t - \frac{V_m}{3\pi}\sin 3\omega_0 t - \cdots.$$

Finding the Fourier coefficients, in general, is tedious. Therefore anything that simplifies the task is beneficial. Fortunately, a periodic function that possesses certain types of symmetry greatly reduces the amount of work involved in finding the coefficients. In Section 17.3 we discuss how symmetry affects the coefficients in a Fourier series.

Use a computer tool to find the Fourier coefficients.

DRILL EXERCISES

17.1 Derive the expressions for a_v, a_k, and b_k for the periodic voltage function shown if $V_m = 60\pi$ V.

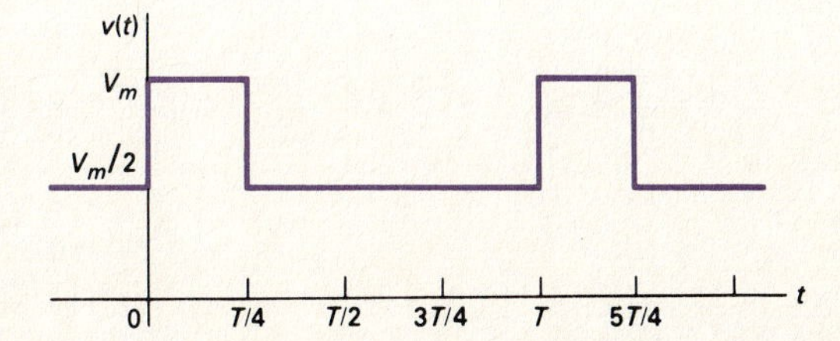

ANSWER: $a_v = 37.5\pi$ V, $a_k = \frac{30}{k}\sin\frac{k\pi}{2}$ V, $b_k = \frac{30}{k}\left(1 - \cos\frac{k\pi}{2}\right)$ V.

17.2 Refer to Drill Exercise 17.1.

a) What is the average value of the periodic voltage?

b) Compute the numerical values of $a_1 - a_5$ and $b_1 - b_5$.

c) If $T = 628.32$ ms, what is the fundamental frequency in radians per second?

d) What is the frequency of the fifth harmonic in hertz?

e) Write the Fourier series up to and including the fifth harmonic.

ANSWER: (a) 117.81 V; (b) 30 V, 0 V, −10 V, 0 V, and 6 V; 30 V, 30 V, 10 V, 0 V, and 6 V; (c) 10 rad/s; (d) 7.96 Hz; (e) $v(t) = 117.81 + 30\cos 10t + 30\sin 10t + 30\sin 20t - 10\cos 30t + 10\sin 30t + 6\cos 50t + 6\sin 50t$ V.

17.3 THE EFFECT OF SYMMETRY ON THE FOURIER COEFFICIENTS

Four types of symmetry may be used to simplify the task of evaluating the Fourier coefficients:

1. even-function symmetry
2. odd-function symmetry

3. half-wave symmetry
4. quarter-wave symmetry

The effect of each type of symmetry on the Fourier coefficients is discussed in the following sections.

Even-Function Symmetry

A function is defined as even if

$$f(t) = f(-t). \tag{17.13}$$

Functions that satisfy Eq. (17.13) are said to be even because polynomial functions with only even exponents possess this characteristic. For even periodic functions, the equations for the Fourier coefficients reduce to

$$a_v = \frac{2}{T}\int_0^{T/2} f(t)\,dt; \tag{17.14}$$

$$a_k = \frac{4}{T}\int_0^{T/2} f(t)\cos k\omega_0 t\,dt; \tag{17.15}$$

$$b_k = 0 \quad \text{for all } k. \tag{17.16}$$

Note that all the b coefficients are zero if the periodic function is even. Figure 17.6 illustrates an even periodic function. The derivations of Eqs. (17.14), (17.15), and (17.16) follow directly from Eqs. (17.3), (17.4), and (17.5). In each derivation, we select $t_0 = -T/2$ and then break the interval of integration into the range from $-T/2$ to 0 and 0 to $T/2$, or

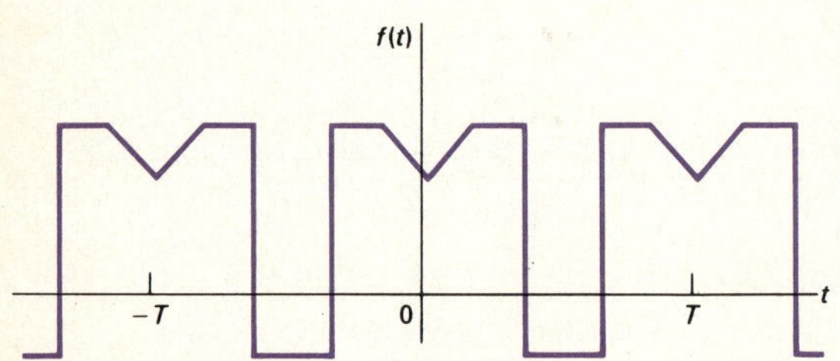

FIGURE 17.6 An even periodic function, $f(t) = f(-t)$.

$$\begin{aligned} a_v &= \frac{1}{T}\int_{-T/2}^{T/2} f(t)\,dt \\ &= \frac{1}{T}\int_{-T/2}^{0} f(t)\,dt + \frac{1}{T}\int_0^{T/2} f(t)\,dt. \end{aligned} \tag{17.17}$$

Now we change the variable of integration in the first integral on the right-hand side of Eq. (17.17). Specifically, we let $t = -x$ and note that $f(t) = f(-x) = f(x)$ because the function is even. We also observe that $x = T/2$ when $t = -T/2$ and $dt = -dx$. Then

$$\int_{-T/2}^{0} f(t)\,dt = \int_{T/2}^{0} f(x)(-dx) = \int_0^{T/2} f(x)\,dx, \tag{17.18}$$

which shows that the integration from $-T/2$ to 0 is identical to that from 0 to $T/2$; therefore Eq. (17.17) is the same as Eq. (17.14). The derivation of Eq. (17.15) proceeds along similar lines. Here,

$$a_k = \frac{2}{T}\int_{-T/2}^{0} f(t)\cos k\omega_0 t\, dt + \frac{2}{T}\int_{0}^{T/2} f(t)\cos k\omega_0 t\, dt, \quad \textbf{(17.19)}$$

but

$$\int_{-T/2}^{0} f(t)\cos k\omega_0 t\, dt = \int_{T/2}^{0} f(x)\cos(-k\omega_0 x)(-dx)$$

$$= \int_{0}^{T/2} f(x)\cos k\omega_0 x\, dx. \quad \textbf{(17.20)}$$

As before, the integration from $-T/2$ to 0 is identical to that from 0 to $T/2$. Combining Eq. (17.20) with Eq. (17.19) yields Eq. (17.15).

All the b coefficients are zero when $f(t)$ is an even periodic function because the integration from $-T/2$ to 0 is the exact negative of the integration from 0 to $T/2$; that is,

$$\int_{-T/2}^{0} f(t)\sin k\omega_0 t\, dt = \int_{T/2}^{0} f(x)\sin(-k\omega_0 x)(-dx)$$

$$= -\int_{0}^{T/2} f(x)\sin k\omega_0 x\, dx. \quad \textbf{(17.21)}$$

When we use Eqs. (17.14) and (17.15) to find the Fourier coefficients, the interval of integration must be between 0 and $T/2$.

ODD-FUNCTION SYMMETRY

A function is defined as odd if

$$f(t) = -f(-t). \quad \textbf{(17.22)}$$

Functions that satisfy Eq. (17.22) are said to be odd because polynomial functions with only odd exponents have this characteristic. The expressions for the Fourier coefficients are

$$a_v = 0; \quad \textbf{(17.23)}$$

$$a_k = 0, \quad \text{for all } k; \quad \textbf{(17.24)}$$

$$b_k = \frac{4}{T}\int_{0}^{T/2} f(t)\sin k\omega_0 t\, dt. \quad \textbf{(17.25)}$$

Note that all the a coefficients are zero if the periodic function is odd. Figure 17.7 shows an odd periodic function.

We use the same process to derive Eqs. (17.23), (17.24), and (17.25) that we used to derive Eqs. (17.14), (17.15), and (17.16). We leave the derivations to you in Problem 17.4.

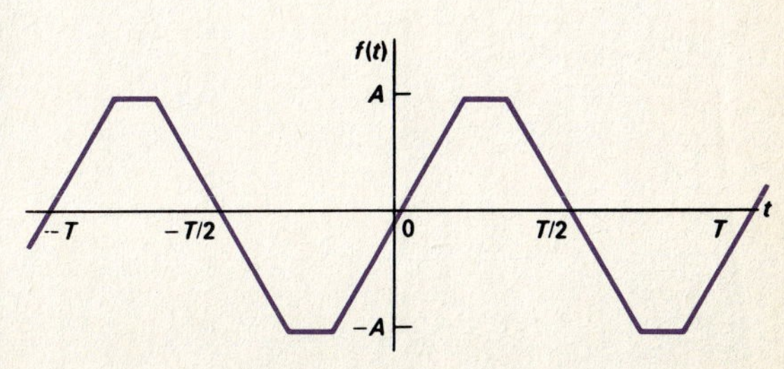

FIGURE 17.7 An odd periodic function $f(t) = -f(-t)$.

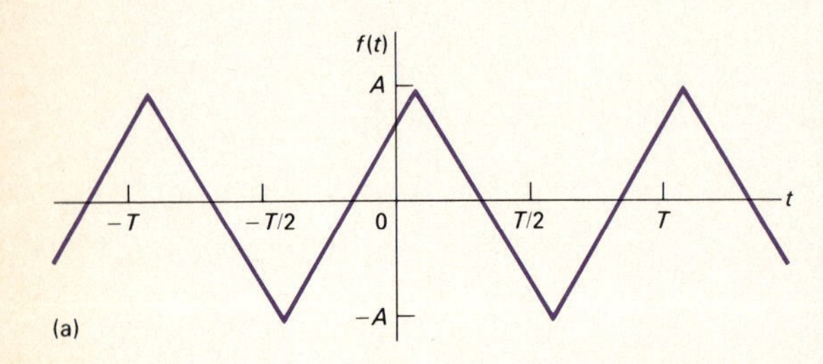

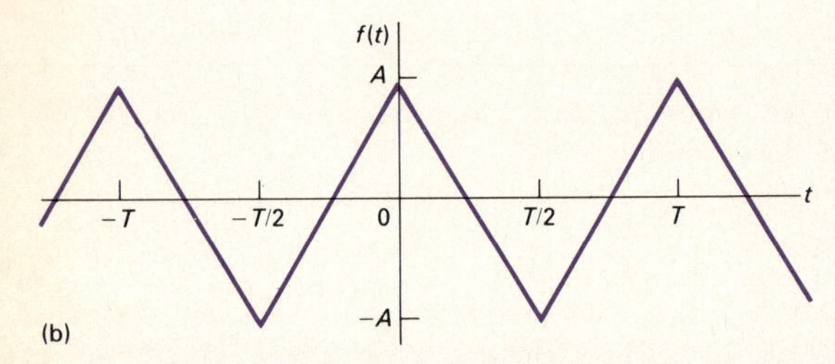

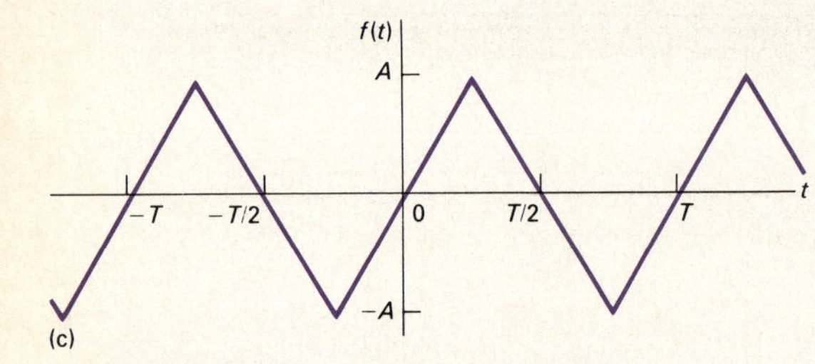

FIGURE 17.8 How the choice of where $t = 0$ can make a periodic function even, odd, or neither: (a) a periodic triangular wave that is neither even nor odd; (b) the triangular wave of (a) made even by shifting the function along the t axis; (c) the triangular wave of (a) made odd by shifting the function along the t axis.

The evenness, or oddness, of a periodic function can be destroyed by shifting the function along the time axis. In other words, the judicious choice of where $t = 0$ may give a periodic function even or odd symmetry. For example, the triangular function shown in Fig. 17.8(a) is neither even nor odd. However, we can make the function even, as shown in Fig. 17.8(b), or odd, as shown in Fig. 17.8(c).

Half-Wave Symmetry

A periodic function possesses half-wave symmetry if it satisfies the constraint

$$f(t) = -f(t - T/2). \tag{17.26}$$

Equation (17.26) states that a periodic function has half-wave symmetry if, after it is shifted one-half period and inverted, it is identical to the original function. For example, the functions shown in Figs. 17.7 and 17.8 have half-wave symmetry, whereas those in Figs. 17.5 and 17.6 do not. Note that half-wave symmetry is not a function of where $t = 0$.

If a periodic function has half-wave symmetry, both a_k and b_k are zero for even values of k. Moreover, a_v also is zero because the average value of a function with half-wave symmetry is zero. The expressions for the Fourier coefficients are

$$a_v = 0; \tag{17.27}$$

$$a_k = 0 \quad \text{for } k \text{ even}; \tag{17.28}$$

$$a_k = \frac{4}{T}\int_0^{T/2} f(t)\cos k\omega_0 t\, dt \quad \text{for } k \text{ odd}; \tag{17.29}$$

$$b_k = 0 \quad \text{for } k \text{ even}; \tag{17.30}$$

$$b_k = \frac{4}{T}\int_0^{T/2} f(t)\sin k\omega_0 t\, dt \quad \text{for } k \text{ odd}. \tag{17.31}$$

We derive Eqs. (17.27)–(17.31) by starting with Eqs. (17.3), (17.4), and (17.5) and choosing the interval of integration as $-T/2$ to $T/2$. We then divide this range into the intervals $-T/2$ to 0 and 0 to $T/2$. For example, the derivation for a_k is

$$\begin{aligned} a_k &= \frac{2}{T}\int_{t_0}^{t_0+T} f(t)\cos k\omega_0 t\, dt \\ &= \frac{2}{T}\int_{-T/2}^{T/2} f(t)\cos k\omega_0 t\, dt \\ &= \frac{2}{T}\int_{-T/2}^{0} f(t)\cos k\omega_0 t\, dt + \frac{2}{T}\int_0^{T/2} f(t)\cos k\omega_0 t\, dt. \end{aligned} \tag{17.32}$$

Now we change a variable in the first integral on the right-hand side of Eq. (17.32). Specifically, we let

$$t = x - T/2.$$

Then

$$x = T/2 \quad \text{when } t = 0;$$

$$x = 0 \quad \text{when } t = -T/2;$$

and

$$dt = dx.$$

We rewrite the first integral as

$$\int_{-T/2}^{0} f(t)\cos k\omega_0 t\,dt = \int_{0}^{T/2} f(x - T/2)\cos k\omega_0(x - T/2)\,dx. \qquad (17.33)$$

Note that

$$\cos k\omega_0(x - T/2) = \cos\,(k\omega_0 x - k\pi) = \cos k\pi\,\cos k\omega_0 x$$

and that, by hypothesis,

$$f(x - T/2) = -f(x).$$

Therefore Eq. (17.33) becomes

$$\int_{-T/2}^{0} f(t)\cos k\omega_0 t\,dt = \int_{0}^{T/2} [-f(x)]\cos k\pi\,\cos k\omega_0 x\,dx. \qquad (17.34)$$

Incorporating Eq. (17.34) into Eq. (17.32) gives

$$a_k = \frac{2}{T}(1 - \cos k\pi)\int_{0}^{T/2} f(t)\cos k\omega_0 t\,dt. \qquad (17.35)$$

But $\cos k\pi$ is 1 when k is even and -1 when k is odd. Therefore Eq. (17.35) generates Eqs. (17.28) and (17.29).

We leave it to you to verify that this same process can be used to derive Eqs. (17.30) and (17.31) (see Problem 17.5).

We summarize our observations by noting that the Fourier series representation of a periodic function with half-wave symmetry has zero average, or dc, value and contains only odd harmonics.

QUARTER-WAVE SYMMETRY

The term **quarter-wave symmetry** describes a periodic function that has half-wave symmetry and, in addition, symmetry about the midpoint of the positive and negative half-cycles. The function illustrated in Fig. 17.9(a) has quarter-wave symmetry about the midpoint of the

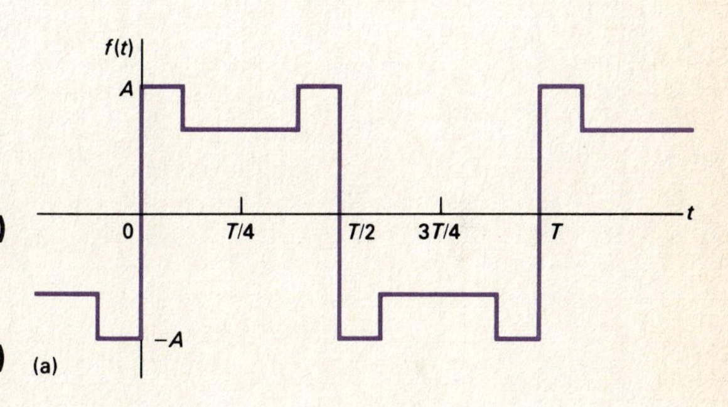

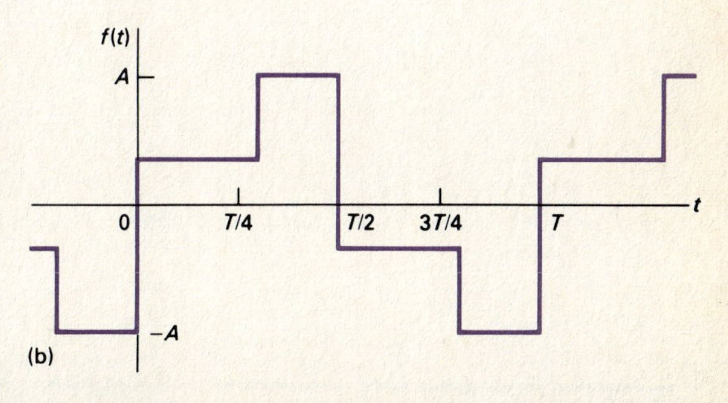

FIGURE 17.9 (a) A function that has quarter-wave symmetry; (b) a function that does not have quarter-wave symmetry.

positive and negative half-cycles. The function in Fig. 17.9(b) does not have quarter-wave symmetry, although it does have half-wave symmetry.

A periodic function that has quarter-wave symmetry can always be made either even or odd by the proper choice of the point where $t = 0$. For example, the function shown in Fig. 17.9(a) is odd and can be made even by shifting the function $T/4$ units either right or left along the t axis. However, the function in Fig. 17.9(b) can never be made either even or odd. To take advantage of quarter-wave symmetry in the calculation of the Fourier coefficients, you must choose the point where $t = 0$ to make the function either even or odd.

If the function is made even, then

$$\begin{aligned} a_v &= 0, \quad \text{because of half-wave symmetry;} \\ a_k &= 0 \quad \text{for } k \text{ even, because of half-wave symmetry;} \\ a_k &= \frac{8}{T}\int_0^{T/4} f(t)\cos k\omega_0 t\,dt \quad \text{for } k \text{ odd;} \\ b_k &= 0 \quad \text{for all } k\text{, because the function is even.} \end{aligned} \tag{17.36}$$

Equations (17.36) result from the function's quarter-wave symmetry in addition to its being even. Recall that quarter-wave symmetry is superimposed on half-wave symmetry, so we can eliminate a_v and a_k for k even. Comparing the expression for a_k, k odd, in Eqs. (17.36) with Eq. (17.29) shows that combining quarter-wave symmetry with evenness allows the shortening of the range of integration from 0 to $T/2$ to 0 to $T/4$. We leave the derivation of Eqs. (17.36) to you in Problem 17.6.

If the quarter-wave symmetric function is made odd,

$$\begin{aligned} a_v &= 0, \quad \text{because the function is odd;} \\ a_k &= 0 \quad \text{for all } k\text{, because the function is odd;} \\ b_k &= 0 \quad \text{for } k \text{ even, because of half-wave symmetry;} \\ b_k &= \frac{8}{T}\int_0^{T/4} f(t)\sin k\omega_0 t\,dt \quad \text{for } k \text{ odd.} \end{aligned} \tag{17.37}$$

Equations (17.37) are a direct consequence of quarter-wave symmetry and oddness. Again, quarter-wave symmetry allows the shortening of the interval of integration from 0 to $T/2$ to 0 to $T/4$. We leave the derivation of Eqs. (17.37) to you in Problem 17.7.

Example 17.2 shows how to use symmetry to simplify the task of finding the Fourier coefficients.

EXAMPLE 17.2

Find the Fourier series representation for the current waveform shown in Fig. 17.10.

SOLUTION

We begin by looking for degrees of symmetry in the waveform. We find that the function is odd and, in addition, has half-wave and quarter-wave symmetry. Because the function is odd, all the a coefficients are zero; that is, $a_v = 0$ and $a_k = 0$ for all k. Because the function has half-wave symmetry, $b_k = 0$ for even values of k. Because the function has quarter-wave symmetry, the expression for b_k for odd values of k is

$$b_k = \frac{8}{T}\int_0^{T/4} i(t)\sin k\omega_0 t\, dt.$$

In the interval $0 \le t \le T/4$, the expression for $i(t)$ is

$$i(t) = \frac{4I_m}{T}t.$$

Thus

$$\begin{aligned} b_k &= \frac{8}{T}\int_0^{T/4}\frac{4I_m}{T}t\sin k\omega_0 t\, dt \\ &= \frac{32I_m}{T^2}\left(\frac{\sin k\omega_0 t}{k^2\omega_0^2} - \frac{t\cos k\omega_0 t}{k\omega_0}\Bigg|_0^{T/4}\right) \\ &= \frac{8I_m}{\pi^2 k^2}\sin\frac{k\pi}{2} \qquad (k \text{ is odd}). \end{aligned}$$

The Fourier series representation of $i(t)$ is

$$\begin{aligned} i(t) &= \frac{8I_m}{\pi^2}\sum_{n=1,3,5,\ldots}^{\infty}\frac{1}{n^2}\sin\frac{n\pi}{2}\sin n\omega_0 t \\ &= \frac{8I_m}{\pi^2}\left(\sin\omega_0 t - \frac{1}{9}\sin 3\omega_0 t + \frac{1}{25}\sin 5\omega_0 t - \frac{1}{49}\sin 7\omega_0 t + \cdots\right). \end{aligned}$$

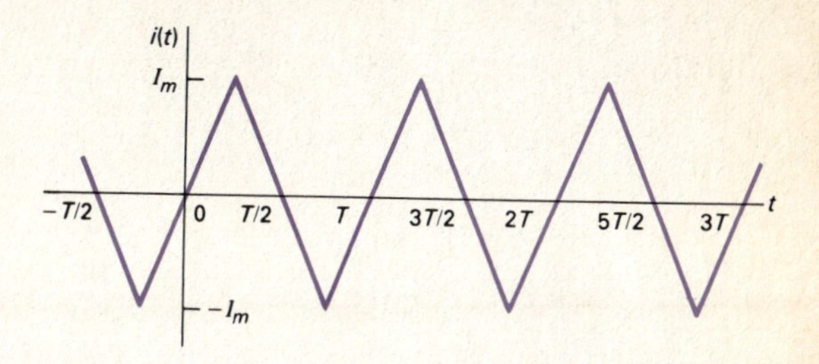

FIGURE 17.10 The periodic waveform for Example 17.2.

DRILL EXERCISE

17.3 Derive the Fourier series for the periodic voltage shown.

ANSWER: $v_g(t) = \dfrac{12V_m}{\pi^2}\displaystyle\sum_{n=1,3,5,\ldots}^{\infty}\frac{\sin(n\pi/3)}{n^2}\sin n\omega_0 t.$

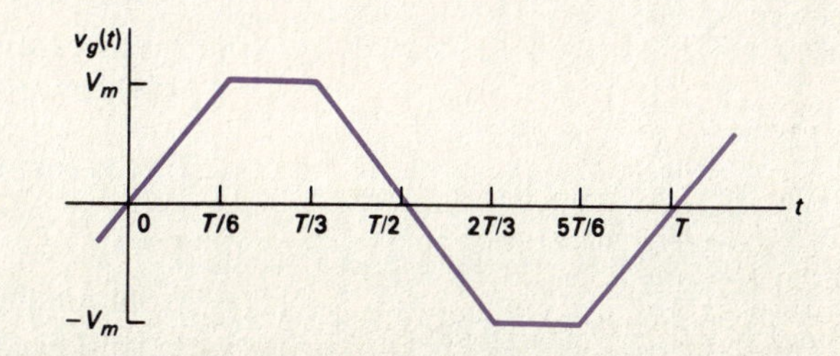

17.4 An Alternative Trigonometric Form of the Fourier Series

In circuit applications of the Fourier series, we combine the cosine and sine terms in the series into a single term for convenience. Doing so allows the representation of each harmonic of $v(t)$ or $i(t)$ as a single phasor quantity. The cosine and sine terms may be merged in either a cosine expression or a sine expression. Because we chose the cosine format in the phasor method of analysis (see Chapter 9), we choose the cosine expression here for the alternative form of the series. Thus we write the Fourier series in Eq. (17.2) as

$$f(t) = a_v + \sum_{n=1}^{\infty} A_n \cos(n\omega_0 t - \theta_n), \tag{17.38}$$

where A_n and θ_n are defined by the complex quantity

$$a_n - jb_n = \sqrt{a_n^2 + b_n^2}\,\underline{/-\theta_n} = A_n\underline{/-\theta_n}. \tag{17.39}$$

We derive Eqs. (17.38) and (17.39) using the phasor method to add the cosine and sine terms in Eq. (17.2). We begin by expressing the sine functions as cosine functions; that is, we rewrite Eq. (17.2) as

$$f(t) = a_v + \sum_{n=1}^{\infty} a_n \cos n\omega_0 t + b_n \cos(n\omega_0 t - 90^\circ). \tag{17.40}$$

Adding the terms under the summation sign by using phasors gives

$$\mathcal{P}\{a_n \cos n\omega_0 t\} = a_n\underline{/0^\circ} \tag{17.41}$$

and

$$\mathcal{P}\{b_n \cos(n\omega_0 t - 90^\circ)\} = b_n\underline{/-90^\circ} = -jb_n. \tag{17.42}$$

Then

$$\begin{aligned}\mathcal{P}\{a_n \cos n\omega_0 t + b_n \cos(n\omega_0 t - 90^\circ)\} &= a_n - jb_n \\ &= \sqrt{a_n^2 + b_n^2}\,\underline{/-\theta_n} \\ &= A_n\underline{/-\theta_n}.\end{aligned} \tag{17.43}$$

When we inverse-transform Eq. (17.43), we get

$$\begin{aligned}a_n \cos n\omega_0 t + b_n \cos(n\omega_0 t - 90^\circ) &= \mathcal{P}^{-1}\{A_n\underline{/-\theta_n}\} \\ &= A_n \cos(n\omega_0 t - \theta_n).\end{aligned} \tag{17.44}$$

Substituting Eq. (17.44) into Eq. (17.40) yields Eq. (17.38). Equation (17.43) corresponds to Eq. (17.39). If the periodic function is either even or odd, A_n reduces to either a_n (even) or b_n (odd), and θ_n is either 0° (even) or 90° (odd).

The derivation of the alternative form of the Fourier series for a given periodic function is illustrated in Example 17.3.

EXAMPLE 17.3

a) Derive the expressions for a_k and b_k for the periodic function shown in Fig. 17.11.

b) Write the first four terms of the Fourier series representation of $v(t)$ using the format of Eq. (17.38).

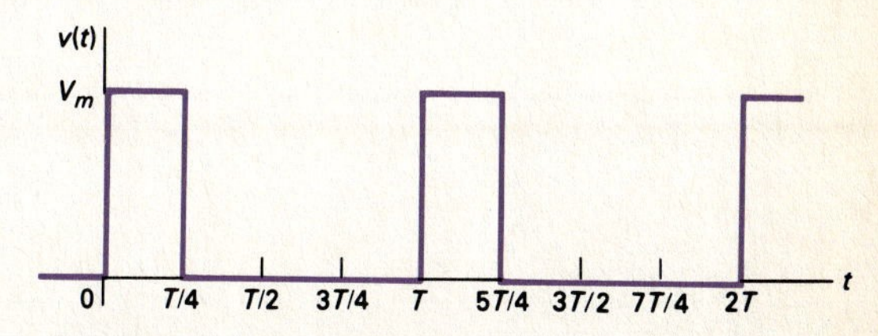

FIGURE 17.11 The periodic function for Example 17.3.

SOLUTION

a) The voltage $v(t)$ is neither even nor odd, nor does it have half-wave symmetry. Therefore we use Eqs. (17.4) and (17.5) to find a_k and b_k. Choosing t_0 as zero, we obtain

$$a_k = \frac{2}{T}\left[\int_0^{T/4} V_m \cos k\omega_0 t\, dt + \int_{T/4}^{T} (0) \cos k\omega_0 t\, dt\right]$$

$$= \frac{2V_m}{T} \left.\frac{\sin k\omega_0 t}{k\omega_0}\right|_0^{T/4} = \frac{V_m}{k\pi} \sin \frac{k\pi}{2}$$

and

$$b_k = \frac{2}{T}\int_0^{T/4} V_m \sin k\omega_0 t\, dt$$

$$= \frac{2V_m}{T}\left(\left.\frac{-\cos k\omega_0 t}{k\omega_0}\right|_0^{T/4}\right)$$

$$= \frac{V_m}{k\pi}\left(1 - \cos\frac{k\pi}{2}\right).$$

b) The average value of $v(t)$ is

$$a_v = \frac{V_m(T/4)}{T} = \frac{V_m}{4}.$$

The values of $a_k - jb_k$ for $k = 1, 2,$ and 3 are

$$a_1 - jb_1 = \frac{V_m}{\pi} - j\frac{V_m}{\pi} = \frac{\sqrt{2}V_m}{\pi}\angle{-45^\circ};$$

$$a_2 - jb_2 = 0 - j\frac{V_m}{\pi} = \frac{V_m}{\pi}\angle{-90^\circ};$$

$$a_3 - jb_3 = \frac{-V_m}{3\pi} - j\frac{V_m}{3\pi} = \frac{\sqrt{2}V_m}{3\pi}\angle{-135^\circ}.$$

Thus the first four terms in the Fourier series representation of $v(t)$ are

$$v(t) = \frac{V_m}{4} + \frac{\sqrt{2}V_m}{\pi}\cos(\omega_0 t - 45^\circ) + \frac{V_m}{\pi}\cos(2\omega_0 t - 90^\circ) + \frac{\sqrt{2}V_m}{3\pi}\cos(3\omega_0 t - 135^\circ) + \cdots.$$

Next we illustrate how to use a Fourier series representation of a periodic excitation function to find the steady-state response of a linear circuit.

DRILL EXERCISE

17.4 a) Compute A_1–A_5 and θ_1–θ_5 for the periodic function shown if $V_m = 60\pi$ volts.

b) Using the format of Eq. (17.38), write the Fourier series for $v(t)$ up to and including the fifth harmonic assuming $T = 628.32$ ms.

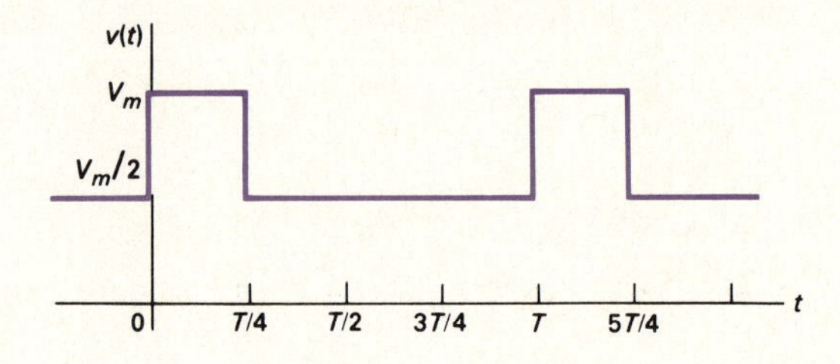

ANSWER: (a) 42.43, 30, 14.14, 0, 8.49 V, and +45°, +90°, +135°, not defined, +45°; (b) $v(t) = 117.81 + 42.43\cos(10t - 45^\circ) + 30\cos(20t - 90^\circ) + 14.14\cos(30t - 135^\circ) + 8.49\cos(50t - 45^\circ) + \cdots$ V.

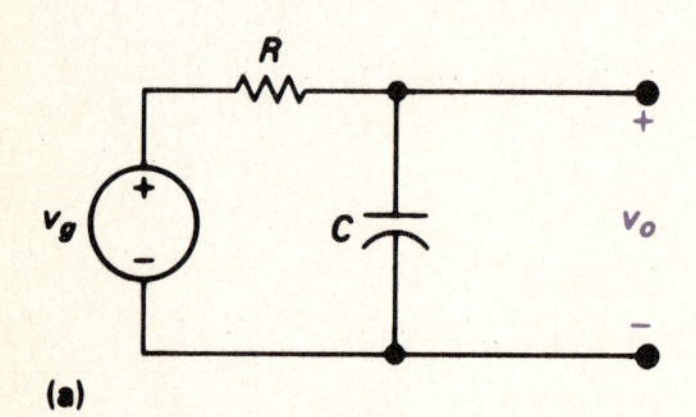

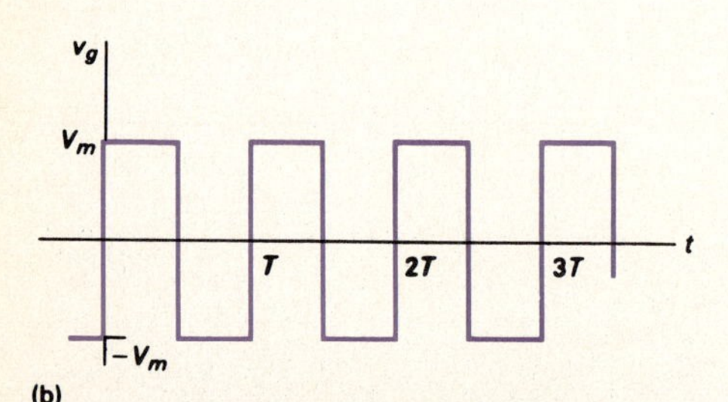

FIGURE 17.12 An *RC* circuit excited by a periodic voltage: (a) the *RC* series circuit; (b) the square-wave voltage.

17.5 AN ILLUSTRATIVE APPLICATION

The *RC* circuit shown in Fig. 17.12(a) illustrates how to use a Fourier series in circuit analysis. The circuit is energized with the periodic square-wave voltage shown in Fig. 17.12(b). The voltage across the capacitor is the desired response, or output, signal.

The first step in finding the steady-state response is to represent the periodic excitation source with its Fourier series. After noting that the source has odd, half-wave, and quarter-wave symmetry, we know that the Fourier coefficients reduce to b_k, with k restricted to odd integer values:

$$b_k = \frac{8}{T}\int_0^{T/4} V_m \sin k\omega_0 t\, dt$$

$$= \frac{4V_m}{\pi k} \quad (k \text{ is odd}). \tag{17.45}$$

Then the Fourier series representation of v_g is

$$v_g = \frac{4V_m}{\pi} \sum_{n=1,3,5,\ldots}^{\infty} \frac{1}{n} \sin n\omega_0 t. \tag{17.46}$$

Writing the series in expanded form, we have

$$v_g = \frac{4V_m}{\pi} \sin \omega_0 t + \frac{4V_m}{3\pi} \sin 3\omega_0 t$$

$$+ \frac{4V_m}{5\pi} \sin 5\omega_0 t + \frac{4V_m}{7\pi} \sin 7\omega_0 t + \cdots. \tag{17.47}$$

The voltage source expressed by Eq. (17.47) is the equivalent of infinitely many series-connected sinusoidal sources, each source having its own amplitude and frequency. To find the contribution of each source to the output voltage, we use the principle of superposition.

For any one of the sinusoidal sources, the phasor-domain expression for the output voltage is

$$\mathbf{V}_o = \frac{\mathbf{V}_g}{1 + j\omega RC}. \tag{17.48}$$

All the voltage sources are expressed as sine functions, so we interpret a phasor in terms of the sine instead of the cosine. In other words, when we go from the phasor domain back to the time domain, we simply write the time-domain expressions as $\sin(\omega t + \theta)$ instead of $\cos(\omega t + \theta)$.

The phasor output voltage owing to the fundamental frequency of the sinusoidal source is

$$\mathbf{V}_{o1} = \frac{(4V_m/\pi)\underline{/0^\circ}}{1 + j\omega_0 RC}. \tag{17.49}$$

Writing $\mathbf{V}_{o1}$ in polar form gives

$$\mathbf{V}_{o1} = \frac{(4V_m)\underline{/-\beta_1}}{\pi\sqrt{1 + \omega_0^2 R^2 C^2}}, \tag{17.50}$$

where

$$\beta_1 = \tan^{-1} \omega_0 RC. \tag{17.51}$$

From Eq. (17.50), the time-domain expression for the fundamental frequency component of v_o is

$$v_{o1} = \frac{4V_m}{\pi\sqrt{1+\omega_0^2R^2C^2}}\sin(\omega_0 t - \beta_1). \quad \textbf{(17.52)}$$

We derive the third-harmonic component of the output voltage in a similar manner. The third-harmonic phasor voltage is

$$\mathbf{V}_{o3} = \frac{(4V_m/3\pi)\angle 0^\circ}{1 + j3\omega_0 RC}$$
$$= \frac{4V_m}{3\pi\sqrt{1+9\omega_0^2R^2C^2}}\angle -\beta_3, \quad \textbf{(17.53)}$$

where

$$\beta_3 = \tan^{-1} 3\omega_0 RC. \quad \textbf{(17.54)}$$

The time-domain expression for the third-harmonic output voltage is

$$v_{o3} = \frac{4V_m}{3\pi\sqrt{1+9\omega_0^2R^2C^2}}\sin(3\omega_0 t - \beta_3). \quad \textbf{(17.55)}$$

Hence the expression for the kth-harmonic component of the output voltage is

$$v_{ok} = \frac{4V_m}{k\pi\sqrt{1+k^2\omega_0^2R^2C^2}}\sin(k\omega_0 t - \beta_k)\ (k \text{ is odd}), \quad \textbf{(17.56)}$$

where

$$\beta_k = \tan^{-1} k\omega_0 RC \quad (k \text{ is odd}). \quad \textbf{(17.57)}$$

We now write down the Fourier series representation of the output voltage:

Use a computer tool to plot as many terms in the sum as you wish.

$$v_o(t) = \frac{4V_m}{\pi}\sum_{n=1,3,5,\ldots}^{\infty}\frac{\sin(n\omega_0 t - \beta_n)}{n\sqrt{1+(n\omega_0 RC)^2}}. \quad \textbf{(17.58)}$$

The derivation of Eq. (17.58) was not difficult. But, although we have an analytic expression for the steady-state output, what $v_o(t)$ looks like is not immediately apparent from Eq. (17.58). As we mentioned earlier, this shortcoming is a problem with the Fourier series approach. Equation (17.58) is not useless, however, because it gives some feel for the steady-state waveform of $v_o(t)$, if we focus on the frequency response of the circuit. For example, if C is large, $1/n\omega_0 C$ is small for the higher order harmonics. Thus the capacitor short circuits the high-frequency components of the input waveform, and the higher order harmonics in Eq. (17.58) are negligible compared to the lower order harmonics. Equation (17.58) reflects this condition in that, for large C,

$$v_o \approx \frac{4V_m}{\pi\omega_0 RC} \sum_{n=1,3,5,\ldots}^{\infty} \frac{1}{n^2} \sin(n\omega_0 t - 90^\circ)$$

$$\approx \frac{-4V_m}{\pi\omega_0 RC} \sum_{n=1,3,5,\ldots}^{\infty} \frac{1}{n^2} \cos n\omega_0 t. \quad \textbf{(17.59)}$$

Equation (17.59) shows that the amplitude of the harmonic in the output is decreasing by $1/n^2$, compared to $1/n$ for the input harmonics. If C is so large that only the fundamental component is significant, then to a first approximation

$$v_o(t) \approx \frac{-4V_m}{\pi\omega_0 RC} \cos \omega_0 t, \quad \textbf{(17.60)}$$

and Fourier analysis tells us that the square-wave input is deformed into a sinusoidal output.

Now let's see what happens as $C \to 0$. The circuit shows that v_o and v_g are the same when $C = 0$, because the capacitive branch looks like an open circuit at all frequencies. Equation (17.58) predicts the same result because, as $C \to 0$,

$$v_o = \frac{4V_m}{\pi} \sum_{n=1,3,5,\ldots}^{\infty} \frac{1}{n} \sin n\omega_0 t. \quad \textbf{(17.61)}$$

But Eq. (17.61) is identical to Eq. (17.46), and therefore $v_o \to v_g$ as $C \to 0$.

Thus Eq. (17.58) has proven useful because it enabled us to predict that the output will be a highly distorted replica of the input waveform if C is large, and a reasonable replica if C is small. In Chapter 14, we looked at the distortion between the input and output in terms of how much memory the system weighting function had. In the frequency domain, we look at the distortion between the steady-state input and output in terms of how the amplitude and phase of the harmonics are altered as they are transmitted through the circuit. When the network significantly alters the amplitude and phase relationships among the harmonics at the output relative to that at the input, the output is a distorted version of the input. Thus in the frequency domain, we speak of amplitude distortion and phase distortion.

For the circuit here, amplitude distortion is present because the amplitudes of the input harmonics decrease as $1/n$, whereas the amplitudes of the output harmonics decrease as

$$\frac{1}{n} \frac{1}{\sqrt{1 + (n\omega_0 RC)^2}}.$$

Use a computer tool to explore the relationship between the system weighting function and the amplitude and phase distortion.

This circuit also exhibits phase distortion because the phase angle of each input harmonic is zero, whereas that of the nth harmonic in the output signal is $-\tan^{-1} n\omega_0 RC$.

THE DIRECT APPROACH TO STEADY-STATE RESPONSE: AN EXAMPLE

For the simple RC circuit shown in Fig. 17.12(a), we can derive the expression for the steady-state response without resorting to the Fourier series representation of the excitation function. Doing this extra analysis here adds to our understanding of the Fourier series approach.

To find the steady-state expression for v_o by straightforward circuit analysis, we reason as follows. The square-wave excitation function alternates between charging the capacitor toward $+V_m$ and $-V_m$. After the circuit reaches steady-state operation, this alternate charging becomes periodic. We know from the analysis of the single time-constant RC circuit (Chapter 7) that the response to abrupt changes in the driving voltage is exponential. Thus the steady-state waveform of the voltage across the capacitor in the circuit shown in Fig. 17.12(a) is as shown in Fig. 17.13.

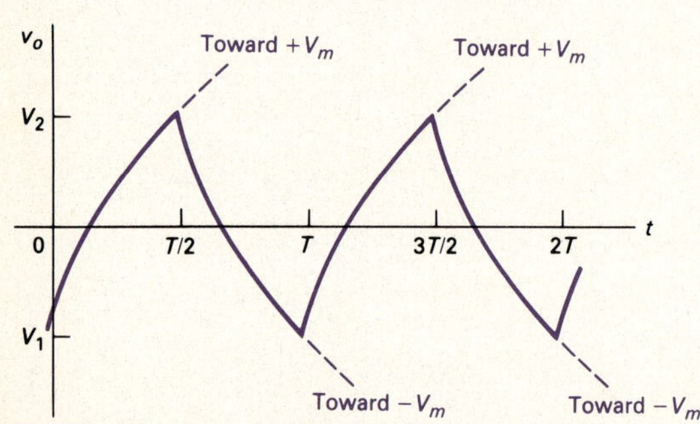

FIGURE 17.13 The steady-state waveform of v_o for the circuit in Fig. 17.12(a).

The analytic expressions for $v_o(t)$ in the time intervals $0 \leq t \leq T/2$ and $T/2 \leq t \leq T$ are

$$v_o = V_m + (V_1 - V_m)e^{-t/RC}, \quad 0 \leq t \leq T/2; \tag{17.62}$$

$$v_o = -V_m + (V_2 + V_m)e^{-[t-(T/2)]/RC}, \quad T/2 \leq t \leq T. \tag{17.63}$$

We derive Eqs. (17.62) and (17.63) by using the methods of Chapter 7, as summarized by Eq. (7.60). We obtain the values of V_1 and V_2 by noting from Eq. (17.62) that

$$V_2 = V_m + (V_1 - V_m)e^{-T/2RC}, \tag{17.64}$$

and from Eq. (17.63) that

$$V_1 = -V_m + (V_2 + V_m)e^{-T/2RC}. \tag{17.65}$$

Solving Eqs. (17.64) and (17.65) for V_1 and V_2 yields

$$V_2 = -V_1 = \frac{V_m(1 - e^{-T/2RC})}{1 + e^{-T/2RC}}. \tag{17.66}$$

Substituting Eq. (17.66) into Eqs. (17.62) and (17.63) gives

$$v_o = V_m - \frac{2V_m}{1 + e^{-T/2RC}}e^{-t/RC}, \quad 0 \leq t \leq T/2, \tag{17.67}$$

and

$$v_o = -V_m + \frac{2V_m}{1 + e^{-T/2RC}}e^{-[t-(T/2)]/RC}, \quad T/2 \leq t \leq T. \tag{17.68}$$

Equations (17.67) and (17.68) indicate that $v_o(t)$ has half-wave symmetry and that therefore the average value of v_o is zero. This result agrees with the Fourier series solution for the steady-state response—namely, that because the excitation function has no zero frequency component, the response can have no such component. Equations (17.67) and (17.68) also show the effect of changing the size of the capacitor. If C is small, the exponential functions quickly vanish, $v_o = V_m$ between 0 and $T/2$, and $v_o = -V_m$ between $T/2$ and T. In other words, $v_o \rightarrow v_g$ as $C \rightarrow 0$. If C is large, the output waveform becomes triangular in shape, as Fig. 17.14 shows. Note that for large C, we may approximate the exponential terms $e^{-t/RC}$ and $e^{-[t-(T/2)]/RC}$ by the linear terms $1-(t/RC)$ and $1-\{[t-(T/2)]/RC\}$, respectively. Equation (17.59) gives the Fourier series of this triangular waveform.

Figure 17.14 summarizes the results. The dashed line in Fig. 17.14 is the input voltage, the solid blue line depicts the output voltage when C is small, and the solid black line depicts the output voltage when C is large.

Use a computer to vary the value of C and plot the output voltage.

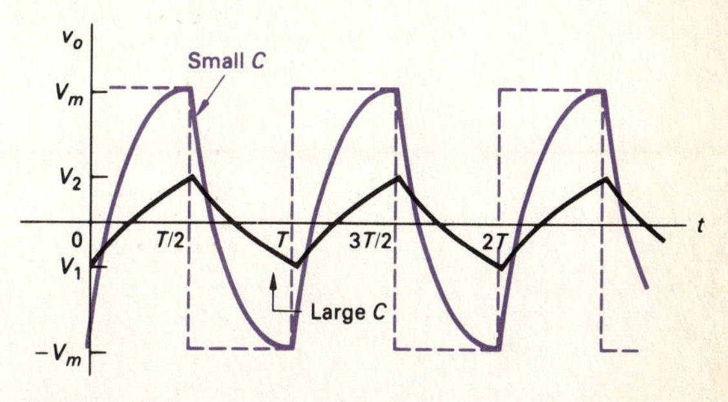

FIGURE 17.14 The effect of capacitor size on the steady-state response.

DRILL EXERCISE

17.5 a) Show that for large values of C, Eq. (17.67) can be approximated by the expression

$$v_o(t) \approx \frac{-V_m T}{4RC} + \frac{V_m}{RC}t.$$

Note that this expression is the equation of the triangular wave for $0 \leq t \leq T/2$. *Hints*: (1) Let $e^{-t/RC} \approx 1 - (t/RC)$ and $e^{-T/2RC} \approx 1 - (T/2RC)$; (2) put the resulting expression over the common denominator $2 - (T/2RC)$; (3) simplify the numerator; and (4) for large C, assume that $T/2RC$ is much less than 2.

b) Substitute the peak value of the triangular wave into the solution for Problem 17.9 (see Fig. P17.9[b]) and show that the result is Eq. (17.59).

ANSWER: (a) Derivation; (b) from Problem 17.9, $a_k = -8V_p/\pi^2k^2$, where $k = 1, 3, 5, \ldots$, and V_p is the peak value of the triangular wave. Therefore, for the triangular wave described in (a),

$$a_k = \frac{-2V_m T}{\pi^2 k^2 RC} = \frac{-4V_m}{\pi \omega_0 RCk^2}.$$

Finally, we verify that the steady-state response of Eqs. (17.67) and (17.68) is equivalent to the Fourier series solution in Eq. (17.58). To do so we simply derive the Fourier series representation of the periodic function described by Eqs. (17.67) and (17.68). We have already noted that the periodic voltage response has half-wave symmetry. Therefore the Fourier series contains only odd harmonics. For k odd,

$$a_k = \frac{4}{T}\int_0^{T/2}\left(V_m - \frac{2V_m e^{-t/RC}}{1+e^{-T/2RC}}\right)\cos k\omega_0 t\,dt$$

$$= \frac{-8RCV_m}{T[1+(k\omega_0 RC)^2]} \quad (k \text{ is odd}), \tag{17.69}$$

$$b_k = \frac{4}{T}\int_0^{T/2}\left(V_m - \frac{2V_m e^{-t/RC}}{1+e^{-T/2RC}}\right)\sin k\omega_0 t\,dt$$

$$= \frac{4V_m}{k\pi} - \frac{8k\omega_0 V_m R^2C^2}{T[1+(k\omega_0 RC)^2]} \quad (k \text{ is odd}). \tag{17.70}$$

To show that the results obtained from Eqs. (17.69) and (17.70) are consistent with Eq. (17.58), we must prove that

$$\sqrt{a_k^2 + b_k^2} = \frac{4V_m}{k\pi}\frac{1}{\sqrt{1+(k\omega_0 RC)^2}} \tag{17.71}$$

and that

$$\frac{a_k}{b_k} = -k\omega_0 RC. \tag{17.72}$$

We leave you to verify Eqs. (17.69)–(17.72) in Problems 17.19 and 17.20. Equations (17.71) and (17.72) are used with Eqs. (17.38) and (17.39) to derive the Fourier series expression in Eq. (17.58); we leave the details to you in Problem 17.21.

With this illustrative circuit, we showed how to use the Fourier series in conjunction with the principle of superposition to obtain the steady-state response to a periodic driving function. Again, the principal shortcoming of the Fourier series approach is the difficulty of ascertaining the waveform of the response. However, by thinking in terms of a circuit's frequency response, we can deduce a reasonable approximation of the steady-state response by using a finite number of appropriate terms in the Fourier series representation. (See Problems 17.25 and 17.27.)

17.6 Average-Power Calculations with Periodic Functions

If we have the Fourier series representation of the voltage and current at a pair of terminals in a linear lumped-parameter circuit, we can easily express the average power at the terminals as a function of the harmonic voltages and currents. Using the trigonometric form of the Fourier series expressed in Eq. (17.38), we write the periodic voltage and current at the terminals of a network as

$$v = V_{\mathrm{dc}} + \sum_{n=1}^{\infty} V_n \cos (n\omega_0 t - \theta_{vn}) \quad \textbf{(17.73)}$$

and

$$i = I_{\mathrm{dc}} + \sum_{n=1}^{\infty} I_n \cos (n\omega_0 t - \theta_{in}). \quad \textbf{(17.74)}$$

The notation used in Eqs. (17.73) and (17.74) is defined as follows:

V_{dc} = the amplitude of the dc voltage component;
V_n = the amplitude of the nth-harmonic voltage;
θ_{vn} = the phase angle of the nth-harmonic voltage;
I_{dc} = the amplitude of the dc current component;
I_n = the amplitude of the nth-harmonic current;
θ_{in} = the phase angle of the nth-harmonic current.

We assume that the current reference is in the direction of the reference voltage drop across the terminals (using the passive sign convention), so that the instantaneous power at the terminals is vi. The average power is

$$P = \frac{1}{T}\int_{t_0}^{t_0+T} p\,dt = \frac{1}{T}\int_{t_0}^{t_0+T} vi\,dt. \quad \textbf{(17.75)}$$

To find the expression for the average power, we substitute Eqs. (17.73) and (17.74) into Eq. (17.75) and integrate. At first glance this appears to be a formidable task, since the product vi requires multiplying two infinite series. However, the only terms to survive integration are the products of voltage and current at the same frequency. A review of Eqs. (17.8), (17.9), and (17.10) should convince you of the validity of this observation. Therefore Eq. (17.75) reduces to

$$P = \frac{1}{T} V_{\mathrm{dc}} I_{\mathrm{dc}} t \Big|_{t_0}^{t_0+T} + \sum_{n=1}^{\infty} \frac{1}{T}\int_{t_0}^{t_0+T} V_n I_n \cos (n\omega_0 t - \theta_{vn}) \times \cos (n\omega_0 t - \theta_{in})\,dt. \quad \textbf{(17.76)}$$

Now, using the trigonometric identity

$$\cos\alpha \cos\beta = \frac{1}{2}\cos(\alpha - \beta) + \frac{1}{2}\cos(\alpha + \beta),$$

we simplify Eq. (17.76) to

$$P = V_{dc} I_{dc} + \frac{1}{T} \sum_{n=1}^{\infty} \frac{V_n I_n}{2} \int_{t_0}^{t_0+T} [\cos(\theta_{vn} - \theta_{in}) + \cos(2n\omega_0 t - \theta_{vn} - \theta_{in})]\, dt. \quad \textbf{(17.77)}$$

The second term under the integral sign integrates to zero, so

$$P = V_{dc} I_{dc} + \sum_{n=1}^{\infty} \frac{V_n I_n}{2} \cos(\theta_{vn} - \theta_{in}). \quad \textbf{(17.78)}$$

Equation (17.78) is particularly important because it states that in the case of an interaction between a periodic voltage and the corresponding periodic current, the total average power is the sum of the average powers obtained from the interaction of currents and voltages of the same frequency. Currents and voltages of different frequencies do not interact to produce average power. Therefore, in average-power calculations involving periodic functions, the total average power is the superposition of the average powers associated with each harmonic voltage and current. Example 17.4 illustrates the computation of average power involving a periodic voltage.

EXAMPLE 17.4

Assume that the periodic square-wave voltage in Example 17.3 is applied across the terminals of a 15 Ω resistor. The value of V_m is 60 V, and that of T is 5 ms.

a) Write the first five nonzero terms of the Fourier series representation of $v(t)$. Use the trigonometric form given in Eq. (17.38).

b) Calculate the average power associated with each term in (a).

c) Calculate the total average power delivered to the 15 Ω resistor.

d) What percentage of the total power is delivered by the first five terms of the Fourier series?

SOLUTION

a) The dc component of $v(t)$ is

$$a_v = \frac{(60)(T/4)}{T} = 15 \text{ V}.$$

From Example 17.3 we have

$$
\begin{aligned}
A_1 &= \sqrt{2}\,60/\pi = 27.01 \text{ V};\\
\theta_1 &= 45^\circ;\\
A_2 &= 60/\pi = 19.10 \text{ V};\\
\theta_2 &= 90^\circ;\\
A_3 &= 20\sqrt{2}/\pi = 9.00 \text{ V};\\
\theta_3 &= 135^\circ;\\
A_4 &= 0;\\
\theta_4 &= 0^\circ;\\
A_5 &= 5.40 \text{ V};\\
\theta_5 &= 45^\circ;\\
\omega_0 &= \frac{2\pi}{T} = \frac{2\pi(1000)}{5} = 400\pi \text{ rad/s}.
\end{aligned}
$$

Thus, using the first five nonzero terms of the Fourier series,

$$
\begin{aligned}
v(t) = 15 &+ 27.01 \cos(400\pi t - 45^\circ)\\
&+ 19.10 \cos(800\pi t - 90^\circ)\\
&+ 9.00 \cos(1200\pi t - 135^\circ)\\
&+ 5.40 \cos(2000\pi t - 45^\circ) + \cdots \text{ V}.
\end{aligned}
$$

b) The voltage is applied to the terminals of a resistor, so we can find the power associated with each term as follows:

$$
\begin{aligned}
P_{\text{dc}} &= \frac{15^2}{15} = 15 \text{ W};\\
P_1 &= \frac{1}{2}\frac{27.01^2}{15} = 24.32 \text{ W};\\
P_2 &= \frac{1}{2}\frac{19.10^2}{15} = 12.16 \text{ W};\\
P_3 &= \frac{1}{2}\frac{9^2}{15} = 2.70 \text{ W};\\
P_5 &= \frac{1}{2}\frac{5.4^2}{15} = 0.97 \text{ W}.
\end{aligned}
$$

c) To obtain the total average power delivered to the 15 Ω resistor, we first calculate the rms value of $v(t)$:

$$V_{\text{rms}} = \sqrt{\frac{(60)^2(T/4)}{T}} = \sqrt{900} = 30 \text{ V}.$$

The total average power delivered to the 15 Ω resistor is

$$P_T = \frac{30^2}{15} = 60 \text{ W}.$$

d) The total power delivered by the first five nonzero terms is

$$P = P_{dc} + P_1 + P_2 + P_3 + P_5 = 55.15 \text{ W}.$$

This is (55.15/60)(100), or 91.92% of the total.

DRILL EXERCISE

17.6 The trapezoidal voltage function in Drill Exercise 17.3 is applied to the circuit shown. If $12V_m = 986.96$ V and $T = 6283.19$ ms, estimate the average power delivered to the 1 Ω resistor.

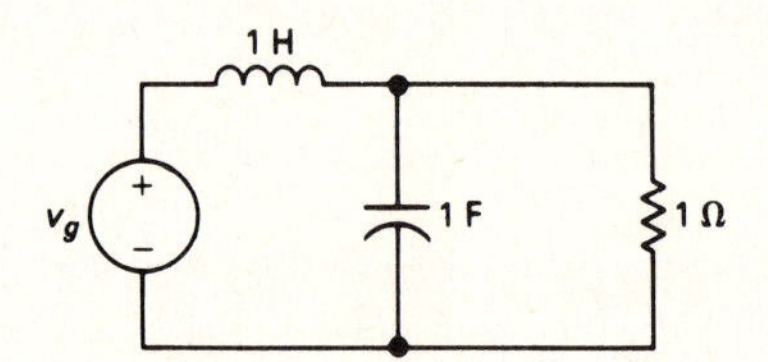

ANSWER: 3750 W.

17.7 The RMS Value of a Periodic Function

The rms value of a periodic function can be expressed in terms of the Fourier coefficients; by definition,

$$F_{rms} = \sqrt{\frac{1}{T}\int_{t_0}^{t_0+T} f(t)^2\, dt}. \quad \textbf{(17.79)}$$

Representing $f(t)$ by its Fourier series yields

$$F_{rms} = \sqrt{\frac{1}{T}\int_{t_0}^{t_0+T}\left[a_v + \sum_{n=1}^{\infty} A_n \cos(n\omega_0 t - \theta_n)\right]^2 dt}. \quad \textbf{(17.80)}$$

The integral of the squared time function simplifies because the only terms to survive integration over a period are the product of the dc term and the harmonic products of the same frequency. All other products integrate to zero. Therefore Eq. (17.80) reduces to

$$F_{\text{rms}} = \sqrt{\frac{1}{T}\left(a_v^2 T + \sum_{n=1}^{\infty} \frac{T}{2} A_n^2\right)}$$

$$= \sqrt{a_v^2 + \sum_{n=1}^{\infty} \frac{A_n^2}{2}}$$

$$= \sqrt{a_v^2 + \sum_{n=1}^{\infty} \left(\frac{A_n}{\sqrt{2}}\right)^2}. \qquad \textbf{(17.81)}$$

Equation (17.81) states that the rms value of a periodic function is the square root of the sum obtained by adding the square of the rms value of each harmonic to the square of the dc value. For example, let's assume that a periodic voltage is represented by the finite series

$$v = 10 + 30\cos(\omega_0 t - \theta_1) + 20\cos(2\omega_0 t - \theta_2) + 5\cos(3\omega_0 t - \theta_3) + 2\cos(5\omega_0 t - \theta_5).$$

The rms value of this voltage is

$$V = \sqrt{10^2 + (30/\sqrt{2})^2 + (20/\sqrt{2})^2 + (5/\sqrt{2})^2 + (2/\sqrt{2})^2} = \sqrt{764.5} = 27.65 \text{ V}.$$

Usually, infinitely many terms are required to represent a periodic function by a Fourier series, and therefore Eq. (17.81) yields an estimate of the true rms value. We illustrate this result in Example 17.5.

EXAMPLE 17.5

Use Eq. (17.81) to estimate the rms value of the voltage in Example 17.4.

SOLUTION

From Example 17.4,

$V_{\text{dc}} = 15$ V;

$V_1 = 27.01/\sqrt{2}$ V, the rms value of the fundamental;

$V_2 = 19.10/\sqrt{2}$ V, the rms value of the second harmonic;

$V_3 = 9.00/\sqrt{2}$ V, the rms value of the third harmonic;

$V_5 = 5.40/\sqrt{2}$ V, the rms value of the fifth harmonic.

Therefore

$$V_{\text{rms}} = \sqrt{15^2 + \left(\frac{27.01}{\sqrt{2}}\right)^2 + \left(\frac{19.10}{\sqrt{2}}\right)^2 + \left(\frac{9.00}{\sqrt{2}}\right)^2 + \left(\frac{5.40}{\sqrt{2}}\right)^2}$$
$$= 28.76 \text{ V}.$$

From Example 17.4, the true rms value is 30 V. We approach this value by including more and more harmonics in Eq. (17.81). For example, if we include the harmonics through $k = 9$, the equation yields a value of 29.32 V.

DRILL EXERCISE

17.7 a) Find the rms value of the voltage shown for $V_m = 100$ V.

b) Estimate the rms value of the voltage, using the first three nonzero terms in the Fourier series representation of $v_g(t)$.

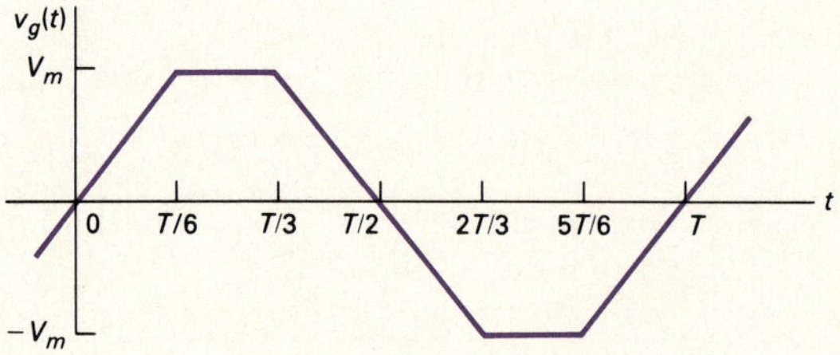

ANSWER: (a) 74.5356 V; (b) 74.5306 V.

17.8 THE EXPONENTIAL FORM OF THE FOURIER SERIES

The exponential form of the Fourier series is of interest because it allows us to express the series concisely. The exponential form of the series is

$$f(t) = \sum_{n=-\infty}^{\infty} C_n e^{jn\omega_0 t}, \quad \textbf{(17.82)}$$

where

$$C_n = \frac{1}{T}\int_{t_0}^{t_0+T} f(t) e^{-jn\omega_0 t}\, dt. \quad \textbf{(17.83)}$$

To derive Eqs. (17.82) and (17.83), we return to Eq. (17.2) and replace the cosine and sine functions with their exponential equivalents:

$$\cos n\omega_0 t = \frac{e^{jn\omega_0 t} + e^{-jn\omega_0 t}}{2}; \tag{17.84}$$

$$\sin n\omega_0 t = \frac{e^{jn\omega_0 t} - e^{-jn\omega_0 t}}{2j}. \tag{17.85}$$

Substituting Eqs. (17.84) and (17.85) into Eq. (17.2) gives

$$f(t) = a_v + \sum_{n=1}^{\infty} \frac{a_n}{2}(e^{jn\omega_0 t} + e^{-jn\omega_0 t}) + \frac{b_n}{2j}(e^{jn\omega_0 t} - e^{-jn\omega_0 t})$$

$$= a_v + \sum_{n=1}^{\infty} \left(\frac{a_n - jb_n}{2}\right) e^{jn\omega_0 t} + \left(\frac{a_n + jb_n}{2}\right) e^{-jn\omega_0 t}. \tag{17.86}$$

Now we define C_n as

$$C_n = \frac{1}{2}(a_n - jb_n) = \frac{A_n}{2}\angle{-\theta_n}, \quad n = 1, 2, 3, \cdots. \tag{17.87}$$

From the definition of C_n,

$$C_n = \frac{1}{2}\left[\frac{2}{T}\int_{t_0}^{t_0+T} f(t)\cos n\omega_0 t\, dt - j\frac{2}{T}\int_{t_0}^{t_0+T} f(t)\sin n\omega_0 t\, dt\right]$$

$$= \frac{1}{T}\int_{t_0}^{t_0+T} f(t)(\cos n\omega_0 t - j\sin n\omega_0 t)\, dt$$

$$= \frac{1}{T}\int_{t_0}^{t_0+T} f(t)e^{-jn\omega_0 t}\, dt\,, \tag{17.88}$$

which completes the derivation of Eq. (17.83). To complete the derivation of Eq. (17.82), we first observe from Eq. (17.88) that

$$C_0 = \frac{1}{T}\int_{t_0}^{t_0+T} f(t)\, dt = a_v. \tag{17.89}$$

Next we note that

$$C_{-n} = \frac{1}{T}\int_{t_0}^{t_0+T} f(t)e^{jn\omega_0 t}\, dt = C_n^* = \frac{1}{2}(a_n + jb_n). \tag{17.90}$$

Substituting Eqs. (17.87), (17.89), and (17.90) into Eq. (17.86) yields

$$f(t) = C_0 + \sum_{n=1}^{\infty}(C_n e^{jn\omega_0 t} + C_n^* e^{-jn\omega_0 t})$$

$$= \sum_{n=0}^{\infty} C_n e^{jn\omega_0 t} + \sum_{n=1}^{\infty} C_n^* e^{-jn\omega_0 t}. \tag{17.91}$$

Note that the second summation on the right-hand side of Eq. (17.91) is equivalent to summing $C_n e^{jn\omega_0 t}$ from -1 to $-\infty$; that is,

$$\sum_{n=1}^{\infty} C_n^* e^{-jn\omega_0 t} = \sum_{n=-1}^{-\infty} C_n e^{jn\omega_0 t}. \quad \textbf{(17.92)}$$

Because the summation from -1 to $-\infty$ is the same as the summation from $-\infty$ to -1, we use Eq. (17.92) to rewrite Eq. (17.91):

$$f(t) = \sum_{n=0}^{\infty} C_n e^{jn\omega_0 t} + \sum_{-\infty}^{-1} C_n e^{jn\omega_0 t}$$

$$= \sum_{-\infty}^{\infty} C_n e^{jn\omega_0 t}, \quad \textbf{(17.93)}$$

which completes the derivation of Eq. (17.82).

Example 17.6 illustrates the process of finding the exponential Fourier series representation of a periodic function.

EXAMPLE 17.6

Find the exponential Fourier series for the periodic voltage shown in Fig. 17.15

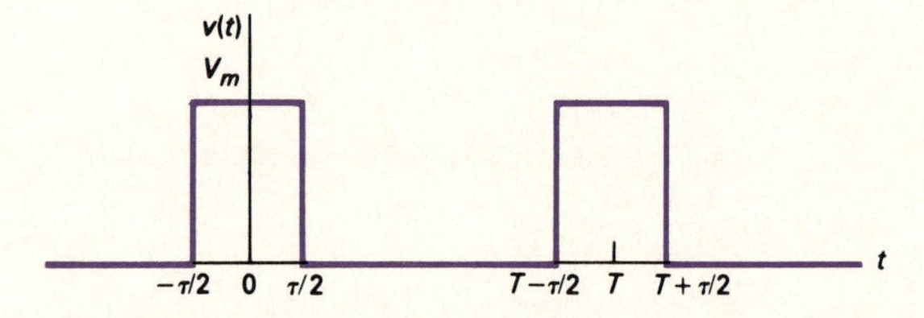

FIGURE 17.15 The periodic voltage for Example 17.6.

SOLUTION

Using $-\tau/2$ as the starting point for the integration, we have, from Eq. (17.83),

$$C_n = \frac{1}{T}\int_{-\tau/2}^{\tau/2} V_m e^{-jn\omega_0 t}\, dt$$

$$= \frac{V_m}{T}\left(\frac{e^{-jn\omega_0 t}}{-jn\omega_0}\right)\Bigg|_{-\tau/2}^{\tau/2}$$

$$= \frac{jV_m}{n\omega_0 T}(e^{-jn\omega_0\tau/2} - e^{jn\omega_0\tau/2})$$

$$= \frac{2V_m}{n\omega_0 T}\sin n\omega_0\tau/2.$$

Here, because $v(t)$ has even symmetry, $b_n = 0$ for all n, and hence we expect C_n to be real. Moreover, the amplitude of C_n follows a $(\sin x)/x$ distribution, as indicated when we rewrite

$$C_n = \frac{V_m\tau}{T}\frac{\sin(n\omega_0\tau/2)}{n\omega_0\tau/2}.$$

We say more about this subject in Section 17.9. The exponential series representation of $v(t)$ is

$$v(t) = \sum_{n=-\infty}^{\infty} \left(\frac{V_m \tau}{T}\right) \frac{\sin(n\omega_0\tau/2)}{n\omega_0\tau/2} e^{jn\omega_0 t}$$

$$= \left(\frac{V_m \tau}{T}\right) \sum_{n=-\infty}^{\infty} \frac{\sin(n\omega_0\tau/2)}{n\omega_0\tau/2} e^{jn\omega_0 t}.$$

We may also express the rms value of a periodic function in terms of the complex Fourier coefficients. From Eqs. (17.81), (17.87), and (17.89),

$$F_{\text{rms}} = \sqrt{a_v^2 + \sum_{n=1}^{\infty} \frac{a_n^2 + b_n^2}{2}}; \quad \textbf{(17.94)}$$

$$|C_n| = \frac{\sqrt{a_n^2 + b_n^2}}{2}; \quad \textbf{(17.95)}$$

$$C_0^2 = a_v^2. \quad \textbf{(17.96)}$$

Substituting Eqs. (17.95) and (17.96) into Eq. (17.94) yields the desired expression:

$$F_{\text{rms}} = \sqrt{C_0^2 + 2\sum_{n=1}^{\infty} |C_n|^2}. \quad \textbf{(17.97)}$$

DRILL EXERCISES

17.8 Derive the expression for the Fourier coefficients C_n for the periodic function shown. *Hint*: Take advantage of symmetry by using the fact that $C_n = (a_n - jb_n)/2$.

ANSWER: $C_n = -j\dfrac{10}{\pi n}\left(1 + \cos\dfrac{n\pi}{3}\right)$, $\quad n$ odd.

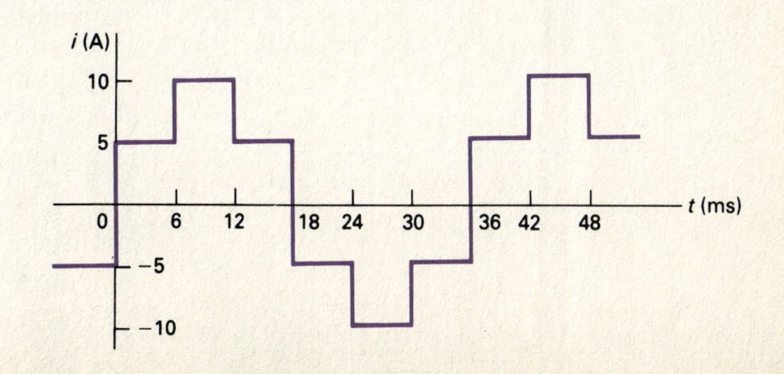

17.9 a) Calculate the rms value of the periodic current in Drill Exercise 17.8.

b) Using C_1–C_{11}, estimate the rms value.

c) What is the percentage of error in the value obtained in (b), based on the true value found in (a)?

d) For this periodic function, how many terms must be used to estimate the rms value before the error is less than 1%?

ANSWER: (a) $\sqrt{50}$ A; (b) 6.980 A; (c) -1.28%; (d) $n = 17$; therefore, the first six nonzero harmonic terms of the series are required.

17.9 AMPLITUDE AND PHASE SPECTRA

A periodic time function is defined by its Fourier coefficients and its period. In other words, when we know a_v, a_n, b_n, and T, we can construct $f(t)$, at least theoretically. When we know a_n and b_n, we also know the amplitude (A_n) and phase angle ($-\theta_n$) of each harmonic. Again, we cannot, in general, visualize what the periodic function looks like in the time domain from a description of the coefficients and phase angles; nevertheless, we recognize that these quantities characterize the periodic function completely. Thus, with sufficient computing time, we can synthesize the time-domain waveform from the amplitude and phase angle data. Also, when a periodic driving function is exciting a circuit that is highly frequency selective, the Fourier series of the steady-state response is dominated by just a few terms. Thus the description of the response in terms of amplitude and phase may provide an understanding of the output waveform.

Use a computer tool to construct these plots from the terms in the Fourier Series.

We can present graphically the description of a periodic function in terms of the amplitude and phase angle of each term in the Fourier series of $f(t)$. The plot of the amplitude of each term versus the frequency is called the **amplitude spectrum** of $f(t)$, and the plot of the phase angle versus the frequency is called the **phase spectrum** of $f(t)$. Because the amplitude and phase angle data occur at discrete values of the frequency (that is, at $\omega_0, 2\omega_0, 3\omega_0, \ldots$), these plots also are referred to as **line spectra**.

AN EXAMPLE

Amplitude and phase spectra plots are based on either Eq. (17.38) (A_n and $-\theta_n$) or Eq. (17.82) (C_n). We focus on Eq. (17.82) and leave the plots based on Eq. (17.38) to Problem 17.45. To illustrate the amplitude and phase spectra, which are based on the exponential form of the Fourier series, we use the periodic voltage of Example 17.6.

To aid the discussion, we assume that $V_m = 5$ V and $\tau = T/5$. From Example 17.6,

$$C_n = \frac{V_m \tau}{T} \frac{\sin (n\omega_0 \tau/2)}{n\omega_0 \tau/2}, \quad (17.98)$$

which for the assumed values of V_m and τ reduces to

$$C_n = 1 \frac{\sin (n\pi/5)}{n\pi/5}. \quad (17.99)$$

Figure 17.16 illustrates the plot of the magnitude of C_n from Eq. (17.99) for values of n ranging from -10 to $+10$. The figure clearly shows that the amplitude spectrum is bounded by the envelope of the $|(\sin x)/x|$ function. We used the order of the harmonic as the frequency scale because the numerical value of T is not specified. When we know T, we also know ω_0 and the frequency corresponding to each harmonic.

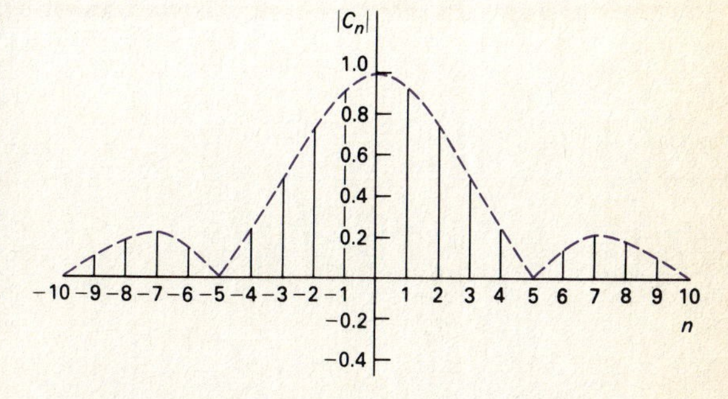

FIGURE 17.16 The plot of $|C_n|$ versus n when $\tau = T/5$, for the periodic voltage for Example 17.6.

Figure 17.17 provides the plot of $|(\sin x)/x|$ versus x, where x is in radians. It shows that the function goes through zero whenever x is an integral multiple of π. From Eq. (17.98),

$$n\omega_0 \left(\frac{\tau}{2}\right) = \frac{n\pi\tau}{T} = \frac{n\pi}{T/\tau}. \quad (17.100)$$

From Eq. (17.100), we deduce that the amplitude spectrum goes through zero whenever $n\tau/T$ is an integer. For example, in the plot, τ/T is 1/5, and therefore the envelope goes through zero at $n = 5, 10, 15$, and so on. In other words, the fifth, tenth, fifteenth, ... harmonics are all zero. As the reciprocal of τ/T becomes an increasingly larger integer, the number of harmonics between every π radians increases. If $n\pi/T$ is not an integer, the amplitude spectrum still follows the $|(\sin x)/x|$ envelope. However, the envelope is not zero at an integral multiple of ω_0.

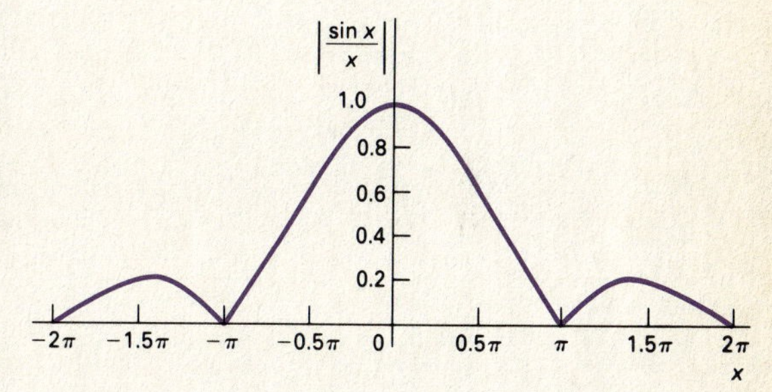

FIGURE 17.17 The plot of $|(\sin x)/x|$ versus x.

Because C_n is real for all n, the phase angle associated with C_n is either zero or 180°, depending on the algebraic sign of $(\sin n\pi/5)/(n\pi/5)$. For example, the phase angle is zero for $n = 0$, ±1, ±2, ±3, and ±4. It is not defined at $n = \pm5$, because $C_{\pm5}$ is zero. The phase angle is 180° at $n = \pm6$, ±7, ±8, and ±9, and it is not defined at ±10. This pattern repeats itself as n takes on larger integer values. Figure 17.18 shows the phase angle of C_n given by Eq. (17.89).

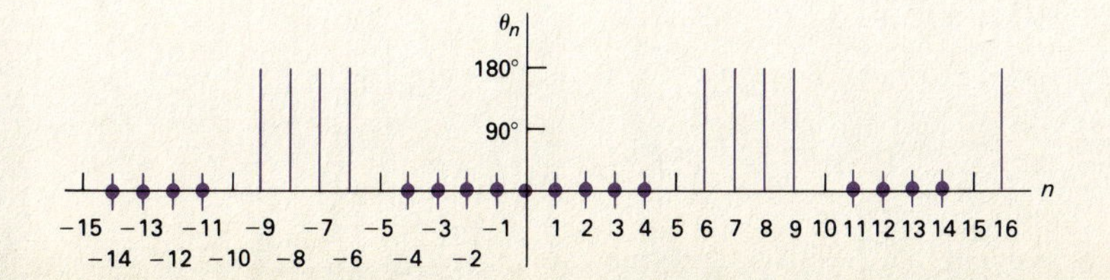

FIGURE 17.18 The phase angle of C_n.

Now, what happens to the amplitude and phase spectra if $f(t)$ is shifted along the time axis? To find out, we shift the periodic voltage in Example 17.6 t_0 units to the right. By hypothesis,

$$v(t) = \sum_{n=-\infty}^{\infty} C_n e^{jn\omega_0 t}; \tag{17.101}$$

therefore

$$v(t - t_0) = \sum_{n=-\infty}^{\infty} C_n e^{jn\omega_0(t-t_0)} = \sum_{n=-\infty}^{\infty} C_n e^{-jn\omega_0 t_0} e^{jn\omega_0 t}, \tag{17.102}$$

which indicates that shifting the origin has no effect on the amplitude spectrum, because

$$|C_n| = |C_n e^{-jn\omega_0 t_0}|. \tag{17.103}$$

However, reference to Eq. (17.87) reveals that the phase spectrum has changed to $-(\theta_n + n\omega_0 t_0)$ radians. For example, let's shift the periodic voltage in Example 17.6 $\tau/2$ units to the right. As before, we assume that $\tau = T/5$; then the new phase angle θ'_n is

$$\theta'_n = -(\theta_n + n\pi/5). \tag{17.104}$$

We have plotted Eq. (17.104) in Fig. 17.19 for n ranging from -8 to $+8$. Note that no phase angle is associated with a zero amplitude coefficient.

You may wonder why we have devoted so much attention to the amplitude spectrum of the periodic pulse in Example 17.6. The reason is that this particular periodic waveform provides an excellent way to illustrate the transition from the Fourier series representation of a periodic function to the Fourier transform representation of a nonperiodic function. We discuss the Fourier transform in Chapter 18.

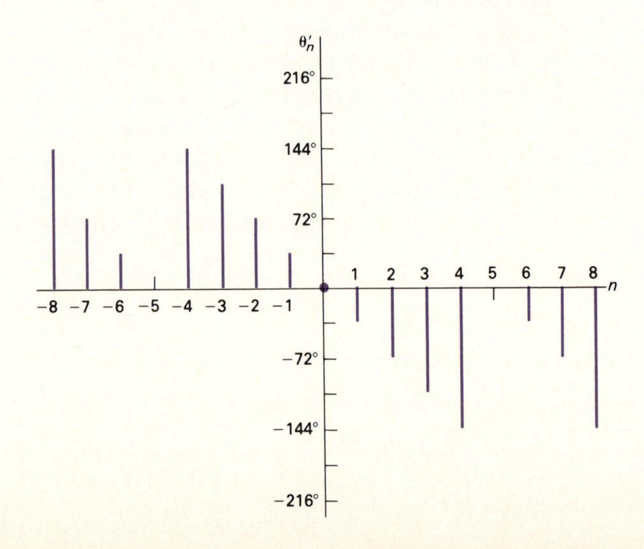

FIGURE 17.19 The plot of θ'_n versus n for Eq. (17.104).

DRILL EXERCISE

17.10 The function in Drill Exercise 17.8 is shifted along the time axis 9 ms to the left. Write the exponential Fourier series for the periodic current.

ANSWER:

$$i(t) = \frac{10}{\pi} \sum_{n=-\infty(\text{odd})}^{\infty} \frac{1}{n}\left(1 + \cos\frac{n\pi}{3}\right) e^{(j\pi/2)(n-1)} e^{jn\omega_0 t} \text{ A.}$$

SUMMARY

- A **periodic function** is a function that repeats itself every T seconds.
- A period is the smallest time interval (T) that a periodic function can be shifted to produce a function identical to itself.
- The Fourier series is an infinite series used to represent a periodic function. The series consists of a constant term and infinitely many harmonically related cosine and sine terms.
- The **fundamental frequency** is the frequency determined by the fundamental period ($f_0 = 1/T$ or $\omega_0 = 2\pi f_0$).
- The **harmonic frequency** is an integer multiple of the fundamental frequency.
- The **Fourier coefficients** are the constant term and the coefficient of each cosine and sine term in the series. (See Eqs. [17.3]–[17.5].)
- Five types of symmetry are used to simplify the computation of the Fourier coefficients (see Section 17.3):
  - *even,* in which all sine terms in the series are zero
  - *odd,* in which all cosine terms and the constant term are zero
  - *half-wave,* in which all even harmonics are zero;
  - *quarter-wave, half-wave, even,* in which the series contains only odd harmonic cosine terms
  - *quarter-wave, half-wave, odd,* in which the series contains only odd harmonic sine terms

- In the alternative form of the Fourier series, each harmonic represented by the sum of a cosine and sine term is combined into a single term of the form $A_n \cos(n\omega_0 t - \theta_n)$.
- For steady-state response, the Fourier series of the response signal is determined by first finding the response to each component of the input signal. The individual responses are added (superimposed) to form the Fourier series of the response signal. The response to the individual terms in the input series is found by either frequency domain or s-domain analysis.
- The waveform of the response signal is difficult to obtain without the aid of a computer. Sometimes the frequency response (or filtering) characteristics of the circuit can be used to ascertain how closely the output waveform matches the input waveform.
- Only harmonics of the same frequency interact to produce average power. The total average power is the sum of the average powers associated with each frequency.
- The rms value of a periodic function can be estimated from the Fourier coefficients. (See Eqs. [17.81], [17.94], and [17.97].)
- The Fourier series may also be written in exponential form by using Euler's identity to replace the cosine and sine terms with their exponential equivalents.
- The Fourier series is used to predict the steady-state response of a system when the system is excited by a periodic signal. The series assists in finding the steady-state response by transferring the analysis from the time domain to the frequency domain.

PROBLEMS

17.1 a) Verify Eqs. (17.6) and (17.7).

b) Verify Eq. (17.8). *Hint:* Use the trigonometric identity $\cos\alpha \sin\beta = \frac{1}{2}\sin(\alpha + \beta) - \frac{1}{2}\sin(\alpha - \beta)$.

c) Verify Eq. (17.9). *Hint:* Use the trigonometric identity $\sin\alpha \sin\beta = \frac{1}{2}\cos(\alpha - \beta) - \frac{1}{2}\cos(\alpha + \beta)$.

d) Verify Eq. (17.10). *Hint:* Use the trigonometric identity $\cos\alpha \cos\beta = \frac{1}{2}\cos(\alpha - \beta) + \frac{1}{2}\cos(\alpha + \beta)$.

17.2 Derive Eq. (17.5).

17.3 Find the Fourier series expressions for the periodic voltage functions shown in Fig. P17.3. Note that Fig. P17.3(a) illustrates the square wave; Fig. P17.3(b) illustrates the full-wave rectified sine wave, where $v(t) = V_m \sin(\pi/T)t, 0 \le t \le T$; and Fig. P17.3(c) illustrates the half-wave rectified sine wave, where $v(t) = V_m \sin(2\pi/T)t$, $0 \le t \le T/2$.

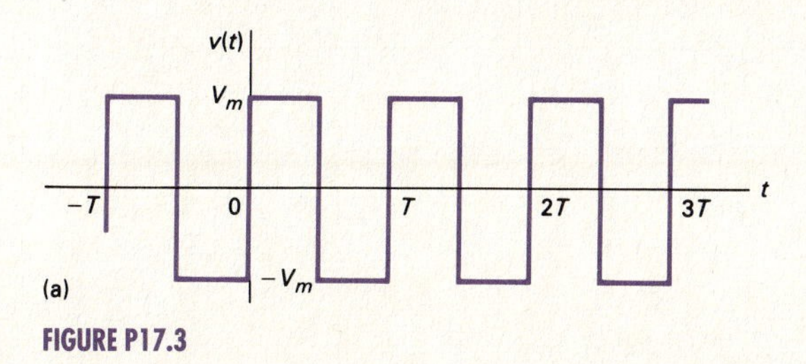

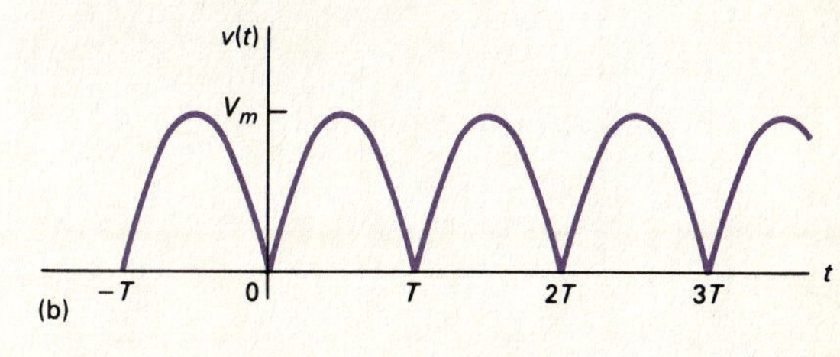

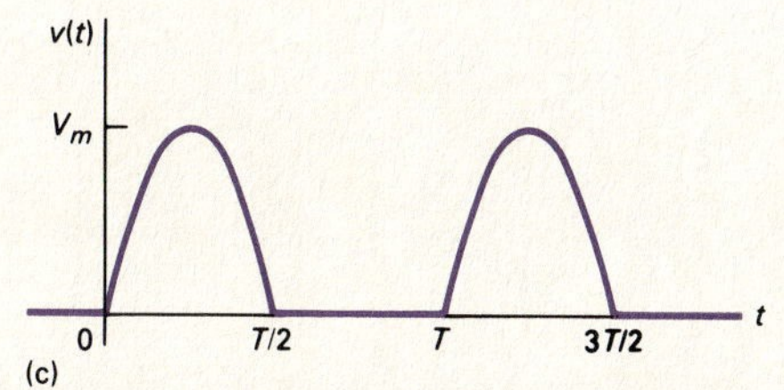

FIGURE P17.3

17.4 Derive the expressions for the Fourier coefficients of an odd periodic function. *Hint:* Use the same technique as used in the text in deriving Eqs. (17.14), (17.15), and (17.16).

17.5 Show that if $f(t) = -f(t - T/2)$, the Fourier coefficients b_k are given by the expressions

$$b_k = 0 \qquad \text{for } k \text{ is even;}$$

$$b_k = \frac{4}{T}\int_0^{T/2} f(t)\sin k\omega_o t\, dt \qquad \text{for } k \text{ odd.}$$

Hint: Use the same technique as used in the text to derive Eqs. (17.28) and (17.29).

17.6 Derive Eqs. (17.36). *Hint:* Start with Eq. (17.29) and divide the interval of integration into 0 to $T/4$ and $T/4$ to $T/2$. Note that because of evenness and quarter-wave symmetry, $f(t) = -f(T/2 - t)$ in the interval $T/4 \le t \le T/2$. Let $x = T/2 - t$ in the second interval and combine the resulting integral with the integration between 0 and $T/4$.

17.7 Derive Eqs. (17.37). Follow the hint given in Problem 17.6 except that because of oddness and quarter-wave symmetry, $f(t) = f(T/2 - t)$ in the interval $T/4 \le t \le T/2$.

17.8 For each of the periodic functions in Fig. P17.8, specify

a) ω_o in radians per second;

b) f_o in hertz;

c) the value of a_v;

d) the equations for a_k and b_k; and

e) $v(t)$ as a Fourier series.

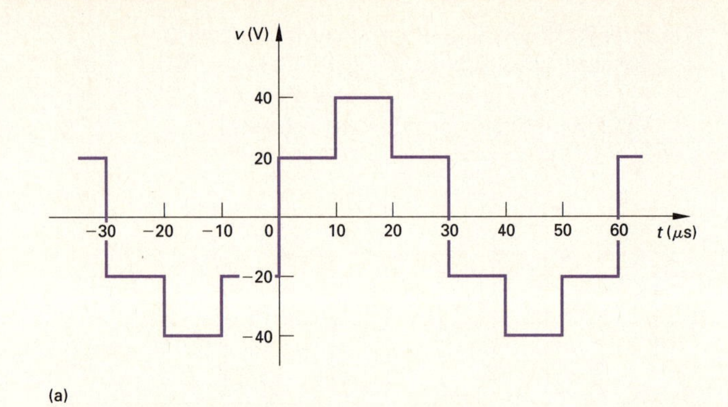

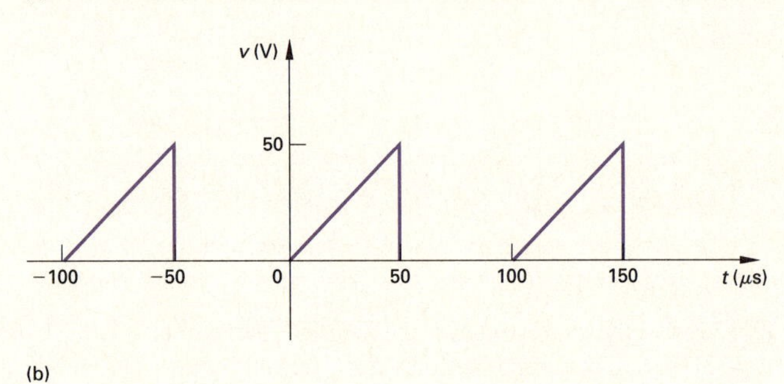

FIGURE P17.8

17.9 Find the Fourier series of each periodic function shown in Fig. P17.9.

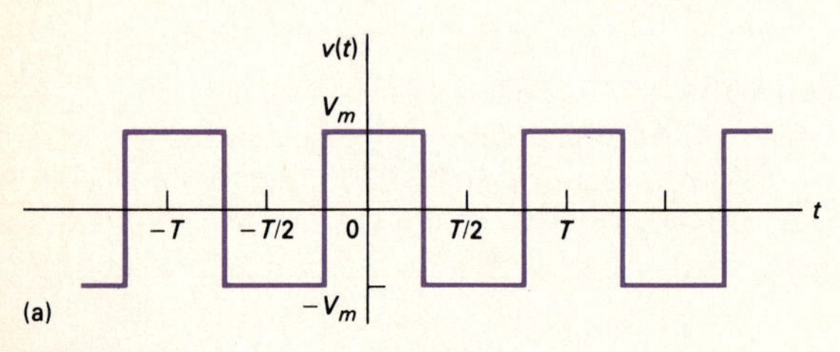

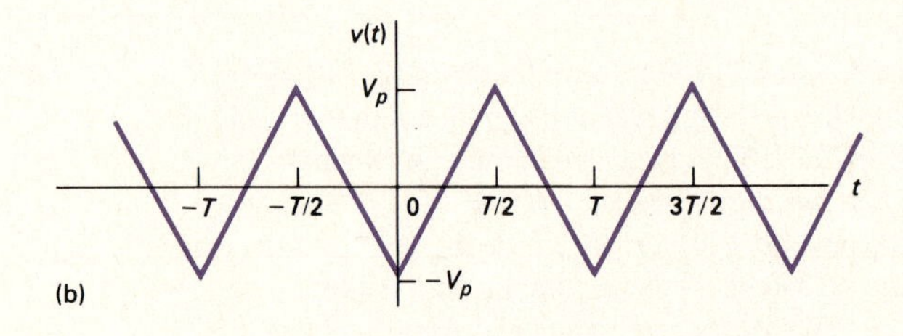

FIGURE P17.9

17.10 Derive the Fourier series for the periodic voltage shown in Fig. P17.10, given that

$$v(t) = 50 \cos \frac{2\pi}{T} t \text{ V}, \qquad -T/4 \le t \le T/4;$$

$$v(t) = -100 \cos \frac{2\pi}{T} t \text{ V}, \qquad T/4 \le t \le 3T/4$$

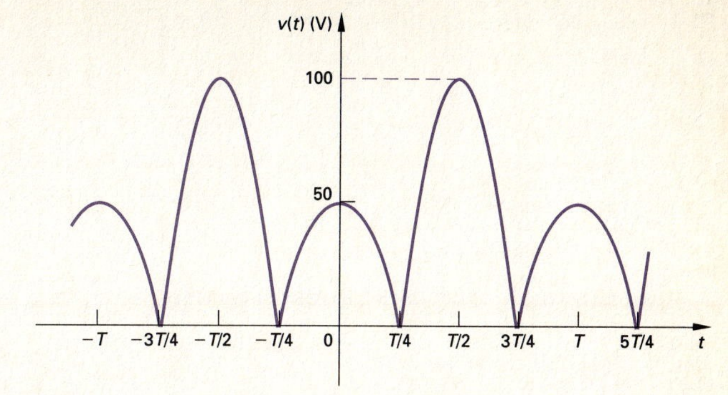

FIGURE P17.10

17.11 a) Derive the Fourier series for the periodic current function shown in Fig. P17.11.

b) Repeat part (a) if the vertical reference axis is shifted $T/2$ units to the right.

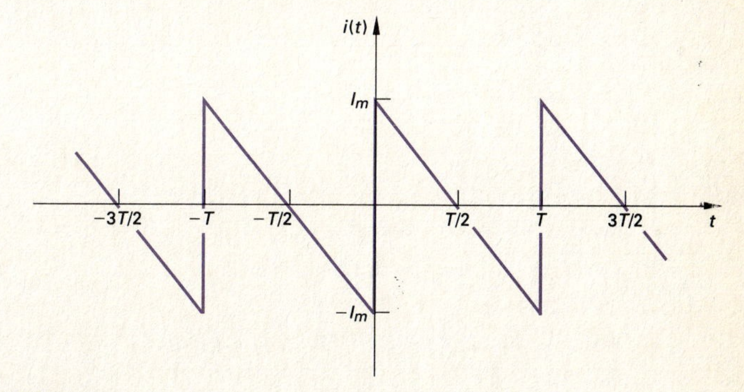

FIGURE P17.11

17.12 It is given that $f(t) = 0.25t^2$ over the interval $-4 < t < 4$ s.

a) Construct a periodic function that satisfies this $f(t)$ between -4 and $+4$ s, has a period of 16 s, and has half-wave symmetry.

b) Is the function even or odd?

c) Does the function have quarter-wave symmetry?

d) Derive the Fourier series for $f(t)$.

e) Write the Fourier series for $f(t)$ if $f(t)$ is shifted 4 s to the right.

17.13 Repeat Problem 17.12 given that $f(t) = 0.25t^3$ over the interval $-4 < t < 4$ s.

17.14 It is given that $v(t) = 10t \sin 0.25\pi t$ V over the interval $-4 \leq t \leq 4$ s. The function then repeats itself.

a) What is the fundamental frequency in radians per second?

b) Is the function even?

c) Is the function odd?

d) Does the function have half-wave symmetry?

17.15 One period of a periodic function is described by the following equations:

$v(t) = 10$ V, $\quad -5$ ms $\leq t \leq 5$ ms;

$v(t) = (20 - 2000t)$ V, $\quad 5$ ms $\leq t \leq 15$ ms;

$v(t) = -10$ V, $\quad 15$ ms $\leq t \leq 25$ ms;

$v(t) = (2000t - 60)$ V, $\quad 25$ ms $\leq t \leq 35$ ms.

a) What is the fundamental frequency in hertz?

b) Is the function even?

c) Is the function odd?

d) Does the function have half-wave symmetry?

e) Does the function have quarter-wave symmetry?

f) Give the numerical expressions for a_v, a_k, and b_k.

17.16 The periodic function shown in Fig. P17.16 is odd and has both half-wave and quarter-wave symmetry.

a) Sketch one full cycle of the function over the interval $-T/4 \leq t \leq 3T/4$.

b) Derive the expression for the Fourier coefficients b_k.

c) Write the first five nonzero terms in the Fourier expansion of $f(t)$.

d) Use the first five nonzero terms to estimate $f(T/4)$.

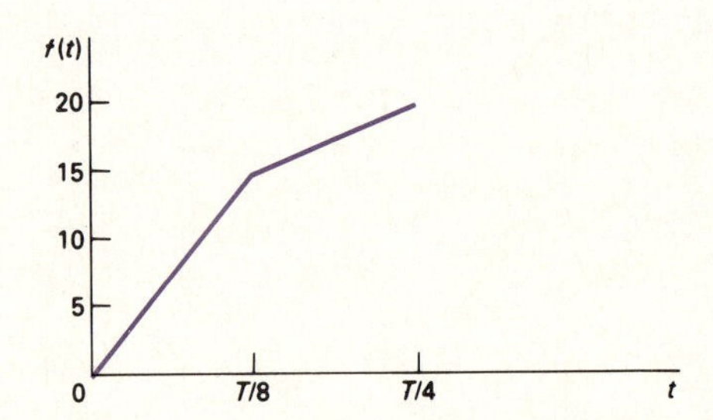

FIGURE P17.16

17.17 It is sometimes possible to use symmetry to find the Fourier coefficients, even though the original function is not symmetrical! With this thought in mind, consider the function in Drill Exercise 17.1. Observe that $v(t)$ can be divided into the two functions illustrated in Figs. P17.17(a) and (b). Furthermore, we can make $v_2(t)$ an even function by shifting it $T/8$ units to the left. This is illustrated in Fig. P17.17(c). At this point we note that $v(t) = v_1(t) + v_2(t)$ and that the Fourier series of $v_1(t)$ is a single-term series consisting of $V_m/2$. To find the Fourier series of $v_2(t)$, we first find the Fourier series of $v_2(t + T/8)$ and then shift this series $T/8$ units to the right. Use the technique just outlined to verify the Fourier series given as the answer to Drill Exercise 17.2(e).

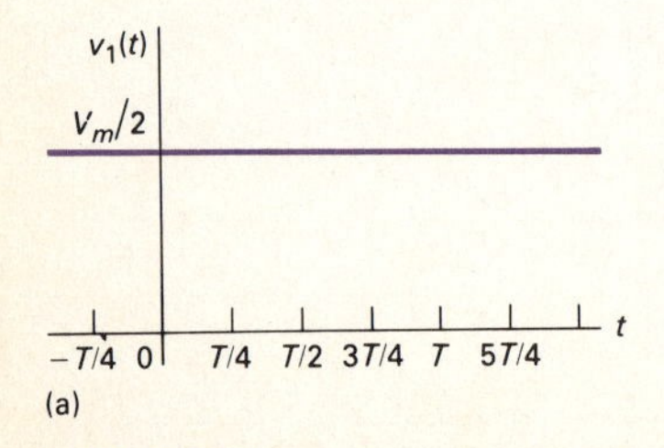

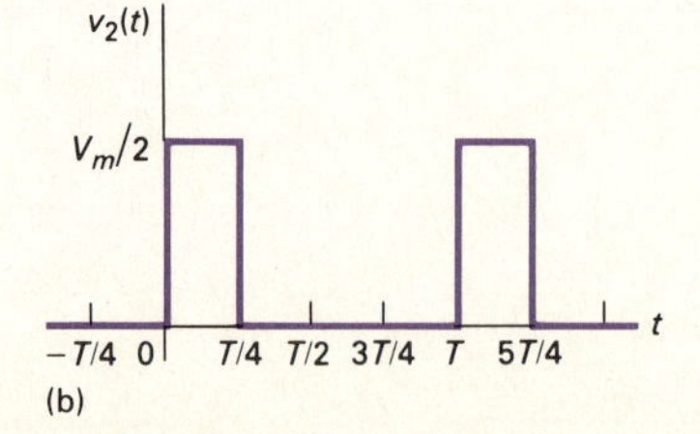

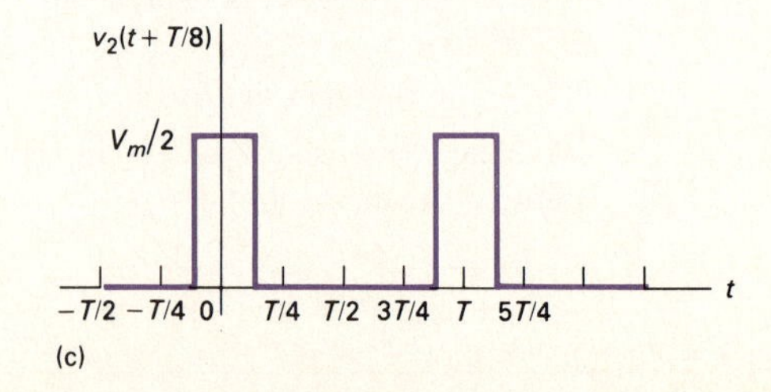

FIGURE P17.17

17.18 a) Derive the Fourier series for the periodic function shown in Fig. P17.18 when $I_m = 9.45\pi$ A. Write the series in the form of Eq. (17.38).

b) Use the first five nonzero terms to estimate $i(T/8)$.

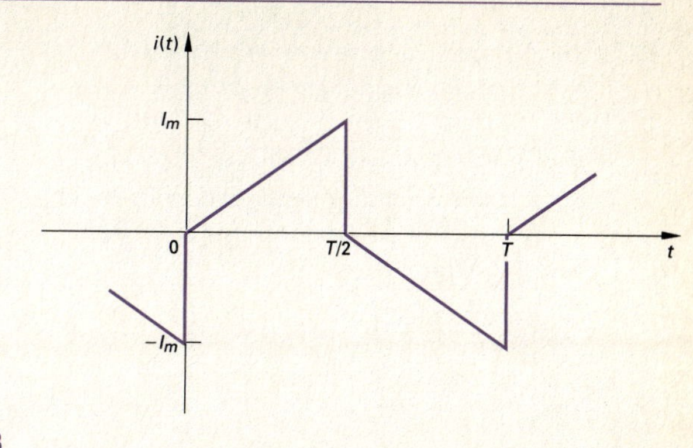

FIGURE P17.18

17.19 Derive Eqs. (17.69) and (17.70).

17.20 a) Derive Eq. (17.71). *Hint:* Note that $b_k = (4V_m/\pi k) + k\omega_o RCa_k$. Use this expression for b_k to find $a_k^2 + b_k^2$ in terms of a_k. Then use the expression for a_k to derive Eq. (17.71).

b) Derive Eq. (17.72).

17.21 Show that when we combine Eqs. (17.71) and (17.72) with Eqs. (17.38) and (17.39), the result is Eq. (17.58). *Hint:* Note from the definition of β_k that

$$\frac{a_k}{b_k} = -\tan \beta_k,$$

and from the definition of θ_k that

$$\tan \theta_k = -\cot \beta_k.$$

Now use the trigonometric identity

$$\tan x = \cot (90 - x)$$

to show that $\theta_k = (90 + \beta_k)$.

17.22 The square-wave voltage shown in Fig. P17.22(a) is applied to the circuit shown in Fig. P17.22(b).

a) Find the Fourier series representation of the steady-state current i.

b) Find the steady-state expression for i by straightforward circuit analysis.

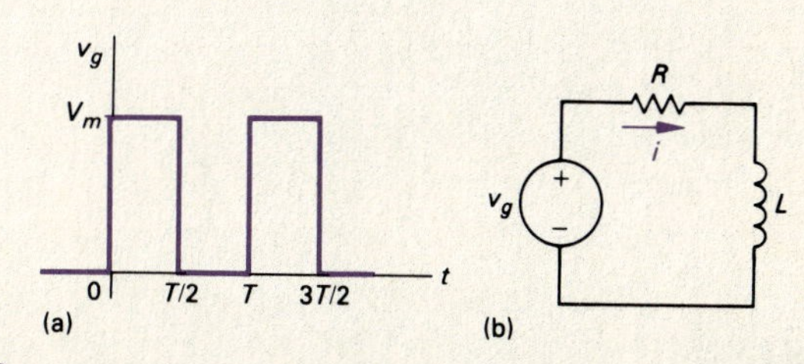

FIGURE P17.22

17.23 The periodic triangular-wave voltage seen in Fig. P17.23(a) is applied to the circuit shown in Fig. P17.23(b).

Derive the first three nonzero terms in the Fourier series that represents the steady-state voltage v_o if $V_m = 281.25\pi^2$ mV and the period of the input voltage is 200π ms.

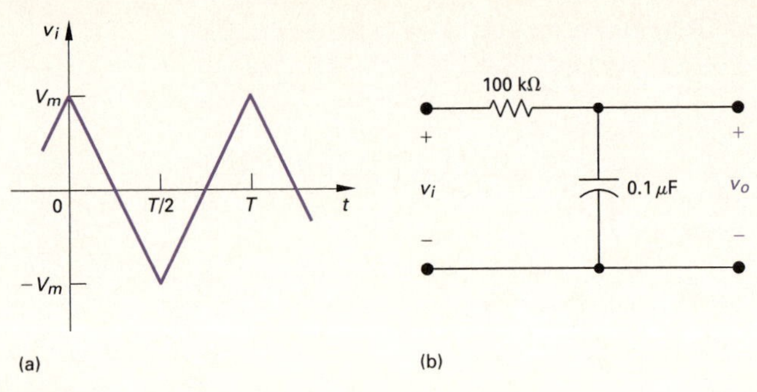

FIGURE P17.23

17.24 The periodic square-wave voltage seen in Fig. P17.24(a) is applied to the circuit shown in Fig. P17.24(b).

Derive the first three nonzero terms in the Fourier series that represents the steady-state voltage v_o if $V_m = 15\pi$ V and the period of the input voltage is 4π ms.

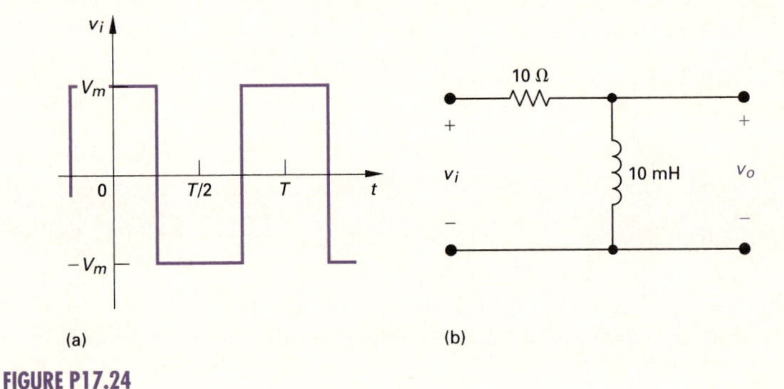

FIGURE P17.24

17.25 The periodic square-wave shown in Fig. P17.25(a) is applied to the circuit shown in Fig. P17.25(b).

a) Derive the first four nonzero terms in the Fourier series that represents the steady-state voltage v_o if $V_m = 10.5\pi$ V and the period of the input voltage is π ms.

b) Which harmonic dominates the output voltage? Explain why.

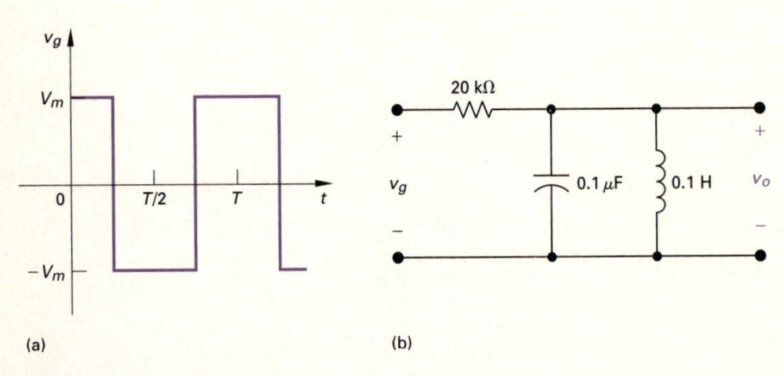

FIGURE P17.25

17.26 The periodic square-wave voltage described in Problem 17.25 is applied to the circuit shown in Fig. P17.26.

a) Derive the first four nonzero terms in the Fourier series that represents the steady-state voltage v_o.

b) Which frequency component in the input voltage is eliminated from the output voltage? Explain why.

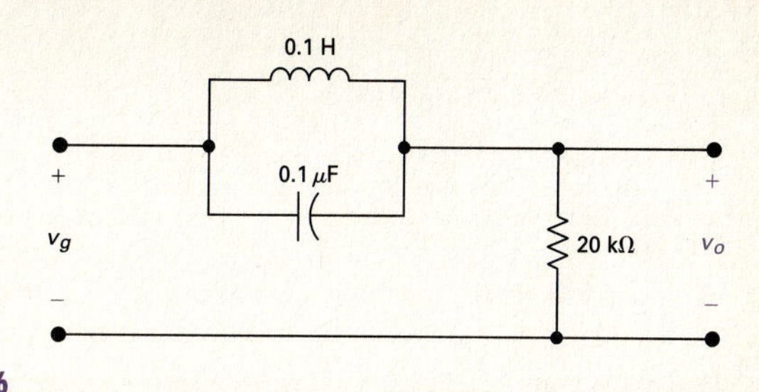

FIGURE P17.26

17.27 The full-wave rectified sine-wave voltage shown in Fig. P17.27(a) is applied to the circuit shown in Fig. P17.27(b).

a) Find the first three nonzero terms in the Fourier series representation of v_o.

b) Does your solution for v_o make sense? Explain.

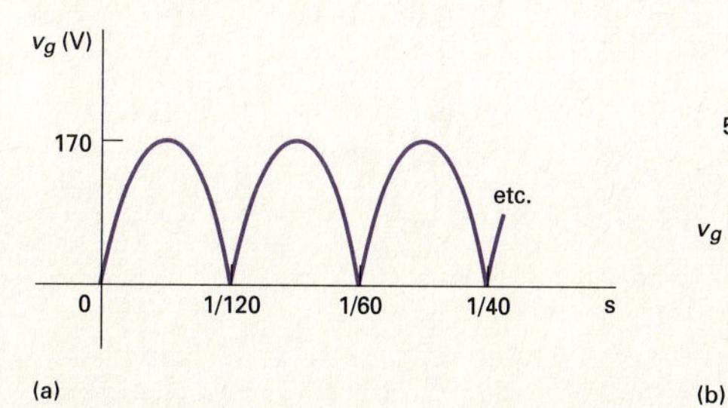

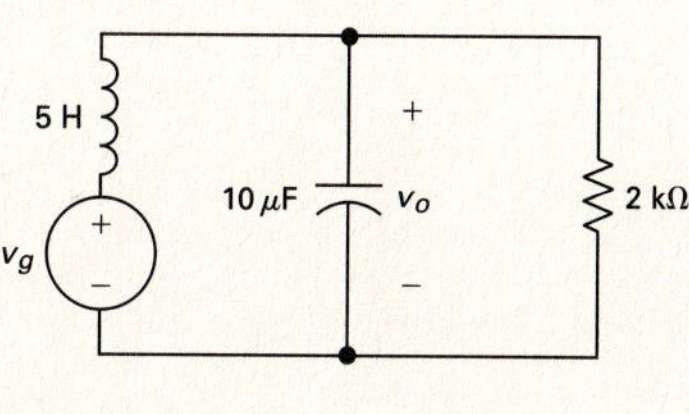

FIGURE P17.27

17.28 The periodic current shown in Fig. P17.28(a) is used to energize the circuit shown in Fig. P17.28(b). Write the time-domain expression for the fifth-harmonic voltage in the expression for v_o.

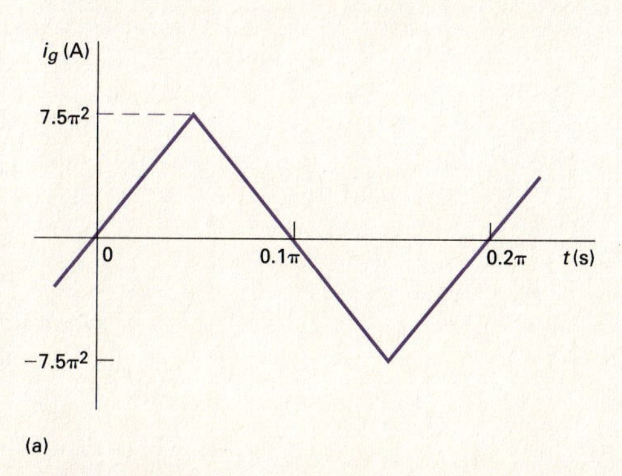

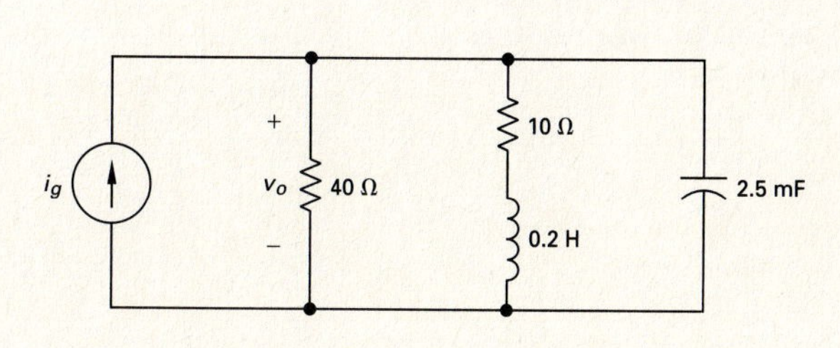

FIGURE P17.28

17.29 A periodic voltage having a period of 0.1π ms is given by the following Fourier series:

$$v_g = 45 \sum_{n=1,3,5,\ldots}^{\infty} \frac{\pi^2 n^2 - 8}{n^3} \sin\frac{n\pi}{2} \cos n\omega_o t \text{ V}.$$

This periodic voltage is applied to the circuit shown in Fig. P17.29. Find the amplitude and phase angle of the component of v_o that has a frequency of 300 krad/s.

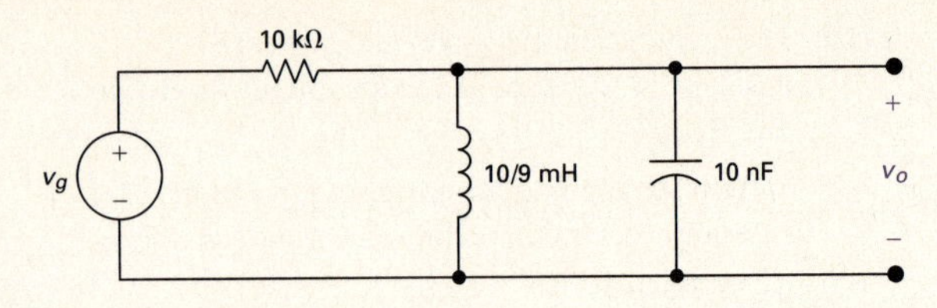

FIGURE P17.29

17.30 a) Estimate the rms value of the periodic square-wave voltage shown in Fig. P17.30(a) by using the first four nonzero terms in the Fourier series representation of $v(t)$.

b) Calculate the percentage of error in the estimation if

$$\% \text{ error} = \left[\frac{\text{estimated value}}{\text{exact value}} - 1\right] \times 100.$$

c) Repeat parts (a) and (b) if the periodic square-wave voltage is replaced by the periodic triangular voltage shown in Fig. P17.30(b).

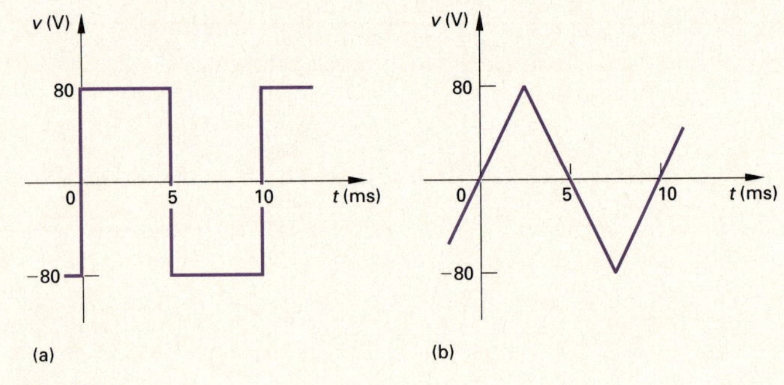

FIGURE P17.30

17.31 a) Estimate the rms value of the full-wave rectified sinusoidal voltage shown in Fig. P17.31(a) by using the first four nonzero terms in the Fourier series representation of $v(t)$.

b) Calculate the percentage of error in the estimation. (See Problem 17.30.)

c) Repeat (a) and (b) if the full-wave rectified sinusoidal voltage is replaced by the half-wave rectified sinusoidal voltage shown in Fig. P17.31(b).

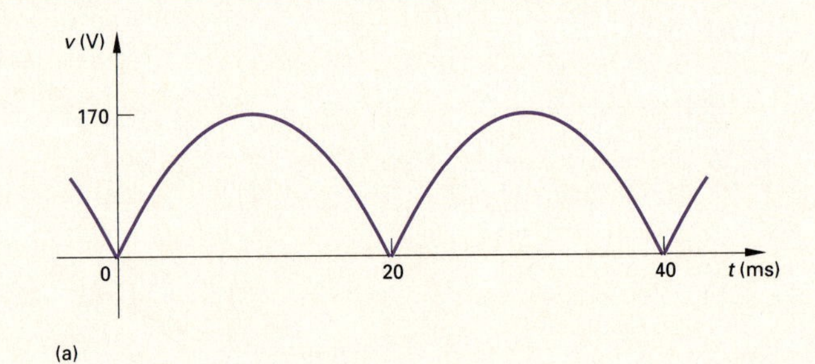

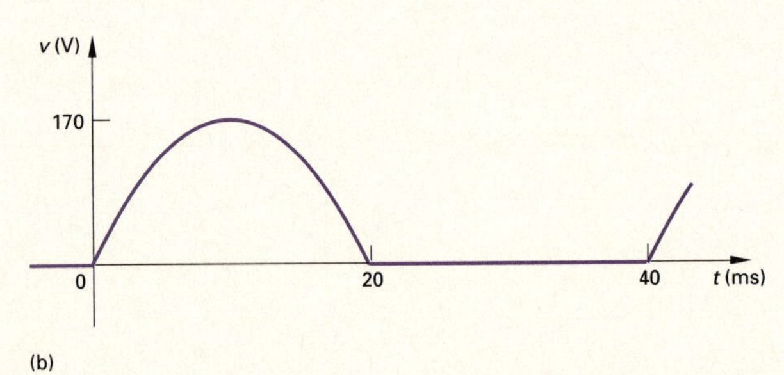

FIGURE P17.31

17.32 a) Derive the expressions for the Fourier coefficients for the periodic current shown in Fig. P17.32.

b) Write the first four nonzero terms of the series using the alternative trigonometric form given by Eq. 17.38.

c) Use the first four nonzero terms of the expression derived in part (b) to estimate the rms value of i_g.

d) Find the exact rms value of i_g.

e) Calculate the percentage of error in the estimated rms value.

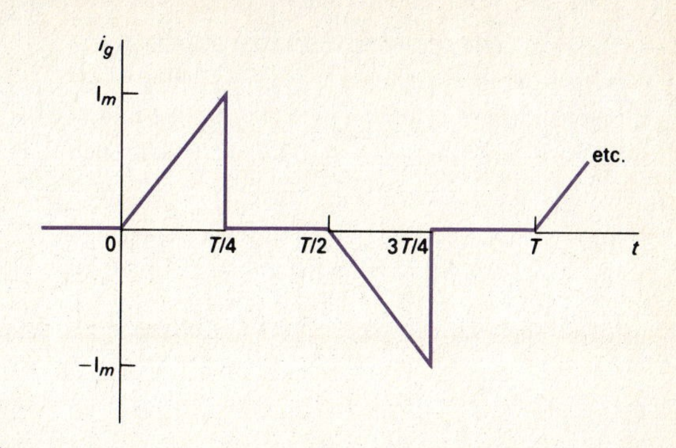

FIGURE P17.32

17.33 a) Use the first four nonzero terms in the Fourier series approximation of the periodic voltage shown in Fig. P17.33 to estimate its rms value.

b) Calculate the true rms value of the voltage.

c) Calculate the percentage of error in the estimated value.

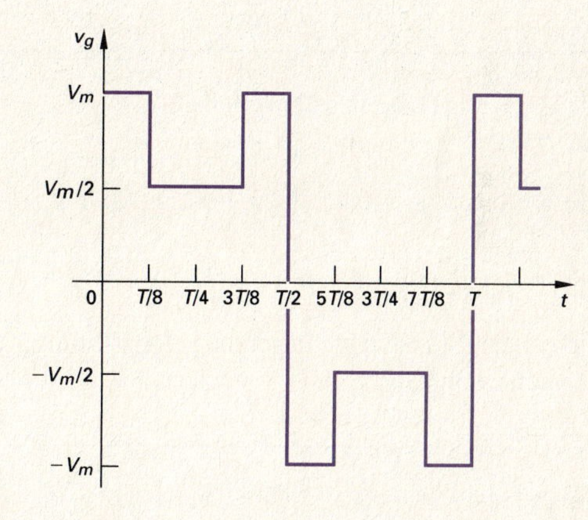

FIGURE P17.33

17.34 The rms value of any periodic triangular wave having the form depicted in Fig. P17.34(a) is independent of t_a and t_b. Note that for the function to be single valued, $t_a \leq t_b$. The rms value is equal to $V_p/\sqrt{3}$.

Verify this observation by finding the rms value of the three waveforms depicted in Figs. P17.34(b), (c), and (d).

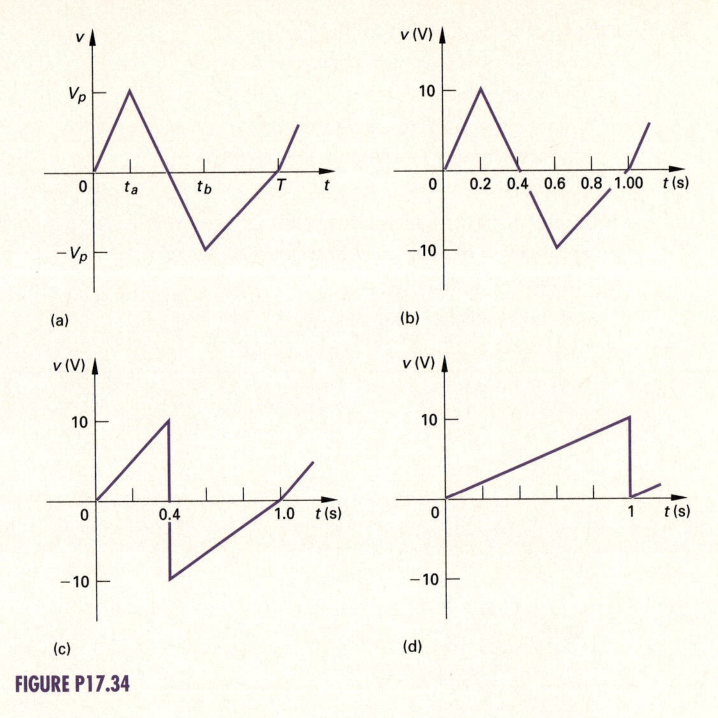

FIGURE P17.34

17.35 The voltage and current at the terminals of a network are

$$v = 40 + 125\cos(400t + 60^\circ) + 50\sin 1200t \text{ V};$$

$$i = 5 + 4\sin(400t + 30^\circ) + 2\cos(1200t + 30^\circ) \text{ A}.$$

The current is in the direction of the voltage drop across the terminals.

a) What is the average power at the terminals?

b) What is the rms value of the voltage?

c) What is the rms value of the current?

17.36 The periodic current in a 1 kΩ resistor is shown in Fig. P17.36.

a) Use the first three nonzero terms in the Fourier series representation of $i(t)$ to estimate the average power dissipated in the resistor.

b) Calculate the exact value of the average power dissipated in the 1 $k\Omega$ resistor.

c) What is the percentage error in the estimated value of the average power dissipated?

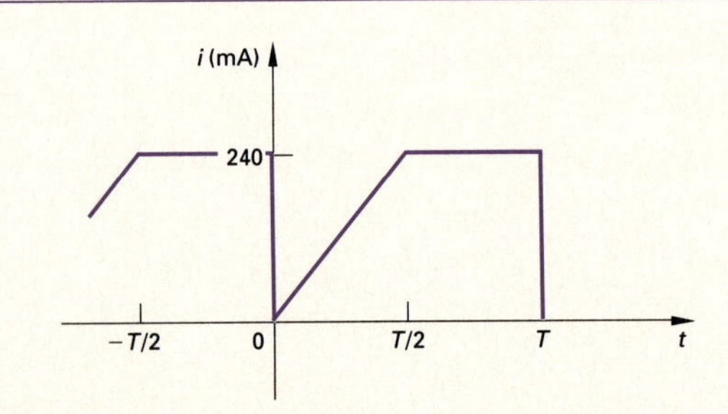

FIGURE P17.36

17.37 The periodic voltage shown in Fig. P17.37 is applied to a 10 Ω resistor.

a) Use the first three nonzero terms in the Fourier series representation of $v(t)$ to estimate the average power dissipated in the 10 Ω resistor.

b) Calculate the exact value of the average power dissipated in the 10 Ω resistor.

c) What is the percentage of error in the estimated value of the average power?

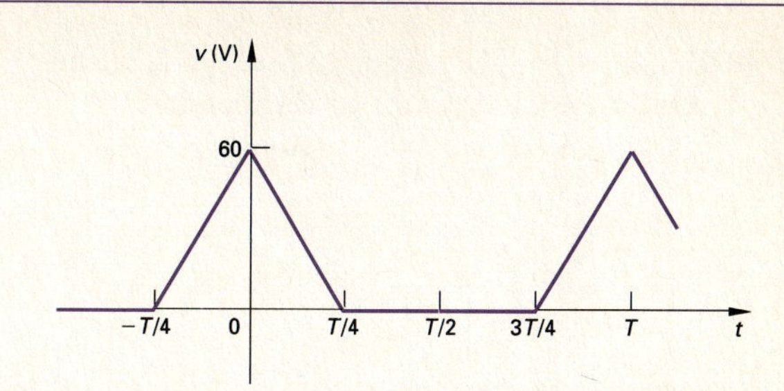

FIGURE P17.37

17.38 The triangular-wave voltage source is applied to the circuit in Fig. P17.38(a). The triangular-wave voltage is shown in Fig. P17.38(b). Estimate the average power delivered to the $50\sqrt{2}$ Ω resistor when the circuit is in steady-state operation.

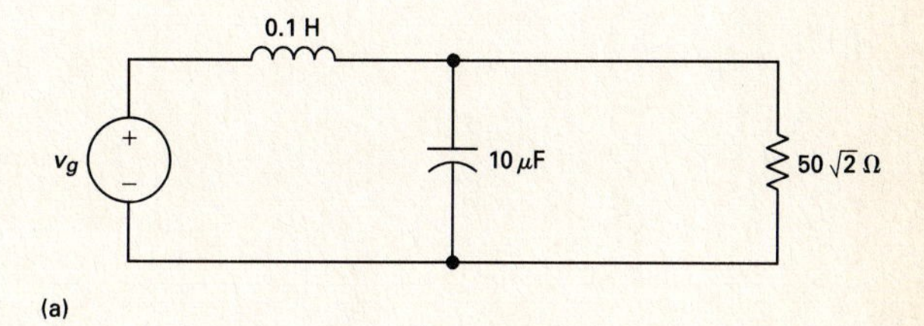

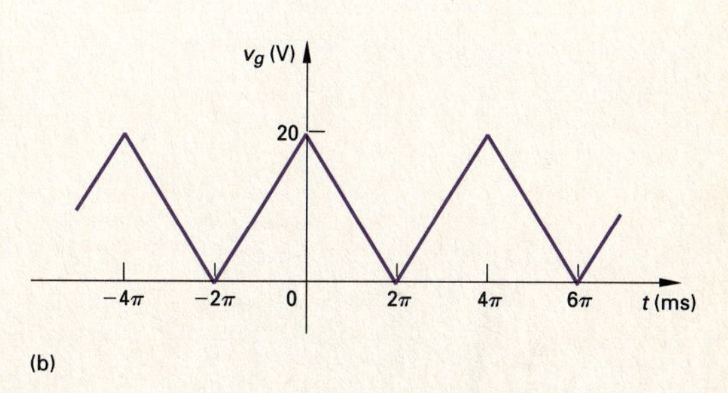

FIGURE P17.38

17.39 Assume the periodic function described in Problem 17.16 is a voltage with a peak amplitude of 20 V.

a) Find the rms value of the voltage.

b) If this voltage is applied to a 15 Ω resistor, what is the average power dissipated in the resistor?

c) If v_g is approximated by using just the fundamental frequency term of its Fourier series, what is the average power delivered to the 15Ω resistor?

d) What is the percentage of error in the estimation of the power dissipated?

17.40 Use the exponential form of the Fourier series to write an expression for the voltage shown in Fig. P17.40.

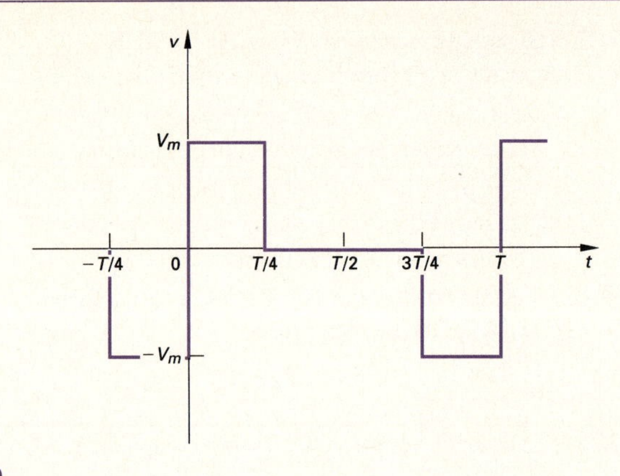

FIGURE P17.40

17.41 Derive the expression for the complex Fourier coefficients for the periodic current shown in Fig. P17.41.

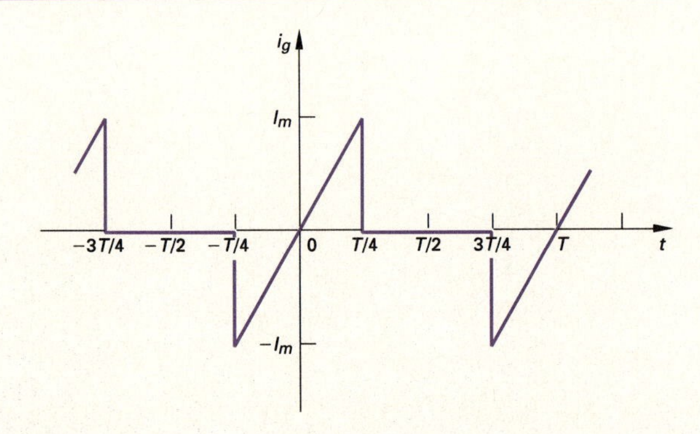

FIGURE P17.41

17.42 a) The periodic current in Problem 17.41 exists in a 24 Ω resistor. If $I_m = 5$ A, what is the average power delivered to the resistor?

b) Assume $i_g(t)$ is approximated by a truncated exponential form of the Fourier series consisting of the first five nonzero terms, that is, $n = 0, 1, 2, 3, 4,$ and 5. What is the rms value of the current, using this approximation?

c) If the approximation in part(b) is used to represent i_g, what is the percentage of error in the calculated power?

17.43 The periodic voltage source in the circuit shown in Fig. P17.43(a) has the waveform shown in Fig. P17.43(b).

a) Derive the expression for C_n.

b) Find the values of the complex coefficients $C_o, C_{-1}, C_1, C_{-2}, C_2, C_{-3}, C_3, C_{-4}$, and C_4 for the input voltage v_g if $V_m = 54$ V and $T = 10\pi\,\mu$s.

c) Repeat part (b) for v_o.

d) Use the complex coefficients found in part (c) to estimate the average power delivered to the 250 Ω resistor.

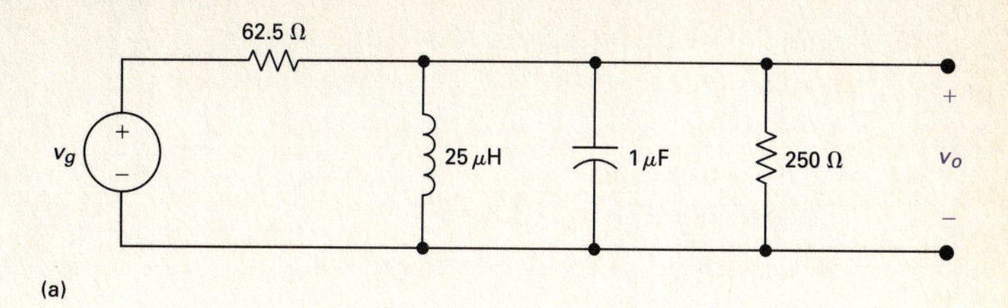

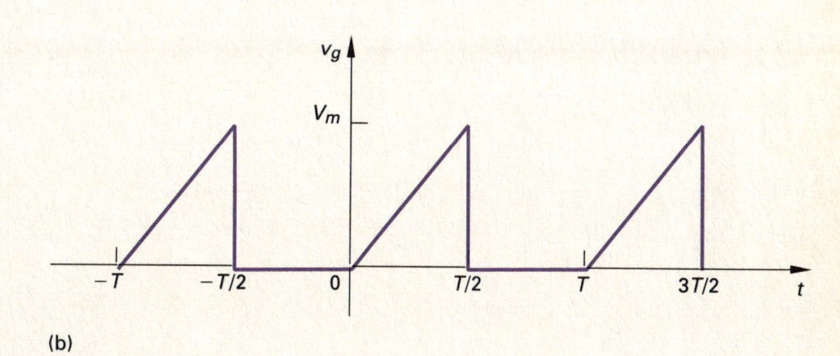

FIGURE P17.43

17.44 a) Find the rms value of the periodic voltage in Fig. P17.43(b).

b) Use the complex coefficients derived in Problem 17.43(b) to estimate the rms value of v_g.

c) What is the percentage of error in the estimated rms value of v_g?

17.45 a) Make an amplitude and phase plot, based on Eq. (17.38), for the periodic voltage in Example 17.3. Assume V_m is 40 V. Plot both amplitude and phase versus $n\omega_o$, where $n = 0, 1, 2, 3, \ldots$

b) Repeat part (a), but base the plots on Eq. (17.82).

17.46 a) Make an amplitude and phase plot, based on Eq. (17.38), for the periodic current in Problem 17.36. Plot both amplitude and phase versus $n\omega_o$ where $n = 0, 1, 2, \ldots$

b) Repeat part (a), but base plots on Eq. (17.82).

17.47 A periodic voltage is represented by a truncated Fourier series. The amplitude and phase spectra are shown in Figs. P17.47(a) and (b), respectively.

a) Write an expression for the periodic voltage using the form given by Eq. (17.38).

b) Is the voltage an even or odd function of t?

c) Does the voltage have half-wave symmetry?

d) Does the voltage have quarter-wave symmetry?

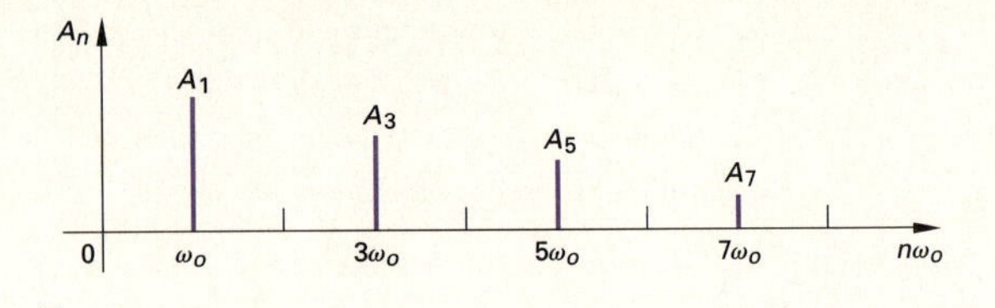

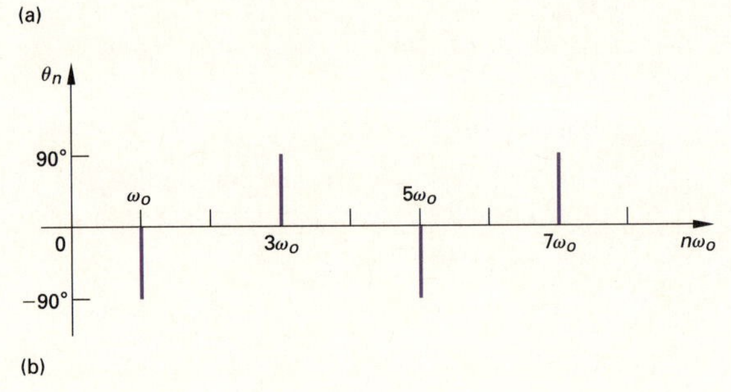

FIGURE P17.47

17.48 A periodic function is represented by a Fourier series that has a finite number of terms. The amplitude and phase spectra are shown in Figs. P17.48(a) and (b), respectively.

a) Write the expression for the periodic current using the form given by Eq. (17.38).

b) Is the current an even or odd function of t?

c) Does the current have half-wave symmetry?

d) Calculate the rms value of the current in milliamperes.

e) Write the exponential form of the Fourier series.

f) Make the amplitude and phase spectra plots on the basis of the exponential series.

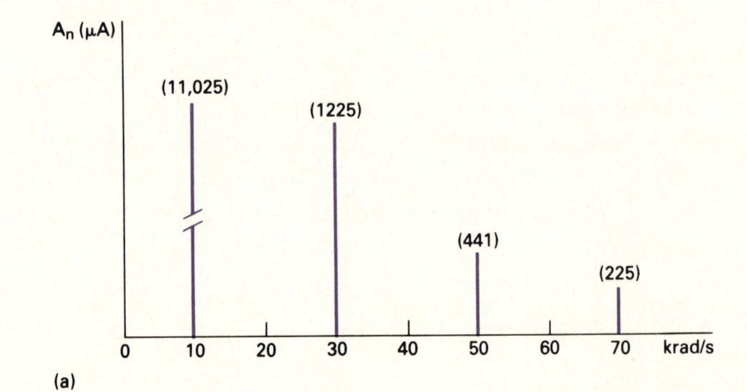

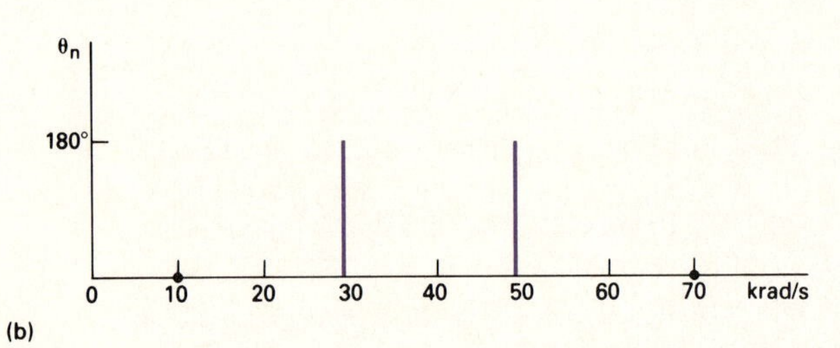

FIGURE P17.48

17.49 The resistors and capacitors in the circuit in Fig. P17.49 have been selected to synthesize a second-order low-pass Butterworth filter having a corner frequency of 1 krad/s. The voltage transfer function is

$$H(s) = \frac{V_o}{V_g} = \frac{10^6}{s^2 + 1000\sqrt{2}s + 10^6}.$$

If the input voltage to this filter is a full-wave rectified sine wave with an amplitude of 2.5π V and a fundamental frequency of 5000 rad/s, write the first two terms of the Fourier series that represents the output voltage.

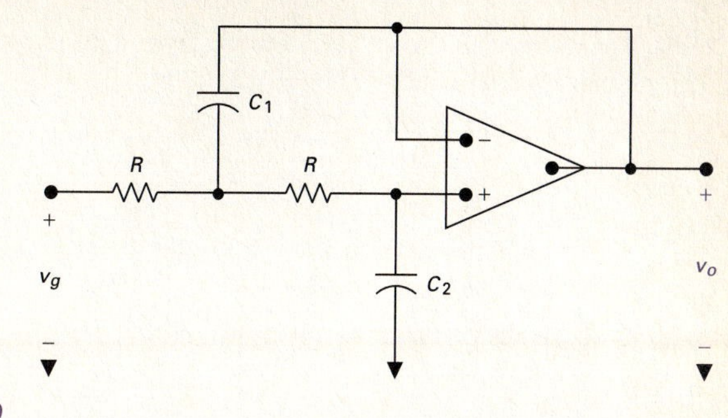

FIGURE P17.49

17.50 The resistors and capacitors in the circuit in Fig. P17.50(a) have been selected to synthesize a third-order low-pass Butterworth filter having a corner frequency of 1 rad/s. The voltage transfer function is

$$H(s) = \frac{V_o}{V_g} = \frac{1}{s^3 + 2s^2 + 2s + 1}.$$

The input to this low-pass filter is the periodic triangular-wave voltage shown in Fig. P17.50(b). Write the first three terms in the Fourier series that represents the output voltage of the filter.

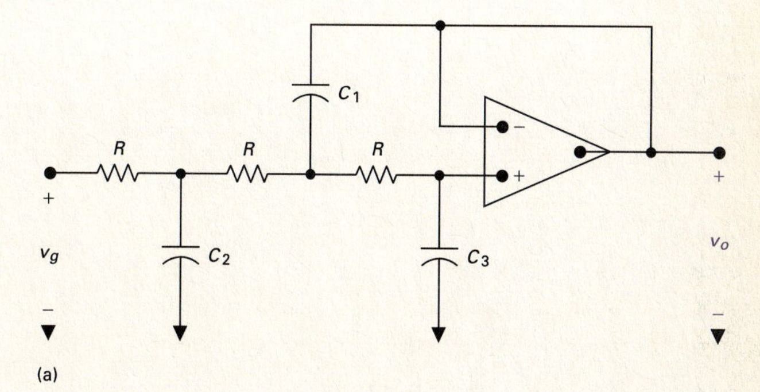

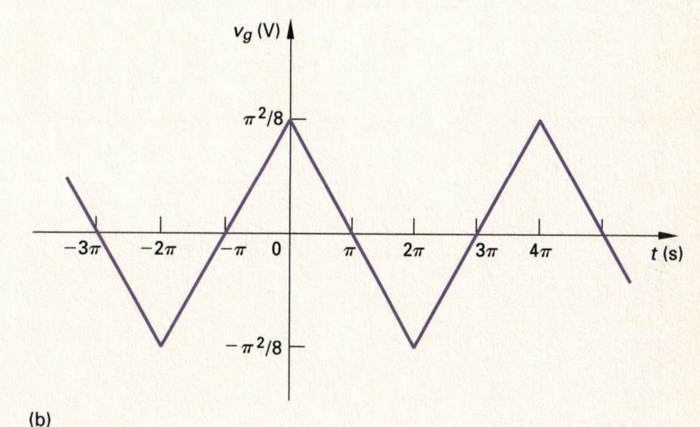

FIGURE P17.50

17.51 The transfer function (V_o/V_g) for the narrow-band bandpass filter circuit in Fig. P17.51(a) is

$$H(s) = \frac{-K_o\beta s}{s^2 + \beta s + \omega_o^2}$$

a) Find K_o, β, and ω_o^2 as functions of the circuit parameters R_1, R_2, R_3, C_1, and C_2.

b) Write the first three terms in the Fourier series that represents v_o if $R_1 = 25\ \text{k}\Omega$, $R_2 = 20.016\ \Omega$, $R_3 = 50\ \text{k}\Omega$, $C_1 = C_2 = 0.10\ \mu\text{F}$, and v_g is the periodic triangular voltage in Fig. 17.51(b).

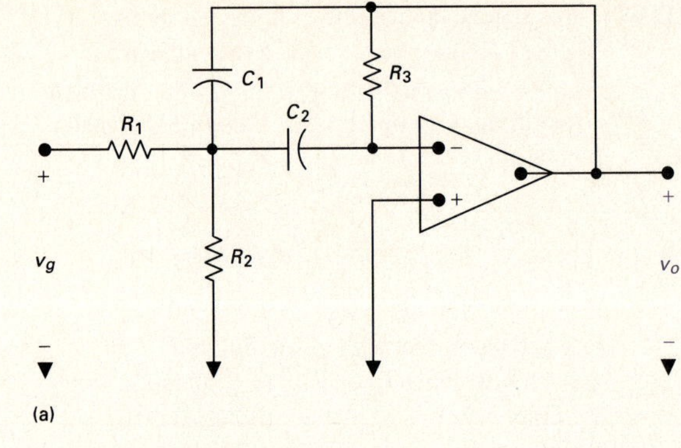

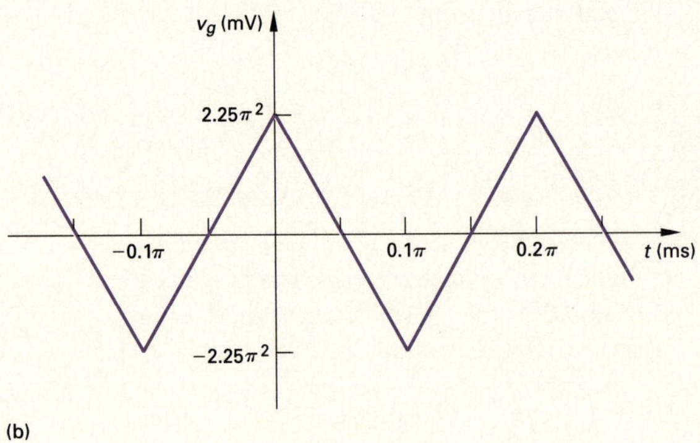

FIGURE P17.51

18 THE FOURIER TRANSFORM

CHAPTER CONTENTS

In Chapter 17 we discussed the representation of a periodic function by means of a Fourier series. This series representation enables us to describe the periodic function in terms of the frequency-domain attributes of amplitude and phase. The Fourier transform extends this frequency-domain description to functions that are not periodic. Through the Laplace transform, we already introduced the idea of transforming an aperiodic function from the time domain to the frequency domain. You may wonder, then, why yet another type of transformation is necessary. Strictly speaking, the Fourier transform is not a new transform. It is a special case of the bilateral Laplace transform, with the real part of the complex frequency set equal to zero. However, in terms of physical interpretation, the Fourier transform is better viewed as a limiting case of a Fourier series. We present this point of view in Section 18.1, where we derive the Fourier transform equations.

The Fourier transform is more useful than the Laplace transform in certain communications theory and signal-processing situations. Although we cannot pursue the Fourier transform in depth, its introduction here seems appropriate while the ideas underlying the Laplace transform and the Fourier series are still fresh in your mind.

18.1 The Derivation of the Fourier Transform

We begin the derivation of the Fourier transform, viewed as a limiting case of a Fourier series, with the exponential form of the series:

$$f(t) = \sum_{n=-\infty}^{\infty} C_n e^{jn\omega_0 t}, \tag{18.1}$$

where

$$C_n = \frac{1}{T} \int_{-T/2}^{T/2} f(t) e^{-jn\omega_0 t}\, dt. \tag{18.2}$$

In Eq. (18.2), we elected to start the integration at $t_0 = -T/2$.

Allowing the fundamental period T to increase without limit accomplishes the transition from a periodic to an aperiodic function. In other words, if T becomes infinite, the function never repeats itself and hence is aperiodic. As T increases, the separation between adjacent harmonic frequencies becomes smaller and smaller. In particular,

$$\Delta\omega = (n+1)\omega_0 - n\omega_0 = \omega_0 = \frac{2\pi}{T}, \tag{18.3}$$

and as T gets larger and larger, the incremental separation $\Delta\omega$ approaches a differential separation $d\omega$. From Eq. (18.3),

$$\frac{1}{T} \to \frac{d\omega}{2\pi} \quad \text{as } T \to \infty. \tag{18.4}$$

As the period increases, the frequency moves from being a discrete variable to becoming a continuous variable, or

$$n\omega_0 \to \omega \quad \text{as } T \to \infty. \tag{18.5}$$

In terms of Eq. (18.2), as the period increases, the Fourier coefficients C_n get smaller. In the limit, $C_n \to 0$ as $T \to \infty$. This result makes sense, because we expect the Fourier coefficients to vanish as the function loses its periodicity. Note, however, the limiting value of the product $C_n T$; that is,

$$C_n T \to \int_{-\infty}^{\infty} f(t)e^{-j\omega t}\,dt \quad \text{as } T \to \infty. \tag{18.6}$$

In writing Eq. (18.6) we took advantage of the relationship in Eq. (18.5). The integral in Eq. (18.6) is the **Fourier transform of *f*(*t*)** and is denoted

$$F(\omega) = \mathscr{F}\{f(t)\} = \int_{-\infty}^{\infty} f(t)e^{-j\omega t}\,dt. \tag{18.7}$$

We obtain an explicit expression for the inverse Fourier transform by investigating the limiting form of Eq. (18.1) as $T \to \infty$. We begin by multiplying and dividing by T:

$$f(t) = \sum_{n=-\infty}^{\infty} (C_n T)e^{jn\omega_0 t}\left(\frac{1}{T}\right). \tag{18.8}$$

As $T \to \infty$, the summation approaches integration, $C_n T \to F(\omega)$, $n\omega_0 \to \omega$, and $1/T \to d\omega/2\pi$. Thus in the limit, Eq. (18.8) becomes

$$f(t) = \frac{1}{2\pi}\int_{-\infty}^{\infty} F(\omega)e^{j\omega t}\,d\omega. \tag{18.9}$$

Equations (18.7) and (18.9) define the Fourier transform. Equation (18.7) transforms the time-domain expression $f(t)$ into its corresponding frequency-domain expression $F(\omega)$. Equation (18.9) defines the inverse operation of transforming $F(\omega)$ into $f(t)$.

Let's now derive the Fourier transform of the pulse shown in Fig. 18.1. Note that this pulse corresponds to the periodic voltage in Example 17.6 if we let $T \to \infty$. The Fourier transform of $v(t)$ comes directly from Eq. (18.7):

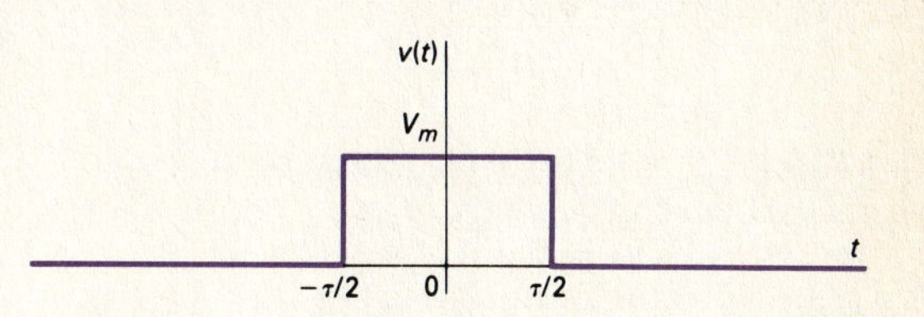

FIGURE 18.1 A voltage pulse.

$$\begin{aligned} F(\omega) &= \int_{-\tau/2}^{\tau/2} V_m e^{-j\omega t}\,dt \\ &= V_m \left.\frac{e^{-j\omega t}}{(-j\omega)}\right|_{-\tau/2}^{\tau/2} \\ &= \frac{V_m}{-j\omega}\left(-2j\sin\frac{\omega\tau}{2}\right), \end{aligned} \tag{18.10}$$

which can be put in the form of $(\sin x)/x$ by multiplying the numerator and denominator by τ. Then,

$$F(\omega) = V_m \tau \frac{\sin \omega\tau/2}{\omega\tau/2}. \tag{18.11}$$

For the periodic train of voltage pulses in Example 17.6, the expression for the Fourier coefficients is

$$C_n = \frac{V_m \tau}{T} \frac{\sin n\omega_0 \tau/2}{n\omega_0 \tau/2}. \tag{18.12}$$

Comparing Eqs. (18.11) and (18.12) clearly shows that, as the time-domain function goes from periodic to aperiodic, the amplitude spectrum goes from a discrete line spectrum to a continuous spectrum. Furthermore, the envelope of the line spectrum has the same shape as the continuous spectrum. Thus as T increases, the spectrum of lines gets denser and the amplitudes get smaller, but the envelope doesn't change shape. The physical interpretation of the Fourier transform $F(\omega)$ is therefore a measure of the frequency content of $f(t)$. Figure 18.2 illustrates these observations. The amplitude spectrum plot is based on the assumption that τ is constant and T is increasing.

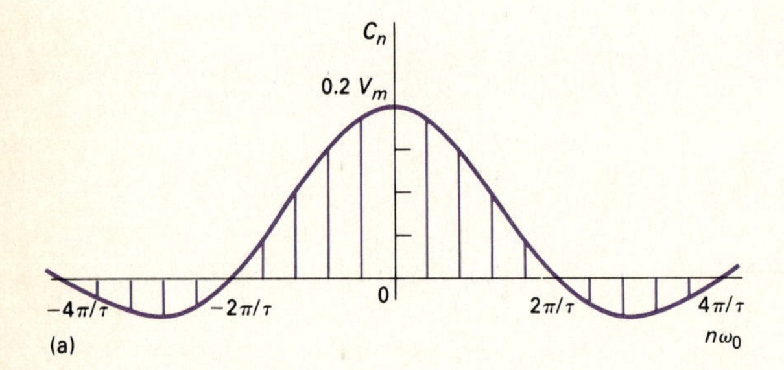

C_n
$0.1\,V_m$
$-4\pi/\tau$ $-2\pi/\tau$ 0 $2\pi/\tau$ $4\pi/\tau$
$n\omega_0$
(b)

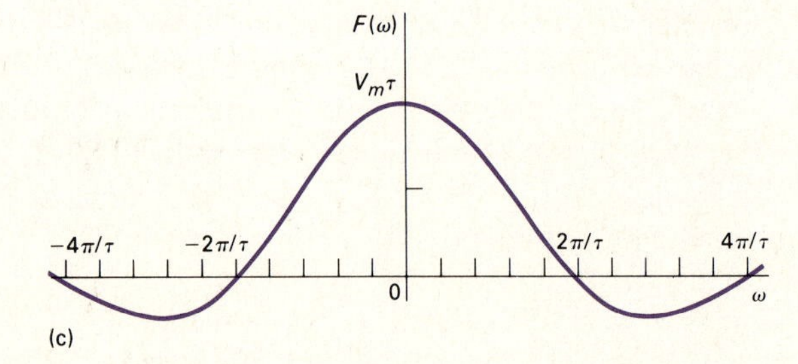

FIGURE 18.2 Transition of the amplitude spectrum as $f(t)$ goes from periodic to aperiodic: (a) C_n versus $n\omega_0$, $T/\tau = 5$; (b) C_n versus $n\omega_0$, $T/\tau = 10$; (c) $F(\omega)$ versus ω.

18.2 The Convergence of the Fourier Integral

A function of time $f(t)$ has a Fourier transform if the integral in Eq. (18.7) converges. If $f(t)$ is a well-behaved function that differs from zero over a finite interval of time, convergence is no problem. *Well-behaved* implies that $f(t)$ is single valued and encloses a finite area over the range of integration. In practical terms, all the pulses of finite duration that interest us are well-behaved functions. The evaluation of the Fourier transform of the rectangular pulse discussed in Section 18.1 illustrates this point.

If $f(t)$ is different from zero over an infinite interval, the convergence of the Fourier integral depends on the behavior of $f(t)$ as $t \to \infty$. A single-valued function that is nonzero over an infinite interval has a Fourier transform if the integral

$$\int_{-\infty}^{\infty} |f(t)|\, dt$$

exists and if any discontinuities in $f(t)$ are finite. An example is the decaying exponential function illustrated in Fig. 18.3. The Fourier transform of $f(t)$ is

$$F(\omega) = \int_{-\infty}^{\infty} f(t)e^{-j\omega t}\, dt = \int_{0}^{\infty} Ke^{-at}e^{-j\omega t}\, dt$$

$$= \frac{Ke^{-(a+j\omega)t}}{-(a+j\omega)}\Bigg|_0^{\infty} = \frac{K}{-(a+j\omega)}(0-1)$$

$$= \frac{K}{a+j\omega}. \quad \textbf{(18.13)}$$

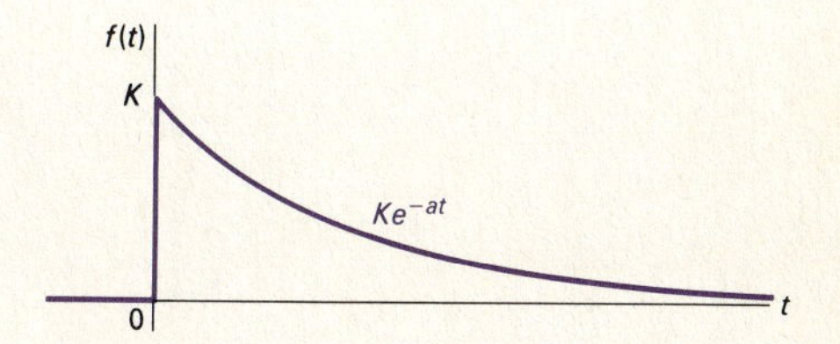

FIGURE 18.3 The decaying exponential function $Ke^{-at}u(t)$.

A third important group of functions have great practical interest but do not in a strict sense have a Fourier transform. For example, the integral in Eq. (18.7) doesn't converge if $f(t)$ is a constant. The same can be said if $f(t)$ is a sinusoidal function—$\cos \omega_0 t$—or a step function—$Ku(t)$. These functions are of great interest in circuit analysis, but, to include them in Fourier analysis, we must resort to some mathematical subterfuge. First, we create a function in the time domain that has a Fourier transform and at the same time can be made arbitrarily close to the function of interest. Next, we find the Fourier transform of the approximating function and then evaluate the limiting value of $F(\omega)$ as this function approaches $f(t)$. Last, we define the limiting value of $F(\omega)$ as the Fourier transform of $f(t)$.

Let's demonstrate this technique by finding the Fourier transform of a constant. We can approximate a constant with the exponential function

$$f(t) = Ae^{-\epsilon|t|}, \quad \epsilon > 0. \quad \textbf{(18.14)}$$

As $\epsilon \to 0$, $f(t) \to A$. Figure 18.4 shows the approximation graphically. The Fourier transform of $f(t)$ is

$$F(\omega) = \int_{-\infty}^{0} Ae^{\epsilon t}e^{-j\omega t}\, dt + \int_{0}^{\infty} Ae^{-\epsilon t}e^{-j\omega t}\, dt. \quad \textbf{(18.15)}$$

Carrying out the integration called for in Eq. (18.15) yields

$$F(\omega) = \frac{A}{\epsilon - j\omega} + \frac{A}{\epsilon + j\omega} = \frac{2\epsilon A}{\epsilon^2 + \omega^2}. \quad \textbf{(18.16)}$$

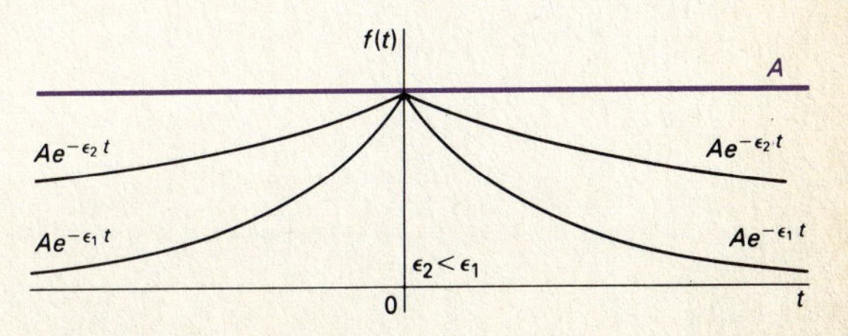

FIGURE 18.4 The approximation of a constant with an exponential function.

The function given by Eq. (18.16) generates an impulse function at $\omega = 0$ as $\epsilon \to 0$. You can verify this result by showing that (1) $F(\omega)$ approaches infinity at $\omega = 0$ as $\epsilon \to 0$; (2) the duration of $F(\omega)$ approaches zero as $\epsilon \to 0$; and (3) the area under $F(\omega)$ is independent of ϵ. The area under $F(\omega)$ is the strength of the impulse and is

$$\int_{-\infty}^{\infty} \frac{2\epsilon A}{\epsilon^2 + \omega^2} d\omega = 4\epsilon A \int_{0}^{\infty} \frac{d\omega}{\epsilon^2 + \omega^2} = 2\pi A. \qquad \textbf{(18.17)}$$

In the limit, $f(t)$ approaches a constant A, and $F(\omega)$ approaches an impulse function $2\pi A\delta(\omega)$. Therefore the Fourier transform of a constant A is defined as $2\pi A\delta(\omega)$, or

$$\mathcal{F}\{A\} = 2\pi A\delta(\omega). \qquad \textbf{(18.18)}$$

In Section 18.4 we say more about Fourier transforms defined through a limit process. Before doing so, in Section 18.3 we show how to take advantage of the Laplace transform to find the Fourier transform of functions for which the Fourier integral converges.

DRILL EXERCISES

18.1 Use the defining integral to find the Fourier transform of the following functions:

a) $f(t) = -A, \quad -\tau/2 \le t < 0;$
$f(t) = A, \quad 0 < t \le \tau/2;$
$f(t) = 0 \quad \text{elsewhere}.$

b) $f(t) = 0, \quad t < 0;$
$f(t) = te^{-at}, \quad t \ge 0, a > 0.$

ANSWER: (a) $-j\left(\frac{2A}{\omega}\right)\left(1 - \cos\frac{\omega\tau}{2}\right)$; (b) $\frac{1}{(a + j\omega)^2}$.

18.2 The Fourier transform of $f(t)$ is given by

$F(\omega) = 0, \quad -\infty \le \omega < -2;$
$F(\omega) = 2, \quad -2 < \omega < -1;$
$F(\omega) = 1, \quad -1 < \omega < 1;$
$F(\omega) = 2, \quad 1 < \omega < 2;$
$F(\omega) = 0, \quad 2 < \omega \le \infty.$

Find $f(t)$.

ANSWER: $f(t) = \frac{1}{\pi t}(2\sin 2t - \sin t)$.

18.3 Using Laplace Transforms to Find Fourier Transforms

We can use a table of unilateral, or one-sided, Laplace transform pairs to find the Fourier transform of functions for which the Fourier integral converges. The Fourier integral converges when all the poles of $F(s)$

lie in the left half of the s plane. Note that if $F(s)$ has poles in the right half of the s plane or along the imaginary axis, $f(t)$ does not satisfy the constraint that $\int_{-\infty}^{\infty} |f(t)|\, dt$ exists.

The following rules apply to the use of Laplace transforms to find the Fourier transforms of such functions.

1. If $f(t)$ is zero for $t \leq 0^-$, we obtain the Fourier transform of $f(t)$ from the Laplace transform of $f(t)$ simply by replacing s by $j\omega$. Thus

$$\mathscr{F}\{f(t)\} = \mathscr{L}\{f(t)\}_{s=j\omega}. \quad \textbf{(18.19)}$$

For example, say that

$$f(t) = 0, \quad t \leq 0^-;$$
$$f(t) = e^{-at}\cos\omega_0 t, \quad t \geq 0^+.$$

Then

$$\mathscr{F}\{f(t)\} = \left.\frac{s+a}{(s+a)^2+\omega_0^2}\right|_{s=j\omega}$$
$$= \frac{j\omega+a}{(j\omega+a)^2+\omega_0^2}.$$

2. Because the range of integration on the Fourier integral goes from $-\infty$ to $+\infty$, the Fourier transform of a negative-time function exists. A negative-time function is nonzero for negative values of time and zero for positive values of time. To find the Fourier transform of such a function, we proceed as follows. First, we reflect the negative-time function over to the positive-time domain and then find its one-sided Laplace transform. We obtain the Fourier transform of the original time function by replacing s with $-j\omega$. Therefore, when $f(t) = 0$ for $t \geq 0^+$,

$$\mathscr{F}\{f(t)\} = \mathscr{L}\{f(-t)\}_{s=-j\omega}. \quad \textbf{(18.20)}$$

For example, if

$$f(t) = 0 \quad (\text{for } t \geq 0^+)$$

and

$$f(t) = e^{at}\cos\omega_0 t \quad (\text{for } t \leq 0^-),$$

then

$$f(-t) = 0 \quad (\text{for } t \leq 0^-)$$

and

$$f(-t) = e^{-at}\cos\omega_0 t \quad (\text{for } t \geq 0^+).$$

Both $f(t)$ and its mirror image are plotted in Fig. 18.5. The

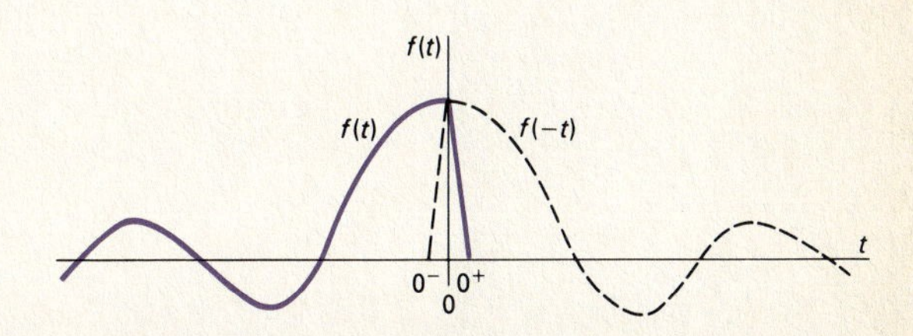

FIGURE 18.5 The reflection of a negative-time function over to the positive-time domain.

Fourier transform of $f(t)$ is

$$\begin{aligned}\mathscr{F}\{f(t)\} &= \mathscr{L}\{f(-t)\}_{s=-j\omega} = \left.\frac{s+a}{(s+a)^2+\omega_0^2}\right|_{s=-j\omega} \\ &= \frac{-j\omega+a}{(-j\omega+a)^2+\omega_0^2}.\end{aligned}$$

3. Functions that are nonzero over all time can be resolved into positive- and negative-time functions. We use Eqs. (18.19) and (18.20) to find the Fourier transform of the positive- and negative-time functions, respectively. The Fourier transform of the original function is the sum of the two transforms. Thus if we let

$$f^+(t) = f(t) \quad (\text{for } t > 0)$$

and

$$f^-(t) = f(t) \quad (\text{for } t < 0)$$

then

$$f(t) = f^+(t) + f^-(t)$$

and

$$\begin{aligned}\mathscr{F}\{f(t)\} &= \mathscr{F}\{f^+(t)\} + \mathscr{F}\{f^-(t)\} \\ &= \mathscr{L}\{f^+(t)\}_{s=j\omega} + \mathscr{L}\{f^-(-t)\}_{s=-j\omega}. \end{aligned} \quad \textbf{(18.21)}$$

An example of using Eq. (18.21) involves finding the Fourier transform of $e^{-a|t|}$. For the original function, the positive- and negative-time functions are

$$f^+(t) = e^{-at} \quad \text{and} \quad f^-(t) = e^{at}.$$

Then

$$\mathscr{L}\{f^+(t)\} = \frac{1}{s+a}$$

and

$$\mathscr{L}\{f^-(-t)\} = \frac{1}{s+a}.$$

Therefore, from Eq. (18.21),

$$\begin{aligned}\mathscr{F}\{e^{-a|t|}\} &= \left.\frac{1}{s+a}\right|_{s=j\omega} + \left.\frac{1}{s+a}\right|_{s=-j\omega} \\ &= \frac{1}{j\omega+a} + \frac{1}{-j\omega+a} \\ &= \frac{2a}{\omega^2+a^2}.\end{aligned}$$

If $f(t)$ is even, Eq. (18.21) reduces to

$$\mathscr{F}\{f(t)\} = \mathscr{L}\{f(t)\}_{s=j\omega} + \mathscr{L}\{f(t)\}_{s=-j\omega}. \quad \textbf{(18.22)}$$

If $f(t)$ is odd, then Eq. (18.21) becomes

$$\mathscr{F}\{f(t)\} = \mathscr{L}\{f(t)\}_{s=j\omega} - \mathscr{L}\{f(t)\}_{s=-j\omega}. \quad \textbf{(18.23)}$$

DRILL EXERCISE

18.3 Find the Fourier transform of each function. In each case, a is a positive real constant.

a) $f(t) = 0, \quad t < 0;$
$f(t) = e^{-at}\sin\omega_0 t, \quad t \geq 0.$

b) $f(t) = 0, \quad t > 0;$
$f(t) = -te^{at}, \quad t \leq 0.$

c) $f(t) = te^{-at}, \quad t \geq 0;$
$f(t) = te^{at}, \quad t \leq 0.$

ANSWER: a) $\dfrac{\omega_0}{(a+j\omega)^2+\omega_0^2}$; b) $\dfrac{1}{(a-j\omega)^2}$; c) $\dfrac{-j4a\omega}{(a^2+\omega^2)^2}$.

18.4 Fourier Transforms in the Limit

As we pointed out in Section 18.2, the Fourier transforms of several practical functions must be defined by a limit process. We now return to these types of functions and develop their transforms.

The Fourier Transform of a Signum Function

We showed that the Fourier transform of a constant A is $2\pi A\delta(\omega)$ in Eq. (18.18). The next function of interest is the signum function, defined as $+1$ for $t > 0$, and -1 for $t < 0$. The signum function is denoted $\text{sgn}(t)$ and can be expressed in terms of unit-step functions, or

$$\text{sgn}(t) = u(t) - u(-t). \quad \textbf{(18.24)}$$

Figure 18.6 shows the function graphically.

To find the Fourier transform of the signum function, we first create a function that approaches the signum function in the limit:

$$\text{sgn}(t) = \lim_{\epsilon \to 0}[e^{-\epsilon t}u(t) - e^{\epsilon t}u(-t)]. \quad \textbf{(18.25)}$$

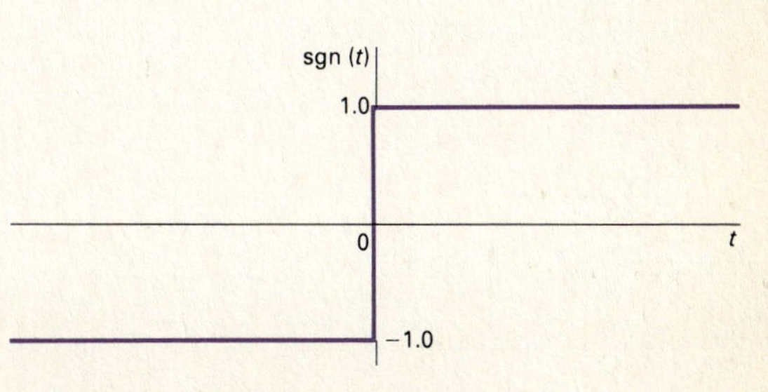

FIGURE 18.6 The signum function.

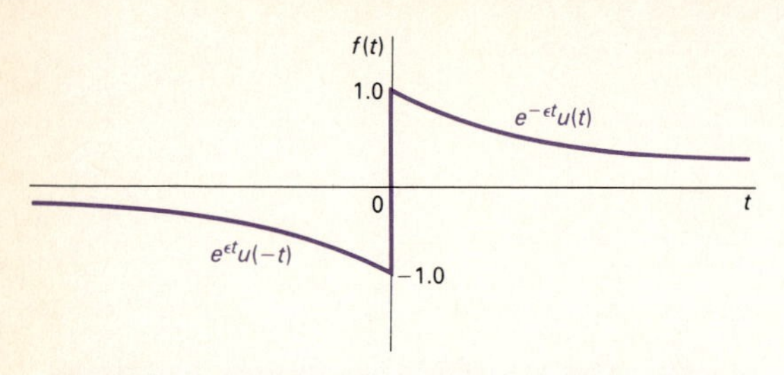

FIGURE 18.7 A function that approaches sgn(t) as ϵ approaches zero.

The function inside the brackets, plotted in Fig. 18.7, has a Fourier transform because the Fourier integral converges. Because $f(t)$ is an odd function, we use Eq. (18.23) to find its Fourier transform:

$$\begin{aligned}\mathscr{F}\{f(t)\} &= \left.\frac{1}{s+\epsilon}\right|_{s=j\omega} - \left.\frac{1}{s+\epsilon}\right|_{s=-j\omega} \\ &= \frac{1}{j\omega+\epsilon} - \frac{1}{-j\omega+\epsilon} \\ &= \frac{-2j\omega}{\omega^2+\epsilon^2}. \end{aligned} \qquad \textbf{(18.26)}$$

As $\epsilon \to 0$, $f(t) \to$ sgn(t), and $\mathscr{F}\{f(t)\} \to 2/j\omega$. Therefore

$$\mathscr{F}\{\text{sgn}(t)\} = \frac{2}{j\omega}. \qquad \textbf{(18.27)}$$

The Fourier Transform of a Unit-step Function

To find the Fourier transform of a unit-step function, we use Eqs. (18.18) and (18.27). We do so by recognizing that the unit-step function can be expressed as

$$u(t) = \frac{1}{2} + \frac{1}{2}\text{sgn}(t). \qquad \textbf{(18.28)}$$

Thus

$$\begin{aligned}\mathscr{F}\{u(t)\} &= \mathscr{F}\left\{\frac{1}{2}\right\} + \mathscr{F}\left\{\frac{1}{2}\text{sgn}(t)\right\} \\ &= \pi\delta(\omega) + \frac{1}{j\omega}. \end{aligned} \qquad \textbf{(18.29)}$$

The Fourier Transform of a Cosine Function

To find the Fourier transform of cos $\omega_0 t$, we return to the inverse-transform integral of Eq. (18.9) and observe that if

$$F(\omega) = 2\pi\delta(\omega - \omega_0), \qquad \textbf{(18.30)}$$

then

$$f(t) = \frac{1}{2\pi}\int_{-\infty}^{\infty}[2\pi\delta(\omega-\omega_0)]e^{j\omega t}\,d\omega. \qquad \textbf{(18.31)}$$

Using the sifting property of the impulse function, we reduce Eq. (18.31) to

$$f(t) = e^{j\omega_0 t}. \qquad \textbf{(18.32)}$$

Then, from Eqs. (18.30) and (18.32),

$$\mathscr{F}\{e^{j\omega_0 t}\} = 2\pi\delta(\omega - \omega_0). \qquad \textbf{(18.33)}$$

We now use Eq. (18.33) to find the Fourier transform of $\cos\omega_0 t$, because

$$\cos\omega_0 t = \frac{e^{j\omega_0 t} + e^{-j\omega_0 t}}{2}. \quad \textbf{(18.34)}$$

Thus

$$\begin{aligned}\mathscr{F}\{\cos\omega_0 t\} &= \frac{1}{2}(\mathscr{F}\{e^{j\omega_0 t}\} + \mathscr{F}\{e^{-j\omega_0 t}\})\\ &= \frac{1}{2}[2\pi\delta(\omega-\omega_0) + 2\pi\delta(\omega+\omega_0)]\\ &= \pi\delta(\omega-\omega_0) + \pi\delta(\omega+\omega_0). \quad \textbf{(18.35)}\end{aligned}$$

The Fourier transform of $\sin\omega_0 t$ involves similar manipulation, which we leave for Drill Exercise 18.4. Table 18.1 presents a summary of the transform pairs of the important elementary functions.

We now turn to the properties of the Fourier transform that enhance our ability to describe aperiodic time-domain behavior in terms of frequency-domain behavior.

TABLE 18.1

FOURIER TRANSFORMS OF ELEMENTARY FUNCTIONS

$f(t)$	$F(\omega)$
$\delta(t)$ (impulse)	1
A (constant)	$2\pi A\delta(\omega)$
$\text{sgn}(t)$ (signum)	$2/j\omega$
$u(t)$ (step)	$\pi\delta(\omega) + 1/j\omega$
$e^{-at}u(t)$ (positive-time exponential)	$1/(a+j\omega)$
$e^{at}u(-t)$ (negative-time exponential)	$1/(a-j\omega)$
$e^{-a\lvert t\rvert}$ (positive- and negative-time exponential)	$2a/(a^2+\omega^2)$
$e^{j\omega_0 t}$ (complex exponential)	$2\pi\delta(\omega-\omega_0)$
$\cos\omega_0 t$ (cosine)	$\pi[\delta(\omega+\omega_0) + \delta(\omega-\omega_0)]$
$\sin\omega_0 t$ (sine)	$j\pi[\delta(\omega+\omega_0) - \delta(\omega-\omega_0)]$

DRILL EXERCISE

18.4 Find $\mathscr{F}\{\sin\omega_0 t\}$.

ANSWER: $j\pi[\delta(\omega+\omega_0) - \delta(\omega-\omega_0)]$.

18.5 Some Mathematical Properties

The first mathematical property we call to your attention is that $F(\omega)$ is a complex quantity and can be expressed in either rectangular or polar form. Thus from the defining integral,

$$\begin{aligned} F(\omega) &= \int_{-\infty}^{\infty} f(t)e^{-j\omega t}\,dt \\ &= \int_{-\infty}^{\infty} f(t)(\cos\omega t - j\sin\omega t)\,dt \\ &= \int_{-\infty}^{\infty} f(t)\cos\omega t\,dt - j\int_{-\infty}^{\infty} f(t)\sin\omega t\,dt. \end{aligned} \tag{18.36}$$

Now we let

$$A(\omega) = \int_{-\infty}^{\infty} f(t)\cos\omega t\,dt \tag{18.37}$$

and

$$B(\omega) = -\int_{-\infty}^{\infty} f(t)\sin\omega t\,dt. \tag{18.38}$$

Thus, using the definitions given by Eqs. (18.37) and (18.38) in Eq. (18.36), we get

$$F(\omega) = A(\omega) + jB(\omega) = |F(\omega)|e^{j\theta(\omega)}. \tag{18.39}$$

The following observations about $F(\omega)$ are pertinent:

1. The real part of $F(\omega)$—that is, $A(\omega)$—is an even function of ω; in other words, $A(\omega) = A(-\omega)$.
2. The imaginary part of $F(\omega)$—that is, $B(\omega)$—is an odd function of ω; in other words, $B(\omega) = -B(-\omega)$.
3. The magnitude of $F(\omega)$,—that is, $\sqrt{A^2(\omega) + B^2(\omega)}$,—is an even function of ω.
4. The phase angle of $F(\omega)$—that is, $\theta(\omega) = \tan^{-1} B(\omega)/A(\omega)$—is an odd function of ω.
5. Replacing ω by $-\omega$ generates the conjugate of $F(\omega)$; in other words, $F(-\omega) = F^*(\omega)$.

Hence, if $f(t)$ is an even function, $F(\omega)$ is real, and if $f(t)$ is an odd function, $F(\omega)$ is imaginary. If $f(t)$ is even, from Eqs. (18.37) and (18.38),

$$A(\omega) = 2\int_{0}^{\infty} f(t)\cos\omega t\,dt \tag{18.40}$$

and

$$B(\omega) = 0. \tag{18.41}$$

If $f(t)$ is an odd function,

$$A(\omega) = 0 \tag{18.42}$$

and

$$B(\omega) = -2\int_0^{\infty} f(t)\sin\omega t\,dt. \tag{18.43}$$

We leave the derivations of Eqs. (18.40) through (18.43) for you as Problems 18.7 and 18.8.

If $f(t)$ is an even function, its Fourier transform is an even function, and if $f(t)$ is an odd function, its Fourier transform is an odd function. Moreover, if $f(t)$ is an even function, from the inverse Fourier integral,

$$\begin{aligned} f(t) &= \frac{1}{2\pi}\int_{-\infty}^{\infty} F(\omega)e^{j\omega t}\,d\omega = \frac{1}{2\pi}\int_{-\infty}^{\infty} A(\omega)e^{j\omega t}\,d\omega \\ &= \frac{1}{2\pi}\int_{-\infty}^{\infty} A(\omega)(\cos\omega t + j\sin\omega t)\,d\omega \\ &= \frac{1}{2\pi}\int_{-\infty}^{\infty} A(\omega)\cos\omega t\,d\omega + 0 \\ &= \frac{2}{2\pi}\int_0^{\infty} A(\omega)\cos\omega t\,d\omega. \end{aligned} \tag{18.44}$$

Now compare Eq. (18.44) with Eq. (18.40). Note that, except for a factor of $1/2\pi$, these two equations have the same form. Thus the waveforms of $A(\omega)$ and $f(t)$ become interchangeable if $f(t)$ is an even function. For example, we have already observed that a rectangular pulse in the time domain produces a frequency spectrum of the form $(\sin\omega)/\omega$. Specifically, Eq. (18.11) expresses the Fourier transform of the voltage pulse shown in Fig. 18.1. Hence a rectangular pulse in the frequency domain must be generated by a time-domain function of the form $(\sin t)/t$. We can illustrate this requirement by finding the time-domain function $f(t)$ corresponding to the frequency spectrum shown in Fig. 18.8. From Eq. (18.44),

$$\begin{aligned} f(t) &= \frac{2}{2\pi}\int_0^{\omega_0/2} M\cos\omega t\,d\omega = \frac{2M}{2\pi}\left(\frac{\sin\omega t}{t}\right)\Bigg|_0^{\omega_0/2} \\ &= \frac{1}{2\pi}\left(2M\frac{\sin\omega_0 t/2}{t/2}\right) \\ &= \frac{1}{2\pi}\left(M\omega_0\frac{\sin\omega_0 t/2}{\omega_0 t/2}\right). \end{aligned} \tag{18.45}$$

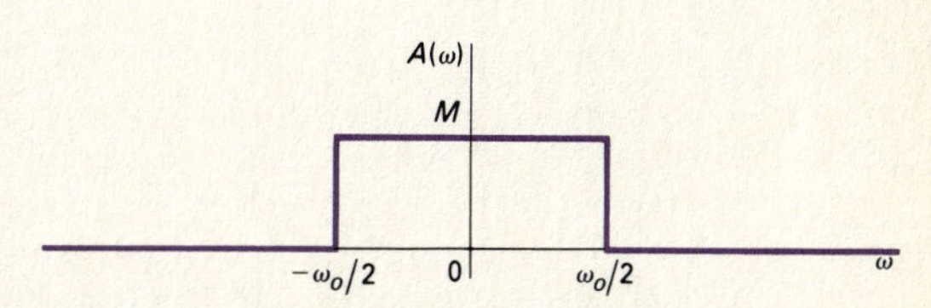

FIGURE 18.8 A rectangular frequency spectrum.

We say more about the frequency spectrum of a rectangular pulse in the time domain versus the rectangular frequency spectrum of $(\sin t)/t$ after we introduce Parseval's theorem.

DRILL EXERCISES

18.5 If $f(t)$ is a real function of t, show that the inversion integral reduces to

$$f(t) = \frac{1}{2\pi}\int_{-\infty}^{\infty}[A(\omega)\cos\omega t - B(\omega)\sin\omega t]\,d\omega.$$

ANSWER: Derivation.

18.6 If $f(t)$ is a real, odd function of t, show that the inversion integral reduces to

$$f(t) = -\frac{1}{2\pi}\int_{-\infty}^{\infty}B(\omega)\sin\omega t\,d\omega.$$

ANSWER: Derivation.

18.6 OPERATIONAL TRANSFORMS

Fourier transforms, like Laplace transforms, can be classified as functional and operational. So far, we have concentrated on the functional transforms. We now discuss some of the important operational transforms. With regard to the Laplace transform, these operational transforms are similar to those discussed in Chapter 13. Hence we leave their proofs to you as Problems 18.9–18.16.

Use a computer tool to assist in these derivations.

1. **Multiplication by a Constant** If $\mathscr{F}\{f(t)\} = F(\omega)$,

$$\mathscr{F}\{Kf(t)\} = KF(\omega). \quad \textbf{(18.46)}$$

2. **Addition (Subtraction)** If

$$\mathscr{F}\{f_1(t)\} = F_1(\omega),$$
$$\mathscr{F}\{f_2(t)\} = F_2(\omega),$$

and

$$\mathscr{F}\{f_3(t)\} = F_3(\omega),$$

then

$$\mathscr{F}\{f_1(t) - f_2(t) + f_3(t)\} = F_1(\omega) - F_2(\omega) + F_3(\omega). \quad \textbf{(18.47)}$$

3. **Differentiation** The Fourier transform of the first derivative of $f(t)$ is

$$\mathscr{F}\left\{\frac{df(t)}{dt}\right\} = j\omega F(\omega). \quad \textbf{(18.48)}$$

The nth derivative of $f(t)$ is

$$\mathscr{F}\left\{\frac{d^n f(t)}{dt^n}\right\} = (j\omega)^n F(\omega). \quad \textbf{(18.49)}$$

Equations (18.48) and (18.49) are valid if $f(t)$ is zero at $\pm\infty$.

4. **Integration** If $g(t) = \int_{-\infty}^{t} f(x)\,dx$,

$$\mathscr{F}\{g(t)\} = \frac{F(\omega)}{j\omega}. \quad \textbf{(18.50)}$$

Equation (18.50) is valid if

$$\int_{-\infty}^{\infty} f(x)\,dx = 0.$$

5. **Scale Change** Dimensionally, time and frequency are reciprocals. Therefore when time is stretched out, frequency is compressed (and vice versa), as reflected in the functional transform

$$\mathscr{F}\{f(at)\} = \frac{1}{a}F\left(\frac{\omega}{a}\right), \quad a > 0. \quad \textbf{(18.51)}$$

Note that when $0 < a < 1.0$, time is stretched out, whereas when $a > 1.0$, time is compressed.

6. **Translation in the Time Domain** The effect of translating a function in the time domain is to alter the phase spectrum and leave the amplitude spectrum untouched. Thus

$$\mathscr{F}\{f(t-a)\} = e^{-j\omega a}F(\omega). \quad \textbf{(18.52)}$$

If a is positive in Eq. (18.52), the time function is delayed, and if a is negative, the time function is advanced.

7. **Translation in the Frequency Domain** Translation in the frequency domain corresponds to multiplication by the complex exponential in the time domain:

$$\mathscr{F}\{e^{j\omega_0 t}f(t)\} = F(\omega - \omega_0). \quad \textbf{(18.53)}$$

8. **Modulation** Amplitude modulation is the process of varying the amplitude of a sinusoidal carrier. If the modulating signal is denoted $f(t)$, the modulated carrier becomes $f(t)\cos\omega_0 t$. The amplitude spectrum of this carrier is one-half the amplitude spectrum of $f(t)$ centered at $\pm\omega_0$, that is,

$$\mathscr{F}\{f(t)\cos\omega_0 t\} = \frac{1}{2}F(\omega - \omega_0) + \frac{1}{2}F(\omega + \omega_0). \quad \textbf{(18.54)}$$

9. **Convolution in the Time Domain** Convolution in the time domain corresponds to multiplication in the frequency domain.

In other words,

$$y(t) = \int_{-\infty}^{\infty} x(\lambda)h(t-\lambda)\,d\lambda$$

becomes

$$\mathcal{F}\{y(t)\} = Y(\omega) = X(\omega)H(\omega). \tag{18.55}$$

Equation (18.55) is important in applications of the Fourier transform, because it states that the transform of the response function $Y(\omega)$ is the product of the input transform $X(\omega)$ and the system function $H(\omega)$. We say more about this relationship in Section 18.7.

10. **Convolution in the Frequency Domain** Convolution in the frequency domain corresponds to finding the Fourier transform of the product of two time functions. Thus if

$$f(t) = f_1(t)f_2(t),$$

then

$$F(\omega) = \frac{1}{2\pi}\int_{-\infty}^{\infty} F_1(u)F_2(\omega-u)\,du. \tag{18.56}$$

Table 18.2 summarizes these 10 operational transforms and another operational transform that we introduce in Problem 18.16.

TABLE 18.2

OPERATIONAL TRANSFORMS

$f(t)$	$F(\omega)$
$Kf(t)$	$KF(\omega)$
$f_1(t) - f_2(t) + f_3(t)$	$F_1(\omega) - F_2(\omega) + F_3(\omega)$
$d^n f(t)/dt^n$	$(j\omega)^n F(\omega)$
$\int_{-\infty}^{t} f(x)\,dx$	$F(\omega)/j\omega$
$f(at)$	$\frac{1}{a}F\left(\frac{\omega}{a}\right),\ a > 0$
$f(t-a)$	$e^{-j\omega a}F(\omega)$
$e^{j\omega_0 t}f(t)$	$F(\omega-\omega_0)$
$f(t)\cos\omega_0 t$	$\frac{1}{2}F(\omega-\omega_0) + \frac{1}{2}F(\omega+\omega_0)$
$\int_{-\infty}^{\infty} x(\lambda)h(t-\lambda)d\lambda$	$X(\omega)H(\omega)$
$f_1(t)f_2(t)$	$\frac{1}{2\pi}\int_{-\infty}^{\infty} F_1(u)F_2(\omega-u)\,du$
$t^n f(t)$	$(j)^n \frac{d^n F(\omega)}{d\omega^n}$

DRILL EXERCISES

18.7 a) Find the second derivative of the function described in Problem 18.1(b).

b) Find the Fourier transform of the second derivative.

c) Use the result obtained in (b) to find the Fourier transform of the function in (a). (*Hint*: Use the operational transform of differentiation.)

ANSWER: (a) $\frac{d^2 f}{dt^2} = \frac{2A}{\tau}\delta\left(t + \frac{\tau}{2}\right) - \frac{4A}{\tau}\delta(t) + \frac{2A}{\tau}\delta\left(t - \frac{\tau}{2}\right)$; (b) $\frac{4A}{\tau}\left(\cos\frac{\omega\tau}{2} - 1\right)$; (c) $\frac{4A}{\omega^2\tau}\left(1 - \cos\frac{\omega\tau}{2}\right)$.

18.8 The rectangular pulse shown can be expressed as the difference between two step voltages; that is,

$$v(t) = V_m u\left(t + \frac{\tau}{2}\right) - V_m u\left(t - \frac{\tau}{2}\right) \text{ V.}$$

Use the operational transform for translation in the time domain to find the Fourier transform of $v(t)$.

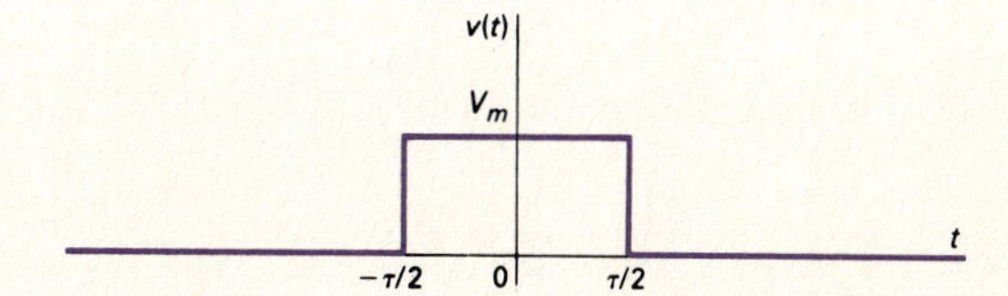

ANSWER: $V(\omega) = V_m\tau\frac{\sin(\omega\tau/2)}{(\omega\tau/2)}$.

18.7 CIRCUIT APPLICATIONS

The Laplace transform is used more widely to find the response of a circuit than is the Fourier transform, for two reasons. First, the Laplace transform integral converges for a wider range of driving functions, and second, it accommodates initial conditions. Despite the advantages of the Laplace transform, we can use the Fourier transform to find the response. The fundamental relationship underlying the use of the Fourier transform in transient analysis is Eq. (18.55), which relates the transform of the response $Y(\omega)$ to the transform of the input $X(\omega)$ and the transfer function $H(\omega)$ of the circuit. Note that $H(\omega)$ is the familiar $H(s)$ with s replaced by $j\omega$.

Example 18.1 illustrates how to use the Fourier transform to find the response of a circuit.

EXAMPLE 18.1

Use the Fourier transform to find $i_o(t)$ in the circuit shown in Fig. 18.9. The current source $i_g(t)$ is the signum function 20 sgn(t) A.

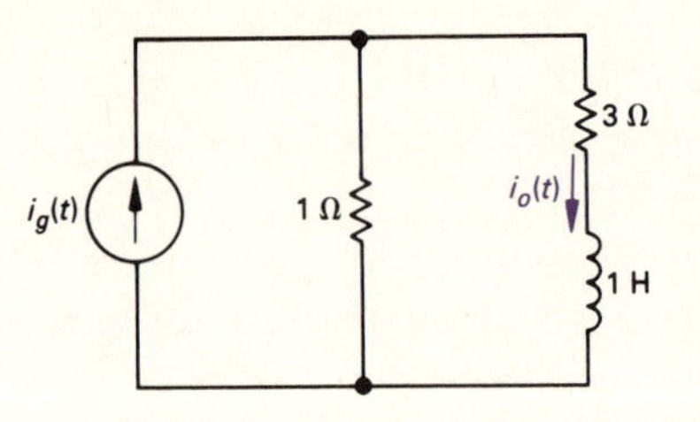

FIGURE 18.9 The circuit for Example 18.1.

SOLUTION

The Fourier transform of the driving source is

$$I_g(\omega) = \mathscr{F}\{20\ \text{sgn}(t)\} = 20\left(\frac{2}{j\omega}\right) = \frac{40}{j\omega}.$$

The transfer function of the circuit is the ratio of $\mathbf{I}_o$ to $\mathbf{I}_g$; so

$$H(\omega) = \frac{\mathbf{I}_o}{\mathbf{I}_g} = \frac{1}{4 + j\omega}.$$

The Fourier transform of $i_o(t)$ is

$$I_o(\omega) = I_g(\omega)H(\omega) = \frac{40}{j\omega(4 + j\omega)}.$$

Expanding $I_o(\omega)$ into a sum of partial fractions yields

$$I_o(\omega) = \frac{K_1}{j\omega} + \frac{K_2}{4 + j\omega}.$$

Evaluating K_1 and K_2 gives

$$K_1 = \frac{40}{4} = 10 \quad \text{and} \quad K_2 = \frac{40}{-4} = -10.$$

Therefore

$$I_o(\omega) = \frac{10}{j\omega} - \frac{10}{4 + j\omega}.$$

The response is

$$i_o(t) = \mathscr{F}^{-1}[I_o(\omega)] = 5\ \text{sgn}(t) - 10e^{-4t}u(t).$$

Figure 18.10 shows the response. Does the solution make sense in terms of known circuit behavior? The answer is yes, for the following reasons. The current source delivers −20 A to the circuit between $-\infty$ and 0. The resistance in each branch governs how the −20 A divides between the two branches. In particular, one-fourth of the −20 A appears in the i_o branch; therefore i_o is −5 for $t < 0$. When the current source jumps from −20 A to +20 A at $t = 0$, i_o approaches its final value of +5 A exponentially with a time constant of $\frac{1}{4}$ s.

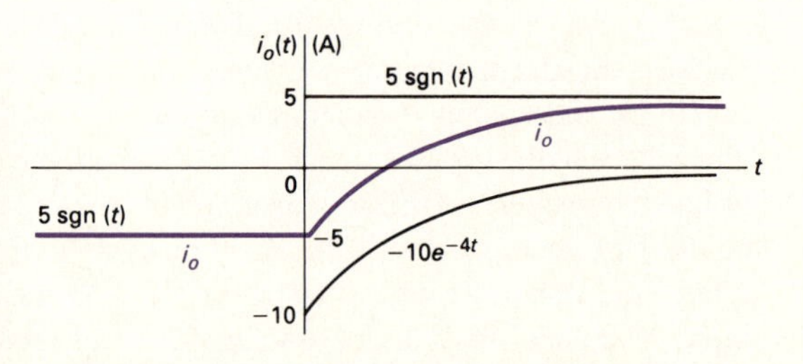

FIGURE 18.10 The plot of $i_o(t)$ versus t.

An important characteristic of the Fourier transform is that it directly yields the steady-state response to a sinusoidal driving function. The reason is that the Fourier transform of $\cos \omega_0 t$ is based on the assumption that the function exists over all time. Example 18.2 illustrates this feature.

EXAMPLE 18.2

The current source in the circuit in Example 18.1 (Fig. 18.9) is changed to a sinusoidal source. The expression for the current is

$$i_g(t) = 50 \cos 3t \text{ A}.$$

Use the Fourier transform method to find $i_o(t)$.

SOLUTION

The transform of the driving function is

$$I_g(\omega) = 50\pi[\delta(\omega - 3) + \delta(\omega + 3)].$$

As before, the transfer function of the circuit is

$$H(\omega) = \frac{1}{4 + j\omega}.$$

The transform of the current response then is

$$I_o(\omega) = 50\pi \frac{\delta(\omega - 3) + \delta(\omega + 3)}{4 + j\omega}.$$

Because of the sifting property of the impulse function, the easiest way to find the inverse transform of $I_o(\omega)$ is by the inversion integral:

$$\begin{aligned} i_o(t) = \mathscr{F}^{-1}\{I_o(\omega)\} &= \frac{50\pi}{2\pi} \int_{-\infty}^{\infty} \left[\frac{\delta(\omega - 3) + \delta(\omega + 3)}{4 + j\omega}\right] e^{j\omega t}\, d\omega \\ &= 25\left(\frac{e^{j3t}}{4 + j3} + \frac{e^{-j3t}}{4 - j3}\right) \\ &= 25\left(\frac{e^{j3t}e^{-j36.87^\circ}}{5} + \frac{e^{-j3t}e^{j36.87^\circ}}{5}\right) \\ &= 5[2\cos(3t - 36.87^\circ)] \\ &= 10\cos(3t - 36.87^\circ). \end{aligned}$$

We leave you to verify that the solution for $i_o(t)$ is identical to that obtained by phasor analysis.

Use a computer tool to assist the verification.

DRILL EXERCISES

18.9 The current source in the circuit shown delivers a current of 10 sgn(t) A. The response is the voltage across the 1 H inductor. Compute (a) $I_g(\omega)$; (b) $H(j\omega)$; (c) $V_o(\omega)$; (d) $v_o(t)$; (e) $i_1(0^-)$; (f) $i_1(0^+)$; (g) $i_2(0^-)$; (h) $i_2(0^+)$; (i) $v_o(0^-)$; and (j) $v_o(0^+)$.

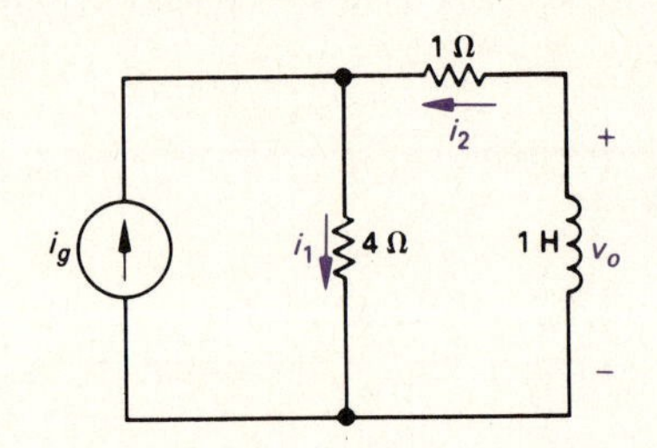

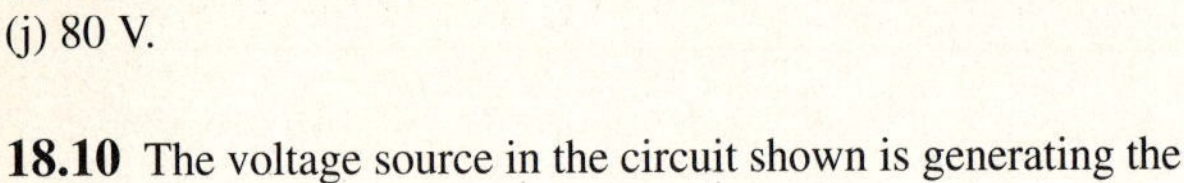

ANSWER: (a) $20/j\omega$; (b) $4j\omega/(5+j\omega)$; (c) $80/(5+j\omega)$; (d) $80e^{-5t}u(t)$ V; (e) −2 A; (f) 18 A; (g) 8 A; (h) 8 A; (i) 0 V; (j) 80 V.

18.10 The voltage source in the circuit shown is generating the voltage

$$v_g = e^t u(-t) + u(t) \text{ V}.$$

a) Use the Fourier transform method to find v_a.

b) Compute $v_a(0^-)$, $v_a(0^+)$, and $v_a(\infty)$.

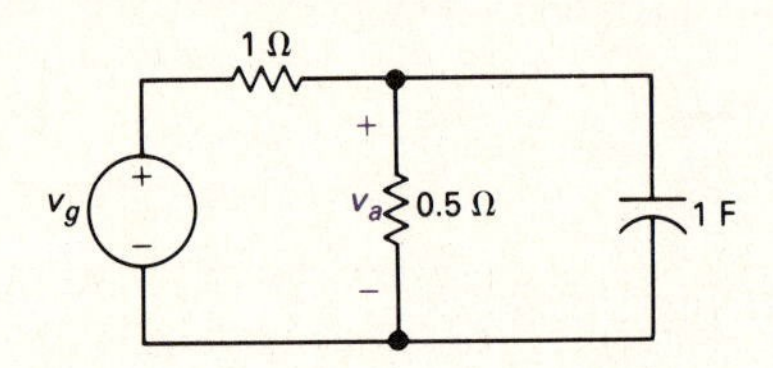

ANSWER: (a) $v_a = \frac{1}{4}e^t u(-t) - \frac{1}{12}e^{-3t}u(t) + \frac{1}{6} + \frac{1}{6}\text{sgn}(t)$ V; (b) $\frac{1}{4}$ V, $\frac{1}{4}$ V, $\frac{1}{3}$ V.

18.8 PARSEVAL'S THEOREM

Parseval's theorem relates the energy associated with a time-domain function of finite energy to the Fourier transform of the function. Imagine that the time-domain function $f(t)$ is either the voltage across or the current in a 1 Ω resistor. The energy associated with this function then is

$$W_{1\Omega} = \int_{-\infty}^{\infty} f^2(t)\,dt. \qquad \textbf{(18.57)}$$

Parseval's theorem holds that this same energy can be calculated by an integration in the frequency domain, or specifically,

$$\int_{-\infty}^{\infty} f^2(t)\,dt = \frac{1}{2\pi}\int_{-\infty}^{\infty} |F(\omega)|^2\,d\omega. \qquad \textbf{(18.58)}$$

Therefore the 1 Ω energy associated with $f(t)$ can be calculated either by integrating the square of $f(t)$ over all time or by integrating $1/2\pi$

times the square of the Fourier transform of $f(t)$ over all frequency. Parseval's theorem is valid if both integrals exist.

The average power associated with time-domain signals of finite energy is zero when averaged over all time. Therefore, when comparing signals of this type, we resort to the energy content of the signals. Using a 1 Ω resistor as the base for the energy calculation is convenient for comparing the energy content of voltage and current signals.

We begin the derivation of Eq. (18.58) by rewriting the kernel of the integral on the left-hand side as $f(t)$ times itself and then expressing one $f(t)$ in terms of the inversion integral:

$$\int_{-\infty}^{\infty} f^2(t)\,dt = \int_{-\infty}^{\infty} f(t)f(t)\,dt$$

$$= \int_{-\infty}^{\infty} f(t)\left[\frac{1}{2\pi}\int_{-\infty}^{\infty} F(\omega)e^{j\omega t}\,d\omega\right]dt. \quad \textbf{(18.59)}$$

We move $f(t)$ inside the interior integral, because the integration is with respect to ω, and then factor the constant $1/2\pi$ outside both integrations. Thus Eq. (18.59) becomes

$$\int_{-\infty}^{\infty} f^2(t)\,dt = \frac{1}{2\pi}\int_{-\infty}^{\infty}\left[\int_{-\infty}^{\infty} F(\omega)f(t)e^{j\omega t}\,d\omega\right]dt. \quad \textbf{(18.60)}$$

We reverse the order of integration and in so doing recognize that $F(\omega)$ can be factored out of the integration with respect to t. Thus

$$\int_{-\infty}^{\infty} f^2(t)\,dt = \frac{1}{2\pi}\int_{-\infty}^{\infty} F(\omega)\left[\int_{-\infty}^{\infty} f(t)e^{j\omega t}\,dt\right]d\omega. \quad \textbf{(18.61)}$$

The interior integral is $F(-\omega)$, so Eq. (18.61) reduces to

$$\int_{-\infty}^{\infty} f^2(t)\,dt = \frac{1}{2\pi}\int_{-\infty}^{\infty} F(\omega)F(-\omega)\,d\omega. \quad \textbf{(18.62)}$$

In Section 18.6, we noted that $F(-\omega) = F^*(\omega)$. Thus the product $F(\omega)F(-\omega)$ is simply the magnitude of $F(\omega)$ squared, and Eq. (18.62) is equivalent to Eq. (18.58). We also noted that $|F(\omega)|$ is an even function of ω. Therefore we can also write Eq. (18.58) as

$$\int_{-\infty}^{\infty} f^2(t)\,dt = \frac{1}{\pi}\int_{0}^{\infty} |F(\omega)|^2\,d\omega. \quad \textbf{(18.63)}$$

An Example Illustrating Parseval's Theorem

We can best demonstrate the validity of Eq. (18.63) with a specific example. If

$$f(t) = e^{-a|t|},$$

the left-hand side of Eq. (18.63) becomes

$$\begin{aligned}\int_{-\infty}^{\infty} e^{-2a|t|}\,dt &= \int_{-\infty}^{0} e^{2at}\,dt + \int_{0}^{\infty} e^{-2at}\,dt\\ &= \frac{e^{2at}}{2a}\bigg|_{-\infty}^{0} + \frac{e^{-2at}}{-2a}\bigg|_{0}^{\infty}\\ &= \frac{1}{2a} + \frac{1}{2a} = \frac{1}{a}.\end{aligned} \qquad (18.64)$$

The Fourier transform of $f(t)$ is

$$F(\omega) = \frac{2a}{a^2 + \omega^2},$$

and therefore the right-hand side of Eq. (18.63) becomes

$$\begin{aligned}\frac{1}{\pi}\int_{0}^{\infty} \frac{4a^2}{(a^2+\omega^2)^2}d\omega &= \frac{4a^2}{\pi}\frac{1}{2a^2}\left(\frac{\omega}{\omega^2+a^2} + \frac{1}{a}\tan^{-1}\frac{\omega}{a}\right)\bigg|_{0}^{\infty}\\ &= \frac{2}{\pi}\left(0 + \frac{\pi}{2a} - 0 - 0\right)\\ &= \frac{1}{a}.\end{aligned} \qquad (18.65)$$

Note that the result given by Eq. (18.65) is the same as that given by Eq. (18.64).

THE INTERPRETATION OF PARSEVAL'S THEOREM

Parseval's theorem gives a physical interpretation that the magnitude of the Fourier transform squared, $|F(\omega)|^2$, is an energy density (in joules per hertz). To see it, we write the right-hand side of Eq. (18.63) as

$$\frac{1}{\pi}\int_{0}^{\infty} |F(2\pi f)|^2 2\pi\ df = 2\int_{0}^{\infty} |F(2\pi f)|^2 df, \qquad (18.66)$$

where $|F(2\pi f)|^2\ df$ is the energy in an infinitesimal band of frequencies (df), and the total 1 Ω energy associated with $f(t)$ is the summation (integration) of $|F(2\pi f)|^2 df$ over all frequency. We can associate a portion of the total energy with a specified band of frequencies. In other words, the 1 Ω energy in the frequency band from ω_1 to ω_2 is

$$W_{1\Omega} = \frac{1}{\pi}\int_{\omega_1}^{\omega_2} |F(\omega)|^2\,d\omega. \qquad (18.67)$$

Note that expressing the integration in the frequency domain as

$$\frac{1}{2\pi}\int_{-\infty}^{\infty} |F(\omega)|^2\,d\omega$$

instead of

$$\frac{1}{\pi}\int_0^{\infty} |F(\omega)|^2 \, d\omega$$

allows Eq. (18.67) to be written in the form

$$W_{1\Omega} = \frac{1}{2\pi}\int_{-\omega_2}^{-\omega_1} |F(\omega)|^2 \, d\omega + \frac{1}{2\pi}\int_{\omega_1}^{\omega_2} |F(\omega)|^2 d\omega. \quad \textbf{(18.68)}$$

Figure 18.11 shows the graphic interpretation of Eq. (18.68).

Examples 18.3, 18.4, and 18.5 illustrate calculations involving Parseval's theorem.

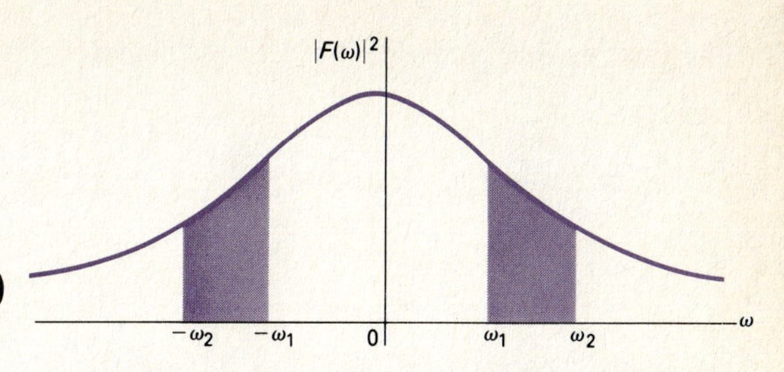

FIGURE 18.11 The graphic interpretation of Eq. (18.68).

EXAMPLE 18.3

The current in a 40 Ω resistor is

$$i = 20e^{-2t}u(t) \text{ A}.$$

What percentage of the total energy dissipated in the resistor can be associated with the frequency band $0 \leq \omega \leq 2\sqrt{3}$ rad/s?

SOLUTION

The total energy dissipated in the 40 Ω resistor is

$$W_{40\Omega} = 40\int_0^{\infty} 400e^{-4t} \, dt$$

$$= 16{,}000 \left.\frac{e^{-4t}}{-4}\right|_0^{\infty}$$

$$= 4000 \text{ J}.$$

We can check this total energy calculation with Parseval's theorem:

$$F(\omega) = \frac{20}{2 + j\omega}.$$

Therefore

$$|F(\omega)| = \frac{20}{\sqrt{4 + \omega^2}}$$

and

$$W_{40\Omega} = \frac{40}{\pi}\int_0^{\infty} \frac{400}{4 + \omega^2} d\omega = \frac{16{,}000}{\pi}\left(\frac{1}{2}\tan^{-1}\left.\frac{\omega}{2}\right|_0^{\infty}\right)$$

$$= \frac{8000}{\pi}\left(\frac{\pi}{2}\right) = 4000 \text{ J}.$$

The energy associated with the frequency band $0 \leq \omega \leq 2\sqrt{3}$ rad/s is

$$W_{40\Omega} = \frac{40}{\pi}\int_0^{2\sqrt{3}} \frac{400}{4+\omega^2}d\omega = \frac{16{,}000}{\pi}\left(\frac{1}{2}\tan^{-1}\frac{\omega}{2}\bigg|_0^{2\sqrt{3}}\right)$$

$$= \frac{8000}{\pi}\left(\frac{\pi}{3}\right) = \frac{8000}{3}\text{ J.}$$

Hence the percentage of the total energy associated with this range of frequencies is

$$\eta = \frac{8000/3}{4000} \times 100 = 66.67\%.$$

Use a computer tool to calculate the energy associated with a different band of frequencies.

EXAMPLE 18.4

The input voltage to an ideal bandpass filter is

$$v(t) = 120e^{-24t}u(t)\text{ V.}$$

The filter passes all frequencies that lie between 24 and 48 rad/s, without attenuation, and completely rejects all frequencies outside this passband.

a) Sketch $|V(\omega)|^2$ for the filter input voltage.

b) Sketch $|V_o(\omega)|^2$ for the filter output voltage.

c) What percentage of the total 1 Ω energy content of the signal at the input of the filter is available at the output?

SOLUTION

a) The Fourier transform of the filter input voltage is

$$V(\omega) = \frac{120}{24 + j\omega}.$$

Therefore

$$|V(\omega)|^2 = \frac{14{,}400}{576 + \omega^2}.$$

Fig. 18.12 shows the sketch of $|V(\omega)|^2$ versus ω.

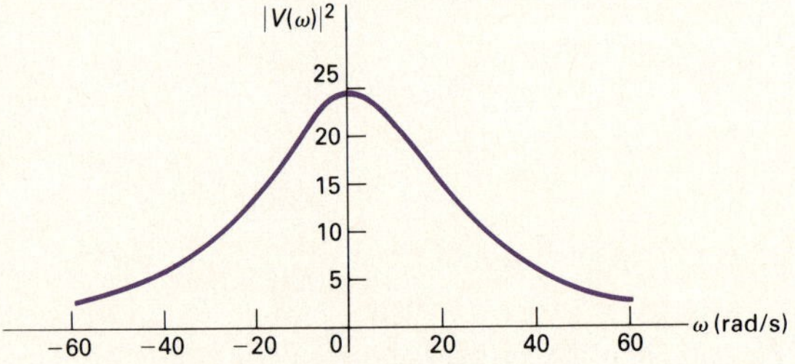

FIGURE 18.12 $|V(\omega)|^2$ versus ω for Example 18.4.

b) The ideal bandpass filter rejects all frequencies outside the passband, so the plot of $|V_0(\omega)|^2$ versus ω appears as shown in Fig. 18.13.

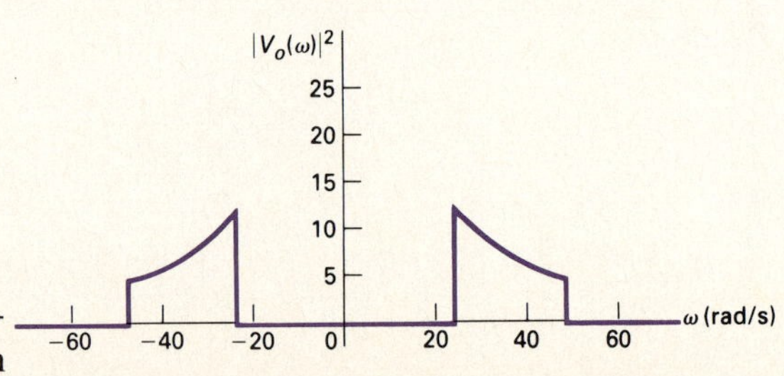

FIGURE 18.13 $|V_o(\omega)|^2$ versus ω for Example 18.4.

c) The total 1 Ω energy available at the input to the filter is

$$W_i = \frac{1}{\pi}\int_0^{\infty} \frac{14{,}400}{576+\omega^2}d\omega = \frac{14{,}400}{\pi}\left(\frac{1}{24}\tan^{-1}\frac{\omega}{24}\Big|_0^{\infty}\right)$$

$$= \frac{600}{\pi}\frac{\pi}{2} = 300 \text{ J.}$$

The total 1 Ω energy available at the output of the filter is

$$W_o = \frac{1}{\pi}\int_{24}^{48} \frac{14{,}400}{576+\omega^2}d\omega = \frac{600}{\pi}\tan^{-1}\frac{\omega}{24}\Big|_{24}^{48}$$

$$= \frac{600}{\pi}(\tan^{-1}2 - \tan^{-1}1) = \frac{600}{\pi}\left(\frac{\pi}{2.84} - \frac{\pi}{4}\right)$$

$$= 61.45 \text{ J.}$$

The percentage of the input energy available at the output is

$$\eta = \frac{61.45}{300} \times 100 = 20.48\%.$$

Parseval's theorem makes it possible to calculate the energy available at the output of the filter even if we don't know the time-domain expression for $v_o(t)$.

EXAMPLE 18.5

The input voltage to the low-pass RC filter circuit shown in Fig. 18.14 is

$$v_i(t) = 15e^{-5t}u(t) \text{ V.}$$

a) What percentage of the 1 Ω energy available in the input signal is available in the output signal?

b) What percentage of the output energy is associated with the frequency range $0 \le \omega \le 10$ rad/s?

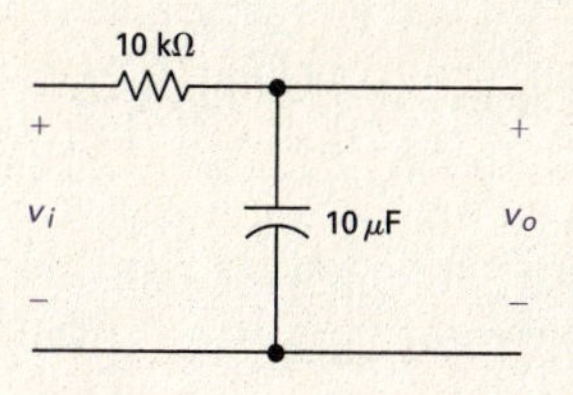

FIGURE 18.14 The low-pass RC filter for Example 18.5.

SOLUTION

a) The 1 Ω energy in the input signal to the filter is

$$W_i = \int_0^{\infty} (15e^{-5t})^2\,dt = 225\frac{e^{-10t}}{-10}\Big|_0^{\infty} = 22.5 \text{ J.}$$

The Fourier transform of the output voltage is

$$V_o(\omega) = V_i(\omega)H(\omega),$$

where

$$V_i(\omega) = \frac{15}{5 + j\omega}$$

and

$$H(\omega) = \frac{1/RC}{1/RC + j\omega} = \frac{10}{10 + j\omega}.$$

Hence

$$V_o(\omega) = \frac{150}{(5 + j\omega)(10 + j\omega)}$$

and

$$|V_o(\omega)|^2 = \frac{22{,}500}{(25 + \omega^2)(100 + \omega^2)}.$$

The 1 Ω energy available in the output signal of the filter is

$$W_o = \frac{1}{\pi}\int_0^\infty \frac{22{,}500}{(25 + \omega^2)(100 + \omega^2)}d\omega.$$

We can easily evaluate the integral by expanding the kernel into a sum of partial fractions:

$$\frac{22{,}500}{(25 + \omega^2)(100 + \omega^2)} = \frac{300}{25 + \omega^2} - \frac{300}{100 + \omega^2}.$$

Then

$$W_o = \frac{300}{\pi}\left\{\int_0^\infty \frac{d\omega}{25 + \omega^2} - \int_0^\infty \frac{d\omega}{100 + \omega^2}\right\}$$

$$= \frac{300}{\pi}\left[\frac{1}{5}\left(\frac{\pi}{2}\right) - \frac{1}{10}\left(\frac{\pi}{2}\right)\right] = 15 \text{ J}.$$

The energy available in the output signal therefore is 66.67% of the energy available in the input signal; that is,

$$\eta = \frac{15}{22.5}(100) = 66.67\%.$$

b) The output energy associated with the frequency range $0 \leq \omega \leq 10$ rad/s is

$$W'_o = \frac{300}{\pi}\left\{\int_0^{10} \frac{d\omega}{25 + \omega^2} - \int_0^{10} \frac{d\omega}{100 + \omega^2}\right\}$$

$$= \frac{300}{\pi}\left(\frac{1}{5}\tan^{-1}\frac{10}{5} - \frac{1}{10}\tan^{-1}\frac{10}{10}\right) = \frac{30}{\pi}\left(\frac{2\pi}{2.84} - \frac{\pi}{4}\right)$$

$$= 13.64 \text{ J}.$$

The total 1 Ω energy in the output signal is 15 J, so the percentage associated with the frequency range 0 to 10 rad/s is 90.97%.

THE ENERGY CONTAINED IN A RECTANGULAR VOLTAGE PULSE

We conclude our discussion of Parseval's theorem by calculating the energy associated with a rectangular voltage pulse. In Section 18.1 we found the Fourier transform of the voltage pulse to be

$$V(\omega) = V_m\tau\frac{\sin\omega\tau/2}{\omega\tau/2}. \tag{18.69}$$

To aid our discussion, we have redrawn the voltage pulse and its Fourier transform in Figs. 18.15(a) and (b), respectively. These figures show that, as the width of the voltage pulse (τ) becomes smaller, the dominant portion of the amplitude spectrum (that is, the spectrum from $-2\pi/\tau$ to $2\pi/\tau$) spreads out over a wider range of frequencies. This result agrees with our earlier comments about the operational transform involving a scale change—in other words, when time is compressed, frequency is stretched out, and vice versa. To transmit a single rectangular pulse with reasonable fidelity, the bandwidth of the system must be at least wide enough to accommodate the dominant portion of the amplitude spectrum. Thus the cutoff frequency should be at least $2\pi/\tau$ rad/s, or $1/\tau$ Hz.

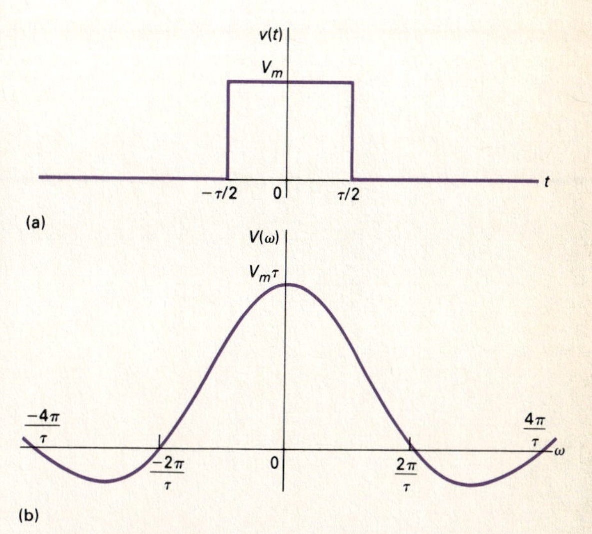

FIGURE 18.15 The rectangular voltage pulse and its Fourier transform: (a) the rectangular voltage pulse; (b) the Fourier transform of $v(t)$.

We can use Parseval's theorem to calculate the fraction of the total energy associated with $v(t)$ that lies in the frequency range $0 \leq \omega \leq 2\pi/\tau$. From Eq. (18.69),

$$W = \frac{1}{\pi}\int_0^{2\pi/\tau} V_m^2\tau^2\frac{\sin^2\omega\tau/2}{(\omega\tau/2)^2}d\omega. \tag{18.70}$$

To carry out the integration called for in Eq. (18.70), we let

$$x = \frac{\omega\tau}{2}, \tag{18.71}$$

noting that

$$dx = \frac{\tau}{2}d\omega \tag{18.72}$$

and that

$$x = \pi, \text{ when } \omega = 2\pi/\tau. \tag{18.73}$$

If we make the substitutions given by Eqs. (18.71) through (18.73), Eq. (18.70) becomes

$$W = \frac{2V_m^2\tau}{\pi}\int_0^{\pi}\frac{\sin^2 x}{x^2}dx. \tag{18.74}$$

We can integrate the integral in Eq. (18.74) by parts. If we let

$$u = \sin^2 x \tag{18.75}$$

and

$$dv = \frac{dx}{x^2}, \tag{18.76}$$

then

$$du = 2\sin x \cos x\, dx = \sin 2x\, dx \tag{18.77}$$

and

$$v = -\frac{1}{x}. \tag{18.78}$$

Hence

$$\int_0^{\pi} \frac{\sin^2 x}{x^2} dx = -\frac{\sin^2 x}{x}\bigg|_0^{\pi} - \int_0^{\pi} -\frac{1}{x}\sin 2x\, dx$$

$$= 0 + \int_0^{\pi} \frac{\sin 2x}{x}\, dx. \tag{18.79}$$

Substituting Eq. (18.79) into Eq. (18.74) yields

$$W = \frac{4V_m^2\tau}{\pi}\int_0^{\pi} \frac{\sin 2x}{2x} dx. \tag{18.80}$$

To evaluate the integral in Eq. (18.80), we must first put it in the form of $\sin y/y$. We do so by letting $y = 2x$ and noting that $dy = 2\,dx$ and $y = 2\pi$ when $x = \pi$. Thus Eq. (18.80) becomes

$$W = \frac{2V_m^2\tau}{\pi}\int_0^{2\pi} \frac{\sin y}{y} dy. \tag{18.81}$$

The value of the integral in Eq. (18.81) can be found in a table of sine integrals.[1] Its value is 1.41815, so

$$W = \frac{2V_m^2\tau}{\pi}(1.41815). \tag{18.82}$$

The total 1 Ω energy associated with $v(t)$ can be calculated either from the time-domain integration or the evaluation of Eq. (18.81) with the upper limit equal to infinity. In either case, the total energy is

$$W_t = V_m^2\tau. \tag{18.83}$$

The fraction of the total energy associated with the band of frequencies between 0 and $2\pi/\tau$ is

$$\eta = \frac{W}{W_t}$$

$$= \frac{2V_m^2\tau(1.41815)}{\pi(V_m^2\tau)}$$

$$= 0.9028. \tag{18.84}$$

[1]M. Abramowitz and I. Stegun, *Handbook of Mathematical Functions* (New York: Dover, 1965), p. 244.

Therefore approximately 90% of the energy associated with $v(t)$ is contained in the dominant portion of the amplitude spectrum.

DRILL EXERCISES

18.11 The voltage across a 50 Ω resistor is

$$v = 4te^{-t}u(t) \text{ V}.$$

What percentage of the total energy dissipated in the resistor can be associated with the frequency band $0 \leq \omega \leq \sqrt{3}$ rad/s?

ANSWER: 94.23%.

18.12 Assume that the magnitude of the Fourier transform of $v(t)$ is as shown. This voltage is applied to a 6 kΩ resistor. Calculate the total energy delivered to the resistor.

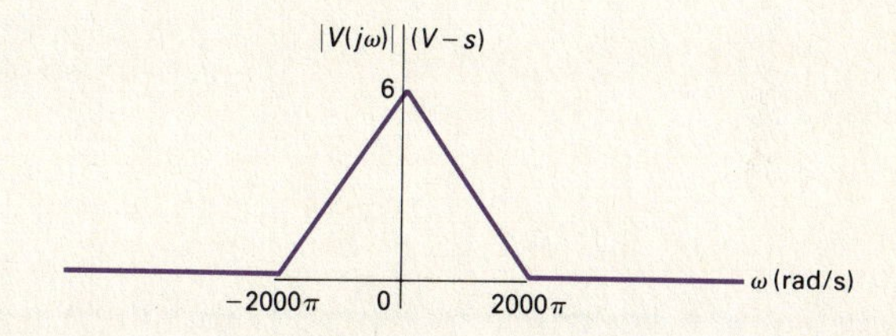

ANSWER: 4 J.

SUMMARY

- The **Fourier transform** gives a frequency-domain description of an aperiodic time-domain function. Depending on the nature of the time-domain signal, one of three approaches to finding its Fourier transform may be used:
  - —If the time-domain signal is a well-behaved pulse of finite duration, the integral that defines the Fourier transform is used.
  - —If the one-sided Laplace transform of $f(t)$ exists and all the poles of $F(s)$ lie in the left half of the s plane, $F(s)$ may be used to find $F(\omega)$.
  - —If $f(t)$ is a constant, a signum function, a step function, or a sinusoidal function, the Fourier transform is found by using a limit process.
- Functional and operational Fourier transforms that are useful in circuit analysis are tabulated in Tables 18.1 and 18.2.

- The Fourier transform of a response signal $y(t)$ is

$$Y(\omega) = X(\omega)H(\omega),$$

where $X(\omega)$ is the Fourier transform of the input signal $x(t)$, and $H(\omega)$ is the transfer function $H(s)$ evaluated at $s = j\omega$.

- The Fourier transform accommodates both negative-time and positive-time functions and therefore is suited to problems described in terms of events that start at $t = -\infty$. In contrast, the unilateral Laplace transform is suited to problems described in terms of initial conditions and events that occur for $t > 0$.

- The magnitude of the Fourier transform squared is a measure of the energy density (joules per hertz) in the frequency domain (Parseval's theorem). Thus the Fourier transform permits us to associate a fraction of the total energy contained in $f(t)$ with a specified band of frequencies.

PROBLEMS

18.1 Use the defining integral to find the Fourier transform of the following functions:

a) $f(t) = A \sin \frac{\pi}{2} t, \quad -2 \leq t \leq 2,$

$f(t) = 0 \quad$ elsewhere;

b) $f(t) = \frac{2A}{\tau} t + A, \quad -\frac{\tau}{2} \leq t \leq 0,$

$f(t) = -\frac{2A}{\tau} t + A, \quad 0 \leq t \leq \frac{\tau}{2},$

$f(t) = 0 \quad$ elsewhere.

18.2 a) Find the Fourier transform of the function shown in Fig. P18.2.

b) Show that the magnitude of $F(\omega)$ is zero when $\omega = 0$.

c) Sketch $|F(\omega)|$ versus ω when $A = 1$ and $\tau = 1$. *Hint:* Evaluate $|F(\omega)|$ at $\omega = 0, 2, 4, 6, 8, 9, 10, 12, 14$, and 15.5. Then use the fact that $|F(\omega)|$ is an even function of ω.

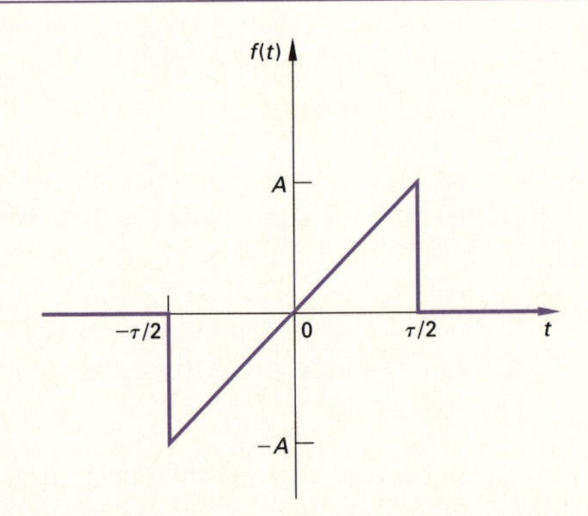

FIGURE P18.2

18.3 The Fourier transform of $f(t)$ is shown in Fig. P18.3.

a) Find $f(t)$.

b) Evaluate $f(0)$.

c) Sketch $f(t)$ for $-10 \leq t \leq 10$ s when $A = 20\pi$ and $\omega_o = 2$ rad/s. *Hint:* Evaluate $f(t)$ at $t = 0, 1, 2, 3, \ldots, 10$ s and then use the fact that $f(t)$ is even.

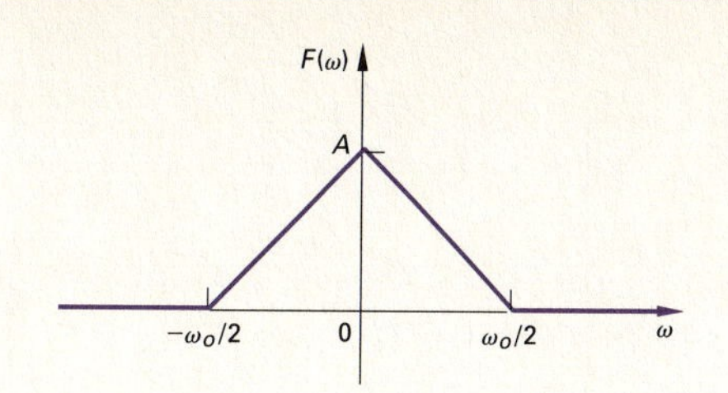

FIGURE P18.3

18.4 Find the Fourier transform of each of the following functions. In all of the functions, a is a positive real constant and $-\infty \leq t \leq \infty$.

a) $f(t) = |t|e^{-a|t|}$;

b) $f(t) = t^3e^{-a|t|}$;

c) $f(t) = e^{-a|t|}\cos\omega_o t$;

d) $f(t) = e^{-a|t|}\sin\omega_o t$;

e) $f(t) = \delta(t - t_o)$.

18.5 Use the inversion integral (Eq. 18.9) to show that $\mathscr{F}^{-1}\{2/j\omega\} = \text{sgn}(t)$. *Hint:* Use Drill Exercise 18.6.

18.6 Find $\mathscr{F}\{\cos\omega_o t\}$ by using the approximating function

$$f(t) = e^{-\epsilon|t|}\cos\omega_o t,$$

where ϵ is a positive real constant.

18.7 Show that if $f(t)$ is an even function,

$$A(\omega) = 2\int_0^\infty f(t)\cos\omega t\,dt$$

and

$$B(\omega) = 0.$$

18.8 Show that if $f(t)$ is an odd function,

$$A(\omega) = 0$$

and

$$B(\omega) = -2\int_0^\infty f(t)\sin\omega t\,dt.$$

18.9 a) Show that $\mathscr{F}\{df(t)/dt\} = j\omega F(\omega)$, where $F(\omega) = \mathscr{F}\{f(t)\}$. *Hint:* Use the defining integral and integrate by parts.

b) What is the restriction on $f(t)$ if the result given in part (a) is valid?

c) Show that $\mathscr{F}\{d^n f(t)/dt^n\} = (j\omega)^n F(\omega)$, where $F(\omega) = \mathscr{F}\{f(t)\}$.

18.10 a) Show that

$$\mathcal{F}\left\{\int_{-\infty}^{t} f(x)dx\right\} = \frac{F(\omega)}{j\omega},$$

where $F(\omega) = \mathcal{F}\{f(x)\}$. *Hint:* Use the defining integral and integrate by parts.

b) What is the restriction on $f(x)$ if the result given in part (a) is valid?

c) If $f(x) = e^{-ax}u(x)$, can the operational transform in part (a) be used? Explain.

18.11 a) Show that

$$\mathcal{F}\{f(at)\} = \frac{1}{a}F\left(\frac{\omega}{a}\right), \quad a > 0.$$

b) Given that $f(at) = e^{-a|t|}$ for $a > 0$, sketch $F(\omega) = \mathcal{F}\{f(at)\}$ for $a = 0.5$, 1.0, and 2.0. Do your sketches reflect the observation that *compression* in the time domain corresponds to *stretching* in the frequency domain?

18.12 Derive each of the following operational transforms:

a) $\mathcal{F}\{f(t-a)\} = e^{-j\omega a}F(\omega)$;

b) $\mathcal{F}\{e^{j\omega_o t}f(t)\} = F(\omega - \omega_o)$;

c) $\mathcal{F}\{f(t)\cos\omega_o t\} = \dfrac{1}{2}F(\omega-\omega_o)+\dfrac{1}{2}F(\omega+\omega_o)$.

18.13 Given $y(t) = \int_{-\infty}^{\infty} x(\lambda)h(t-\lambda)\,d\lambda$, show that $Y(\omega) = \mathcal{F}\{y(t)\} = X(\omega)H(\omega)$, where $X(\omega) = \mathcal{F}\{x(t)\}$ and $H(\omega) = \mathcal{F}\{h(t)\}$. *Hint:* Use the defining integral to write $\mathcal{F}\{y(t)\} = \int_{-\infty}^{\infty}\left[\int_{-\infty}^{\infty} x(\lambda)h(t-\lambda)\,d\lambda\right]e^{-j\omega t}\,dt$. Next, reverse the order of integration and then make a change in the variable of integration, that is, let $u = t - \lambda$.

18.14 Given $f(t) = f_1(t)f_2(t)$, show that $F(\omega) = (1/2\pi)\int_{-\infty}^{\infty} F_1(u)F_2(\omega-u)du$. *Hint:* First, use the defining integral to express $F(\omega)$ as

$$F(\omega) = \int_{-\infty}^{\infty} f_1(t)f_2(t)e^{-j\omega t}dt.$$

Second, use the inversion integral to write

$$f_1(t) = \frac{1}{2\pi}\int_{-\infty}^{\infty} F_1(u)e^{j\omega t}du.$$

Third, substitute the expression for $f_1(t)$ into the defining integral and then interchange the order of integration.

18.15 Suppose that $f(t) = f_1(t)f_2(t)$, where

$$f_1(t) = \cos\omega_o t,$$
$$f_2(t) = 1, \quad -\tau/2 < t < \tau/2,$$
$$f_2(t) = 0, \quad \text{elsewhere}.$$

a) Use convolution in the frequency domain to find $F(\omega)$.

b) What happens to $F(\omega)$ as the width of $f_2(t)$ increases so that $f(t)$ includes more and more cycles on $f_1(t)$?

18.16 a) Show that

$$(j)^n \left[\frac{d^n F(\omega)}{d\omega^n}\right] = \mathscr{F}\{t^n f(t)\}.$$

b) Use the result of part (a) to find each of the following Fourier transforms:

$$\mathscr{F}\{te^{-at}u(t)\};$$

$$\mathscr{F}\{|t|e^{-a|t|}\};$$

$$\mathscr{F}\{te^{-a|t|}\}.$$

18.17 a) Use the Fourier transform method to find $v_o(t)$ in the circuit shown in Fig. P18.17. The initial value of $v_o(t)$ is zero, and the source voltage is $100u(t)$ V.

b) Sketch $v_o(t)$ versus t.

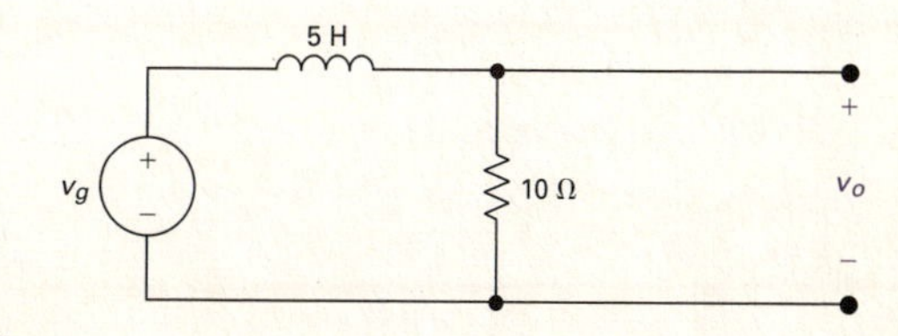

FIGURE P18.17

18.18 Repeat Problem 18.17 if the input voltage (v_g) is changed to $100\ \text{sgn}(t)$.

18.19 a) Use the Fourier transform method to find $i_o(t)$ in the circuit shown in Fig. P18.19 if $i_g = 100\ \text{sgn}(t)\ \mu$A.

b) Does your solution make sense in terms of known circuit behavior? Explain.

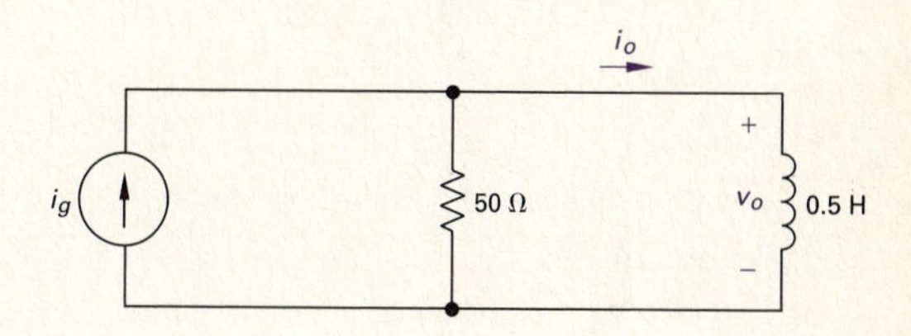

FIGURE P18.19

18.20 Repeat Problem 18.19 except replace $i_o(t)$ with $v_o(t)$.

18.21 a) Use the Fourier transform to find v_o in the circuit in Fig. P18.21 if $v_g = 20\ \text{sgn}(t)$.

b) Does your solution make sense in terms of known circuit behavior? Explain.

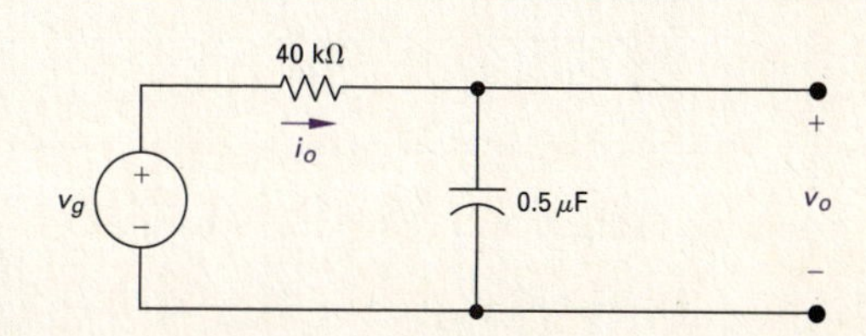

FIGURE P18.21

18.22 Repeat Problem 18.21 except replace v_o with i_o.

18.23 The voltage source in the circuit in Fig. P18.23 is given by the expression

$$v_g = 15\,\text{sgn}(t)\ \text{V}.$$

a) Find $v_o(t)$.

b) What is the value of $v_o(0^-)$?

c) What is the value of $v_o(0^+)$?

d) Use the Laplace transform method to find $v_o(t)$ for $t > 0^+$.

e) Does the solution obtained in part (d) agree with $v_o(t)$ for $t > 0^+$ from part (a)?

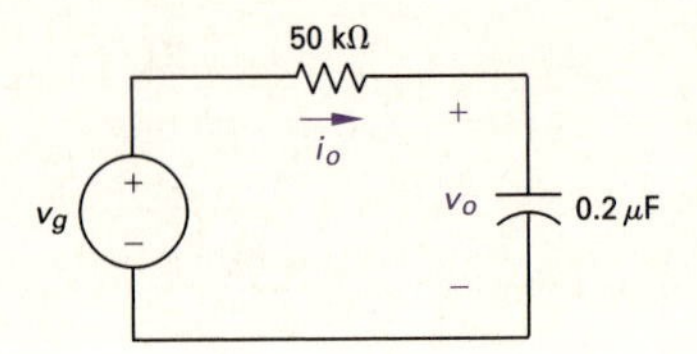

FIGURE P18.23

18.24 Repeat Problem 18.23 except replace $v_o(t)$ with $i_o(t)$.

18.25 a) Use the Fourier transform to find v_o in the circuit in Fig. P18.25 if i_g equals $3e^{-5|t|}$ A.

b) Find $v_o(0^-)$.

c) Find $v_o(0^+)$.

d) Use the Laplace transform method to find v_o for $t \geq 0$.

e) Does the solution obtained in part (d) agree with v_o for $t > 0^+$ from part (a)?

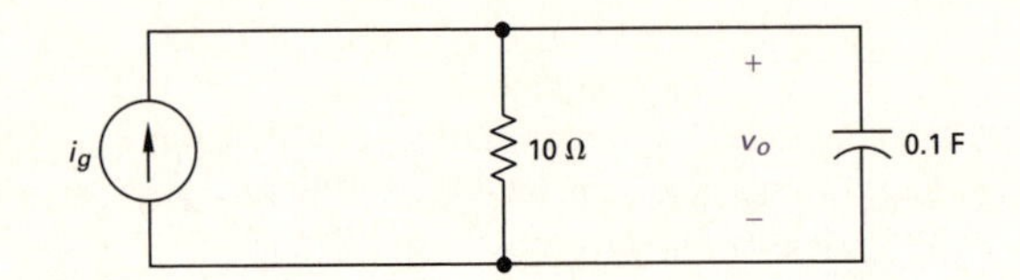

FIGURE P18.25

18.26 Use the Fourier transform method to find v_o in the circuit in Fig. P18.26 if $v_g = 125\cos 75t$ V.

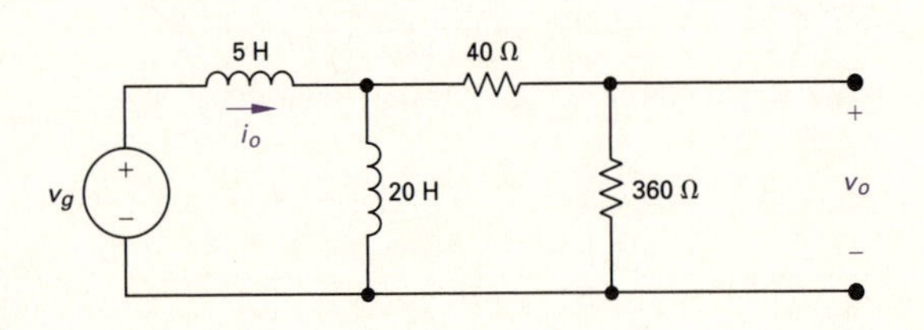

FIGURE P18.26

18.27 Assume v_g in the circuit in Fig. P18.26 changes to

$$v_g = -75e^{50t}u(-t) + 75e^{-50t}u(t)\ \text{V}.$$

Use the Fourier transform method to find i_o.

18.28 a) Use the Fourier transform method to find v_o in the circuit shown in Fig. P18.28. The current source generates the current

$$i_g = \text{sgn}(t) \text{ A}.$$

b) Calculate $v_o(0^-)$, $v_o(0^+)$, and $v_o(\infty)$.

c) Do the results in part (b) make sense in terms of known circuit behavior? Explain.

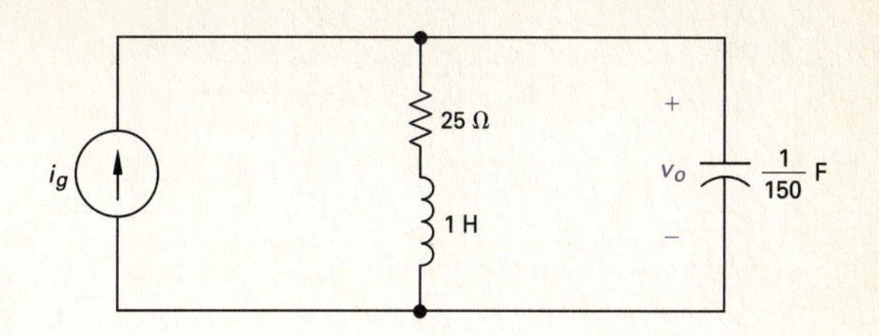

FIGURE P18.28

18.29 a) Use the Fourier transform method to find v_o in the circuit in Fig. P18.29 when

$$i_g = 18e^{10t}u(-t) - 18e^{-10t}u(t) \text{ A}.$$

b) Find $v_o(0^-)$.

c) Find $v_o(0^+)$.

d) Do the answers obtained in parts (b) and (c) make sense in terms of known circuit behavior? Explain.

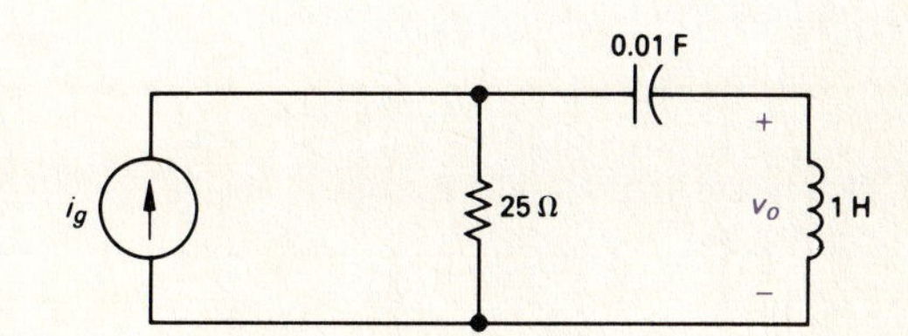

FIGURE P18.29

18.30 The voltage source in the circuit in Fig. P18.30 is generating the signal

$$v_g = 5 \text{ sgn}(t) - 5 + 30e^{-5t}u(t) \text{ V}.$$

a) Find $v_o(0^-)$ and $v_o(0^+)$.

b) Find $i_o(0^-)$ and $i_o(0^+)$.

c) Find v_o.

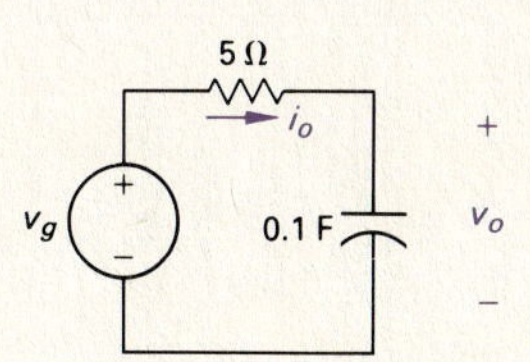

FIGURE P18.30

18.31 a) Use the Fourier transform method to find v_o in the circuit in Fig. P18.31 when

$$v_g = 60e^{5t}u(-t) + 900te^{-5t}u(t) \text{ V}.$$

b) Find $v_o(0^-)$.

c) Find $v_o(0^+)$.

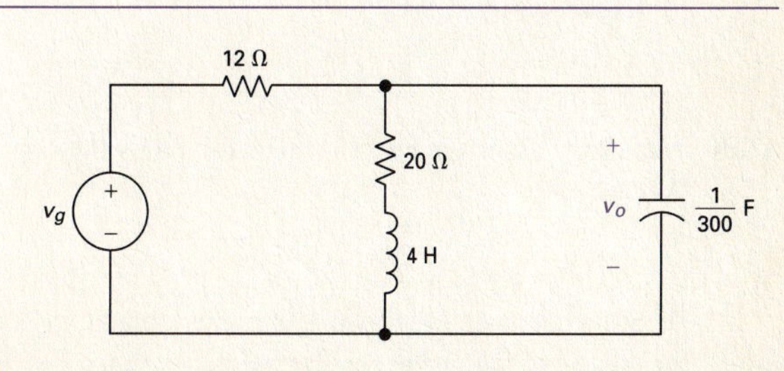

FIGURE P18.31

18.32 When the input voltage to the system shown in Fig. P18.32 is $15u(t)$ V, the output voltage is

$$v_o = \left[10 + 30e^{-20t} - 40e^{-30t}\right]u(t) \text{ V}.$$

What is the output voltage if $v_i = 15\,\text{sgn}(t)$ V?

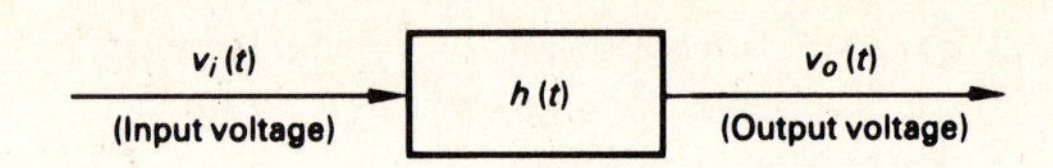

FIGURE P18.32

18.33 It is given that $F(\omega) = e^{\omega}u(-\omega) + e^{-\omega}u(\omega)$.

a) Find $f(t)$.

b) Find the 1 Ω energy associated with $f(t)$ via time-domain integration.

c) Repeat part (b) using frequency-domain integration.

d) Find the value of ω_1 if $f(t)$ has 90% of the energy in the frequency band $0 \le \omega \le \omega_1$.

18.34 The input current signal in the circuit seen in Fig. P18.34 is

$$i_g = 30e^{-2t}\,\mu\text{A},\, t \ge 0^+.$$

What percentage of the total 1 Ω energy content in the output signal lies in the frequency range 0 to 4 rad/s?

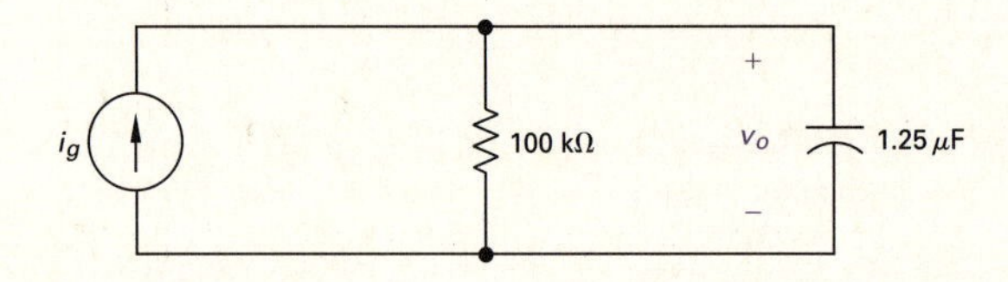

FIGURE P18.34

18.35 The input voltage in the circuit in Fig. P18.35 is $v_g = 30e^{-|t|}$ V.

a) Find $v_o(t)$.

b) Sketch $|V_g(\omega)|$ for $-5 \le \omega \le 5$ rad/s.

c) Sketch $|V_o(\omega)|$ for $-5 \le \omega \le 5$ rad/s.

d) Calculate the 1 Ω energy content of v_g.

e) Calculate the 1 Ω energy content of v_o.

f) What percentage of the 1 Ω energy content in v_g lies in the frequency range $0 \le \omega \le 2$ rad/s?

g) Repeat part (f) for v_o.

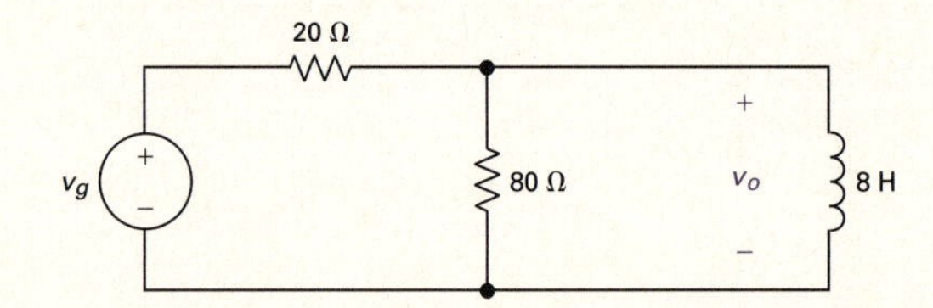

FIGURE P18.35

18.36 The circuit shown in Fig. P18.36 is driven by the voltage

$$v_1 = 60e^{-5t}u(t) \text{ V}.$$

What percentage of the total 1 Ω energy content in the output voltage v_2 lies in the frequency range $0 \le \omega \le 10$ rad/s?

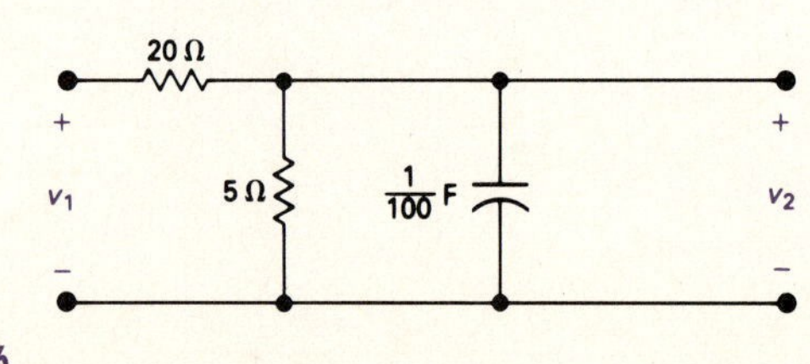

FIGURE P18.36

18.37 The amplitude spectrum of the input voltage to the high-pass RC filter in Fig. P18.37 is

$$V_i(\omega) = \frac{200}{|\omega|}, \quad 100 \leq |\omega| \leq 200 \text{ rad/s},$$

$$V_i(\omega) = 0, \qquad \text{elsewhere.}$$

a) Sketch $|V_i(\omega)|^2$ for $-300 \leq \omega \leq 300$ rad/s.

b) Sketch $|V_o(\omega)|^2$ for $-300 \leq \omega \leq 300$ rad/s.

c) Calculate the 1 Ω energy in the signal at the input of the filter.

d) Calculate the 1 Ω energy in the signal at the output of the filter.

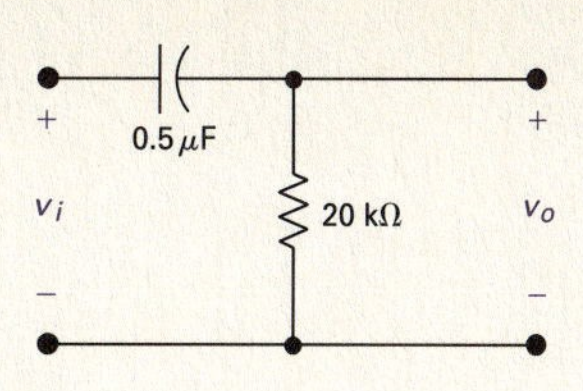

FIGURE P18.37

18.38 The input voltage to the high-pass RC filter circuit in Fig. P18.38 is

$$v_i(t) = Ae^{-at}u(t).$$

Let α denote the corner frequency of the filter, that is, $\alpha = 1/RC$.

a) What percentage of the energy in the signal at the output of the filter is associated with the frequency band $0 \leq \omega \leq \alpha$ if $\alpha = a$?

b) Repeat part (a), given that $\alpha = \sqrt{3}\,a$.

c) Repeat part (a), given that $\alpha = a/\sqrt{3}$.

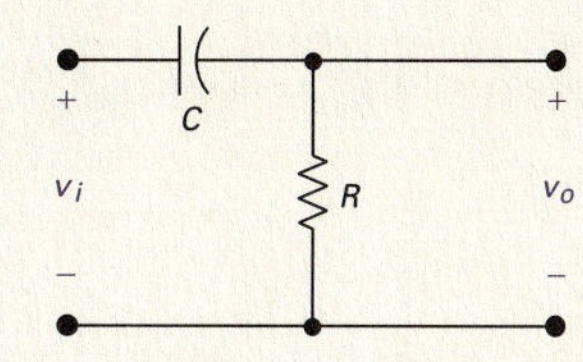

FIGURE P18.38

19 Two-Port Circuits

CHAPTER CONTENTS

So far, we have frequently focused on the behavior of a circuit at a specified pair of terminals. Recall that we introduced the Thévenin and Norton equivalent circuits solely to simplify circuit analysis relative to a pair of terminals. In analyzing some electrical systems, focusing on two pairs of terminals is also convenient. In particular, this is helpful when a signal is fed into one pair of terminals and then, after being processed by the system, is extracted at a second pair of terminals. Because the terminal pairs represent the points where signals are either fed in or extracted, they are referred to as the **ports** of the system. In this chapter, we limit the discussion to circuits that have one input and one output port.

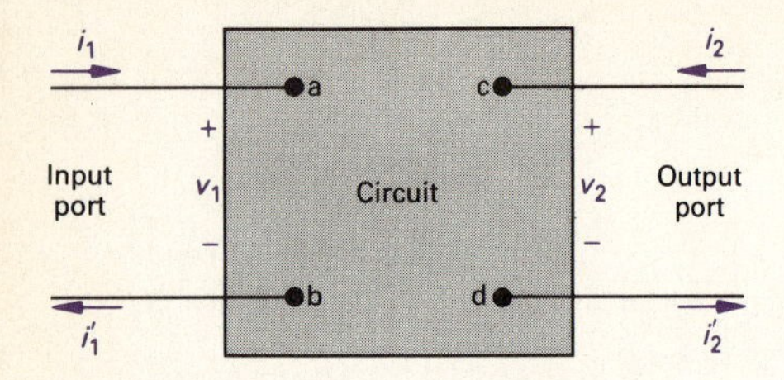

FIGURE 19.1 The two-port building block.

Figure 19.1 illustrates the basic two-port building block. Use of this building block is subject to several restrictions. First, there can be no energy stored within the circuit. Second, there can be no independent sources within the circuit; dependent sources, however, are permitted. Third, the current into the port must equal the current out of the port; that is, $i_1 = i_1'$ and $i_2 = i_2'$. Fourth, all external connections must be made to either the input port or the output port; no such connections are allowed between ports, that is, between terminals a and c, a and d, b and c, or b and d. These restrictions simply limit the range of circuit problems to which the two-port formulation is applicable.

The fundamental principle underlying two-port modeling of a system is that only the terminal variables (i_1, v_1, i_2, and v_2) are of interest. We have no interest in calculating the currents and voltages inside the circuit. We have already stressed terminal behavior in the analysis of operational amplifier circuits. In this chapter we formalize that approach by introducing the two-port parameters.

19.1 THE TERMINAL EQUATIONS

In viewing a circuit as a two-port network, we are interested in relating the current and voltage at one port to the current and voltage at the other port. Figure 19.1 shows the reference polarities of the terminal voltages and the reference directions of the terminal currents. The references at each port are symmetric with respect to each other; that is, at each port the current is directed into the upper terminal, and each port voltage is a rise from the lower to the upper terminal. This symmetry makes it easier to generalize the analysis of a two-port network and is the reason for its universal use in the literature.

The most general description of the two-port network is carried out in the s domain. For purely resistive networks, the analysis reduces to solving resistive circuits. Sinusoidal steady-state problems can be solved either by first finding the appropriate s-domain expressions and then replacing s with $j\omega$, or by direct analysis in the frequency domain. Here, we write all equations in the s domain; resistive networks and sinusoidal steady-state solutions become special cases. Figure 19.2 shows the basic building block in terms of the s-domain variables I_1, V_1, I_2, and V_2.

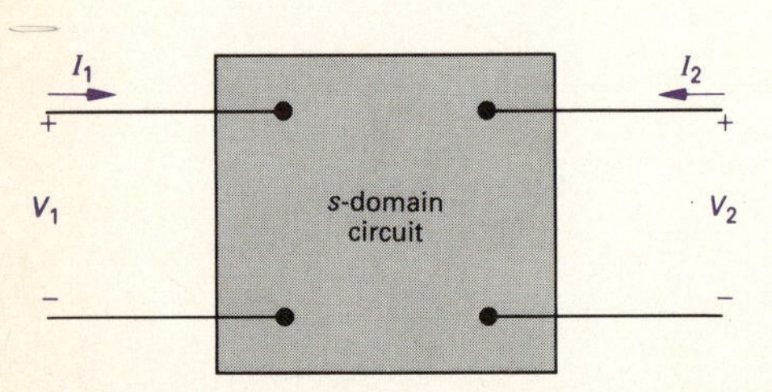

FIGURE 19.2 The s-domain two-port basic building block.

Of these four terminal variables, only two are independent. Thus for any circuit, once we specify two of the variables, we can find the two remaining unknowns. For example, knowing V_1 and V_2 and the circuit within the box, we can determine I_1 and I_2. Thus we can describe a two-port network with just two simultaneous equations.

However, there are six different ways in which to combine the four variables:

$$V_1 = z_{11}I_1 + z_{12}I_2,$$
$$V_2 = z_{21}I_1 + z_{22}I_2; \quad (19.1)$$

$$I_1 = y_{11}V_1 + y_{12}V_2,$$
$$I_2 = y_{21}V_1 + y_{22}V_2; \quad (19.2)$$

$$V_1 = a_{11}V_2 - a_{12}I_2,$$
$$I_1 = a_{21}V_2 - a_{22}I_2; \quad (19.3)$$

$$V_2 = b_{11}V_1 - b_{12}I_1,$$
$$I_2 = b_{21}V_1 - b_{22}I_1; \quad (19.4)$$

$$V_1 = h_{11}I_1 + h_{12}V_2,$$
$$I_2 = h_{21}I_1 + h_{22}V_2; \quad (19.5)$$

$$I_1 = g_{11}V_1 + g_{12}I_2,$$
$$V_2 = g_{21}V_1 + g_{22}I_2. \quad (19.6)$$

These six sets of equations may also be considered as three pairs of mutually inverse relations. The first set, Eqs. (19.1), gives the input and output voltages as functions of the input and output currents. The second set, Eqs. (19.2), gives the inverse relationship, that is, the input and output currents as functions of the input and output voltages. Equations (19.3) and (19.4) are inverse relations, as are Eqs. (19.5) and (19.6).

The coefficients of the current and/or voltage variables on the right-hand side of Eqs. (19.1) to (19.6) are called the **parameters** of the two-port circuit. Thus when using Eqs. (19.1), we refer to the z parameters of the circuit. Similarly, we refer to the y parameters, the a parameters, the b parameters, the h parameters, and the g parameters of the network.

19.2 THE TWO-PORT PARAMETERS

We can determine the parameters for any circuit by computation or measurement. The computation or measurement to be made comes directly from the parameter equations. For example, suppose that the problem is to find the z parameters for a circuit. From Eqs. (19.1),

$$z_{11} = \left.\frac{V_1}{I_1}\right|_{I_2=0}, \quad (19.7)$$

$$z_{12} = \left.\frac{V_1}{I_2}\right|_{I_1=0}, \tag{19.8}$$

$$z_{21} = \left.\frac{V_2}{I_1}\right|_{I_2=0}, \tag{19.9}$$

and

$$z_{22} = \left.\frac{V_2}{I_2}\right|_{I_1=0}. \tag{19.10}$$

Equations (19.7)–(19.10) reveal that the four z parameters can be described as follows:

1. z_{11} is the impedance seen looking into port 1 when port 2 is open.
2. z_{12} is a transfer impedance. It is the ratio of the port 1 voltage to the port 2 current when port 1 is open.
3. z_{21} is a transfer impedance. It is the ratio of the port 2 voltage to the port 1 current when port 2 is open.
4. z_{22} is the impedance seen looking into port 2 when port 1 is open.

Therefore the impedance parameters may be either calculated or measured by first opening port 2 and determining the ratios V_1/I_1 and V_2/I_1, and then opening port 1 and determining the ratios V_1/I_2 and V_2/I_2. Example 19.1 illustrates the determination of the z parameters for a resistive circuit.

EXAMPLE 19.1

Find the z parameters for the circuit shown in Fig. 19.3.

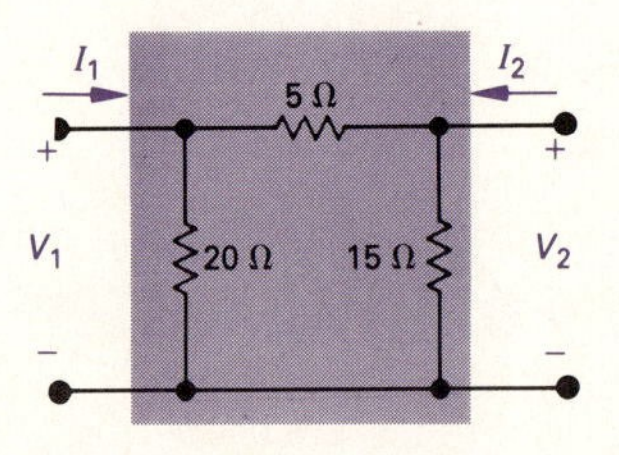

FIGURE 19.3 The circuit for Example 19.1.

SOLUTION

The circuit is purely resistive, so the s-domain circuit is also purely resistive. With port 2 open, that is, $I_2 = 0$, the resistance seen looking into port 1 is the 20 Ω resistor in parallel with the series combination of the 5 and 15 Ω resistors. Therefore

$$z_{11} = \left.\frac{V_1}{I_1}\right|_{I_2=0} = \frac{(20)(20)}{40} = 10\,\Omega.$$

When I_2 is zero, V_2 is

$$V_2 = \frac{V_1}{15+5}(15) = 0.75V_1,$$

and therefore

$$z_{21} = \left.\frac{V_2}{I_1}\right|_{I_2=0} = \frac{0.75V_1}{V_1/10} = 7.5\,\Omega.$$

When I_1 is zero, the resistance seen looking into port 2 is the 15 Ω resistor in parallel with the series combination of the 5 and 20 Ω resistors. Therefore

$$z_{22} = \left.\frac{V_2}{I_2}\right|_{I_1=0} = \frac{(15)(25)}{40} = 9.375\,\Omega.$$

When port 1 is open, I_1 is zero and the voltage V_1 is

$$V_1 = \frac{V_2}{5+20}(20) = 0.8V_2.$$

With port 1 open, the current into port 2 is

$$I_2 = \frac{V_2}{9.375}.$$

Hence

$$z_{12} = \left.\frac{V_1}{I_2}\right|_{I_1=0} = \frac{0.8V_2}{V_2/9.375} = 7.5\,\Omega.$$

Use a circuit simulator to confirm these results.

Equations (19.7)–(19.10) and Example 19.1 show why the parameters in Eqs. (19.1) are called the z parameters. Each parameter is the ratio of a voltage to a current and therefore is an impedance with the dimension of ohms.

We use the same process to determine the remaining port parameters, which are either calculated or measured. A port parameter is obtained by either opening or shorting a port. Moreover, a port parameter is an impedance, an admittance, or a dimensionless ratio. The dimensionless ratio is the ratio of either two voltages or two currents. Equations (19.11)–(19.15) summarize these observations.

$$y_{11} = \left.\frac{I_1}{V_1}\right|_{V_2=0}\ \text{S};$$
$$y_{12} = \left.\frac{I_1}{V_2}\right|_{V_1=0}\ \text{S};$$
$$y_{21} = \left.\frac{I_2}{V_1}\right|_{V_2=0}\ \text{S};$$
$$y_{22} = \left.\frac{I_2}{V_2}\right|_{V_1=0}\ \text{S}. \qquad \textbf{(19.11)}$$

$$a_{11} = \left.\frac{V_1}{V_2}\right|_{I_2=0};$$
$$a_{12} = -\left.\frac{V_1}{I_2}\right|_{V_2=0}\ \Omega;$$
$$a_{21} = \left.\frac{I_1}{V_2}\right|_{I_2=0}\ \text{S};$$
$$a_{22} = -\left.\frac{I_1}{I_2}\right|_{V_2=0}. \qquad \textbf{(19.12)}$$

$$b_{11} = \left.\frac{V_2}{V_1}\right|_{I_1=0};$$

$$b_{12} = -\left.\frac{V_2}{I_1}\right|_{V_1=0} \ \Omega;$$

$$b_{21} = \left.\frac{I_2}{V_1}\right|_{I_1=0} \ \text{S};$$

$$b_{22} = -\left.\frac{I_2}{I_1}\right|_{V_1=0}. \quad \textbf{(19.13)}$$

$$h_{11} = \left.\frac{V_1}{I_1}\right|_{V_2=0} \ \Omega;$$

$$h_{12} = \left.\frac{V_1}{V_2}\right|_{I_1=0};$$

$$h_{21} = \left.\frac{I_2}{I_1}\right|_{V_2=0};$$

$$h_{22} = \left.\frac{I_2}{V_2}\right|_{I_1=0} \ \text{S}. \quad \textbf{(19.14)}$$

$$g_{11} = \left.\frac{I_1}{V_1}\right|_{I_2=0} \ \text{S};$$

$$g_{12} = \left.\frac{I_1}{I_2}\right|_{V_1=0};$$

$$g_{21} = \left.\frac{V_2}{V_1}\right|_{I_2=0};$$

$$g_{22} = \left.\frac{V_2}{I_2}\right|_{V_1=0} \ \Omega. \quad \textbf{(19.15)}$$

DRILL EXERCISES

19.1 Find the y parameters for the circuit in Fig. 19.3.

ANSWER: $y_{11} = 0.25$ S; $y_{12} = y_{21} = -0.2$ S; $y_{22} = \frac{4}{15}$ S.

19.2 Find the a and b parameters for the circuit in Fig. 19.3.

ANSWER: $a_{11} = \frac{4}{3}$; $a_{12} = 5\ \Omega$; $a_{21} = \frac{2}{15}$ S; $a_{22} = 1.25$; $b_{11} = 1.25$; $b_{12} = 5\ \Omega$; $b_{21} = \frac{2}{15}$ S; $b_{22} = \frac{4}{3}$.

The two-port parameters are also described in relation to the reciprocal sets of equations. The impedance and admittance parameters are grouped into the immittance parameters. The term **immittance** denotes a quantity that is either an impedance or an admittance. The a and b parameters are called the **transmission** parameters because they describe the voltage and current at one end of the two-port network in terms of the voltage and current at the other end. The immittance and transmission parameters are the natural choices for relating the port variables. In other words, they relate either voltage to current variables or input to output variables. The h and g parameters relate cross-variables, that is, an input voltage and output current to an output voltage and input current. Therefore the h and g parameters are called **hybrid** parameters.

Example 19.2 illustrates how a set of measurements made at the terminals of a two-port circuit can be used to calculate the a parameters.

EXAMPLE 19.2

The following measurements pertain to a two-port circuit operating in the sinusoidal steady state. With port 2 open, a voltage equal to $150\cos 4000t$ V is applied to port 1. The current into port 1 is $25\cos(4000t - 45^\circ)$ A, and the port 2 voltage is $100\cos(4000t + 15^\circ)$ V. With port 2 short-circuited, a voltage equal to $30\cos 4000t$ V is applied to port 1. The current into port 1 is $1.5\cos(4000t + 30^\circ)$ A, and the current into port 2 is $0.25\cos(4000t + 150^\circ)$ A. Find the a parameters that can describe the sinusoidal steady-state behavior of the circuit.

SOLUTION

The first set of measurements gives

$$\mathbf{V}_1 = 150\angle 0^\circ \text{ V}; \quad \mathbf{I}_1 = 25\angle -45^\circ \text{ A};$$
$$\mathbf{V}_2 = 100\angle 15^\circ \text{ V}; \quad \mathbf{I}_2 = 0 \text{ A}.$$

From Eqs. (19.12),

$$a_{11} = \left.\frac{\mathbf{V}_1}{\mathbf{V}_2}\right|_{I_2=0} = \frac{150\angle 0^\circ}{100\angle 15^\circ} = 1.5\angle -15^\circ$$

and

$$a_{21} = \left.\frac{\mathbf{I}_1}{\mathbf{V}_2}\right|_{I_2=0} = \frac{25\angle -45^\circ}{100\angle 15^\circ} = 0.25\angle -60^\circ \text{ S}.$$

The second set of measurements gives

$$\mathbf{V}_1 = 30\angle 0^\circ \text{ V}; \quad \mathbf{I}_1 = 1.5\angle 30^\circ \text{ A};$$
$$\mathbf{V}_2 = 0 \text{ V}; \quad \mathbf{I}_2 = 0.25\angle 150^\circ \text{ A}.$$

Therefore

$$a_{12} = -\left.\frac{\mathbf{V}_1}{\mathbf{I}_2}\right|_{V_2=0} = \frac{-30\angle 0^\circ}{0.25\angle 150^\circ} = 120\angle 30^\circ \ \Omega$$

and

$$a_{21} = -\frac{\mathbf{I}_1}{\mathbf{I}_2}\bigg|_{V_2=0} = \frac{-1.5\angle 30^\circ}{0.25\angle 150^\circ} = 6\angle 60^\circ.$$

DRILL EXERCISE

19.3 The following measurements were made on a two-port resistive circuit. With 10 mV applied to port 2 and port 1 open, the current into port 2 is 0.25 μA, and the voltage across port 1 is 5 μV. With port 2 short-circuited and 50 mV applied to port 1, the current into port 1 is 50 μA, and the current into port 2 is 2 mA. Find the h parameters of the network.

ANSWER: $h_{11} = 1000\,\Omega$; $h_{12} = 5 \times 10^{-4}$; $h_{21} = 40$; $h_{22} = 25\,\mu$S.

RELATIONSHIPS AMONG THE TWO-PORT PARAMETERS

Because the six sets of equations relate to the same variables, the parameters associated with any pair of equations must be related to the parameters of all the other pairs. In other words, if we know one set of parameters, we can derive all the other sets from the known set. Because of the amount of algebra involved in these derivations, we merely list the results in Table 19.1.

Although we do not here derive all the relationships listed in Table 19.1, we do derive those between the z and y parameters and between the z and a parameters. These derivations illustrate the general process involved in relating one set of parameters to another. To find the z parameters as functions of the y parameters, we first solve Eqs. (19.2) for V_1 and V_2. We then compare the coefficients of I_1 and I_2 in the resulting expressions to the coefficients of I_1 and I_2 in Eqs. (19.1). From Eqs. (19.2),

$$V_1 = \frac{\begin{vmatrix} I_1 & y_{12} \\ I_2 & y_{22} \end{vmatrix}}{\begin{vmatrix} y_{11} & y_{12} \\ y_{21} & y_{22} \end{vmatrix}} = \frac{y_{22}}{\Delta y} I_1 - \frac{y_{12}}{\Delta y} I_2 \qquad \textbf{(19.16)}$$

and

$$V_2 = \frac{\begin{vmatrix} y_{11} & I_1 \\ y_{21} & I_2 \end{vmatrix}}{\Delta y} = -\frac{y_{21}}{\Delta y} I_1 + \frac{y_{11}}{\Delta y} I_2. \qquad \textbf{(19.17)}$$

TABLE 19.1

PARAMETER CONVERSION TABLE

$$z_{11} = \frac{y_{22}}{\Delta y} = \frac{a_{11}}{a_{21}} = \frac{b_{22}}{b_{21}} = \frac{\Delta h}{h_{22}} = \frac{1}{g_{11}}$$

$$z_{12} = -\frac{y_{12}}{\Delta y} = \frac{\Delta a}{a_{21}} = \frac{1}{b_{21}} = \frac{h_{12}}{h_{22}} = -\frac{g_{12}}{g_{11}}$$

$$z_{21} = -\frac{y_{21}}{\Delta y} = \frac{1}{a_{21}} = \frac{\Delta b}{b_{21}} = -\frac{h_{21}}{h_{22}} = \frac{g_{21}}{g_{11}}$$

$$z_{22} = \frac{y_{11}}{\Delta y} = \frac{a_{22}}{a_{21}} = \frac{b_{11}}{b_{21}} = \frac{1}{h_{22}} = \frac{\Delta g}{g_{11}}$$

$$y_{11} = \frac{z_{22}}{\Delta z} = \frac{a_{22}}{a_{12}} = \frac{b_{11}}{b_{12}} = \frac{1}{h_{11}} = \frac{\Delta g}{g_{22}}$$

$$y_{12} = -\frac{z_{12}}{\Delta z} = -\frac{\Delta a}{a_{12}} = -\frac{1}{b_{12}} = -\frac{h_{12}}{h_{11}} = \frac{g_{12}}{g_{22}}$$

$$y_{21} = -\frac{z_{21}}{\Delta z} = -\frac{1}{a_{12}} = -\frac{\Delta b}{b_{12}} = \frac{h_{21}}{h_{11}} = -\frac{g_{21}}{g_{22}}$$

$$y_{22} = \frac{z_{11}}{\Delta z} = \frac{a_{11}}{a_{12}} = \frac{b_{22}}{b_{12}} = \frac{\Delta h}{h_{11}} = \frac{1}{g_{22}}$$

$$a_{11} = \frac{z_{11}}{z_{21}} = -\frac{y_{22}}{y_{21}} = \frac{b_{22}}{\Delta b} = -\frac{\Delta h}{h_{21}} = \frac{1}{g_{21}}$$

$$a_{12} = \frac{\Delta z}{z_{21}} = -\frac{1}{y_{21}} = \frac{b_{12}}{\Delta b} = \frac{-h_{11}}{h_{21}} = \frac{g_{22}}{g_{21}}$$

$$a_{21} = \frac{1}{z_{21}} = -\frac{\Delta y}{y_{21}} = \frac{b_{21}}{\Delta b} = -\frac{h_{22}}{h_{21}} = \frac{g_{11}}{g_{21}}$$

$$a_{22} = \frac{z_{22}}{z_{21}} = -\frac{y_{11}}{y_{21}} = \frac{b_{11}}{\Delta b} = -\frac{1}{h_{21}} = \frac{\Delta g}{g_{21}}$$

$$b_{11} = \frac{z_{22}}{z_{12}} = -\frac{y_{11}}{y_{12}} = \frac{a_{22}}{\Delta a} = \frac{1}{h_{12}} = -\frac{\Delta g}{g_{12}}$$

$$b_{12} = \frac{\Delta z}{z_{12}} = -\frac{1}{y_{12}} = \frac{a_{12}}{\Delta a} = \frac{h_{11}}{h_{12}} = -\frac{g_{22}}{g_{12}}$$

$$b_{21} = \frac{1}{z_{12}} = -\frac{\Delta y}{y_{12}} = \frac{a_{21}}{\Delta a} = \frac{h_{22}}{h_{12}} = -\frac{g_{11}}{g_{12}}$$

$$b_{22} = \frac{z_{11}}{z_{12}} = \frac{y_{22}}{y_{12}} = \frac{a_{11}}{\Delta a} = \frac{\Delta h}{h_{12}} = -\frac{1}{g_{12}}$$

$$h_{11} = \frac{\Delta z}{z_{22}} = \frac{1}{y_{11}} = \frac{a_{12}}{a_{22}} = \frac{b_{12}}{b_{11}} = \frac{g_{22}}{\Delta g}$$

$$h_{12} = \frac{z_{12}}{z_{22}} = -\frac{y_{12}}{y_{11}} = \frac{\Delta a}{a_{22}} = \frac{1}{b_{11}} = -\frac{g_{12}}{\Delta g}$$

$$h_{21} = -\frac{z_{21}}{z_{22}} = \frac{y_{21}}{y_{11}} = -\frac{1}{a_{22}} = -\frac{\Delta b}{b_{11}} = -\frac{g_{21}}{\Delta g}$$

$$h_{22} = \frac{1}{z_{22}} = \frac{\Delta y}{y_{11}} = \frac{a_{21}}{a_{22}} = \frac{b_{21}}{b_{11}} = \frac{g_{11}}{\Delta g}$$

$$g_{11} = \frac{1}{z_{11}} = \frac{\Delta y}{y_{22}} = \frac{a_{21}}{a_{11}} = \frac{b_{21}}{b_{22}} = \frac{h_{22}}{\Delta h}$$

$$g_{12} = -\frac{z_{12}}{z_{11}} = \frac{y_{12}}{y_{22}} = -\frac{\Delta a}{a_{11}} = -\frac{1}{b_{22}} = -\frac{h_{12}}{\Delta h}$$

$$g_{21} = \frac{z_{21}}{z_{11}} = -\frac{y_{21}}{y_{22}} = \frac{1}{a_{11}} = \frac{\Delta b}{b_{22}} = -\frac{h_{21}}{\Delta h}$$

$$g_{22} = \frac{\Delta z}{z_{11}} = \frac{1}{y_{22}} = \frac{a_{12}}{a_{11}} = \frac{b_{12}}{b_{22}} = \frac{h_{11}}{\Delta h}$$

$$\Delta z = z_{11}z_{22} - z_{12}z_{21}$$
$$\Delta y = y_{11}y_{22} - y_{12}y_{21}$$
$$\Delta a = a_{11}a_{22} - a_{12}a_{21}$$
$$\Delta b = b_{11}b_{22} - b_{12}b_{21}$$
$$\Delta h = h_{11}h_{22} - h_{12}h_{21}$$
$$\Delta g = g_{11}g_{22} - g_{12}g_{21}$$

Comparing Eqs. (19.16) and (19.17) with Eqs. (19.1) shows

$$z_{11} = \frac{y_{22}}{\Delta y}, \quad (19.18)$$

$$z_{12} = -\frac{y_{12}}{\Delta y}, \quad (19.19)$$

$$z_{21} = -\frac{y_{21}}{\Delta y}, \quad (19.20)$$

and

$$z_{22} = \frac{y_{11}}{\Delta y}. \quad (19.21)$$

To find the z parameters as functions of the a parameters, we rearrange Eqs. (19.3) in the form of Eqs. (19.1) and then compare coefficients. From the second equation in Eqs. (19.3),

$$V_2 = \frac{1}{a_{21}} I_1 + \frac{a_{22}}{a_{21}} I_2. \quad \textbf{(19.22)}$$

Therefore, substituting Eq. (19.22) into the first equation of Eqs. (19.3) yields

$$V_1 = \frac{a_{11}}{a_{21}} I_1 + \left(\frac{a_{11}a_{22}}{a_{21}} - a_{12} \right) I_2. \quad \textbf{(19.23)}$$

From Eq. (19.23),

$$z_{11} = \frac{a_{11}}{a_{21}}; \quad \textbf{(19.24)}$$

$$z_{12} = \frac{\Delta a}{a_{21}}. \quad \textbf{(19.25)}$$

From Eq. (19.22),

$$z_{21} = \frac{1}{a_{21}}; \quad \textbf{(19.26)}$$

$$z_{22} = \frac{a_{22}}{a_{21}}. \quad \textbf{(19.27)}$$

Example 19.3 illustrates the usefulness of the parameter conversion table.

EXAMPLE 19.3

Two sets of measurements are made on a two-port resistive circuit. The first set is made with port 2 open, and the second set is made with port 2 short-circuited. The results are as follows:

Port 2 Open	**Port 2 Short-Circuited**
$V_1 = 10$ mV	$V_1 = 24$ mV
$I_1 = 10\ \mu$A	$I_1 = 20\ \mu$A
$V_2 = -40$ V	$I_2 = 1$ mA

Find the h parameters of the circuit.

SOLUTION

We can find h_{11} and h_{21} directly from the short-circuit test:

$$h_{11} = \left. \frac{V_1}{I_1} \right|_{V_2=0} = \frac{24 \times 10^{-3}}{20 \times 10^{-6}} = 1.2 \text{ k}\Omega$$

and

$$h_{21} = \left.\frac{I_2}{I_1}\right|_{V_2=0} = \frac{10^{-3}}{20 \times 10^{-6}} = 50.$$

The parameters h_{12} and h_{22} cannot be obtained directly from the open-circuit test. However, a check of Eqs. (19.7) through (19.15) indicates that the four a parameters can be derived from the test data. Therefore h_{12} and h_{22} can be obtained through the conversion table. Specifically,

$$h_{12} = \frac{\Delta a}{a_{22}} \quad \text{and} \quad h_{22} = \frac{a_{21}}{a_{22}}.$$

The a parameters are

$$a_{11} = \left.\frac{V_1}{V_2}\right|_{I_2=0} = \frac{10 \times 10^{-3}}{-40} = -0.25 \times 10^{-3};$$

$$a_{21} = \left.\frac{I_1}{V_2}\right|_{I_2=0} = \frac{10 \times 10^{-6}}{-40} = -0.25 \times 10^{-6}\text{ S};$$

$$a_{12} = -\left.\frac{V_1}{I_2}\right|_{V_2=0} = -\frac{24 \times 10^{-3}}{10^{-3}} = -24\,\Omega;$$

$$a_{22} = -\left.\frac{I_1}{I_2}\right|_{V_2=0} = -\frac{20 \times 10^{-6}}{10^{-3}} = -20 \times 10^{-3}.$$

The numerical value of Δa is

$$\begin{aligned}\Delta a &= a_{11}a_{22} - a_{12}a_{21} \\ &= 5 \times 10^{-6} - 6 \times 10^{-6} = -10^{-6}.\end{aligned}$$

Thus

$$h_{12} = \frac{\Delta a}{a_{22}} = \frac{-10^{-6}}{-20 \times 10^{-3}} = 5 \times 10^{-5}$$

and

$$h_{22} = \frac{a_{21}}{a_{22}} = \frac{-0.25 \times 10^{-6}}{-20 \times 10^{-3}} = 12.5\,\mu\text{S}.$$

DRILL EXERCISE

19.4 The following measurements were made on a two-port resistive circuit: With port 1 open, $V_2 = 15$ V, $V_1 = 10$ V, and $I_2 = 30$ A; with port 1 short-circuited, $V_2 = 10$ V, $I_2 = 4$ A, and $I_1 = -5$ A. Calculate the y parameters.

ANSWER: $y_{11} = 0.75$ S; $y_{12} = -0.5$ S; $y_{21} = 2.4$ S; $y_{22} = 0.4$ S.

RECIPROCAL TWO-PORT CIRCUITS

If a two-port circuit is **reciprocal**, the following relationships exist among the port parameters:

$$z_{12} = z_{21}; \quad (19.28)$$

$$y_{12} = y_{21}; \quad (19.29)$$

$$a_{11}a_{22} - a_{12}a_{21} = \Delta a = 1; \quad (19.30)$$

$$b_{11}b_{22} - b_{12}b_{21} = \Delta b = 1; \quad (19.31)$$

$$h_{12} = -h_{21}; \quad (19.32)$$

$$g_{12} = -g_{21}. \quad (19.33)$$

A two-port circuit is reciprocal if the interchange of an ideal voltage source at one port with an ideal ammeter at the other port produces the same ammeter reading. Consider, for example, the resistive circuit shown in Fig. 19.4. When a voltage source of 15 V is applied to the port ad, it produces a current of 1.75 A in the ammeter at port cd. The ammeter current is easily determined once we know the voltage V_{bd}. Thus

$$\frac{V_{bd}}{60} + \frac{V_{bd} - 15}{30} + \frac{V_{bd}}{20} = 0, \quad (19.34)$$

and $V_{bd} = 5$ V. Therefore

$$I = \frac{5}{20} + \frac{15}{10} = 1.75 \text{ A}. \quad (19.35)$$

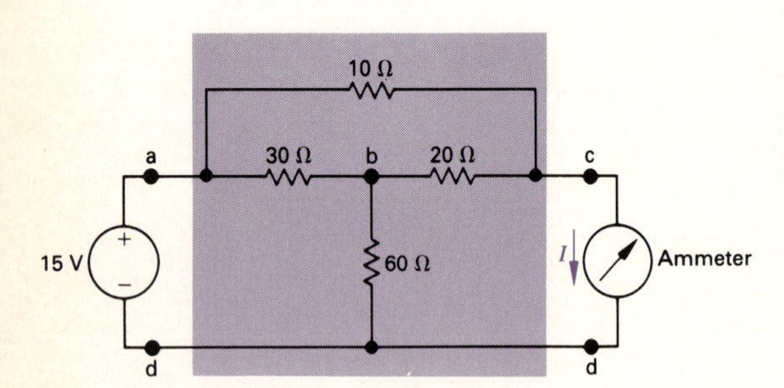

FIGURE 19.4 A reciprocal two-port circuit.

If the voltage source and ammeter are interchanged, the ammeter will still read 1.75 A. We verify this by solving the circuit shown in Fig. 19.5:

$$\frac{V_{bd}}{60} + \frac{V_{bd}}{30} + \frac{V_{bd} - 15}{20} = 0. \quad (19.36)$$

From Eq. (19.36), $V_{bd} = 7.5$ V. The current I_{ad} equals

$$I_{ad} = \frac{7.5}{30} + \frac{15}{10} = 1.75 \text{ A}. \quad (19.37)$$

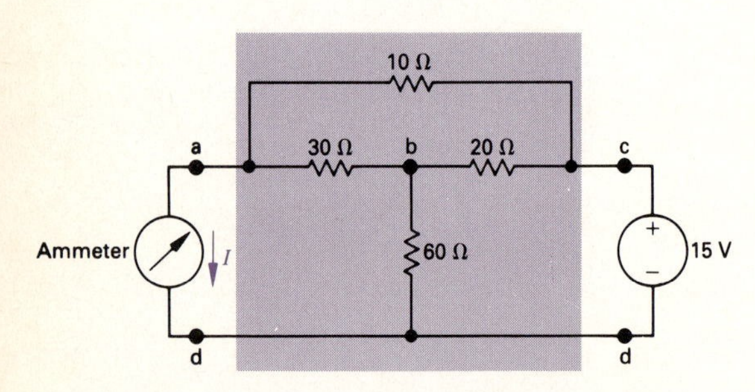

FIGURE 19.5 The circuit shown in Fig. 19.4, with the voltage source and ammeter interchanged.

A two-port circuit is also reciprocal if the interchange of an ideal current source at one port with an ideal voltmeter at the other port produces the same voltmeter reading.

DRILL EXERCISE

19.5 a) Calculate the reading of the ideal voltmeter in the circuit shown.

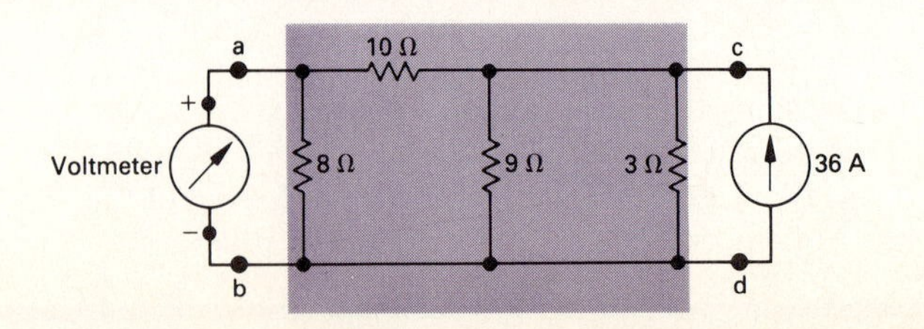

b) Interchange the voltmeter and the ideal current source. Calculate the voltmeter reading.

ANSWER: (a) 32 V; (b) 32 V.

For a reciprocal two-port circuit, only three calculations or measurements are needed to determine a set of parameters.

A reciprocal two-port circuit is **symmetric** if its ports can be interchanged without disturbing the values of the terminal currents and voltages. Figure 19.6 shows four examples of symmetric two-port circuits. In such circuits, the following additional relationships exist among the port parameters:

$$z_{11} = z_{22}; \quad \textbf{(19.38)}$$

$$y_{11} = y_{22}; \quad \textbf{(19.39)}$$

$$a_{11} = a_{22}; \quad \textbf{(19.40)}$$

$$b_{11} = b_{22}; \quad \textbf{(19.41)}$$

$$h_{11}h_{22} - h_{12}h_{21} = \Delta h = 1; \quad \textbf{(19.42)}$$

$$g_{11}g_{22} - g_{12}g_{21} = \Delta g = 1. \quad \textbf{(19.43)}$$

For a symmetric reciprocal network, only two calculations or measurements are necessary to determine all the two-port parameters.

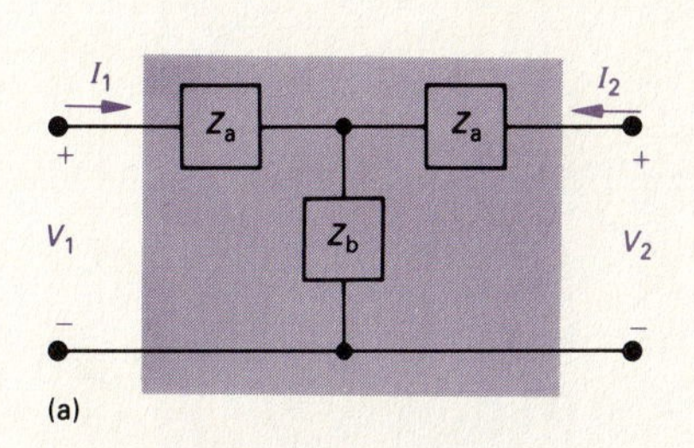

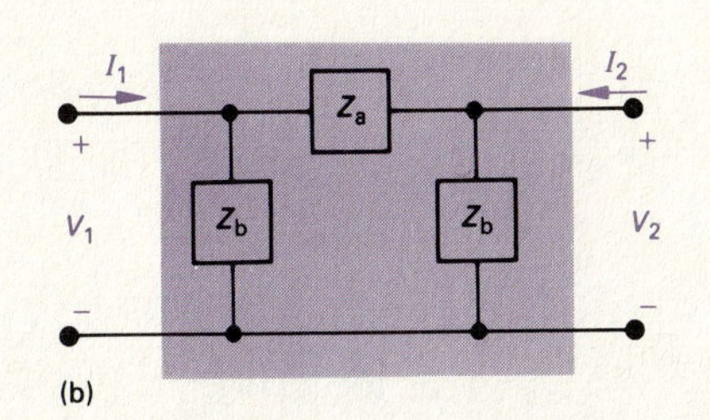

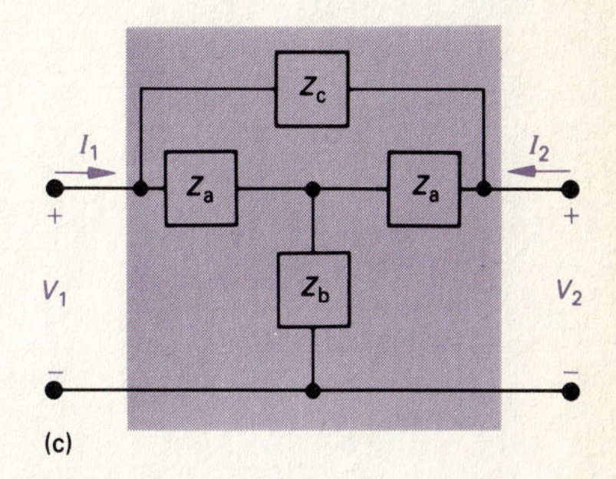

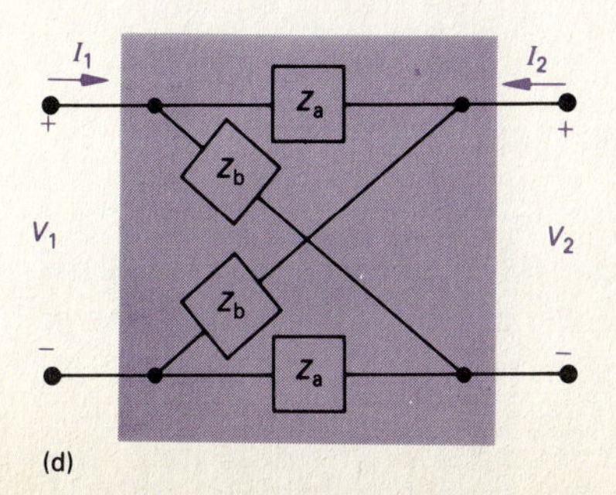

FIGURE 19.6 Four examples of symmetric two-port circuits: (a) a symmetric tee; (b) a symmetric pi; (c) a symmetric bridged tee; (d) a symmetric lattice.

DRILL EXERCISE

19.6 The following measurements were made on a resistive two-port network that is symmetric and reciprocal: With port 2 open, $V_1 = 95$ V and $I_1 = 5$ A; with a short circuit across port 2, $V_1 = 11.52$ V and $I_2 = -2.72$ A. Calculate the z parameters of the two-port network.

ANSWER: $z_{11} = z_{22} = 19\ \Omega$, $z_{12} = z_{21} = 17\ \Omega$.

19.3 ANALYSIS OF THE TERMINATED TWO-PORT CIRCUIT

In the typical application of a two-port model, the circuit is driven at port 1 and loaded at port 2. Figure 19.7 shows the s-domain circuit diagram for a typically terminated two-port model. Here, Z_g represents the internal impedance of the source, V_g the internal voltage of the source, and Z_L the load impedance. Analysis of this circuit involves expressing the terminal currents and voltages as functions of the two-port parameters, V_g, Z_g, and Z_L.

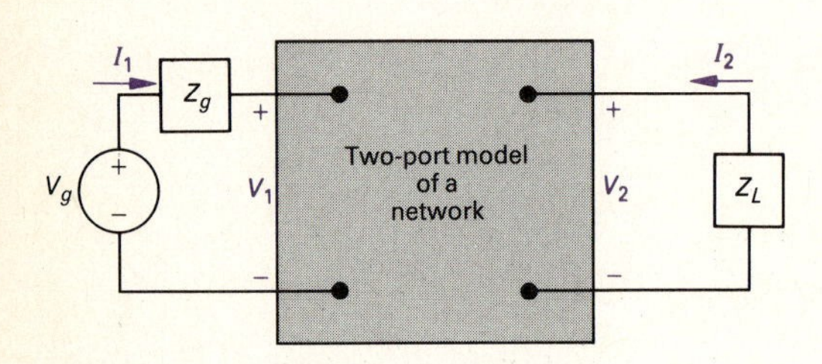

FIGURE 19.7 A terminated two-port model.

Six characteristics of the terminated two-port circuit define its terminal behavior:

1. the input impedance $Z_{in} = V_1/I_1$, or the admittance $Y_{in} = I_1/V_1$;
2. the output current I_2;
3. the Thévenin voltage and impedance (V_{Th}, Z_{Th}) with respect to port 2;
4. the current gain I_2/I_1;
5. the voltage gain V_2/V_1; and
6. the voltage gain V_2/V_g.

THE SIX CHARACTERISTICS IN TERMS OF THE Z PARAMETERS

To illustrate how these six characteristics are derived, we develop the expressions using the z parameters to model the two-port portion of the circuit. Table 19.2 summarizes the expressions involving the y, a, b, h, and g parameters.

TABLE 19.2

TERMINATED TWO-PORT EQUATIONS

***z* PARAMETERS**

$$Z_{\text{in}} = z_{11} - \frac{z_{12}z_{21}}{z_{22} + Z_L}$$

$$I_2 = \frac{-z_{21}V_g}{(z_{11} + Z_g)(z_{22} + Z_L) - z_{12}z_{21}}$$

$$V_{\text{Th}} = \frac{z_{21}}{z_{11} + Z_g}V_g$$

$$Z_{\text{Th}} = z_{22} - \frac{z_{12}z_{21}}{z_{11} + Z_g}$$

$$\frac{I_2}{I_1} = \frac{-z_{21}}{z_{22} + Z_L}$$

$$\frac{V_2}{V_1} = \frac{z_{21}Z_L}{z_{11}Z_L + \Delta z}$$

$$\frac{V_2}{V_g} = \frac{z_{21}Z_L}{(z_{11} + Z_g)(z_{22} + Z_L) - z_{12}z_{21}}$$

***y* PARAMETERS**

$$Y_{\text{in}} = y_{11} - \frac{y_{12}y_{21}Z_L}{1 + y_{22}Z_L}$$

$$I_2 = \frac{y_{21}V_g}{1 + y_{22}Z_L + y_{11}Z_g + \Delta y Z_g Z_L}$$

$$V_{\text{Th}} = \frac{-y_{21}V_g}{y_{22} + \Delta y Z_g}$$

$$Z_{\text{Th}} = \frac{1 + y_{11}Z_g}{y_{22} + \Delta y Z_g}$$

$$\frac{I_2}{I_1} = \frac{y_{21}}{y_{11} + \Delta y Z_L}$$

$$\frac{V_2}{V_1} = \frac{-y_{21}Z_L}{1 + y_{22}Z_L}$$

$$\frac{V_2}{V_g} = \frac{y_{21}Z_L}{y_{12}y_{21}Z_g Z_L - (1 + y_{11}Z_g)(1 + y_{22}Z_L)}$$

***a* PARAMETERS**

$$Z_{\text{in}} = \frac{a_{11}Z_L + a_{12}}{a_{21}Z_L + a_{22}}$$

$$I_2 = \frac{-V_g}{a_{11}Z_L + a_{12} + a_{21}Z_g Z_L + a_{22}Z_g}$$

$$V_{\text{Th}} = \frac{V_g}{a_{11} + a_{21}Z_g}$$

$$Z_{\text{Th}} = \frac{a_{12} + a_{22}Z_g}{a_{11} + a_{21}Z_g}$$

$$\frac{I_2}{I_1} = \frac{-1}{a_{21}Z_L + a_{22}}$$

$$\frac{V_2}{V_1} = \frac{Z_L}{a_{11}Z_L + a_{12}}$$

$$\frac{V_2}{V_g} = \frac{Z_L}{(a_{11} + a_{21}Z_g)Z_L + a_{12} + a_{22}Z_g}$$

***b* PARAMETERS**

$$Z_{\text{in}} = \frac{b_{22}Z_L + b_{12}}{b_{21}Z_L + b_{11}}$$

$$I_2 = \frac{-V_g \Delta b}{b_{11}Z_g + b_{21}Z_g Z_L + b_{22}Z_L + b_{12}}$$

$$V_{\text{Th}} = \frac{V_g \Delta b}{b_{22} + b_{21}Z_g}$$

$$Z_{\text{Th}} = \frac{b_{11}Z_g + b_{12}}{b_{21}Z_g + b_{22}}$$

$$\frac{I_2}{I_1} = \frac{-\Delta b}{b_{11} + b_{21}Z_L}$$

$$\frac{V_2}{V_1} = \frac{\Delta b Z_L}{b_{12} + b_{22}Z_L}$$

$$\frac{V_2}{V_g} = \frac{\Delta b Z_L}{b_{12} + b_{11}Z_g + b_{22}Z_L + b_{21}Z_g Z_L}$$

***h* PARAMETERS**

$$Z_{\text{in}} = h_{11} - \frac{h_{12}h_{21}Z_L}{1 + h_{22}Z_L}$$

$$I_2 = \frac{h_{21}V_g}{(1 + h_{22}Z_L)(h_{11} + Z_g) - h_{12}h_{21}Z_L}$$

$$V_{\text{Th}} = \frac{-h_{21}V_g}{h_{22}Z_g + \Delta h}$$

$$Z_{\text{Th}} = \frac{Z_g + h_{11}}{h_{22}Z_g + \Delta h}$$

***g* PARAMETERS**

$$Y_{\text{in}} = g_{11} - \frac{g_{12}g_{21}}{g_{22} + Z_L}$$

$$I_2 = \frac{-g_{21}V_g}{(1 + g_{11}Z_g)(g_{22} + Z_L) - g_{12}g_{21}Z_g}$$

$$V_{\text{Th}} = \frac{g_{21}V_g}{1 + g_{11}Z_g}$$

$$Z_{\text{Th}} = g_{22} - \frac{g_{12}g_{21}Z_g}{1 + g_{11}Z_g}$$

TABLE 19.2 (continued)

***h* PARAMETERS (continued)**	***g* PARAMETERS (continued)**
$\frac{I_2}{I_1} = \frac{h_{21}}{1 + h_{22}Z_L}$	$\frac{I_2}{I_1} = \frac{-g_{21}}{g_{11}Z_L + \Delta g}$
$\frac{V_2}{V_1} = \frac{-h_{21}Z_L}{\Delta h Z_L + h_{11}}$	$\frac{V_2}{V_1} = \frac{g_{21}Z_L}{g_{22} + Z_L}$
$\frac{V_2}{V_g} = \frac{-h_{21}Z_L}{(h_{11} + Z_g)(1 + h_{22}Z_L) - h_{12}h_{21}Z_L}$	$\frac{V_2}{V_g} = \frac{g_{21}Z_L}{(1 + g_{11}Z_g)(g_{22} + Z_L) - g_{12}g_{21}Z_g}$

The derivation of any one of the desired expressions involves the algebraic manipulation of the two-port equations along with the two constraint equations imposed by the terminations. If we use the z-parameter equations, the four that describe the circuit in Fig. 19.7 are

$$V_1 = z_{11}I_1 + z_{12}I_2; \quad \textbf{(19.44)}$$

$$V_2 = z_{21}I_1 + z_{22}I_2; \quad \textbf{(19.45)}$$

$$V_1 = V_g - I_1Z_g; \quad \textbf{(19.46)}$$

$$V_2 = -I_2Z_L. \quad \textbf{(19.47)}$$

Equations (19.46) and (19.47) describe the constraints imposed by the terminations.

To find the impedance seen looking into port 1, that is, $Z_{\text{in}} = V_1/I_1$, we proceed as follows. In Eq. (19.45) we replace V_2 with $-I_2Z_L$ and solve the resulting expression for I_2:

$$I_2 = \frac{-z_{21}I_1}{Z_L + z_{22}}, \quad \textbf{(19.48)}$$

We then substitute this equation into Eq. (19.44) and solve for Z_{in}:

$$Z_{\text{in}} = z_{11} - \frac{z_{12}z_{21}}{z_{22} + Z_L}. \quad \textbf{(19.49)}$$

To find the terminal current I_2, we first solve Eq. (19.44) for I_1 after replacing V_1 with the right-hand side of Eq. (19.46). The result is

$$I_1 = \frac{V_g - z_{12}I_2}{z_{11} + Z_g}. \quad \textbf{(19.50)}$$

We now substitute Eq. (19.50) into Eq. (19.48) and solve the resulting equation for I_2:

$$I_2 = \frac{-z_{21}V_g}{(z_{11} + Z_g)(z_{22} + Z_L) - z_{12}z_{21}}. \quad \textbf{(19.51)}$$

The Thévenin voltage with respect to port 2 equals V_2 when $I_2 = 0$. With $I_2 = 0$, Eqs. (19.44) and (19.45) combine to yield

$$V_2\Big|_{I_2=0} = z_{21} I_1 = z_{21}\frac{V_1}{z_{11}}. \quad (19.52)$$

But $V_1 = V_g - I_1 Z_g$, and $I_1 = V_g/(Z_g + z_{11})$; therefore substituting the results into Eq. (19.52) yields the open-circuit value of V_2:

$$V_2\Big|_{I_2=0} = V_{\mathrm{Th}} = \frac{z_{21}}{Z_g + z_{11}} V_g. \quad (19.53)$$

The Thévenin, or output, impedance is the ratio V_2/I_2 when V_g is replaced by a short circuit. When V_g is zero, Eq. (19.46) reduces to

$$V_1 = -I_1 Z_g. \quad (19.54)$$

Substituting Eq. (19.54) into Eq. (19.44) gives

$$I_1 = \frac{-z_{12} I_2}{z_{11} + Z_g}. \quad (19.55)$$

We now use Eq. (19.55) to replace I_1 in Eq. (19.45), with the result that

$$\frac{V_2}{I_2}\Big|_{V_g=0} = Z_{\mathrm{Th}} = z_{22} - \frac{z_{12} z_{21}}{z_{11} + Z_g}. \quad (19.56)$$

The current gain I_2/I_1 comes directly from Eq. (19.48):

$$\frac{I_2}{I_1} = \frac{-z_{21}}{Z_L + z_{22}}. \quad (19.57)$$

To derive the expression for the voltage gain V_2/V_1, we start by replacing I_2 in Eq. (19.45) with its value from Eq. (19.47); thus

$$V_2 = z_{21} I_1 + z_{22}\left(\frac{-V_2}{Z_L}\right). \quad (19.58)$$

Next we solve Eq. (19.44) for I_1 as a function of V_1 and V_2:

$$z_{11} I_1 = V_1 - z_{12}\left(\frac{-V_2}{Z_L}\right)$$

or

$$I_1 = \frac{V_1}{z_{11}} + \frac{z_{12} V_2}{z_{11} Z_L}. \quad (19.59)$$

We now replace I_1 in Eq. (19.58) with Eq. (19.59) and solve the resulting expression for V_2/V_1:

$$\frac{V_2}{V_1} = \frac{z_{21} Z_L}{z_{11} Z_L + z_{11} z_{22} - z_{12} z_{21}}$$

$$= \frac{z_{21} Z_L}{z_{11} Z_L + \Delta z}. \quad (19.60)$$

To derive the voltage ratio V_2/V_g, we first combine Eqs. (19.44), (19.46), and (19.47) to find I_1 as a function of V_2 and V_g:

$$I_1 = \frac{z_{12}V_2}{Z_L(z_{11}+Z_g)} + \frac{V_g}{z_{11}+Z_g}. \quad \textbf{(19.61)}$$

We now use Eqs. (19.61) and (19.47) in conjunction with Eq. (19.45) to derive an expression involving only V_2 and V_g; that is,

$$V_2 = \frac{z_{21}z_{12}V_2}{Z_L(z_{11}+Z_g)} + \frac{z_{21}V_g}{z_{11}+Z_g} - \frac{z_{22}}{Z_L}V_2, \quad \textbf{(19.62)}$$

which we can manipulate to get the desired voltage ratio:

$$\frac{V_2}{V_g} = \frac{z_{21}Z_L}{(z_{11}+Z_g)(z_{22}+Z_L) - z_{12}z_{21}}. \quad \textbf{(19.63)}$$

The first entries in Table 19.2 summarize the expressions for these six attributes of the terminated two-port circuit. Also listed are the corresponding expressions in terms of the y, a, b, h, and g parameters.

Example 19.4 illustrates the usefulness of the relationships listed in Table 19.2.

EXAMPLE 19.4

The two-port circuit shown in Fig. 19.8 is described in terms of its b parameters, the values of which are

$$b_{11} = -20, \; b_{12} = -3000\,\Omega, \; b_{21} = -2\text{ mS}, \text{ and } b_{22} = -0.2.$$

a) Find the phasor voltage $\mathbf{V}_2$.

b) Find the average power delivered to the 5 kΩ load.

c) Find the average power delivered to the input port.

d) Find the load impedance for maximum average power transfer.

e) Find the maximum average power delivered to the load in (d).

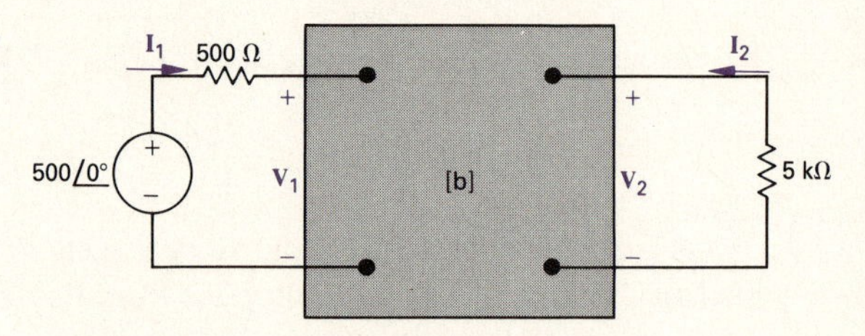

FIGURE 19.8 The circuit for Example 19.4.

SOLUTION

a) To find $\mathbf{V}_2$, we have two choices from the entries in Table 19.2. We may choose to find $\mathbf{I}_2$ and then find $\mathbf{V}_2$ from the relationship $\mathbf{V}_2 = -\mathbf{I}_2 Z_L$, or we may find the voltage gain $\mathbf{V}_2/\mathbf{V}_g$ and calculate $\mathbf{V}_2$ from the gain. Let's use the latter approach. For the b-parameter values given, we have

$$\begin{aligned}\Delta b &= (-20)(-0.2) - (-3000)(-2\times 10^{-3})\\ &= 4 - 6\\ &= -2.\end{aligned}$$

From Table 19.2,

$$\frac{\mathbf{V}_2}{\mathbf{V}_g} = \frac{\Delta b Z_L}{b_{12} + b_{11}Z_g + b_{22}Z_L + b_{21}Z_gZ_L}$$

$$= \frac{(-2)(5000)}{-3000 + (-20)500 + (-0.2)5000 + [-2 \times 10^{-3}(500)(5000)]}$$

$$= \frac{10}{19}.$$

Then,

$$\mathbf{V}_2 = \left(\frac{10}{19}\right) 500 = 263.16\underline{/0^\circ} \text{ V}.$$

b) The average power delivered to the 5000 Ω load is

$$P_2 = \frac{263.16^2}{2(5000)} = 6.93 \text{ W}.$$

c) To find the average power delivered to the input port, we first find the input impedance Z_{in}. From Table 19.2,

$$Z_{\text{in}} = \frac{b_{22}Z_L + b_{12}}{b_{21}Z_L + b_{11}}$$

$$= \frac{(-0.2)(5000) - 3000}{-2 \times 10^{-3}(5000) - 20}$$

$$= \frac{400}{3} = 133.33\ \Omega.$$

Now $\mathbf{I}_1$ follows directly:

$$\mathbf{I}_1 = \frac{500}{500 + 133.33} = 789.47 \text{ mA}.$$

The average power delivered to the input port is

$$P_1 = \frac{0.78947^2}{2}(133.33) = 41.55 \text{ W}.$$

d) The load impedance for maximum power transfer equals the conjugate of the Thévenin impedance seen looking into port 2. From Table 19.2,

$$Z_{\text{Th}} = \frac{b_{11}Z_g + b_{12}}{b_{21}Z_g + b_{22}}$$

$$= \frac{(-20)(500) - 3000}{(-2 \times 10^{-3})(500) - 0.2}$$

$$= \frac{13{,}000}{1.2} = 10{,}833.33\ \Omega.$$

Therefore $Z_L = Z_{\text{Th}}^* = 10{,}833.33\ \Omega$.

e) To find the maximum average power delivered to Z_L, we first find $\mathbf{V}_2$ from the voltage-gain expression $\mathbf{V}_2/\mathbf{V}_g$. When Z_L is 10,833.33 Ω, this gain is

$$\frac{\mathbf{V}_2}{\mathbf{V}_g} = 0.8333.$$

Thus

$$\mathbf{V}_2 = (0.8333)(500) = 416.67 \text{ V},$$

and

$$P_2(\text{maximum}) = \frac{1}{2}\frac{416.67^2}{10{,}833.33} = 8.01 \text{ W}.$$

DRILL EXERCISE

19.7 The b parameters of the two-port network shown are $b_{11} = 2000/3$, $b_{12} = 2/3$ MΩ, $b_{21} = 1/15$ S, and $b_{22} = -100/3$. The network is driven by a sinusoidal current source having a maximum amplitude of 100 μA and an internal impedance of $1000 + j0$ Ω. It is terminated in a resistive load of 10 kΩ.

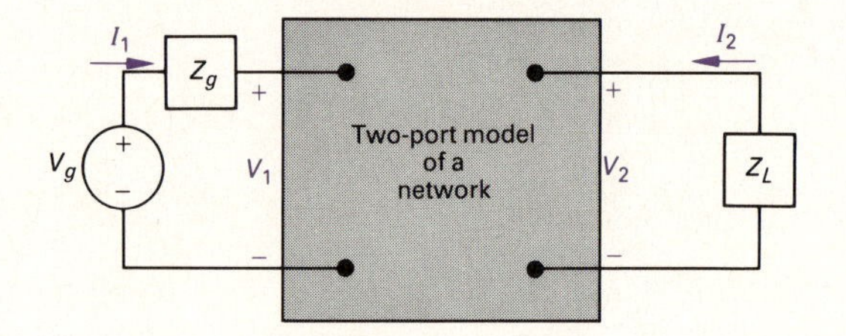

a) Calculate the average power delivered to the load resistor.

b) Calculate the load resistance for maximum average power.

c) Calculate the maximum average power delivered to the resistor in (b).

ANSWER: (a) 80 mW; (b) 40 kΩ; (c) 125 mW.

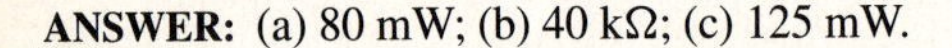

19.4 INTERCONNECTED TWO-PORT CIRCUITS

Synthesizing a large, complex system is usually made easier by first designing subsections of the system. Interconnecting these simpler, easier-to-design units then completes the system. If the subsections are modeled by two-port circuits, synthesis involves the analysis of interconnected two-port circuits.

Two-port circuits may be interconnected five ways: (1) in cascade, (2) in series, (3) in parallel, (4) in series-parallel, and (5) in parallel-series. Figure 19.9 depicts these five basic interconnections.

We analyze and illustrate only the cascade connection in this section. However, if the four other connections meet certain requirements, we can obtain the parameters that describe the interconnected circuits by simply adding the individual network parameters. In particular, the z parameters describe the series connection, the y parameters the parallel connection, the h parameters the series-parallel connection, and the g parameters the parallel-series connection.[1]

The cascade connection is important because it occurs frequently in the modeling of large systems. Unlike the other four basic interconnections, there are no restrictions on using the parameters of the individual two-port circuits to obtain the parameters of the interconnected circuits. The a parameters are best suited for describing the cascade connection.

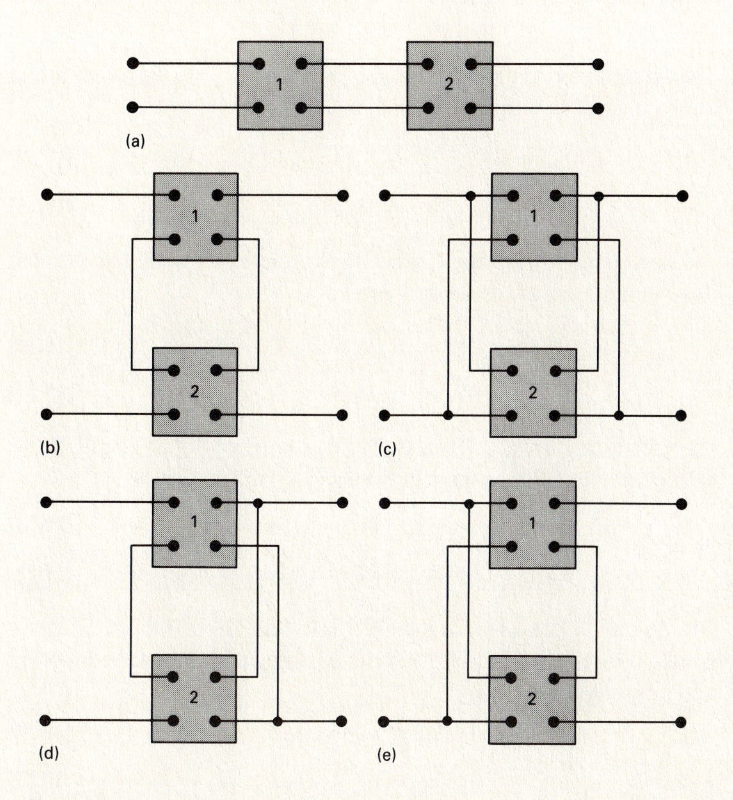

FIGURE 19.9 The five basic interconnections of two-port circuits: (a) cascade; (b) series; (c) parallel; (d) series-parallel; (e) parallel-series.

[1] A detailed discussion of these four interconnections is presented in Henry Ruston and Joseph Bordogna, *Electric Networks: Functions, Filters, Analysis* (New York: McGraw-Hill, 1966), chap. 4.

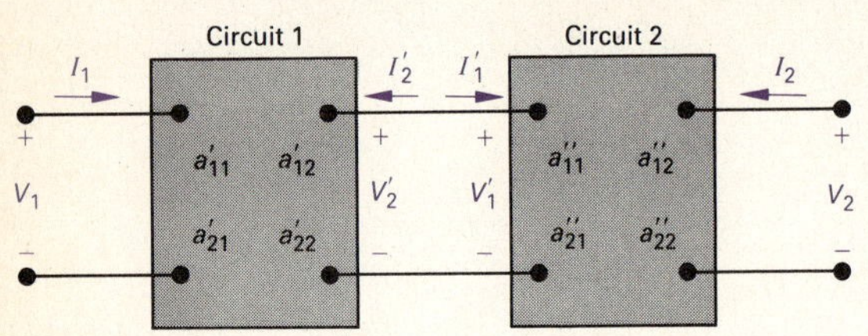

FIGURE 19.10 A cascade connection.

We analyze the cascade connection by using the circuit shown in Fig. 19.10, where a single prime denotes a parameters in the first circuit and a double prime denotes a parameters in the second circuit. The output voltage and current of the first circuit are labeled V_2' and I_2', and the input voltage and current of the second circuit are labeled V_1' and I_1'. The problem is to derive the a-parameter equations that relate V_2 and I_2 to V_1 and I_1. In other words, we seek the pair of equations

$$V_1 = a_{11}V_2 - a_{12}I_2 \quad \textbf{(19.64)}$$

and

$$I_1 = a_{21}V_2 - a_{22}I_2, \quad \textbf{(19.65)}$$

where the a parameters are given explicitly in terms of the a parameters of the individual circuits.

We begin the derivation by noting from Fig. 19.10 that

$$V_1 = a_{11}'V_2' - a_{12}'I_2'; \quad \textbf{(19.66)}$$

$$I_1 = a_{21}'V_2' - a_{22}'I_2'. \quad \textbf{(19.67)}$$

The interconnection means that $V_2' = V_1'$ and $I_2' = -I_1'$. Substituting these constraints into Eqs. (19.66) and (19.67) yields

$$V_1 = a_{11}'V_1' + a_{12}'I_1'; \quad \textbf{(19.68)}$$

$$I_1 = a_{21}'V_1' + a_{22}'I_1'. \quad \textbf{(19.69)}$$

The voltage V_1' and the current I_1' are related to V_2 and I_2 through the a parameters of the second circuit:

$$V_1' = a_{11}''V_2 - a_{12}''I_2; \quad \textbf{(19.70)}$$

$$I_1' = a_{21}''V_2 - a_{22}''I_2. \quad \textbf{(19.71)}$$

We substitute Eqs. (19.70) and (19.71) into Eqs. (19.68) and (19.69) to generate the relationships between V_1, I_1 and V_2, I_2:

$$V_1 = (a_{11}'a_{11}'' + a_{12}'a_{21}'')V_2 - (a_{11}'a_{12}'' + a_{12}'a_{22}'')I_2 \quad \textbf{(19.72)}$$

$$I_1 = (a_{21}'a_{11}'' + a_{22}'a_{21}'')V_2 - (a_{21}'a_{12}'' + a_{22}'a_{22}'')I_2. \quad \textbf{(19.73)}$$

By comparing Eqs. (19.72) and (19.73) to Eqs. (19.64) and (19.65), we get the desired expressions for the a parameters of the interconnected networks, namely,

$$a_{11} = a_{11}'a_{11}'' + a_{12}'a_{21}''; \quad \textbf{(19.74)}$$

$$a_{12} = a_{11}'a_{12}'' + a_{12}'a_{22}''; \quad \textbf{(19.75)}$$

$$a_{21} = a_{21}'a_{11}'' + a_{22}'a_{21}''; \quad \textbf{(19.76)}$$

$$a_{22} = a_{21}'a_{12}'' + a_{22}'a_{22}''. \quad \textbf{(19.77)}$$

Example 19.5 illustrates how to use Eqs. (19.74)–(19.77) to analyze a cascade connection with two amplifier circuits.

EXAMPLE 19.5

Two identical amplifiers are connected in cascade, as shown in Fig. 19.11. Each amplifier is described in terms of its h parameters. The values are $h_{11} = 1000\,\Omega$, $h_{12} = 0.0015$, $h_{21} = 100$, and $h_{22} = 100\,\mu\text{S}$. Find the voltage gain V_2/V_g.

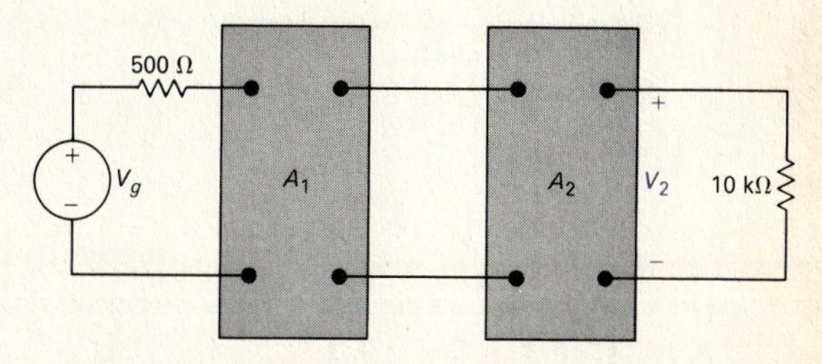

FIGURE 19.11 The circuit for Example 19.5.

SOLUTION

The first step in finding V_2/V_g is to convert from h parameters to a parameters. The amplifiers are identical, so one set of a parameters describes the amplifiers:

$$a'_{11} = \frac{-\Delta h}{h_{21}} = \frac{+0.05}{100} = 5 \times 10^{-4},$$

$$a'_{12} = \frac{-h_{11}}{h_{21}} = \frac{-1000}{100} = -10\,\Omega,$$

$$a'_{21} = \frac{-h_{22}}{h_{21}} = \frac{-100 \times 10^{-6}}{100} = -10^{-6}\,\text{S},$$

$$a'_{22} = \frac{-1}{h_{21}} = \frac{-1}{100} = -10^{-2}.$$

Next we use Eqs. (19.74)–(19.77) to compute the a parameters of the cascaded amplifiers:

$$\begin{aligned}
a_{11} &= a'_{11}a'_{11} + a'_{12}a'_{21} \\
&= 25 \times 10^{-8} + (-10)(-10^{-6}) = 10.25 \times 10^{-6}; \\
a_{12} &= a'_{11}a'_{12} + a'_{12}a'_{22} \\
&= (5 \times 10^{-4})(-10) + (-10)(-10^{-2}) = 0.095\,\Omega; \\
a_{21} &= a'_{21}a'_{11} + a'_{22}a'_{21} \\
&= (-10^{-6})(5 \times 10^{-4}) + (-10^{-6})(-0.01) \\
&= 0.0095 \times 10^{-6}\,\text{S}; \\
a_{22} &= a'_{21}a'_{12} + a'_{22}a'_{22} \\
&= (-10)(-10^{-6}) + (-10^{-2})^2 = 1.1 \times 10^{-4}.
\end{aligned}$$

From Table 19.2,

$$\frac{V_2}{V_g} = \frac{Z_L}{(a_{11} + a_{21}Z_g)Z_L + a_{12} + a_{22}Z_g}$$

$$= \frac{10^4}{[10.25 \times 10^{-6} + 0.0095 \times 10^{-6}(500)]10^4 + 0.095 + 1.1 \times 10^{-4}(500)}$$

$$= \frac{10^4}{0.15 + 0.095 + 0.055} = \frac{10^5}{3} = 33{,}333.33.$$

Thus an input signal of 150 μV is amplified to an output signal of 5 V. For an alternative approach to finding the voltage gain V_2/V_g, see Problem 19.41.

If more than two units are connected in cascade, the a parameters of the equivalent two-port circuit can be found by successively reducing the original set of two-port circuits one pair at a time.

DRILL EXERCISE

19.8 Each element in the symmetric bridged-tee circuit shown is a 15 Ω resistor. Two of these bridged tees are connected in cascade between a dc voltage source and a resistive load. The dc voltage source has a no-load voltage of 100 V and an internal resistance of 8 Ω. The load resistor is adjusted until maximum power is delivered to the load. Calculate (a) the load resistance, (b) the load voltage, and (c) the load power.

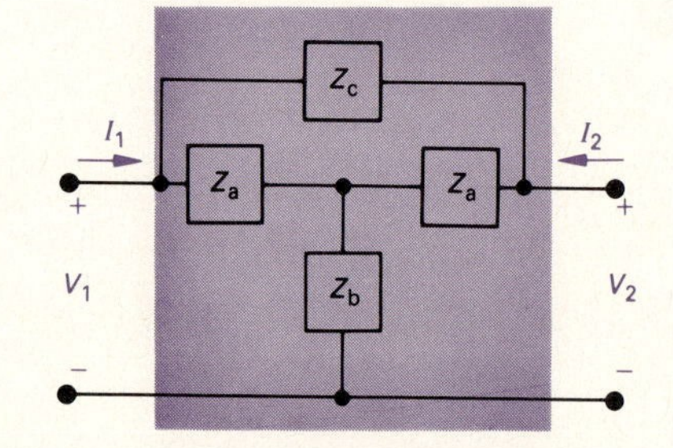

ANSWER: (a) 14.44 Ω; (b) 16 V; (c) 17.73 W.

SUMMARY

- The **two-port model** is used to describe the performance of a circuit in terms of the voltage and current at its input and output ports.
- The model is limited to circuits in which
  - — no independent sources are inside the circuit between the ports;
  - — no energy is stored inside the circuit between the ports;
  - — the current into the port is equal to the current out of the port; and
  - — no external connections exist between the input and output ports.

- Two of the four terminal variables (V_1, I_1, V_2, I_2) are independent; therefore only two simultaneous equations involving the four variables are needed to describe the circuit.
- The six possible sets of simultaneous equations involving the four terminal variables are called the z-, y-, a-, b-, h-, and g-parameter equations. See Eqs. (19.1)–(19.6).
- The parameter equations are written in the s domain. The dc values of the parameters are obtained by setting $s = 0$, and the sinusoidal steady-state values are obtained by setting $s = j\omega$.
- The relationships among the six sets of parameters are given in Table 19.1.
- Any set of parameters may be calculated or measured by invoking appropriate short-circuit and open-circuit conditions at the input and output ports. See Eqs. (19.7)–(19.15).
- A two-port circuit is **reciprocal** if the interchange of an ideal voltage source at one port with an ideal ammeter at the other port produces the same ammeter reading. The effect of reciprocity on the two-port parameters is given by Eqs. (19.28) through (19.33).
- A reciprocal two-port circuit is **symmetric** if its ports can be interchanged without disturbing the values of the terminal currents and voltages. The added effect of symmetry on the two-port parameters is given by Eqs. (19.38) through (19.43).
- The performance of a two-port circuit connected to a Thévenin equivalent source and a load is summarized by the relationships given in Table 19.2.
- Large networks can be divided into subnetworks by means of interconnected two-port models. The cascade connection was used in this chapter to illustrate the analysis of interconnected two-port circuits.

PROBLEMS

19.1 Find the h and g parameters for the circuit in Example 19.1.

19.2 Find the z parameters for the circuit shown in Fig. P19.2.

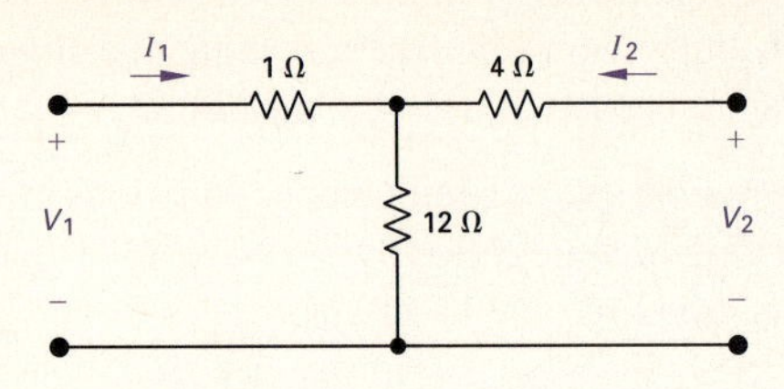

FIGURE P19.2

19.3 Find the b parameters for the circuit shown in Fig. P19.3.

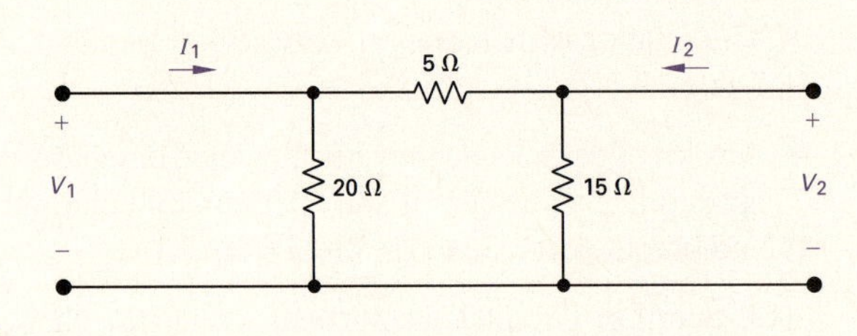

FIGURE P19.3

19.4 Find the g parameters for the circuit in Fig. P19.4.

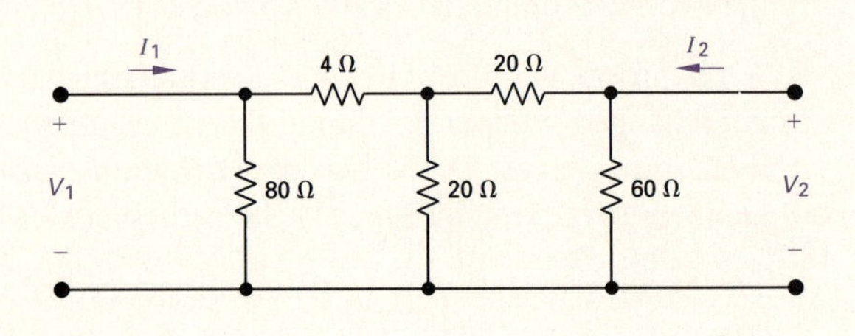

FIGURE P19.4

19.5 Find the a parameters for the circuit in Fig. P19.5.

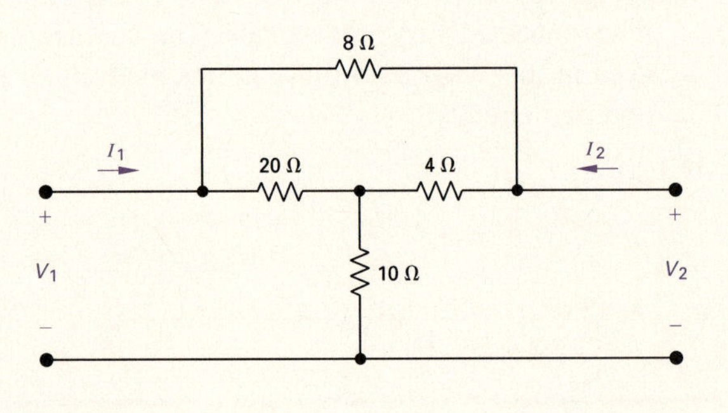

FIGURE P19.5

19.6 Use the results obtained in Problem 19.5 to calculate the h parameters of the circuit in Fig. P19.5.

19.7 Find the y parameters for the circuit in Fig. P19.7.

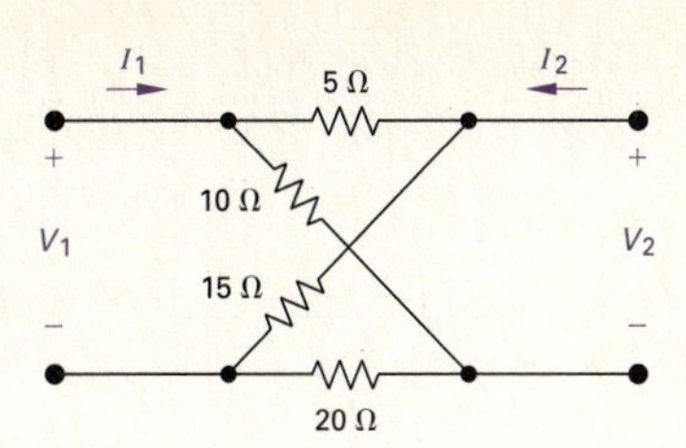

FIGURE P19.7

19.8 Use the results obtained in Problem 19.7 to calculate the z parameters for the circuit in Fig. P19.7.

❖ **19.9** Select the values of R_1, R_2, and R_3 in the circuit in Fig. P19.9 so that $b_{11} = 5$, $b_{12} = 140\ \Omega$, $b_{21} = 100$ mS, and $b_{22} = 3$.

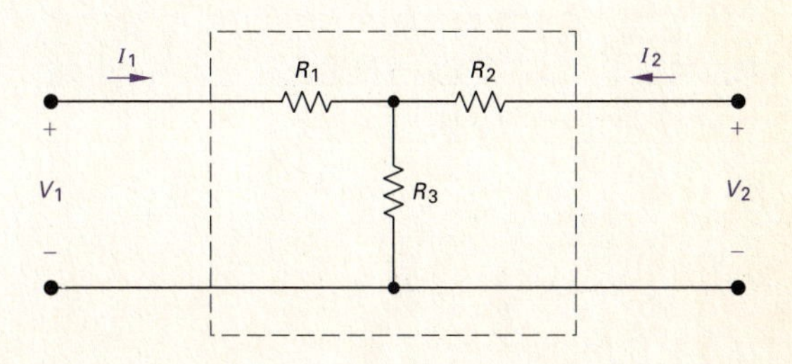

FIGURE P19.9

19.10 Find the g parameters of the two-port circuit shown in Fig. P19.10.

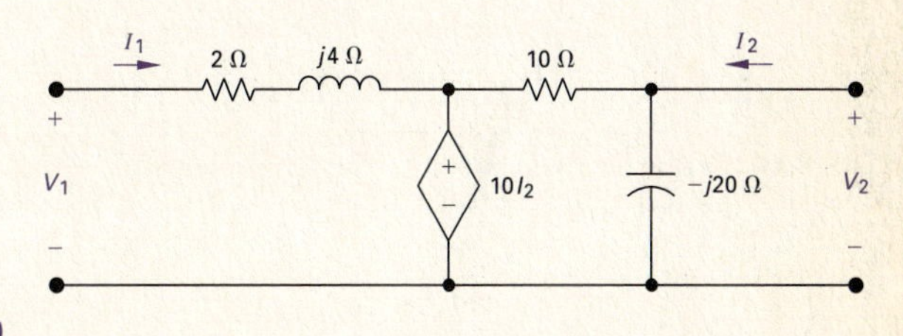

FIGURE P19.10

19.11 The following direct-current measurements were made on the two-port network shown in Fig. P19.11.

Port 2 Open	Port 2 Short-Circuited
$V_1 = 8$ mV	$V_1 = 5$ V
$I_1 = 4\ \mu$A	$I_1 = 5$ mA
$V_2 = -8$ V	$I_2 = 250$ mA

Calculate the h parameters for the network.

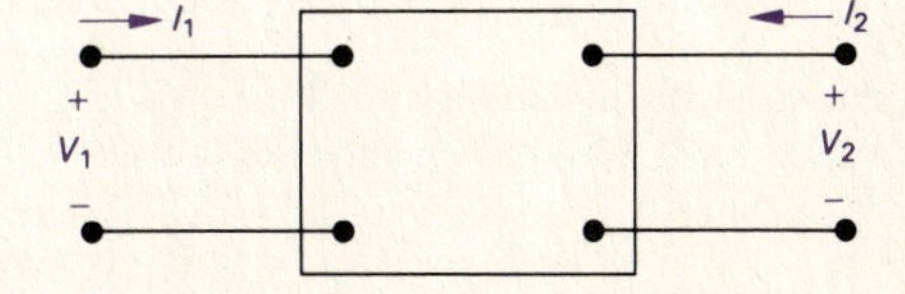

FIGURE P19.11

19.12 a) Use the measurements given in Problem 19.11 to find the a parameters for the network.

b) Check your calculations by finding the a parameters directly from the h parameters found in Problem 19.11.

19.13 Find the frequency-domain values of the a parameters for the two-port circuit shown in Fig. P19.13.

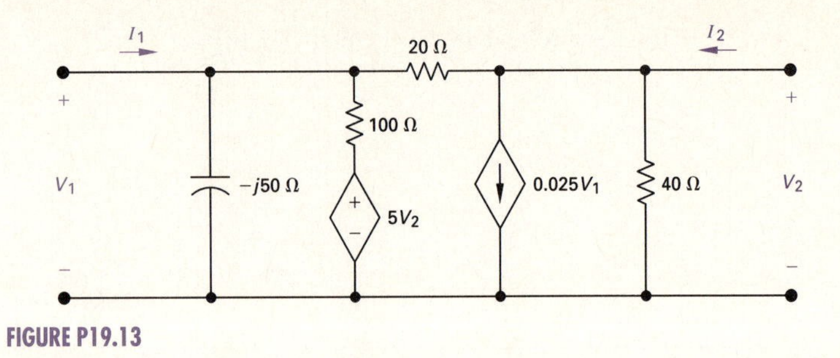

FIGURE P19.13

19.14 Find the h parameters for the two-port circuit shown in Fig. P19.13.

19.15 Find the g parameters for the operational amplifier circuit shown in Fig. P19.15.

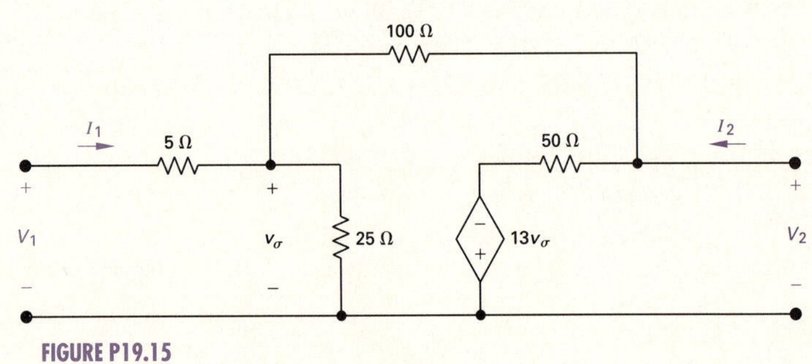

FIGURE P19.15

19.16 Derive the expressions for the h parameters as functions of the a parameters.

19.17 Derive the expressions for the y parameters as functions of the b parameters.

19.18 Derive the expressions for the g parameters as functions of the z parameters.

19.19 The operational amplifier in the circuit shown in Fig. P19.19 is ideal. Find the h parameters of the circuit.

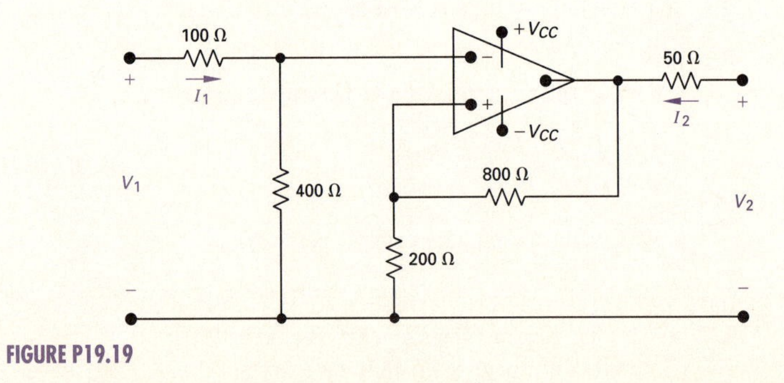

FIGURE P19.19

19.20 Find the s-domain expressions for the a parameters of the two-port circuit shown in Fig. P19.20.

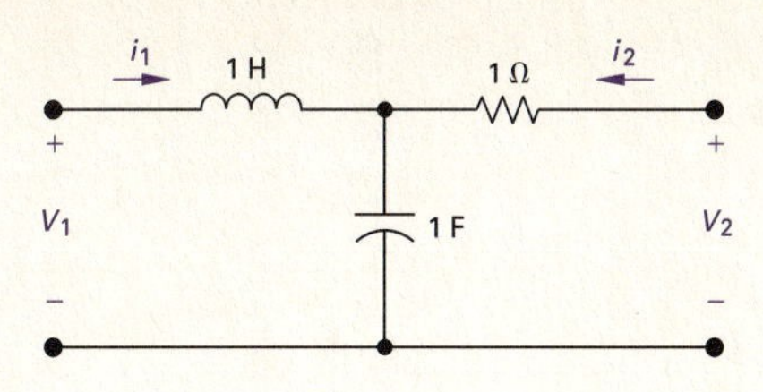

FIGURE P19.20

19.21 Find the s-domain expressions for the y parameters of the two-port circuit shown in Fig. P19.21.

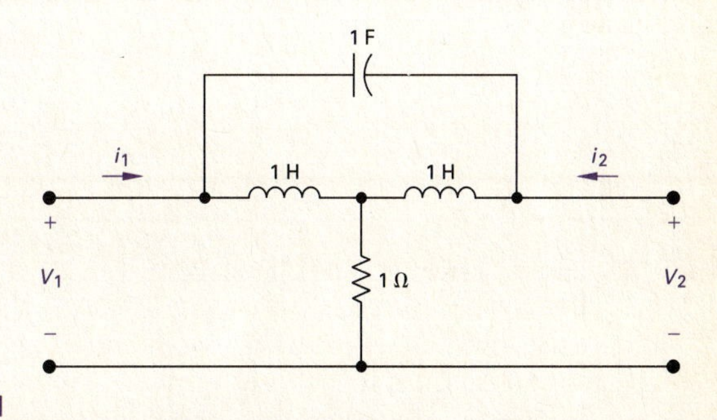

FIGURE P19.21

19.22 Is the two-port circuit shown in Fig. P19.22 symmetric? Justify your answer.

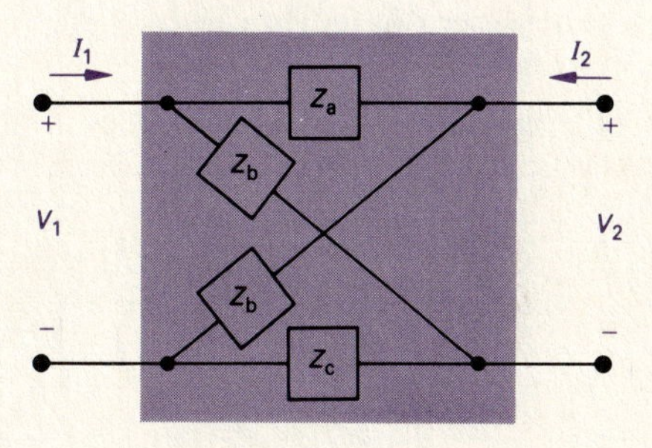

FIGURE P19.22

19.23 a) Use the defining equations to find the s-domain expressions for the h parameters for the circuit in Fig. P19.23.

b) Show that the results obtained in part (a) agree with the h-parameter relationships for a reciprocal symmetric network.

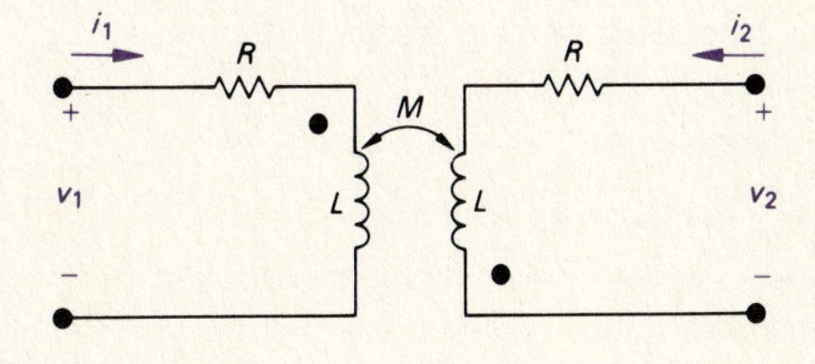

FIGURE P19.23

19.24 Derive the expression for the input impedance ($Z_{in} = V_1/I_1$) of the circuit if Fig. 19.7 in terms of the b parameters.

19.25 Derive the expression for the current gain I_2/I_1 of the circuit in Fig. 19.7 in terms of the g parameters.

19.26 Derive the expression for the voltage gain V_2/V_1 of the circuit in Fig. 19.7 in terms of the y parameters.

19.27 Derive the expression for the voltage gain V_2/V_g of the circuit in Fig. 19.7 in terms of the h parameters.

19.28 Find the Thévenin equivalent circuit with respect to port 2 of the circuit in Fig. 19.7 in terms of the z parameters.

19.29 The linear transformer in the circuit shown in Fig. P19.29 has a coefficient of coupling of 0.80. The transformer is driven by a sinusoidal voltage source whose internal voltage is $v_g = 125 \cos 5000t$ V. The internal impedance of the source is $70 + j0\ \Omega$.

a) Find the frequency-domain z parameters of the linear transformer.

b) Use the z parameters to derive the Thévenin equivalent circuit with respect to the terminals of the load.

c) Derive the steady-state time-domain expression for v_2.

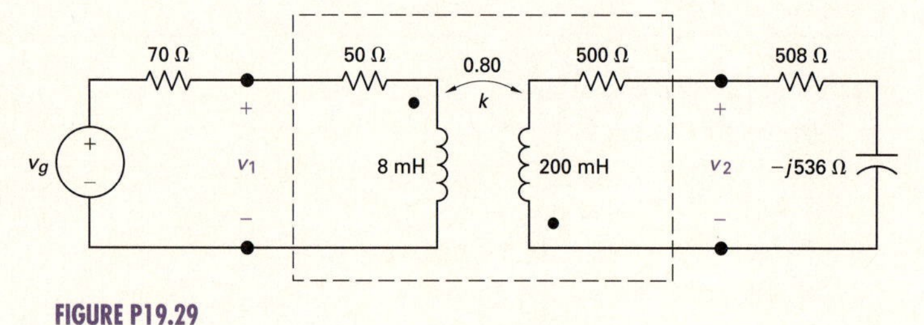

FIGURE P19.29

19.30 The a parameters of the amplifier in the circuit shown in Fig. P19.30 are

$a_{11} = a_{22} = 0.50;$

$a_{12} = j10\ \Omega;\ a_{21} = j75$ mS.

Find the ratio of the output power to that supplied by the ideal voltage source.

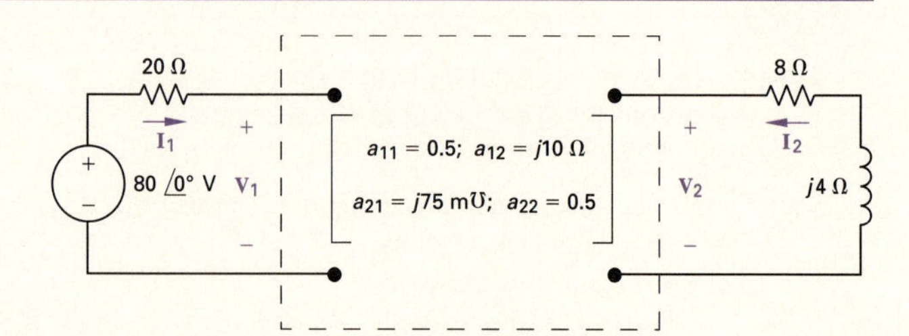

FIGURE P19.30

19.31 The b parameters for the two-port circuit in Fig. P19.31 are

$$b_{11} = \frac{3 + j1}{3}; b_{12} = -1 + j4\ \Omega$$

$$b_{21} = \frac{1}{3}\text{S}; b_{22} = 1 + j1$$

The load impedance Z_L is adjusted for maximum average power transfer to Z_L. The ideal voltage source is generating a sinusoidal voltage of

$$v_g = 90 \cos 8000t \text{ V}.$$

a) Find the rms value of V_2.

b) Find the average power delivered to Z_L.

c) What percentage of the average power developed by the ideal voltage source is delivered by Z_L?

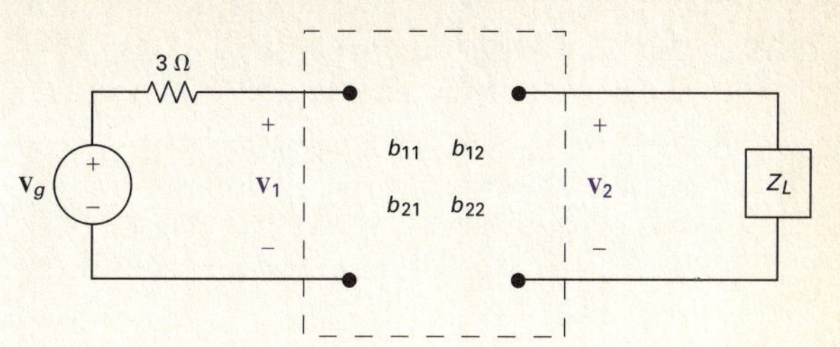

FIGURE P19.31

19.32 The h parameters for the two-port power amplifier circuit in Fig. P19.32 are

$h_{11} = 500\ \Omega; h_{12} = 10^{-3};$

$h_{21} = 50;$ and $h_{22} = 50\ \mu\text{S}.$

The internal impedance of the source is $1500 + j0\ \Omega$, and the load impedance is $10{,}000 + j0\ \Omega$. The ideal voltage source is generating a voltage

$$v_g = 250 \cos 40{,}000t \text{ mV}.$$

a) Find the rms value of V_2.

b) Find the average power delivered to Z_L.

c) Find the average power developed by the ideal voltage source.

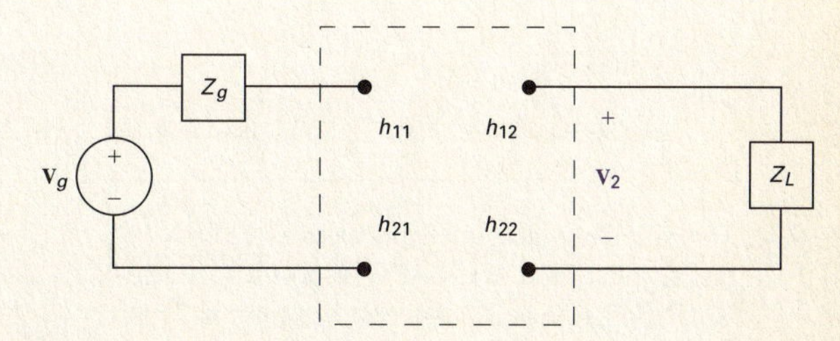

FIGURE P19.32

19.33 For the terminated two-port amplifier circuit in Fig. P19.32, find

a) the value of Z_L for maximum average power transfer to Z_L;

b) the maximum average power delivered to Z_L; and

c) the average power developed by the ideal voltage source when maximum power is delivered to Z_L.

19.34 a) Find the s-domain expressions for the h parameters of the circuit in Fig. P19.34.

b) Port 2 in Fig. P19.34 is terminated in a resistance of 800Ω, and port 1 is driven by a step voltage source $v_1(t) = 45u(t)$ V. Find $v_2(t)$ for $t > 0$ if $C = 0.1\ \mu$F and $L = 400$ mH.

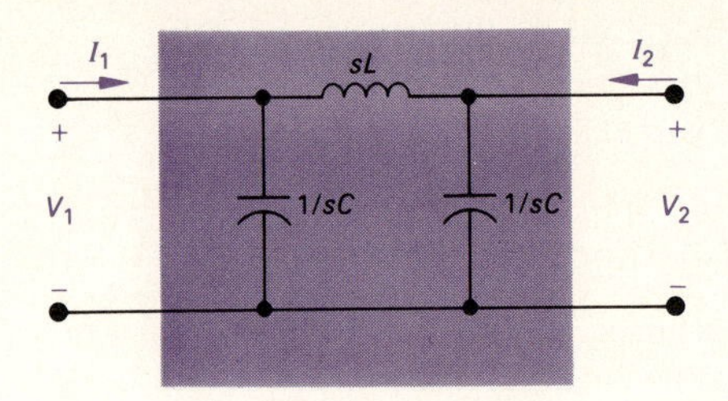

FIGURE P19.34

19.35 a) Find the z parameters for the two-port network in Fig. P19.35.

b) Find v_2 for $t > 0$ when $v_g = 50u(t)$ V.

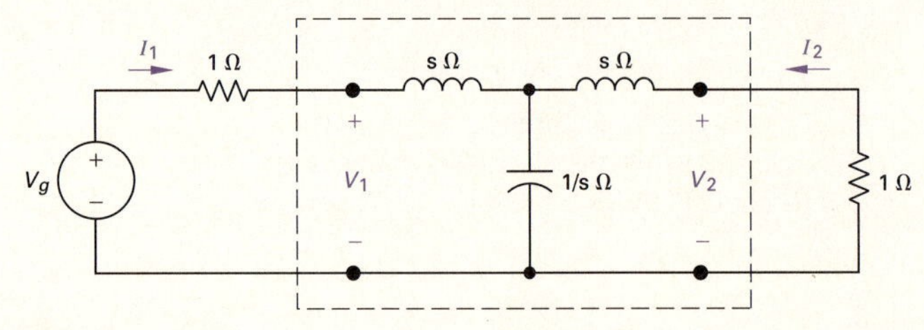

FIGURE P19.35

19.36 The following measurements were made on a resistive two-port network. With port 2 open and 90 V applied to port 1, the port 1 current is 2.5 A, and the port 2 voltage is 50 V. With port 2 short-circuited and 195.2 mV applied to port 1, the port 1 current is 8.2 mA, and the port 2 current is −5 mA. Find the maximum power (milliwatts) that this two-port circuit can deliver to a resistive load at port 2 when port 1 is driven by a dc voltage source with an internal resistance of 4 Ω and an internal voltage of 22.8 V.

19.37 The following dc measurements were made on the resistive network shown in Fig. P19.37.

Measurement 1	Measurement 2
$V_1 = 4$ V	$V_1 = 20$ mV
$I_1 = 5$ mA	$I_1 = 20\ \mu$A
$V_2 = 0$ V	$V_2 = 40$ V
$I_2 = -200$ mA	$I_2 = 0$ A

A variable resistor R_o is connected across port 2 and adjusted for maximum power transfer to R_o. Find the maximum power.

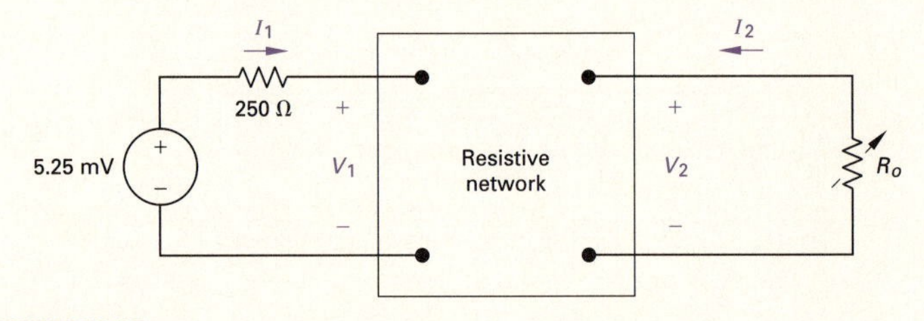

FIGURE P19.37

19.38 The two networks shown in Fig. P19.38 are identical. The g parameters of each network are

$$g_{11} = 1 \text{ mS}; g_{12} = -4 \times 10^{-3};$$

$$g_{21} = -4000; g_{22} = 96 \text{ k}\Omega.$$

Calculate the output voltage v_o if $v_g = 4.8$ mV dc.

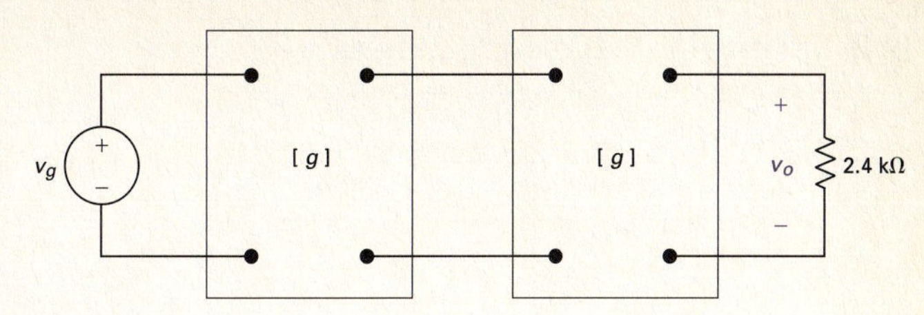

FIGURE P19.38

19.39 The h parameters of the first two-port circuit in Fig. P19.39(a) are

$$h_{11} = 1000 \ \Omega; h_{12} = 5 \times 10^{-4};$$

$$h_{21} = 40; \text{ and } h_{22} = 25 \ \mu\text{S}.$$

The circuit in the second two-port circuit is shown in Fig. P19.39(b), where $R = 72$ kΩ. Find v_o if $v_g = 9$ mV dc.

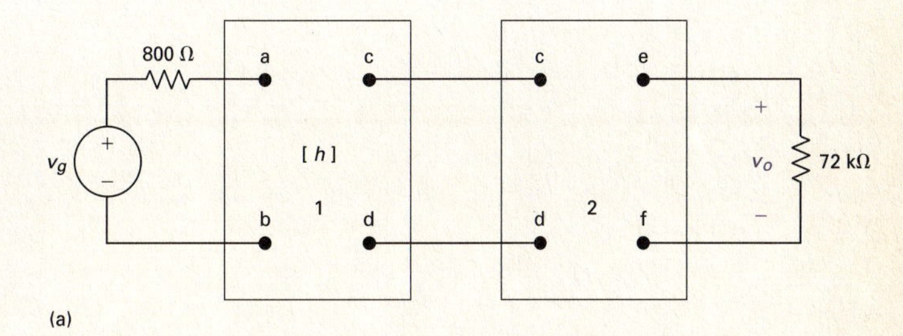

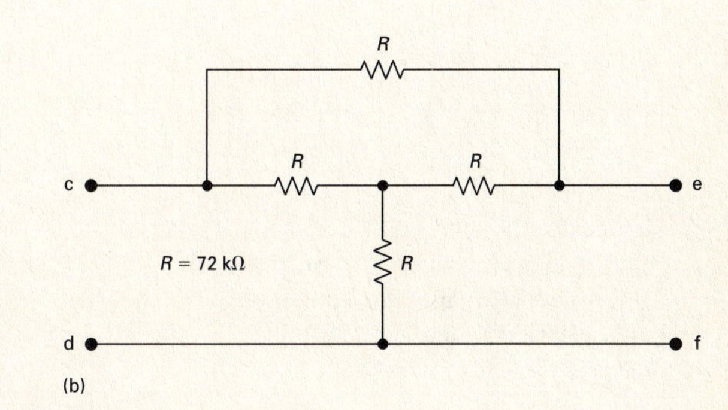

FIGURE P19.39

19.40 The networks D and E in the circuit in Fig. P19.40 are reciprocal and symmetric. For network D, it is known that $a'_{11} = 5$ and $a'_{12} = 24\Omega$. The impedance Z_o is adjusted for maximum average power transfer to Z_o. Find Z_o if $Z_g = (5 + j0)\ \Omega$.

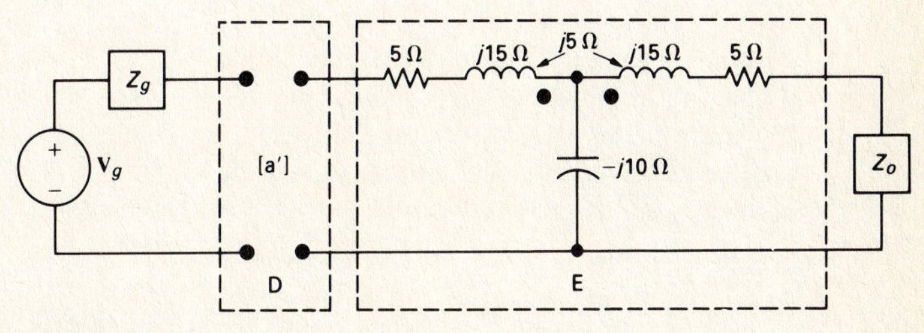

FIGURE P19.40

19.41 a) Show that the circuit in Fig. P19.41 is an equivalent circuit satisfied by the h-parameter equations.

b) Use the h-parameter equivalent circuit of part (a) to find the voltage gain V_2/V_g in the circuit in Fig. 19.11.

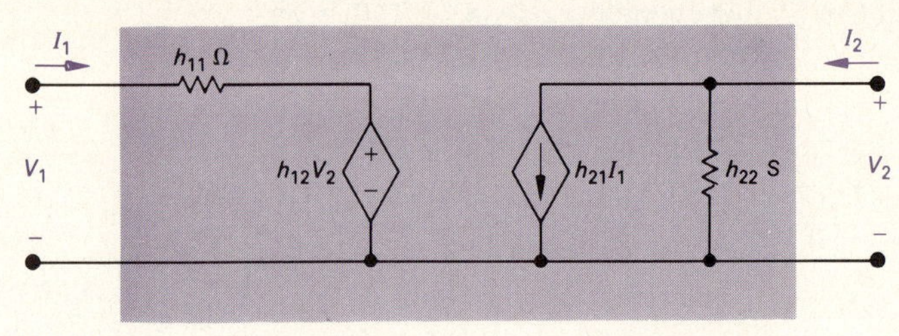

FIGURE P19.41

19.42 a) Show that the circuit in Fig. P19.42 is an equivalent circuit satisfied by the z-parameter equations.

b) Assume that the equivalent circuit in Fig. P19.42 is driven by a voltage source having an internal impedance of Z_g ohms. Calculate the Thévenin equivalent circuit with respect to port 2. Check your results against the appropriate entries in Table 19.2.

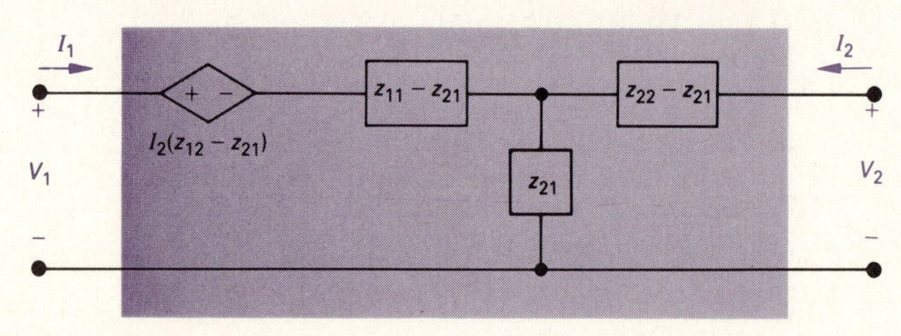

FIGURE P19.42

19.43 a) Show that the circuit in Fig. P19.43 is also an equivalent circuit satisfied by the z-parameter equations.

b) Assume that the equivalent circuit in Fig. P19.43 is terminated in an impedance of Z_L ohms at port 2. Find the input impedance V_1/I_1. Check your results against the appropriate entry in Table 19.2.

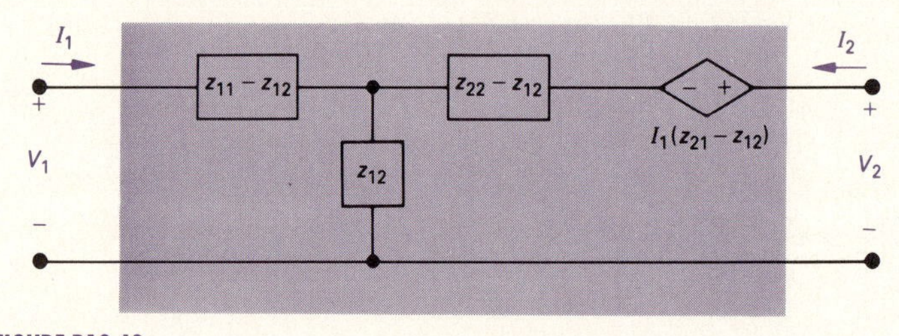

FIGURE P19.43

19.44 a) Derive two equivalent circuits that are satisfied by the y-parameter equations. *Hint:* Start with Eqs. (19.2). Add and subtract $y_{21}V_2$ to the first equation of the set. Construct a circuit that satisfies the resulting set of equations, by thinking in terms of node voltages. Derive an alternative equivalent circuit by first altering the second equation in Eqs. (19.2).

b) Assume that port 1 is driven by a voltage source having an internal impedance Z_g, and port 2 is loaded with an impedance Z_L. Find the current gain I_2/I_1. Check your results against the appropriate entry in Table 19.2.

19.45 a) Derive the equivalent circuit satisfied by the g-parameter equations.

b) Use the g-parameter equivalent circuit derived in part (a) to solve for the output voltage in Problem 19.39. *Hint:* Use Problem 3.60 to simplify the second two-port circuit in Problem 19.39.

The Solution of Linear Simultaneous Equations

Circuit analysis frequently involves the solution of linear simultaneous equations. Our purpose here is to review the use of determinants to solve such a set of equations. The theory of determinants (with applications) can be found in most intermediate-level algebra texts. (A particularly good reference for engineering students is Chapter 1 of E.A. Guillemin's *The Mathematics of Circuit Analysis* [New York: Wiley, 1949]. In our review here, we will limit our discussion to the mechanics of solving simultaneous equations with determinants.

A.1 Preliminary Steps

The first step in solving a set of simultaneous equations by determinants is to write the equations in a rectangular (square) format. In other words, we arrange the equations in a vertical stack such that each variable occupies the same horizontal position in every equation. For example, in Eqs. (A.1), the variables i_1, i_2, and i_3 occupy the first, second, and third position, respectively, on the left-hand side of each equation:

$$\begin{aligned} 21i_1 - 9i_2 - 12i_3 &= -33, \\ -3i_1 + 6i_2 - 2i_3 &= 3, \\ -8i_1 - 4i_2 + 22i_3 &= 50. \end{aligned} \tag{A.1}$$

Alternatively, one can describe this set of equations by saying that i_1 occupies the first column in the array, i_2 the second column, and i_3 the third column.

If one or more variables are missing from a given equation, they can be inserted by simply making their coefficient zero. Thus Eqs. (A.2) can be "squared up" as shown by Eqs. (A.3):

$$\begin{aligned} 2v_1 - v_2 &= 4, \\ 4v_2 + 3v_3 &= 16, \\ 7v_1 + 2v_3 &= 5; \end{aligned} \tag{A.2}$$

$$\begin{aligned} 2v_1 - v_2 + 0v_3 &= 4, \\ 0v_1 + 4v_2 + 3v_3 &= 16, \\ 7v_1 + 0v_2 + 2v_3 &= 5. \end{aligned} \tag{A.3}$$

A.2 Cramer's Method

The value of each unknown variable in the set of equations is expressed as the ratio of two determinants. If we let N, with an appropriate subscript, represent the numerator determinant and Δ represent the denominator determinant, then the kth unknown x_k is

$$x_k = \frac{N_k}{\Delta}. \tag{A.4}$$

The denominator determinant Δ is the same for every unknown variable and is called the **characteristic determinant** of the set of equations. The numerator determinant N_k varies with each unknown. Equation (A.4) is referred to as **Cramer's method** for solving simultaneous equations.

A.3 The Characteristic Determinant

Once we have organized the set of simultaneous equations into an ordered array, as illustrated by Eqs. (A.1) and (A.3), it is a simple matter to form the characteristic determinant. This determinant is the square array made up from the coefficients of the unknown variables.

For example, the characteristic determinants of Eqs. (A.1) and (A.3) are

$$\Delta = \begin{vmatrix} 21 & -9 & -12 \\ -3 & 6 & -2 \\ -8 & -4 & 22 \end{vmatrix} \tag{A.5}$$

and

$$\Delta = \begin{vmatrix} 2 & -1 & 0 \\ 0 & 4 & 3 \\ 7 & 0 & 2 \end{vmatrix}, \tag{A.6}$$

respectively.

A.4 THE NUMERATOR DETERMINANT

The numerator determinant N_k is formed from the characteristic determinant by replacing the kth column in the characteristic determinant with the column of values appearing on the right-hand side of the equations. For example, the numerator determinants for evaluating i_1, i_2, and i_3 in Eqs. (A.1) are

$$N_1 = \begin{vmatrix} -33 & -9 & -12 \\ 3 & 6 & -2 \\ 50 & -4 & 22 \end{vmatrix}, \tag{A.7}$$

$$N_2 = \begin{vmatrix} 21 & -33 & -12 \\ -3 & 3 & -2 \\ -8 & 50 & 22 \end{vmatrix}, \tag{A.8}$$

and

$$N_3 = \begin{vmatrix} 21 & -9 & -33 \\ -3 & 6 & 3 \\ -8 & -4 & 50 \end{vmatrix}. \tag{A.9}$$

The numerator determinants for the evaluation of v_1, v_2, and v_3 in Eqs. (A.3) are

$$N_1 = \begin{vmatrix} 4 & -1 & 0 \\ 16 & 4 & 3 \\ 5 & 0 & 2 \end{vmatrix}, \tag{A.10}$$

$$N_2 = \begin{vmatrix} 2 & 4 & 0 \\ 0 & 16 & 3 \\ 7 & 5 & 2 \end{vmatrix}, \tag{A.11}$$

and

$$N_3 = \begin{vmatrix} 2 & -1 & 4 \\ 0 & 4 & 16 \\ 7 & 0 & 5 \end{vmatrix}. \quad \text{(A.12)}$$

A.5 The Evaluation of a Determinant

The value of a determinant is found by expanding it in terms of its minors. The **minor** of any element in a determinant is the determinant that remains after the row and column occupied by the element have been deleted. For example, the minor of the element 6 in Eq. (A.7) is

$$\begin{vmatrix} -33 & -12 \\ 50 & 22 \end{vmatrix},$$

while the minor of the element 22 in Eq. (A.7) is

$$\begin{vmatrix} -33 & -9 \\ 3 & 6 \end{vmatrix}.$$

The **cofactor** of an element is its minor multiplied by the sign-controlling factor

$$-1^{(i+j)},$$

where i and j denote the row and column, respectively, occupied by the element. Thus the cofactor of the element 6 in Eq. (A.7) is

$$-1^{(2+2)}\begin{vmatrix} -33 & -12 \\ 50 & 22 \end{vmatrix},$$

and the cofactor of the element 22 is

$$-1^{(3+3)}\begin{vmatrix} -33 & -9 \\ 3 & 6 \end{vmatrix}.$$

The cofactor of an element is also referred to as its **signed minor**.

The sign-controlling factor $-1^{(i+j)}$ will equal $+1$ or -1 depending on whether $i + j$ is an even or odd integer. Thus the algebraic sign of a cofactor alternates between $+1$ and -1 as we move along a row or column. For a 3×3 determinant, the plus and minus signs form the checkerboard pattern illustrated here:

$$\begin{vmatrix} + & - & + \\ - & + & - \\ + & - & + \end{vmatrix}$$

A determinant can be expanded along any row or column. Thus the first step in making an expansion is to select a row i or a column j. Once a row or column has been selected, each element in that row or column is multiplied by its signed minor, or cofactor. The value of the determinant is the sum of these products. As an example, let us evaluate the determinant in Eq. (A.5) by expanding it along its first column. Following the rules just explained, we write the expansion as

$$\Delta = 21(1)\begin{vmatrix} 6 & -2 \\ -4 & 22 \end{vmatrix} - 3(-1)\begin{vmatrix} -9 & -12 \\ -4 & 22 \end{vmatrix} - 8(1)\begin{vmatrix} -9 & -12 \\ 6 & -2 \end{vmatrix}. \tag{A.13}$$

The 2×2 determinants in Eq. (A.13) can also be expanded by minors. The minor of an element in a 2×2 determinant is a single element. It follows that the expansion reduces to multiplying the upper-left element by the lower-right element and then subtracting from this product the product of the lower-left element times the upper-right element. Using this observation, we evaluate Eq. (A.13) to

$$\begin{aligned} \Delta &= 21(132 - 8) + 3(-198 - 48) - 8(18 + 72) \\ &= 2604 - 738 - 720 = 1146. \end{aligned} \tag{A.14}$$

Had we elected to expand the determinant along the second row of elements, we would have written

$$\begin{aligned} \Delta &= -3(-1)\begin{vmatrix} -9 & -12 \\ -4 & 22 \end{vmatrix} + 6(+1)\begin{vmatrix} 21 & -12 \\ -8 & 22 \end{vmatrix} - 2(-1)\begin{vmatrix} 21 & -9 \\ -8 & -4 \end{vmatrix} \\ &= 3(-198 - 48) + 6(462 - 96) + 2(-84 - 72) \\ &= -738 + 2196 - 312 = 1146. \end{aligned} \tag{A.15}$$

The numerical values of the determinants N_1, N_2, and N_3 given by Eqs. (A.7), (A.8), and (A.9) are

$$N_1 = 1146, \tag{A.16}$$

$$N_2 = 2292, \tag{A.17}$$

and

$$N_3 = 3438. \tag{A.18}$$

It follows from Eqs. (A.15) through (A.18) that the solutions for i_1, i_2, and i_3 in Eq. (A.1) are

$$\begin{aligned} i_1 &= \frac{N_1}{\Delta} = 1\text{ A}, \\ i_2 &= \frac{N_2}{\Delta} = 2\text{ A}, \end{aligned} \tag{A.19}$$

and

$$i_3 = \frac{N_3}{\Delta} = 3 \text{ A}.$$

We leave you to verify that the solutions for v_1, v_2, and v_3 in Eqs. (A.3) are

$$v_1 = \frac{49}{-5} = -9.8 \text{ V},$$

$$v_2 = \frac{118}{-5} = -23.6 \text{ V}, \tag{A.20}$$

and

$$v_3 = \frac{-184}{-5} = 36.8 \text{ V}.$$

A.6 MATRICES

A system of simultaneous linear equations can also be solved using matrices. In what follows, we briefly review matrix notation, algebra, and terminology.[1]

A **matrix** is by definition a rectangular array of elements; thus

$$\mathbf{A} = \begin{bmatrix} a_{11} & a_{12} & a_{13} & \cdots & a_{1n} \\ a_{21} & a_{22} & a_{23} & \cdots & a_{2n} \\ \cdots & \cdots & \cdots & \cdots & \cdots \\ a_{m1} & a_{m2} & a_{m3} & \cdots & a_{mn} \end{bmatrix} \tag{A.21}$$

is a matrix with m rows and n columns. We describe A as being a matrix of order m by n, or $m \times n$, where m equals the number of rows and n the number of columns. We always specify the rows first and the columns second. The elements of the matrix—$a_{11}, a_{12}, a_{13}, \ldots$—can be real numbers, complex numbers, or functions. We denote a matrix with a boldface capital letter.

The array in Eq. (A.21) is frequently abbreviated by writing

$$\mathbf{A} = [a_{ij}]_{mn}, \tag{A.22}$$

where a_{ij} is the element in the ith row and the jth column.

If $m = 1$, **A** is called a **row matrix**, that is,

$$\mathbf{A} = [\ a_{11} \quad a_{12} \quad a_{13} \quad \cdots \quad a_{1n}\ ]. \tag{A.23}$$

---

[1]An excellent introductory-level text in matrix applications to circuit analysis is Lawrence P. Huelsman, *Circuits, Matrices, and Linear Vector Spaces* (New York: McGraw-Hill, 1963).

If $n = 1$, **A** is called a **column matrix**, that is,

$$\mathbf{A} = \begin{bmatrix} a_{11} \\ a_{21} \\ a_{31} \\ \vdots \\ a_{m1} \end{bmatrix}. \qquad \textbf{(A.24)}$$

If $m = n$, **A** is called a **square matrix**. For example, if $m = n = 3$, the square 3 by 3 matrix is

$$\mathbf{A} = \begin{bmatrix} a_{11} & a_{12} & a_{13} \\ a_{21} & a_{22} & a_{23} \\ a_{31} & a_{32} & a_{33} \end{bmatrix}. \qquad \textbf{(A.25)}$$

Also note that we use brackets [ ] to denote a matrix, whereas we use vertical lines | | to denote a determinant. It is important to know the difference. A matrix is a rectangular array of elements. A **determinant** is a function of a square array of elements. Thus if a matrix **A** is square, we can define the determinant of **A**. For example, if

$$\mathbf{A} = \begin{bmatrix} 2 & 1 \\ 6 & 15 \end{bmatrix},$$

then

$$\det \mathbf{A} = \begin{vmatrix} 2 & 1 \\ 6 & 15 \end{vmatrix} = 30 - 6 = 24.$$

A.7 MATRIX ALGEBRA

The equality, addition, and subtraction of matrices apply only to matrices of the same order. Two matrices are equal if, and only if, their corresponding elements are equal. In other words, $\mathbf{A} = \mathbf{B}$ if, and only if, $a_{ij} = b_{ij}$ for all i and j. For example, the two matrices in Eqs. (A.26) and (A.27) are equal because $a_{11} = b_{11}$, $a_{12} = b_{12}$, $a_{21} = b_{21}$, and $a_{22} = b_{22}$:

$$\mathbf{A} = \begin{bmatrix} 36 & -20 \\ 4 & 16 \end{bmatrix}, \qquad \textbf{(A.26)}$$

$$\mathbf{B} = \begin{bmatrix} 36 & -20 \\ 4 & 16 \end{bmatrix}. \qquad \textbf{(A.27)}$$

If **A** and **B** are of the same order, then

$$\mathbf{C} = \mathbf{A} + \mathbf{B} \qquad \textbf{(A.28)}$$

implies

$$c_{ij} = a_{ij} + b_{ij}. \tag{A.29}$$

For example, if

$$\mathbf{A} = \begin{bmatrix} 4 & -6 & 10 \\ 8 & 12 & -4 \end{bmatrix} \tag{A.30}$$

and

$$\mathbf{B} = \begin{bmatrix} 16 & 10 & -30 \\ -20 & 8 & 15 \end{bmatrix}, \tag{A.31}$$

then

$$\mathbf{C} = \begin{bmatrix} 20 & 4 & -20 \\ -12 & 20 & 11 \end{bmatrix}. \tag{A.32}$$

The equation

$$\mathbf{D} = \mathbf{A} - \mathbf{B} \tag{A.33}$$

implies

$$d_{ij} = a_{ij} - b_{ij}. \tag{A.34}$$

For the matrices in Eqs. (A.30) and (A.31), we would have

$$\mathbf{D} = \begin{bmatrix} -12 & -16 & 40 \\ 28 & 4 & -19 \end{bmatrix}. \tag{A.35}$$

Matrices of the same order are said to be **conformable** for addition and subtraction.

Multiplying a matrix by a scalar k is equivalent to multiplying each element by the scalar. Thus $\mathbf{A} = k\mathbf{B}$ if, and only if, $a_{ij} = kb_{ij}$. It should be noted that k may be real or complex. As an example, we will multiply the matrix **D** in Eq. (A.35) by 5. The result is

$$5\mathbf{D} = \begin{bmatrix} -60 & -80 & 200 \\ 140 & 20 & -95 \end{bmatrix}. \tag{A.36}$$

Matrix multiplication can be performed only if the number of columns in the first matrix is equal to the number of rows in the second matrix. In other words, the product **AB** requires the number of columns in **A** to equal the number of rows in **B**. The order of the resulting matrix will be the number of rows in **A** by the number of columns in **B**. Thus if $\mathbf{C} = \mathbf{AB}$, where **A** is of order $m \times p$ and **B** is of order $p \times n$, then **C** will be a matrix of order $m \times n$. When the number of columns in **A** equals the number of rows in **B**, we say **A** is conformable to **B** for multiplication.

An element in **C** is given by the formula

$$c_{ij} = \sum_{k=1}^{p} a_{ik}b_{kj}. \tag{A.37}$$

The formula given by Eq. (A.37) is easy to use if one remembers that matrix multiplication is a row-by-column operation. Hence to get the ith, jth term in **C**, each element in the ith row of **A** is multiplied by the corresponding element in the jth column of **B**, and the resulting products are summed. The following example illustrates the procedure. We are asked to find the matrix **C** when

$$\mathbf{A} = \begin{bmatrix} 6 & 3 & 2 \\ 1 & 4 & 6 \end{bmatrix} \quad \textbf{(A.38)}$$

and

$$\mathbf{B} = \begin{bmatrix} 4 & 2 \\ 0 & 3 \\ 1 & -2 \end{bmatrix}. \quad \textbf{(A.39)}$$

First we note that **C** will be a 2×2 matrix and that each element in **C** will require summing three products.

To find C_{11} we multiply the corresponding elements in row 1 of matrix **A** with the elements in column 1 of matrix **B** and then sum the products. We can visualize this multiplication and summing process by extracting the corresponding row and column from each matrix and then lining them up element by element. So to find C_{11} we have

$$\begin{array}{l|c|c|c} \text{Row 1 of } \mathbf{A} & 6 & 3 & 2 \\ \hline \text{Column 1 of } \mathbf{B} & 4 & 0 & 1 \end{array};$$

therefore

$$C_{11} = 6 \times 4 + 3 \times 0 + 2 \times 1 = 26.$$

To find C_{12} we visualize

$$\begin{array}{l|c|c|c} \text{Row 1 of } \mathbf{A} & 6 & 3 & 2 \\ \hline \text{Column 2 of } \mathbf{B} & 2 & 3 & -2 \end{array};$$

thus

$$C_{12} = 6 \times 2 + 3 \times 3 + 2 \times (-2) = 17.$$

For C_{21} we have

$$\begin{array}{l|c|c|c} \text{Row 2 of } \mathbf{A} & 1 & 4 & 6 \\ \hline \text{Column 1 of } \mathbf{B} & 4 & 0 & 1 \end{array}$$

and

$$C_{21} = 1 \times 4 + 4 \times 0 + 6 \times 1 = 10.$$

Finally, for C_{22} we have

$$\begin{array}{l|c|c|c} \text{Row 2 of } \mathbf{A} & 1 & 4 & 6 \\ \hline \text{Column 2 of } \mathbf{B} & 2 & 3 & -2 \end{array},$$

from which

$$C_{22} = 1 \times 2 + 4 \times 3 + 6 \times (-2) = 2.$$

It follows that

$$\mathbf{C} = \mathbf{AB} = \begin{bmatrix} 26 & 17 \\ 10 & 2 \end{bmatrix}. \quad \textbf{(A.40)}$$

In general, matrix multiplication is not commutative, that is, $\mathbf{AB} \neq \mathbf{BA}$. As an example, consider the product **BA** for the matrices in Eqs. (A.38) and (A.39). The matrix generated by this multiplication is of order 3×3, and each term in the resulting matrix requires adding two products. Therefore if $\mathbf{D} = \mathbf{BA}$, we have

$$\mathbf{D} = \begin{bmatrix} 26 & 20 & 20 \\ 3 & 12 & 18 \\ 4 & -5 & -10 \end{bmatrix}. \quad \textbf{(A.41)}$$

Obviously, $\mathbf{C} \neq \mathbf{D}$. We leave you to verify the elements in Eq. (A.41).

Matrix multiplication is associative and distributive. Thus

$$(\mathbf{AB})\mathbf{C} = \mathbf{A}(\mathbf{BC}), \quad \textbf{(A.42)}$$

$$\mathbf{A}(\mathbf{B} + \mathbf{C}) = \mathbf{AB} + \mathbf{AC}, \quad \textbf{(A.43)}$$

and

$$(\mathbf{A} + \mathbf{B})\mathbf{C} = \mathbf{AC} + \mathbf{BC}. \quad \textbf{(A.44)}$$

In Eqs. (A.42), (A.43), and (A.44), we assume that the matrices are conformable for addition and multiplication.

We have already noted that matrix multiplication is not commutative. There are two other properties of multiplication in scalar algebra that do not carry over to matrix algebra.

First, the matrix product $\mathbf{AB} = 0$ does not imply either $\mathbf{A} = 0$ or $\mathbf{B} = 0$. (*Note:* A matrix is equal to zero when all its elements are zero.) For example, if

$$\mathbf{A} = \begin{bmatrix} 1 & 0 \\ 2 & 0 \end{bmatrix} \quad \text{and} \quad \mathbf{B} = \begin{bmatrix} 0 & 0 \\ 4 & 8 \end{bmatrix},$$

then

$$\mathbf{AB} = \begin{bmatrix} 0 & 0 \\ 0 & 0 \end{bmatrix} = 0.$$

Hence the product is zero, but neither **A** nor **B** is zero.

Second, the matrix equation $\mathbf{AB} = \mathbf{AC}$ does not imply $\mathbf{B} = \mathbf{C}$. For example, if

$$\mathbf{A} = \begin{bmatrix} 1 & 0 \\ 2 & 0 \end{bmatrix}, \quad \mathbf{B} = \begin{bmatrix} 3 & 4 \\ 7 & 8 \end{bmatrix}, \quad \text{and} \quad \mathbf{C} = \begin{bmatrix} 3 & 4 \\ 5 & 6 \end{bmatrix},$$

then

$$\mathbf{AB} = \mathbf{AC} = \begin{bmatrix} 3 & 4 \\ 6 & 8 \end{bmatrix}, \quad \text{but } \mathbf{B} \neq \mathbf{C}.$$

The **transpose** of a matrix is formed by interchanging the rows and columns. For example, if

$$\mathbf{A} = \begin{bmatrix} 1 & 2 & 3 \\ 4 & 5 & 6 \\ 7 & 8 & 9 \end{bmatrix}, \quad \text{then} \quad \mathbf{A}^T = \begin{bmatrix} 1 & 4 & 7 \\ 2 & 5 & 8 \\ 3 & 6 & 9 \end{bmatrix}.$$

The transpose of the sum of two matrices is equal to the sum of the transposes, that is,

$$(\mathbf{A} + \mathbf{B})^T = \mathbf{A}^T + \mathbf{B}^T. \tag{A.45}$$

The transpose of the product of two matrices is equal to the product of the transposes taken in reverse order. In other words,

$$[\mathbf{AB}]^T = \mathbf{B}^T\mathbf{A}^T. \tag{A.46}$$

Equation (A.46) can be extended to a product of any number of matrices. For example,

$$[\mathbf{ABCD}]^T = \mathbf{D}^T\mathbf{C}^T\mathbf{B}^T\mathbf{A}^T. \tag{A.47}$$

If $\mathbf{A} = \mathbf{A}^T$, the matrix is said to be **symmetric**. Only square matrices can be symmetric.

A.8 IDENTITY, ADJOINT, AND INVERSE MATRICES

An **identity matrix** is a square matrix where $a_{ij} = 0$ for $i \neq j$, and $a_{ij} = 1$ for $i = j$. In other words, all the elements in an identity matrix are zero except those along the main diagonal, where they are equal to 1. Thus

$$\begin{bmatrix} 1 & 0 \\ 0 & 1 \end{bmatrix}, \quad \begin{bmatrix} 1 & 0 & 0 \\ 0 & 1 & 0 \\ 0 & 0 & 1 \end{bmatrix}, \quad \text{and} \quad \begin{bmatrix} 1 & 0 & 0 & 0 \\ 0 & 1 & 0 & 0 \\ 0 & 0 & 1 & 0 \\ 0 & 0 & 0 & 1 \end{bmatrix}$$

are all identity matrices. Note that identity matrices are always square. We will use the symbol **U** for an identity matrix.

The **adjoint** of a matrix **A** of order $n \times n$ is defined as

$$\text{adj } \mathbf{A} = [\Delta_{ji}]_{n \times n}, \tag{A.48}$$

where Δ_{ij} is the cofactor of a_{ij}. (See Section A.5 for the definition of a cofactor.) It follows from Eq. (A.48) that one can think of finding the adjoint of a square matrix as a two-step process. First construct a matrix made up of the cofactors of **A**, and then transpose the matrix of cofactors. As an example we will find the adjoint of the 3×3 matrix

$$\mathbf{A} = \begin{bmatrix} 1 & 2 & 3 \\ 3 & 2 & 1 \\ -1 & 1 & 5 \end{bmatrix}.$$

The cofactors of the elements in **A** are

$$\begin{aligned}
\Delta_{11} &= 1(10 - 1) = 9, \\
\Delta_{12} &= -1(15 + 1) = -16, \\
\Delta_{13} &= 1(3 + 2) = 5, \\
\Delta_{21} &= -1(10 - 3) = -7, \\
\Delta_{22} &= 1(5 + 3) = 8, \\
\Delta_{23} &= -1(1 + 2) = -3, \\
\Delta_{31} &= 1(2 - 6) = -4, \\
\Delta_{32} &= -1(1 - 9) = 8, \\
\Delta_{33} &= 1(2 - 6) = -4.
\end{aligned}$$

The matrix of cofactors is

$$\mathbf{B} = \begin{bmatrix} 9 & -16 & 5 \\ -7 & 8 & -3 \\ -4 & 8 & -4 \end{bmatrix}.$$

It follows that the adjoint of **A** is

$$\text{adj } \mathbf{A} = \mathbf{B}^T = \begin{bmatrix} 9 & -7 & -4 \\ -16 & 8 & 8 \\ 5 & -3 & -4 \end{bmatrix}.$$

One can check the arithmetic of finding the adjoint of a matrix by using the theorem

$$\text{adj } \mathbf{A} \cdot \mathbf{A} = \det \mathbf{A} \cdot \mathbf{U}. \qquad \textbf{(A.49)}$$

Equation (A.49) tells us that the adjoint of **A** times **A** equals the determinant of **A** times the identity matrix, or for our example,

$$\det \mathbf{A} = 1(9) + 3(-7) - 1(-4) = -8.$$

If we let $\mathbf{C} = \text{adj } \mathbf{A} \cdot \mathbf{A}$ and use the technique illustrated in Section A.7, we find the elements of **C** to be

$$\begin{aligned}
c_{11} &= 9 - 21 + 4 = -8, \\
c_{12} &= 18 - 14 - 4 = 0,
\end{aligned}$$

$$
\begin{aligned}
c_{13} &= 27 - 7 - 20 = 0, \\
c_{21} &= -16 + 24 - 8 = 0, \\
c_{22} &= -32 + 16 + 8 = -8, \\
c_{23} &= -48 + 8 + 40 = 0, \\
c_{31} &= 5 - 9 + 4 = 0, \\
c_{32} &= 10 - 6 - 4 = 0, \\
c_{33} &= 15 - 3 - 20 = -8.
\end{aligned}
$$

Therefore

$$
\begin{aligned}
\mathbf{C} &= \begin{bmatrix} -8 & 0 & 0 \\ 0 & -8 & 0 \\ 0 & 0 & -8 \end{bmatrix} = -8 \begin{bmatrix} 1 & 0 & 0 \\ 0 & 1 & 0 \\ 0 & 0 & 1 \end{bmatrix} \\
&= \det \mathbf{A} \cdot \mathbf{U}.
\end{aligned}
$$

A square matrix $\mathbf{A}$ has an **inverse**, denoted as $\mathbf{A}^{-1}$, if

$$\mathbf{A}^{-1}\mathbf{A} = \mathbf{A}\mathbf{A}^{-1} = \mathbf{U}. \tag{A.50}$$

Equation (A.50) tells us that a matrix either premultiplied or postmultiplied by its inverse generates the identity matrix $\mathbf{U}$. For the inverse matrix to exist, it is necessary that the determinant of $\mathbf{A}$ not equal zero. Only square matrices have inverses, and the inverse is also square.

A formula for finding the inverse of a matrix is

$$\mathbf{A}^{-1} = \frac{\operatorname{adj} \mathbf{A}}{\det \mathbf{A}}. \tag{A.51}$$

The formula in Eq. (A.51) becomes very cumbersome if $\mathbf{A}$ is of an order larger than 3 by 3.[2] Today the digital computer eliminates the drudgery of having to find the inverse of a matrix in numerical applications of matrix algebra.

It follows from Eq. (A.51) that the inverse of the matrix $\mathbf{A}$ in the previous example is

$$
\begin{aligned}
\mathbf{A}^{-1} &= -1/8 \begin{bmatrix} 9 & -7 & -4 \\ -16 & 8 & 8 \\ 5 & -3 & -4 \end{bmatrix} \\
&= \begin{bmatrix} -1.125 & 0.875 & 0.5 \\ 2 & -1 & -1 \\ -0.625 & 0.375 & 0.5 \end{bmatrix}.
\end{aligned}
$$

You should verify that $\mathbf{A}^{-1}\mathbf{A} = \mathbf{A}\mathbf{A}^{-1} = \mathbf{U}$.

---

[2] You can learn alternative methods for finding the inverse in any introductory text on matrix theory. See, for example, Franz E. Hohn, *Elementary Matrix Algebra* (New York: Macmillan, 1973).

A.9 PARTITIONED MATRICES

It is often convenient in matrix manipulations to partition a given matrix into submatrices. The original algebraic operations are then carried out in terms of the submatrices. In partitioning a matrix, the placement of the partitions is completely arbitrary, with the one restriction that a partition must dissect the entire matrix. In selecting the partitions, it is also necessary to make sure the submatrices are conformable to the mathematical operations in which they are involved.

For example, consider using submatrices to find the product $\mathbf{C} = \mathbf{AB}$, where

$$\mathbf{A} = \begin{bmatrix} 1 & 2 & 3 & 4 & 5 \\ 5 & 4 & 3 & 2 & 1 \\ -1 & 0 & 2 & -3 & 1 \\ 0 & 1 & -1 & 0 & 1 \\ 0 & 2 & 1 & -2 & 0 \end{bmatrix}$$

and

$$\mathbf{B} = \begin{bmatrix} 2 \\ 0 \\ -1 \\ 3 \\ 0 \end{bmatrix}.$$

Assume that we decide to partition $\mathbf{B}$ into two submatrices, $\mathbf{B}_{11}$ and $\mathbf{B}_{21}$; thus

$$\mathbf{B} = \begin{bmatrix} \mathbf{B}_{11} \\ \mathbf{B}_{21} \end{bmatrix}.$$

Now since $\mathbf{B}$ has been partitioned into a two-row column matrix, $\mathbf{A}$ must be partitioned into at least a two-column matrix; otherwise the multiplication cannot be performed. The location of the vertical partitions of the $\mathbf{A}$ matrix will depend on the definitions of $\mathbf{B}_{11}$ and $\mathbf{B}_{21}$. For example, if

$$\mathbf{B}_{11} = \begin{bmatrix} 2 \\ 0 \\ -1 \end{bmatrix} \quad \text{and} \quad \mathbf{B}_{21} = \begin{bmatrix} 3 \\ 0 \end{bmatrix},$$

then $\mathbf{A}_{11}$ must contain three columns, and $\mathbf{A}_{12}$ must contain two columns. Thus the partitioning shown in Eq. (A.52) would be acceptable for executing the product $\mathbf{AB}$:

$$\mathbf{C} = \left[\begin{array}{ccc:cc} 1 & 2 & 3 & 4 & 5 \\ 5 & 4 & 3 & 2 & 1 \\ -1 & 0 & 2 & -3 & 1 \\ 0 & 1 & -1 & 0 & 1 \\ 0 & 2 & 1 & -2 & 0 \end{array}\right] \left[\begin{array}{c} 2 \\ 0 \\ -1 \\ \hdashline 3 \\ 0 \end{array}\right]. \tag{A.52}$$

If, on the other hand, we partition the **B** matrix so that

$$\mathbf{B}_{11} = \begin{bmatrix} 2 \\ 0 \end{bmatrix} \quad \text{and} \quad \mathbf{B}_{21} = \begin{bmatrix} -1 \\ 3 \\ 0 \end{bmatrix},$$

then $\mathbf{A}_{11}$ must contain two columns, and $\mathbf{A}_{12}$ must contain three columns. In this case the partitioning shown in Eq. (A.53) would be acceptable in executing the product $\mathbf{C} = \mathbf{AB}$:

$$\mathbf{C} = \left[\begin{array}{cc:ccc} 1 & 2 & 3 & 4 & 5 \\ 5 & 4 & 3 & 2 & 1 \\ -1 & 0 & 2 & -3 & 1 \\ 0 & 1 & -1 & 0 & 1 \\ 0 & 2 & 1 & -2 & 0 \end{array}\right] \left[\begin{array}{c} 2 \\ 0 \\ \hdashline -1 \\ 3 \\ 0 \end{array}\right]. \tag{A.53}$$

For purposes of discussion, we will focus on the partitioning given in Eq. (A.52) and leave you to verify that the partitioning in Eq. (A.53) leads to the same result.

From Eq. (A.52) we can write

$$\mathbf{C} = [\ \mathbf{A}_{11} \quad \mathbf{A}_{12}\ ] \begin{bmatrix} \mathbf{B}_{11} \\ \mathbf{B}_{21} \end{bmatrix} = \mathbf{A}_{11}\mathbf{B}_{11} + \mathbf{A}_{12}\mathbf{B}_{21}. \tag{A.54}$$

It follows from Eqs. (A.52) and (A.54) that

$$\mathbf{A}_{11}\mathbf{B}_{11} = \begin{bmatrix} 1 & 2 & 3 \\ 5 & 4 & 3 \\ -1 & 0 & 2 \\ 0 & 1 & -1 \\ 0 & 2 & 1 \end{bmatrix} \begin{bmatrix} 2 \\ 0 \\ -1 \end{bmatrix} = \begin{bmatrix} -1 \\ 7 \\ -4 \\ 1 \\ -1 \end{bmatrix},$$

$$\mathbf{A}_{12}\mathbf{B}_{21} = \begin{bmatrix} 4 & 5 \\ 2 & 1 \\ -3 & 1 \\ 0 & 1 \\ -2 & 0 \end{bmatrix} \begin{bmatrix} 3 \\ 0 \end{bmatrix} = \begin{bmatrix} 12 \\ 6 \\ -9 \\ 0 \\ -6 \end{bmatrix},$$

and

$$\mathbf{C} = \begin{bmatrix} 11 \\ 13 \\ -13 \\ 1 \\ -7 \end{bmatrix}.$$

The **A** matrix could also be partitioned horizontally once the vertical partitioning is made consistent with the multiplication operation. In this simple problem, the horizontal partitions can be made at the discretion of the analyst. Therefore **C** could also be evaluated using the partitioning shown in Eq. (A.55):

$$\mathbf{C} = \left[\begin{array}{ccc:cc} 1 & 2 & 3 & 4 & 5 \\ 5 & 4 & 3 & 2 & 1 \\ \hdashline -1 & 0 & 2 & -3 & 1 \\ 0 & 1 & -1 & 0 & 1 \\ 0 & 2 & 1 & -2 & 0 \end{array}\right] \left[\begin{array}{c} 2 \\ 0 \\ -1 \\ \hdashline 3 \\ 0 \end{array}\right]. \tag{A.55}$$

From Eq. (A.55) it follows that

$$\mathbf{C} = \begin{bmatrix} \mathbf{A}_{11} & \mathbf{A}_{12} \\ \mathbf{A}_{21} & \mathbf{A}_{22} \end{bmatrix} \begin{bmatrix} \mathbf{B}_{11} \\ \mathbf{B}_{21} \end{bmatrix} = \begin{bmatrix} \mathbf{C}_{11} \\ \mathbf{C}_{21} \end{bmatrix}, \tag{A.56}$$

where

$$\mathbf{C}_{11} = \mathbf{A}_{11}\mathbf{B}_{11} + \mathbf{A}_{12}\mathbf{B}_{21},$$
$$\mathbf{C}_{21} = \mathbf{A}_{21}\mathbf{B}_{11} + \mathbf{A}_{22}\mathbf{B}_{21}.$$

You should verify that

$$\mathbf{C}_{11} = \begin{bmatrix} 1 & 2 & 3 \\ 5 & 4 & 3 \end{bmatrix} \begin{bmatrix} 2 \\ 0 \\ -1 \end{bmatrix} + \begin{bmatrix} 4 & 5 \\ 2 & 1 \end{bmatrix} \begin{bmatrix} 3 \\ 0 \end{bmatrix}$$
$$= \begin{bmatrix} -1 \\ 7 \end{bmatrix} + \begin{bmatrix} 12 \\ 6 \end{bmatrix} = \begin{bmatrix} 11 \\ 13 \end{bmatrix},$$

$$\mathbf{C}_{21} = \begin{bmatrix} -1 & 0 & 2 \\ 0 & 1 & -1 \\ 0 & 2 & 1 \end{bmatrix} \begin{bmatrix} 2 \\ 0 \\ -1 \end{bmatrix} + \begin{bmatrix} -3 & 1 \\ 0 & 1 \\ -2 & 0 \end{bmatrix} \begin{bmatrix} 3 \\ 0 \end{bmatrix}$$
$$= \begin{bmatrix} -4 \\ 1 \\ -1 \end{bmatrix} + \begin{bmatrix} -9 \\ 0 \\ -6 \end{bmatrix} = \begin{bmatrix} -13 \\ 1 \\ -7 \end{bmatrix},$$

and

$$\mathbf{C} = \begin{bmatrix} 11 \\ 13 \\ -13 \\ 1 \\ -7 \end{bmatrix}.$$

We note in passing that the partitioning in Eqs. (A.52) and (A.55) is conformable with respect to addition.

A.10 Applications

The following examples demonstrate some applications of matrix algebra in circuit analysis.

EXAMPLE A.1

Use the matrix method to solve for the node voltages v_1 and v_2 in Eqs. (4.5) and (4.6).

SOLUTION

The first step is to rewrite Eqs. (4.5) and (4.6) in matrix notation. Collecting the coefficients of v_1 and v_2 and at the same time shifting the constant terms to the right-hand side of the equations gives us

$$1.7v_1 - 0.5v_2 = 10, \qquad \text{(A.57)}$$
$$-0.5v_1 + 0.6v_2 = 2.$$

It follows that in matrix notation, Eq. (A.57) becomes

$$\begin{bmatrix} 1.7 & -0.5 \\ -0.5 & 0.6 \end{bmatrix} \begin{bmatrix} v_1 \\ v_2 \end{bmatrix} = \begin{bmatrix} 10 \\ 2 \end{bmatrix}, \qquad \text{(A.58)}$$

or

$$\mathbf{AV} = \mathbf{I}, \qquad \text{(A.59)}$$

where

$$\mathbf{A} = \begin{bmatrix} 1.7 & -0.5 \\ -0.5 & 0.6 \end{bmatrix},$$

$$\mathbf{V} = \begin{bmatrix} v_1 \\ v_2 \end{bmatrix},$$

and

$$\mathbf{I} = \begin{bmatrix} 10 \\ 2 \end{bmatrix}.$$

To find the elements of the V matrix, we premultiply both sides of Eq. (A.59) by the inverse of A; thus

$$\mathbf{A}^{-1}\mathbf{AV} = \mathbf{A}^{-1}\mathbf{I}. \qquad \text{(A.60)}$$

Equation (A.60) reduces to

$$\mathbf{UV} = \mathbf{A}^{-1}\mathbf{I}, \qquad \text{(A.61)}$$

or

$$\mathbf{V} = \mathbf{A}^{-1}\mathbf{I}. \tag{A.62}$$

It follows from Eq. (A.62) that the solutions for v_1 and v_2 are obtained by solving for the matrix product $\mathbf{A}^{-1}\mathbf{I}$.

To find the inverse of $\mathbf{A}$, we first find the cofactors of $\mathbf{A}$. Thus

$$\begin{aligned} \Delta_{11} &= (-1)^2(0.6) = 0.6, \\ \Delta_{12} &= (-1)^3(-0.5) = 0.5, \\ \Delta_{21} &= (-1)^3(-0.5) = 0.5, \\ \Delta_{22} &= (-1)^4(1.7) = 1.7. \end{aligned} \tag{A.63}$$

The matrix of cofactors is

$$\mathbf{B} = \begin{bmatrix} 0.6 & 0.5 \\ 0.5 & 1.7 \end{bmatrix}, \tag{A.64}$$

and the adjoint of $\mathbf{A}$ is

$$\text{adj } \mathbf{A} = \mathbf{B}^T = \begin{bmatrix} 0.6 & 0.5 \\ 0.5 & 1.7 \end{bmatrix}. \tag{A.65}$$

The determinant of $\mathbf{A}$ is

$$\det \mathbf{A} = \begin{vmatrix} 1.7 & -0.5 \\ -0.5 & 0.6 \end{vmatrix} = (1.7)(0.6) - (0.25) = 0.77. \tag{A.66}$$

From Eqs. (A.65) and (A.66), we can write the inverse of the coefficient matrix, that is,

$$\mathbf{A}^{-1} = \frac{1}{0.77}\begin{bmatrix} 0.6 & 0.5 \\ 0.5 & 1.7 \end{bmatrix}. \tag{A.67}$$

Now the product $\mathbf{A}^{-1}\mathbf{I}$ is found:

$$\begin{aligned} \mathbf{A}^{-1}\mathbf{I} &= \frac{100}{77}\begin{bmatrix} 0.6 & 0.5 \\ 0.5 & 1.7 \end{bmatrix}\begin{bmatrix} 10 \\ 2 \end{bmatrix} \\ &= \frac{100}{77}\begin{bmatrix} 7 \\ 8.4 \end{bmatrix} = \begin{bmatrix} 9.09 \\ 10.91 \end{bmatrix}. \end{aligned} \tag{A.68}$$

It follows directly that

$$\begin{bmatrix} v_1 \\ v_2 \end{bmatrix} = \begin{bmatrix} 9.09 \\ 10.91 \end{bmatrix}, \tag{A.69}$$

or $v_1 = 9.09$ V and $v_2 = 10.91$ V.

EXAMPLE A.2

Use the matrix method to find the three mesh currents in the circuit in Fig. 4.24.

SOLUTION

The mesh-current equations that describe the circuit in Fig. 4.24 are given in Eq. (4.34). The constraint equation imposed by the current-controlled voltage source is given in Eq. (4.35). When Eq. (4.35) is substituted into Eq. (4.34), the following set of equations evolves:

$$\begin{aligned} 25i_i - 5i_2 - 20i_3 &= 50, \\ -5i_i + 10i_2 - 4i_3 &= 0, \\ -5i_1 - 4i_2 + 9i_3 &= 0. \end{aligned} \tag{A.70}$$

In matrix notation, Eqs. (A.70) reduce to

$$\mathbf{AI} = \mathbf{V}, \tag{A.71}$$

where

$$\mathbf{A} = \begin{bmatrix} 25 & -5 & -20 \\ -5 & 10 & -4 \\ -5 & -4 & 9 \end{bmatrix},$$

$$\mathbf{I} = \begin{bmatrix} i_1 \\ i_2 \\ i_3 \end{bmatrix},$$

and

$$\mathbf{V} = \begin{bmatrix} 50 \\ 0 \\ 0 \end{bmatrix}.$$

It follows from Eq. (A.71) that the solution for **I** is

$$\mathbf{I} = \mathbf{A}^{-1}\mathbf{V}. \tag{A.72}$$

We find the inverse of **A** by using the relationship

$$\mathbf{A}^{-1} = \frac{\text{adj } \mathbf{A}}{\det \mathbf{A}}. \tag{A.73}$$

To find the adjoint of **A**, we first calculate the cofactors of **A**. Thus

$$\begin{aligned} \Delta_{11} &= (-1)^2(90 - 16) = 74, \\ \Delta_{12} &= (-1)^3(-45 - 20) = 65, \\ \Delta_{13} &= (-1)^4(20 + 50) = 70, \\ \Delta_{21} &= (-1)^3(-45 - 80) = 125, \\ \Delta_{22} &= (-1)^4(225 - 100) = 125, \\ \Delta_{23} &= (-1)^5(-100 - 25) = 125, \\ \Delta_{31} &= (-1)^4(20 + 200) = 220, \\ \Delta_{32} &= (-1)^5(-100 - 100) = 200, \\ \Delta_{33} &= (-1)^6(250 - 25) = 225. \end{aligned}$$

The cofactor matrix is

$$\mathbf{B} = \begin{bmatrix} 74 & 65 & 70 \\ 125 & 125 & 125 \\ 220 & 200 & 225 \end{bmatrix}, \qquad \text{(A.74)}$$

from which we can write the adjoint of **A**:

$$\text{adj } \mathbf{A} = \mathbf{B}^T = \begin{bmatrix} 74 & 125 & 220 \\ 65 & 125 & 200 \\ 70 & 125 & 225 \end{bmatrix}. \qquad \text{(A.75)}$$

The determinant of **A** is

$$\det \mathbf{A} = \begin{vmatrix} 25 & -5 & -20 \\ -5 & 10 & -4 \\ -5 & -4 & 9 \end{vmatrix}$$

$$= 25(90 - 16) + 5(-45 - 80) - 5(20 + 200) = 125.$$

It follows from Eq. (A.73) that

$$\mathbf{A}^{-1} = \frac{1}{125} \begin{bmatrix} 74 & 125 & 220 \\ 65 & 125 & 200 \\ 70 & 125 & 225 \end{bmatrix}. \qquad \text{(A.76)}$$

The solution for **I** is

$$\mathbf{I} = \frac{1}{125} \begin{bmatrix} 74 & 125 & 220 \\ 65 & 125 & 200 \\ 70 & 125 & 225 \end{bmatrix} \begin{bmatrix} 50 \\ 0 \\ 0 \end{bmatrix} = \begin{bmatrix} 29.60 \\ 26.00 \\ 28.00 \end{bmatrix}. \qquad \text{(A.77)}$$

The mesh currents follow directly from Eq. (A.77). Thus

$$\begin{bmatrix} i_i \\ i_2 \\ i_3 \end{bmatrix} = \begin{bmatrix} 29.6 \\ 26.0 \\ 28.0 \end{bmatrix} \qquad \text{(A.78)}$$

or $i_1 = 29.6$ A, $i_2 = 26$ A, and $i_3 = 28$ A.

Example A.3 illustrates the application of the matrix method when the elements of the matrix are complex numbers.

EXAMPLE A.3

Use the matrix method to find the phasor mesh currents $\mathbf{I}_1$ and $\mathbf{I}_2$ in the circuit in Fig. 9.36.

SOLUTION

Summing the voltages around mesh 1 generates the equation

$$(1 + j2)\mathbf{I}_1 + (12 - j16)(\mathbf{I}_1 - \mathbf{I}_2) = 150\underline{/0^\circ}. \quad \text{(A.79)}$$

Summing the voltages around mesh 2 produces the equation

$$(12 - j16)(\mathbf{I}_2 - \mathbf{I}_1) + (1 + j3)\mathbf{I}_2 + 39\mathbf{I}_x = 0. \quad \text{(A.80)}$$

The current controlling the dependent voltage source is

$$\mathbf{I}_x = (\mathbf{I}_1 - \mathbf{I}_2). \quad \text{(A.81)}$$

After substituting Eq. (A.81) into Eq. (A.80), the equations are put into a matrix format by first collecting, in each equation, the coefficients of $\mathbf{I}_1$ and $\mathbf{I}_2$; thus

$$\begin{aligned} (13 - j14)\mathbf{I}_1 - (12 - j16)\mathbf{I}_2 &= 150\underline{/0^\circ}, \\ (27 + j16)\mathbf{I}_1 - (26 + j13)\mathbf{I}_2 &= 0. \end{aligned} \quad \text{(A.82)}$$

Now, using matrix notation, Eq. (A.82) is written

$$\mathbf{AI} = \mathbf{V}, \quad \text{(A.83)}$$

where

$$\mathbf{A} = \begin{bmatrix} 13 - j14 & -(12 - j16) \\ 27 + j16 & -(26 + j13) \end{bmatrix},$$

$$\mathbf{I} = \begin{bmatrix} \mathbf{I}_1 \\ \mathbf{I}_2 \end{bmatrix}, \quad \text{and} \quad \mathbf{V} = \begin{bmatrix} 150\underline{/0^\circ} \\ 0 \end{bmatrix}.$$

It follows from Eq. (A.83) that

$$\mathbf{I} = \mathbf{A}^{-1}\mathbf{V}. \quad \text{(A.84)}$$

The inverse of the coefficient matrix **A** is found using Eq. (A.73). In this case, the cofactors of **A** are

$$\begin{aligned} \Delta_{11} &= (-1)^2(-26 - j13) = -26 - j13, \\ \Delta_{12} &= (-1)^3(27 + j16) = -27 - j16, \\ \Delta_{21} &= (-1)^3(-12 + j16) = 12 - j16, \\ \Delta_{22} &= (-1)^4(13 - j14) = 13 - j14. \end{aligned}$$

The cofactor matrix **B** is

$$\mathbf{B} = \begin{bmatrix} (-26 - j13) & (-27 - j16) \\ (12 - j16) & (13 - j14) \end{bmatrix}. \quad \text{(A.85)}$$

The adjoint of **A** is

$$\text{adj } \mathbf{A} = \mathbf{B}^T = \begin{bmatrix} (-26 - j13) & (12 - j16) \\ (-27 - j16) & (13 - j14) \end{bmatrix}. \tag{A.86}$$

The determinant of **A** is

$$\begin{aligned} \det \mathbf{A} &= \begin{vmatrix} (13 - j14) & -(12 - j16) \\ (27 + j16) & -(26 + j13) \end{vmatrix} \\ &= -(13 - j14)(26 + j13) + (12 - j16)(27 + j16) \\ &= 60 - j45. \end{aligned} \tag{A.87}$$

The inverse of the coefficient matrix is

$$\mathbf{A}^{-1} = \frac{\begin{bmatrix} (-26 - j13) & (12 - j16) \\ (-27 - j16) & (13 - j14) \end{bmatrix}}{(60 - j45)}. \tag{A.88}$$

Equation (A.88) can be simplified to

$$\begin{aligned} \mathbf{A}^{-1} &= \frac{60 + j45}{5625} \begin{bmatrix} (-26 - j13) & (12 - j16) \\ (-27 - j16) & (13 - j14) \end{bmatrix} \\ &= \frac{1}{375} \begin{bmatrix} -65 - j130 & 96 - j28 \\ -60 - j145 & 94 - j17 \end{bmatrix}. \end{aligned} \tag{A.89}$$

Substituting Eq. (A.89) into (A.84) gives us

$$\begin{aligned} \begin{bmatrix} \mathbf{I}_1 \\ \mathbf{I}_2 \end{bmatrix} &= \frac{1}{375} \begin{bmatrix} (-65 - j130) & (96 - j28) \\ (-60 - j145) & (94 - j17) \end{bmatrix} \begin{bmatrix} 150\angle 0^\circ \\ 0 \end{bmatrix} \\ &= \begin{bmatrix} (-26 - j52) \\ (-24 - j58) \end{bmatrix}. \end{aligned} \tag{A.90}$$

It follows from Eq. (A.90) that

$$\begin{aligned} \mathbf{I}_1 &= (-26 - j52) = 58.14\angle -116.57^\circ \text{ A}, \\ \mathbf{I}_2 &= (-24 - j58) = 62.77\angle -122.48^\circ \text{ A}. \end{aligned} \tag{A.91}$$

In the first three examples, the matrix elements have been numbers—real numbers in Examples A.1 and A.2, and complex numbers in Example A.3. It is also possible for the elements to be functions. Example A.4 illustrates the use of matrix algebra in a circuit problem where the elements in the coefficient matrix are functions.

EXAMPLE A.4

Use the matrix method to derive expressions for the node voltages V_1 and V_2 in the circuit in Fig. A.1.

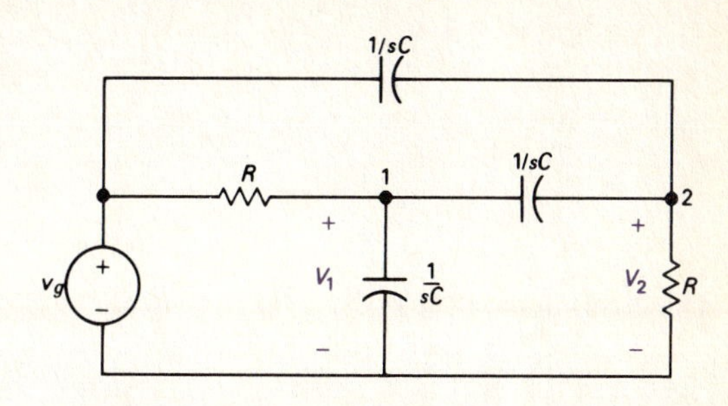

FIGURE A.1 The circuit for Example A.4.

SOLUTION

Summing the currents away from nodes 1 and 2 generates the following set of equations:

$$\frac{V_1 - V_g}{R} + V_1 sC + (V_1 - V_2)sC = 0,$$
$$\frac{V_2}{R} + (V_2 - V_1)sC + (V_2 - V_g)sC = 0. \quad \text{(A.92)}$$

Letting $G = 1/R$ and collecting the coefficients of V_1 and V_2 gives us

$$(G + 2sC)V_1 - sCV_2 = GV_g,$$
$$-sCV_1 + (G + 2sC)V_2 = sCV_g. \quad \text{(A.93)}$$

Writing Eq. (A.93) in matrix notation yields

$$\mathbf{AV} = \mathbf{I}, \quad \text{(A.94)}$$

where

$$\mathbf{A} = \begin{bmatrix} G + 2sC & -sC \\ -sC & G + 2sC \end{bmatrix},$$

$$\mathbf{V} = \begin{bmatrix} V_1 \\ V_2 \end{bmatrix}, \quad \text{and} \quad \mathbf{I} = \begin{bmatrix} GV_g \\ sCV_g \end{bmatrix}.$$

It follows from Eq. (A.94) that

$$\mathbf{V} = \mathbf{A}^{-1}\mathbf{I}. \quad \text{(A.95)}$$

As before, we find the inverse of the coefficient matrix by first finding the adjoint of **A** and the determinant of **A**. The cofactors of **A** are

$$\Delta_{11} = (-1)^2[G + 2sC] = G + 2sC,$$
$$\Delta_{12} = (-1)^3(-sC) = sC,$$
$$\Delta_{21} = (-1)^3(-sC) = sC,$$
$$\Delta_{22} = (-1)^4[G + 2sC] = G + 2sC.$$

The cofactor matrix is

$$\mathbf{B} = \begin{bmatrix} G+2sC & sC \\ sC & G+2sC \end{bmatrix}, \tag{A.96}$$

and therefore the adjoint of the coefficient matrix is

$$\text{adj } \mathbf{A} = \mathbf{B}^T = \begin{bmatrix} G+2sC & sC \\ sC & G+2sC \end{bmatrix}. \tag{A.97}$$

The determinant of **A** is

$$\det \mathbf{A} = \begin{vmatrix} G+2sC & sC \\ sC & G+2sC \end{vmatrix} = G^2 + 4sCG + 3s^2C^2. \tag{A.98}$$

The inverse of the coefficient matrix is

$$\mathbf{A}^{-1} = \frac{\begin{bmatrix} G+2sC & sC \\ sC & G+2sC \end{bmatrix}}{(G^2 + 4sCG + 3s^2C^2)}. \tag{A.99}$$

It follows from Eq. (A.95) that

$$\begin{bmatrix} V_1 \\ V_2 \end{bmatrix} = \frac{\begin{bmatrix} G+2sC & sC \\ sC & G+2sC \end{bmatrix} \begin{bmatrix} GV_g \\ sCV_g \end{bmatrix}}{(G^2 + 4sCG + 3s^2C^2)}. \tag{A.100}$$

Carrying out the matrix multiplication called for in Eq. (A.100) gives

$$\begin{bmatrix} V_1 \\ V_2 \end{bmatrix} = \frac{1}{(G^2 + 4sCG + 3s^2C^2)} \begin{bmatrix} (G^2 + 2sCG + s^2C^2)V_g \\ (2sCG + 2s^2C^2)V_g \end{bmatrix}. \tag{A.101}$$

Now the expressions for V_1 and V_2 can be written directly from Eq. (A.101); thus

$$V_1 = \frac{(G^2 + 2sCG + s^2C^2)V_g}{(G^2 + 4sCG + 3s^2C^2)}, \tag{A.102}$$

and

$$V_2 = \frac{2(sCG + s^2C^2)V_g}{(G^2 + 4sCG + 3s^2C^2)}. \tag{A.103}$$

The interested reader can test the validity of Eqs. (A.102) and (A.103) by using them to find $v_1(t)$ and $v_2(t)$ in Problem 14.28.

In our final example, we illustrate how matrix algebra can be used to analyze the cascade connection of two two-port circuits.

EXAMPLE A.5

Show by means of matrix algebra how the input variables V_1 and I_1 can be described as functions of the output variables V_2 and I_2 in the cascade connection shown in Fig. 19.10.

SOLUTION

We begin by expressing, in matrix notation, the relationship between the input and output variables of each two-port circuit. Thus

$$\begin{bmatrix} V_1 \\ I_1 \end{bmatrix} = \begin{bmatrix} a'_{11} & -a'_{12} \\ a'_{21} & -a'_{22} \end{bmatrix} \begin{bmatrix} V'_2 \\ I'_2 \end{bmatrix}, \tag{A.104}$$

and

$$\begin{bmatrix} V'_1 \\ I'_1 \end{bmatrix} = \begin{bmatrix} a''_{11} & -a''_{12} \\ a''_{21} & -a''_{22} \end{bmatrix} \begin{bmatrix} V_2 \\ I_2 \end{bmatrix}, \tag{A.105}$$

Now the cascade connection imposes the constraints

$$V'_2 = V'_1 \quad \text{and} \quad I'_2 = -I'_1. \tag{A.106}$$

These constraint relationships are substituted into Eq. (A.104). Thus

$$\begin{aligned} \begin{bmatrix} V_1 \\ I_1 \end{bmatrix} &= \begin{bmatrix} a'_{11} & -a'_{12} \\ a'_{21} & -a'_{22} \end{bmatrix} \begin{bmatrix} V'_1 \\ -I'_1 \end{bmatrix} \\ &= \begin{bmatrix} a'_{11} & a'_{12} \\ a'_{21} & a'_{22} \end{bmatrix} \begin{bmatrix} V'_1 \\ I'_1 \end{bmatrix}. \end{aligned} \tag{A.107}$$

The relationship between the input variables (V_1, I_1) and the output variables (V_2, I_2) is obtained by substituting Eq. (A.105) into Eq. (A.107). The result is

$$\begin{bmatrix} V_1 \\ I_1 \end{bmatrix} = \begin{bmatrix} a'_{11} & a'_{12} \\ a'_{21} & a'_{22} \end{bmatrix} \begin{bmatrix} a''_{11} & -a''_{12} \\ a''_{21} & -a''_{22} \end{bmatrix} \begin{bmatrix} V_2 \\ I_2 \end{bmatrix}. \tag{A.108}$$

After multiplying the coefficient matrices, we have

$$\begin{bmatrix} V_1 \\ I_1 \end{bmatrix} = \begin{bmatrix} (a'_{11}a''_{11} + a'_{12}a''_{21}) & -(a'_{11}a''_{12} + a'_{12}a''_{22}) \\ (a'_{21}a''_{11} + a'_{22}a''_{21}) & -(a'_{21}a''_{12} + a'_{22}a''_{22}) \end{bmatrix} \begin{bmatrix} V_2 \\ I_2 \end{bmatrix}. \tag{A.109}$$

Note that Eq. (A.109) corresponds to writing Eqs. (19.72) and (19.73) in matrix form.

COMPLEX NUMBERS

Complex numbers were invented to permit the extraction of the square roots of negative numbers. Complex numbers simplify the solution of problems that would otherwise be very difficult. The equation $x^2 + 8x + 41 = 0$, for example, has no solution in a number system that excludes complex numbers. These numbers, and the ability to manipulate them algebraically, are extremely useful in circuit analysis.

B.1 NOTATION

There are two ways to designate a complex number: with the cartesian, or rectangular, form or with the polar, or trigonometric, form. In the **rectangular form**, a complex number is written in terms of its real and imaginary components; hence

$$n = a + jb, \tag{B.1}$$

where a is the real component, b is the imaginary component, and j is by definition $\sqrt{-1}$.[1]

In the **polar form**, a complex number is written in terms of its magnitude (or modulus) and angle (or argument); hence

$$n = ce^{j\theta}, \tag{B.2}$$

[1] You may be more familiar with the notation $i = \sqrt{-1}$. In electrical engineering, i is used as the symbol for current, and hence in electrical engineering literature, j is used to denote $\sqrt{-1}$.

where c is the magnitude, θ is the angle, e is the base of the natural logarithm, and, as before, $j = \sqrt{-1}$. In the literature, the symbol $\underline{/\theta}$ is frequently used in place of $e^{j\theta}$; that is, the polar form is written

$$n = c\underline{/\theta}. \tag{B.3}$$

Although Eq. (B.3) is more convenient in printing text material, Eq. (B.2) is of primary importance in mathematical operations because the rules for manipulating an exponential quantity are well known. For example, because $(y^x)^n = y^{xn}$, then $(e^{j\theta})^n = e^{jn\theta}$; because $y^{-x} = 1/y^x$, then $e^{-j\theta} = 1/e^{j\theta}$; and so forth.

Because there are two ways of expressing the same complex number, we need to relate one form to the other. The transition from the polar to the rectangular form makes use of Euler's identity:

$$e^{\pm j\theta} = \cos\theta \pm j\sin\theta. \tag{B.4}$$

A complex number in polar form can be put in rectangular form by writing

$$\begin{aligned} ce^{j\theta} &= c(\cos\theta + j\sin\theta) \\ &= c\cos\theta + jc\sin\theta \\ &= a + jb. \end{aligned} \tag{B.5}$$

The transition from rectangular to polar form makes use of the geometry of the right triangle, namely,

$$\begin{aligned} a + jb &= (\sqrt{a^2 + b^2})e^{j\theta} \\ &= ce^{j\theta}, \end{aligned} \tag{B.6}$$

where

$$\tan\theta = b/a. \tag{B.7}$$

It is not obvious from Eq. (B.7) in which quadrant the angle θ lies. The ambiguity can be resolved by a graphical representation of the complex number.

B.2 The Graphical Representation of a Complex Number

A complex number is represented graphically on a complex-number plane, which uses the horizontal axis for plotting the real component and the vertical axis for plotting the imaginary component. The angle of the complex number is measured counterclockwise from the positive real axis. The graphical plot of the complex number $n = a + jb = c\angle\theta$, if we assume that a and b are both positive, is shown in Fig. B.1. This plot makes very clear the relationship between the rectangular and polar forms. Any point in the complex-number plane is uniquely defined by giving either its distance from each axis (that is, a and b) or its radial distance from the origin (c) and the angle of the radial measurement θ.

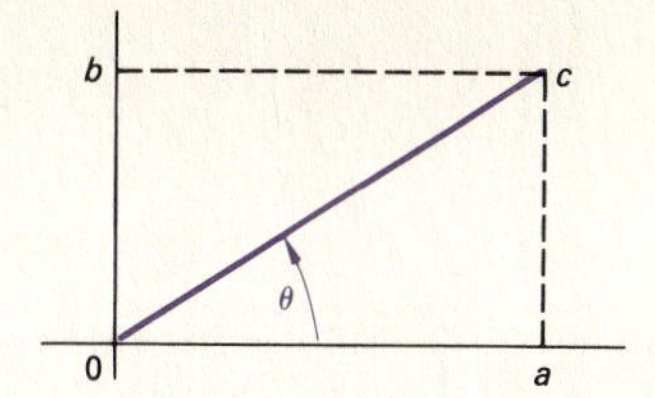

FIGURE B.1 The graphical representation of $a + jb$ when a and b are both positive.

It follows from Fig. B.1 that θ is in the first quadrant when a and b are both positive, in the second quadrant when a is negative and b is positive, in the third quadrant when a and b are both negative, and in the fourth quadrant when a is positive and b is negative. These observations are illustrated in Fig. B.2, where we have plotted $4 + j3$, $-4 + j3$, $-4 - j3$, and $4 - j3$.

Note that we can also specify θ as a clockwise angle from the positive real axis. Thus in Fig. B.2(c) we could also designate $-4 - j3$ as $5\angle{-143.13^\circ}$. In Fig. B.2(d) we observe that $5\angle 323.13^\circ = 5\angle{-36.87^\circ}$. It is customary to express θ in terms of negative values when θ lies in the third or fourth quadrant.

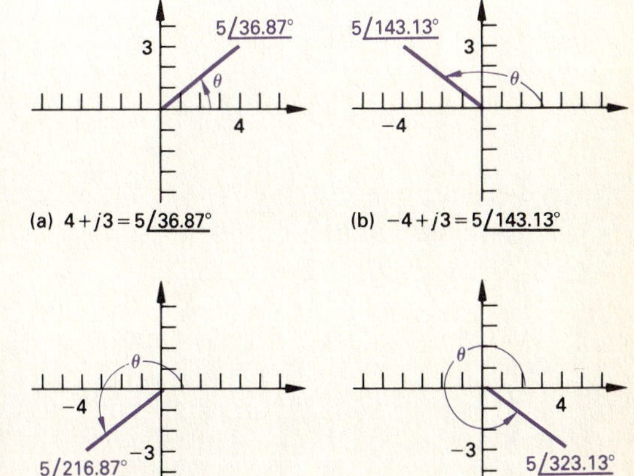

FIGURE B.2 The graphical representation of four complex numbers.

The graphical interpretation of a complex number also shows the relationship between a complex number and its conjugate. The **conjugate of a complex number** is formed by reversing the sign of its imaginary component. Thus the conjugate of $a + jb$ is $a - jb$, and the conjugate of $-a + jb$ is $-a - jb$. When we write a complex number in polar form, we form its conjugate simply by reversing the sign of the angle θ. Therefore the conjugate of $c\angle\theta$ is $c\angle{-\theta}$. The conjugate of a complex number is designated with an asterisk. In other words, n^* is understood to be the conjugate of n. Figure B.3 shows two complex numbers and their conjugates plotted on the complex-number plane. Note that conjugation simply reflects the complex numbers about the real axis.

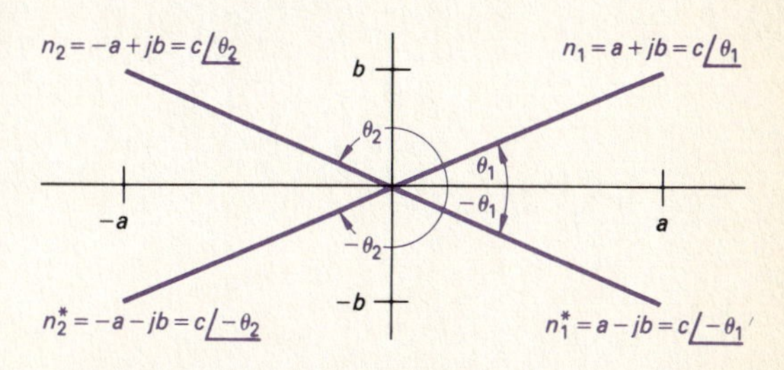

FIGURE B.3 The complex numbers n_1 and n_2 and their conjugates n_1^* and n_2^*.

B.3 ARITHMETIC OPERATIONS

ADDITION (SUBTRACTION)

To add or subtract complex numbers, we must express the numbers in rectangular form. Addition involves adding the real parts of the complex numbers to form the real part of the sum, and the imaginary parts to form the imaginary part of the sum. Thus, if we are given

$$n_1 = 8 + j16$$

and

$$n_2 = 12 - j3,$$

then

$$n_1 + n_2 = (8 + 12) + j(16 - 3) = 20 + j13.$$

Subtraction follows the same rule. Thus

$$n_2 - n_1 = (12 - 8) + j(-3 - 16) = 4 - j19.$$

If the numbers to be added or subtracted are given in polar form, they are first converted to rectangular form. For example, if

$$n_1 = 10\angle 53.13^\circ$$

and

$$n_2 = 5\angle -135^\circ,$$

then

$$\begin{aligned} n_1 + n_2 &= 6 + j8 - 3.535 - j3.535 \\ &= (6 - 3.535) + j(8 - 3.535) \\ &= 2.465 + j4.465 = 5.10\angle 61.10^\circ, \end{aligned}$$

and

$$\begin{aligned} n_1 - n_2 &= 6 + j8 - (-3.535 - j3.535) \\ &= 9.535 + j11.535 \\ &= 14.966\angle 50.42^\circ. \end{aligned}$$

MULTIPLICATION (DIVISION)

Multiplication or division of complex numbers can be carried out with the numbers written in either rectangular or polar form. However, in most cases, the polar form is more convenient. As an example, let's find the product $n_1 n_2$ when $n_1 = 8 + j10$ and $n_2 = 5 - j4$. Using the rectangular form, we have

$$\begin{aligned} n_1 n_2 &= (8 + j10)(5 - j4) = 40 - j32 + j50 + 40 \\ &= 80 + j18 \\ &= 82\angle 12.68^\circ. \end{aligned}$$

If we use the polar form, the multiplication $n_1 n_2$ becomes

$$\begin{aligned} n_1 n_2 &= (12.81\angle 51.34^\circ)(6.40\angle -38.66^\circ) \\ &= 82\angle 12.68^\circ \\ &= 80 + j18. \end{aligned}$$

The first step in dividing two complex numbers in rectangular form is to multiply the numerator and denominator by the conjugate of the denominator. This reduces the denominator to a real number. We then divide the real number into the new numerator. As an example, let's find the value of n_1/n_2, where $n_1 = 6 + j3$ and $n_2 = 3 - j1$. We have

$$\begin{aligned}\frac{n_1}{n_2} &= \frac{6 + j3}{3 - j1} = \frac{(6 + j3)(3 + j1)}{(3 - j1)(3 + j1)} \\ &= \frac{18 + j6 + j9 - 3}{9 + 1} \\ &= \frac{15 + j15}{10} = 1.5 + j1.5 \\ &= 2.12\underline{/45^\circ}.\end{aligned}$$

In polar form, the division of n_1 by n_2 is

$$\begin{aligned}\frac{n_1}{n_2} &= \frac{6.71\underline{/26.57^\circ}}{3.16\underline{/-18.43^\circ}} = 2.12\underline{/45^\circ} \\ &= 1.5 + j1.5.\end{aligned}$$

B.4 USEFUL IDENTITIES

In working with complex numbers and quantities, the following identities are very useful:

$$\pm j^2 = \mp 1, \qquad \text{(B.8)}$$

$$(-j)(j) = 1, \qquad \text{(B.9)}$$

$$j = \frac{1}{-j}, \qquad \text{(B.10)}$$

$$e^{\pm j\pi} = -1, \qquad \text{(B.11)}$$

$$e^{\pm j\pi/2} = \pm j. \qquad \text{(B.12)}$$

Given that $n = a + jb = c\underline{/\theta}$, it follows that

$$nn^* = a^2 + b^2 = c^2, \qquad \text{(B.13)}$$

$$n + n^* = 2a, \qquad \text{(B.14)}$$

$$n - n^* = j2b, \qquad \text{(B.15)}$$

$$n/n^* = 1\underline{/2\theta}. \qquad \text{(B.16)}$$

B.5 The Integer Power of a Complex Number

To raise a complex number to an integer power k, it is easier to first write the complex number in polar form. Thus

$$\begin{aligned} n^k &= (a + jb)^k \\ &= (ce^{j\theta})^k = c^k e^{jk\theta} \\ &= c^k(\cos k\theta + j \sin k\theta). \end{aligned}$$

For example,

$$\begin{aligned} (2e^{j12^\circ})^5 &= 2^5 e^{j60^\circ} = 32e^{j60^\circ} \\ &= 16 + j27.71, \end{aligned}$$

and

$$\begin{aligned} (3 + j4)^4 &= (5e^{j53.13^\circ})^4 = 5^4 e^{j212.52^\circ} \\ &= 625e^{j212.52^\circ} \\ &= -527 - j336. \end{aligned}$$

B.6 The Roots of a Complex Number

To find the kth root of a complex number, we must recognize that we are solving the equation

$$x^k - ce^{j\theta} = 0, \tag{B.17}$$

which is an equation of the kth degree and therefore has k roots.

To find the k roots, we first note that

$$ce^{j\theta} = ce^{j(\theta+2\pi)} = ce^{j(\theta+4\pi)} = \cdots. \tag{B.18}$$

It follows from Eqs. (B.17) and (B.18) that

$$x_1 = (ce^{j\theta})^{1/k} = c^{1/k}e^{j\theta/k}, \tag{B.19}$$

$$x_2 = [ce^{j(\theta+2\pi)}]^{1/k} = c^{1/k}e^{j(\theta+2\pi)/k}, \tag{B.20}$$

$$x_3 = [ce^{j(\theta+4\pi)}]^{1/k} = c^{1/k}e^{j(\theta+4\pi)/k}, \tag{B.21}$$

$$\vdots\ .$$

We continue the process outlined by Eqs. (B.19), (B.20), and (B.21) until the roots start repeating. This will happen when the multiple of

π is equal to $2k$. For example, let's find the four roots of $81e^{j60^\circ}$. We have

$$x_1 = 81^{1/4}e^{j60/4} = 3e^{j15^\circ},$$
$$x_2 = 81^{1/4}e^{j(60+360)/4} = 3e^{j105^\circ},$$
$$x_3 = 81^{1/4}e^{j(60+720)/4} = 3e^{j195^\circ},$$
$$x_4 = 81^{1/4}e^{j(60+1080)/4} = 3e^{j285^\circ},$$
$$x_5 = 81^{1/4}e^{j(60+1440)/4} = 3e^{j375^\circ} = 3e^{j15^\circ}.$$

Here, x_5 is the same as x_1, so the roots have started to repeat. Therefore we know the four roots of $81e^{j60^\circ}$ are the values given by x_1, x_2, x_3, and x_4.

It is worth noting that the roots of a complex number lie on a circle in the complex-number plane. The radius of the circle is $c^{1/k}$. The roots are uniformly distributed around the circle, the angle between adjacent roots being equal to $2\pi/k$ radians, or $360/k$ degrees. The four roots of $81e^{j60^\circ}$ are shown plotted in Fig. B.4.

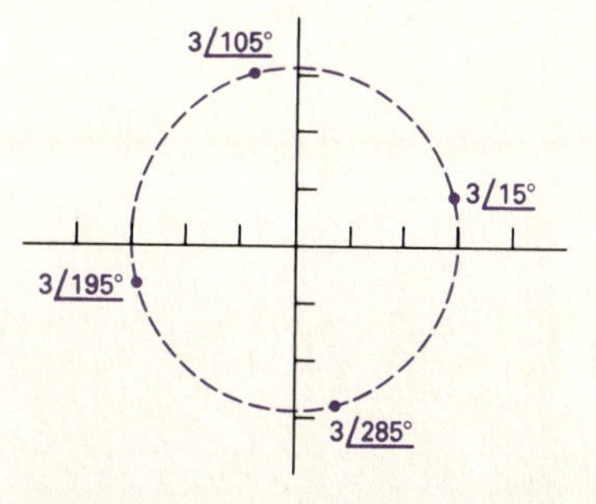

FIGURE B.4 The four roots of $81e^{j60^\circ}$.

Topology in Circuit Analysis

Here we introduce the topological properties of a circuit that allow us to prove the validity of the node-voltage and mesh-current methods of circuit analysis. This introduction and discussion of topological concepts also adds some depth of understanding to circuit analysis in general.

Circuit analysis revolves around two sets of constraints. The first is the set of constraints imposed on the circuit variables by the elements themselves—for example, Ohm's law. The second is the set of constraints forced on the variables by the interconnections of the elements—for instance, Kirchhoff's laws. By focusing on the constraints imposed by interconnections, circuit or network topology leads us to write the minimum number of independent equations that describe the circuit.

To show how to use topology to derive the node-voltage and mesh-current methods of analysis, we need to expand on the vocabulary introduced in Chapter 4. Thus in addition to the terms *node, branch, path, loop, mesh, planar,* and *nonplanar,* we must add the concepts of *graph, tree, cotree, link, cut set, fundamental cut set,* and *fundamental loop*.

C.1 Topological Concepts

The first step in studying the topological properties of a network is to suppress the nature of the circuit elements. We do so by constructing a graph of the circuit. The **graph** consists of redrawing the circuit, with a line representing each branch of the network. Obviously, the line conceals the nature of the element in a branch. Figure C.1 shows a circuit and its graph.

(a)

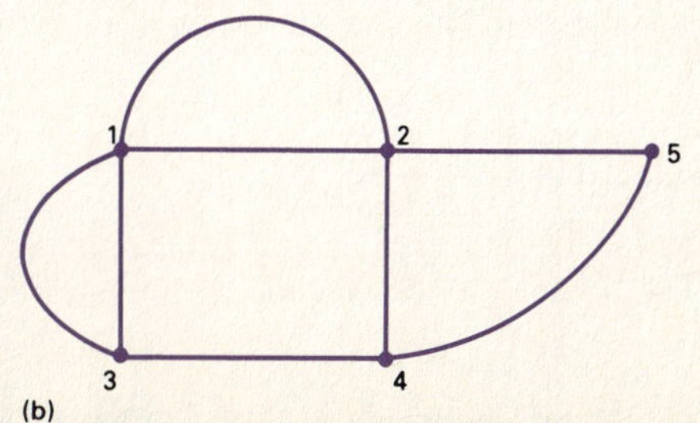

FIGURE C.1 (a) A circuit; (b) its graph.

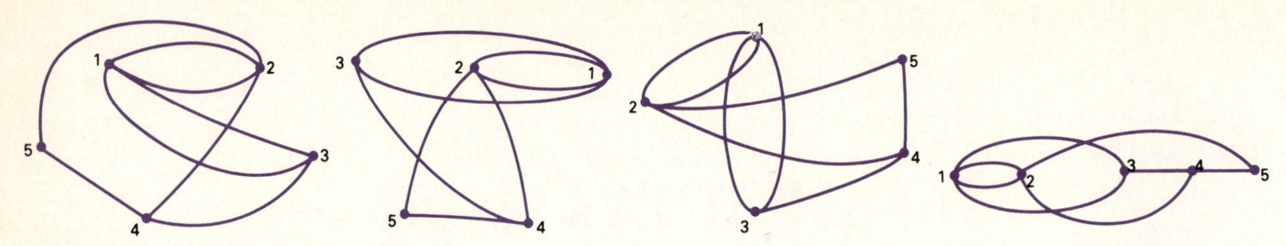

FIGURE C.2 Four graphs that are topologically equivalent to the graph shown in Fig. C.1(b).

The graph of a circuit is a visual aid that contains the pertinent information about the interconnections of nodes and branches. It completely specifies the circuit's topological character.

Before we show how to use topology to derive node-voltage or mesh-current equations, a brief explanation of this important branch of mathematics is in order. The word **topology** refers to the science of place. In mathematics, topology is a branch of geometry in which figures are considered perfectly elastic. Their flexible nature allows elastic, as opposed to rigid, motions. Thus figures can be stretched, twisted, squeezed, pulled, and bent. Two figures are topologically equivalent only if one figure can be made to coincide with the other by an elastic deformation. Thus a sphere and a cube are topologically equivalent, as are a circle and a square.

The topological properties of a figure are those that are invariant under elastic deformation. The topological properties of a circuit are those that are invariant with the stretching, squeezing, bending, or twisting of the circuit graph. For example, all the graphs in Fig. C.2 are topologically equivalent to the graph in Fig. C.1(b), because they all depict the same interconnections between the five nodes and eight branches.

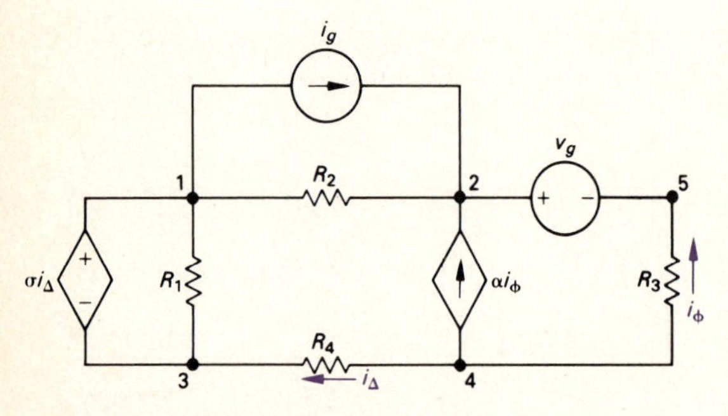

FIGURE C.3 A circuit that is topologically the same as the circuit shown in Fig. C.1(a).

Two networks are also topologically equivalent if they differ only in the elements that make up their branches. Thus the circuit shown in Fig. C.3 is the same, topologically, as the circuit shown in Fig. C.1(a).

A **tree** is defined as any set of connecting branches that connects every node to every other node without forming any closed paths or loops. Consider, for example, the graph shown in Fig. C.4, where we have labeled the branches and nodes for purposes of discussion. One of the trees in this graph consists of the branches a, d, and e; these three connected branches connect all four nodes of the graph without forming any loops. Figure C.5 shows this tree of the graph, with the branches of interest highlighted by heavy lines.

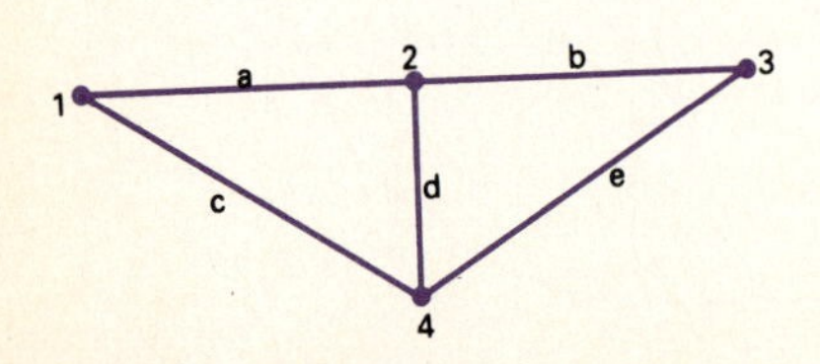

FIGURE C.4 A graph.

In general, a graph contains multiple trees. The graph shown in Fig. C.4 has eight trees. The seven trees other than the one shown in Fig. C.5 are b,d,c; c,a,b; e,b,a; a,c,e; b,e,c; a,b,d; and c,d,e.

The number of branches in any selected tree is always one less than the number of nodes, that is, $n - 1$. This condition follows directly from the definition of a tree. Once a tree for a graph has been defined, the remaining branches are referred to as the **links**, or **chords**. The collection of links is called a **complementary tree**, or **cotree**. Thus the cotree of the tree shown in Fig. C.5 consists of the branches b and c. The branches of a cotree may or may not be connected, whereas the branches of a tree are always connected. Also, cotrees may have loops; by definition, a tree contains no loops.

FIGURE C.5 A tree of the graph shown in Fig. C.4.

The last topological property germane to our discussion is the **cut set**. A cut set is a minimal set of branches that, when cut, divides the graph into two groups of nodes. The adjective *minimal* means that a cut set of a graph cannot itself contain a cut set that would divide the original graph into the same two groups of nodes. For example, in the graph shown in Fig. C.6, the branches a, c, f, and h are a cut set that divides the graph into the two groups of nodes shown in Fig. C.7. Cutting the branches a, c, f, h, and e would also divide the original graph into two graphs, one containing nodes 1, 2, and 4, and the other nodes 3 and 5. However, this configuration is not a cut set, because it contains the cut set a,c,f,h. Thus a cut set is the smallest number of branches needed to cut the graph into two specified groups of nodes.

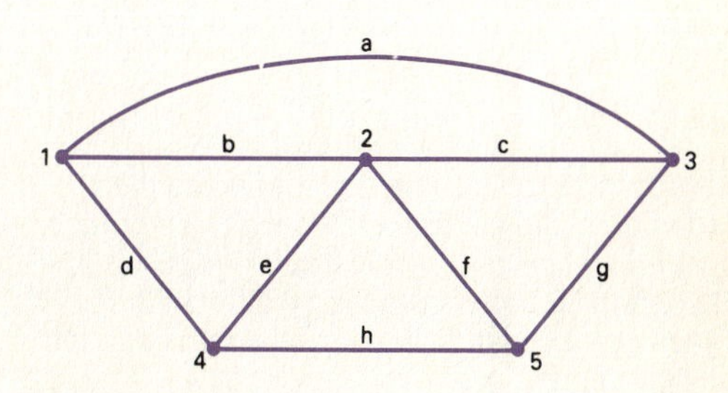

FIGURE C.6 A graph

When a graph is cut into two subgraphs,[1] an isolated node may form one subgraph. Therefore the cut sets in Fig. C.6 are d,e,h; a,b,d; b,c,e,f; a,c,g; and f,g,h. Just as there are many trees in a given graph, there are also many cut sets.

Keep in mind that the algebraic sum of the currents in the cut-set branches is zero. This characteristic becomes apparent when you view the cut set as the branches tying one portion of a circuit to another. For example, consider stretching the graph shown in Fig. C.6 so that the cut-set branches a, c, f, and h clearly form the tie between the nodes 1,2,4 and 3,5. Figure C.8 shows the resulting graph. Kirchhoff's current law requires that the algebraic sum of the currents in the cut-set branches a, c, f, and h equal zero; otherwise, a charge would accumulate in either the A or B portion of the circuit. Although a graph may contain many loops and many cut sets, to derive an independent set of circuit equations we need to focus on a subset of these loops and cut sets, called the fundamental loops and fundamental cut sets.

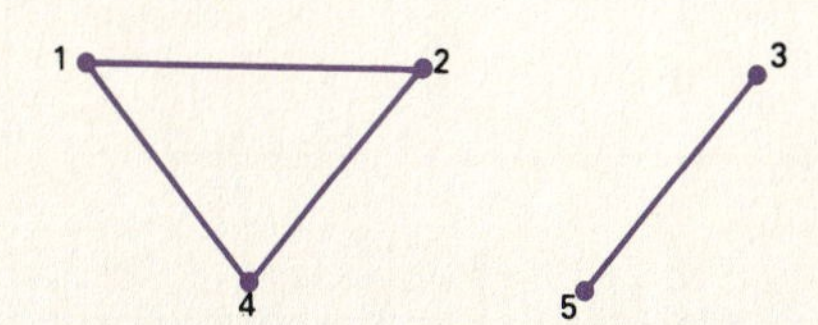

FIGURE C.7 The two subgraphs of the graph shown in Fig. C.6, created by the cut set a,c,f,h.

A **fundamental loop** is a loop that contains one and only one link. Returning to the graph in Fig. C.6, we define a tree as the branches d, e, f, and g. Then the four fundamental loops are d,$\underline{\text{b}}$,e; f,$\underline{\text{c}}$,g; e,f,$\underline{\text{h}}$; and d,$\underline{\text{a}}$, g,f,e, with the underlined letter denoting the link in each loop. The set of fundamental loops in a circuit is not unique, because the chosen tree is not unique. The number of fundamental loops, however, is unique and equals the number of links—that is, $b' - n + 1$, where b' is the total number of branches in the graph.

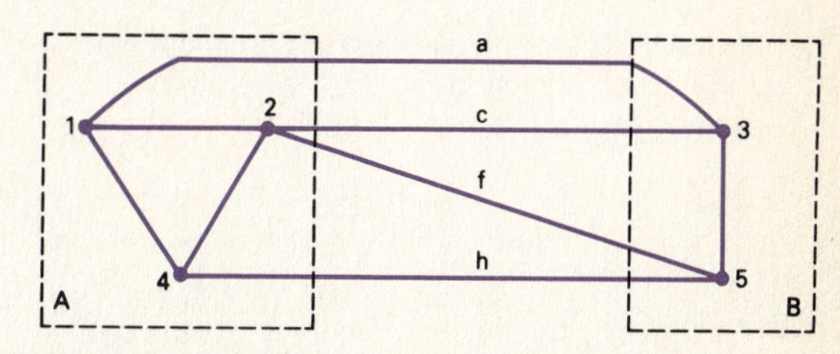

FIGURE C.8 An elastic deformation of the graph shown in Fig. C.6.

A **fundamental cut set** is a cut set that contains one and only one tree branch. To find a fundamental cut set associated with a tree, follow these steps: (1) select a tree branch; (2) divide the graph into

[1] A graph g is said to be a subgraph of the graph G if all the nodes and all the branches in g are also in G and if each branch in g has the same end nodes as in G.

two subgraphs, one containing all the tree branches connected to one node of the selected tree branch, and the other containing all the tree branches connected to the other node; (3) separate these two subgraphs by stretching the selected tree branch; and (4) draw in just those links that span the two subgraphs. These links and the selected tree branch form a fundamental cut set.

To illustrate this procedure, we return to the graph shown in Fig. C.6 and, as before, select a tree as the set of branches d, e, f, and g. Next we find the fundamental cut set associated with the tree branch labeled e. In accordance with step 2, we construct the graph in Fig. C.9(a). Figure C.9(b) shows the result of stretching the graph in Fig. C.9(a), and Fig. C.9(c) shows the construction of the links that span the two subgraphs. From Fig. C.9(c), we identify the fundamental cut set associated with tree branch e as the branches a, b, h, and $\boxed{\text{e}}$, with the boxed letter denoting the tree branch. The three remaining fundamental cut sets are a,b,$\boxed{\text{d}}$; a,c,h,$\boxed{\text{f}}$; and a,c,$\boxed{\text{g}}$. The graphs associated with these cut sets appear in Figs. C.10(a), (b), and (c), respectively.

FIGURE C.9 A graphic technique for finding the fundamental cut set associated with tree branch e: (a) the subgraphs associated with branch e; (b) separating the subgraphs by stretching branch e; (c) constructing the links that span the two subgraphs.

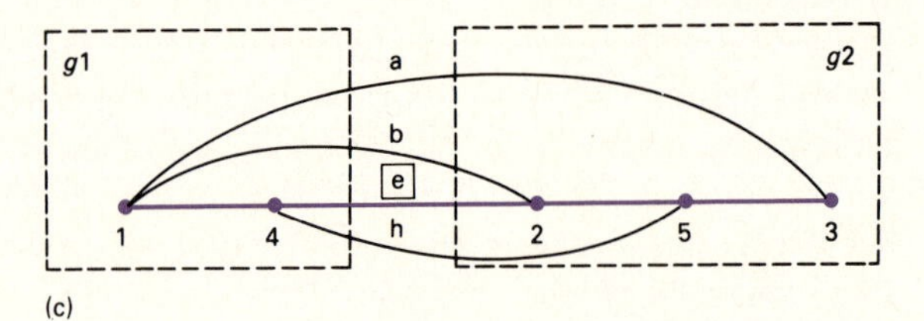

FIGURE C.10 The remaining fundamental cut sets of the graph shown in Fig. C.6, when the selected tree is the set of branches d, e, f, and g.

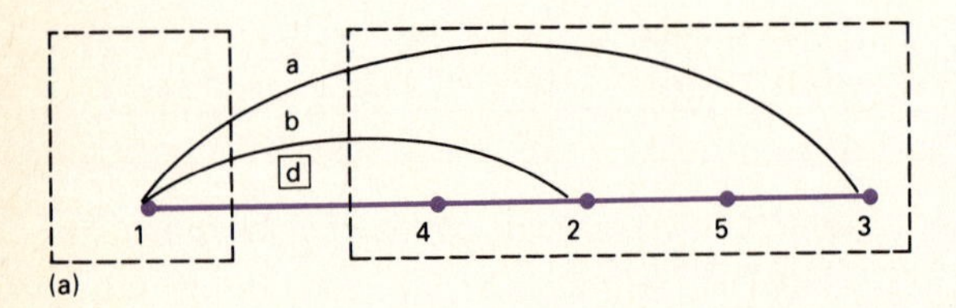

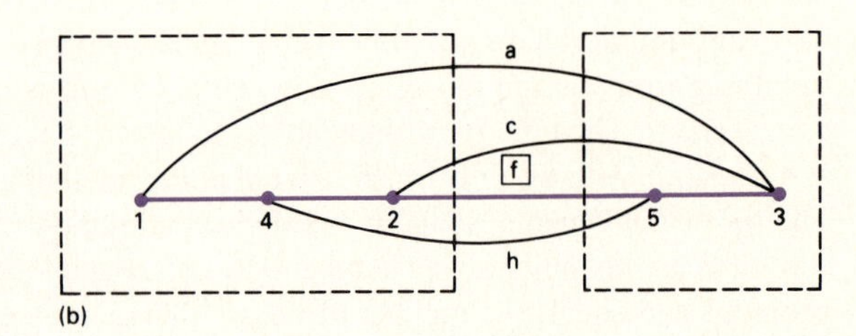

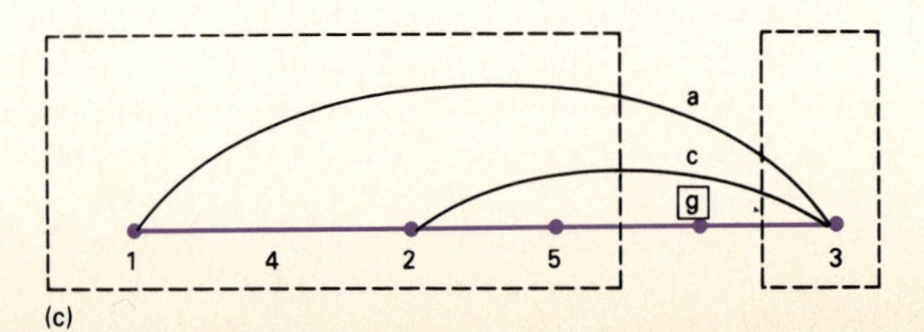

We now show how to use the topological properties of trees, links, fundamental loops, and fundamental cut sets to prove the validity of the node-voltage, loop-current, and mesh-current methods of analysis. In Section C.2 we validate the node-voltage method, and in Section C.3 we justify the loop-current method. Section C.4 discusses how the mesh-current method can be used to replace the loop-current method in planar circuits.

C.2 A Topological Approach to the Node-Voltage Method

The derivation of the node-voltage method is rooted in two fundamental characteristics. First, once the tree-branch voltages are known, the link voltages are also known. Second, an independent set of equations involving the tree-branch voltages can be derived by summing the currents in every fundamental cut set. The cut-set currents must add to zero, in accordance with Kirchhoff's current law. These equations are guaranteed to be an independent set, because each fundamental cut set introduces a tree branch that is not in any other fundamental cut set. The circuit shown in Fig. C.11 illustrates these characteristics.

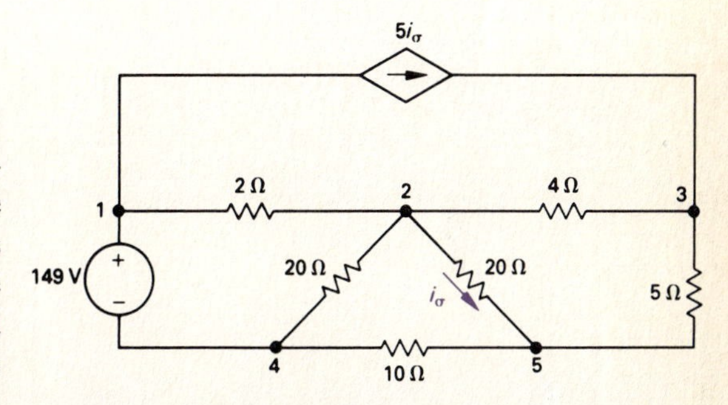

FIGURE C.11 A circuit illustrating a topological approach to analysis.

The first step in the topological approach is to construct a circuit graph and assign a voltage and a current to each branch in the graph. For convenience, we always match the direction of the branch current to the direction of the voltage drop across the branch. Figure C.12 presents the graph with the assigned variables for the circuit shown in Fig. C.11.

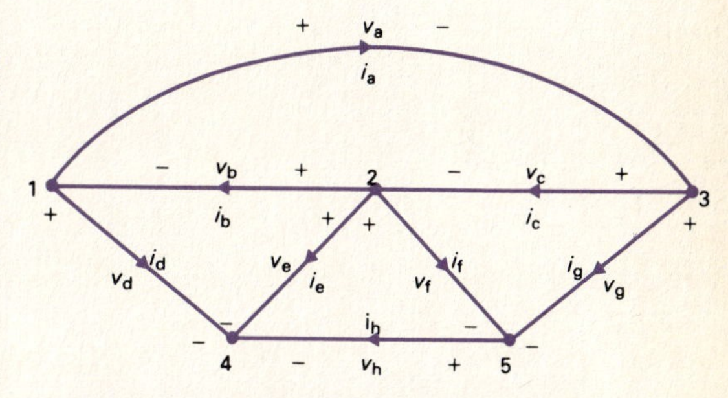

FIGURE C.12 A graph, with assigned branch variables, for the circuit shown in Fig. C.11.

Now we select a tree. Although there are in general many trees to choose from, we can facilitate our analysis by adhering, whenever possible, to the following guidelines:

- Place the branch in a tree if it contains a voltage source or if the branch voltage controls dependent sources.
- Place the branch in a cotree if it contains a current source or if the branch current controls dependent sources.

Hence we select a tree consisting of the branches d, e, c, and g, as shown in Fig. C.13. By defining the tree, we also define the fundamental cut sets. Here, the four fundamental cut sets are (1) a,b,d; (2) a,b,e,h; (3) a,c,f,h; and (4) g,f,h. For this circuit, we know the tree-branch voltage v_d, and therefore we have only three unknown tree-branch voltages—namely, v_e, v_c, and v_g. Thus, we have to sum only the currents in cut sets (2) through (4). Figure C.14 depicts these three cut sets.

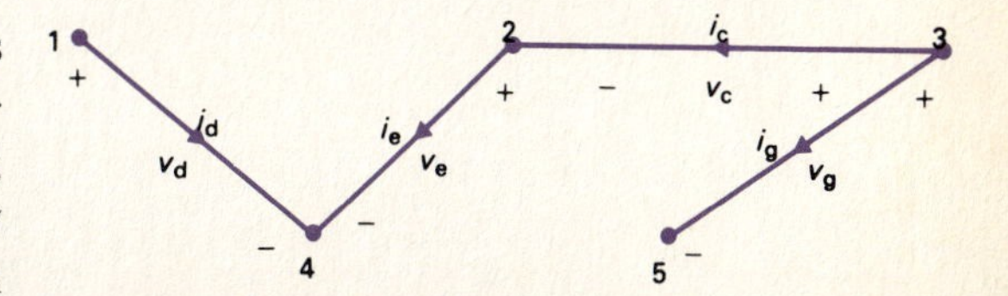

FIGURE C.13 A tree of the graph shown in Fig. C.12.

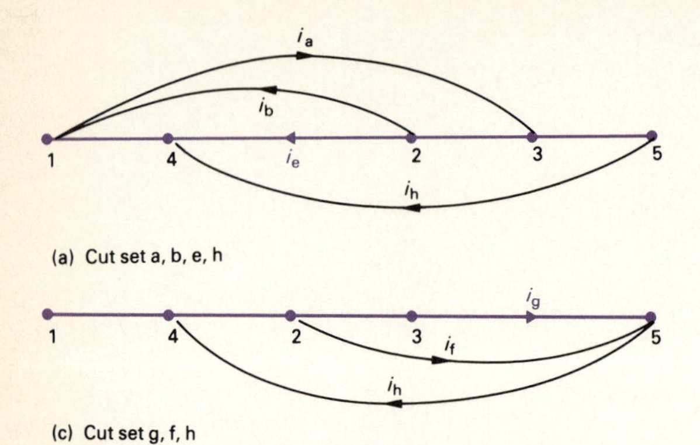

(a) Cut set a, b, e, h

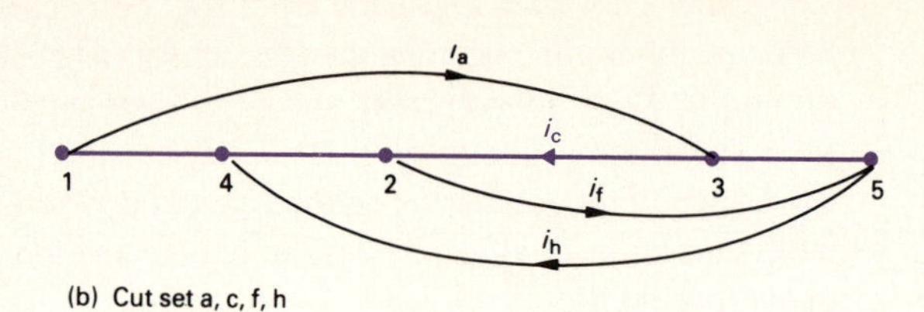

(b) Cut set a, c, f, h

(c) Cut set g, f, h

FIGURE C.14 A graphic representation of the pertinent cut sets.

If the cut-set currents oriented toward the right are made positive, then those oriented toward the left are negative. Thus the three cut-set current equations are

$$\begin{aligned} i_a - i_b - i_e - i_h &= 0; \\ i_a - i_c + i_f - i_h &= 0; \\ i_g + i_f - i_h &= 0. \end{aligned} \tag{C.1}$$

To find the tree-branch voltages, we simply express the cut-set currents as functions of these voltages. Reference to both the circuit diagram and its graph yields the set of equations:

$$\begin{aligned} &i_a = 5i_\sigma = 5i_f = \frac{5(v_g - v_c)}{20}; \quad i_b = \frac{v_e - 149}{2}; \\ &i_c = \frac{v_c}{4}; \quad i_e = \frac{v_e}{20}; \quad i_f = \frac{v_g - v_c}{20}; \\ &i_g = \frac{v_g}{5}; \quad \text{and} \quad i_h = \frac{v_e + v_c - v_g}{10}. \end{aligned} \tag{C.2}$$

Substituting Eqs. (C.2) into Eqs. (C.1) and multiplying the resulting equations by 20 generates the set of independent equations:

$$\begin{aligned} 7v_g - 7v_c - 13v_e &= -1490; \\ 8v_g - 13v_c - 2v_e &= 0; \\ 7v_g - 3v_c - 2v_e &= 0. \end{aligned} \tag{C.3}$$

The solutions for the tree-branch voltages are

$$\begin{aligned} v_g &= 40 \text{ V}; \\ v_c &= 4 \text{ V}; \\ v_e &= 134 \text{ V}. \end{aligned} \tag{C.4}$$

You should verify that once the tree-branch voltages are known, the current, voltage, and power associated with any branch in the circuit can be found.

The transition from tree-branch voltages to node voltages is based on the fact that, if one node is selected as a reference, the resulting $n-1$ node voltages are always expressible in terms of $n-1$ tree-branch voltages. (Remember that, by definition, every tree consists of a set of branches that connects all the nodes without forming loops.) We have already shown that the $n-1$ tree-branch voltages constitute an independent set of variables that can be used to describe a circuit; thus the node voltages also form an independent set of $n-1$ variables. The node voltages are chosen more often than the tree-branch voltages because they can be identified without specifying a tree.

For example, in the circuit shown in Fig. C.11, we arbitrarily chose node 4 as the reference and let v_1, v_2, v_3, and v_5 denote the resulting node voltages. If we now select the tree depicted in Fig. C.13, the node voltages as functions of the tree-branch voltages are

$$v_1 = v_d; \quad v_2 = v_e;$$
$$v_3 = v_e + v_c; \quad \text{and} \quad v_5 = v_e + v_c - v_g. \qquad \text{(C.5)}$$

Because v_c, v_d, v_e, and v_g are a set of independent variables, so are v_1, v_2, v_3, and v_5.

Now suppose that we retain the same reference node but change the tree to consist of the branches d, h, f, and c. Then, from the graph shown in Fig. C.12,

$$v_1 = v_d; \quad v_2 = v_h + v_f;$$
$$v_3 = v_h + v_f + v_c; \quad \text{and} \quad v_5 = v_h. \qquad \text{(C.6)}$$

But v_c, v_d, v_f, and v_h are independent variables, because they are a set of tree-branch voltages. Thus, as before, we conclude that v_1, v_2, v_3, and v_5 also form a set of independent variables.

We conclude our discussion of the topological approach to the node-voltage method with the following summary:

1. Once the tree-branch voltages are determined, the link voltages can be calculated, and therefore every branch voltage is known.

2. The tree-branch voltages form a set of independent variables that can be used to describe the circuit.

3. A set of independent equations involving the tree-branch voltages can be derived by summing the branch currents in the $n-1$ fundamental cut sets in accordance with Kirchhoff's current law.

4. Once a reference node has been selected, the $n-1$ node voltages can always be expressed as combinations of $n-1$ tree-branch voltages, and therefore the node voltages can also be used as a set of independent variables to describe the circuit.

5. Node voltages are used more often than tree-branch voltages, because they can be defined without reference to a specific tree.

C.3 A Topological Approach to the Loop-Current Method

The loop-current method is valid for both planar and nonplanar circuits. After introducing this method, we offer a numerical example of how to use it to analyze a nonplanar circuit. Then in Section C.4 we show how the mesh-current method evolves from the loop-current method when the circuit is planar.

The key to the loop-current method is to recognize that, if we know the $b' - n + 1$ link currents, we can calculate the $n - 1$ tree-branch currents simply by summing the currents in the $n - 1$ fundamental cut sets. Thus the $b' - n + 1$ link currents form a set of independent variables that we can use to describe a circuit. We derive a set of independent equations, using the link currents as variables, by defining a loop current for each of the $b' - n + 1$ fundamental loops in the circuit and then summing the voltages around each loop in accordance with Kirchhoff's voltage law. Note that, because each fundamental loop contains only one link, each loop current is identical to a link current.

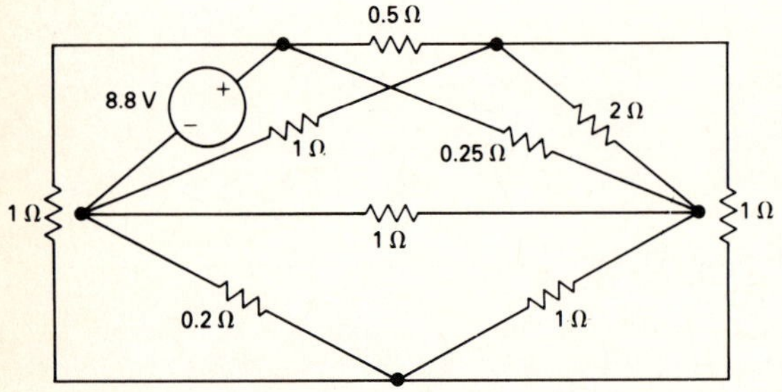

FIGURE C.15 A nonplanar circuit used to illustrate the loop-current method.

We use the nonplanar circuit in Fig. C.15 to illustrate the loop-current method. The circuit contains five nodes and ten branches, so we have to derive and solve six $(10 - 5 + 1)$ simultaneous equations. Figure C.16 shows a graph of the circuit, with the nodes and branches labeled to aid the discussion.

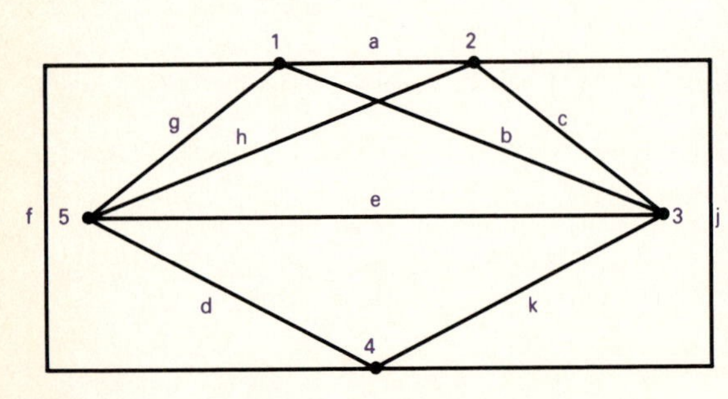

FIGURE C.16 A labeled graph of the circuit shown in Fig. C.15.

Next, we select a tree and at the same time assign reference directions for all the branch currents. Figure C.17 shows the results of these decisions. Note that the tree branches are g, h, j, and k and that the six link currents are i_a–i_f.

Now that we have selected the tree, we can identify the six fundamental loops. Recall that each loop current coincides with a link current. The six fundamental loops are as follows:

Loop	Branches
a	a, h, g
b	b, k, j, h, g
c	c, k, j
d	d, h, j
e	e, h, j, k
f	f, j, h, g

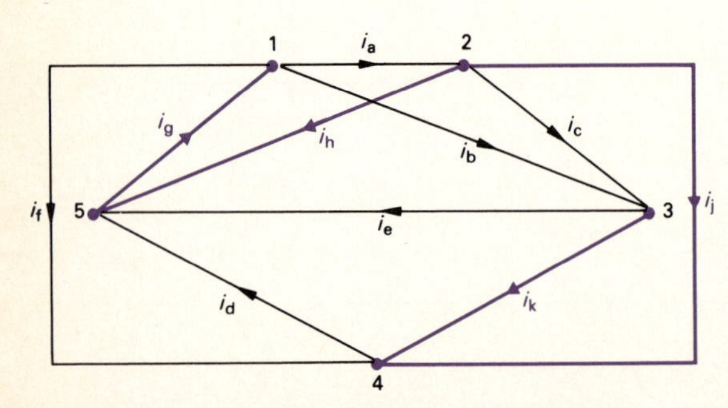

FIGURE C.17 The graph shown in Fig. C.16, with the selected tree and branch current references.

If we attempted to draw the six loop currents on the graph shown in Fig. C.17, the resulting figure would be difficult to decipher. However, the information of primary interest at this point is the identification and orientation of the loop currents in each tree branch. Hence we depict only this information in Fig. C.18, where each tree branch is isolated and the loop currents, along with their directions, are shown next to each branch.

We derive the following set of loop-current equations by summing the voltages around each fundamental loop. Each equation starts with the voltage across a link and then proceeds around the fundamental loop containing that link. The tracing direction corresponds to the direction of the link current. The equations start with loop a and proceed in alphabetical order:

$$
\begin{aligned}
0 &= 0.5i_a + 1(i_a + i_b + i_f - i_d - i_e) - 8.8;\\
0 &= 0.25i_b + 1(i_b + i_c - i_e) + 1(i_b + i_c + i_f - i_d - i_e)\\
&\quad + 1(i_a + i_b + i_f - i_d - i_e) - 8.8;\\
0 &= 2i_c + 1(i_b + i_c - i_e) + 1(i_b + i_c + i_f - i_d - i_e);\\
0 &= 0.2i_d + 1(i_d + i_e - i_a - i_b - i_f)\\
&\quad + 1(i_d + i_e - i_b - i_c - i_f);\\
0 &= 1i_e + 1(i_d + i_e - i_a - i_b - i_f)\\
&\quad + 1(i_d + i_e - i_b - i_c - i_f) + 1(i_e - i_b - i_c);\\
0 &= 1i_f + 1(i_b + i_c + i_f - i_d - i_e)\\
&\quad + 1(i_a + i_b + i_f - i_d - i_e) - 8.8.
\end{aligned}
\tag{C.7}
$$

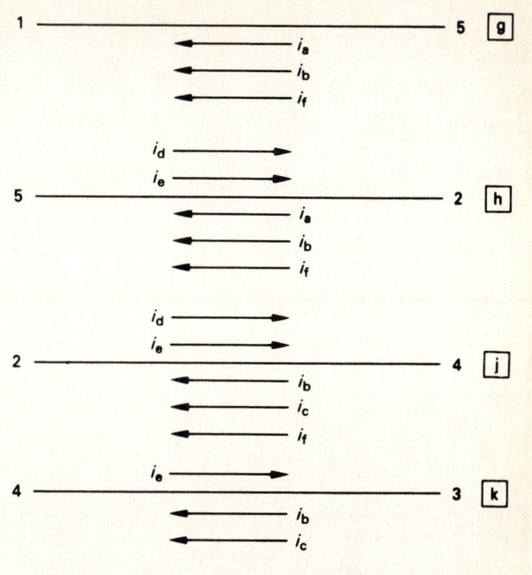

FIGURE C.18 The identification and orientation of the loop currents in each tree branch.

Rearranging Eq. (C.7) to facilitate solution by a computer gives

$$
\begin{aligned}
8.8 &= 1.5i_a + i_b + 0i_c - i_d - i_e + i_f;\\
8.8 &= i_a + 3.25i_b + 2i_c - 2i_d - 3i_e + 2i_f;\\
0 &= 0i_a + 2i_b + 4i_c - i_d - 2i_e + i_f;\\
0 &= i_a - 2i_b - i_c + 2.2i_d + 2i_e - 2i_f;\\
0 &= -i_a - 3i_b - 2i_c + 2i_d + 4i_e - 2i_f;\\
8.8 &= i_a + 2i_b + i_c - 2i_d - 2i_e + 3i_f.
\end{aligned}
\tag{C.8}
$$

Solving for the six loop (link) currents yields

$$
\begin{aligned}
i_a &= 7.28 \text{ A};\\
i_b &= 10.4 \text{ A};\\
i_c &= -0.52 \text{ A};\\
i_d &= 12.6 \text{ A};\\
i_e &= 6.20 \text{ A};\\
i_f &= 6.28 \text{ A}.
\end{aligned}
\tag{C.9}
$$

To find the tree-branch currents, we sum the currents in each fundamental cut set. The four fundamental cut sets are (1) g,b,a,f; (2) h,a,b,e,d,f; (3) j,c,b,e,d,f; and (4) k,e,b,c. The cut-set equations then are

$$
\begin{aligned}
i_g &= i_a + i_b + i_f;\\
i_h &= i_a + i_b + i_f - i_d - i_e;\\
i_j &= i_d + i_e - i_b - i_c - i_f;\\
i_k &= i_b + i_c - i_e.
\end{aligned}
\tag{C.10}
$$

The numerical values of the tree-branch currents are

$$\begin{aligned} i_g &= 23.96 \text{ A}; \\ i_h &= 5.16 \text{ A}; \\ i_j &= 2.64 \text{ A}; \\ i_k &= 3.68 \text{ A}. \end{aligned} \tag{C.11}$$

Now that we know all the branch currents, we have enough information to calculate any branch voltage or power that may be of interest.

C.4 The Mesh-Current Method (Planar Circuits)

Having demonstrated the loop-current method, we now show that in planar circuits, the mesh-current method can be used to replace the loop-current method. For planar circuits where every window forms part of the boundary of the graph,[2] the mesh-current method is equivalent to the loop-current method, because we can always construct a tree so that the cotree is the set of perimeter links. Then the fundamental loops and meshes are identical.

For example, consider the graph shown in Fig. C.19, where, as before, the heavy lines represent the tree branches. Note that the meshes are identical to the fundamental loops. Because the number of fundamental loops is $b' - n + 1$, the number of meshes also equals $b' - n + 1$.

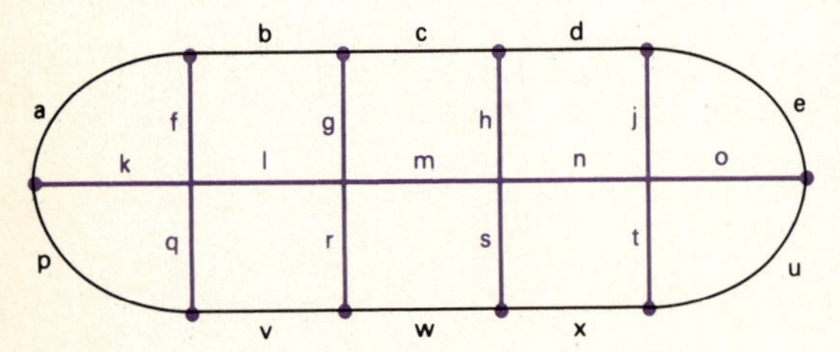

FIGURE C.19 A planar graph where every window forms part of the boundary.

What happens if the circuit graph has one or more interior windows? We show (1) that the number of meshes still equals $b' - n + 1$ and (2) that the mesh equations form an independent set. To aid the discussion of item (1), we use the circuit graph shown in Fig. C.20, which has one interior window. The graph has 20 branches and 12 nodes, so we expect to count 20 − 12 + 1, or 9, meshes. The graph does indeed contain 9 meshes.

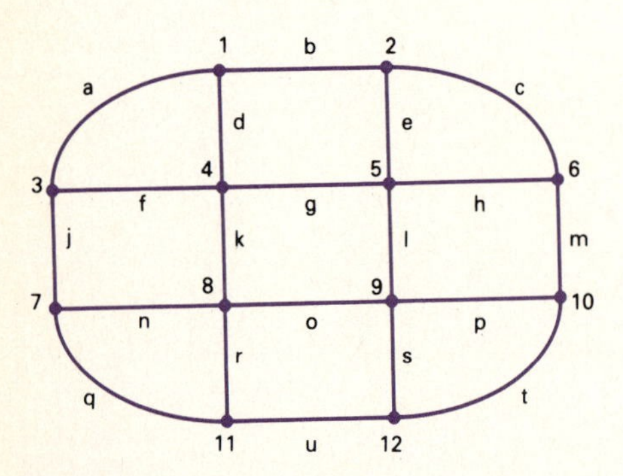

FIGURE C.20 A planar graph with one interior window.

We deduce that the number of meshes always equals the number of links by reasoning as follows. First, we select a tree and then add the links to the graph one at a time. Each time we add a link, it either forms a new loop or divides an existing loop in two. Hence the addition of each link corresponds to the addition of new loop. After the last link has been added, the loops correspond to the windows or meshes of the graph; that is, the number of meshes equals the number

[2]The **boundary** of a planar graph is made up of the branches that form the perimeter of the graph. For example, in Fig. C.12, the boundary consists of the branches a, g, h, and d.

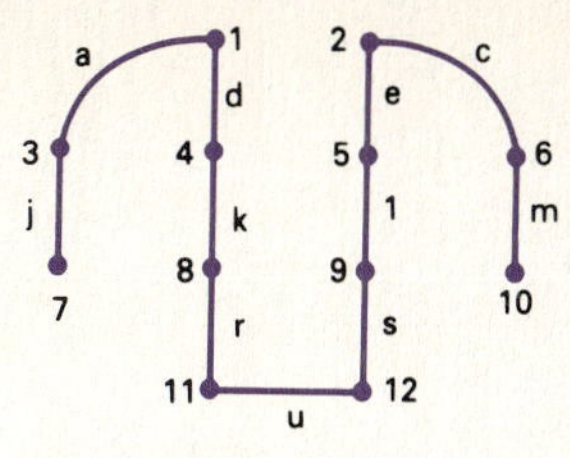

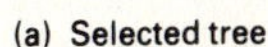
(a) Selected tree

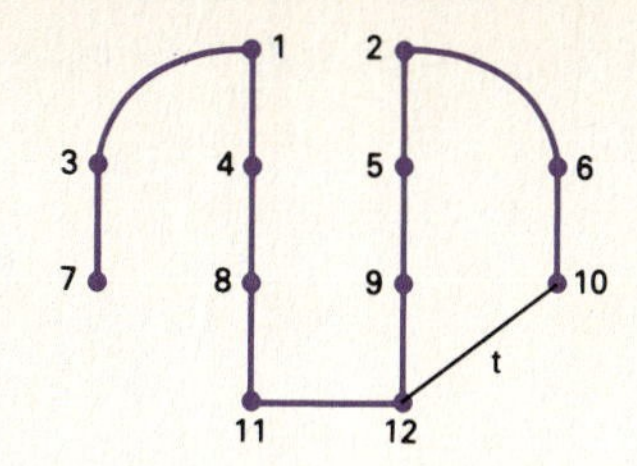

(b) 1 link, 1 loop

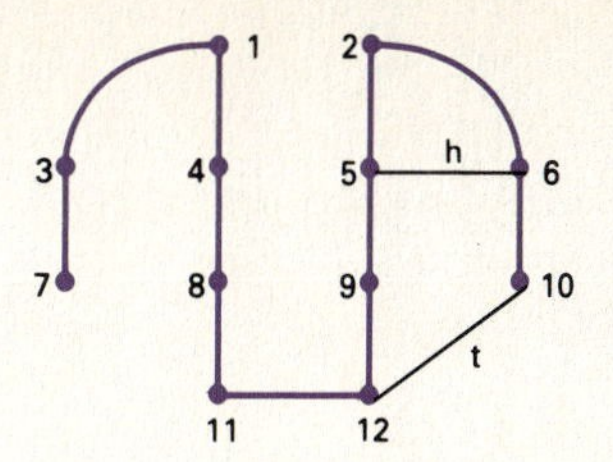

(c) 2 links, 2 loops

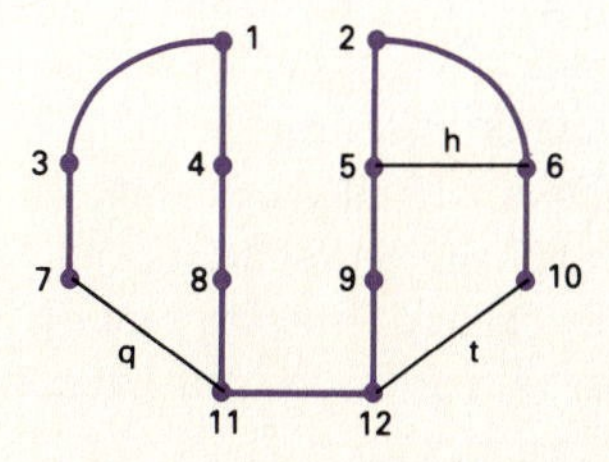

(d) 3 links, 3 loops

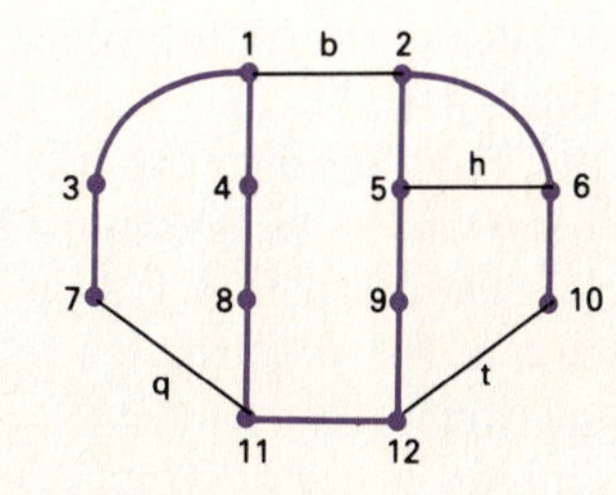

(e) 4 links, 4 loops

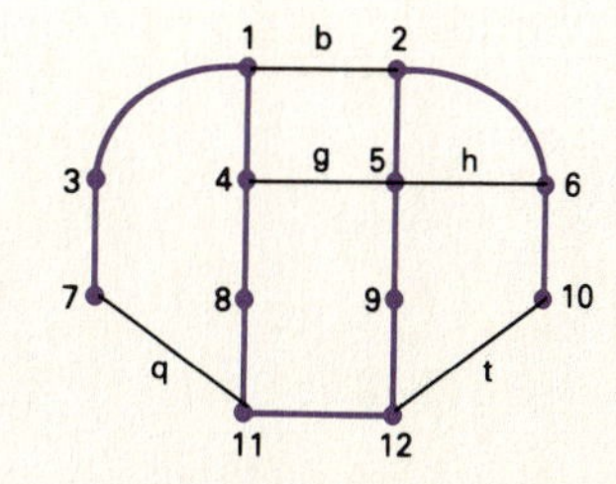

(f) 5 links, 5 loops

FIGURE C.21 An illustration showing how the addition of a link adds a loop to the graph.

of links. Figure C.21 shows the result of adding the first five links in this process for the graph shown in Fig. C.20. Here, we selected branches j, a, d, k, r, u, s, l, e, c, and m to form the tree.

We have established that the number of meshes in a planar circuit equals $b' - n + 1$, so the remaining task is to show that the system of mesh equations is independent. Recall from algebra that a set of linear equations is independent if no equation in the set can be derived from a linear combination of the remaining equations. If one of the equations can be derived from a linear combination of the others, we say the equations are dependent. We can state this criterion for dependence in a set of mesh equations as follows. Let $m_1, m_2, \ldots, (m'_b - n + 1)$ represent each mesh equation in the set of equations. Let $c_1, c_2, \ldots, (c'_b - n + 1)$ be a set of constants, not all of which are zero. Then, if the equations are dependent, we can find values for the c's so that

$$c_1 m_1 + c_2 m_2 + \cdots + (c'_b - n + 1)(m'_b - n + 1) = 0. \qquad \textbf{(C.12)}$$

If the sum in Eq. (C.12) cannot be made equal to zero, the equations are independent.

To show that the mesh equations form an independent set, we assume that the equations satisfy Eq. (C.12), and we then demonstrate that we always get a contradiction. Before proceeding to our general case, however, let's consider a circuit with no interior meshes, because in this case it is easy to verify that the mesh equations form an independent set. In the circuit graph shown in Fig. C.22, we select a tree so that each mesh contains a perimeter branch. The branch voltages with their reference polarities are also given.

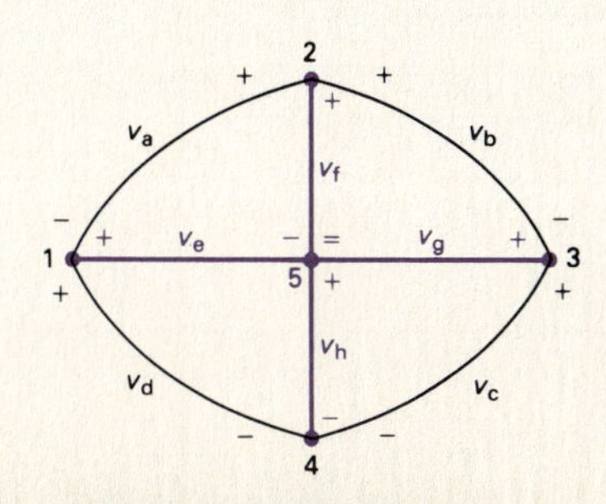

FIGURE C.22 A circuit graph.

If we sum clockwise around each mesh and assign a plus sign to a voltage drop, we get four mesh equations:

$$\begin{aligned} m_1: &\quad -v_a + v_f - v_e = 0; \\ m_2: &\quad v_b + v_g - v_f = 0; \\ m_3: &\quad v_c - v_h - v_g = 0; \\ m_4: &\quad -v_d + v_e + v_h = 0. \end{aligned} \tag{C.13}$$

Note that each equation contains a voltage that does not appear in any other equation—each mesh is identical to a fundamental loop and hence contains only one link voltage. Thus we can never find a set of constants c_1, c_2, c_3, and c_4 of which some are not zero, to satisfy Eq. (C.12). Therefore the mesh equations are independent.

We now return to our general case where the circuit may contain one or more interior meshes. First, we acknowledge that some, but not all, of the constants c_1 to $(c'_b - n + 1)$ may be zero. For purposes of discussion, we number the meshes so that the first k constants—that is, c_1 to c_k—are not zero. Then Eq. (C.12) reduces to

$$c_1 m_1 + c_2 m_2 + \cdots + c_k m_k = 0. \tag{C.14}$$

By hypothesis, we can eliminate the summing of the voltages around the meshes from $k + 1$ to $b' - n + 1$. In this group of meshes, at least one mesh always has a branch common with at least one other mesh in the group 1 to k. If we denote the voltage across this common branch as v_x, in the 1 to k mesh equations only one equation will contain v_x. Hence, finding a set of nonzero coefficients (c_1 to c_k) to eliminate this voltage is impossible. Therefore Eq. (C.14) cannot be satisfied, and the set of $b' - n + 1$ mesh equations must be independent.

For example, consider the circuit graph shown in Fig. C.23 and assume that c_1–c_9 are nonzero constants and that $c_{10} = 0$. Therefore we do not sum around mesh 10. Hence, v_x appears only in equation m_6, making impossible either

$$\sum_{k=1}^{9} c_k m_k = 0 \tag{C.15}$$

or

$$\sum_{k=1}^{10} c_k m_k = 0. \tag{C.16}$$

Therefore the 10 mesh equations are independent.

We summarize the topological approach to the loop-current and mesh-current methods of analysis as follows:

1. The loop-current method is based on summing the voltages around each fundamental loop of the circuit in accordance with Kirchhoff's voltage law.

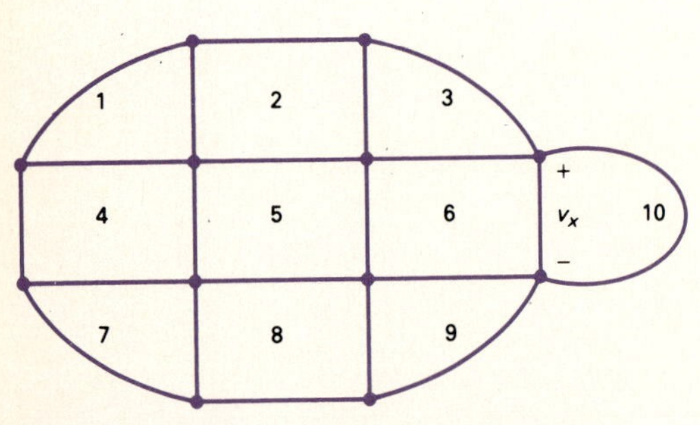

FIGURE C.23 A circuit graph used to illustrate the independence of the mesh equations.

2. The loop-current method generates a set of independent equations because each fundamental loop contains a link-branch voltage that appears in no other fundamental loop.
3. The loop-current method can be used for either planar or nonplanar circuits.
4. If a circuit is planar, the loop-current method can be replaced with the simpler mesh-current method.
5. The mesh-current method is simpler because, in general, identifying the meshes in a circuit is easier than identifying the fundamental loops.
6. The number of meshes in a circuit equals the number of links, that is, $b' - n + 1$.
7. The $b' - n + 1$ mesh equations are also independent.

THE DECIBEL

We defined the amplitude of $H(j\omega)$ in decibels in Eq. (15.66). However, the original definition of *decibel* involves power ratios. Tracing back to this definition is worthwhile because it is still a widely accepted use of the term.

Telephone engineers who were concerned with the power loss across the cascaded circuits used to transmit telephone signals introduced the decibel. Figure D.1 defines the problem. There, p_i is the power input to the system, p_1 is the power output of circuit A, p_2 is the power output of circuit B, and p_o is the power output of the system. The power gain of each circuit is the ratio of the power out to the power in. Thus

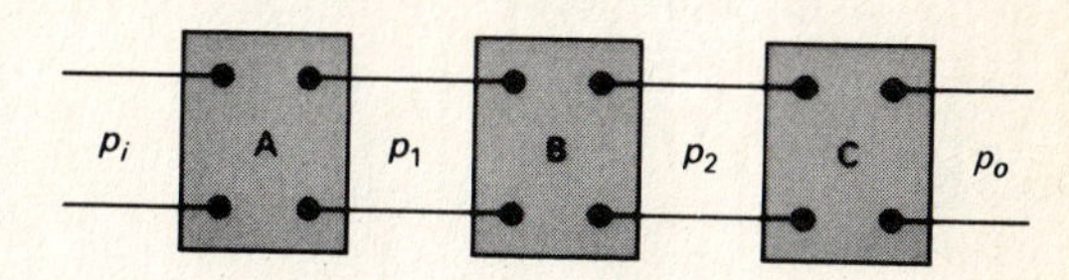

FIGURE D.1 Three cascaded circuits.

$$\sigma_A = \frac{p_1}{p_i}, \quad \sigma_B = \frac{p_2}{p_1}, \quad \text{and} \quad \sigma_C = \frac{p_o}{p_2}.$$

The overall power gain of the system is simply the product of the individual gains, or

$$\frac{p_o}{p_i} = \frac{p_1}{p_i}\frac{p_2}{p_1}\frac{p_o}{p_2} = \sigma_A\sigma_B\sigma_C.$$

The multiplication of power ratios is converted to addition by means of the logarithm; that is,

$$\log_{10}\frac{p_o}{p_i} = \log_{10}\sigma_A + \log_{10}\sigma_B + \log_{10}\sigma_C.$$

This log ratio of the powers was named the **bel**, in honor of Alexander Graham Bell. Thus we calculate the overall power gain, in bels, simply by summing the power gains, also in bels, of each segment of the transmission system. In practice, the bel is an inconveniently large quantity. One-tenth of a bel is a more useful measure of power gain; hence the **decibel**. The number of decibels equals 10 times the number of bels, so

$$\text{Number of decibels} = 10\log_{10}\frac{p_o}{p_i}.$$

When we use the decibel as a measure of power ratios, in some situations the resistance seen looking into the circuit equals the resistance loading the circuit, as illustrated in Fig. D.2. When the input resistance equals the load resistance, we can convert the power ratio to either a voltage ratio or a current ratio:

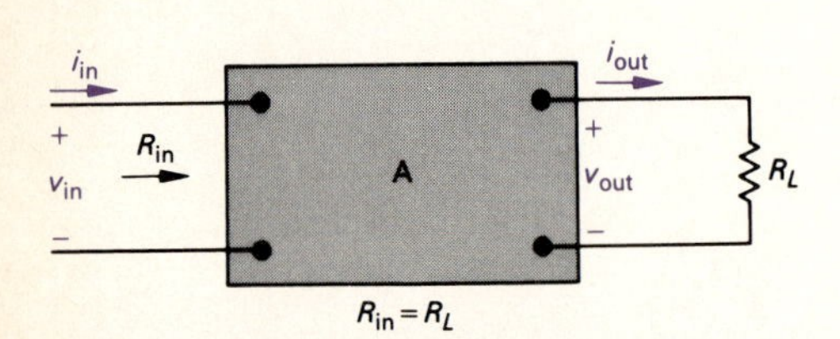

FIGURE D.2 A circuit in which the input resistance equals the load resistance.

$$\frac{p_o}{p_i} = \frac{v_{out}^2/R_L}{v_{in}^2/R_{in}} = \left(\frac{v_{out}}{v_{in}}\right)^2$$

or

$$\frac{p_o}{p_i} = \frac{i_{out}^2 R_L}{i_{in}^2 R_L} = \left(\frac{i_{out}}{i_{in}}\right)^2.$$

These equations show that the number of decibels becomes

$$\begin{aligned}\text{Number of decibels} &= 20\log_{10}\frac{v_{out}}{v_{in}}\\ &= 20\log_{10}\frac{i_{out}}{i_{in}}.\end{aligned} \quad \textbf{(D.1)}$$

The definition of the decibel used in Bode diagrams—that is, Eq. (15.66)—is borrowed from the results expressed by Eq. (D.1), since these results apply to any transfer function involving a voltage ratio, a current ratio, a voltage-to-current ratio, or a current-to-voltage ratio. You should keep the original definition of the decibel firmly in mind because it is of fundamental importance in many engineering applications.

When you are working with transfer function amplitudes expressed in decibels, having a table that translates the decibel value to the actual value of the output/input ratio is helpful. Table D.1 gives some useful pairs. The ratio corresponding to a negative decibel value is the reciprocal of the positive ratio. For example, −3 dB corresponds to an output/input ratio of 1/1.41, or 0.707. Interestingly, −3 dB corresponds to the half-power frequencies of the filter circuits discussed in Chapters 15 and 16.

The decibel is also used as a unit of power when it expresses the ratio of a known power to a reference power. Usually the reference power is 1 mW and the power unit is written dBm, which stands for

TABLE D.1

SOME dB-RATIO PAIRS

dB	RATIO	dB	RATIO
0	1.00	30	31.62
3	1.41	40	100.00
6	2.00	60	10^3
10	3.16	80	10^4
15	5.62	100	10^5
20	10.00	120	10^6

"decibels relative to one milliwatt." For example, a power of 20 mW corresponds to ± 13 dBm.

AC voltmeters commonly provide dBm readings that assume not only a 1 mW reference power but also a 600 Ω reference resistance (a value commonly used in telephone systems). Since a power of 1 mW in 600 Ω corresponds to 0.7746 V (rms), that voltage is read as 0 dBm on the meter. For analog meters, there usually is exactly a 10 dB difference between adjacent ranges. Although the scales may be marked 0.1, 0.3, 1, 3, 10, and so on, in fact 3.16 V on the 3 V scale lines up with 1 V on the 1 V scale.

Some voltmeters provide a switch to choose a reference resistance (50, 135, 600, or 900 Ω) or to select dBm or dBV (decibels relative to one volt).

An Abbreviated Table of Trigonometric Identities

1. $\sin(\alpha \pm \beta) = \sin\alpha\cos\beta \pm \cos\alpha\sin\beta$

2. $\cos(\alpha \pm \beta) = \cos\alpha\cos\beta \mp \sin\alpha\sin\beta$

3. $\sin\alpha + \sin\beta = 2\sin\frac{\alpha+\beta}{2}\cos\frac{\alpha+\beta}{2}$

4. $\sin\alpha - \sin\beta = 2\cos\left(\frac{\alpha+\beta}{2}\right)\sin\left(\frac{\alpha-\beta}{2}\right)$

5. $\cos\alpha + \cos\beta = 2\cos\left(\frac{\alpha+\beta}{2}\right)\cos\left(\frac{\alpha-\beta}{2}\right)$

6. $\cos\alpha - \cos\beta = -2\sin\left(\frac{\alpha+\beta}{2}\right)\sin\left(\frac{\alpha-\beta}{2}\right)$

7. $2\sin\alpha\sin\beta = \cos(\alpha-\beta) - \cos(\alpha+\beta)$

8. $2\cos\alpha\cos\beta = \cos(\alpha-\beta) + \cos(\alpha+\beta)$

9. $2\sin\alpha\cos\beta = \sin(\alpha+\beta) + \sin(\alpha-\beta)$

10. $\sin 2\alpha = 2\sin\alpha\cos\alpha$

11. $\cos 2\alpha = 2\cos^2\alpha - 1 = 1 - 2\sin^2\alpha$

12. $\cos^2\alpha = \frac{1}{2} + \frac{1}{2}\cos 2\alpha$

13. $\sin^2\alpha = \frac{1}{2} - \frac{1}{2}\cos 2\alpha$

14. $\tan(\alpha \pm \beta) = \dfrac{\tan\alpha \pm \tan\beta}{1 \mp \tan\alpha\tan\beta}$

15. $\tan 2\alpha = \dfrac{2\tan\alpha}{1 - \tan^2\alpha}$

An Abbreviated Table of Integrals

1. $\displaystyle\int xe^{ax}\,dx = \frac{e^{ax}}{a^2}(ax-1)$

2. $\displaystyle\int x^2e^{ax}\,dx = \frac{e^{ax}}{a^3}(a^2x^2-2ax+2)$

3. $\displaystyle\int x\sin ax\,dx = \frac{1}{a^2}\sin ax - \frac{x}{a}\cos ax$

4. $\displaystyle\int x\cos ax\,dx = \frac{1}{a^2}\cos ax + \frac{x}{a}\sin ax$

5. $\displaystyle\int e^{ax}\sin bx\,dx = \frac{e^{ax}}{a^2+b^2}(a\sin bx - b\cos bx)$

6. $\displaystyle\int e^{ax}\cos bx\,dx = \frac{e^{ax}}{a^2+b^2}(a\cos bx + b\sin bx)$

7. $\displaystyle\int \frac{dx}{x^2+a^2} = \frac{1}{a}\tan^{-1}\frac{x}{a}$

8. $\displaystyle\int \frac{dx}{(x^2+a^2)^2} = \frac{1}{2a^2}\left(\frac{x}{x^2+a^2}+\frac{1}{a}\tan^{-1}\frac{x}{a}\right)$

9. $\displaystyle\int \sin ax\sin bx\,dx = \frac{\sin(a-b)x}{2(a-b)} - \frac{\sin(a+b)x}{2(a+b)},\quad a^2\neq b^2$

10. $\displaystyle\int \cos ax\cos bx\,dx = \frac{\sin(a-b)x}{2(a-b)} + \frac{\sin(a+b)x}{2(a+b)},\quad a^2\neq b^2$

11. $\displaystyle\int \sin ax\cos bx\,dx = -\frac{\cos(a-b)x}{2(a-b)} - \frac{\cos(a+b)x}{2(a+b)},\quad a^2\neq b^2$

12. $\displaystyle\int \sin^2 ax\,dx = \frac{x}{2} - \frac{\sin 2ax}{4a}$

13. $\displaystyle\int \cos^2 ax\,dx = \frac{x}{2} + \frac{\sin 2ax}{4a}$

14. $\displaystyle\int_0^\infty \frac{a\,dx}{a^2 + x^2} = \frac{\pi}{2}, \quad a > 0;$

$\displaystyle = 0, \quad a = 0;$

$\displaystyle = \frac{-\pi}{2}, \quad a < 0$

15. $\displaystyle\int_0^\infty \frac{\sin ax}{x}\,dx = \frac{\pi}{2}, \quad a > 0;$

$\displaystyle = \frac{-\pi}{2}, \quad a < 0$

16. $\displaystyle\int x^2 \sin ax\,dx = \frac{2x}{a^2}\sin ax - \frac{a^2x^2 - 2}{a^3}\cos ax$

17. $\displaystyle\int x^2 \cos ax\,dx = \frac{2x}{a^2}\cos ax + \frac{a^2x^2 - 2}{a^3}\sin ax$

18. $\displaystyle\int e^{ax}\sin^2 bx\,dx = \frac{e^{ax}}{a^2 + 4b^2}\left[(a\sin bx - 2b\cos bx)\sin bx + \frac{2b^2}{a}\right]$

19. $\displaystyle\int e^{ax}\cos^2 bx\,dx = \frac{e^{ax}}{a^2 + 4b^2}\left[(a\cos bx - 2b\sin bx)\cos bx + \frac{2b^2}{a}\right]$

List of Examples

Chapter 2

Chapter 3

CHAPTER 4

CHAPTER 5

CHAPTER 6

CHAPTER 7

CHAPTER 8

CHAPTER 9

CHAPTER 10

CHAPTER 11

CHAPTER 12

CHAPTER 13

CHAPTER 14

CHAPTER 15

Chapter 16

Chapter 17

Chapter 18

CHAPTER 19

Answers to Selected Problems

Chapter 1

1.1 2001.10 m/s

1.3 6.14 s

1.5 6 sin 4000t mC

1.7 156.04 μm/s

1.9 a) 400 W, delivering
b) entering
c) gain

1.11 a) 3000 W, B to A
b) 1200 W, A to B
c) 1080 W, A to B
d) 9600 W, B to A

1.13 a) 215,000 C
b) 2036.67 kJ

1.15 a) 2 ms
b) 649.61 mW
c) 2400 μJ

1.17 a) $t = 0$
b) 3.2 W
c) 5 mJ

1.20 a) car B
b) 72 kJ

1.22 a) 5.39 W
b) 9 J

1.25 yes, $p_{\text{del}} = p_{\text{abs}} = 40$ kW

Chapter 2

2.3 a) 5 A
b) 19 A
c) 6650 W

2.5 no, violates Kirchhoff's current law

2.9 a) Yes, no violation of Kirchhoff's laws
b) no, because the voltages across the independent and dependent current sources are indeterminate.

2.12 a) 2 A
b) $p_{5\Omega} = 320$ W; $p_{25\Omega} = 400$ W; $p_{70\Omega} = 280$ W; $p_{10\Omega} = 360$ W; $p_{8\Omega} = 800$ W
c) $p_{\text{diss}} = p_{\text{dev}} = 2160$ W

2.20 $R = 15\ \Omega$

2.25 a) 0 A
b) -3.2 A
c) -12.8 A

2.28 a) $i_\Delta = 5$ A; $v_o = 80$ V
b) $p_{\text{dev}} = p_{\text{diss}} = 4230$ W

Chapter 3

3.3 a) 15 Ω
b) 20 Ω

3.6 126 V

3.9 $i_o = 1.92$ A; $i_g = 12$ A

3.13 -2.7 V

3.17 a) $R_1 = 60$ kΩ; $R_2 = 15$ kΩ
b) 250 mW

3.23 $R_1 = 257.5\ \Omega$; $R_2 = 1030\ \Omega$; $R_3 = 8240\ \Omega$; $R_4 = 41{,}200\ \Omega$

3.27 6.25 A

3.31 -5 A

3.35 a) $R_1 = 29.95$ kΩ; $R_2 = 120$ kΩ; $R_3 = 150$ kΩ
b) 316.8 V
c) 330 V

3.39 -20%

3.44 324 W

3.49 3 A

3.50 9 A, 388.80 W

3.54 a) 4 A
b) 7.84 A
c) -19.84 V
d) 6.48 kW

3.57 a) 50 Ω
b) 936 W

CHAPTER 4

4.2 50 V

4.5 $v_1 = 100$ V; $v_2 = 50$ V

4.8 1816 W

4.11 20 V, 180 W

4.15 3.2 V

4.17 a) 1250 W
b) 21.85 kW

4.20 a) 8.8 kW
b) 8.8 kW

4.25 1084.50 W

4.27 a) $i_a = 9.8$ A; $i_b = -0.2$ A; $i_c = -10$ A
b) $i_a = -1.72$ A; $i_b = 1.08$ A; $i_c = 2.80$ A

4.30 740 W

4.35 2016 W

4.37 a) -5.2 mA
b) 200 mW
c) 2.912 mW

4.40 180 W

4.42 530 W

4.46 a) $i_a = -7$ A; $i_b = -8$ A; $i_c = 8$ A; $i_d = 10$ A; $i_e = 18$ A
b) $p_{\text{dev}} = p_{\text{diss}} = 8240$ W

4.49 10.8 A

4.52 a) 1 A
b) 1 A

4.55 $V_{\text{Th}} = 60$ V, $R_{\text{Th}} = 10\ \Omega$

4.58 a) $V_{\text{Th}} = 30$ V; $R_{\text{Th}} = 20\ \Omega$
b) $R_{\text{Th}} = 20\ \Omega$

4.61 a) -16.67 V
b) -7.41%

4.64 $V_{\text{Th}} = 280$ V; $R_{\text{Th}} = 20$ kΩ

4.67 $V_{\text{Th}} = 0$ V; $R_{\text{Th}} = 25\ \Omega$

4.69 $V_{\text{Th}} = 25$ V; $R_{\text{Th}} = 5$ kΩ

4.73 a) 5 kΩ
b) 45 mW

4.76 a) 6.4 Ω
b) 90 W

4.79 3.77%

4.83 $v_o = 270$ V; $i_o = 2.55$ A

4.86 10 V

4.90 40 V

4.96 23.09 W

4.98 0 A

CHAPTER 5

5.1 a) -10 V
b) -12 V
c) 8 V
d) -8 V
e) 12 V
f) $3.2 \le v_a \le 5.6$ V

5.4 -3.26 mA

5.7 $-250\ \mu$A

5.10 a) -4 V
b) $-2.5 \le v_b \le -1.3$ V

5.13 $-360\ \mu$A

5.18 $0 \le R_f \le 60$ kΩ

5.20 a) 10.54 V
b) $-4.55\text{ V} \le v_g \le 4.55$ V
c) 181.76 kΩ

5.23 a) 8.03 V
b) $i_a = -100\ \mu$A; $i_b = 100\ \mu$A
c) 6.4625 for v_a; 4.375 for v_b

5.27 $R_a = 200\ \text{k}\Omega$; $R_b = 1\ \text{M}\Omega$; $R_f = 16.5\ \text{k}\Omega$

5.31 a) -15.95 V
b) $-638\ \text{mV} \le v_b \le 962\ \text{mV}$

5.34 a) 13.49
b) $v_n = 999.45$ mV; $v_p = 999.83$ mV
c) 387.78 μV
d) 692.47 pA
e) 13.5, 1 V, 1 V, 0 V, 0 A

5.38 $v_{o1} = 12.4$ V; $v_{o2} = 7$ V

5.41 $v_o = -5$ V; $i_o = 400\ \mu$A

6.20 15 H

6.23 183.26 ms

6.26 3.2 nF charged to 20 V, positive at terminal a

6.29 a) $10e^{-500t}$ V
b) $8e^{-500t} + 7$ V
c) $2e^{-500t} - 7$ V
d) 400 nJ
e) 1625 nJ
f) 1225 nJ
g) yes

CHAPTER 6

6.2 a) $i = 0, t < 0$
$i = 50t$ A, $0 \le t \le 10$ ms
$i = 1 - 50t$ A, 10 ms $\le t \le$ 20 ms
$i = 0$, 20 ms $\le t \le \infty$
b) $v = 0, t < 0$
$v = 2$ V, $0 < t < 10$ ms
$v = -2$ V, 10 ms $< t <$ 20 ms
$v = 0$, 20 ms $< t < \infty$
$p = 0, t \le 0$
$p = 100t$ W, $0 < t < 10$ ms
$p = -2 + 100t$ W, 10 ms $< t <$ 20 ms
$p = 0$, 20 ms $\le t \le \infty$
$w = 0, t \le 0$
$w = 50t^2$ J, $0 \le t \le 10$ ms
$w = 0.02 - 2t + 50t^2$ J, 10 ms $\le t \le$ 20 ms
$w = 0$, 20 ms $\le t \le \infty$

6.6 a) 6.30 ms
b) 94 mV

6.7 a) $(2 - 10t)e^{-5t}$ mV
b) 735.76 μW
c) absorbing
d) 73.58 μJ
e) 108.27 μJ, 200 ms

6.9 1007 W, absorbing

6.12 21 V

6.14 a) $i = 0, t < 0$
b) $i = 5(\cos 3000t + 13 \sin 3000t)e^{-1000t}$ mA
c) no
d) yes, current jumps instantaneously from 0 to 5 mA at $t = 0$
e) 25 μJ

6.17 a) 33.67 μJ
b) 225 μJ

CHAPTER 7

7.1 a) 16 Ω
b) 50 ms
c) 0.8 H
d) 62.5 J
e) 12.77 ms

7.5 $32e^{-50t}$ V, $t \ge 0^+$

7.8 33.33%

7.11 133.89 Ω

7.15 a) 40 nJ
b) 50 nJ

7.18 33.33%

7.21 a) $9.9e^{-1000t}$ mA
b) 42.14%

7.24 a) 132.07 μJ
b) 9.16 ms

7.27 a) 160×10^{-6} A/V
b) $5.4e^{-40t}$ V, $t \ge 0^+$

7.30 a) $10e^{-1000t}$ V
b) 11.11%
c) $(10/3)e^{-1000t} - (40/3)$ V, $t \ge 0$
d) $(20/3)e^{-1000t} + (40/3)$ V, $t \ge 0$
e) 80 μJ

7.33 a) 200 V, 20 Ω, 40 mH
b) 1.39 ms

7.36 $8 + 2e^{-125{,}000t}$ mA, $t \ge 0$
$-50e^{-125{,}000t}$ V, $t \ge 0^+$

7.40 17.33 ms

7.45 a) 50 mA, 1.6 kΩ, 250 nF, 400 μs
b) 643.78 μs

7.48 a) 4.5 mA
b) 0 A
c) 250 μs
d) $4.5e^{-4000t}$ mA, $t \geq 0^+$
e) $45 - 40.5e^{-4000t}$ V, $t \geq 0^+$

7.51 $-45 + 90e^{-800t}$ V, $t \geq 0$

7.56 a) 2.88 μJ
b) 5.76 μJ
c) 8.64 μJ

7.59 a) 4 A
b) 2.68 A
c) 986.39 mA
d) -192.76 V
e) -48.19 V

7.62 408.15 μs

7.66 232.56 nA, left to right

7.71 a) $v_o(t) = 0, t < 0$
$v_o(t) = 80e^{-20,000t}$ V, $0 < t < 80\ \mu$s
$v_o(t) = -63.85e^{-20,000(t-80\times10^{-6})}$ V,
$80\ \mu\text{s} < t < \infty$
b) $v_o(80^-\ \mu\text{s}) = 16.15$ V
$v_o(80^+\ \mu\text{s}) = -63.85$ V
c) $i_o(80^-\ \mu\text{s}) = i_o(80^+\ \mu\text{s}) = 15.96$ mA

7.76 12.43 ms

7.80 30 kΩ

7.83 25 ms

7.86 $v_o = 25{,}000t - 10$ mV
$v_p = 40 - 35e^{-500t}$ mV
$v_f = 25{,}000t - 50 + 35e^{-500t}$ mV

CHAPTER 8

8.1 a) $s_1 = -10{,}000$ rad/s, $s_2 = -40{,}000$ rad/s
b) overdamped
c) 3125 Ω
d) $s_1 = -16{,}000 + j12{,}000$ rad/s
$s_2 = -16{,}000 - j12{,}000$ rad/s
e) 2500 Ω

8.5 $v = (75{,}000t + 15)e^{-10,000t}$ V, $t \geq 0$

8.7 a) 1250 Ω, 0.4 μF, $-25{,}000$ V/s, 5 V
b) $(10{,}000t - 12)e^{-1000t}$ mA, $t \geq 0$

8.12 a) 2500 Ω
b) $(80{,}000t - 15)e^{-2000t}$ V, $t \geq 0$
c) 10.11 V
d) 60.17%

8.15 $5(\sqrt{3}\sin 2000\sqrt{3}t + 3\cos 2000\sqrt{3}t)e^{-2000t}$ V, $t \geq 0$

8.17 $-2e^{-2000t} + 8e^{-8000t}$ V, $t \geq 0$

8.21 $(4 - 12e^{-40t} + 3e^{-160t})$ mA, $t \geq 0$

8.24 $60 - 750{,}000te^{-10,000t} - 105e^{-10,000t}$ mA, $t \geq 0$

8.27 $16e^{-400t} - 4e^{-16,000t}$ V, $t \geq 0$

8.30 a) $(16{,}000t + 20)e^{-800t}$ V, $t \geq 0$
b) $160 - 160(400t + 1)e^{-800t}$ mA, $t \geq 0$

8.33 a) 25,000 Ω, 5 H
b) 0, 12 A/s
c) $4e^{-1000t} - 4e^{-4000t}$ mA, $t \geq 0$
d) 462.10 μs
e) 1.89 mA
f) $-20e^{-1000t} + 80e^{-4000t}$ V, $t \geq 0$

8.36 a) 1250 Ω
b) 6 mA, 60 A/s
c) $(56{,}250t + 15)e^{-5000t}$ V, $t \geq 0$

8.39 $(108\cos 10t - 12\sin 10t)e^{-10t}$ V, $t \geq 0$

8.42 $52 + 78e^{-2000t} - 34e^{-6000t}$ V, $t \geq 0$

8.46 399.96 Ω, 0.20 H

8.49 a) $0 \leq t \leq 0.5$ s
$v_{o1} = -1.6t$ V
$v_o = 10t^2$ V
$0.5^+\text{s} < t < t_{\text{sat}}$
$v_{o1} = 0.8t - 1.2$ V
$v_o = 15t - 5t^2 - 3.75$ V
b) $t_{\text{sat}} = 3.5$ s

8.51 a) $\dfrac{d^2v_o}{dt^2} = -\dfrac{v_g}{R^2C^2}$
b) $\dfrac{d^2v_o}{dt^2} = \dfrac{v_g}{R^2C^2}$, same except for a reversal in sign
c) two integrations of the input signal with one op amp

CHAPTER 9

9.1 a) 80 V
b) 500 Hz
c) 3141.59 rad/s
d) -0.5236 rad
e) $-30°$
f) 2 ms
g) (1/6) ms
h) $80\cos(1000\pi t + \pi/2)$ V
i) (11/6) ms
j) (5/3) ms

9.4 a) 400π rad/s
b) $20\cos(400\pi t - 45°)$ A

9.8 169.71 V

9.10 a) 502.65 krad/s
b) 90°
c) $-39.79\ \Omega$
d) $0.05\ \mu F$
e) $-j39.79\ \Omega$

9.13 $500\cos(4000t - 53.13°)\ \mu A$

9.16 $200\angle 53.13°$ mS
$120 + j160$ mS

9.19 a) 0.40 H, 0.10 H
b) $5\cos 50{,}000t$ mA
$12.5\cos 50{,}000t$ mA

9.22 a) 1273.24 Hz
b) $1.5\cos 8000t$ mA

9.25 $111.80\cos(8000t - 100.30°)$ V

9.27 a) $6\angle 22.5°\ \Omega$
b) i_g lags v_g by 25 μs

9.32 a) $\mathbf{I}_b = 2.50\angle 90°$ A, $\mathbf{I}_c = 10\angle 36.87°$ A
$\mathbf{V}_g = 358.47\angle 67.01°$ V
b) $i_b = 2.5\cos(800t + 90°)$ A
$i_c = 10\cos(800t + 36.87°)$ A
$v_g = 358.47\cos(800t + 67.01°)$ V

9.35 $44.72\cos(5000t + 26.57°)$ V

9.38 $188.43\angle -42.88°$ V

9.42 $36\cos 2000t$ V

9.45 $\mathbf{I}_a = 30\angle 0°$ A
$\mathbf{I}_b = 30 - j20$ A
$\mathbf{I}_c = 30 + j10$ A
$\mathbf{I}_d = 30\angle -90°$ A

9.48 $i_a = 1.58\cos(10^6 t + 108.43°)$ A
$i_b = 1.12\cos(10^6 t - 26.57°)$ A
$i_c = 2.5\cos(10^6 t + 180°)$ A

9.51 $\mathbf{V}_{\text{Th}} = 72 - j24$ V
$Z_{\text{Th}} = 1.8 + j5.4\ \Omega$

9.55 $\mathbf{V}_{\text{Th}} = 80 - j80$ V, $Z_{\text{Th}} = 16 - j12\ \Omega$

9.58 $\mathbf{I}_{\text{N}} = 6\angle -90°$ mA; $Z_{\text{N}} = (4 + j3)$ kΩ

9.60 $v_o(t) = 36\cos 2000t$ V

9.63 a) $4.24\cos(2 \times 10^5 t - 45°)$ V
b) 3.54 V

9.66 $6.71\cos(2500t + 153.43°)$ V

9.68 a) $Z_{ab} = [Z/(1 + K)]$
b) $C_{ab} = C(1 + K)$

CHAPTER 10

10.1 a) 1044.42 W (absorbing)
731.31 VAR (absorbing)
b) 85.51 W (absorbing)
234.92 VAR (delivering)
c) 573.58 W (delivering)
819.15 VAR (delivering)
d) 225 W (delivering)
389.71 VAR (absorbing)

10.4 0.90 lagging, 0.43; 0.43 leading, −0.90; 0.57 leading, −0.82

10.7 a) 15.81 V
b) 62.5 W

10.10 16 W, −32 VAR, 35.78 VA

10.15 600 W

10.18 0.8587

10.21 $g1$: 6200 W, 1490 VAR
$g2$: 5960 W, 970 VAR

10.24 a) 133.48 V
b) 256 W
c) 1788.59 μF
d) 126.83 V
e) 184.96 W

10.27 a) 313.60 W
b) $-625\ \Omega$
c) $468.75 + j0\ \Omega$
d) 205.40 W

10.30 a) 2.94 W
b) 200 Ω, 1 μF
c) 4.58 W, yes
d) 5 W
e) 250 Ω, 0.5 μF
f) Yes

10.34 a) 100 Ω
b) 6400 W

10.38 2644.4 V (rms), 2369.99 V (rms)

10.40 12.5 mW

CHAPTER 11

11.3 $v_{AB} = 13{,}198.23\cos\omega t$ V
$v_{BC} = 13{,}198.23\cos(\omega t + 120°)$ V
$v_{CA} = 13{,}198.23\cos(\omega t - 120°)$ V

11.6 a) $\mathbf{I}_o = 0$
b) $\mathbf{V}_{BN} = 1117.03\angle 117.54^\circ$ V
c) $\mathbf{V}_{BC} = 1909.14\angle 87.54^\circ$ V
d) unbalanced

11.9 a) 15.24 A
b) 6583.94 V

11.12 a) $\mathbf{I}_{AB} = 176\angle 16.26^\circ$ A
$\mathbf{I}_{BC} = 44\angle -156.87^\circ$ A
$\mathbf{I}_{CA} = 22\angle 120^\circ$ A
b) $\mathbf{I}_{aA} = 182.48\angle 9.53^\circ$ A
$\mathbf{I}_{bB} = 219.75\angle -162.37^\circ$ A
$\mathbf{I}_{cC} = 46.78\angle 50.96^\circ$ A
c) $\mathbf{I}_{ba} = \mathbf{I}_{AB} = 176\angle 16.26^\circ$ A
$\mathbf{I}_{cb} = \mathbf{I}_{BC} = 44\angle -156.26^\circ$ A
$\mathbf{I}_{ac} = \mathbf{I}_{CA} = 22\angle 120^\circ$ A

11.13 a) $\mathbf{I}_{AB} = 105.60\angle 36.87^\circ$ A
$\mathbf{I}_{BC} = 105.60\angle 156.87^\circ$ A
$\mathbf{I}_{CA} = 105.60\angle -83.13^\circ$ A
b) $\mathbf{I}_{aA} = 182.90\angle 66.87^\circ$ A
$\mathbf{I}_{bB} = 182.90\angle 186.87^\circ$ A
$\mathbf{I}_{cC} = 182.90\angle -53.13^\circ$ A
c) $\mathbf{I}_{ba} = \mathbf{I}_{AB} = 105.60\angle 36.87^\circ$ A
$\mathbf{I}_{cb} = \mathbf{I}_{BC} = 105.60\angle 156.87^\circ$ A
$\mathbf{I}_{ac} = \mathbf{I}_{CA} = 105.60\angle -83.13^\circ$ A

11.17 a) $8.22\angle -71.57^\circ$ A
b) 99.76%

11.20 a) $905{,}883 + j474{,}564$ VA
b) 99.35%

11.23 a) 138.46 A (rms)
b) 0.892 lagging

11.26 a) 9081.94 V
b) 8394.89 V
c) 97.88%
d) 99.22%
e) 15.35 μF

11.29 a) 242.58 A (rms)
b) 41,510.12 VAR

CHAPTER 12

12.1 a) 0.75
b) 26 mH
c) 2

12.4 upper terminal

12.8 a) $14.4 - j19.2\ \Omega$
b) $30 + j52.8\ \Omega$
c) 35.78 V
d) 32%

12.11 a) 0.559
b) 3.16 A

12.14 a) $\mathbf{V}_{\text{Th}} = 850 + j850$ V
$Z_{\text{Th}} = 85 + j85\ \Omega$
b) 18,062.5 W

12.17 a) 117 V
b) 152.10 W
c) 43.14%

12.20 a) 104 V
b) 135.20 W
c) 66.02%

12.23 $10 + j0\ \Omega$

12.28 a) 156.25 W
b) $\sum P_{\text{dev}} = \sum P_{\text{diss}} = 250$ W

12.31 a) $750 - j250\ \Omega$
b) $60\angle 0^\circ$ mA
$47.43\angle -18.43^\circ$ V

12.33 a) 54 W
b) 1.02%
c) $\sum P_{\text{dev}} = \sum P_{\text{diss}} = 5292$ W

12.36 a) $a_1 = 5, a_2 = 50$
b) 250 mW
c) $800\angle 0^\circ$ mV

12.40 a) $L_1 = 60$ mH, $L_2 = 80$ mH, $M = 20$ mH
b) 0.2887

12.43 $86.10\angle -41.99^\circ$ V

CHAPTER 13

13.4 As $\epsilon \to 0$, the amplitude $\to \infty$, the duration $\to 0$, and the area is independent of ϵ (i.e., $A = 1$).

13.8 a) $1/(s + a)^2$
b) $s/(s^2 + \omega^2)$
c) $(s \cos\theta - \omega \sin\theta)/(s^2 + \omega^2)$
d) $1/(s^2 - 1)$
e) $[\cosh\theta + s(\sinh\theta)]/(s^2 - 1)$

13.12 a) $s\omega/(s^2 + \omega^2)$
b) $s^2/(s^2 + \omega^2)$
c) 2

13.16 a) $40e^{-3s}/(s+8)$
b) $5(e^{-2s} - 2e^{-4s} + 2e^{-8s} - e^{-10s})/s^2$

13.23 a) $320(e^{-1000t} - e^{-4000t})$ mA, $t \geq 0$
b) $(48 - 64e^{-1000t} + 16e^{-4000t})u(t)$ V

13.25 a) $10e^{-1000t}\sin 7000t$ V, $t \geq 0$
b) $[7 + 7.07e^{-1000t}\cos(7000t + 171.87°)]u(t)$ mA

13.27 a) $(20t - 4 + 4e^{-5t})u(t)$
b) $(250 - 200te^{-t} - 250e^{-t})u(t)$
c) $[30t - 8 + 10e^{-3t}\cos(t + 36.87°)]u(t)$
d) $(20 - 2.5t^2e^{-t} - 15te^{-t} - 20e^{-t})u(t)$
e) $[16 + 89.44te^{-2t}\cos(t + 26.57°) + 113.14e^{-2t}\cos(t + 98.13°)]u(t)$

13.30 $(20{,}000t + 5)e^{-4000t}u(t)$ mA

13.32 $(10e^{-1000t} - 10e^{-4000t})u(t)$ V

13.40 a) $300u(t) = 60i_1 + 10\dfrac{di_1}{dt} + 5\dfrac{di_2}{dt}$

$0 = 5\dfrac{di_1}{dt} + 40i_2 + 5\dfrac{di_2}{dt}$

b) $\dfrac{300}{s} = (60 + 10s)I_1 + 5sI_2$

$0 = 5sI_1 + (40 + 5s)I_2$

c) $I_1(s) = \dfrac{60(s+8)}{s(s+4)(s+24)}$

$I_2(s) = \dfrac{-60}{(s+4)(s+24)}$

d) $i_1(t) = (5 - 3e^{-4t} - 2e^{-24t})u(t)$ A
$i_2(t) = (-3e^{-4t} + 3e^{-24t})u(t)$ A

e) $i_1(\infty) = 5$ A, $i_2(\infty) = 0$

f) Yes. $i_1(0^+) = i_2(0^+) = 0$, since there is no energy stored at $t = 0^+$. At $t = \infty$, $i_1 = 300/60 = 5/$A, and $i_2 = 0$, since the 40Ω resistor is shorted.

CHAPTER 14

14.4 a) $(s^2 + 8000s + 25 \times 10^6)/s$ Ω
b) $-p_1 = 0$
$-z_1 = -4000 + j3000$ rad/s
$-z_2 = -4000 - j3000$ rad/s

14.7 $-p_1 = -1 + j1$ rad/s
$-p_2 = -1 - j1$ rad/s
$-z_1 = -0.5 + j0.866$ rad/s
$-z_2 = -0.5 - j0.866$ rad/s

14.11 a) $90{,}000/(s^2 + 8000s + 25 \times 10^6)$
b) $v_o(\infty) = 0$, $v_o(0^+) = 0$
c) $(30e^{-4000t}\sin 3000t)u(t)$ V

14.14 $v_o = \dfrac{30s^2 + 420{,}000s + 720 \times 10^6}{s(s+4000)(s+10{,}000)}$

$v_o = (18 + 20e^{-4000t} - 8e^{-10{,}000t})u(t)$ V

14.17 $130.21e^{-875t}\cos(3000t + 57.48°)u(t)$ V

14.20 $(10^4t + 1)e^{-10{,}000t}u(t)$ V

14.25 $-25e^{-1000t}\cos 2000t\,u(t)$ mA

14.28 a) $I_a = \dfrac{8}{s(s+3)}$

$I_b = \dfrac{9(s+1)}{s(s+3)}$

b) $i_a = (6 - 6e^{-3t})u(t)$ A
$i_b = (3 + 6e^{-3t})u(t)$ A

c) $v_a = \dfrac{30}{s^2} + \dfrac{60}{s(s+3)}$

$v_b = \dfrac{30}{s^2} - \dfrac{120}{s(s+3)}$

$v_c = \dfrac{30}{s^2} + \dfrac{60}{s(s+3)}$

d) $v_a = [30t + 20 - 20e^{-3t}]u(t)$ V
$v_b = [30t - 40 + 40e^{-3t}]u(t)$ V
$v_c = [30t + 20 - 20e^{-3t}]u(t)$ V

e) 32.67

14.30 $(10 - 10e^{-0.5t}\cos 0.5t)u(t)$ A

14.34 $(60 - 3 \times 10^5te^{-5000t} - 60e^{-5000t})u(t)$ mA

14.37 a) $240(s+40)/s(s+20)(s+80)$
b) $i_1(0^+) = 0$, $i_1(\infty) = 6$ A
c) $(6 - 4e^{-20t} - 2e^{-80t})u(t)$ A

14.39 $(10e^{-t} + 5e^{-10t})u(t)$ A

14.43 $-[40e^{-5000t}\sin 10{,}000t]u(t)$ mV

14.46 56.5 ms

14.49 a) $250/(s+250)$, $-p_1 = -250$ rad/s
b) $s/(s+250)$, $-z_1 = 0$, $-p_1 = -250$ rad/s
c) $s/(s+20{,}000)$, $-z_1 = 0$, $-p_1 = -20{,}000$ rad/s
d) $20{,}000/(s+20{,}000)$, $-p_1 = -20{,}000$ rad/s
e) $500/(s+625)$, $-p_1 = -625$ rad/s

14.51 a) $-40{,}000s/(s+1000)(s+5000)$
b) $-z_1 = 0$, $-p_1 = -1000$ rad/s, $-p_2 = -5000$ rad/s

14.56 $70{,}000(s+8000)/(s+10{,}000)(s+40{,}000)$
$-z_1 = -8000$ rad/s
$-p_1 = -10{,}000$ rad/s
$-p_2 = -40{,}000$ rad/s

14.63 $15e^{-10t}$ V, $0 \leq t \leq \infty$

14.66 a) $5(1 - e^{-50t})$ V, $0 \le t \le 0.1s$
$[10e^{-50(t-0.1)} - 5e^{-50t} - 5]$ V, $0.1s \le t \le 0.2s$
$[10e^{-50(t-0.1)} - 5e^{-50(t-0.2)} - 5e^{-50t}]$ V, $0.2s \le t \le \infty$
b) $50e^{-50t}$ μA, $0 < t < 0.1s$
$50e^{-50t} - 100e^{-50(t-0.1)}$ μA, $0.1s < t < 0.2s$
$50e^{-50t} - 100e^{-50(t-0.1)} + 50e^{-50(t-0.2)}$ μA, $0.2s < t \le \infty$
c) $i_o(0.1^-) = 0.34$ μA
$i_o(0.1^+) = -99.66$ μA
$i_o(0.2^-) = -0.67$ μA
$i_o(0.2^+) = 49.33$ μA

14.70 $12.30\cos(10{,}000t - 26.57°)$ V

14.73 $61.84\cos(6000t + 66.37°)$ V

14.78 a) $(8 - 6e^{-50{,}000t})u(t)$ V
b) $40\delta(t) + (2 + 4.5e^{-50{,}000t})10^6 u(t)$ pA
c) $v_o = 8u(t)$ V
$i_o = 160\delta(t) + 2 \times 10^6 u(t)$ pA

14.81 a) $4\delta(t) - 40{,}000e^{-10^4 t}u(t)$ V
b) The impulsive voltage will show up across the two inductors at $t = 0^+$ and establish a current at 0^+ equal to

$$i_L = \frac{1000}{20}\int_{0^-}^{0^+} 5\delta(t)dt = 250 \text{ A}$$

therefore, $i_L = 250e^{-10^4 t}u(t)$ A

$$v_o = 16 \times 10^{-3}\frac{di_L}{dt} = -40{,}000e^{-10^4 t}u(t) \text{ V}.$$

At $t = 0$, $\frac{di_L}{dt} = 250\delta(t)$

therefore, $v_o(0) = (16 \times 10^{-3})[250\delta(t)] = 4\delta(t)$. Hence the solution makes sense in terms of known circuit behavior.

CHAPTER 15

15.1 a) 2021.27 Hz
b) $H(j\omega_c) = 0.71\angle -45°$
$H(j0.2\omega_c) = 0.98\angle -11.31°$
$H(j5\omega_c) = 0.20\angle -78.69°$
c) $v_o(\omega_c) = 7.07\cos(12{,}700t - 45°)$ V
$v_o(0.2\omega_c) = 9.81\cos(2540t - 11.31°)$ V
$v_o(5\omega_c) = 1.96\cos(63{,}500t - 78.69°)$ V

15.5 a) $(1/RC)/[s + (R + R_L)/RR_LC]$
b) 0
c) $R_L/(R + R_L)$
d) $(1/RC)[1 + (R/R_L)]$
e) 11 krad/s; $0.9091\angle 0°$; $0.6428\angle -45°$; $0.9046\angle -5.71°$; $0.0905\angle -84.29°$

15.9 a) 5.31 kΩ
b) 333.86 Hz

15.12 $\omega_o = 110$ krad/s; $\beta = 21$ krad/s; $Q = 5.24$

15.15 a) 100 krad/s
b) 15.92 kHz
c) 8
d) 93.95 krad/s
e) 14.96 kHz
f) 106.45 krad/s
g) 16.94 kHz
h) 12.5 krad/s

15.18 a) $450\cos 25{,}000t$ mV
b) $318.20\cos(22{,}624.69t + 45°)$ mV
c) $318.20\cos(27{,}624.69t - 45°)$ mV

15.21 a) 1 Mrad/s
b) 62.5 krad/s
c) 16
d) $200\cos 10^6 t$ mV

15.24 a) 1 Mrad/s
b) 159.15 kHz
c) 15
d) 967.22 krad/s
e) 153.94 kHz
f) 1.03 Mrad/s
g) 164.55 kHz
h) 10.61 kHz

15.27 a) $\frac{K(s^2 + \omega_o^2)}{s^2 + \beta s + \omega_o^2}$, $K = \frac{R_L}{R + R_L}$; $\beta = \left(\frac{RR_L}{R + R_L}\right)\frac{1}{L}$; $\omega_0^2 = \frac{1}{LC}$
b) $\frac{1}{\sqrt{LC}}$
c) $\left(\frac{RR_L}{R + R_L}\right)\frac{1}{L}$
d) $\frac{\omega_o L(R + R_L)}{RR_L}$
e) 0
f) K
g) K

15.29 a) $R = 1.5$ kΩ, $L = 2.5$ mH
b) 1.74

15.34 a) 0, 132.29 krad/s
b) 93.54 krad/s
c) 2.07

15.36 a) 50 Ω
b) −10.88 dB

15.39 63.7 times as large as the DTMF tones

CHAPTER 16

16.2 a) $-\dfrac{C_1\,[s+(1/R_1C_1)]}{C_2\,[s+(1/R_2C_2)]}$

b) $-R_2/R_1$

c) $-C_1/C_2$

d) Yes, as $\omega \to 0$, the capacitor branches behave as open circuits, and as $\omega \to \infty$, the capacitor branches approach short circuits, making the resistive branches negligible.

16.9 a) $R = 0.05\ \Omega$, $L = 1$ H; $C = 1$ F

b) $R = 5000\ \Omega$, $L = 2.5$ H, $C = 0.25$ nF

c) See Fig. P16.8.

16.16 a) 200 mH

b) $(10 - 10e^{-2500t}\cos 2500t)u(t)$ mA

16.18 a) $k_m = 25$, $k_f = 1000$

b) $(1250 + 5250e^{-500t} + 750e^{-1000t})u(t)$ V

16.26 a) 3

b) −32.65 dB

16.28 a) $H(s) = \dfrac{1}{(s+1)(s^2+s+1)}$

b) $H'(s) = \dfrac{(4000\pi)^3}{(s+4000\pi)[s^2+4000\pi s+(4000\pi)^2]}$

c) −32.65 dB

16.34 a) bandpass

b) 2 kHz, 8 kHz, 4 kHz, 0.67

c) $H'(s) = \dfrac{65{,}536 \times 10^{12}\pi^4 s^4}{D_1D_2D_3D_4}$,

where

$D_1 = s^2 + 3060\pi s + 16 \times 10^6\pi^2$

$D_2 = s^2 + 7392\pi s + 16 \times 10^6\pi^2$

$D_3 = s^2 + 12{,}240\pi s + 256 \times 10^6\pi^2$

$D_4 = s^2 + 29{,}568\pi s + 256 \times 10^6\pi^2$

d) $0.996\angle 0^\circ$

16.38 a) $H'(s) = \dfrac{729 \times 10^{10}\pi^3 s^3}{D_1D_2D_3D_4}$,

where

$D_1 = s + 1000\pi$

$D_2 = s + 9000\pi$

$D_3 = s^2 + 1000\pi s + 10^6\pi^2$

$D_4 = s^2 + 9000\pi s + 81 \times 10^6\pi^2$

b) −3.89 dB, 19.99 dB

c) yes

16.43 a) unity-gain low-pass filter

b) $H(s) = \dfrac{\left(\dfrac{1}{R_1R_2C_1C_2}\right)}{s^2 + \left(\dfrac{1}{R_1C_1}\right)s + \left(\dfrac{1}{R_1R_2C_1C_2}\right)}$

c) two

d) $G_1 = b_1C_1$

$G_2 = \dfrac{b_o}{b_1}C_2$

e) no

f) first section: $R_1 = 1.307\ \Omega$, $R_2 = 0.765\ \Omega$
second section: $R_1 = 0.541\ \Omega$, $R_2 = 1.848\ \Omega$

16.45 $\dfrac{s^6}{(s^2+0.518s+1)(s^2+\sqrt{2}s+1)(s^2+1.932s+1)}$

16.46 a) $\dfrac{s^6}{D_1D_2D_3}$,

where

$D_1 = s^2 + 12{,}950s + 625 \times 10^6$

$D_2 = s^2 + 35{,}350s + 625 \times 10^6$

$D_3 = s^2 + 48{,}300s + 625 \times 10^6$

b) −3.01 dB

16.47 a) unity-gain high-pass filter

b) $\dfrac{s^2}{s^2 + \dfrac{1}{R_2C_2}s + \dfrac{1}{R_1R_2C_1C_2}}$

c) two

d) $R_1 = \dfrac{b_1}{b_oC_1}$, $R_2 = \dfrac{1}{b_1C_2}$

e) no

f) 1 Ω, 1 Ω

16.51 a) $H'(s) = -\left[H'_{lp}(s) + H'_{hp}(s)\right]$

$H'_{lp} =$

$$\frac{4096 \times 10^8\pi^4}{[s^2+612\pi s+(800\pi)^2][s^2+1478.4\pi s+(800\pi)^2]}$$

$H'_{hp} =$

$$\frac{s^4}{[s^2+9792\pi s+(12{,}800\pi)^2][s^2+23{,}654.4\pi s+(12{,}800\pi)^2]}$$

b) −44.19 dB

CHAPTER 17

17.3 a) $\dfrac{4V_m}{\pi}\displaystyle\sum_{n=1,3,5}^{\infty}\frac{1}{n}\sin n\omega_0 t$

b) $\dfrac{2V_m}{\pi}\left[1 + 2\displaystyle\sum_{n=1}^{\infty}\frac{\cos n\omega_0 t}{(1-4n^2)}\right]$

c) $\dfrac{V_m}{\pi} + \dfrac{V_m}{2}\sin\omega_{0t} + \dfrac{2V_m}{\pi}\displaystyle\sum_{n=2,4,6}^{\infty}\frac{\cos n\omega_0 t}{(1-n^2)}$

17.8 a) $\omega_{0a} = 104.72$ krad/s
$\omega_{0b} = 62.83$ krad/s
b) $f_{0a} = 16.67$ kHz
$f_{0b} = 10$ kHz
c) $a_{va} = 0$; $a_{vb} = 12.5$ V
d) $a_{ka} = 0$ for all k
$$b_{ka} = \frac{80}{\pi k}\left[1 + \cos\frac{k\pi}{3}\right],\ k \text{ odd}$$
$b_{ka} = 0$, k even
$$a_{kb} = \frac{50}{\pi^2 k^2}(\cos k\pi - 1)$$
$$b_{kb} = -\frac{50}{\pi k}\cos k\pi$$
e) $$v_a(t) = \frac{80}{\pi}\sum_{n=1,3,5,\cdots}^{\infty}\frac{[1+\cos(n\pi/3)]}{n}\sin n\omega_0 t$$
$v_b(t) =$
$$\frac{50}{\pi}\left\{\sum_{n=1}^{\infty}\frac{1}{n^2\pi}(\cos n\pi - 1)\cos n\omega_0 t + \frac{1}{n}\cos n\pi \sin n\omega_0 t\right\}$$

17.11 a) $$\frac{2I_m}{\pi}\sum_{n=1}^{\infty}\frac{\sin(n\omega_0 t)}{n}$$
b) $$\frac{2I_m}{\pi}\sum_{n=1}^{\infty}\frac{\cos n\pi}{n}\sin n\omega_0 t$$

17.14 a) $\pi/4$ rad/s
b) yes
c) no
d) no

17.18 a) $$18.90\sum_{n=1,3,5,\cdots}^{\infty}\frac{1}{n}\sqrt{1+(4/n^2\pi^2)}\cos(n\omega_0 t - \theta_n),$$
where $\theta_n = 180° - \tan^{-1}(n\pi/2)$
b) 7.22 A

17.23 $2238.83\cos(10t - 5.71°) + 239.46\cos(30t - 16.70°) + 80.50\cos(50t - 26.57°) + \cdots$ mV

17.26 a) $42\cos(2000t - 0.60°) + 13.99\cos(6000t + 177.32°) + 5.98\cos(14{,}000t + 184.17°) + \cdots$ V
b) fifth harmonic, because the circuit is a bandreject filter with a center frequency of 10 krad/s

17.29 $29.50\cos(300{,}000t + 180°)$ V

17.33 a) $0.755 V_m$
b) $0.7906 V_m$
c) -4.5 %

17.36 a) 37.23 W
b) 38.40 W
c) -3.05 %

17.40 $$\sum_{n=-\infty}^{n=\infty} -j\frac{V_m}{\pi n}\left[1-\cos\left(\frac{n\pi}{2}\right)\right]e^{jn\omega_0 t}$$

17.42 a) 100 W
b) 1.9144 A
c) -12.04%

17.44 a) 22.05 V
b) 21.29 V
c) -3.44%

17.47 a) $A_1 \sin\omega_0 t - A_3 \sin 3\omega_0 t + A_5 \sin 5\omega_0 t - A_7 \sin 7\omega_0 t$
b) odd
c) yes
d) yes

17.50 $0.9923\cos(0.5t - 60.26°) + 0.0316\cos(1.5t + 173.88°) + 0.0026\cos(2.5t + 137.26°) + \ldots$ V

CHAPTER 18

18.1 a) $$\frac{-j4\pi A}{\pi^2 - 4\omega^2}\sin 2\omega$$
b) $$\frac{4A}{\omega^2\tau}\left[1-\cos\left(\frac{\omega\tau}{2}\right)\right]$$

18.4 a) $$\frac{2(a^2-\omega^2)}{(a^2+\omega^2)^2}$$
b) $$\frac{-j48a\omega(a^2-\omega^2)}{(a^2+\omega^2)^4}$$
c) $$\frac{a}{a^2+(\omega-\omega_0)^2} + \frac{a}{a^2+(\omega+\omega_0)^2}$$
d) $$\frac{-ja}{a^2+(\omega-\omega_0)^2} + \frac{ja}{a^2+(\omega+\omega_0)^2}$$
e) $e^{-j\omega t_0}$

18.15 a) $$\frac{\tau}{2}\left\{\frac{\sin[(\omega+\omega_0)(\tau/2)]}{(\omega+\omega_0)(\tau/2)} + \frac{\sin[(\omega-\omega_0)(\tau/2)]}{(\omega-\omega_0)(\tau/2)}\right\}$$
b) $F(\omega) \to \pi[\delta(\omega+\omega_0) + \delta(\omega-\omega_0)]$

18.23 a) $15\,\text{sgn}(t) - 30e^{-100t}u(t)$ V
b) -15 V
c) -15 V
d) $(15 - 30e^{-100t})u(t)$ V
e) yes

18.25 a) $(12.5e^{-t} - 7.5e^{-5t})u(t) + 5e^{5t}u(-t)$ V
b) 5 V
c) 5 V
d) $(12.5e^{-t} - 7.5e^{-5t})u(t)$ V
e) yes

18.30 a) $v_o(0^-) = v_o(0^+) = -10$ V
b) $i_o(0^-) = 0$ A, $i_o(0^+) = 8$ A
c) $5\,\text{sgn}(t) + (10e^{-2t} - 20e^{-5t})u(t) - 5$ V

18.33 a) $\dfrac{(1/\pi)}{1+t^2}$

b) $\dfrac{1}{2\pi}$ J

c) $\dfrac{1}{2\pi}$ J

d) 1.15 rad/s

18.36 82.05%

18.38 a) 18.17%

b) 27.23%

c) 10.57%

CHAPTER 19

19.2 $z_{11} = 13\ \Omega$, $z_{12} = 12\ \Omega$, $z_{21} = 12\ \Omega$, $z_{22} = 16\ \Omega$

19.5 $a_{11} = 1.40$, $a_{12} = 6.4\ \Omega$, $a_{21} = 80$ mS, $a_{22} = 1.08$

19.8 $z_{11} = 12\ \Omega$, $z_{12} = 1\ \Omega$, $z_{21} = 1\ \Omega$, $z_{22} = 10.5\ \Omega$

19.11 $h_{11} = 1000\ \Omega$, $h_{12} = -5 \times 10^{-4}$, $h_{21} = 50$, $h_{22} = 25\ \mu$S

19.14 $h_{11} = 15 - j5\ \Omega$, $h_{12} = 1.5 - j0.50$, $h_{21} = (-375 + j125)10^{-3}$, $h_{22} = 37.5 + j12.5$ mS

19.19 $h_{11} = 500\ \Omega$, $h_{12} = 0$, $h_{21} = -40$, $h_{22} = 20$ mS

19.29 a) $z_{11} = 50 + j40\ \Omega$, $z_{12} = -j160\ \Omega$, $z_{21} = -j160\ \Omega$, $z_{22} = 500 + j1000\ \Omega$

b) $V_{Th} = 158.11\angle{-108.43^\circ}$ V, $Z_{Th} = 692 + j936\ \Omega$

c) $92.31\cos(5000t - 173.41^\circ)$ V

19.32 a) 35.36 V

b) 125 mW

c) 18.75 μW

19.35 a) $z_{11} = \dfrac{s^2+1}{s}\ \Omega$, $z_{12} = \dfrac{1}{s}\ \Omega$, $z_{21} = \dfrac{1}{s}\ \Omega$, $z_{22} = \dfrac{s^2+1}{s}$

b) $\left[25 - 25e^{-t} + 18.9e^{-0.5t}\cos(1.32t + 90^\circ)\right]u(t)$ V

19.38 23.01 V

19.40 $7.01 - j3.98\ \Omega$

INDEX